DATE DUE

MR 2 0 06			

DEMCO 38-296

HANDBOOK OF
AGRICULTURE

MARCEL

® *HANDBOOK OF AGRICULTURE*
© **IDEA BOOKS, S.A.**
Rosellón, 186, 1º 4ª
08008 Barcelona - Spain
Tel. 93 453 30 02
Fax 93 454 18 95
http://www.ideabooks.es
e-mail: ideabooks@fonocom.es

© For this English Edition:
MARCEL DEKKER, Inc.
270 Madison Ave.
New York, NY 10016
Tel. 212 696 9000
Fax 212 685 4540
http://www.dekker.com

... *Technical Agricultural Engineer*
AGRICULTURAL TECHNIQUES IN EXTENSIVE CROPS
DEFENSE OF CULTIVATED PLANTS
SOILS, FERTILIZERS AND ORGANIC MATTER
Mª PAZ YUSTE PÉREZ / *Technical Agricultural Engineer*
HORTICULTURE
FRUIT TREES
GREENHOUSE CULTIVATION

TRANSLATION
abc TRADUCCIONS

GRAPHIC DESIGN AND DRAWINGS
LLUÍS LLADÓ TEXIDÓ

LITERARY REVISION
CARMEN VILASECA

LITERARY ENGLISH REVISION
MARCEL DEKKER, Inc.

PRODUCTION OF GRAPHICS
ALEX CHIFONI

PHOTOGRAPHS
We gratefully acknowledge the collaboration of the manufacturers and producers who
have sent us material for use in this encyclopedia.
The company's own archive, Alfa Omega and Estudio Barambio.

PRE-PRODUCTION
ESTUDIO CHIFONI

PRINTING
ROL-PRESS

PRINTED IN SPAIN

LEGAL DEPOSIT: B-15.996-99

ISBN: 84-8236-089-2 Spanish edition

ISBN: 0-8247-7914-2 English edition
1999 Edition

This six-part volume deals with the options open to those developing and maintaining an agricultural operation. Sections are dedicated to soil preparation, fertilizer application, the defense of cultivated plants from diseases and pests, and greenhouse cultivation.

This is a thorough work for the professional agriculturalist, a practical reference to guide in choosing the optimal soil-preparation procedure for a given crop, the type of seed to buy, or the best machinery to perform an agricultural task, in light of climatic factors, the site, the size of the area cultivated and the potential profitability of the investment.

Progress in agriculture, like other aspects of the modern world, is speeding up. Nowadays, those involved in industrial agriculture must

Pruning with compressor pruning shears

Rainfall-type watering system

Machinery for scraping in extensive crops

have a clear idea of the type of crop that is most appropiate and profitable for them, bearing in mind market variations, changes in eating habits, the possible uses of their property, and the storage and transport facilities available.

Since it is now hard to find specialized labor, mechanizing and automating the work may be advantageous, but the profitability of these potentially very large investments should be studied carefully. This means that it is essential to select the best seeds, the most modern machinery and the most suitable agricultural chemicals, while taking into account the regulations governing products authorized in different countries, as well as packing and labelling regulations. You need good information to get good harvests.

Bunch of Matilde-type grapes

Tractor with trench-digging accessory

Conveyor belt and packer

Laboratory of "In vitro" cultivation

Corn ready for harvest

Many authors have written books on agriculture, ranging from technical handbooks and monographs to guides for enthusiasts. However, the lack on the market of an up-to-date and informative work aimed exclusively at the professional led us to conceive a comprehensive volume written in practical language and aiming to spread current knowledge of agricultural science.

A work of this type might easily be so large as to be unwieldy, since so many subjects are treated. When creating this work, however, our intention was to summarize all the general subjects that may be of interest when growing crops, and to create a practical reference guide for professional farmers.

This work deals with different areas of agriculture, such as horticulture, fruit tree cultivation, plant pathology, extensive crops, greenhouses, soils, and fertilizers.

It has been produced with the help of many of the companies in the field who have informed us of the latest cultivation techniques and methods, as well as the range of new products now on the market.

This work is aimed at all professional farmers. It is intended to be interesting and easy to understand, but with the rigor needed for it to be useful. It will also be of use for those studying at agricultural training centers.

We hope that the reading public this work is aimed at will receive it with as much enthusiasm as we have had in producing it.

Machinery for cultivating small fields

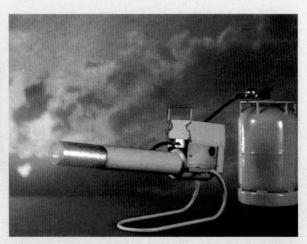

Device to scare away birds and animals

Mari Paz Yuste Pérez
Agricultural
Technical
Engineer

The power of 10 notation

The metric system makes great use of multiplication and division by factors of ten. For example, converting fertilizer doses from grams per square meter to kilograms per hectare is simply a question of multiplying by the right factors of ten.

10^3 is 10 x 10 x 10, one thousand, and is a one followed by three zeros. 10^{10} is 1 followed by 10 zeros.

10^{-3} is 1 divided by 10 x 10 x 10 = 1/1000 = 0.001.

To multiply factors of ten, it is only necessary to add the power of ten. Thus 10^3 x 10^3 = 10^6.
To divide by factors of ten, the powers of ten are subtracted: $10^7/10^5 = 10^2$.

The most important metric prefixes, with examples:

deci = 10^{-1} = one tenth (deciliter)
centi = 10^{-2} = one hundredth (centimeter)
milli = 10^{-3} = one thousandth (milligram)
micro = 10^{-6} = one millionth (microgram)
nano = 10^{-9} = one thousand-millionth (nanometer)

kilo = 10^3 = one thousand (kilogram)
mega = 10^6 = one million (megaWatt)

1 metric ton = 10^3 kg = 10^6 g; 1 metric ton is a thousand kilograms and a million grams.

Length
The metric system is very useful because it is possible to measure distance, weight, and other factors on a single continuous scale. Though it is now defined in terms of atoms, the meter was originally intended to be exactly one ten millionth of the distance between the equator and the North Pole. The meter is divided into smaller fractions, such as centimeters and millimeters, etc., that are all factors of ten.

If you are transplanting seedlings at a distance of 13 cm, how many plants are needed for a single row 17.20 m long?
17.20 m = 17.2 x 100 cm = 1,720 cm
1,720/13 = 132 plants

1 meter is roughly 39.4 inches (almost exactly 1.1 yards, 1.094 to be exact)
1 yard is 0.9144 meters.

A distance of 1000 m is thus approximately 1,100 yards.
1 mile is 1.609 kilometers.
To convert miles to kilometers, multiply by five and divide by eight.
To convert kilometers to miles, multiply by eight and divide by five.

Area
Area is length x width.

The unit of area most used in this book is the hectare, which is a square
100 m x 100 m = 10,000 m². There are 100 hectares in a square kilometer.

1 acre is 0.405 hectares. A 32 acre site is thus 32 x 0.405 = 12.96 hectares.
1 hectare is thus almost exactly $2^{1/2}$ (2.5) acres (to be precise, 2.471 hectares). An 8 ha site is thus 8 x 2.471 = 19.768 acres

To convert grams per square meter to kilograms per hectare: multiply by the number of square meters in a hectare (10,000) and divide by the number of grams in a kilogram (1,000).
Thus 7 grams per square meter corresponds to 7 x 10,000 g/ha = 7 x 10 kg/ha = 70 kg/ha.

Volume
Volume is length x width x height.
1 cm x 1 cm x 1 cm is one cubic centimeter.
1 m x 1 m x 1 m is one cubic meter. A cubic meter contains 100 x 100 x 100 (= 10^6 = 1 million) cubic centimeters. A liter is a cube of water 10 cm x 10 cm x 10 cm.

Weight
The gram was originally defined as the weight of one cubic centimeter of water. A liter is 1,000 cubic centimeters of water, and thus weighs one kilogram. A cubic meter of water (10^6 cubic centimeters) weighs exactly one metric ton.

One pound is 453.6 grams (0.4536 kg).
One kilo is 2.205 pounds
One ounce is 28.35 grams.

Liquid measures
The measures for liquids have the same names in US English and UK English, but different values.

US values
1 fluid ounce = 0.0296 liters
1 pint = 0.4732 liters
1 gallon = 3.7854 liters

UK values
1 fluid ounce = 0.0283 liters
1 pint = 0.5683 liters
1 gallon = 4.5461 liters

Weight per unit area
One ounce per square yard is equal to 28.35 gm per square yard. A yard is equal to 0.9144 meters, so a square yard is 0.9144 x 0.9144 square meters (= 0.836 square meters). One square meter is thus equal to 1/0.836 square yards = 1.1961 square yards.
To obtain the quantity per square meter, multiply the quantity in grams by the conversion factor 1.1961. In this case, 28.35 gm x 1.1961 = 33.911 gm.

1 ounce per square yard = 33.911 g/m²

2 oz./sq. yd = 67.822 g/m^2
3 oz./sq. yd = 101.733 g/m^2
4 oz./sq. yd = 135.644 g/m^2
5 oz./sq. yd = 169.555 g/m^2

Density is the weight per unit volume.
Density is expressed as a dimensionless value, such as 0.918, and gives the density of the substance in relation to water, which has a density of 1. A density of 0.918 is equal to 0.918 grams per cubic centimeter, or 918 kilograms per cubic meter.

Rainfall

Rainfall is an example of the type of calculation that is easy in the metric system. Suppose you have a 0.5 hectare site with 17 trees, and 7 millimeters of rain fall, how much rain has the site received?

For 1 square meter of soil, 1 mm of rainfall corresponds to a volume of water of 1 mm x 1 m x 1 m. This can be calculated in millimeters:
1 mm x 1,000 mm x 1,000 mm = 1,000,000 cubic millimeters = 1,000 cubic centimeters = 1 liter. It can also be calculated in meters:
0.001 m x 1 m x 1 m = 0.001 cubic meters = 1 liter.

0.5 hectares is 5,000 square meters, and if each receives 1 liter the site receives a total of 5,000 liters = 5 cubic meters = 5 metric tons. If there are 17 trees, then 5,000/17 liters has fallen for each tree = 294 liters.

TABLE OF EQUIVALENTS AND CONVERSIONS

LENGTHS

Centimeters	x	0,3937	=	inches	x	2,5400	= Centimeters
Meters	x	3,2808	=	feet	x	0,3048	= Meters
Meters	x	1,0936	=	yards	x	0,9144	= Meters
Meters	x	0,5468	=	fathoms	x	1,8288	= Meters
Kilometers	x	0,6214	=	miles	x	1,6093	= Kilometers
Kilometers	x	0,5396	=	miles nautical (U.K.)	x	1,8532	= Kilometers
Kilometers	x	0,5399	=	miles nautical (U.S.A.)	x	1,8520	= Kilometers

SURFACE

Sq centimeters	x	0,1550	=	sq inches	x	6,4516	= Sq centimeters
Sq meters	x	10,7639	=	sq feet	x	0,0929	= Sq meters
Hectare	x	2,4710	=	acres	x	0.4047	= Hectares
Sq kilometer	x	0,3861	=	sq miles	x	2,5900	= Sq kilometers

VOLUME

Cubic centimeters	x	0,0610	=	cubic inches	x	16,3873	= Cubic centimeters
Cubic meters	x	35,3145	=	cubic feet	x	0,0283	= Cubic meters
Cubic meters	x	1,3080	=	cubic yards	x	0,7646	= Cubic meters
Liters	x	0,2200	=	gallons (U.K.)	x	4,5461	= Liters
Liters	x	0,2642	=	gallons (U.S.A.)	x	3,7850	= Liters
Liters	x	1,7596	=	pints (U.K.)	x	0,5683	= Liters
Hectoliters	x	2,7497	=	bushels (U.K.)	x	0,3637	= Hectoliters
Hectoliters	x	2,8378	=	bushels (U.S.A.)	x	0,3524	= Hectoliters

WEIGHT

Grams	x	0,0353	=	ounces (Av)	x	28,3500	= Grams
Grams	x	0,0321	=	ounces (Troy)	x	31,1526	= Grams
Kilograms	x	2,2046	=	pound	x	0,4536	= Kilograms
Metric ton	x	0,9842	=	ton (U.K.)	x	1,0160	= Metric ton
Metric ton	x	1,1023	=	ton (U.S.A.)	x	0,9072	= Metric ton

1

Soils and Fertilizers

SOILS

1. INTRODUCTION

1.1. GENERAL POINTS

The word soil comes from the Latin word *solum*, which means base or bottom. The most general definition is that soil is a layer of weathered parent material covering most of the Earth's surface.

This layer, varying from a few centimeters thick to several meters, is where animals and plants can interact with the mineral world and establish a dynamic relationship with it. Plants obtain water and essential nutrients from the soil, and animal life depends on plants.

Mankind has always been in close contact with the soil, and for just that reason the concept and definition of soil is so universal that everybody has their own idea of what soil is. To an architect, soil is the foundation on which to construct buildings. To a mining engineer, the soil is the inconvenient layer that has to be removed (with all the inopportune costs)

in order to extract the desired metal. For us, and for the soil scientist, the soil, considered as a cultivation medium, is the mixture of mineral and organic materials that supports plant life, and which has formed from weathered rock by the action of the climate and living organisms.

Pedology, soil science, is the name given to the science that studies soils, considered as natural entities, in all their aspects, such as their physical structure, morphology, internal organization, their physical, chemical, mineral and biological characteristics, their fertility, as well as their origins, classification, development, taxonomy, geographical development, mapping, usage, improvement and conservation.

Considered from the perspective of agricultural management, one branch of soil science (or **pedology**) studies the soil as the basis of plant growth, that is to say, the soil's fertility and its water economy as factors in plant growth.

1.2. DESCRIPTION OF AGRICULTURAL SOIL

When analyzing a plot of agricultural soil, we can distinguish two aspects, its **physiography** and its **morphology**.

The physiography is the external part, the surface, what can be seen, and essentially consists of the distinguishing factors of the ground surface, such as its slope, how stony it is, the plant cover, etc. The morphology is the hidden part, what we cannot see without excavating the soil. From an exclusively agricultural point of view, we are interested in both aspects. The physiography gives us an idea of the agricultural tasks that need to be carried out on the surface, while the morphology gives us an idea of the soil material we are dealing with, including its physical and chemical properties, and thus how they can be corrected, if necessary.

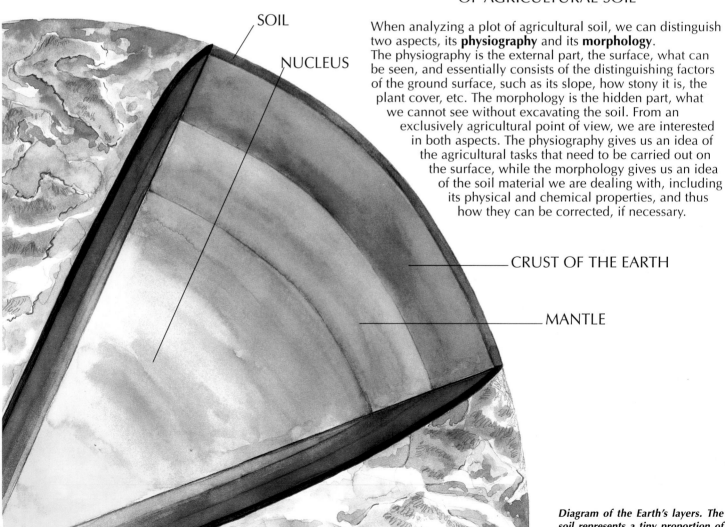

SOIL

NUCLEUS

CRUST OF THE EARTH

MANTLE

Diagram of the Earth's layers. The soil represents a tiny proportion of the Earth's crust.

1.2.1. Soil sampling

If what we are trying to do is describe the soil surface, we are restricting ourselves to the discipline of physiography. Characterizing the surface of agricultural soil is done by taking samples, by extracting part of the surface soil. This extraction is known as **sampling**.

To remove some of the surface soil, we must use sampling techniques, dealt with in more detail in specialized works. Broadly speaking, on sites that are physiographically homogeneous, samples should be taken in a grid, collecting material of all sizes equally. All the soil obtained is then mixed together to make a sample of about 1 kg.

author to author and depend on which school of soil science they prefer. Finally, there is the C horizon, generally formed by the parent material of the soil.

Example of the distribution of samples on the surface of a plot

Below: diagrammatic soil profile

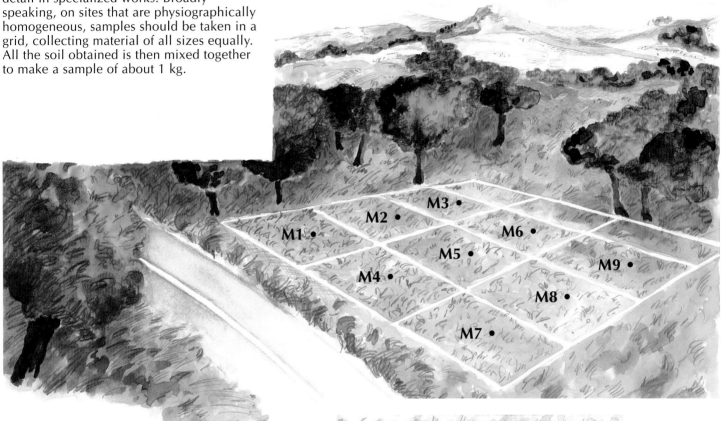

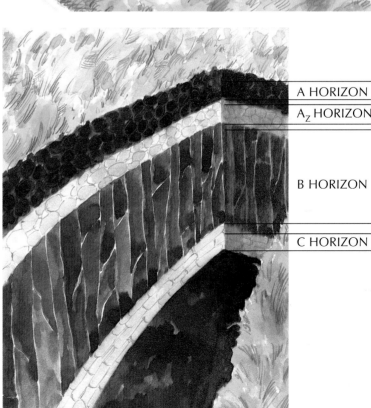

A HORIZON

A_Z HORIZON

B HORIZON

C HORIZON

1.2.2. The soil profile

When cultivating deep-rooting crops, we are also interested in the soil's morphology, but to study it we have to dig quite deep, often more than 1 m. We will consider sampling, but in this case sampling the different horizons of the soil **profile**.

Soils develop into different layers at different depths. Digging a vertical section of the soil lets us see what is normally hidden, meaning we can look closely at these different layers.

The different layers in the soil profile are known as **horizons**, and are given names starting at the beginning of the alphabet. The top horizon is thus the A horizon, and is usually the richest in organic material. The middle part of the soil profile is usually richer in clay particles and lighter in color than the upper horizon; this is the B horizon, or deep soil.

The A and B horizons often show sub-horizons (A_{00}, A_0, B_1, B_2, etc.), special forms of the horizon in question. The names given to them vary greatly from

The parent material lies under the *solum* and goes down to the bedrock. This horizon may be very deep, thin or totally absent. The soil profile consists of the A and B horizons and at least the top of the C horizon, when this is present.

1.3. QUALITATIVE EVALUATION

In the same site where we took the samples (whether physiographically on the surface or at depth by cutting a soil profile), a careful inspection reveals many of the soil's physical properties. By looking at it and feeling it, we can compare its physical properties and measure them against some sort of scale, such as size, texture, intensity, etc. Each soil has its own set of physical properties, and these depend of the composition of its components, the relative quantity of each component and the way the are bound together.

1.3.1. Thickness

The soil's thickness, its depth, varies from one part of the Earth's surface to another (between a few centimeters and a few meters). Cutting a soil profile shows us how deep the soil is. If the soil is deep, we will have far fewer problems than if we are cultivating a soil only a few centimeters deep. To give a relevant example, if we cut a soil profile and find that the top of the soil is only 2-3 cm above the parent material, we can rule out agriculture, as increasing the depth of soil, by landfill or mining the parent material with explosives, would be too expensive.

1.3.2. Color

A soil's color is one of its easiest features to identify, and it is very useful because it tells us a lot about factors like the soil's organic matter content, the climate, drainage and the soil's mineral composition.
Most of the minerals that make up soils, mainly in the surface (A) horizons, vary in color from white to pale grey. There are a few black minerals, some red ones and a few of other colors. The reddish, brownish or greyish color of most normal soils is due to the presence of two materials that are powerful coloring agents, the humus fraction of the organic matter and the different compounds of iron. Organic matter is broken down by the action of microbes into humus. Humus consists of very small particles,

almost black in color, and is a very effective coloring agent. If 5% of the soil is organic matter, this is enough to turn the soil black or almost black.
Iron has two different oxidation states (ferrous oxide, FeO, and ferric oxide, Fe_2O_3), which change in color depending on their degree of hydration, their location, their distribution, their oxidation state, etc.

1.3.2.1. Assessing soil color

Assessing soil color is done using the Munsell color system. This system is based on the three variables of a color: hue, chroma and value. The drawing below shows a page of the Munsell color determination code (The Munsell Book of Colors).
The **hue** refers to the dominant wavelength of the light reflected by an object, defined in terms of five primary colors and their mixtures; blue, green, yellow, red and purple. The **chroma** is a measure of the degree of saturation of a color, its purity. The **value** is the measure of the brightness, how bright or dark a color is.

Mineral compound	Chemical formula	Color
Ferrous oxide	FeO	Bluish grey
Hydrated ferric oxide (limonite)	$Fe_2O_3 * xH_2O$	Yellowish brown
Ferric oxide (hematite)	Fe_2O_3	Red

Iron compounds that influence soil color, according to Thomson (1988)

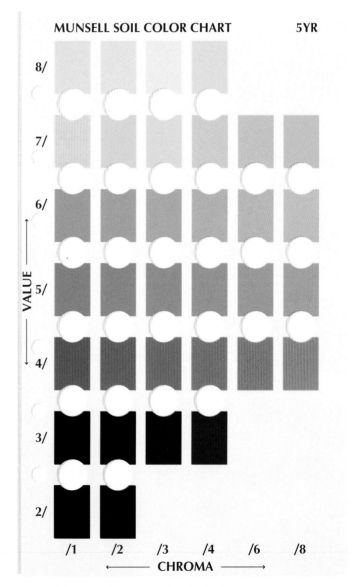

A page of the Munsell code for color classification of soils

2. SOLID SOIL COMPONENTS

Soils are made up of solids, liquid and gases, mixed together in variable proportions. The relative quantities of air and water present depend to a large extent on the intensity of the binding between the solid particles. Aggregates consisting of small particles tend to be very different from those made up of large particles. Both a soil's texture (one way of assessing the size of its particles) and its structure (the way the particles bind together) influence the total volume of soil pores and how it is distributed.

We advise those who have not read much on agricultural management to make an effort to read this chapter on the *Solid Components of the Soil*, the *Liquid Phase of the Soil* and *Soil Chemistry*. As the soil is a "Solid-Water-Air" continuum, it requires understanding of many concepts that are discussed in all three sections.

In the next few sections we will define the solid soil components, i.e., the organic and mineral parts, as well as their main characteristics, and look briefly at ways of analyzing them and then characterizing them, in order to give the reader some valid parameters, so that if they have to undertake a soil analysis, they can understand what the results mean.

2.1. ORGANIC MATTER (O.M.)

It is well known that the organic matter in the soil is derived from the remains of higher plants, animals, and in general the remains of any organic matter input into the soil. These materials all gradually decompose in the soil. Decomposition is usually started by earthworms and insects chewing the material, digesting part of it and breaking the rest down into fragments. Different types of microbe break down these fragments, the body wastes and eventually the dead bodies of the insects, worms, etc. Breakdown is not only due to the action of plants and animals, but is often also partly performed by the oxidizing or reducing conditions within the soil.

The soil's abundant population consists of many plants and animals, varying in size from microscopic, such as bacteria, to multicellular organisms, such as burrowing mammals or large tree roots. A normal soil contains, including plants and animals, 2-3 kg/m^2 of living matter, about 20,000-30,000 kg/ha of animals and plants.

Animal and plant wastes and remains return to the soil where they are processed by insects, worms, fungi, bacteria and other living organisms. The cycle is eventually completed when the carbon dioxide and mineral nutrients are released and become available to plants once more. If this did not happen, soil fertility would decline, all the plants would die, and everything else would die soon afterwards. Measuring the percentage of organic matter in a soil sample is usually done in a specialized laboratory, and quantifying the amount requires use of complex analytical procedures. Briefly, analysis is based on the oxidation of the organic matter (the O.M.'s unit structure is long chains of hydrogenated carbon atoms), which is then corrected (using the empirically derived value 1.724) to determine the percentage of O.M.

$$\% \text{ O.M.} = \% \text{ C} * 1.724$$

Most soils contain between 1% and 6% O.M. Of course, in very arid soils (deserts) the percentage will be less than 1%, while in tropical jungles, where organic input into the soil is high, it may be more than 6%.

2.1.1. Components

The organic matter present in the soil can be classified by its origin or chemical composition.

	kg/ha	Organisms per hectarea
Living macroorganisms		
Roots	15,000	
Insects	1,000	20,000,000
Earthworms	500	1,000,000
Nematodes	50	200,000,000
Crustaceans	40	400,000
Snails, slugs	20	10,000
Rodents, snakes, etc.	20	200
Dead but identifiable remains of macroorganisms	4,000	
Living microorganisms		
Bacteria	3,000	2 X 10^{18}
Fungi	3,000	2 X 10^{14}
Actinomycetes	1,500	5 X 10^{16}
Protozoa	100	5 X 10^{12}
Algae	100	1 X 10^{10}
Dead and finely divided organic matter	150,000	

The average estimated content of organic matter per hectare of soil formed in prairies in a sub-humid temperate region. Based on Thomson (1988)

2.1.1.1. Classification by origin

We can define four different sources:

• **Living macroorganisms**. Multicellular plants and animals make a very important contribution to the soil structure (plant roots modify the soil structure) and percentage of O.M. Some animals, like earthworms, have an effect similar to roots, as they open new paths that air can flow along, and they also represent a large input of organic matter into the soil (an abundant population of earthworms eats and excretes many tonnes of soil per hectare per year).

• **Dead but identifiable remains of macroorganisms**. The remains of dead multicellular organisms represent an input of organic matter into the soil in the form of roots, leaves, dead animals, etc.

*Fertilizing a field
with dung (O.M)*

• **Living microorganisms**. They are directly responsible for decomposing fresh organic matter. There are a large number of types of microorganism, including bacteria, actinomycetes, fungi (both the "true" fungi and the slime molds), algae, nematodes and protozoa. Each type acts in a different way, but their description (effects, habits, way of life, etc.) belongs not here but in the next section, which deals with the dynamics of organic matter in the soil. Note that microbial activity is essential to release nutrients from the O.M. into the soil, and without their activity soils would become sterile and all life would cease.

• **Dead organic matter and finely divided organic matter**. Dead and finely divided materials are the main components of humus. Humus is defined as the **colloidal** organic products derived from the decomposition of fresh organic matter and of the synthetic activities of soil microorganisms. Humus tends to form a layer covering the soil's mineral particles and is closely associated with clays.

*2.1.1.2. Classification
by chemical composition*

Decomposing organic matter, from its incorporation into the soil to its complete breakdown into minerals, passes through several stages that can be classified chemically. We can distinguish three types:

• **Fresh organic matter**. Fresh organic matter that has only just entered the soil, and is beginning to decompose.
• **Humus**. This represents 10-15% of the soil organic matter. Humus consists of colloidal organic products derived from the decomposition of plant remains and synthesis by soil microorganisms.
• **Humic substances**. These are considered the true components of the humus, as they are the

A **colloid**, or a colloidal solution, is the name given to a substance that is dispersed (not dissolved) in a medium, that diffuses slowly and cannot diffuse across a dialysis membrane, unlike true solutions. The particles can only be seen with an electron microscope as their spherical diameter (Ø) is roughly 0.1 to 0.001 μm, and their essential properties are due to their dispersed phase not their chemical composition.

organic materials that last in the soil without being broken down. Some authors consider that they are the true humus. These compounds, dark or black in color, have a relatively high molecular weight and are formed by secondary synthesis. The term humic substances is used generically for this finely divided dark material, which can be divided on the basis of its solubility characteristics. There are three subdivisions of humic substances based on their behavior when dissolved colloidally in a weakly alkaline solution of NaOH or NH_4OH. They are:

- *Humin*. This is the part of the soil organic matter that does not dissolve in an NaOH or NH_4OH solution.

- *Humic acids*. This is the most important fraction, and has received most study. It is a dark organic material that dissolves in NaOH or NH_4OH, and which precipitates on acidification to pH 1 or 2.

- *Fulvic acids*. These are the remaining organic substances that do not precipitate when a soil solution is acidified to pH 1 or 2 (i.e., they remain in colloidal suspension).

2.1.2. Distribution within the soil

The distribution of organic matter within the soil is usually very irregular, and depends on the type of soil, the area's climate, its minerals, the site's greater or lesser plant cover, etc.
Within a single profile, the surface (A) horizons are usually richer in humus than the deeper horizons, because organic remains fall directly on the surface and rarely reach the lower horizons, except in the form of humus (dead and finely divided material).

When talking about cultivated ground, the amount of organic matter present and its distribution may vary greatly due to past human disturbance. This means that if, for example, we cultivate a plot of ground with lettuces, and harvest them when ready, we must return to the soil the organic matter and nutrients that the crop plants had to extract for their growth. If we don't do this, the soil will be impoverished within a few years.

The O.M. content of any soil horizon depends partly on the size of the annual input of organic remains, and partly on the percentage of organic matter that is mineralized every year. When the two processes are in balance (within limits of tolerance that allow a margin of variation due to eluviation in some highly washed soils), the content of organic matter stabilizes. Even so, the distribution of the organic matter within the soil depends on a large number of causal factors. Let's consider some of them.

the left hand graph. When talking about the temperature, however, the higher the average annual temperature, the lower the amount of organic matter present, because microbial decomposition is faster, and thus the percentage of O.M. in the soil is reduced.

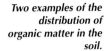

Two examples of the distribution of organic matter in the soil.
The first case (A) is a prairie soil in a sub-humid temperate region.

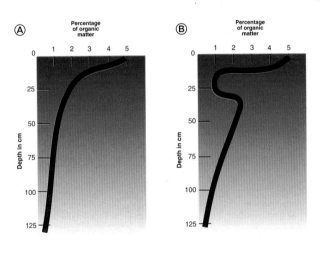

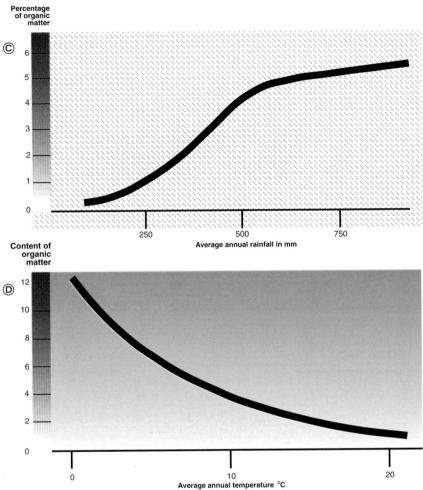

• **The effect of the vegetation**. As previously mentioned, the distribution of O.M. in the soil depends on the vegetation. In the example shown in the accompanying graph, we can see the distribution of the O.M. in a prairie soil and in a forest soil. Graph A shows the distribution of O.M. in a typical temperate zone soil supporting a prairie (the organic matter decreases with depth). Graph B shows a forest soil with an accumulation of organic matter about 25 cm below the surface. This soil shows illuvial accumulation of humus. This occurs in some wet sites with a large input of organic matter (leaves, twigs, etc.), because rain or irrigation water washes the humus (finely divided dead material) down into the deeper B horizons.

• **The effect of the climate**. In terms of the effect of rainfall, the greater the rainfall, the greater the percentage of organic matter in the soil. So the greater the annual rainfall, the greater the plant growth and the greater the input of organic matter to the soil, as is clearly shown by

• **The effect of topography**. In soils on steep slopes, typical soil erosion by weather tends to reduce soil depth in the higher points of hills and mountains, and thus to increase the depth of the soils in lower areas by settling. As the upper areas are left without soil, there is less vegetation and the annual input of organic matter is also lower. In valleys, however, the deeper soils support much more vegetation, and thus they contain a higher percentage of O.M., as shown in the accompanying graph.

• **The effect of time**. In terms of soil formation, the process of change in organic matter over time probably takes several thousand years. Rocks supply most of the nutrients needed for plant life, except for nitrogen. Over time, the nitrogen content of a virgin soil increases due to rainfall and the nitrogen-fixing activities of azotobacter and rhizobium bacteria (which fix atmospheric nitrogen). The increase in the soil's nitrogen concentration allows more and more plants to

The second case (B) deals with a forest soil. Note the illuvial accumulation of humus at a depth of between 25 and 50 cm. Thomson & Troc (1988).
Two examples of the influence of climate on the distribution of organic matter in the soil. The upper graph (C) represents the different O.M. content in prairie soils with a different annual average rainfall. The lower graph (D) shows that in prairie soils, higher average annual temperatures cause the percentage of O.M. to decrease.

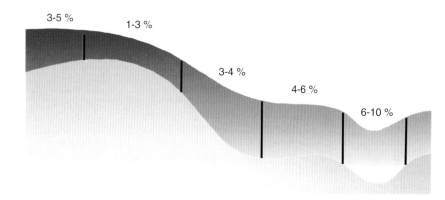

3-5 % 1-3 % 3-4 % 4-6 % 6-10 %

The relationship between the topographical position, the thickness of the A horizon and the percentage of O.M. in a sub-humid temperate climate

grow on the soil, leading to an increase in its organic matter content. This is followed by a long period of time when the percentage of organic matter remains stable for many years. Finally, there is a fifth phase when the soil's organic matter content declines, and the soil gradually becomes less and less fertile.

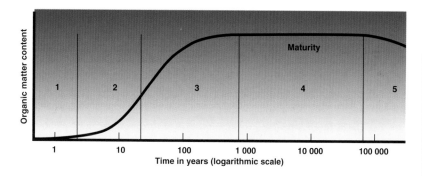

Fictitious graph trying to show the changes in the percentage of O.M. in a soil over time. The time scale, which is logarithmic, is only indicative and varies greatly from one soil to another.

2.1.3. Functions

The characteristics of organic matter and the functions it plays in agricultural soil are the most important part of this whole section. Organic matter greatly affects the soil, providing it with features it would otherwise lack. These features are used in agriculture to correct underlying soil deficiencies, the most common ones being the inability of sandy soils to retain water and nutrients, and the waterlogging and compaction of clay soils. Let's consider some of the soil factors determined by its organic matter content:

• **Color**. The dark color typical of organic or very organic soils may help to raise their temperatures. Just 5% O.M. in a soil is enough to make it dark, or even black. It is well known that dark colors absorb more radiation than light ones, and that dark things warm up more.

• **Retaining water**. Organic matter can retain up to 20 times its own weight of water. This physical property helps to keep the soil moist, thus avoiding soil desiccation and contraction, and improving moisture retention in sandy soils.

On the right: soil prepared for tilling

• **Combination with clay minerals**. O.M. joins soil particles together into structural units

(aggregates), which improve soil aeration. In clay soils, where waterlogging may be a terrible problem, organic matter allows gaseous exchange, stabilizes the structure and increases permeability.

• **Chelation**. O.M. forms stable complexes with Cu^{2+}, Mn^{2+}, Zn^{2+} and other divalent cations. When O.M. chelates the trace elements (i.e., when it binds very tightly to them), it reduces the plants' ability to take them up.

• **Solubility in water**. The insolubility of organic matter is partly due to its association with clays. The salts of divalent and trivalent cations with organic matter are also insoluble. Isolated O.M. is also partly insoluble. Yet part of the organic matter is soluble in water, and is lost to the water table by leaching.

• **Relation to pH**. Organic matter buffers the soil's pH. That is to say, the pH remains stable

when an acid or base is added.

• **Cation exchange**. Organic matter increases the soil's cation exchange capacity (C.E.C.) by 20-70%. The total acidity of the isolated humus fractions is between 3,000 and 14,000 mmol kg^{-1}.

• **Mineralization**. Organic materials break down to CO_2, NH_4^+, NH_3, PO_4^{3-} and SO_4^{2-}. These are the source of the nutrient elements needed for plant growth.

• **Combination with organic molecules**. Organic matters influence the biological activity, persistence and biodegradability of pesticides and herbicides. It modifies the amount of pesticides to be applied for effective control.

2.1.4. The dynamics of organic matter

Now that we have considered the components of the organic matter, its distribution and its role in agricultural, we must turn to the dynamic changes that organic matter undergoes. Before becoming humus, as already mentioned, animal and plant remains undergo chemical changes, which are extremely hard to identify in the laboratory. These transformations are generically known as *soil organic matter dynamics*. The microorganisms responsible for breakdown, the distinctive features of each type of organic matter, and the characteristics of the soil environment all affect the mineralization of O.M. The subjects dealt with in this section include quantitative parameters used to assess humus formation and mineralization, such as the K1 and K2 coefficients and the C/N ratio.

2.1.4.1. Microorganisms

Soil microorganisms are responsible for the breakdown of organic matter. The term microorganisms refers to a large number of microscopic animals (microfauna) and plants (microbiota). The total mass of living microorganisms per unit soil volume is usually called the **microbial biomass**. The bacteria are usually present in the largest numbers, but fungi usually dominate when considered by weight. Let's consider the bacteria in a little detail.

• **Bacteria are the simplest living organisms**. They vary greatly but their DNA is always a circular loop in the cytoplasm; they are called prokaryotes. Other organisms have cells with a nucleus and their DNA is on chromosomes, and they are called eukaryotes. Some bacteria may form colonies. According to Clark's 1954 study, a fertile soil with fresh organic matter may contain around a thousand million bacteria per gram. Some bacteria play very defined roles, such as obtaining energy from oxidizing the nitrogen in ammonia to nitrates. Others simply participate in the general process of decomposition of organic materials. Bacteria can be divided into aerobic and anaerobic bacteria, depending on whether they need oxygen to survive. Aerobic bacteria obtain energy by oxidizing, for example, a sugar to carbon dioxide, while anaerobic bacteria oxidize some of the sugar molecules to carbon dioxide and reduce the rest of the sugar molecules to produce, for example, alcohol. Some anaerobic bacteria may reduce inorganic ions, like NO_3^- or SO_4^{2-}. Some bacteria, called facultative anaerobes, can grow aerobically when there is free oxygen gas (O_2) in the soil, but if oxygen is not present, they can grow anaerobically. Anaerobic methods of obtaining energy produce less energy than aerobic methods. Aerobic bacteria obtain a greater yield for their "labor" than anaerobic bacteria. Anaerobic bacteria usually grow in waterlogged soil, where no oxygen gas is available for their respiration. Both types of bacteria are often present in the same soil.

Bacteria can also be divided, on the basis of how they obtain carbon, into autotrophs and heterotrophs. Autotrophs obtain the carbon they need to synthesize proteins, etc., from inorganic sources, such as CO_2 and CO_3^{2-}. Heterotrophs have to obtain carbon in the form of organic molecules. The most important autotrophs are photoautotrophs, which use the energy of sunlight to power synthesis, but some bacteria are chemo-autotrophs, and use energy from transformation of minerals to power their synthesis of organic compounds from inorganic carbon. The bacteria that oxidize ammonium ions to nitrite and those that oxidize nitrite to nitrate are chemo-autotrophs. Finally, note that bacteria grow better in alkaline environments, where calcium is abundant; in acidic environments, bacterial populations are lower.

Organic matter can retain up to 20 times its own weight of water. This characteristic of O.M. may be of great importance for crops traditionally grown in very dry summer conditions. Courtesy of Antonio Climato (Instituto Propagazione Specie Legnose), Florence, 1989.

Main components of the soil microflora

GROUP	POPULATION/ AVERAGE	GENERAL CHARACTERISTICS	CONDITIONS OF THE SOIL ENVIRONMENT	CHARACTERISTIC ACTIVITIES	CONSEQUENCES
Virus	Variable 50/100μ	DNA or RNA molecule, bacteriophages	Require a host cell	Obligate parasites of bacteria, actinomycetes, nematodes, fungi	Viral diseases in infected plants
Bacteria	0-10⁹ ind/g Very variable	Variable morphology Fixed/mobile Aerobic/anaerobic Heterotrophic/autotrophic	Aeration. moisture. T. 21 to 38°C pH 6-8 Ca²⁺	$NH_4^+ \longrightarrow NO_2^- \longrightarrow NO_3^-$ $NO_3^- \longrightarrow$ N org. $NO_3^- \longrightarrow NH_4^+$ $S^{2-} \longrightarrow SO_4^{2-}$ $SO_4^{2-} \longrightarrow S^{2-}$ $Fe^{2+} \longrightarrow Fe^{3+}$ Biodegradation	Fixation of atmospheric nitrogen Nitrogen fixation Mineral alteration Rhizosphere (nodules) Alcoholic fermentation Cellulose breakdown
Actinomycetes	≈700 kg/ha Ø mycelia 1,5 mμ	Complex bacteria with morphology similar to fungi, with mycelia	pH 6,5 - 7 Some are resistant to high temperatures	Parasites, and in some cases, in symbiosis	Fermentation Nitrogen supply Antibiotic production
Fungi	1,000 - 1,500 kg/ha Length 100 m Ø < 5 mμ		Not very harsh conditions. Favoured by acid environments.	Parasites, saprophytes and in symbiosis (mycorrhiza)	Alcoholic fermentation, breakdown of lignin and chitin. Cellulose breakdown in acid environments. Pre-nitrifying
Algae	10⁵ ind/g	With photosynthetic pigments	Limited depth (0 - 2 cm) Autotrophic	Can degrade O.M. Produce slime	Cultivation medium for colonizing bacteria

• **Actinomycetes**. Actinomycetes are bacteria that form filamentous colonies. The filaments are the same diameter as the bacteria (and as coarse clay) and they are often interwoven and branched, making them difficult to count. They are not as abundant as other bacteria, but there are still a lot of them (several million per gram of soil). Like bacteria, they grow better in alkaline environments with abundant calcium. The typical smell of newly plowed soil is due to actinomycetes.
As they are bacterial colonies, they are much larger than single bacteria, but they are smaller than fungi. They play a role in the soil similar to those of other microorganisms, and they have been thoroughly studied as some strains are the sources of antibiotics used in medicine.

• **The algae**. This is a very broad term for organisms with chlorophyll that can synthesize organic matter from carbon dioxide and water,

Compounds of soil fauna

GROUP	SIZE	SPECIES
Macrofauna	2 to 10 mm	Insect larvae Arthropods Earthworms
Mesofauna	0.2 to 2 mm	Nematodes, oligochete worms small insect larvae and microarthropods
Microfauna	0 to 0.2 mm	Microscopic protozoa, nematodes and arthropods

with the energy of sunlight, but which are not plants. Algae may be bacteria, single cell eukaryotes, large eukaryotes, such as seaweeds, and some flowering plants are even known loosely as algae. Like some facultative anaerobic bacteria, the algae present within the soil where there is no sunlight can obtain energy by oxidizing other organic materials, just like other microorganisms. Soil algae vary in size from single cell algae to multicellular ones, which are often visible to the naked eye as filaments on puddles. The most familiar ones are the slimy stringy scum that grows in stagnant water. Algae tolerate a wide range of humidity conditions, thriving in boggy soils, and they are also very important in desert conditions. Some algae, such as the cyanobacteria (blue-green algae), are able to fix atmospheric nitrogen; the nitrogen input to crops from these algae is not very important in typical agricultural soils, but in some special crops, such as rice, it may be very important.
It is well known that a symbiotic relation of great importance can be formed between algae and

some fungi, composite organisms known as lichens. The alga supplies the fungus with carbohydrates produced by photosynthesis, and in return, the fungus helps by absorbing mineral elements, water and anchoring it to the rock. Lichens are often the first organisms to colonize bare rock, and their weathering of the parent rock is very slow, but may take place over many years.

• **The protozoans**. The protozoans are unicellular eukaryotes that are less numerous than the bacteria, and need a more constant watery environment. Biologically speaking, some protozoans appear to be half plant and half animal, because they can take in solid particles like animals, yet at the same time they can photosynthesize. Protozoans eat bacteria (among other things) and it has been observed that the more protozoans there are, the less bacteria there are.

• **Nematodes**. Nematodes are well known for the damage they can cause. Many nematodes are parasites of the roots of crops such as fruit trees, sugar beet, corn, soybean, tobacco, etc. Some are even parasites of human beings. Most live in decomposing organic matter. They are between 0.2 and 2 mm long.

2.1.4.2. Characteristics of the environment

The ease of decomposition of organic matter depends mainly on the microclimate of the soil, as this affects the activity of the microorganisms discussed in the previous section. These microclimatic characteristics are the soil's aeration, its temperature and its pH. Microorganisms, like all living things, need a certain minimum set of environmental conditions in order to grow. These minimum requirements, which are strictest for the simplest organisms, are usually moisture and warmth, and in the case of soil microorganisms, a suitable soil pH too.
In dry or very dry soil conditions, microorganisms are totally inactive, but in wet conditions there

are, as mentioned, two possibilities: 1) when the soil is at F.C. (field capacity: the micropores are saturated with water and the macropores are full of air), microbial activity is aerobic, but 2) when the soil is completely saturated with water, waterlogged (the macropores and micropores are full of water), the microorganisms can only obtain energy anaerobically.

Temperature is another factor limiting microbial activity. For example, in the Arctic tundra microbial activity is restricted to the few months of summer, and even so their activity is minimal. The moderately cold winters in temperate climates greatly reduce microbial activity. In tropical areas, where temperatures are consistently high all year round, microbial activity is much greater.

The pH, as we shall see when we discuss soil chemistry, measures how acid or alkaline the environment is, something that greatly affects microbial activity. Many microorganisms are active in alkaline environments but only a few types of microbes can decompose organic matter in acid environments.

The table immediately below shows the effects of moisture, temperature and pH on the decomposition of organic matter and on humus formation.

Composition	Funcional groups in meq/100 g of product after kiln-drying at 550°	
C 45-65 %	COOH	300-600
O 48-30 %	OH phenol	280-360
N 2-6 %	OH alcohol	300-500
H 5 %	C = O	300
	O - CH₃	20-50
	Total acidity	500-1,400

Common chemical characteristics of all the humic substances. Saña & Soliva (1987)

idea of decomposition and humus formation. To recap:

1) Fresh organic matter is gradually incorporated within the soil over the course of time, and as organisms die or shed body parts.

To the left: the relationship between environmental factors and the rate of breakdown of organic matter

2) Microorganisms decompose this fresh organic matter, at a faster or slower rate depending on their environmental conditions. Decomposition, or mineralization, is defined as the break down of large complex molecules to smaller, simpler, ones (either organic molecules or minerals).

General cycle of transformation of fresh organic matter into humus. Josa &

		ENVIRONMENTAL CHARACTERISTICS				
Aeration	Moisture	T	pH	Decom-position	Humus formation
Aerobic	Moderately wet	High	Basic	Fast	Abundant
Anaerobic	Very wet	Low	Acid	Slow	Low

2.1.4.3. Mineralization and humus formation

The chemical processes involved in mineralization and humus formation by microorganisms are, as already mentioned, very complex. In fact, the study of how organic matter is mineralized is only just beginning. The many analytical chemists studying these processes disagree on which pathways of synthesis lead to which results.

The reader should now understand the general

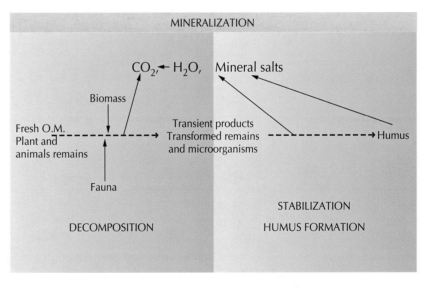

Hereter (1995)

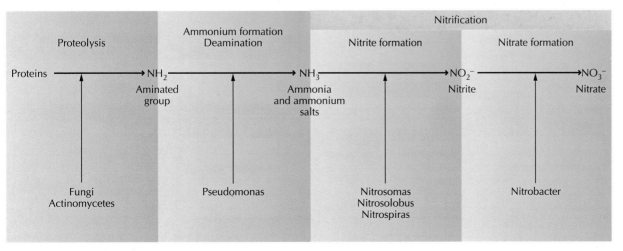

Decomposition and mineralization of nitrogen-containing materials. Josa & Hereter

Breakdown of soluble sugars and starch.
Saña & Soliva (1987)

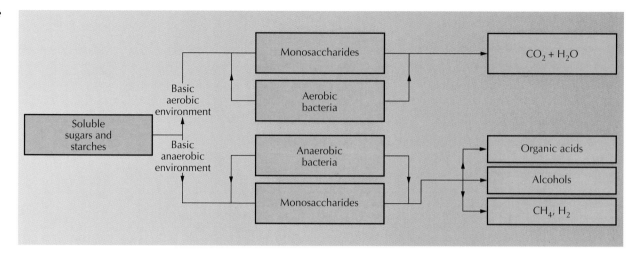

3) This decomposition includes a first stage in which most of the fresh organic matter is mineralized.

4) A small amount of the fresh organic matter is not rapidly mineralized, and undergoes a very complex series of chemical transformations and is broken down to very small organic particles. This finely divided material is known as humus, and the transformation of organic matter to *humus* is known as *humus formation.*

5) *Humus formation* is the process by which the carbon in organic remains is changed into humus by biochemical and/or chemical processes. Chemically, this is the formation of new, more complex, molecules (polymer formation). The nitrogen incorporated into these molecules is of great importance.

6) The *humic substances* are the most stable part of the humus. There are three humic types of humic substances, divided on the basis of their physical properties when they form colloidal dispersals; humic acids, fulvic acids and humins. Humic substances are acidic polymers with a high molecular weight, more-or-less aromatic properties, that are formed by the condensation of a large number of substances whose chemical

characteristics are listed in the table (previous page). The general structure is a compound with a polycyclic aromatic core with side chains and phenolic acids, carbohydrates and polypeptides.

7) Over time the humic substances are also mineralized, and their inorganic nutrients released into the soil.

Some of these biochemical and chemical processes are outlined in the accompanying charts. They are based on recent studies, and we can expect new insights from the study of humus formation and mineralization in the near future. Pay special attention to the chart showing: the decomposition of proteins (nitrogen-containing organic compounds) into nitrites and then nitrates; humus formation from organic sources of carbon, such as sugars and lignin, and sources of nitrogen such as plant proteins, soil minerals or nitrogen fixed from the atmosphere; the breakdown of soluble sugars and starch into carbon dioxide (CO_2), water (H_2O), methane (CH_4), hydrogen (H_2), alcohols and organic acids; the breakdown of cellulose to humus, unaltered cellulose and gases, such as carbon dioxide, hydrogen, methane, etc.; and finally the transformation of lignin into humus, polyphenols and unaltered lignin.

Humus formation from lignin.
Saña & Soliva (1987)

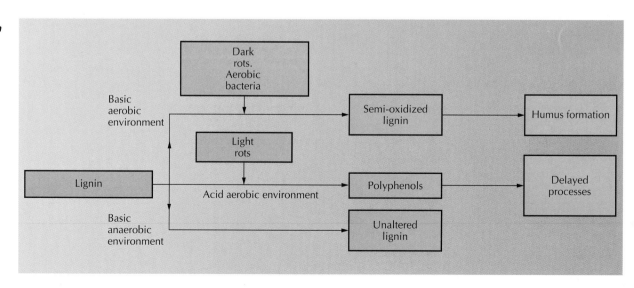

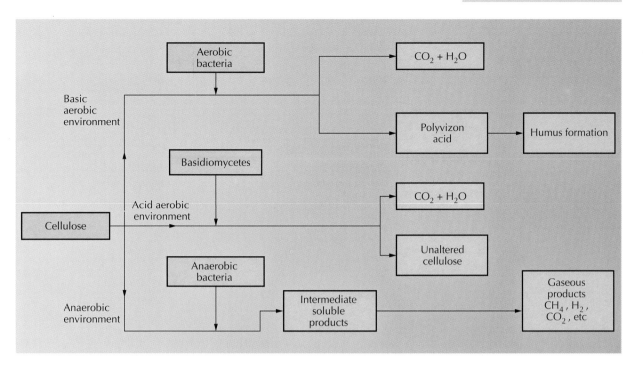

*Humus formation
from cellulose.
Saña & Soliva (1987)*

2.1.4.4. K1 and K2 coefficients

These coefficients, related to mineralization and humus formation, could have been included within the previous section, but the fact that they are recent and important led us to dedicate some space to them.

• **K1 coefficient**. This is also known as the *isohumic coefficient*, and is the amount of humus formed from a unit weight of dry organic matter input to the soil. Bearing in mind that most organic remains decompose without turning into humus, this coefficient is calculated in each case on the basis of the carbon/nitrogen ratio discussed in the next section.

• **K2 coefficient**. The K2 coefficient, or humus *mineralization coefficient*, is a measurement of the percentage of stable humus that is mineralized every year. In most soils, average annual decomposition of humus is between 2,000 and 4,000 kg/ha (this is only a small proportion of the input of organic remains into the soil every year, but remember that most of the remains are broken down without being turned into humus). This amount may represent less than 1% of the humus present in a cold region, or more than 25% of the humus content of a tropical soil. The K2 coefficient thus depends on climatic conditions and on the presence of ions that stabilize the humus, but in general for temperate regions an average value of K2 ≈ 0.02 (about 2% of the humus is mineralized every year).

2.1.4.5. The C/N ratio

Organisms consist mainly of molecules of carbon, hydrogen, oxygen and nitrogen, with lesser quantities of phosphorus and sulfur.

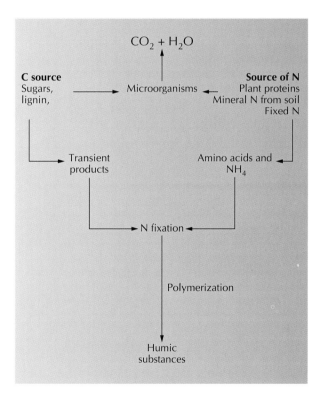

*Diagram summarizing
the synthesis of humic
substances.
Josa & Hereter (1995)*

Humus contains all the elements that have been taken up by plants, but not in the same proportions as in plant tissues.
The percentage of carbon measured in a laboratory belongs exclusively to the organic C, as does the organic nitrogen, which is also measured by **oxidation of the organic matter**. Soil microorganisms decompose fresh organic matter and obtain energy by oxidizing the organic carbon to carbon dioxide gas and water.

Fresh organic matter + O_2 = Energy + CO_2 + H_2O

In the laboratory analysis of the total nitrogen of a quantity of soil, the percentages of organic nitrogen and ammonium are measured. Most of the nitrogen in the soil is organic; only a small part of the nitrogen in the soil is present in a mineral form, such as nitrates, nitrites or free ammonium.

Decomposition and changes in N, based on the C/N ratio of the plant material. Fuentes (1989)

After obtaining energy, the microorganisms reproduce (generally by division), increasing their nitrogen requirements (because building new organisms requires proteins). This nitrogen can come from two sources, either directly from organic remains or directly from soil minerals. Some bacteria, azotobacteria, can even fix atmospheric nitrogen (N_2) directly.

Thus nitrogen can be considered a limiting factor in the decomposition of organic matter. Soil scientists derive a ratio known as the **carbon/nitrogen ratio**, or C/N ratio, from the joint action of all the microorganisms, the carbon and organic nitrogen they need to reproduce.

The ratio of the carbon content of the soil organic matter to its nitrogen content is a rough guide to how far the organic matter present has broken down.

Depending on its C/N value, we can tell whether a sample of organic matter is highly decomposed or only slightly. If the C/N value is 50-80, there is a lot of fresh organic matter and not very much microbial activity. If the value is between 15 and 40, breakdown is close to equilibrium, and part of the nitrogen released is incorporated within the soil. If the C/N value is around 10, the decomposition of organic matter is considered to have reached equilibrium, meaning that the amounts of carbon and nitrogen are adequate to ensure the process neither slows down or speeds up.

2.2. MINERALS AND ROCKS

The solid components of the soil include organic matter (discussed in the preceding section) and the inorganic materials derived from the rocks and minerals, and these are the inorganic and solid constituents of agricultural soils.

Rocks and minerals are dealt with in this chapter. Our aim is not to produce a compendium of soil geology, but an overview to acquaint the reader with geological terminology, so they can understand a soil analysis, and grasp the difference, for example, between a carbonated soil and a silicated one.

Because, in the final analysis, soil is essentially nothing but weathered rock, the composition of the minerals in the soil is in theory the same as that of the rocks making up the earth's crust. However, the relative abundance of the elements in the crust differs from that of the Earth as a whole, due to the sorting effect of gravity.

In the Earth's formation, large quantities of light elements, such as hydrogen and helium, rose into the atmosphere and escaped into space. The heaviest metals, such as iron and nickel tended to sink towards the center of the Earth. As a result, only the nine elements listed in the accompanying table are present in real abundance in the Earths crust. The atoms and ions of these nine elements account for a total of 99% of the elements making up the Earth's crust.

Representation, in percentages, of the elements (their atoms and ions) present in the Earth's crust

Organic matter	C/N ratio	Rate of decomposition	Fate of nitrogen
Cereal straw	50-80	Slow	Consumed by microorganisms
Dung with straw	20-40	Slow	Near to equilibrium
Rotted dung	15-20	Medium	Part of the nitrogen released is incorporated into the soil
Green mulches of legumes, liquid dung	10-20	Fast	Much is incorporated
Stabilized humus	9-10	Slow	Slow incorporation

The tenth most important element (not shown in the table) is *titanium,* and it only accounts for 0.2% of the Earth's crust.

These differences in percentages based on volume are even greater when considered in terms of the number of ions and atoms. This difference is due to the fact that ions are different sizes. An ion is an atom or group of atoms with a positive or negative electric charge. Anions are ions with a negative charge, and cations are ions with a positive charge. A cation is formed when an atom loses one or more of its electrons, and thus acquires one or more positive charges. When this happens, the remaining electrons are attracted closer to the nucleus and the cation's size effectively diminishes. A cation is thus smaller than the atom it is derived from. However, when an atom gains one or more electrons and becomes an anion, its size increases, meaning that the resulting anion is larger than the original atom.

Oxygen is not only the most abundant element in the soil minerals, but it is also the only one that can form anions. Due to their large size and abundant numbers, oxygen anions occupy more than 90% of the volume of the Earth's crust. Thus most of the minerals and rocks in a soil can be considered to consist of a structure of oxygen atoms, with a variety of cations (mainly silicate and aluminium ions) filling the gaps. Minerals based on oxygen and silicon are thus the most abundant in the Earth's crust, and deserve a chapter to themselves. These minerals consisting of combined oxygen and silicon are called *silicates,* and those that also contain aluminium are known as *aluminosilicates.*

Element	Chemical simbol	Ions	Percentage of total number of atoms and ions
Oxygen	O	O^{2-}	60
Silicon	Si	Si^{4+}	20
Aluminium	Al	Al^{3+}	6
Hydrogen	H	H^+	3
Sodium	Na	Na^+	3
Calcium	Ca	Ca^{2+}	2
Iron	Fe	Fe^{2+} and Fe^{3+}	2
Magnesium	Mg	Mg^{2+}	2
Potassium	K	K^+	1

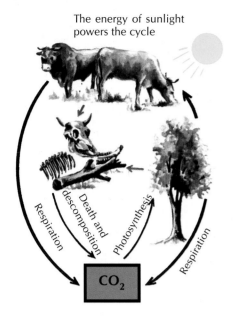

The energy of sunlight powers the cycle

two of these square-based pyramids joined together at their bases, with one tip pointing up and the other pointing down. The cube is the simplest possible arrangement.

• **Alteration**. This is the change in the rock's mineral and chemical composition due to the action of the atmosphere, underground water or hot springs. The alteration depends on the temperature of the water, the composition of the rock and its degree of fragmentation. The main chemical processes involved in alteration are hydrolysis, oxidation, hydration, dissolving away and carbonation.

• **Weathering**. Weathering is the alteration and later breakdown of rocks by physical or chemical processes, the most important of them being the action of water, though factors related to the weather, such as temperature and wind, are also important.

2.2.1. The classification of minerals

Minerals are classified on the basis of their chemical composition and their crystal structure. The following is a list of the types of mineral that exist, according to their chemical composition, and examples of each type of mineral:

• Native elements: native silver, gold, etc.
• Sulfides and sulfur minerals: pyrites and chalcopyrites, etc.
• Fe oxides and hydroxides: hematite, goethite, etc.
• Al oxides and hydroxides: gibbsite, etc.
• Halogen salts: halite (rock salt), carnallite, etc
• Ca carbonates: calcite, etc.
• Mg carbonates: dolomite, etc.
• Phosphates: apatite, etc.
• Sulfates: gypsum, epsomite, etc.
• Silicates: quartz, micas, clays, etc.
• Substances of organic origin: coal, oil, etc.

Almost all the minerals in the soil consist of silicates and carbonates. We could almost say that the solid matter of the soil consists entirely of silicates, carbonates and organic matter.
As a result, unless stated otherwise, we tend to be discussing silicates and carbonates, their different types, their structures, etc.
Of all the elements of the periodic table, the nine listed above account for 99% of all the components of the soil. These elements and their combinations are combined together to form minerals. Many different processes can cause minerals to combine and form rocks. In the soil we can find rocks consisting of an almost pure single element (nuggets of gold, for example), but rocks usually consist of combinations of different minerals.
The essential terms you need to understand rocks and minerals are:

• **Rocks**. Rocks are the materials forming the Earth's crust. Rocks are defined as an association of minerals. Rocks may also consist of fragments of other rocks, or petrified organic remains. In general they show some degree of statistical uniformity in terms of their hardness and consistency, but rocks can sometimes be soft and deformable, mobile or even liquid or gaseous.

• **Minerals**. Minerals are natural inorganic chemical compounds, and are almost always present as a crystalline solid. Primary minerals formed during the crystallization of the rock. Secondary minerals arose after the rock formed, and are generally derived from primary minerals.

• **Crystals**. Crystals are solids whose atoms (either ions or molecules) are ordered regularly following a basic distribution (unit cell) that is repeated again and again to form the mineral's crystal lattice. The three basic, and most important, structural units are *tetrahedral, octahedral* and *cubic*. A tetrahedron is a three-dimensional figure with four equal triangular faces, and can be represented as a pyramid with a triangular base. An octahedron can be thought of as

The table on the next page shows the minerals that make up most of the soil. Minerals can be classified by their stability in the environment, placing them on a scale ranging from those that are easily altered to those that are hardest to alter. Quartz, at the hard end of the list, is a very stable mineral and is barely altered or weathered. The two right hand columns show the elements these minerals consist of, divided into those with the greatest presence (main constituents) and those with the least presence (lesser constituents).

Nutrients elements and the percentages of them released when some minerals are altered

The table to the right lists the minerals that are sources of some nutrient elements, and the percentage released from each one on breakdown. The nutrients and the minerals that can supply them will be dealt with in more detail in the last chapter of this book, the one on fertilizers. (Though many fertilizers are now chemically synthesized, minerals have been applied to the earth for many centuries to correct specific deficiencies in agricultural soils.)

POTASSIUM RELEASED (K)	MAGNESIUM RELEASED (Mg)	CALCIUM INPUT (Ca)
Orthoclase 15-17 %	Dolomite 21 %	Feldspars
Microcline 20 %	Olivine 30-40 %	Gypsum
Moscovite 11 %	Pyroxenes 40 %	Calcite
Biotite 8-11 %	Biotite 17 %	Anphiboles
Ilite	Glauconite	

Main soil minerals and their component elements, arranged from easily altered to hard to alter

NAME OF MINERAL	CONSTITUTION	
	GREATER	LESSER
Olivine	Si, Mg, Fe	Mn, Zn, Cu, No, Ni, Co
Garnet	Si, Ca, Mg, Fe, Al	Mn
Augite	Si, Ca, Mg, Fe, Al	Mn, Zn, Cu, Ni, Co, V
Hornblende	Si, Ca, Mg, Fe, Al	Mn, Zn, Cu, Ni, Co, V
Biotite	Si, K, Mg, Fe, Al	Mn, Zn, Cu, Ni, Co, V
Apatite	P, Ca, O	(F)
Anorthite	Si, Ca, Al	Cu, Mn
Andesite	Si, Na, Ca, Al	Cu, Mn
Oligoclase	Si, Na, Ca, Al	Cu
Albite	Si, Na, Al	Cu
Orthoclase	Si, K, Al	Cu
Ilmenite	Fe, Ti	Co, Ni, V
Magnetite	Fe	Zn, Co, Ni, V
Tourmaline	Si, Ca, Mg, Fe, B, Al	
Zircon	Si, Zr	
Quartz	Si	

2.2.1.1. Silicate minerals

These minerals are based on silicon and share one thing in common, they are all based on the tetrahedral structure of silica $(SiO_4)^{4-}$, four oxygen atoms at the four corners of a regular tetrahedron, reflecting the alignment of the covalent bonds between the silicon and the oxygen. The silicon atom occupies the center of the tetrahedron (covalent radius of silicate = 0.042 nm and of oxygen = 0.135 nm). The oxygen atoms are shared by consecutive tetrahedra, meaning that the formula can be expressed as SiO_2.

On the basis of the distribution and arrangement of the tetrahedra, we can classify silicates into the following types:

Classification of silicates on the basis of the type of aggregation of their tetrahedra

Class	Tetrahedra grouping	Si:O ratio	External charge per Si tetrahedron	Example
Nesosilicates	Independent	1:4	−4	Olivine
Sorosilicates	Pairs	2:7	−3	Hemimorphite
Cyclosilicates	Rings	1:3	−2	Beryl
Inosilicates	Single chains	1:3	−2	Augite
	Double chains	4:11	−1,5	Hornblende
Phyllosilicates	Sheets or layers	2:5	−1	Mica and clay minerals
Tectosilicates	Three-dimensional structures	1:2	0	Quartz and feldspars

• **Nesosilicates.** These are isolated tetrahedra joined by cations that act as bridges between the tetrahedra. This type of silicate mineral includes olivine, garnet, andalusite, etc.
• **Sorosilicates.** The tetrahedra are in pairs joined to each other by a shared oxygen atom, and have cations that compensate the charges of the oxygen

atoms that are not shared, and the pairs are joined to other pairs by cations.
• **Cyclosilicates.** The tetrahedra are in groups of three, or four but usually six, and are arranged in flat rings. Tourmaline is an example of a cyclosilicate.
• **Inosilicates.** These are chains of tetrahedra that share one or two of their corners. They can be simple chains (pyroxenes such as augite) or double chains (amphiboles such as hornblende).
• **Tectosilicates.** These are minerals consisting of a three-dimensional grouping of tetrahedra where all the corners are shared by adjacent tetrahedra. Logically, their three-dimensional structure makes these silicates more resistant to weathering; the best known tectosilicates are quartz (SiO_2) and the feldspars, such as orthoclase $(KAlSi_3O_8)$ and plagioclase $(CaAl_2Si_2O_8)$.
• **Phyllosilicates.** These are minerals with a structure in the form of layers of tetrahedra that share the three oxygen atoms in the same plane, and their common empirical formula is $(Si_2O_5)^{2-}$. Examples of phyllosilicates include white mica or muscovite, biotite and the **clay minerals**.

The clay minerals, or the clay mineral group, is the name for the phyllosilicates whose physical and chemical properties in the retention of nutrients are so important that, together with humus, they are largely responsible for a soil's fertility.
The phyllosilicates, or layer silicates, are especially important in the soil because they include the silicated clay minerals and the micas. Both these groups are important in the chemistry and fertility of the soil. Though they should have been described before the tectosilicates, because their crystalline structure is simpler, we have kept them till last because they are so important for agriculture, the reason why they are also treated at greater length.

Phyllosilicates consist of layers that have very strong internal bonds, but are only weakly joined to other planes. These sheets are only three or four oxygen atoms thick, and explain mica's well-known property of breaking into extremely thin sheets. The internal sheets of phyllosilicates can be thought of as a layer cake of tetrahedra and octahedra. In some phyllosilicates, the 1:1 silicates, these layers alternate, while in the 2:1 silicates each unit layer consists of tetrahedral layers above and below a octahedral layer. The octahedral layers may contain trivalent cations, such as Al^{3+}, or divalent ions, such as Mg^{2+} and Fe^{2+}. The octahedral layers formed with hydroxyl anions (OH^-) instead of oxygen, form stable minerals such as gibbsite ($Al_2(OH)_6$), and brucite ($Mg_3(OH)_6$).

To finish, look at the accompanying table listing the different types of clay minerals, belonging to the phyllosilicates, divided by the type of layers that alternate, either 1:1 (tetrahedron, octahedron) or 1:2 (tetrahedron, octahedron, tetrahedron).

2.2.1.2. Nonsilicate minerals

Only a few nonsilicate minerals are important in the soil. The rest are only present in small amounts, or are too soluble to remain long in most soils. Some are too inert to be of functional importance. Only calcite and some very stable oxides need to be considered. **Calcite** is the most abundant form of calcium carbonate present in the soil. Calcium carbonate is only slightly soluble in water, and the depth to which the $CaCO_3$ has been washed in the soil can be used as an indicator of how effective washing is and the thickness of the *solum*. Thus, for example, in arid soils, the low rainfall does not wash the calcium carbonate beyond the solum and gives rise to sub-surface carbonated horizons. Carbonate ions are triangular in structure, with three oxygen atoms surrounding a central carbon atom (the carbon atom is smaller than the silicon atom). Equal numbers of Ca^{2+} and CO_3^{2-} are packed together to form calcite and other forms of calcium carbonate.

Gibbsite is an aluminium-rich mineral that accumulates in weathered soils. It consists of layers of Al^{3+} in octahedral coordination between two layers of OH^- anions. The adjacent layers show little attraction to each other; as a result, gibbsite crystals tend to be flaky and are frequently too small for us to see that they are crystalline.

The chemical formula of gibbsite is $Al(OH)_3$ and is usually present with two related aluminium minerals, boehmite, $AlO(OH)$, and diaspore, $HAlO_2$. These minerals are considered to be hydroxides, or are simply grouped together with the oxides, such as hematite, anatase, etc. The **ore** of aluminium that smells of earth is called **bauxite** and consists of these three minerals.

Hematite (Fe_2O_3) is an oxide of ferric (3^+) iron. The Fe^{3+} ions occupy octahedral spaces between closely packed oxygen ions. This close packing and its high iron content mean that hematite has a density (5.26 g/cm³) twice that of most soil minerals. Yet the density of hematite particles in the soil never reaches this level, partly due to the presence of silicates and aluminium minerals, but also because much of the iron is in the hydrated form with large amounts of water. One iron mineral that contains hydrogen is **goethite**, FeO(OH).

Hematite is the most frequent cause of a reddish color in soil. Goethite and similar compounds are known by the inclusive name of **limonite**, and are frequently responsible for yellowish-brown soil colors. They and the aluminium minerals occur in almost all soils in small quantities. Both groups are formed by weathering processes and are highly insoluble. They tend to accumulate in highly weathered soils, and are particularly abundant in many tropical environments. Veins of iron are probably the end results of the weathering, at some time in the past, of iron rich rocks.

Types of phyllosilicate clay minerals, divided by the type of alternation of layers

TYPE	GROUP	SUBGROUP	FORM
1:1 te:oc	Kaolinite-Serpentine	Dioctahedral	Kaolinite
1:2 te:oc:te	Smectites	Dioctahedral	Montmorillonite
	Vermiculites	Dioctahedral	Dioctahedral
		Trioctahedral	Trioctahedral
	Micas and Illites	Dioctahedral	Muscovite
		Trioctahedral	Biotite
	Chlorites	Trioctahedral	Chlorite

Anatase is a form of titanium oxide, TiO_2. Titanium occupies a site within the octahedron of oxygen, forming a very stable structure. Anatase particles are usually very small, but they are even more resistant to weathering than oxides of iron, aluminium and silicon. Anatase is sometimes used as a mineral of reference to estimate the quantity of minerals that a soil has lost through weathering.

In their fine clay fraction, all soils contain some material that has no identifiable structure. These materials are known as **allophane**. Its constituents are too small to be identified specifically, but they probably include the blocks that intervene in the construction of the minerals; tetrahedra containing silicon, and octahedra housing aluminium, magnesium and iron.

Allophane is dominant in some young soils, formed on volcanic ashes. This is probably because the liquid ejected from the volcano into the atmosphere cools instantaneously, and does not have enough time to form mineral crystals large enough to be identified. Soils formed from these materials are rich in fine clay, and in general they are very fertile.

An **ore** is a mineral used as a raw material to extract a metal. The term is used for the minerals that can be commercially exploited. An exploitable mineral contains the ore, which is the useful part, and the earth it was in, the non-usable part.

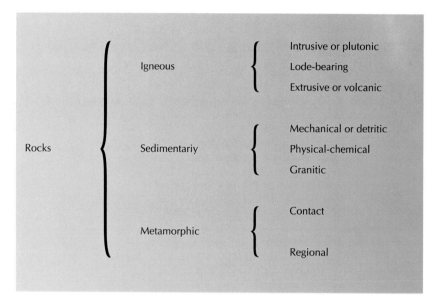

Classification of rocks by how they formed

2.2.2. The classification of rocks

Rocks are classified by they way they were formed. The diagram above shows a classification of rocks by their origins. Rocks can be divided into three main groups by their origins, igneous, sedimentary and metamorphic rocks.

Silicon-containing rocks used to be called *acid rocks*, and those containing little silicon were called *basic rocks*. These names have nothing to do with their pH, but many books use these terms.

2.2.2.1. Igneous rocks

Igneous rocks form when volcanic **magma** solidifies as it cools. Magma is a mixture of rocky material in a liquid state, due to the high temperatures within the Earth, below its crust. Magma may also contain solid materials, as well as crystals and rock fragments, and even gases.

Phosphorus cycle

These rocks are mainly of the silicate type. They are usually divided into **intrusive** (or **plutonic**) rocks, which are formed at great depth, and **extrusive** (or **volcanic**) rocks, which solidify when they are ejected from within the Earth's crust by volcanic eruptions. The main component of igneous rocks is quartz (SiO_2), and this gives rise to a preliminary classification of these rocks:

• Acid rocks. More than 66% quartz (eg. granite)
• Intermediary rocks. From 66% to 52% quartz (eg. andesite)
• Basic rocks. Between 52% and 45% quartz (eg. gabbro)
• Ultra-basic rocks. Less than 45% quartz (eg. peridotite)

The main intrusive rocks, classified by whether they are acid or basic (following the old terminology)

• **Granite**. Acid composition, consisting of quartz, potassium feldspar and to a lesser extent plagioclase and biotite or muscovite.

• **Granodiorite**. Acid composition, consisting mainly of quartz, plagioclase and to a lesser extent,

potassium feldspar, biotite and often hornblende.

• **Tonalite**. Acid composition, consisting mainly of plagioclase, quartz, biotite and hornblende.

• **Diorite**. Intermediary composition, consisting mainly of alkaline feldspar (usually accounting for 50-75% of the rock's mass), and iron and magnesium minerals, normally hornblende and biotite.

• **Syenite**. Intermediate composition, consisting of alkaline feldspar (orthoclase and/or albite), subordinate plagioclase and some minerals largely consisting of iron and magnesium (biotite, hornblende and occasionally augite), that may sometimes contain quartz and sometimes feldspathoids.

• **Gabbro**. Basic composition, consisting of plagioclase, pyroxenes and more often olivine.

• **Peridotite**. Ultrabasic, consisting mainly of olivine.

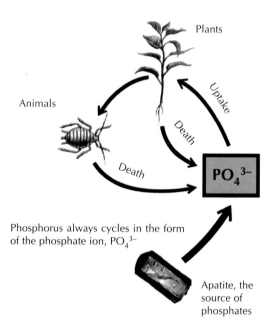

Phosphorus always cycles in the form of the phosphate ion, PO_4^{3-}

Apatite, the source of phosphates

The main extrusive (volcanic) rocks, classified on the basis of their acid or basic nature are:

• **Rhyolite**. Acid composition, consisting of quartz, alkaline feldspar and/or sodium plagioclase, and to a lesser extent, minerals dominated by iron and magnesium, such as biotite. It can be considered the volcanic equivalent of granite.

• **Dacite**. Acid composition, consisting basically of plagioclase, quartz, hornblende, biotite and occasionally pyroxene. It is the volcanic equivalent of granodiorite and tonalite.

• **Andesite**. Intermediate in composition, consisting mainly of plagioclase and minerals dominated by iron and magnesium (pyroxenes, hornblende and, less commonly, biotite). It is the volcanic equivalent of diorite.

Aragonite (macle)

• **Trachyte**. Intermediate in composition, consisting mainly of alkaline feldspar with some iron-magnesium minerals (pyroxene, biotite or hornblende). It is the volcanic equivalent of syenite.

• **Basalt**. Basic in composition, consisting of plagioclases and pyroxenes, and in some cases, olivine. Basalt is the most abundant volcanic rock. It is the volcanic equivalent of gabbro.

2.2.2.2. Sedimentary rocks

Sedimentary rocks were formed by the physical, chemical and biological processes that take place on the Earth's surface. They consist of particles derived from pre-existing rocks that were deposited after transport by an erosive agent (wind, ice, seas, currents or rivers). They are classified on the basis of the particle size of the original grains, or by their chemical composition.

The table below classifies sedimentary rocks by their particle size, with the name of the particle of origin, its size, and the name of the sediment. We can thus divide the sedimentary rocks into: **conglomerates** and **breccia**; **sandstones**; **limonites**; and **argillites**, **marls** and **slates**. Sandstones can be subdivided on the basis of the nature of the grains of sand.

Calcarenite, **gypsarenite**, **quartzarenite** and **arkose** are the sandstones whose generic sand contains more than 25% feldspar, **litarenite** has sand grains with more than 75% quartz and is dominated by fragments of stone, and **greywackes** are sandstones with a high percentage of silt or clay between the sand grains. Marls, unlike argillite and slates, contain 35-65% $CaCO_3$.

Sedimentary rocks can be divided into three groups on the basis of their composition; carbonate minerals, evaporites and carbonaceous rocks. Let's take a quick look at them.

Carbonated sedimentary rocks have been enriched in carbonates or magnesites by some process that occurred after their formation:

• **Limestone**. A sedimentary rock whose mineral composition is based on calcite and aragonite, two mineral forms of $CaCO_3$. Aragonite turns into calcite on heating to 400°C. The rock can contain up to 10% mineral dolomite ($MgCO_3$).
• **Dolomitic limestone**. Sedimentary rock whose mineral composition is based on calcite and aragonite. The rock can contain up to 50% mineral dolomite ($MgCO_3$).

Natural rhombohedron of calcite

Dolomite on magnesite

Limonite (a pseudomorph of pyrites)

• **Calcareous dolomite**. Sedimentary rock whose mineral composition is based on dolomite. It can contain up to 50% calcite.
• **Dolomite**. Sedimentary rock whose mineral composition is based on dolomite. It may contain up to 10% calcite.

Classification of the sedimentary rocks by particle size

Name of the particle	Particle size in mm	Name of sediment	Name of the rock
Rudite	more than 2	Gravel	Conglomerates Breccias
Arenite	from 2 to 1/16	Sand	Sandstones
Lutite	from 1/16 to 1/256	Silt	Limonites
Argillite	less than 1/256	Clay	Argillites marls and slates

A list of minerals going from the most easily weathered to the least. Those with bonds in three dimension, such as quartz, are much harder to degrade than those that dissolve in water, such as gypsum and calcite. Taken from Jackson & Sherman (1953)

Evaporitic sedimentary rocks form due to the precipitation of mineral salts by evaporation. The most important ones (with their chemical formula) are: **gypsum** ($CaSO_4 * 2H_2O$), **anhydrite** ($CaSO_4$), **halite** (NaCl), **carnallite** (KCl) and **epsomite** ($MgSO_4 * 7H_2O$).

Carbonaceous sedimentary rocks were formed by a process in which the plant organic matter was turned into stone. This accumulation of plant matter took place in basins on land and in the sea; coal formation took place after sedimentation and consisted mainly of enrichment with carbon and loss of oxygen. This group includes **peat**, **lignite**, **coal** and **anthracite**.

1. Gypsum	$CaSO_4 \cdot 2H_2O$ and other more soluble salts
2. Calcite	$CaCO_3$
3. Hornblende	$NaCa_2(Mg, Fe, Al)_5(Si, Al)_8O_{22}(OH)_2$
4. Biotite mica	$K(Mg, Fe)_3(AlSi_3)O_{10}(OH)_2$
5. Feldspars	
Plagioclase	$CaAl_2Si_2O_8$ -$NaAlSi_3O_8$
Orthoclase	$KALSi_3O_8$
6. Quartz	SiO_2
7. Muscovite	$KAL_2(AlSi_3)O_{10}(OH)_2$
8. Clay mica intermediates	
(including illite, vermiculite and chlorite)	
9. Montmorillonite	$(Mg, Al)_2Si_4O_{10}(OH)_2 \cdot H_2O$
10. Kaolinite	$Al_4Si_4O_{10}(OH)_8$
11. Gibbsite	$Al(OH)_3$
12. Hematite	Fe_2O_3
13. Anatase	TiO_2

2.2.2.3. Metamorphic rocks

Metamorphic rocks have undergone some type of transformation in their structure, mineral composition and usually in their chemical

Many companies sell, wholesale and retail, grains of different particles size to be mixed with other substrates.

compositions due to the action of external agents. The most important modifying agents are pressure and temperature. The pressure can be of two different types, **static pressure** due to the weight of the overlying column of rock, which exerts a pressure of approximately 300 bars per kilometer of depth, and the pressure due to tectonic forces, called **stress**. The lowest temperature at which metamorphism occurs is about 200°C (the temperature at which clays lose their water) and the upper limit is 800°C to 1,000°C. This variation in temperature is due to the geothermal gradient, that is to say, the heat produced mechanically in tectonic deformations or by magmatic intrusions, which can lead to large temperature increases in the rocks receiving the magma.

The time factor is also important in metamorphic activity, as how much the rock is modified depends to a large extent on how long the external forces act upon it. In general, the longer the time, the greater the mineralogical and structural adaptation to the new conditions determined by external agents.

Metamorphic rocks can be classified by their mineralogical and chemical composition, by the type and degree of the metamorphism they have undergone, by the type of rock they are derived from, etc. The main metamorphic rocks are:

• **Slates**. Phyllites, schists and paragneisses, in order of greater metamorphism and particle size, are all derived from lutitic rocks.

• **Quartzites**. Derived from quartzoarenite (sandstone). Quartz is the most abundant of all minerals.

• **Marbles**. These are metamorphic rocks derived from calcareous sediments. Calcite is the most frequent mineral.

• **Gneiss**. A rock generally formed from intrusive igneous rocks. It consists mainly of quartz and feldspar. Its grain size is large and its foliation is irregular.

2.2.3. Weathering

Weathering is the overall term for a series of mechanical and chemical processes that break down the rock, its minerals, and thus, the soils. Almost all these processes are regulated by the climate. Rainfall and temperature are key factors and, together with the hardness and chemical composition of the rocks, minerals and soil, they determine how the weathering process develops. The passing of time is accompanied by a progressive change in the soil's mineral composition. Some differences in the rate of weathering can be easily explained; calcite and gypsum simply dissolve away, without needing any sort of structural change. The silicon-oxygen bonds of hornblende lie in just one dimension, those of biotite in two dimensions and feldspars and quartz have bonds in three dimensions. The strongest bonds in the silicates are those between

silicon and oxygen, and they are thus the ones that make minerals most resistant to weathering. A list of minerals ordered from the most easily weathered to the most resistant to weathering is shown in the accompanying table.
Weathering includes chemical processes that are very complex to describe, not our purpose here, but it is worth distinguishing between two types of weathering, mechanical weathering and chemical weathering.

2.2.3.1. Mechanical weathering

In cold climates, the action of ice is the most common form of weathering. Changes in temperature mean that the low moisture content of the rocks and soils alternately freezes solid and melts. The action of ice will only occur where total temperatures vary above or below zero

Mechanical weathering; in rocky areas, subject to large temperature variations, the action of mechanical weathering produces screes, large areas consisting largely of small pebbles and other small stones, the result of the ice breaking the rocks.

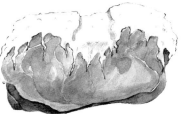

Water filters into the rocks through pores and cracks.

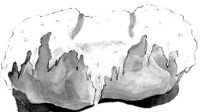

When it freezes, water occupies a volume 9% greater.

Thus, when the water freezes it can break or even split the parent material and other rocks.

degrees and where there is liquid water.
The pressure of the ice crystals that form may shatter rock outcrops and the parent material. Other types of crystal can also have the same effect. All surface and underground waters contain dissolved salts derived from the leaching of minerals and soils. When the moisture evaporates, salt crystals grow and may split porous rocks apart. This destructive action of salt occurs especially in sandstones and in desert regions where evaporation is exceptionally high. Even in the most arid desert conditions, rocks contain enough moisture for these cracks to form.

Extreme temperature changes between the day and night can also break down rocks by a continued process of contraction and expansion. This phenomenon is common in the desert, where the difference between daytime and nighttime temperatures is extreme.
Weathering can also be caused by living organisms, such as the roots of growing plants.

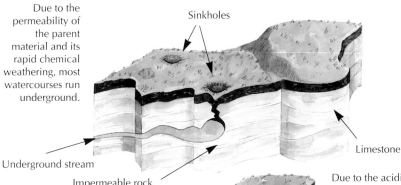

Due to the permeability of the parent material and its rapid chemical weathering, most watercourses run underground.

Sinkholes

Underground stream

Impermeable rock

Limestone

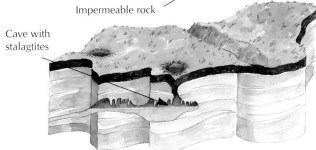

Cave with stalagtites

Due to the acidity of the water, the calcareous caverns get larger and larger. Calcareous water falling drop by drop forms stalactites, hanging down from the roof, while stalagmites form from the ground upwards.

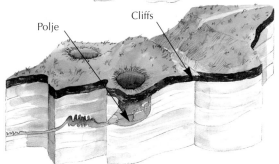

Polje

Cliffs

The sinkholes get bigger and bigger. The system of caves collapses, forming bowl-shaped poljes. Karstic landscapes may also have cliffs and exposed limestones potholes.

2.2.3.2. Chemical weathering

Chemical weathering mainly occurs when weak acids in surface or underground water dissolve different minerals. When carbon dioxide dissolves in water, a part combines with the water to form carbonic acid, H_2CO_3. This acid dissolves the calcium carbonate of limestone, and the flow of water through this permeable rock may give rise to caves and tunnels. When the calcareous water evaporates, the calcium carbonate is precipitated in limestone formations. In regions with limestone parent material and high rainfall, this type of weathering is associated with **karstic landscapes** (named after the former province of Karst, on the Italian-Slovenian frontier), with potholes, sinkholes, **poljes** and caves. Slovenia's magnificent karstic caverns are famous for their size and for the beauty of their stalactites and stalagmites.

Polje is an internationally used Slovenian word, used to define a closed depression of karstic origin.

Some views of the Slovene region of Karst, showing excellent calcareous formations in the form of stalactites and stalagmites

When talking about the weathering of minerals and the parent material, we must not forget the action of the organic acids produced by humus. These acids turn feldspar and mica into hydroxides of iron and aluminium, and into hydrated aluminium silicates, clay minerals that form kaolin, and bauxite. All that is left of granite is a few grains of quartz. Weathered granite is usually known as **granitic sand** and it is widely used in gardening.

2.2.4. Particle size and texture

So far, we have looked at the two main solid components of the soil, organic matter and the minerals and rocks. In this section, we are going to look at weathered minerals and rocks. The minerals and weathered rocks become the particles that make up the soil. The form and size of these particles (their granulometry), the percentage of each one of them (and thus the texture) and the different types of aggregates (the structure) they form with each other, all greatly affect the soil's characteristics.

2.2.4.1. Particle size

When classifying the particles of a soil we must think back to the beginning of the chapter on soils (sampling and profiles), which briefly explained the types of extraction taken for later analysis. When we have a sample of soil, we can determine the percentage corresponding to each type of particle, and then define the texture we are dealing with.
The total weight of the sample must be divided by its **particle size**, that is to say, determining the percentages of each size of soil particle.
By an initial sieving with a sieve with holes greater than 2 mm, we can separate the coarse fraction. Below is a table showing the large elements by particle size, as well as the diameters of the fine soil, whose particles determine the soil's texture.

Division by particle size. Size intervals for the different soil particles

SOIL FRACTION	SUBFRACTIONS		SIZE IN MILLIMETERS
Coarse elements		Blocks	> 200 mm
		Stones	20-200 mm
		Gravel	2-20 mm
Fine earth	Sand	Very coarse sand	2 - 0.1 mm
		Coarse sand	1-0.5 mm
		Average sand	0.5-0.25 mm
		Fine sand	0.25-0.10 mm
		Very fine sand	0.10-0.05 mm
		Loam	0.05-0.002 mm
		Clay	< 0.002 mm

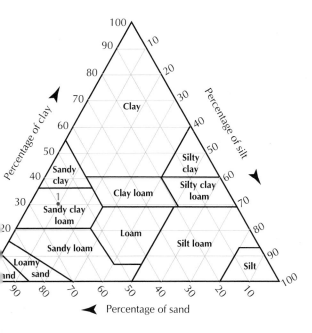

2.2.4.3. Interpreting soil texture

Every particle present makes its contribution to the nature of the soil as a whole. Clays and organic matter are important because they can store nutrients and water. The smallest particles may also help other larger particles to join together to form aggregates. The largest particles (usually sand) form the soil's basic structure. They account for most of its weight, and help to ensure good aeration and permeability. Soils with large amounts of coarse sand can usually withstand heavy weights without becoming compacted.

2.2.4.2. Soil class

With the rest of the sample (particles smaller than 2 mm), the soil's class can be determined in a laboratory. Analysis of the **class** of a soil sample is expensive because it takes a long time, and has to be carried out in a specialized laboratory. Soil consists of particles of three different size categories, **sand**, **silt** and **clay**, and its class depends on the amount of each fraction present. Analysis of the class requires finding out how much of the total sample weight consists of sand (between 2 mm and 0.05 mm), silt (0.05 mm and 0.002 mm) and clay (less than 0.002 mm). Once the percentages of sand, silt and clay have been measured in the laboratory, we can locate the values on the **soil-class triangle**. This triangle is divided into twelve areas containing all the possible proportions of sand, silt and clay. The numbers on the three scales are at an angle to indicate which of the faint blue lines they correspond to. Thus in the soil-class triangle above, the intersection of lines labelled 1 (in red), within the sandy clay loam fraction, represents 30% clay (following the horizontal line), 10% silt (following the line parallel to the left side of the triangle) and 60% sand (following the line parallel to the right hand side of the triangle). So follow the lines on the diagram to show yourself that the following mixtures correspond to the names given:

60% sand, 25% silt and 15% clay = sandy loam
25% sand, 45% silt and 30% clay = clay loam
28% sand, 54% silt and 18% clay = silt loam

As the reader will have found out when referring to the triangle of textures, or soil-class triangle, a new term has been introduced, **loam**. Loamy soil, soil with a loamy texture, consists of a balanced mix of sand, silt and clay, with the properties of all three fractions. Soils with a loamy texture are the most

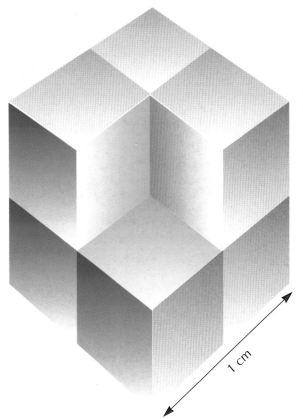

0.2 cm

1 cm

The surface area of a gram of soil is inversely proportional to the diameter of its particles. When a particle is divided into smaller particles, the total surface area increases. If we divide a block with sides 1 cm long, and a surface area of 6 cm², into small blocks with sides 0.2 cm long, we get 125 blocks with a total surface area of 30 cm². If we cut the same block into even smaller particles, with sides 0.001 cm long, we get a total surface area of 6,000 cm².

appreciated by farmers, as they have the good aeration and open texture of sandy soils, the good nutrient retention of clays and the good water retention of silty clay soils.

Sandy soils are generally highly permeable to air, water and roots, but have two major limitations. The first is their low water retention capacity, and the second is their insufficient ability to store nutrients. To achieve high levels of production, frequent additions of water and nutrients are required. The presence of a high percentage of organic matter would help to compensate the deficiency of clay, but most sandy soils are very poor in organic matter. Of course, the limitations of these sandy soils can be remedied if one has access to fertilizers and irrigation water, but they are expensive. If excess water and fertilizer are applied, the fertilizer may be lost by washing.

A/ Soil profiles are often exposed during the construction of roads and highways.
B/ In uniform soils, the texture applies to the entire profile.

Sandy soil's limited ability to retain water and nutrients is related to the low total surface area of its particles as a whole. The smaller the diameter of its particles, the greater the total surface area of a gram of soil.

Clay particles not only have a large surface area, but are also electrically charged, and this means that clays can retain nutrients on their surfaces. Sands, however, lack this property. So nutrient loss by washing from clay soils is very small compared to losses if the same quantity of nutrients was present in a sandy soil.

Clays retain much more water than sands, essentially because they have a much larger surface area that can retain a film of water. A quantity of water that would cause heavy washing in a sandy soil may not even wet a clay soil far enough down to cause any washing at all. The dissolved nutrients are considered to be lost from the soil when the water reaches beyond the volume exploited by the roots and drains away.

Soils that contain too much clay have a high capacity to retain water, but usually show inadequate aeration. Surprisingly, a high organic matter content helps to improve both waterlogging in clay soils and water shortage in sandy soils. The organic matter helps the clay particles to stick to each other, forming aggregates between which there is space for the air.

Loams and silt loams are highly desirable for most purposes. They have enough clay to retain adequate levels of water and nutrients to ensure excellent plant growth, but not so much as to

Ⓐ

Ⓑ

present aeration problems or problems during cultivation operations. They contain sufficient loam to gradually form more clay (to replace what is lost by eluviation and erosion) and to liberate nutrients when it is weathered. A soil containing about 7% to 27% clay and roughly similar quantities of silt and sand, has a loam texture. Loam soils, with several types of organic matter, are very good for most purposes.

Unless stated otherwise, when we say a soil has this texture or that texture, we are referring to its topsoil, its A horizon. Of course, in uniform soils, the texture refers to the entire profile, but in most cases there are variations in texture with increasing depth. Many soils show sufficient differences for the horizons to be placed in different soil classes. These variations in texture may be very important, especially when the B horizon contains a lot of clay, as this greatly reduces penetration by water, air and roots. In general, any abrupt change in texture causes a delay in the movement of water. This can be very favourable, if for example it occurs at a reasonable depth in the profile, so that it favors the retention of additional water for a relatively long time, meaning it can be used by plants in the dry season.

C/ Some soil profiles consist entirely of the parent material.
D/ Some surface ("A") horizons hardly contain enough soil for plants to grow.
E/ In most cases, texture varies with depth.

Ⓒ

Ⓓ

Ⓔ

2.3 ORGANIZATION OF THE SOLID COMPONENTS

We have defined all the solid components of the soil, the organic matter in different stages of decomposition, rocks, minerals and all their different stages of weathering. Now we need to see how the solid components bind together to make up the soil.

2.3.1. Structure

The **structure** is defined as the arrangement of the individual particles of the soil. The smallest particles, such as clays and humus, tend to join together to form aggregates or structural units, consisting of natural aggregations of primary particles (sand, silt and clay), that occur in the soil and persist. Their natural origin and their persistence distinguishes them from clods, aggregates formed by disturbances such as plowing.

2.3.1.1. Classification

Aggregates occur in several forms. By size, they can be coarse, medium or fine, and their degree of development may be weak, moderate or strong. The broadest general classification of aggregate is by their geometry. Aggregates may be granular,

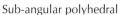

Granular

Laminar

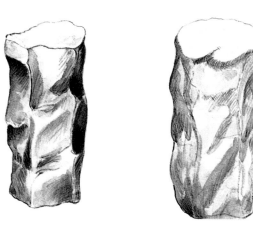

Sub-angular polyhedral   Angular polyhedral

Prismatic   Columnar

laminated, angular polyhedral and sub-angular polyhedral, prismatic or columnar.

Careful observation is required to detect weak aggregates, but strong aggregates are clearly visible and can easily be told apart. Some soils, especially those poor in clay, totally lack aggregates, as do very sandy soils lacking humus. The presence of aggregates is very important in clay soils, as they improve its permeability. Air, water and roots move far more quickly between aggregates than through them, and a structured soil is much more permeable than an unstructured soil.

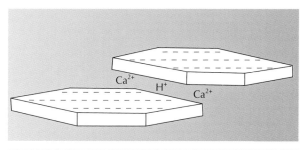

Diagram of two electrically charged clay sheets joining together because of the cations between them

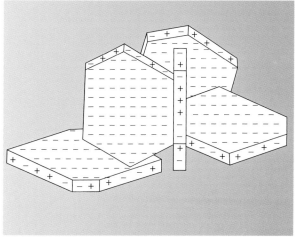

Several clay sheets after joining together in the way shown in the diagram above.

2.3.1.2. Formation

Aggregates may form in two ways. The first is that, because the clay particles have negative electric charges that attract positively charged ions (cations, such as Ca^{2+}, H^+, Fe^{3+},) to which further sheets of clay then bind. The second way is that the presence of organic matter (humus) in the soil, consisting of long chains of atoms with many side branches and different types of functional groups, which may be ionized, can give rise to positively or negatively charged sites. The stability of aggregates has more to do with the runoff of rainwater and soil erosion than with crop yields. Thus, on soils dominated by clays, conserving the soil structure is of great importance. If we do some task with heavy machinery on a waterlogged soil, we compact it and partly destroy its structure. Partial destruction of the soil structure causes its destabilization, and this leads to later problems. When soil structure is non-existent, soils get waterlogged easily (puddled soil), and then, when they dry out, they form an impenetrable crust.

Types of soil structure

2.3.2 Soil density

Its **density** is an important property of any material, but in the case of soils, the density is extremely important. A soil's density gives us an idea of the amount of micropores and macropores it contains. Soil porosity is of great importance in agriculture because it determines the amount of water and air that can accumulate in the soil for plants to use. Soil has an apparent density and a real density, its bulk density and its particle density.

2.3.2.1. The bulk density

The **bulk density** (Bd), or apparent density, is the weight of the soil solids per unit of the total soil volume. The Bd data must be expressed in units of weight and volume, usually in grams per cubic centimeter (g/cm^3). It is measured in two stages, in the field and in the laboratory.
Its formula is:

$$Bd = \frac{\text{Sample weight in g}}{\text{Unaltered sample volume in } cm^3}$$

where: the unaltered sample volume is determined in the field by extracting a portion of soil using a metal cylinder of a known volume (usually 100 cm^3), taking care not to alter the soil structure. The soil removed must then be dried in a laboratory (to drive off the water) at 105°C and weighed, giving the dry weight in grams.
The bulk density of the A horizons in mineral soils may vary between 1 and 1.6 g/cm^3 (in organic soils,

The relationship between the bulk density, the percentage of solids and percentage of pore volume in soils with a real density of 2.65 g/cm^3

Bulk density		% of volume solids	% of volume pores
g/cm^3	kg/m^3		
1.0	1,000	38	62
1.1	1,100	42	58
1.2	1,200	45	55
1.3	1,300	49	51
1.4	1,400	53	47
1.5	1,500	57	43
1.6	1,600	60	40

it is lower and may be as low as 0.1 g/cm^3 in sphagnum peat). Most of the variations are due to differences in the total volume of pores. In general, fine-textured soils show greater porosity and a lower apparent density than sandy soils. Of course the apparent density of a soil varies with its degree of compaction. Compressing the soil causes its pore volume to decrease, increasing its weight per unit volume. Overloading a soil tends to compact the lower (B) horizons, giving them a greater bulk density than the topmost (A) horizons.
The presence of organic matter in a soil lowers its bulk density. This is because its density is lower than that of the soil and because, when it forms aggregates, it maintains the soil's structure and

The bulk density of an artificial substrate is very important; it determines their structure, and their structures determines their use.

porosity, and this also lowers its bulk density.
To show how to calculate the bulk density and to see how useful it is in practice, let's look at these two examples.

1) We wish to remove the earth to a depth of one meter from a 1 ha site covered by a homogeneous soil with an apparent density of 1.3 g/cm^3. What weight of soil has to be removed?
Let's look at this as an example of using the metric system
A hectare (1 ha) is 100 m x 100 m = 10,000 m^2 (10^2 m x 10^2 m = 10^4 m^2).
To find the volume, multiply the area by the height; 10,000 m^2 x 1 m = 10,000 m^3 (10^4 m^2 x 1 m = 10^4 m^3).
To find the weight, multiply the volume by the density.
(As a guideline: the density of water is 1 g/cm^3, and this corresponds to 1 kg per liter, and 1 metric ton per cubic meter.)
The total weight is the volume multiplied by the density = 10,000 m^3 x 1.3 t/m^3 = 13,000 tons.
Or, a cubic meter is 10^2 x 10^2 x 10^2 cm = 10^6 cm^3, and so 10^4 m^3 = 10^{10} cm^3. A kilo is equal to a thousand grams (10^3) and a ton is a thousand kilos or a million (10^6) grams, so the the full calculation is 10^{10} cm^3 x 1.3gcm^{-3} = 1.3 x 10^{10} g = 1.3 x 10^4 T.
2) What weight in kg of organic substrate, whose density is 0.9 g/cm^3, are we going to need to fill 24 circular containers of diameter 19 cm and 10 cm tall (accepting $\pi = 3.14$)?

VOLUME OF CONTAINER
$3.14 *(19/2)^2*10 = 2{,}833.8\ cm^3$

WEIGHT OF SUBSTRATE PER CONTAINER
$2{,}833.8\ cm^3*0.9\ g/cm^3 = 2550.4\ g$

TOTAL WEIGHT, IN KG, FOR 24 CONTAINERS
$2550.4/1000*24 = 61.2\ kg$

Ⓐ

Ⓒ

A/ Fine structure (0-10 mm, without fibers); recommended for seed beds and rooting cuttings of ornamental plants, and ideal for crops requiring acid conditions.

B/ Medium structure (0-21 mm, with fibers); for pricking out all sorts of plants and cultivation in their final pot.

C/ Coarse structure (0-35 mm, with fibers); recommended for long-lived plants in large containers.

2.3.2.2. The particle density

The particle density (Pd) is the weight of the solids in the soil per unit volume of soil after discounting the volume of the pores. So, unlike the bulk density, the particle density of the soil is calculated from a volume that excludes the volume occupied by the macropores and micropores. The volume of the sample without the pores will always be smaller than with them, and so the value of its particle density will always be greater than its bulk density. The values of the Pd are also expressed in units of weight and volume, usually in grams per cubic centimeter (g/cm^3).

The reader may find that other books on soils use different terms to express the same ideas. Terms like "**specific weight**" and "**volumetric weight**" were formerly used for the real density, and though not expressed in units, they usually referred to grams per cubic centimeter.

When discussing the Pd value, we exclude the volume of pores from the calculation, and we find that most soils' real density varies within the narrow margin of $2.60 \ g/cm^3$ to $2.75 \ g/cm^3$. This is because quartz, feldspar and silicate colloids, whose densities are all in this range, generally account for most of the minerals in the soil. Yet, when abnormally large amounts of heavy minerals are present, such as magnetite, garnets, epidote, zircon, tourmaline and hornblende, the real density of a soil may exceed $2.75 \ g/cm^3$. It is important to grasp that particle density has nothing to do with the smallness of the particles of a given mineral or the arrangement of the solids in the soil.

Due to the fact that organic matter weighs much less than the same volume of solid minerals, the amount of organic matter in a soil greatly affects its particle density. As a result, the topsoil has a lower particle density than the subsoil. The highest density, in these conditions, is usually $2.4 \ g/cm^3$ or less.

Yet for more general calculations, the average density of the particles in an arable topsoil can be accepted to be about $2.65 \ g/cm^3$.

Real density can be measured in the laboratory using a small flask known as a **pycnometer**. To obtain precise results, the operation has to be

Field of sunflowers

carried out with great care. But, as already mentioned, for most calculations involving real density, we can use the average value of 2.65 g/cm³ and feel sure that any error will be minimal.

2.3.3. Porosity

The process that transforms rock into soil consists largely of the hollowing out and breaking down of

the material, and the formation of pores within the solid mass. These are often referred to as empty spaces, but they in fact contain air and water. If there were no pores, the soil wold not be suitable for plant life.

The relative volumes of the three states of matter (solid, liquid and gas) present in the soil can be represented by a generalized diagram like the pie chart below. In general terms, roughly half the volume of the A horizon is occupied by solid matter, and the other half is pore space. Most of the solids are minerals, but a small proportion, as shown in the pie chart, is organic matter (in organic soils, this proportion is much higher and the mineral part is lower). The pore volume occupied by air and by water varies with changes in the soil's moisture content. In general, the larger pores are full of air unless the soil is completely waterlogged, while the smaller pores are full of water unless the soil is totally desiccated. Air and water can enter and leave the intermediate-sized pores when the soil's water content varies.

Sandy soils usually have a lower volume of pores than fine-textured soils, but they are almost always well aerated (unless there is some sub-surface factor to prevent water movement). Their good aeration is partly due to the fact that most of their soil pores are large enough to allow the water that enters within them to drain away afterwards. This ensures air circulation is good, as air only fails to reach a small number of isolated pores. Of course the fact that sandy soils lose water so easily means they have a very limited capacity to retain water for use by plants.

Densities of some common soil minerals

MINERAL		DENSITY IN g/cm³
Quartz		2.65
Feldspars	Orthoclase	2.56
	Plagioclase	2.60-2.76
Micas		2.76-3.00
Silicates	of the clay group	2.00-2.70
Hydroxides of Fe and Al		2.40-4.30

Pie chart below right. Standard volumes of the soil components in a typical "A" horizon. The solid part of the soil (mineral and organic) usually occupies about 50% of the volume.

In volcanic areas, such as the Canary Islands, soils are rich in nutrients, but O.M. must be added to ensure crops can make use of the abundant minerals ejected from the bowels of the Earth.

Volcanic areas are often very mountainous and vehicles cannot be used, so cultivation requires use of the traditional method, draft animals.

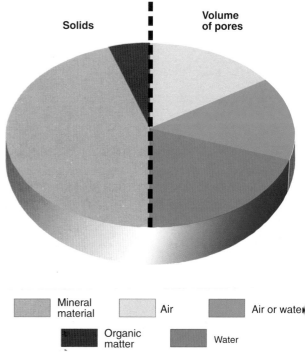

| Mineral material | Air | Air or water |
| Organic matter | Water | |

Clay loams and clays usually have a very large total number of pores, but they retain a lot of water, even when there are no subsurface factors restricting drainage. Their pores are abundant, but very small. Unless the soil has a good structure, most pores in these soils are smaller than the thickness of the water film that is retained around a soil particle. The few large pores that exist may even be cut off from the circulation of air, because they are connected by small pores that remain full of water for much of the time.

We can distinguish two general types of pores, micropores and macropores, which give rise to types of porosity, aeration porosity and capillary porosity. As a reference, macropores are those with a diameter greater than 8μ, and all those less than 8 μ are considered micropores.

2.3.3.1. Percentage porosity

A soil's **porosity**, as discussed in the preceding section, represents the sum of all the gaps between the soil particles, the sum of the macropores and the micropores. This total porosity can be measured numerically. The porosity, a soil's total percentage porosity (% P), is calculated on the basis of its particle density and bulk density values. Consider the following formulas:

$$\% \ P = \frac{Pd - Bd}{Pd} * 100$$

From the % porosity, we can determine the % solids by subtraction:

$$\% \ solids = 100 - \% \ P$$

When planting vineyards or orchards it is often necessary to open up the soil with the traditional hoe, to ensure a porous soil for the new rootstock. Photograph courtesy of FORD

3. THE LIQUID PHASE OF THE SOIL

So far, we have considered the solid components of soils, organic matter and minerals. When discussing organic matter, we pointed out that a soil is usually about half solid matter and about half pores. At the end of the outline of the solid soil components, we were introduced to the ideas of porosity and density, which are nothing but different ways of quantitatively evaluating the pore space.

The pore space is vital for plants, as it serves as a reserve of air and water. Air, specifically oxygen, is essential for root respiration; if for any reason (such as prolonged waterlogging) the plant roots have no access to gaseous oxygen (O_2), they asphyxiate and die.

Water is essential for plant life, and is greatly affected by the soil's porosity. The purpose of this chapter is to describe the role of the water retained in the soil by porosity.

3.1. WATER

Water is vital for plant growth because their physiological processes require water, but water is also vital because it contains dissolved nutrients. Rainfall and the other forms of precipitation are the main forms of water input, but they would be of little use to plants if the soil did not retain water for them to use from one rainfall to the next. The soil's capacity to store water depends on its depth, texture, structure and other fundamental properties.

We can distinguish between two general types of pores, micropores and macropores, and they are responsible for different types of porosity, aeration porosity and capillary porosity. When we say that the soil is at field capacity (F.C.), we mean the soil's micropores are saturated with water and its macropores are full of air, or to put it another way, its capillary porosity is saturated with water and its aeration porosity contains air. This water reserve contained in the micropores after drainage is known as the useful **water retention capacity**. Most studies give the diameter of 8 µ as the minimum size necessary for water to drain freely.

3.1.1. Water in the soil

The soil does not always contain the same quantity of water. The soil's water content varies over time, due to the inputs/outputs, some of them external (gains from rainfall and/or irrigation), others due to evaporation (losses due to high temperatures) and the water taken up by plants (losses).

To explain how plants obtain their nutrients from the soil botanists use extremely complex descriptions of *osmotic potential*. But for our purposes it is sufficient to consider that the plant obtains the nutrients it needs from the soil water

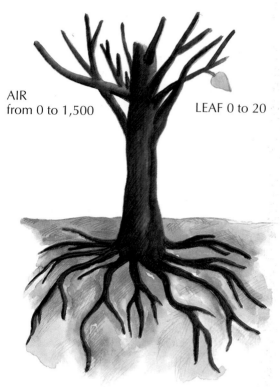

AIR
from 0 to 1,500

LEAF 0 to 20

SOIL 0 to 15

solution by a pressure difference. That is to say, when the plant opens its **stomata**, a difference in water potential is created between the soil or soil water solution and the exterior of the plant. This difference in potential means the nutrient-bearing soil water solution enters the plant root cells and travels up to the leaves (the site of photosynthesis). Most of the water evaporates into the atmosphere, but the nutrients and some water are retained by the leaf, which synthesizes organic matter from carbon dioxide, water and nutrients with energy from sunlight. This organic matter is put to two purposes: part is used in the plant's structural growth and part as a source of energy. It is important to grasp that the plant has to expend energy to absorb these nutrients.

Let's consider what happens in a soil after irrigation or abundant rainfall. All the pore space (the macropores and the micropores) is saturated with water; when the soil is full of water, it is usually said to be **waterlogged**.

Over the course of two or three days, the **gravitational water** (the water occupying the macropores) gradually percolates down to the underlying **water table**. When the macropores have lost their water and the soil retains water in its micropores, we say the soil is at **field capacity** (F.C.).

Field capacity is considered to be the optimum for plants. The soil water is easily accessible, or to put it another way, the plants have to expend little or no energy to obtain water.

The water occupying the micropores is gradually lost over time, either through evaporation or because the plants use it.

As water is lost from the soil, less and less water is

Division of the water potentials of air (in atm. cm³) between the leaf and the soil. Due to the difference in potential, the water absorbed by the roots rises up to the stomata, where it is used by the leaf, and part is lost by evaporation.

Osmotic pressure is the pressure present on one side of a semipermeable membrane separating two solutions of different concentrations when the system is in equilibrium. Semipermeable membranes, which can be natural or artificial, let some types of molecules (the solvent) pass through freely but not others (the solute molecules).

A *stoma* (plural: stomata) is a small opening in the epidermis of the underside of the leaf that can be closed and opened by the two surrounding kidney-shaped guard cells. This means that the cells below the stomata are in contact with the outside air.

The *water table* is the name of the set of insoluble minerals or parent material present below the soil surface. This layer varies in depth, depending on the morphology of the soil. It often forms a water reserve in the subsoil, where the rainwater percolating from the surface accumulates, and this can be exploited by means of artesian wells.

available to the plant. The less water remains, the more energy must be expended in extracting it from the soil. After a certain point, plants cannot obtain any more water, because too much energy is needed to obtain it. This point has two names: the **temporary wilting point** (T.W.P.) and the **permanent wilting point** (P.W.P.). In both cases the plant cannot use the little water left in the soil, but in the first case the plant will recover if it is watered, while in the second case its death is inevitable.

The amount of water held between field capacity and the permanent wilting point is the **plant-available water**.

The permanent wilting point of a soil does not mean that the soil contains no more water, but that what is left is not available to the plants. The water available in the soil but which cannot be used by plants is called unavailable water. Water **not available to plants** can be divided into two fractions. The **hygroscopic water** is the percentage of water that is left in air-dried soil, at 98% relative humidity and 25°C (atmospheric evaporation extracts more water than plants can take up), and the remaining portion is the **chemically combined water,** which is tightly bonded to the molecules of the soil particles, and which can only be extracted and measured by complicated laboratory analysis.

3.2. WATER ENERGY

As we already know, the plant has to expend energy to extract water from the soil, because water is subject to forces retaining it in the capillary micropores. The more tightly the water is held in the soil, the more energy the plant has to expend. The amount of force with which the water is retained by the soil is usually called the **water potential** and is discussed in this section.

3.2.1. Total water potential

We define the total water potential as the amount of work needed by a unit of chemically pure water to transfer an infinitesimally small portion of water from a **pool** located at a given height and at atmospheric pressure, until it acquires the same unit energy of water as the water in the point in question. This work has to be done in a way that is reversible and without changes in temperature.

The free energy of the soil water is expressed as its **water potential**, the sum of forces that retain or move the water within the soil. The total water potential is represented by the Greek letter psi, with the symbol ψ. Pure water on the soil surface has a potential of 0. The total water potential is calculated using the formula:

$$P\text{ total} = \text{matrix P} + \text{osmotic P} + \text{hydrostatic P} + \text{gravitational P}$$

$$\Psi_w = \Psi_m + \Psi_o + \Psi_h + \Psi_g$$

A "**pool**" of water is a deposit of chemically pure free water that is in equilibrium with atmospheric pressure. This pool can be located anywhere as long as it is common to the entire system being studied.

Before the plant reaches the T.W.P. (temporary wilting point), it must be watered. If it isn't watered, vegetative growth stops, and this greatly reduces the final yield.

• Ψ_m: Known as the matrix potential, it is the sum of the forces adsorbing the water onto the surface of the soil particles and of the capillary forces due to surface tension. Its value is negative and is expressed in units of pressure.

• Ψ_s: This is the osmotic potential due to osmotic pressure of the substances dissolved in the soil water. Its value is negative and expressed in pressure units.

• Ψ_h: The hydrostatic pressure, due to the hydrostatic pressure of a column of water to be supported at a given point.
Its value is positive and it is also measured in units of pressure.

• Ψ_g: The gravitational potential. It is the possible potential between two points due to their different heights, due to gravity. Its value is also positive and it is expressed in units of pressure.

The values of the hydrostatic pressure and the gravitational potential are usually so small that they are of little or no importance.

3.2.2. The water potential units

Water potential can be expressed in different types of units. We can consider that water potential is comparable to suction or negative pressure, so that any unit suitable for expressing pressure is suitable for water potential. Atmospheres and bars are the most widely used units. Other less widely used units are the height of a column of water or of mercury, or the weight per unit surface area. This variety of units means that conversion factors are needed to convert units. Below these lines is a table of the main units and their conversion factors.

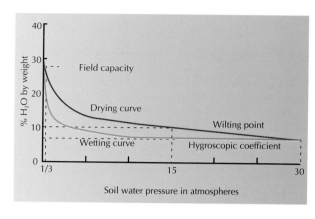

% H$_2$O by weight — Field capacity — Drying curve — Wilting point — Wetting curve — Hygroscopic coefficient

Soil water pressure in atmospheres

For example, the pF value of a column of water 1,000 cm tall, is -log 1,000 = -3, and is generally written pF = -3.
Confusion may arise when we measure the soil water content and its water potential in centimeters. So, it is usually recommended to use bars, pF or atmospheres to calculate pressure.
The subdivisions of the soil water are based on how tightly it is bound, and are thus very closely related to the water potential, and are frequently expressed in atmospheres and pF. The following rough values are used:

Field capacity = 1/3 atm. pF = 2.5

Wilting point 15 atm. pF = 4.2

Hygroscopic coefficient = 30 atm. pF = 4.5

3.3. MOVEMENT OF WATER

Soil water moves as both liquid water and as water vapor. The movement of liquid water is controlled by the water potential: water movements are much greater when the soil is wet than when it is dry. In dry soil, the water is so tightly bound to the soil particles that there is almost no movement of liquid water. The extent of movement of water vapor between different parts of the soil depends greatly on their temperatures. The presence of a temperature gradient leads to the creation of a vapor gradient, and this gives rise to the movement of water vapor through the water-filled pores from the warmest part of the soil to the coldest. When heavy rain falls, part infiltrates into the soil, and part is lost by surface runoff. **Percolation** can be defined as the downward movement of water in the soil, and is usually applied to the change from soil saturation to field capacity, while **infiltration** is when water enters the soil from the surface and filters downwards. The **rate of infiltration** is the speed that water filters down through the soil. This rate depends a lot on the soil and its physical condition, but in general it is something like 1 or 2 cm per hour.

EQUIVALENCES BETWEEN UNITS OF WATER POTENTIAL

BAR	ATMOSPHERES (ATM.)	TORRICELLI (mm Hg)	pF	HEIGHT OF A COLUMN OF WATER (h) (in cm) H$_2$O
0.001	0.000987	0.752		1.02
0.01	0.00987	7.52	1	10.22
0.3	0,297	225.7	2.5	340
1	1	760	3	1,000
15	14.85	11,286	4.2	15,320

As the reader will notice when comparing the different units and values in the table, they include a new unit of pressure, the pF value. This unit is only used in soil science and agriculture for the water potential. The pF value is logarithmic, and is the negative of the value of the water pressure, expressed as a height (h) of a column of water, expressed in centimeters:

$$pF = -\log h \text{ (en cm)}$$

3.4. THE PROPERTIES OF WATER IN THE SOIL

Water gives soil its consistency. The soil's consistency means we can talk of its plasticity or adhesiveness when the soil contains water, and its firmness or hardness when it is dry. The soil's water retention capacity depends on its intrinsic properties, its texture, clay types, structure, organic matter content, etc.

3.4.1. Factors that influence water retention

The soil's ability to retain useful or available water is one of its most important characteristics. In dry farming, retaining useful water is vital, as the crops depend entirely on the soil water reserves. The factors affecting water retention are the soil's texture, its clay types, its structure, its organic matter content, as well as the thickness and sequence of the horizons in its profile. As already pointed out when discussing soil texture, the clays greatly affect water retention. Their small particles retain more water than coarse- or sandy-textured soil, both at field capacity and at the permanent wilting point.

Logically, the depth of the soil also affects how much water it can retain. Thus a deep soil will have greater water reserves than one where the bedrock is close to the surface. A well-structured soil retains more water than an unstructured one, as a well-structured soil has macropores and micropores that help to maintain the soil's water capacity. Organic matter can retain up to 20 times its own weight of water. Of course, the reason why organic matter can absorb so much water is that it has a very low density and is very porous.

The sequence of layers in a soil profile may have a considerable influence on the soil's water retention capacity, not only because of the inherent differences between the different textured horizons, but also because of their influence on the movement of water. There is generally a delay in the downwards movement of the water when there is a drastic change in the soil texture. This delay is sometimes so large that it increases the water retention capacity of the upper layer. This water is loosely retained and is easily accessible to plants unless there are aeration problems.

3.4.2. Mechanical behaviour

The cohesion binding together the particles to form aggregates is known as the soil consistency. The soil's consistency has to be measured using dry, moist and wet soil. It is assessed qualitatively or semi-qualitatively rather than quantitatively. Depending on the soil's water content, the consistency may be expressed in terms of plasticity, firmness, hardness and adhesiveness.

Depending on the consistency of the soil to be cultivated, we have to choose light or heavy machinery. (Courtesy of LARDINI).

3.4.2.1. Plasticity

The plasticity of a soil is defined when the soil is almost saturated with water. Plasticity is the soil's ability to acquire and maintain a new form when pressure is applied and then removed. A soil's plasticity can be assessed as slightly plastic, plastic or very plastic. It is the soil's clay content that makes it plastic, and the more clay it contains, the more plastic it is.

3.4.2.2. Adhesiveness

A soil's adhesiveness, like its plasticity, is defined when the soil is saturated with water. It is defined as the soil's ability to adhere to other objects.
Like plasticity, we divide soils into slightly adhesive, adhesive or very adhesive.
Like plasticity, the adhesiveness is directly proportional to the clay content.
Thus a sandy soil shows little or no adhesiveness.

3.4.2.3. Firmness

Firmness is the name given to a soil's ability to deform when a given pressure is applied. A soil may be loose, very friable, friable, firm, very firm or extremely firm.
This qualitative assessment is performed when the soil has about 50% of its pores saturated.

3.4.2.4. Hardness

Hardness is assessed when the soil is dry. It is defined as the resistance of a dry soil to being broken down.
A soil can be loose, weak, slightly hard, hard, very hard or extremely hard.
These terms do not only describe the difficulty of breaking up a given aggregate, but also express the resistance to root penetration and the effort required to plow or to dig a hole.

is the medium in which all the processes of photosynthesis take place. To put it another way, thanks to water, the plants can take up chemical ions from the soil solution, and with CO_2 and the energy of sunlight, they can carry out photosynthesis in the leaves, producing organic matter and releasing oxygen gas (O_2). Soil chemistry includes aspects of the chemistry of solutions and of solids (mineralogy). The zone of contact between the solid and liquid phases is of great importance in soil chemistry. It is, in many aspects, related to the chemistry of colloids, in which surface effects play a major role. In most soils many more ions are adsorbed onto the surface of solid particles than are truly dissolved. The ions adsorbed are in a slow-acting equilibrium with the ions absorbed within mineral particles, and a rapid-acting equilibrium with the ions dissolved in the liquid phase.

4.1 THE pH

Water consists of molecules whose formula is H_2O, indicating that each molecule consists of one oxygen atom and two hydrogen atoms. In the immense majority of the molecules of a sample of water, a single oxygen atom is linked by covalent bonds to two hydrogen bonds. But in a small number of the water molecules (one in every 600 million – 1 in 6×10^8), one of the bonds is absent, and the molecules are present as two ions, with opposing electrical charges, H^+ and OH^-.

In a balanced ecosystem, the amount of organic matter fixed by photosynthesis is the same as that returned to the pool by the decomposition of organisms, and the energy that powers this cycle comes from the sun.

Ratio of the pH and the pOH to the molarity of alkaline and acid solutions

4. SOIL CHEMISTRY

The chemical properties of soils, or soil chemistry, is the area of soil science that joins together the soil's previously discussed properties and characteristics with the fertilizers dealt with in the next chapter.

It is well known that plants "feed" on chemicals that they obtain from the soil, and together with light and carbon dioxide (CO_2), they are used to make organic matter. It is also known that water

$$H_2O \rightleftharpoons H^+ + OH^-$$

In pure water, the number of the two types of ions will be the same, as each water molecule will form one of each. The water is said to be neutral. In a liter of water there will only be 0.1 microgram of H^+ ions, representing a concentration of 0.0000001 grams per liter, written 10^{-7} g/l. If we add a substance that supplies H^+ ions (an acid), their concentration will increase (if it is ten times greater, it will be 0.000001 g/l or 10^{-6} g/l), and the water becomes more acidic. But if we then add a substance that forms OH^- anions (a base, or an alkali), the concentration of H^+ ions will decline, as some of the H^+ ions react with the added OH^- ions, forming molecules of water (for example if it is reduced by a factor of 100, the solution will have 0.00000001 g/l (10^{-8} g/l).

To avoid using all these awkward numbers with so many zeros and negative powers of ten, the idea of pH was introduced. This is a measure of the concentration of H^+ ions in a solution, and thus its acidity or alkalinity. Instead of expressing the quantity of hydrogen ions as 10^{-7}, 10^{-6}, 10^{-8}, they

pH	Acidity (molarity of H^+)	Alkalinity (molarity of OH^-)	pOH
0	1.0	0.00000000000001	14
1	0.1	0.0000000000001	13
2	0.01	0.000000000001	12
3	0.001	0.00000000001	11
4	0.0001	0.0000000001	10
5	0.00001	0.000000001	9
6	0.000001	0.00000001	8
7	0.0000001	0.0000001	7
8	0.00000001	0.000001	6
9	0.000000001	0.00001	5
10	0.0000000001	0.0001	4
11	0.00000000001	0.001	3
12	0.000000000001	0.01	2
13	0.0000000000001	0.1	1
14	0.00000000000001	1.0	0

are given as 7 (which is neutral), 6 (acidic), 8 (basic), etc.

Thus a solution with a pH of 5 has 10 times as many hydrogen ions as a solution with pH 6, and 100 (=10^2) as many as one with pH 7. The increase is thus logarithmic. The more acidic a solution is, the lower its pH, and the more basic a solution is, the higher its pH.

Water does not present any resistance to these changes in pH, as its pH changes when we add any acid or base that provides H^+ or OH^- ions. Solutions of some substances make water more resistant to these changes, so that small additions of acids or bases do not changes the solution's pH. They are known as *buffer solutions*, because they buffer the changes in pH due to increased concentrations of the H^+ or OH^- ions. If two buffer solutions are mixed together, the final pH of the solution will depend on which substance has the greater buffering power.

4.1.1. Measuring the pH

The pH is measured in a laboratory by dissolving an **aliquot** of soil in distilled water. Using a pH-meter, the concentration of H^+ and OH^- ions is measured at a standard temperature of 25°C. There is a brief table to help interpret the results of soil pH analyses. The table below shows the terms used to describe different pH values.

pH values	Denomination
Below 4.0	Extremely acid
4.0 to 5.0	Very strongly acidic
5.0 to 5.5	Strongly acidic
5.5 to 6.0	Moderately acidic
6.0 to 6.7	Slightly acidic
6.7 to 7.3	Range of neutral
7.0	Neutral
7.3 to 8.0	Weakly alkaline
8.0 to 8.5	Moderately alkaline
8.5 to 9.0	Highly alkaline
9.0 to 10.0	Very highly alkaline
Greater than 10	Extremely alkaline

4.1.2. The soil pH

Look at the figure showing the pH values of some typical soils, and notice that most of them are between pH 4 and pH 8. Almost all soils with a pH of 8 or more have problems of salinity or excess Na^+ ions in their cation exchange sites. Soils with a pH of around 4 are acidified by sulfate ions.

Washing processes remove bases from the soil, and thus tend to cause the pH to fall in the long term. This process of natural decrease in pH is particularly important in young soils and is of less importance in senile soils, where weathering processes have eliminated most of the clays with a 2:1 type structure.

Typical soils with their corresponding pH values. Taken from Thomson (1988)

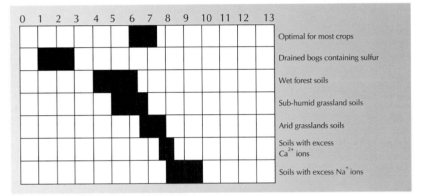

The input of organic matter gives rise to the formation of organic acids that displace the bases from the cation exchange complex, and thus acidify, the site because they lower the percentage saturation of bases.

Washing processes are more intense in acid conditions, because the dominant form of weathering releases more cations and less are retained at the cation exchange sites. Forests located in wet climates usually give rise to soils that are more acid than those of grass meadows, as they lose the bases from their cation exchange complexes by illuviation, causing acidification.

An **aliquot** is a known weight (any known weight) of a sample. The sample may be of any type of matter, whether solid, liquid or gas. The word is often used by analytical laboratories.

pH-meter: a device to measure the pH of a solution.

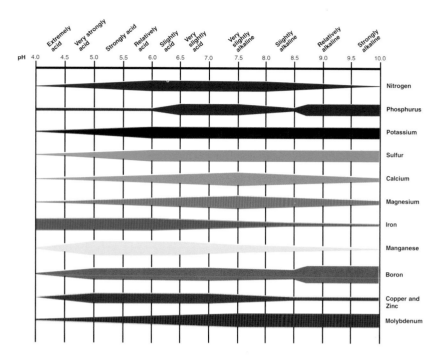

Extremely acid Very strongly acid Strongly acid Relatively acid Slightly acid Very slightly acid Very slightly alkaline Slightly alkaline Relatively alkaline Strongly alkaline

Nitrogen
Phosphorus
Potassium
Sulfur
Calcium
Magnesium
Iron
Manganese
Boron
Copper and Zinc
Molybdenum

Nutrient availability to plants in function of soil pH. Taken from Troug (1946)

Electrical conductivity values and their corresponding salinity assessment

Conductometer: a device to measure electrical conductivity

The siemens (S, from the scientist W. von Siemens) is the unit of electrical conductance. It is the reciprocal of the value in ohms (ohm) and is represented by the letter "S". Thus $1/S = \Omega$. The siemens is the accepted S.I. unit and replaced the mho (ohm spelt backwards). The *millisiemens*, or mS (one thousandth of a siemens), and the *microsiemens*, µS (one millionth of a siemens) are also widely used.

4.1.3. The effect of pH on nutrients

pH affects plant growth through their uptake of nutrients. The soil pH influences the rate of nutrient release through weathering, the solubility of all the materials in the soil, and the quantity of nutrient ions stored in the cation exchange complexes. The pH is thus a good guideline indicating which ions may be deficient. The diagram above shows the relative solubility of different nutrients at different pH values. When faced with pH values that prevent nutrient uptake, the farmer has two choices; either apply something to change the pH to a more favorable value, or apply sufficient fertilizer to remedy the deficiency, despite the pH. Nutrient requirements vary from crop species to crop species, and so the optimum pH also varies. The optimum pH is usually between 6.0 and 7.5, because all the nutrients are reasonably available to plants in this range of pH values, though some acid-loving plants prefer acidic soils with pH values of 5 or 6, such as azaleas, rhododendrons and strawberries.

4.2. SALINITY

Whenever soil drainage is disrupted for whatever reason (and especially in arid and semi-arid regions where rainfall is low and evaporation is intense), this favours the accumulation of salts on the soil surface or just below it. This phenomenon is due to the evaporation of water that has been borne up to the surface by rising capillary action, causing salts to accumulate gradually in the surface horizon.
In general, salination occurs when the presence of salts starts to modify the soil's chemical structure. Saline soils contain many sulfate and chloride salts of Na, Ca, Mg and K concentrated in the "A" horizon of arid soils. In extremely dry conditions, large amounts of these sulfates and chlorides may form a white crust in the topsoil. They are known as **solonchǎk**, or **white alkaline**, soils.
The carbonates of the common metals, especially sodium and potassium, dissolve the soil organic matter (forming soluble humates), and this darkens the color of the salt solution and crust (**forming solonetz, or black alkaline soils**). These black alkaline soils may also be due to calcium chloride or excess sodium nitrate. This black alkali soil solution is toxic to plants at much lower levels than white alkaline soil solution, as it is more highly alkaline.
It is easy to understand that a high concentration of salts in the surface "A" horizons prevents the plant taking up nutrients by the osmotic processes described above: the pressure gradient in the soil is too great, due to the presence of the salts, for the plant to absorb the cations it needs.

4.2.1. Electrical conductivity

A soil's electrical conductivity, or E.C., is used to measure its salinity in the laboratory. This easy test consists of dissolving an aliquot (a sample of known weight) in distilled water, and using a **conductometer** to measure the electrical conductivity in **milliSiemens** at 25°C. The conductometer reading is multiplied by the empirically obtained factor 0.64 to give the total quantity of salts in milligrams per liter. This measurement is based on the fact that distilled water has a conductivity of 0, and thus the value for the solution's conductivity must be due to the salts dissolved in it. Underneath these lines there is a small table to help interpret the results of analytical measurements of soil conductivity.

E.C. mS cm^{-1} (milliSiemens per cm)	Assessment
< 0.6	Not saline
0.6-1.2	Slightly saline
1.2-2.4	Saline
2.4-6.0	Very saline
> 6.0	Hypersaline

4.2.2. The origins of salinity

Specialized bibliographies deal with four hypotheses for the origin of salinity. The first is that some soils were saline before the intervention of human beings, for example due to deposition at sea or on land. Soil salinity may also be increased by human action, such as using excessively saline water to irrigate crops on initially "healthy" soils. If a soil receives rainfall and irrigation water that is exactly the same as the sum of what the plants use and what is lost by evaporation from the soil, then the salts not taken up by the vegetation will accumulate in the soil, and of course irrigation water from any source will contain some minerals, however low their concentration. This does not occur in wet areas, but does occur in arid and semi-arid areas, because in wet areas the excess rainfall washes the salts through the soil, illuviating them down towards the water table.

A third mechanism is that where there is a saline layer below the soil surface, the salts will reach the cultivable topsoil. The fourth and least reason is that excess fertilizer that plants are unable to absorb will lead in the long-term to soil salination.

4.2.3. The effects of salinity

Soil salinity seems to have two negative impacts on plants. One is that the increase in the salts dissolved in the soil water solution makes it harder to take up nutrients (due to osmosis, water tends to leave the roots instead of entering them bearing nutrients). The second reason is that excessive concentrations of some salts are **phytotoxic**, toxic to plants.

A plant's sensitivity to soil salinity depends on the species it belongs to. Plants from coastal habitats often resist salinity much better than plants that evolved in the wet forests of the interior of the continents. The table below lists plants by their tolerance of salinity, assessed as high, medium and low.

High tolerance	Medium tolerance	Low tolerance
Date palm	Olive	Pear
Beetroot	Vine	Apple
Cabbage	Melon	Orange
Asparagus	Tomato	Grapefruit
Spinach	Cabbage	Plum
Esparto grass	Cauliflower	Almond
Bermuda grass	Lettuce	Apricot
Rhodes grass	Potato	Peach
Canada wild rye	Carrot	Strawberry
Grama grass	Pea	Lemon
Barley	Squash	Avocado
Trefoil	Sweet clover	Radish
Sugar beet	Mountain bromegrass	Celery
Turnip	Strawberry clover	Runner bean
Cotton	Dallis grass	White clover
	Sudan grass	Meadow foxtail
	Alfalfa	Hybrid clover
	Tall fescue grass	Red clover
	Rye	Ladino clover
	Wheat	Broad bean
	Oats	
	Orchard grass	
	Meadow fescue	
	Canary grass	
	Foxtail brome	
	Rice	
	Sorghum	
	Corn	
	Sunflower	

4.3. CARBONATES

This brief section deals with an explanation of the importance of carbonates ($CaCO_3$) and how to assess them. As already explained in the section on density and porosity, the solid part of the soil consists essentially of silicates, carbonates and organic matter. Leaving out the organic matter, we are left with the mineral component, the silicates and carbonates. Some soils consist of just silicates and organic matter, while others may also contain substantial amounts of carbonates. Carbonates greatly influence soil fertility, because, as we shall see later when discussing with the nutrient elements, the calcium cation (Ca^{2+}) is present in a far larger percentage than the other individual cations (Mg^{2+}, Na^+, K^+) and often there is more calcium than all the other ions put together.

4.3.1. The measurement of carbonates

There are two sides to the measurement of carbonates, *in situ* measurements (in the field), and *in vitro* measurements (in the laboratory). A first approach to estimate the amount of carbonates in the field is to sprinkle a few drops of hydrochloric acid on a sample of the soil. If there is any carbonate for the acid to react with, effervescence occurs (it froths) and carbon dioxide is given off:

$$2\ HCl + CaCO_3 = CO_2 \nearrow + CaCl_2 + H_2O$$

If this preliminary field test shows no effervescence at all, the soil is siliceous and no carbonates are present. If effervescence occurs, we will have to measure the percentage of carbonate in the laboratory using Bernard's calcimeter, which is based, like the field test, on the release of carbon dioxide, which is then measured. By finding the volume of CO_2 released, we can calculate the amount of carbonate in the soil, and thus the percentage of $CaCO_3$.

4.3.2. Assessment of results

Below this paragraph there is a small table to help interpret the results. Note that 5% carbonates usually ensures that a good content of Ca^{2+} cations is adsorbed onto the clay-humus complex, enough to meet the plant's nutritional requirements for this element. If carbonate content reaches 10%, the soil's physics and chemistry are dominated by the carbonates, and due to the high percentage of Ca^{2+} ions (which prevent other elements from occupying the negative sites on the clay-humus complex) you should avoid trying to cultivate **calcifuge plants**.

Percentages	Assessment
0-1	Very low
1-10	Low
10-30	Medium
30-60	High
> 60	Very high

4.4. THE CATION EXCHANGE COMPLEX

The silicate clay minerals, allophane and humus, are known as the **clay-humus complex**. These solid components of the soil possess all the major characteristics known as the cation exchange capacity, or C.E.C.

4.4.1. Adsorption

All these materials (clay, allophane and humus) possess negative charges, meaning that they attract cations. This attraction and retention of the cations dissolved in the liquid phase of the soil by the clay-humus complex (or the solid phase) is known as **adsorption**.

A **phytotoxic** substance is poisonous to plants.

List of plants in relation to their sensitivity to soil salinity. Taken from the U.S. Salinity Laboratory Staff (1954)

Percentage carbonate content, $CaCO_3$, in a soil and their corresponding evaluation

Calcifuges are plants that cannot grow in limy soils, soils rich in calcium carbonate.

The (adsorbed) cations on the clay-humus complex can now be taken up by plants. The forces involved in **absorption** by the plant are probably stronger than adsorption onto the complex.

Sorption is the name often given to the combination of the two processes. Nutrients with positive charges are adsorbed onto the clay-humus complex, at the same time as they are taken up by plants for their nutrition. And by extension, we give the name **desorption** to the displacement of ions from the solid phase of the soil to the liquid phase, the soil solution, due to the action of another, stronger (more electropositive), ion which displaces it.

4.4.2. Characterizing the C.E.C.

The silicate clay minerals, the allophane and the humus all possess negative charges, and these attract cations. These cations are exchangeable if they can be replaced by others that are also dissolved in the liquid surrounding the particles. Replacement is possible if the bonds are not very strong and the sites are accessible to the soil solution. These cations absorbed by the clay-humus complex are the source of the nutrients needed for plant growth. The soil's cation exchange capacity is of vital importance to plant growth. Very sandy soils lacking humus and clay, where exchange capacity is very low, have very poor plant cover.

There is a precise analytical procedure to determine the soil's C.E.C., but is very complicated, being time-consuming rather than complex. We can estimate the C.E.C. if we know the amount of organic matter and clay.

Refer to the table below, which shows the value of the cation exchange capacity in terms of **_milliequivalents_** per 100 g of soil, and try to calculate the C.E.C. of a given soil.

*A **milliequivalent** is the quantity of material that combines with, or replaces, one milligram of hydrogen. The number of milligrams in a milliequivalent is calculated by dividing the atomic weight (or molecular weight or ionic weight) by its valency. Avogadro's number ($6.022045 * 10^{23}$ mol^{-1}) of reactive charges correspond to one equivalent or 1,000 milliequivalents.*

C.E.C. (cation exchange capacity) of humus and some clays. Thomson (1988)

	Cation exchange capacity, in meq/100 g	
	Representative value	Representative range
Humus	200	100-300
Vermiculite	150	100-200
Allophane	100	50-200
Montmorillonite	80	60-100
Illite	30	20- 40
Chlorite	30	20- 40
Peat	20	10- 30
Kaolinite	8	3- 15

On the right: a table of C.E.C. values and an evaluation

Supposing that when we started managing a farm or plot, we had commissioned an agricultural laboratory to perform a soil texture analysis, we would have available to us data like the following: LOAMY soil with 3% organic matter (humus), with a TEXTURE, based on the percentages of particles of different sizes, of 20% clay, 40% silt and 40% sand, with the

following proportions of different types of clay (as % of total weight) of 3% vermiculite, 7% montmorillonite, 9% illite and 1% kaolinite. From there we would perform the following calculation:

3% humus	0.03*200 = 6.00
3% vermiculite	0.03*150 = 4.50
7% montmorillonite	0.07*80 = 5.60
9% illite	0.09*30 = 2.70
1% kaolinite	0.01*8 = 0.08
	18.88 meq

So the soil will have a C.E.C. of 18.88 meq per 100 g of soil.

4.4.3. Interpretation of the results

In reality, the C.E.C. is a measure of the soil's fertility, its ability to store cations. When the C.E.C. is measured in the laboratory, it should be approximately the same as the sum of the value in milliequivalents for each ion. The quantity of each cation can be measured in the laboratory by different procedures, and the results then expressed in terms of milliequivalents per 100 g of soil. Thus the sum of the milliequivalents of Ca^{2+}, Mg^{2+}, K^+, Na^+ should be close to the value of the C.E.C. measured in the laboratory.

We can think of the C.E.C. as a store, whose size depends on the amount of organic matter and clay, and from which the plants can take the nutrients that they need.

Refer to the table immediately below to compare the value we have calculated for the soil's cation exchange capacity to see if this soil can retain cations or not.

C.E.C. in meq/100 g	Evaluation
< 5	Poor
5-10	Low
10-15	Low normal
15-25	High normal
25-40	High
> 40	Very high

4.5. NUTRIENT ELEMENTS

We know that plants have to take up nutrient ions, which enter the plant due to the difference in water potential along the soil-plant-atmosphere continuum, and these inorganic ions are used, together with energy from the sun, in the synthesis of organic matter in photosynthesis.

We also know that these ions are derived from the soil minerals in their various forms. These ions derived from weathered minerals ionize (meaning they have either a positive or negative charge), when they dissolve in the soil solution.

Most minerals are very insoluble, but a small quantity will always ionize when it comes into contact with the soil solution.
As we have already seen, these dissolved ions, if they are positively charged, are retained on the clay-humus complex, as this possesses negative charges. The main cations, in order of decreasing abundance, are Ca^{2+}, Mg^{2+}, Na^+, K^+ and NH_4^+. This occurs in carbonated soils or ones with a basic pH, but in acidic soils, ones with an acidic pH, Al^{3+}, Fe^{3+} and H^+ ions are usually present. The concentration of cations adsorbed onto the C.E.C. is far greater than the quantity of free cations in solution, mainly because these ions are lost by washing.
These four cations (calcium, magnesium, sodium and potassium [Ca^{2+}, Mg^{2+}, Na^+, K^+]) in the table below are the most common and account for 99% of the ions adsorbed by clays and humus. Other ions, such as ammonium, iron, cobalt, copper, manganese and zinc (NH_4^+, Fe^{2+}, CO^{2+}, Cu^+, Mn^{2+}, Zn^{2+}), only represent 1% of the cations present in the soil, though this very low presence is usually sufficient to meet the plants' requirements for these vital elements.
The cation exchange complex also has some positive charges, known as the **anion exchange complex** (A.E.C.), where anions, ions with negative charges, are adsorbed. Anion exchange sites may be due to the presence of amine groups (nitrogen-containing radicals) in the humus, or with bonds ending in a cation on the edge of a clay mineral, or with a hydroxyl group (OH^-) that ionizes with materials such as $Al(OH)_3$ or $Fe(OH)_3$.
The probability of the ionization of large quantities of OH^- ions is partly due to the abundance of OH^- containing minerals and partly on the pH. This probability increases in highly weathered and washed soils, because the Al and Fe compounds tend to accumulate in them, and because in acidic conditions, most of the OH^- ions formed then combine with H^+ ions to produce water.

Component	Iones
Calcite	$(Ca^{2+})(CO_3^{--})$
Calcium sulfate	$(Ca^{2+})(SO_4^{--})$
Hidroxylapatite	$(Ca^{2+})^5 (PO_4^{---})^3 (OH^-)$
Aluminium hydroxide	$(Al^{3+})(OH^-)^3$
Iron (Fe3+) hydroxide	$(Fe^{3+})(OH^-)^3$
Variscite	$(Al^{3+})(OH^-)^2 (H_2PO_4^-)$

Base	Ca^{2+}	Mg^{2+}	K^+	Na^+	Other bases
Percentage of total exchangeable bases	75-85	12-18	1-5	1	1

Dean & Rubins confirmed in 1947 that part of the exchange capacity of cations and anions arises at the edge of clay crystals. The number of positive and negative charges on the edge of the crystalline structures may be altered by variations in the pH. The anion exchange capacity increases at low pH values (in acidic soils), while the cation exchange capacity increases with high pH values (basic soils). Specialist articles on soils often dismiss anion exchange to the background, possibly because it is frequently masked by cation exchange, or because it has only recently begun to be studied, or because most soils are dominated by cations, and because highly anionic soils are only found in highly weathered and extremely acid tropical soils. The most frequent anions are trivalent, divalent and monovalent phosphate ions (PO_4^{3-}, HPO_4^{2-}, $H_2PO_4^-$), carbonates and bicarbonates (CO_3^{2-}, HCO_3^-) sulfates, bisulfates and nitrates (SO_4^{2-}, HSO_4^-, NO_3^-) hydroxyl groups and chlorides (OH^-, Cl^-).
Plant roots can take up ions from the soil solution or from the clay-humus complex. The plant needs to spend less energy to take up nutrients from the soil solution than from the clay-humus complex, but the availability of the nutrients is much greater in the clay-humus fraction because the dissolved ions are washed towards the water table, decreasing their concentration in the soil solution. In the following discussion of fertility, we will learn about modifying the concentration of the ions in the soil solution or in the clay-humus complex by means of chemical or organic fertilizers.

Some soil minerals and the ions they form. Most minerals are very insoluble, but a small amount always ionizes when it comes into contact with the soil solution.

Average percentages of cations retained in the clay-humus complex. Note that the Ca^{2+} ion accounts for an overwhelming percentage compared to all the other cations put together.

5. ARTIFICIAL SOILS, SUBSTRATES

Mineral soil is the universal culture medium for plant growth, though in plants cultivated in pots or containers, it has been gradually replaced by substrates with a higher percentage of organic matter.

We use the term substrate to mean an artificial soil, whether of organic origin or not, that is used to grow plants, especially ornamental plants cultivated in greenhouses. What is discussed here could equally well be included in the section on greenhouses in this encyclopedia, because artificial soils or substrates are so closely linked to greenhouse cultivation. However, as on many occasions artificial substrates are used to correct the physical, chemical, water retention or nutrient properties of the mineral soil in the field, we considered it relevant to locate it within the section on soils, given their increasing importance in agriculture as a replacement for the conventional mineral soil.

5.1. GENERAL ASPECTS OF SUBSTRATES

In addition to supporting and anchoring the plant, an artificial soil or substrate must supply the plant with suitable amounts of air, water and mineral nutrients, just like the mineral soil.

If the proportions of these components are not suitable, plant growth may be affected, giving rise to a number of **pathologies**, including:

* **Asphyxia** due to the lack of oxygen, which prevents root respiration and the aerobic respiration of the soil organisms.
* **Dehydration** due to lack of water, which may even lead to the plant's death.
* **Excess or deficiency of nutrients**, an imbalance between their relative concentrations, which limits plant growth.
* **Diseases**, which may be produced indirectly by the above causes, as the plants become more vulnerable to attack by virus, bacteria, fungi, etc.

*The physical
properties of some
substrates. Taken
from Pagés &
Matallana (1984)*

*General properties of
some properties of
substrates. Taken
from
Richardson (1992)*

Material	Aeration	Water retention	Nutrient content
Peat	Good	Good	Poor
Sand	Poor/good	Poor	None
Perlite	Very good	Poor/good	None
Polyspan	Very good	None	None
Vermiculite	Poor/good	Good	Poor/good
Rock wool	Very good	Poor	Poor
Absorbent rock wool	Poor	Very good	Poor
Shredded bark	Very good	Good	Poor

The study of a substrate is performed from the same perspective as that of the mineral soil. Thus, the organic matter, the minerals (though present in much lower percentages), the substrate's range of

particle size, its density, porosity, structure, water and water dynamics, as well as the chemistry of the substrates, and its pH, salinity, and C.E.C. can all be understood in the terms used to discuss ordinary mineral soils.

In the next sections, in which we define the physical and chemical characteristics of the substrates, we shall only make a series of general points dealing with the clearest difference between mineral soils and artificial substrates.

5.1.1. Physical properties

When discussing the composition of a mineral soil we compared the different amounts of its different components, with roughly 50% solid matter, but in artificial substrates there is much less mineral material, which is replaced by the organic matter. The diagrams on the following page show the differences between the percentage composition of a mineral soil and that of an artificial and organic substrate after saturating them with water and letting them drain freely, that is to say when both are at field capacity. The proportion of the solid, liquid and gaseous phases in a growth medium vary with the nature of the medium and with the external conditions (drainage, temperature, humidity, etc.).

Substrate	O.M. %	Bd_a g/cm^3	Pd_r g/cm^3	p %
Light peat	87.2	0.076	1.35	94.3
Dark peats	55.5	0.296	1.83	83.7
Pine bark	69.6	0.286	1.64	82.6
Forest topsoil	62.3	0.303	1,75	82.6
Cork	—	0.145	0.922	84.3
Grapeseed cake	91.0	0.156	1.33	88.3
Rice husks	86.9	0.103	1.39	92.6
Wool clippings	79.0	0.153	1.50	89.5
Volcanic earth	<1.00	0.682	2.65	74.2
Perlite	1.37	1.13	2.63	95.2
Vermiculite	8.96	0.146	2.52	94.2

The first thing to notice is that the artificial substrate has a much lower content of solids than the mineral soil (because O.M. is very porous). This means that in a given volume of substrate, there will be more space available for air and water than in the same volume of mineral soil. This means that plants can grow in a small volume of substrate, such as that contained in a plant pot.

In general, if a substrate does not possess adequate fertility, it can be remedied by applying fertilizers, or by washing with water to

*Physical properties
of substrates. The
particle densities
of different types
of substrates.
Taken from Wilson
(1984)*

SUBSTRATE	Pd g/cm^3
Light peat	1.55
Sand	2.62
Vermiculite	2.61
Mineral soil	2.54
Perlite	2.37
Bark	2.00
Pine needles	1.90

eliminate excess salts. But if its physical structure is unsuitable, it is difficult to improve. The fact that it is impossible to improve the substrate in a container means that more attention must be paid to its physical structure than to its chemical properties.

The reduced volume of cultivation medium in a container compared to natural soil in the field means that the substrate must possess much better physical properties, such as aeration and water retention. To begin with, it must have 85% or more porosity so that it can store large amounts of air and water within the limited space of the container.

the depth, the more water is retained per unit volume of substrate.

Even if a substrate's water retention is high, the forces retaining it within the small pores may be greater than the suction pressure the plant can exert, and the water is thus not available to the plant. It is thus very useful to know the amount of available water, which will depend on the size of the smallest pores and on the concentration of salts in the soil solution. The greater the concentration of salts, the greater the suction the plant has to apply, and in extreme cases, it may even lead to the plant losing water.

The amount of water available in a substrate is

Comparison of the composition of a mineral soil and an organic substrate

MINERAL SOIL

ORGANIC SUBSTRATE

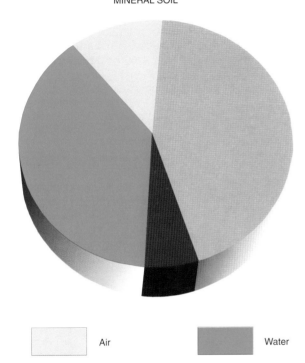
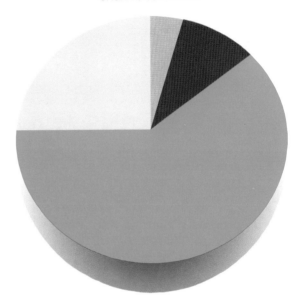

| | Air | | Water | | Organic matter | | Mineral fraction |

A substrate must also have a good distribution of pore sizes, as if it has mostly macropores, it will have good aeration (plenty of oxygen), but it will be poor at retaining water. If there are too many micropores, this will result in poor aeration (even though there is an adequate water reserve) and the roots may suffer from lack of oxygen. The total porosity of a substrate can be calculated easily (as shown for a mineral soil) if we know its bulk density and its particle density. The porosity will allow us to measure some important values of any substrate, such as the quantity of substrate contained in a given volume, the degree of weathering of some soil components, the amount of mineral material included, and also to check the degree of compaction. If the substrate contains additives with a unit structure of closed cells, such as perlite or polyspan, it is necessary to determine the particle density by using a pycnometer to calculate the effective or open porosity.

The total quantity of water retained by a substrate depends on the proportion of small-sized pores and on the thickness, i.e. the depth, of the substrate within the container. The greater

considered to be the quantity extracted between vacuum pressures corresponding to 10 and 100 cm columns of water, and this can only be measured in the laboratory, using vacuum equipment. If the complete graph of suction can be plotted, it is also possible to find out the amount of reserve water and the macroporosity. Knowing these values is important because it affects how often the plants need to be irrigated.

The bulk density expressed in grams dry matter per milliliter of substrate (g dry matter/ml substrate) of some substrates. Note that the smaller the particle size of a substrate, the higher its bulk density. Taken from Thompson (1988)

Component	Size, mm	Bulk density, g/ml
Moss peat		0.03-0.14 (mostly around 0.1)
Pine bark	2-5	0.12
	0.5-1	0.21
	< 0.5	0.30
	mix	0.25-0.27
Eucalyptus chippings		0.23
Sand	0.5-1	1.28
Clay		1.2
Diatomite		0.42
Calcined attapulgite	0.5-1	0.53
Lignite	0.5-1	0.47
	0.1-0.5	0.46
Exfoliated vermiculite	1-2	0.11
Clay pellets		0.3-0.6
Perlite	2-5	0.21
Slag	0.5-2	0.85

The porosity to air is the most important physical property of a substrate, and it can be measured by simple methods, some of them feasible for a farmer to carry out.

If a substrate has a low porosity to air, irrigation must be limited, especially in greenhouses, where losses due to evapotranspiration are low, so as not to saturate the air-filled pores (waterlogging). And to the contrary, a substrate with a high porosity to air requires frequent watering in summer, in order to replace the losses of water.

In general, the physical properties of a substrate cannot be predicted simply on the basis of knowing the components and their proportions, as these components vary greatly from one area to another and when they are mixed together, the components interact, and this means that the physical properties of the final mixture is not just a weighted average of all its components. So, in each case it is necessary to measure the properties of the components, or mixtures, used and this can in some cases be performed on the farm, but may have to be performed in a laboratory.

5.1.2. Chemical properties

A substrate's pH, its acidity or alkalinity, is one of its most important parameters, as the following factors all depend on the pH value:

• The possible presence of aluminium or manganese compounds poisonous to plants that limit their growth.
• The ease of assimilation of mineral nutrients, as their availability to plant roots depends largely on the substrate pH.
• The quantity of nutrients retained as a reserve in the exchange complex, as the capacity of the organic matter is much greater at higher pH values. This is why it is so important to know the value of the substrate's C.E.C. and pH.

The pH also affects the solubility of phosphorus, which is higher at lower pH values, meaning that low pH values increase the risks of losses by leaching and of toxicity at excessively high concentrations.

SUBSTRATE	C.E.C. meq./100 g
Young sphagnum moss	140-160
Eutrophic peat	70-80
Bark	70-80
Inert	0.1-1.0

Salinity, excess dissolved salts in the soil solution in the cultivation medium, is one of the most important nutritional problems when cultivating plants in containers. The effects are similar to dehydration due to lack of water, and it can be corrected by washing out the salts with excess water. It is easy to find out the degree of salinity by measuring the electrical conductivity.

Heavy metal contamination of substrates is a cause for concern among environmentalists, due to their ability to pollute the environment permanently. If you suspect, or know, that a substrate contains sewage sludge, slag, rubbish or other residues or byproducts that might contain heavy metals, it is necessary to check their concentration. This is because heavy metals are not only toxic to plants but can also enter the food chain leading to human beings when these substrates are used to cultivate vegetables. Many authorities recommend using substrates that are chemically inert (peat, perlite, vermiculite, etc.) when cultivating plants for human consumption.

The methods used to measure the fertility of organic substrates are different from those used for mineral soils. These differences affect all the different stages of chemical analysis, from sample preparation to the expression of the results, as well as the composition of the solution used to extract the available nutrients. As the results of the analysis of pH, conductivity and available nutrients depend to a large extent on the method used, it is essential to know which method was used in order to interpret the results correctly. As a general rule, one should choose to buy the substrates with the labels with the most complete details of composition, as this is to some extent a guarantee of how reliable the producer is. As long, of course, as they meet your needs.

Sustrate	pH	Conduct. µS/cm	Na mg/l	K mg/l	Ca mg/l	Mg mg/l	N-NH₄ mg/l(N)	N-NO₃ mg/l(N)	PO₄ mg/l(P)	Cl mg/l	SO₄ mg/l (SO₄)
Dark peat	6.5	363	40	119	204	24	54	78	19	73	15
Light peat (LP)	6.1	326	36	124	173	30	26	85	18	57	340
Peat mixture	5.2	478	59	147	231	56	70	125	35	117	560
LP + perlite	7.0	315	33	110	135	21	34	98	<5	30	306
LP + vermiculite	6.1	475	32	159	330	47	<5	200	27	56	387
LP + Pine bark (PB)	5.4	113	33	44	61	13	<5	28	12	57	49
LP + PB + wood fiber (WF)	6.6	110	26	61	51	8	<5	5	<5	46	40
LP + PB + sand	5.2	76	19	11	41	8	<5	6	<5	44	54
LP + PB + earth	6.7	310	31	113	209	34	14	23	<5	138	415
LP + compost	6.8	373	96	364	67	13	<5	6	26	330	36
LP + earthworm humus	7.4	3,360	910	3,422	860	237	20	766	<5	2,667	2,631
LP + pozzalana + WF	7.2	321	67	229	127	25	<5	54	<5	137	219
Pine bark (PB)	4.8	106	28	53	52	11	<5	15	13	49	72
Ground PB	7.2	241	35	88	167	20	5	5	5	71	350
PB + purines	7.2	816	96	606	65	19	155	66	<5	189	1,223
Ground PB + purines	7.1	839	128	682	177	21	183	119	119	344	1,086
Coco fiber	5.6	222	126	230	28	6	<5	14	14	206	51

5.2. TYPES OF SUBSTRATES AND THEIR CHARACTERISTICS

Substrates can be divided into organic and inorganic substrates. Organic substrates usually consist of peat or some other type of organic remains, such as pine bark, and they behave in a special way, because they tend to decompose to minerals (as they are organic). Inorganic substrates are made up of several different inert inorganic materials and are usually a product, or by-product, of some industrial process.
It is often worth mixing together different types of inorganic substrate, as then you combine the properties of the components making up the mix. As already pointed out, mixes do not show properties corresponding to the proportions of their components, and in fact each mix is like a unique substrate with its own special properties. Before mixing components, it is worth consulting in some specialist center, or the manufacturer, which mixes are best for the crop we are going to cultivate, and the properties of the mixture. Legislation in many countries obliges the manufacturers to sell these products by volume, and not by weight; this is a measure to prevent fraud, because these substrates can retain so much water that unscrupulous manufacturers could supersaturate it with water and sell the water at the price of the substrate.

	Composition (mg/l)				
	Nitrogen	Phosphorus	Potassium	Calcium	Magnesium
Pine bark	310	25	120	395	25
Fir bark	440	70	340	1,200	110
Sphagnum moss	450	2	14	150	20

Total nutrient content of organic substrates derived from peat and pine bark. Taken from Solbraa (1974)

5.2.1. Organic substrates

Organic substrates are derived from plant materials that have undergone some degree of humus formation. The best known and most thoroughly studied is peat, but now some conifer remains, such as pine bark and pine needles, are being used with some success. Every organic substrate tends to turn into humus, just like organic matter in the soil, and during its mineralization it releases nutrient ions, and so it is important to know its chemical composition, in order to take it into account when calculating the dose of fertilizer to be applied or when preparing a nutrient solution. The table shows the total nutrient content of the most common organic substrates.

5.2.1.1. Peat

Peat is defined as the broken down remains of the vegetation of a bog, which has not decomposed completely due to the excess of water and lack of oxygen, and deposited over long periods of time, favouring the formation of relatively dense layers of organic matter (Penningsfield & Kurzmann, 1975). Other authorities (such as Strasburger et al., 1977) have pointed out that this natural substrate consists of deposits of the remains of mosses and higher plants that are in a state of slow carbonization, in the absence of oxygen, meaning that they maintain their anatomical structure for a long time.

Depending on the site of origin of the peat, peats are classified into low, transitional and high.

Valley bogs (soligenous mires) are eutrophic and produce *fen peat*, highly decomposed peats that are not suitable for agriculture, as they show low porosity, deficient water retention and air retention, and they may even contain materials toxic to plants on their exchange complex.

The most important characteristics of the origin, formation and components of different types of peat

Raised bogs (ombrogenous bogs) are oligotrophic and produce *raised bog peat* from peat bogs that form in cold regions with high rainfall and relatively high moisture (Canada, former USSR, Finland, Poland and Ireland). They consist mainly of ***Sphagnum spp.***, about 90% by weight. These peats retain large amounts of water, and the living topmost layer gradually buries the lower dead layers. Some of these peat bogs are up to 10 m deep, and started to form more than 10,000 years ago. Depending on the degree of humus formation, there are two types; **slightly decomposed peat** (SLDP), or light peat, is widely used in agriculture as it has excellent physical properties, such as a very open structure and a high capacity to retain water and air; and strongly decomposed peat (STDP) or **black peat**, which is a very dark color. Dark peat is not so highly appreciated as its high degree of decomposition means that it has lost many of its properties. Finally, there are **transition peats**, which are

The influence of differences in plant composition on the water retention capacity of peats. Data expressed in grams of water per 100 g dry peat. Taken from Penningsfield & Kurzmann (1975)

Light sphagnum peat distributed by Químicas Sicosa, S.A.

Sphagnum is a moss. Sphagnum is the name of the genus, and spp. means any of the species of the genus Sphagnum. It is used for both plants and animals.

Sphagnum

COUNTRY	ANNUAL PRODUCTION (millions of m^3)	ESTIMATED AREA (thousands of km^2)
Former USSR	300.0	1,500.0
Germany	6.0	11.1
China	4.0	34.8
USA	1.6	402.0
UK	1.5	15.8
Canada	1.1	1,700.0
Ireland	1.1	11.8
Sweden	0.8	70.0
Poland	0.8	13.5
Finland	0.7	104.0

FORMATION			
ORIGIN	PROCESSES	ENVIRONMENTAL FACTORS	COMPONENTS
Plant remains (mosses and other plants)	Slow carbonization + stratification	Absence of O$_2$. Excess moisture. Low temperatures	Partially decomposed organic matter

PEAT	RETENTION CAPACITY
Sphagnum	1,000-1,500
Carex	700-800
Eriophorum	500-600
Wood pantanosa	400-500

typical of central Europe (France, Germany) with characteristics intermediate between valley bogs and raised bogs.

Because peats are organic matter, they show the same properties as the organic matter of mineral soils, which we have already discussed. They usually have a high water retention capacity, and generally have an acid pH. Their C.E.C. is usually very high as is their porosity, or ability to retain air. These characteristics are present to a lesser or greater extent depending on the type of peat and the extent it has turned into humus.

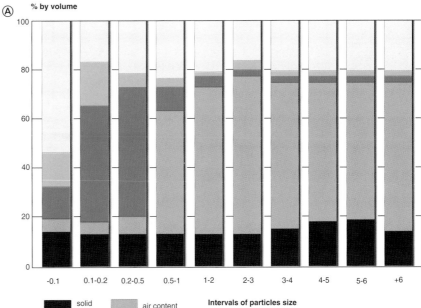

A

% by volume

5.2.1.2. Conifer remains

In the last thirty years, many substrates have been studied for use in agriculture. Many of them, due to their poor physical and chemical properties, are no longer in use. One of the artificial soils that has given the best results is the remains of several types of plant. Two of the most important are the bark and the needles of different species of pine (*Pinus spp.*). The bark and needles have a very high particle density, something like 2.00 for bark and 1.90 for needles, ensuring good retention of water and air.

Refer to the table above, which shows a study of the percentages of solid material and the air and water content, and note that if we want the substrate to have a good retention of water available to plants, we must choose a particle size between 0.1 and 0.5 cm. If, however, we want the substrate to have good aeration, then we must choose larger particles (1 cm or larger). The pine barks with a larger particle size will require regular inputs of water, as particles of this size retain little water available to the plant. They basically behave like the particles in a soil; the smaller they are, the better their water retention and the worse their air retention; while the larger the particles, the better the retention

of air and the worse the retention of water. In chemical terms, it is worth pointing out that their exchange capacities range between 70-80 meq/100 g of substrate, far higher than the normal values for a mineral soil.

5.2.2. Inert substrates

Inert substrates are used for hydroponic cultivation in a greenhouse. As hydroponic cultivation uses nutrient solutions, the substrate has to be chemically unreactive, i.e., it should not absorb or supply any elements at all. In the case of hydroponic cultivation, it is worth thoroughly washing substrates like gravel and sand, in order to remove any remaining soil, because this might change the nutrient solution. Substrates for hydroponics must also be biologically inert, free of pests and diseases, as they may bring latent diseases into the crop.

A/ Study of the pine bark substrate and its capacity to retain water and air. Within the different grades of pine bark, choose grades that have a small particle size (0.1-0.5 cm) as they are the ones that retain the most water. B/, C/ and D/ Different aspects of peat processing in Germany. Photos by courtesy of NV VAN ISRAEL

D

5.2.2.1. Gravel

There are three types of gravel on the market, derived from different sources. There are **quartz gravels**, derived from siliceous or "acidic" rocks. Make sure the granules are not very large and that the edges are not sharp. Its water retention capacity is poor, meaning it has to be watered regularly. Its chemical behaviour, however, is excellent, as it neither supplies nor absorbs any element. Its price is low, but transport is expensive. **Pumice stone gravel** is derived from basaltic, or "basic", rocks (with a low silicon content). Unlike quartz gravel, it has excellent physical properties. When the particle size is in the range 2-15 mm, the volume of pores accounts for 85% of the total volume. **River gravel** can also be used as a substrate, but it has the same low porosity problems as quartz gravel.

5.2.2.2. Sands

Like gravel, sand is a natural substrate. The only sands that are suitable for cultivation are siliceous sands or ones that consist mainly of quartz (calcareous sands are not usually suitable). The sands used in agricultural are usually from rivers (siliceous sands), because in many countries extracting sand from beaches (calcareous sands) is against the law.
The only difference between siliceous sands and the gravels described above is their particle size. The particle size of sands is between 2 mm and 0.05 mm. Over time, the sand is weathered and loses its good aeration properties, though it usually lasts several years.
Sand is expensive, and so it can only be used in highly profitable crops. Sand is now very widely used in the construction of sports fields, mixed approximately 50%/50% with peat.

Crops on volcanic ash. These soils are very fertile, rich in all types of nutrients.

The green of a golf course and the turf of a football pitch must have extraordinarily high capacity of aeration (provided by the sand), but they must also have a high capacity to retain water and nutrients (provided by the peat).

5.2.2.3. Volcanic earth

Like gravel and sand, volcanic earth (tuff, or volcanic tuff) is a natural substrate, but it is of volcanic origin. The particle size varies from a few millimeters to 1.5 cm. Volcanic earth is reddish in color and is very porous, giving the substrate very good aeration. Its large pores make it a poor substrate in terms of water retention. It is often used as a surface decoration, scattering a thin layer over other substrates, in plant pots, containers and planters.

5.2.2.4. Perlite

Perlite is a binary compound of ferrite and cementite, which are both obtained through metallurgical processes. There are two types of perlite, which differ in their microscopic structure, which may be laminar or granular. When granular perlite is heated to 1,000°C, it expands, giving rise to very light spheroid forms, whose bulk density is only 130-180 kg/m^3 (0.13-0.18 g/cm^3).
This expanded material is used in agriculture either on its own or mixed with other substrates, for soilless cultivation or in containers. It cannot be used in the open air, because it is so light it would be blown away by the wind.
It is an artificial inert substrate, white, and with roughly spherical particles 2-6 mm in diameter. It is chemically inert at a pH of 7-7.5, but at a very low (acid) pH values it may liberate

aluminium, one of its components.
Peat is often mixed with perlite in order to improve the peat's drainage and aeration. It is easily deformed when gently squashed between the fingers, meaning that its useful life in cultivation is limited, as the mechanical action of the roots eventually ruins it.

5.2.2.5. Vermiculite

Vermiculite is a magnesium-containing hydrated sheet silicate mineral ($Mg_3Si_4O_{10}(OH)_2 \times H_2O$). Like perlite, it expands when it is heated rapidly (to 300°C), reaching four times its original volume.
In agricultural terms, vermiculite is an expanded and exfoliated clay. Its particle size is between 5 and 10 mm. It is thus a low density material, with a good water retention capacity. The fact that it is a clay means that it retains the ion-adsorbing properties of clays.
It is sometimes alkaline, due to its possible magnesium content. Despite all these advantages, it also has its disadvantages, as with time it compacts and loses its water retention capacity. It is often included in formulations mixed with other substrates, thus improving the water retention properties of the resulting substrate.

5.2.2.6. Rock wool

Rock wool is an inorganic material obtained by mixing diabase (60%), limestone rock (20%) and coal (20%) in solution at 1,600°C. It is an artificial substrate, but is not totally inert as it supplies small quantities of iron, magnesium, manganese and especially calcium. Its pH value is slightly alkaline, ranging between 7 and 9, though over time it tends to fall to neutral.
It is sold commercially in the granulated form. Its bulk density is low, giving it a high water retention capacity. It has a high water retention capacity at low water potential values, and in addition the retained water gradually flows from the top of the container to the bottom. It is usually mixed with other substrates to combine their different properties.

5.2.2.7. Polystyrene

This is a thermoplastic material obtained by polymerizing styrene. It is obtained by heating an artificial compound consisting of round white particles, 4-12 mm in diameter. Its weight is low, as is its water retention capacity, but its aeration is good. Its pH is 6 to 6.5.

5.2.2.8. Polyurethane

This is the generic name for several synthetic polymers that contain the urethane group. When heated they expand to form a foam. It is totally inert, very light, with a very stable structure and a very high porosity (98%), and so its air retention capacity is very high. Its great disadvantage is that is has no water retention capacity. It is often used as a seed bed for seed germination.

Vermiculite, before and after expansion ("popping")

6. SOIL DISINFECTION

Often, the repeated cultivation of the same species leads to an increase in the crop's pathogens in soil. We say that the soil is "tired" or "exhausted". This exhaustion may also be due to a lack of nutrients, and this question is dealt with in the chapter of fertilizers. If you think about it, it is logical that the populations of pathogenic microorganisms will increase when the same crop is permanently cultivated in the same soil. The root diseases due to these microorganisms increase exponentially, eventually making it impossible to continue cultivating the same species on the site.

Soil disinfection is an agricultural practice consisting of applying pesticides or steam in order to eliminate, or at least reduce, the crop parasites present in the soil.

Within the section on organic matter, we mentioned the organisms living in the soil. Many of them are beneficial to crops, but some raise major problems for plants. It is necessary to distinguish here between the multicellular animals, or higher animals, and the microorganisms, or microscopic organisms, which are usually unicellular. The animals include the insects (beetles, butterflies, flies, homopteran bugs, etc.), millipedes and the nematodes. All these organisms damage the roots and crown of plants. The microorganisms include small nematodes, fungi, actinomycetes, bacteria and viruses. The seeds of weeds may also accumulate in the soil. Some disinfectant products act as herbicides and are more active and more dangerous, and must be used with great care.

Basically, disinfectants meet the farmer's need to control two groups of organisms that inflict great damage on plants, nematodes and fungi. Nematodes are usually very abundant in soil, and seriously damage plant roots. Fungi, such as *Fusarium spp., Verticillium spp.* and *Phytophthora spp.* are responsible, respectively, for fusarium wilt, verticillium wilt

Typical symptoms of gummosis produced by Phytophthora *on the trunk of an orange tree. This soil fungus attacks the roots, neck and lower trunk. One product effective against soil* Phytophthora *is Fosetil-Al at 80%. The active material is sold as ALERTE by AgrEvo and is manufactured by RHÔNE-POULENC.*

and blight, as well as many mildews, rots and blights of plant roots and stems.

Obviously, disinfection must be carried out before the crop is planted, because even the least herbicidal disinfectants would cause serious damage to the plants. And of course, after applying the product you must respect the manufacturer's instruction on how long you must wait before performing any operation (especially sowing) on the disinfected ground.

Soil temperature and humidity are important factors to bear in mind when applying disinfectants. Most products are ineffective at temperatures below about 10°C. At higher temperatures, the gas is better distributed and disinfection is more effective. The soil moisture

should not be any higher than the field capacity, as excessive water in the soil prevents the product's penetration. Even for products that are applied to the soil as solutions, it is better for the soil to be as dry as possible, as the greater the pore volume full of air, the better the product will penetrate.

It is also important to take into account the percentage of organic matter in a soil, as it retains the disinfectants. In extremely organic soils, it is better to increase the period of time stipulated by the manufacturer by a few days. Clays also show an ability to retain pesticides. The price of these products and the cost of specialized personnel to apply them mean they are too expensive for use in extensive agriculture. However, they are used in intensive horticulture, especially among greenhouse crops.

Some of these products are extremely toxic, and absolutely must be applied by highly trained staff. In many countries, there is very strict legislation governing the use, handling and storage of these products.

Every year new disinfectant products are launched onto the market. The following pages contain a list of the ones most widely used in Spain, and how they should be applied. Anyway, the farmer should visit several companies and get information on the different products on offer, the microorganisms they kill, how they must be applied, their chemical composition, their concentration, the follow-up safety period, their toxicology, etc.

To finish, note that very strong disinfectants, the ones that even kill weed seeds, greatly destabilize the soil. This means that they leave it without life, and this leads to an imbalance between the treated plot and untreated neighboring ones. This happens even inside a greenhouse, as the disinfectant gases travel within the porous space of the soil.

6.1. DISINFECTING SOIL WITH STEAM

One of the first systems to be applied, steam disinfection consists of introducing steam into the soil through a nozzle. This is usually done at a depth of 10 to 30 cm and requires specialized machinery. Steam disinfection is effective against insects and some nematodes, and has some effect against fungi, though its only effect on the most detrimental fungi, such as *Fusarium spp.*, is to reduce their population. Use of steam, though it has little destabilizing effect on the soil, has fallen into disuse.

In those soils without problems with fungi, bacteria or nematodes, removing weeds mechanically is cheaper than any other method.

6.2 DISINFECTING THE SOIL WITH CHEMICAL PRODUCTS

These products, produced by the agricultural chemicals industry, are often known as pesticides. They are usually classified according to the microorganism to be controlled, or by their formulation, either solid, liquid or gas. They can also be classified by how they are applied. There are three main methods of application; injection into the soil, dissolved in irrigation water or by covering the soil with plastic sheeting and applying the product as a gas.

In general the (chemically) weakest disinfectants are the ones whose action is least effective, but they are the ones that destabilize the soil least. The most effective disinfectants, the most toxic ones with the broadest spectrum of activity, usually have a highly destabilizing effect on the soil. We will briefly look at some of the most common.

authors suggest the formulation with more chloropicrin gives better results against fungi. It is applied by injecting the gas under a plastic sheeting at a rate of 500-800 kg/ha. After three days, the plastic sheeting can be removed, but you must wait two weeks before handling the soil. In order for the treatment to be effective against weed seeds, the highest dose has to be applied. The time that must elapse before you can plant varies depending on a large number of factors. After using methyl bromide, before you sow you should carry out the "lettuce test", which consists of planting two or three lettuce seedlings in different places, and then observing how they react.

6.2.2. Dichloropropane

Dichloropropane (or **DD**, as it is sold in Spain) is basically used to kill nematodes, and is not very effective at controlling other organisms. It is

Polyvalent disinfectants and their effectiveness against different soil organisms ("x" represents moderately effective, and "xx" represents highly effective)

ACTIVE MATERIAL	APPLICATION	EFFECTIVENESS			
		Fungicide	Nematicide	Herbicide	Insecticide
Methyl bromide 98 % 2 % Chloropicrin	Under plastic	x/xx	xx	xx	xx
Methyl bromide 67 % 33 % Chloropicrin	Under plastic	xx	xx	xx	xx
Dichloropropene	By injection and in irrigation water		xx		
Dichloropropane	By injection and in irrigation water		xx		
Dazomet	Plowing in granules	xx	x/xx	x	xx
Metam-sodi	By injection and in irrigation water	xx	x/xx	x	xx

6.2.1. Methyl bromide and chloropicrin

Methyl bromide is a liquid bromine compound that evaporates to gas at a temperature of 3.6°C, and is highly toxic. Chloropicrin, another gas, is extremely irritant. The two gases are very effective at disinfecting soil.

Methyl bromide is an extremely toxic gas, and because it is also odorless, there is legislation obliging producers to include at least 2% chloropicrin, because this gas is tear-producing and has an intense smell (as well as being a fumigant), and is thus easily detected.

It is sold commercially as a liquid, in flattened green metal canisters, similar to the smallest sizes of butane gas canister. There are two formulations of these gases on the market in Spain; methyl bromide 67%:33% chloropicrin, and methyl bromide 98%:02% chloropicrin. Both formulations are highly effective against fungi, nematodes, weed seeds, and as insecticides, though some

applied by injection into the soil (at a depth of 15-20 cm) at a dose of 400 to 800 kg/ha. Its safety period is from four to eight weeks.

6.2.3. Dichloropropene

Like dichloropropane, dichloropropene is excellent against nematodes, but is not very effective against other organisms. It is applied by injection into the soil, or dissolved in irrigation water. It is applied at a rate of 600 to 700 kg/ha, and the safety period that must elapse is 4 weeks.

6.2.4. Dazomet

Dazomet is a granular product that has to be mixed into the soil by plowing. It is an acceptable soil disinfectant, with a good action against insects and fungi, but its effectiveness against nematodes and weeds is only moderate. The dose is between 600 and 700 kg/ha, and the safety period is about 3 weeks.

6.2.5. Metam-sodi

Metam-sodi, or VAPAM, is an acceptable soil disinfectant, as it is an effective insecticide and fungicide, but its effectiveness against nematodes and weeds is only moderate. It can be applied by injection into the soil or in irrigation water. The dose is usually between 800 and 1,200 kg/ha. Its safety period depends on the soil temperature; at a temperature of 12-18°C, you must wait about 3 weeks, but at a temperature of only 8°C, the waiting period is 6 weeks.

6.3. THE BEHAVIOR OF DISINFECTED SOILS

It is important to realize that a disinfected soil, especially one disinfected with methyl bromide, is defenceless. By eliminating every trace of plant and animal life from the soil, it has become a virgin site, where microorganisms can arrive very easily, because there is no competition. The disinfected soil thus behaves as an empty receiver that will eventually recover its animal and plant life.

After disinfecting the soil and waiting the necessary safety period, it is worth cultivating the species most vulnerable to the organisms we have eliminated, and less sensitive crops in the second planting.

The whole subject of soil disinfectants, especially methyl bromide, is of great concern to environmentalists. Methyl bromide contains the element **bromine**, which is a liquid that evaporates very readily. Bromine gas is extremely irritating to the eyes, nose, etc. As it is a disinfectant, it is obviously highly toxic to living things. Some percentage of the bromine is retained on the clay-humus complex and it is eliminated from the soil so slowly that if disinfections are frequent, it may even poison the plants due to accumulation of bromine. Bromine also reacts with metals, leading to formation of metallic bromine compounds, and these contaminate the soil and water table.

Bromine: an element, whose symbol is Br, that is a member of the halogen group of elements

Cultivated fields

7. CULTIVATION WITHOUT SOIL, HYDROPONICS

As agriculture has gradually turned from an art into a science, it has become possible to greatly increase control over the environmental conditions that plants grow in.

In terms of the soil, one way of increasing this control is to replace the soil with a less complex medium that fulfils the same functions. Science's response to this challenge has been the development of hydroponic cultivation.

The artificial cultivation mediums available have mainly been created for plants whose high market price makes them more profitable (flowers, ornamental plants, etc.).

Practising this cultivation method requires some basic technical knowledge, and great dedication from the producer, and without them reliable production is very difficult. Nowadays, hydroponic cultivation is widespread, and this has made it much more precise, though it is important to realize that the initial start-up investment is very high.

You might imagine that hydroponic cultivation is a recent technique, but surprisingly this technique goes back to the 17th century. The first recorded scientific investigation was in 1868, when Van Helmont planted a cutting of willow in a special container. After five years of watering it with nothing but rainwater, he confirmed that it had grown. The illustration is a drawing of a container designed for cultivation in water by Sachs (1868). Controversy as to whether it was the earth or the water that provided the nutrients needed for its growth led the scientific community to undertake further research into hydroponic cultivation.

7.1. HYDROPONIC SYSTEMS

Hydroponic cultivation consists of replacing the soil with a natural or artificial substrate that may be solid or liquid. This type of cultivation is not restricted to cultivation in water, but also refers to cultivation in inert media, such as perlite, volcanic earth and expanded clay.

The plant's nutrition is based on the nutrient solution that is supplied, and this has to meet all the plant's nutrient requirements. Thus, only the mineral nutrients needed for optimum plant growth are added to the water. As you can imagine, the nutrient solutions consists of dissolved ions (cations and anions).

There are two ways of preparing the nutrient solution. The first method is to prepare a concentrated stock solution for each chemical element, and the second method is to add the salts directly to water, as long as the compounds used are known to be compatible with each other. We must always remember that the nutrient solution's pH must be suitable for the specific crop.

The most important environmental factors affecting hydroponic cultivation are temperature, light, carbon dioxide supply, moisture and oxygen levels around the roots.

These factors must all be studied in detail and interrelated on the basis of the expected production and the species we want to cultivate. Within the system of hydroponic cultivation, two different techniques can be distinguished, cultivation without soil and cultivation using an inert substrate. We can further classify different types of cultivation technique. Cultivation without a substrate includes cultivation in a tank of nutrient solution, the nutrient film system and horizontal or vertical aeroponics. Cultivation methods with an inert substrate include hydroponic cultivation in gravel or sand or rock wool.

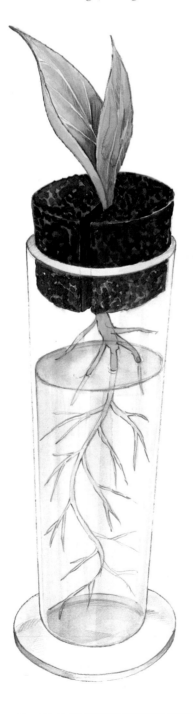

Recipient for water cultivation, acording to Sachs (1868)

With modern "in vitro" cultivation techniques, plant growth occurs on a nutrient gel that is also the plant's support.

7.2. CONCLUSIONS

The scarcity of arable soil that affects humanity on a world scale suggests that in the not-so-distant future hydroponic cultivation will be a great success. Furthermore, hydroponic cultivation makes production possible in sites where the soil is not suitable for cultivation, because hydroponic cultivation does not need to use the soil.

The great disadvantage of hydroponic cultivation is its high cost, meaning that it can not be recommended for vegetable crops that do not have a high profit margin. But in the future, cheaper hydroponic systems will be introduced, at the same time as governments will create grants to encourage agricultural production, given the increasing need for food of the world's increasing population.

This section of cultivation without soil and hydroponics has been a brief overview, as hydroponic cultivation is really most important in greenhouse crops. In the section on greenhouse cultivation, the reader will find another section on hydroponics, which describes the chemicals that have to be used, how to make up the solutions, and the advantages and disadvantages of hydroponics.

FERTILIZERS

1. GENERAL CONCEPTS

In the previous chapter, which discussed soils, we were introduced to the importance of the cation exchange complex (C.E.C.) and the nutrient elements that are absorbed onto it. We studied the ions, the form in which plants take up the nutrients they need. As we have seen, the plant uses these ions, together with water and carbon dioxide, to produce organic matter for their energy requirements and to build up their body in the process called photosynthesis, which means "synthesis using light".

To the right: Feeding well-nourished grass to livestock is also good for people

Applying fertilizer to plants is one way that human beings artificially modify the concentration of ions in the soil solution, in order to increase crop yields. This modification is usually achieved by adding something to the soil, and the products used include animal dung, mixes of synthesized chemicals, and inputs of nutrient-rich minerals from other sites.

Diagram of the nutrition of plants

As we have already seen, the weathered parent material and its constituent minerals are the origin of soils and the source of nutrients for the plants. Some places on the planet's surface are very rich in nutrients, such as the nitrate mining areas of Chile. Chilean nitrate was one of the first fertilizers to be used, after animal dung. Human beings are thought to have started agriculture in the Neolithic period. It is important to realize that the first books on farming date from roughly 400 BC-300 AD. These early books all mention that dung is good for crops. Human beings learned long ago that incorporating some types of organic matter into the soil led to increased harvests.

The 19th century Industrial Revolution caused a boom in research and development of chemical fertilizers, inorganic fertilizers based on chemicals synthesized in laboratories.

Around 1840, Liebig and the German school of agricultural chemists started to manufacture the first artificial fertilizers, especially ones supplying potassium. In 1898, William Crookes started to search for chemical methods of supplying nitrogen to the soil. In 1906, however, F. Haber worked out how to use atmospheric nitrogen to make a synthetic fertilizer, and with the help of K. Bosch the process was brought into commercial production.

Fertilizer production and consumption has grown spectacularly in industrialized countries since the Second World War. As the world's population is increasing, the need for food is also increasing. These needs can be met by feeding our plants with the correct amount of nutrients in the form of fertilizer. This does not only apply to those plants grown for human consumption, but also those to be fed to livestock, because if we supply nutrients correctly to our plants, we ensure that our animals are well fed, and thus that human beings are well fed.

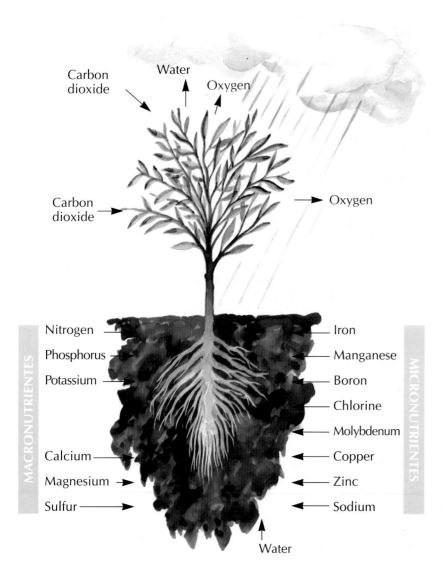

Carbon dioxide
Water
Oxygen
Carbon dioxide
Oxygen

MACRONUTRIENTES

Nitrogen
Phosphorus
Potassium

Calcium
Magnesium
Sulfur

Water

MICRONUTRIENTES

Iron
Manganese
Boron
Chlorine
Molybdenum
Copper
Zinc
Sodium

1.1. GENERAL POINTS

As we have already pointed out plants require nutrients in the form of the chemical elements or their compounds. These elements must be absorbed in the form of ions, or to put it another way, they must be dissolved in the soil solution in order for the plant to take them up.

The composition of plants is similar to that of animals. 80-90% is water, leaving 10-20% dry matter.

Like animals, the most active tissues are the plant parts with the highest water content, the meristems, or growing tips, which contain about 93% water, while this percentage decreases to 55% in the most highly lignified tree trunks.

The analysis of the average chemical composition of plant dry matter (see the accompanying table) shows that the most important elements in its composition are carbon, oxygen and hydrogen. These three elements account for 95% of the dry weight of plants. They are abundant and their sources cannot be exhausted, so that they are not a source of concern from the point of view of plant nutrition. From the point of view of plant nutrition, the main nutrients required are nitrogen, phosphorus and potassium, called the macroelements. The secondary elements, sulfur, calcium and magnesium are needed in lesser amounts than the macronutrients. The micronutrients are essential for plants but they only require them in very small amounts. They are also known as trace elements, because when analyzing plant materials, traces of these elements are found. Micronutrients are essential to the plants, but they are only present in tiny quantities in plant tissues. The most important ones are iron, manganese, copper, zinc, boron, molybdenum and chlorine.

1.2. FERTILIZER UNITS

A fertilizer contains chemical substances that, when they come into contact with the soil solution, can be transformed into ions. These ions must be in a form suitable for the plant to take up and use for its nutrition. These fertilizer elements are quantified as fertilizer units and vary with the richness of the fertilizer.

Fertilizers may contain all the elements listed above (macroelements - nitrogen, phosphorus and potassium; secondary elements - sulfur, calcium and magnesium; microelements - iron, manganese, copper, zinc, boron, molybdenum and chlorine), some of them, or often just one of them. The fertilizer's richness of each element is expressed in a standardized way. Nitrogen for example, is expressed as **total N**, phosphorus is expressed in terms of **P_2O_5**, and potassium is expressed as **K_2O**. Each secondary element and each of the remaining microelements is expressed in its own way, as we shall see over the course of this chapter.

The phosphorus content and potassium content of fertilizers have traditionally been expressed in terms of their oxides. The terms *phosphoric acid* and *potash* mean that they are measured in terms of their oxides, and this is a simple mathematical calculation based on the percentage values. Fertilizers in fact do not contain P_2O_5 or K_2O. Some manufacturers indicate the content in both ways. Thus the fertilizer's label may contain information in terms of total N, P_2O_5 and K_2O or the values expressed as the chemical elements N-P-K.

We may need the factors to convert these values for oxides to the value for the element itself (which we are really interested in knowing). The conversion factors are calculated on the basis of the atomic weight, and are:

$$\%P *2.29 = \%P_2O_5 \qquad \%P_2O_5 * 0.44 = \% P$$

$$\%K *1.20 = \%K_2O \qquad \%K_2O * 0.83 = \% K$$

Fertilizer packets do not usually list the chemical compounds they are made of. This information is generally not necessary, because the guaranteed analysis shows the purchaser the quantity of each nutrient that it contains. One fertilizer unit is equal to 1 kg of nitrogen, of phosphoric acid or of potash (or by extension a kilo of any of the remaining elements). The exact calculation of the basis of the type of fertilizer is dealt with in the next section, which deals with the richness of fertilizers and how it is calculated.

Element	Symbol	Percentage
Carbon	C	40-50 %
Oxygen	O	42-44 %
Hydrogen	H	6-7 %
Nitrogen	N	1-3 %
Phosphorus	P	0.05-1 %
Potassium	K	0.3-3 %
Calcium	Ca	0.5-3.5 %
Sulfur	S	0.1-0.5 %
Magnesium	Mg	0.03-0.08 %
Sodium	Na	0.001-3.5 %
Silcon	Si	0.005-1.35 %
Chlorine	Cl	0.15-0.25 %
Iron	Fe	Traces
Manganese	Mn	Traces
Cooper	Cu	Traces
Zinc	Zn	Traces
Boron	B	Traces
Molybdenum	Mo	Traces
Cobalt	Co	Traces
Aluminium	Al	Traces
Fluorine	F	Traces
Selenium	Se	Traces
Bromine	Br	Traces
Iodine	I	Traces

Percentage of the elements present in the largest amounts in plants. The twelve most important elements account for more than 99% of the dry weight. Taken from Javillier

1.3. THE RICHNESS OF FERTILIZERS

As has already been pointed out, fertilizers may contain one or several fertilizer elements. The most usual way of describing a fertilizer is by using a series of three numbers that represent the percentages of N-P-K (nitrogen, phosphorus, potassium). So if the label of a commercial fertilizer says 20-5-10, it means the fertilizer contains:

20%	nitrogen (total N)
5%	phosphoric acid (P soluble in citrate in the form of P_2O_5)
10%	potassium (K soluble in water and in the form of K_2O)

If we wished to calculate the exact content of each element, we have to use the factors of conversion in the above table

20%	N total
5%	P_2O_5 * 0.44 = 2.2% P
10%	K_2O * 0.83 = 8.3% K

Right: A diagram representing Liebig's Law of the Minimum for fertilizer elements. If we try to fill the barrel in the diagram with water, this will flow out through the smallest stave, which in the example is the element phosphorus (P). The maximum yield of a crop will be limited by the element that is limiting in the soil.

Another way of calculating the richness of a fertilizer is to work out the number of fertilizer units of each element that it contains. The problem is approached from the opposite point of view. Suppose we have a 25 kg sack of fertilizer whose N-P-K richness is 20-5-10. 1st - How many kg of fertilizer do we need to have one fertilizer unit of nitrogen? 2nd - How many kg for one fertilizer unit of phosphoric acid? 3rd - And for one fertilizer unit of potassium oxide? Supposing that the contents of the sack correspond to 100%, that the total fertilizer content is 20% + 5% + 10% = 35%, and that the other 65% consists of other ingredients that do not contribute to the fertilizer, then:
we would need

1) 100/20 = **5 kg** of 20-5-10 to make one fertilizer unit of nitrogen
2) 100/5 = **20 kg** of 20-5-10 to make one fertilizer unit of phosphoric acid
3) 100/10 = **10 kg** of 20-5-10 to make one fertilizer unit of potassium oxide.

Calculation of the fertilizer units is especially useful for farmers who prefer to mix their own fertilizers, and buy several monovalent fertilizers (ones that are sources of a single element). This has the advantage that each fertilizer unit costs much less (monovalent fertilizers always work out cheaper than mixtures). The farmer can thus change, depending on the requirements of the crop, the amount of each element, and the best way to do this is on the basis of the fertilizer units, because the number of kilos of each fertilizer to be used is known.

Co Mn Na Ca   S   N   P   K   B Fe Mg Cu

1.4. THE PROPORTIONS OF NUTRIENTS

To make good compost with grass, it is worth using a granulated acid conservation agent, such as Euro Sil, produced by TIMAC, S.A.

Calculation fertilizer richness in terms of fertilizer units may seem rather irrelevant, but knowing how to do this is very useful, because specialist articles often discuss a crop's needs in terms of fertilizer units of nitrogen, phosphorus and potassium. Using these calculations the farmer can use any type of fertilizer by converting the figures given into equivalent units that are suitable for their crop.
Some fertilizers only contain one or two of the three elements N-P-K. These fertilizers usually put a zero instead of a value for the missing element(s). The formula 0-20-0 indicates a fertilizer containing 20% phosphorus (as P_2O_5) and no nitrogen or potassium.

The calculation of the percentage P would be:

$$20\%P_2O_5 * 0.44 = 8.8\% P$$

And the calculation of the fertilizer unit:

100/20 = **5 kg** of 0-20-0 make one fertilizer unit of phosphoric acid.

The proportions of nutrients are only useful for ternary fertilizers consisting of the three nutrient elements N-P-K. These proportions are calculated from the ratios of the three elements in the following way:
In the previous example of a 20-5-10 ternary fertilizer, we divide the smallest value (5) by itself, and the others by the same value. Thus 20/5 = 4, 5/5 = 1, and 10/5 = 2, and so the proportion of N-P-K in a 20-5-10 fertilizer is 4:1:2.
This means that for every four parts of nitrogen, we are supplying one unit of phosphorus and two of potassium. This proportion is especially useful as in many articles the application of fertilizers is referred to using these ratios. The study of plant nutrition shows that a balanced supply of all the fertilizer elements is more important than supplying minerals separately.

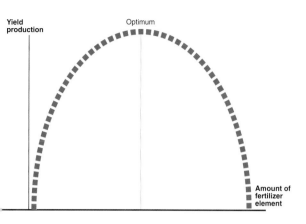

This question is clarified in the next section, which states the Law of the Minimum. Plants, like all organisms, show nutritional patterns in common. Just as a person cannot survive on proteins, fats or sugars alone, plants need a balanced nutrition with all the different ions they require.

1.5. LIMITING FACTORS

The Law of the Minimum, or the Law of Limiting Factors, was formulated in the 19th century by Liebig. It says, more or less, "*The size of the yield obtained is determined by the element that is present in the lowest amount in relation to the needs of the crop*". This law reveals that "fertilizers show solidarity", that is to say, that the shortage of just one essential element affects production as a whole, even if the other elements are all present in sufficient quantities. In fact, it would be more precise to say that each production factor acts better the closer the other factors are to their optimum values. The optimum value for each factor can be considered to be independent.

The Law of the Minimum does not only apply to nutrient elements, but also affects all the other factors influencing growth that may limit yields. Thus all the growth factors are independent and any one may be limiting.

Rational cultivation forms a whole that must harmoniously combine all the growth factors, so that each one is acting in the best conditions, and not restricted by a deficiency of another factor.

Obviously, intensive fertilizer application may be considerably less effective if a low productivity variety is planted or if weeds are abundant. And likewise a high yield variety will not be able to attain its potential if fertilizer application is insufficient, or if there are too many weeds that compete with the crop and restrict its growth.

The factors affecting growth are very varied, and include genetic, cultivation, climatic and nutritional factors. In terms of nutritional factors, one can estimate that fertilizer application leads to harvest increases of 25-50% when compared to an unfertilized control plot of ground. This law is normally represented as a barrel of

water (see drawing) in which each stave represents one of the elements. The thickness of each stave represents the proportion, with respect to the others, of the element required by the plant. A shortage of one of them limits the capacity of the others and sets the maximum possible yield. If the presence of the other elements was optimal (all the staves of equal height), the crop's yield would be the optimum, i.e., the highest.

The Law of the Minimum has another aspect, the Law of the Maximum, which postulates that increases beyond a certain quantity of fertilizer do not increase production, but in fact cause it to decline. This maximum depends on the needs of each species, and on other factors, such as the climate, soil, etc.

Valuable conclusions can be drawn from the Law of the Minimum and the Law of the

The law of decreasing yields, or the Law of the Maximum. As we increase the amount of fertilizer applied, we increase production, until we reach the optimum, after which production decreases.

Maximum. The purpose of a fertilizer is to supply nutrients to the plant and to prevent deficiencies. The fertilizer should therefore be as complete and balanced as possible, because it is of no use at all if it supplies many fertilizer units of nitrogen but supplies no units of the other elements. Excessive fertilizer application, above the crop's requirements, apart from being a waste of money, reduces the production from the optimum point, as shown in the graphic representation of the Law of the Maximum (above).

Furthermore, excessive fertilizer use contaminates the water table, as the most mobile and soluble elements, such as nitrogen, are lost by washing.

Using the right dosage of fertilizer and applying it when the plant needs it, increases production and leads to lower production costs and a lower final price.

A factory producing spherical granulated fertilizers, containing organic matter and minerals, super potassium fertilizers, and ternary mixes. Photo by permission of ETS. PLANTIN.

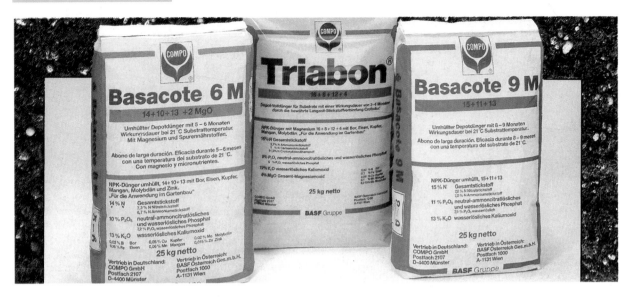

2. CLASSIFICATION OF FERTILIZERS

Fertilizers can be classified in several different ways. One criterion is their nutrient content, and they can also be classified by whether their components are of mineral or organic origin. In the case of manufactured fertilizer, the manufacturing process is very important, as this gives rise to different types of fertilizer, and the following are the three main ways of classifying fertilizers.

2.1. BY THEIR PHYSICAL STATE

We can divide fertilizers by the physical state they are sold in, that is to say, whether they are a solid, a liquid or a gas.

Solid fertilizers come in a huge number of different formulations. Some consist of just one compound, while others contain two or three. Some are intended for application to the soil surface while others are for deep application. They are usually granules with a relatively small particle size, and they can be applied by incorporation into the soil (at depth) using machinery or by hand (on the surface). These granules are sometimes sold as soluble fertilizers so the farmer can dissolve them for application with irrigation water.

Liquid fertilizers are relatively stable solutions of nutrient ions. Solutions can be either true solutions or colloids. There is a huge number of fertilizer formulations, but most consist of the macronutrients N-P or N-P-K. Microelements are often added. They are usually used as leaf feeds and in combined fertilizer application with irrigation.

The most common form of gaseous fertilizer is bottles of carbon dioxide. In an enclosed site, such as a greenhouse, the concentration of CO_2 in the air may vary, as the greenhouse is a closed system with a controlled atmosphere. If you have suitable measurement and dosage equipment, increasing the carbon dioxide level during the day leads to an increase in production. Increasing the concentration of carbon dioxide in a greenhouse using cylinders of CO_2 is known as *carbon dioxide fertilization*.

2.2. BY THEIR COMPOSITION

Fertilizers can be organic in composition or mineral. Perhaps the best known organic fertilizer is animal dung, though there are many others, such as peat, composts and other commercial products, such as guano, fish meal, molasses, etc.

When they weather, all the natural **minerals** release elements (mainly as anions) that are nutrients for plants. Some places are very rich in minerals containing nutrients, such as the deposits of sodium nitrate in Chile, and deposits of minerals containing potassium chloride, such as sylvite.

Chemically synthesized fertilizers are formulated on the basis of natural fertilizer minerals or their primary modifications, the byproducts of the coal industry or organic waste products. All these components, physically and chemically mixed together, form synthetic chemical fertilizers, which have become more and more widely used since their invention in the second half of the last century. The increase in their production, sale and use is mainly due to past over-exploitation of natural mineral deposits, which led to shortage of the fertilizers made from them.

2.3. BY THEIR FORMULATION

Fertilizers can also be classified on the basis of the nutrient elements they contain. There are two main types, **simple** and **compound** fertilizers. Simple, or **monovalent**, fertilizers, only contain a single fertilizer element, while compound fertilizers contain more than one. Fertilizers with two elements are called **binary** fertilizers, and those with three elements are called **ternary**. Depending on how they are manufactured, fertilizers may be **mixes** (physically mixed together) or **complexes** (chemically combined). Compound fertilizers may also contain secondary nutrients and/or trace elements.

Many of the fertilizers that contain a single element, and some containing two, consist of a single compound. In these cases, the name of the compound figures on the label. For example ammonium nitrate is 33.5-0-0 and potassium nitrate is 13.5-0-38

2.3.1. Simple fertilizers

Simple, monovalent or single-compound fertilizers, contain a single fertilizer element. Examples include ammonium nitrate ($NH_4^+NO_3^-$) and potassium chloride (K^+Cl^-). Ammonium nitrate fertilizers usually contain about 33.5% total N. Potassium chloride fertilizers usually contain about 50% K_2O. Note that the ammonium nitrate and the potassium oxide fertilizers contain other elements that contribute nothing to the number of fertilizer units, in this case, oxygen, hydrogen and chlorine. These elements represent the filler in the fertilizer.

Ammonium nitrate fertilizer with 33.5% total N has 66.5% (100-33.5%) filler. This filler is equivalent to the fertilizer oxygen, nitrogen and hydrogen, together with its intrinsic impurities. Its standard three figure notation is 33.5-0-0.
Potassium chloride fertilizer with a richness of 50% contains 50% K_2O ($K_2O * 83 = 41.5\%$ K) and 58.5% chlorine, oxygen and other impurities. In standard notation, it would be 0-0-50.

2.3.2 Compound fertilizers

Compound fertilizers contain at least two of the three main fertilizers, N-P-K. Most of the fertilizers manufactured belong to this group, which is the most widely used and consumed. Within this group there are two subgroups, fertilizer mixes and chemical complexes or compounds. Both subgroups can be divided into binary and ternary fertilizers. All this notation is based on the three macronutrients, but many compound fertilizers also contain secondary elements and micronutrients.

2.3.2.1 Fertilizer mixes

Fertilizer mixes are fertilizers that contain two or three elements and have been obtained by mechanically mixing simple compounds together, sometimes with a little water so they form granules. They are sold as relatively small granules, and may be binary (P-K combinations) or ternary (N-P-K combinations).
There is a huge number of different fertilizers on the market that vary in richness and consist of binary or ternary mixes. This is because the fertilizer companies, due to competition or in response to customer demand, launch new formulations onto the market every year that are supposedly better than the old ones.
They are called fertilizer mixes because of the way they are manufactured. These fertilizers are produced by bulk blending, and their special characteristic is that they are made by simply mechanically mixing simple or binary granulated fertilizers of the same density and with a similar particle size. The simple fertilizers used in mixtures are usually as concentrated as possible in order to ensure the greatest fertilizing power in the smallest volume possible (with the consequent savings in transport, handling and sacks). The fertilizers most commonly used in these mixes include ammonium nitrate, ammonium phosphate, concentrated superphosphate, potassium chloride and urea.

Binary fertilizer mixes are essentially different formulations of P-K. Most are mixes of natural phosphates, **slag** and potassium chloride or sulfate. Potassium sulfate is more expensive than potassium chloride and is more appreciated, mainly because the chloride ion can be toxic to plants if the soil concentration becomes too high. The notation used for fertilizers whose potassium is derived from potassium sulfate is, for example, 0-12-20S. Usage and sales of binary fertilizers are rising. The most common commercial formulations include 0-18-18, 0-19-19, 0-24-11, 0-12-12, 0-13-13, 0-13-7, 0-12-18 and 0-12-20S. Some binary compounds are enriched with microelements, such as boron and zinc.
Ternary fertilizer mixes, unlike binary mixes, are increasingly uncommon, as they are being displaced from the marketplace by the ternary complex mixes discussed next.

Previous page and below: The combination of one of two different types of Basacote with Triabon makes it possible to optimize fertilizer application to the crop's needs, in a way that is flexible and profitable. (Courtesy of BASF)

As we shall see when we discuss the role of phosphorus as a fertilizer, *slag* (or Thomas slag) is a by-product of smelting iron, and contains a lot of phosphoric acid.

Left: General view of a workshop producing blended granulated fertilizers. The fertilizers manufactured here include organo-mineral, superphosphate and ternary mineral fertilizers. (Installations of ETS. PLANTIN)

As they are physical and mechanical mixes, some components tend to be heavier, with a density different from the others. So the farmer must take great care when handling and applying them, especially if they are going to be applied mechanically, as these differences in weight and density of the components of the mix prevent them from being evenly distributed. This means that some parts of the crop receive more of one of the components of the mix, and less of others, while in other zones, the opposite occurs.

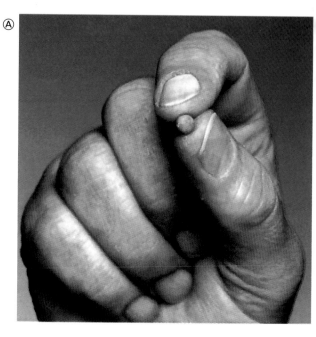

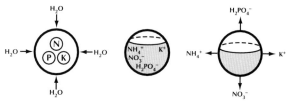

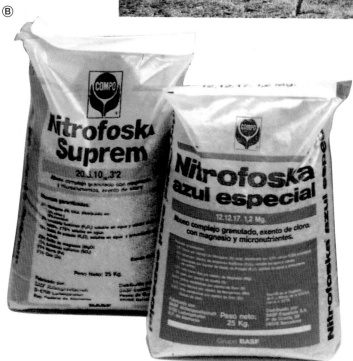

The great advantage of complex fertilizers over mixes is that every grain contains the same amount of nutrients as the label on the packaging. Thus a small grain of a complex fertilizer whose richness is 12-12-17S, contains 12% total nitrogen, 12% P_2O_5 and 17% K_2O derived from sulfate. This is very useful when the fertilizer is applied by machinery, as each nutrient is distributed equally.

Complex binary compound fertilizers are compounds of N-K and N-P. The best example of an N-K binary is potassium nitrate (13-0-44). Important examples of N-P fertilizers include monoammonium phosphate ($PO_4H_2NH_4$), usually sold as 11-48-0 and 10-51-0, and diammonium phosphate ($PO_4H(NH_4)_2$), usually sold as 18-46-0 and 18-50-0 formulations. Many chemical reactions give rise to intermediate products that can be considered as binary fertilizers, but are normally used for the synthesis of complex ternary fertilizers.

2.3.2.2. Complex compound fertilizers

Complex chemical fertilizers, or chemical combination fertilizers, are obtained by chemically reacting the raw materials used with the intermediate products that form in the mixture. The first chemical transformation performed to make a fertilizer was to dissolve the tricalcium phosphate contained in natural phosphates in sulfuric acid to make superphosphate.

The main fertilizers produced this way are ammonium and potassium nitrates, ammonium and potassium phosphates, monocalcium and bicalcium phosphates, ammonium sulfate, potassium sulfate, calcium sulfate, ammonium chloride, potassium chloride, etc.

Ternary complex compound fertilizers are N-P-K compounds produced by chemical synthesis. In the last few years these ternary fertilizers have become overwhelmingly important and there are many different formulations.

To compare the many different ternary complex fertilizers on the market, it is worth calculating the proportions of nutrients, as explained in section 1.3. Comparing a 12-12-24 with an 18-18-36 shows that they have the same proportions, 1-1-2. The only difference between the two is that we are going to need more kilograms of the first than the second for the same effect. This system of calculation of proportions is a simple way of assessing the price of fertilizers, because we can always work out the cost of each fertilizer unit.

Many of these ternary fertilizers contain microelements, and these microelements are expensive as their manufacture is costly. You cannot compare the prices of two ternary compounds if they are not the same (for example, one containing microelements with one that does not). Another factor that may increase the cost of the fertilizer is whether the potassium is in the form of potassium chloride or sulfate. The same formulation (i.e., the same proportions of N-P-K) will always be more expensive if the potassium is in the form of potassium sulfate. In general, the units of nitrogen and phosphoric acid are usually more expensive than those of potassium, so we cannot compare two fertilizers with formulas 10-10-20 and 20-10-10 (though both have 40% total richness). The complex compound fertilizers include important new synthetic fertilizers that represent major advances in fertilizer application. The latest generation is exemplified by *slow release* fertilizers. This should not be confused with fertilizers that only release nitrogen slowly, important fertilizers that are discussed in the section on nitrogen.

Slow release complete fertilizers are usually very complete fertilizers with all the N-P-K macronutrients (with potassium sulfate), secondary elements (magnesium and calcium) and most of the micronutrients the plant needs. The nutrients are contained within a resin covering that expands when it comes into contact with the moisture of the soil. When the resin expands, pores form that allow soil water to enter and dissolve the nutrients inside, which are then released. They are called slow release because the nutrients are released gradually when the soil is wetted by irrigation or rain.

A good crop requires correct fertilizer application.
Photo courtesy of
INDUSTRIAS QUÍMICAS SICOSA, S.A.

Chlorophyll is the green pigment in plants. It is present in the chloroplast and is responsible for photosynthesis.

Symbiosis is a state of balanced physiological interdependence between two or more organisms of different species. Both of them benefit from the symbiosis.

Estimates of the quantity of N₂ fixed in a year by Rhizobium bacteria in legume root nodules.

3. CHEMICAL FERTILIZERS

The three elements that account for most of the composition of plants, carbon, hydrogen and oxygen, need not concern us, as they are present in such large quantities in the soil, water and air that plants have no problem obtaining them.
The three basic elements required for plant nutrition, the macronutrients nitrogen, phosphorus and potassium, are discussed in the following chapters.
Sulfur, calcium and magnesium are *secondary elements* in plant nutrition. Plants need less of them than of the macronutrients but more than of the microelements.
The essential microelements, iron, manganese, copper, zinc, boron, molybdenum and chlorine are required by plants in tiny quantities, but a deficiency may cause serious disorders of the plant's metabolism.

3.1. MACROELEMENTS: NITROGEN

Nitrogen is the only of the nutrient elements that is not present in the soil's bedrock.
Nitrogen is present in nature in two states, as a gas in the atmosphere (80% of the atmosphere is N_2), and in a combined state, either mineral or organic. Organic nitrogen cannot be used by plants, as plants cannot digest living organisms, and nitrogen is one of the building blocks of the proteins of living things. Only the mineral forms of nitrogen can be taken up by plants.
Nitrogen is responsible for the green parts of the plant, whose growth, leaves, vigor and foliage are all intimately related to nitrogen. Plants are often an intense green color if they have enough nitrogen, because nitrogen (together with magnesium) is an essential component of **chlorophyll**.

The farmer must be careful when applying nitrogen fertilizers. Excess nitrogen supply, or nitrogen alone, may lead to very lush leaf growth, and this may be detrimental to production of other parts. If we apply only nitrogen fertilizers, we obtain very vigorous plants but production will be little and late (the Law of the Minimum).

3.1.1. The nitrogen cycle

The drawing illustrates the cycle of nitrogen in the soil, plant, living organisms and the atmosphere, and all their possible interactions. Atmospheric nitrogen (N_2) is fixed by two pathways, either by the direct fixation from the air by Azotobacteria or by the action of **symbiotic** bacteria that live in nodules on the roots of leguminous plants (legumes). These bacteria, *Rhizobium spp.*, form a symbiotic relationship with legumes; they provide the plant with nitrogen fixed from the atmosphere, while they receive organic substances produced by photosynthesis from the plant.
The nitrogen derived from the organic matter incorporated into the soil (excrement, dead bodies, body parts) is degraded by microorganisms. The nitrogen passes through several stages until it is completely mineralized. Studying and measuring this is very complicated.

Crop	Nitrogen fixed, kg/ha
Alfalfa	150
Sweet clover	120
Red clover	90
Soybean	60
Broad bean	25
Pea	25

We need only say that proteins (nitrogen-containing organic compounds) are first broken down to ammonium, and this is converted into amine groups (NH_2), which are broken down to ammonia and ammonia salts and these are oxidized, first to nitrite (NO_2^-) and then to nitrate (NO_3^-).

The nitrogen input into agricultural soils is derived from the remains of organic matter from plants and animals, the nitrogen fixed by azotobacteria and *Rhizobium* and the fertilizers applied by the farmer. These inputs of nitrogen are rapidly broken down by microorganisms. The final form of inorganic nitrogen, nitrate (NO_3^-) is extremely soluble, and is easily lost from the soil by leaching or washing, and then flows to the sea or into the water table. Small amounts of inorganic nitrogen are evaporated together with seawater and return to the atmosphere as gaseous nitrogen (N_2).

A nitrate mine in the Antofagasta area of Chile, producing "Nitrato de Chile" (sodium nitrate, known as "Chilean saltpeter")

The nitrogen cycle in the soil-water-air continuum. Note the importance of N_2 fixation by Rhizobium spp. and azotobacteria.

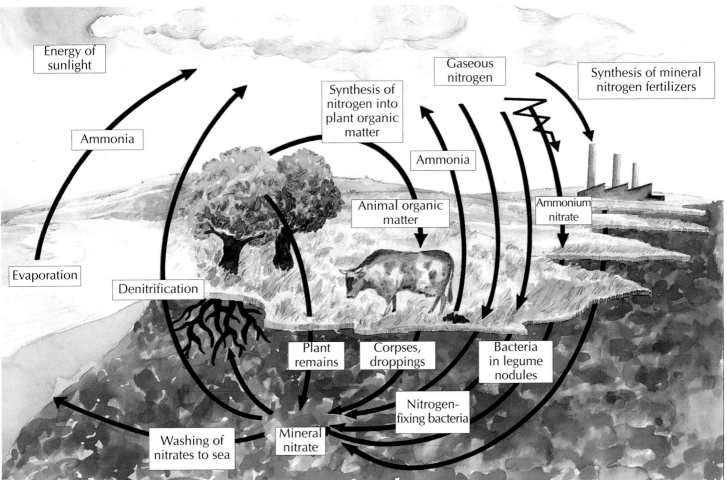

3.1.2. Nitrogen fertilizers

For a long time, the only nitrogen fertilizers available to farmers were the fossil deposits of sodium nitrate in Chile, and the ammonium sulfate obtained as a byproduct of distilling coal. Nowadays, the high demand for nitrogen fertilizers is largely met by chemically synthesized fertilizers manufactured from the nitrogen in the atmosphere. Ammonium is the main intermediate product, and is the basis of the manufacture of most nitrogen fertilizers.

The many simple nitrogen fertilizers can be divided on the basis of the type of nitrogen they contain. All the complex ternary fertilizers (with or without microelements) are produced from simple nitrogen compounds.

3.1.2.1. Organic nitrogen fertilizers

There are several organic nitrogen fertilizers on the market, all containing at least 3% organic nitrogen.

The most important are cake, dried blood, horn meal, desiccated meat, tanned hides, wool residues, fish meal, etc. These fertilizers are useful because they create a reserve of nitrogen in the soil that is gradually released over time.

3.1.2.2. Synthesized organic fertilizers

These fertilizers contain organic nitrogen obtained by chemical synthesis. They are gradual or slow release fertilizers, whose main feature is that they release the nitrogen slowly in order to reduce losses due to washing and to adapt to the plant's ability to absorb the nitrogen, thanks to the combination of urea with different aldehydes.

The most common products are urea-formaldehyde or urea-formol compounds Crotonil idendiurea® and Isobutyl idendiurea®. These two different urea-formaldehyde formulations are registered trademarks of BASF, and can be considered environment friendly fertilizers because they release nitrogen slowly in accordance with the plant's needs, thus preventing contamination of the water table by the nitrates in the drainage water.

These fertilizers are expensive, and their main market is among flower growers,

An example of a complex ternary fertilizer with part of its nitrogen as CRODOTUR®. Manufactured by BASF, S.A.

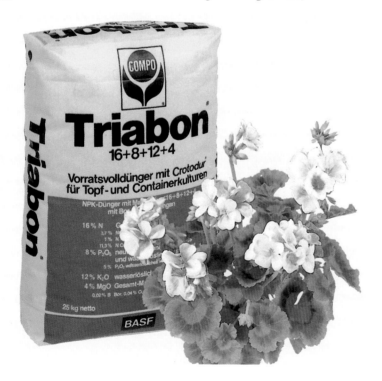

horticulturalists and turf specialists.
These fertilizers are sold as solid granulated formulations, with several different nitrogen-containing compounds. Apart from Crotodur® and Isodur®, they may also contain nitrogen in the form of ammonium or nitrate. They are often high-tech complex ternary fertilizers, containing phosphoric acid and potassium, usually present as sulfate. These fertilizers usually contain calcium, magnesium and microelements.

3.1.2.3. Nitrogen in the form of urea

Urea is an amide, and by weight contains 46% nitrogen as urea. Depending on its degree of compaction, urea has a low density, about 0.8 g/cm³. The action of a special enzyme (urease, which is produced by some bacteria), hydrolyzes urea in the soil to ammonium which is then nitrified.

The most important thing to remember is that until the urea is hydrolyzed it descends through the soil like a nitrate, and is not retained on the clay-humus complex. But once it has been hydrolyzed, it behaves like an ammonium fertilizer. Thus for the plant to use the urea, it must be broken down by the bacterial enzyme urease.

Thus high microbial activity and a high humus content will favour hydrolysis of urea. In soils with a normal biological activity, this hydrolysis is rapid, taking three or four days in a soil with a suitable content of organic matter.

In calcareous and alkaline soils, urea is a good nitrogen fertilizer that rapidly supplies nitrogen to the plant. But in acid soils with a high rainfall, the transformation of urea into ammonia is slower, due to the lower levels of **urease**, and you should use other types of nitrogen fertilizer instead.

Nitrogen in the form of ammonia is slightly volatile, and because urea is converted into ammonia before being nitrified, it is worth burying this fertilizer slightly below the surface immediately after it is applied.

Because it is highly soluble, urea can be formulated as a liquid feed that is applied to the leaves (foliar feeding). At times of stress when the plant requires an additional input of nitrogen, it can be applied as a foliar feed for the plant to absorb through its leaves. Urea is also sold in other solid formulations either on its own or associated with other sources of nitrogen, mainly nitrate and ammonium. The more technically advanced ternary complex fertilizers are those whose total N content is in the form of urea, nitrate and ammonium, together with sufficient phosphoric acid and whose potassium is derived from potassium sulfate. These fertilizers usually also contain calcium, magnesium and microelements.

3.1.2.4. Nitrogen in the form of ammonium: ammonium sulfate

Ammonium sulfate is sold as small crystals and contains 20-21% N in the form of ammonium. Depending on their origin, there are three types of sulfate, namely synthesized sulfates, sulfates from cokers, and sulfates recovered from several different industries (such as artificial textiles and plastics).

Ammonium sulfate contains 23-24% sulfur, which is normally an additional benefit. Plants of the cress family (Cruciferae) require a lot of sulfur, and for them it is essential. This fertilizer makes the soil slightly more acid, meaning that it is especially suitable for soils with an excessively high (alkaline) pH value.

As in the case of the different formulations of urea, the higher the soil's temperature and alkalinity, the faster the nitrogen in the ammonium is oxidized to nitrate. The ammonium in fertilizers are adsorbed well onto the clay-humus complex before they are oxidised to nitrates. It is worth making an effort to cover the fertilizer with a little soil to avoid loss by volatilization.

We should also mention anhydrous ammonium (N content 82%) and nitrogen-containing liquid fertilizers whose richness varies between 18.2% and 30%. This type of liquid fertilizer often contains various form of N, such as urea, nitrate and ammonium. There are also liquid ternary complex fertilizers whose nutrient elements are in suspension. Part of their total N is in the form of ammonium and they also contain phosphoric acid, potassium, calcium, magnesium and microelements.

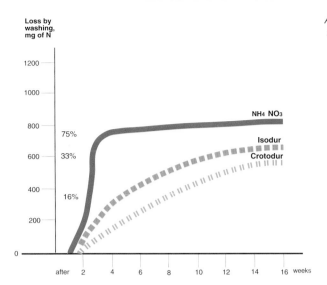

A typical experiment to show the washing of nitrogen from three different compounds. Taken from Mung & Dressel

(Left) Chilean saltpeter (sodium nitrate)

3.1.2.5. Nitrogen in the form of nitrates

Nitrate fertilizers are highly soluble in water and are not effectively absorbed onto the clay-humus complex, and so they are easily lost by washing. The plant can only absorb them when they are dissolved in the soil solution, but they are among the ions most rapidly absorbed by the plant when they are dissolved in the soil.

Nitrate fertilizers are recommended for use when we require a rapid response from the plant, at the end of winter, if we applied fertilizer late in the fall, at times of *stress*, etc.

Sodium nitrate, or Chilean saltpeter, is the oldest of the nitrogen-containing fertilizers, and contains 16% N and 25% sodium. It is no longer widely used, except by those growing beetroot, as Chilean saltpeter contains a relatively high amount of boron, which these plants require a lot of.

There are other nitrates such as calcium nitrate (25% CaO), containing 15% nitrogen as nitrate, and nitrate of calcium and magnesium, which contains 15% nitrogen.

3.1.3. Characteristics and properties

When considering the forms of nitrogen in the preceding section discussing the decomposition of organic nitrogen, it is worth pointing out the ones that are most readily used by plants. Nitrogen in the form of the ammonium ion (NH_4^+) can be adsorbed onto the clay-humus complex and taken up by plants, but it has the disadvantage that it is rapidly nitrified by microorganisms into nitrate, meaning that its effective concentration in the soil is very low.

Nitrogen in the form of nitrate (NO_3^-) is extremely soluble. It is not retained by the clay-humus complex, and is thus easily lost by leaching (washing). Plants can take nitrates up from the soil solution before they are lost.

Using nitrogen in the form of nitrates has to be well thought out, because they are so highly soluble. When nitrate fertilizers are applied, they are washed down through the soil to the water table, and this has two effects; the first is that the expensive fertilizer is wasted, as the plants cannot use the nitrogen that is lost, and the second is that the water table and aquifers are contaminated with nitrites and nitrates. Recent studies analyzing underground waters in the Almería region of Spain found excessive levels of nitrites and nitrates. This contamination by nitrogen salts is due to excess application of nitrate fertilizers by local farmers.

Examples of ternary fertilizers with part of their nitrogen in the form of ISODUR®. These fertilizers are very suitable for lawns, as the slow release of the nitrogen means it does not "burn" the grass. Manufactured by BASF, S.A.

Sports ground fertilized with ISODUR® slow-release nitrogen fertilizer. Manufactured by BASF, S.A.

One way to partly relieve the problem of nitrate pollution of underground water is to use fertilizers whose total N is partly as nitrates and partly as ammonium. Thus, in the first days the plant can take up the nitrogen in the form of nitrates, and when they are lost by washing, they start to take up the ammonium.

The total N of the most technically advanced fertilizers is in the form of urea, ammonium sulfate, nitrates and Crotonil idendiurea® or Isobutyl idendiurea® (though one or more of them components may not be present). Note this supplies the plant with an initial supply of nitrogen (urea and nitrates), a second input when the first has finished (ammonium) and finally the plant receives the slow-release nitrogen from the formaldehyde compounds in Crotodur® and Isodur®.

Remember that legumes have symbiotic *Rhizobium* bacteria in their root nodules that can fix nitrogen from the atmosphere. This is important because it means that legumes require less nitrogen than other plants. Referring to specialist reference works or asking local suppliers with good reputations are both good ways of establishing how much fertilizer to apply to each crop and when. Finally, remember that humus is mineralized in the soil, and as it is broken down it releases nitrogen, which becomes available to the plant, and so it is important to know roughly how much humus is present in a soil in order to take into account when calculating the nitrogen input. Clearly a sandy soil containing 0.5% organic matter is not going to have the same amount of available nitrogen as one with roughly 10% organic matter.

3.2 THE MACROELEMENTS: PHOSPHORIC ACID

What we normally refer to as phosphoric acid is in fact phosphorus pentoxide (or phosphoric anhydride), which is phosphorus (P) combined with oxygen (O). We have already listed the conversion factors necessary to convert the % P into % P_2O_5 and vice versa. We have also already seen that the plant can absorb phosphorus in the form of the anions PO_4^{3-}, PO_4H^{2-}, $PO_4H_2^{-}$. For convenience, when talking about phosphoric acid and the anions it produces we shall write them as PO_4^{3-}.

3.2.1. The phosphorus cycle

The phosphate in the soil is mainly derived from the parent material and mineralized organic matter. The mineral phosphate forming part of the parent material is highly insoluble and is unavailable to the plant. Mineral phosphate is mainly present in the form of orthophosphates, the most representative being tricalcium phosphate (apatite), magnesium phosphates and the phosphates of iron and aluminium. The phosphorus that is unavailable to plants is called *retrograde* P_2O_5.

In acid soils, it is possible that the phosphates may precipitate as their iron or aluminium salts, and in very alkaline soils, they may precipitate as calcium phosphate. These highly insoluble precipitates are comparable to organic phosphate, which is also highly insoluble. Because of slight variations in pH, the effects of organic matter, microbial activity and the fact that some plants can absorb them directly, these two forms of phosphorus (organic and inorganic) means that the phosphorus can be absorbed onto the clay-humus complex and then be absorbed by the plants.

Part of the phosphorus fraction retained on the C.E.C. may dissolve into the soil solution. The phosphorus dissolved in the soil solution is the most easily taken up by the plant, though plants may also absorb some of the phosphorus adsorbed onto the clay-humus complex.

In order to understand this process better, the reader should carefully study the diagram to the right which shows the changes phosphorus undergoes in the soil.

3.2.2. Phosphate fertilizers

Depending on their origin, phosphate fertilizers can be divided into two groups. The first group consists of byproducts of some metal refinery industries. Among other impurities, the iron minerals of Lorraine and Normandy contain small quantities of phosphorus which must be removed during refining, in Bessemer convertors, in order to produce high quality steel. These impurities are heated to a high temperature in the molten mass and calcium is added, and this basic slag, which is a mixture of calcium, phosphates, calcium silicates, removes the phosphorus from the iron. This is dephosphorizing basic slag, or Thomas slag, named after the British engineer who perfected this improvement of the Bessemer converter.

The second group of phosphate fertilizers consists of the natural deposits of phosphate minerals that occur in sites around the world. These mineral phosphates are aluminium and calcium phosphates salts, sedimentary tricalcium phosphate or crystallized tricalcium phosphate. These minerals contain about 65% to 77% tricalcium phosphate, which is equal to about 30-33% P_2O_5 and 48-52% CaO.

These natural phosphates are treated industrially in a variety of physical and chemical processes to obtain superphosphates, orthophosphoric acids, calcium phosphates, bicalcium phosphates, etc. These different forms are the basis of all the different simple phosphate fertilizers, whether solid or liquid. All the binary, ternary and complex liquid fertilizers, with or without microelements, that are on the market are made from these simple phosphate fertilizers.

3.2.2.1. Calcium superphosphates

Superphosphate minerals are obtained by reacting natural phosphates with sulfuric and phosphoric acid. The quantity of acid is dosed so that almost all the tricalcium phosphate is transformed into the water-soluble monocalcium phosphate. The phosphate content of a superphosphate is limited by this industrial process; it is impossible to obtain a phosphorus content greater than one quarter of the total phosphorus content of tricalcium phosphate.

Simple superphosphate is a 50% mixture of monocalcium phosphate and calcium sulfate (gypsum). It usually contains 16-24% P_2O_5 that is *soluble in a citrate solution*. It also contains between 9% and 12% sulfur, 28% CaO and small quantities of microelements (Fe, Zn, Mn, B, Mo). The most commonly sold superphosphate is P_2O_5 superphosphate, containing 18% P.

Enriched superphosphate is obtained by reacting

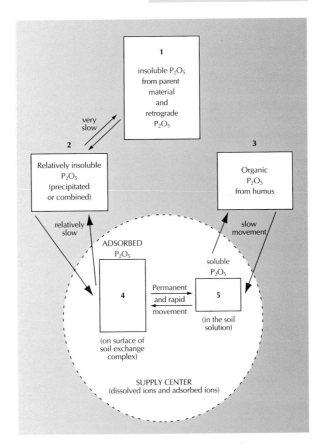

natural phosphates with a mixture (which may vary in proportions) of sulfuric and phosphoric acid, and this gives a content of 25-35% phosphoric acid. This also produces a significant amount of calcium sulfate, giving it a 5-9% sulfur content.

Triple superphosphate, obtained by reacting phosphoric acid with natural phosphates, contains 38-48% P_2O_5.

Superphosphates can be applied to all crops, both on the surface and at depth. They are also indicated for all alkaline soils with normal supplies of calcium. Superphosphates may be sold as a fine powder or granules, granules being more common.

3.2.2.2. Calcium phosphates or precipitated phosphate

Bicalcium phosphate, Ca_2HPO_4, contains 38-42% phosphoric acid. Bicalcium phosphate is mainly used in animal feeds, in mineral supplements or making compost from corn, mixed with urea at a dose of 2-3 kg/T of forage.

3.2.2.3. Phosphal

Phosphal is the name of a fertilizer produced from aluminium-calcium phosphates imported from Senegal, which are made soluble by heating to 600°C and are then milled. This fertilizer has a 34% acid phosphate content and is sold finely ground with a diameter of 0.16 mm. It is effective in alkaline soils and acid soils, though its behavior is better in basic soils. It is also more effective if the soil clay-humus complex is abundant. It should be buried fast, preferably a few weeks before starting cultivation, so it can dissolve and occupy sites on the C.E.C.

*The **solubility** of soluble forms of phosphorus is usually expressed in terms of a solution of citric acid in water, because the analytical method employed uses ammonium citrate to measure the phosphorus.*

3.2.2.4. Dephosphorizing slags

Dephosphorizing basic slags, Thomas slags, are byproducts of certain metallurgical processes, as discussed above. Thomas slags are sold as a very heavy black powder (100 kg occupies less than 50 liters) containing between 12% and 20% phosphoric acid, usually in the range 15-17%. The phosphoric acid in these slags is present in a special form, complex combinations of phosphates and silicophosphates. They also contain 45-55 kg of lime, and so they are very suitable as fertilizers for acid soils and grasslands. These slags also contain microelements, such as magnesium, manganese, zinc, copper, cobalt, molybdenum, silica and iron oxides.

Perhaps the most inconvenient feature of these slags is the large amount of dust they give off during application. This can be diminished by wetting the mass at the moment of distribution, by pouring water directly into the center of the hopper applying the slag, at a rate of 2 to 3 liters of water per sack.

The element sulfur is abundant in nature, either in a pure form (sulfur deposits in Italy, deposits in the U.S.), or combined with metals (pyrites, for example) or gas (oil or steam).
Courtesy of SANDOZ

3.2.2.5. Milled natural phosphates

Laboratory analysis is essential to measure a soil's phosphorus content.
Photograph ceded by ETS. PLANTIN

Milled natural phosphates usually consist of 60-77% of tricalcium phosphate, together with calcium carbonate and calcium fluoride. Natural phosphates vary in their phosphorus content depending on their origin, and contain 26-33% total P_2O_5. They used to be sold as very fine powders, but now they are sold as a granulate. It seems that some crops can make better use of milled natural phosphates; this is true of fodder crops, legumes and members of the cress family, because the fertilizer has a direct effect on the roots.

It is worth covering the fertilizer with soil after applying it to the surface. It is also not recommended to use milled phosphates in excessively limy soils, where the movement of PO_4^{3-} is slow.

3.2.2.6. Condensed phosphates

Ammonium polyphosphate is a condensed phosphate, the only one that is used in agriculture. Raising the concentration of the phosphoric acids yields metaphosphoric acid, two salts of which are occasionally used as fertilizers, though they are not very important.

Monopotassium phosphate is sometimes used in irrigation, and contains 51% P_2O_5 and almost 34% K_2O, meaning it contains almost 85% of the main fertilizer elements.

3.2.2.7. Binary, ternary and liquid phosphate fertilizers

So far we have considered the simple phosphate fertilizers, from which all the binary, ternary and complex fertilizers on the market are manufactured. We need not go into this in great detail, but remember the binary fertilizers that we have already discussed, such as monoammonium phosphates, diammonium phosphates, and ammonium polyphosphates, as well as nitrophosphates, ammonium phosphates, etc. It is worth pointing out that recently introduced ternary fertilizers consist mainly of the simple phosphates already discussed, most contain nitrogen in the form of urea, ammonium or nitrate, the potassium is in the form of potassium sulfate, and they contain microelements. Modern liquid fertilizers show the same characteristics, with the added advantage that they can be applied in the irrigation water, and as a foliar feed. Liquid fertilizers consisting exclusively of N-P-K are usually clear solutions, that is to say true solutions. To the contrary, if microelements are added, they are in a colloid, not true solutions, as the elements are in suspension and are not dissolved.

3.2.3. Characteristics and properties

Phosphorus is the least mobile element in the soil. A crop's phosphorus supply comes mostly from the soil's reserves. The yield of the fertilizers applied in each season is very low, and thus the crops may absorb between 5% and 30% of the phosphorus applied as fertilizer. The best policy for phosphorus application is to program it in the medium and long term, in order to maintain the soil's fertility in phosphorus.

Thus, the dose of phosphorus should be calculated over the long term and should be equal to the net quantity taken up by the crops + losses by washing + losses by retrograde fixation - the assessed input from organic inputs. Laboratory analysis is required to measure the assimilable phosphorus in a soil. On the basis of this analysis and quantitative measurement, we can decide whether the doses of phosphate fertilizer need only replace the phosphorus that has been extracted or lost, or whether we have to remedy the soil's low phosphorus levels.

To find out the amount of phosphorus needed by the plants, we have to consider their real needs, as every species has different needs and these change over the course of the plant's growth. These needs must be combined with the measured soil fertility. All these parameters are dealt with in more detail in specialized articles on different crops. They explain the needs of each crop and the critical moments when the plant needs an additional supply.

As a guideline, we can accept that a 40 **ppm** of phosphorus is high, and that the soil has good reserves of this element. If, however, levels are only 8-10 ppm the soil has few phosphorus reserves, though the assessment of phosphorus levels in a plot of ground should be performed by a specialist who is aware of the details of the farm, and also the general soil features of the region.

3.3. MACROELEMENTS: POTASSIUM

Normally, the potassium (or potash) in fertilizers is measured as K_2O. Most countries use this notation to express the potassium (K) content of their fertilizers. Potassium oxide contains 83% by weight of potassium. We dealt (on page 75) with the conversion factors to convert potassium

to its oxide and vice versa.

Potassium is mainly present in the plant as the dissolved K^+ cation. Potassium is concentrated in the youngest tissues and plays an important role regulating the plant's functions. Potassium also increases the plant's resistance to fungal diseases. It is also important in photosynthesis. It is needed for the formation of sugars, and the plants that are grown for their sugar and starch reserves, such as potatoes, beetroot and grapes, take up a lot of potassium. Some studies relate this directly to the quantity of light. Thus, in areas where sunshine is very intense, plants absorb less potassium than the same plants in areas where sunshine is less intense.

Nitrogen and potassium seem to be closely interrelated. Some articles say that potassium plays a role in synthesis of proteins (which contain nitrogen), and they thus argue that the proportion of the two elements has to be precisely calculated to ensure the highest plant yield. A plant with sufficient potassium will have a greater concentration of minerals in its sap, and this increases its resistance to frost. Finally, potassium plays a role in plant transpiration, meaning that a good supply of potassium improves the plant's water economy.

3.1.1. The potassium cycle

Crystalline and volcanic rock formations are usually rich in potassium (from 2-7% in granite feldspars) but this potassium is in the form of highly insoluble silicates, and can not be taken up by the plant. Weathering turns part of this insoluble potassium into cations that are dissolved in the soil solution. In calcareous soils, natural potassium reserves are very low, as their constituents do not include any potassium-containing minerals.

The potassium liberated by natural minerals

ppm means parts per million, which is the same as milligrams per kilogram, or grams per ton (or milliliters per liter, etc.).

The mobile and immobile phases of potassium in the soil

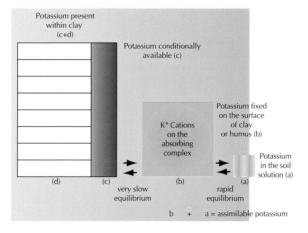

or supplied in the form of fertilizer can enter into the (1:2) clays and be trapped there, meaning the plant cannot make use of it. This trapped fraction is known as retrograde potassium or non-exchangeable potassium. A small part of the K^+ cations may return to the surface of the clay forming part of the cations adsorbed onto the clay-humus complex.

An equilibrium is reached between the potassium of the C.E.C. and the ionized potassium in the soil solution, so that when the plant absorbs potassium from the soil solution, the clay-humus complex releases ions to the soil solution and makes up for the loss. The dynamics of potassium in the soil are summed up in the diagram on the preceding page.

Some devices allow sowing and fertilizer application at the same time.

Schists are slate minerals, and in general any laminar rock.

3.3.2. Potassium fertilizers

The natural mineral containing potassium is sylvite. This contains potassium chloride (28%) and sodium chloride (56%), together with **schists** and other sediments (16%), that is to say it contains 16-17% of potassium measured as K_2O. Sylvine is not used directly as a potassium fertilizer, but has to be industrially transformed, in order to eliminate almost all the schists and common salt, raising the concentration of potassium chloride to 60%. To avoid the low toxicity to plants of the chloride ion, the potassium chloride is treated with sulfuric acid to obtain potassium sulfate, with a content equivalent to 50% K_2O.
The solubility of potassium fertilizers does not vary greatly from one to another. What is really important is whether the potassium is the chloride or sulfate salt, as some species of plant are sensitive to chloride ions. As potassium sulfate has to undergo a second transformation, fertilizers with potassium sulfate are usually more expensive than ones with potassium chloride. The most important potassium fertilizers are discussed in detail in the following sections.

3.3.2.1. Potassium chloride

Potassium chloride usually contains the equivalent of 60-61% K_2O. It is the best, most widely used and cheapest of all. It is on the market as potassium chloride powder (60% K_2O), granulated potassium chloride (60% K_2O) and pearled potassium chloride (61% K_2O). As it contains almost no sodium, it is suitable for all crops (except those that are highly sensitive to chloride ions, such as runner beans, tobacco, flax, etc.).

3.3.2.2. Potassium sulfate

Potassium sulfate, which contains the equivalent of 50% K_2O, also contains 18% sulfur (S). Its chloride content is very low, and there are two commercial qualities of potassium sulfates, normal potassium sulfate (with 2.5% Cl) and high quality potassium sulfate (with only 0.5% Cl), which is slightly more expensive. High quality potassium sulfate fertilizer is suitable for plants with high sulfur requirements, and where the agricultural produce is of high quality, such as vineyards producing high quality wines, flax, flower cultivation, tobacco, etc.

3.3.2.3. Patentkali®

Patentkali® is a registered trademark of BASF and is imported from Germany. It is a fertilizer mix of potassium sulfate and magnesium sulfate. It contains the equivalent of 28% K_2O, 8% Mg, and 18% S. It is recommended for crops that require a lot of magnesium, and which are sensitive to chloride ions, and is applied at a dose of 200-1,000 kg/ha.

3.3.2.4. Binary, ternary and liquid potassium fertilizers

So far we have dealt with the simple potassium fertilizers, from which all the binary, ternary and liquid fertilizers on the market are made. We need say no more on them. The discussion of binary compounds dealt with P-K compounds, such as combined phosphorus and potassium fertilizers, superpotash, etc. Binary N-K compounds, such as potassium nitrate, were also discussed.
Complex ternary fertilizers are increasingly formulated with potassium sulfate instead of potassium chloride. These high-tech ternary fertilizers are expensive, but they are of very high quality. Their formulation often includes secondary elements, such as calcium and magnesium, as well as microelements.
There are many different liquid formulations for applying to the leaves or in irrigation water. As we pointed out when talking of phosphoric acid, there are two types of foliar feed, depending on whether they are true solutions or colloidal. Potassium chloride raises a problem of solubility for the manufacturers when manufacturing ternary fertilizers. The potassium is often in the form of sulfate, which avoids the problem of insolubility. This fertilizer thus tends to make the soil a little more acid.
There are ternary liquid fertilizers on the market whose potassium is in the form of potassium chloride. They are supersaturated solutions, which use a physical system to maintain the crystals of potassium chloride in suspension.
Liquid fertilizers may also contain microelements, and are sold as colloidal solutions or fertilizer solutions that contain N-P-K + microelements. The problem with colloids is that their nutrient content cannot exceed a given limit, because this would cause the nutrients to precipitate. As a result, true solutions usually last for a longer time than false or colloidal solutions, which tend to precipitate.

3.3.3. Characteristics and properties

Potassium is a relatively immobile element, though not as immobile as phosphorus, and so much of the potassium used by crops comes from the soil reserve. However, the contribution made by fertilizers is more important in the case of potassium. As occurs with phosphorus, the behaviour of potassium in the soil is not consistent, and depends on factors, such as soil texture, the characteristics of the crop, etc. Calculating the dose of potassium fertilizer to apply to the crop, just as in the case of phosphorus, is determined by the amount taken up by the crop plus the losses by washing. Potassium is lost in a different way to the way phosphorus is lost.

Potassium losses by washing are much greater than those of phosphate, and may be very large in sandy soils. The less clay and organic matter there is in the soil, the greater the losses by washing. Some authors calculate the losses of potassium using the following formula:

$$\text{Losses of } K_2O \text{ by washing} = \frac{400}{\text{Clay} + \text{O.M.}}$$

On the other hand, the soil's capacity to adsorb the potassium cations (K^+) depends largely on its content of double-layer (2:1) clays, which trap cations in sites that are not accessible to plants, meaning that the retention of potassium cations depends on the amount and type of clays present. We can leave the calculation of the doses of potassium fertilizers to the specialists, as they involve many factors, such as analysis of the amount of assimilable potassium in the soil, analysis of the plants (to find out how much potassium they take up), and the practical and agronomic data of the plot of ground in question. As a guideline, a level of 47 ppm of exchangeable potassium, for example a sandy soil, are considered very poor. As a contrast, 665 ppm of assimilable potassium in a clay is considered very high.

It is briefly worth mentioning the concept of luxury consumption. Unlike phosphorus, the plant absorbs all the potassium available to it in the soil. But this consumption leads to an increase in the productivity, as far as an optimum (the Law of the Maximum) where the productivity is greatest, and after this production does not increase, but in fact often declines. This reduction is explained as an imbalance between an excessive quantity of potassium with respect to the other elements (the Law of the Minimum). This decrease is explained as an imbalance between the excessive amount of potassium with respect to the other elements (the Law of the Minimum).

3.4 SECONDARY ELEMENTS

As a result of technical advances, the list of elements that are considered to be essential for plants continues to increase. Of course, merely finding an element among the ashes of a plant does not mean it is essential for the plant. Nor is there any reason why the elements that plants require should be the same as those that animals require.

European Union (EU) legislation has a category for the secondary elements, namely sulfur, calcium and magnesium. The difference between the N-P-K elements and the secondary elements is simply a question of quantity. Laboratory analysis of plant ash shows sulfur, calcium and magnesium are present in lesser amounts than nitrogen, phosphorus and potassium. The study of the secondary elements and microelements is still young, and the EU legislation on them is very recent.

3.4.1. Sulfur

In terms of fertilizers, sulfur is expressed in terms of sulfur trioxide (SO_3); the conversion factor between S and SO_3 is 2.5. So,

$$\%S * 2.5 = SO_3 \qquad \%SO_3 * 0.4 = \% S$$

Sulfur is a component of many proteins, as are nitrogen and phosphorus. The plant absorbs sulfur from the soil in the form of the sulfate ion, SO_4^{2-} in order to synthesize organic matter. Many plants require a lot of sulfur, such as legumes, members of

A/ Packaging plant
B/ Quality control of fertilizers is necessary to ensure their nutrient content is consistent
C/ Field tests are essential to assess if a new formulation works.

Ⓐ　　　　Ⓑ　　　　Ⓒ

the cress family, onions, artificial meadows, etc. Sulfur in the soil is derived from sulfate minerals but also from organic matter (75-90%). Sulfur is released from organic matter by microbial activity, as is nitrogen, and this makes it available to plants, either adsorbed onto the clay-humus complex or dissolved in the soil solution.

The soil loses its fertility in sulfates because they are removed within the crops (and the more sulfur the crop requires, the more is lost) and by washing of the anions. Sulfur returns to the soil (ignoring fertilizers) from the atmosphere when sulfur dioxide is washed out of the air, the mineralization of humus, irrigation water and dung, which contains about 0.5 kg S per ton.

The N-P-K fertilizers previously discussed, which have their potassium in the form of potassium sulfate, provide significant amounts of sulfate to the soil. But if the laboratory analysis shows there is a serious sulfur deficiency, basic sulfur can be supplied to the soil and the microorganisms will convert it into sulfuric acid (H_2SO_4), which ionizes and becomes available in the soil solution.

Schematic structural formula of chlorophyll. Taken from A. Finci

Sulfur cycle

SO_2

Animals

Residues

Assimilation

Organic sulfur

SO_4^{2-}

Washing

Sulfur minerals

3.4.2. Calcium

To be a calcium fertilizer, a fertilizer must have more than 3% calcium, calculated as CaO. Often, in basic and carbonated soils ($CaCO_3$), the calcium concentration is more than sufficient for the plant's nutrition. In fact, all the salts described in the N-P-K section may be salts of calcium.

Calcium plays a major role in all phases of plant life. Its presence in the cell sap is essential for plant growth at all moments from germination to seed ripening. Calcium also makes plant tissues more resistant. We will deal with calcium (lime) again when we discuss treatments for acid soils.

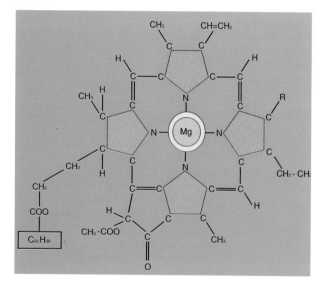

3.4.3. Magnesium

The current regulations are that magnesium must be stated as its equivalent in magnesium oxide (MgO). The factors to convert Mg to MgO are

$$\% \text{ Mg} * 1.66 \text{ } \% \text{ MgO} \quad \% \text{ MgO} * 0.6 = \% \text{ Mg}$$

Magnesium is essential to the plant because a magnesium atom is at the heart of the chlorophyll molecule. The drawing above is a schematic diagram of the chlorophyll molecule. Magnesium deficiency in the plant leads to a reduction in photosynthetic activity and thus of the agricultural yield. Mineral magnesium is usually in the form of calcareous minerals, such as dolomite, that contain variable proportions of calcium and magnesium. The rocks with the highest magnesium content are eruptive rocks and limestone sedimentary rocks. The magnesium that is of interest to the farmer is the usable or exchangeable magnesium. This is thought to be 2-10% of the total soil magnesium.

It is worth pointing out the idea of the antagonism of ions, the competition between the soil ions to occupy the sites on the C.E.C.

This competition is especially clear in the case of the Mg^{2+} ion, because high levels of other ions, such as H^+ (in acid soils), or excessively high levels of exchangeable potassium (K^+), and even high levels of Ca^{2+} may relegate magnesium to a secondary position, and though it is present in sufficient quantities for the plant in the soil, it is hard to take up.

The extraction of Mg by plants is about 30 kg/ha per year. Some plants, such as legumes and fruit trees, require a lot of magnesium. In addition to what is taken up by the crops, the soil is considered to lose 15-40 kg/ha per year by leaching. The Mg may be supplied in the form of natural minerals, such as dolomite (18-20% MgO), or magnesium-containing fertilizers, such as calcium and magnesium nitrate, Patentkali® (8% MgO), and magnesium sulfate (16% MgO), etc.

3.4.4. Sodium, chlorine and aluminium

These three elements, which we include as secondary elements, are not normally considered secondary elements, but they are of such importance, for a variety of reasons, that they deserve special attention.

Sodium (Na) is not considered to be essential for plants, though large quantities of sodium are found in plant ash. Some crops, such as beetroot, respond well to sodium, even if they have received plenty of potassium fertilizer; their ash contains a lot of sodium. It seems that sodium, in the special conditions of potassium deficiency, can take over some of potassium's functions within the plant. As we shall see in the chapter on improvement of saline soils, sodium has a negative impact on soil structure.

Chlorine (Cl) is present everywhere and is highly mobile in the soil. The presence of chlorides is relatively negative, especially in the form of common salt (sodium chloride). As is well known, excessive soil salinity is toxic to plants. Therefore, excess chlorides are more likely to be a problem than a deficiency. The plant needs a small amount of chlorine, but as it is so abundant in nature, it is often present in excess. There is antagonism between chloride ions and sulfates. Fertilizers with potassium in the form of potassium chloride supply more than enough chloride to meet the plant's needs.

Aluminium (Al) is not essential for plants, but because it is present in large quantities in the soil, it is worth pointing out that it is toxic to plant roots. In acidic soils, exchangeable aluminium (Al^{3+}) blocks the action of other elements like calcium, magnesium, manganese, phosphates, etc. This gives rise to the paradox that, though the nutrients are available in the soil, they are not available on the clay-humus complex because the aluminium has occupied the exchangeable sites. The presence of large amounts of the aluminium ion in the soil may lead to plant deficiencies of P_2O_5, even if there a reasonable amount of phosphorus in the soil, as the aluminium reacts with the phosphate to form aluminium phosphate, a highly insoluble molecule that blocks the assimilable phosphorus.

3.5. THE MICROELEMENTS

The number and importance of the microelements varies depending on the sources consulted. Advances in biochemistry and analytical chemistry make it likely that the number of microelements known to be essential for plant growth will increase in the future, and that the precise requirements for each one will be established.

Microelement	g/Ha
Manganese, iron	500 g
Zinc, boron	200 g
Copper	100 g
Molybdenum	10 g
Cobalt	1 g
Selenium	0.02 g

Uptake of microelements by plants. These values are very low, but they are vital.

It is known that plants only need very small quantities of these microelements, but that they are vital. Most of them form part of enzymes involved in synthesis of organic matter (such as copper, zinc, molybdenum) or as co-enzymes in this synthesis. It is thought that many of them can be replaced by other elements with similar chemical properties. The seven microelements considered to be essential for plant growth are: iron, manganese, zinc, copper, boron, molybdenum and chlorine (Fe, Mn, Zn, Cu, B, Mo, Cl), and the following is a brief description of their characteristics.

3.5.1. Iron

Iron forms part of many enzymes and is essential for chlorophyll synthesis. Iron is present in the soil in adequate amounts to meet plant needs, which are roughly 1-2 kg/ha per year. Iron deficiency in plants causes yellowing of the leaves, etc., as they have less chlorophyll. This means that photosynthesis is not fully functional. Iron deficiency in plants is often due not to a shortage in the soil, but to excess calcium (lime), which immobilizes the iron (nutrient antagonism).

Application of iron sulfate to the soil does not usually give the desired results, as the iron is mobilized by the soil $CaCO_3$. As we shall see in the next section, on fertilizers with microelements, this problem can be partly resolved by using chelate fertilizers.

Sites near the sea often show problems of salination due to excess sodium.

3.5.2. Boron

Boron is not a well-known element, but it is known to play a role in the synthesis of the plant cell wall. In general, the concentration of boron (B) in the soil is 1-2 ppm, and deficiencies may occur if this descends to 0.6 ppm. Boron uptake, as often happens with iron, may be blocked in excessively limy soils with too high a pH. Boron is present in four different forms in the soil; water soluble boron, boron bound to organic matter, boron in clay minerals and borosilicates.

Wet regions suffer loss of boron through washing. These losses, together with the quantity removed in the crops, mean that a large and increasing number of soils in wet regions now show a requirement for boron fertilizers.

A/ Damage produced by boron deficiency in an apple (internal cork). Photo ceded by BORAX ESPAÑA, S.A.
B/ Boron-deficient apples (left) compared with healthy ones (right). SOLUBOR and FERTIBOR are the leading brands of agricultural boron. They are manufactured by BORAX ESPAÑA, S.A.
C/ Corky tissue in pear. Note that the fruit is externally deformed. Photo ceded by BORAX ESPAÑA, S.A.
D/ Rough bark ("toadskin") of apples with papery bark. Photo ceded by BORAX ESPAÑA, S.A.
E/ A change in the color of the leaves is the first sign of boron deficiency in olives. Photo ceded by BORAX ESPAÑA, S.A.
F/ Chlorosis begins at the tips and then spreads to the rest of the plant. BORAX ESPAÑA, S.A.

(A)

(B)

(C)

(D)

(E)
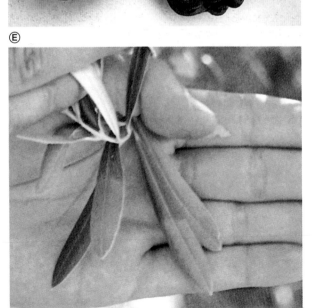

(F)

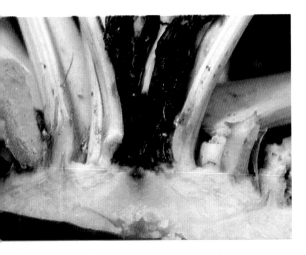

3.5.4. Copper

It was recognized around 1900 that copper stimulated plant growth. In the areas where **Bordeaux mixture** was sprayed on fruit trees and vegetables, their growth was much greater. Most soils have copper as Cu^{2+} ions, but if the oxidation status is low, copper may be present as Cu^+ ions. The cation exchange complex retains copper ions so firmly that they are even less mobile than Ca^{2+} ions. The concentration of copper ions in the soil solution is a few parts per million. Copper is most soluble in acidic soils, and its solubility decreases with higher pH values. Copper is an important co-enzyme needed to activate several enzymes in plants. It is also involved in chlorophyll production. Like iron and copper, iron and manganese are closely related. Thus excess of copper gives rise to chlorotic symptoms similar to those indicating a deficiency of iron. Copper shows low mobility within the plant, and so the clearest deficiency symptoms are seen in new organs and new growth, as this is where copper normally accumulates.

3.5.5. Zinc

Zinc is present in almost all soils in small quantities, that are however sufficient in most soils and for most plants. In addition to being a catalyst and regulator of plant growth, it is involved in the production of the auxins that control growth, and in corn a deficiency of zinc leads to shorter internodes. A soil deficiency of zinc may be due to soils with a high pH, as this makes zinc insoluble, or to a high concentration of PO_4^{3-} ions (ion antagonism). Analysis of zinc concentrations in different plants seems to show very large differences between species, even between species growing in the same soil, and thus with comparable fertility.

3.5.6. Molybdenum

Molybdenum is the only trace element that is more soluble in basic environments than acidic ones, meaning that molybdenum deficiencies are resolved by making the soil more alkaline by liming (applying $CaCO_3$). The most important fact about molybdenum is that it is essential for legumes, as it plays a role in the fixation of atmospheric nitrogen by the *Rhizobium* bacteria in their root nodules. The first symptoms of molybdenum deficiency appear similar to nitrogen deficiency.

3.5.7. Fertilizers with micronutrients

Great care should be taken when applying fertilizers with micronutrients, especially if the fertilizers contain elements, like boron, that may be toxic to plants. The problem lies in that, given that the plant's need for microelements is so small, the difference between deficiency, the difference between the correct dose and an excess is very small, meaning that it is often easy to apply too much.

3.5.3. Manganese

Manganese (Mn) is needed for chlorophyll production and in the plant's enzyme systems. Manganese is washed from well-drained acid soils because oxidation and acidity increase its solubility. Dissolved manganese migrates to wetter and/or more alkaline conditions where it precipitates as small hard dark particles, called *nodules* or *concretions*. The ratio of the manganese in the plant to that in the soil is very high. The similarity of manganese to iron means they are antagonistic; the symptoms of iron toxicity are similar to those of manganese deficiency and vice versa.

Boron deficiency causes heart rot in beetroot. Photo ceded by BORAX ESPAÑA, S.A.

Bordeaux mixture is a classic fungicide consisting of copper sulfate and lime.

A mineral form of boron: borax

A mineral form of boron: boronatrocalcite

1

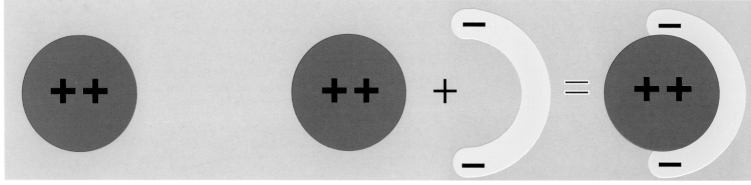

1/ The metallic micronutrients (Fe, Mn, Cu, Zn) are combined with certain substances to form chelates. When a chelate forms, the organic molecules encloses the metal cation and forms a complex with no electrical charge. Chelation prevents microelements from being prematurely fixed, makes their absorbtion easier for both root and leaf, and increases their mobility within the plant.

2/ The difference between a shortage and an excess of a microelement may be very small. The photo shows symptoms of excessive boron in a plant of Euphorbia pulcherrima. To deal with this problem, chlorine free formulations are often used, such as 20-10-20 or 15-5-25 fertilizers from GRACE-SIERRA INTERNATIONAL, B.V.

3/ Examples of chelated microelements. Hortrilon® is a product with several micronutrients and Fetrilon® 13% is a single EDTA chelate of iron. Both are manufactured by BASF, S.A.

Ferric chlorosis *is a plant disorder due to iron deficiency.*

So you must take care to apply the correct dose and take great care to ensure the microelements are evenly distributed.

Unfortunately, micronutrient deficiency can seriously reduce the quantity and quality of your harvest, but without being so severe as to produce clearly recognizable symptoms of deficiency. Thus all the different fertilizer elements may show two types of deficiency: with symptoms or symptomless. A healthy looking plant may well not show optimum production because of this latent (hidden) deficiency. This point is also true of the macroelements and secondary elements.

When a crop shows a deficiency of a given microelement, it is not enough to apply it directly to the soil, because in most cases it is an **induced deficiency**. That is to say, the soil contains an adequate amount of the microelement, but it cannot be used by the plant because the pH is too high (see 4.1.3. The effect of pH on nutrients) or because of ion antagonism (see 3.4.3. Magnesium). To get around this problem, fertilizers to correct deficiencies of microelements are usually formulated as chelates.

Chelate is word derived from the Greek word for claw. Chelates are organic compounds, soluble in water, and able to bind with metallic cations and immobilize them. These cations can exchange with other cations because they are very weakly ionized, due to the organic chelating agent.

The first microelement to be produced in the form of a synthesized chelate was iron. Chelates of iron exemplify how this type of fertilizer acts. Iron supplied in the form of ferrous (Fe^{2+}) iron is soluble in water and rapidly becomes available to plants, because it dissolves and ionizes. The

ferrous (Fe^{2+}) ions are rapidly oxidized to ferric (Fe^{3+}) ions, and they precipitate as ferric (Fe^{3+}) oxide, or some other equally insoluble Fe^{3+} compound. The chelated iron is also soluble in water, but it does not ionize, and the iron remains in a soluble form that the roots can absorb easily. One of the best known chelating agents is ethylene diamine tetra-acetic acid (**EDTA**). The ionizable H^+ of the acetic acid part of the molecule can be replaced by metallic cations. These replacements are indicated by a prefix indicating the element that is chelated, such as Fe-EDTA or Zn-EDTA. In the soil, Fe-EDTA resists both microbial attack and hydrolysis. EDTA's stability and effectiveness against *ferric chlorosis* is greater when the soil pH is slightly acid, but there are other chelating agents, such as EDDHA (ethylene diamine di (O-hydroxy xiphenyl acetic acid), that are stable even in basic soils (with a high pH). The more modern chelating agents, such as EDDHA, are usually more expensive than older ones, such as EDTA, but they are worth using on really alkaline soils because they are highly effective.

There are products on the market to correct deficiencies of a single metallic microelement, such as the brands (registered by BASF) Fetrilon® 13%, Mantrilon® 9% and Zitrilon® 10% (with a 13%, 9% and 10% content of Fe, Mn and Zn, respectively), or with several metallic microelements, such as Hortrilon®, which contains 4.8% MgO, 0.5% B, 5% Fe, 2% Mn, 0.5% Mo and 0.5% Zn.

hese microelements are chelated with
:DTA, but there are others, such as Basafer®
vith a 6% iron content as the EDDHA
·helate.

Other fertilizers, such as those containing
boron, may be sold, in addition to chelates,
such as sodium borate or borax (11.3% B),
boracite or sodium borate (14.5% B), and
products based on sodium pentaborate for
leaf sprays (19% B).
If you wish to apply magnesium in a form
that is less expensive, magnesium sulfate can
be applied as a foliar feed twice a year, thus
avoiding the cost of chelated magnesium. In
terms of copper, it is possible to alternate
applications of chelates to the soil with foliar
application of copper oxychloride, which has
the advantage of also being a fungicide.
Another copper compound is copper sulfate
($CuSO_4 * 5\ H_2O$), which like copper
oxychloride is a fungicide. With reference to
application of molybdenum fertilizers, most
phosphate fertilizers contain small quantities
of Mo, which is usually enough, as the crop's
requirements are so small.
Microelements are not only sold alone, but
are also sold as binary and ternary
combinations, for application to leaves, in
irrigation water or into the soil, with the
potassium in the form of potassium sulfate
and with one or more chelated metal
microelements. They can be sold as solid or
liquid fertilizers.

*Part of the factory of
INDUSTRIAS
QUÍMICAS DEL
VALLÉS, which
manufactures copper
compounds such as
copper oxychloride,
which is not only a
fungicide but is also
effective against
copper deficiency.*

Ⓐ   Ⓑ   Ⓒ

*Nutrient deficiency
negatively affects
plants and their
production.*

*A/ Necrotic patches
and malformations
appear at the base of
the leaf.
B/ Deficient
pollination and seed
set.
C/ Irregular seed set
with the obvious loss
of production.
D/ and E/ Splits in
the stalk and
breakage of the
capitulum (the
seed head).*

d

e

*The company "20
MULE TEAM" sells a
wide range of borate
fertilizers to meet all
requirements.*

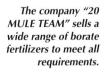

4. ORGANIC FERTILIZERS

In the chapter on the solid part of the soil, we stressed the importance of the organic matter. We dealt with the origin and classification of the organic matter in the soil, the biological and climatic factors influencing the soil's organic matter content, its physical and chemical properties, and the role it plays in the soil.

Over time, the organic matter in the soil is mineralized by microorganisms, which return the nutrient elements (H, C, O, N, P, K, etc.) to the soil. When the soil's organic matter content is lowered, the exchange capacity decreases because there is less humus, and this diminishes fertility. As we have already seen, organic matter contributes to the soil's water retention capacity, its porosity (aeration), etc.

We have already looked at the importance of the K1 coefficient, which refers to the amount of humus formed from one unit weight of dry organic matter input into the soil, and the K2 coefficient, the percentage of stable humus that is mineralized every year (K2 $\approx$ 0.002). If we repeatedly cultivate the same plot of ground, its organic matter is gradually mineralized and it declines in quality, and this is even truer for horticultural crops, such as lettuce, in which the entire plant is removed from the field, meaning that almost no organic matter is returned to the soil.

It is necessary to replace the organic matter so that the soil does not lose its capacity to retain water (structure), nutrients (C.E.C.) and air (porosity).

Any sort of organic remain can become incorporated into the soil. Each one has its advantages and disadvantages, and the farmer has to decide which one to use, taking into account their advantages, price, ease of application, how long they last, etc.

4.1. THE ORIGIN OF ORGANIC MATTER

Returning organic matter is perhaps the oldest agricultural activity still practised. Treatises on agriculture written between 2,400 and 1,700 years ago discuss applying dung to crops.

Dung long ago ceased to be as important as it was in the distant past. The amount of dung has declined, there is much less horse dung and much more cattle dung, and it is much harder to obtain. The irregular distribution of stockraising means that some agricultural areas have an excess of dung, while other areas have almost none.

As a result, alternatives to dung have been sought in order to continue returning organic matter to the soil to replace that lost by mineralization every year.

Cultivated fields

Dung, organic wastes from crops, such as stems, roots, stubble, residues from fodder crops, and in general any sort of organic matter, are considered in the next sections.

4.1.1. Dung

As is well known, dung is the droppings of different farm animals, after being rotted for a while in the stable or dungheap, which often contains part of the animals' bedding (mainly straw).

Dung, like all organic matter, gives the soil structure, the ability to retain water and nutrients, as well as the fertilizer units released when it mineralizes. It also helps to maintain an acceptable population of soil microorganisms (a soil without microbial life would be dead).

product and increases its weight, it helps us to retain the liquid excrements and thus make best use of their high levels of N and P.

Note that horse dung is considerably richer than that of cattle, and sheep dung is richer than horse dung. Bird droppings are five times richer than cattle dung, especially in terms of phosphates and calcium.

Nutrient	kg/t
Nitrogen	4.0
P_2O_5	2.5
K_2O	5.5
Sulfur	0.5
Magnesium	2.5
Calcium	5.0
Manganese	0.04
Boron	0.004
Copper	0.002

Average composition of dung in kg per ton, calculated as fresh product containing 20-25% dry weight

Type of livestock	kg per day 1,000 kg live weight	% N		% P		% K	
		Solid	Liquid	Solid	Liquid	Solid	Liquid
Cows	70-100	0.5	0.25	0.11	0.06	0.41	0.21
Hogs	70	0.5	0.1	0.13	0.42	0.37	0.09
Chickens	60	1.5		0.43		0.41	
Bedding		0.5		0.125		0.4	

Dung production and composition

4.1.1.1. Composition

The composition of dung varies greatly, as it depends on many factors like the species and age of the livestock, the bedding used, the inclusion or not of liquid excrement, and the extent of the decomposition and washing that may have taken place during storage or composting. The feed the animals have eaten is also important, as is the proportion of straw to droppings, how the livestock are managed, etc. Above there is a table showing the average amounts of nutrients in dung, calculated on the basis of 25% dry weight and in kilograms per hectare.

In general, cattle and hog dung contains more nitrogen and potassium than phosphorus. Their solid excrement shows similar proportions, but their liquid excrements (urine) are different, as cattle urine contains more N and K than hog urine, which however contains more P. Above there is a table showing the amount of dung produced per day per thousand kilos live weight and the proportion of nutrients each one contains.

As you can see in the second table, half of the nitrogen and more than half of the potassium are in the liquid excrements (urine). So it is often convenient to add straw or some other type of organic bedding to the dung, because although it dilutes the

Dung also contains the other essential nutrients in varying proportions, but almost always in proportions roughly similar to what the plant needs. As dung's phosphorus content is normally relatively low, it is generally worthwhile complementing dung with a phosphate fertilizer, which is mixed with the dung, making what is called fortified dung.

Photo of machine packaging of an organic substrate to supply organic matter to soil. Photo courtesy of GRENA S.R.L.

4.1.1.2. Composting

Dung contains valuable nutrients that become accessible to plants when it is buried in the ground. But when the dung is rotted in the open air, many of its nutrients are lost by washing or evaporation.

Fermented dung can be bought. HUMUS VITA SUPER is fermented cattle and horse dung with chicken droppings that has been rotted for at least 12 months. Manufactured by FOMET, S.A.

the field as soon as possible, and then buried to avoid the loss of gases and the washing of nutrients (especially nitrogen).

In some farms with livestock, the dung is collected in ditches, where it is washed, and the washings are channelled to a collector. Once there, the bacteria decompose much of the organic matter in the same way as in the soil, releasing its nutrients.

4.1.2. Other organic remains

The current scarcity of dung in some areas has led to the study and use of other organic compounds. The best known are remains after harvesting, stubble, corn stems, potato remains, the green parts of sugarbeet, etc. Some plants are often cultivated so that they can be buried into the ground as fertilizer. Most of the fast-growing fodder plants are examples of this type of green manure. Compost consisting of plant remains fermented in the same way as dung is widely used in gardening. Recently, there has been research into composts made of seaweeds, crushed grape skins and ground vine prunings, peat, and compost made from urban rubbish.

4.1.2.1. Characteristics

Harvest remains represent the plant parts that are not used by human beings, such as leaves, stems, roots and other aboveground or underground parts. These residues should not be ignored as they represent an annual input of humus of 500-800 kg/ha. Remember that the annual loss of humus is 700-1,000 kg/ha. Bear in mind that easily rotted materials (burial of short cycle forage crops) produces very little humus.

Many gaseous nutrients that are produced in the first decomposition, such as CO_2, NH_3 and H_2S, escape into the atmosphere, while other byproducts of decomposition, such as nitrogen, potassium, some of the phosphorus, and the micronutrients are easily lost by washing.

The best way to compost dung is in a highly compacted ditch or trench, so that it undergoes anaerobic decomposition and is protected from washing of nutrients by the rain. Once it has been composted, the dung has to be applied to

There is machinery on the market that makes it possible to harvest the crop and then leave the crop plant's remains in the soil, in order to incorporate its organic matter into the soil. (Photo courtesy of MASSEY FERGUSON)

Annual supply of
humus from crop
residues

Crop	dry weight t/ha	Humus kg/ha
Wheat (stubble)	3-4	450-600
Wheat (stubble + buried straw)	5-7	1,200-1,500
Barley (stubble)	4-6	300-450
Corn (buried stems)	8-10	1,200-1,500
Green parts of sugarbeet	4-6	600-900
Potato wastes	1	casi nulo
Green parts of mustard	3	100
Alfalfa (two years)	8-10	800-1,000
Temporary meadows	15-18	750-900

No till, or minimum tilling, techniques rely on using machinery to plant the crop with the minimum number of journeys by the machinery. The photo shows the same field as on the preceding page, but from a different angle. On the dried remains of the previous crop, the seeds are sown, fertilizer is applied, and the seed is then buried, all in a single journey. (Courtesy of MASSEY FERGUSON)

To the contrary, lignified materials, such as straw produce a lot of humus, but require additional nitrogen in order to rot.

We have provided a table of the most common plant remains that, on burial in soil, produce large amounts of humus. All the values in the table are close to the quantities of humus obtained from dung, and are usually about 100 kg/t per year.

Several species are cultivated just to be buried in the ground, as this produces large amounts of humus. Different authors disagree on the amount of humus obtained by *green manuring*, but they agree that fresh organic matter from green plants is very effective, and that it has a large and immediate effect on microbial activity and on the soil's physical properties and fertility.

Compost is mainly used in gardens and intensive horticulture. The fermentation method used is similar to the rotting of dung. Several different types of plant remains are piled together and covered with earth, sometimes with the addition of ammonia fertilizers to boost the microorganisms. A well rotted compost can be obtained within a few months, though it is better to wait for a year. It is then scattered on the soil and then buried, providing humus and nutrients, as well as improving the soil structure.

Peat is also used as a source of organic matter, but its high price means it is too expensive for extensive agriculture, leaving aside whether it is a good source of humus. Peat has its advantages. It is often used as bedding in stables because it absorbs a lot of water and is thus easily enriched with fertilizers in urine.

Composted city rubbish is obtained from sorted domestic rubbish, which is ground and heated, after a phase of hot fermentation in industrial plants to avoid pollution and environmental impacts. Its average nutrient content, in terms of dry weight, is 0.8-1% N, 0.4-0.7% P_2O_5, 0.25-0.4% K_2O, 2.5-5% Ca and 0.15-0.4% Mg.

5. APPLICATION OF CHEMICAL FERTILIZERS

We have already seen the main limiting factors affecting the crop's ability to make use of fertilizer. In the first place, the fertilizer needs a suitable pH value if it is to be available to the plant. In the second place, there must be a good cation exchange capacity so that the ions of the fertilizer are adsorbed onto the C.E.C. and are not lost by washing. In the third place, the soil solution must be abundant enough to dissolve the ions. It is also important to bear in mind that there may be antagonism between some ions, because sometimes, especially with microelements, the plants may show deficiency even though there is enough of the nutrient in the soil.

There are other causes conditioning the crop's use of nutrients. The genetic features of the crop species and of the variety all affect their use of nutrients, and for example a crop of alfalfa (which fixes atmospheric nitrogen) that does not receive enough phosphorus and potassium, will not reach its full potential. The plant variety is also important. If a hybrid (genetically selected) variety is not supplied with the nutrients it requires, its yield will be far lower than a normal (non-hybrid) plant.

The soil's morphology also conditions the effectiveness of the fertilizers. The yield will not be the same in a shallow impermeable soil as in a deep permeable soil, event though the same dose of the same fertilizer is applied. The climate is also important. It seems clear that the same dose of the same fertilizer will not have the same effect if applied to irrigated plants as if applied to crops grown using dry farming techniques. To finish, recall the Law of the Maximum and the Law of the Minimum, which postulate that nutrients must be combined in the right proportions and that there is a maximum or optimum point of fertilizer units, above which production does not increase but decreases.

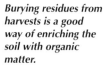

Burying residues from harvests is a good way of enriching the soil with organic matter.

The uptake of nutrient elements by the crop is by means of a water potential along the soil-plant-air continuum, as we saw in the section on the liquid phase of the soil, the soil solution. The movement of the fertilizer within the soil has been dealt with in the sections of the individual nutrients. We saw that nitrogen is extremely mobile within the soil and often shows large losses due to washing, and that phosphorus and potassium are the least mobile of the soil macronutrients, because they both are relatively insoluble.

5.1. APPLICATION METHODS

Fertilizers can be applied before, during or after sowing the seed. When to apply basically depends on the quantity of fertilizer you want to apply. If the quantity is small, it can be applied at the moment of sowing. If the quantity is large, it can be applied before or after sowing, or even divided into two or more separate doses.

You must take into account the fact that nitrogen is highly mobile. It is usual to recommend application of nitrogen fertilizers just before sowing, so that it is not lost due to washing. Applications of the more immobile elements, such as phosphoric acid and potassium, can be programmed in the long term, and phosphorus and potassium fertilizers can be applied before or after sowing as an investment in the future. An initial input of phosphorus and potassium fertilizer into the soil can be made to fertilize the first crop and as an investment in future crops. In parallel, nitrogen is applied several times (staggered), just before sowing and during the crop's germination.

The fertilizer is scattered on the soil, or worked into the soil, so that the roots can absorb it more effectively at the right moment. The application method varies with the fertilizer's texture, the fertilizer element's mobility within the soil, and the layout of the root system.

Regardless of whether the fertilizer is liquid or solid, there are two ways of applying it, surface fertilizing or deep fertilizing. Surface fertilizing, when the fertilizer is applied to the soil surface, either the entire surface or part of it, while deep fertilizing, it is buried in the soil by plowing or by using special machinery. If the fertilizer is buried by plowing, this is known as deep fertilizer application to the entre site. If it is applied using machinery, the fertilizer will be located in defined strips or bands.

5.1.1. Location

The fertilizer that is distributed in strips, localized, either on the surface or at depth, can be divided into three groups depending on its location.

Starter fertilizer is the application of small doses of fertilizer (generally nitrogen and phosphates) near the seed to stimulate the young plant's growth. This makes use of the favorable action of phosphoric acid on the root growth of young plants, and the N*P interaction, which is very effective in the early stages of growth (for example, in corn, early potato).

Strip fertilizing is applying fertilizer along the line of seeds. It is normal to apply a complete N-P-K fertilizer. It can be left on the surface, next to the furrow, or it can be buried using specialized machinery. There are very complex machines that can carry out three tasks at once: planting the seed, applying the fertilizer and often some sort of antiparasitic product.

Deep fertilizing during planting is carried out in crops that grow in a site for many years, such as fruit trees and vineyards. Once or twice a year (in spring and fall) fruit trees are fertilized with N-P-K. These applications of fertilizer to an established crop are carried out with specialized machinery called *deep fertilizers*. The deep fertilizer consists basically of one or more rows of teeth, behind which special conduits channel the fertilizer into the base of the furrow.

5.2. CALCULATING FERTILIZER DOSES

The problem of calculating fertilizer dosage is often one of the hardest the farmer has to perform. What usually happens is that the confused farmers generally ends up following the dosage indications given by the suppliers. The manufacturers usually indicate on the label the dosage to apply and how to apply it, and the manufacturers often recommend doses greater than the farmer need apply. The dosage instructions are always guidelines, and the manufacturers often state this on the label. These dosages are normally calculated supposing that no fertilizer has been applied to the soil.

5.2.1. A theoretical ideal case

A theoretically complete calculation of the required dosage of fertilizer would be extremely complex, expensive and time-consuming. To begin with, a thorough laboratory analysis of the plant tissues would have to be performed, in order to determine the amounts of each of the fertilizer components in the plant. This approach has the disadvantage that we usually need to know the dose of fertilizer before cultivating the crop, not during its growth. One alternative method is to consult the more specialized books containing the results of these analyses. These percentages are called the **total extraction** values of the nutrient elements and they are quantified for almost all important crops. But knowledge of the total extraction should not be confused with the plant's real needs. It should not be forgotten that knowledge of plant nutrition is incomplete, and many parameters are unknown.

The second step would be thorough soil sampling and its later laboratory analysis to determine the soil's physical and chemical characteristics, soil class, porosity, density, C.E.C., N-P-K nutrient status, secondary

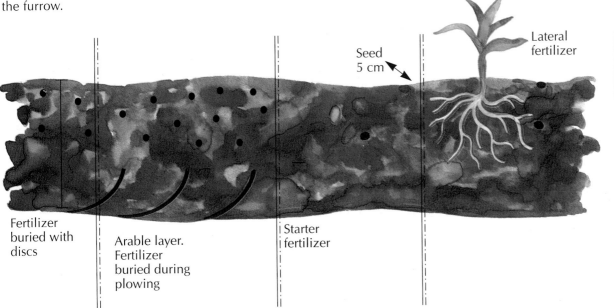

Seed 5 cm

Lateral fertilizer

Fertilizer buried with discs

Arable layer. Fertilizer buried during plowing

Starter fertilizer

Diagram showing the different types of fertilizer application. Fertilizer spread on the surface would be located in a thin film on top of the soil.

elements, microelements, etc. The results of the analysis of nutrient levels seek to provide the value of the nutrient elements available in the soil (not those that are immobilized). These values are usually expressed in terms of high, medium or low, in comparison with reference soils.

At this stage, the technician on the farm should bring together the analyses of the plants and the soil with the special features of the farm in question, and on this basis calculate a "made to measure" dosage for the case in question. Even so, there are unknowable factors, such as the future weather, that may mean that too much or too little fertilizer is applied.

Machinery for applying fertilizer in extensive cultivation. (Manufactured by GILLES)

Organoleptic qualities are the qualities that affect the human senses, such as the feel, taste or smell.

5.2.2. A more realistic case

This section seeks offers a solution halfway between the ideal approach to plant nutrition and complete simplification of this highly complex problem. This proposal starts from the basis that there are two types of nutrient element: nitrogen, which is mobile, washable and therefore scarce, and the pair phosphorus and potassium, which are immobile, non-washable and often present in sufficient quantities in the soil to ensure plant nutrient supply, greatly simplifying the problem.

Nitrogen, as it is present in very low levels in the soil, will limit the maximum yield (the Law of the Minimum, the smallest "stave in the barrel" is the factor limiting production. The doses of nitrogen fertilizer must equal the amount extracted by the crop in question, as there can never be an excess of nitrogen in the soil (as a guideline follow the manufacturer's dosage instructions). In normal soils, the doses of phosphorus and potassium must be maintenance doses, in the order of 60-80 fertilizer units of P_2O_5 and 60-100 fertilizer units of K_2O per hectare. The soil should be regularly analyzed to check that the maintenance doses are sufficient, or if the doses should be increased to replace the losses. To finish, note that it is possible in some cases to satisfy simultaneously with a single application of a ternary fertilizer, all or part of the requirements of the plant and the soil for nitrogen, phosphorus and potassium.

5.3. PLANNING FERTILIZER APPLICATION

It is worth planning fertilizer application at the beginning of the growing season. This means that you can order the fertilizer well in advance, and this usually means that it works out cheaper. The most widely used technique is to establish the total need of P-K for each plot for a year, and the N required for the next crop.

It is necessary to consult the nutrient requirements of each crop, especially if the crop has a high requirement for a given nutrient. After certain crops that require a lot of nutrients, such as maize, the fertilizer applied for the next crop has to be supplemented. Legume crops require less nitrogen than other crops, as the *Rhizobium* bacteria in their root nodules can fix atmospheric nitrogen.

In the cultivation of established fruit trees, you can draw up a plan to apply fertilizer three times a year. One balanced N-P-K application in spring, the next one a month before harvest with a large dose of K to favor sugar formation in the fruit, and a final application in fall to prepare the plant to produce vigorous buds the next spring.

In horticultural crops with a short cycle, a deep application can be made that is not excessively loaded with P-K and two or three applications of N can be made during the plant's growth. You should take great care when applying nitrogen fertilizer to crop species grown for their green parts, such as lettuces, as if there is insufficient P-K in the soil, excess nitrogen will lead to poor head formation, with very green heads that are also misshapen.

In general, the plant extracts give us an idea of its nutritional needs, and on this basis the fertilizer needed can be divided into two or three applications over the course of the plant's life cycle, the first application at depth and the others on the surface.

To sum up, nitrogen is responsible for the green parts of the plant, its overall growth. Phosphorus plays a very important role in seed germination, while potassium is of great importance in ensuring the fruit is of high **organoleptic** quality.

5.4. FERTILIZER APPLICATION SYSTEMS

Fertilizer application systems are the direct consequence of the different commercial preparations of different fertilizers. We can classify the fertilizers into solids (crystallized, crushed, pearled and powdered), liquids (solutions or suspensions) and those that are gases at room temperature.

The most important and widely accepted solid fertilizers are the granulates, because this presentation is convenient to handle and easy to apply and store. Nor does it give off lots of dust when applied, a major reason for preferring granulates to powders. In general, fertilizers are hygroscopic, meaning that they absorb water from the atmosphere. When the temperature increases, the water is restored and they stick together. This is why fertilizers are usually sold in plastic sacks, as this prevents them from

retractile teeth, the **special teeth** with a rigid cutter, and for the deepest applications, **the subsoil cutter** which can reach a depth of 30 cm. Some farmers use sacks of soluble fertilizer incorrectly by placing them in the flow of irrigation water. The sacks have holes pierced in them, and the water gets in, dissolves the fertilizer and transports it to the plants. This method, partly comprehensible in small plots, has the disadvantage that the distribution of fertilizer within the plot is extremely irregular.

5.4.2. Foliar feeds

Fertilizers applied to the leaves, foliar feeds, are sold already dissolved. As we have already seen, there are two types of foliar feed, those that form true solutions, clear solutions, and those that form nutrient suspensions, or colloidal solutions. The fertilizers based on solutions consist of simple compounds or binary N-P fertilizers. Colloidal solutions normally contain N-P-K and usually contain microelements.
It is well known that the green parts of plants,

Gathering beetroot leaves for conversion into foliar feed. (Machinery manufactured by GILLES)

Different types of teeth for burying liquid fertilizers

absorbing water, and means they can be stored without problems. Solid fertilizers can be applied by hand or with machines, on the surface or at depth, and the applications vary with how concentrated the fertilizer is, the dose and the proportion of nutrients.

5.4.1. Fertilizer application in irrigation water

Some solid fertilizers with a very small particle size are sold as soluble fertilizers, and they are the ones most suitable for application in irrigation water. Dissolving fertilizer in irrigation water is one way of applying it, and this requires the use of mixing vessels, pumps, channels and sprays to supply the fertilizer as a liquid. It is essentially applying fertilizers in the irrigation water. The commercial argument underlying the sale of these fertilizers is that plants absorb nutrients much better if they are dissolved in irrigation water. This is true, but they have the disadvantage that the ions are much more vulnerable to washing once they are in the soil. This means that they need to be more frequently applied (regular applications) than granulates. Combined irrigation and fertilizer application is important in enclosed environments, where it is possible to monitor nutrient inputs and losses very closely. This system has the advantage that once the expensive application system has been installed, the labor costs of fertilizer application are greatly reduced, as a single person can mix the solutions and turn on the pump to distribute the dissolved fertilizer through the combined fertilizer-irrigation system.
Soluble fertilizers intended for application in irrigation water may also be applied locally with specialized machinery. The accompanying drawing shows several types of teeth used in local application, depending on the depth at which we wish to apply the fertilizer. There are **vibratory teeth, flexible cultivator teeth, rigid**

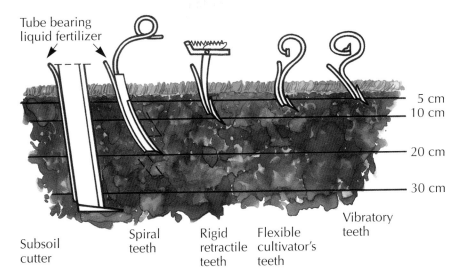

Tube bearing liquid fertilizer

5 cm
10 cm
20 cm
30 cm

Subsoil cutter

Spiral teeth

Rigid retractile teeth

Flexible cultivator's teeth

Vibratory teeth

especially their leaves, are capable of absorbing dissolved nutrients through their stomata. Foliar application reduces the losses of elements as much as possible, as they are rapidly absorbed. They have the further advantage that they can be applied together with the many plant protection products on the market. Foliar feeding should be reserved for moments when the plant is under stress, and for whatever reason requires a nutrient supplement. Urgent examples might be: detection of a given deficiency, during a very dry period, after a period of extreme cold, after hail damage, etc. The best known foliar application of N is application of dissolved urea. This nitrogen containing compound is highly soluble and has a surprisingly strong effect on plants. The doses for foliar application have to be precisely calculated, as the nutrients enter the plant rapidly through the leaves and may cause toxic effects due to excess nutrients.

6. APPLICATION OF ORGANIC FERTILIZERS

Technical experts recommend a policy of combining application of organic fertilizers with chemical fertilizers. Mineral fertilizers are essential to complement the dung; they supply, from outside the farm, large quantities of the elements N-P-K, that are summed to those already present in the farm and help to correct the soil's nutritional deficiencies.

In terms of fertilizer units, application of organic matter is totally insufficient, mainly with respect to the fertilizer nutrients, mainly because the nutrients, especially phosphorus, potassium and the microelements are released slowly and often do not meet the crop's immediate needs. Application of organic matter is, however, essential to maintain and improve the soil's structure, and its capacity to retain nutrients, water and air.

Application of organic matter should be considered an investment in the medium and long term future. The incorporation of dung into the soil should be carried out in fall-winter, so that when the crop is planted in the spring, the organic matter has thoroughly decomposed. Dung is usually applied at a ratio of 30 t/ha, but higher doses (40-50 t/ha) can be used if it is done to improve the soil's physical structure. As already pointed out, the dung should be taken from the dungheap and buried as quickly as possible, in order to prevent losses due to evaporation.

Burial of plant remains, as well as dung, implies the incorporation of weed seeds and also favours the spread of fungal diseases (*Fusarium, Verticillium*, etc.). If a crop is profitable enough, it is better to use organic matter that is free of weed seeds, as the cost of weeding them out is high.

All organic fertilizers must contain levels of toxic heavy metals that are below the legal limits. When cultivating sensitive species, such as legumes, it is necessary to monitor the levels of copper, cadmium, mercury, etc., especially when these residues come from sewage treatment plants or are contaminated with industrial residues.

Applying dung to a field.

7. CORRECTING AND IMPROVING SOILS

The previous chapters have dealt with the application of organic or inorganic fertilizers to restore soil fertility, and of course this fertility diminishes with each season of plant growth. In certain cases, however, such as when turning a woodland into cropland, or improving saline ground near the coast, a general change of the soil's characteristics may be required.

Soil correction is when we are trying to change the soil's physical and/or chemical properties. These improvements imply application of several tons per hectare of some material in order to modify the soil's properties. These operations are expensive, and are only worthwhile for highly profitable crops.

Chemical corrections aim to change the soil's pH, in order to make the nutrient elements present more available to the plants. This may be **correction of acid soils** or **correction of alkaline soils**. It may also be necessary to modify the soil's physical properties (texture, density, porosity, etc.).

decrease in fertility and a soil environment that is unfavorable for growth of most plants. Raising the pH to a value nearer neutral may well be worthwhile, despite its high cost. This operation is called **liming**.

If liming is carried out in a well organized cultivation system, it leads to a long term increase in fertility. Liming is a tricky operation, because harvests immediately after liming are usually abundant, as by raising the pH we have favored adsorption of exchangeable bases and their later uptake by plants. Making these bases available gradually impoverishes the soil if it does not receive regular inputs of organic matter and fertilizers to maintain its fertility.

7.1.1. Materials used in liming

Logically, the material used should raise the alkalinity gently. To avoid drastic changes that may have very negative effects on the plants, liming should be gradual. If we have to raise the pH from say 5 to 7, then it is better to apply lime two or three times over a reasonable period, than to try and raise the pH all at once. This is true in general of

When ground is recovered for agriculture, in addition to removing the stones and levelling the site, it is worth carrying out a physical and chemical analysis to see if any soil improvements are necessary. In the case shown in the photograph, uncultivated ground is being transformed to introduce vineyards.

This is **correction of loose soils** or **correction of heavy soils**.

7.1. CORRECTION OF ACID SOILS

In regions where rainfall is high enough to ensure good crops, most soils are acidic. This acidity is due to washing of bases in the water that percolates through the soil. This leads to a

agricultural questions: it is better to solve a problem little by little than to try some miracle instant cure that may lead to more expenses than benefits in the long term.

The cations applied should be mainly Ca^{2+} and Mg^{2+}, with little or no Na^+, due to the well known salination effects of sodium ions.

A major project like raising the soil pH may be extremely expensive if we are talking about many hectares, and so the product used for liming should be relatively cheap.

Calcium carbonate (CaCO$_3$, lime) is the product most often used for liming. The material used is ground limestone rock, an excellent source of lime that is abundant and cheap. We are supplying desirable cations such as calcium, and also anions, carbonates, that do not have any toxic properties. The effect of lime is to raise alkalinity gently, but effectively, and it is a very cheap material as there are many deposits of limestone all over the world.

Other materials that can be used to lime a soil are CaCl$_2$ (calcium chloride), and calcium sulfate (CaSO$_4$ * 2H$_2$O, gypsum). These materials all have problems, such as their high price, toxic effects on plants, or the fact that they contain calcium but are neutral salts and thus do not raise alkalinity.

7.1.2. How much lime is needed

Minerals containing calcium carbonate usually have impurities, and this reduces the effectiveness of liming. Measuring the percentage of CaCO$_3$ is essential to calculate the number of kg needed to carry out the liming we are planning. The greater the percentage of calcium carbonate and the more finely milled the material, the less product will be needed for the same effect. The **effective calcium carbonate equivalent** (**ECCE**) is the measurement of a calcium carbonate's effectiveness.

There are two ways of calculating the exact amount of lime needed to raise a soil's pH, the pH-base saturation method, and the buffer solution method. They are very complicated, both in theory and in practice, as they involve use of measurements like the pH, the C.E.C., base saturation, and the ECCE of the calcium carbonate, meaning that they must be measured and calculated by a technician with the help of a specialist laboratory.

7.2. CORRECTING ALKALINE SOILS

Sometimes to cultivate certain plant species it is desirable to make the soil more acid. This is frequent in gardening because some flowers, such as azaleas and rhododendrons, and berries, such as bilberries, grow on acid soils. Some potato cultivators also prefer to acidify the soil in order to prevent potato scab.

Applications of elemental sulfur (S) or sulfur compounds are the most common method of lowering soil pH. The sulfur should be applied long enough in advance for it to be oxidized to sulfate. Agricultural sulfur has the great advantage of being very cheap, and the disadvantage that it acts in the medium and long term. If sulfuric acid was available as an industrial byproduct (sulfuric acid from copper smelting), we would have a very cheap alternative to agricultural sulfur, that would be equally effective, and faster acting.

Machinery produced by GILLES to incorporate the materials needed to improve the soil

Concentrated H_2SO_4 is applied to the cleared soil or in a 3% solution on Bermuda grass (*Cynodon dactylon*). Surprisingly, tests carried out by Ryan, Stroehlein and Miyamoto (1995) showed there were no adverse effects when concentrated sulfuric acid was applied directly to cleared soil. When cultivating plants with high requirements for iron (such as *Hydrangea macrophylla*), it is worth using ferrous (Fe^{2+}) sulfate and/or aluminium sulfate (alum). Nitrogen fertilizers that contain ammonium substantially reduce pH over time. Urea acidifies neutral or acid soils, but has no effect when there are abundant free carbonates, as they prevent hydrolysis of urea.

7.2.1. Problems of alkaline soils

When we were talking about the effect of pH on nutrients, we saw the drawing showing the solubility of each nutrient at different pH values. The chemical problems of alkaline soils are caused by the reduced availability of phosphorus, potassium and most micronutrients. Iron deficiencies are especially frequent in alkaline soils and cause ferric chlorosis in many species of plant.

In extremely alkaline, or basic, soils, however, the amount of salts dissolved in the soil solution is so high that plants have difficulties in water uptake. The osmotic pressure of the soil solution may be greater than that in the cell sap of normal plants.

7.2.2. Different types of saline soils

There are four different types of saline soils. All four have alkaline or very alkaline pH values, and their main problem is that due to the large amount of calcium and or sodium, the cation exchange capacity of other nutrients is in some way or another blocked.

Excessively limy (calcium-rich) soils are usually present as patches in wet regions. They are young soils, with an excess of calcium carbonate. They often show waterlogging, as the water table is very near the soil surface. They show low fertility because the Ca^{2+} ion occupies the exchange sites and prevents other ions from occupying them. These soils can be improved by better drainage, dunging to acidify the soil, and applying iron and magnesium sulfates to the foliage if the plants show clear ferric chlorosis.

Saline soils have a high concentration of soluble salts, which prevents plants from taking up nutrients (due to the difference in osmotic pressure). One can obtain an idea of the concentration of soluble salts in a soil from its electrical conductivity value. A soil is considered to be saline when its conductivity exceeds 2.4 millisiemens/cm.

A product for treating sodic soils. It acts by displacing the sodium ions in the exchange sites and replacing them with calcium ions. Manufactured by PROMISOL, S.A.

Saline soils can often be recovered by washing the salts out with abundant irrigation water. If the irrigation water is extremely saline, irrigation should be even more frequent and abundant to wash out as much of the salt as possible. Obviously intensive washing to remove salts only works if there is a good drainage system.

Sodic soils have more than 15% of their exchange sites occupied by Na^+ ions. For this

reason, they are poor in other soluble salts. Their pH value is around 8.5 and they are the most difficult soils to correct. It is worth ensuring drainage is good and practising frequent and abundant washings. There are products on the market that act by displacing the sodium ions adsorbed onto the clay-humus complex. They often consist of massive applications of calcium ions, which due to their high ionic strength, displace the Na^+ ions from the complex. The Na^+ ions in the soil solution can easily be washed by irrigation water or rain.

Sodic-saline soils combine the properties of sodic soils with those of saline ones. They have both a high concentration of soluble salts and 15% or more of their exchange sites occupied by the sodium ion. Recovery of these soils is in two stages. The first is to displace the sodium ion, using the products available on the market, and then to apply as much irrigation water as possible in order to wash out the sodium salts. Logically, this only works if drainage is excellent.

7.3. CORRECTION OF VERY LIGHT, OPEN SOILS

Normally, very open soils lack structure. We have already seen that structure is due to the

aggregates, and that these form due to organic matter. Very open soils are usually dominated by sand, and have a low level of organic matter. This gives rise to very light soils, whose main problems are the poor retention of water and nutrients. These soils often show high porosity values meaning they are suitable for cultivating any plant, as long as the water supply is constant and they receive regular feeding. The characteristics of these soils would make them partially suitable as substrates for hydroponic cultivation in a greenhouse, but in the open air, these soils represent a major problem for the farmer.

It is partly possible to correct these soils with regular applications of organic matter. In general, an application of well rotted dung before each crop will gradually reduce the soil's lightness. Normal applications of dung are about 30 t/ha, and these should be doubled in order to improve the soil gradually. It is also possible, though far more expensive, to import a quantity of clay in order to lower the average particle size and make the soil less sandy. Or perhaps the intermediate solution would be an application of organic matter and clay, which would improve the soil structure and increase its fertility.

7.4. CORRECTING VERY HEAVY SOILS

Very heavy soils are considered the opposite of very light soils. Their clay texture makes them highly compact and they tend to show waterlogging. If a soil suffers both compaction and waterlogging, there is the risk of frequent problems of root asphyxiation.

In very heavy soils infiltration of water is difficult and slow, meaning that it is hard to fertilizer nutrients to penetrate the soil. Very compact soils often show problems of aeration, and this reduces aerobic microbial action, which may be replaced by anaerobic respiration, causing a slowdown in the rate of decomposition of organic matter.

Surprisingly, the treatment to improve heavy soils is the same as the treatment for light soils, increasing its organic matter content. The organic matter in a clay soil acts like a sponge, creating pore space that can be occupied by air or water, thus reducing compaction.

In general, applying well rotted dung before planting each crop will reduce the soil's compactness. Normal applications of dung are about 30 t/ha, and these should be doubled in order to improve the soil structure gradually. It is also possible to import sand in order to increase the average particle size, and make the soil less clay. Or perhaps the answer would be an intermediate application of organic matter and sand (preferably siliceous river sand) which would increase the porosity (aeration) when it was mixed in.

*The typical symptoms in **Euphorbia pulcherrima** of excess soluble salts in the soil. The problem can be solved by frequent and abundant irrigation to wash the salts from the soil. Photo ceded by GRACE-SIERRA INTERNATIONAL, B.V.*

8. DEFICIENCIES OF NUTRIENT ELEMENTS

A deficiency of an element is when it does not reach optimum levels in the plant. Deficiencies may be **latent** (hidden) or **external** (visible). Insufficient uptake of nutrients causes nutritional irregularities in the plant, and these are accompanied (in external deficiencies) by pathological symptoms called **deficiency diseases** that may adopt many forms, depending on the element that is lacking (chlorosis, necrosis, leaf deformations, variations in coloring, inadequate growth, etc). It is important for the farmer to realize that a hidden deficiency may be responsible for lower than expected production, and that it is quantitatively much easier to correct a latent deficiency than one that is externally visible, because a visible deficiency is much worse than a hidden one, as the deficiency is more severe.

Clearly, quantifying a deficiency is based on a purely agricultural point of view, and more precisely in function of the optimum for plant production. Nutritional tests can be carried out by anybody, and you should try to perform them regularly, time permitting.

Suppose we wish to see the effect of P and K fertilizer application at a dose of 0 or 67 kg. To do this, there must be 2 x 2 = 4 plots —a total of 4 separate plots, in order to include all the combination of 2 fertilizers (P + K) and 2 doses (0 + 67 kg). The table underneath shows that the first plot, the control plot, is not fertilized. The second plot is only treated with 67 kg of K. The third plot only receives 67 kg of P, while the fourth plot receives 67 kg of both P + K. The effects on production are clear, the highest production is in the fourth plot, with fodder production of 9.79 t/ha.

The trials can be as extensive as you wish. If you wished to test the effects of three different fertilizers (N, P, K) at three different doses of 0, 50 and 100 kg/ha, then you would need 3 x 3 x 3 = 27 different plots to test all the different combinations.

8.1. CAUSES OF DEFICIENCY

From the agricultural point of view there are two different types of deficiency. **Absolute deficiencies**, or **primary deficiencies**, are caused by a deficiency of a given nutrient element in the soil, and it need only be applied for the deficiency to be resolved. In **induced deficiencies**, the nutrient elements are present in the soil but cannot be absorbed because it they are not in an assimilable state, and generally they are blocked because the pH is unsuitable or due to ion antagonism. This means that the deficiency is indirect because it is induced by unfavorable circumstances. It is not enough just to apply the element to the soil, as the deficiency will not be resolved. The unfavorable conditions must be changed, or the nutrient must be applied directly to the leaves. As already pointed out, study of human nutrition has advanced much further than plant nutrition, and the interactions of nutrients in the plant are still not fully understood. It is known that some interactions are positive and that others are negative. Positive interactions between two or more elements often mean they reinforce each other, and the result of their joint action is greater than the sum of their individual effects.

Negative interactions, or ion antagonism, are often the cause of induced nutrient deficiencies, some of which are considered next.

When phosphorus is applied, the nitrogen content of the plant declines if the plant is suffering a nitrogen deficiency, but it increases if nitrogen is abundant. Increased absorption of calcium and magnesium is accompanied by decreased absorption of potassium. One of the signs of potassium deficiency appears if the calcium content exceeds that of potassium more than eight-fold. This antagonism shows that these ions tend to replace each other. The same antagonism exists between magnesium and calcium, and between magnesium and potassium.

It is well known that the ions of calcium and iron show antagonism. Excess calcium in the soil, together with a high pH value, blocks the absorption of iron, causing the typical symptoms of ferric chlorosis. There is a similar antagonism between iron and manganese, causing a sort of competition between them, so that the symptoms of toxicity due to excess iron correspond to manganese deficiency and vice versa.

Copper deficiency leads to an accumulation of iron in the leaves, while excess copper gives rise to chlorotic symptoms that appear to suggest iron deficiency. As copper shows low mobility within the plant, the symptoms of the deficiency are clearest in the new growth and new organs, where the copper would normally accumulate.

A 2 x 2 experimental scheme showing the influence of applying N and P fertilizers on alfalfa production

Control plot No P No K 5.48 t/ha fodder	Treatment K No P 67 kg K 7.57 t/ha fodder (increase of 2.09 t/ha)
Treatment P 67 kg P No K 8.04 t/ha fodder (increase of 2.56 t/ha)	P+K treatment 67 kg P 67 kg K 9.79 t/ha fodder (increase of 4.31 t/ha)

The insolubility of iron phosphates means that iron contributes to phosphorus deficiency and vice versa. High levels of phosphorus may lower availability of iron and contribute to iron deficiency. High levels of phosphorus in the soil tends to lower the availability of zinc, and when zinc content is high, phosphorus absorbtion is lowered.

Molybdenum is essential for fixation of atmospheric nitrogen in the root nodules of legumes. It is also required for the enzyme that turns nitrates into forms the plant can use. As a result molybdenum deficiency can cause nitrogen deficiency in the plant, and nitrogen deficiency is usually the first symptom when molybdenum is present in insufficient amounts.

8.2. SYMPTOMS

The diagnosis of latent and/or external deficiencies is difficult. It is done by contrasting the results of three methods of investigation, soil analysis, analysis of the plant, and when the deficiency is visible, visual examination of the symptoms.

Farmers do not normally have the time or money to go to a laboratory specialized in the qualitative and quantitative assessment of plant deficiencies. We will give the reader two ways of assessing deficiencies that can be used directly in the field and based on a simple visual examination. The first a list of deficiencies and their most visible effects on the plants. The second is a key, similar in concept to the list, that the reader can work through to find out which nutrient is deficient. For those who are not used to working through a key, the idea is that they are based on elimination: if something is A, then it is not AA and viceversa. After the first question, you must chose between B and BB and then between C and CC and so on, until you reach the deficiency in question.

8.2.1. Description of the symptoms

Some elements are mobile within the plant, while others are immobile. This is of great importance when assessing deficiencies. Mobile elements, such as nitrogen, can be translocated to the growing tips (the receptors of the nutrients taken up from the soil). Immobile nutrients, such as sulfur, are found in the young parts and cannot be translocated to old leaves.

Nitrogen. Nitrogen is mobile within the plant, and deficiency symptoms are shown by the lowest leaves (the oldest). The oldest leaves turn yellow (chlorotic) and when they dry they turn greenish brown or black. If nitrogen is lacking in later stages of development, the stems are short and thin. In some plants, chlorotic leaves may develop orange or reddish colors. The leaves fall prematurely, and shoot and root growth is restricted. Branching, flowering, fruit and seed production all decrease.

Phosphorus. Phosphorus is also mobile within the plant, so deficiency symptoms are shown by old leaves. The leaves turn a dull dark green, and stem, petioles and leaf veins turn reddish-purple (because of the synthesis of **anthocyanins** due to the deficiency).

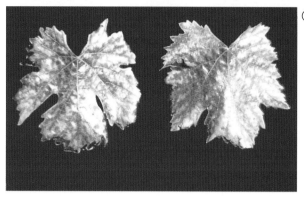

A/ *Ferric chlorosis in a peach tree. Kelamix plus, produced by SICOSA, S.A., is an EDDHA chelate of iron that is a very effective treatment for chlorosis.*
B/ *Yellow mottling in vine leaves caused by a lack of boron. Photo ceded by BORAX ESPAÑA, S.A.*
C/ *Calcium deficiency is normally shown by* Euphorbia pulcherrima *during its growing phase, especially when it has been exposed to excessive light (40,000 lux) and irrigation water with high levels of sodium. This can be controlled by applying 15-0-15 fertilizer from GRACE-SIERRA INTERNATIONAL, B.V.*
D/ *The characteristic symptoms of magnesium deficiency in* Euphorbia pulcherrima. *The poinsettia requires a lot of magnesium during its growing stage. A 15-5-25 fertilizer with a high magnesium content, such as the one produced by GRACE-SIERRA INTERNATIONAL, B.V., is suitable for correcting this problem.*

Anthocyanin is the generic name for the group of plant pigments responsible for the reddish, purple or blue color of many flowers, fruit, barks and roots.

A/ *Characteristic symptoms of molybdenum deficiency. The poinsettia requires a lot of molybdenum. If it is deficient, molybdenum must be applied gradually. Photo ceded by GRACE-SIERRA INTERNATIONAL, B.V.*

B/ *Characteristic symptoms of zinc deficiency. The problem is shown by young leaves in their early growth, and can be mistaken for manganese deficiency. A 15-5-25 fertilizer with a high Zn content should be applied to remedy the deficiency. Manufactured by GRACE-SIERRA INTERNATIONAL, B.V.*

C/ *A thick skin with inadequately developed pulp and little juice are the characteristic symptoms of boron deficiency in oranges. BORAX ESPAÑA, S.A. produces a wide range of boron-containing fertilizers.*

D/ *and* **E/** *Boron deficiency in the leaves of the Navelina and Navelate varieties of orange. Photo ceded by BORAX ESPAÑA, S.A.*

Marginal necrosis may appear in leaves, petioles and fruit. In terms of general growth, the effects of phosphorus deficiency are very similar to those of nitrogen deficiency: restricted growth of the roots and aboveground parts, short, thin stems, lowered production of fruit and seeds, etc.

• **Potassium**. Potassium is a mobile element, and thus potassium deficiency is most clearly shown in old leaves. The old leaves present pale chlorotic patches with the appearance of "burns" (necrosis) at

the leaf tips and edges. These areas of dead tissue progress from the tip to the base and from the leaf margin towards the intervein area. The leaf tip tends to curve downwards. The root system is poorly developed, the internodes are slightly shorter and the stems are weaker. Seed production is greatly diminished.

• **Sulfur.** Sulfur is a highly immobile element, so that sulfur deficiency, unlike phosphorus and nitrogen deficiency, is located in the younger shoots and parts. The young leaves show a general chlorosis, both along the veins and between the veins, and the younger the leaf, the yellower it is. The leaves grow less than in control plants, and the entire shoot is pale. In some plants (apple trees and strawberries), anthocyanin colorings may be present. There is never necrosis in these young leaves.

• **Calcium.** Calcium is an immobile element, and it deficiency is shown in young leaves, shoots and the growth zone of the root. The young leaves on the shoots, especially those around the apical bud, are malformed, showing curling, and curl around the growing tip and the apical meristem eventually dies. If the axillary buds develop, they also die in the end. In the already developed young leaves large chlorotic patches appear at the edges. The roots turn mushy and die.

• **Magnesium.** Magnesium is mobile and its deficiency is shown in the old leaves. It produces

typical intervein chlorosis on old leaves, which very rarely show necrosis.
The leaf tip and margins may curl up. The leaves do not dry up. In some cases, anthocyanin colorations may occur, for example in cotton and cherry trees.

Iron. Iron is immobile, and so deficiency is shown by the shoots and young leaves. Severe chlorosis appears between the veins in the young leaves and shots, while the main veins and the secondary veins stand out in green against a yellow background. In extreme conditions, necrosis and burns are present on the leaf margins and tips, and the leaves that are forming may show malformations.

• **Boron.** Boron is not very mobile, and so deficiency appears in the young leaves. The leaves surrounding the terminal bud turn light green at their base, and eventually fall off. Later growth shows distorted leaves that are stunted and fragile. Eventually the growing tips suffer necrosis and the apical meristem dies along with the tip of the shoot. Fleshy organs may rot internally (internal necrosis).

• **Manganese.** Manganese is immobile within the plant, and so deficiency is first shown in new leaves. Chlorotic patches appear all over the leaf and develop into intervein necrosis.

• **Zinc.** Zinc is relatively immobile, and the deficiency is shown by adult leaves. Intervein chlorosis occurs, and rapidly growing patches occupy the spaces between the veins, sometimes invading the nerves. Because auxin synthesis declines, the internodes are shorter, and the leaves may be small or thickened. In the final stages, the leaves may necrose on the edges and at the tip.

• **Molybdenum.** Molybdenum deficiency leads to nitrogen deficiency. This generally appears in the lower leaves, with intervein mottling and curving of the leaf. The entire edge of the leaf may dry up, and the leaf is narrow and ribbon-like ("whiptail").

• **Copper.** Copper is immobile and so deficiency is visible in the new leaves and shoots. They look bleached (apical bleaching), turning grey and looking dry and soft. The leaves located immediately below the tip are often unable to hold themselves upright.

The peach tree is very sensitive to ferric chlorosis. The photo shows a healthy specimen of Prunus persica with deep green leaves. It is clearly very different to the chlorotic peach (page 113). Photo courtesy of AGREVO

8.2.2. Key

Key to identifying deficiency symptoms in plant nutrition

SYMPTOMS	Deficient element
A) The plant's oldest (lowest) leaves are most affected; localized or general effects.	
B) Effects almost always generalized; more or less marked drying out of lower leaves; the plant is light or dark green.	
C) Plant is light green; the lower leaves are yellow, and when they dry out, they turn a light brown color; if the shortage is in the later stages of growth, the stems are short and thin.	**Nitrogen**
CC) Plant is dark green; it frequently shows red or purple coloration; the lower leaves are yellow and on drying turn a dark greenish or black color; the stems are short and thin if the element is scarce during later phases of development.	**Phosphorus**
BB) Effects are almost always localized; mottling or chlorosis in the lower leave, with or without areas of dead tissue; little or no drying out of these tissues.	
C) The mottled or chlorotic leaves may typically turn red, as happens in the cotton plant; dead zones sometimes appear; the leaf tip and margins twist, concave upwards; stems thin.	**Magnesium**
CC) Leaves chlorotic or mottled, with large or small areas of dead tissue.	
D) Small areas of dead tissue, generally at the tip and between the veins, and more pronounced at the edge of the leaves; thin stems.	**Potassium**
DD) Rapidly growing patches, generally between the veins, eventually invading the secondary leaf veins and even the main veins; thick stems, shortening of stem internodes.	**Zinc**
AA) The youngest leaves, including those around the buds, are affected; localized symptoms.	

SYMPTOMS	Deficient element
B) The terminal bud dies and distortions appear at the leaf tip or base of the young leaves.	
C) The young leaves around the terminal bud are usually curved at the beginning and eventually die at the tip and margins, so later growth is characterized by discontinuous growth at these points; the stems die, beginning at the terminal bud.	**Calcium**
CC) The young leaves of the terminal bud turn light green at the base, and are eventually shed. Later growth gives rise to twisted leaves; the stem eventually dies near the terminal bud.	**Boron**
BB) The terminal bud generally remains alive; chlorosis or bleaching of the youngest leaves, around the bud; veins are light or dark green.	
C) The young leaves are permanently bleached (apical bleaching), without patches or clear chlorosis; new shoots, as well as the branches and stems (in the area immediately below the tip) are flabby and often cannot hold themselves upright in the later stages of severe deficiency of the element.	**Copper**
CC) The young leaves are not bleached; chlorosis, with or without patches of dead tissue scattered over the leaves.	
D) Areas of dead tissue scattered throughout the leaf; the smallest veins tend to remain green, giving the leaf a "grid" appearance.	**Manganese**
DD) Generally, without dead area; the chlorosis may or may not attack the veins, meaning they may appear light or dark.	**Zinc**
E) Young leaves with light green veins and intervein tissue.	**Sulfur**
EE) Young leaves are chlorotic; the main veins are usually green; the stems are short and thin.	**Iron**

8.3. CORRECTIONS

Correcting non-induced deficiencies is simple. All that is needed is to apply the element lacking to the soil over time. In urgent cases, when a season's crop depends on it, foliar applications can be used, and these are surprisingly rapid and affective. Induced deficiencies are more complicated, as corrective applications are not enough, and the unfavorable soil conditions have to be changed.

Soil improvement is essentially modifying its pH and texture. These modifications are often enough to eliminate a deficiency disease, as soil improvements often require expensive long-term investments, and the problem is often how to find a rapid solution to an urgent deficiency.

In these cases, the experts recommend applying the nutrients to the leaves; this type of application must be carried out with great care, as excess nutrients could cause the opposite problem, excess fertilizer. Great care is especially important in the case of nitrogen leaf feeds with urea and microelements. If you think you are dealing with an induced deficiency, it really is worthwhile employing the services of an expert or laboratory to assess the scale of the problem and plan a suitable program to deal with it.

BIBLIOGRAPHY

ABAD, M. & NOGUERA, V.
Las turbas Revista Agricultura, 716-722
Madrid, March, 1992

ANSORENA, J.
Sustratos, Propiedades y Caracterización
Madrid: Mundi Prensa, 1994.

BUCKMANN, H.O & BRADY, N.C.
Naturaleza y propiedades de los suelos
3rd edition
Barcelona: Hispano Americana, 1985.

EL PAÍS NEWSPAPER
Atlas El País-Aguilar
Madrid, Aguilar, 1991

ENCICLOPÈDIA CATALANA
Gran Enciclopèdia Catalana
Reprint of second edition
Barcelona, 1990

GENERALITAT OF CATALONIA
Desinfecció de sòls amb bromur de metil
Barcelona, 1985

GROS, A. & DOMÍNGUEZ VIVANCOS, A.
Abonos. Guía práctica de la fertilización.

8th edition, revised and extended.
Madrid: Mundi Prensa, 1992

HERETER, A. & JOSA, R.
Edafología Barcelona Higher School of Agriculture,
1995-1996

HERETER, A. & JOSA, R.
Interpretació de resultats analítics
Barcelona Higher School of Agriculture
1991-1992

LIÑÁN, C.
Vademecum de productos fitosanitarios
Madrid, Ediciones Agrotécnicas, 1995

PENNINGSFIELD, F. & KURZMANN, P.
Cultivos hidropónicas en turba
Madrid: Mundi Press, 2nd edition, 1983.

STRASBURGER, F., NOLL, F., SCHENCK, H. & SCHIMPER,
A.F.W.
Tratado de botánica
Barcelona: Marín, 6th edition, 1981

THOMPSON, L.M., & TROEH, F.R.
Los suelos y su fertilidad
Barcelona: Reverté 4th edition, 1988

2

Fruit trees

1. INTRODUCTION

Arboriculture is the application of different agricultural techniques to trees, and fruit cultivation is the cultivation of fruit-bearing plants.

Many trees are of agricultural importance, but we are going to deal mainly with fruit trees, a complex group mostly cultivated for their fleshy fruit. They are from many different origins; temperate-cold zones (the apple), hot-temperate zones (the apricot), subtropical zones, (oranges and loquats) and tropical areas (pineapples and bananas).

The fruit tree, like any other tree, consists of two main parts, the aboveground parts, which we can see, and the root system, consisting of the roots as a whole.

The root system fulfils many roles, including:

• Anchoring the tree in the soil
• Uptake of water and nutrients from the soil
• Transporting organic substances throughout the plant
• Secreting certain organic substances
• Respiration
• Growth in length and thickness, and branching of the roots.

The aboveground part of the tree is the entire structure above the soil. It consists of the skeleton, formed by the trunk and the branches, and the crown, formed by the tree's most active elements, such as the buds, flower buds, twigs, flowers and fruit.

The aboveground part fulfils a series of functions. We can distinguish between those performed by the skeleton and by the crown.

• **THE SKELETON.** This mainly performs mechanical functions, but also has some physiological functions.

• *Mechanical functions.* It supports the leaves, flowers and fruit. Mechanical resistance to climatic adversity, e.g., wind or storms.
• *Physiological functions.* Circulation of nutrients along the plant's system of vessels. It has reserves in its wood and lignified tissue.
• *Biological functions.* Respiration and growth in thickness.

• **THE CROWN.** This has many functions.

• *Mechanical functions.* It shades the wood, screening it from the sun. It also protects the flower buds, flowers and fruit.
• *Physiological functions.* Storage of reserves and circulation of nutrients, like the trunk and branches.
• *Biological functions.* The most important are:

Photosynthesis. This is the process by which the green parts of the plant synthesize carbohydrates from the carbon dioxide in the air and the water supplied by the roots, and is powered by the energy of sunlight. All the sugars and reserves the tree produces are the result of photosynthesis.

Respiration. Plant respiration is the oxidation of carbohydrates produced in photosynthesis to obtain the energy required for the tree's growth, development and reproduction.
Transpiration. Transpiration is the process by which the plant emits the water absorbed by the roots. When this water evaporates, it cools the leaves, eliminating the heat produced in respiration and keeping the tree's temperature stable.

• *Reproductive functions.* These are performed by the flowers and fruit.

It should be pointed out that the growth organ of trees is the bud. The bud is cone shaped and consists of a meristem, a growing point, that is protected within protective hairs, and externally by bracts or scales. Buds are classified by their type, their position or how they develop.

• **By their type:**

• *Vegetative buds.* These give rise to a shoot when they develop.
• *Flower buds.* These develop into a flower or inflorescence.
• *Mixed buds.* These develop into shoots and flowers.

• **By their position:**

• *Terminal buds.* Located at the tip of the twig or branch. They may be vegetative or flower buds.
• *Axillary buds.* An axillary bud is produced in the axil of every leaf along the branch. They may be vegetative, flower or mixed buds.
• *Stipule buds, or replacement buds.* They occur on both sides of the axillary bud and replace it if it does not develop normally.
• *Basal buds.* These buds are located at the base of the shoot or branch.

• **By their development:**

• *Normal buds.* These buds develop normally, sprouting the year after their formation.
• *Latent, or dormant, buds.* These buds' development is inhibited and they remain dormant within the branch for several years.
• *Adventitious buds.* These are new buds that form spontaneously in old wood.

In almost all tree species, buds form during the growth period and then differentiate in a way that makes them clearly visible during the resting period.

When spring arrives, the differentiated buds start to grow. Their growth starts with an increase in the bud's size, and the opening of its protective scales, releasing the hairs, followed a few days later by the first leaves. This first stage is called sprouting and is the first stage after the bud breaks.

Growth continues with the development of the leaves and the axillary buds, creating a partly lignified association called the twig. When the vegetative growth period starts, the shoots gradually lignify and the terminal and the axillary buds become more prominent. At this stage, the twig becomes a **shoot**. Next spring, the buds on the branch will grow into new shoots, while the shoot will lignify and become a branch.

The main stem is vertical and forms the tree's trunk. The later branches that arise from the trunk are the *primary* branches, those that grow from the primary branches are s*econdary,* and so on.

Vegetative stems only produce vegetative buds, fruiting stems bear flower buds, and mixed shoots produce both types of bud.

1.1. PLANT GROWTH CYCLES

In temperate zones, environmental conditions vary considerably over the course of the year, and fruit trees are adapted to this.

Plant growth is mainly conditioned by the ambient temperature, and they respond by alternating phases of active growth (when conditions are favorable) with resting phases (when conditions are unfavorable).

Therefore, they show two relatively clear distinct periods.

• Winter rest period

In the winter rest period the tree does not show vegetative activity; photosynthesis is reduced to a minimum and neither growth nor flowering occurs.

Other processes, such as respiration, transpiration, uptake and circulation of nutrients by the roots, are slowest at the beginning and end of the rest period, or throughout the rest period in hot zones.

The tree ceases activity and starts its rest period coinciding with the low temperature period, and this cold period is necessary for the normal physiology of species from temperature zones.

• Active period

This period extends from the beginning of biological activity till the rest period starts. This activity includes all the tree's physiological and biological processes. This takes the form of the growth and development of shoots and branches, and the appearance of flowers and fruit.

The buds, shoots, flowers and fruit all change in appearance over the period of activity. These changes in appearance are **phenological changes**.

The first external sign of the start of vegetative growth is when the buds increase in size. This is the beginning of the vegetative cycle. The scale leaves protecting the bud separate and the inner down is visible. This is called bud **break**.

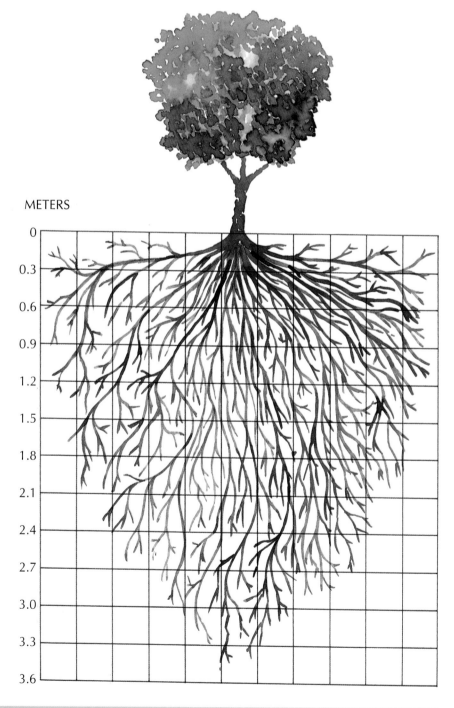

Diagram representing the growth in the soil of the roots of an isolated adult apple tree, in a loamy fertile soil without competition from other trees

METERS

0

0.3

0.6

0.9

1.2

1.5

1.8

2.1

2.4

2.7

3.0

3.3

3.6

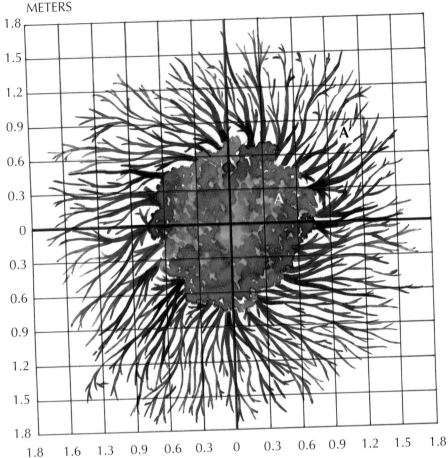

METERS

A Area covered by crown
A′ Area occupied by root system

Diagram showing the growth of the roots of an adult apple tree (seen from above), in a loamy fertile soil

To the right: Annual cycles of a fruit tree

After formation of flower buds, the flower primordia are formed, and the bud then enters a resting period. After the winter rest, the bud sprouts and then flowering occurs. The flowering process includes the pollination and fertilization of the flower. Flowering lasts from 10 to 25 days, depending on the species. Once the flower has been pollinated and fertilized, it grows into a fruit. The fruit grows, changes color and then undergoes ripening. Fruit ripening can occur in late spring or early summer in early fruit, in high summer for high season

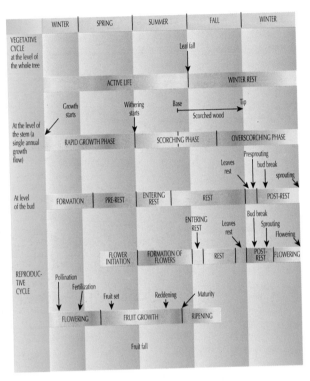

Soon afterwards, the first leaves appear. This phenomenon is called **sprouting**. Growth becomes intense, as long as the environmental conditions are appropriate. This is known as **spring growth** and lasts until temperatures reach 35-40°C. When temperatures are high, the trees enters the summer growth halt, which does not occur in some areas and may last several weeks in others, especially if there is water shortage or drought. When temperatures decline, fall sprouting occurs, which starts the second growth period which lasts until growth stops with the winter halt in growth. In deciduous species, this rest starts when the leaves are shed. This is the end of the growth cycle.

It should be pointed out that the activity cycle of the roots is longer. The roots become active 2-4 weeks before the aboveground parts, and cease growth 2-3 weeks after them.

The reproductive cycle is one of the most important parts of the vegetative cycle, and begins with flower bud formation and finishes the following year when the fruit ripens.

Flower induction is the process in which a vegetative meristem undergoes an irreversible modification into a flower bud. This change is caused by several very different external stimuli.

fruits, or in fall-winter for late fruit, and this depends on the crop species.

1.2. PHASES IN THE LIFE OF A TREE

The number of years that, on average, a tree lives in its natural environment before dying of natural causes is known as its life expectancy. This life expectancy varies greatly, mainly depending on the species. Some fruit trees, such as the peach, rarely live more than 25 years, while others may live for centuries, such as olives. By relating the growth of a fruit tree to its production, we can establish five periods or phases in the life of a tree.

• **Youth**. This includes the tree's first years of life. It typically shows vegetative growth with little flowering or fruiting. Depending on the species, this phase may last 2-7 years.
• **Starting production**. Flower and fruit production gradually increase. Depending on the species, this phase may last 3-10 years.

• **Adult stage**. This coincides with peak production. In this phase, the tree manages to maintain a balance, maintaining normal growth as well as full flowering and fruiting. This is the longest phase, and *may* last 10-40 years, depending on the species and the cultivation techniques.

• **Aging**. Growth declines considerably, though flowering may continue to be abundant. Fruit formation also declines. This phase is long and slow, depending not only on the species, but also on good cultivation techniques, as these may maintain acceptable levels of production for a longer time.

• **Old age**. This is the last few years of the tree's life. Growth is almost non-existent, and flowering and fruiting decline to nothing.

Left: classification of buds

The structure of a vegetative bud

TYPE OF SHOOT	NAME	CHARACTERISTICS	LENGTH	TYPICAL OF
VEGETA-TIVE	WOOD	The normal shoot with vegetative buds	0.5 to 2 m	Stone and seed fruits
	SUCKER	Excessively developed shoot with vegetative buds	Over 3 m	Stone and seed fruits
	SHORT	Weak underdeveloped shoot with vegetative buds	Less than 0.4 m	Seed fruits
	VERY SHORT	Very short with only terminal bud	0.5 cm	Seed fruits
FRUITING	MIXED SHOOT	Like a normal vegetative shoot but with some flower buds	0.5 to 2m	Stone fruits
	VEGETATIVE FLOWERING	Like the short. The terminal bud is vegetative and the rest are flower buds	Less than 0.4 m	Stone fruits
	SHORT	Short in which the terminal bud has turned into a flower bud	Less than 0.4 m	Seed fruits
	SHORT FLOWERING	Like a short shoot. The flower buds are bunched on a single axis	3-5 cm	Stone fruits
	VERY SHORT FLOWERING	Very short in which the terminal bud has turned into a flower bud	0.5 cm	Seed fruits
	VERY SHORT SHOOT	Very short shoot elongated by 2 or more years of vegetative growth and in which the terminal bud is a flower bud	5-10 cm	Seed fruits

Ⓐ

TERMINAL BUD

AXILLARY BUDS

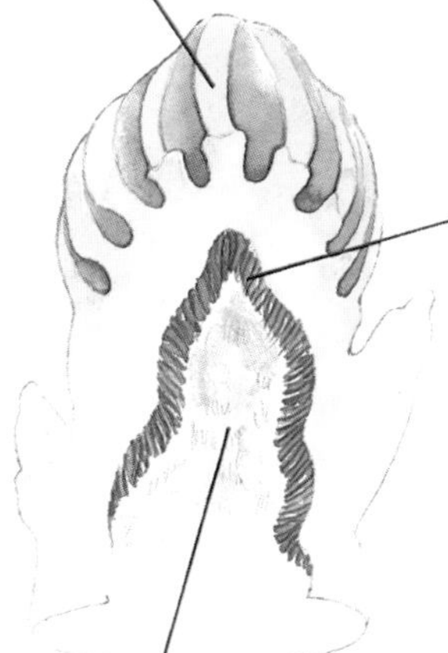

SCALE LEAVES

PROTECTIVE HAIRS

A/ Organs of seed fruits

B/ Organs of stone fruits

APICAL CONE

1. VEGETATIVE BUD

2. VERY SHORT

3. ELONGATED SHORT SHOOT

4. SHORT

5. VERY SHORT WITH FLOWER

6. WOODY BRANCH

7. STUMP

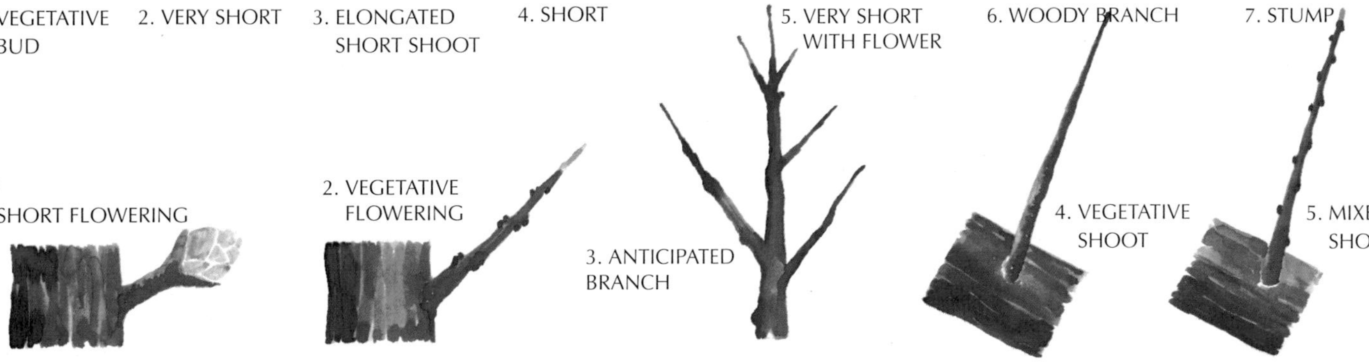

Ⓑ

1. SHORT FLOWERING

2. VEGETATIVE FLOWERING

3. ANTICIPATED BRANCH

4. VEGETATIVE SHOOT

5. MIXED SHOOT

2. REPRODUCTION AND PROPAGATION OF FRUIT TREES

2.1. SEEDS

Plants grown from seed are robust, well rooted in the soil, and do not require large investments. But bear in mind that plants grown from seeds are not genetically identical to each other, and this leads to differences in behavior that are incompatible with modern cultivation methods.

It should also be pointed out that the modern cultivars obtained through hybridization and sports do not breed true from seed.

Sowing is one of the most important tasks when it comes to planning a crop. To sow well, you must taken into account the following factors:

A/ Seed sowing in members of the Rose family bearing pomes, "seed fruit" (apple, pear, etc.)
B/ Seed sowing in members of the Rose family bearing drupes, "stone fruit" (peach, etc.)

In the second case, the plant will work out more expansive for the farmer, but it will be firmly grafted and growing.

• Use selected seeds
• Prepare the soil correctly
• Distribute the seeds correctly

Production depends on the environmental conditions and on the plant's productive capacity. This second factor depends in turn upon the high quality of the seed and on its genetic characteristics.

2.1.1. Seed requirements and storage

To get good results in a crop, seeds with the following characteristics must be chosen:

• They must belong to the species or variety you want
• They must be free of impurities
• They should be easy to store
• They must germinate to give healthy plants

The origin of the seeds is important in relation to the climatic characteristics of the site of origin, as this factor may influence their resistance to low temperatures.

The genetic characteristics of the seeds are also important, as are the techniques used to separate the seed from the fruit. It is important to know the treatments the seed has undergone during the industrial separation process, to ensure it has not negatively affected their viability.

Bearing in mind that it is almost impossible to recognize a variety on the basis of its seed alone, the seeds bought should be certified, as this guarantees that the entire production of the seed has been thoroughly controlled.

Certification is carried out by the technicians of

A) Duration of sowing 1 year + 1 year in nursery

YEAR	January	February	March	April	May	June	July	August	September	October	November	December.	LOCATION
1			Sowing		Germination	Growth and development							Seedbed in nursery
2	Transplanting			Growth and development									Transplant in nursery
3		Transplanting						Graft scion					Transplant into the field
4		Grafting of scion		Sprouting		Growth and development					1 st pruning		Field

B) Duration of sowing 1 year + 2 years in nursery

YEAR	January	February	March	April	May	June	July	August	September	October	November	December	LOCATION
1			Sowing		Germination	Growth and development							Seedbed in nursery
2	Transplanting			Growth and development					Grafting of scoin				Transplant to nursery
3		Grafting of scion		Sprouting		Growth and development							Nursery
4	1 st pruning and transplant												Transplanting to field

the body authorized to do so, and they control all the phases of seed production and gathering.
Some varieties are better than others in terms of producing seed for cultivation.

Good seed producing varieties include:

• **Apples**

Ben Davis
Annurca

• **Peaches**

Lovell
Elberta
GF-305
Nemaguard
Stribling
Okinawa
Rancho Resistent

• **Plums**

St. Julien Hybrid 1
St. Julien Hybrid 2

As seeds are the origin of all crops, it is important to determine their qualities.

• **Purity**. This is how clean the seeds are. It is important to distinguish between the whole seeds of the desired species, the seeds that belong to unwanted species, and the impurities (the stones, earth or broken seeds that may be present). The purity is expressed as a percentage, by weight or in parts per thousand.

• **Germination**. The germination is the percentage of pure seeds that in favorable conditions will grow into normal plants. The seeds of many trees, such as citrus fruits, persimmons, loquats, poplar, elm and willow, lose their ability to germinate in a few days or weeks if they are not correctly stored.
Other plants, such as the olive, retain their ability to germinate normally for many years.

Other factors to be taken into account are:

• *Health*. They should be disinfected in order to prevent certain diseases and pests.
• *Moisture*. It is important to know the seed's moisture content when it comes to storing the seeds.
• *Quality*. Factors, such as size, shape, color and shine all reflect the seed's quality.

Seeds are stored for a variety of reasons. One of them is to maintain their germination, and the other is to induce faster and easier germination.
Seeds that rapidly lose their ability to germinate are usually left on the tree, within the fruit, for as long as possible, almost until the moment they are to be sown.

In general, seeds that show good germination accompanied by a long dormant period are usually treated by **stratification**.

Stratification is storage of seeds between alternating layers of wet sand at a temperature of 2-10°C, depending on the species.

The seeds may be submerged in water for 12 hours before stratification.

Depending on the species, stratification lasts for 1-4 months.

• Almond and apricot: 3-4 weeks
• Apple: 2 months
• Pear, peach and cherry: 3 months
• Persimmon: 6 months

2.1.2. Treatments to favor germination

Before the seed is sown, it may require treatment to encourage it to germinate. This preparation may be physical or chemical.

• **Physical preparation**
In general, this is the drying, selection and clasification carried out by the seed producer.

• **Chemical preparation**
This is done to kill the pests and diseases that can be transmitted by seeds, and to protect them from the insects that may attack the earliest stages of germination.

In general, to encourage germination in seeds with a hard stone it is worth softening the join between the two parts of the stone.

This is done by stratifying the stone. As explained above, stratification is storing the stones between layers of wet sand. The best time to carry out this operation is late fall/early winter. The process lasts roughly three months, until the stone starts to open.

Other treatments applied to the seeds to favor germination are:

• **Application of hormones**
Applying gibberellins to dormant seeds may increase their germination and the growth of recently germinated plants. This has been demonstrated in peaches, apples and vines.

• **Immersion in hot water**
This is carried out on seeds with very hard or impermeable seed cases in order to soften the outer layers, and to wash out germination inhibitors.
Immersion is for 10-12 hours.

• **Mechanical scarification**
This is carried out to break the seed case. Care must be taken not to damage the embryo and leave the seed inviable.

*Inside of a nursery
with irrigation system*

*Detail of interior of
shaded area*

*A variety of plants
forming roots*

2.1.3. Sowing times and methods

Sowing is especially important and a good crop
depends on it.
The seeds of trees with hard stones are sown
after stratification in late winter/early spring,
when the root begins to grow out.
The seeds should be sown in seed beds arranged
in lines about 70-80 cm apart. Germination
does not occur until a month after sowing.
The seedling is grown for use as a **rootstock**, and
it will later be grafted with the desired variety.
The rootstocks will remain in the seedbeds until
the following winter, when they can be
transplanted into the field to be grafted and
grown.
Seed fruit should be sown into a seedbed, either
by scattering or with a seed drill in late
winter/early spring.
At the beginning of the next winter, 9-10 months
after sowing, the plants obtained in the seed bed
are transplanted into straight lines 60-80 cm
apart with a distance between the plants of 20 to
30 cm.
A year after the first transplant, they are
definitively planted into the orchard.
Preparation for sowing should include preparing
the soil, in order make the seedling's
germination easier and to ensure it roots well.
The seed must find everything it needs for its
germination in the soil, that is to say, warmth, air
and moisture.

2.2. VEGETATIVE PROPAGATION

Vegetative propagation is based on the ability of
some parts of the plant to produce new shoots
and roots, or the ability of a grafted bud to grow
into a new plant.
There are several vegetative propagation
techniques, such as taking cuttings, layering and
grafting, and we shall now discuss them.

Vegetative propagation has two main advantages
over reproduction by seeds:

• The offspring plants are all identical to each
other and to the parent plant;
• It gives rise to plants that are in a short
juvenile phase, meaning that the start of fruit
production will not take so long.

When carrying out vegetative propagation it is
very important to respect the polarity of the
different plant parts to be used as
reproductive material. That is to say, the part
that was nearest to the roots must be
orientated so that it will produce the roots,
while the topmost part must be orientated so
it can produce shoots. They must conserve
the same orientation (polarity) as they had on
the mother plant. If the polarity is inverted,
the cutting's ability to root is greatly
diminished.

One way of propagating using shoots produced
by the roots is to use the suckers that some trees,
such as the plum and the hazel, produce
spontaneously from their roots and crown.
Using these shoots has gone out of fashion, as
the new plant inherits this tendency to
produce suckers and this leads to a greater
number of operations to get rid of them.
These suckers are also prone to degenerate
and to transmit diseases, especially virus
diseases.

2.2.1. Cuttings

This is a vegetative propagation technique based on using the parts of the plant, such as the shoots, branches, leaves and roots. Once these parts have been removed from the mother plant and placed in the right conditions, they can develop roots and give rise to a complete new plant.

In general, woody plants are propagated with techniques of this type, using parts of the branch or shoot. In the case of branches, the cutting is a **woody cutting**, while cuttings from shoots are **semi-woody** or **softwood cuttings**, and normally bear leaves.

FACTORS		
PLANT	— Nutritional status	
	— Age	
	— Vegetative status	
	— Polarity	
EXTERNAL FACTORS	— Atmospheric humidity	
	— Temperature	
	— Light intensity	
	— Tipe of substrate	
	— Substrate moisture and humidity	
CULTIVATION	— Hormone application	
	— Foliar feed	
	— Incisions	
	— Warmth from below	
	— Misting - Spraying	

SUBSTRATES	ADVANTAGES	DISADVANTAGES	TYPES
STERILE	— Free of disease — Allow good aeration — Retain moisture	— Do not supply any nutrients — Are normally used as mixes	— Sand. Retains little irrigation water — Vermiculite — Perlite
NON STERILE	— Supply nutrients		— Light peat. Is not attacked by fungi and parasitic bacteri. Rich in rootforming substances. pH 4-4,5 — Dark peat. Cannot be used on its own as it creates an anoxic environment that causes the cutting to rot. pH 5-7. — Moorland soil. Risk of nematode invasion.

TYPE OF CUTTING	ATMOSPHERIC CONDITIONS	SUBSTRATE	TIME OF YEAR AND OPERATIONS
Woody cutting without leaves	— Geenhouse or cold tunnel — Misting to maintain humidity at 60-70% — Intense light	— Sand or fine vermiculite + 10% (by volume) moorland soil	— End of February — If cutting is taken in Dec/Jan, ensure air temperature is 13 - 15 °C
Herbaceous cutting with leaves	— Greenhouse or tunnel — Misting or irrigation to maintain humidity at 90% — Weak light	— Sand or vermiculite + 10% (by volume) of light peat	— July and August until mid September — Treatment with hormones — Nitrogen based leaf feed

2.2.1.1. Types of cutting

We can distinguish five types of cutting:

• **Simple cutting**. The most widely used. It consists of a portion 25-35 cm long of a vegetative branch. In vines, cuttings may be up to 50 cm long.

• **Cutting with "heel"**. This is the same as a simple cutting, but has a "heel" of wood from the branch it was taken from.

• **Stump cutting**. The same as the preceding two, but has 6-10 cm of the branch it was taken from.

• **Shoot cutting**. Branch or part of a branch, 1 or 2 years old, and 1-2 m long. This is used in figs, olives, poplars, etc.

• **Semi-woody cutting**. A portion of the shoot with the current year's leaves. This technique is used in stone fruit, which are very hard to root using woody cuttings, and also in evergreens.

Upper left:
Factors that influence the rooting of cuttings

Below left:
Substrates used for cuttings. General characteristics

Above:
Mixture of substrates for cuttings of fruit trees and vines

Different types of cuttings

SIMPLE CUTTING

CUTTING WITH HEEL

STUMP CUTTING

SHOOT CUTTING

SEMI-WOODY CUTTING

2.2.1.2. Propagation by cuttings

Woody cuttings are taken during the winter period. They are stored stratified, or in a refrigerated environment, until the moment they are going to be planted, at the end of the winter period.

The cutting is normally 15-20 cm long, but in some special cases, such as the vine, it may be 50 cm long, and in the olive it may be up to 2 m long.

The cuttings are planted in the nursery where the soil has been suitably prepared and fertilized. Planting can be performed by pushing the cutting into the substrate, or they can be planted by opening up furrows and leaning the cuttings on the wall of each furrow, then covering it and pressing down the earth around the roots.

Semi-woody cuttings of deciduous species are taken in late spring, when the sprouts have reached maturity and the leaves are fully developed.

Semi-woody cuttings of evergreen species can be taken in the fall-winter period. Cuttings of some species can only be taken during a relatively limited period.

Planting cuttings

Stratified while awaiting planting.

Dig a hole 12-15 cm deep.

Place the cuttings vertically 12-15 cm apart.

Firm the soil and leave the top 2-3 cm of each cutting exposed.

Remove the cuttings, taking care not to damage the roots.

2.2.1.3. Additional techniques for taking cuttings

There are several ways of encouraging and promoting root growth in cuttings, including:

• **Hormone application**. There is a wide range of chemical products, similar to naturally occurring plant growth regulators, that favor root growth. This group of hormones is known as **auxins**.

- *Indole Acetic Acid, IAA*
- *Indole Butyric Acid, IBA*
- *Alfa-Naphthalene Acetic Acid, ANA*

The hormone can be applied as a liquid, by submerging the base of the cutting in a solution of the product, at a concentration of 20-200 ppm for 24 hours, or of 500-10,000 ppm for 5 seconds.
It is also possible to treat cuttings with hormone rooting powders, by wetting the base of the cuttings and then dusting them with the powder.

• **Misting and high humidity**. They act to help root formation in semi-woody cuttings or cuttings with leaves.
Maintaining the ambient humidity at a high level helps cuttings to root. This can only be done in greenhouses. The idea is to reduce transpiration as much as possible, meaning that the cuttings live longer, and have more time to root.
Misting is applying water as a mist directly onto the cuttings placed in lines to root.
The water can also contain things like foliar feeds at very low doses.
Misting devices can be installed in greenhouses and in the open air. The water vapor has a double effect on the cuttings; it reduces transpiration and acts to buffer temperatures, ensuring the cuttings are not negatively affected by high temperatures.
The substrate must be well drained to prevent root asphyxia. The water used should not be calcium-rich, and must be free of impurities, as they would block the nozzles of the misters.

• **Warmth from below**. The cuttings are taken at the beginning of the winter rest period and are placed in rows or previously prepared boxes. The substrate within is kept at a temperature of 21°C by electric heater elements regulated by a thermostat.
The basal heating means that the cuttings, which have been treated with hormones, quickly form root callus and then roots, which are followed by the opening of the buds and sprouting.

2.2.1.4. Aftercare of cuttings

Plants grown from woody cuttings are dug up during the winter, in their rest period.

They can be dug up by hand or with special machinery, special plows that cut below the roots and lift up the rooted cuttings.
When dealing with cuttings of evergreen plants, they should be dug up with a root ball, thus avoiding the shock of transplanting.
In both cases, they can be transplanted directly into the field, or to other areas of the nursery for later grafting.

2.2.2. Layering

Layering is the production of a new plant by causing roots to grow on a branch that is still joined to the parent plant, and which is later separated when it has grown enough roots to live independently.
The branch is brought into contact with an open, moist, substrate to encourage root formation.
Except for the hazel and some varieties of fruit tree, layering is no longer widely used, because it requires a large area of ground, several years for propagation to occur, and reduces the yield of the parent plant.

2.2.2.1. Types of layering

The following layering methods exist:

• **Simple layering**. This is carried out in shrubby trees that have branches that can be bent down to the ground. There are two methods, multiple layering (with several simple bends on a single branch) and tip layering (where the tip of the branch is buried in the soil).

Simple layering

A/ Bending the branch down to the ground in winter
B/ Spring/summer. Rooting and sprouting
C/ Winter. Young plant after separation

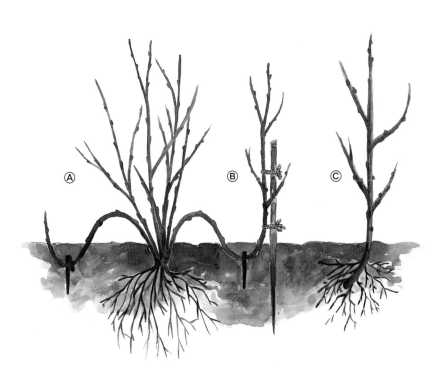

• **Branch layering.** The entire length of the branch is buried.

• **Rootstock layering.** This method is used in fields for producing graft stocks, especially of apples, quinces and plums.

Branch layering. On the right, rootstock layering

Ⓐ

In winter the tree is planted at 35-40° with respect to the soil. It will sprout in spring.

Ⓑ

A ditch is dug in winter and the bent stem is buried.

Ⓒ

During the spring and summer, it will produce shoots that will root.

Ⓓ

Separation is carried out in winter.

Ⓐ

Cut the stem to ground level in winter.

Ⓑ

In spring, when the sprouts have reached 30 cm, earth up around the stem.

Ⓒ

In summer the roots form.

Ⓓ

The earth is removed and the new plants separated in winter.

Ⓔ

The shoots after separation.

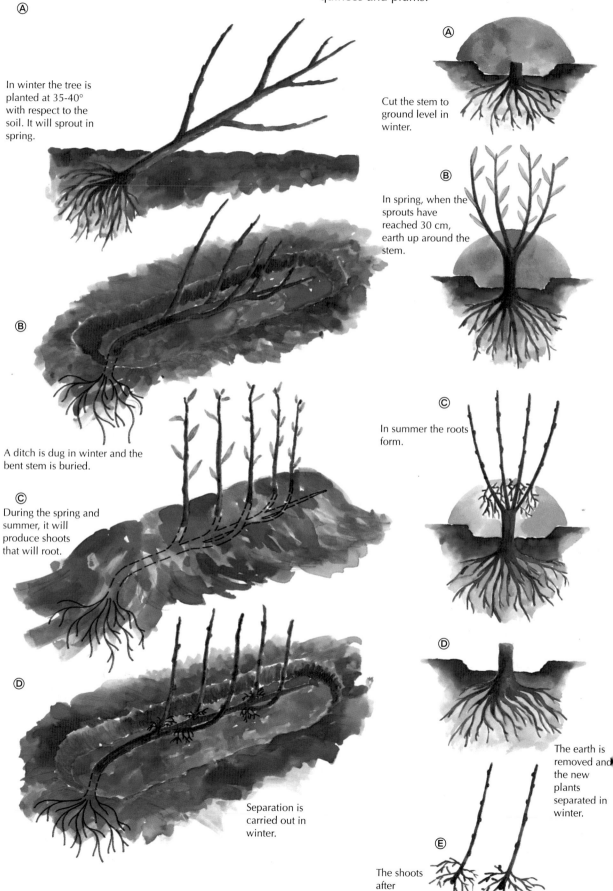

- **Air layering** This technique is used when the tree to be propagated does not have branches that can be bent down to the ground.

2.2.2.2. Layering propagation techniques

- **Simple layering**. During the winter rest period, dig a trench about 25-30 cm long near the tree's trunk.
The selected branch is bent down and buried leaving the tip exposed. The tip is tied to a guide. In spring, the buried buds will sprout and then develop roots, while the aboveground buds will sprout. It is separated in the next winter, giving rise to a new plant.

- **Branch layering**. The tree to be propagated is planted almost parallel to the soil, at an angle of 35-40°, and it is kept down with a "U"-shaped device.
A series of sprouts will appear along the stem. In the following winter dig a trench 30 cm deep and bury the layered stem.
The following spring and summer, the shoots will grow and the buried parts will produce roots.
When winter arrives, the cuttings are separated from the parent plant, thus obtaining new plants after a period of two years.
The plant nearest to the roots of the mother plant will be the tree to propagate, and thus the cycle can be performed again following the process mentioned above.

- **Rootstock layering**. During the winter, the trunk is cut at ground level. In spring, when the shoots are about 30 cm tall, it is earthed up to leave only 10 cm of the shoots above the earth. The shoots will grow in the summer, and the buried parts will form roots. The next winter, the earth is carefully removed and the plants are separated to obtain the new plants.

- **Air layering**. In winter a wedge-shaped lateral incision is made in the branch to be layered, to provoke the appearance of adventitious roots.
The area where the cut is made is then covered with a mixture of substrates, and is all bound together and to the tree with fabric or plastic with holes pierced in it.
In the following winter, the cutting is separated below the root ball.

2.2.3. Grafting

Grafting is a vegetative propagation technique that consists of joining together parts of two different plants, so that they bind together and the sap can flow, thus forming a single new plant that can grow and develop.

A graft consists of two parts:

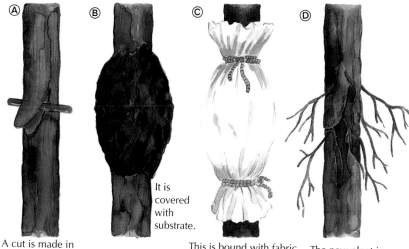

Ⓐ Ⓑ Ⓒ Ⓓ

It is covered with substrate.

A cut is made in winter.

This is bound with fabric or plastic with holes pierced in it.

The new plant is separated in winter.

Air layering

- The lower part, the **stock** or **rootstock**. This essentially consists of the root system, the part already growing on the site, and a portion of stem.
- The upper part, the **scion**, the **graft**, or the **variety**. This gives rise to the aerial and productive part.

The rootstock can be obtained by sexual or asexual reproduction.

Crown graft in a cherry tree

The main varieties used as stocks for grafting varieties of fruit trees. The columns on a green background, headed ROOTSTOCKS, is a list of the main stocks used. The columns headed ORIGIN indicates the biological origin of the stocks in the column on the right - APR (apricot), ALM (almond), PLU (plum), ALMxA (almond x apple), PLUxA (plum x apple), APP (apple), QUI (quince), PEA (pear) and CHE (cherry). The columns with titles in blue shows the scions that can be grafted on the stock.

Examples:
1. The stock "Missour" (green column) is biologically a peach (PEA, in the origin column) and is biologically close enough to accept grafts of varieties of almond, peach and nectarine; it is not, however, compatible with varieties of apricot or plum.

2. The Kirchensaller stock is biologically a pear, and only varieties of pear can be grafted successfully onto it.

Nectarines	Peach	Plum	Almond	Apricot	ORIGIN	ROOTSTOCKS
X	X	X	X	X	APR	Apricot
X	X	X	X	X	ALM	Almond
X	X	X			PLU	Brompton
X	X	X			PLU	Plum INRA GF-43, GF-2038, GF-2037
X	X	X			PLU	Damson C
X	X				PLU	Damson P-12
	X		X		PLU	Damson P-1869
X	X		X		ALMxAPP	Hybrid GF-557
X	X		X	X	APP	Hybrid GF-677, IS-5/18, IS-5/23
X	X	X	X	X	APP	Common peach
X	X				APP	Bangour
X	X		X	X	APP	GF-305
X	X				APP	GF-278, GF-763
X	X		X		APP	Missour, PS-A3, PS-A5, PS-A6, PS-92, PS-C14
X	X	X	X	X	APP	Nemaguard
X	X				APP	Okinawa
X	X			X	APP	Rancho R
X	X			X	APP	Stribling S-37 and S-60
X	X			X	APP	Harrow Blood, Siberian Ccrabapple, Bokhara Shahil, Tzin Pee tac, Yunnan, Pl 36436
X	X				APP	Higama, Rubira, Rutger's red leaf
		X	X	X	PLU	Marianna 2624, Marianna GF-8-1
X	X				PLU	Marianna P-10-2
		X		X	PLU	Myrobalan
		X		X	PLU	BM-8, INRA GF-31, P-12
		X			PLU	INRA GF-31-6, P-1030
				X	PLU	29-C
			X		PLU	P-34-16
		X			PLU	Prunus tomentosa
		X			PLU	Prunus besseyi
		X			PLU	Prunus hortulata (Fla-1-1)
				X	PLU	Reine Claude GF-1380
X	X	X			PLU	St. Julien A & St. Julie of Orleans
X	X	X			PLU	St. Julien GF-655-2
X	X				PLU	St. Julien Hybrid 1 & Hybrid 2
		X			PLU	Selec. Pixy (S. Julián Orleans)
					PLU	Prunus institia
		X			PLUxAPP	Rigotti-1, P-1609, S-2729, P322x5 1058, S 749 x S 1490, P 322 x P 871 and S1 and S2

Cherry	Pear	Apple	ORIGIN	ROOTSTOCKS
		X	APP	Wild apple
		X	APP	M-1, M-13, M-16, M-25
		X	APP	MM-109, MM-104, MM-111, MI-793, M-2, M-4, M-7, MM-106
		X	APP	M-9, M-26, M-27
		X	APP	Bittenfelder
		X	APP	Grahams Gubilaum
		X	APP	Skierniewice P-1, P-2, y P-22
		X	APP	Alnarp 2 (A$_2$)
		X	APP	Budagowski-9 (B-9)
		X	APP	B-119, B-146, B-491
		X	APP	MAC (1, 4, 9, 25, 39, 46, 16, 24, 30)
		X	APP	Pillnitz (PIR-0, PIR-900)
		X	APP	Dab
	X		QUI	Angers, Sydo and Adams quinces
	X		QUI	Fontenay quince
	X		QUI	Provence quince, INRA BA-29
	X		QUI	EM-A, EM-C
	X		QUI	Malling A quince
	X		QUI	INRA quince C-85-1
	X		QUI	INRA quince C-98-4
	X		PEA	Wild pear
	X		PEA	Wild pear clones; 2267, 2268, 2269, 2270, 2271, 2272, 2273, 2274, 2275, 2276, 2277, 2278
	X		PEA	Pyrus communis (class Old Home) and clone OH x F
	X		PEA	Fieudiere
	X		PEA	Kirchensaller
X			CHE	Prunus avium
X			CHE	Prunus avium F-12-1
X			CHE	Prunus mahaleb (St. Lucie cherry)
X			CHE	Prunus cerasus (acid cherries)
X			CHE	Colt (P. avium x P. pseudocerasus hybrid)
X			CHE	Prunus betulaefolia
X			CHE	Prunus ussuriensis
X			CHE	Prunus calleryana
X			CHE	GM (Grand Manil) 1, 8, 9, 17, 61, 65, 79, 85 and 156

There are several reasons for carrying out grafting.

• Main aims
- To propagate a profitable variety.
- To spread a variety. Most commercially profitable varieties do not grow well on their own roots.

• Secondary aims:
- To adapt a species to specific climatic and soil conditions (negative conditions, such as root asphyxia, drought, tired soil, chlorosis and low temperatures).
- To induce greater or lesser growth and vigor, and a longer- or shorter-lived tree.
- To bring production forward in the weaker rootstocks.
- To improve the general quality of the fruit; its size, color, taste, etc.
- To increase resistance to certain pests in a given area. This is one of the most effective and profitable ways of fighting pests.
- To introduce an effective pollinator variety in orchards where this need was not foreseen in advance.
- To renew plantations of varieties that are no longer marketable, and thus need to be replaced.

The last two points are really regrafts, that is to say, grafting onto a tree that is already a graft.

Fig	Lemon	Mandarin	Orange	Loquat	Persimmon	Walnut	Hazel	ROOTSTOCKS
					X			Diospyros kaki
					X			Diospyros lotus (Loto italico)
					X			Diospyros virginiana (Common or American, persimmon)
						X		Juglans regia
						X		Juglans nigra
						X		Juglans hindsii
						X		Paranox (J. regia x J. hindsii)
							X	Corylus tubulosa (C. maxima)
							X	Corylus colurna (Turkish hazel)
							X	Corylus avellana
				X				Wild loquat (Eriobotrya japonica)
				X				Quince
				X				Hawthorn
X								Fig (cutting or shoot)
	X	X	X					Bitter orange (Citrus aurantium)
		X	X					Sweet orange (C. sinensis)
	X♦	X	X					Mandarin orange (C. reticulata)
	X	X	X					Common mandarin
	X	X						King Mandarin
	X♦	X	X					Parcirus trifoliata
	X	X	X					Citrange Carrizo
	X♦	X	X					Citrange Troyer (P. trifoliata x W. Nare)
	X	X						Citrus taiwanica
		X						Ragspur lime
		X						Rough lemon (C. jambhiri)
	X							Citrus macrophylla
	X							Citrus volkameriana
	X							Sampson Tangelo
	X							Orangelo 4475
	X							Citrus amolicarpa
	X							Citrus depressa
	X							Citrus junos
	X							Citrus penniversiculata

(♦ only successful with some varieties)

2.2.3.1. Preconditions for grafting

Before grafting, the following points should be taken into account.

The affinity between the stock and the variety
The cambium of the two plants must come into direct contact, so the plants can join
You must respect the stem-root polarity when inserting the scion.
• Use a grafting method appropriate for the stock and the variety to be grafted.
• Use grafting tools that are clean and sharp.
• Carry out the graft corrrectly. The cuts should be clean and the cambium zones must be in direct contact, and they may also be immobilized and fixed together using ties or putty.
• Correct aftercare. Remove any gum exuded in some stone fruits, and break the ties around the graft if they are strangling it.

following steps should be taken to store it:

• Bundle into groups of 30-40 units and store vertically in dry sand or in chambers 1-2°C above zero.
• Wet the sand to make it easier to remove the bundles and then wash them to remove all sand.
• Cut the tips off each variety and it will then be ready to graft.

Observations for performing the graft
When a woody graft is performed in late winter, the stock or rootstock should be in a later stage of growth than the scion. This is important in order to ensure the stock can nourish the scion. The best time to take material from the parent plant is in late summer or early fall, when the tree is entering its winter rest period. This process can be carried out until the end of winter.

Placing the scion in the stock

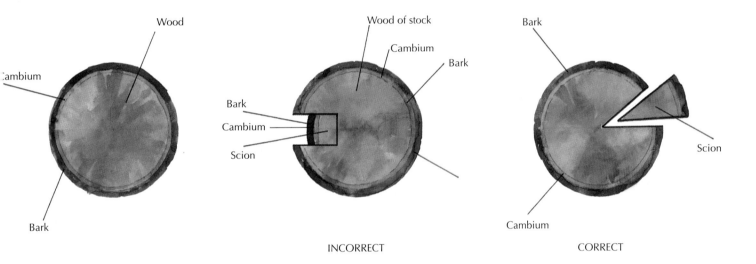

INCORRECT   CORRECT

2.2.3.2. Requirements for the plant material to be grafted

Choosing the material to be grafted is of great importance, and will determine whether the graft succeeds or fails. The scion's characteristics determine whether the resulting plant is better than the rootstock or not.

• The material must be taken from highly productive plants that show all the best features of the variety in question.
• The parent plant has to be healthy, well nourished and of productive age, and the scion should be collected in good weather, and never during frosts.
• The material should be one year old, with a diameter of about 1 cm, well lignified and vigorous, and preferably from the outermost shoots of the crown.

If the material is collected long in advance, the

2.2.3.3. Types of graft

There are three different types of graft, depending on the material to be grafted:

• Bud grafts
In bud grafts the scion consists of a bud with a little bark and wood.

This is called **vegetative budding** when grafting is performed at a time when the grafted bud can develop immediately, i.e., it is carried out in spring or early summer. This is a Californian technique that has not become very widespread, as it has drawbacks, such as the inadequate growth of the grafted plant, which usually does not grow more than a meter tall.

Dormant bud grafting is carried out in late summer, and so the grafted bud does not sprout until the following spring. To perform this type of graft, the bud has to be dormant, and the stock must be actively growing in order to separate the bark easily from the wood.

The main types of budding are:

• *T-budding*
A T-shaped cut, with a length of about 30 mm, is made in the bark of the stock, the sides are raised and the scion is inserted beneath them. The scion consists of a bud and a little bark and wood.
The T-shaped cut is sometimes made upside down (inverted T), especially in stone fruit that ooze gum when wounded, as this would asphyxiate the bud.

• *Ring graft*
A ring of bark several centimeters long is removed from the stock and replaced by a similar ring from the chosen variety, bearing at least one bud.
The stock and the branch the bud is taken from must be more or less the same size to ensure success.
This budding method is suitable for hardwood species such as walnut.

• *Striated ring graft*
This is similar to ring grafting, but instead of removing a ring of bark from the stock, the stock's bark is cut into strips that are used to cover the bud

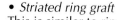

• *Tip ring graft*
This is similar to the two previous methods, but it is inserted just below the tip of the stock's stem, which has been cut back.

Mallorcan graft
A cut is made in the stock that is shaped so the scion fits neatly into it.

The cut may be angled at one end, simple, or double, when it is shaped at both ends.

It is difficult to carry out because the cuts have to be almost perfect. It is mainly used on vines.

• **Approach grafting**

In this method of grafting, the graft is only separated once it has fused with the stock, and until then both stock and scion remain on their own roots.

The main types of approach grafting are:

• *Simple approach grafting*

A portion of bark and wood are removed and the branches are tied together. The top of the stock and the base of the scion are not removed until the following year.

A/ When the stems differ in diameter
B/ When the stems are the same diameter

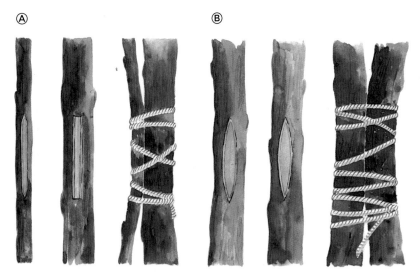

• *Patch graft*

A portion of the stock's bark is removed and replaced with a similar patch from the variety to be grafted, bearing at least one bud.

This bud graft method is used when the stock has thicker bark than the variety to be grafted.

• *Cleft graft*

This is a variation on the preceding method, but a V-shaped portion of bark and wood is cut from the stock, into which the scion (V-shaped on one side) fits, as it is the same shape.

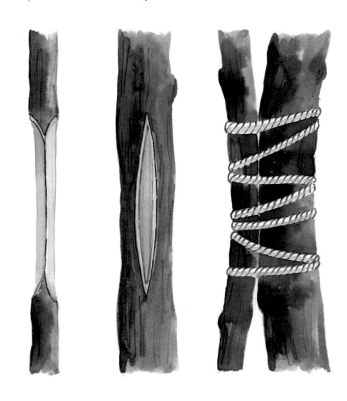

• *Notch graft*

A notch is cut in the stock, removing a piece of wood that is replaced by a piece of the same shape from the variety to be grafted.

• *Whip and tongue approach graft*
The notches cut are carefully shaped to ensure a close fit.

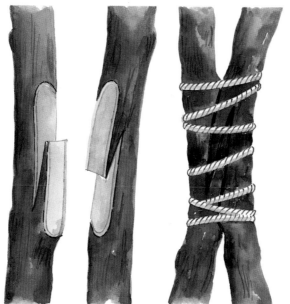

• **Scion graft**
In this case, the graft consists of a portion of shoot with one or more buds. It is performed in spring, when the stock has renewed vegetative growth, and it is carried out when one wants to graft onto adult trees.
Like budding, the stock should be in a later stage of growth than the buds grafted onto it.

The main forms of bud graft are:

• *Cleft grafting*
The scion is inserted into a cut down the center of the limb which has been cut back, taking care to ensure that the cambium zones of the two are in close contact.

The cut may be:

A/ Radial with a single scion
B/ Side to side with two scions
C/ Goatsfoot

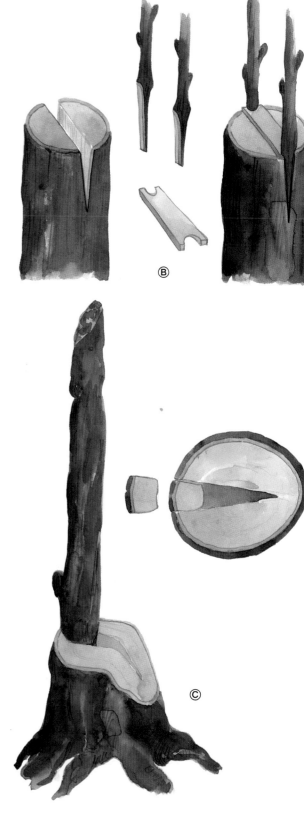

- Radial with a single scion
- Side to side with two scions
- Double or cross, with two side to side cuts
- Goatsfoot

The goatsfoot is similar to the radial cut, but differs in that the stock is cut back, and half the stem is cut to leave a slant.

Saddle graft
The stock is cut and shaped to form a saddle, and the scion is shaped to sit on it, so the cut surfaces are in full contact.

Whip and tongue
There are two variants, the simple and double.
The simple form is a slanting cut, made in the scion and in the stock, so that the two cut surfaces fit smoothly together. The double whip and tongue is similar to the single, but the cut is slightly slanted to form a heel, which acts to fit the scion and stock together.

Side graft
The cut made in the stock is lateral. It is not necessary to cut the stock back.

Spur graft
Like the preceding method, but the graft is sited near the base of a branch, which is cut back above the site of the graft.

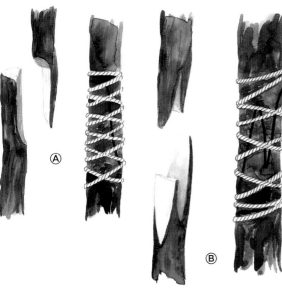

A/ Saddle graft
B/ Whip and tongue graft
C/ Side graft
D/ Spur graft
E/ Tip graft

• Tip graft
This is performed in the tip of the stock, which is not cut back. It is mainly used in the walnut.

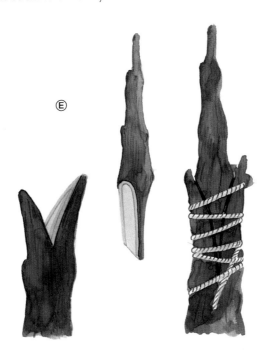

• *Empty graft*
A wedge is removed from the stock that is the same shape as the bottom of the inserted graft.

• *Gaillard*
A graft with a double cleft graft in which the stock is not cut back, but is curved towards the side opposite to where the scions are inserted.

• *Wedge grafting*
Like the preceding method, but the wedge is cut from the scion, not from the stock.

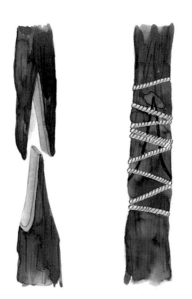

• *Bridge grafting*
The two tips of a scion are grafted into the trunk to form a bridge. This grafting method is used to bypass damaged areas of bark.

• *Arch*
The tip of a branch is grafted onto the same stock. This grafting method is also used to bypass damaged bark.

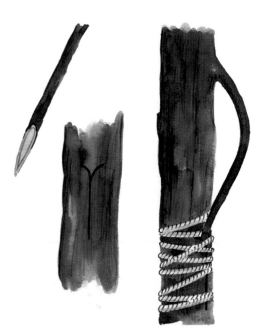

• Crown grafting
In this grafting method, the scion is introduced between the wood and the bark of the stock. Different forms of crown graft include:

• Classic crown graft
The base of the scion is cut at a slant and the bark of the stock is cut longitudinally so the scion fits into the notch.
One, two, or more scions can be grafted into the stock, and these are known as single, double and multiple crown grafts.

• Matching graft
This is the name for the techniques that remove a portion of wood from the stock to make a cavity that the scion fits into.
The best known is the *triangular,* in which the base of the scion has two cuts, meaning that it is triangular in cross-section.

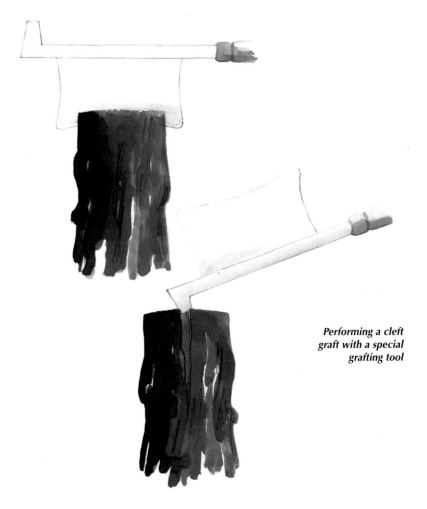

Performing a cleft graft with a special grafting tool

• Clarin
The base of the scion has 2 slanted cuts, while the stock which has been cut at a slant, has no longitudinal cut.

Thickening of graft site due to incompatibility

2.2.3.4. Ties

The purpose of the tie is to prevent the callus that forms from separating the scion from the stock, and also to favor the growth of vascular tissue, prevent dirt from entering and ensure close contact between the cambium of the scion and the stock.

The most widely used materials are raffia or cellophane film. The main advantage of these materials is that they are degradable, and so do not need to be cut off later. Rubber bands and tacks are also used.

Another way of keeping the scion in place is using **putty** or wax, which is melted by warming and then spread on the exposed parts of the stock and scion. The putty is applied with a spatula, and its purpose is to prevent oxidation and dehydration of the joint.

Using small tacks to fix the scions in place.

2.2.3.5. Affinity with graft stock

The scion and the stock are said to be compatible when they can form an effective and lasting joint.

The greater the number of physiological, anatomical and nutritional similarities, the greater the affinity. In general, the closer the botanical relationship, the greater the affinity.

The most important causes of incompatibility are:

- Different rates of transpiration in stock and scion
- Different rates of sap transport in the two vascular systems
- Accumulation of toxic substances in the graft zone
- Necrosis of the vascular system caused by biochemical interactions between stock and scion
- Toxicity due to proteins released by one or the other
- Transmission of viral disease by the scion.

Symptoms of incompatibility

- Slow callus formation
- Unequal growth
- Death of the scion bud or shoot
- Reddish or purplish color in the leaves in late summer or prematurely

Formula and recipe for grafting putty

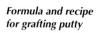

COLD PUTTIES	
White pitch 500 g Black pitch 600 g Tallow 500 g Yellow beeswax 260 g	The ingredients are melted together over a gentle heat, and left to cool until is lukewarm, and then 250 g of alcohol are added. If the putty is too dense, add more alcohol or a little tallow; if it is too runny, add more pitch or wax.
Rosin (colophony) 1,000 g Wax 500 g Alcohol 200 g	As above.
Yellow beeswax 65 g Turpentine 65 g White pitch 32 g Tallow 16 g	Everything is melted together, at the same time. (Lhomme-Lefort formula).

HOT PUTTIES	
Rosin 830 g Black pitch 100 g Tallow 100 g Sieved ash 40 g	Melt it all together, and then add the ash.
Rosin 1,000 g Black pitch 1,000 g Tallow 100-200 g	Dissolve the rosin and the black pitch in hot water, and then gradually add the tallow until the cooled putty is no longer glassy (Pieri formula).
Rosin 500 g Linseed oil 1,000 g Paraffin 2,500 g	Adriance and Brison formula.

Premature leaf fall

Early differentiation of buds and rapid entry into production

Discontinuities in the graft, which is weakly structured and may break.

The appearance of small buds in the stock above the graft area

Accumulation of starch above the point of grafting

2.2.3.6. Grafting methods

Two grafting methods are of special interest:

Regrafting

Regrafting is carried out on adult trees in order to replace the variety forming the tree's crown. Regrafting is done to correct errors such as the lack of pollinating varieties in an orchard, or to replace varieties that are out of favor with more fashionable ones.

Regrafting is only carried out in plantations that are in good condition with many years of future production.

It can be carried out on main branches or secondary branches, or both, but after heavy pruning.

The grafting method used is budding, and as many branches are budded as possible.

One major factor to bear in mind is to treat the pruning damage and seal it with fungicidal pastes to avoid later problems.

• Reinvigoration

A good stock variety is planted next to the tree to be reinvigorated, and it is then grafted, either by a T-graft or approach grafting. After three years, the root mass has been substantially increased.

2.2.3.7. Aftercare

Grafts must be protected from winter frosts by covering the plants before winter arrives, especially in cold areas, and they also need to be protected against late frosts in spring.

To favour the development of a main stem, the terminal bud must be trained, and any other shoots that may compete with the main stem should be removed. To form an espalier the stem should be pinched out below the third well-formed leaf. This will bring the buds located in the leaf axils into growth, causing greater branching, and above all makes it possible to remove any sprouts arising from the stock.

A/ Reinvigorating a tree by approach grafting.

B/ Reinvigorating a tree by "T" grafting a separate stem.

3. THE CLIMATE IN FRUIT CULTIVATION

Every species or variety grows and produces best in a given set of conditions. All fruit trees are conditioned in their cultivation by the different factors forming the climate.

Fruit trees can be classified on the basis of their climatic requirements:

- Temperate species
- Hot-temperate species
- Sub-tropical species
- Tropical species

Factors affecting a tree's growth and production

It should be pointed out that species show a high ability to adapt to new climates as a result of genetic selection and improvement.

	GROWTH OF TREE	AGRONOMIC VALUE (OWN CAPACITY)	VIGOR AND FERTILITY HEALTH STATUS ROBUSTNESS RESISTANCE TO PESTS AND DISEASES ADAPTABILITY
		POSSIBILITIES OF ENVIRONMENT	LIMITING FACTORS CONDITIONING FACTORS
		CLIMATIC FACTORS SOIL FACTORS AGRONOMIC FACTORS	ECOLOGICAL OPTIMUM
	PRODUCTIVE POTENTIAL	PLANTATION SYSTEM	
		CULTIVATION TECHNIQUES	

Intrinsic factors	Extrinsic factors			
The tree	Ecological factors			Economic and commercial
	Climatic	Soil	Agronomic	
	Temperature	Depth	Relief	Land capital
	Rainfall		Access	Amount invested
Variety	Wind	Permeability	Comunications	Financing
	Moisture		Infrastructure	General expenses
	Sunshine	Alkalinity	Availability of work force	Fixed expenses
Rootstock	Storms		Training work force	Financial costs
	Hail	Fertility	Previous experience	Commercial orientation
	Hailstones		Availability of water	Cash flow
Selection	Snow	Salinity	Machinery available	Accounting
	Fog		etc.	Management
	Altitude			Contributions
Health status	Orientation/Aspect			Taxes
	Microclimate			etc.

• Temperate species

Their main feature is that they need a winter rest, coinciding with the low temperature period. They can withstand temperatures as low as –15°C. These species are also sensitive to hot summer temperatures.

Examples: apple, pear, plum and cherry.

• Hot-temperate species

They have a weaker requirement for winter rest than temperate species. They are more sensitive to low winter temperatures, but resist high summer temperatures better.

Examples: peach, apricot, Japanese plum, vine, olive, almond.

• Subtropical species

They are sensitive to low temperatures and grow better in high temperatures. There are three groups:

- Moderate heat requirement: fig, pecan, pistachio, persimmon.
- Heat-requiring: citrus fruit, avocado, loquat, custard apple.
- High heat requirement: palms, date palms.

• Tropical species

These species cannot tolerate low temperatures.
Examples: banana, guava, mango.

In general, when talking of fruit trees in temperate zones, this includes temperate, hot-temperate and subtropical fruits. The study of climate is very complex, as it deals with many different factors, all of them interrelated. We will deal with these factors individually first, and then discuss them in a more integrated way. The main features of the climate are:

- Temperature
- Moisture and rainfall
- Luminosity
- Wind
- Storms, hail, hailstones
- Snow

3.1. THE TEMPERATURE

The temperature is one of the most important features of a given climate. It is important to distinguish between the temperatures during the winter rest period and those in the growing period.

3.1.1. Winter temperatures

These are the temperatures during the tree's winter rest period, when it is not growing. The average temperatures in this period are around 10°C, but may fall as low as –10°C, which causes serious problems.

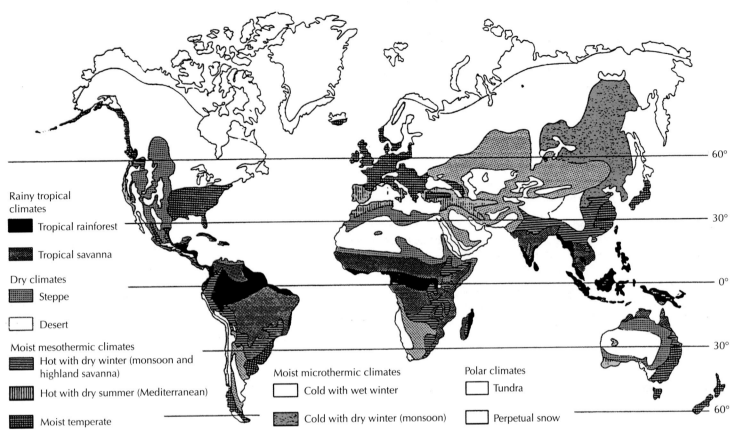

Rainy tropical climates
- ■ Tropical rainforest
- ▨ Tropical savanna

Dry climates
- ▦ Steppe
- ☐ Desert

Moist mesothermic climates
- ▤ Hot with dry winter (monsoon and highland savanna)
- ▥ Hot with dry summer (Mediterranean)
- ▦ Moist temperate

Moist microthermic climates
- ☐ Cold with wet winter
- ▨ Cold with dry winter (monsoon)

Polar climates
- ☐ Tundra
- ☐ Perpetual snow

60°
30°
0°
30°
60°

When the temperature falls below 0°C, this is called a frost. How serious a frost is depends not only on the temperature but also on how long it lasts and when it occurs.

A tree's resistance to these periods of low temperature is mainly genetically determined, but is influenced by the tree's nutritional and nutritional status and general health.

The root system is the part that is least resistant to the cold. A temperature at the roots of –5°C or –10°C can cause the tree to die. This is unlikely as the soil would have to freeze to a depth of 40-50 cm, as a result of much lower air temperatures over a long period.

One defence against low root temperatures is earthing up, which is done before the arrival of winter. This is commonly done in vineyards and young plantations.

Buds are also sensitive to low temperatures, especially flower buds, which are damaged or killed by temperatures of –10°C. Branches are also sensitive to low temperatures, fruiting structures and young branches that are not highly lignified.

These all damage the tree but do not endanger its life. Damage to the wood can, however, kill the tree.

If the low temperatures occur early, before the wood has hardened, it may affect the bark and the sites of branching.

The damage takes the form of the death of patches of the bark on the south-facing parts of the tree.

The world's main climatic zones. The world's most important fruit-producing areas are between the 30th and 50th parallels in the north and south hemispheres.

Cold-hour requirements of different species

Minimum temperatures tolerated for half an hour by different fruit-producing species (Saunier, 1960)

Cold-hour requirements chart (hours from 0 to 2000):

Species	Cold-hours (range)
Fig	
Vine	
Persimmon	
Almond	
Quince	
Blackberry	
Apricot	
Peach	
Bilberry	
Cherry	
Sour cherry	
Pecan	
Japanese plum	
Walnut	
Redcurrant	
Pear	
Hazel	
Raspberry	
Apple	
European plum	
American plum	

Species	State of development		
	Closed flower buds showing color	Full flower	Young fruit
Peach	–3.9°	–2.5°	–1.6°
Apple	–3.9°	–2.2°	–1.6°
Cherry	–3.9°	–2.2°	–1.1°
Pear (1)	–3.9°	–1.7°	–1.1°
Pear (2)	–4.4°	–2.2°	–1.1°
Japanese cherry	–3.9°	–2.2°	–1.1°
Cherry	–5.0°	–2.8°	–1.1°
Apricot	–3.9°	–2.2°	–0.5°
Almond	–3.3°	–2.7°	–1.1°
Vine	–1.1°	–1.1°	–0.5°
Walnut	–1.1°	–1.1°	–1.1°

(1) Sensitive varieties: Butter (or beurré) Bosc, Butter Anjou, Conference
(2) Resistant varieties: Butter Clairgeau, Butter Hardy, Passe Crasanne, Williams, Duchess of Angoulême

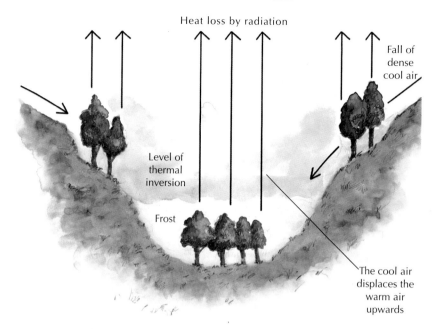

Heat loss by radiation

Fall of dense cool air

Level of thermal inversion

Frost

The cool air displaces the warm air upwards

Frost formation by heat loss
In a valley planted with fruit trees, heat is lost by radiation and this cools the air, and the cold air falls to the valley bottom because of it is denser, while the warmer air is displaced upwards. This leads to a frost in the valley bottom, while its sides are above 0°C.

If the temperature falls sharply, the damage may reach as far as the stem's pith. This causes zones of necrosis within the stem that greatly diminish its mechanical resistance to, for example, the wind.
If the necrotic area is large, the affected branch, or even the entire tree, dies above the affected part.
The part of the tree most at risk is the crown, where the temperature is lowest.
To heal the tree, first let all the effects of frosting appear, so that you can cut all the damaged wood away, back to healthy wood.
The wound must then be disinfected and sealed to prevent further necrosis. Putty should be used to seal the cut, as this favors healing.
If the necrotic area has led to formation of a hollow, this should also be sealed, using an inert filler, such as clay, plaster or expanded polyurethane. This improves the mechanical resistance of the branch or tree.

Right: The optimal values of average temperatures during the summer period for several fruits

Another aspect of winter temperatures is precisely the opposite to that discussed above, namely that they may not be low enough.

If an area has mild winters, and one year temperatures do not descend enough, the trees may show three types of symptom in the next growing season:

• Delay bud break and sprouting, delaying the normal flowering period, and changes in the order that different varieties flower.
• Irregular and scattered sprouting, meaning that some buds have not broken when others are already in full flower, which extends the flowering period and thus the fruit production period.
• Shedding the flower buds, which can be caused by high as well as low temperatures.

The length of the winter rest period depends on the individual tree's genetic characteristics and its physiological and nutritional status. Trees can be divided into three groups on the basis of their requirement for cold-hours.

• High cold-hour requirement (more than 700 hours): apple, pear, plum, cherry, sweet chestnut and walnut
• Medium cold-hour requirement (400-700 hours): some varieties of pear, hazel, olive, Japanese plum and peach.
• Low cold-hour requirement (400 hours): some varieties of peach, African apricot, almond, fig and quince.

Species	Optimum temperature (°C) during the summer period
Apple	18 - 24°C
Pear	20 - 25°C
Peach	22 - 26°C
Apricot	20 - 24°C
European plum	18 - 20°C
Japanese plum	20 - 24°C
Sweet cherry	18 - 22°C
Sour cherry	16 - 18°C
Almond	20 - 26°C
Walnut	20 - 22°C
Hazel	18 - 24°C

Groups of species (temperate zone with 500 mm annual average rainfall)

A. Drought resistant species that can be dry farmed	B. Less drought-resistant, complementary irrigation recommended	C. Sensitive to drought, and require irrigation
Oil-producing olive	Olive for harvesting green	Citrus fruit in general
Vine for wine production	Table grape	Apple on medium vigor stocks
Almond	Apple on wild stock	
Fig	Apple on vigorous stock	Apple on weak stocks
Pistachio	Pear on wild stock	Pear on quince
Apricot	Peach (early varieties)	Peach
St. Lucie cherry	Plum (early varieties)	Plum
Caper	Cherry on P. avium stock	Hazel
	Cherry on P. cerasus stock	Kiwi
	Walnut	Raspberry
	Pomegranate	Redcurrant
	Loquat	Bilberry

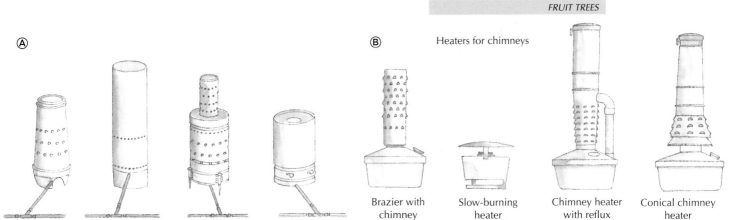

Brazier with chimney · Slow-burning heater · Chimney heater with reflux · Conical chimney heater

3.1.2. Spring temperatures

The beginning of spring marks the start of vegetative growth in fruit trees, accompanied by flowering, pollination and fertilization, and at this time the tree is sensitive to variations in temperature.

The most cold-sensitive parts of the flower are the ovary, the ovules and the base of the style, which freeze and die if the temperature within the flower reaches –1°C or –2°C for more than half an hour.

In general, a dormant bud is more resistant than one that is opening, and this is more resistant than the open flower, which is in turn more resistant than the recently set fruit. After fruit-set, the fruit becomes more and more resistant.

It is also true in general that trees from cold-temperate areas, such as apple, pear, plum and cherry, are more resistant to cold temperatures than typically Mediterranean species such as the peach, almond and apricot.

There are sensitive trees that avoid possible frosts by flowering late in the season. These include the walnut, olive and vine.
If the frost was intense, the bud or flower bud turns black, dies and falls from the plant. If the frost was not so intense, the flower bud may open, but the flower will not form fruit if the ovary or style has suffered damage.
To assess the extent of damage caused by a frost, the bud or flower bud should be cut longitudinally 48 hours after the frost and inspected.

This damage may also occur during flowering, and the flower may be left sterile by damage to the ovary and style.
During fruit-set, the damage depends not only on the lowest temperature reached, but also on the size of the fruit.

If the frosts kills the embryo, the fruit stops growing and is shed immediately, or after 2-3 weeks, if the fruit is relatively large.
The damage can be assessed by cutting across a fruit about 48 hours after the frost.

If the frost was mild, the damage may be restricted to the fruit's peel, causing cracks that will later turn into brown patches that reduce its market quality.
As we have seen, spring frosts can have an important effect on both flowering and fruit-set. It is thus very important to understand the spring frost regime of the site you cultivate, including details like the probable dates of frost, the period of time when they may occur, and their intensity and length.

There are 4 types of spring frost.

• **Convection frosts**. These are caused by masses of cold air below 0°C, accompanied by winds from the north. Temperatures fall sharply and this causes much damage.

• **Radiation frosts**. These are caused by the heat loss by the ground and plants during the night. Heat loss is greatest on a clear cloudless night, when there is no wind and humidity is low.

• **Frosts due to thermal inversions**. These occur because cold air is denser than warm air and falls to ground level, meaning that the air at ground level is the coldest.

• **Frosts caused by evaporation**. These occur in cold, dry environments, and are due to the cooling caused by the evaporation of water. The heat needed to evaporate the water comes from the plant, cooling it. When wind accompanies these cold, dry, conditions, the risk of freezing due to evaporation increases.

3.1.2.1. Defenses against frost

Defenses against frost may be direct or indirect, but the best results are obtained by combining the two.

• **Indirect methods**

Indirect methods seek to avoid damage to the tree by ensuring that the temperatures do not get so low that the tree is damaged.

• By choosing species and varieties that are resistant to low temperatures.

• Selecting the right site for cultivation, taking into account the relief, and the microclimate. Avoid sites that are boxed in (where air cannot circulate freely) and in large open spaces.
The best areas to plant a crop are well-ventilated mountain slopes that face the sun, and where cold air can flow freely downhill.

Moist areas located in the shade, with poor air flow and close to watercourses and pools, all favor evaporation frosts, which are usually not intense, but are dangerous.

• Using the correct cultivation techniques. A tree's condition, adequate irrigation, fertilization, pruning and good general health, all favour the plant's vigor, and therefore its resistance to frosts. Trees that are not vigorous are much more likely to be affected by frosts.

Flat, compacted sites irradiate much less heat than plowed areas or sites with plant cover. Moist soils also irradiate much less heat than dry soils, though moist soils may increase the risk of evaporation frosts if the environment is dry.

• Direct methods

Direct methods try to maintain the environmental temperature above the frost damage temperature.

This can be done in several ways:

• Protective shields

These shields consist of smoke, and together with the water vapor of the air, this creates a thick smog, or an artificial mist can be created directly.
The screens are located slightly above the soil, and their purpose is to reduce radiation by the soil.
For greatest effectiveness, there should be no wind to blow them away, and they should not be used near towns or highways, because they are a hazard.
They are generally used in mild frosts, or as a complement to other defense systems.

• Fans

Fans work by creating a draft of air to mix the cold air at ground level with the overlying layer of warm air. This raises the temperature of the colder layer. Helicopters can also be used to do this, but it is more expensive.
Fans are powered by diesel or electric motors, and can be placed at ground level or above the ground.
This method is used in mild frosts.

• Heating
Heating is creating heat by burning a source of energy. The heat may come from heaters or burners. There are many commercial models of both types, which can use a wide range of fuels.
This system is very effective, but is expensive to install and maintain.
In general, devices with large chimneys that radiate heat are more effective than ones with open flames, especially if they are connected to each other, rather than in isolation.
Defense is more effective if there are more heaters, even if they are smaller.
The number of devices should be doubled at the edge of the plot, especially on the northern edge.
This system has the disadvantage in that space is needed for storage, distributing and collecting the devices, and the fuel is expensive.

• Spray irrigation

This is based on the heat released when water freezes solid (changes from the liquid state to the solid). The idea is to water the tree's surface by spraying while the frost lasts. The water freezes on the vegetation, giving up heat.
This gives rise to secondary problems, such as excess weight of ice, which can cause branch breakage, and waterlogging.
In general, this is the most reliable, easiest and most versatile method, despite the high cost of the equipment and its installation, because it can be used in normal irrigation of the crop, meaning that it pays for itself more quickly.

3.1.3. Summer temperatures

This refers to the temperatures between the end of spring to the beginning of fall, the temperatures in the period of active growth, after flowering.
In general, fruit trees can grow vegetatively over a wide range of temperatures, though the temperatures they are best adapted to lie within a narrower range.
In the summer period, it is very rare for temperatures to fall below 0°C.
Temperatures may, however, occasionally fall well below normal for the season. The most serious consequence is that the final fruit is smaller and of lesser market value, and in addition the fruit takes longer to ripen and the tree as a whole produces less vegetative growth.

High summer temperatures are a much more frequent problem, when temperatures exceed 30°C in dry sunny environments.

This may cause the following symptoms:

• Lessening of photosynthetic activity above 30°C
• Summer growth halt at temperatures of 32-36°C

At higher temperatures, **heat waves**, the leaves and buds are scorched, dehydrated by the loss of tissue water, and then yellow and die, causing the tree to shed all its leaves. Scorching, or burning, may be due to excessive sunshine or a very dry environment.

Near harvest time, high temperatures are unfavorable for fruit coloration, which is the result of the contrast between the daytime and nighttime temperatures.

3.2. MOISTURE AND AVERAGE ANNUAL RAINFALL

To support normal vegetative growth and peak production, fruit trees need adequate water in the soil they are growing in.

The water needed to maintain this level of humidity is derived mainly from rainfall. All this means that average annual rainfall is critically important when cultivating fruit trees.

A soil water shortage leads to less photosynthesis, and this may cause the tree's growth to decrease. It may provoke an even longer summer growth halt, and even the death of the tree if the water shortage is severe and prolonged.

Other effects of water shortage include:

• Decrease in flower production

• Lowered fruit production, with fewer, smaller, fruit
• Lowered quality, with less attractive, poorly colored, fruit

The restrictions imposed by rainfall are not just due to the average annual rainfall, but also how it is distributed over the course of the year. The needs of fruit trees can be considered to be met by an annual average rainfall of 700 mm per year.

The water requirements of fruit trees vary greatly, depending on the variety and what stage of growth it is in. In general, early ripening fruit require less water than late ripening fruit, and the tree's water requirements decline after the fruit is harvested.

There are two types of farming, depending on whether they use irrigation or not.

• **Dry farming**: when fruit trees can grow vegetatively and produce a reasonable crop using only the rain that falls.

• **Irrigated farming**: when additional water is supplied by irrigation.

Note that damage may also be caused by rain, mainly in the flowering period. If flowering coincides with a period of persistent heavy rain, pollination and fruit-set may be affected.

Distribution of rainfall over the world. Note that some desert areas have been transformed into fruit producing regions by the spread of irrigation. The different areas are defined on the basis of average annual rainfall in millimeters.

Average annual rainfall in mm

- > 2.000
- 1.500 a 2.000
- 1.000 a 1.500
- 500 a 1.000
- 250 a 500
- < 250

This is mainly because:

• Insects, especially bees, do not fly in rainy weather.
• The pollen is washed to the ground, and is also washed from the stigmas by the rain.
• The temperature is lower, meaning that pollination and fruit-set are slower.
• Flowers may be destroyed by intense rain.

Rain can also damage the fruit. After a dry period, intense water uptake may cause cracking because the epidermis does not stretch enough. These cracks allow entry of fungi, which cause the fruit to rot. If these cracks are very small, they may heal by forming corky tissue without any further damage other than forming corky brown patches. This makes the skin look like a potato, and is known as **russeting**.
Other causes may lead to russeting, such as mild frosts in the early phases of fruit development, viral diseases, certain nutrient deficiencies and some agricultural chemicals used to treat the plants.
Intense rain may cause fruit-fall, or waterlogging of the ground, leading to problems of root asphyxia. How serious these problems are depends on the tree's resistance.
High levels of environmental moisture favors the appearance and spread of fungal and bacterial diseases.
In wet periods, fungi become a serious problem. In dry periods, parasitic mites ("scabs") are the main problem.

Obviously, light intensity affects photosynthesis and therefore the tree's vegetative growth, flower production, and the size, color and composition of the fruit. These factors in turn determine the quantity and quality of production.
When light intensity does not reach the minimum required by the fruit tree, photosynthesis declines, leading to a reduction in its growth and development. Insufficient light also affects flower induction and differentiation, flowering and fruit formation. The harvest is reduced within the tree's crown, the fruit are smaller and poorly colored, as pigment formation is low due to the lack of light. Low light intensity affects your choice of variety to cultivate, and varieties with green or yellow fruit are recommended. It also affects pruning, and a flat shape is desirable to make the most of the light.

Excess sunshine normally coincides with hot dry periods, making the damage worse:

• The **fruit** pigments may be destroyed, making the fruit darker in color. One characteristic effect is rouging on the side of the fruit exposed to the most sun, due to excess sunshine. It is mainly observed in plums, apples and pears.

• The **leaves** wither, dry out, die and are shed. This is known as scorching. It also occurs mainly in plums, apples and pears.

• The **wood** is very badly affected. Large, sometimes deep, ulcers form that heal badly. Their treatment is the same as for damage caused by frost. One effective way of protecting the trunk is by liming it. This protects the trunk from the sun, and also from pests and diseases.

3.4. CLIMATIC EVENTS

The main climatic events are wind, hail, hailstones and snow.

3.4.1. Wind

Intense winds have two types of negative effect:

• **Mechanical effects**. Winds can damage flowers, fruit, leaves and branches. They may be torn off, beaten off or battered by intense winds.
Another mechanical effect is noted in young plantations in zones with dominant winds from one direction. The trunk is at an angle and the crown is deformed, leaving the tree unbalanced.
In terms of cultivation techniques, the wind makes application of sanitary treatments and sprinkler irrigation more difficult.

Better annual renewal in lobulate forms. Lobulate forms:
A/ are better illuminated than simple globose shapes B/, meaning their production is higher and better quality. Their vegetative renewal will also be better.

Ⓐ   Ⓑ

3.3. LIGHT INTENSITY

Fruit trees can grow vegetatively and flower in a wide range of light intensities. If the light is weaker or stronger than this range it has negative effects on the tree. Their light requirements vary with the time of year, but fruit trees normally require a lot of sunlight, and thus grow better in climates with few clouds and strong sunshine.

Permeable barrier

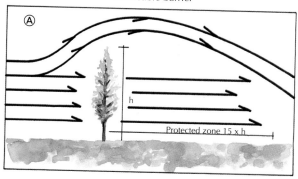

Impermeable barrier

Barrier open at base

Windbreaks vary in height and permeability to airflow, the combination of which determines the size of the area protected:

• **Height**. The higher the windbreak, the more effective it is. This is why living windbreaks are more effective, because inert ones are rarely more than 3 m tall.

• **Permeability**. Experiments have shown that semipermeable windbreaks are more effective than impermeable ones, as these create eddies behind them.

Disadvantages of windbreaks

• The loss of useful agricultural soil. This loss is reduced by planting spindle-shaped species and locating them on access routes, irrigation ditches or the edges of the field.
• Root competition between the fruit tree and the windbreak, generally because the windbreak is planted with a robust variety. To reduce this competition, prune the roots parallel to the windbreak, at a depth of 1 m. This will reduce the growth of roots into the cultivated ground. This is accompanied by irrigation and fertilizer application similar to that of the fruit trees.
• Shading of the fruit trees by the windbreak. This obviously depends on the orientation of the windbreak relative to the course of the sun.
• An increase in fungal disease, due to the higher humidity, and of pests as they can shelter in the

Different defensive effects of different types of windbreaks
A/ Permeable barriers filter the wind, reducing its speed, and effectively defend a strip 15 to 20 times the height of the windbreak.
B/ Impermeable barriers cause the appearance of eddies. Defense is less effective.
C/ Windbreaks that are open at the base do not slow the wind down effectively. They are not an effective defense.

• **Physiological effects**. During flowering, strong winds prevent bees and other insects from flying, making pollination impossible. One of the most important causes of scorching is hot dry winds. Another physiological affect they cause is a rapid increase in transpiration; the tree loses water rapidly and dehydrates, causing desiccation, leaf fall and general weakening. If the plantation is close to the sea, the main problem is salty winds, which are toxic to the trees, and reduce their growth.
The main defense against the mechanical effects of wind are windbreaks. These also reduce evapotranspiration, saving irrigation water. They have to be installed at right angles to the dominant wind direction, in order to filter the wind and thus slow it down.

There are two types of windbreak:

• **Inert**. These consist of a non-living material, such as brickwork, cane or plastic sheeting.

• **Living**. These generally consist of trees with an upright growth form.

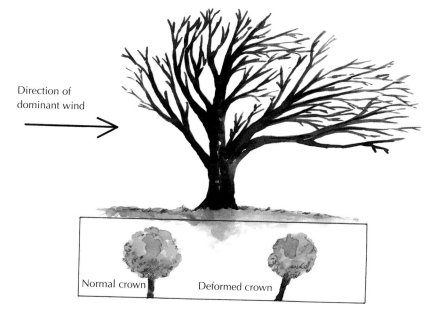

Direction of dominant wind

Normal crown   Deformed crown

windbreak.
• The plantation needs the protection of windbreaks most in the years after planting, when the windbreak is young too. Thus, windbreaks should be planted several years in advance or relatively large trees should be planted.
• Windbreaks increase the risk of radiation frosts in spring and increase temperatures in summer.

Hot winds are a special case. When hot winds blow, placing windbreaks may increase damage. If the winds are salty, the best barriers are impermeable.

Mechanical effects of wind
The dominant wind deforms the crowns of the trees, leading to an unbalanced distribution of branches and a trunk that is not vertical.

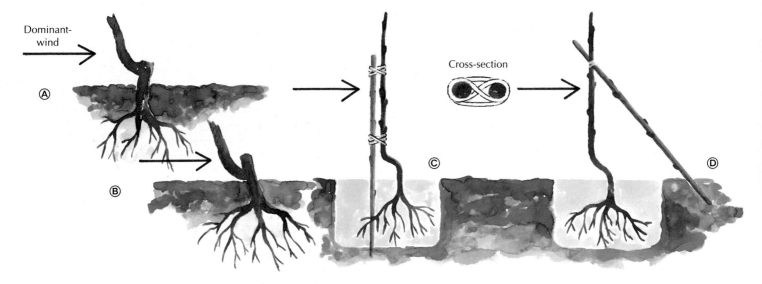

Dominant-wind

Ⓐ

Ⓑ

Cross-section

Ⓒ

Ⓓ

A/ and B/ In new plantations subject to frequent winds, the "crook" of the graft should point into the wind to avoid breakages, and it should be slightly angled to compensate for the effect of the wind.
C/ and D/ Different forms of "staking" against the wind, in young trees.

Species used as windbreaks have to possess certain features, namely:

• Rapid vertical growth
• Tallness
• Spindle shape
• Robust, vigorous and adapt well
• Non-invasive root system
• Not very dense, evergreen, foliage
• Wood that can be used and which does not break easily

The following species may be suitable as windbreaks:

• **Poplars**. They usually give good results, and their wood can be used. They are however, sensitive to cold conditions and require a lot of water.
• **Birch**. They are used in wet areas with a cool climate. The birch does not resist cold, heat or drought.
• **Eucalyptus**. Tall and fast-growing. The wood can be used, but they are highly invasive and their wood is easily broken.
• **Tamarisk**. They are resistant to cold and salinity. They are widely used near the sea. They grow well on any soil, but do not grow very tall.
• **Cypress** (*Cupressus spp.*). They are widely used. They are evergreen and reach a good height, but are slow-growing.
• **Chamaecyparis**. Used in wet areas. Does not resist drought or hot environments. Good height and dense foliage.
• **Cupressocyparis**. Not very resistant to drought and hot environments. More vigorous than the above conifers and faster growing.
• **Thuja** (arborvitae). More resistant to heat and salinity than *Chamaecyparis*, but shorter and slow-growing. It forms very compact hedges.
• **Privet**. Robust, sensitive to drought and cold-resistant. Forms barriers that are fast-growing but low and compact.
• **Cherry laurel**. Robust and cold-resistant. Low but compact hedge.
• **Laurel**. Compact and requires humidity.

• **Cane**. Robust and drought resistant, but does not form very high barriers. It is invasive and creates a risk of fires in fall.
• **Bamboo.** Does not require much moisture, but sensitive to cold and drought. Forms taller barriers than cane, but it is slower growing. It is also invasive and hard to eliminate.

3.4.2. Hail and hailstones

Hail and hailstones are climatic phenomena generally associated with storms, normally in spring and summer, after high temperatures on a calm day. These storms are often accompanied by strong winds and intense rain, which aggravate the damage and lead to irreparable damage.

Hail is precipitation in the form of spherical grains of ice 2 to 5 mm in diameter. When the diameter is larger and the stones are irregular, they are called hailstones.

When a plantation is affected by hail and hailstones, the consequences are generally very serious.

The severity of the damage depends on the size of the hail, their descent velocity and how long the hailstorm lasts.

There may be impact wounds on the leaves and fruit, and defoliation and fruit loss may be total. The tree's bark is also negatively affected by the impacts, and the plant is generally weakened.

The affected fruit loses its market value and the affected wood must be pruned out to renew it. A secondary problem is fungal diseases, as the high humidity and wounds to the tree mean they can gain easy access to the tree.

Once the damage has been done, the measures that can be taken are of little use, as the damage cannot be made good. Thus the best thing to do is to try to prevent hailstorm damage.

This seeks mainly to reduce the damage that may be caused. The idea is to try and cause a larger number of hail particles to form, that are smaller in diameter and have a lower descent velocity.

The idea relies on changing the normal development of a storm cloud by introducing a large number of possible nuclei for freezing. The nuclei are solid particles of any of many types, but their importance is that they induce the water of the cloud to solidify on them. This means that there are many nuclei, which are therefore smaller and less energetic.

The solid particles used for this purpose are
• *Silver iodide.* The most widely used.
• *Lead iodide.* Cheaper than silver iodide, but more polluting.
• *Chlorosulfonic acid.*

There are several ways of sowing the clouds with the particles:

• **Coal burners**. The burner consists of two concentric metallic tubes. The coked coal is introduced into the inner tube. This coal has been impregnated with a 2% solution of silver iodide.
When the coal is burnt, the inner cylinder gets red hot, causing the silver iodide to sublime, forming tens of thousands of freezing nuclei that are emitted into the air. This is the cheapest method known and each burner covers a zone of 10-20 km^2.

• **Silver iodide generators**. The generator is similar to the burner described above, but the inner tube is a chimney into which a nozzle injects a solution of silver iodide in acetone. The freezing nuclei are propelled through the chimney and into the atmosphere by a fan. The area covered is greater than burners, about 50 km^2, but installation and maintenance are expensive.

• **Rockets**. Two sorts of rocket can be used:

• *Granifuge rockets.* They explode in the center of the cloud, emitting a shock wave that breaks up the hail grains that are forming, causing them to fall as rain or sleet.
• *Chemical rockets.* These explode and break the ampoules they contain at a given height. The compound they contain forms freezing nuclei that are dispersed within the selected part of the cloud.

The first type of rocket is only effective in minor storms, and the second, in addition to being expensive, is only effective as a complement to the use of generators on the ground.

• **Airplanes**. The use of airplanes is the most sophisticated method known. This method requires a weather radar network to locate and measure the characteristics of the clouds that are forming and the size of the storm.

The information obtained is analyzed and the necessary measures are taken to avert the hailstorm, using generators or rockets installed on the airplane.

This method is very expensive, and requires specialized equipment and staff.

• **Use of plastic sheets or mesh**. These nets are placed on fixed structures and cover the crops during the period of risk of hailstorms.
The sheets prevent the direct impact of the hail against the tree, and thus limit damage.
The method is expensive, as the areas of fruit trees to be covered are large and many supports are needed. Setting the screens up and taking them down makes these structures even more expensive.
Sheeting is only used on small areas of highly profitable crops.

3.4.3. Snow

In general, except in specific cases, snow is a beneficial climatic factor, as it alleviates severe cold spells by supplying moisture to the soil and preventing it from freezing solid. The problem is when there are early frosts in fall, before leaf fall, and also in areas where crops with evergreen leaves are cultivated, such as olives and citrus fruit.
The problem for trees with evergreen leaves is that the weight of snow overloads the crown, causing the branches to break, or even the entire tree if the weight is too great.
The damage caused may be greater if, after snowing, the cold spell continues. This delays thawing, and lengthens the time the tree must bear the added load.
Another major problem is when intense snowfall is followed by very cold weather, as this causes the snow to freeze solid on the ground. This causes the tree's crown to freeze, causing the lower part of the tree to die off and the almost certain death of the tree.
The way to avoid major overloading is to prune them in advance. In zones that are at risk of heavy snowfall, when pruning you should aim for more lobulate shapes, and stout pyramidal skeletons.
To prevent the crown of the tree from freezing, the stem is earthed up in fall before the snow period. This protects both the lower part of the tree and the roots.

4. THE SOIL

The soil provides the fruit tree with all the mineral nutrients and water that it needs to grow. It also acts to support and anchor the root system.

The soil is a complex mixture of minerals, organic matter and living organisms, and is in constant activity (See Section 1: Soils, fertilizers and organic matter).

The soil's components are:

Next page, below: Hydraulic beater to harvest olives

• **Minerals** that vary greatly is composition and particle size. They account for 45-50% of the soil's total volume.
• **Organic matter**, consisting of animal and plant remains in different stages of decomposition. This is a much smaller proportion of the soil, roughly 0.5-5%.
• **Air** with a composition similar to the atmosphere as a whole, and **water**, both in proportions varying with the type of soil, but occupying roughly half the total volume.
• **Living things**, such as bacteria, algae, fungi, grubs, earthworms, and insects that live in the soil.

The world's main soil types
A wide variety of deciduous fruit trees can be cultivated if the soil pH and depth are suitable, and internal drainage is good. The soil of many deserts is suitable, with irrigation, as long as its salt content is not excessive (taken from USDA)

The type of soil is of great importance when deciding which species to plant. The main features of soils that affect crops and have to be taken into account when selecting are:

• Depth
• Permeability
• Lime content and pH value
• Fertility
• Salinity

All these features are determined by the physical, chemical and biological characteristics of each soil.

4.1. DEPTH

Depth is defined here as the thickness of soil that the fruit tree's roots can exploit without any restriction.

In general, it is considered that a soil less than 50 cm deep is inadequate for the growth of fruit trees. The lower limit is a soil 1 m deep, as long as it is sufficiently fertile.

In general, the deeper the soil, the better the conditions are for the roots. Root growth can be restricted by several factors:

• **Mechanical factors**, such as the presence of bedrock or a very compact soil horizon, that the roots cannot grow through.

• **Chemical factors**, such as the presence of a horizon so saline that it is toxic to plants, or a very alkaline horizon that roots can not grow in.

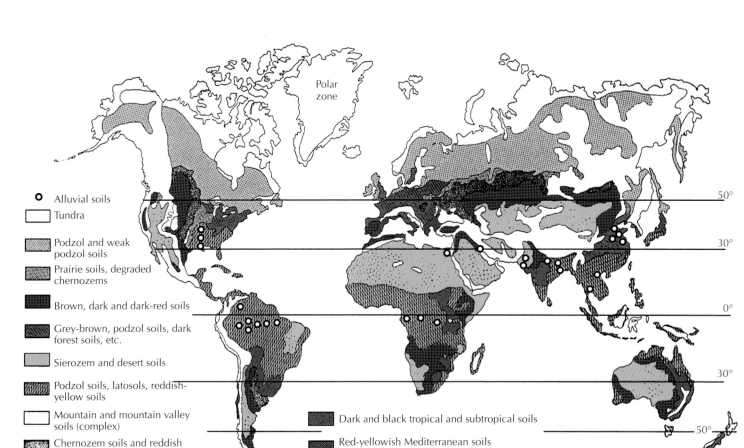

Polar zone

○ Alluvial soils

☐ Tundra

▨ Podzol and weak podzol soils

▨ Prairie soils, degraded chernozems

▨ Brown, dark and dark-red soils

▨ Grey-brown, podzol soils, dark forest soils, etc.

▨ Sierozem and desert soils

▨ Podzol soils, latosols, reddish-yellow soils

☐ Mountain and mountain valley soils (complex)

▨ Chernozem soils and reddish brown soils

▨ Dark and black tropical and subtropical soils

▨ Red-yellowish Mediterranean soils (including *terra rossa*), mainly in mountains.

50°
30°
0°
30°
50°

• **Physiological factors**, such as the presence of a very wet soil horizon or water table that lead to poor aeration and root asphyxia.

These problems are hard to resolve, and if the bedrock is near the surface or there are saline or lime horizons, they cannot be resolved. When the limitation is due to a compacted horizon, it is possible to break it up, depending on the horizon's thickness and hardness.
If there is a very wet horizon or the water table is near the surface, one solution is to install a drainage system, but this is usually so expensive that the investment is not worthwhile.

4.2. PERMEABILITY

Permeability conditions the movement of water in the soil, and thus the amount of oxygen available to the roots.
Permeability is a measure of the speed of infiltration of water into the soil in question. Values are usually 5-15 cm/h. Values of less than 5 cm/h are typical of heavy clay soils, with problems of root asphyxia. Values above 25 cm/h indicate soils that are excessively sandy and are infertile, due to the continuous washing of nutrients.

Poor permeability may be due to:

• *The presence of an impermeable horizon*
• *The presence of a hardened layer underground due to plowing*
• *An excessively silty or clay soil*
• *A continuous structure*

Root asphyxia is caused by shortage of O_2 in the soil. The root hairs die first, and then the larger roots are affected, which may damage the entire root system and lead to the tree's later death. If damage is not very serious, the root system may recover, especially if the soil condition that caused asphyxia is remedied.
The aboveground part of the tree may show chlorosis and gradually wither, shedding its leaves and fruit, though it may completely die in a few days if root asphyxia is total. Problems caused by root asphyxia are always serious, and often irreversible.

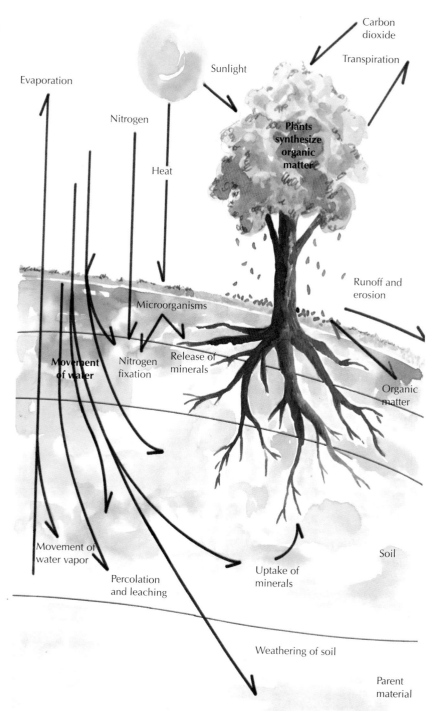

The processes taking place in the soil and their importance for the tree.
Plants absorb water and minerals from the soil and, with carbon fixed by photosynthesis, they are used to synthesize organic substances. Most of this organic matter is returned to the soil when it decomposes, releasing the nutrients that are returned so a new cycle can begin.

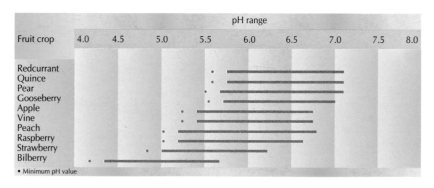

Fruit crop	pH range								
	4.0	4.5	5.0	5.5	6.0	6.5	7.0	7.5	8.0
Redcurrant									
Quince									
Pear									
Gooseberry									
Apple									
Vine									
Peach									
Raspberry									
Strawberry									
Bilberry									

• Minimum pH value

pH values for several fruit-bearing species

On the next page.: Top: sensitivity to salinity of different fruit species. Bottom: Soil salinity values. These are expressed as electrical conductivity (of a soil solution in water) in msiemens/cm.

4.3. LIME CONTENT AND pH

Calcium is an important element in the nutrition of fruit trees, and reaches high concentrations in the woody organs and leaves. In general, all fruit trees require at least 2-6% active calcium in the soil.

High levels of calcium give the soil bad physical properties, diminishing its stability and structure, and favoring the appearance of a surface crust and a hardened layer underground.

In the tree, excess calcium leads to the appearance of ferric chlorosis, with leaf yellowing, although the veins remain green, and in the worst cases this leads to leaf death and fall. This weakens the tree and may, in the worst cases, lead to its death.

Ferric chlorosis occurs because the plant does not have enough iron for synthesis of chlorophyll, and so much less chlorophyll is produced.

In Limy soils, those with a high pH, and worst of all, soils combining both, even though iron may be present, it is insoluble, and is thus unavailable to the plant.

In these cases, what is important is not the total amount of lime in the soil, but the amount of active calcium.

Active calcium is calcium in a soluble state within the soil, and it immobilizes the iron within the soil.

The level of active calcium bears no relation to the total amount of calcium. In soils with high levels of total carbonates (more than 40%), there may be low levels of active calcium, meaning there is no risk of ferric chlorosis.

It has already been pointed out that iron may be immobilized in soils that do not contain a lot of calcium but which are alkaline, with pH values greater than 7.5, and the problem is worst in very limy soils with high pH values. Fruit-producing species show different tolerances to the problem of ferric chlorosis. The almond, olive and vine are among the species that best resist limy soils, and can grow in soils with 30% active calcium.

The most sensitive fruit trees are pears grafted onto a quince stock, and peach grafted on wild stock, in which chlorosis appears at an active calcium level of 6% and pH value of 7.5. The symptoms of calcium deficiency are flabby fruit, branch death, poor flowering, low resistance to cankers and bitter pits.

The pH range that most trees can adapt to is quite wide, between 6 and 7.8.
pH values below 6 are not favorable for root activity. The availability of calcium, magnesium and potassium is very low, and microbial activity declines, as does nitrogen fixation.

4.4. FERTILITY

Soil fertility is a difficult concept to define. It is taken to be the set of soil characteristics that allow maximum yields to be attained, as long as climate and farming practices are suitable. In practice, however, the influence of other factors is so important that it is difficult to define the characteristics that determine fertility.

But in general, the factors that best define soil fertility are:

• **Organic matter content**
Fruit trees can grow in soils with very different content of organic matter. The most suitable levels are 2-4% organic matter in irrigated crops, and 1-2% in dry farming. In soils with more than 4% organic matter, some species of fruit tree may be affected by fungal diseases of the root system, or by the insolubility of some nutrients.

• **Nutrient content**
The tree takes up water from the soil and carbon dioxide from the air. For the tree's nutrition, other mineral elements are required, and these must be taken up from the soil. There are two main groups, which depend on the amount of the nutrient required:

• *Macroelements*, of which relatively large amounts are required: nitrogen, phosphorus, potassium, sulfur, calcium and magnesium.
• *Microelements*, or *oligoelements* of which much smaller amounts are needed: iron, zinc, copper, manganese, molybdenum, boron and chlorine.

Sulfur and calcium do not present problems as they are usually present in the soil in adequate amounts.

Problems of fertility are usually the easiest to solve once they have been defined.

Applying remedies and fertilizers is the most logical way to resolve these problems, although its economic cost must be taken into account, as it may be very expensive in very poor soils (see Section 1: Fertilizers).

4.5. SALINITY

It is important to be aware of the danger of salination, which may be due to soil salinity or irrigation with saline water.

In general, fruit-bearing species are very sensitive to salinity, and so it is an important limiting factor.

Salt tolerant plants (2 g/l of NaCl)	
Carob	(Ceratonia silicua)
Date palm	(Phoenix dactylifera)
Pistachio	(Pistacia vera)

Relatively tolerant (1 y 2 g/l of NaCl)	
Olive	(Olea europaea)
Vine	(Vitis vinifera)
Fig	(Ficus carica)
Pomegranate	(Punica granatum)

Salt sensitive species (< 1 g/l of NaCl)	
Apricot	(Prunus armeniaca)
Almond	(Prunus amygdalus)
Quince	(Cydonia japonica)

Very sensitive species (<0.5 g/l of NaCl)	
Peach	(Prunus persica)
Pear	(Pyrus communis)
Apple	(Malus pumila)
Plum	(Prunus domestica)
Citrus fruit	(Citrus spp.)
Loquat	(Eriobotrya japonica)
Pecan	(Carya illinoensis)
Walnut	(Juglans regia)
Avocado	(Persea americana)
Pomelo	(Citrus grandis)

Electrical conductivity of a dissolved soil extract. Expressed in millisiemens/cm at 25°C.	Soil salinity	Crop growth
0 - 2	None	Normal for all crops.
2 - 4	Low	Only very sensitive crops are affected.
4 - 8	Moderate	Most crops are affected. Only tolerant plants can grow.
8 - 16	High	Only very tolerant plants can grow.

A simple sum allows us to calculate the risk as a function of the sodium chloride content in the soil extract expressed in g/l.
It is not worth cultivating fruit trees in areas of high salinity, but in some cases certain stocks adapt better and are more resistant, meaning that some species can be cultivated in soils where they would not normally be viable.

The stocks include:
• **Peach**: Moussour wild, *Prunus davidiana*.
• **Pear**: *Pyrus betulaefolia*
• **Orange**: *Citrus macrophylla*, Rangpur lime, orange mandarina

In general, trees, affected by problems of salinity show the symptoms of chlorosis, and the leaves are small, pale and leathery, growth comes to a halt, and the scarce fruit are smaller than normal.

4.6. SOIL STUDY AND FRUIT TREES

To carry out an *in situ* examination of the soil, a pit 1 m deep has to be dug, that is wide enough for you to get inside it.
Depending on how heterogeneous the site is, more pits may have to be dug.
The edges of the pit should be sharp, gently removing the compaction produced when digging. This allows us to observe the different soil horizons directly.

The soil study should define the following characteristics:

Different models of spiked roller and their uses

• The unobstructed depth of soil
• The presence of obstacles, such as bedrock or the water table.
• Average permeability. Signs of waterlogging or poor aeration.

FRUIT SPECIES	Fertility	Depth	Moisture	Compaction	Limestone	Chlorides
Avocado	◐	◐	●	⊕		●
Apricot	○	○	◐	⊕	⊕	⊙
Caper	⊖	⊖	⊖	⊕		
Carob	⊙	⊖	⊖		⊙	⊙
Hazel	⊖	○	⊖	⊕	⊙	⊙
Jujube		⊕	◐	⊕		
Sweet chestnut	⊖		●			⊖
Castaño	⊕	○	○	●		
Cherry	◐	○	◐	●		
Sour cherry	⊙		◐	⊕		
Custard apple						
Prickly pear	⊕	⊕	◐	⊕		⊖
European plum		●	◐	⊕	⊕	◐
Japanese plum		●	◐	⊕	◐	◐
Evergreen oak	⊕		⊙	⊕	⊕	
Raspberry		○	⊕	◐		
Pomegranate	○	⊕	○	●		⊖

FRUIT SPECIES	Fertility	Depth	Moisture	Compaction	Limestone	Chlorides
Redcurrant			○			●
Fig			○		⊙	⊙
Persimmon			●			
Apple	○	○	○	⊕		●
Peach	◐	○	◐	⊕	◐	
Quince	◐	⊖	●	●	◐	⊙
Mulberry	◐	○	◐			
Loquat			◐			●
Walnut	○	○	◐	⊕	⊕	●
Olive	⊕	◐	⊖			⊙
Pecan	○	●	●	●		
Date palm		○	●			◐
Pear	◐	◐	◐	◐	◐	⊙
Nut pine	⊕		⊙			
Pistachio	⊖		⊖		⊖	⊖
Banana	●	○	●			◐
Vine	⊕	⊕	⊕		⊕	⊙
Blackberry	○	●	◐		◐	

Key:

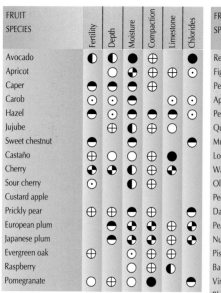

SOIL	FERTILITY	DEPTH	MOISTURE
●	Requires a very rich soil	Requires a very deep soil	Irrigation needed
◐	Needs a rich soil	Requires a deep soil	Moist soil
○	Prefers a rich soil	Prefers a deep soil	Relatively moist soil
⊕	Tolerates a relatively poor soil	Tolerates a relatively shallow soil	Relatively drought resistant
⊙	Tolerates a poor soil	Tolerates shallow soil	Drought resistant
⊖	Tolerates a very poor soil	Tolerates very shallow soil	Very drought resistant

SOIL	COMPACTION	LIME	CHLORIDES
●	Tolerates heavy soils	Lime intolerant	Very sensitive
◐		Limited dose	
○		Prefers a non-lime soil	
⊕	Prefers a light soil	Tolerates lime	
⊙		Needs lime	Quite resistant
⊖	Requires a very light soil	Must have lime	Very resistant

❂ For grafts, adaptation to the soil depends on the stock
(the absence of symbols shows the plant is indifferent or information is not available).

Soil requirements of different fruit species

• Compacted soil. Presence or not of a hardened layer due to plowing.
• The soil's structure at depth.
• Behavior of the root system of wild plants, both how well they grow and their density and the depth they reach.
• Abundance of life. Galleries and presence of earthworms and insects.

All these factors will affect tree growth if an orchard is planted.
It is worth noting that the fruit trees that are most demanding in terms of soils are peaches and apricots, which are the most sensitive to compact soils with poor permeability.

Good drainage is important for healthy tree growth, but it is desirable that the water table should be at least 120-150 cm below the soil surface.
In soils with a high pH value and high lime content, chlorosis symptoms can be expected due to the immobilization of iron. This mainly affects the peach and the pear.

5. PLANTING FRUIT TREES

Choosing which fruit species to cultivate is conditioned by the area's climatic factors and by economic concerns.
The many varieties available means that choosing is hard, and it should be borne in mind that a mistaken choice is very difficult to put right. To simplify the choice, it is necessary to define the production aims in advance. For example, fruit to be eaten fresh is different from fruit grown for canning or industrial use.

One can also define the ripening period that is desired, and so fruit can be classified into:

• *Extra-early*
• *Early*
• *In season*
• *Late*

Doing this simplifies the final choice of variety to cultivate. The characteristics the chosen variety must possess are known as its **agronomic value** and this is the sum of two values, the cultural and commercial value.

• **Cultural value:**
Ecology
Cultivation requirements
Fertility
Sanitary status
Resistance to pests and diseases

• **Commercial value:**
Production cost
Handling
Storage
Acceptability
Sale price

The optimal agronomic value is shown by an undemanding variety, that is fertile, healthy, pest-resistant, with low production costs, that is easy to handle and store, well accepted and has a high price.
When choosing the stock, this depends mainly on the soil's characteristics, although the stock cannot always be chosen independently on the variety to be grafted, as there may be problems of incompatibility between the two. Thus, once the stock has been chosen, you should check that it is fully compatible with the selected variety.

5.1. PLANTATION LAYOUT

The layout of the plantation depends on the site's relief. Flat sites with gentle slopes use a regular layout, while sites with steep slopes and a rugged relief use irregular, non-geometric, layouts.

The most widely used regular, or geometric planting patterns are:

THE MOST WIDELY USED FRUIT
STOCKS (IN SPAIN)

Pear
Wild
Wild "Kirchensaller"*
Wild Betulaefolia*
Quince E.M.A.
Quince "Angers"
Quince of Provence
Quince B.A. 29
Quince E.M.C.

Cherry
Prunus avium (Reboldo)
P. cerasus (Masto)
St. Lucie
St. Lucie 64
F 12/1
Colt

Apple
Wild*
E.M. II
E.M. VII
E.M. IX
M.M. 106
M.M. 109
M.M. 111
M. 26
M. XXV

Apricot
Wild*
St. Julien (Pollizo)
Myrobalan B

Peach
Wild*
Wild GF-305*
Wild Missour*
Wild Nemaguard*
St. Julien A
St. Julien GF655-2
St. Julien (Pollizo)
Damson 1869
Brompton
Hybrid GF-677
Hybrid Adafuel

Plum
Myrobalan B
Myrobalan
Mariana GF 8/1

Citrus fruit (Oranges)
Bitter orange
Sweet orange
Common mandarin
Orange mandarin
Troyer citrange
Carrizo citrange
Poncirus trifoliata
Citrus macrophylla

Almond
Wild bitter almond
St. Julien

Vine
161-49 Couderc
41 B Millardet
110 Ritcher
99 Ritcher
3309 Couderc
Rupestas de Lot
420 Millardet
196-17 Castell
SO4

* = grown from seed

FRUIT VARIETIES MOST
CULTIVATED IN SPAIN

Pear
Limonera
Ercolini
Butter Early Morettini
Butter Giffard
Blanquilla
Castell (San Juan)
William's
Max Red Bartlett
Conference
Decana Congreso
Passe Crasanne
Roma (Aragón)

Cherry
Bing
Ambrunesa
Napoleón (Monzón)
Burlat
Ramón Oliva (Jaboulay)
Red cherry
Black cherry
Mollar
Cristobalina

Mandarin
Satsumas:
Satsuma Owari
Clausellina
Clementines:
Fina
Oroval

Nules clementine
Tomatera
Esbal
Hernandina
Clementard
Marisol
Arrufatina
Other mandarins and
hybrids:
Common
Fortune
Kara
Nova
Wilking

Apple
Golden Delicious (group)
Starking Delicious (group)
Rome Beauty
King of Pippins
White Pippin
Red Pippin
Grey Pippin
Green Doncella
Jonathan
Granny Smith
Mingan

Almond
Long Desmayo
Red Desmayo
Marcona
Rof
Ramillete
Planeta
Garrigues
Ferragnes
Ferraduel

Orange
Sweet orange
Navel group:
Washington Navel
Thomson
Navelina
Newhall
Navelate
White group:
Common
Cadenera
Salustiana
Castellana
Berne
Valencia Late
Blood oranges:
Double fine
Entrefina
Sanguinelli
Bitter oranges:
Bitter orange

Peach
Springtime
Armgold
Dixired
Cardinal
Redgloba
Maruja
Jerónimo
Sudanell
San Lorenzo
Zaragozano
Yellow September
Springcrest
Armking
Maygrand
Crimson Gold
Nectared (series)
Babygold (series)
Merril (series)

Plum
Golden Japan
Metheey
Santa Rosa
Green Queen Claude
Oullins Queen Claude
Toulouse Queen Claude

Grapefruit
Marsh
Redblush

Apricot
Bulida
Canino
Moniqui
Galta Rocha
White Murcia
Paviot

Vine
Airen
Tempranillo
Garnacha
Palomino
Bobal
Verdejo
Viura
Monastrell
Xarello
Parellada

Lemon
Verna
Fino
Eureka
Lisbon

• **Square frame**. The trees are planted at the corners of a square, whose side is the measure called the *planting frame*. A site might be planted using a 5 m square frame.
• **Rectangular frame**. The trees are planted at the corner of a rectangle, whose longest side is called the row and whose shortest side is the line. The trees could, for example, be planted in a 6 m x 4 m rectangle.
• **Staggered rows**. The trees are placed at the corners of an equilateral triangle, the side length of which is the planting frame.
• **Five of diamonds**. The trees are planted in a square or rectangular frame, but with a fifth tree at the centre of the frame. The trees could thus be planted in a five of diamonds of 5 m x 5m or 4 m x 6 m.
• **Paired rows**. The trees are planted in 2 or 3 staggered rows with separating rows. The trees might be planted in 3 staggered rows 1 m apart with rows 3 m wide.
• **Blocks**. The trees are planted in several staggered rows, forming masses with separating rows.

The most widely used fruit stocks (in Spain)

Types of distributions

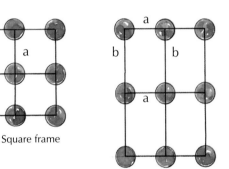
Square frame

Rectangular frame

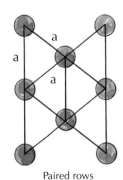
Paired rows

Five of Diamonds

Contour planting

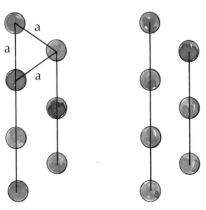
Straggered rows

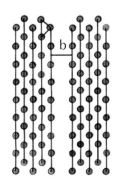
Blocks

5.2. PLANTING DISTANCE

The distance between consecutive trees is known as the **planting frame**. The planting frame allows us to find out how much ground area is occupied by each tree, and thus calculate the number of trees in a hectare (planting density). All this is of great importance to make best use of the entire ground area suitable for agriculture.

The planting frame varies, depending on the species being planted, and also on the how the plot is managed. In dry farming, for example, rainfall is a major limiting factor, and gives rise to competition for soil water between neighboring trees. This means larger planting frames are needed, even though there is little crown growth and part of the ground appears to be underused.

In irrigated plantations, water is not a limiting factor, meaning there is no root competition for water, and the planting frame is calculated on the basis of the greatest development of the crown, to avoid competition for light.

The advantages of large planting frames is that they allow cultivation to be completely mechanized, thus making it cheaper. But if the crop cannot be mechanized, then large frames are expensive and have many drawbacks. The trend is to cultivate smaller fruit trees, thereby increasing the planting density and thus production per hectare.

Thus to decide on the planting frame we must not only foresee the future growth of the trees and the type of crop, but also to define in advance the cultivation techniques to be used, the training system, the machinery to be used, the irrigation system, the harvesting method, and to sum up, all the operations that will have to be performed over the course of the year.

compacted soil. This will favor root growth. Together with the soil preparation, fertilizers and other corrective measures should be applied at depth. This is called **deep fertilization**.

The soil can be prepared manually, without machinery, or using machinery.

5.3.1. Preparing the site by hand

This is the traditional method, and it is carried out in small plots or sites where mechanization is not feasible.

A depth of soil up to 60 cm is dug up, using hoes, picks and similar tools.

This is the best possible type of preparation, as the soil is uniformly treated and broken up, but it requires a lot of physical labor and is not really worthwhile.

5.3.2. Preparing the site mechanically

There are several techniques to prepare the entire area of the site. One of them is deep plowing, plowing the soil at depth and turning over the top layer of soil, to a depth of as much as 80 cm. This requires use of a moldboard plow, with a single furrow or double furrow, pulled by a sufficiently powerful tractor. Turning the soil over makes it possible to bury the fertilizer at depth, together with any plant remains present on the surface. However, deep plowing has the disadvantage that it can only be applied where the soil is homogeneous with depth.

This must be done when the soil is ready, i.e., when it is adequately moist, but not waterlogged. This greatly reduces the soil's resistance and makes the work easier.

Number of trees/ha (planting density) with different planting frames

| Distance between trees (m) | Square frame or rectangular frame | | | | | | | | | | | |
| | Distance between rows (m) | | | | | | | | | | | |
	2.0	2.5	3.0	3.5	4.0	4.5	5.0	5.5	6.0	6.5	7.0	7.5
1.0	5,000	4,000	3,333	2,857	2,500	2,222	2,000	1,818	1,667	1,538	1,429	1,333
1.5	3,333	2,670	2,222	1,905	1,667	1,481	1,333	1,212	1,111	1,026	952	889
2.0	2,500	2,000	1,667	1,428	1,250	1,111	1,000	909	833	769	714	667
2.5	2,000	1,600	1,333	1,143	1,000	889	800	727	667	615	571	533
3.0	1,667	1,333	1,111	952	833	741	667	606	556	513	476	444
3.5	1,428	1,143	952	816	714	635	571	519	476	440	408	381
4.0	1,250	1,000	833	714	625	556	500	455	417	385	357	333
4.5	1,111	889	741	635	556	494	444	404	370	342	317	296
5.0	1,000	800	667	571	500	444	400	364	333	308	286	267
5.5	909	727	606	519	455	404	368	331	303	280	260	242
6.0	833	667	556	476	417	370	333	303	278	256	238	222
6.5	769	615	513	440	385	342	308	280	256	237	220	205

| Staggered rows | | | | | | | | | | | | | |
| Distance between trees (m) | | | | | | | | | | | | | |
1	1.5	2.0	2.5	3.0	3.5	4.0	4.5	5.0	5.5	6.0	6.5	7.0	7.5
23,094	10,264	5,773	3,695	2,566	1,885	1,443	1,140	923	763	641	546	471	410

5.3. PREPARING THE SITE

The aim of preparing the site is to leave the soil in the best conditions for the fruit tree's growth.

As a whole, these operations seek to dig up, loosen and aerate the site in order to improve its water retention capacity, and get rid of any possible

Once the site has been plowed, a further series of tasks should be carried out, i.e., breaking up clods, levelling, in order to leave the soil in the best condition.

Another technique to prepare the soil is subsoil plowing. This is performed by driving a subsoil spike into the ground, which is then pulled by the tractor.

The depth plowed can be greater than deep plowing, as it does not turn the earth over, but opens it and breaks it up. You need not wait until the soil is ready, as the effect is greater if the soil is dry, as long as it is possible to drive the spike into the ground. Subsoil plowing does not leave the soil as loose as deep plowing, but it can be very useful in soils where you do not want to mix the different layers.

To thoroughly loosen the soil, plowing should be carried out a second time, at right angles to the first, forming a grid pattern. After the subsoil plowing, complementary tasks should be carried out on the surface in order to ensure incorporation of fertilizer applied at depth and to level the surface.

If the site to be prepared is very large, or if the planting frame is very large (more than 8 m), and in robust species of little value, these tasks need only be carried out along the lines where the trees will be planted, saving time and money.

The strip worked can be between 1 and 4 m wide, and can be deep plowed or subsoil plowed.

5.3.3. Additional tasks

After performing jobs like deep plowing or subsoil plowing the soil tends to be in clods and rather hard on the surface. In order to bury the fertilizers and eliminate the tracks of the previous plowing, complementary tasks have to be performed to leave the soil flat and smooth for later planting.

This surface plowing is only to a depth of 30 cm, and can be done with different types of disk harrows or spike harrows, depending on the soil.

5.3.4. Timetable

The site should be prepared in fall. You should wait until the first rains and the soil is in good condition before starting these tasks.

You should wait for at least 20 days after deep plowing before carrying out the first complementary plowing, so that the rain and temperature changes can soften the clods of earth and make harrowing easier. Complementary plowing should be carried soon afterwards, in order to have the site ready for planting as soon as possible, even if planting has to be delayed.

If subsoil plowing was performed instead of deep plowing, the same is true, though you need not wait until the soil is in good condition.

5.4. PLANTING THE TREES

Once the site has been prepared, and is flat and in good tilth, it is ready for planting.

Planting consists of three phases. The first is **marking the ground**, that is to say physically marking where each tree will be planted. The second is **digging holes**, and the third phase is **planting the tree**.

5.4.1. Marking the ground

As already pointed out, marking the ground is simply transferring your planting design onto the site. You will need the following tools:

• **Poles or range poles**. To mark straight lines. They should be 1 or 2 m tall and brightly painted so they are clearly visible at a distance.
• **Twine**. To mark the lines. Normally smooth hemp twine, of diameter 6-8 mm, weighing little and easy to tense, is used.
• **Tape measure**. To measure distances.
• **Canes or stakes**. To mark the exact site where the tree is to be planted. They have to be at least 30-40 cm tall and should be resistant.
• **Marking frame**. This is used to mark the frames along the contours of the site. It consists of a wooden or metal framework, of the same length as the planting frame (or a fraction of its length, such as 2.5 m in a frame of 5 m) and a height of 1 m. It should also have two spirit levels.

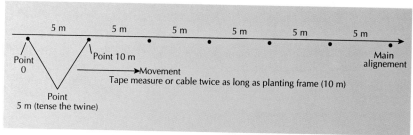

Marking staggered rows (5 m)

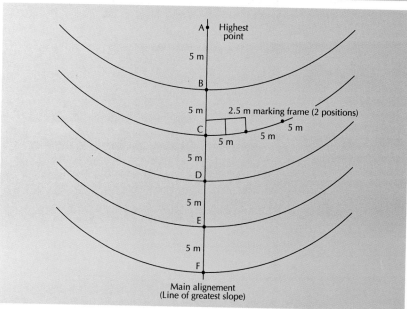

Marking the site along contours (5 m)

To start marking the site, the first thing is to choose the best orientation for the plantation. This should take into account both the amount of sunshine and the need to use machinery later.

To make best use of the sunshine, the lines of the plantation should run north-south. If the sunshine is very strong, the lines should be aligned east-west.

For mechanization, it is worth aligning the lines in the same direction as the longest axis of the plot.

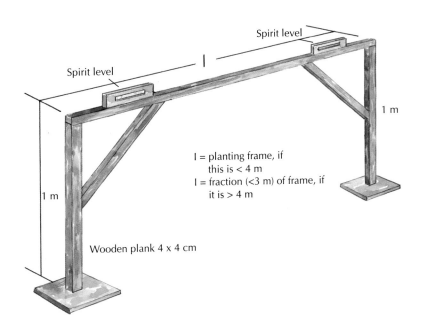

Spirit level

Spirit level

1 m

I = planting frame, if
   this is < 4 m
I = fraction (<3 m) of frame, if
   it is > 4 m

1 m

Wooden plank 4 x 4 cm

Marking frame for sites on a slope

The irrigation system also has to be taken into account. In plots where irrigation water is distributed by gravity, the direction of the slope determines the orientation of the plantation, while if irrigation is by sprinkler, the layout of the piping must be followed. Marking begins at one of the corners of the plot, at a distance of 3-5 m from the edge, and this will be a peripheral service lane.

Starting from point 0, the basic alignment is marked in the chosen direction. To do this, a string is tied from the initial range pole to a second pole at the end of the string. To prolong the alignment, ropes are tied to successive poles sited by eye.

To avoid mistakes in the base alignment when this is very large, a third range pole is positioned 4 m before the final range pole, and the prolongation of the second twine is tied to it.

Following the already prepared base alignment using a tape measure, the planting sites are marked with canes or sticks.

Steps to mark a site for planting in a regular geometrical arrangement

The canes are all placed on the same side of the string, and so they are solidly in position and upright. Once the base alignment has been marked, the vertical alignments must be marked. To do this a right-angled triangle is constructed from point 0, with sides 6 and 8 m long and a hypotenuse of 10 m. This is done to mark the vertical alignment at point 0.

To obtain the second vertical alignment, a simple isosceles triangle made with the measuring tape is used. The vertical alignments must also be marked at the other end of the base alignment, as other intermediate ones should also be marked. The number of vertical intermediates must be equal to the highest multiple of the planting frame that is less than 25 m. For example, for a planting frame of 5 m, a vertical alignment should be made every 25 m along base alignment.

Once the vertical alignments have been marked with twine, the planting sites should be marked with canes or stakes in accordance with the planting frame you have chosen.

Marking the areas between two vertical alignments is performed as follows: First, use the tape measure to mark from one alignment to the next and place

the canes, checking visually that they are aligned correctly in every direction. Once they have been placed, the equipment is moved in parallel and the same operation is repeated.

In plantations in staggered rows, it is not necessary to obtain vertical alignments or mark the area between them. All you need do is mark equilateral triangles from the base alignment and move from cane to cane.

If the site is to be planted along contour lines, the base alignment should coincide with the line of the steepest slope. The plot should be divided into more or less uniform sectors and the greatest slope in each one is marked.

The points will be marked on this line at a distance equal to the planting frame, and from them, the other planting points at the same contour level are marked using the marking frame.

If two curves move too far apart, another line is included between them, and if they come too close, one of the two is interrupted.

There is another series of important points, such as the location of the pollinators or the posts for the support structures if there are any, as well as water points and transverse access lanes. These points must be clearly marked to distinguish them from the planting points. Paint the canes or stakes a different color to distinguish them.

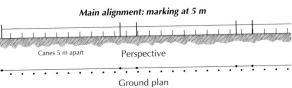

Phase 1. Replanting operations (rectangular frame 5x4 m)
Main alignment: layout

1st pole   3rd pole   2nd pole   5th post   4th post

Perspective view

1st twine   2nd twine   3rd twine

cords   Ground plan   cords

Main alignment: marking at 5 m

Canes 5 m apart   Perspective

Ground plan

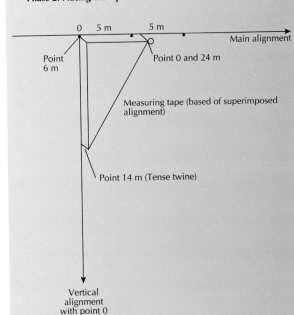

Phase 2. Placing the tape measure to obtain vertical alignments

0   5 m   5 m

Main alignment

Point 6 m

Point 0 and 24 m

Measuring tape (based of superimposed alignment)

Point 14 m (Tense twine)

Vertical alignment with point 0

Phase 3. Layout of vertical alignments and marking at a distance of 4 m

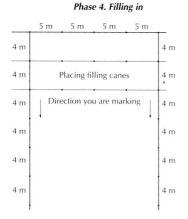

Canes 5 m apart
Elevation

Ground plan

Canes 4 m apart

Vertical alignments (25 m apart)

Phase 4. Filling in

5 m	5 m	5 m	5 m
4 m			4 m
4 m	Placing filling canes		4 m *
4 m	Direction you are marking		4 m
4 m			4 m
4 m			4 m
4 m			4 m

5.4.2. Digging the holes

The holes can be dug manually or mechanically.
Digging by hand is only done on small plots or where machinery cannot reach.
The hole is dug with a spade or a shovel and should be about 60 x 40 cm and about 50 cm deep.
Mechanical digging is done with special

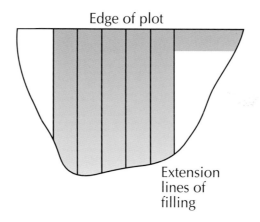

Edge of plot

Extension lines of filling

machines, which make holes about 20-30 cm in diameter and 50 cm deep.
These machines consist of a helical axis in the form of a drill and terminating in a drill bit, which is fitted onto a gear system protected by a shield and it is all powered by the tractor.
Digging holes can raise certain problems. In very sandy light soils, the walls of the hole fall in, and in very heavy clay soils, the **glass effect** occurs; the walls of the hole, which have been compacted, act as a container, and so if it rains before the trees are planted, water accumulates and creates a problem.
Furthermore, after planting the growing roots take a long time to grow beyond these walls to explore the rest of the soil.
Then, you must remove the protective covering around the root ball, and place the plant in the hole and then fill it with soil.
When pressing the earth down to firm it, take great care not to tread on the root ball, as this could break the roots. So only press down the surrounding earth. If the root ball is protected by peat or cardboard, it need not be removed before planting.

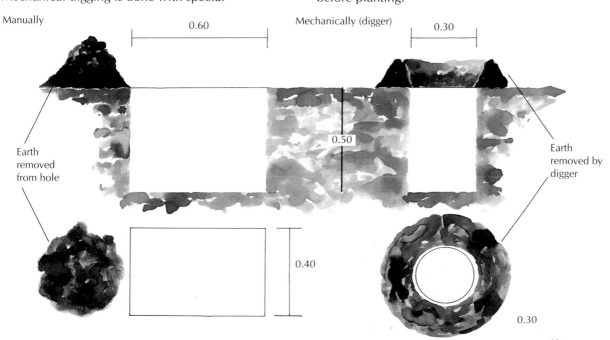

Manually

Mechanically (digger)

0.60

0.30

0.50

Earth removed from hole

Earth removed by digger

0.40

0.30

Ways of digging a hole

could cause the tree to die.

Reception of the trees at their destination must be rapid, and their general health should be checked, as should their labelling. Failures in new plantings are often due to damage during transport.

Once the trees have been unloaded, the best thing is to plant them immediately, but this is not always possible, meaning that they must be stored for a while.

If the plant comes as bare roots, it must be protected from direct sun, and from dehydration, and must be planted before it starts sprouting. To ensure this the bundles should be placed in shaded ditches ("heeled in") covering them with loose earth or sand.

If the plant comes as a root ball or in a container, and cannot be planted immediately, something similar should be done. Another way of keeping deciduous trees during their rest period, both bare-root and root-ball trees, is storage in a cold chamber, at 2-4°C and a relative humidity of 70% and in the dark. In this way, the trees can be stored long enough to replace any planting failures.

When it is time to plant the trees, they should be taken out of storage and placed in the holes dug in the ground.

Before bare-root trees are placed in the hole, the roots should be pruned, removing those that are dead or damaged. They are then daubed; that is to say, the roots are submerged in a runny mixture of soil and water.

This is done to keep the roots moist and improve their contact with the earth.

A fungicide can be added to this mix, especially for trees that are sensitive to fungal disease.

Once the trees have been placed in the holes, planting is completed. The planting technique depends on the tree's presentation, and there are several methods.

• **Bare-root trees**. Start by placing loosened soil in the bottom of the hole to the desired height. The plant is placed inside and supported in the correct position, so that the roots spread out over the bottom of the hole, without forcing them. Then soil is added until the roots are completely covered. Tamp the soil down with a tamping beetle or your foot, and completely fill the hole.

The depth of planting is important, as if the roots are too near the surface they may be damaged by cold weather, and if they are too deep they may suffer root asphyxiation. Therefore you must take into account that the first irrigation water or rainfall will cause the earth in the hole to settle and its level will fall.

• **Root-ball trees**. Plant as above, but place earth in the bottom of the hole until the top of the root ball is level with the soil surface.

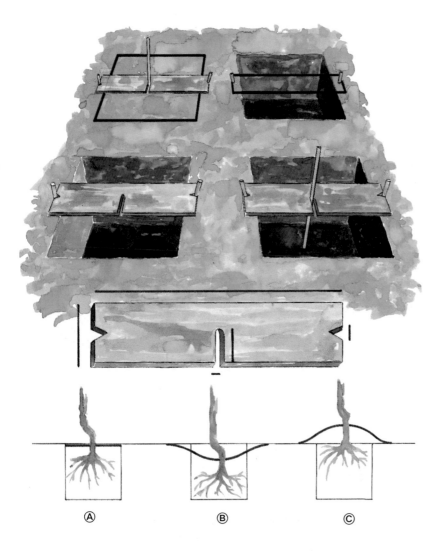

Above: Marking board

To mark exactly where the tree should be planted once the hole has been dug

Planting the tree

A/ Correct planting: once the earth in the hole has settled, the tree is at the same depth as it was in the nursery.

B/ Too deeply planted: the tree will "revert" by growing roots above the site of the graft. The trees may sometimes die of asphyxia.

C/ Raised planting: this is done on wet soils or where the water table is very high. The trees are badly anchored and frequently fall over.

5.4.3. Planting

Planting is placing the plant in the site where it is going to grow and develop. Plants ready for planting usually come in one of two forms: either bare roots, in which the roots lack any protective material, or as a root ball with some earth or in some type of container.

Bare root planting is used in deciduous trees when they are in the winter rest period, or in very robust species or very young plants. Root-balled plants are used in evergreen species, or deciduous species if the species is very delicate or expensive, in order to ensure the planting is successful.

In general, the trees to be planted are bought at a producer nursery, and so they have to be transported to the field.

To do this, they are first removed from the ground and prepared for sale, in late fall.

The bare-root trees are grouped together in bundles of 10 to 100 trees, depending on their size, and they are packed in bunches, with straw, plastic or burlap.

When transporting root-balled trees, it is important that the root ball should not dry out, break up or receive heavy blows, as this

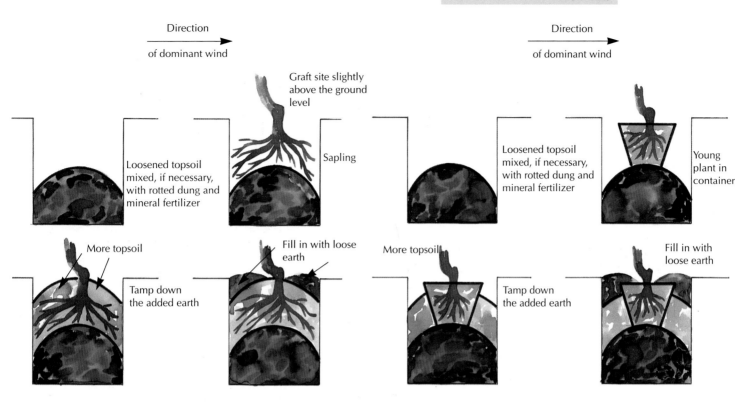

Direction → of dominant wind

Graft site slightly above the ground level

Loosened topsoil mixed, if necessary, with rotted dung and mineral fertilizer

Sapling

More topsoil

Fill in with loose earth

Tamp down the added earth

Direction → of dominant wind

Loosened topsoil mixed, if necessary, with rotted dung and mineral fertilizer

Young plant in container

More topsoil

Fill in with loose earth

Tamp down the added earth

5.5. AFTERCARE

Planting has finished once the plant has taken root in the soil. So aftercare to ensure rooting is important. The first thing to do is to water the plant, and this should be done as soon as possible after planting. The purpose is to bring the roots into contact with the soil, ensuring there are no gaps or pockets of air that may desiccate the roots. Watering is repeated every 15-30 days.

Other aftercare includes straightening trees that are not upright, covering roots that are exposed after watering, or making an irrigation basin around the plant, which is of great importance in areas that are dry farmed. These basins run around the base of the tree and their purpose is to collect more water when rain falls.

Training pruning and staking, if they are necessary, also have to be done after planting. One problem arising after planting is the replacement of trees that have not taken root. It is normal for there to be some failures when planting, but these should not exceed 2-3%. If the failure rate is higher, you can suppose you made some mistake in planting.

Failures when planting are directly related to the quality of the plants, the techniques used, and the prevailing climatic conditions, especially water shortage or harsh frosts.

The hardest thing about replacement is recognizing which trees have failed before they sprout in spring, as the failures have to be replaced as quickly as possible. If this is not done during the winter, it should be left until the following winter, meaning that a year has been lost. If the tree is withered, with wrinkled bark and dry buds, it has definitely not taken root.

Planting is considered to have concluded successfully when the trees sprout in spring. After this, a series of aftercare measures are required to ensure the plant continues sprouting normally. The most important aftercare in the first months is watering, as the plant does not yet have a well developed root system. Watering should be frequent, but not intense, to ensure the plant does not suffer water shortage.

Another aftercare to consider is maintaining the soil loose and free of weeds, to avoid competition for water and nutrients.

5.6. WHEN TO PLANT

The time chosen for planting is important. Planting must be carried out on mild, sunny, days, with no wind and some moisture. Planting is performed during the tree's rest period, when transpiration is lowest, and the danger of dehydration is therefore lowest.

In temperate areas that are not very rainy and where winters are mild, trees can be planted in late fall, even though they have the entire winter in front of them. This means they can take advantage of the winter rains. But if the local climate is cold, with the danger of severe frosts, planting is performed at the end of winter to avoid damage by low temperatures.

On the left:
Planting bare roots.
On the right:
Planting a root ball

Pruning almonds

6. PRUNING AND TRAINING FRUIT TREES

The tree can be considered as a fruit-producing machine, and the farmer's main interest is to do what he can to increase the plantation's yield.

Pruning is one of the main factors influencing production, together with fertilizer application, irrigation, working the soil, and prevention of disease. Thus pruning cannot be considered in isolation as a production factor.

Pruning orders and regularizes the tree's productive potential, but in no way creates it. To sum up, the level of productivity, is the result of several parameters, including pruning.

6.1. GENERAL PRINCIPLES

The tree's laboratory is located in its leaves, and transforms raw materials into complex products.

A fruit tree's productivity is directly proportional to the amount of sugars it photosynthesizes, which in turn is directly proportional to the photosynthetically active leaf area.

The main purpose of pruning is to act on the aerial parts of the tree so that the largest possible number of leaves are exposed to the sunshine. The shape of the tree's crown is very important, and you should aim to extend the area of leaves rather than aim for a dense compact shape.

A tree's life consists of three development phases:

Schematic diagram showing what to aim for when training fruit trees.

A/ Natural development of the crown
- creation of an unproductive zone
- productive space is unoccupied

B/ Guided development of the crown
- disappearance of the unproductive zone
- all space is productive

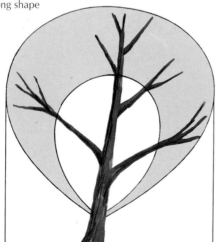

Ⓐ Wrong shape

Natural development of the crown

Productive zones

Unproductive zones

Ⓑ Right shape

Productive zones

Guided development of the crown

• **Youth**, when vegetative growth occurs, but no flowers are produced. The tree grows a lot but does not produce flower buds or flowers. Your aim in this period is to obtain the greatest volume of crown in the shortest time.

• **Adolescence and maturity**, when both vegetative growth and flower formation occur. Your aim in this period is to maintain the tree in an equilibrium between growth and maximum production, for as long a period as possible.
Pruning is an important technique in this period, as you must ensure young branches are produced in order to obtain more active leaves.

• **Old age**, when flower formation occurs, but there is little or no vegetative growth. In this period, you should aim to cut the crown back, in order to induce a second flush of youth in the tree.

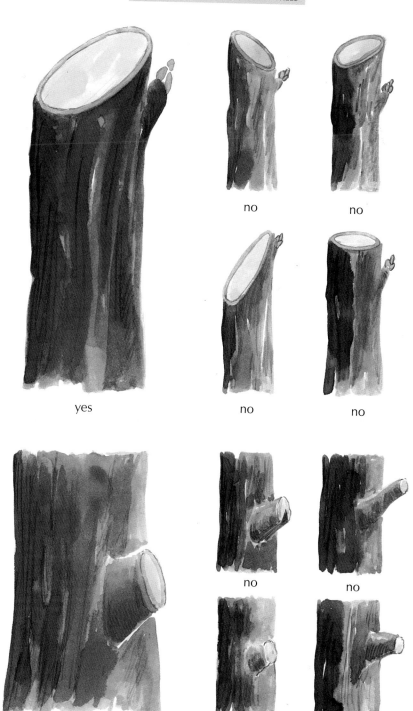

6.1.1. Technical aspects of pruning

A fruit tree is not a typical tree or a shrub or a herbaceous perennial. Their intermediate nature means they can be treated in a wide variety of ways to increase production.
The aerial part's natural appearance is determined by its growth form. This is characteristic of each species and variety, and greatly influences the tree's behavior.

The best, or most rational, way to prune the tree does not go against its natural growth form.
A fruit tree that has never been pruned tends to grow upwards and outwards, leaving its base without leaves.
This means that the productive areas of the plant are harder to reach, making collecting the fruit more expensive, while creating a non-productive zone at the center of the crown.
Pruning can reduce unproductive areas to a minimum, and guide the aerial parts towards more worthwhile spaces.
To achieve this, the laws governing plant growth have to be grasped and respected.

Cutting the shoot must be done near a bud, and the cut should be at a slight angle. The cut should not be too close to, or too far from, the bud. Cutting branches should be done close to the trunk, and at a slight angle. The branch should not be cut down to the trunk, nor should a stump be left, and the bark should not be broken.

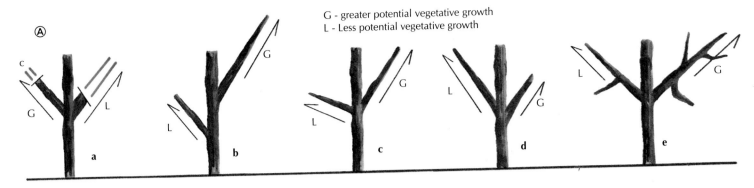

G - greater potential vegetative growth
L - Less potential vegetative growth

Ⓐ

a   b   c   d   e

Ⓐ

A/ Interactions between 2 branches

Rules on interactions to be taken into account in order to ensure the crown is balanced:

• **Between 2 branches**:
A) 2 branches of the same size, at the same height, at the same angle. If we leave one longer than the other, we strengthen the first and weaken the second.
B) 2 branches of the same size and at the same angle, but at different levels. The upper branch is always favored over the lower branch.
C) 2 branches located at the same level, but at different angles. The branch forming the smallest angle from the stem dominates the other.
D) 2 branches at the same height and the same angle, but of different sizes: The largest (in cross-section) will be dominant.
E) 2 branches located at the same level, with the same angle and size: The one with most secondary branches will be dominant.

• **Between 2 parts of the crown**
A) Between symmetrical elements of the framework. You should try to maintain a balance between the symmetrical elements of the tree's crown, that is to say, the two main leaders should be balanced and have the same vegetative potential.
B) Between different elements of the crown.

B/ Interaction between vegetation on two different parts of the crown

This occurs in specimens with a central axis in which not all the elements of the framework are at the same angle. The central axis, because it is growing vertically, dominates the leader branches (at an angle), and to keep it in balance with the leaders, it has to be severely pruned.
When a branch is removed, the plant loses the reserves stored in it as well as its leaf area. This always weakens the tree, so that the intensity of pruning should be in keeping with the tree's vigor.
After heavy pruning, many vigorous shoots sprout below the cut. This is because the sap that flowed into the branch does not flow elsewhere, but accumulates near the cut, causing these shoots to sprout.
Sap tends to rise to the highest parts of the tree and to the tips of the vertical branches, to the detriment of the lower part of the tree. The lower branches of the trees should be left longer than the upper ones so that plenty of light can reach them and ensure they are vigorous.
Sugar-bearing sap is distributed better in the horizontal branches, the ones that bear most flowers and fruit. To induce a branch to fruit, it has to be bent downwards.
In general, you should aim for simple, easily trained forms with evenly distributed vegetation that covers a given area.

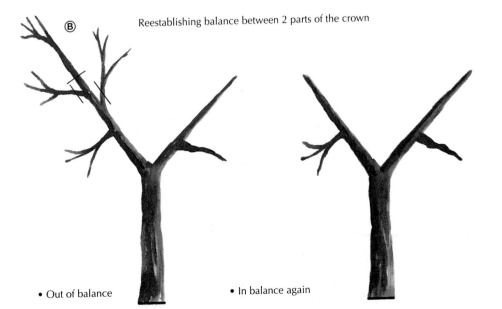

Reestablishing balance between 2 parts of the crown

Ⓑ   Reestablishing the framework between 2 different elements of the framework

Ratio of the lengths to be pruned between a and b when the two are balanced.

Vegetative potential
a > b

• Out of balance   • In balance again

• Out of balance   • In balance again

Whatever the shape, the following are all necessary:

Air and light. All parts of the tree must have air and light. The tree has to be well illuminated to ensure maximum leaf activity and fruit coloration.

The framework or **frame** should be restricted to what is strictly necessary, because if it is too large, it reduces the space for the fruit-bearing wood.

Minimum pruning during the tree's training, so that it grows rapidly and forms its frame as soon as possible, without worrying too much about the leaf cover, which will grow quickly.

Suitable angles in the main branches. The branches are all trained at the same angle in order to maintain balance and divide the tree's vigor equally.

To achieve the greatest amount of productive wood, as near as possible to the tree's framework.

Do not remove the tip buds from the one-year-old branches, as this disturbs the balance of vegetative growth between the different parts of the tree, and also delays fruiting and slows down the development if the tree is in the training period.

The tree's size and volume depend on the vigor of the rootstock, on the variety grafted and the site's fertility.

The training and pruning should be the best for each variety, and you should not try to adapt the plant to the training method.

6.1.2. Basic rules for shaping the tree's skeleton

When structuring the crown, you have to try to establish a hierarchy between all the elements forming the framework.

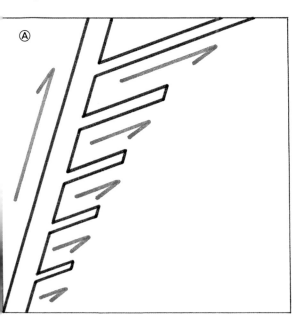

The relations to be borne in mind are:

• Relations of thickness
The purpose is to ensure that all the year's shoots should be equally vigorous everywhere in the crown, i.e., those near the ground should be as vigorous as the higher ones.

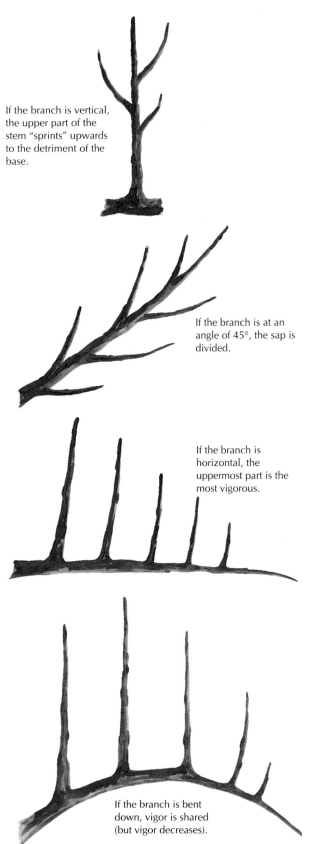

If the branch is vertical, the upper part of the stem "sprints" upwards to the detriment of the base.

If the branch is at an angle of 45°, the sap is divided.

If the branch is horizontal, the uppermost part is the most vigorous.

A/ Incorrect: Imbalance between the elements of the framework. The vegetation grows upwards and the base of the crown is left bare.

If the branch is bent down, vigor is shared (but vigor decreases).

The way the crown's vegetative energy is distributed can be compared to a fluid. The bigger the cross-section, the greater the flow.
Forcing the tree to branch from the base means that the tree acquires a conical shape,

B/ Correct form: a suitable ratio between the elements of the framework. Equal vigor of the shoots in all the different areas of the crown.

C/ a) Schematic relation between the length ratios of the elements of the framework.
b) Distance ratios between the elements of the framework.

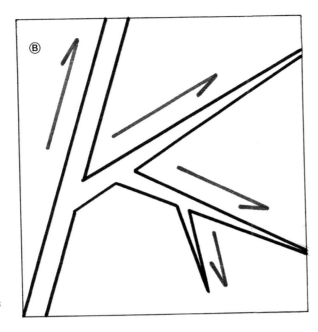

so that the vegetation cannot grow upwards, as the sap has to follow the route imposed on it.

• Length relations

The crown's framework consists of primary, secondary, tertiary, branches, etc. This hierarchy of branches becomes more important as the tree becomes more vigorous.
There should be a constant length ratio between the branches of different degree, whatever the form of framework chosen.

• Relations of distance

The stronger and longer the leaders, the more space should be left between them.

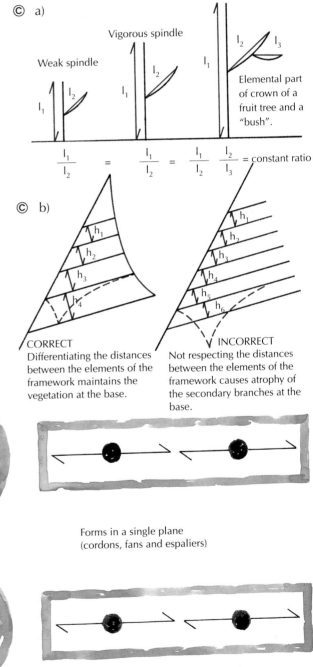

Ⓒ a)

Weak spindle

Vigorous spindle

Elemental part of crown of a fruit tree and a "bush".

$$\frac{l_1}{l_2} = \frac{l_1}{l_2} = \frac{l_1}{l_2} \quad \frac{l_2}{l_3} = \text{constant ratio}$$

Ⓒ b)

CORRECT
Differentiating the distances between the elements of the framework maintains the vegetation at the base.

INCORRECT
Not respecting the distances between the elements of the framework causes atrophy of the secondary branches at the base.

The height of the crown is determined by the plant's vigor.

The more vigorous the tree, the higher its crown.

Ⓓ

High trunk   Medium trunk   Low trunk

D/ Graphic representation of the crown's growth as a function of its height.
E/ Volume and area of different crown shapes

Ⓔ

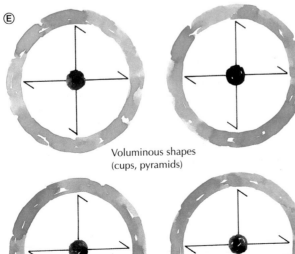

Voluminous shapes
(cups, pyramids)

Forms in a single plane
(cordons, fans and espaliers)

6.1.3. Basic rules for shaping the crown

• Height of the crown

The development of the crown is inversely proportional to the height of the base of the crown.

From a technical point of view, a space must be left between the crown and the soil to avoid damage to the vegetation and for machinery to pass.

• Crown height and distance between trees

There has to be a constant ratio between the height of the crown and the planting frame. The minimum distance is determined by the width needed for machinery to pass and the optimal distance should allow sufficient sunshine to reach the lower levels of the crown. Otherwise, the base of the crown loses its leaves and atrophies.

The taller the crown, the greater the distance necessary between trees.

In general, we can accept that the optimum distance is 1.5 times the crown's height. That is to say, if the crown is 2.5 m tall, the optimum distance between rows would be 1.5 x 2.5 m = 3.75 m.

• Choosing the shape of the crown

The ideal form is the one giving the maximum active leaf surface with the minimum volume.

There are two types of crown shape:

• *Circular shapes, volumes.* The framework of the crown develops around a real axis, or an imaginary one if the center is open, and the vegetation grows out in every direction.
• *Shapes in one plane, flat shapes.* The crown's framework develops along a line.

For circular shapes, the ideal arrangement is planting in squares, as this ensures sunshine reaches every point in the crown. This has the technical and economic disadvantage that an excessive distance is needed between trees for machinery to pass.

Flat shapes have a series of advantages:

• It is easier to mechanize work in the field
• Harvesting can be rationalized
• Pruning and clearing tasks can be rationalized
• Application of sanitary products is easier
• The crown can start nearer the ground
• Greater economy in herbicide use
• Makes better use of sunshine.

6.2. TYPES OF PRUNING

Over the course of the year, the tree passes through different vegetative stages, and pruning must adapt to them.

There are two main types of pruning:

Pruning an almond tree.

• Summer pruning

This done in late spring, when the tree is growing actively. It consists of removing the tip of excessively vigorous shoots in order to favor the growth of other shoots.

Summer pruning can also be carried out in the summer. This is mainly done to remove the shoots that emerge beyond the ends of the leaders and main branches that form the framework, as well as the shoots that are growing inwards into the crown.

Summer pruning also serves to remove all the very vigorous shoots that are not worth keeping in order to renew an older part of the crown.

In early harvesting, summer pruning cannot be carried out in the three preceding weeks, to avoid a negative effect on fruit growth.

SPECIES	VARIETY	TRAINING	DENSITY
APPLE	All	Cup	Low
	All	Reg. fan	Medium
	All	Irreg. fan*	Medium
	Golden	Drapeau	Medium
	Golden, Idared	Spindle	High
PEAR	All	Cup	Low
	All	Reg. fan.	Medium
	All	Irreg fan*	Medium
	All	Spindle	High
PEACH	All	Cup	Low
AND	All	Irreg. fan*	Medium-low
NECTARINE	All	Free fan	Medium
	All	Spindle	Medium-high
CHERRY	All	Cup	Low
	All	Irreg. fan*	Medium

All fruit trees adapt well to training in a cup shape.
* Irregular fan: this is the same as the regular or cassic fan, except that the branches at the same height leave the trunk at the same point; in the irregular fan the leaders are located randomly on one side or the other, with no points where two branches leave the trunk.

Machine prepared for pruning after the harvesting equipment has been removed

*On the right:
How some species adapt to different training systems*

In this case, summer pruning is carried out after harvesting has finished.

• Winter pruning
This is carried out when the tree is in its winter rest period. Depending on the aim, there are several different types.

• *Regulated pruning*. This pruning aims to train the tree's frame and is of great importance in the first years of the tree's life.
• *Production pruning*. This aims to increase production, i.e., to provoke or speed up the production of flower-producing organs.
• *Maintenance pruning*. This allows harmonious balance of the tree's vigor and fertility. An excessively vigorous tree only produces long shoots, while an excessively fertile tree produces too many flowers and fruit, and is rapidly exhausted and then dies.
• *Renewal pruning*. This aims to stimulate sprouting in order to rejuvenate senescent trees. Renewal requires experience and knowledge of how to treat the different parts of the crown correctly and decide the correct lengths.

• Other types of pruning
• *Pruning for fruiting*. This is carried out in summer and consists of pinching out the tips of the productive branches, but not the structural branches, to induce them to form fruit.
• *Healing pruning*. This consists of the different operations carried out to treat sick trees, by eliminating the affected woody parts.

6.3. SHAPING SYSTEMS

6.3.1. Helicoidal cup

The tree's framework consists of three main leaders, or arms, i.e., the main branches, at different heights on the stem but of equal length, at an acute angle of 35-40° to the vertical.
These leaders will be the base of the secondary branches that will grow from them. The tree as a whole will have a conical shape.

Forming a helicoidal cup shape takes 4 years.

1st year
The one-year-old tree should be cut back to 70-100 cm above the ground when it is planted, depending on the desired height.
If the rootstock is grafted with a dormant bud, it should be allowed to grow to a height of 60 cm, and then the leader is pinched out to make it branch.
In spring of the same year, the three branches that will form the base of the framework are chosen, ensuring they are as well distributed as possible, with an angle of 120° between them, and a difference in height of 10-20 cm.
During the winter, the main branches chosen as leaders should be confirmed, and all the shoots below them are removed. The shoots at the tips of the main branches are also removed, as are all the buds that will grow inwards into the crown. The other buds are left, as long as they are not very vigorous and there are not very many of them.

2nd year
In spring, all the shoots growing into the center of the crown are pinched out, as are any vigorous shoots at the tips of the main branches. Excessively vigorous shoots are removed, as are any fruit that may form.

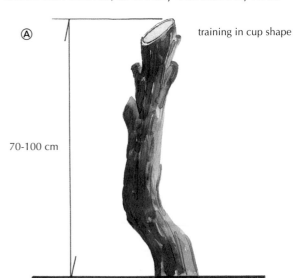

Ⓐ

70-100 cm

training in cup shape

Ⓑ

45°

45°

45°

120°

45°

After pruning
in winter of
first year

Branch at acute angle

Ⓒ

Correcting the angle of insertion of a leader

Branches too close together

Correcting the position of two of the leaders

Ⓓ

45° 45°

120°

45° 45°

45° 45°

After pruning in
spring in the
second year

After winter pruning in
second year

*Pruning to form a
helicoidal cup*
A/ 1st year in spring
B/ Winter
*C/ Different
corrections*
D/ Second year
E/ 3rd year

Ⓔ

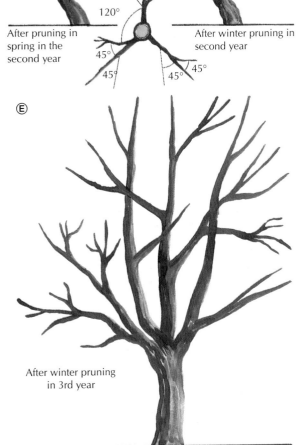

After winter pruning
in 3rd year

**Training pruning:
Classic or regular fan**

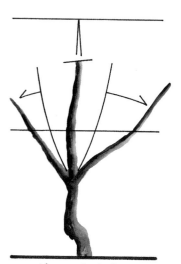

After pruning in 1st year

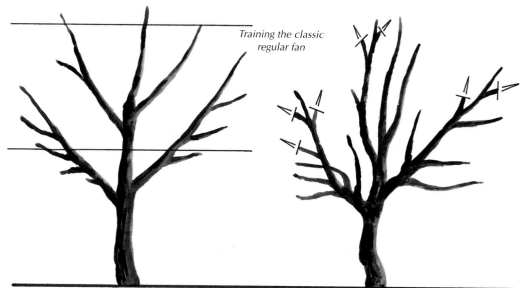

*Training the classic
regular fan*

After winter 2nd year

*After growth in 2nd year, when
summer pruning has been performed*

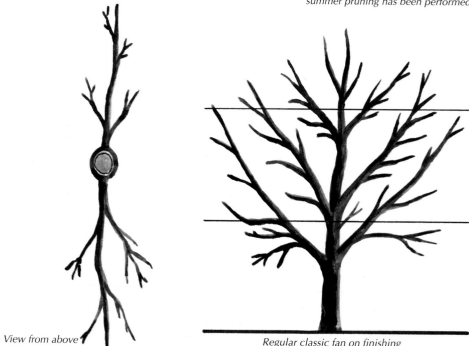

View from above

Regular classic fan on finishing

winter, an extension shoot will be chosen on
ach main branch that is at the tip and is not at an
ngle to the main branch. All the other shoots on
e top 20-30 cm of the stem are removed, as are
l those that will grow inwards into the crown and
l those shoots on the trunk below the leaders.
one of the leaders is more vigorous, it should be
ut back to maintain its height in balance with the
thers, and another shoot should be used as the
xtension shoot.
elect the secondary branches located on the side
pposite to the ones chosen the year before. As
e secondary branches are located alternately,
ere is no competition between the two, or
etween them and the leader.
he very vertical branches are removed, and tip
uds on the secondary branches are not
emoved.

rd year
runing is carried out during active growth in
pring, and is the same as the year before. In
vinter, a third secondary branch is chosen using
e same criteria as the previous years.

th year
he leaders are cut back to a vigorous shoot that
cts as an extension, and does not have its tip bud
emoved.
he secondary shoots on the first level may start to
ear fruit, as they have finished their formation.
When training has finished, the tree should
ave three leaders, all of them with the same
mount of vegetation, and be conical in
hape.

6.3.2. Regular classic fan

This is based on a central axis bearing the 4 or 5
evels forming the fan.
ach one of these levels consists of two branches
on opposite sides of the trunk at an angle of 45-
50° to the central axis. The distance between the
ayers is 50-100 cm.
To perform this type of training, staking is
necessary. The staking consists of posts and
support wires to hold the branches in place.
Training a fan takes 4-5 years, depending on the
number of layers.

1st year
The tree should be cut back to 50-70 cm above
the ground when it is planted.
In the spring, the three best branches are chosen to
form the framework. One will be the central axis,
and the other two will be aligned with the
plantation.
In winter, the 2 chosen side branches are bent
down and tied to the wires, and their tip buds
removed.
The central axis is cut back to the height you want
the second layer to form at.
If the 2 lateral branches are not sufficiently
vigorous, the axis is cut 20 cm above the initial
cut. This strengthens the lateral branches, and
despite losing a year, the tree's framework is
ensured.

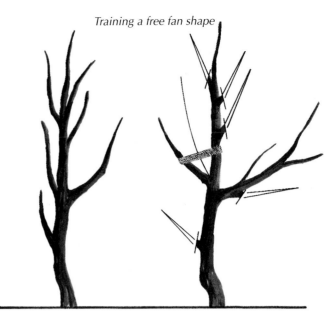

Training a free fan shape

Summer pruning in 1st year   After winter pruning in 2nd year

Summer pruning in 2nd year   After winter pruning in 2nd year

2nd year
In spring, the only shoots to be removed are
vigorous shoots that grow vertically upwards from
the branches of the first layer or at the tip of the
leader.
In winter, the second layer is formed in the same
way as the first.
Remove the shoots at the tips of the branches of
the first layer and the highly vertical branches, and
leave the ones growing outwards.
The central axis is pinched out at the height you
want the third level to form.

3rd year
Each further level making up the fan is formed in
the same way as above.
Once the desired height has been reached, the
central axis should be cut back to a shoot acting
as an extension, to avoid it growing any further.

*Training a free fan
shape (continued
on next page).*

After winter pruning in the 3rd year: formed free fan

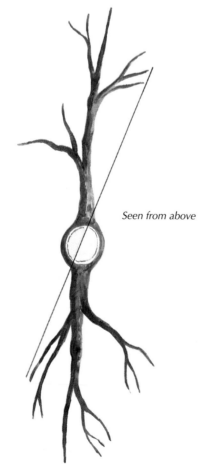

Seen from above

A fan is well formed when:

• Every layer is longer than the layer above it.
• The 2 arms at the same layer are equally vigorous
• The secondary branches on each arm follow a strict hierarchy.

An important variation that should be mentioned is the irregular fan. The difference is that the arms forming the layers are inserted at random on both sides, and two arms are never inserted at the same point.

6.3.3. Free fan

The difference between this and the regular fan is that it allows the tree to grow freely, thus reducing the number of prunings to be carried out.
Its shape is less symmetrical and rigid than the regular fan, but its general appearance is similar. Another important difference is that it is not staked and wired. To achieve the same effect, canes are tied to the branches to ensure they adopt the angles desired.
Training takes 3-5 years, depending on the number of layers desired.

1st year
The tree is not cut back when planted, but any excessively vigorous shoots are removed.
In the spring 2 branches are chosen that will form the first layer and the others are pinched out, but not the central axis. If the tree is very vigorous, the first 2 stories can be chosen at this stage.
In winter, the branches are bent with canes so they adopt the desired angle.

2nd year
In spring, the buds are pinched out from the tips of the main branches, except for the terminal bud, and any excessively vigorous shoots are removed.
In winter, the same criterion is followed as in the regular fan, except that the side branches on the arms are allowed to grow, as long as they do not get in the way of cultivation tasks.

3rd year
The same procedure as the previous year is followed. When the tree's vegetative growth is short or if it is beginning to lose the branches at the base, it is cut back. This consists of cutting back the central axis, cutting just above a well-formed shoot at an acute angle, to avoid it growing too high.
The primary and secondary branches should also be cut back to favor the vegetation along the entire branch. Excessively vigorous branches are removed, as are very old fruiting branches.

Pruning on planting

End of growth in 1st year, after pruning in spring

After winter pruning in 1st year

6.3.4. Spindlebush

This is a very simple method of training that is based on a central axis bearing 7-8 branches that are distributed in no particular order, but are a good distance apart and alternate around the stem to avoid shading each other and ensure the light reaches all branches equally.
The branches at the base are longer than the higher ones, and the further up the central axis they are borne the shorter they are.

1st year
If the sapling is vigorous and has produced shoots at the base, these are respected, but if they are further up the plant, they should be cut back to 2 buds at the time of planting to favor sprouting of the lower buds.
In spring, the shoots are removed from the upper tips of the branches chosen to be the leaders and from the vigorous branches that might compete with them.
The axis is always left untouched.
In winter, the shoots are removed from the end of the central axis, as are all the very vigorous buds, except on the main branches.

2nd year
Follow the same criteria as the previous year.

Training spindlebush

End of growth in 2nd year

7. CULTIVATION TECHNIQUES

Cultivation techniques include all the tasks performed on the site and the tree to ensure good production.

One of the most important is soil maintenance. This is mainly removing weeds, but it serves other purposes, such as maintaining good soil structure by preventing compaction, as well as improving fertility and preventing erosion.

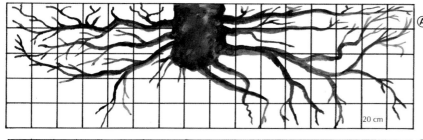

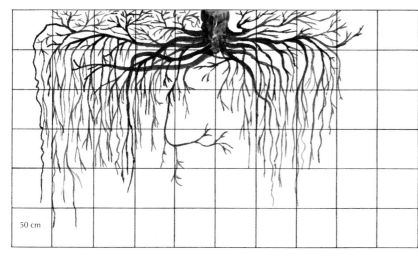

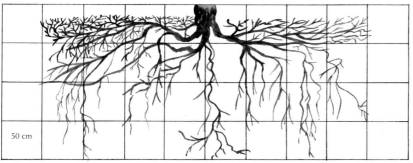

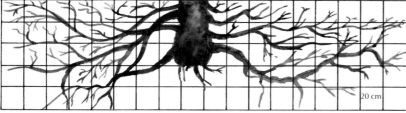

There are several soil maintenance techniques. Some leave the spontaneous vegetation alone, while others remove it by continuous weeding or herbicide applications.

A) Site free of weeds

• Weeding by plow

Advantages:

• There is no competition for nitrogen or water
• There are no host plants for fungus and pests, thus reducing the risk of attacks.
• It reduces the number of rats and moles.
• Application of organic matter and fertilizer to the soil is easier.

Disadvantages:

• Low mobility of phosphorus and potassium
• Lower content of organic matter
• Higher risk of ferric chlorosis
• Possible damage to soil by heavy machinery
• Degradation of soil structure
• Compaction and damage by root suffocation
• Damage if frosts occur soon afterwards
• The cost of carrying out the weeding

• Using herbicides

Advantages:

• Reduces the amount of plowing to be done
• As the ground is not plowed, its structure improves and the roots grow better
• Consumption of fertilizer and water is less than when plant cover is present, and the same as when plowing is performed, though the water losses are lower with plant cover
• Damage due to low temperatures is diminished
• Less ferric chlorosis than with plowing

Disadvantages:

• The cost of the herbicides
• Accumulation of herbicides in the soil over the years
• Soil compaction due to heavy machinery, though this is less than that caused by plowing

B) Site with plant cover

• Temporary plant cover

This consists of plant cover, either spontaneous or seeded, covering the entire surface of the plantation or just in bands. The bands can be sited underneath the fruit trees, leaving the center of the plots clean, or by clearing the zone under the fruit trees and leaving the plant cover in the center of the plot.

It is called temporary because the plant cover is only allowed to grow from spring to fall, and in late fall the plants are buried in the ground, together with the fertilizer and organic matter if necessary.

Advantages:

• It improves the mobility of phosphorus and potassium
• Reduces compaction and thus root asphyxia

- Increases the content of organic matter
- Saves plowing and improves the soil structure
- Roots can grow into the surface layer
- Protection from erosion and cracking after irrigation or rain
- Reduction of the soil's temperature range
- Less ferric chlorosis
- Fertilizers and organic matter can be applied in the winter tasks.

Disadvantages:

- Greater need for nitrogen and water
- Avoids plowing, but not reaping
- More moles and thus damage to the root system, especially if the plant cover is growing directly under the trees.
- More fungi and pests in general
- Greater damage by frosts in spring
- If the plant cover reaches up to the trunk, damage due to rotting may appear on the collar.

- **Permanent plant cover**

The same as temporary plant cover, but the vegetation is not buried in the winter, which is why it is called permanent. It too may cover the entire plot or be in bands.
The advantages and disadvantages are the same as above, but the advantages are greater, as are the disadvantages, and it also makes it harder to apply organic matter and fertilizers.
When the cut grass is not removed from the plot, it is called a **grass-mulch**.

C) Site with other types of cover

- **Mulching**

No plowing at all is performed, and the soil is totally covered with straw, organic matter or plastic material. The cover may be complete or in strips.

Advantages:

- Saves a lot of water
- No plowing
- Potassium usage improves
- Average texture improves

Disadvantages:

- Root asphyxia is made worse in compacted soils
- The cost of the mulch
- Harder to apply fertilizers
- Increases risk from frosts
- Increases the number of moles
- If straw is used, the risk of fire increases.
- It is irreversible, and you cannot change to other techniques without causing damage to the superficial root system.

The applications of these techniques are:

- *Site weeded by plowing*
Recommended for fruit trees in their first 3-4 years in most cases, and especially for dry farming of fruit trees where rainfall is not high.

- *Weeds cleared by herbicides*
Only in trees 3 or more years old, and you must also take into account the variety's tolerance to the herbicide used.

- *Temporary plant cover*
This can be recommended, in general, for fruit trees more than 3-4 years old, as the disadvantages decline. The fruit trees that require little water, such as peaches and other stone fruit (apricot, almond and cherry) do not tolerate plant cover well.
In dry farming plantations it cannot be recommended, as competition for water is a problem.

- *Permanent plant cover*
For fruit trees more than 4 years old

- *Mulching*
Can only be recommended in loose or sandy soils, or where water availability is limited.

7.1. HERBICIDES IN FRUIT GROWING

Herbicides are products used to kill weeds on a site, taking "weed" to mean any unwanted plants.
The best herbicide is not one that kills everything on the site, but one that does a good job, destroying the plants that cause the most problems, and not leaving residues that will be a problem for the crop.

When choosing a herbicide and deciding when to apply it, you should bear in mind the following general rules:

Left, several cultivation techniques.
A/ Diagram of the root system of a peach tree in an unplowed site without grass; only herbicides have been applied. The roots can develop very well in the topmost layers.
B/ Diagram of the development of the root of a peach tree: the right side has been subject to regular plowing to a depth of 20 cm, while the left side has had plant cover.
C/ Diagram of the root system of a peach tree, in a well structured plot of soil and regularly mulched (with straw). There are abundant small roots near the surface because this area is better aerated and more fertile.
D/ Roots of a peach after three years of "non-cultivation"; the roots are restricted to the level below the layer of soil invaded by the roots of the weeds.

Ⓐ

	ATA (aminotriazol)	Atrazine	Dalapon	2,4-D	Diuron	Simazine
Agropyron repens (quack grass)	S	SS-SR	S	R	SR-R	SR-R
Cynodon dactylum (Bermuda grass)	SS	R	S	R	R	R
Cirsium (thistle)	SS	SR	R	S	SR	R
Convulvulus (bindweed)	SR	SR	R	S	R	R
Equisetum (horsetail)	SR	SR	R	S	R	R
Rumex (sorrel)	S	S-SR	R	SR	SR-R	—
Ranunculus	SS	S-SR	—	S-SS	SS-SR	SR
Sonchus (sow-thistle)	S	S	—	SS	SS	SS
Tussilago (coltsfoot)	—	SS	—	R	—	R

S = Sensitive   SS = Semi-sensitive   SR = Semi-resistant   R = Resistant

Some points on weeds

- Young weeds are easier to kill than old ones;
- In general, weeds are easier to destroy in conditions that favor their germination and rapid growth, except if you are using pre-emergence herbicides.

In relation to the climatic conditions and soil

- Unsuitable temperatures or rain during or after herbicide application may lead to negative results;
- Fine-textured or clay soils need larger doses of herbicide than coarse-textured or sandy soils.

Ⓑ

Herbicide	Herbicidal action				Applied to weeds		Apple	Pear on wild stock	Pear on quince stock	Peach	Cherry	Plum	Apricot
	Absorption		Contact	Pre-emergence	Post-emergence								
	Roots	Leaves			Young	Adult							
Aminotriazol (ATA)	X	X X	—	X X	X X	X X	Yes	Yes	No	Yes	Yes	Yes	Yes
Altrazine	X X	X X	—	X X	X X	—	Yes 4	No	No	No	No	No	No
Bromacil (A)	X X	—	—	X X	X	—	No	No	No	Yes 2	No	Yes 2	Yes 2
Carbethamide	—	X	—	X X	X	—	Yes	Yes	Yes	Yes	Yes	Yes	Yes
Clorthiamide (B)	X X	—	—	X X	X	—	Yes	Yes 3	No	Yes 3	No	Yes 3	Yes 3
Dalapon	X	X X	—	—	—	X X	Yes 4	Yes 4	Yes 4	Yes 4	Yes 4	Yes 4	Yes 4
Diuron	X X	X	—	X X	X	—	Yes 3	Yes 3	No	No	No	No	No
DNOC and DNBP	—	—	X X	—	X X	X X	Yes	Yes	Yes	Yes	Yes	Yes	Yes
Diquat	—	X	X X	—	X X	X X	Yes	Yes	Yes	Yes	Yes	Yes	Yes
Glyphosate	—	X X	—	—	X	X X	Yes	Yes	Yes	Yes	Yes	Yes	Yes
Methabenzthiazuron++	X X	X X	—	X X	X X	X	Yes	Yes	Yes	Yes	Yes	Yes	
2,4-D (C)	—	X X	—	—	X X	X	Yes	Yes	Yes	Yes	Yes	Yes	Yes
Oxadiazon	—	X X	X	X X	X X	X	Yes	Yes	Yes	Yes	Yes	Yes	Yes
Paraquat	—	X	X X	—	X X	X X	Yes	Yes	Yes	Yes	Yes	Yes	Yes
Propyzamide + Diuron	X X	X	—	X X	X X	X	Yes 3	Yes 3	No	No	No	No	No
Simazine (D)	X X	X	—	X X	X	—	Yes 3	Yes 3	Yes 3	No	No	No	Yes D
Terbacyl	X X	X	—	X X	X —	Yes	Yes ?	Yes ?	Yes ?	Yes	No	? ?	
Mixtures:													
Benzuride-ATA-Dalapon	X X	X X	—	—	X X	X X	Yes 3	Yes 3	Yes 4	Yes 4	Yes 4	Yes 4	? ?
ATA-Diuron	X X	X X	—	X	X X	X	Yes 3	Yes 3	No	No	No	No	No
ATA-Simazine	X X	X X	—	X X	X X	X	Yes 3	Yes 3	Yes 3	No	No	No	No
Diuron-Paraquat	X X	X X	X X	X	X X	X X	Yes 3	Yes 3	No	No	No	No	No
Metabenzthiazuron-ATA-MCPA	X X	X X	—	X X	X X	X	Yes 3	Yes 3	Yes 3	No	Yes	Yes	Yes
Phenobenzuron-ATA-Dalapon	X X	X X	—	—	X X	X	Yes 4	Yes 4	? ?	Yes 4	Yes 4	Yes 4	No

A   Treat only well developed plantations and on soil that is not very loose
B   Better not to use on very open soils, and use low doses on apricots
C   Use non volatile amine salt and treat under guidance
D   Use maximum of 1.5 kg of Simazine

Numbers: Age at which it can be applied XX Most suitable or most effective period X Useful action, but less than above. — Unsuitable or ineffective.

A/ Effectiveness of a major group of herbicides against frequent weeds of fruit plantations
B/ Summary of the application characteristics of different herbicides
C/ Effectiveness of some herbicides against dicotyledonous weeds

Machinery to apply agricultural chemicals to plants

Ⓒ

	Diuron	Metabenzthiazuron	Simazine
Amaranthus (pigweed)	S	S	SR
Capsella bursa pastoris (Shepherd's purse)	S	S	S
Chenopodium (goosefoot)	S	S	S
Cirsium (thistle)	SR	R	R
Papaver (poppy)	S	S	S
Euphorbia	—	—	S
Fumaria	SR	SR	S
Sonchus	SS	SR	SS
Convulvulus (bindweed)	R	SR	R
Portulaca	S	S	SR
Raphanus (wild radish)	S	S	S
Ranunculus arvensis	SS	SS	SR
Polygonum spp.	SS	SS	S
Veronica	R	S	S

In relation to the product and its applications

- You should not apply the same product more than three years in a row. Or better still, you should only use it two years in a row. The continued application of the same herbicide eliminates some weeds, but selects in favor of the forms that are more resistant.
- Apply the correct dose. The maximum dosage should only be applied the first year, after which lower doses should be administered.
- Never mix herbicides together, unless specifically instructed to do so on the label.
- Distribute the herbicide as evenly as possible over the site.
- Get to know the characteristics of each product, and fully respect each and every one of the safety norms recommended by the producer, and their recommendations for each type of site and crop.
- Store them under lock and key in places that are clearly labelled.

7.1.1. The properties of the most widely used herbicides

• Aminotriazol or ATA

This is a systemic herbicide that mainly penetrates through the leaves. It prevents chlorophyll formation and it takes a few weeks to act. It does not remain in the soil for long, 3 to 5 weeks.
It is effective against many species of annual and perennial plants. Very effective against grasses.
It is applied after emergence on actively growing plants at a dose of 4.5 kg of active material per hectare, though it is normally used combined with longer-lasting products, such as Diuron or Simazine.
Take care not to wet the green parts of the fruit tree when applying the herbicide.

• Atrazine

This is mainly absorbed by the roots. It acts by blocking the plant's respiration. It can remain in the soil for more than 6 months.
It is effective against most young plants, except crabgrass (*Digitaria*).
It can be applied before or after weed germination, at a dose of 2.5-3 kg of active material per hectare.
The only fruit trees it can be used on are apple trees more than four years old.

• Chlorthiamide

This agent is only active as a herbicide when it comes into contact with the ground, when it is transformed into another compound dichlobenyl.
It is absorbed slowly by the roots, but it shows good persistence in the soil.
It is effective against a large number of grasses and annual and perennial plants, but inactive against *Mercurialis* and *Sorghum*.
It is applied before emergence, though it can also be applied after emergence, at a dose of 6-8 kg active material per hectare.
It can not be used on pears, cherries or quinces. In apples, peaches, apricots and plums, it can only be used when the tree is more than three years old.

• Dalapon

This is a systemic herbicide, and is absorbed by the leaves.
It is effective against actively growing grasses.
It is applied after emergence, at a dose of 8 kg of active material per hectare. In general, it can be applied to all fruit trees, but only to cherries when they are more than 4 years old. This product is not widely used.

• DNOC, Dinitro-ortho-cresol

This systemic herbicide is absorbed through the leaves. It is fast acting and does not remain in the soil.
It acts on many annual and perennial plants, and is effective against *Convulvulus*.

Application of insecticides. The photo shows application in a pear orchard.

It is applied after emergence, at a dose of 1 kg active material per hectare, and when the temperature is below 25°C.
In general, it can be applied to all fruit trees without age restrictions.

• Diquat

This is a fast-acting contact herbicide. It destroys the chlorophyll of the green parts, and does not remain in the ground.
It is effective against young plants, though it is less effective against grasses than Paraquat.
It is applied after emergence, at a dose of 0.7-1.0%, to all types of fruit trees without age restrictions.

• Simazine

This is mainly absorbed through the leaves, and it blocks the plant's respiration. It is fast-acting and lasts in the soil for 6-12 weeks.

It is effective against many grasses, with an effect similar to Atrazine, but has the advantage of being more selective.

It is applied before emergence, though it can be used in early treatment after emergence, mixed with ATA. For best results, the soil should be reasonably moist.

The dose to be applied varies between 2 and 4 kg, depending on the tree's species and age.

• Bromacyl

This is mainly absorbed by the roots and acts by inhibiting the process of photosynthesis. It is short-lived in the soil, as it is broken down by microorganisms.

It is effective against a large number of plants and grasses.

It can be applied before or shortly after emergence, at a dose of 1.6 kg, or 2 kg if the tree is more than 4 years old.

It is only used on apples and pears on wild stock.

• Metabenzthiazuron, or Tribunil

This is absorbed by both leaves and roots. It only remains in the soil for a short period, 3-4 months.

It can be applied before emergence or shortly after emergence, at a dose of 1.8-2.8 kg of active material per hectare, depending on the type of weeds.

It can be used on all species and ages of tree.

• Paraquat

This is a contact herbicide that destroys the chlorophyll in the green parts of the plants. It is fast-acting and only remains a short time in the soil.

It is effective against young annual grasses. It is applied after germination, at a dose of 0.7-1.2 liters of active material per hectare. A wetting agent and a high volume of water should be added.

7.1.2. Mixtures of herbicides

The most common mixtures of herbicides are:

- ATA + Simazine
- ATA + Diuron
- Diuron + Paraquat
- Benzuride + ATA + Dalapon
- Metabenzthiazuron + ATA
- Metabenzthiazuron + Diuron
- Metabenzthiazuron + Simazine

These mixtures are effective against a very large number of weeds, but if any escape

their action, you should consider using a complementary herbicide.

Weeds showing resistance include:

- *Convulvulus* (bindweed). Can be eliminated by 2,4-D and MCPA.

- *Cirsium* (thistle). Can be eliminated by 2,4-D and MCPA.

- *Cynodon dactylon* (Bermuda grass) and *Sorghum.* They can be eliminated by treating the shoots with Dalopon.

- *Lolium,* or rye-grass. Before emergence, treat with Metabenzthiazuron, or ATA once it has emerged.

7.1.3. Other herbicides

Compounds of more recent origin are also important herbicides, but their properties are not so well known.

• Glyphosate

Absorbed by the leaves. Fast-acting and only remains in soil a short period.

Very effective against grasses and annual weeds.

Applied after emergence, after harvesting or during vegetative active growth, at a dose of 1.1-1.2 kg of active material per hectare. Suitable for all types and ages of fruit tree.

• Oxydiazon

Contact herbicide used before and after germination against many annual grasses. Effective against *Convulvulus* (bindweed).

• Carbethamide

Absorbed by the roots and effective against annual grasses. Application is in winter, at a dose of 3 kg of active material per hectare.

• Propyzamide

Absorbed by the roots and effective against annual and perennial grasses. Its activity depends on the soil humidity and temperature, which has to be medium-low. Applied before or after emergence, together with Diuron or Simazine.

• Terbacil

Absorbed by the roots and acts to inhibit photosynthesis.

It is compatible with stone fruits and is effective against many annual grasses and plants.

Phenobenzuron

)sorbed through the roots. Can be applied before
 soon after germination, and is effective against
nnual grasses. It is used mixed with ATA and
alapon.
 has little effect on stone fruit such as the plum,
•erry and peach and against some fruit trees, such
 apples and pears.

7.2. THINNING THE FRUIT

•ost fruit species produce more fruit than required
•r a good harvest.
•inning the fruit is done to prevent the branches
•eaking, to increase the size of the fruit, to improve
•, color and quality, and to stimulate the flower
•itiation that will produce the following year's
•arvest.
•creasing the ratio of leaves to fruit, by removing
•me of the fruit, means that the remaining fruit are
•rger. This gives rise to a decrease in production,
•it the fruit are larger.
•n appropriate balance between fruit size and
•oduction is 20-40 leaves per fruit.

INCREASE THINNING	— Young trees — Rain — High humidity — High maximum temperatures — Frosty nights — Soft water for spraying — Slow drought conditions — High fruit concentration — Low vigor — Narrow frame — Weakly pruned — Abundant flowering — Poor pollination — Addition of wetting agents — Abundant previous harvest
DECREASE THINNING	— Adult trees — Dry environment — Low humidity — Long periods of maximum temperature — Nights without frost — Hard water for spraying — Fast drought conditions — Low concentration — Moderate vigor — Large frame — Heavily pruned — Weak flowering — Good pollination — Non-addition of wetting agents — Poor previous harvest

•rly thinning stimulates flower production the
•llowing year in the varieties that tend to produce a
•w crop the year following a good crop. This
•ehavior is due to the absence of flower induction in
•e bearing years.
•e later the fruit is thinned, the smaller the resulting
•crease in the other fruit, and the intensity of
•inning is always dependent on the size desired by
•e market, and how abundant fruit-set was in the
•st place.

There are three methods of thinning.

• Manual thinning
This is removal of the fruit by hand. All small or
weak fruit should be removed, regardless of the
space they leave, as long as those left are not so
close together that they squash each other when
they grow.

• Mechanical thinning
- A high pressure jet of water shortly after flowering
- A stiff hair brush to knock off the small fruit
- A shaker of the same type as used for harvesting.
This has the disadvantage that it also shakes off the
larger fruit.

• Chemical thinning
Chemical thinning has the following advantages
over the preceding methods: lower costs, greater
fruit size and quality, and better regulation of
production in varieties that alternate.
Its disadvantages are: the risk of frost after early
application, over-thinning, damage to the leaves and
variable results, depending on the age and vigor of
the individual tree.

You are advised to try out the product on a small
number of trees before performing the first full-scale
application.

*Factors affecting
thinning*

• DNOC 4,6-dinitro-ortho-cresol. This destroys the
pollen and pistils and slows down growth of the
pollen tube.
• NAA (Naphthalene Acetic Acid), NAAm
(Naphthalene Acetamide), NPA (Naphthylplalamic
acid) and 3-CPA (2(3 chlorophenoxy)-propionamide.
They halt embryonic development, and cause fruit
to be shed.
• Sevin (1-naphthyl n-methyl carbonate). Prevents
movement of the growth compounds.
• Ethrel (2-chloroethyl phosphoric acid). Acts by
causing ethylene release in the tissues, which
stimulates abscission.

7.3 PLANT GROWTH REGULATORS IN FRUIT TREE CULTIVATION

The most direct action on a plant is to manipulate its
hormonal balance to achieve a given response.
A plant's functioning does not depend only on
the concentration of natural hormones, but
also on the balance between the different
hormones.
This equilibrium varies over the course of the fruit's
development. Thus the application of NAA shortly
after flowering causes thinning, while if applied
later, it prevents fruit fall in the period before
gathering. Not only is the time of application
important, but also its concentration.

7.3.1. The different growth regulators

The plant growth regulators, or plant hormones, are
organic compounds of natural or synthetic origin,
that in small concentrations accelerate,

modify or inhibit physiological processes within the plant.

Growth regulators, whether natural or not, can be divided into 5 groups that differ in their chemistry and effects on plants.

• Auxins

The auxins are the growth regulators that control the rate of elongation of the cells in the shoots.

They can cause or delay the abscission (shedding) of young fruit or delay the abscission of ripe fruit.

They can also stimulate the synthesis of ethylene in fruit, speeding up their ripening. Some auxins stimulate rooting in cuttings of many species.

• Gibberellins

On the next page: Growth regulators in pomology

All the gibberellins are natural products of the fungus *Gibberella fujikuroi*. They are derivatives of gibberellic acid (GA_3).

They act on cell division and elongation, break dormancy in seeds and buds, and it seems that, together with auxins, they prevent the abscission of young fruit. More than 50 compounds with gibberellin activity are now known. The most widely used ones include GA_3, GA_7, and Gaq (the subindex denotes different forms with minor structural differences).

• Cytokinins

Cytokinins are purine derivatives that stimulate cell division, apical dominance, branching and bud induction, accelerate seed germination, and prevent the senescence and abscission of flowers, fruit and leaves.

The following have cytokinin activity:
- BA (6-benzylamine purine)
- Kinetin (6-furfurylamino purine)
- 2 ip (6-benzylamino tetrahydroiranyl purine)
- Zeatin (6-hydroxymethyl butyrylamino purine)

• Ethylene

Ethylene has several interesting properties: it accelerates fruit ripening and color development, it promotes leaf and fruit abscission, its stimulates flower induction, and breaks dormancy in buds and seeds.

In fruit cultivation it is applied to the trees as products that then release ethylene. The most widely used is etefon (2-chloroethyl phosphoric acid), also known as Ethrel or CEPA.

• Growth inhibitors

Growth inhibitors inhibit or delay cell division and elongation in tissues.

- *Inhibitors* suppress cell division, meaning that growth stops.
- *Retardants* reduce growth without causing malformations and increase the green color of leaves, and flower induction.

7.3.2. The use of growth regulators

Plant growth regulators are used for a variety of purposes in fruit cultivation:

• Propagation of fruit trees

Synthetic auxins are good at stimulating rooting in cuttings and transplants, as well as stimulating grafts.

The application of gibberellins stimulates the germination of seeds for wild stocks in some specie such as citrus fruit.

• Dormancy

In some cases, they control dormancy. Spraying gibberellins onto peach trees, in late winter, advances sprouting, even if there have not been enough cold-hours.

• Flower induction

In seed fruit, the application of growth retardants or inhibitors produces an increase in flowering. Treatments with gibberellin during flowering provoke weaker flowering in the following spring.

• Fruit development

Auxins are used to stimulate fruit-set and to prevent shedding of flowers, and they also prevent fruit fall Gibberellins favor fruit-set and development, reducing shedding and parthenocarpic induction.

• Ripening and senescence

Gibberellins delay maturity and prevent skin aging in citrus fruit.

Some auxins are used to stimulate vegetative grow and delay shedding of fruit.

• Thinning

Products like synthetic auxins, ethrel, sevin and DNOC are used with good results in chemical thinning.

• Tree growth

Alor, CCC and etefon are used to control excessive growth.

To avoid the appearance of suckers when cutting back, paints with auxins (ANA) are used.

Use	Crop	Concentrations or quantity	Period of treatment
Gibberellins (GA³)			
Reduces effects of yellows virus	Sour cherry	15-25 ppm	10-15 days after petal fall
Delays ripening			
Larger and better-textured fruit	Cherry	5-10 ppm	3 weeks before harvest
Reduces cracking caused by rain			
Improves fruit size and shape	Apple	5-25 ppm	First petal fall
Improves fruit-set	Pear (some cultivars)	10-20 ppm	During flowering or on petal fall
Prevents premature ripening	Pear (Williams)	100 ppm	4 weeks before harvest
Improves fruit quality	Plum (Italian)	20-50 ppm	4-5 weeks before harvest
Improves fruit-set and size	Bilberry	10-50 ppm	Flowering-petal fall
Improves fruit and bunch size	Vine (Black Corinth)	2.5-5 ppm	Shortly after flowering
Increases fruit size	Vine (Sultanina)	2.5-20 ppm	Flowering
Produces looser bunches	Vine (Sultanina)	20-40 ppm	On fruit-set
Induces seedlessness	Vine (Delaware)	100 ppm	Before full flower
Increases size, advances ripening	Vine (Delaware)	100 ppm	On fruit-set
Produces looser bunches, less rot	Strains with compact bunches	1-10 ppm	2-3 weeks before full flowering
Increases seed germination	Apple, pear, cherry, hazel	5-100 ppm	Before germination
Delays bud break	Cherry, plum, peach	200-800 ppm	Start of fall
Accelerates defoliation	Nurseries	2,000 ppm	Before growth starts
Increases seed germination	Several species	100 - 500 ppm (24 hours immersión)	
Growth retardants (SADH, CCC)			
Increase fruit-set	Vine	100-1,000 ppm CCC	Leaf spray
Flower initiation	Pear	1,000 ppm CCC	40-50 days before full flowering
Flower initiation	Pear, apple	500-1,000 ppm SADH	30-40 days before full flowering
Control of growth	Apple	1,000-2,000 ppm SADH	30 days before full flowering
Prevents pre-harvest fall and improves quality	Apple	1,000-2,000 ppm SADH	45-60 days before full flowering
Delays flowering, increases fruit-set	Apple	4,000 ppm SADH	Spray in fall
Advances ripening, color and promotes fruit fall	Cherry, peach and plum	500-2,000 ppm SADH	2-5 weeks before full flowering
Delays flowering	Almond	2,000 - 4,000 ppm SADH	June, September, October
Increases fruit-set	Vine	2,000 ppm SADH	Start of flowering
Prevents premature ripening	Pear	SADH	30 days before harvest
Chemical pruning: branching	Pear	500 ppm SADH	Early summer
Auxins (ANA, NAD, AIB, 2,4-D, 2,4,5-TP)			
Chemical thinning	Apple	10-20 ppm ANA	15-25 days before full flowering
Chemical thinning	Pear	10-15 ppm ANA	15-21 days before full flowering
Chemical thinning	Apple	20-50 ppm NAD	7-14 days before full flowering
Increases fruit-set	Pear	2-7.5 ppm 2,4,5-TP	After harvest
Prevents early fruit fall	Pear	10 ppm ANA	3 weeks before harvest
Prevents early fruit fall	Apple	20 ppm ANA	3-4 weeks before harvest
Prevents early fruit fall	Apple	10 ppm 2,4,5-TP	5-6 weeks before harvest
Prevents early fruit fall	Apricot, plum (Italian)	5-20 ppm 2,4,5-TP	Before stone hardens
Reduces cracking caused by rain	Cherry	1 ppm ANA	35 days before harvest
Reduces production of suckers	Apple, pear, plum, cherry and hazel	1,000 ppm 2,4-D or ANA	Early summer
Increases fruit-set	Blackberry	β-NOAA++ 50-100 ppm	When fruit half size
Rooting of cuttings	Several species	Lavado con 20-200 ppm AIB++	Before callus formation
Rooting of cuttings	Several species	Inmersión rápida 500-5,000 ppm AIB	Before callus formation
Cytokinins (BA, Kinetin)			
Increases branching	Several species	100-200 ppm	Early summer
Increases fruit length	Apple	BA 25 ppm	Within 10 days of full flowering
Increases seed germination	Several species	100-500 ppm (24 hours immersión)	
Ethylene releasers (etefon)			
Flower initiation	Many species	100-1,000 ppm	Early summer
Chemical pruning	Plum (Italian)	200-500 ppm	Early summer
Chemical thinning	Peach, plum, apple	20-200 ppm	4-8 weeks after full flowering
Promotes ripening and color	Apple, fig	250-500 ppm	1-2 weeks before harvest
Accelerates husk shedding	Walnut	400-500 ppm	When husk starts to split
Accelerates husk splitting	Hazel	900-1,000 ppm	When fruits starts to open
Cause abscission and makes collection easier	Bilberry and gooseberry	500-2,000 ppm	10 days before harvest
Cause abscission and makes collection easier	Peach, cherry, plum, pear, apple	500-2,000 ppm	10 days before harvest
Cause abscission and makes collection easier	Vine	250 ppm	2 weeks before harvest

ppm = parts per million
The months are for the northern hemisphere

8. FERTILIZER APPLICATION AND IRRIGATION

The availability to the plant of nutrients in the soil depends on a series of factors, such as the climate, the type of soil, its moisture content, its pH, its nutrient and humus content, the species cultivated and the stock it is grown on.

In order for the plant to absorb the minerals from the soil, they have to be taken up from the soil solution into the cells within the roots.

The nutrients have to cross the **plasma membrane** into the xylem vessels, and from here the minerals are translocated to the rest of the plant.

In this process of absorption, the plant expends energy, which it obtains from respiration. This energy is required because the minerals are entering the root against the concentration gradient, because the concentration of minerals in the root is greater than that in the soil. This is why it is so important that the soil should have enough oxygen for the roots to respire.

8.1. ESSENTIAL ELEMENTS

A plant's needs are determined by the quantity of each element present in it. Thus the levels found on analysis of the plant should be taken as within a defined range of concentrations. Below this, there is a deficiency, and above it there is an excess, possibly even toxicity.

Analysis of the soil is important in establishing nutrient availability and its acidity or alkalinity, and analysis of the leaf is important because it shows the plant's nutritional status.

On the following page: Nutritional status of the leaves of several fruit species in high summer showing the levels of deficiency, normality and excess

Always remember that the idea is to feed the plant, not the soil. Analysis of the leaves gives a rough idea of the fertilizer requirement, and also helps to identify any disorder that might have occurred, whether toxicity or deficiency, and the corrective measures to take.

Note that the level of the different elements in the leaf varies over the course of the year, and that these standard values are only valid for leaves collected at the appropriate time and treated correctly.

An element is considered essential if it meets the following requirements:

• The lack of an essential element prevents full growth of the plant
• Lack or deficiency can only be corrected by supplying the element, which can not be replaced by any other.
• It must fulfil the two preceding points for a wide range of plants.

The following 16 elements are considered essential for plants: carbon, oxygen, hydrogen, nitrogen, phosphorus, potassium, calcium, magnesium, sulfur, iron, manganese, zinc, copper, molybdenum, boron and chlorine.

These elements are divided into 2 groups, depending on the amount used by the plant.

• **Macroelements** are required by plants in large amounts.
Carbon, hydrogen and oxygen are obtained from the air and water, while the rest are obtained from the soil. The other macroelements are nitrogen, phosphorus, potassium, calcium, magnesium and sulfur.
• **Microelements** are required by plants in very small amounts. Excess may be toxic, especially in the case of boron and copper. This group consists of iron, manganese, zinc, copper, molybdenum, boron, chlorine and sodium.

• **Characteristics of the main macroelements**

Nitrogen. This is generally found in the soil in organic form, because it is not a product of the weathering of soil minerals.
When organic matter is mineralized, nitrogen and ammonia are released, and this ammonia is rapidly oxidized to the nitrate ion (NO_3^-). This ion is easily taken up by plants, but may be lost by leaching.

• **Phosphorus**. Phosphorus is present in the soil as organic and inorganic compounds. Most of the inorganic compounds are effectively insoluble.
The amount of assimilable phosphorus in the soil solution is small in proportion to what the plant takes up. As phosphates are taken up from the soil solution, further phosphates in the soil dissolve into the soil solution. Phosphate is mainly present as calcium phosphate, which is slightly soluble.

Nutrient availability in soils. The broader the horizontal band representing each element, the more soluble it is. The element's solubility varies with the soil pH.

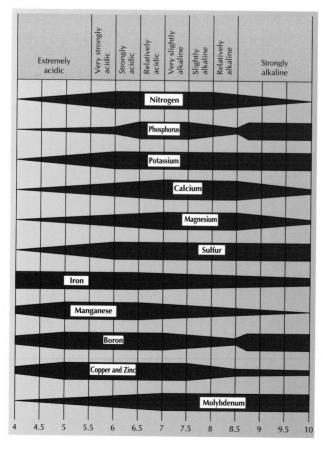

	Extremely acidic	Very strongly acidic	Strongly acidic	Relatively acidic	Very slightly alkaline	Slightly alkaline	Relatively alkaline	Strongly alkaline

Nitrogen
Phosphorus
Potassium
Calcium
Magnesium
Sulfur
Iron
Manganese
Boron
Copper and Zinc
Molybdenum

4   4.5   5   5.5   6   6.5   7   7.5   8   8.5   9   9.5   10

Potassium. Even if the soil contains large amounts of potassium, only a small part is taken up by the plant.

Minerals such as mica and feldspar release potassium slowly over time.

A small amount of potassium is present (dissolved in the soil solution), and this is absorbed by the plant. As this potassium is withdrawn from the soil solution, exchangeable potassium that was bound to clays and organic matters enters the soil solution.

Calcium. Calcium deficiency is unusual, except in acid soils, which are sometimes limed to raise their pH.

Magnesium. Magnesium deficiency is frequent in some zones, and this is usually remedied by applying magnesium sulfate at a rate of 150-500 kg/ha, depending on the crop and the variety.

Sulfur. There are several sources of sulfur in the soil: irrigation water, residues from sanitary treatments, organic matter and sulfur-containing fertilizers.

• **Characteristics of the main microelements**

Boron. The available boron in the soil can decline due to the losses caused by extraction by the crops, as well as losses due to leaching and reversion to forms that can not be assimilated by the plant. Its availability is diminished in soils with low humidity, as well as in over-limed soils.

Sodium. Production is negatively affected by a high content of soluble salts. If there is a high ratio of exchangeable sodium to the other exchangeable bases, the physical status of the soil becomes unsuitable.

In general, fruit trees have a very low tolerance of high salt levels.

Excess salinity can be washed from the soil with irrigation water, if this is of good quality, as long as enough is applied and drainage is adequate.

To eliminate excess exchangeable sodium, gypsum (calcium sulfate) is added, as this increases the percentage base saturation and improves the soil structure.

8.2. ORGANIC MATTER

The organic matter of the soil consists of all the plant and animal remains that have been decomposed and changed by microbial activity to a greater or lesser extent.

The soil's organic matter is derived from

• Remains of crops or weeds
• Input of dung or organic fertilizers
• Remains of living things, such as fungi, algae or bacteria, that live in the soil

The residues are decomposed by the actions of microorganisms and gradually transformed, and may follow two different processes.

Nutritional status[1]	N	K	P	Ca	Mg	Mn	Fe	Cu	B	Zn
						ppm dry weight				
Apple										
LN	1.5	0.9	0.08	0.20	0.18	20	40	1	30	10
N	2.0	1.2	0.12	1.0	0.24	25	50	4	35	18
AN	2.3	3.0	0.30	2.5	1.0	200	400	50	80	100
EX	3.5	4.0	0.70	3.0	2.0	450	500	100	100	200
Pear										
LN	1.9	0.4	0.08	0.20	0.18	20	40	1	30	10
N	2.2	0.7	0.12	1.0	0.24	25	50	4	35	18
AN	2.4	3.0	0.30	2.5	1.0	200	400	50	80	100
EX	3.5	4.0	0.70	3.0	2.0	450	500	100	100	200
Cherry										
LN	1.7	1.0	0.08	0.20	0.18	20	40	1	30	10
N	2.3	1.2	0.12	1.0	0.24	25	50	4	35	18
AN	2.6	3.0	0.30	2.5	1.0	200	400	50	80	100
EX	4.0	4.0	0.70	3.0	2.0	450	500	100	100	200
Peach										
LN	2.0	1.0	0.08	0.20	0.18	20	40	1	30	10
N	2.8	1.5	0.12	1.0	0.24	25	50	4	55	18
AN	3.8	3.0	0.30	2.5	1.0	200	400	50	80	100
EX	4.5	4.0	0.70	3.0	2.0	450	500	100	100	200
Plum										
LN	1.7	1.0	0.08	0.20	0.18	20	40	1	30	10
N	2.2	1.4	0.12	1.0	0.24	25	50	4	35	18
AN	2.5	3.0	0.30	2.5	1.0	200	400	50	80	100
EX	3.5	4.0	0.70	3.0	2.0	450	500	100	100	200
Hazel										
LN	1.8	0.4	0.08	0.20	0.18	20	40	1	30	10
N	2.2	0.7	0.12	1.0	0.24	25	50	4	35	18
AN	2.5	2.0	0.30	2.5	1.0	200	400	50	80	100
EX	3.5	3.0	0.70	3.0	2.0	450	500	100	100	200
Walnut										
LN	2.0	0.9	0.08	0.20	0.18	20	40	1	75	10
N	2.3	1.2	0.12	1.0	0.24	25	50	4	90	18
AN	2.8	2.0	0.30	2.5	1.0	200	400	50	100	100
EX	4.5	3.0	0.70	3.0	2.0	450	500	100	150	200
Pecan										
LN	1.6	0.4	0.08	0.20	0.18	140	40	1	40	10
N	2.3	1.0	0.12	0.7	0.30	200	75	4	60	18
AN	2.8	1.5	0.30	1.5	1.0	500	150	50	100	100
EX	3.0	2.0	0.70	3.0	2.0	1,000	300	100	600	200
Almond										
LN	1.5	1.0	0.08	0.20	0.25	20		2	30	10
N	2.4	1.5	0.12	1.0	0.50	75		10	35	25
AN	3.0	3.0	0.30	2.5	1.0	200		50	80	100
EX	4.0	4.0	0.70	4.0	2.0	450		100	100	200

LN = Less than normal
N = Normal
EN = Above normal
EX = Excess

Diagram of the decomposition of organic matter in the soil

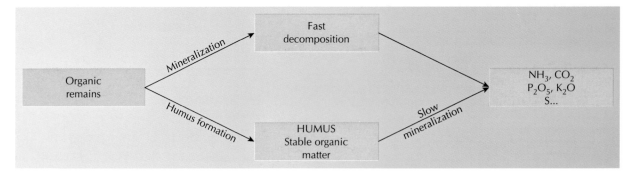

• **Mineralization**, when they are completely decomposed to mineral compounds like carbon dioxide, water, ammonia, phosphates, sulfates, etc.

• **Humus formation**, when they are transformed into organic complexes that are more stable and which decompose slowly. The stable portion of the organic mater is known as **humus**.
The nitrogen content of the humus seems to be related with complexes that decompose slowly. This is an advantage, because humus decomposition leads to slow release of nitrogen in a form that can be taken up by plant roots. Humus contains 3-6% nitrogen and 55-58% carbon.

Irrigation channel in citrus plantation

Fresh organic matter, green compost, incorporated into the soil is broken down by microbes, partly to minerals and partly to humus. It thus adds nutrients to the soil that can be more rapidly absorbed.
Most of the organic matter input, 60% to 70%, is actively mineralized within about 2 years, while the rest becomes humus, and mineralizes very slowly, depending on the climatic and soil conditions.

The conditions that favour mineralization rather than humus formation are:

• Good aeration
• High temperature
• Sufficient moisture
• Plant remains rich in nitrogen are easily broken down by microorganisms.
This is why wet and cold areas have more organic matter in their soils than hot zones.

8.3. INTERACTIONS BETWEEN ESSENTIAL ELEMENTS

An essential element may affect the use or uptake of another. There are two main types of interaction:

• **Antagonistic**, when one element acts against, or diminishes the effect of, another.
• **Synergistic**, when one element acts to increase the effect of another.

Antagonistic interactions may not be reciprocal. For example, an increase in the level of nitrogen in the soil decreases phosphorus uptake, and an increase in phosphorus likewise reduces uptake of nitrogen; an increase in nitrogen in a soil poor in boron seriously reduces boron uptake, but a lot of boron reduces use of nitrogen.
Some antagonistic interactions may have beneficial effects, such as the application of excess sulfur (sulfate, SO_4^{2-}) to young trees. This excess decreases the uptake of arsenic, which is poisonous to the plant.
Another example is the example of adding calcium or magnesium to irrigation water in order to decrease the absorption of soluble copper.
Manganese is an antagonist of iron. Excess manganese in the soil converts iron into a form that cannot be absorbed by the plant, leading to chlorosis caused by iron deficiency.

Average contents of the organic materials most frequently used as fertilizers

MATERIAL	N	P_2O_5	K_2O	Organic matter	Reaction
Stable dung	0.4	0.2	0.40	30	A
Corral dung	0.7	0.34	0.65	60	A
Sheep dung	1.0	0.3	1.0	60	A
Pig dung	0.5	0.3	0.65	60	A
Chicken, etc., droppings	1.6	1.25	0.9	50	B
Alfalfa (dry)	2.5	0.5	2.1	85	—
Alfalfa straw	1.5	0.3	1.5	82	—
Cereal straw	0.6	0.2	1.1	80	—
Wool clippings	0.8	1.2	—	—	—
Dried blood	13.0	2.0	1.0	80	A
Hoof and horn	7.15	—	—	—	—
Dry sewage	2.0	2.0	—	—	A
Cotton seedcake	7.0	3.0	2.0	80	A
Air-dried seaweed	1.5	0.5	2.0	—	—
Dried peat	2.0	—	—	—	—

(Nutrient elements %)

(A) Acidic
(B) Basic

However, if there is insufficient manganese in the soil, the amount of iron that can be assimilated increases to levels that may even be toxic. The plant also suffers from a manganese deficiency, which leads to chlorosis.

A high level of manganese also reduces the absorption of nitrogen.

You should realize that the different types of chlorosis that may appear in the plant as a result of a lack of any one of the elements iron, manganese, magnesium, copper or potassium, may be due to a lack of the element in the soil, but it may also be due to its blockage by another element present in the soil.

Thus, iron (Fe^{2+}) is precipitated by a high concentration of phosphorus (PO_4^{3-}). The plant's roots are thus surrounded by iron they can not use. There may be interactions between more than two elements. Thus, potassium negatively affects magnesium uptake when there are high levels of nitrogen, but not if the levels are low.

8.4. FERTILIZERS

In theory, fertilizer is applied when the plant needs more elements than it can take up from the soil. Nitrogen is the element most applied as a fertilizer. The other elements only need to be applied to restore what has been extracted in the harvest, or when there is a clear need for them.

Because complex N-P-K fertilizers are in general much more expensive than simple fertilizers, ones with a single element, they should only be used when all three elements are necessary. Even so, in most cases you can just mix together three simple fertilizers.

In the case of phosphorus, the normal process of soil weathering means that small quantities of phosphorus are available.

Yet if you have to apply phosphorus, the best form is as calcium superphosphates. If you need to apply potassium, the best form is usually as potassium sulfate.

• Superphosphates

There are three formulations:
• Normal calcium superphosphate, with 16-18% P_2O_5.
• Double, or enriched, calcium superphosphate, with 25% P_2O_5.
• Triple, or concentrated, calcium superphosphate, with 36% P_2O_5.

They are all soluble fertilizers that are very readily taken up and which do not affect the soil if they are applied in limy soils.

When choosing from a set of nitrogen fertilizers, you should base your choice on an important factor; the price per fertilizer unit.

In addition, the effects of nitrogen fertilization on soil pH and on ion balance should be taken into account.

In general, fertilizers that reduce the pH should not be used, especially in soils with a pH value of 5.5 or less. If you want to raise the soil pH a little, use $Ca(NO_3)_2$ (lime saltpeter).

Soils with a pH value of 7.5 or greater respond well to acidifying fertilizers such as $(NH_4)_2SO_4$, which neutralize part of the soil alkalinity.

Some soils show deficiencies in some of the different elements that are essential for plant growth. There may be deficiencies of nitrogen, boron, zinc, iron, potassium, magnesium, manganese and in a few cases, phosphorus and sulfur.

Every plantation requires different elements and in different proportions, so that previous mixtures of different fertilizers should not be used. The different elements should be applied separately as required. You are advised not to use a mineral element unless there is convincing proof that it is necessary.

Apart from nitrogen, phosphorus and potassium, the other elements are not applied in deep fertilization, but you should wait until the fruit trees show deficiency symptoms and then correct them by foliar application of fertilizers. This is always cheaper than applying them in the initial deep fertilization.

Organic inputs may provide sufficient amounts of some elements, such as sulfur, zinc, copper, manganese and molybdenum.

There are three elements that are however, worth including in the deep fertilization, magnesium, boron and iron. These deep inputs are best carried out separately from the deep fertilization mentioned above.

• Magnesium deficiency

This is frequent in acid soils or those that have received large applications of potassium fertilizers. Some species of fruit trees, such as apples, are especially sensitive to magnesium deficiency. The deep input recommended is of the order of 100 kg/ha of MgO using magnesium sulfate (16% MgO). A different answer, if you are going to lime as a remedy anyway, is to apply dolomite or magnesium-rich limestone.

• Boron deficiency

This is frequent on acid soils and on soils that have been irrigated for a long time to grow crops that require a lot of boron. The deep input should be in the form of sodium borate (borax, 11.4% boron) at a dose of 20-40 kg/ha, or boracite (14.2% boron) at a dose of 15-30 kg/ha.

• Iron deficiency or ferric chlorosis

This is frequent in fruit tree plantations. One of the causes is usually the lack of soluble iron in the soil. To remedy this, iron compounds are applied in deep fertilization. Ferrous (Fe^{2+}) sulfate can be used in the form of large crystals to slow down its oxidation (to Fe^{3+}), at a dose of 300-500 kg/ha, though it may cause toxicity.

Using iron chelates is much more effective, as they are rapidly fixed in the soil.

The recommended form of application is to spray the chelates onto the soil as a powder and then irrigate. The doses used are roughly 1.5-2.0 kg iron/ha, and this gives better results than applying the solid to the soil.

8.4.1. Corrections

Treatment seeks to correct the initial characteristics of the soil. The two most common treatments in our soils are:

Application of potassium

• **Adding organic matter**, to improve the level of organic matter in the soil. To cultivate fruit trees, there should be roughly 2-3% organic matter in the soil.

However, achieving these levels may require large inputs, which are expensive.

Normal applications of dung range from 30-60 t/ha up to 80-100 t/ha in small plots that are irrigated.

The dung or organic matter most frequently used is cattle dung or sheep droppings, with a greater or lesser proportion of straw bedding in different areas. Hog dung is the cheapest, but it produces the least humus and is relatively acid. Chicken droppings, and bird droppings in general, tend to increase alkalinity and decompose slowly.

• **Liming for acid soils with a pH below 6**. In soils with a pH between 6 and 7, symptoms of calcium and magnesium deficiency may appear, making treatment necessary.

Good results are obtained with crushed limestone (calcium carbonate), dolomite (a natural mixture of calcium and magnesium carbonates), and quicklime (calcium oxide) or slaked lime (calcium hydroxide).

You should not try to raise the pH value by more than 1.0 in a year, if it is below 6.

Liming should be performed immediately after dunging, once the dung has been buried. Limestone is buried separately. In order to ensure good distribution, it should be crushed and very dry.

8.4.2. Deep fertilization

Fertilizer application at depth completes the process of planting the fruit tree. It is carried out before planting, when it is easier to apply the fertilizers, spread them and distribute the nutrients, and bury it all at a greater depth. All this is much harder once the trees have been planted.

The main aims of this application are:
• To correct the soil's possible deficiencies.
• To establish an appropriate level of fertility.
• To create a reserve of nutrients to ensure the tree grows well during its first years.

Application of phosphorus

This fertilizer application should consist of

P in reserve ppm	pH value	Cultivation	Fertilization at depth Kg P_2O_5/Ha	Observation
> 5	Any	Any	Unnecessary	Monitoring with analysis every three years
	< 5.5	Dry farming	150	Lime to pH 5.5
		Irrigation	300	or greater
<5	5.5-7.8	Dry farming	150	—
		Irrigation	300	—
	> 7.8	Dry farming	400	—
		Irrigation	600	—

For phosphorus, the norm to follow might be:

K in reserve ppm	Cultivation	Deeo fertilization Kg K_2O/Ha	Observations
> 20	Any	Unnecessary	—
10-20	Dry farming	Unnecessary	Check, analysis every 2/3 years
	Irrigation	200	
< 10	Dry farming	100	—
	Irrigation	400	—

For potassium, the norm to follow might be:

phosphate and potassium fertilizers. The organic matter previously applied acts as a source of nitrogen. There is no reason to apply nitrogen fertilizers, as the time of year when this application is performed means that the rain would wash it down into the soil, where it would be lost.

The amount of P_2O_5 and K_2O that has to be applied in this deep fertilization varies from soil to soil, depending on its reserves and its pH value.

In general, preliminary fertilization at depth in a fruit plantation should be calculated on the basis of 200-600 kg/ha of P_2O_5 and 200-400 kg/ha K_2O, depending on what is needed.

Greater inputs are not recommended, nor are they worthwhile.

8.4.3. Fertilization of some fruit species

• **Applying fertilizer to stone and seed fruit**

Their requirements vary from species to species. The amount of a given mineral element extracted from the soil is hard to calculate, as this varies with the species and variety, and also with the management and cultivation techniques used.

The time and method of fertilization application in these crops can be as follows:

• Phosphorus-potassium fertilizer, together with part of the nitrogen fertilizer, is applied in fall/spring, at a depth of 15-50 cm, in strips 80-90 cm from the tree's trunk in young plantations, and at a greater distance and depth in established plantations.
• The rest of the nitrogen fertilizer (complementary fertilizer) is applied in the periods of maximum growth in young plantations, and before sprouting, and after harvesting in established plantations.

• **Applying fertilizer to citrus trees**

Citrus fruit require intensive fertilization with nitrogen (up to 400-500 kg/ha).

Citrus fruit frequently show deficiencies of elements like iron, zinc, manganese and copper. They may also show symptoms of magnesium deficiency due to excess potassium.

In general, uptake of soil nutrients by citrus fruit occurs all year long, depending on the temperature.

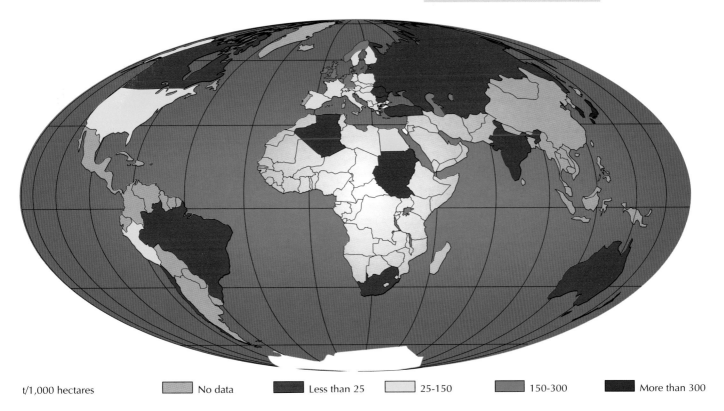

t/1,000 hectares   ☐ No data   ■ Less than 25   ☐ 25-150   ■ 150-300   ■ More than 300

he dose of fertilizer might vary between:

300-400 kg/ha N
50-100 kg/ha P_2O_5
50-150 kg/ha K_2O

he time of application depends on the different species and varieties:

Early variety orange. Application is in late winter/early spring, with 60-70% of the ertilizer and the rest in mid summer.
Mid variety orange. The first application is carried out after harvesting and the second at he end of summer, following the same schedule.
Late variety orange. Fertilizer can be applied three or four times, the first time in ate winter/early spring, using 20% of the ertilizer, the second time after harvest, with 40-50%, and the rest divided between late summer and mid/late winter.
Medium variety lemon. Fertilizer is applied twice, once in late winter/early spring, and again in mid summer, with a larger application in spring.
Late variety lemon. Fertilizer is applied 3 times, the first two of them as above, and the hird in mid fall.

• Applying fertilizer to almonds

Fertilizer is applied after harvest, and is complemented in spring, after flowering, with an application of nitrogen.

The doses are calculated as a function of the soil's richness and vary between:

80-120 kg/ha N
50-150 kg/ha P_2O_5
80-160 kg/ha K_2O

• Applying fertilizer to olives

The hardening of the stone is a critical moment when the tree's reserves are exhausted in the production of the current crop, and accumulation for reserves for the following season starts.
The factor limiting this crop is water, and so the amount of fertilizer applied has to be established in accordance with the possible production permitted by the rainfall.
The doses of fertilizer depend on the production and are calculated as follows:

• For medium to low production of 20-50 kg (calculated per tree):

0.4-0.6 kg N
0.3-0.5 kg P_2O_5
0.3-0.6 kg K_2O

• For production higher than 50 kg (per tree):

1.0-1.5 kg N
0.5-1.0 kg P_2O_5
0.7-1.5 kg K_2O

Fertilization at depth, with half the nitrogen and all the phosphorus and potassium, should be carried out in fall, and be located at a depth of 20-30 cm as close as possible to the trees without damaging their main roots. The rest of the nitrogen is applied at the end of winter to meet the tree's needs in spring.

Zones of humidity produced by drip irrigation

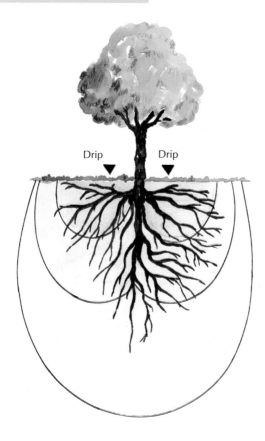

8.5. IRRIGATING FRUIT TREES

The tree's behaviour in relation to the soil water, depends mainly on the site's characteristics, the frequency and intensity of irrigation, the depth the water penetrates the soil and the spread of the root system.

The most widely used irrigation systems in fruit trees are irrigation in furrows, sprays and drip irrigation.

The following points should be borne in mind:

• **Irrigation along furrows** is not suitable on sites with steep slopes, unless it follows the contours.

• Applying fertilizer to vineyards

Water is the main factor limiting vines. Pruning restricts the vine's growth to adapt it to local water availability. This may be why vines show little response to fertilizer application.
Vines accumulate reserves in their stems, roots and the shoots of the stem, and these determine the next year's growth and harvest. Fertilization application should thus not only meet the needs of the crop, but also allow for the formation of reserves.
The vine shows two peaks in the uptake of mineral elements, the first after flowering, and the second before the winter growth halt. Nitrogen directly affects the plant's vigor and production, and excess is detrimental to quality, meaning that it should be applied with caution. Nitrogen deficiency, however, may lead to flower fall. Potassium shows peak absorption after flowering, and can be absorbed in excess without affecting production or quality.

The fertilizer dose depends on the area's characteristics and, of course, on the soil's fertility. In general, the dose per hectare might vary between:

$$30\text{-}70 \text{ kg N}$$
$$30\text{-}80 \text{ kg } P_2O_5$$
$$40\text{-}120 \text{ kg } K_2O$$

Drip irrigation system in apples and carob trees

The fertilizer should be buried between the lanes, at a depth of roughly 20-25 cm, as most of the roots are at a depth of 20-40 cm. Application is in late winter/early spring.

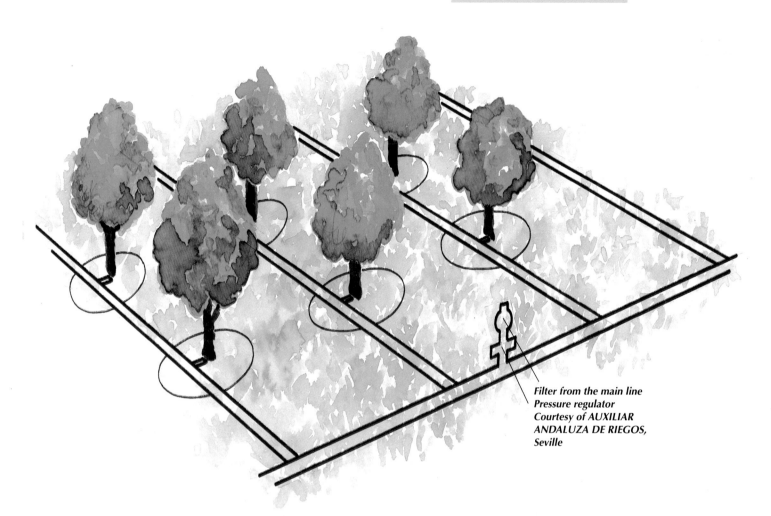

Filter from the main line
Pressure regulator
Courtesy of AUXILIAR
ANDALUZA DE RIEGOS,
Seville

• **Spray irrigation**, is suitable for a wide range of soils and reliefs. It is especially useful in abrupt sites that can not be levelled or are very steep, that have shallow soil and are easily eroded.
One of the advantages of spraying is that it also protects the plant against excessive heat or cold. The spray can be fixed, use perforated tubing or may be rotatory (a sprinkler).

• **Drip irrigation**, or **trickle system**, maintains continuous humidity in the zone around the root system. It supplies water at low pressure at a dose of 3-8 l/h per plant, a large saving of water.
A drip irrigation system consists of a pump, filters, a solenoid valve controlled by a timer, a pressure regulator, plastic tubing for the main distribution, capillary tubing with individual tricklers and devices to measure soil moisture.

Drip irrigation has a series of major advantages:

• Lower water consumption
• Uniform distribution of the water in sites with abrupt relief
• Makes use of machinery easier
• Makes weed control easier

Design of an irrigation system. The main line, equipped with a filter and a pressure regulator, supplies the distribution branch which supplies the lateral feeders, which have one or two tricklers per tree.

Different spray systems

9. FLOWERING, POLLINATION AND FRUIT GROWTH

The main aim of every fruit plantation is to obtain fruit. To ensure production is profitable, it must meet minimum requirements of quantity and quality.

The fruit is the result of the growth of the flower, meaning that production of many fruit requires production of many flowers.

The farmer's intervention seeks to obtain favorable conditions for the tree to produce as many flowers as possible, in accordance with its fruiting habits. This ensures good production.

Flower induction is a physiological change that occurs at a given moment and causes the bud to develop into a flower instead of a vegetative shoot. This physiological change is controlled by a series of internal and external stimuli.

Flower induction is followed by morphological differentiation leading to flower formation, and the primordia that will give rise to the sepals; then the petals, stamens and pistils all develop within the flower bud.

Flower induction is thus clearly very important, as are ways of influencing it and increasing the number of flower buds.

There has been a great deal of research into flower induction, but there are still no definitive conclusions.

Flowering
A/ Pistil
B/ Flower
C/ Stamen

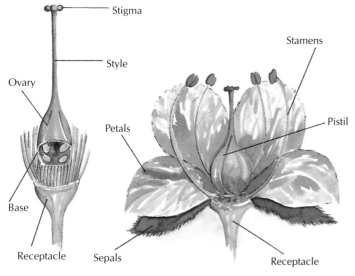

Parts of a complete flower

Ⓐ

Ⓑ

On the right:
Pollination
Pollen grain germination on the stigma

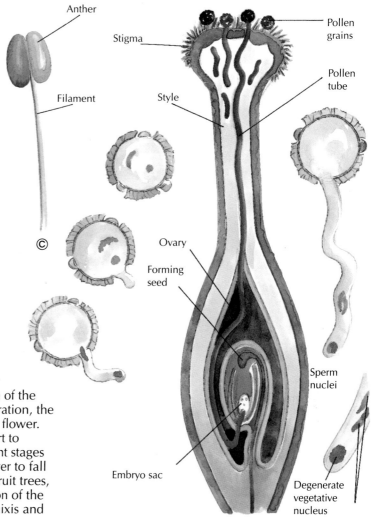

Ⓒ

9.1. FLOWERING

The tree's flowering cycle starts with flower induction and is followed by differentiation of the flower bud. After a period of rest and maturation, the bud sprouts the next spring, giving rise to a flower. Flowering begins when the flower buds start to open. This is followed by a series of different stages that lead to **fruit-set**, which causes the flower to fall and starts the fruit's development. In most fruit trees, fruit formation is the result of the fertilization of the flower, though there are also cases of apomixis and parthenocarpy.

The flowering process consists of the following stages

- Pollen formation
- Pollination
- Pollen germination
- Growth of the pollen tube
- Fertilization

9.1.1. Flower induction

The formation of a flower bud starts when flowering is induced.

• Hypotheses to explain flower induction

The first hypotheses considered that the flower formation depends on the presence in the plant of unknown substances produced in the leaf (J. Sachs, H. Müller-Thurgau, O. Loen and others). Later, flower induction was thought to be due to the a given balance between the level of carbohydrates and mineral salts (G. Klebs, Kraus and Kraybill). They thought that if carbohydrates dominate, then flower formation is abundant, and if mineral salts dominate, then vegetative growth dominates.

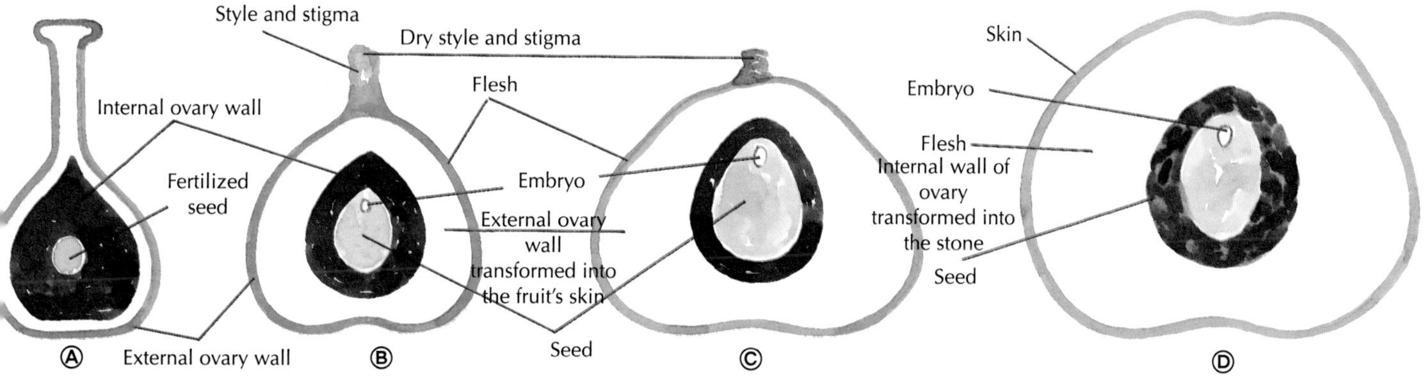

Diagram labels (A–D):

A: Style and stigma / Internal ovary wall / Fertilized seed / External ovary wall

B: Dry style and stigma / Flesh / Embryo / External ovary wall transformed into the fruit's skin / Seed

C: (fruit cross-section)

D: Skin / Embryo / Flesh / Internal wall of ovary transformed into the stone / Seed

Fruit development. How a fleshy fruit forms

ter work by Hooker, Petter and Phillips ontradicted the previous theory, and postulated the esence of a hormonal substance produced in the af.
ae currently accepted theory is that flower duction is based on a given hormonal balance ithin the bud. This balance is determined by a set interacting factors (environmental, nutritional, nysiological and genetic).

Flower induction period

principle, an endogenous balance is quired that is only achieved when the tree as accumulated a certain level of reserves, at is to say, when the tree has reached a ertain age.
deciduous fruit trees, induction occurs at the end the spring growth period of the year before owering.
evergreen fruit trees, induction occurs after the inter cold has passed, in the year before the flower produced.
nere are special cases, such as the fig, which as two flower induction periods. The first is in te spring and gives rise to flowers in the same ear (which will form true figs), and the second at the end of summer and gives rise to owers in the following spring (early figs).

Factors that influence flower induction

ower induction is the result of the action of a set of ctors of different natures that act upon the ormonal balance within the bud.

Nutritional factors

any species flower more if they receive an pplication of nutrients such as nitrogen or nosphorus, potassium, trace elements, and this also epends on the time of application.
is important to maintain a high level of arbohydrates in order to obtain a large number of ower buds, though this does not determine the oundance of flowering.

Environmental factors

ertain environmental factors may influence ower induction. Thus, lack of sunshine leads to ss flower buds, and a period of drought llowed by heavy irrigation favours the oduction of flower buds.

In relation to the temperature, it is known that for good flower induction some deciduous species need to accumulate a given number of cold-hours in winter.

• Cultivation factors

In a young plant, fertilizers rich in nitrogen favor vegetative growth and diminish flower induction, while in the adult plant, if the application is split over the year, production of flower buds is increased.
Lack of sunshine affects flower induction, and so pruning should aim to increase the light levels throughout the tree. Pruning is also carried out to reduce excessive leaf area, as this has a negative effect on flower induction.
Other complementary techniques, such as bark-ringing and bending the branch downwards favor the accumulation of carbohydrates, and thus greater production of flower buds.
The use of growth retardants also increases flower production. Both CCC and Alar (SADH) increase flower induction in apples, pears and citrus fruits, but to a variable extent in peaches.
Other growth regulators, such as auxins or cytokinins, do not have effects on flower induction, while the gibberellins inhibit flower induction.

9.1.2. The flowering period

Flowering time may determine the variety you should select, especially in zones with a frost risk period.
The flowering time is characteristic of each variety, though it may be affected by environmental and cultural factors.
It is important to be aware of the flowering time of the different varieties in a plantation when it comes to choosing pollinator varieties.

The length of flowering is taken to be 10-25 days, and full flowering is when 50-90% of the flowers are open.

9.1.3. Pollination

Woody species may have one of two forms of sexuality:

• **Species with separate male flowers and female flowers**
- *On the same tree (monoecious)*, such as the walnut and hazel.

Pollinator varieties

	Good pollinator varieties	Bad pollinator varieties
Pear	Butter Gifford, Dr. Jules Guyot, Butter Early-Moretini Castell, William, Butter Hardy, Doyenne du Congress, Conference, Passe Crassane	Ercolini White Aranjuez Roma (Aragón)
Apple	Golden Delicious, Rome Beauty, Macintosh, King of Pippins, Green Doncella, Starking Delicious	Canadian White Pippin Stayman Winesap
Apricot		Usually there are no problems.
Almond		All varieties should be considered totally or partially self-incompatible. Cross-pollination is recommended.
Cherry		Almost all varieties should be considered to be self-sterile. Cross-pollination is recommended.
Plum		Cross-pollination is recommended.
Peach		There are no problems in fertilization. The varieties are self-fertile.

- *On different trees (dioecious)*, such as the pistachio, carob, kiwi fruit, date palm and palm.

• **Species with hermaphroditic flowers**. These are the large majority and include almond, cherry, apple, peach, pear, orange, olive, and vine.
It is important to realize that pollination is a different process to fertilization.
Pollination is the transfer of the pollen grain from the anthers of the stamens to the stigma, while fertilization is the fusion of two gametes, a female gamete located in the embryo sac and a male gamete within the growing pollen tube. The pollination process starts when the mature anthers open and expose the pollen grains, and it finalizes when the pollen grains reach the stigma.

There are several types of pollination:

• **Depending on the source of the pollen**:
- *Allogamous fertilization*, or *outbreeding*, when the pollen is from another variety of the same species. These species may be self-sterile, that is to say, they can not fertilize themselves with their own pollen. They may, however, be partially self-sterile or partially self-fertile.
- *Autogamous pollination*, or *self-pollination*, when they are pollinated with their own pollen. These species are self-fertile.

• **Depending on the way that pollen is transferred:**
- *Wind pollination*, when the pollen is transferred by the wind, as in hazel, walnut, olive, sweet chestnut, pistachio and vine.

Planting pollinators
A/ Excess to pollination requirements, and only recommended in very unfavorable conditions (climate, variety, number of insects.
B/, C/, D/, E/ Recommended in normal conditions.
C/ The pollinators are planted in complete rows. This does not give rise to problems due to differences in vigor and longevity within a single line, and facilitates harvesting and the application of treatments.
F/ Percentage considered insufficient, unless the conditions mentioned above are ideal. The third variant offers a better distribution, as it reduces the distances between pollinators and the other trees.

1 — With a single pollinator variety

A. 50% pollinators

```
X  O  X  O  X  O
X  O  X  O  X  O
X  O  X  O  X  O
X  O  X  O  X  O
X  O  X  O  X  O
X  O  X  O  X  O
X  O  X  O  X  O
```

B. 33% pollinators

```
X  O  O  X  O  O
X  O  O  X  O  O
X  O  O  X  O  O
X  O  O  X  O  O
X  O  O  X  O  O
X  O  O  X  O  O
X  O  O  X  O  O
```

C. 25% pollinators in complete rows

```
O  O  O  X  O  O  O  X
O  O  O  X  O  O  O  X
O  O  O  X  O  O  O  X
O  O  O  X  O  O  O  X
O  O  O  X  O  O  O  X
O  O  O  X  O  O  O  X
O  O  O  X  O  O  O  X
```

D. 25% pollinators

```
O  X  O  X  O  X
O  O  O  O  O  O
O  X  O  X  O  X
O  O  O  O  O  O
O  X  O  X  O  X
O  O  O  O  O  O
O  X  O  X  O  X
O  O  O  O  O  O
```

E. 16% pollinators

```
O  O  X  O  O  X
O  O  O  O  O  O
O  O  X  O  O  X
O  O  O  O  O  O
O  O  X  O  O  X
O  O  O  O  O  O
O  O  X  O  O  X
O  O  O  O  O  O
```

F. 10% pollinators (three different arrangements of pollinator trees)

```
O  O  X  O  O      O  X  O  X      O  X  O  O
O  O  O  X  O      O  O  O  O      O  O  O  O
O  O  X  O  O      O  O  O  O      O  O  O  X
O  O  O  X  O      O  O  O  O      O  O  O  O
O  O  X  O  O      O  O  O  O      O  O  O  O
O  O  O  X  O      O  X  O  X      O  X  O  O
O  O  X  O  O      O  O  O  O      O  O  O  O
O  O  O  X  O      O  O  O  O      O  O  O  X
                   O  O  O  O      O  O  O  O
                   O  O  O  O      O  O  O  O
```

2 — With two pollinator varieties

A. 33% pollinators (16.5% + 16.5%) with two different arrangements of pollinator varieties

```
X  V  V  O  V  V  X      X  V  V  X  V  V
X  V  V  O  V  V  X      O  V  V  O  V  V
X  V  V  O  V  V  X      X  V  V  X  V  V
X  V  V  O  V  V  X      O  V  V  O  V  V
X  V  V  O  V  V  X      X  V  V  X  V  V
X  V  V  O  V  V  X      O  V  V  O  V  V
X  V  V  O  V  V  X      X  V  V  X  V  V
X  V  V  O  V  V  X      O  V  V  O  V  V
```

B. 50% pollinators (25% + 25%) with two different arrangements of pollinator varieties

```
O  Y  X  Y  O  Y  X      O  Y  O  Y
O  Y  X  Y  O  Y  X      X  Y  X  Y
O  Y  X  Y  O  Y  X      O  Y  O  Y
O  Y  X  Y  O  Y  X      X  Y  X  Y
O  Y  X  Y  O  Y  X      O  Y  O  Y
O  Y  X  Y  O  Y  X      X  Y  X  Y
O  Y  X  Y  O  Y  X      O  Y  O  Y
O  Y  X  Y  O  Y  X      X  Y  X  Y
```

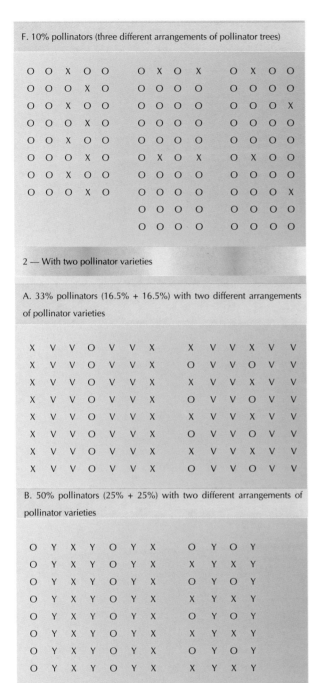

- *Insect pollination*, when insects, mainly bees and wasps, transfer the pollen, as occurs in apricot, almond, cherry, plum, apple, peach and pear.

These external agents that transfer the pollen are called **pollinators**.
The range of pollinator insects is much less than the wind, which may be up to 10 km.

Bees show the following characteristics:

• *Constancy*. If they frequent the flowers of a given variety, they will do so until it finishes flowering.
• *Flights between plants of 10-18 m*, meaning this should be the maximum distance between pollinator trees.
• *Distance from the hive*. They are most active in a radius of 250 m, so two hives per hectare are recommended.
• *Hive conditions*. Bees do not fly if there is a strong wind, rain or temperatures below 10°C on clear days, and have an optimum temperature of 25°C.

It should be kept in mind that many necessary insecticides are also toxic to bees, meaning that you must take great caution and use the correct dose of the active material that is least dangerous to bees but which ensures adequate pest control, or use biological methods, whenever this is possible. The best policy is to avoid any type of treatment during the flowering period, as they are usually not effective and waste time and diminish production.

Pesticides and products relatively non-toxic to bees and other pollinating insects

FUNGICIDES	Cihexastan	ANA (Rodofix Phymone)
Sulfur	Dicofol	NAD (Amid-thin)
Benomyl	Fenbutestan	Deminozide (Alar)
Binapacril	Tetradiphon	Chlorocholine chloride
Bordeaux mix	Triciclestan	(Cycocel)
Captafol	Tetradifon Dicofol	
Captan		**HERBICIDES**
Carbendazime	**INSECTICIDES**	Bromacil
Chlortalonil	Mineral oils	Dichlobenyl
Dichlofuamide	Bacillus thuringiensis	Diquat
Dinocap	Bromophos	2, 4-D
Ditalinfos	Endosulfan	Diuron
Dithianone	Ethiophencarb	EPIC
Dodin	Ethión	Glyphorate
Folpet	Fosalone	Linuron
Glycophene	Menazon (Azidition)	Oxidiazón
	Synergized Pyrethrum	Paraquat
ACARICIDES	Pirimicarb	Promethrine
Amitraz		Simazine
Benzoxymate	**PLANT HORMONES**	Terbacil
Bromopropilate	Gibberellic acid	Trifluraline

Bigger harvests

**Flowering habits of
some fruit trees**

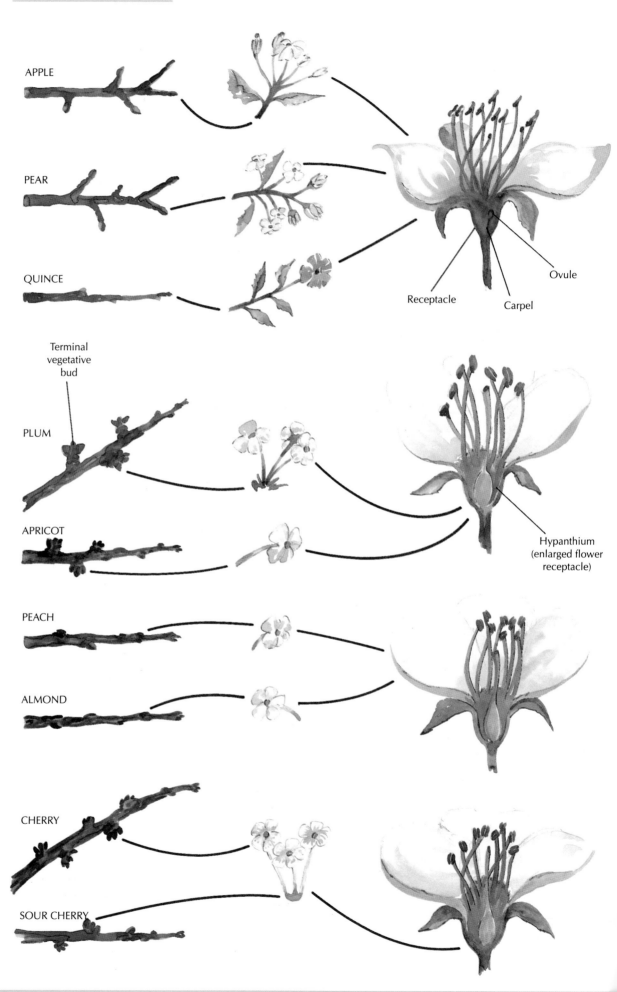

APPLE

PEAR

QUINCE

Receptacle

Carpel

Ovule

Terminal
vegetative
bud

PLUM

APRICOT

Hypanthium
(enlarged flower
receptacle)

PEACH

ALMOND

CHERRY

SOUR CHERRY

It is also important to understand that **pollinator** trees may be required. These trees are of different varieties and produce pollen to ensure cross-fertilization of the main variety planted. The framework for planting the pollinators is one out of every three rows. This ensures a suitable proportion.

If the fruit is small in size, such as cherries and almonds, it is worth increasing this proportion of pollinator trees, in order to ensure good production.

To ensure optimum pollination, the flowering period of the pollinator should start shortly before that of the main variety. Wild varieties that are good pollen producers can be used. The most important thing is that the two varieties should be compatible, and start flower production at the same age.

Two pollinator varieties may be planted, because:

• Pollination is better, as there are two varieties
• To ensure full coverage of the flowering period of the main variety. One pollinator variety is in full flower when the variety to be pollinated is starting to flower, while the second pollinator reaches full flowering when the variety to be pollinated is finishing.

9.1.4. Fertilization

Once the pollen grain has reached the stigma, it swells and starts to germinate, which takes place 12-36 hours after pollination.

The pollen grain breaks open and the pollen tube grows out and down through the style to the ovary, where it fertilizes the ovule.

Several pollen tubes may enter the style. Each one is derived from a separate pollen grain, but only a single pollen tube can enter each ovule and fertilize it.

The ambient temperature greatly affects fertilization. The best temperature is 20-25°C, and fertilization does not occur below 5°C or above 35°C.

Moist periods with mild temperatures favor fertilization, unlike dry periods.

Heavy rain and strong wind are unfavorable, as they cause the pollen to clump together.

Pollination is the end of the flowering process. It is followed by cell division and growth within the ovary. When the ovary is visible, the fruit has set, and fruit growth starts.

9.1.5. Sterility and its causes

Sterility can be caused by phenomena acting at any moment from flower formation to fruit-set. The causes of sterility can be divided into external, due to factors outside the tree, or internal, when they are due to the tree. These causes are almost always inter-related, but we are going to look at them separately.

• **External causes**

A) Climatic causes

In practice, climatic effects are usually most clearly seen in plantations of fruit trees.

Both flowering and fertilization are complex processes that are influenced negatively by any deviation from normal environmental factors. The most important climatic factor is the temperature. The optimum temperature is 18-25°C. Below 10°C, the process is slowed down, as it is above 25-30°C.

Temperatures below 0°C are dangerous, as the flower is sensitive to cold and may die.

Rain is another major climatic factor. Wet, rainy, springs cause bad harvests due to the phenomenon called **flower shedding**, which causes inadequate fertilization.

This is due to several reasons, such as clumping of pollen grains and limitations on the flight of pollinating insects.

Low humidity accompanied by high temperatures also has a negative effect during the period the stigmas are receptive, as they desiccate and the pollen does not stick to them.

The final climatic factor to consider is the wind. If the wind speed is greater than 10 km/h bees have difficulty flying, and this negatively affects pollination.

If the wind is dry, it dehydrates the stigmas, and if it is intense it may damage the flower and cause it to be shed.

B) Nutritional causes

Nutritional imbalance or deficiency may lead to morphological and physiological sterility. Resistance to climatic causes of sterility is also linked to good nutritional status.

It is hard to say which mineral elements are most important. The most important deficiencies can be due to nitrogen, boron and magnesium.

Excess levels of elements may also affect flowering, as plants on soils that are excessively fertile or that have received too much nitrogen fertilizer tend to produce abundant vegetative growth but few flowers.

C) Cultivation-related causes

Some inappropriate cultivation techniques may have a very negative effect, including:

• *Pruning*. Inadequate pruning favors abundant production in alternating years ("biennial production"), deprives the tree of the light and warmth needed for the flower buds, encourages fungal attacks and deprives the reproductive cells of the food supplies needed for their later growth.
Late pruning is not a good idea.

• *Irrigation*
You should not irrigate during flowering.

• *Plowing*. Inappropriate plowing may damage surface roots, causing water loss, which is not a good idea during flowering.

The same is true of thinning, bark-ringing, pinching out, and in general, of all complementary practices, as depending on how they are performed, they may delay the normal vegetative growth essential for the following harvest and for the accumulation of reserves for the following flowering period, and thus elongate the productive life of the tree.

D) Accidental causes

Accidental causes occasionally have negative effects on fruit-set, or on the flowers and fruit.

For example:

Parasitic attacks or toxic effects of the chemical products used. Any action that causes a shock to the tree during flowering.

• Intrinsic causes of sterility

A) Genetic causes

Some characteristics, such as the viability of the pollen grains (their percentage germination), ovule degeneration and pollen tube growth, are genetically regulated. Occasionally, this regulation does not function properly and gives rise to specific cases of total or partial self-sterility.
Totally self-sterile trees are usually removed, and partly self-sterile ones require the presence of varieties capable of pollinating them.

B) Physiological

Closely linked to genetic causes, physiological causes of sterility are often the symptoms of genetic problems.
Premature degeneration of the ovules or slow growth of the pollen tube are examples of physiological problems of genetic origin.

C) Morphological causes

Many morphological causes of sterility are also of genetic origin, but these causes are clearly visible. Pistil abortion, macrostylia (longer style than stamens), little or no production of pollen grains, lack of synchronization between the ripening of stamens and pistils, are all morphological causes that limit pollination and cause sterility.

9.1.6. Parthenocarpy and apomixis

Production of seedless fruits, which were not fertilized or in which the embryo died, is known as **parthenocarpy**, and is frequent in the banana, pineapple, some figs and some citrus trees and vines. Parthenocarpy may occasionally occur in pears, apples and some other stone fruit.

There are several types of parthenocarpy:

• Natural parthenocarpy

This is related to the tree's auxin content and is of genetic origin: it may be:

• *Stimulatory*, when fruit-set and growth occur with pollination, even though fertilization does not take place.

• *Vegetative*, when fruit-set and fruit growth occur without the stimulation of pollination.

• Accidental parthenocarpy

This is due to climatic factors, such as high or low temperatures at the time of fruit-set.

• Induced parthenocarpy

Parthenocarpy is caused by application of plant growth regulators. The purpose of this treatment is to give the right dose of hormones to trigger fruit development. The results of this treatment are not very consistent.

• *In seed fruit*, synthetic auxins (ANA, 2-4D, 2-4-5T, 2-4-5TP, etc.) are used even though they have other effects, such as the reduction of fruit-set and increased fruit fall. In general, gibberellins give better results, though the fruit are smaller and deformed.

Growth retardants, such as CCC and Alar give good results if they are applied during full flowering.

• *In stone fruit*, they are mainly applied to prevent pollen incompatibility or sterility.

Apomixis is the formation of fruit and seeds without fertilization. It occurs relatively frequently in some fruit crops, such as the walnut, the hazel and citrus fruit. It is not agronomically profitable as the seeds are of poor quality, even though seeds are formed, and production cannot be controlled.

9.1.7. Basic principles for good pollination and fertilization

The problems that can appear during pollination and fertilization are many and complex. It is hard to define any basic principles, but these are some recommendations.

• It is recommended to plant pollinators in almost all fruit species, except varieties that are known to be self-fertile.

• You are advised to plant two different varieties of pollinator, 50% of each.

Flowering habits of some fruit trees

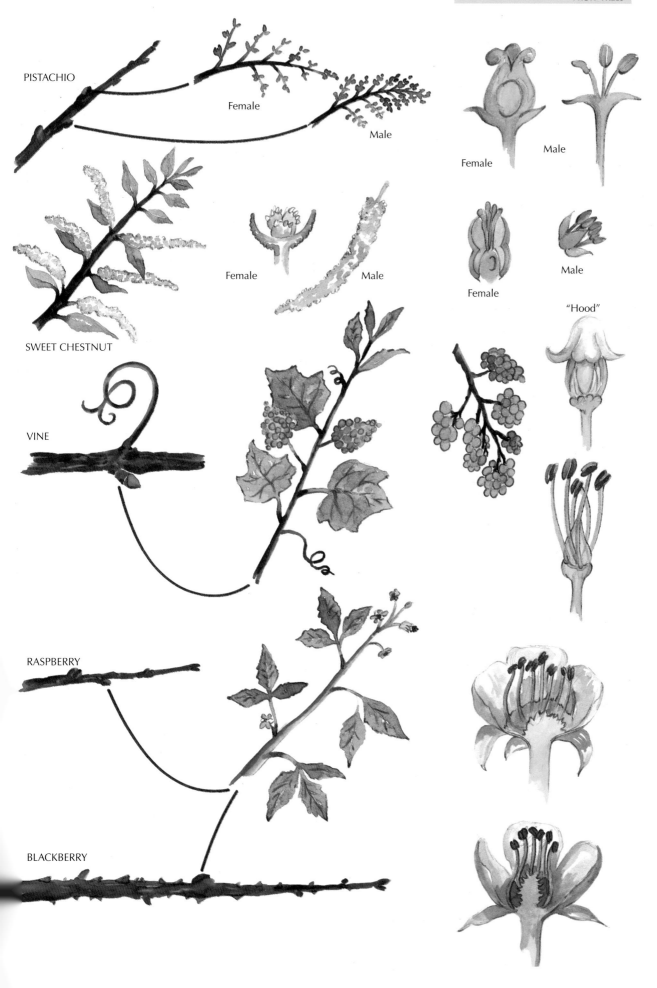

PISTACHIO

Female

Male

Female

Male

SWEET CHESTNUT

Female

Male

Female

Male

VINE

"Hood"

RASPBERRY

BLACKBERRY

Flowering habits of some fruit trees

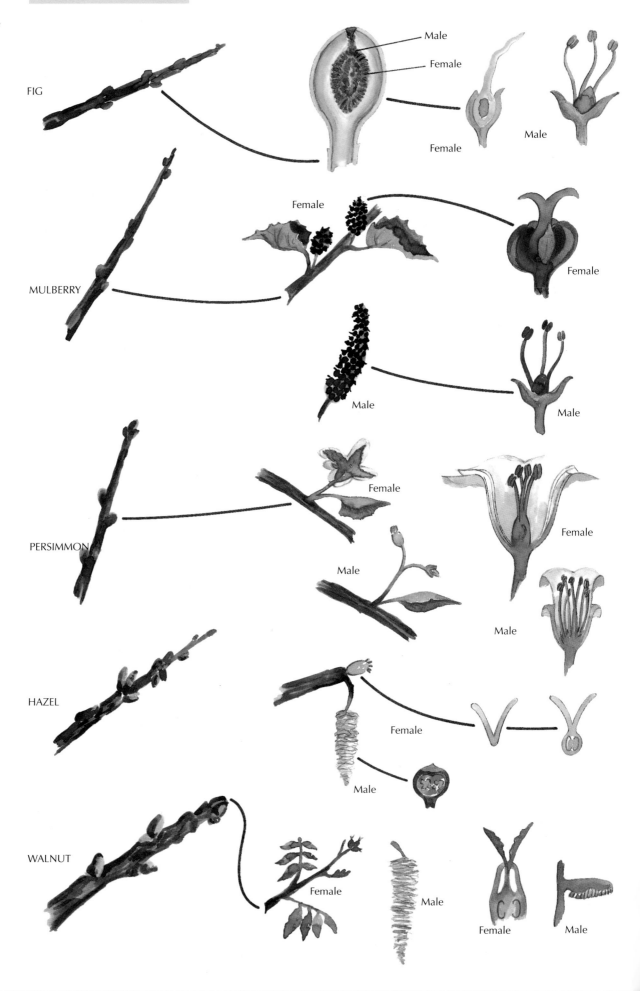

FIG

Male

Female

Female

Male

MULBERRY

Female

Female

Male

Male

PERSIMMON

Female

Male

Female

Male

HAZEL

Female

Male

WALNUT

Female

Male

Female

Male

• The pollinator trees should also produce high-value fruit.
• Complete flowering of the pollinator varieties should fully coincide with that of the main variety.
• The pollinator trees should account for 10-20% of the total number of trees, with a maximum distance of 30-40 m between a tree and its nearest pollinator.
• The arrangement of the pollinators within the plantation is important, and it should be regular and homogeneous, thus avoiding the problems during harvest, irrigation and treatments.
• The lists of varieties showing infertility with each other given in bibliographies are only guidelines. In every different zone you should check whether the varieties are infertile.
• It is worth renting beehives during the flowering period, placing between 1 and 6 hives per hectare.
• Chemicals should not be sprayed during the flowering period, and if they have to be applied, use products that are not toxic to insects, and apply as a mist.
• Cultivation techniques should aim for the trees to be in peak condition at flowering time.

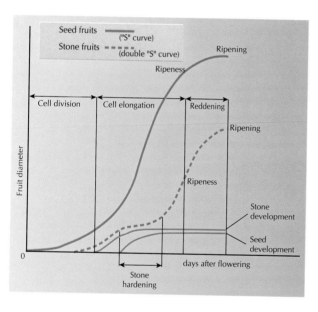

9.2. FRUIT GROWTH

Once fertilization has finished and the fruit has set, the fruit begins to grow.
This growth finishes when the fruit is mature, after about 70-80 days in early varieties of cherry and apricot, but 250-300 days in late varieties of apple and pear.

The growth of the fruit can be divided into four phases:

• *Fruit-set, the start of growth*
• *Active growth*, with a large increase in the fruit's size and weight to almost the typical full size.

• *Ripening*. The fruit increases slightly in size, but changes dramatically in color, and attains its characteristic taste. Reddening marks the end of active growth and the onset of ripeness.
• *Senescence or aging*. The fruit withers.

9.2.1. Factors influencing fruit growth

The factors conditioning fruit growth are water availability and the reserves within the fruit.
Fruit consists mainly of water, accounting for 50-90% of the weight of the ripe fruit.
Water is also the vehicle for the supply of mineral elements and the translocation of carbohydrates.
All this affects the fruit tree's requirements for

Growth of banana fruit

soil water during the fruit's development period. If the tree can not obtain enough water from the soil, the fruit may be affected; it is smaller, and may dehydrate and form wrinkles, and the fruit may even be shed.
Fruit formation also requires a lot of nutrients. The most important component in the first phases of growth is nitrogen.
One climatic factor that most affects fruit development is the temperature. Fruit growth is favoured by medium to high temperatures without sharp nighttime falls.
One physiological factor conditioning the development of the fruit is the number of seeds present, compared to the normal number. This is a consequence of the circumstances of pollination and fertilization. The number of seeds conditions the quantity of the harvest, its volume and quality, as well as the date the fruit ripens. Fruits with less seeds than normal are shed more easily, have odd shapes, and are smaller and of poorer quality, and they also take longer to mature.

Type-curves of fruit growth

9.2.2. Fruit fall

Not all the fruit that has set reaches maturity. Fruits fall off the tree throughout fruit growth.

Distribution of fruit fall over time

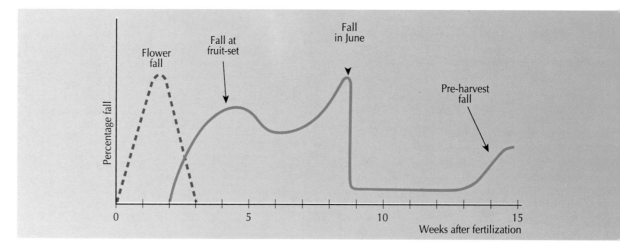

These falls may be divided into *physiological falls*, when they occur to a greater or lesser extent, and *accidental falls*, when they occur due to accident, etc.

• **Physiological falls**

In general, this is due to one of the following four reasons:

• *Flowers with abnormal or over-developed pistils.*
• *Fall on fruit-set.* This is the shedding of residual flowers, unfertilized flowers or fruit that have not set well. In abundant flowering, 75-90% of the flowers may be shed, but this does not endanger the harvest.
• *June fall.* This is due to the competition between fruit during the growth phase in early june.
The fruit with the least seeds, deficient nutrition or delayed growth, are shed, and this benefits those that are left, which are bigger and better quality than they otherwise would be.
This shedding may, in normal conditions, represent 10-30% of the fruit and reduces the thinning necessary later.
• *Mature shedding or pre-harvest fall.* This is caused by the growth of the abscission layer. Some varieties produce this layer prematurely and the fruit falls before it can be harvested.

• **Accidental falls**

These are due to several causes and may occur at any moment. They may be caused by parasitic attacks, or climatic reasons, such as late irrigation after a dry period, treatment with copper compounds after flowering, nutrient deficiencies, deep plowing during flowering, damage by tools, or bird damage.

Cocoa tree, checking fruit ripeness by touch.

9.2.3. Alternation or biennial production

Alternation is the tendency of fruit trees to alternate years with heavy fruit production with years of poor production.
Production is thus biennial and is caused by factors within the tree. In different cases, the following

causes may be involved:
• *Interference between vegetative growth and flower induction.* Shoots appear to control flower induction by means of hormones.
• *Interference between fruit growth and vegetative growth.* The fruit compete with the growing tips. These develop less, meaning that next year's harvest will be reduced.
• *Exhaustion of carbohydrate reserves in the roots.*

Several cultivation techniques are effective at controlling biennial production, including:

• *Fruit thinning.* This is carried out in the first stages of fruit development.
• *Scoring.* This is normal practice in citrus fruits, and seeks to increase flowering in the areas above the scored bark.
• *Harvesting early.*
• *Pruning.* Regular moderate pruning reduces alternation in many species.
• *Hormonal flower induction treatments.* Treatment with growth retardants, such is CCC or Alar, is limited as few experiments have been performed.

9.2.4. Fruit ripening

The fruit ripening process consists of a set of physical and chemical changes. These changes determine the color, taste, smell and texture of the fruit.

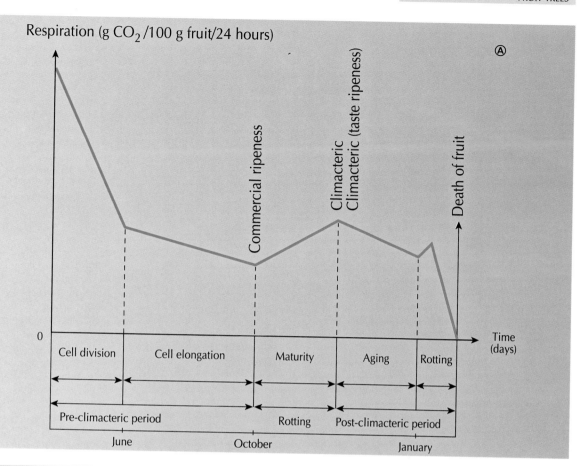

Respiration (g CO_2/100 g fruit/24 hours)

Ⓐ

Commercial ripeness

Climacteric
Climacteric (taste ripeness)

Death of fruit

0

Time (days)

| Cell division | Cell elongation | Maturity | Aging | Rotting |

Pre-climacteric period

Rotting

Post-climacteric period

June

October

January

Graph of respiration over the course of fruit development
A/ Fruits with climacteric
B/ Fruits without climacteric

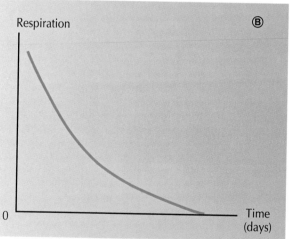

Respiration

Ⓑ

0

Time (days)

The fruit's maturity can be considered in three ways:

• *Consumer ripeness* or *taste ripeness*, when the fruit is ready to be eaten. This point depends on the particular taste of the consumer.
• *Commercial maturity*, or *harvest ripeness*, when the fruit is harvested early so it can reach the sale point when it is ready for consumption. The fruit is collected and separated from the tree, and continues to ripen until taste ripeness.
• *Physiological maturity*, when the seeds are ripe and can germinate successfully.

Ripening is a consequence of biochemical activity within the fruit, which is activated by a series of physiological processes.

Fruit	Edible portion (analyzed)	Water	Protein	Fiber	Sugars	Ascorbic acid (vitamin C) (mg)	MOA*
POMES Apple	Pome without peel or core	84	0.25	2.2	11.4	5	m
PEAR	Pome without peel or core	83	0.25	2.5	10.8	4	m**
HESPERIDIA (CITRUS) Orange	Pulp, with no peel or seeds	86	0.82	2.0	8.5	50	c
Lemon	Juice	91	0.32	—	1.6	50	c
Grapefruit	Pulp, without rind or seeds	90	0.06	0.6	5.3	40	c
DRUPES Peach	Drupe with skin without stone	86	0.06	1.4	9.1	8	m**
Apricot	Drupe with skin without stone	86	0.6	2.1	6.7	7	m
Plum	Drupe with skin without stone	84	0.6	2.1	9.6	3	m
Cherry	Drupe with skin without stone	81	0.6	1.7	11.9	5	m
BERRIES Grape	No seeds, with skin	79	0.6	0.9	16.1	4	m
Banana	Without skin	71	1.2	3.4	16.2	10	m

Chemical composition of fleshy fruits (g per 100 g)
*MOA = Main organic acid;
m = malic acid;
c = citric acid
** In some varieties there is more citric acid than malic acid.

**Flowering habits of
some fruit trees**

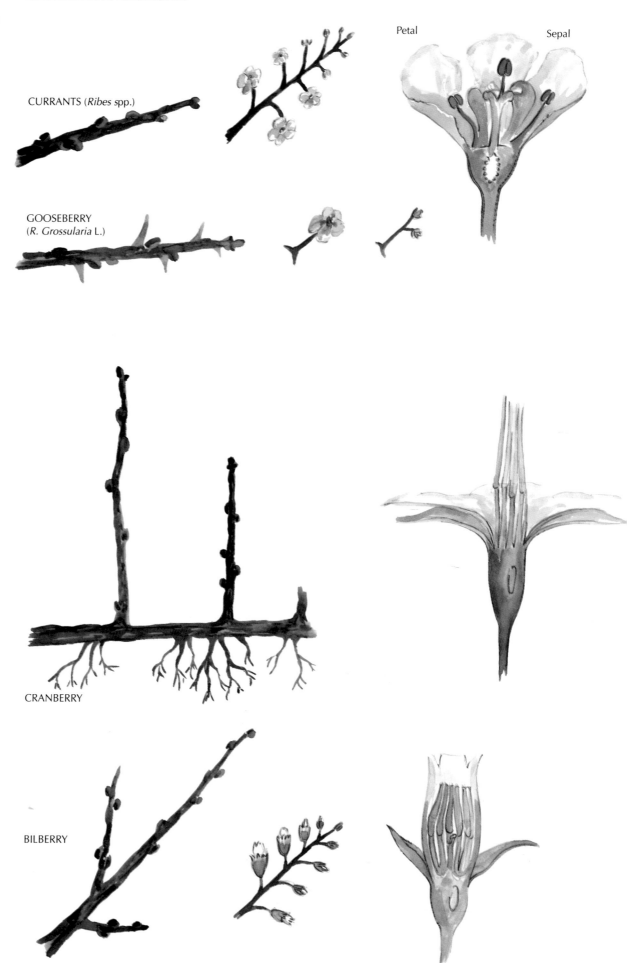

CURRANTS (*Ribes* spp.)

Petal   Sepal

GOOSEBERRY
(*R. Grossularia* L.)

CRANBERRY

BILBERRY

• *Transpiration.* The process by which the fruit loses some of the water that arrives through the vessels.
• *Respiration.* Respiration consumes oxygen from the air and gives off carbon dioxide.
• *Photosynthesis.* The opposite of respiration, in photosynthesis the fruit synthesizes carbohydrates from CO_2 absorbed from the air. This continues until the fruit changes color and loses its chlorophyll.
• *Fermentation.* This occurs when the fruit ages or if it lacks oxygen, and it emits CO_2 and forms alcohol.

These physiological processes set off several activities within the fruit.

• *Emission of volatile substances.* These are mainly ethylene and aroma-producing agents, such as the esters, alcohols, aldehydes and ketones that give the fruit its typical smell. Emission increases with temperature and maturity.
• *Fruit constituents*, which may accumulate, disappear or be transformed.
• *Sugar formation* from starches. The starch gradually disappears and when it has disappeared the concentration of soluble sugars peaks and fruit growth ends. The main sugars present in cell sap are fructose, sucrose and glucose.
• *Changes in acidity.* The organic acids decrease as ripening occurs, but they are still present at maturity. The main organic acids are malic acid, the main acid in stone and seed fruits, and citric acid, which is found in citrus fruit.
Other fruit acids include ascorbic acid (vitamin C) and succinic acid. This change is speeded up by higher temperatures.
• *Appearance of vitamins.* Vitamins, especially vitamin C, increase during ripening, while tannins decrease and disappear.
• *Changes in pigmentation.* Pigments increase during the ripening process, requiring oxygen and light. The main pigments are carotenes, which give rise to yellow and orange colors, and the anthocyanins, which form red and blue colors. The chlorophyll gradually disappears, and is replaced by these pigments.

The fruit's texture also changes as it matures. It softens, due to water accumulation and the weakening of the cell walls.
To follow the fruit's ripening, we can measure its respiration. This is the volume of CO_2 given off by the fruit's respiration in a given time and at a constant temperature.

There are three periods in the fruit's respiration:

• *Pre-climacteric.* The intensity of respiration decreases.

• *Climacteric.* This is when respiration increases to a peak. This peak is the climacteric and marks the end of the ripening process.
• *Post-climacteric.* The end of the climacteric marks the beginning of the fruit's senescence. The fruit ages, rots and dies, and its respiration declines to nothing.

Fruits show two main types of ripening, as revealed by graphs of their respiration.

• *Climacteric fruit* undergo a dramatic increase in respiration during ripening. Fruit with a climacteric include:
Apples, pears, plums, peaches, nectarines, apricots, avocadoes, bananas, custard apples, mangoes, papaya, kiwi, figs and persimmons.
Harvest ripeness coincides with the start of the climacteric, and taste ripeness with the climacteric.
• *Non-climacteric fruit* do not have a climacteric. Respiration declines evenly throughout ripening. They include:
Cherries, grapes, grapefruit, oranges, lemons, mandarins, melons, pineapples, strawberries, olives and mandarins.

• **The influence of hormones on the ripening process**

• *Ethylene.* Also known as the ripening hormone. Ethylene is produced by ripening fruit, but production is greater in fruit with a climacteric. Applying ethylene to pre-climacteric fruit induces ripening in fruit that show a climacteric, causing a displacement of the respiratory curve in time, but without affecting its general outline.
In non-climacteric fruit that have been separated from the tree, respiration can be stimulated with ethylene at any moment.

High temperatures favor ethylene production in fruit.

• *Auxins* delay maturity. They favor ethylene production, but make the plant's tissues insensitive to the ethylene.

• *Cytokinins* delay the senescence of the fruit's peel.

• *Gibberellins* delay loss of chlorophyll and the accumulation of carotenes in the peel. They are used to keep the skin firm and stimulate regreening in citrus fruit.
• *Abscissic acid*, ABA, causes fruit senescence and accelerates the ripening process.

Fruit should be harvested at taste ripeness, as ripening does not occur once the fruit is separated from the mother plant.

Underside of a citrus leaf showing aphid infestation.

10. DISORDERS OF FRUIT TREES

10. 1. MAIN PESTS AND DISEASES

This large group consists of organisms that affect in one way or another the plant's health, and thus the quality of its production.

Pests belong to several groups:

- Insects
- Acarid (mites)
- Fungi
- Bacteria
- Virus
- Mycoplasmas

You must understand a pest's life cycle and feeding habits in order to find the best way to combat it. Pests vary greatly in their life cycles, and can cause damage at more than one developmental stage. The details of individual pest species are considered in more detail in another part of this work.
The following are some of the most common pests and diseases of fruit trees.

10.1.1. Insects

The most important insects pests are:

- *Thrips*, including pear thrips, which attack stone fruit and the flowers and fruit of the pear.
- *Tiger moth.* Attacks pears and almonds.
- *Leafhoppers.* Attack fruit trees and forestry plantations.
- *Rust flies.* Attack the leaves of the pear and quince.
- *Phylloxera.* A very important pest of vines.
- *Aphids*, etc. There are many different types of aphid (or greenfly, blackfly, etc.) and they may attack different fruit trees, such as the cherry, plum, apple, peach, pear and walnut.
- *Scale insects.* They attack the leaves and fruit of seed fruits, such as the vine, and the walnut.
- *Soft scale insects.* They attack the shoots of stone and seed fruit, as well as the young fruits.
- *Red scale and San José scale insects.* They attack fruit and seed fruits.
- *Apple red bug.* Attacks the bark and wood of fruit trees.
- *Plum weevil.* This attacks the fruit of stone and seed fruits, as well as bilberries, gooseberries, persimmons and vines.
- *Fruit flies.* There are many fruit flies that vary with the fruit tree they attack, such as cherry, apple or

Misting. System by AGROTÉCNICA.

walnut. There is also the Mediterranean fruit fly, which attacks several different fruit trees.
• *Orange walnut bug.*
• *The peach twig borer* also attacks almonds and apricots.
• *The peach tree borer* and the *apple borer*. The peach tree borer also attacks plums and apricots, and the apple borer also attacks pears.
• *Apple moth.* Attacks stone and seed fruit.
• *Coddling moth.* Attacks apples, pears and walnuts.
• *Fruit leaf-roller.* Attacks the leaves of stone and seed fruit.

10.1.2. Mites

The most important are:

• *Yellow spider mite.* Attacks the leaves of apples and pears.
• *Red spider mite.* Attacks the leaves and fruits of several fruit trees and the vine.
• *Eriophyes vitis infestation.* Attacks pears, vines, hazels and walnuts.
• *Mite infestation.* Attacks pears, apples, peaches, plums and figs.

10.1.3. Fungi

The most important are:

• *Apple speckling*
• *Mildew.* Attacks several fruit species.
• *Peach leaf blister (leaf curl).* It also attacks nectarines and almonds.
• *Dark rot.* Attacks stone fruit.

10.1.4. Bacteria

• *Fire-blight (Erwinia).* Affect seed fruit of the rose family.
• *Bacterial canker.* Affects stone fruit, as well as pears and occasionally apples and citrus fruit.
• *Agrobacterium* (crown gall). Attacks the crown and roots of apples, pears, cherries, plums, apricots, peaches, almonds and vines.
• *Phytophthora.* Attacks the root and crown of many fruit trees.
• *Verticillium wilt.* Attacks the root and crown of many fruit trees.

10.1.5. Viruses

Viral diseases are classified by the symptoms they cause in the host plant, and the most important are:

• *Rough skin virus*
• *Stony pit virus*
• *Star cracks virus*
• *Chat fruit virus*
• *Vein yellow virus*
• *Mosaic virus* (different forms attack many fruit trees).

10.1.6. Mycoplasma bacteria

These bacteria give rise to mycoplasma infections, which include:

• *Pear decline*
• *Cherry albinism*
• *Peach mosaic*

10.2 NON-PARASITIC DISORDERS

These are caused when the plant's normal functioning is disturbed by external physical, chemical or mechanical agents.
These agents are mainly from the two media in which the plant grows, the air and the soil. Human beings, however, are a further cause of disturbance, for example by using inappropriate cultivation techniques.

10.2.1. Atmospheric disturbances

• **The effect of lack of sunlight**
Lack of light causes **legginess**, in which the stems are excessively long, with very weak leaves that have lost their green color.

• **The effect of heat**
Excess heat greatly increases transpiration. The plant has to replace this with water from the roots, as long as there is enough water in the soil.

• The effect of temperatures

Low temperatures have a variety of effects on fruit trees: freeze burns to fine bark, blackening of branches, the death of resting buds, flower and fruit fall, and root freezing. The type of damage will depend on the time of year that the low temperatures occur, and how long and how low they are.

One effect of low temperatures is what is known as **sunstroke**. This causes damage on the southern side of the trunk, branch or fruit. This is caused when tissue heating occurs on a sunny winter's day due to intense sunshine, followed by rapid cooling in the shade or after sunset.

• The action of snow

Snow protects the plant from intense cold, and when it melts, it provides water for the soil. Snow can cause mechanical damage to branches by overloading them.

• The effect of rain

Excess rain can cause mechanical damage to leaves, flowers and fruits, causing them to break and fall. It can also make pollen inviable and thus totally prevent fertilization. Continuous rain for long periods creates excess moisture which leads to abnormal plant growth and also favors the spread of fungal diseases.
The fruit have less taste, and some may rot and others may crack.

• The effect of fog.

Fog deprives trees of light and heat while increasing humidity, thus favoring fungal diseases.

• The effect of wind

The damage depends on wind intensity. Strong winds tear off flowers and fruit, and damage leaves and branches. Gentle winds favor fertilization of the flowers and transpiration, meaning the plant is more active. If these gentle winds are prolonged, this transpiration eventually exhausts the tree, and may cause the leaves to wilt. Salt-bearing winds are very negative, because they contain (in suspension) salts that damage the leaves and "burn" them.

Other disturbances caused by atmospheric agents include:

• Vine blush

Vine blush is the sudden reddening of the leaves in summer, due to the effect of strong winds and sharp falls in temperature.

• Mealy fruit

Fruit ripening is a fermentation process in which sugars are turned into alcohol, which combines with fruit acids to give esters, which are responsible for the fruit's smell. If this process occurs during hot weather, the acids volatilize before they can react with the alcohol, and this means the fruit's flesh is floury and tasteless.
Mealiness occurs in early varieties of pears and apples when the summer is very hot.

• Honeydew

This is a sugary, sticky, liquid that appears on the leaf surface. This occurs in May, when the sun shines after a few days of rain. This is accompanied by infestations of aphids, which feed on the honeydew.

• Gummosis

Gummosis is a disease of citrus trees. It is a degenerative process of the bark cells in which their starch turns into slime, which is exuded by the branches and trunk. Gummosis leads to progressive weakening of the tree, whose leaves yellow and fall.
There are several causes. It may be the tree's reaction to unfavorable circumstances (not usually parasites), such as an impermeable subsoil, excess moisture, lack or excess of a mineral nutrient or a frost.
Gummosis may also be accompanied by microbes like *Bacteria gummis*.
The remedy is not to use excess organic fertilizers or excess irrigation, especially in impermeable sites, and airing the soil by deep plowing. As a curative technique, you can scrape clean the parts affected by the gum, treat the wound with acid iron sulfate, and then putty to prevent contact with the air. Gummosis also occurs in cherries, peaches and apricots.

• Mite infestation

This is due to excess moisture and a large amount of nutrient elements in the soil, accompanied by a lack of calcium. It takes the form of cracks in the bark of the branches, which then heal producing corky growths, and the branch starts to look "scabby".

• Corkiness

This is growth of corky tissue over the lenticels (air pores) in the bark of the stem, so that when you touch them you can detect a yellow powder. Suberosis, like mite infestation, is due to excess moisture and nutrients in the soil.
The most affected plants are: pear, peach, apple, cherry, plum and vine.

• Scalding of grapes

This affects grapes on hot days in summer or fall. The heat of the direct sun can cause withering and desiccation of the bunches not protected by leaves.

• Bitter pit

Bitter pit is the appearance of greenish depressions on the peel, which turn into dark circular patches, especially near the tip. The causes are complex. It occurs in young trees with an excessive supply of nitrogen, d is made worse by excessively frequent irigation, severe pruning and fruit that are o big. Alternating wet and dry periods, as ell as high transpiration in hot dry eather, can also trigger bitter pit. It mainly fects apples.

Stony fruit

ony pears have a hard flesh, which is tough nd lacks juice. The causes are unfavorable imatic conditions, especially lack of oisture.

Fruit cracking

his mainly occurs when heavy rains fall after dry period. It is an osmotic phenomenon aused by the rainwater that adheres to the uit.
 prevent this disorder in sensitive fruit, such cherries and peaches, you should apply owdered copper sulfate at the beginning of pening.

• Mobilization of food reserves to heal the wounds and replace the damaged organs.
• Faster transpiration.
• Tissues rot if the wound does not cure properly, leading to diseases like gangrene, gummosis and other rots.
• They are also an entrance for pathogenic organisms and they increase the plant's susceptibility to insect attack.

10.2.3. Disorders due to soil conditions

• **Site fatigue**

Site fatigue is caused by several factors, and occurs when a site is replanted with the same species or a closely related one.
Planting the same species of fruit tree is due to market demand. The best way to solve soil fatigue is to alternate crop species, as long as the economic expectations are favorable.

Root system of a peach tree planted at different depths.
The deeper the planting depth and the more compacted the soil, the worse the roots grow. The roots grow upwards trying to escape the asphyxiating conditions.

10.2.2. Disorders due to wounds

he importance of a wound depends on its osition, how deep it is and the organ that is lamaged.
Damage can be caused by cuts, breakages, lows, or abrasion caused by pruning, animal ites, and by atmospheric agents such as vind, hail and frosts.

hese wounds may cause:

  Disordered growth in the plant due to the absence of the part lost.
• Lower harvest quality and quantity.
• Formation of woody calluses as the tree's response o try and cure the wound.

The symptoms of fatigue are:

• Decreased and irregular growth
• Fewer shoots and branches
• Diminished production
• Smaller leaves
• Intervein chlorosis
• Poor response to intense fertilization
• Shorter life-span
• Poor nutrient absorption from soil
• Small, shorter, roots

The main causes of fatigue are:

• Deterioration of the physical characteristics of the soil, including compaction and degradation by excessive plowing.
• Nutrient deficiencies. Consumption of a given

element by one crop provokes an imbalance or deficiency for the next crop. It may also be caused by incorrect fertilizer application or excessive washing of the soil.
• Salinity, linked to excessive fertilizer application.
• Secretion of toxic substances. Fruit trees secrete substances that may damage the next crop, and even the soil microflora. The fruit tree that is most affected is the peach.
• Soil pH. The pH influences the breakdown of the structure, as well as the solubility of the mineral elements.
Fungal and bacterial attacks on the roots are also reasons why the fruit tree may show fatigue.
• Nematodes are one of the main causes of fatigue. They cause damage that encourages attacks by fungi, bacteria and viruses.
• Accumulation of chemical products in the soil, such as residues of herbicides and treatment chemicals.

Fatigue may be due to one of these factors or a combination, and varies from one area to another.

The techniques used to prevent site fatigue are:

• Rotation of fruit trees with herbaceous species.
• If you grow fruit trees in the same site, grow a different species or use more vigorous rootstock varieties.
• Supply organic matter to improve the soil structure.
• Balanced application of fertilizers to meet the crop's needs.
• When uprooting the plantation, try to remove all the remains of the previous plants.
• Before replanting, turn the soil thoroughly in order to expose the soil microorganisms to the sun.
• Replace all the soil in the planting holes. This technique is not effective against nematodes.
• If chemicals have accumulated in the soil, you are advised not to use them abusively and to follow the manufacturer's instructions.
• Apply treatments to eliminate fungi, harmful bacteria or nematodes, when present, or if repeating the crop.

Nutrient deficiencies in fruit trees and citrus fruit

• Measures should be taken against reinfestations, especially of nematodes. They may be spread by tools or by infected saplings. The new attack is usually faster and more virulent.

• Root asphyxia

Root asphyxia occurs when the soil's oxygen content is low, and this means that some microorganisms can grow within the root by anaerobic respiration, that is to say, without air. Anaerobic respiration generates products like lactic acid, ethyl alcohol and carbon dioxide, which are all toxic to the roots.

Oxygen may not be reaching the soil for several reasons:

• The most important reason is excess water around the roots (waterlogging). When the soil is waterlogged, the water displaces the air.
• Soil with a high percentage of clays and silt. Clay and silt can give rise to a very compact soil structure, with few large air-filled pores.
• Compaction due to other causes, such as heavy machinery, can cause the same effects.
• Deterioration of soil structure due to unsuitable cultivation techniques.

The signs that the plant is suffering from root asphyxia are:

• Dehydration of the aboveground parts and general weakness.
• Leaf yellowing and fall.
• When spring arrives, the tree flowers and produces shoots, but once the reserves have been exhausted, the symptoms of asphyxia appear. If asphyxia is only mild, the tree may even develop fruits, but these fall before ripening or develop irregularly.
• The tree becomes increasingly susceptible to parasite attack, crown gall canker and root rot due to honey fungus (*Armillaria*).
• Asphyxiation is accompanied by a characteristic rotting smell and a bluish coloration of the soil near the roots.
• The external root tissues are necrosed and the inner ones develop dark patches.

KEY TO DEFICIENCIES IN FRUIT TREES	Deficient element
A. The symptoms occur in old leaves, which fall when the young leaves become affected.	
1. Intervein chlorotic or necrotic zones. The leaf edges are not affected to begin with. Leaf size may not be affected.	Magnesium
2. The leaf edges are affected first, both the tip and sides. The leaf curl upwards and are smaller than normal.	Potassium
B. The symptoms appear in the youngest leaves:	
1. The young leaves are yellow with green veins.	
a) The shoot internodes are short, forming rosettes of yellowing leaves. The older leaves turn a bronze color, and fall easily.	Zinc
b) The internodes are almost normal length. The leaves are yellow, except for the veins, which form a green grid on a yellow background. The youngest growing leaves may have almost no green color. As they age, they may turn slightly green.	Iron
c) The leaf tip turns yellow. The main veins are green. The growing tip of the branch dies and new lateral buds grow.	Copper

	Deficient element
2. Young leaves green or only slightly yellow.	
a) Young green leaves bent upwards in a "boat" shape. The terminal buds abort and new side growth is produced, which then dies.	Boron
b) The main leaf vein is cut short, and tips are rounded.	Molybdenum
C. Symptoms in any part of the plant, or the entire plant.	
1. Small leaves, uniform pale green, and reduced growth:	
a) The petiole (leaf stalk), the lower part of the central leaf vein and the young shoots all have a purplish tinge in their early growth. This color may disappear afterwards, and the color improves.	Phosphorus
b) The leaves are pale, and this gets worse as they get older.	Nitrogen
2. Normal leaf size. The areas between the main veins are pale. The lateral veins stand out like green bands. The fine veins can not be distinguished. The young growing leaves do not show symptoms. Appears halfway between magnesium deficiency and iron deficiency.	Manganese
D. Symptoms mainly in the fruit. Corkiness in young fruit. Internal necrotic patches in fruit or corkiness.	Boron

• The poorly developed roots tend to grow upwards towards the surface in search of the air they need.
• The plant's root hairs, which absorb water from the soil are dry and the collar shows a corky layer.

The best ways to counteract root asphyxia are:

• To choose rootstocks that are resistant to root asphyxia.
• Prevent the problem of root waterlogging by ensuring good drainage conditions.
• Aerate the soil thoroughly before planting, by plowing as deep as possible.
• Incorporate organic matter into the soil to improve its structure.
• Ensure the variety planted, and its planting and cultivation, are appropriate.
• On sites at risk of root asphyxia, plow as little as possible, so you do not cause compaction. Avoid use of rotatory machinery, and use spring-toothed harrows instead. This prevents deterioration of the soil structure.

If symptoms of root asphyxia appear, a series of operations should be carried out to prevent it getting worse.

• Remove the topmost soil around the trees so the roots are closer to the surface and more air can reach them.
• Apply abundant iron sulfate and feed the tree with leaf feeds.
• If the tree is bearing abundant fruit, part must be removed if the tree is to support the rest to ripeness.
• Change fertilizer applications.

The plant's nutrition requires the presence in the soil of a number of nutrient elements that are essential for their normal growth.
Lack of one of these elements causes decreased vegetative growth, even if all the other elements are present.
Nutritional disorders are complex. They may be the result of the lack of one or more elements in the soil. They may also be due to the antagonistic action of one element preventing another element from being taken up by the roots, or by physical or chemical soil conditions that make one element insoluble.

The main deficiencies of each crop species are:

• *Vines*: Ferric chlorosis, nitrogen deficiency, magnesium deficiency.
• *Apples and pears*: Ferric chlorosis, deficiency of nitrogen, magnesium, manganese, zinc, potassium and boron.
• *Cherry* : sensitive to magnesium deficiency.
• *Plum*: sensitive to deficiencies of nitrogen, potassium and manganese.
• *Peach*: sensitive to deficiencies of manganese and potassium.
• *Strawberry*: sensitive to deficiencies of iron and manganese.

• **Ferric chlorosis**

This chlorosis is caused by decreased chlorophyll synthesis in the leaf, because the plant does not have enough iron.
The symptoms are that the leaves turn yellow while the leaf veins remain green. It starts with the youngest leaves, which in the worst cases may necrose and die. There is a general decrease in tree growth and longevity, and both productivity and fruit quality are lower.

Ferric chlorosis can be due to several causes:

• Shortage of iron, or soluble forms of iron, in the soil.
• Problems in uptake due to root damage.
• Internal problems within the tree, such as damage to the vessels that prevents translocation.
• The iron in the soil is present in insoluble forms, because the soil is alkaline.

Iron shortage in the soil is a nutritional problem that can be resolved by supplying the soil with iron. The most effective way of doing this is with chelates, organic compounds consisting of iron bound tightly to the chelating agent. These chelates maintain the iron in a form that plants can assimilate, and prevent it from being precipitated within the soil.

KEY TO DEFICIENCIES IN FRUIT TREES	Deficient element		Deficient element
I. Symptoms in young leaves or shoots.		II. Symptoms in mature leaves:	
A. Uniform leaf color in each area. Decreased growth:		A. Loss of green color, initially locally and then spreading gradually:	
1. New leaves are pale green or yellow. Little fruit, which is pale in color.	Nitrogen	1. Parallel to the main vein. The leaf base remains green. Premature leaf shed.	Magnesium
2. New leaves are yellowy green or brighter yellow than above. Sulfur.	Sulfur	2. Along the leaf edge, affecting zones between the veins. Calcium.	Potassium
3. Shoots are very short. Foliage lacks shine. Fruit has gum patches. Seeds abort. Excessive fruit fall.	Boron	3. In groups near the tip or outer half of the leaf. The color changes from pale yellow to bronze. Excessive wilting.	
4. Leaves are almost normal. The fruit is small, with thin peel, and falls early.	Potassium	4. Intervein mottling, with yellow or orange borders. Dark brown patches on the underside.	Molybdenum
5. Large, very dark, leaves. Gum deposits on petioles. Gum oozes out of shoots and fruit. Shoots and secondary shoots die.	Copper		
B. Leaves with non-uniform appearance.		B. Loss of color, not localized to begin with.	
1. Very small leaves, sharp-tipped, with bright yellow mottling that contrasts with the general green of the leaf. Fruit are small and pale.	Zinc	1. Yellow-green, or even yellow, leaves with whitish nerves.	Nitrogen
2. Rather small leaves. Mottled pale green, or greyish in horseshoe shape opening out from the central vein.	Manganese	2. Matt green and occasionally yellowish orange. In severe cases, the leafs suffer necrosis. The fruit is thick and spongy, with an empty center and a very acid taste.	Phosphorus
3. The network of fine green veins against a background of a pale green, yellow or whitish leaf. The shoots show reduced growth and then die.	Iron		

11. HARVESTING THE FRUIT

The fruit must ripen correctly to obtain high quality produce, and ripening must be understood if it is to be stored for any period of time.

Therefore, harvesting at the right time is of great importance. Commercial planning, the fruit's end use (immediate consumption, short or long-term storage, etc.) should suggest the precise moment of harvesting when the fruit fulfils the relevant commercial requirements. It is important to grasp the difference between harvest ripeness and market ripeness. Harvest ripeness is the state when fruit for long-term storage should be harvested, while market ripeness is when the fruit is almost ready to be eaten.

Harvest ripeness is also known as pre-maturity or physiological ripeness, and it is of most relevance to us. Maturity indicators have been selected, which are signs that can be easily measured in order to define the status of the fruit.

In principle, there is no measure that defines the maturity of the fruit, so two or three are used together. Personal experience is also essential in choosing the time to harvest.

Recommended values for hardness of pears at the time of harvest (5/16" tip = 8 mm)

Variety	Hardness (pounds)	Variety	Hardness (pounds)
Anjou	13-15	Limonera	15-17
Blanquilla	14-16	M. Hardy	11-12
B. Luisa	15-16	M. Red Bartlett	14-15
Conference	13-15	Passe Crasane	12-14
D. Comice	13-14	Williams	17-19
G. Leclerc	13-14		

Variety	Hardness (pounds)	Variety	Hardness (pounds)
Golden Delicious	15-17	Granny Smith	14-16
Starking Delicious	16-17	Jonathan	16-18
Rome Beauty	16-18	Wellspur	16-17
Red Delicious	16-17	Stayman	15-17

Most recommended values for apple hardness at the time of harvest, for apples for long-term storage (7/16" tip = 11 mm)

11.1. INDICATORS OF RIPENESS

There are three types of indicators of ripeness:

• Climatic indicators

• *Heat units.* This takes the temperature into account, and is thus more reliable than the fruit's age.

The heat units are summed on the basis of the average monthly temperatures. The norm, for seed fruit, is an increase of 7.2°C between full flowering and harvest.

• Physiological measures

• *The fruit's age.* This is the number of days between full flowering and ripening. This period varies from variety to variety and from site to site, and even from year to year, and so it is not a very precise index.
• *Flesh color.* The flesh changes color as it ripens. It is not widely used.

• *Peel color.* This is an important measure. Bear in mind that some fruits turn a red color, while others do not, and that this color does not always mean ripeness. The disappearance of chlorophyll is just as important as the appearance of color. What is assessed is the background color and how it changes.

There are standard color charts for yellow fruit, and devices called colorimeters to measure the intensity of the color.
• *Seed color.* This is quite reliable, especially in pears. As the fruit ripens, the seeds change color from whitish to dark, or even black.
• *How easily fruit is detached.* Not very widely used, due to its variability.
• *Fruit size and weight* are useful guides, but not sufficient.
• *Pulp hardness.* This is a precise value measured with a penetrometer. It is based on the fact that the pulp becomes softer as the fruit ripens. The resistance of the pulp is shown on a scale, either in pounds or kilos. The diameter of the probe is standardized.
• *Respiration.* This relies on measuring the fruit's respiration, using a respirometer, to calculate the CO_2 released and the O_2 consumed. As the fruit matures, its respiration declines to a minimum, which is at harvest ripeness. After this, respiration increases until market ripeness.

• Chemical measures

• *Starch content.* The starch content declines as the fruit matures, as it is turned into sugar. The right time to harvest coincides with the disappearance of the last starch from the fruit. When a portion of the fruit is placed in a solution of iodine in potassium iodide, the presence of starch is shown by the appearance of a deep violet-blue color. The solution consists of 2.5 g of iodine + 10 g potassium iodide per liter of water.
• *Acid content.* The concentration of acids declines as the fruit ripens.
• *Sugar content.* As maturity arrives, complex sugars (polysaccharides, etc.) are turned into simple ones (fructose, etc.). This change in sugar content can be measured using a refractometer.
• *Sugar/acid ratio.* This ratio varies as the fruit ripens, because the acid decreases and the sugar increases.

Some of these measures can be calculated by the cultivator, but others have to be assessed by technicians or specialist centers. Its importance means that every farm should have tools to work out how mature the fruit is in 3 or 4 different ways.

11.2. HARVESTING UNRIPE FRUIT

Fruits harvested before maturity, when they are still green, wither, and even if they turn yellow, the

Large-scale manual collection: (A, B) cross and longitudinal sections of a harvesting cart with boxes. *Platforms are used for the manual collection of fruit from trees in rows. The lower fruit are collected from the ground, while the rest are picked from a platform, with a single level. The collecting tubes replace sacks, as they permit great freedom of movement to the operator. Getting rid of steps and collection sacks increases the productivity of the labor force.*

Collection tubes

Fork lift truck

Rollers

Ⓐ

Ⓑ

Tree trunk

Pinch grip

Type C grip on a branch

Ⓐ

Ⓑ

Apple

YES

NO

Pear

(Left)
Manual harvesting:
Picking apples and pears is done by lifting and simultaneously twisting the fruit, rather than simply pulling it hard. This principle is true for most other species.

(Right)
Mechanical harvesting:
Two general types of grips used to collect fruit by shaking the branch
A/ Type C grip on a branch
B/ Pinch grip on the trunk of a tree

flesh remains hard and has an insipid acidic taste, and does not last as long.
Fruit that is collected late lacks flavor, has a floury texture and is extremely susceptible to disease.

The advantages and disadvantages of harvesting unripe fruit are:

• In excessively early harvests

- Lack of color and loss of 10-20% of weight.
- Do not ripen well in refrigerated chambers.
- Less smell and taste, as they have had less time to accumulate reserves.
- The fruit is more likely to be scorched, the longer the period it is stored and the lower the temperature.
- Given that transpiration is greater in the unripe fruit, it is more likely to wither.
- Greater tendency to bitter pit.
- The only advantage is that the fruit resists rot, and also handling, better.

• Excessively late harvests

- Fruit fall is abundant.
- Greater susceptibility to glassiness.
- Shorter storage life and more attacks of mold.
- Greater sensitivity to manipulation, mechanical damage, and internal rotting.
- Greater sensitivity to low temperatures and CO_2.
The only advantage is that coloring is more complete and the fruit is larger.

11.3. RECOMMENDATIONS FOR HARVESTING

- To harvest correctly, follow these basic guidelines.
- Harvest the fruit with its stalk, the petiole, but without leaves.
- Harvest the fruit at harvest ripeness, or close to it.
- Take the greatest care when handling, avoiding blows and bruises.
- Reduce handling operations as much as possible.

The ideal is for the fruit to go straight from the tree to the final packing for storage.

• Do not harvest wet or very moist fruit.
• Do not leave harvested fruit exposed to the sun.
• Put fruit picked at market ripeness into boxes of a single layer, as with melons.
• Large fruit is normally near the top of the tree, as is fruit on unproductive trees, and this should be sold first as it does not store for long.

In relation to the packaging, note that:

• The best packing is plastic, as it does not store moisture, it is light-weight and easy to clean. It also causes less damage to the fruit.

• Weak or flexible packing cause the most damage to the fruit.
• Wooden boxes should have smoothed edges.

Conditions that packing should fulfil:
• Side opening (at least 15% of surface area)
• Wooden laths with smoothed edges
• Separation between laths of no more than 6 mm
• Rigid structure
• Standard size to make paletting easier

Side protectors are not necessary in plastic packing but are needed in wooden ones, whose surface is discontinuous. These side protectors are recommended for transport, but not when the fruit is placed in refrigerated chambers, as it prevents the cold air from coming into full contact with the fruit. This is resolved by holes in the protectors, representing about 25% of the surface area.
If the fruit must be transported a long distance, or over poor roads, side and bottom protection in the lorry are essential.
In long journeys by lorry, the boxes of fruit on the top and at the rear must have something soft between them and the roof, so the fruit is not violently shaken. If there is no padding, great mechanical damage may be caused.

11.4. VARIETIES OF PEARS

Agronomic characteristics

1/ KAISER " C. BOSC". Semi-vigorous. Poor affinity with quince stock. Very productive. Attacked by pear suckers (jumping plant lice). Very little cultivated in Spain. High quality fruit and not sensitive to fungi in the field. Medium to large sized fruit.
2/ BLANQUILLA DE ARANJUEZ. Very vigorous. Requires a quince stock and requires hormones to induce branching and ensure fruit-set: responds very well to hormones. Recommended to prune for a shape that weakens the tree and ensures good aeration, as this improves production and controls the mottling this variety is prone to.
3/ BUENA LUISA. Very productive. On calcareous and/or infertile sites, grows best on a wild stock. A quince stock favors vigor. Fruit thinning improves size and coloring. Strong tendency to alternate production. It seems to be one of the varieties least affected by fire blight.
4/ CONFERENCE. Not very vigorous. On infertile soils, a wild stock is recommended. It is very sensitive to leaf scorch (Brussone). Resistant to mottling. Prefers wet sites and environments. Sometimes, some fruit cracks in May, but they heal very well. Better to prune aiming for an open shape.
5/ Passe Crasanne. Not a very vigorous tree. On infertile sites, better on a wild stock. In some areas, it has problems entering production. Trees with abundant reserves at the end of a season show better fruit-set the following flowering season. Medium quality.
6/ DOYENNE DU COMICE. Pear from the Pyrenean area of Puigcerdà.
SUPER COMICE. Very vigorous tree. Irregular and erratic production. Pruning and pinching out are necessary to force it to start production. Prune hard.
7/ ERCOLINI. Medium to good vigor. Very productive. Sometimes requires thinning. Pollinates many other early and semi-early flowering varieties. Very commercial variety: medium quality.
8/ GENERAL LECLERC. Very vigorous. Productive. Usually produces large, unevenly colored, fruit. The fruit are very susceptible to rots and diseases in the period immediately before harvest. When young shows strong tendency to shed fruit. Be prudent with this variety.
9/ JULES GUYOT. Low affinity with quince stock. Low vigor, very productive (parthenocarpic production). Recommended for areas with long and/or harsh winters. Very robust plant and fruit in the field.
10/ ABBOT FETEL Irregular and erratic production. Good quality. It is being planted in Italy again. Weak to medium vigor. Medium to large fruit.
11/ BUTTER AREMBERG. Its main virtue is its resistance to spring frosts.
12/ BUTTER HARDY. Vigorous to very vigorous. Slow to start production. Fruit medium to large. Good quality. The time of harvest must be just right if it is to be stored for 2.5 to 3 months. Requires long branches. Very resistant to calcium.
13/ ALEX. DROUILLARD. "Countess". Good vigor. Rather sensitive to chlorosis and leaf scorch. Medium to large fruit. Productive to very productive (tendency to alternate). Medium to low quality.
14/ EPINE DU MAS. Duke of Bordeaux. Medium vigor. Low affinity to quince stock. Medium size fruit. Productive. Sometimes needs thinning.
15/ HIGHLAND. Good vigor. Very productive. Tends to shed fruit if crop is heavy. Medium to large fruit. Medium quality. This variety should be monitored. Also suitable for preserving, etc.
16/ PR. DROUARD. Medium vigor. Very productive. Best grown on wild stock.
17/ PR. HERON. Medium vigor. Production starts young and is abundant. Rather sensitive to spring frosts. Should be monitored. Taste is inconsistent.

11.5. VARIETIES OF PLUMS

	1st month of summer		2nd month of summer		3rd month of summer	
	10	20	10	20	10	20
	Black Star					
	Strival					
		Black Gold				
			Black Diamond			
						Sungold
						Angeleno

Variety	Ripening	Fruit size	Skin color	Shape	Pollination
Black Star	− 0.5	Large	Shiny black	Round	Bl. Gold, Bl. Diamond
Strival	− 0.4	Large	Violet red	Spherical	Sorriso, Pr. Golden Jap.
Black Gold	+ 0.2	Large	Deep blue	Round	Angeleno, Bl. Diamond
Black Diamond	+ 13	Very large	Dark violet	Flattened	Angeleno, Bl. Gold
Angeleno	+ 68	Large	Dark violet	Round	Bl. Gold, Bl. Diamond
Sungold	+ 68	Large	Greenish-yellow	Round	Friar, Laroda, Fortune

1/ *Sungold*
2/ *B. Stark*
3/ *S. Black Gold*
4/ *B. Diamond*
5/ *Angeleno*
6/ *Strival*

11.6. VARIETIES OF APPLES

VARIETIES ILLUSTRATED: Standard Red Delicious: 1/ RED Delicious: **2/** STARKING Delicious: **3/** RICHARD Delicious: **4/** TOPRED Delicious: **5/** ROYAL RED Delicious: **6/** EARLY RED ONE (*)
(*) Though it is a standard variety (a sport of Red King) it has very intense coloring and diffuse, as if it was a Spur.
VARIETIES NOT SHOWN IN THE ILLUSTRATIONS: STARK Delicious: HI-EARLY: SHARP RED Delicious: SHOTWELL Delicious.
AGRONOMIC CHARACTERISTICS: All these varieties are vigorous to very vigorous; they must be grafted on stocks that are not very vigorous and/or trained in a way that reduces vigor, both to shape them and to ensure an early start to production.
They are sensitive to low spring temperatures (more than Golden Delicious).
Once they have started to produce, production is consistent and high, and for the same shape and stock, show less tendency to biennial production than Golden Delicious. They require pollinator trees (in cold areas, at least 25%). They are sensitive to mottling, and also to canker if they are forced when shaping. Little or no sensitivity to powdery mildew.
The most recommended varieties at the moment are Topred (unbeatable colour) and Starking. Though it may be better quality, the variety Red Delicious is no longer grown because of its slow and inadequate coloring.

VARIETIES ILLUSTRATED: Red Delicious "SPURS": 7/ OREGON; **8/** STARKRIMSON; **9/** WELLSPUR; **10/** RED SPUR; **11/** MILLER STURDY; **12/** ELITE; **13/** REDCHIEF; **14/** TOPCROP; **15/** CLEAR RED.
VARIETIES NOT SHOWN IN THE ILLUSTRATIONS: STARKSPUR RED; SPUR; STARKSPUR PRIME; RED; STARKSPUR; SUPREME; "SEMI-SPURS"; EARLY RED ONE (*).
(*) See Table Standard Red Delicious
AGRONOMIC CHARACTERISTICS: Most Spur varieties, though they show good vigor, show less growth than the standard varieties. Some are, however, equally vigorous. Graft onto medium to very vigorous stocks.
These varieties respond very badly to bending the branches flat or down (they try to grow back up), and this may lead to attacks or severe attacks of papery *canker*
The natural development of the Spur is in the form of a closed clump and this should be respected as far as possible.
Very productive and with intensely colored fruit (red), that may sometimes be excessive. The varieties Oregon, Elite, and Clear Red are remarkably red which makes them very commercial. They are sensitive to mottling, papery canker, and preferred (especially Starkrimson) by the red spider mite. Initial quality (not very high) improves with storage.

VARIETIES ILLUSTRATED: 16/ GOLDEN; **17/** SMOOTHEE; **18/** GOLDEN 972
AGRONOMIC CHARACTERISTICS: Smoothee is a Golden cross, with little or no russetting. Golden INRA-972: less affected by russetting than Golden Standard.
They are the most productive Golden crosses or selections. They are very easy to shape. Generally sensitive to powdery mildew and leaf-fall (water, heat and/or nutritional problems). In situations of risk of russetting, the varieties 972 and Smoothee should be chosen.

VARIETIES ILLUSTRATED: 19/ BELGOLDEN (Goldensheen); **20/** LYSGOLDEN (Goldenir)

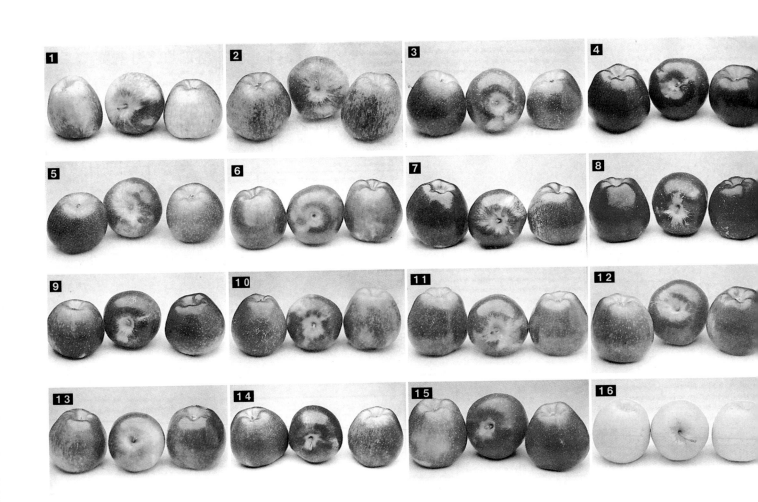

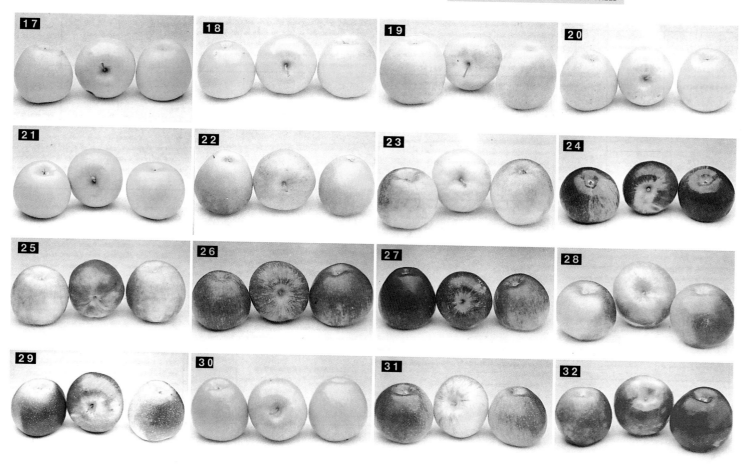

AGRONOMIC CHARACTERISTICS: Both varieties are less vigorous than Golden. Their productivity is also a little lower. The fruit shape is slightly longer (especially in Lysgolden) and the fruit size is very good (larger than Golden and 972). They show slight pink coloration on the side exposed to most sun (more pronounced in Belgolden). More sensitive to powdery mildew than Golden and Golden 972.

VARIETIES ILLUSTRATED: 21/ MUTSU (Crispin)
AGRONOMIC CHARACTERISTICS: Golden cross of Japanese origin (Golden x Indo). Its greatest, and perhaps only, interest is that it is resistant, or shows little susceptibility, to russetting. Very vigorous tree. The fruit are longer and greener in color than Golden Standard. Large fruit. Quality inferior to Golden.

VARIETIES NOT SHOWN IN THE ILLUSTRATIONS, "SPURS": GOLDEN AUVIL; STARK GOLDEN.
AGRONOMIC CHARACTERISTICS: These varieties show the typical features of "spurs" (less vigor, closed growth form and rapid start to production and elongated fruit), and also show a tendency to alternation, greater susceptibility to russetting, and clearly inferior fruit quality. Not recommended.

VARIETIES ILLUSTRATED: 22/ BLUSHING GOLDEN (Grifer)
AGRONOMIC CHARACTERISTICS: This is a partly colored Golden cross (Jonathan x Golden). It shows good vigor, a good and rapid start to production. Fruit is longer than Golden, with the sun-exposed side totally colored, and totally free from russetting.

VARIETIES ILLUSTRATED: 23/ JONOGOLD (Golden x Jonathan).
AGRONOMIC CHARACTERISTICS: Medium to high vigor. Very productive. Fruit size is good. It is very regular. The color is better in hilly areas. In areas where the color can develop fully, the fruit is very attractive. The quality is excellent. This variety's set of characteristics make it one to take into consideration.

VARIETIES ILLUSTRATED: Jonathan group: 24/ JONEE sport of Blackjon; **25/** GOLDJON sport of Blackjon
VARIETIES NOT SHOWN IN THE ILLUSTRATIONS: BLACKJON; NURED JONATHAN sport of Jonathan
AGRONOMIC CHARACTERISTICS: Good vigor and high production. Well-colored, very attractive fruit: medium size. Within the acid varieties, it is of excellent quality. Highly sensitive to powdery mildew. Blackjon is the most susceptible. Gradual cooling is recommended. Applying calcium chloride to soil may improve fruit quality and storage.

VARIETIES ILLUSTRATED: Stayman group: 26/ IMPROVED BLACKSTAYMAN (Nured Stayman); **27/** STAYMAN RED
AGRONOMIC CHARACTERISTICS: High vigor. Average growth form, regular productivity. Fruit size is good. Deep red color, but may be too dark when removed from refrigeration, making them less attractive. Rather sensitive to powdery mildew and mottling; harvested early, it is rather susceptible to bitter pit.

VARIETIES ILLUSTRATED: 28/ IDARED (Jonathan x Wagener)
AGRONOMIC CHARACTERISTICS: Medium vigor, rapid start to production and very productive. Good fruit size, pale red coloring of half the fruit (insufficient). The quality is excellent, especially for Central European tastes. Sensitive to powdery mildew. Fruit ripens over a long period.

VARIETIES ILLUSTRATED: 29/ GLOSTER (Richared x Cloche)
AGRONOMIC CHARACTERISTICS: Vigorous and well branched. Good productivity. The fruit vary quite a lot in size, shape and color. In favorable zones, the fruit's color and appearance is attractive.
The *Cemiostoma* leaf miner seems to be highly attracted to this variety. Requires careful monitoring.

VARIETIES ILLUSTRATED: 30/ GRANNY SMITH
AGRONOMIC CHARACTERISTICS: Good vigor, but branches tend to be left without leaves at base. Begins production at early age, and productivity is medium. Fruit size is medium to good. It is sensitive to powdery mildew and to (mosaic) virus.
As it is late-ripening (it is the latest variety), it should not be planted in areas with a short fall. As long as it is stored well, its quality (slightly acid, juicy and crunch⬤ is excellent, and especially after March-April, when other varieties are starting to deteriorate.

VARIETIES ILLUSTRATED: 31/ ROME BEAUTY; **32/** ROME RED BEAUTY; **33/** COOPER RB-1 (Spur of Beauty).
AGRONOMIC CHARACTERISTICS: Variety with good vigor, medium productivity, with a tendency to biennial production. Good fruit size. Average quality fruit. The Cooper RB-1 Spur is more productive and its fruit show more color. This variety's main, and perhaps only, interest is its resistance to cold winters (it flowers very late).

VARIETIES ILLUSTRATED: 34/ BOSKOOP BEAUTY; **35/** GREY PIPPIN
VARIETIES NOT SHOWN IN THE ILLUSTRATIONS: WHITE PIPPIN and other European varieties, such as Cox's Orange, Crimson Cox's, Mans Pippin, Clochard Pippin.
AGRONOMIC CHARACTERISTICS: These varieties, especially Boskoop Beauty, generally show a tendency to biennial production. Although the prices they attair are the highest on the market, it should be remembered that this is for a minority taste. Only a few markets, and with restrictions, would accept small quantities o⬤ this variety.

VARIETIES ILLUSTRATED: 36/ BELCHARD (Chantecler) (Golden x Clochard Pippin)
AGRONOMIC CHARACTERISTICS: Good vigor and tendency to "denude" the bases of the branches. Starts fruiting rapidly. Its productivity is medium, as is the fruit size and quality.

VARIETIES ILLUSTRATED: 37/ CHARDEN (Golden x Clochard Pippin)
AGRONOMIC CHARACTERISTICS: Very vigorous plant. Rapid start to production and good productivity. Its fruiting shape means that almost no thinning is required. In flat hot areas, the fruit are large and tend to mealiness. On hillsides the fruit is small, and they do not tend to mealiness. As they are very prone to vir⬤ attack, propagation must from with healthy material.

VARIETIES ILLUSTRATED: 38/ GOLCHARD (Cloden) (Golden x Clochard Pippin)
AGRONOMIC CHARACTERISTICS: Semi-vigorous (it is a vigorous Spur). Very productive, with very uniform medium-size fruit. Good quality. It tends to fruit on young wood, at the tip of the branches.

VARIETIES ILLUSTRATED: 39/ BELRENE (sport of King of Pippins)
AGRONOMIC CHARACTERISTICS: Very similar to King of Pippins, of which it is a sport. The fruits of Belrene have the most color, and their ripening is highly grouped.

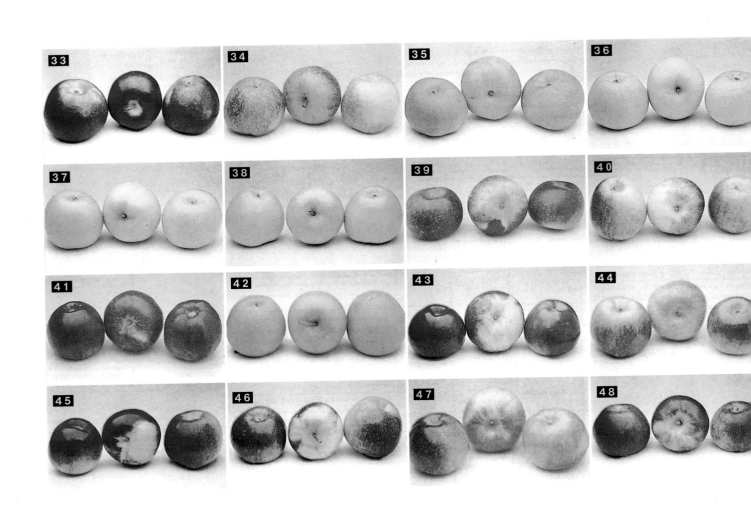

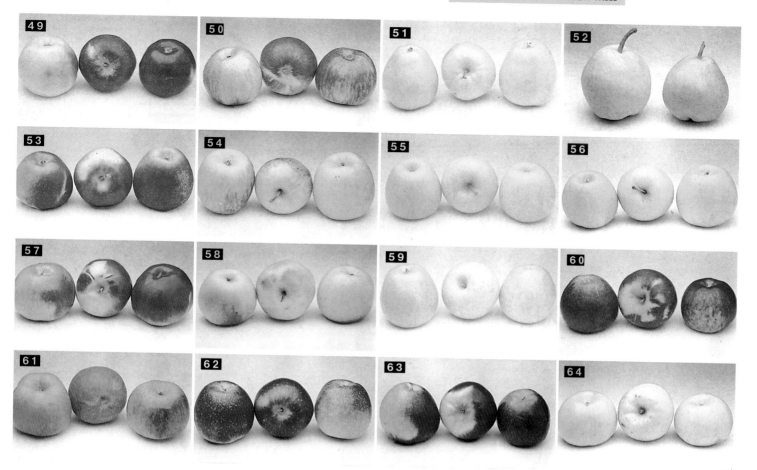

VARIETIES ILLUSTRATED: 40 and 41/ GALA and ROYAL GALA (Kidd's Orange Red x Golden)
AGRONOMIC CHARACTERISTICS: Behavior and vigor similar to Golden. Rapid start to production, and very high productivity. Medium to large fruit and excellent quality (very juicy fruit, crunchy and good smell). Royal Gala is a sport that is totally red.

VARIETIES ILLUSTRATED: 42/ BERTANE
AGRONOMIC CHARACTERISTICS: A mutation of Golden. The fruit is 100% protected from russetting. Quality good to very good.

VARIETIES ILLUSTRATED: 43/ SPARTAN (McIntosh x Newton)
AGRONOMIC CHARACTERISTICS: Medium vigor tree that takes a long time to enter production. Medium productivity. Medium quality and fruit size. Prone to drop fruit, but less than McIntosh.

VARIETIES ILLUSTRATED: 44/ QUERINA
AGRONOMIC CHARACTERISTICS: Quite vigorous plant with "weeping" branching. Starts production rapidly and its productivity is good. This variety's main interest is that it is resistant to mottling. The high quality fruit are very attractive.

VARIETIES ILLUSTRATED: 45/ PRIMA
AGRONOMIC CHARACTERISTICS: Medium vigor: tendency to "bare" patches on branches. Average productivity, which is higher in areas with harsh winters. Medium fruit size. Medium quality. Its best point, like the variety Querina, is that it is resistant to mottling. As a whole (fruit size and color), QUERINA is better.

VARIETIES ILLUSTRATED: 46/ PAULA RED
AGRONOMIC CHARACTERISTICS: Average vigor and rapid start to production. An excellent summer variety with medium-large fruit, good appearance and taste. Prone to fruit-fall.

VARIETIES ILLUSTRATED: 47/ MELROSE (Jonathan x Delicious)
AGRONOMIC CHARACTERISTICS: Semi-vigorous tree. Slow to start production. Medium productivity. Good fruit size. Good quality (very firm flesh, which is appreciated by the Nordic and Central European market). Requires sites that favor coloring, or otherwise it colors very badly.

Illustrations without text: 48/ AKANE; **49/** BEN DAVIS; **50/** CARDINAL; **51/** CLOCHE; **52/** WINTER DEACON (Flor d'hivern); **53/** DELBARD JUBILEE; **54/** DELCORF; **55/** GOLDEN B; **56/** GOLDEN HYB.; **57/** JERSEYMAC; **58/** OZARK GOLD; **59/** PERMAIN; **60/** RED WINESAP; **61/** SPIGOLD; **62/** SUMMERRED; **63/** TYDEMAN'S; **64/** GREEN DONCELLA.

*Photos and text
courtesy of
VIVAS COOPERATIVI
RAUSCEDO*

11.7. WINE GRAPES

Characteristics

AIREN

White variety cultivated especially in central Spain. It is a vigorous variety, with large bunches with wings.
The grapes are cylindrical, golden in color, large with waxy bloom and soft flesh. Not very sensitive to spring frosts because it sprouts late.
Excellent resistance to drought and very robust.
Clearly more fertile at the base, and thus adapted to a very short "Gobelet" type pruning.

ALBARIÑO

Sprouts early (approximately the same time as Chasselas) and rather late to ripen (two weeks after Chasselas). It is a vigorous variety and also very fertile, with grape production of 10-12 t/ha in the most common forms of cultivation.
The bunches are small, generally with a wing, which is sometimes so big that the bunch appears to be double. The berries are spherical and rather small, yellowish green in color, with a slightly muscatel-like flavour.
The most suitable shape to prune for is the traditional vine about 2 m tall, with very low planting densities, less than 1,000 rootstocks/ha and a planting frame of 3 x 4 or even 4 x 4, with mixed pruning back to leave some spurs (2-3 buds) and some rods (6-7 buds).

CABERNET SAUVIGNON

The bunch is medium-small, cylindrical, normally with a larger wing, rather compact with a medium size grape, spherical, blue-violet skins, firm pulp that is fleshy with a slightly herby taste.
Quite a vigorous variety, it sprouts medium to late, and the vegetation is relatively erect with medium-short internodes.
It adapts well to temperate climates and is best in dry or well ventilated sites.
In the north it prefers zones that are exposed to good sunlight, or hills and light soils, especially in valleys.
It does not do well on excessively fertile, wet, soils which lead to great vigor and problems in stem lignification.
It adapts well to different pruning shapes taking into account soil and climatic conditions. Production is regular and constant.
Ripens in the third period.

CARIÑENA

A variety that originated in Spain, and is now mainly cultivated in Aragon, Catalonia and the "la Rioja" region. In Italy, it is only grown on the island of Sardinia.
It is a vigorous variety, with an upright habit, and thus adapts well to pruning into a vase shape, but in windy areas, this may lead to problems due to stem breakage.
Sprouting and ripening are late (third period).
The bunch is large, cylindrical-conical and compact. The berry is spherical, medium size, and black with thick astringent skin.
Productive capacity is high, almost excessive, as this may compromise ripening in zones with a heat deficit.

CHARDONNAY

This variety is relatively homogeneous. The main differences between forms are in the greater or lesser seed size, and especially the grape's organoleptic qualities. The plant is vigorous and sprouts very early, meaning it should not be cultivated in areas at risk of late frosts. Vigorous shoots with a short internode, the vegetative growth tends to be balanced.
It can be trained in several ways, and adapts well to different planting frames, as long as they are not very narrow.
It can be pruned hard in the south or medium-long in the north as long as the buds left are in keeping with the vigor.
Production is average and constant for all forms of training. If pruning is mild, it may become abundant, but this decreases fruit quality.

GARNACHA

Variety that originated in Spain, and was later cultivated throughout the French Midi region (Grenache). In Italy, it is known as Tocai Rosso and Cannonau.
A vigorous variety that sprouts medium-late and produces stout shoots with short internodes, it adapts well to different areas. It adapts well to different methods of training, as long as they allow some room for expansion, and you should aim to leave few but long shoots. It raises problems for complete mechanization.

Garnacha (Tocai Rosso) VCR 3
An excellent and constant producing clone. Bunch is not excessively compact, and thus more disease resistant. Good sugar content and fixed acidity.

GRACIANO

This variety is originally from the La Rioja region of Spain. Its vigor is good, it sprouts late and ripens in mid October.
Its disease resistance is good. The bunches have two short cylindrical "shoulders", less prominent than in the Tempranillo variety, and they do not hang down.
The grape is round, dark black in color, relatively small in size, with thin skin, firm colorless flesh and very thick seeds. From afar, the new shoots as a whole have a reddish tinge. This makes it possible to recognize this variety in spring.

MACABEO

White variety. Seems to have originated in Spain, from where it spread to the French Midi region. It is cultivated in Catalonia, Aragon and the Upper Ebro. It is now widely grown in Castilla-La Mancha.
It is a vigorous variety, with a large compact bunch; the grape is round, medium size, and has a thin skin.
It adapts well to all climatic conditions and altitudes, but better if cultivated in cool fertile soils in heat regions II and III.
Grows well on most stocks, but best on less vigorous ones. Excess production causes quality to decline greatly. Grows well when pruned hard, though it can be mildly pruned depending on the soil and climatic conditions. This variety is prone to attacks of grey mold (*Botrytis*).

MERLOT

Relatively vigorous, average sprouting, normal shoots with short internodes, with balanced vegetative growth.
The average bunch is pyramidal with wings, and relatively loose. It adapts well to different methods of training and pruning. It is thus easily cultivated, and even totally free shapes can be completely mechanized, and it prefers medium pruning. The planting frame can also vary depending on the site, especially its fertility. Production is abundant and constant. With an equal load of shoots, it is more productive when they are left long rather than cut short.

MONASTRELL

A black Spanish variety that is widely grown on the east coast of Spain, especially in Murcia and Alicante.
The bunches are of medium size, and bluish in color, and the grapes are round with thick skin and juicy flesh.
It is sensitive to mildew and grey mould.
It ripens late, and does not adapt to all situations, as it may not ripen sufficiently.

SMALL GRAPE MUSCATEL (Moscato blanco)

The variety is relatively homogeneous. Its distinguishing characteristics are related to the shape of the bunch, its productivity and its smell, which are often linked to the cultivation environment.
The raceme is of medium size, semi-compact or semi-loose, cylindrical-pyramidal, ellipsoidal, amber yellow in color, easily separated, consistent skin, and fleshy pulp with a clear muscatel taste.
The plant is relatively vigorous, with medium-early sprouting, stout shoots with medium-short internodes, and relatively balanced vegetative growth. Grows well on calcareous sites, but not excessively clay soils or on sites exposed to the north. It adapts well to different pruning and training methods, as long as there are not too many long shoots.

PARELLADA

A white variety from Catalonia, mainly cultivated in the Alt Penedès area. It is a vigorous variety, with large bunches that are a little squashed if production is high. The grape is medium-large and has a hard skin, with a characteristic golden pink color.
It sprouts late, and ripens from September 20 onwards.
All the buds are very productive. They should be pruned hard so as not to "exhaust" the rootstock.
This variety is sensitive to drying out, and thus has to be grafted on stocks that make good use of the soil.
It should be cultivated at an altitude of more than 300 meters.

PINOT BLANC

A medium vigor variety that sprouts medium-early, with long shoots and lax vegetation with relatively long internodes. It does not adapt well to wet sites that tend to produce chlorosis. It prefers dry climates, or at least sites that are well exposed. It adapts well to different training techniques and planting densities. It prefers pruning back to medium or long shoots. Ripening is medium-early. Harvesting is early if the grape is for use in sparkling wines, and to avoid attacks of *Botrytis*. Its disease resistance is normal, except for its sensitivity to *Botrytis* and chlorosis.

PINOT NOIR

This variety is not very uniform. There are two different biotypes, the result of different aims in selection, that vary in leaf shape, bunch size and shape, and the quantity and quality of production. This description refers to the best quality Pinot Noir, cultivated in Burgundy.

The bunch is small, compact, cylindrical, normally with a single clear "wing", a short thick stalk for the bunch (peduncle), and a medium-small grape that tends to crack easily, and has black-violet skin, and clear, juicy, sweet flesh with a simple taste.

A medium-vigor variety, it is a medium sprouter, with branched shoots and medium-short internodes. It adapts well to a variety of soils, but it is better if they are not excessively fertile and wet. It prefers temperate climates that are not excessively hot, and good exposure.

RHINE RIESLING

This variety is relatively uniform. There are some differences in the size of the bunch and the organoleptic characteristics of the wine it produces.

The bunch is small, compact, with medium-small spheroidal grapes that are yellow with a firm skin, and juicy flesh with a delicate aromatic taste.

It prefers training forms at low densities, and pruning back to medium-large shoots. Production is good and constant. Medium ripening season. There are some problems in mechanical harvesting.

Fruit-set may be a problem in areas where it does not adapt well. It requires well exposed and ventilated areas to avoid damage from *Botrytis*.

SAUVIGNON

This variety consists of different biotypes that vary in their bunch size, and especially in the smell of the final product.

A vigorous variety, with medium-early sprouting, dense vegetative growth, upright shoots with short internodes, with many secondary buds that sprout. It adapts well to different sites, but prefers sites that are not very fertile or wet, or tend to induce chlorosis. It prefers well exposed sites and dry and temperate climates.

It adapts well to different training methods, with high densities. Pruning back to relatively many medium or long shoots. In espalier forms, manual or machine intervention is needed to correct the positions of the shoots, and summer pruning, especially before fruit-set and before harvest.

TEMPRANILLO

This is a vigorous variety with a medium size bunch that is compact, long and has a "wing".
The grape is of medium size, round and has normal skin.
It sprouts early, and matures in mid September.
It adapts well to all types of soils, and prefers south-facing sun sites.
It is not very sensitive to spring frosts, but is sensitive to hot wi in spring.
Sensitive to *Botrytis*.
Depending on the cultivation zone, it can be pruned hard or mildly, respecting the balance between production and quality and taking into account that it is a medium producing variety.

TREIXADURA

It is a vigorous variety, with good productivity, that prefers a broad training shape, and slightly cool climates. The leaf is of average size, round and entire or divided into three lobes, and bright green in color. The shoot is vigorous, spindle-shaped and upright.
The bunch is of medium size, cylindrical to conical, with a "wing", and very compact.
The stalk of the bunch is short and slightly lignified. The grape is medium to large, uniform and elliptical, with a thin waxy skin.
The flesh is not very sugary and it has a special taste.
Its spread is restricted because of its tendency to produce unbalanced wines. This variety is not particularly sensitive to any disease, though the bunch is so compact that it may be affected by *Botrytis cinerea*.

VERDEJO

This variety is the most cultivated in the Rueda. D.O. district. It moderately vigorous, with a strong and highly branched shoot. The vegetation's growth form is prostrate, with small leaves. Th bunch is small, loose, and the grapes are yellow and spherical with a waxy bloom and thick skin and a neutral taste.

XAREL·LO *(as spelled in the catalonian alphabet)*

This white variety is from Catalonia and was introduced by Greek sailors.
It is a robust variety, of medium vigor, that produces not very compact bunches. The grape is round and has a thick skin. It ripens after September 15.
It adapts well to most soils up to an altitude of 400 m.
Pruning has to leave long shoots, as the basal buds are not very productive.
Production is good, but depends on climatic conditions at the time of flowering, because of its tendency to drop its flowers. It shows excellent affinity with all stocks.

ALPHONSE LAVALLÉE (Ribier)

This variety was obtained in France in the last century, by crossing "Bellino" with "Lady Dawres Seedling". The bunch is medium to large, cylindrical to conical, with a "wing", not very compact, and with an average weight of 500-600 g.
The grape is spherical, medium to large, and weighs 7 to 9 g. The skin has a waxy bloom and is firm, with a uniform dark blue color, a relatively crunchy and juicy flesh, and a simple taste.
The fertility is 1.5 and productivity is good. It ripens 30-35 days after Cardinal.
Good vigor, and adapts well to all training shapes, but the planting frame should not be too big.

CARDINAL

This variety has rather large bunch, cylindrical to conical and elongated, relatively loose, sometimes with a "wing", and an average weight of 500-600 g. Medium large grape, round, relatively firm skin, red-violet color, not very uniform. The flesh is crunchy, sweet and pleasant, with a neutral taste, 2 to 3 seeds per grape, and an average weight of 7-9 g. The grape ripens early (around 15-20 July).
This variety is quite vigorous. It prefers vigorous stocks. Does not do well if pruning leaves long shoots. Its real fertility is approximately 1.5. It adapts well to both espalier and cordons.

ITALY

This is now grown in many countries, especially Italy, Spain, France, Northern Africa and Greece. The bunch is large, cylindrical to conical, with one or two "wings", not very compact, and with an average weight of 600-700 g. The grape is large and oval, with thick firm skin and a waxy bloom, and is golden yellow in color. The flesh is crunchy with a delicate Muscatel taste. Grape weight: 8-10 g.
Real fertility is 1.20. High production. Matures in the third period.

MATILDE

The bunch is very large, cylindrical to conical, elongated, not very compact, with an average weight of 700-800 g.
Large grape, average weight 6-7 g, juicy flesh with a slightly aromatic taste, and yellow in color.
Optimum real fertility of nearly 1.80. High production. Matures 5-10 days after Cardinal.
Vigorous variety. Needs large planting frames, but pruning should not leave excessively long shoots.
It shows normal affinity with most graft stocks, but the most vigorous stocks are the most useful (1103 Paulsen, Kober 5BB).

*Storage of fruit
and vegetables*

12. STORAGE

Successful storage is based on three points:

• **Field conditions**, such as the site, the climatic conditions, the rootstock, the variety, fertilization, pruning, thinning, pollination, harvesting time and the care taken during harvesting.
• **Handling of the fruit**.
• **Optimal functioning of the refrigeration chambers**. They should be insulated, thermostat controlled, high evaporator yield, effective seal for a controlled atmosphere, defrosting, automatic and with recirculation of air.

12.1. STOREHOUSE

Operations to be carried out:

• **Pre-ripening**. This consists of storing the fruit a few days in the storehouse at ambient temperature. This operation is negative, as the fruit loses weight due to evaporation of water, the risk of fungal attack increases, the storage life decreases and the flesh becomes less juicy and more mealy (floury).
No more than 24 hours should elapse between harvesting and the fruit entering the refrigeration chamber, and this period is only 12 hours for pears. The only advantage of pre-ripening is that it rectifies the negative effects of an excessively early harvest. Yet even so, it is much more negative than positive.

• **Classification by quality and size**. This is performed before placing the fruit in the chamber, and has a series of drawbacks:
• The time it takes is equivalent to a pre-ripening period.
• The correct removal of all fruit that is damaged.

• The damage caused by this classification operation.
• The time and space needed.
You are advised to carry out a preliminary rough classification in the field, and mechanical classification by quality and weight when the fruit is removed from the refrigerator chamber.

12.2. TREATMENTS

The most important treatments are those seeking to prevent scalding, especially in red apples, and rotting due to *glosporium*. Treatment for bitter pit is less common.
Ethoxyquin is used to prevent scalding. The most important thing in preventing *Glosporium* rot is to get the fruit inside the chamber as soon as possible

Recommended storage conditions, storage time, color production and the physical characteristics of the fruit of different deciduous species

Fruit	Storage temperature (°C)		Relative humidity (%)	Approximate storage time (days)	Upper freezing limit (° C)	Water content (%)	Specific heat capacity (Kcal/Kg/°C)	Heat production (Kcal/t/day) when the fruit is stored at		
	Minimum	Maximum						0°C	4-5°C	20-21°C
Avocado	4	13	85-90	14-28	− 0.28	65.4	0.72	—	1,222-1,833	4,500-21,194
Apricot	− 0.6	0	90	7-14	− 1.06	85.4	0.88	—	500-2,306	3,667-7,639
Bilberry	− 0.6	0	90-95	14	− 1.28	82.3	0.86	139-639	556-750	3,167-5,333
Cranberry	− 2	4	90-95	60-120	− 0.89	87.4	0.90	167-194	250-278	667-1.111
Persimmon	− 1		90	90-120	− 2.17	78.2	0.83			1,222-1,472
Cherry	− 1	− 0.6	90-95	14-21	− 1.78	80.4	0.84	250-333	583-861	1,722-1,944
Plum	− 0.6	0	90-95	14-28	− 0.83	85.7	0.89	111-194	250-556	1,028-1,583
Raspberry	− 0.6	0	90-95	2-3	− 1.11	80.6	0.85	1,083-1,528	1,889-2,361	
Strawberry	0		90-95	5-7	− 0.78	89.9	0.92	750-1,083	1,000-2,028	6,250-11,972
Pomegranate	0		90	14-28	− 3.00	82.3	0.86			
Currants (*Ribes Sp.*)	− 0.6	0	90-95	7-14	− 1.00	84.7	0.88			
Gooseberry (*R. grossularia*)	− 0.6	0	90-95	14-28	− 1.11	88.9	0.91	417-528	750-833	
Sour cherry	0		90-95	3-7	− 1.67	83.7	0.87	361-806	778-806	2,389-3,056
Fig	− 0.6	0	85-90	7-10	− 2.44	78.0	0.82	—	667-806	3,472-5,806
Apple	− 1	4	90	90-240.	− 1.50	84.1	0.87	139-250	306-444	1,028-2,139
Peach	− 0.6	0	90	14-28	− 0.94	89.1	0.91	250-389	389-556	3,611-6,250
Quince	− 0.6	0	90	60-90	− 2.00	85.3	0.88			
Blackberry (*Rubus ursinus*) «Blackberry»	− 0.6	0	90-95	2-3	− 0.83	84.8	0.88	1,083-1,194	1,917-2,500	9,528-11,778
«Dewberry»	− 0.6	0	90-95	2-3	− 1.28	84.5	0.88			
«Loganberry»	− 0.6	0	90-95	2-3	− 1.28	83.0	0.86			
Nectarine	− 0.6	0	90	14-28	− 0.89	81.8	0.85			
Pear	− 2	− 0.6	90-95	60-210	− 0.156	82.7	0.86	194-417	306-611	1,833-4,278
Elderberry	− 0.6	0	90-95	7-14	—	79.8	0.84			
Grape (wine)	− 1	− 0.6	90-95	90-180	− 2.17	81.6	0.85	83-139	194-361	
Grape (American)	− 0.6	0	85	14-56	− 1.28	81.9	0.86	167	333	2,000

without pre-ripening and treat the fruit on the tree with benomyl, methyltrophanate or dichlofluanide. In the case of bitter pit, they should be washed in a 3% solution of calcium chloride, an excellent fungicide. Mixing it with ethoxyquin is effective against bitter pit, rots and scalding.

12.3. PACKING

There are three types of packing:
• Totally closed boxes
• Crates (boxes with plastic or wooden laths)
• Wooden pallets
The best material is plastic, as it is light, easily cleaned, does not damage the fruit and does not take up moisture, though it has the disadvantage of being expensive.
If wooden packing is used, the laths should have smoothed edges and the distance between the lathes should be less than 6 mm. The sides should be at least 15% open. It is worth putting a protector with holes at the bottom of the box in order for air to circulate. This is optional on the sides.
Wooden packing should be moistened in advance, because if it is very dry it will absorb moisture within the chamber.
It is worth pointing out that the number of fruit damaged by bruising is much greater in one-way, non-reusable boxes than in reusable packing.

12.4. LOADING

When loading, the following factors should be considered to ensure satisfactory storage:
• *Intensity*. This is the number of kilos that enter the refrigeration chamber every day. This is calculated on the basis of the chamber's refrigeration capacity.
• *Arrangement and stacking*. The packages must be arranged so that their longest axis is parallel to the flow of air. This creates the largest number of passageways to ensure good air flow.
• *Density*. The chamber should be loaded so that 10% of its total volume is left free. You must leave a gap between the load and the roof, between the load and the ventilator, and between the load and the sides of the chamber, in order to ensure correct air circulation and cooling.

The basic guidelines to obey are:

• There should be a pallet at the base and between every 6-7 layers of packing.
• There must be a distance of at least 5 cm between the load and the walls of the chamber.
• On the side facing the ventilators, the distance between the wall and the load should be at least 10 cm.
• If the evaporator and ventilators are at one end of the chamber, a strip should be left unloaded that is as wide as the evaporator and approximately half a meter lower.
• In relation to the height of the load, stacking should be stepped so that in the first quarter of the chamber, the load is no higher than the height of the base of the ventilators. From this point, the height in

the stack can increase until the gap is only 25-30 cm.
• The minimum distances mentioned above have to be increased in bigger chambers, leaving 15-20% of the total space free.
• It is essential to leave at least one passage to make monitoring easier, if the fruit is going to be stored for any period of time.

Multi-use fruit and vegetable grader. Manufactured by Calibrex (France)

• Load in accordance with the fruit's storage potential. This depends on a series of factors, such as the age of the plant, the rootstock, the quantity of the harvest, the fruit size, the site, fertilization with nitrogen, climate and production.
Storage potential refers to how well the fruit will store, as well as the appearance of possible physiological disorders.
• Loading should be fast. This is closely related to what was said on pre-ripening.
• Take the thermostat into account when loading. It is important not to put hot fruit next to the thermostat, as the temperature is not representative of the temperature in the chamber as a whole, but it can still turn the refrigeration on, thus causing damage by freezing fruit that is already cold.
If boxes with hot fruit are placed in the chamber, and they can only be placed near the thermostat, the problem can be resolved in two ways:
- Place boxes that are already cooled on top of them.
- Stop the compressor for half an hour and turn on the fans, so that the cold air lowers the temperature of the new fruit.

12.5. STORAGE CONDITIONS

12.5.1. Conventional cold-storage

• **Temperature**

You should always start with fruit that is a suitable condition for long-term storage. In pears, the initial temperature to reach is 0°C, and this is then lowered to –0.6°C. The temperature should never exceed a maximum of 0.5°C or a minium of –1°C.
For Starking apples the storage temperature is between 0°C and 1°C. For Golden apples the storage temperature is between 0.5°C and 1.5°C. European varieties are stored between 2.5°C and 3.5°C.

A high storage temperature shortens the length of the storage period, and low temperatures may cause problems such as *storage scald*, which is when the fruit has trouble ripening properly once it has left the chamber. One important idea is cooling at a decreasing temperature. In fruit that is too green this ensures a slight increase in the quality of the finish (ripeness and color) and also reduces the refrigeration necessary during loading.

The application of this decreasing temperature is variable, as it depends on the variety and the degree of ripeness, as well as the fruit's storage potential. It is usually between 15 days and 6-7 weeks.

• Moisture

The relative humidity should be high, roughly 88-92%. At lower levels, the fruit loses weight due to the loss of water. The fruit wrinkles, loses its juiciness and becomes less attractive.

Higher relative humidity values are hard to achieve and are not recommended, as they favor the attack of fungi and also give the fruit an unpleasant taste and smell.

To achieve the right level of humidity, 5 points should be taken into account:

• Wooden packing should not be very dry when it enters the chamber, as it can absorb 20% of its weight in water vapor.
• The fruit should not have been pre-ripened, as it loses a lot of moisture during this process.
• In chambers that are not totally full, it is hard to achieve the right levels of humidity.
• The temperature difference should be as little as possible, and try to avoid differences of more than 6-7°C. The temperature difference is the difference between the storage temperature of the chamber and that in the evaporator, the gas expansion temperature. To ensure this temperature difference is as small as possible, between 1.5-1.8 m^2 of evaporating surface is needed per tonne capacity.
• To carry out defrosting carefully.

12.5.2. In a controlled atmosphere

This system makes it possible to work with temperatures that are not so low, and gives better results.

The humidity in this system, all other things being equal, is always higher than in normal storage, with levels of 90-94%.

These levels are attained because the system is hermetically sealed and because the higher temperatures reduce the temperature difference and due to the fact that the concentrations of the different gases leads to decreased metabolism and transpiration.

The basis of cold-storage in a controlled atmosphere is controlling the concentrations of oxygen and carbon dioxide, which can vary between 2% and 5%, although the optimum is a level of 3% for both gases.

The following points should be borne in mind:

• The O_2 concentration should not decrease below

2%, as this would provoke anaerobic respiration (fermentation).
• The best levels are 3-5% O_2 and 2-4% CO_2, as these levels reduce scald caused by high levels of CO_2.
• Exceptions: the Passe Crasanne pear tolerates CO_2 levels up to 10%.
• Ripe fruit is sensitive to CO_2.
• The color and maturity should be those desired for the moment of sale, as the fruit hardly changes at all during storage.

12.6. CHECKING AND CORRECTIVE OPERATIONS

12.6.1 Conventional cold storage

Controls to be carried out in storage:

• Defrosting

This operation is essential, because if the ice is not removed from the evaporator, it does not refrigerate effectively.

There are 4 ways of doing this:

• *Using air.* This is done with the compressor at rest and the ventilators in operation. This method is slow but is excellent for incorporating moisture into the environment. The disadvantage is that when the defrosting has finished, the air is dry, meaning that may cause water loss. Even so, this is the best system and the most widely used.
• *Using water.* This methods consists of spraying water on the evaporator. This is a rapid system and hardly raises the temperature, but the humidity is not reabsorbed into the environment.
• *Using electric heater elements.* This method is rapid but the moisture is not incorporated into the environment. It also leads to quite large power consumption and raises the temperature to some extent.
• *Using hot gas.* The system is rapid but decreases the humidity levels within the chamber. It also causes a large increase (3°C) in the temperature within the chamber.

• Turnover of air
This operation is no longer used, as the air that enters in the different controls is more than enough.

• Temperature
The temperature should be checked regularly, especially the minimum temperatures in the coldest points within the chamber.
After an automatic stoppage of the compressor, the lowest temperatures in the chamber are near the ground, about 30 cm above the floor.
In winter the refrigeration is often off for several hours, and this causes the air to stratify, the coldest air accumulating at the lowest levels. In order to avoid this stratification, the installations must be turned on for 2-3 minutes every hour.

• Humidity
The relative humidity is hard to measure, as it varies depending on whether the refrigeration is in operation or not, and it is not the same in the passageways as within the packing.

• Airflow
This is a characteristic of the installation, and good airflow is necessary to ensure that the cold air reaches everywhere in the chamber and to prevent the formation of warm pockets or sites that are not accessible.
There should be no problems in a chamber with refrigerators that allow an adequate rate of recirculation and an airspeed of between 0.25 and 0.4 m/s. If problems occur, they are probably due to incorrect loading, or an excessive or badly distributed load.
The coefficient of recirculation is the number of times per hour the ventilators inject or move a volume of air equal to the volume of the empty chamber.
The normal coefficient is 22-24, though in overloaded chambers it should be raised to 27-30.

• Monitoring the fruit
This depends on the fruit. If it is very long-lasting, the controls are infrequent, once every 3-4 weeks starting 2.5-3 months after the fruit is placed in the chamber. Fruit that is susceptible to deterioration should be checked every 15 days, starting 6-7 weeks after the fruit is placed in the chamber.

• Checking for soft scald
It is important to check for soft scald if the fruit was riper than desirable when collected.
If you detect the first symptoms of soft scald when performing a check, you must do the following:

• Pick out all the affected fruit, as this will rot and affect the healthy fruit.
• Partially unload the chamber, leaving no more than 70% of the total capacity, to ensure maximum air flow.
• Raise the storage temperature by about 2°C relative to the temperature at the time of the check.
• Ventilate intensely with the doors open and the ventilators in operation, for at leat two hours.

If you do this, you will probably stop the soft scald progressing and prevent more fruit being affected.
If the scald is normal, losses will be caused, especially because the fruit is less attractive.

In other disorders, such as bitter pit, the process is irreversible once it has started, and the fruit should be sold rapidly. If rot appears, you must act quickly to save what you can.

• Damage due to low temperatures
If by error the fruit is damaged by excessively low temperatures, but has not frozen solid, what you must do is turn off the refrigeration and open the chamber so the temperature can rise a few degrees. Never take the fruit outside to the ambient temperature, as rapid thawing has a dreadful effect.

12.6.2. Cold storage in a controlled atmosphere

Checks to be carried out during storage:

• *Defrosting.* All that was said in the preceding section on defrosting applies fully.
• *Humidity.* This does not show any problems.
• *Temperature.* Check the control devices work properly and regulate the temperature effectively before you close the store with the fruit inside it.
• *Gas concentration.* O_2 and CO_2 are checked by measuring their concentrations.
• *If O_2 is missing,* air is injected using a turbine until the desired level is reached.
• *If there is excess O_2,* it is burnt off.
• *If there in not enough CO_2,* it is worth reducing the time the decarbonizer is in operation.
• *If there is excess CO_2,* increase the time the decarbonizer is turned on.
• *Airtightness.* It is essential to check that the chamber is airtight, because if it is not it will be difficult to maintain the gas levels and temperature, etc.
• *Checking on the fruit.* It is worth carrying out monthly checks after mid January.

12.7. UNLOADING THE COLD CHAMBER

In general, there are few problems when it comes to unloading the fruit. Apples, for example, raise no problems. The longer pears are kept, however, the more problematic they become.
If the storage period is too long, when they are removed the fruit show problems in ripening.
To avoid external softening or mechanical bruising, the fruit should not be handled when it is still cold, shortly after it has come out of the cold chamber, especially after more than 4 months of storage.
One of the solutions to the difficulty of ripening the fruit after it has left the chamber, is to raise the temperature a couple of degrees a week or two before removing it.
This ensures correct, simultaneous and uniform ripening.

The chamber should be unloaded on the basis of two criteria

• Unloading should be done as quickly as possible, as water losses increase as the chamber is emptied.
• Unloading should start at the sides, lowering the height of the stacks, and opening passageways, to improve the conditions during this final phase of conservation.

13. FRUIT SPECIES

In order to correctly apply the principles established in the other sections, it is necessary to know the general characteristics of the main species of cultivated fruit-bearing plants.

13.1. SEED FRUIT

The fruit is a fleshy pome. The fleshy part of the fruit is the receptacle of the tissue of the calyx. This is why fruit can sometimes be produced without fertilization.

pollinator varieties are planted in the plantation.
All the varieties show rootstock compatibility. This is why some varieties are used as rootstocks in certain site conditions or because of their resistance to certain diseases.
It grows well in temperate climates, favoring the wet side. It is less affected by cold than by heat. It is damaged by abundant moisture and humidity during flowering.
From flowering to maturity takes 100 days on average in early varieties and 147 days in medium varieties.
The fruit do not need strong light to color well, but do need moderate airing.

Apple, and section showing seeds

Cultivar	Size of tree	Days from flowering to harvest	Flovering period[1]	Fruit size[2]	Fruit color[3]	Use[4]	Days of cold storage (1)	Need for pollination	General productivity[5]
Giffard	M	100-120	M	M-L	Y-P	F	Short	Yes	G
Early Morettini	—	100-125	E	—	—	F	—	Yes	G
Limonera	M	105-125	M	L	Y	F	Short	Yes	G
Clapp Favorite	L	105-130	M	L	Y-P	F	50-70	Yes	H-G
Williams	M	110-135	M	M	Y-P	F, T	70-85	Yes	H-G
Seckel	M-L	120-140	M	S	R-P	F	90-100	Yes	H-G
Butter Hardy	M	130-150	M	M	R	F, T	75-140	Yes	G
Eldorado	M	140-160	M	M	G-Y	F	180-220	Yes	G
Anjou	L	140-165	E-M	M-L	G-P	F	175-185	Yes	L-G
Bosc	M	150-165	L	M-L	R	F, C	90-100	Yes	G
Packham's Triumph	M	150-165	M	M-L	G	F	170-190	Yes	G
Doyenné du Comice	L	150-170	L	L	G-P	F	90-105	Yes	L-G
Angouleme	M	150-170	E	L	Y	F	—	Yes	G
Flemish Beauty	M	160-180	L	M	Y-P	F	—	Yes	G
Conferencie	M	160-180	—	M-L	G	F	—	Yes	H-G
Easter	M	160-185	—	M	Y-P	—	90-100	Yes	—
Winter Nelis	M	160-185	L	S	G	F	175-230	Yes	H-G
Forelle	M	160-190	E	S	G-P	F	—	Yes	H-G
Kieffer	M	170-190	E	L	Y-P	F, T	90-120	Yes	H-G
Glou Morceau	M	170-200	—	L	G-Y	F	Long	Yes	G
Clairgeau	M	170-200	M	L	Y-P	F	—	Yes	G
Passe Crasanne	M	180-210	M	M-L	G-R	F	Long	Yes	—
Nijiscike	M	—	L	M-L	Y	F	Long	Yes	H-G
Ya Li	M	—	E	L	Y-G	F	Long	Yes	

KEYS: 1 E = early, M = medium, L = late
2 S = small, M = medium, L = large
3 Y = yellow, G = green, R = reddish, P = pink
4 F = Fresh consumption, T = tinned, C = compote
5 G = good, H = high, L = low
(1) The storage time is several weeks longer in a controlled atmosphere

Pear cultivars of the world and their main features

The most important seed fruit are:

• Pear
• Apple
• Quince

Pear, showing ripe fruit on branch, and section

13.1.1. Pear

Pyrus communis. Rose family (Rosaceae)

The pear is a tree with a round pyramidal growth form when it is young, and can live for 75 years. It is generally deciduous and sometimes produces thorns.
The leaves are oval, finely toothed or entire, glabrous, occasionally downy, and shiny above.
The pear flowers in April, at a temperature of 10°C. Its flowers are white or sometimes pink.
The fruit is a globose or pyriform (pear-shaped) pome that ripens between July and October.
The seeds are within stringy cells and are matt black. In certain climates, some cultivars can produce fruit without seeds, though the best production is when

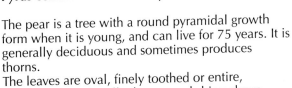

Quince. Detail of flower

n terms of sites, it grows well in deep, cool, soils,
hat are clay-siliceous-limestone and with quite a lot
of humus.

13.1.2. Apple

Pyrus malus. Rose family (Rosaceae)

The apple is a deciduous tree with thornless
branches. The crown is round and the tree can reach
an age of 80 years.
The leaves are oval, serrate, soft, and are glabrous or
downy depending on the variety. The
hermaphroditic flowers are white, pink or crimson,
and open in the middle of May.
The fruit is a globose (spherical) pome that contains
shiny dark seeds. Some varieties ripen in just 70
days, while others take 180 days.
It is one of the most widespread fruit trees in the
temperate zones because it adapts well to a variety
of climates.
In general, it requires a temperate climate, cool and
rather moist, and does not adapt well to hot dry
regions.
Its great genetic variety means it is impossible to
generalize about the apple's climate requirements.
Some varieties need a long cold period while others
only require a short one.
The apple tree can produce in sites that are not very
sunny, where a pear could not produce, but they do
require a well-aired site.
In terms of the soil, the apple requires permeable
clay-limestone or clay-siliceous soils. It is less
demanding than the pear in the depth of soil it
requires.

13.1.3. Quince

Cydonia oblonga. Rose family (Rosaceae)

Shrub or small tree, deciduous and thornless.
The leaves are oval, alternate, entire, dark green
above and downy on the underside.
The large solitary flowers are white or pink.

Cultivar	Ripening (days before or after variety Elberta)	Flesh color[1]	Fruit size	Adherence to stone	Attractiveness	Firmness of flesh	Table quality	Industrial quality	Resistance to shot hole	Cold hours	Main use[2]
							Scale (from 0 to 10, where 10 = the best				
Marcus	−63	Y	5	4	6	4	6	4	5	800	FLC
Mayflower	−58	Y	3	3	6	3	5	3	7	1.200	FLC
Earlired	−50	Y	6	5	8	6	7	4	7	850	FLC
Collins	−49	Y	5	5	8	6	7	4	—	—	FLC
Cardinal	−46	Y	6	4	8	6	7	4	7	950	LC
Early Redhaven	−44	Y	6	5	8	7	7	—	—	—	LC
Dixirred	−42	Y	6	4	7	6	7	4	8	1.000	LC
Redcap	−42	Y	6	4	8	6	8	4	7	750	LC
Erly-Red-Fre	−40	W	7	6	7	6	8	4	8	900	FL
Sunhaven	−38	Y	7	6	7	7	8	5	8	900	FLC
Merrill Gemfree	−38	Y	7	6	6	7	7	5	6	—	LC
Early East	−37	Y	7	3	6	5	5	3	—	—	FL
Jerseyland	−33	Y	8	8	6	8	8	7	4	850	LC
Dixigem	−32	Y	7	7	7	7	9	7	7	850	LC
Arp Beauty	−32	Y	8	4	8	7	7	3	—	—	FL
Prairie Dawn	−32	Y	7	5	8	7	7	6	—	—	FL
Redhaven	−30	Y	7	7	8	8	8	7	8	950	FLC
Raritan Rose	−27	W	8	8	8	6	9	7	9	950	FL
Golden Jubilee	−25	Y	8	8	7	6	8	8	8	850	FL
Prairie Daybreak	−25	Y	9	6	6	6	5	—	—	—	FL
Ranger	−25	Y	8	8	7	8	8	8	9	950	LC
Newday	−25	Y	8	6	6	6	6	6	—	—	FL
Washington	−24	Y	8	8	8	8	8	8	—	—	LC
Triogem	−22	Y	8	8	7	8	8	7	6	850	FLC
Fairhaven	−19	Y	8	8	7	7	8	8	7	850	L
Glohaven	−19	Y	8	8	7	8	8	6	—	—	LC
Western Pride	−19	Y	8	6	6	7	6	—	—	—	FL
Sunhigh	−17	Y	9	8	8	8	9	9	4	750	C
Vedette	−17	Y	8	5	6	6	6	3	—	—	FL
Richhaven	−16	Y	9	9	9	9	9	9	4	1.000	LC
July Elberta	−15	Y	8	9	7	8	9	9	5	750	FL
Southland	−14	Y	9	9	9	9	9	9	6	750	FLC
Halehaven	−14	Y	9	9	6	7	9	8	7	850	LC
Redglobe	−14	Y	9	9	10	10	9	9	7	850	FLC
Loring	−11	Y	9	9	9	9	9	9	8	800	LC
Veteran	−11	Y	8	8	8	7	9	8	8	1.100	FL
Slappey	−11	Y	8	9	6	6	6	7	—	—	FL
Delight	−11	Y	7	9	7	6	7	—	—	—	FL
Gene Elberta	−11	Y	7	8	8	7	9	—	—	—	FLC
Goldeneast	−10	Y	8	6	6	6	6	—	—	—	FLC
Belle	−8	W	8	9	6	7	9	8	9	850	FL
Redelberta	−8	Y	9	9	7	9	9	—	—	900	LC
Suncrest	−4	Y	9	9	9	10	9	9	4	850	LC
Sullivans Early Elberta	−4	Y	9	9	7	8	8	8	6	900	C
Early Elberta	−3	Y	9	9	8	9	9	9	8	850	LC
Merril 49'er	−3	Y	8	7	7	7	6	6	—	—	LC
Elberta	0	Y	9	9	8	9	8	8	7	900	C
Redskin	0	Y	9	9	9	9	9	8	9	650	FLC
H.H. Brilliant	0	Y	9	9	9	9	9	8	5	750	LC
Dixiland	0	Y	9	9	9	9	9	—	7	750	C
Madison	0	Y	9	9	8	8	9	—	—	—	FL
J. H. Hale	+1	Y	10	9	9	10	9	7	6	900	C
Halberta Giant	+2	Y	10	7	7	8	6	—	—	—	FL
Gold Medal	+2	Y	7	6	7	7	6	—	—	—	FLC
Afterglow	+5	Y	9	9	7	9	8	7	7	750	C
Alamar	+5	Y	10	9	9	8	7	—	—	—	FLC
Rio Oso Gem	+6	Y	10	9	8	10	9	8	7	900	C
Constitution	+10	Y	9	9	7	8	8	7	7	750	C
Autumn	+12	Y	9	9	7	8	7	7	8	850	C
Late Elberta	+12	Y	8	6	5	7	6	6	—	—	—
Krummel	+27	Y	9	9	7	9	7	7	7	900	C

SOURCE: USDA Handbook 280 by H. W. Fogle et al., 1965.
1. W = white, Y = yellow
2. F = family, L = Local, C = commercial.

The fruit is round, thick and covered in down, and is
green before ripening and yellow after ripening.
When it ripens it develops a very characteristic
pleasant smell.
The quince is originally from the hottest regions of
southeast Europe and Asia Minor, and thus requires
a very hot climate, though it does resist low
temperatures. It should not be grown on a site
exposed to the north.

*Characteristics of the
cultivars of peach in
order of ripening*

*Detail of apple shoot
opening*

Cultivars of European plum

Cultivar	Ripening period	Needs pollination	Fruit shape	Fruit size	Loose stone	Skin color[1]	Flesh color[1]	Use[2]
Tragedy	Early	Yes	Oval	Large	Yes	Y-P	G-Y	F
California Blue	Early	No	Oval	Large	Yes	A	Y	F
Iroquois	Early	No	Oval	Medium	Yes	A	G-Y	FC
Early Italian	Early	No	Oval	Medium	Yes	A	Y	FC
Lombard	Early	No	Oval	Large	Yes	R	Y	F
Queenston	Early	—	Oval	Medium	Yes	A	Y	F
Parsons	Early	Yes	Oval	Medium	Yes	R-A	Y-A	FP
Stanley	Medium	No	Obovate	Medium	Yes	A	Y	FC
Bluefre	Medium	No[3]	Oval	Large	Yes	A	Y	F
Damson	Medium	No	Oval	Small	Yes	A	Y	FC
Bluebell	Medium	Yes	Oval	Large	Yes	A	Y	FC
Grand Duke	Medium	Yes	Oval	Large	No	A	Y	F
Agen	Medium	No	Obovate	Medium	Yes	A	Y	FCP
German	Medium	Yes	Oval	Medium	Yes	A	Y	FCP
Italian	Medium	No	Oval	Medium	Yes	A	Y-A	FCP
Imperial	Medium	Yes	Oval	Medium	Yes	A	Y	FP
Brooks	Medium	No	Oval	Very large	Yes	A	Y	FP
Sugar	Medium	No	Oval	Medium	Yes	A	Y	FP
Sargeant	Medium	—	Oval	Medium	Yes	A	Y	FP
Reina Claudia	Late	No	Oval	Medium	Yes	G-A	G-Y	FC
President	Late	Yes	Oval	Large	Yes	P	Y	F
Pozegaca	Late	—	Oval	Small	—	A	—	L
Standard	Late	Yes	Oval	Large	Yes	A-P	G-Y	F
Vision	Late	Yes	Oval	Large	Yes	A	Y	F
Victoria	Late	No	Oval	Medium	Yes	R	Y	F
Moyer Perfecto	Late	No	Oval	Large	Yes	A	Y	FP

1. Y = Yellow, B = blue, A = ambar, P = purple, R = red, G = green
2. F = fresh, P = prunes (drying), C = conserves, L = liqueur
3. Cross-pollination may increase fruit-set in some areas

Detail of almond fruit

It is not very demanding in its soil requirements. It grows well on cool limestone soils and along the banks of watercourses.

It is mainly used as a dwarfing stock for pears, and the fruit also provides quince jam and jelly. The main cultivars are: De Angers, Naranjo, Bereczky, Champion, De Rea Mammoth, De Metz, Meech, Smyrna, Pineapple and Van Deman.

13.2. STONE FRUIT

The fruit is a drupe, normally with a single seed. The most widespread genus is *Prunus*, which includes the almond, peach, nectarine, plum, cherry, apricot as well as many species that are only cultivated as rootstocks.

Pollination and fertilization of the ovule are necessary because, unlike seed fruit, stone fruit never develop parthenocarpically.

13.2.1. Almond

***Prunus amygdalus*. Rose family (Rosaceae)**

A deciduous tree, originally from the hot regions of western Asia.

The leaves are entire, lanceolate and serrated. The hermaphroditic flowers are white or pink. The fruit is a green fleshy drupe, more oval in shape than elongated. It contains a woody shell (endocarp) holding one or two almonds, which bear a dark brown skin.

Almost all varieties of almond are self-sterile and thus require pollinator varieties. Bearing in mind that pollination occurs during a cold, wet period of the year, it is important to rent bees to obtain good production.

Soft-shelled varieties: Princesa, Infausta, Mollar canal, Mollar cartagena, Blanqueta, Mollar fita, Re caracola, Mollar trincheta, Mollar blanca.

Hard-shelled varieties: Fanareta, Poteta, Fina del Alto, Planeta, Desmayo, Largueta, Marcona, Pelad.

It is a robust tree that tolerates winter frosts. It likes well-aired zones that are not subject to spring frost during the flowering period.

In terms of soil, it prefers a dry, light, stony soil tha is deep and permeable.

13.2.2. Peach

***Prunus persica*. Rose family (Rosaceae)**

The peach is originally from China. It is a medium small tree with a relatively short life of 20 to 50 years.

Cultivar	Somatic number of chromosomes	Size of tree	Days from flowering to harvest	Fruit size	Fruit color[1]	Use[2]	Maximum number of days storage	Earliness	Biennial production	Global productivity[3]	Self-compatibility
Yellow Transparent	34	S-M	70-100	S	Y	C	80	Good	Yes	Mod.	Partly
Gravenstein	51	L	110-130	L	RS	F, C	90	Bad	Yes	Mod.	No
James Grieve	—	M	110-130	L	YRS	F	100	Good	No	Mod.	—
Antonovka	—	M	110-130	M	Y	C	100	Good	Yes	Mod.	—
Wealthy	34	M	120-135	M	R	F, C	80	Good	Yes	Good	Partly
Winter Banana	34	M	150-165	L	YW	C,F	150	Good	No	Good	No
Cortland	34	M	125-140	S	R	F, C	150	Good	No	Good	Weakly
McIntosh	34	M-L	124-145	M	R	F, C	130	Good	No	Good	Weakly
Cox's Orange Pippin	34	M	130-160	M	YW	F	130	Good	Moderate	Medium	No
Rhode Island Greening	51	L	130-155	M-L	G	C	180	Bad	Moderate	Medium	No
Bramley	51	L	135-155	L	G	C	190	Medium	Yes	Mod.	Partly
Ralls	—	M	—	M	R	F	185	Bad	Yes	Medium	—
Jonathan	34	M	135-150	S	R	C, F	120	Good	No	Good	Weakly
Grimes Golden	34	M	140-150	M	Y	F, C	120	Good	Moderate	Good	Partly
Golden Delicious	34	M	140-160	M-L	Y	F, C	160	Very good	Sí	Very good	No
Delicious (and sports)	34	M-L	140-160	M-L	R	F, C	180	Medium	Moderate	Medium	No
Spur Delicious	34	M	145-165	M-L	R	F, C	180	Good	Moderate	Good	No
Boskoop	51	G	145-165	L	GYR	F, C	160	Bad	Moderate	Medium	—
Northern Spy	34	G	145-170	L	W	F, C	180	Very bad	Moderate	Bad	No
Mutsu	51	G	145-170	L	GY	F, C	190	Good	No	Good	No
York Imperial	34	G	155-175	M-L	R	C	180	Bad	Yes	Medium	Partly
Rome Beauty	34	S-M	160-175	L	R	C	240	Very good	No	Very good	Weakly
Newtown	34	M	160-175	M	GY	C, F	200	Bad	Yes	Medium	Partly
Winesap	34	M	160-180	S-M	R	C, F	240	Medium	Moderate	Mod.	No
Stayman	51	L	160-175	M-L	R	F, C	180	Good	No	Good	Weakly
Sturmer Pippin	—	M	160-175	M-L	GY	F	210	—	—	—	—
Granny Smith	34	L	180-210	M-L	G	F, C	210	Medium	No	Good	—

NOTE: the dashes indicate lack of data. 1 Y = Yellow, (W = washed). S= striped, R = Red, G = green
2 F = Fresh, C = Culinary
3 V = Very.

Apple cultivars and their important features

Peach

Cherry

The leaves are alternate, lanceolate and serrated, and light green in color.
The flowers appear before the leaves, and are pink, or occasionally white.
The fruit is spherical, with a more-or-less clear longitudinal groove. The skin is glabrous or downy, and green or yellow in color. The flesh is juicy and sweet-smelling, and can be white, yellow or reddish. The stone is elongated, hard, with very clear sinuous grooves, and contains the seed. Most peach cultivars are self-fertile. The self-sterile varieties include *Halberta, Candoka, Alamar* and *Mikado*.
The **nectarine** is simply a peach with recessive genes that produces fruit that lack the downy hairs, i.e., they are glabrous.
The peach tree is sensitive to the climate. It requires a lot of heat and abundant light for the fruit to ripen and color well. Brusque temperature changes in spring, as well as frosts and cold breezes, seriously prejudice flowering. The peach is also prejudiced by rapid alternation of moisture and sunshine, prolonged rain and late frosts.
The best sites are in well-aired valleys, not exposed to frosts and sheltered from winds.
The best soils are sandy, siliceous-calcareous, deep, and above all cool and light.

13.2.3. Apricot

Prunus armeniaca. Rose family (Rosaceae)

The apricot tree is originally from China.
The leaves are oval, slightly heart-shaped, irregularly toothed, smooth, shiny, and dark green above and lighter on the underside.
The large solitary white or pink flowers are produced in spring. The fruit is globose, with a thick groove. The skin is more or less orange, and is covered with very fine hairs. The flesh is juicy and perfumed.
The best-known cultivars are: Yellow Alexandrian, Shiny Alexandrian, Roman, Liabaud, Luizet, Nancy, Real, De Tours, Boncarande, Común del Lac and Early Sardinian.
To produce fruit, the apricot needs abundant air, heat and light. In valleys it hardly produces any flowers, and it grows better in higher areas, even if they are dry and battered by the wind. If it is sheltered, these conditions even increase production.
It requires light soils that are warm, permeable, sandy, stony and not very fertile.

13.2.4 Cherry

Prunus avium. Rose family (Rosaceae)

A tall tree originally from the forests of Europe.

Right: Hazel fruits

Right: Detail of chestnut, showing catkins and female flowers

Characteristics of cultivars of cherry

The leaves are large, alternate, oval, doubly toothed and a dark green color.
The large scented flowers are white or pink.
The fruit is a red globose heart-shaped drupe with sweet juicy flesh.
The main cultivars are: Cristobalina, Temprana de Sol, Garrafal burlat, Garrafal Moreau, Corazón de Palomo, Van, Wing, Garrafal Napoleon and Picota.

Cultivar	Flowering date[2]	Requires polli-nators	Incompa-tibility group[3]	Fruit size	Fruit color[4]	Tendency to splitting	Pulp texture[5]	Taste[5]	Use[6]
Very early									
Seneca	M	Yes	10	S	B	Low	D	D	F
Vista	E	Yes	11	M	B	Low	G	G	F
Burlat Precoz[1]	M	Yes	—	L	B-R	High	G	G	FE
Early Purple	M	Yes	—	M	B	High	M	M	FEC
Bigarreau de Schrecken	L	Yes	—	M	B	High	G	G	FEC
Early									
Black Tartarian	M	Yes	1	S-M	B	Low	D	M	F
Viva	—	Yes	4	M	B-R	Very low	M	G	F
Vega	M	Yes	12	L	W	Medium	VG	M	CE
Venus	E	Yes	2	M	B	High	G	G	FE
Chinook	E	Yes	9	L	B	High	G	G	FE
Corum[1]	E	Yes	—	L	W	Medium	M-G	G	CFE
Nona[1]	M	Yes	—	L	B	—	—	—	—
Macmar	M	Yes	—	L	W	High	G	G	CFE
Knight's Early Black	L	Yes	1	M	B	High	G	M	ECF
Bada[1]	M-L	Yes	—	L	W	—	G	M	CE
Larian[1]	M	Yes	—	M	B	—	D	M	F
Mid season									
Merton Bigarreau	M	Yes	2	L	B	Low	G	G	FEC
Bing	M	Yes	2	L	B	High	VG	VG	FE
Rainier	M	Yes	9	L	W	Medium	G	G	FE
Napoleón	M	Yes	3	L	W	Medium	G	G	CEF
Sam[1]	L	Yes	—	M	B	Low	—	G	FE
Sue	M	Yes	4	L	W	High	VG	G	CEF
Schmidt	M	Yes	8	M	B	Low	G	G	FE
Vernon	L	Yes	3	L	B	Low	G	G	CEF
Vic	M	Yes	13	M	B	—	G	G	EC
Stella[1]	M	No	—	L	B	Medium	M	M	F
Emperor Francis	—	Yes	3	M	W-R	Low	G	G	—
Berryessa	M	Yes	—	L	W	—	G	M	CE
Late									
Windsor	L	Yes	2	M	B-R	Low	G	G	C
Gold	—	Yes	6	S	Y	Very low	G	M	C
Lamida	L	Yes	—	L	B	Medium	G	M	FEC
Spalding	L	Yes	—	M	B	Medium	G	M	F
Van	M	Yes	2	L	B	Medium	VG	G	FEC
Jubilee[1]	M	Yes	—	L	B	Low	M	G	FE
Hudson	M	Yes	9	M	R-B	Low	VG	G	FC
Ulster	M	Yes	13	M	B	Low	G	G	FE
Hedelfingen	M	Yes	7	L	B	Low	G	G	FEC
Lambert	L	Yes	3	L	B	High	G	G	FE
Black Republican	E	Yes	—	M	B	High	VG	G	FE

1 Cultivar compatible with Bing, Lambert and Napoleón
2 M = mid season, E = early, L = late
3 The cultivars with the same number are inter-compatible, and can not pollinate each other.
4 W = white, B = black, R = Red, Y = Yellow.
5 D = deficient, M = medium,. G = good, VG = very good.
6 F = fresh, E = conserve, C = confectionary j = jam

There is another species, *Prunus cerasus* or sour cherry. This is a smaller tree than the common cherry. Its fruit is red, with skin that separates from the flesh, which is juicy and sour.
In general, the cherry is not very sensitive to temperature variations, and thus adapts well to different climates.
It is resistant to winds. It needs well-aired bright sites.
It is not very demanding in the soils it requires, and it can be grown in limestone, stony and sandy soils. It is prejudiced by excessively wet or impermeable sites.

13.2.5. Plum

Prunus domestica. Rose family (Rosaceae)

The plum is a medium-sized tree, originally from Anatolia and Persia.
The leaves are oblong, serrated, smooth above and hairy on the underside. The flowers are solitary and white in color.
The fruit is a grooved drupe, round or oval, glabrous, and generally with waxy skin. The fruit may be yellow, red or violet.

There are 2 closely related species:

• Japanese plum (*Prunus salicina*). The fruit is conical or heart-shaped, with a sharper tip than in the other species. It is mainly eaten fresh. The main cultivars are Santa Rosa, El Dorado, Duarte, Queen Anne and Sinka.
• Sloe, or blackthorn. *Prunus spinosa.* Spiny with a round acidic fruit.

In general, the plum is a tree of warm climates, though there are some cold-resistant varieties. It is damaged by moisture and by frosts in spring, during its flowering period.
It is the least demanding of all fruit trees in its soils requirements. It grows well in all soils, as long as they are not excessively wet or clay.

13.2.6. Olive

Olea europaea. Olive family (Oleaceae)

The olive tree is probably from Asia Minor. There are two sub-species, the wild olive and the domestic, or cultivated, olive.

The olive is a very tough tree, and long-lived, and can grow very large.

The leaves are evergreen, simple, entire, oval, lanceolate, and finish in a sharp tip. They are light green on the upper side of the leaf and silvery on the underside, because of the many hairs present.

The flowers are small, whitish-green in color, and in long clusters.

The fruit, the olive, is an oval drupe that is fleshy and rich in lipids. It is light green and as it ripens it turns purplish, reddish and eventually black. The fruit contains a stone that protects the seed.

The most widespread cultivars for oil production are: Picual, Hojiblanca, Arbequina, Cornicabra, Verdial Real, Negral and Empeltre.

The most widely cultivated table olive cultivars are: Manzanilla, Gordal and Sevillana.

The olive is very closely associated with the zone that has a Mediterranean climate, consisting of mild winters and hot summers with almost no rain.

The olive does not like the cold. Frosts may be dangerous, especially during the flowering period. It tolerates high temperatures in summer very well. It does, however, need a winter rest period to flower and fruit.

It requires a site with deep, fertile, soil that is permeable and cool.

13.3. NUTS

13.3.1. Walnut

Juglans regia. Walnut family (Juglandaceae)

The walnut is a vigorous deciduous tree, originally from Persia.

The leaves are alternate, large, glabrous, serrated or entire, and are deep green in color with a sharp smell.

The flowers are monoecious. The male flowers are borne in catkins on shoots produced the previous year. The female flowers are grouped in terminal racemes on the branches produced that year.

The fruit is a large drupe called a walnut, which has large divisions with 2 or 4 incomplete cells.

Juglans species can hybridize with each other when they are cultivated in the same site. All the species of the genus Juglans are edible, but Juglans regia (common, or English, walnut) is the most important, while the black walnut (Juglans nigra) is of less importance for its fruit, though its wood is highly appreciated.

The main cultivars are: Franquette, Eureka, Placentia, Grenoble, Mayette, Parisienne, Santa Barbara and Concord.

Most specimens of the common walnut are

self-fertile, but many cultivars shed their pollen before the female flowers open, meaning that plants of pollinator cultivars must be planted.

It is sensitive to low temperatures and to high ones, and is easily damaged by spring frosts. Walnuts do well on mountainous sites, and if grown as isolated trees, they are robust and vigorous.

In terms of the site, it is a delicate tree that requires limy soils, that are permeable, deep and cool, and it is damaged by standing water.

13.3.2. Hazel

Corylus avellana. Birch family (Betulaceae)

Deciduous shrub, originally from Asia Minor. The yellowish green leaves are oval, doubly toothed, rough, large, with hairs on the underside.

The hazel is monoecious, so a single plant bears both female flowers and male flowers. The male flower is a pendulous yellow catkin. The female flowers are surrounded by persistent hairy bracteoles.

The fruit is a globose ovoid nut with a woody pericarp, or shell, surrounded by an involucre of leaves and grouped in bunches at the end of the shoots. The hazel is almost round, with a tip that is first green and hairy, and then turns reddish.

Nut pine

Pecan: fruit

Characteristics of the main varieties of almonds

Cultivar	Almond (% of total weight of fruit)	Shape	Size	As pollinator of Barcelona	Productivity	Almond
Barcelona	43	Round	M-L	—	Moderate	Smooth
Daviana	52	Oval	M	M-G	Weak	Smooth
DuChilly	44	Elongated	L	G	High	Wrinkled
White Aveline	50	Oval-flat	S	M-G	Moderate	Smooth
Montebello	42	Round	M	Null	Moderate	Smooth
Brixnut	42	Round	L-M	Null	High	A little wrinkled
Halls Giant	46	Round	M	G	Moderte	Rather rough
Royal	Low	Oval	L	Bad	Moderate	Rather rough
Nooksack	43	Elongated	M	Bad	Moderate	Rough

The flowers appear before the leaves, in early spring, and the fruit matures in fall. Hazels are self-sterile so that varieties capable of pollination have to be planted at a rate of one pollinator to fourteen productive plants.
The most important variety in Spain is the Negreta. Other varieties include: Piñolenca, Gironella, Grifoll, Morell, Tereneta, Grossal, Ribet, Baccilara, Panuttara, Trevisond Imperial, Giant cob and Santa Anna.

13.3.3. Pistachio

Pistacia vera. Cashew family (Anacardiaceae)

A small dioecious tree, originally from Syria. The leaves are alternate, simple, trifoliate or pinnate, and are smaller and darker green in the male tree.
The small unisexual flowers are purple.
The pistachio is resistant to dry summer conditions and prefers sunny positions. It is also resistant to low temperatures, though the flowers freeze at a temperature of -2°C.
The fruit is an ovoid elongated dry drupe, and divided into two equal valves.
The pistachio is a member of the cashew family, like the cashew nut, the mango, the terebinth (*P. terebinthus*) and the mastic tree (*P. lentiscus*).

Shoots on a strawberry tree

As the tree is dioecious, one male must be planted for every 6 female plants. Pistachios are wind-pollinated, and the male plants produce large amounts of pollen, generally before the female pistils are receptive. A new technique to resolve this problem is to collect, store and then disperse the pollen when the female flowers are receptive.

13.3.4. Sweet chestnut

Castanea sativa. Beech family (Fagaceae)

The sweet chestnut, or Spanish chestnut, is a large deciduous tree from the Mediterranean region. The serrate leaves are entire, in two rows, with numerous parallel veins and glabrous leaves above and below. The male flowers are straw yellow catkins. The female flowers are borne close to the catkins.
The chestnuts are large, ovoid in shape, brown, with a large pale basal scar, and 1 to 3 (sometime 5 or 7) are borne in a prickly capsule, which opens on maturity into 4 valves or divisions.
Fresh chestnuts contain about 50% carbohydrates, and can be stored at 4.4°C for 8 weeks, as long as they do not have mold. An hour in water at 68°C eliminates the mold without damaging the chestnut. Chestnuts can be stored at 4.5°C for a year, if they are dried out to 10% humidity. Dried chestnuts are rehydrated by soaking in water or by steaming for half an hour before use. Unlike most nuts, the chestnut contains little oil but abundant starch. They have to be boiled or roasted to make them digestible.
It is a tree of temperate climates, and does not tolerate great heat and prolonged dry conditions. Frosts and prolonged frosts seriously damage it.
Chestnuts prefer deep loose soils that are cool and rich in organic matter. It is also highly calcifugous (lime-hating), and is not recommended for sites with more than 1% calcium carbonate.

13.3.5. Carob

Ceratonia siliqua. Bean family (Leguminosae)

The carob tree is originally from Syria, and is of medium size.
The leaves are compound with round leathery leaflets that are a shiny dark green.
The small reddish purple flowers are borne in racemes.
The fruit is a brown pod, 10 to 20 cm long and 2 to 3 cm broad, that contains the carob bean. It is adapted to a hot maritime climate, and is damaged by low temperatures. It prefers rocky and calcareous soils, and does not thrive on wet clay soils.

13.3.6. Nut pine

Pinus pinea. Pine family (Pinaceae)

The stone pine is a large tree, whose characteristic umbrella-shaped crown is the reason for its other name, the umbrella pine.
The leaves are rigid needles up to 15 cm long by 1 mm wide and light green.
The male flowers are yellow catkins 10 mm long.
The female flowers are oval reddish-green cones 8-10 cm wide and 10-15 cm long.

The fruit is a cone, a pine cone consisting of woody scales, and red to brown in color. Each scale leaf has 2 seeds or pine nuts on the upper surface.

The seeds have a hard dark shell.

The nut pine grows in the coastal region. It is adapted to a hot or temperate climate. It needs a sunny site, and is sensitive to frosts and sharp changes in temperature.

It prefers loose, moist and deep soils.

13.4. SMALL FRUIT

13.4.1. Raspberry

Rubus. Rose family (Rosaceae)

A shrub that may be deciduous or evergreen, and which grows in stony sites on mountains.

The stems are erect or trailing. Most of the stem bears thorns, and the stems are generally short-lived.

The leaves are alternate, entire, oval and serrate, and have a rachis (leaf-stalk).

The flowers are small, white or tinged pink, and are borne in a terminal raceme.

The fruit is an aggregate of many small drupes, or drupelets.

Different types:

- *Rubus idaeus* (raspberry)
- *Rubus occidentalis* (black raspberry)
- *Rubus ursinus* (western blackberry)

It grows better in temperate climates than hot ones. It can not withstand great heat. It prefers deep, cool, fertile, soils that trap heat.

13.4.2. Bilberry or cranberry

Vaccinium. Heath family (Ericaceae)

A shrub of forests and mountain meadows. The leaves are alternate, entire or serrate, and deep green in color. The solitary flowers are white. The fruit is a round berry, with a sweet taste, containing several seeds, and topped by the persistent remains of the calyx.

Several species of *Vaccinium* are cultivated:
- American cranberry - *Vaccinium macrocarpum*
A North American species, very productive, that requires open sunny sites.
- Bilberry - *Vaccinium myrtillus*
The best known of these fruits. It occurs in mountains, under the shade of large trees, such as pines, firs, beeches, on sandy loose soils.

13.4.3. Currants

Ribes. Currant family (Grossulariaceae)

A thornless shrub, occasionally with spines and bristly stipules, palmate leaves, small, serrate smooth above and downy on the underside.

The flowers are small in racemes. The fruit retains the persistent calyx of the flower, and may be red, white, yellow or black.

There are 4 main species of *Ribes* with edible fruit.
- *Ribes sativum*
- *Ribes rubrum*
- *Ribes nigrum* (black currant)

The black currant is susceptible to mildew, and the white-pine blister rust, which is most unfortunate as its vitamin C content is higher than most other fruit, and it is used as a source of vitamin C in Europe.

The best white variety is White Imperial.
- *Ribes grossularia*. The gooseberry.
A shrub reaching a height of up to 1 m, with branches that sometimes bear stout spines, which may be up to 1 cm long and mainly with three points.

The leaves are heart-shaped or wedge-shaped, with 3 to 5 lobules, with blunt tip and with scalloped teeth that may be glabrous or pubescent.

The flowers may be single or in a group of two, and are greenish in color. The fruit is globose or ovoid, and may be smooth or hairy, and may be red, yellow or green.

The gooseberry is very robust, and prefers a temperate climate and a semi-shady site. It prefers light soils that are not too cool and with limestone.

13.4.4. Strawberry tree

Arbutus unedo. Heath family (Ericaceae)

A shrub with alternating oval leaves, that are coriaceous (leathery), shiny and dark green. The petioles are short and reddish. The bell-shaped flowers are white, and the fruit are globose berries, that are red when ripe.

This shrub is very widespread in the Mediterranean Basin. It grows in dry arid sites in hot zones.

13.4.5. Vine

Vitis vinifera. Vine family (Vitaceae)

The vine is the most widely cultivated deciduous broadleaf species. It is originally from the regions south of the Caspian Sea.

Viñedo

Fruit of the white currant, red currant and black currant

Raspberry

Summary of the characteristics of parthenocarpic varieties of figs in Spain

The vine is a sarmentose shrub, a climber that attaches itself with tendrils. If the vine is cultivated, the stem is cut back to a rootstock, so it only bears the current year's shoots, the only ones that can produce fruit.

The leaves are simple and toothed.

The flowers, borne in racemes, are small and greenish. The fruit is a fleshy berry, juicy, with 2 to 4 seeds.

- *Wine-making varieties*: Palomino, Moscatel, Pedro Ximénez, Listar, Garrido and Garnacha, Ribote, Miguel del Arco, Batista, Gorgollosa, Vinater, Malvasía, Albillo, Negramoll, Tintilla, Vermepuela, Forastera, Albarín, Negrín, Verdeja, Blanquilla, Doradilla, Tempranilla, Picapoll, Xarel·lo, Macabeo and Parellada.

- *Table grapes*: Albillo, Chasselas dorado, Batiller de Bayreuth, Italia, Moscatel, Rosaki, San Jaime, Sultanina, Tefa de Vaca and Valencia.

The best grapes are grown in hot dry climates. The vine is sensitive to rapid falls in temperature, to co winds from the north and prolonged rains.

Variety	Production	Harvest time	Fruit color	Production aim	Flesh	Fruit size
Maella White	Figs	September-October	Green	Fresh and dry	Very sweet	Large
Colar	Early figs-Figs	June/July and August/September	Very black, striped	Fresh	Very sweet	Very large
Goine	Early figs-Figs	Mid June and August/September	Black with reddish neck	Fresh	Sweet	Medium
Early white	Early figs-Figs	June and August	Light green	Fresh and dry	Sweet	Very large
Ñoral	Early figs-figs	June/July and August/September	Greenish white	Fresh	Sweet	Medium
Verdal	Figs	October/November	Green	Fresh	Very sweet	Large
Moscactel	Early figs-figs	August and October	Light green	Fresh and dry	Very sweet	Medium
Broad-leaf or Florancha	Early figs-Figs	June and August	Violet-black, striped	Fresh	Very sweet	Large
White	Figs	September	Pale green-white	Fresh and dry	Sweet	Large
Piel de toro	Figs	September/October	Black	Fresh	Sweet	Large
Pajareros	Early figs-Figs	June and August	Green-yellow	Fresh	Sweet	Medium
Black Neapolitan	Early figs-Figs	June and September	Black-violet	Fresh	Very sweet	Very large
Burjasot	Figs	September	Green	Fresh and dry	Very sweet and fine	Large
Cuello de dama	Figs	October	Black	Fresh	Very sweet	Large
Perolasos	Early figs-Figs	Mid June and August/September	Bluish-black	Fresh	Sweet	Medium
Parejal	Early figs-Figs	June and August	Black	Fresh	Sweet	Large
Date figs	Figs	August/September	Greenish white	Fresh and dry	Very sweet	Medium
White rose	Early figs-Figs	June and August	Reddish black	Fresh	Very sweet	Medium
Rojisca	Early figs-Figs	June and August/September	Black	Fresh	Sweet	Small
Black Bordisot	Early figs-Figs	Late June and September	Black	Fresh and dry	Sweet	Medium
Valencia and Murcia Black	Figs	September/October	Black	Fresh	Very sweet	Medium
Common or wild	Early figs-Figs	Early June and August	Striped black	Fresh	Very sweet	Medium
Pacueca or Lampaga	Early figs-Figs	June and August	Black-green-violet	Fresh	Sweet	Large

n terms of soil, the vine is not very demanding. It prefers light soils, permeable, siliceous or stony, not very humid, that dry easily and warm up rapidly.

13.5. COMPOUND FRUIT

13.5.1. Mulberry

Morus. Mulberry family (Moraceae).

Dioecious trees, deciduous and without thorns. The leaves are entire, toothed, very rough to the touch, and dark green in color.
The fruit is an ovoid compressed achene, that may be red, pink, white or shiny black (mulberry), and superficially similar to a blackberry.
The fruit is sweet and tasty, but excessively soft and vulnerable to handling. It has a weak abscission layer, and often falls when touched, even before ripening. Most mulberry trees are dioecious, so that the male trees can be cultivated as ornamentals.

There are two main species:
• *Morus alba*, cultivated to feed its leaves to silkworms. Originally from Central Asia.
• *Morus nigra*, cultivated for its edible fruit. Originally from Persia.

The ideal climate for mulberries is temperate or sub-tropical. Avoid low, moist sites that are exposed to late hoar frost or frosts. The best sites are sheltered, with deep, open, permeable soils that warm up easily.

13.5.2. Fig

Ficus carica. Mulberry family (Moraceae)

The fig is from western Asia. In the wild, this species tends to remain as a shrub. It grows fast and its roots are highly penetrating.
The leaves are large, divided into 3 or 5 lobules, and heart-shaped at the base. The leaves are palmate and are rough above and with tough hairs on the underside.
The flowers of the fig usually produce pollen, but many varieties produce parthenocarpic fruit.
The fig is not a simple fruit but a swollen empty flower receptacle that contains many flowers within its cavity.
This receptacle may be greenish or darkish violet, and the edible part is the interior, which consists of a large number of drupes with a soft, juicy, sugary pulp.
Two types of figs appear over the course of the year:
• *Early figs* appear in spring, and form the first harvest.
• *Figs*, or true figs, form in summer-fall, and are the second crop. They are sweeter and have a stronger smell than the early figs.
Well-known varieties of fig include: White, Brianzolo, White Violada, Black Violada, Corazón, Datto, Dal Ábate, Dottato, Gentil, Granado, Mónaco, Celeste, Paraíso, Reina, Black Early Fig, Smyrna, Magnolia and Cuello de Dama.

In hot climates, production is abundant, and the figs are sweeter and easier to preserve.
Autumn rains are very prejudicial meant, and figs need well aired sites.
The fig grows well in all soils, except clay soils. It prefers deep permeable soils.

13.6. CITRUS FRUIT

These trees are all originally from southern Asia and belong to the genus *Citrus* in the rue family (Rutaceae).
One of the features of all the species of this genus is the presence in all the plant's organs of an essential oil that gives it a distinctive smell. The leaves are arranged in a spiral and are entire and elliptical.
The flowers are regular, white or pink, and appear in spring and in fall, though in repeat-flowering species it is difficult to fix the flowering time.
The fruit is a berry with 7 to 12 *segments*. Each segment contains one or more seeds. The fruit ripens from November onwards.

Fig branch with fruit

Detail of the leaves and flowers of the bitter orange.

	Minimum	Optimum	Maximum
Sweet orange	12-14	23-24	36-39
Bitter orange	12.8	23-26	38-39
Lime	17.6	34	35.6
Grapefruit	17	31-34	43-44

Above: Temperature requirements for some citrus fruit

Sweet orange

Persimmon

The tree starts fruiting 5-6 years after planting. Citrus trees require a hot-temperate climate, moist and free of winds. They are sensitive to low temperatures and prefer fertile, deep and permeable soils.
The most widely cultivated species are
• *Citrus aurantium*, the bitter orange
• *Citrus sinensis*, the sweet orange
• *Citrus deliciosa*, the mandarin orange
• *Citrus paradisis*, the grapefruit
• *Citrus limon*, the lemon
• *Citrus aurantiifolia*, the lime

13.6.1. Bitter orange

In general, it is smaller than the sweet orange. The fruit is similar but the pulp contains an acid juice that makes it bitter. It is the most cold-resistant species in the genus *Citrus* and is the species most frequently used as a rootstock.

13.6.2. Sweet orange

A spiny tree with leaves with winged petioles. The fruit is round, yellow-orange and the skin is covered with convex oil glands. The pulp is juicy, sweet and sugary.
The most widely cultivated varieties are: Valencia, Navelina, Salustiana, Común, Castellana, Doblefina, Entrefine, Berna, Sanguinelli, Washington, Thomson and Moro.

13.6.3. Mandarin

The fruit is small, round and flattened, and orange-red.

It is more robust and productive than the common orange, but is shorter-lived.
The most widely cultivated varieties are: Satsuma and Clementine.

13.6.4. Grapefruit

Less robust than the orange. The large fruit is round and yellow-orange in color. The pulp is greenish and spongy, with an almost bitter sugary juice.
Cultivated varieties: Duncan, Ruby and Shambar.

13.6.5. Lemon

The fruit is ovoid, with yellow-green or yellow skin, either smooth or rough. It contains a lot of pulp, with an aromatic acidic juice.
Cultivated varieties: Fino, Verna, Eureka, Verdelli, Monachello, Feminello and Lisbon.

13.6.6. Lime

Large fruit with yellow-green smooth skin.
Varieties: Key and Mexicano.

13.7. EXOTIC FRUITS

13.7.1. Persimmon

Diospyros. Ebony family (Ebenaceae)

This tree is originally from China and Japan.
The leaves are large, green, and turn reddish before they are shed in fall. The female flowers are solitary and the male flowers are borne in cymes.
The fruit is a large juicy berry, usually sold with the elongated calyx still attached at the base. The flesh is soft, juicy and very sweet, and orange in color.
The most important fruit species are:
• *D. virginiana*. The American persimmon, an American species whose fruit is not very palatable
• *D. lotus*. The date plum. It is originally from Asia
• *D. kaki*. The Japanese persimmon. Originally from China. The best of the edible species. It is the most widely used as a rootstock for its fruit.

Left: Banana plants

Right: Date palm with dates

Pollination is by insects, which can transport pollen several hundred meters from a tree with male flowers. Some fruit produce parthenocarpic fruit if there is no local source of pollen.
It is a tree of hot climates, though it can grow in temperate climates as long as it is not planted in moist areas. It is very drought resistant. It prefers a deep, fresh, loose clay soil.

13.7.2. Custard apples

Annona cherimola. **Custard apple family (Annonaceae)**

The custard apple is a small tree originally from Peru.
The leaves are oval, entire, slightly aromatic and a deep green color.
The sweet-smelling flowers are white and appear in summer.
In summer the fruit is as large as a pear, with a scaly skin that turns grey when ripe.
The pulp is buttery, juicy, sweet-tasting and perfumed.
Custard apples have to be grown in a hot climate in a site that is cool and can be irrigated.

13.7.3. Banana

Musa. **Banana family (Musaceae)**

The banana is a herbaceous perennial plant, whose aerial parts die after fruiting, but are replaced by new shoots that grow up from the base. The true stem of the plant is an underground organ that only emerges above the ground when the flower stem is produced.
The leaves are very large and when old are frequently torn by the wind.
The yellow flowers are borne in a complex hanging raceme that may be up to 2 m long, gathered within bracts. The individual bunches, or hands, make up a *spathe*. The entire spathe may consist of 3 or 4 hands, and in high-fruiting varieties, up to 12.
To begin with the fruit is green and turns yellow on ripening.
3 main species are cultivated:

• *Musa paradisiaca.* Thick fruit.
• *Musa cavendishii.* Elongated fruit. This species is cultivated on the Mediterranean coastline of Spain and the Canary islands.
• *Musa sapientium.* Small fruit. Cultivated in Guinea.
The banana needs a warm climate, with constant moisture in the air.
It prefers soils rich in potassium, rich in lime, that do not retain water in winter.

13.7.4. Date palm

Phoenix dactylifera. **Palm family (Palmae)**

The largest cultivation areas are in the Nile valley and in Persia.

The date can reach a height of 20 m. It is a dioecious plant, meaning some trees produce male flowers and other trees produce female flowers. The female flowers are borne on large hanging inflorescences that form a spathe.
The fruit is yellow or dark drupe containing a single elongated seed.
The date requires a climate with high temperatures and dry air, especially in the period between flowering and the ripening of the date. It requires soils with constant moisture, cool, deep, loose and fertile.

13.7.5. Avocado

Persea americana. **Laurel family (Lauraceae)**

The avocado is a fast growing tree originally from the American tropics.
The leaves are perennial, entire, oval, leathery and dark green in color.
The small flowers are grouped in terminal panicles. The fruit are pear-shaped drupes and deep green or purple.
The calory-rich flesh is firm, buttery, with a delicate flavor, and contains a single hard round seed.
Cultivated varieties: Bacon, Zutano, Fuerte, Pinkerton, Hass and Reed.
It needs a hot site that is sheltered from the winds. It adapts to a wide range of rainfall. It prefers a deep fertile soil with frequent irrigation during the summer.

13.7.6. Guava

Psidium guajava. **Myrtle family (Myrtaceae)**

Originally from Central America. It can reach a height of 6 m.
The leaves are opposite, entire, oval, thick, and green with transparent pores.
The flowers are solitary, and white or pink. The fruit is a yellow or red berry, with a strong smell. The flesh is juicy, perfumed, sugary and sour.
It is a very robust tree, but grows best in hot areas with moderate rainfall. It prefers a loose, hot and fertile soil.

13.7.7. Kiwi fruit

Actinidia chinensis. **(Actinidiaceae)**

The kiwi, or Chinese gooseberry, is a climbing shrub originally from China.
The deciduous leaves are large, entire with a light layer of down.
It is a dioecious shrub, meaning there are male plants and female plants. The creamy white flowers are large and the different sexes are very similar, though the female flowers are slightly larger.
The fruit is an elongated oval dark green berry, covered by a dark-greenish skin with abundant short hairs.
The flesh is emerald green and is highly nutritious.

Avocado

Lime

Lemon

Guava

Papaya. Detail of fruit

The female cultivated varieties include; Abbott, Allison, Bruno, Hayward, Monty.
The male cultivated varieties include: Matua, Tomuri and M-3.
The kiwi prefers climates with mild winters and temperate and moist summers. It is sensitive to low temperatures and drought.
It requires light, cool soils that are rich in organic matter and contain little calcium carbonate.

13.7.8. Mango

Kiwi. Detail of fruit

Mangifera indica. Cashew family (Anacardiaceae)

The mango is a large perennial tree that is originally from northern India.
The leaves are alternate, simple, entire, rather leathery, and lanceolate.
The flowers are grouped in a terminal panicle. A single panicle contains both male and female flowers. The flowers vary in color, and may be yellow, green or red.
The fruit is a large ovoid drupe, yellow green in color, with pink, red and violet tones.
Cultivated varieties: Haden, Irwin, Keiff, Kent, Lippens, Maya, Palmer and Sensation.
It is essentially a tropical fruit but adapts to hot temperate climates. Its limiting factor is low temperatures. It has no problem with high temperatures. It is sensitive to wind.
In terms of soil, it is a robust tree that grows well on rocky and calcareous soils, though it prefers deep, light and well-drained soils.

13.7.9. Lychee (litchi)

Litchi chinensis. Soapberry family (Sapindaceae)

The litchi, or lychee, is a small tree that is originally from south China.
The leaves are alternate and compound, and sprout several times a year.
The flowers are small, and grouped in a terminal panicle. There are three different types of flower that open one after another in the panicle. They differ in their degree of sexual development.
The fruit is a round, ovoid, drupe. The peel is bright red, with small, angular, bumps. The flesh is white, translucent, juicy, sweet and pleasant smelling.
The litchi is a tree from the moist subtropical zone and is highly sensitive to environmental conditions. It is more resistant to the cold than the mango and the avocado, but less than the citrus fruits.
It tolerates high temperatures well, though if they are accompanied by dry periods, they may shed their flowers or fruit.
It is not very demanding, but it prefers cool deep soils with good drainage, that are rich in organic matter and slightly acid.

The fruit of the pomegranate

13.7.10 Pineapple

Ananas comosus. Pineapple family (Bromeliaceae)

The pineapple is a herbaceous perennial originally from South America.
The stem only flowers once, and dies after fruiting; a side shoot then grows to replace the mother plant.
The leaves are long and narrow, growing in a spiral on a short stem and forming a rosette. The toothed leaves are green above with a silvery underside due to the abundant hairs.
The flowers, 100 to 200, are hermaphrodite and are borne in the axil of a bract. Humming birds are the main pollinators, and although bees come, they are not effective pollinators.
The compound fruit is the result of the fusion of small individual fruit with the bracts and central axis of the inflorescence. Ripening takes 5 to 6 months.
The top of the fruit is crowned by a tuft of leaves, which continues growing until the pineapple is ripe.
The pineapple is very sensitive to frosts. It needs a sunny climate and adapts very well to dry conditions.
It needs an open soil that is cool, well-aired and permeable.

13.7.11. Papaya

Carica papaya. Papaya family (Caricaceae)

The papaya is a small fast-growing tree that is originally from tropical America.
The leaves are peltate, large and divided into lobes and form a spiral.
The plant is dioecious. It produces female flowers in small groups, and male flowers in large hanging panicles.
The fruit is large berry, fleshy and hollow, and a spherical-oblong shape. The peel is green and turns yellow at the base when it ripens. The flesh is yellow-orange, with a pleasant taste.
The tree is delicate, sensitive to frosts and high air humidity. It requires a lot of sunshine during the ripening of the fruit and should be protected from strong winds. It should have a fertile loamy soil, that is fertile and permeable, well aired and rich in organic matter.

13.8. OTHER FRUIT-PRODUCING PLANTS

13.8.1. Pomegranate

Punica granatum. Pomegranate family (Punicaceae)

Small deciduous trees with spiny branches.

The leaves are opposite, entire, reddish when young, glabrous and shiny above.
The flowers are scarlet or purple.
The fruit is a thick spherical berry crowned by a calyx, and is 6 to 8 cm in diameter and may be yellow, red or pink. The fruit is apple-shaped and covered by a leathery rind that encloses many seeds. The edible part is the juicy pulp around the seed.
The pomegranate is often planted as an ornamental in patios.
If it is cultivated in temperate climates, it should be planted in a sheltered site that is exposed to the sun. It resists drought. It is not very demanding, but prefers a soil that is of average texture, rich and not very moist.

13.8.2. Jujube

Zizyphus jujuba. Buckthorn family (Rhamnaceae)

This is a small tree, deciduous, originally from northern China, with glabrous leaves with stipules and bearing spines up to 3 cm long.
The leaves are simple, arranged in two rows, oblong-oval, with three veins at the base, serrate, smooth, leathery, brilliant, glabrous and firm.
The flowers are small and greenish yellow.
The fruit is an oval or round drupe that is 1.5 cm long and is green when young, turning dark red. The flesh is whitish, sweet and sugary, and is rich in vitamin C.
It requires a temperate climate and resists frosts. It needs a lot of heat to ripen well and supports dry conditions very well. The soil must be deep and cool.

13.8.3. Service tree (sorb)

Sorbus domestica. Rose family (Rosaceae)

The fruit is grouped together (5 to 10 fruit in a group, and are apple- or pear-shaped, and 1.5 to 3 cm long. They are generally green and ripen to a red brown color.
The fruit, the sorb apple, has a very sour taste when naturally ripe, and it is only edible after bletting. Even so, it is indigestible. Sorb apples are used to make a cider that is in great demand in Germany.
It resists wind and cold weather, but suffers from hot and dry conditions. It requires deep, rich, soils that are not very moist.

13.8.4. Medlar

Mespilus germanica. Rose family (Rosaceae)

This is a small tree originally from Europe.
The leaves are large, entire, oval, serrated and downy on both sides.
The flower is large and white.
The fruit is almost round, downy when young and an orangish color. Its taste is astringent and bitter, even when it is totally bletted.
It requires a temperate-cold climate, preferably in cool shady sites that are not too wet.

13.8.5. Loquat (Japanese medlar)

Eriobotrya japonica. Rose family (Rosaceae)

This small tree is originally from eastern China.
The leaves are large, long, lanceolate, leathery, shiny above and velvety on the underside.
The flowers are grouped at the tip of the plant and their sweet smell attracts bees.
The fruit are pomes, usually borne in groups of 4 or 5, and orange-yellow in color. The pulp is soft, sugary and very pleasant. Inside there is a "core" consisting of 1 to 3 seeds.
Cultivated varieties: Palermo, Concha de Oro, Monreal, Santa Rosalía, Olivier, Vanille and Tanaka.
It is more sensitive to the cold than the medlar and can tolerate dry conditions.

13.8.6. Prickly pear

Opuntia ficus-indica. Cactus family (Cactaceae)

This a succulent plant from tropical America. It consists of oval pads, greenish blue with a fleshy texture, and bears spines.
The flower is deep yellow.
The fruit, the prickly pear, is an ovoid berry with many spines, and varies in color from light yellow to deep red. The pulp is soft, sweet and aromatic, with many seeds.
It requires hot climates, and is very resistant to dry conditions but does not resist winds or cold. The soil has to be loose and limy, though it can also grow in infertile soils and between rocks or even on walls.

Loquat

Pineapple

14. FORESTRY

A forest is a group of trees that occupies a large area. A forest is divided into patches. Patches are areas of the forest that differ in their composition, status or age.

Silviculture is the science dealing with the creation, conservation and regeneration of forests.
A forest plays many roles, and these are of increasing importance, not only for ecological and production reasons, but also for the large area occupied by forests throughout the world. A forest produces direct benefits, such as wood and its derivatives (paper, cardboard, lacquer, firewood, charcoal and resins) and indirect benefits. These indirect benefits include the regulation of rivers, protection from erosion and the wood's effect on rainfall. The main element of the forest is the tree. The tree is a woody plant structure that, depending on the species, can reach a height of 90 m and 3 m in diameter. The tree consists of three main parts; the crown, the trunk and the roots.

Parts of the trunk
The sapwood is the functional wood where the sap rises. It consists of colorless living tissues, meaning it is light in color. It consists of the last 10 to 15 annual growth rings the tree has produced.

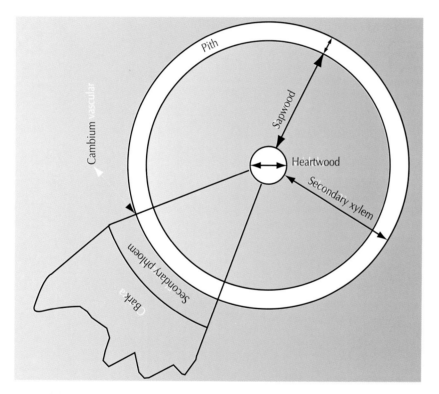

The crown consists of all the branches, leaves, flowers and fruit. Its main function is the production of carbohydrates through photosynthesis. To do this, it needs sunlight, and this is the reason for the competition between the crowns of trees in a forest.
The trunk is a woody column that supports the crown whose main function is to keep it in contact with the root system for transport of water, minerals and nutrients. The outside trunk is protected by the bark.
The roots form the underground part of the tree and play several roles. One role is to anchor the tree in the soil, and another is to supply the tree with water and minerals absorbed from the soil.

• Direct forest products

Forests produce materials that are consumed by people and by industry. These include timber, firewood, resins, nuts, grazing, hunting and fishing, tanning agents, mushrooms, essences, plants and undergrowth as bedding for livestock, stone, earth, etc.
Timber is the most important product extracted from forests, and the timber produced depends on the trees felled. Timber species include the following: cedar, jacaranda, ebony, mahogany, beech and oak. Firewood, however, is used less and less, because firewood use is decreasing as it is replaced by other fuels. The problem of low demand leads to the accumulation of pruning remains, leaf litter and dead branches in the forest. This accumulates because it is not economically worthwhile collecting it.
This accumulation of branches also shelters many pests, and is highly inflammable in a forest fire.
Plant resins are obtained mainly from pines, and are used to produce colophony, used to make varnishes and paints, and turpentine. These plant resins have been almost totally replaced by synthetic resins derived from petroleum, which are easier to produce and cheaper.
The most important nuts produced in forests are pine nuts, chestnuts, walnuts and acorns.
The pine nut comes form the stone pine (*Pinus pinea*), and like the chestnut and walnut, is for human consumption. The acorn, which may come from deciduous or evergreen oaks, has traditionally been fed to livestock, especially pigs.
Much of the world's area of forests is also used for grazing. In some countries more than 50% of grazing is under the shade of forest trees.
The moist environment typical of areas with many plants favours the growth of grass, which is eaten fresh, or as straw, by livestock.
Other important sources of wealth present in many forests are fish and game. Big game species include deer, wild boars, bears, tapirs, ocelots, sloths, jaguars and pumas. Small game species include badgers, hares, peccaries, iguanas, pigeons, pheasants and partridges, etc., and waterfowl.
The rivers contain many fish, which are of interest as sport (angling) and economically. The best tasting or nutritious ones include eels, perch, skates, atherines, edible turtles, etc.

• Indirect benefits of forests

The indirect benefits of the forest are essentially its influence on the climate, soil and water.

• **Heartwood**, is the non-living wood at the center, which gives the tree its strength. It consists of the tree's oldest growth rings, and does not start to form until the trunk is 12-15 years old.

• **Climate**
The forest creates its own climate. The moisture, temperature, rainfall and evaporation of the forest microclimate are different to those of open fields.

• **Temperature**
The crowns of the trees act as barriers, and buffer changes in temperature. Thus, in the summer the temperature within the forest is lower than outside, and in winter, it is higher. It thus tends to reduce variations in temperature.

• **Moisture**
Moisture is created by the transpiration of the trees and it is greatest during the growing period.

• **Wind**
The forest reduces the wind speed by 60-80%. This depends on the density of the crowns, the spacing between them and the size of the forest. It protects crops and livestock from the effects of the wind.

• **Rain**
The amount of rainfall reaching the ground is less within the forest, as the leaves intercept the water before it reaches the soil.

• **Evaporation**
The forest's evaporation is the sum of the evaporation of the soil, transpiration by plants and that of the intercepted rainfall.
Evaporation is determined by the temperature, the relative humidity, the wind, and the atmospheric pressure.

• **Soil**
Forests play a very important role protecting the soil from erosion.
Forests also enrich the soil by incorporating organic matter.

• **Water**
The forest's influence on water resources depends on the climate, the soil, the type of forest, and the site's relief.
The quantity of water produced by a deforested watershed is greater than that of a forested watershed. This is because the forest is a large consumer of water.
In terms of water availability over the course of the year, a deforested watershed produces large amounts of water during the rainy season, but may dry up in the dry season. Rivers leaving a forested watershed produce water throughout the year.
The sediments borne by the rivers also differ. A river leaving a deforested watershed bears more sediment, and this may lead to silting up of reservoirs and cause problems in hydroelectric dams.

14.1. REFORESTATION

Grinding forest remains.

Reforestation is the replacement of one forest mass by a different one. The composition and continuity of the forest depend on this.

Reforestation may be natural or artificial, depending on whether human beings intervene directly or not.

• **Natural method**

Seed dispersal and establishment take place without human intervention, which is limited to favoring the conditions for the germination and growth of the plants, with different tasks, such as:

• Getting rid of old or diseased specimens
• Reducing the density of trees
• Conditioning the soil to favor germination
• Protection from animals
• Shelter, irrigation and fertilization after germination.

Seeds for reforestation arrive spontaneously from the same site or neighboring sites, and are dispersed by the wind.

• **Artificial method**

The seeds and plants are distributed by people, who choose the site, method and species to be planted.

Reforestation may be by sowing or by planting saplings.

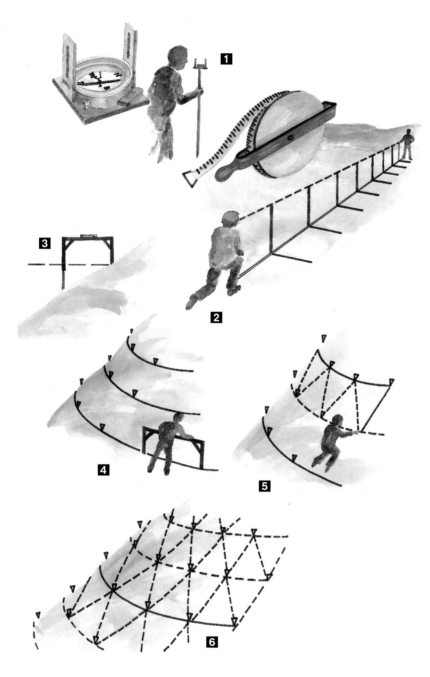

Artificial reforestation

1/ The alignment of the first rows in a rectangular or square distribution, is established by using a compass with a viewer.

2/ The site of the holes is determined using a tape measure. The site of the hole can be marked with a stake, or you can dig the hole immediately.

3/ Marking out the triangles starts by marking the distance between the rows on the same contour level. (Continued on next page)

too slowly and burn the soil, rendering it unsuitable for germination.

The second operation to prepare the soil is to loosen it, to prepare a tilth. This is done by digging holes or ditches (if done partially) or by digging the soil deeply (if it is done thoroughly). This ensures the soil is looser and suitable for germination.

The seed can be sown in several ways:

• *Scattering.* The seed is simply scattered by hand over the soil, and then covered with sieved earth and raked over. It is cheap to perform, but a lot of seeds are needed. The saplings also tend to be in clumps. This is why it is not usually practised.

• *In trenches.* Dig parallel trenches and sow the seed in them. The trenches should be 30-50 cm wide, 1-1.5 m apart.

• *In holes.* Dig holes like those for a plantation. If the seed is not sown in circular spaces but in square ones, it is known as sowing in boxes.

• *In groups.* This technique hardly requires soil preparation. It is done in soils that are not very compact and with a moist climate. Make a hole in the soil with a hoe, place 2 or 3 seeds in it and cover them with the earth that was removed.

• *In ridges.* This is done on wet or compact ground, in order to sow above ground level. The ridges are left to air for one winter before sowing.

The sowing period depends on how long the seed retains its ability to germinate. If the seed only lasts a few days, like the elm, birch, poplar and willow, it should be sown when harvested. The other trees can be sown in fall or spring. The fall is recommended if the climate is hot and dry, as the plant can make use of the rains and ensure it reaches the hot season with a good root system, while spring is recommended in cold climates with risks of winter frosts or strong rains.

• **Direct sowing**

Direct sowing does not usually give very good results, due to seed predation by animals and unfavorable environmental conditions.

Seed is sown if the tree is difficult to cultivate in a nursery, such as holm oaks, and is performed by simply placing the seed in the soil, aiming for a given number of trees in an area after successful germination.

Two things must be done to prepare the soil for sowing. The first is to clear the vegetation, which is only necessary when there is troublesome vegetation that raises problems when sowing. This can be done mechanically or by a **running fire**, burning the plant cover.

This is cheap but dangerous. It can not be performed on days with intense wind, or on days without wind, as the fire would advance

• **Planting**

This is planting saplings of the desired species in previously prepared soil.

Artificial reforestation uses fast-growing species and seeks to optimize use of space, to create a wood that produces more timber than in natural reforestation.

The shrubs and grasses that compete with the growing saplings also have to be removed.

The success of planting depends on:

• The choice of species. It should come from an area with a climate similar to the area where it will be planted.

• Healthy, vigorous plants with a good root system.
• Planting density and methods appropriate for the soil and climatic conditions. Mechanical methods can be used where the site's relief allows.

dunes without offering excessively rigid resistance.
The plantation can be linear or in staggered rows, and is done by opening a slit in the soil, where the plant is placed, and is held in position by the walls of the slit.

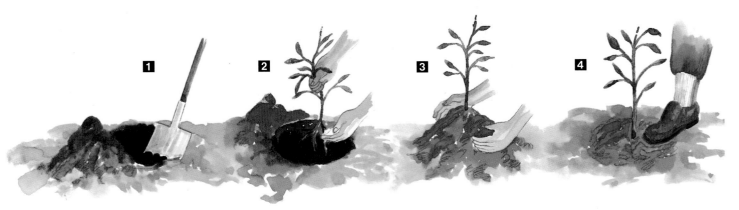

It is worth pointing out that high planting densities lead to higher wood production per hectare than low planting densities.
The diameter of the trunks of the individual trees is, however, less.
The plants used in reforestation may be taken from forested areas, using excess saplings. This removes the excess saplings and thins the forest.
Plants may also come from commercial nurseries, and these give the best results.
Plants from nurseries usually come bare-root, with a root ball or in a container (which may be reusable). The containers should be removed before planting.
If the plants are bare roots, they can only be removed from the site of origin during the rest period.
It is worth preparing the site before planting.

Depending on the forest's characteristics, the operations to be performed are:

• **Clearing**. Consists of burning the invasive plant cover.

• **Uprooting**. This is uprooting the shrubs, together with the removal of the grass layer and the remains of any former vegetation.

• **Drainage**. Improving drainage is essential if the site gets waterlogged. This is done by digging a network of shallow ditches deep enough to ensure the soil drains well. Another method is to lay drains, using stones or other materials.

• **Soil stabilization**. On sites with steep slopes, or at risk from heavy rainfall or gulley erosion, the soil should be consolidated by stakes, dikes, etc., or by sowing appropriate species that will not be washed away by the water or the earth it bears.
In dunes, first construct an artificial dune with canes, to halt the advance of the

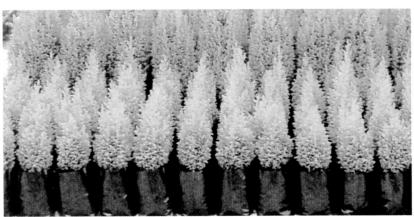

14.2. TREE NURSERIES

A forest nursery is a site intended to produce plants for use in artificial reforestation.
Before creating a nursery for this purpose, it is worth considering whether it would not be cheaper to buy the plants. To do this, take into account the number of saplings to be produced every year, their quality, and the distance between the nursery and the planting site.
There are two types of nursery, temporary and permanent ones.
• **Temporary nurseries** are intended to supply plants for reforestation of a given zone. This type of nursery is sited near to the zone to be planted, and they normally cultivate local species of trees, as they are acclimatized to the soil and climate of the area of forest where they are going to be planted, and will not be subject to the problems arising from transport.
This type of nursery is a simple installation, not requiring buildings or major investment. They do not have artificial irrigation, so the operations must coincide with the rainy periods.
• **Permanent nurseries** are created to produce more than 30,000 plants a year.
The investment required is greater, as they require a site, equipment, irrigation systems and labor.

Planting:
1/ Dig the hole.
2/ Place the plant upright.
3/ Fill in the hole.
4/ Tread down the soil around the plant, so the roots are in contact with the soil.

In the center:
Plants for reforestation

Artificial reforestation (continued):
4/ The distance between the plants is marked along the rows.
5/ In order to mark the equilateral triangles in the staggered rows, the sites for the next row are marked using a cord.
6/ The complete planting following staggered rows.

Setting up a forest nursery

If you have to set up a nursery on a slope, you must terrace it. These make handling easier and prevent erosion.
To protect the plants from winds, windbreaks are needed, or you can plant the nursery leeward (in the wind shade) of a forest.

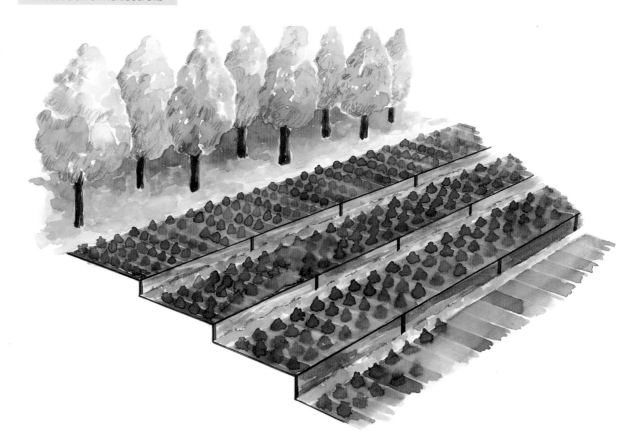

Irrigation systems in a tree nursery:
1/ Underground watering using perforated tubing
2/ Irrigation by channels leading to the plots
3/ Sprinklers (the most frequently used)

When choosing a site to set up a nursery, many factors, such as climate, soil, the relief, communications, access and the availability of labor all have to be taken into account.

The nursery should meet the following requirements: a climate similar to where the trees will be planted, a flat soil with good drainage and good structure, fertile, not stony, and with plentiful water for irrigation.

The size of the nursery depends on the number of plants to be produced every year, the species of tree and the desired size of sapling.

The surface of the nursery should be divided into sections separated by passageways. Each section is dedicated to plants at different stages.

The sections present might be: seedbeds, first transplant seedbeds, nursery for deciduous plants, pot-grown plants, etc. Each section is in turn divided into rectangular plots or beds, which are separated by paths or ridges of earth.

The tasks to be carried out include:

• Sowing
The seed can be sown into seed boxes or directly into previously prepared ridges. Both have to be shaded during the germination phase. The seed is scattered or sown in lines and at a low density.

The seeds used are collected from the trees, using different methods depending on the tree.

The seed should be collected before it is dispersed naturally, and only collect seed from healthy, vigorous, trees.

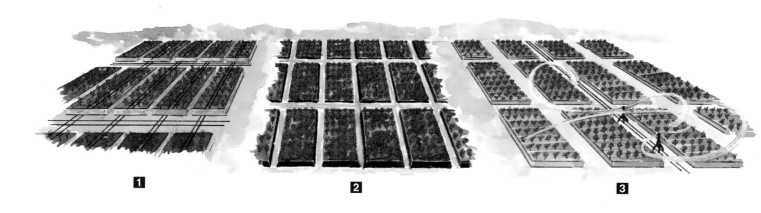

1   2   3

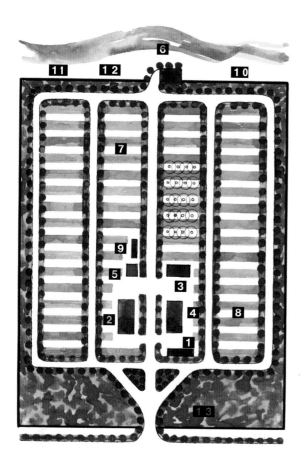

Left: 1/ House for the supervisor 2/ Building with store for tools; storehouse for fertilizers, herbicides, insecticides and seeds; and an office for the supervisor 3/ A shaded area for seedling germination. 4/ Garage for tractors and transport equipment 5/ Water deposit 6/ Pump from water supply 7/ Irrigation system 8/ Beds and areas for plants in pots 9/ Deposit of earth and other organic matter 10/ Windbreak 11/ Hedge 12/ Paths and rows 13/ Demonstration woods

• Pricking out

7-14 days after germination, the seedlings can be pricked out.

This is the plant's first transplantation after germination, moved from the seed bed to the nursery area, and this may mean planting in rows or in pots. This is done to ensure the plant has enough room to grow.

When the seedling is pricked out, the main root should be pinched out to avoid it growing too deep and to encourage growth of the root system.

After pricking out, the seedlings should be watered by sprinkler.

• By cuttings

Some trees can be propagated vegetatively by cuttings rather than seeds.

This is done in plants that root easily, such as willows and poplars.

This is simply planting a young stem, after removing its leaves, which grows adventitious roots and gives rise to a new plant.

• Container cultivation

The most widely used system is to grow the plants in containers.

The size and composition of the container can vary greatly, ranging from polythene balls, to pots made of plastic, clay, cardboard or pressed peat.

The advantage of using container-grown plants for reforestation is that they can grow using the soil in the ball until they root in the site.

14.3. TREE SPECIES

Woody trees are divided into conifers and broadleaves.

• *Conifers* belong to the botanical group gymnosperms, which bear naked seeds in cones.
• *Broadleaves* belong to the angiosperms, the flowering plants, which bear seeds fully enclosed within an ovary.

The main forestry species are:

• Pine

• *Pinus canariensis*, the Canary pine. 3 needles in each bunch. Resinous wood, yellowish-white in color and of high quality.
• *Pinus radiata*, the Monterey pine. 3 needles in each bunch. The wood is soft, light and spongy, and easily bleached. It is less hard than other pines. This species produces excellent pulp for paper.
• *Pinus uncinata*, or black pine. Small timber, white and resinous, and valued for making guitar frames.
• *Pinus sylvestris*, or Scots pine. The wood is compact and resinous, and is of excellent quality as the trunk is straight and free of knots.

*Right 1/ To make the holes in the earth in the pots, you can use a thick nail.
2/ Pines should be pricked out before the cotyledons (seed-leaves) have expanded.
3/ Hold broadleaf seedlings by the cotyledons.
4/ Very long roots should be pinched back with your thumb to about 3 cm.
5/ Firm the soil around the root.
6/ Recently pricked out pine seedlings.*

It is widely used in carpentry and construction. It is the best pine for firewood.

- *Pinus nigra*, or black pine. The high quality wood is suitable for construction or the sawmill. It is similar to the wood of *P. uncinata*. The resin can be used, though the yield is less than that of the cluster pine.
- *Pinus pinaster*, or cluster pine. Its wood is light and very resinous. Its main use is for production of turpentine, though it is also used in construction.
- *Pinus pinea* or stone pine. The wood is relatively resinous, heavy and with many knots. Together with the Aleppo pine, it is the cheapest of the pine timbers, as it is hard to work. It burns rapidly as firewood. It is mainly grown for the pine nuts it bears.
- *Pinus halepensis*, or Aleppo pine. The wood is hard, light in color, and resinous. It is considered to be of poor quality because of the distorted growth form of the trunk. It is used to make packing boxes, railway sleepers or for heating installations.

• Fir
- *Abies alba*, or common silver fir. The wood is white, resinous, lightweight, and easy to work, though it is of lower quality than pine. It is used in cabinet-making.
- *Abies pinsapo*, or Spanish fir. The wood is lightweight, weak and not very resinous. It is mainly used in landscaping.
- *Pseudotsuga menziesi*, or Douglas fir. The wood is very good, hard, resistant, and easy to work. It is used as firewood, for beams and railway sleepers, and to make paper.

• Cedar
- *Cedrus atlantica*, or Atlas cedar, or Atlantic cedar. The good quality wood is soft, aromatic and easy to work. It is used for beams, posts and high quality carpentry.
- *Cedrus deodara*, or Himalayan cedar. Similar to the above.
- *Cedrus libani*, or cedar of Lebanon. Similar to the above.

• Cypress
- *Cupressus sempervirens*, or Italian cypress. The smooth-textured wood is not very resinous, and is aromatic, resistant and easy to work. It is used in cabinet-making and carpentry.
- *Cupressus arizonica*. Similar to the above.

The best known **broadleaf forest trees** are:

• Fagus sylvatica, or beech
The wood is hard, heavy, with a smooth, uniform texture, white or light brown, and easy to work but not very resistant to changes in humidity. The wood is widely used in cabinet-making, but the tree is now mainly planted for landscaping purposes.

• Castanea sativa, or sweet chestnut
The wood is hard and heavy, but elastic, meaning that it is easy to work and is very long-lasting. It is widely used in carpentry, cabinet-making, in casks and barrels, and in construction. It is poor quality firewood, as it is hard to light and does not release much heat. It is also planted for the fruit it produces (chestnuts).

• Oaks
- *Quercus robur*, or English oak. The excellent wood is hard and heavy, resistant to moisture and weatherproof, meaning that it is suitable for shipbuilding. It is also excellent firewood.
- *Quercus petraea* or Durmast oak. The wood is similar to the above. It is used in cabinet-making and railway sleepers.
- *Quercus canariensis* or Canary oak. The wood is of good quality and very porous, used for railway sleepers and building. The acorns are also fed to livestock.
- *Quercus faginea* or beech oak. Good wood, though large pieces are hard to obtain. The acorns are fed to livestock.
- *Quercus suber*, or cork oak. The wood is hard and heavy. It is normally used as firewood, and also to make tools and barrels. Its main industrial use is as the source of cork. The bark is rich in tannins. This tree is of great value in landscaping.
- *Quercus rubra*, or red oak. Very good wood.
- *Quercus ilex*, or holm oak, or evergreen oak. The wood is white or pinkish, dense, homogenous and compact. It is not used in construction because it is too heavy, but it is used in road building and making tools. The wood is highly valued for the large amount of heat it produces.

• Poplars and aspens
- *Populus alba* or white poplar. The wood is of low quality, and has an unpleasant smell when dried.
- *Populus tremula* or aspen. The wood is not valued.
- *Populus nigra* or black poplar. The wood is light and of better quality than the white poplar.

• Eucalyptus
- *Eucalyptus globulus* and *E. camaldulensis*. Their main industrial use is to produce cellulose.
- *Eucalyptus camaldulensis* is also used to make parquet, because of the wood's reddish color.

• Alnus glutinosa, or alder
The wood is hard, fibrous, and hard to saw. It is very resistant to water, and is thus often used in hydraulic works.

• Acer spp., or maples
White wood, light and easy to work. It is used in cabinet-making and carpentry.

• Betula pubescens, or birch
The wood is almost white, and is not dense and does not last long. It is used to produce veneers and to obtain paper.

• Celtis australis, or hackberry
The wood is elastic, flexible and compact, and is used to make oars, carts and vats.

• Ceratonia siliqua, or carob
The wood is hard, heavy and homogenous. It is used in cabinet-making and as a high quality firewood.

• **Corylus avellana**, or hazel
The wood is white and is not very resistant. It is used in basket-making.

• **Fraxinus angustifolia**, or ash
The wood is very good, resistant, elastic, and easy to work. It is used in cabinet making and as firewood.

• **Ilex aquifolium**, or holly
The wood is resistant, heavy, hard and difficult to work. When died black, it resembles ebony. It is used in cabinet-making, construction and as firewood.

• **Walnuts**
• *Juglans regia*, or walnut
The wood is hard, excellent quality and easy to work. It is used in high quality cabinet-making. It is also cultivated for its fruit, the walnut.
• *Juglans nigra*, or black walnut. Similar to the above.

• **Hybrid plane**, or **Platanus hibrida**, or London plane tree.
The wood is hard, resistant, similar to the wood of the beech, but lighter in weight. It is used in construction, carpentry and cabinet-making.

• **Prunus avium**, or wild cherry
The hard dark red wood is easy to work, but has the disadvantage of tending to split. Used in cabinet-making.

• **Robinia pseudoacacia**, or false acacia
The wood is hard, heavy, difficult to work, and also scratches easily. It is used as firewood, in carts and for posts.

• **Salix spp.**, or willows
The wood is soft, light, and not very valuable. It is used in basket-making and to make cellulose.

• **Sophora japonica**, or pagoda tree
The dense wood does not withstand humidity and splits. It is used in cabinet-making.

• **Taxus baccata**, or yew
The good quality wood is reddish, uniform, compact, resistant and elastic. It is used in cabinet-making, bows, lances, veneers and craft work.

• **Linden**, or **lime**
• *Tilia platyphyllos* or large-leafed linden. The wood is light, soft, uniform, and easy to work. It is used in carving.
• *Tilia cordata*, or small-leafed linden. The wood is white, soft, porous, better than that of the large-leafed linden, though it does not withstand the weather. It is used in paper production.

• **Elms**
• *Ulmus minor*, the common elm.
The wood is dense, rot-proof, elastic, soft, and easy to work. It is used in road alignments and in construction.
• *Ulmus glabra* or wych elm. The wood is hard, heavy, elastic, and worse quality than the preceding wood. It is used in carts and shipbuilding.

Knots forming in wood:
1/ Axial section of a trunk with a loose knot.
2/ Axial section of a trunk with healthy knot.
3/ Wooden plank with a loose knot.
4/ Wooden plank with a healthy knot.

Pruning and sawing with pneumatic cutter and saw.

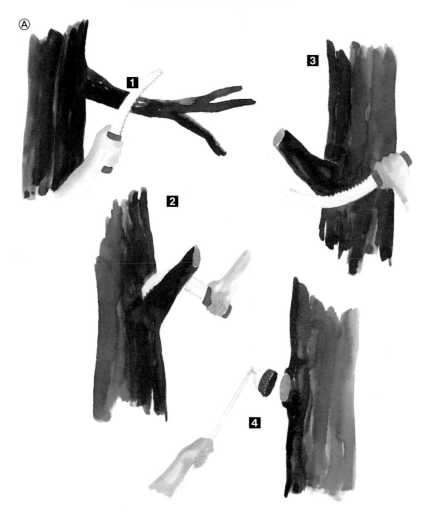

Ⓐ

Ⓒ

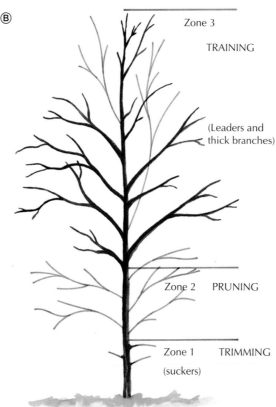

Ⓑ

A

1/ First cut the thick branches about 15 cm from the trunk.
2/ Then cut upwards from below, to prevent the bark from splitting.
3/ The branch is then cut off by cutting down from above.
4/ The wound is covered with some type of sealing putty to prevent rotting and insect attack.

B

The three classic types of operations, which can all be carried out at the same time: training, pruning and trimming.

C

Trimming for a straight stem (above) and pruning back (below).

Zone 3

TRAINING

(Leaders and thick branches)

Zone 2   PRUNING

Zone 1   TRIMMING

(suckers)

14.4. PRUNING FORESTS

Forest pruning consists of influencing the formation of the trees' crowns and eliminating the dead or mutilated branches that might give rise to pests and disease.

The branches are inserted in the woody tissues of the tree. The zone of insertion is called the **knot**, and is normally harder and darker.

A knot is healthy when the surrounding wood is joined to it. If not, the knot is said to be loose.

A healthy knot is formed when the branch was alive, while a loose knot is formed when the branch was dead. This is the reason why loose knots may cause rots in the stem.

The number of loose knots can be reduced by correct pruning, and this achieves cleaner and healthier wood.

There are two types of pruning, natural and artificial:

• **Natural pruning**. This leaves the death and fall of the branch to nature. This death may be due to lack of light or by disease in the branch. The branches that do not receive light are weakened and are more easily attacked by insects and diseases.

• **Artificial pruning**. This is carried out to obtain high quality wood without loose knots. Pruning dead branches has no effect on the tree's growth. In addition to pruning aimed at shaping the stem, artificial pruning has other aims, such as healing the tree, promoting fruiting, maintenance and renewal.

The purpose of shaping is to obtain a straight upright tree. This requires removing the double and triple shoots and slowing the branches that are growing excessively. Suckers and the lower levels of branches are also removed.

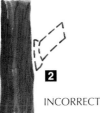

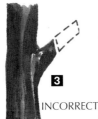

CORRECT   INCORRECT   INCORRECT

Ⓐ

Ⓑ

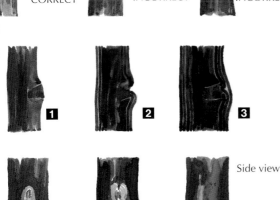

Side view

Front view

To obtain high quality wood, remember that tree wood is not produced until wood has grown over the scar produced by pruning a branch.

The cut should be clean, trying to avoid splitting the wood, or splintering the wood to be removed. The branch should be cut back as close to the trunk as possible, when the tree is young, so that the knot remains within the tree and layers of quality wood can grow over it.

If you wish to obtain wood free of knots, you should never prune once the trunk diameter exceeds 10 cm where the branch is inserted. The wound left by pruning should be as small as possible, preferably less than 5 cm in diameter, so that it can heal well within a short time.

Badly healed wounds allow access by fungi and insects which eat the wood and shorten the tree's life.

High quality wood for carpentry and veneers require correct pruning of the trees. It does not, however, make much sense to prune in plantations that seek to produce firewood or to obtain cellulose.

The best pruning time is during the tree's rest period, preferably towards the end in order to accelerate wound healing.

Pruning of thin, sick or dead branches in resinous trees should preferably be carried out in late fall, in order to reduce sap loss to a minimum, and so the wound can cure over the winter.

One variant of pruning is **twig trimming**. This consists of cutting off twigs so that they can be fed to livestock. This type of trimming is obviously not very good for the tree, as its main aim is to feed the animals.

This trimming may be intense and almost total.

14.5. EXPLOITATION OF SCRUBLAND

The most important exploitation the forest experiences is the extraction of wood, which may be used as wood or made into cellulose, paper, etc., though there are other uses, such as extraction of cork, resins and fruit.

A
Cutting a branch from the trunk should be done leaving the ring to heal. (1). It is counterproductive to cut right back (2) or to leave an excessively large stump (3).

B
Growth over a pruning wound
1/ The year of pruning
2/ After two years, it is beginning to grow over.
3/ After four years, it has totally grown over.

The transport of trunks and the grinding of branches, etc.

Removal by dragging

Removal by lifting the load

Trailer

14.5.1 For wood

To obtain timber, the tree must be felled and transformed into a condition suitable for use by the timber industry. Three points should be taken into account.

• Felling time

The best time to fell is in the rest period, in fall/winter. This is because the wood deforms less as it contains less moisture, and it has less rot, because fungi need heat, and it suffers less damage by caterpillars and woodworm, because it contains less starch.

• The age to fell the tree

The time until the trees are old enough to be felle known as the turnover.

This can be delayed for a year or two, depending market conditions or other conditions.

There are 4 factors closely linked to the felling ag that must all be taken into account because they influence each other. They are: the species and variety of tree, the soil quality, the planting densit and the application the wood is for.

• *The species* and the *variety* indicate the best ag for felling.

• On *good soils*, the age for felling is usually less than on poor quality soils.

• The *application* and *planting density*: for the production of thin wood (building, cellulose factories, etc.) planting should be dense, as the tr will not grow very big and their roots and crowns will not compete. If what is desired is timber, the spacing must be considerably greater.

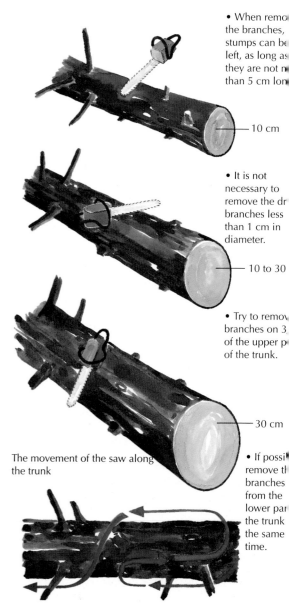

• When remo
the branches,
stumps can be
left, as long as
they are not n
than 5 cm lon

— 10 cm

• It is not
necessary to
remove the dr
branches less
than 1 cm in
diameter.

— 10 to 30

• Try to remov
branches on 3
of the upper p
of the trunk.

— 30 cm

The movement of the saw along
the trunk

• If possi
remove th
branches
from the
lower par
the trunk
the same
time.

• Felling technique

The trees to be cut in a patch of forest are normally marked in advance. Felling should take into account the direction of the fall, removal from the forest and the need to damage young trees as little as possible.

When choosing the direction the tree is to fall, you must take these points into account:

• The natural inclination of the trunk and crown
• The location of the nearby trees
• The direction of the prevailing wind
• The site where the felled trunks are to be piled

To fell the tree correctly, you must do the following:

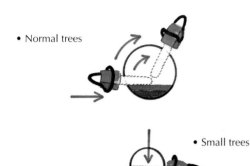

• Normal trees

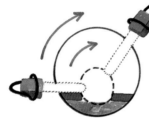

• Small trees

• Large trees
(a groove must be cut)

• Clean around the base, removing the shrubs near the tree and an area for the following operations.
• Clean the trunk to shoulder height by removing the branches.
• First make a groove of 1/5 of the diameter of the trunk at the lowest height possible, making the upper cut first to ensure it coincides with the lower cut. The groove is the cut that is made at the base of the trunk with a horizontal lower face and an upper cut at an angle of 30-40°.
• Then make the transverse cut, the felling cut, which should be made slightly higher than the horizontal cut of the groove and on the opposite side of the trunk. The felling cut is parallel, and leaving a strip of wood of 3 to 4 cm uncut. This is called the **felling hinge**. The groove cut prevents the tree from splitting, by making the fall slower and more directed.
• If the direction of fall does not coincide with the natural fall of the tree, the **felling hinge** is left wider on the other side.
• The preparation of the wood includes cleaning, debranching and debarking the tree and cutting it into logs. Removal from the forest is performed

• Make the upper cut first to ensure it coincides with the lower cut.

• Make the horizontal cut as low as possible in order to obtain as much wood as possible.

• The felling cut has to be large enough (1/5 of the trunk's diameter)

mechanically with dragging equipment or trailers, or using draft animals.

14.5.2. For cork

This is only performed on cork oaks, early in summer, when it is hot and moist. These environmental conditions increase the tree's vegetative activity and mean the cork can be removed easily, without damaging the living layers.
The first extraction is carried out when the tree is 60 cm in circumference and at a height of 1.3 m. The cork obtained from this first stripping is of low quality. After 9 years, the normal rotation period in cork extraction, the second stripping is performed, which produces cork that is still of low quality.
The third stripping starts to produce better quality cork. Over its life the tree can be stripped 9 times. The length of the strip should be twice the circumference of the tree at a height of 1.3 m in the first stripping, at a height of 2.5 m in the second stripping, and at 3 m in the others.

14.5.3. For resin

Resin is extracted from the resin pine and the black pine. The product extracted is pine turpentine and consists of water, resin, and impurities, such as shavings, bark and insects. The resin obtained consists of oil of turpentine and colophony, which can be purified and distilled.
Each tree can produce 2 to 3 kg a year of raw product.
Cheaper industrial alternatives mean that resin extraction no longer covers its cost, and so the future of this activity is not bright.
To produce resin, the forest is thinned early, in order to obtain the large diameters that are best for resin production as soon as possible. This diameter is 30 cm at a height of 1.3 m above the soil.
When the trees have reached a diameter of 18-22 cm, those that are to be thinned out are bled to death, by making as many cuts in them as possible.

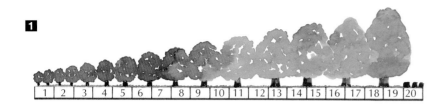

| 1 | 2 | 3 | 4 | 5 | 6 | 7 | 8 | 9 | 10 | 11 | 12 | 13 | 14 | 15 | 16 | 17 | 18 | 19 | 20 |

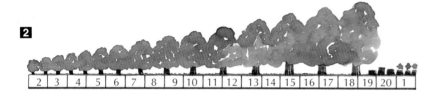

| 2 | 3 | 4 | 5 | 6 | 7 | 8 | 9 | 10 | 11 | 12 | 13 | 14 | 15 | 16 | 17 | 18 | 19 | 20 | 1 |

Rotation of a forest
1/ The plots, one year after regularizing the forest. The figures show the age of the trees in the plot.
2/ The plots, two years after the regularizing the forest.

Resin extraction from a forest starts when the trees have reached a diameter of more than 32 cm. This means that six grooves can be cut, leaving six strips of bark about 3 cm wide.
Each groove will be cut 6 times a year, each cut about 500 cm long and 12 cm wide. Each groove will have a variable number of cuts made.
A metal sheet nailed to the base of the annual groove gathers the resin and channels it towards a collection vessel.
The resin extraction period is during the growing period. Before cutting the grooves, the tree's bark is removed but not to the wood.

14.5.4. For nuts

The nuts that are collected

Diseases produced by fungi
1/ Fungi on a trunk and a stump
2/ Damage caused by a fungus in a trunk
3/ Fungus canker of the stem
4/ Axial section of a stem showing streaks full of gum

• **Acorns**, which are fed to livestock and game.
• **Pine nuts, sweet chestnuts, hazel** and **walnut**, for human consumption.

To produce nuts, the trees must be adults, and large and vigorous enough to flower and fruit. The forest has to be open, and free of scrub and pests.

14.5.5. Poplar cultivation

The plants are hybrid poplar clones. The most widely planted is **Campeador**.
Trees 1 or 2 years old are planted in a 4 x 4 frame, if they are going to be cut after 6 to 8 years for non-thick wood, or in a 6 x 6 frame if they are going to be cut after 10-12 years for thicker wood, better in quality and fetching a higher price.
It is essential to have access to irrigation water or a permanent water table.
Pruning is performed in winter during the third, fourth and fifth years of the sapling's life. The lower whorl of branches is removed, and branches that are excessively dominant and vertical are removed, to prevent them from causing the stem to divide.

14.6. FOREST ROTATION

The main difference between growing crops and growing trees is the time you must wait until the product is ready. The farmer usually harvests once a year, but rotation of trees in a forest may take 10-120 years.
The rotation period depends mainly on what purpose the wood is for. For example, to obtain pulp for paper and cellulose, rotation is 10 to 25 years, while to obtain wood for veneers, it may be as long as 120 years.
It is important that the site's production should show a sustained increase, that is to say that the annual harvest of wood should never be greater than the forest's annual increment.
The following is one example of a rotation:

a) The site is divided into 20 equal plots
b) Every year the trees are harvested from one plot, so that after 20 years there will be a new forest with 20 plots of different ages, with a difference of one year between them.
c) Each plot yields the same volume of wood in the next rotation.

This is a theoretical plan for the forest, as it supposes ideal climatic conditions, such as a balanced forest, similar climatic and growing conditions over the cultivation period, and homogeneous sites. As these circumstances do not exist, this scheme is only theoretical.

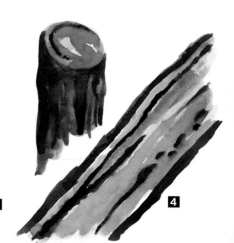

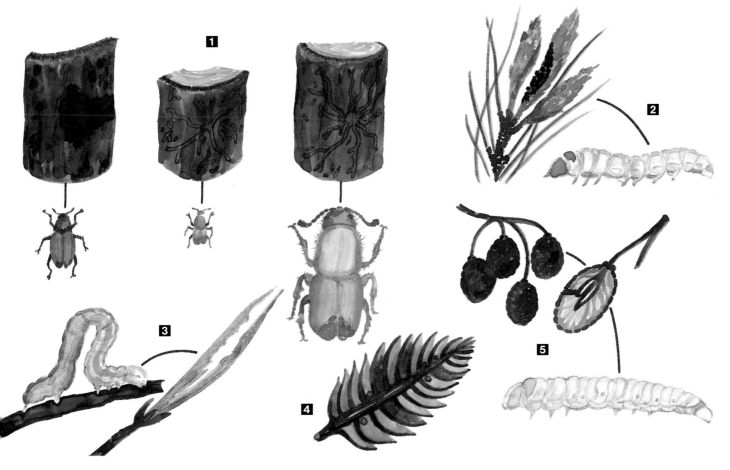

Insect damage
1/ Bark weevils. The damage caused often allows identification of the species of weevil.
2/ Damage caused by a caterpillar in the buds of a pine.
3/ Caterpillar that defoliates pines.
4/ Pine cone attacked by caterpillars.
5/ Fruit and seeds of cedar attacked by a caterpillar.

14.7. FOREST PROTECTION

In general, young trees are more susceptible to attacks by parasites than mature trees, as it is easier for them to be damaged or to die.

Forests that have regenerated naturally are less susceptible to disease and pest attacks than artificial forests.

In turn, a forest with a variety of tree species of different ages, is less susceptible to the spread of fire, to disease and to pests.

Other factors, such as poor soils, high tree density, lack of thinning and pruning, or over-exploitation of the forest all favor the spread of diseases and pests.

14.7.1. Non-parasitic disorders

These non-parasitic disorders are not only important for the direct damage they cause to the tree, but also because they weaken the tree, favoring attacks by parasites.

When considering which tree species to plant, you should choose one that is totally adapted to the climate and soil of the forest where it is to be planted.

Climatic causes

• *Temperature*
Low winter temperatures cause damage. Frosts in spring and fall are dangerous, as they affect the tree outside its rest period. Damage by cold may be reversible or irreversible, and are generally worse if the temperature change is rapid. Many factors affect the seriousness of the damage: the species, the variety, its vegetative state, the soil type, the tree's age and its orientation.

• *Drought*
Drought damage is reversible if it is slight, or irreversible, causing the tree to die. The seriousness of the damage is influenced by factors like the depth and nature of the soils and the distribution of rainfall over the course of the year.

• *Excess water*
This may cause root asphyxia due to lack of oxygen.

• *Snow*
Snow damage varies from species to species. Sleet is also dangerous as this accumulates on the branches and causes breakages and malformations.

• *Wind*
Very strong winds may cause branches to break or even uproot the tree, especially if they are accompanied by rain.
Constant winds in one direction may alter the tree's growth and its shape.

• **Soil**

Plant species are adapted to a definite physical and chemical soil composition. Any variation in this composition may cause disorders in the tree.

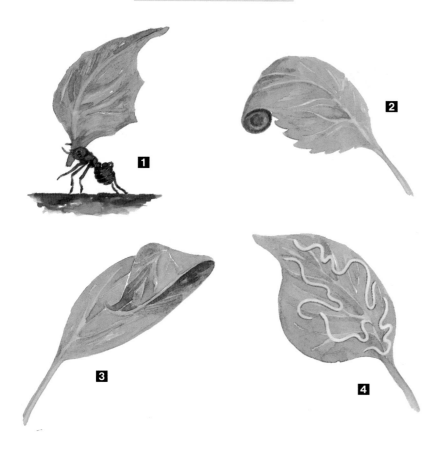

Damage produced by insects
1/ A leaf attacked by a defoliator, in this cases a leaf-cutter ant
2/ A leaf attacked by a leaf curler caterpillar
3/ A leaf attacked by a leaf folder
4/ A leaf attacked by a leaf miner

• Forest fires

Fire is the most important agent destroying forests, with 90% of them caused by humans. The prevention of fires starts from the education of the public to obey the legal regulations. The law controls the use of fire to burn stubble, clear land, etc.
The public must be informed of the degree of danger of forest fires. This depends on the atmospheric conditions at the time, the moisture content of the vegetation, the wind speed and the material in the forest.
Firebreaks are one way of restricting the spread of a fire once it has started. Firebreaks are open strips, cut in the middle of the forest, which prevent the fire from advancing.
Watchtowers help to locate the fires, and there are generally several in an area of forest.

• Wild animals

The forest is a shelter for wild animals. They are in a natural balance with the forest that can be altered by factors like hunting, fire and adverse climatic conditions.
Any environmental imbalance may cause an increase in damage to the trees when the animals search for food, and they may then become pests.
To prevent damage, mesh fencing can be installed to prevent the entry of rabbits, porcupines and wild boars. The fencing should reach down to a depth of at least 25 cm below the soil surface.

If rodent populations are very high, use traps with poisoned bait or remove the ground cover from the soil, as this protects them. The damage produced concentrated on the fruits, seeds, bark, terminal twigs and roots.

14.7.2. Parasitic disorders

In general, insect or fungus attacks do not cause the sudden death of the tree, but they do check growth and cause wood quality to decline.

• Pests

Pests can be classified by their different methods of attack:

• Defoliating insects
- Of the holm oak and cork oak
Lymantria dispar
Tortrix viridiana
Malacosoma neustria

- Of elms and poplars
Euproctis chrysorroea
Leucoma salicis
Galerucella luteola

- Of pines
Lymantria monacha
Thaumetpea pityocampa, the processionary caterpillar.
Dendrolinis pini
Diprion pini
Neodiprion sertifer
Melolontha vulgaris
Anoxia villosa
Amphimallus pini

• Sucking insects
- The most dangerous aphids are *Cedrobium lapo* and *Cinara cedri*.
- The most dangerous scale insect is *Neucaspis pi*.

• Miners and borers
The following attack conifers:
- *Rhiaczonia buoliana* and *Rhiaczonia duplana*. Both attack the shoots of pines.
- *Hylobius abietis*. Causes serious damage to the bark and cambium of trees.
- *Pissodes notatus* or *Dyorictria splendidella*. Both attack the trunks of young pines.

The following affect broadleaf trees:
- *Scolytus scolytus*
- *Scolytus multistriatus*
- *Gypsonoma aceriana*
- *Zeuzera pyrina*
- *Cryptorrhynchos lapathi*, or weevil.
- *Paranthrene tabaniformis. Melanophilla picta* (leaf miners of poplars).
- *Saperda carcharias*.

Pest control

Pest control is carried by an integrated set of 3 techniques:

Pine. Detail of nest of processionary caterpillar

• *Mechanical.* Removing the affected areas of trees or diseased trees.
• *Chemical.* By using insecticides. The problem is that insecticides also eliminate the predators of the pests you wish to control.
• *Biological.* This includes several methods:
- *Breeding and releasing non-damaging predators.*
- *Dispersal of hormones* produced by the insect at critical moments to disrupt its normal development.
- *Releasing sterile males* which compete with the fertile males, thus reducing the pest population.
- *Attracting the insects* using volatile insect hormones or baits. This prevents the males from finding the females, thus reducing the population. These substances are also used in sticky traps to catch the insects.

• **Diseases**

In general, the diseases that affect trees are hard to diagnose. Most are caused by fungi.

The main fungi that attack trees in Spain, and can cause major damage in reforestation, are:

• In nurseries: *Fusarium* and *Alternaria.*

• In poplars and aspens: *Dothichiza popules* (causes black patches on poplar bark), *Cytospora chrysosperma, Venturia pupulina, Melampsora allii-populina* (poplar rust) and *Taphrina aurea.*

• In chestnuts: *Phytophthora cinnamoni* and *Phytophthora cambivora,* the causes of ink disease. *Endothia parasitica,* which causes chestnut blight, and *Mycosphaerella maculiformis.*

• In beech: Beech red heart, which is caused by several species of fungus, *Ungulina marginata, Ganoderma applanatum, Fomes connatus.*
• In elms: *Ceratocystis ulmi* (releases a toxin that poisons the elm's sap), *Dothidella ulmi.*

• In pines: *Armillaria mellea* and *Critocybe tabescens.* These totally destroy the tissues of the neck of the root and the main roots. *Fomes annosus, Cronartium flaccidum, Cenangium ferruginosum, Melampsora pinitarquia, Diplodia pinea, Coleosporium senecionis, Lophodermium pinastri.*

• In holm oaks: *Taphrina kruchii.* This gives rise to "witch's brooms" in the branches of holm oaks.

• In deciduous oaks: *Hypoxylon mediterraneum, Taphrina kruchii, Microsphaera alphitoides.*

• Diseases of other broadleaf trees and conifers:
Botrytis cinerea
Coryneum cardinale
Gnomonia veneta
Rhytisma acerinum
Nectria cinnabarina

Control of diseases

Forest tree species must be planted in zones where the climatic conditions coincide with those of their place of origin.
Moisture and temperature determine the germination of fungus spores. High tree population densities favors high humidity levels. To prevent diseases, it is worth thinning and pruning.
Pruning cuts must be clean, avoiding stumps that can act as entry points for fungi that can attack the tree.
Wounds in the bark caused by tools or fire are also entry routes for pathogenic organisms.
To control epidemics, sick trees should be cut down and burnt; apply suitable fungicides.

BIBLIOGRAPHY

AMORÓS CASTAÑER, M.
Producción de agrios
Madrid: Mundi-Prensa, 1995

BASCOÑANA CASASÚS, M.
Cultivo de la actinida
Barcelona: Aedos, 1989

BAZIN, P.
Repoblación forestal de tierras agrícolas
Madrid: Mundi-Prensa, 1995

BONFIGLIOLI, O.
El injerto en los árboles frutales y la vid
Barcelona: CEAC, 1987

BOVEY, R.
Defensa de las plantas cultivadas
Barcelona: Omega, 1984

BRÉTAUDEAU, J.
Creación de formas frutales
Madrid: Mundi-Prensa, 1982
— *Poda e injerto de frutales*
Madrid: Mundi-Prensa, 1987

CAPDEVILA BATLLES, J.
Frutales y hortalizas. Erradicación de elementos hostiles
Barcelona: Aedos, 1981

COBIANCHI, D.
El ciruelo
Madrid: Mundi-Prensa, 1989

COLETO MARTÍNEZ, J.M.
Crecimiento y desarrollo de las especies frutales
Madrid: Mundi-Prensa, 1989

COUTANCEAU, M.
Fruticultura
Barcelona: Oikos-Tau, 1971

DÍAZ QUERALTO, F.
Práctica de la defensa contra heladas
Lérida: Diladro, 1971

DOMÍNGUEZ, F.
Plagas y enfermedades de las plantas cultivadas
Madrid: Mundi-Prensa, 1993

DURÁN, S.
Replantación de frutales
Barcelona: Aedos, 1976

EDICIONES TÉCNICAS EUROPEAS
El melocotonero. Referencias y técnicas
Lérida: 1989

FERNÁNDEZ ESCOBAR, R.
Planificación y diseño de plantaciones frutales
Madrid: Mundi-Prensa, 1988

FIDEGHELLI, C.
El melocotonero
Madrid: Mundi-Prensa, 1987

FLORES DOMÍNGUEZ, A.
La higuera. Frutal mediterráneo para climas cálidos Madrid: Mundi-Prensa, 1990

GALÁN SAUCO, V.
Los frutales tropicales en los subtrópicos
Madrid: Mundi-Prensa. Vol. I, 1990

GALÁN SAUCO, V.
Los frutales tropicales en los subtrópicos
Madrid: Mundi-Prensa. Vol. II, 1992

GARCÍA FAGREGA
El cerezo
Lérida: Dilagro, 1974

GENERALIDAD DE CATALUÑA
Dept. of Agriculture, Fisheries and Food
Apunts de silvicultura

GIL-ALBERT VELARDE, F.
Tratado de arboricultura frutal
Volumes I, II y III
Ministry of Agriculture
Madrid: Mundi-Prensa, 1989

GRISVARD, P.
La poda de los árboles frutales
Madrid: Mundi-Prensa, 1975

GUT, N.
El albaricoquero
Madrid: Mundi-Prensa, 1963

HAWLEY, R.C. Y SMITH, D.M.
Silvicultura práctica
Barcelona: Omega, 1972

HUBERT, M. Y COURRAUD, R.
Poda y formación de los árboles forestales
Madrid: Mundi-Prensa, 1989

JAVRLARITZA, E.
Manual de manejo forestal
Dept. of Agriculture.
Basque Government.
Vizcaya: 1984

LOOSE, H.
La poda de los árboles frutales
Barcelona: Omega, 1983

LOUSSERT, R.
El olivo
Madrid: Mundi-Prensa, 1980
— *Los agrios*
Madrid: Mundi-Prensa, 1992

MARTÍ CONDEMINAS, P.
El nogal
Barcelona: Sintes, 1986

MARTÍNEZ ZAPORTA, F.
Fruticultura, fundamentos y prácticas
Ed. Ministerio de Agricultura.
Instituto Nacional de Investigaciones Agrarias. Madrid: 1964

MESÓN, M. Y MONTOYA M.
Silvicultura mediterránea
Madrid: Mundi-Prensa, 1993

MINISTRY OF AGRICULTURE
El almendro
Madrid: 1988

MINISTRY OF AGRICULTURE
Frutos secos
Madrid: 1968

MINISTRY OF AGRICULTURE
Plagas e insectos en las masas forestales españolas
Madrid: 1992

MOLINA NOVOA, T.
El avellano. Guía práctica de cultivo
Lérida: Dilagro, 1973

MONTOYA OLIVER, J.M.
La poda de los árboles forestales
Madrid: Mundi-Prensa, 1988
— *Los alcornocales*
Madrid: Mundi-Prensa, 1988

NAVARRO GARNICA, M.
Técnicas de reforestación
Madrid: Icona, 1975

PESSON, P.
Ecología forestal
Madrid: Mundi-Prensa, 1978

PIETER GRIJPMA, IR.
Producción forestal
Manuales para educación agropecuaria
México: Trillas, 1984

PY, CLAUDE
La piña tropical
Madrid: Blume, 1968

RAGAZZINI, D.
El kaki
Madrid: Mundi-Prensa, 1985

REBOUR, H.
Frutales mediterráneos
Madrid: Mundi-Prensa, 1971

SPINA, P.
El Aalgarrobo
Madrid: Mundi-Prensa, 1989

STRASBURGER, E.
Tratado de botánica
Barcelona: Manuel Marín, 1953

TAMARO, D.
Fruticultura
Barcelona: Gustavo Gili, 1979

THOMAS, A.
Hormonas reguladoras del crecimiento vegetal
Barcelona: Omega, 1977

TORRES JUAN, J.
Patología forestal
Madrid: Mundi-Prensa, 1993

TROCME, S. Y GRASS, R.
Suelo y fertilización en fruticultura
Madrid: Mundi-Prensa, 1979

WESTWOOD, M.N.
Fruticultura de zonas templadas
Madrid: Mundi-Prensa, 1982

3

Defense of cultivated plants

Olive trees in Martos
(Jaén - Spain)

1. INTRODUCTION

Like all living organisms, plants are attacked by agents that damage them. These may be physiological disturbances, damage by parasitic microorganisms, viruses etc., or attack by animal parasites. In this chapter, our aim is to describe the agents that cause damage to plants, explain how to assess the damage, and to explain the methods available to prevent or cure them. The branch of agricultural science that studies the damage to plants, their causal agents and the possible remedies, is known as **plant pathology**.

A wide range of chemical products on the market can now treat many pests and diseases that could not be treated 50 years ago. This is the main reason why the world's agricultural production has grown almost exponentially in terms of quantity and quality. Continued study and laboratory research will lead to more and better solutions for plant health in the near future, advances that are essential to meet the growing demand for food from a growing population.

In general, damage by non-living elements are known as accidents. Damage by living agents, such as micro-organisms, viruses, bacteria and fungi are called **diseases**. Multicellular animals, such as insects and mites, are usually called **pests**.

1.1. PHYSIOLOGICAL DISORDERS

Physiological disorders in plants are due to non-organic agents. These disorders are divided into 3 main groups: **physical disturbances** or **climatic disturbances**, produced by non-living agents, damage by human disturbance, or soil disorders. Let's look at some of them.

1.1.1. Physical disturbances or climatic damage

Damage to plants by accidents or the weather cause mechanical wounds in the leaves, trunk, branches or roots. These include damage by weather, such as hail, wind, rain, etc. Climatic agents also include effects that produce no external wounds, but produce damage, such as excess or deficiency of light, and these may lead to physiological disorders, which are often very serious.

1.1.1.1. Wounds

Wounds vary in importance, depending on their size, depth and position on the plant, as wounds have different repercussions depending on the organ affected. They are mainly produced by grazing animals and climatic events, such as hail, lightning, frosts, etc.

Physical damage caused by human action, such as those caused by pruning, creates an entry point for parasitic organisms.

Snow can often cause mechanical damage through its weight, breaking or deforming woody plants.

These wounds cause numerous disturbances to plants, such as:

• Decreased harvest quantity and quality

• Creation of entry points for the organisms that parasitize wounds in plants, such as white rot of grape, mummification of fruit, damage caused by fungi that attack wood, different types of bacteria and fungus, etc.

• Inhibition of the circulation of sap, caused by disorders of the vascular tissues

• Decreased photosynthesis and respiration due to a total or partial loss of the photosynthetic organs. The leaves that are lost affect the growth and production of the plant as a whole. A loss of foliage also leads to decreased evaporation and transpiration.

• Loss of ability to absorb nutrients after damage to the roots.

• Mobilization of the plant's reserves, with the obvious energetic costs, in order to heal the wounds and to replace the organs destroyed.

Each type of plant responds differently to the loss of its parts. Herbaceous plants, when they loose part of their body mass, respond by growing new stems and new foliage. The plant must spend a lot of time on this reconstruction, 6 months to a year, or more (depending on the damage). This period of time may be decisive and cause the loss of a season's harvest. In woody plants, the wounded organs produce a special **corky tissue** over the wound, which gradually covers the wound, forming a resistant callus over the wound. The aerial stems, underground stems, tubers (such as potatoes) and roots all form corky tissue. This corky

tissue limits transpiration and prevents the penetration of infectious microorganisms.

1.1.1.2. Wind

The action of the wind on plants is mainly due to its mechanical effect. Once the wind reaches a certain speed, it can break the branches and stems of woody species. The wind also has a dehydrating effect, as it accelerates transpiration and dehydrates the epidermis and the tissues of the bark. Some plants that grow in a windy habitat are morphologically and physiologically adapted to save as much water as possible; their sunken stomata are thus sheltered from the wind, in order to reduce transpiration as much as possible. This is clearly shown by the members of the cactus family (Cactaceae).

1.1.1.3. Snow, frosts and hail

• Snows

In general, snow is good for plants. A layer of snow protects planted seeds from the external cold, preventing the seed's temperature from falling below 0°C, and bodes well for seed germination in the spring. However, the weight of snow can often cause mechanical damage, crushing herbaceous plants and deforming or breaking woody plants.

• Frosts

A frost is when the temperature falls below 0°C. Each plant, and each organ, requires a given temperature for their growth. Increases or decreases with respect to this optimum point lead to a decrease in production, and if the variation is very large, growth halts, and the plant or part of it may die if the temperatures are extreme. There is thus a temperature that favors germination, a different temperature that favors flowering and another that favors fruit ripening.
In temperate climates, the optimum temperatures for plant growth vary between 10°C and 25°C. Below 0°C, plant growth halts, but if the temperature is far below 0°C, the tissues die and the plant dies. Obviously, each plant species tolerates different

minimum temperatures. Conifers from high mountains, for example, can withstand winter temperatures of –20°C, which would kill a tomato (*Lycopersicon esculentum*) outright.
It is important to realize that plant sap contains dissolved substances that lower its freezing point to below 0°C. The more dissolved material there is, the lower the sap's freezing point. Low temperatures cause part of the water of the plant's tissues to leave the cells and crystallize as ice in the intercellular spaces, meaning that the rest of the sap remaining within the cells becomes more concentrated, and this further lowers its freezing point. Plants from high mountain habitats, such as conifers, box, ivy and holly, resist this concentration of their sap and the freezing and thawing of water in their intercellular spaces without dying.
In temperate latitudes, cold damage can be classified into four groups: **cold spells**, **early frosts** in fall, **winter frosts**, and **late frosts** in spring. Cold spells occur when the temperature suddenly falls below 0°C. The plant's metabolism slows down sharply. The circulation of sap ceases, as does photosynthesis. If this fall in temperature occurs during flowering or fruit ripening, it causes typical disorders, such as shedding of flowers, and deformation, splitting and abnormal fall of the fruit.

Furthermore, when the plant suffers stress of this type, it facilitates the entry of disease-causing viruses and bacteria, and so the viral and bacterial damage can be mistaken for frost damage. Some plants react to cold by producing abnormally large amounts of anthocyanins, meaning that they turn a reddish color.

In the fall, deciduous plants withdraw reserves from their leaves into the main plant. Thus, when the plant sheds its leaves, it does not lose the reserves accumulated in the leaf. Early frosts in fall prevent this process as they disrupt vegetative growth too early. Early frosts cause premature leaf fall and poor ripening of the wood.

Winter frosts, typically a very sharp fall in the temperature, are especially negative when the soil and plants are not covered with snow. Plants from temperate areas that suffer greatly from harsh winter frosts include the walnut (*Juglans regia*), quince (*Cydonia vulgaris*), laurel (*Laurus nobilis*) and fig (*Ficus carica*). Winter colds are generally well tolerated by deciduous trees from temperate climates, as long as the temperatures fall gradually. The negative effects of frosts are greatest when low temperatures last for a long time: deep cracks form in the tree, as well as radial splits that penetrate to a greater or lesser extent into the trunk, and are caused by the differential contraction of the layers of the trunk due to the low temperatures. Damage to trees is much worse when there is a severe frost after a very mild winter. If January and February have been especially warm, the plants start to sprout, and if the cold returns suddenly they are very vulnerable to serious damage in the inner bark and the cambium growth zone.

The trees that are damaged by winter cold generally do not die immediately, because the damage is normally limited to some branches or the trunk. The buds sprout, they sometimes continue growing until flowering, and then when the first hot period arrives, they suddenly dry up. After a severe frost, strong sunshine is especially harmful: sunshine on the frozen tissue can cause damage on the south or southeast of the trunk. The bark, and sometimes the cambium, dies in a relatively uniform broad strip. The bark becomes wrinkled and in older trees it splits, loosens and falls off.

It is very uncommon for the roots to die due to cold weather, but it is more common in sandy soils, where the cold penetrates further. Frost damage usually leads to (bacterial) cankers or parasitic fungi, as these microorganisms can enter through the frost damage.

Late frosts in spring cause the worst damage to cultivated plants in the temperate zones. Vines, fruit trees, strawberries and early potatoes suffer most, and especially the young tissues of the shoots and flowers, which are fully turgid with water. The sensitivity of fruit trees to spring frosts is closely related to the species and variety, as well as the tree's state of growth. Buds that are still closed can withstand low temperatures relatively well, but once they have started to grow they are extremely sensitive, especially the sexual organs. It is common to see ovaries and stigmas that have turned black and died due to the cold, while the corolla remains intact.

The leaves of several plants show very characteristic damage due to cold. The limb is wrinkled, bruised on the upper side, while the lower epidermis is tight and often split. In apples and pears, frost may only affect the surface tissues and cause corky zones to form, often in the form of a corky band running around the fruit.

• Hail

This causes tremendous damage to plants, especially to flowers and fruits, and also to leaves and branches, thus greatly decreasing harvests. This damage, especially the mechanical damage,

Fruit trees of early species and varieties are particularly sensitive to spring frosts, especially the flowers and the young tissues of the shoots.

A layer of snow on pastures and cereal fields protects the seeds from the low external temperatures.

is especially important in vines, fruit trees and vegetables. The most common damage includes damage to fruit and its shedding, loss of foliage, splitting of the stems, shedding of patches of bark, and the breakage of shoots and branches.

1.1.1.4. Lightning

Lightning damage is uncommon and mainly affects large isolated trees. The upper parts of the tree, which are full of sap, conduct the electricity and divide the charge. The splits and clefts caused by the electric discharge are usually only seen below the crown, in the thick branches and the trunk, which are relatively bad conductors. Sometimes vineyards are hit by lightning, and 20 to 100 of the rootstock are affected, but not in any particular order. In fruit or vine plantations cultivated in espalier, the lightning may run

along the iron cables and burn all the plants in an entire row. The symptoms of damage by lightning vary greatly but are unmistakeable.

1.1.1.5. Light and drought

Light and heat, whether too much or too little, can damage plants. Excessive light is usually accompanied by excessive heat. When both light and heat are excessive, this usually causes physiological disorders in plants. Light is necessary for green plants, as it provides the energy they need for photosynthesis. Tissues that form in the dark are yellowish-white. Etiolated plants typically show elongated stems with long internodes, while the leaves are small and may be reduced to scales. This phenomenon is exemplified by potatoes that sprout in a dark chamber.

Many plant disorders are due to inadequate light, for example the spindly growth of seedlings sown too close together, the feeble and weak branches of very leafy trees that have not been pruned, the poor growth of plants under the trees, etc. Flattening of cereals is frequently due to very dense sowing, because the young plants shade each other when their leaves grow. This leads to excessive elongation of the stems (due to lack of light) and weak cell walls, which do not attain the necessary strength. Later, the influence of the wind or rain may cause the heavy ear, in the case of corn, to break at the base. Excess light and heat can also be harmful. In general, plants can withstand high temperatures and strong sunlight relatively well, as long as the increase is gradual. Most of the damage observed is not due to an excessively high temperature, but to an excessively sharp change from cold to hot

In deserts, plants can not grow because of the lack of water, and the large temperature difference between the night and day makes them even more inhospitable for plants.

conditions, from shade or semi-shade into full sunshine. Damage of this type can cause the drying out of delicate plants recently removed from the nursery and placed directly into strong sunshine. One of the clearest examples of this effect is shown by the ornamental plants from the tropics, whose natural habitat is the understorey of the tropical rainforest. They can be cultivated in greenhouses with suitable shading, but if they are directly exposed to the sun on a summer's day, they can be severely damaged.

This damage is called sunscald of the trunk or bark. Thus, when young fruit trees or transplanted vines are taken out of the nursery (where they were shading each other) and transplanted into their definitive site, they may be damaged. This is because they are exposed to high temperatures and strong sunshine when their bark is too thin to protect the cambium from the strong sunshine, because it formed in the shade. Herbaceous plants may suffer even more damage by light. The foliage may be scorched or burnt, grapes may be scalded, and apples, pears, tomatoes, etc., may all be damaged. This damage affects the organs that are normally shaded and which are suddenly exposed to the sun after pruning, trimming or clearing. The leaves may also be scorched after a storm or a strong wind, which may change the orientation of the leaves, meaning that the sunlight reaches parts that were previously shaded.

Damage caused by excess light or heat is made worse if the plants are suffering from water shortage. Strong sunshine in high temperatures causes high levels of transpiration. The water lost by the leaves has to be replaced by water absorbed by the roots from the soil, but in dry conditions, the roots can not get this water, and this causes the plant to dry out, then leaf death and if the drought continues the plant eventually dies.

1.1.1.6. Salty winds

The sensitivity of plants to the salty winds associated with seaside sites depends on the plant's adaptation. Plants whose natural habitat is close to the coast are less sensitive to salty winds and they adapt better to sites near the sea. Examples of this include aloes, palms, cacti, etc., which grow well right on the seafront, as well as other plants, such as pines, carob, oleander, olive and others, which grow perfectly well near the sea, but not right on the coast itself. Plants from continental (inland) locations, such as deciduous forest trees, suffer greatly from airborne salt. The section on Soils in this book contains a list of plants divided into groups (high, medium and low tolerance) on the basis of how well they tolerate salinity.

1.1.1.7. Air pollution

The gases, fumes and dust emitted by factories and the pollution of the large cities frequently damages the surrounding vegetation, for example by depositing a layer of grime on the leaves, and by scorching the leaves and sprouts, and this may lead to a total failure to thrive. Studies of air pollution and its effect on plants are relatively recent. Research by different authors shows which plants are most sensitive to a given pollutant. This plant sensitivity provides us with indicators (the plants) that are a reliable and cheap method of determining whether certain pollutants have reached toxic levels in the air. There are two tables. One (next page) is a list of the plants that are most sensitive to a given pollutant. The other shows the standard composition of dry air and the percentages of the different components. The following paragraphs deal with the most important pollutants, and their direct and indirect effects on plants.

• **Industrial dust**. When the dust is alkaline or neutral it is usually not very toxic for plants. Its negative action is usually mechanical, as the dust is so fine that it can block the stomata and reduce respiration, transpiration and photosynthesis.

• **Fluorine-containing gases**. Fluoride emissions are among the air pollutants that cause the most serious damage in crops and forests. The factories responsible for these emissions are aluminium smelters, ironworks, ceramics factories and phosphate fertilizer producers. Young tissues are most affected by fluoride pollution, which mainly affects the leaves at the tips of the branches. The toxic effect is most clearly shown when a dry period is followed by fine rain, as the gases dissolve in the water, and can enter the plant through the stomata, causing burns within the tissues that spread with time.

Different sulfur-containing pollutants and their dynamics in the atmosphere. Note the formation of ozone due to the reaction of SO_2 with the O_2 of the atmosphere.

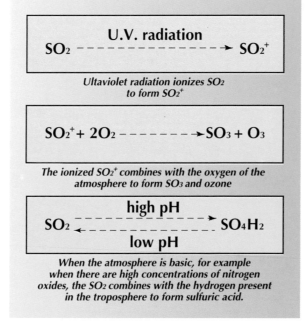

U.V. radiation

$$SO_2 \dashrightarrow SO_2^+$$

Ultaviolet radiation ionizes SO_2 to form SO_2^+

$$SO_2^+ + 2O_2 \dashrightarrow SO_3 + O_3$$

The ionized SO_2^+ combines with the oxygen of the atmosphere to form SO_3 and ozone

high pH
$$SO_2 \leftdashrightarrow SO_4H_2$$
low pH

When the atmosphere is basic, for example when there are high concentrations of nitrogen oxides, the SO_2 combines with the hydrogen present in the troposphere to form sulfuric acid.

Plant species	Pollutant	Reference
Chicory (*Cichorium endivia*) Barley (*Hordeum vulgare*) Alfalfa (*Medicago sativa*) Clover (*Trifolium pratense*)	Sulfur dioxide SO_2	Ormond and Adepipe, 1974 Posthumus, 1976
Tobacco (*Nicotiana tabacum*) Spinach (*Spinacia oleracea*)	Ozone O_3	Heggestad and Darley, 1969 Posthumus, 1976
Stinging nettle (*Urtica urens*) Annual bluegrass (*Poa annua*) Windsor bean (*Vicia faba*)	Peroxyacetylene nitrates (PAN)	Heggestad and Darley, 1969 Posthumus, 1976
Gladiolus (*Gladiolus gandavensis*) Tulip (*Tulipa gesneriana*) Freesia	Hydrofluoric acid (HF)	Reinert, 1975 Posthumus, 1976 Van Ray, 1969
Petunia (*Petunia nyctaginiflora*) Potato (*Solanum tuberosum*)	Ethylene (C_2H_2)	Posthumus, 1976
Spinach (*Spinacia oleracea*) Tobacco (*Nicotiana tabacum*)	Nitrogen dioxide (NO_2)	Posthumus, 1976

List of the plants that are especially sensitive to each pollutant, according to several authors

Some species are more sensitive, such as the pine, the vine, the iris, the gladiolus, the tulip and fodder oats while others are less sensitive, such as the apple, willow, dandelion, white clover and the chrysanthemum. These species, and ones whose leaves are glabrous or waxy withstand relatively high levels of fluoride relatively well.

• **Sulfur-containing gases**. Sulfur dioxide (SO_2) is formed by the combustion of sulfur or sulfur-containing products, such as gasoline, oil, and other petroleum products. Sulfur-containing gases are produced by everything that burns fuel, and by factories producing sulfuric acid, refineries, as well as cars and lorries, etc. Conditions that are favorable for plants, such as high relative humidity, high light intensity and a moderate temperature all favor the gas's entry into the plant through the stomata. It is generally accepted that plants can withstand an average concentration of 0.2 cm^3 of sulfur dioxide per m^3 of air ($\approx$ 0.2 ppm). When the concentration exceeds this level, visible damage occurs. The species most sensitive to sulfur dioxide include alfalfa, barley, oats, wheat, lettuce, endive, spinach and tobacco, while the most resistant species include the apple, the apricot, the vine, the gladiolus, and the lilac. The effect of sulfur dioxide is more intense when exposure is for a long period of time than if exposure is to a high concentration for a short period of time. Its effects on the plant's tissues are usually formation of whitish necrotic patches or the loss of color in the peripheral tissues. There may sometimes be yellowish or brownish coloring. These symptoms often resemble those of nutrient deficiency, though in this case the symptoms are more symmetrical. The sulfur compounds released into the atmosphere, such as sulfur dioxide (SO_2), usually undergo a series of changes until, in some conditions, they are converted into sulfuric acid (H_2SO_4). Sulfuric acid in the air is the cause of "acid rain". Large areas of forest in Europe are already suffering the effects of acid rain. The damage to the plants first takes the form of intervein yellow chlorotic patches that then turn white, dry up and die.

• **Chlorine and hydrochloric acid**. The chlorine in the atmosphere mainly comes from factories producing soda. It may be present as chlorine gas (Cl_2) or combined with hydrogen (H) to form hydrochloric acid (HCl). Hydrochloric acid is three times more toxic to plants than sulfuric acid, though it is unusual for damage to occur, as high atmospheric concentrations of hydrochloric acid are usually only caused by accidents. The elder, cherry, plum, peach, vine, beans, clover and alfalfa are among the plants considered to be most sensitive to chlorine and hydrochloric acid.

Constituent	Chemical symbol	Percentage by volume
Nitrogen	N_2	78.0840000
Oxygen	O_2	20.9460000
Argon	Ar	0.9340000
Carbon dioxide	CO_2	0.0330000
Neon	Ne	0.0018180
Helium	He	0.0005240
Methane	CH_4	0.0002000
Krypton	Kr	0.0001140
Hydrogen	H	0.0000500
Nitrous oxide	N_2O	0.0000500
Xenon	Xe	0.0000087

Percentage composition of the elements of dry air (Taken from Queney, 1974)

• **Tar and tarmac**. Fumes from tar are toxic to plants due to the volatile phenolic compounds they contain. These compounds cause surface lesions in young leaves, which appear to have been varnished, with a relatively clear curling up of the edges. These fumes are emitted from factories that produce electrodes and installations that impregnate wood, and possibly gas factories. The fumes emitted when city streets are tarmacked are similar.

• **Mains gas**. Gas that escapes from breakages or splits in gas mains may damage plants by asphyxiating their roots. This leads to leaf fall, the drying up of the branches and the death of the tree. A bluish tinge in the tree's roots is often a sign of gassing; this bluish color is a symptom of the root's asphyxiation due to lack of oxygen.

1.1.2. Human disturbances

Human disturbances can be divided in two main groups: the mechanical damage done by humans directly, such as pruning, the movement of machinery, etc., and those due to improper usage of chemical products.

When dosage of a pesticide is too high, it may kill the plants and destroy the layer of grass that protected the fruit tree from frosts, as is shown in the photo.

1.1.2.1. Mechanical damage

The damage caused by the action of man has similar consequences to the previously described damage by animals or the weather. Thus, the direct consequence of these human wounds include decreased harvest quality and quantity, creation of entry points for parasites, inhibition of sap circulation due to the damage to the vascular tissues, decreased photosynthesis and respiration due to total or partial loss of the photosynthetic organs, and decreased nutrient uptake.

In general, small wounds, such as pruning cuts, do not need any treatment. Transversal or longitudinal wounds to the trunk or the main branches, as well as those produced when branches break, should be carefully treated. It is necessary to cut back the damaged surface, smooth it, and cover it with a suitable putty or balsam.

In stone fruit, wounds often cause gum production (gummosis) and tylosis, and these block the vessels that the sap flows in. This damage, if widespread, may lead to the tree's death.

1.1.2.2. Incorrect use of pesticides

Most plant damage after pest treatment is due to human carelessness. A mistake when choosing a product, an error in the dosage, incorrect application, inadequate cleaning of the recipients where it is prepared, certain mixtures of insecticides and fungicides that are chemically incompatible, application in strong winds (especially highly volatile herbicides, which may be blown onto adjoining crops), etc., are clear examples of human error causing physiological damage to plants.

Treatment with chemicals represents exposing the plant to a new chemical. In general, most plants can withstand treatment at the doses recommended by the manufacturer. Their guidelines are based on experiments carried out in advance and then checked by the official research stations, but despite this, accidents do happen. This is because there are unknowable factors that can not be taken into account in the tests, such as temperature, wind speed, relative humidity, soil class, the rootstock, the plant's age, etc. All these factors may increase or decrease the plant's sensitivity to pesticide products.

The oldest pesticides have been studied most, and their possible toxic effects on plants, and especially on vines and fruit trees, are well defined. Products that have only come onto the market relatively recently may, however, cause physiological disturbances that are not well defined, and so the farmer is strongly recommended to try out a series of preliminary tests before large-scale use. This is especially true for organic fungicides and synthetic insecticides, whose secondary effects have only recently been observed. In the relevant chapters, the reader will find descriptions of these products and how to apply them. This section deals with the possible damage caused by pesticides. As a general rule, the farmer should read the label carefully to find out which plant species are most sensitive to the product, and the possible toxic effects on each plant.

• **Fungicide toxicity to plants**. There are two main families of fungicide, copper-containing compounds and sulfur-containing ones. In vines and fruit trees, and especially in hybrid varieties, an excessive dose of fungicide causes a halt in growth and burns the vegetative organs. If the tree's leaves have been damaged or the fungicide is applied just before rain, the product enters the plant through the cuticle and stomata, causing relatively severe burns that lead to necrotic patches on the leaves that have clearly defined edges, and to the formation of a rough, corky, skin on the fruit, together with splitting.

• **Insecticide toxicity to plants**. Chlorine-containing insecticides are toxic to plants at doses greater than those stated in the instructions. This toxicity leads to severe chlorosis accompanied by epinasty in the most sensitive plants, such as members of the cucumber family (Cucurbitaceae). One of the chlorine-containing compounds, lindane, acts on the cell nuclei of growing tissues, causing polyploidy (an increase in the number of sets of chromosomes in the cell nucleus). Phosphate-containing insecticides, such as the different forms of parathion and many systemic insecticides, cause deformations in plants similar to those caused by hormone weedkillers such as 2,4-D. These anomalies occur in forced crops under glass (lettuces, tomatoes, ornamentals, etc.) though they may occur in plants grown outdoors, such as tobacco, flax and beetroot. Like lindane, organophosphate insecticides disrupt cell division, especially in growing tissues.

• **Toxicity to plants due to mixing products**. In most cases, where mixing products is appropriate, for example a fungicide and insecticide, the manufacturer's technical specifications will provide guidelines on which types of products can be mixed together, and which can not. The recommendations to carry out these mixtures usually indicate that you must add a wetting agent. A **wetting agent** is an organic solvent with properties similar to a soap, which ensures the product adheres better to the plant or insect. If the wetting power of the solution is too high (because of excess wetting agent), the solution is easily washed away by the rain and the active material is not uniformly distributed on the vegetation. This means that some parts are inadequately protected, while other parts are burnt due to accumulation of the product on a small area. Insecticide-fungicide mixtures cause more accidents when the insecticide is in the form of an emulsion than when it is in the form of a suspension. If an insecticide emulsion is added, the fungicide becomes more toxic and the burns it causes are more dangerous. This does not only apply to organophosphates but also to chlorine-containing insecticides and acaricides.

• **Toxic effects of herbicides on plants**. Herbicides are chemicals intended to kill plants. As we shall see in the chapter on weeds and their elimination, herbicides are very toxic to plants. They can be divided into two main groups, those that act on contact and those that enter the plant. When this type of herbicide enters the plants, it disrupts its metabolism. Some herbicides are totally effective, as they kill all plants, while others are selective herbicides and only kill some weeds, while leaving the crop untouched. The farmer exposes the crops and neighboring crops to the herbicide, so the main precaution to take is not to apply when there is a strong wind. These products are described in detail, together with the precautions necessary to prevent accidents, in the corresponding chapter. Damage caused by incorrect use of a herbicide, depending on the dose applied, may cause leaf and stem deformations,

Apple tree. Variety Golden Delicious. Rose family (Rosaceae)

shoots to wither and flowers to fall and possibly even the fall of the fruit.

1.1.3. Soil disturbances

Soil disturbance causes damage to the plant due to soil defects. The soil's structure, fertility, its excess or lack of moisture, etc., are some of the soil-related causes of damage to plants. There is a description of this problem in the section on Soils in this book, which discusses in detail the different types of nutrient deficiency that a soil may suffer due to unsuitable structure and its consequences, such as dry conditions, excessive moisture, etc.

1.1.3.1. The soil's physical structure

An excessively sandy soil or one that is very rich in clay causes problems for the plant's roots, and thus the entire plant. An excessively sandy soil retains few nutrients and little water, thus frequently giving rise to physiological drought and low nutrient availability. Clay soils are good at retaining water and nutrients, but they often become waterlogged, and this causes the roots to asphyxiate due to lack of oxygen. Adding organic matter is the way to resolve the problems of soils with chronic structural problems. In clay soils organic matter acts like a sponge, providing the soil with enough porous space to contain the oxygen needed for the root's respiration. But in sandy soils the organic matter acts to form aggregates that increase the soil's fertility and water retention, thus ensuring that the plant does not die of water shortage.

1.1.3.2. Excess moisture

Excessively clay soils often suffer from problems related to water due to compaction of the soil around the roots, root asphyxiation due to compaction. Yet a clay soil also retains too much irrigation water or rain throughout its pore space, giving rise to oxygen shortage in the soil, and thus asphyxiation of the roots because they cannot respire. To learn more about the problems of heavy soils and how to correct them, consult the section on Soils in this book. Excess moisture, waterlogging of the soil, causes the foliage to turn shades of yellow, red or purple, the same colors as those caused by a deficiency of nitrogen or phosphorus, and also chlorosis similar to those produced by iron or manganese deficiency. There are also marginal necrotic patches, similar to that caused by potassium deficiency.

In excessively wet soils, fruit trees and vines wilt and droop, showing clear signs of chlorosis. The roots die of asphyxia, and the shortage of oxygen leads to anaerobic fermentation which can give rise to toxic sulfur-containing compounds (such as hydrogen sulfide). In roots with a high glucose content, such as the roots of the apple, anaerobic respiration gives rise to alcohols.

If heavy rain falls after a long dry period, the fruit that is nearly ripe may crack or split because it absorbs a lot of water, swells and bursts the fruit's skin.

1.1.3.3. Drought

Like waterlogging, drought is also a physiological problem related to water, but in this case the problem is a shortage of water. Dry soil does not affect all plants equally, and those with surface roots are the most susceptible. Plants with tap roots, which exploit deeper layers of the soil, resist drought longer. It is well known that some very old trees may resist long periods of drought. The opposite is also true, that plants with a very shallow root system, such as grasses, show little resistance to drought. It seems clear that the soil depth and structure are determining factors in whether plants can resist a given length of drought. In shallow sandy soils, the symptoms will be much worse than in deep loam-clay soils. In very dry years, the fruit of vines and fruit trees are small and frequently fall prematurely. Yellow, red or purple colorations, similar to those produced by deficiencies of nitrogen and phosphorus, are the most common symptoms of damage due to drought, and this may even lead to necrosis of the leaves if the drought continues.

An extreme drought allows certain animal parasites to attack the plant when it is defenseless and infest it. This may be caused by aphids, *Psylla*, scale insects, etc., and some fungi that live on the honeydew secreted by these insects. The best known is the olive scale (*Saissetia oleae* Bern.), a parasite of some plants, mainly the olive, citrus fruit and oleanders, which is associated with a fungus commonly known as sooty mold (*Capnodium spp.*).

1.1.3.4. Nutrient deficiencies

It is well known that plants feed by absorbing dissolved ions from the soil solution, and then

Photographs showing the different stages of the life cycle of the olive scale (Saissetia oleae Bern):
1/ Adult with eggs
2/ Eggs and larvae
3/ Developing adults
4/ Leaves attacked by sooty mold
(Capnodium sp.)
5/ Stem with larvae and adults (note the slight attack of sooty mold).
Photos used with permission of the Department of Agriculture and Stockraising and Fisheries of the Generalitat of Catalonia

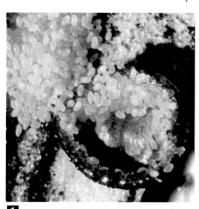

1

2

3

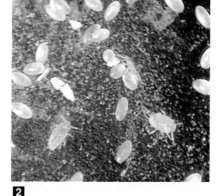

4

5

they transport them to the leaves, where the ions are incorporated into the organic matter photosynthesized from the carbon dioxide of the air, using the energy of sunlight. This organic matter is used by the plant to meet its energy needs and to build its own body parts. The section on Soils in this book lists the nutrients that plants require and the external symptoms of their deficiencies. As a reminder, the main elements involved in plant nutrition are carbon dioxide gas, oxygen gas (both abundant in the atmosphere) and hydrogen (as water), followed by the macroelements nitrogen, phosphorus and potassium (which have to be taken up from the soil solution), and the secondary elements calcium, magnesium and sulfur, and the microelements, needed in small quantities but essential to the plant, such as boron, copper, iron, manganese, molybdenum and zinc. Remember that deficiency of a given element may be due to soil deficiency or to nutrient antagonism. Thus the plants show ferric chlorosis, which in principle is a symptom of iron deficiency in the soil, although there may in fact be enough iron for the plant. This iron is, however, blocked by an excess of calcium or an excessively alkaline soil pH value. To finish, note that the external symptoms of nutrient deficiencies can be distinguished from disturbances due to pests because deficiencies have symmetrical effects on the leaves but parasites do not.

1.1.3.5. Salinity

The opposite, disorders due to excessive nutrient concentrations, may also occur. An excess of one element may not only produce the blockage of another nutrient in the soil, but may also be clearly toxic to the plant in its own right. This mainly occurs with the micronutrients. Boron is an element that causes symptoms of deficiency if it is present in low concentrations but which may be toxic at high levels, and the difference between them is very narrow. In the vine, excess boron clearly reduces the plant's vegetative activity. The terminal leaves are smaller and the edges of the limbs darken and turn downwards, intervein necrotic patches appear that join together and the leaves eventually completely die.
The most frequent case, and the worst for the farmer, is excess salinity due to high levels of sodium or chloride ions. There is a huge amount of material on the study and correction of saline soils. The reader can refer to the part of the section on Soils that deals with saline, sodic and sodic-saline soils. Excess sodium and chloride ions cause burns on the roots and necrosis of the leaf edges. The amount of damage will depend on the quantity of salts present in the soil and the intrinsic sensitivity of the plant in question.

1.1.3.6. Soil fatigue

Soil fatigue is the name given to several different causes which we have already partially described, as well as disorders of organic origin which we will deal with later, and these may occur together or separately. Soil is suffering fatigue when the plants are stunted, the foliage is clearly impoverished and production is lowered in quantity and quality. If it is impossible to obtain the same harvest as in previous years for no apparent reason, this leads the farmer to an understandable state of confusion. The methods available for the recovery of fatigued soils are explained in chapters five and six of this section, which describe the preventive and curative measures to deal with the problems mentioned above. If the same crop is cultivated year after year, it may have exhausted the soil's reserves, or perhaps it shows a special requirement for one element, and has just exhausted this one. The soil may also have a high population of a microorganism that is especially destructive to a given crop. It is also possible that the soil's structure has deteriorated, meaning that it has lost its nutrients, its water retention capacity and its natural drainage. These difficulties are present to an even greater extent in artificial soils. In container cultivation, in greenhouses, soils often lose their structure and their ability to supply nutrients to the plant, mainly due to the fact that the soils used are artificial and rapidly lose their good physical and chemical properties.

1.1.3.7. Acidity and alkalinity

Acidity and alkalinity have been discussed in detail in the section on Soils of this book, which discuss acidic and basic (alkaline) soils, their effects on plants and possible solutions. To recap, some soils are acidic, such as forests in continental climates and high mountain meadows, while others are basic or alkaline, and these are more frequent in coastal areas. The nature of the bedrock from which the soil is derived often determines the acidity or alkalinity of the resulting soil. Yet other factors are also important, such as the quantity of organic matter in the soil, and the influence of rainfall, etc.
A soil's acidity or alkalinity determines the nutrient status of the plants growing on it. In acidic soils, some elements are easier for the plant to use than others. This is true of iron and most other microelements. In alkaline soils, however, some elements, such as calcium and magnesium, occupy the sites on the cation exchange complex and often prevent the plant from taking up iron. To finish, there are some acid-loving plants, namely azaleas and rhododendrons. Most horticultural species, such as tomatoes, onions, leeks, melons, etc., grow better in alkaline soils.

2. VIRAL PARASITES

Some viruses, perhaps the most studied, are parasites of bacteria. Schematic diagram of bacteriophage T2:
1/ Capsid 950Å ∗ 650Å)
2/ Tail (950Å)
3/ Tail nucleus (80Å in diameter)
4/ Base plate (200Å)
5/ Tail fibers (1,500Å)

Viruses are the simplest known form of life. They show some of the properties of living organisms but some of the properties of non-living matter. In general, their structure is extremely simple, an external covering consisting mainly of proteins and lipids, within which there is genetic information in the form of a chain of DNA or RNA, a chain consisting of nitrogen-containing bases, phosphates and sugars (deoxyribose and ribose), arranged in a helix with another chain, a

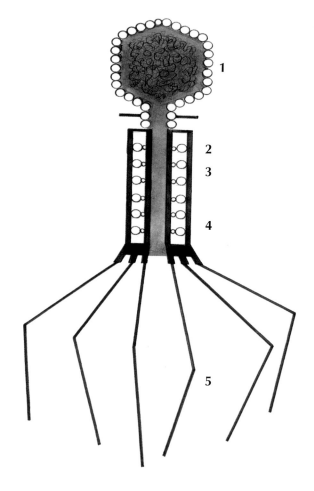

Schematic diagram of DNA (deoxyribose nucleic acid), according to Watson & Crick. Each of the chains of the double helix consists of alternating deoxyribose and phosphate groups. The molecules of deoxyribose are joined to the nitrogen-containing bases. Each base forms hydrogen bonds with a base forming part of the other chain. The four nitrogen-containing bases present are adenine, cytosine, guanine and thymine. Adenine can only pair with thymine, and cytosine can only pair with guanine.

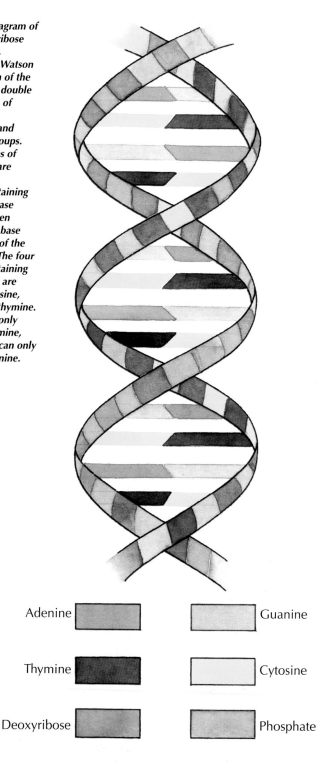

Adenine				Guanine
Thymine				Cytosine
Deoxyribose				Phosphate

double helix. Viruses can only survive as parasites of other living things. Some viruses, bacteriophages, are parasites of bacteria, such as *T2, T4, T6*, etc., which infect *Escherichia coli*, and others cause disease in plants, such as tobacco mosaic virus, or in animals, such as the *Myxomatosis cuniculi* which causes the disease myxomatosis in rabbits and hares. Some viruses infect human beings, such as the flu virus, and the HIV virus responsible for the sadly famous acquired immune deficiency syndrome (AIDS). In general terms, viruses are totally incapable of any metabolism on their own, and show no reaction to external stimuli. They do nothing more than reproduce (though it would be more correct to use to use the word replicate) thanks to the organs and compounds of the host cell (its ribosomes and enzymes). Though they do not have any metabolism of their own, they show the three basic characteristics of genes: identical reproduction within the host cell, transmission of characteristics and the ability to mutate. Another factor that all viruses show is that they are specific. In general, a virus is specific to a species and does not usually infect other species. This might appear to be advantageous, but there is the consequent drawback that mutations occur in the virus's genetic information with some ease. The flu virus is a clear example of this type of viral mutation. The great problem about the flu is that over the course of each year it undergoes so many genetic mutations that last year's vaccines are no longer effective.

To finish, when a virus enters an animal it reacts by producing a protein called **interferon** that specifically inhibits the replication of other viruses that may reach the cell afterwards, though it often does not prevent the virus already inside the cell from replicating. The interferon acts as a sort of natural vaccine to prevent further attacks. So far, because viruses do not have a metabolism of their own, the methods of fighting them are very limited.

2.1. THE NATURE OF VIRUSES

The viruses that infect plants multiply within the host plant cell and undergo mutations in their genetic material and divide into different strains. Some form crystals and others form particles, but they are all visible in the electron microscope (they are invisible in the optical microscope) as large protein-like molecules. Electron microscopes can multiply by a factor of 500,000 to 2,000,000 and make it possible to classify these viral forms into filaments, spherical particles or rods. The vary in size from a diameter of 15 to 30 mμ in the spherical viral particles, which are in fact polyhedra, up to a length of 200 to 700 mμ in a rods and filaments.

Because viruses lack a metabolism they can not "survive" in the open. They can only replicate within their plant hosts and they must a be transmitted by other organisms called **vectors**. Vector organisms are responsible for the spread of viral infections. Insect stings or bites of other animals, and occasionally pollen, are all transmission agents or vectors of viruses. Once the virus is within the plant cell, the virus injects its genetic material into the cell nucleus, which then starts to replicate the virus. As the population of virus increases, they spread from one cell to another through the intercellular channels, the plasmodesmata, and circulate freely through the vessels of the inner bark and sometimes those of the wood, and spread throughout the plant.

Many studies have confirmed that viruses are specific. Thus, in general, each virus only infects a single species of plant. This is because of their method of replicating. When the virus's DNA or RNA penetrates the nucleus it can only combine with the DNA of the plant if their bases pair up.

2.1.1. Mutations

As we already pointed out when discussing the general features of viruses, they show a great tendency to mutate. To put it another way, the chains of DNA or RNA in the viral capsid undergo genetic mutations. Often some plants in a crop infected by a virus show symptoms that are different to the rest. If this difference is transmitted to successive infected plants, this is a new strain of the virus. Genetic engineering has made it possible to select and isolate the new virus and study it. Viruses, like all living things, adapt to the environment in which they live, the host plant, and affect the chances of transmission from one plant to another.

In a crop infected by a virus, we may also find plants that do not appear to be affected. These plants can be considered to be resistant to the virus and they too can be selected for study and propagation. These plants will be resistant to a specific virus and can be sold as resistant forms. This method of combatting viruses, discussed later in more detail, is known as **genetic selection** or genetic **improvement**.

Geneticists can obtain varieties that are free of virus.

Plants can be selected in the laboratory that are free of virus.

2.1.2. Transmission

As the reader will see in the chapter dealing with the prevention of viral diseases, it is essential to understand how viral diseases are transmitted and spread.

Pruning wounds can be entry points for viral diseases.

• **Transmission by mechanical inoculation.** Some viruses are transmitted by contact from

one plant to another. When plants brush against each other they cause superficial grazes that are often enough for the virus to be transmitted from one plant to another. This is true of the tobacco mosaic virus in tobacco and tomato, and of the potato virus.

• **Transmission by insects and other vectors.** Most viruses cannot survive exposure to the air, as they have no metabolism of their own. Thus many viruses would die out if they were not transmitted and propagated by different types of vectors, animals belonging to a variety of zoological groups. The most relevant, in order of increasing importance, are nematodes, acarids (mites), and insects. The insects are the most abundant vectors and the ones that have the most effect. The following orders of insects include vector species: Homoptera (greenfly, blackfly, etc.), Coleoptera (beetles), Orthoptera (locusts and crickets) and Thysanoptera (thrips).
In general, when the vector insect, or other animal, feeds on an infected plant, some of the infected sap is retained in its mouthparts and is later released when the insect feeds on another, healthy, plant. This type of virus is called **non-persistent**, and soon dies outside the plant, and the vector ceases to be infectious. Other viruses, **persistent viruses**, are ingested by the insect along with its food, and then cross the insect's stomach wall and throughout its body, and eventually reach the salivary glands. The insect remains infectious until its death, and will infect all the plants it feeds on.

• **Transmission through the seed and pollen.** It is uncommon for an infected mother plant to transmit a virus to its offspring, but this happens with some plants, such as the mosaic virus in bean and lettuce mosaic in lettuce. Virus transmission by pollen is very rare and only occurs in a few plants and with a few species of virus that are not of agricultural importance.

• **Transmission by vegetative propagation.** Graft scions, cuttings, layers, suckers, stolons, tubers, bulbs, rhizomes and other vegetative reproduction organs can all transmit viruses, and this is of great importance when growing vines, fruit trees, strawberries, raspberries, potatoes and onions. In many cases the farmer is the main agent spreading the virus by taking propagation material from infected plants.

2.1.3. Identification

The farmer can detect and identify viruses simply by observing the symptoms the virus causes in the plant. Unlike other parasites, which can be seen with a hand lens, viruses can only be seen with electron microscopes. Most agricultural laboratories do not have such sophisticated equipment, and use other more practical methods to identify them.

To detect viruses with symptoms that are not clearly visible, specially selected host plants are cultivated in a greenhouse and then inoculated with the virus. These are **test plants** or **indicator plants**. Experimental transmission of the virus to these plants is a **diagnostic** method.

2.2. CLASSIFICATION

Viruses were discovered at the beginning of this century, with the invention of the electron microscope, as these devices made it possible to determine the parasite's shape, polyhedral structure, etc. There is still no agreement on precisely which criteria should be used to classify them. The binomial system of classification is not universally accepted for viruses. In this book, we have used their common names, as these are the most widely accepted.

Many types of virus infection cause yellow patches, recalling a mosaic, to form on the leaves. This characteristic effect is why many plant viruses have names like tobacco mosaic virus, vine mosaic virus, radish yellow mosaic virus, etc.

2.3. LIFE CYCLES

Unlike animals, plants do not produce interferon, that is to say, they do not produce special antibodies against viruses. This is why an infected plant is rarely cured and remains a carrier of the virus until it dies. All the plants propagated vegetatively from it will also be infected. Thus the tubers, rhizomes, bulbs, stolons, layers, cuttings, and graft scions will all be infected with the virus. Some climatic conditions are decisive for the action of virus. Low temperatures, for example, depress the effect of the virus. The potato virus only acts at about 16°C and above, and its activity halts above 20°C, and the symptoms produced disappear. The virus is then latent or hidden. Variations in temperature are closely linked to variations in the intensity of sunshine. This explains why the action of a virus is much stronger at some times of year than at others. At times of average temperatures, such as spring and fall, viral activity is more intense, but it is hardly noticeable in summer and winter. The best known case is the virus that causes mosaic of vine leaves; the mosaic appears intensely in spring, and disappears in August, but returns in the fall.

Other conditions, such as the general health status of the crop (preventing the presence of vectors), optimal nutrition for the plant, a soil suitable for the crop, the absence of competing weeds, etc., help the plant to withstand the action of the virus.

Photo of a plant produced by in vitro cultivation techniques. Genetic selection of virus resistant plants is the most effective preventive measure against viruses. (Photo courtesy of RHÔNE-POULENC)

2.4. SYMPTOMS

Symptoms produced by viruses vary greatly, depending on the plant, the virus and the environmental conditions. The main effects are anomalies in the formation and growth of the plant's organs, and in their functioning. Other symptoms include general or localized malformations, or necrosis or chlorosis, which may be localized or generalized throughout the plant. The leaves of the plant are especially badly affected, and turn reddish and show epinasty, in addition to general disorders of the plant's growth, such as dwarfism. Yellow or white patches often appear distributed like the tiles of a mosaic on the plant's leaves.

The symptoms caused by viruses in plants can be classified by their effect on the plant. This effect depends on the plant's degree of resistance to the virus. Within a single species, we may find plants that resist a particular virus, while others are not resistant. There may even be cases of total immunity, **immune plants**. There are also plants that the virus hardly affects (**tolerant plants**). Finally, when the plant (or its parts) dies before the virus has spread throughout the plant, they are said to be **hypersensitive**. When the virus spreads throughout the plant, this is **systemic infection**, but when the infection is localized, the damage is in the form of **local lesions**.

3. PARASITES OF PLANTS

Cultivated plants can be damaged by the effects of weeds, and the action of parasitic bacteria and fungi. Weeds simply compete with the crop for the elements they need and they are studied separately (see section 7). This chapter deals with the **fungi** and **bacteria**. Fungi and bacteria live totally or partially at the expense of plants. These parasitic species are unable to photosynthesize, meaning that they must grow and reproduce at the cost of plants. Parasitism is essentially a question of nutrition. Plants that can synthesize organic compounds from carbon dioxide (for example in photosynthesis) are called **autotrophs**, while organisms that can not, and thus have to obtain their organically combined carbon from plants or even animals, are **heterotrophs**. There are some flowering plants that are parasitic, such as dodder (*Cuscuta europaea*), some orchids (*Corallorhiza trifida*) and mistletoe (*Viscum album*). Mistletoe has green leaves and is a hemiparasite, that is to say it has specialized in living at the expense of other plants. It can photosynthesize, but takes water and nutrients from the host by means of **haustoria** which grow into the vascular system of the host plant. These plants, together with the carnivorous plants, are extremely unusual, and lie outside the scope of this book. The following sections deal with fungi and bacteria, the parasites that can cause the most damage to cultivated plants.

3.1. BACTERIA

Bacteria are very small **prokaryotes** and some lack chlorophyll and cannot perform photosynthesis. They do not have a cell nucleus, and they are normally 1 to 3.4 μ long and 0.5 to 1 μ wide. They may be spherical (**cocci**), rods (**bacilli**) or twisted spirals. Some bacteria have a flagellum (plural: flagella) or cilia with which they can move through liquid media. In culture, they usually form colonies, which can be identified visually, forming an opaque mass, sometimes yellowish and sometimes pinkish, and generally rather viscous. Bacteria that lack chlorophyll and can not photosynthesize and live at the expense of another organism are **obligate parasites**, and those that grow on dead organic matter are called **saprophytes**. There is, however, a third type of relationship that is of great importance in agriculture, a mutually beneficial relationship between a plant and a bacteria, called **symbiosis**. This is clearly exemplified by the *rhizobium* bacteria that live on members of the bean family (*Leguminosae*). The bacteria obtain organic matter from the plant, while the plant obtains the nitrogen fixed by the bacteria, and many legumes will not grow successfully in the absence of their strain of bacteria. This type of symbiosis is very special, as most bacterial attacks are very negative for plants. Exclusively saprophytic bacteria decompose the dead organic matter and return its mineral elements to the soil, recycling them.

Bacteria are also attacked by viruses, bacteriophages, which cause lysis (splitting) of the bacterial cell, and this releases the contents, thus spreading the bacteriophage.

The shapes of several different bacteria. For comparison, the bottom right drawing shows the ascospore of apple scab (Venturia inaequalis Wint.) at the same scale.

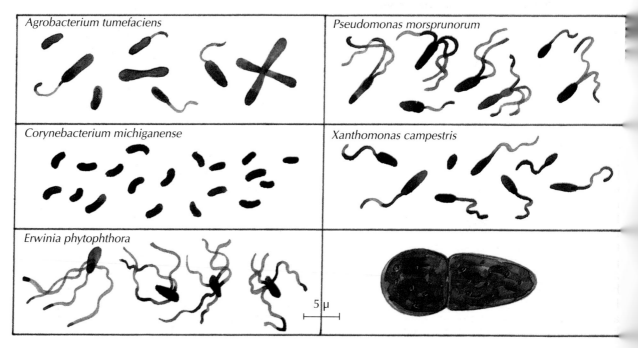

Agrobacterium tumefaciens

Pseudomonas morsprunorum

Corynebacterium michiganense

Xanthomonas campestris

Erwinia phytophthora

5 μ

As viruses can not usually parasitize more than one species of bacteria, they may be useful in identifying bacterial species and strains.

3.1.1. Classification

Bacteria are classified on the basis of their morphology and their nutritional needs for nitrogen and carbohydrates. From the point of view of plant pathology, we are only interested in the rod-shaped bacteria, the bacilli, as they cause most of the diseases of agricultural plants. The main bacilli are:

• **Agrobacterium**. These are small mobile rods, frequently with a single flagellum. *Agrobacterium tumefaciens* of vine and fruit trees is the most important.

• **Corynebacterium**. These are small curved immobile rods, such as the bacteria responsible for tomato wilt (*Corynebacterium michiganense*).

• **Erwinia**. These are small mobile rods with many flagella. The best known example is *Erwinia phytophthora*, the cause of potato "blight".

• **Pseudomonas**. Small straight rods, mobile or immobile, that may or may not have a flagellum. They frequently produce a fluorescent pigment. These include the bacteria responsible for tobacco angular leaf spot (*Pseudomonas tabaci*).

• **Xanthomonas**. Small mobile rods with flagellum, such as *Xanthomonas campestris*, the cause of black rot of cabbage.

3.1.2. Life cycles

The optimal temperature for bacterial growth is quite high, between 25 and 37°C. They are usually present in large numbers in all sorts of decomposing organic matter. Some bacteria can form resistance **spores** to survive adverse climatic conditions. They spend the winter on infected seeds, tubers and litter from infected plants.
Bacteria show two types of reproduction, the simplest being replication, when the bacterial cell divides in two. This gives rise to two identical daughter cells. In adverse conditions bacteria can undergo a form of sexual reproduction, with the exchange of genetic material. **Transformation** is the name given to the transfer of genes to a receptor bacteria in DNA from a donor bacteria. This change in the genetic material helps the bacterial population to adapt to its environment, in accordance with Darwin's theory of evolution by natural selection.
In comparison with fungi, only a few diseases are caused directly by bacteria, mainly because they cannot perforate the epidermis

covering the plant, and have to enter the plant through a wound or through the stomata, and this greatly reduces their possible activity.

3.1.3. Symptoms

Bacterial infections are spread to other plants in drops full of bacteria that ooze out of the stomata or through cracks in the affected tissues. These exudates are spread by climatic agents, such as the rain and wind, or by direct contact, and also through the activities of insect vectors, such as insects and slugs and others. Human activity can transfer bacteria from one plant to another in operations like pinching out, removal of buds, transporting infected plants or seeds, and pruning.

Photo of a peach leaf affected by peach leaf curl (Taphrina deformans). The rough texture and reddish color are very typical. (Photo courtesy of SHELL)

The symptoms produced in plants by bacterial diseases can be divided into three groups:

• **Galls** or **tumors**. The bacteria causes the plant's metabolism to go out of control, and the cells grow in a disorderly way to form growths or cankers. This type of growth is commonly compared to a human cancer, and the most common one is the canker in the roots and at the base of the stem of the vine and fruit trees caused by the bacteria *Agrobacterium tumefaciens*.

• **Infections of the vascular system**. When bacteria invade the plant's vessels, they slow down the circulation of the sap, causing the plant to droop. This is what happens in bacterial diseases of the tomato and cabbage and ring bacteriosis in the potato.

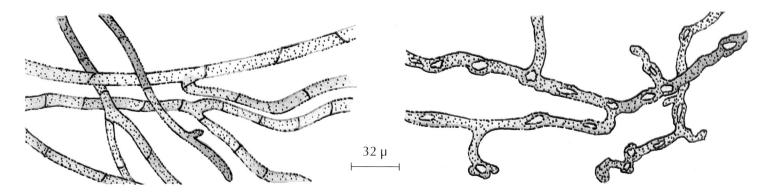

32 μ

The two types of fungal mycelia. On the left septate mycelia (each cell has a nucleus). The mycelium on the right has several nuclei but does not have separate cells.

• **Oily patches**. These are patches with an oily appearance that appear when bacteria grow within the cells of the parenchyma or between them. These symptoms are characteristic of the bacterial diseases of tobacco, beans, plums, potatoes and carrots, all diseases in which the bacteria release enzymes that cause a wet rot of the invaded tissue and release an unpleasant smell.

3.2. FUNGI

The fungi form a group of heterotrophic eukaryotic organisms, and may be saprophytic or parasites of living things, and are clearly very different from plants. The most important fungi in terms of plant pathology are the parasites. Fungi lack chlorophyll and can not photosynthesize, meaning that they are heterotrophic and must depend on carbon fixed by other organisms. Many fungi can alternate between parasitism and feeding saprophytically, such as apple scab and the fungus causing wheat stem rot. They start by living parasitically on the living plant. When the plant dies they continue living on its dead remains, living as a saprophyte. They are **facultative parasites** (meaning they can be parasites).
In addition to the parasitic and saprophytic fungi, there are other forms that have a very important way of feeding themselves, living in **symbiosis**. Each of the two partners obtains some benefit from the association. This is shown by the mycorrhizal fungi that live in association with the roots of some orchids. The mycorrhiza obtain organic compounds (sugars, etc.) from the orchid, while spreading the effective range of the orchid's roots, allowing them to take up more water. The most important example of symbiosis in fungi is the formation of *mycorrhiza* with the roots of trees, especially trees in forests. The mycorrhiza allow the tree to obtain nutrients more effectively, and some trees, such as the pines, are incapable of normal growth in sterilized soil, because they cannot absorb nutrients without their mycorrhizal partner. Most fungi are multicellular, but the most primitive forms are unicellular. Their vegetative apparatus consists of *hyphae*, which are microscopic filaments. The hyphae as a whole are known as the *mycelium*. The mycelia may contain divisions called *septa* (singular: *septum*)

or be continuous; the higher fungi have septate hyphae while the lower fungi do not have septa. In septate mycelia, each compartment is a cell, but primitive fungi, lacking septa, can be thought of as a single cell with many nuclei, all in a chain. A small piece of mycelium is enough to start a new fungus.
Fungal hyphae occur in different forms, depending on the type of fungus and the environmental conditions. They may look like roots, **rhizomorphic**, which cause white rot of roots. In unfavorable environmental conditions, they may group together into compact groups and turn a darkish color. These agglomerations are known as **sclerotia** and allow the fungus to survive during long periods of drought or low temperatures.

3.2.1. Parasitism

There are several factors influencing whether a fungus is able to grow at the plant's expense, such as the plant's sensitivity, the virulence of the fungus and other external factors.

• **Host sensitivity**. The health of the plant, its genetic makeup, and its stage of growth all influence whether the fungus will successfully attack the plant. Often different varieties of a single plant species show different sensitivity; for example certain varieties of vine from America are less sensitive to mildew than their European counterparts. In general terms, the plant's sensitivity may be anywhere between total susceptibility and total resistance.
• **The virulence of the fungus**. The fungus must be potentially pathogenic, i.e., it must be virulent, if it is to successfully attack the living tissue and cause disease despite the defensive measures of the host.
• **External factors**. Climatic factors such as the temperature and the humidity are factors limiting the growth of fungi. Most fungi grow abundant hyphae and mycelia if temperatures and humidity are high. These conditions are most frequent in spring in temperate countries, though they also occur in autumn. This is when the mildews in vines are most virulent. In especially dry years, when there is little or no dew, there may be little or no fungal growth and thus damage. Some species do not fit into this scheme, such as *Fusarium nivale*, which is most active during the winter.

3.2.2. General life cycles

Fungi reproduce by **spores**, which are released from the parent hypha and dispersed by the wind and rain. Spores are tiny organs consisting of one or more cells, and can be produced sexually or asexually. Asexual spore formation is a form of vegetative reproduction, and the spores formed are called conidia or conidiospores. Fungi producing only asexual spores are said to be imperfect, while the perfect state of the fungus is when it produces sexual spores. Vegetative reproduction may also occur by cell division to give rise to a new identical cell. In sexual reproduction, the resulting cell or

can withstand harsh environmental conditions, and which will fruit when favorable conditions arrive. In spring, the spores are released and cause the first infections. The disease then spreads vegetatively, growing rapidly and producing abundant spores. Some spores may go through their entire life cycle on a single species of host plant, while others have two different host species, one for the sexual form and one for the asexual form. In order to germinate, the spore requires precise temperature and humidity conditions. When these conditions occur, the spore germinates and emits a germination tube, which enters the host tissues through the stomata, wounds, lenticels or even by perforating the

Different types of fruiting body in the sexual and asexual phase of basidiomycetes and ascomycetes

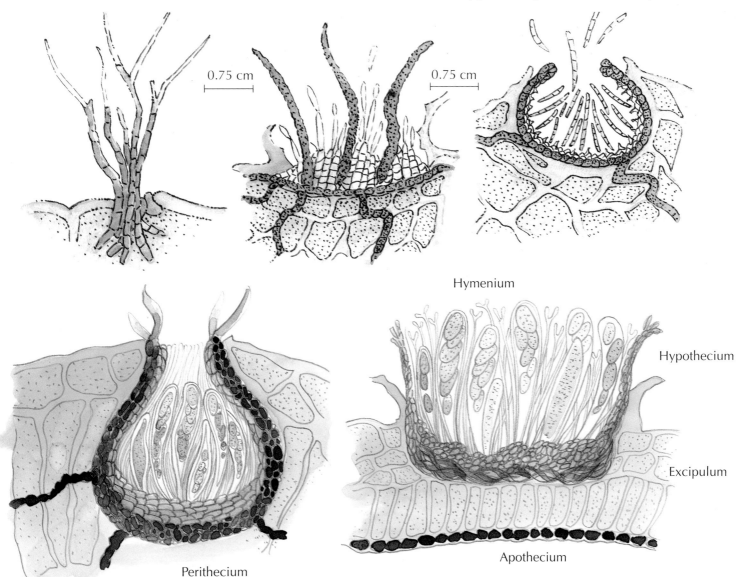

0.75 cm

0.75 cm

Hymenium

Hypothecium

Excipulum

Perithecium

Apothecium

Fleshy fruiting bodies of the sexual phase of a higher fungus. The spores ripen within these fruiting bodies produced by ascomycetes and basidiomycetes.

organism is the product of the genetic fusion of two different nuclei from the fungal mycelium.
The two types of reproduction are, however, associated (in temperate climates) with the climate. Asexual reproduction normally occurs in summer and its role is basically to spread the organism. Sexual reproduction normally occurs at the end of fall and perpetuates the species in especially adverse circumstances. The **spore** or **zygote** is a hard capsule, a resistance stage that

epidermis. From this moment onwards, if the environment is favorable, the mycelium grows and invades the cells or intercellular spaces and feeds on the host. In the early incubation phase there are no external symptoms, but as the infection develops symptoms appear if the weather conditions are suitable.
Sexual reproduction in fungi is a remarkably complicated process. In the lower fungi (with polynucleate hyphae), fertilization

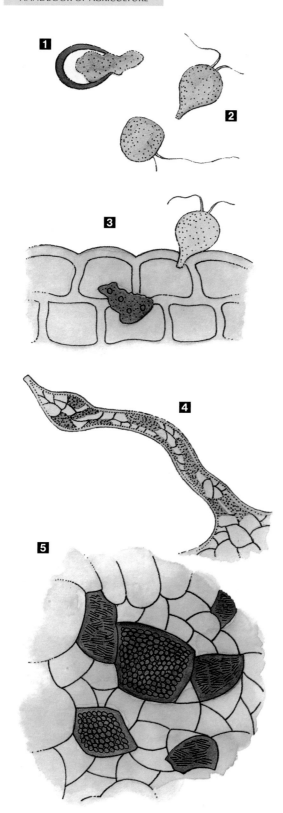

Details of the different developmental stages of the most primitive fungi, the archaemycetes. The drawings show clubroot which affects the cabbage and its relatives (Plasmodiophora brassicae).
1/ Spore
2/ Zoospores
3/ Entry of a zoospore
4/ Root hair with young plasmodia
5/ Cabbage root with cells full of spores (Plasmopara viticola)
According to Bovey, R.

The best known edible fungi, such as the milk caps (the pictures shows Lactarius sangifluus) are basidiomycetes. (Courtesy of the Agriculture, Stockraising and Fisheries Department of the Generalitat of Catalonia).

occurs within the mycelium and the spores produced are free but within a thick membrane, such as the winter oospores produced by mildews. In the most complex fungi, sexual reproduction is accompanied by the formation of relatively fleshy complex organs, called perithecia, apothecia and carpophores, within which the sexually produced spores ripen. Ascomycetes produce ascospores, while basidiomycetes produce basidiospores.

The perithecium of the apple scab and the apothecium of vine leaf spot are examples of the complex sexual organs formed by higher fungi. Most of the mushrooms of culinary importance are basidiomycetes, and the mushroom (or toadstool) is the structure dispersing the ripe sexually produced spores. These include *Lactarius sangifluus* (a milkcap), *Boletus edulis* (the edible boletus) and *Cantharellus cibarius* (the chantarelle).

Asexual, or vegetative, reproduction gives rise to spores in the summer. These are borne in conidia which are generally at the tips of fertile hyphae called conidiophores. In some groups, the conidia form within an open fruiting body (the acervulus) or a closed flask-shaped fruiting body (the pycnidium), which produces pycnidiospores. To resist cold or desiccation, some fungi produce spores with a thick covering, called chlamydospores. In favorable climatic conditions, fungi can produce huge numbers of summer spores. Tobacco mildew can produce up to a million conidiospores per cm^2, while each of the 200,000 spores of black-stem rust of wheat can form a pustule within 10 days that will in turn release about 200,000 spores.

3.2.3. Classification

Fungi are classified on the basis of their method of reproduction into the following groups.

• **Archaemycetes**. These are lower fungi which develop in a very special way. These fungi typically produce zoospores, which are mobile. These flagella-bearing spores are formed by a splitting of the mass of the membraneless protoplasm

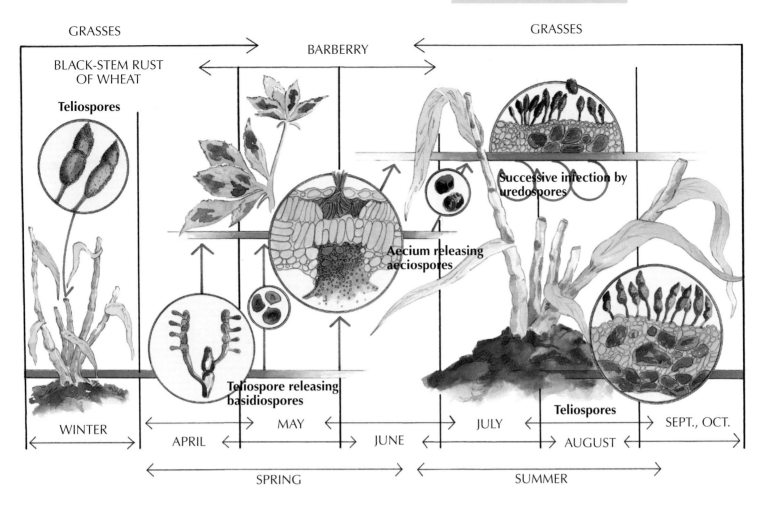

GRASSES

BARBERRY

GRASSES

BLACK-STEM RUST OF WHEAT

Teliospores

Successive infection by uredospores

Aecium releasing aeciospores

Teliospore releasing basidiospores

Teliospores

WINTER | APRIL | MAY | JUNE | JULY | AUGUST | SEPT., OCT.

SPRING | SUMMER

and then manage to enter the host's cells. They multiply rapidly causing increased growth of the infected parts of the plant. Reproduction occurs within the host by means of the formation of a spore mass surrounded by a resistant covering (a cyst). The spores are eventually released when the plant's tissues decompose (club root, and potato warty scab).

• **Phycomycetes**. These fungi have mycelia that do not have divisions (septa). The winter spore is known as the oospore and grows within the plant when good weather arrives. Asexual or vegetative reproduction occurs later, and the mycelium reproduces by producing conidia and conidiospores. This type of reproduction is found in mildews and white rusts.

• **Ascomycetes**. The higher fungi all have septate mycelia. The sexual filaments fuse and form empty receptacles called perithecia, such as apple scap and powdery mildew or in apothecia, such as vine leaf spot. The spores ripen within the asci (singular: ascus) containing the zygotes or ascospores. Asexual reproduction occurs in spring when the summer spores germinate; they may mature on the surface, such as cercospores, within special receptacles such as pycnidia (*Septoria*) or in sclerotia (*Botrytis*). The ascomycetes include the ergot disease of rye, mildew, anthracnosie and powdery mildews.

Life cycle of a basidiomycete. Basidiomycetes are considered to be the most highly evolved fungi, and the diagram shows the complex life cycle of black-stem rust of wheat (Puccinia graminis). According to Bovey, R.

Powdery mildew on a stone fruit

Quality and productivity are the key factors in modern agriculture.

• **Basidiomycetes**. They are also called higher fungi, and like the ascomycetes they have septate mycelia. In their sexual phase, the basidiospores ripen within the basidia, which are comparable to the asci of ascomycetes. In addition to mushrooms like the milk caps and boletus, this group also contains some rusts, loose smuts, and stinking smuts.

Some rusts belonging to the basidiomycete group of fungi require two different hosts to complete their life cycle. These rusts are said to be **heteroecious** and the best known forms are black-stem rust of wheat, pear rust, blackcurrant rust and plum rust. Other rusts live on a single host and are said to be **autoecious**. These include the rusts of sugarbeet, beans and raspberries. The heteroecious rusts may have as many as five different types of fruiting body. The best known of all these basidiomycetes is black-stem rust of wheat (*Puccinia graminis*), which we shall describe as a representative example. During the summer, the rust grows vegetatively on the wheat and produces a series of spores called **uredospores**. These form dark powdery eruptions on the wheat, which shed a dark powder. They germinate and grow into a mycelium that enters the plant through the stomata, and then form new patches of rust and new spores. When the wheat starts to die in late August, black fruiting bodies appear in the pustules, **teliospores**, which are bicellular spores. They are highly resistant and can survive the winter cold. Teliospores are in fact the zygote formed by sexual reproduction. In spring, each cell of the teliospore emits a **promycelium**,

Example of the life cycle of a phycomycete fungus, the downy mildew of vines (Plasmopara viticola). According to Bovey, R.

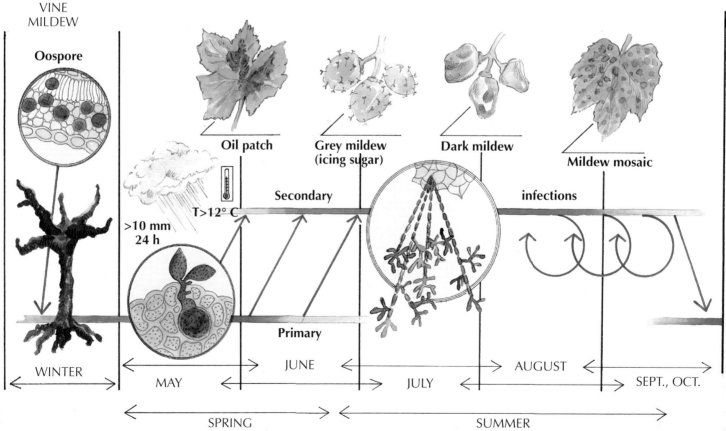

VINE MILDEW

Oospore

Oil patch

Grey mildew (icing sugar)

Dark mildew

Mildew mosaic

Secondary

infections

T>12° C

>10 mm
24 h

Primary

WINTER

MAY

JUNE

JULY

AUGUST

SEPT., OCT.

SPRING

SUMMER

or **basidium** which expands along one edge and contains 4 basidiospores. Once the basidiospores are ripe, they are shed, but they can not penetrate into the plant. The fruiting body can only be produced on a second host, the **barberry** (*Berberis vulgaris*), where two further types of fruiting body are produced. On the upper surface of the leaf, they produce **spermogonia** with **spermatia** which play a major role in the sexual reproduction of the fungus, and on the underside of the leaf they produce **aecia** with **aeciospores**. The germination tube of the aeciospore is unable to penetrate into the barberry, but it can attack wheat and produce new uredospores. This is the life cycle of wheat black-stem rust.

Other basidiomycetes are said to be imperfect, as their entire life cycle is not yet known, because their sexual phase (the "perfect" phase) has not yet been observed.

3.2.4. Symptoms

In fungal diseases of cultivated plants the vegetative parts invaded by the fungal mycelium rapidly turn yellow. These are areas of diseased cells, and they turn darkish when they die. In especially harmful fungi, such as *Fusarium spp.* and *Verticillium spp.*, obstruction of the vascular system prevents the flow of sap and this causes the plant to wither and die soon afterwards. Some species of fungus secrete cytolytic substances (which cause the host cells to break open) and the tissue dies. This is exemplified by the wood fungi, or lignicolous fungi, which rot wood, and those that cause dry or wet rots of fruit. Some fungi cause diseases with symptoms similar to those of some

bacterial diseases. The infected plant then reacts locally with excessive tissue growth, leading to the formation of cankers or galls.

With the exception of vascular fungi, attacks by most fungi are clearly localized on the plant, producing clearly visible fruiting bodies, such as perithecia, spores, apothecia, conidia, etc. The plant then reacts by laying down a cork or lignin barrier to stop the parasite spreading. Sometimes cultivated plants react by producing gum (gummosis) or resin. This is true of forest and fruit trees. Some plants produce toxic substances to slow down the advance of the fungus, though the action of these natural fungicides is highly localized and short-lived. The plant's best defense is a thick cuticle or a waxy covering. These are **external resistance factors**. As with bacteria, plants may show

The product SAPROL, sold by SHELL, contains 19% triphorine. This preventive fungicide is very effective against Moniliosis in apples. (Photo courtesy of SHELL)

greater or lesser resistance to these parasites, depending on their sensitivity. Paradoxically, it is often the plants that are most susceptible that suffer the least damage. This is because the infected cells die so quickly that the fungus can not reproduce and infect the rest of the plant.

4. ANIMAL PARASITES

A/ Taxonomic classification of the animal kingdom. Animals are classified hierarchically into phylum, class, order, suborder, family, genus, species and variety. The name of each organism is denoted by the species, the variety and in many cases by the abbreviation of the name of the researcher who gave it that name.

B/ Diagram of the evolutionary classification of living things, from the most primitive (virus and bacteria) to the most advanced, such as the vertebrates.

We have already looked at viruses, parasitic bacteria, and this chapter deals with the animals that attack crops. Charles Darwin's theory of evolution, published in 1859 in his book "On the Origin of Species by means of Natural Selection or the Preservation of Favoured Races in the Struggle for Life", explained that since the beginning of life, the different species present have been subject to natural selection, and this has led to their adaptation to their environment, due to the inheritance of genetic variations, while those that have not adapted have not survived. Asexually-reproducing unicellular microscopic organisms adapted to their environment due to mutations in their genes which created variation that differentiated them from each other. Over time, organisms evolved that reproduced sexually, which allowed them to show greater diversity, due to genetic recombination. This was followed by the evolution of more complex multicellular organisms, the fungi, plants and animals. Immediately below, there are two reference diagrams. The first is an outline of the classification of all living things, from the simplest living things, the bacteria, to the most complex, such as the vertebrates, including human beings. The second diagram is an outline of the classification of the animal kingdom. On the basis of their characteristics animals are classified into a **phylum**, **class**, **order**, **sub-order**, **family**, **genus** and **species**. The scientific name of an animal consists of two words, the first being the name of the genus it belongs to and the second being the name of the species it belongs to. Thus the scientific name of the insect commonly known as the San José scale is *Quadraspidiotus perniciosus Comst.*, and *Quadraspidiotus* is the name of the genus, *perniciosus* is the name of the species, and *Comst.* is the abbreviated name of the researcher who first gave it this name. It is common to see *Quadraspidiotus sp.* or *Quadraspidiotus spp.* as a way of saying all the species of the genus *Quadraspidiotus*.

Of the different phyla (singular: phylum) shown in the diagram, the most important are the mollusks, nematodes, arthropods and vertebrates. The following section is an analysis of these phyla, in order of their increasing evolutionary complexity

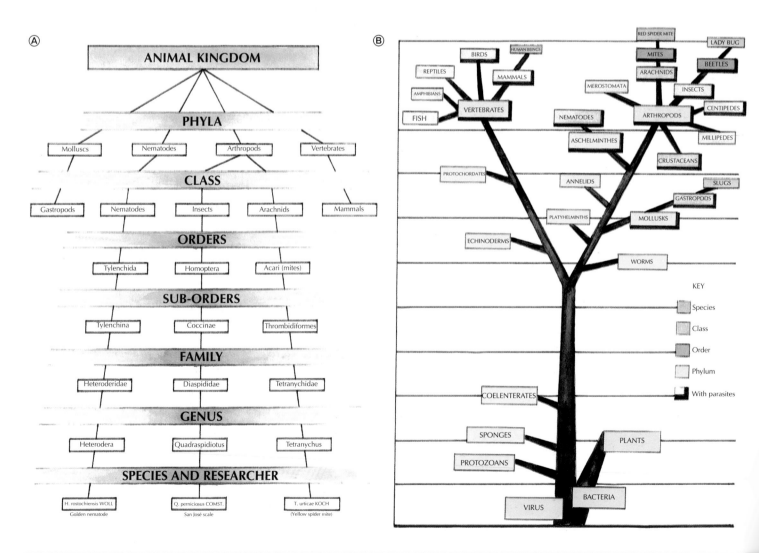

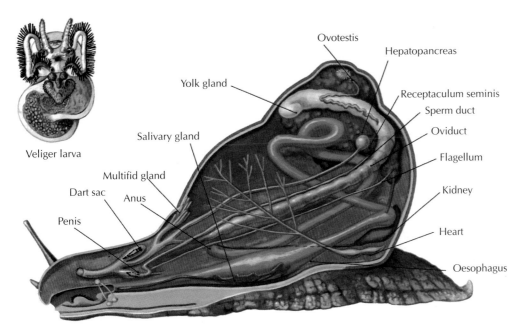

Anatomy of land snail

Veliger larva

Yolk gland

Salivary gland

Multifid gland

Dart sac

Anus

Penis

Ovotestis

Hepatopancreas

Receptaculum seminis

Sperm duct

Oviduct

Flagellum

Kidney

Heart

Oesophagus

but you should realize that the most important parasitic animals on plants are either insects or arachnids.

4.1. MOLLUSKS
(Phylum Mollusca)

The mollusks include the class Gastropoda (gastropods, such as snails, slugs, etc.), which contains many species that can be very damaging if they become abundant in crops. These include the **snails** (*Helix* and *Cepaea*) and slugs (*Limax, Agriolimax, Milax, Daroceras, Limnea,* and *Arion*). The two groups differ in whether they have an external shell or not. The body of a gastropod consists of three parts: the **foot**, a muscular mass used for locomotion; the head, which bears 4 retractile eye stalks, and a mouth with a maxilla and an organ for scraping at the plants, the **radula**, and the **visceral mass**, which is covered by the mantle, a sort of grooved muscular tunic and, in the snails, protected partially by an external shell. Slugs have a rudimentary shell below their skin.

Dry conditions led to a noticeable reduction of the damage caused by slugs and snails. Their movement seems to be limited by the presence or absence of a thin film of water on the soil surface. A slight rainfall may allow large numbers of snails to move around, though they are more active after rainfall than during it. High temperatures do not appear to inhibit their movements, only influencing their movements if the environment is very dry due to high evaporation.

4.1.1. Classification

The class Gastropoda includes the following families; *Testacellidae, Limacidae, Milacidae, Arionidae, Lymnaeidae, Ferussacidae* and *Helicidae.* The families *Limacidae, Milacidae,*

Arionidae and *Lymnaeidae* include all the types of slug that occur in temperate climates. The snails are represented by the family *Helicidae.* The small elongated snails that also damage cultivated plants are members of the family *Ferussaciidae.*

4.1.2. Life cycles

Mollusks are **hermaphroditic**, meaning that each individual is both male and female. The eggs are laid in the soil and hatch within two to four weeks. The snails grow rapidly, depending on the species, taking between 6 weeks and a year to reach adult size.

Snails are more active in some seasons of the year (almost all species showing peaks in spring and especially fall), and also varies over the course of the day. Most snails feed from about two hours after sunset until about two hours before the sunrise. At temperatures below 4.5°C, most snails and slugs cease activity and do not start again until temperatures increase. They then feed and reproduce until the next winter arrives.

4.1.3. Symptoms

Snails and slugs attack all sorts of plants, but cause the most severe economic damage in leaf vegetables such as lettuce, cabbage, broccoli, etc. In other plants, such as fruit trees, attacks by gastropods may cause a lower harvest, by reducing the leaf area (and thus photosynthesis), and by directly attacking the leaves and some fruit, causing the fruit's quality to decline. The product can then only be used by the food processing industry.

The symptoms of snail attack are unmistakeable. Large patches of the leaves are gnawed away by the radula of the gastropods, partially or totally destroying the leaves, and thus reducing the leaf area, and reducing the quantity and quality of the agricultural product.

Their calcareous shell allows land snails to survive dry conditions.

The different species in the family *Ferussaciidae* characteristically attack the fleshy roots of cultivated plants like asparagus, and their bites cause damage that allows entry of saprophytic fungi, bacteria and nematodes, causing the plant's growth to slow down or halt, and may even cause death of the aboveground parts. It is often possible to detect a trail of shiny dry slime on the plants and in the surrounding area, which was left by the animal as it moved. In general, they cause little more than mechanical damage, though they may also be vectors or viral and bacterial diseases. Populations of slugs and snails are restricted to acceptable levels when climatic conditions are normal, but if the spring is very rainy, the populations may grow almost exponentially and cause very serious damage to young crops.

4.2. NEMATODES
(*Nematoda*)

The nematodes includes many species that are damaging to cultivated plants. Nematodes are spindle-shaped (i.e., long, thin and cylindrical), showing bilateral symmetry and they are generally very small in size. They are normally spindle-shaped and have a circular cross-section. They usually have a **mouth** surrounded by **lips** bearing the sensorial organs below which is the buccal cavity, an oesophagus, the intestine, and the rectum which terminates in a ventral anus. The body is covered by a cuticle, and almost never has any sort of external appendage. In general the sexes are separate, and show definite sexual dimorphism (the females are usually larger than the males).
There are more than 40,000 species of

nematode, though only about 15% have been studied and thoroughly described. Most of them live in the sea, though some are parasites of invertebrates (e.g., *Mermithidae*) or of vertebrates (e.g., *Trichinella, Filariodea, Oxyurida, Strongylida,* etc.). Some nematodes are saprophytic and feed on decomposing organic matter. Those that live in the soil are called free-living or plant-eating, nematodes. These are the best known and most studied, as they cause damage to crop plants. Unlike other nematodes, they have a **stylet** activated by muscles, which the nematodes use to penetrate the walls of the root cells. They are generally between 1 and 3 mm long, and usually feed on plant roots, and their abundance in the soil depends on how moist the soil is. In normal conditions, nematode density is about 50 to 400 million per hectare, though in very moist conditions, they may reach 6 million per m^2.

• **Nutrition**. Nematodes parasitic on plants feed on the contents of the cells, whose membranes they perforate with their stylet. They also inject their saliva which liquefies the contents of the cell to be digested. In this way, they gradually destroy the tissues of the plant, though the worst damage caused to the plant by nematodes are usually deformations caused by their toxic saliva, which causes necrosis and deformation of the root tissues. Nematodes are also the most active vectors for the transmissions of viral and bacterial diseases.

• **Parasites of nematodes**. A few nematodes are predators of other nematodes. The most common enemies of nematodes are fungi, whose mycelia close like a trap around nematodes and imprison them, and some bacteria that live on them.

• **Mobility**. Nematodes have a limited ability to move within the soil. This is why plants infected by nematodes are often restricted to clearly defined patches. Nematodes are sometimes spread by the mechanical action of other factors, such as the wind, the hooves of animals, transport machinery, plowing tools, and the transplantation of roots, etc, such as bulbs, rhizomes, stolons, etc.

• **Specificity**. Some species of parasites are able to grow on a large number of plant species, often belonging to families that are taxonomically distant. In other circumstances, certain nematodes that are parasites of plants in a single family show a genetic inability to attack other species of the same family. This is shown by the nematode *Ditylenchus dispsaci*, which attacks wheat, rye and oats but normally has no effect on barley. It is also generally accepted that the highly polyphagous nematodes, such as *Ditylenchus dipsaci* and *Heterodera schachtii*, consist of three different categories:

the **polyphagous** strains, which can attack a large number of plant species, **oligophagous** strains, which can only attack a small number of plants, and **monophagous** strains, which only attack one or two species of plant. In any case, it is not easy to identify and study each species, as each plant species may be host to more than one species of nematode.

4.2.1. Classification

From the point of view of plant health, the phytophagous nematodes can be divided into groups on the basis of their type of parasitism. Nematodes may attack the plant internally (**endoparasitic nematodes**) or externally (**ectoparasitic nematodes**).

• **Endoparasitic nematodes**. This term includes all the nematodes that spend most of their life within the host plant. The nematodes that attack stems, bulbs and roots are represented by two species, *Dytilenchus dipsaci* and *D. destructor*. *D. dipsaci* is highly polyphagous and can attack about 150 species of plant (both wild and cultivated). This species is also divided into a number of biological strains characterized by their affinity for certain plants.
The nematodes that attack leaves belong to the genus *Aphelenchoides*. Morphologically, they resemble *Dytilenchus* but differ in their biology and the type of damage they cause. The nematodes that attack seeds belong to the genus *Anguina*. The nematodes that form cysts on roots belong to the genus *Heterodera*, a genus characterized by the shape of the adult female, whose inflated body becomes spherical or ovoid. The root-gall forming nematodes are species of the genus *Meloidogyne*. The root-eating nematodes are mainly represented by the genus *Pratylenchus*. They spend most of their life within roots, only leaving them when they start to rot.

• **Ectoparasitic nematodes**. These phytophagous species live in the soil, outside the plant tissues. They also move within the area near the roots, the **rhizosphere** and are **free-living nematodes**. This group contains many species and most of them are animal vectors that spread bacterial and viral diseases of plants.

4.2.2. Life cycles

Nematodes reproduce sexually, and generally have to be fertilized before they can produce eggs. Some species, however, show **parthenogenesis**, or are **hermaphroditic**, with both sexes present in the same individual. The eggs are oval or rounded and generally possess a resistant membrane. The egg's embryonic development starts immediately after it is laid, either in the organs of the infected plant or within the soil. The larvae develop within the egg and are the same shape as the adult, but less than a tenth of a millimeter long. They grow by a series of **molts**, when the animal sheds its cuticle. Up to five different stages have been observed between the four moults, though in some classes of nematodes growth is more complicated because they show metamorphosis.

When climatic conditions are dry, soil humidity decreases greatly and the host plants dry out. When this happens, the larvae (whatever stage they are in) can form cysts, and are protected by their cuticle. This type of protection against adverse conditions, called **anabiosis**, often allows them to survive in the soil for many years. When moisture

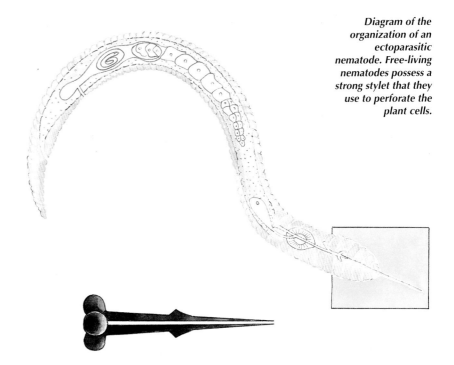

Diagram of the organization of an ectoparasitic nematode. Free-living nematodes possess a strong stylet that they use to perforate the plant cells.

conditions are once more ideal (or the same plant is cultivated again in the same site), the encysted nematodes once more become metabolically active. This water requirement of the nematodes explains why they generally occur at some depth within the soil and generally do not occur in the top 5 centimeters or so.

4.2.3. Symptoms

The symptoms caused by nematode attacks vary greatly, depending on the plant species attacked and the type of parasite. They are often called eelworms or roundworms. In general, the external symptoms of attack take the form of necrosis, deformations and rots of the roots, stem and leaves. The phenomenon of soil "fatigue" described in the first chapter of this book may be largely due to nematode infestation.

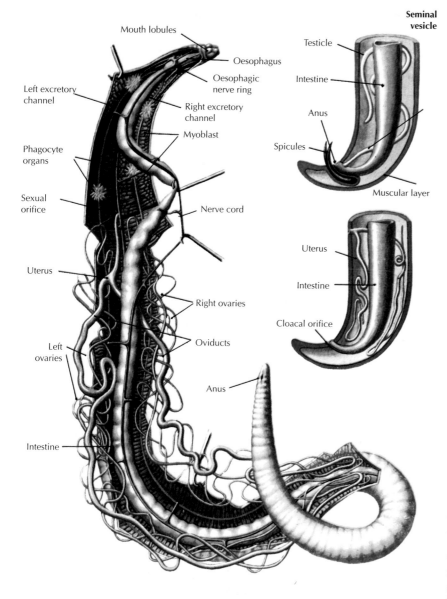

Mouth lobules

Oesophagus

Oesophagic nerve ring

Left excretory channel

Right excretory channel

Myoblast

Phagocyte organs

Sexual orifice

Nerve cord

Uterus

Right ovaries

Left ovaries

Oviducts

Intestine

Anus

Seminal vesicle

Testicle

Intestine

Anus

Spicules

Muscular layer

Uterus

Intestine

Cloacal orifice

Anatomy of the nematode Ascaris lumbricoides

Soil "fatigue" is often caused by nematodes, which parasitize generation after generation of the same plant, and thus increase considerably, rendering the site incapable of continued production. The following sections deal with the most widely known diseases caused by nematodes that attack plants.

• *Pratylenchus pratensis* attacks many species, including members of the following families: the poppy family (Papaveraceae), cress family (Cruciferae), rose family (Rosaceae), legumes (Leguminosae), carrot family (Umbelliferae), potato family (Solanaceae) and many members of the grass family (Gramineae). The roots of young plants attacked by these nematodes rapidly turn dark, the invaded areas turn red and die, the leaves turn yellow, then whitish and soft. Most of the plants die or halt growth.

• *Pratylenchus penetrans.* This nematode causes lily-of-the-valley worm lesion. In addition to attacking the lily-of-the-valley

(*Convallaria*) this nematode attacks many bulbs, such as hyacinths, gladioli, tulips, etc. Growth is halted and the plants dry out when the nematodes are very numerous. The damage they cause allows entry of a fungus (*Cylindrocarpon radicola*) which causes reddening of the roots.

• *Dytilenchus dipsaci.* This species can feed on many different crop plants, and is a severe problem for farmers. It causes leaf deformations in cereals and broadening of the stem, meaning that the plant remains short; the ears develop badly and sometimes abort. Oats are particularly badly attacked, and the base swells. Young plants of rye that are parasitized show unusual growth; the leaf parenchyma grows excessively and the leaf's sides grow up and inwards (like a boat seen head on); the stems remain short and the plant grows many stems. Corn is heavily attacked at the base, which swells and turns soft, causing the plant to fall over. Young plants of buckwheat attacked by this species turn a characteristic purplish color due to anthocyanin formation.

This nematode also attacks onions, garlic and leeks. Onions can be parasitized as they germinate, and they swell at the base and eventually burst. Many ornamental plants are also affected, such as hydrangeas, hyacinths, gladioli, etc. In hyacinths, the parasites cause the appearance in the bulb of concentric blackish circles alternating with healthy zones. In tobacco, the first stages of the disease take the form of small circular bumps a few millimetres in diameter that are lighter in color than normal; these later rise towards the top of the plant. In plants less than two months old, the leaves dry up and the plant dies, but when an older plant is attacked, it falls over.

• *Meloidogyne incognita.* This species attacks the potato, tomato, endive, carrot, lettuce, etc., but it especially attacks vines (*Vitis vinifera*). Many varieties of vine are very sensitive to this nematode. The plants that are attacked produce few branches which are thin and short, the main roots are destroyed, surface roots are then produced that are in turn attacked, and which show small knots. The vine eventually dies.

4.3. ARTHROPODS
(*Phylum Arthropoda*)

The arthropods are all bilaterally symmetrical invertebrates with a segmented body. The different classes of arthropods, despite showing large differences, all show a series of features in common. Their body consists of interdependent segments and has jointed limbs (the meaning of the word "arthropoda"). In general, their body consists of a distinct head region, a thorax and an abdomen.

These three body regions vary greatly in the different classes of arthropods, and have different names. The insects have a clear **head**, **thorax** and **abdomen**. The arachnids (spiders, etc.) have a fused head and thorax, the **cephalothorax** (or **prosoma**) and the **opisthosoma**. In many arthropods, there is a fourth type of body region that lack appendages and is sited behind the anus. It is laminar or needle-shaped and is called a **telson**.

As a rule, each segment has a single pair of appendages. Early in their evolution, these segments were presumably similar, but natural selection has led to their specialization by means of morphological differentiation due the different functions they play. The **cephalic appendages** are sensorial or mouthparts or for gripping. The **thoracic appendages** are locomotory and are adapted to walking, jumping or swimming, depending on the species. The **abdominal appendages**, when present, are usually highly transformed, for example, the organs used in copulation. One of the characteristic features of arthropods is that they are covered by a chitinous external skeleton or **exoskeleton**. The cuticle consists of two parts, a thin **epicuticle** on the outside and the thicker **endocuticle** on the inside. This exoskeleton often contains substances other than chitin, such as calcium carbonate, to make it harder. The endocuticle is the site of attachment of the internal muscles that allow arthropods to move. The arthropods' exoskeleton is rigid and does not allow the animal to grow, and so they have to grow by molting their skin and growing a new one. Each time an arthropod molts its exoskeleton, it produces a new, larger one so it can continue growing. Depending on the class of arthropod, the molts may be simple, when each stage is like the one before, but bigger, and the adults have wigs and reproduce. Other arthropods have a complex series of molts, best shown by the butterflies (Lepidoptera) which undergo **metamorphosis** from caterpillars to butterflies. The development of an arthropod (and of most invertebrates and some vertebrates) implies growth through a series of stages from the egg to the adult. In insects showing metamorphosis, the animal's life cycle starts with the **egg**, which hatches into the **larva**, then the **pupa** and finally the **adult**, also known as the **imago**. Many species have larval stages that are much longer than the adult stage, because the imago is only the sexually mature form, while the larval phase is associated with active growth.

The respiratory system of the arthropods varies greatly from class to class. Aquatic arthropods (such as crustaceans) have **gills**, while terrestrial arthropods may have **cutaneous respiration** and tracheal or pseudotracheal respiration. The nervous system consists of a **brain**, a **sympathetic** nervous system and a nerve tube located ventrally with respect to the digestive tube. Some senses are well developed in arthropods, for example touch, located in the antennae and the sensory hairs found all over the body, especially at the tips of the appendages. Their visual organs vary greatly from class to class. Some have simple eyes and others have compound eyes. They have a hormone system that regulates important functions like growth, molting, metamorphosis and reproduction.

Most arthropods are either male or female (the sexes are distinct and separate), and many show very clear **sexual dimorphism** (the females are different from the males). **Parthenogenesis** is common among arthropods, and **hermaphrodites** are very rare. Most are **oviparous**, meaning they lay eggs that hatch to produce **larvae**. If the larvae is similar to the adult, it passes through a series of molts. If the larvae is totally unlike the adult, it will have to undergo **metamorphosis**.

The most important arthropods in terms of plant pathology are the arachnids, centipedes and millipedes and the insects, though the insects are by far the most important.

Morphology and anatomy of the honey-bee (Apis mellifica), an animal representing the insects.

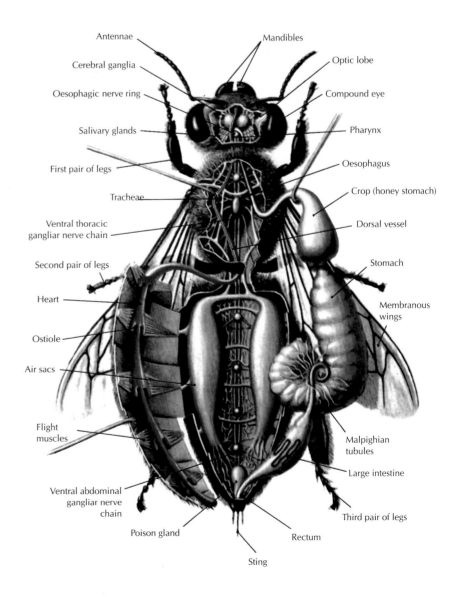

Antennae

Cerebral ganglia

Oesophagic nerve ring

Salivary glands

First pair of legs

Tracheae

Ventral thoracic gangliar nerve chain

Second pair of legs

Heart

Ostiole

Air sacs

Flight muscles

Ventral abdominal gangliar nerve chain

Poison gland

Sting

Mandibles

Optic lobe

Compound eye

Pharynx

Oesophagus

Crop (honey stomach)

Dorsal vessel

Stomach

Membranous wings

Malpighian tubules

Large intestine

Third pair of legs

Rectum

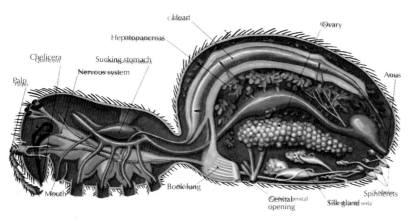

Chelicera
cuticle
Palp
Palpo
Hepatopancreas
Heart
Ovary
Chelicera
Sucking stomach
Nervous system
Anus
Mouth
Book-lung
Genital opening
genital
Silk gland
seda
Spinnerets

Vertical longitudinal section of a spider

Tetranychus urticae
Class Arachnida

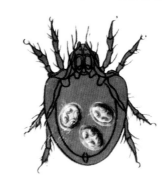

Orobatid mite with cistercus of an anoplocephalid

4.3.1. Arachnids (Class *Arachnida*)

This class includes **spiders**, **scorpions** and the tiny spider-like animals called mites (or acarid mites). The arachnids are terrestrial arthropods with bodies divided into two distinct parts, the **prosoma** and the **opisthosoma**. The prosoma is basically a fused head and thorax, and the opisthosoma is the posterior part. In the adults, the prosoma has simple eyes and six pair of appendages: a pair of **chelicerae** located in front of the mouth opening and which terminate in pincers (chelae) or hooks, a pair of **pedipalps**, usually on either side of the mouth, and four pairs of walking legs. The opisthosoma or thorax houses the genital orifice (near the front) and the anus (at the rear). The body usually shows some segmentation, but this is difficult to detect in mites.

Arachnids are mainly terrestrial, though some mites are semi-aquatic. Respiration is normally by tracheae; the **spiracles** are located in the abdomen, but they are sometimes present in the thorax or near the mouth. In the vast majority of mites, **respiration** is through the **skin**. The arachnids feed in the ways we have already discussed; most feed on live prey (carnivores), others feed on dead bodies (saprophytes) and many of them are parasites of plants. Except for some mites that reproduce parthenogenetically, most arachnids have clearly different male and female sexes. They are **oviparous**, meaning that after copulation the female lays eggs. The class *Arachnida* contains nine orders, the **scorpions**, **pseudoscorpions**, **solifuges**, **palpigrades**, **pedipalps**, **spiders**, **ricinuleids** and **mites**. By far the most important from the point of view of plant physiology are the mites.

The mites are very small arachnids, rarely exceeding 1 mm in width. In general, the nymph and adult have four pairs of legs, but the larva only has three pairs. The body is normally very short, as broad as it is long, and a single piece with an ovoid or wormlike shape. There is often a transversal groove that passes behind the second pair of legs. The entire front part, including the mouth parts and the first two pairs of legs, form the **proterosome**. The posterior part, with two pairs of legs and the abdomen, forms the **histerosome**. Depending on the species, the proterosome may have up to 5 ocular organs, together with one or three pairs of sensory hairs. Mites show such variation that it is difficult to define general characteristics that encompass them all. Mites vary greatly in size and shape, color, and have very different reproductive cycles.

4.3.1.1. Classification

The diagram on the following page shows a schematic classification of the mites (order Acari). This order is divided into the sub-orders *Onychopalpida, Mesostigmata, Ixodidae, Trombididae* and *Sarcoptidae*. Each order is divided into families, each family into genera, and each genus into species. The most important phytophagous mites are members of

External anatomy of the body of a mite. Ventral view:
Prot: *proterosoma*
Gn: *gnathosoma*
Pr: *propodosoma*
Hist: *histerosoma*
M: *metapodosoma*
O: *opisthosoma*
Ch: *chelicerae*

Hist

Prot

O

M

Pr

Gn

ch

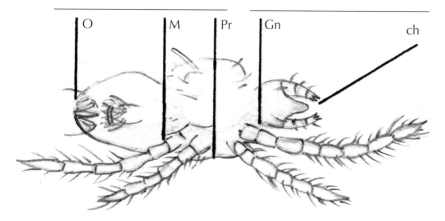

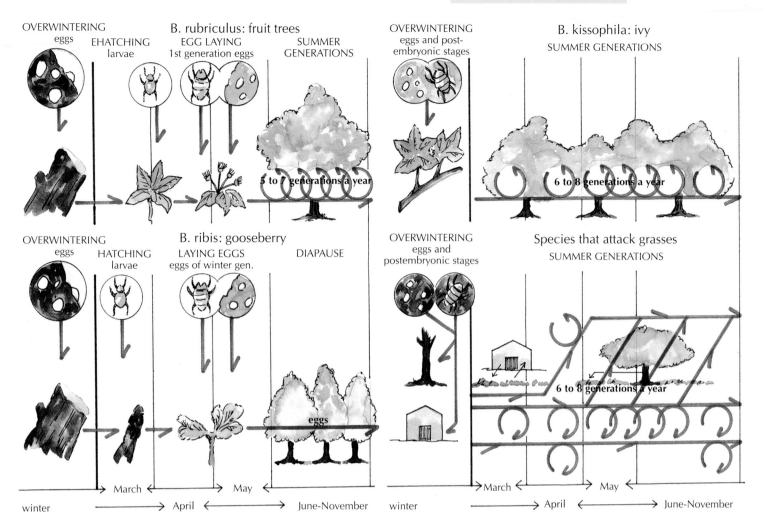

OVERWINTERING eggs — EHATCHING larvae — B. rubriculus: fruit trees EGG LAYING 1st generation eggs — SUMMER GENERATIONS — 5 to 7 generations a year

OVERWINTERING eggs and post-embryonic stages — B. kissophila: ivy SUMMER GENERATIONS — 6 to 8 generations a year

OVERWINTERING eggs — HATCHING larvae — B. ribis: gooseberry LAYING EGGS eggs of winter gen. — DIAPAUSE — eggs

OVERWINTERING eggs and postembryonic stages — Species that attack grasses SUMMER GENERATIONS — 6 to 8 generations a year

winter → March ← April ← May ← June-November   winter → March ← April ← May ← June-November

Different life cycles of the mites of the species of Bryobia *(According to Mathys)*

the sub-orders *Trombidiformes* and *Sarcoptiformes*, which include the following species.

• *Trombidiformes.* The ticks or thrombidiform mites possess a single pair of spiracles, which are sometimes absent, either on or near the proterosome. The palps, generally free and well-developed, are modified into pincers or sensory organs. The non-phytophagous species that are predators have chelicerae that are modified to grasp or immobilize their prey. The thrombidiform mites are divided into several families that include the phytophagous species. The most important species, and the best studied, are: *Eriophyes vitis*, the cause of vine leaf blister, *Eriophyes piri*, the pear leaf blister mite, linden and lilac leaf blister (*Eriophyes tiliae* and *E. lowi*), vine acarosis (*Phyllocoptes vitis*), dry acarosis of tomatoes (*Vasates lycopersici*) and

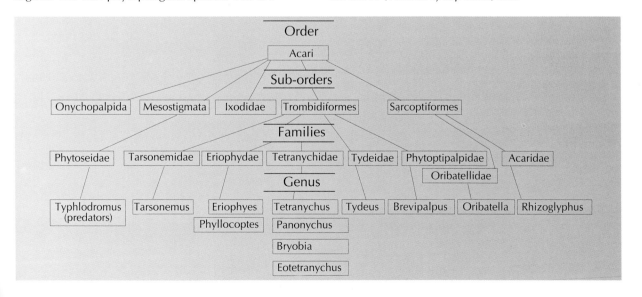

Order — Acari

Sub-orders — Onychopalpida | Mesostigmata | Ixodidae | Trombidiformes | Sarcoptiformes

Families — Phytoseidae | Tarsonemidae | Eriophydae | Tetranychidae | Tydeidae | Phytoptipalpidae | Acaridae | Oribatellidae

Genus — Typhlodromus (predators) | Tarsonemus | Eriophyes | Phyllocoptes | Tetranychus | Panonychus | Bryobia | Eotetranychus | Tydeus | Brevipalpus | Oribatella | Rhizoglyphus

Taxonomy of the mites

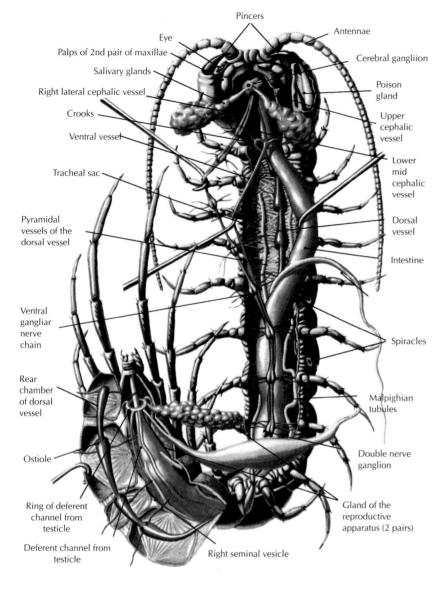

Pincers
Eye
Palps of 2nd pair of maxillae
Salivary glands
Right lateral cephalic vessel
Crooks
Ventral vessel
Tracheal sac
Pyramidal vessels of the dorsal vessel
Ventral gangliar nerve chain
Rear chamber of dorsal vessel
Ostiole
Ring of deferent channel from testicle
Deferent channel from testicle
Right seminal vesicle

Antennae
Cerebral gangliion
Poison gland
Upper cephalic vessel
Lower mid cephalic vessel
Dorsal vessel
Intestine
Spiracles
Malpighian tubules
Double nerve ganglion
Gland of the reproductive apparatus (2 pairs)

Morphology and anatomy of Lithobius forfictus, a typical centipede

4.3.1.2. Life cycles

Mites can reproduce sexually or parthenogenetically. Most are **oviparous**, though some species are **ovoviviparous**, and some are even **viviparous**. The eggs may be spherical or ovoid and hatch to produce larvae with three pairs of legs, or just two pairs in some phytophagous groups. They then develop through a series of molts and mobile stages called **nymphs**. There are four mobile stages after hatching, the **larva**, two **nymph** stages and the **adult** (or imago). Their development may be complicated by a third nymph stage, which may be absent. In phytophagous mites, the number of generations in a year is variable, but there are usually many generations, and together with their high potential rate of population growth, this means that these harmful species can multiply very quickly. Different stages overwinter in different species.

The great variations in reproduction in mites means that we cannot offer a simple life cycle. To illustrate mite life cycles, we shall discuss a mite that has been widely studied, that mainly attacks fruit trees, the red spider mite. This mite mainly affects apple trees, but affects a wide range of other fruit-producing plants, such as peaches, gooseberries, vines, strawberries, etc. It overwinters in the form of **winter eggs**, which are sometimes extraordinarily abundant, and are laid under the bark of the trunk and branches, especially in sheltered sites. They start to hatch in mid April and all hatch within about 20 days. The larva are about 0.2 mm long, eat the underside of the leaves and take 18 to 20 days to grow. The successive generations take a shorter or longer time to grow depending on the temperature. A generation may grow to maturity in just 7 or 8 days in the summer, but may take 20 to 25 days in fall. The first **adults** appear in early May. The females are fertilized on hatching and start laying eggs 3 days later. There is thus a short period when there are no eggs on the plant, when the winter eggs have hatched but the first summer eggs have not yet been laid.

4.3.1.3. Symptoms

The larvae, nymphs and adults all feed by sucking the sap from cells perforated by their **stylet**, mouthparts adapted for piercing and sucking. The effect of all these bites disturbs the plant's metabolism and the palisade cells are destroyed. This causes a halt in growth, deformations, blackening or darkening of the leaves, chlorosis and different types of growths (galls) that may cause the leaves to fall, and depress the plant's growth. Like nematodes and mollusks, mites are vectors of viruses and bacteria because they pierce the cell, and so in addition to the damage they cause in their own right, they may also spread viral and bacterial diseases.

strawberries (*Tarsonemus pallidus*). The previous page has diagrams of the life cycles of several mites of the sub-family *Bryobiinae*, which mainly attack seed fruit (*Bryobia rubrioculus*), gooseberries (*Bryobia ribis*), ivy (*Bryobia kissophila*) and grasses (*Bryobia praetiosa, B. cristata* and *B. graminum*). Another sub-family of importance in plant pathology includes the genera *Panonychus* and *Tetranychus*; the red spider mite of fruit trees (*Panonychus ulmi*) and the polyphagous yellow spider mite (*Tetranychus urticae*) are both highly damaging to cultivated plants. A detailed description of the mites is beyond the scope of this book, and the reader will have to consult reference works to find more on the taxonomy of mites.

• *Sarcoptiformes*. They do not possess spiracles, or they have a tracheal system leading to spiracles and porous areas located in different parts of the body. The mandibles are almost always in the form of chelicera. Simple palps and anal suckers are often present. This sub-order includes damaging species such as the flour mite (*Acarus siro*), the cheese mite (*Tyrolychus casei*) and the bulb mite (*Rhyzoglyphus echinopus*).

Due to the large number of different mites, the symptoms of mite attacks vary greatly. Thus, grape leaf blister mite (*Eriophyes vitis*) causes formation of reddish or greenish galls on the upper face of the leaf and white ones that later turn brownish or reddish on the underside. Thus galls are initially isolated, but may grow together if the mites are present in large numbers. Pear leaf blister, which is caused by a different mite, invades the young leaves before they have unrolled, and produces small bumps. These bumps are light green to begin with and then turn reddish or brown, and their lower surface is covered with overgrown hairs that shelter the adult forms. The affected tissues suffer necrosis, and the leaves dry out and die. The fruit are often attacked, too, and this causes the fruit to be shed prematurely. The damage caused by this family of mites (*Eriophyniae*), such as the erinose mite of the leaves of lime, raspberry, tulip, walnut, hazel, gooseberry, and chrysanthemums, all show similar symptoms.

Another family, the *Tetranychinae*, contains species that mainly affect the apple, pear, plum, peach and cherry, as well as the vine, gooseberry and strawberry, and several ornamental shrubs and trees. The symptoms are similar in all the species attacked: the mite is found on the underside of the leaf, which acquires a typical greyish satiny appearance. The harvest is dramatically reduced as a consequence of the decreased photosynthesis and leaf fall. In general, hot dry summer weather, especially in August, causes the mite to population to boom; heavy rainfall, however, tends to wash the animal off the plant, which usually recovers. The different species of the family *Acaridae* include mites that eat agricultural products, such as flour, cheese, cereal grains, bulbs, etc. The bulb mite (*Rhyzoglyphus echinopus*) can attack the bulbs of tulips, hyacinths, gladioli, daffodils, etc. The flowers are deformed, and wounded bulbs are the most liable to attack. This mite is also an effective vector, transmitting fungal spores as well as bacteria.

4.3.2. Centipedes and millipedes

The **centipedes** and **millipedes** are terrestrial arthropods that possess a single pair of antennae and whose body consists of many similar segments, each bearing one or two pairs of legs. The adults do not have wings and the sexes are on separate individuals. There are four similar orders, the **centipedes (Chilopoda)**, the **Symphyla**, the **Pauropoda** and the **millipedes (Diplopoda)**. Their characteristics vary greatly, making it impossible to provide a description encompassing them all.

In general, the members of this group are neither numerous or prejudicial except in wet, or at least moist, sites. They burrow

down into the earth when the ground is dry. Depending on the species, they may be **carnivorous** or **eat plants**. The herbivores only attack plant tissues that are rich in water, and the millipedes and the members of the Symphyla cause the worst damage when beet and cereals sown in the fall do not germinate because of cold or wet weather.

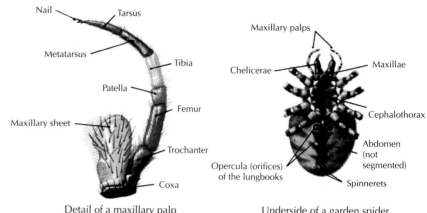

Detail of a maxillary palp and the names of its parts

Underside of a garden spider

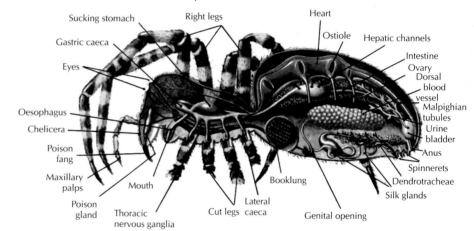

Anatomy of *Epeira diademata* the garden spider, in a longitudinal section.

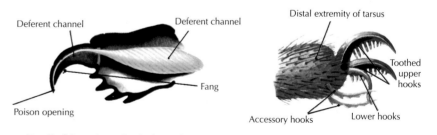

Detail of the poison gland of a spider

Tip of a spider's leg

4.3.2.1. Classification

• **Order Chilopoda (centipedes)** They have filiform (thread-like) antennae, a more or less flattened body consisting of many segments, each one with a single pair of appendages. The mouthparts are strong and, behind the labium, consist of a pair of powerful toothed mandibles and two pairs of maxillae. The legs of the first body segment have evolved into powerful hooks, pincers, that inject a poison that paralyzes the prey. The genital orifice is near the

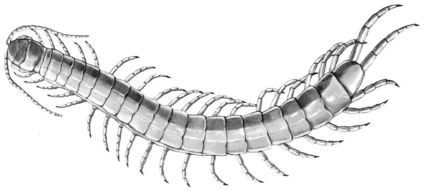

Scolopendra can grow quite large and the eggs are incubated by the female.

Scutigerella immaculata, lower surface

Iulus, showing the many legs

Polydesmus, whose body is fragile

Decapauropus cuenoti

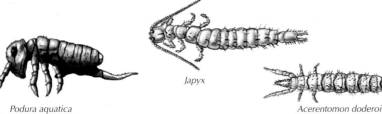

Podura aquatica

Japyx

Acerentomon doderoi

Centipedes, millipedes, etc.

rear of the body. The most common form is the centipede *Scolopendra* which lives in temperate, preferably arid, climates and whose bite is usually painful.

• **Order Diplopoda** (millipedes). They have rather short antennae, with 7 or 8 distinct parts. The body is more or less cylindrical and may consist of 11 to more than 100 segments, and the cuticle is frequently impregnated with calcium salts. The mouthparts are reduced to a pair of slightly thickened mandibles, protected by the labrum and a lower lip. The first body segment has no legs, but the other segments (except for the tip of the tail, which is two fused segments) bear two pairs of legs.

The two genital orifices are on the third segment, after the second pair of legs. Most millipedes have a series of repellant pores that open into the lateral part of the dorsal arch of the segments.

• **Order Symphyla**. These animals are very small and whitish in color. They have a low number of body segments and the genital orifice is located near the front.

• **Order Pauropoda**. They are less than 2 mm long and they have forked antennae: they have few segments and legs; the genital orifice is located at the base of the second pair of legs.

4.3.2.2. Life cycles

This group consists of unisexual animals, that are oviparous, with internal fertilization and simple molting. Their life cycles differ from order to order. As an example, we shall describe the life cycle of *Scutigera inmaculada*, a phytophagous centipede that causes significant damage in a wide range of plants.
The eggs are pearly white and are covered in tiny ridges. The eggs are laid in packets containing 4 to 12 eggs, sometimes up to 25, from March to August, peaking in May and June, and the eggs have an incubation period of 15 days. The individuals pass through seven larval stages. The individuals can live for four years and counting all the different larval stages, they may molt as many as fifty times. The number and frequency of molts, depends on genetic factors, nutrition, temperature and humidity. The adults can withstand very low temperatures for a long time. It has been shown that some individuals can withstand temperatures as low as 2°C for a long time.

4.3.2.3. Symptoms

Centipedes feed on insects, arachnids and grubs, though a few also attack plants. *Geophilus longicornis* burrows galleries in potato tubers. Other planteating centipedes such as *G. carpophagus* burrow into the fallen fruit of plums, apricots and peaches. The ones that eat vegetables, such as *Halophilus subterraneus*, which attacks the neck of lettuces, celery and onions, are also very common.
The order **Symphyla**, after the centipedes, contains most of the species that are damaging to plants. *Scutigera inmaculada* causes serious damage in newly germinated maize, though it often also attacks beans, vetch, cucumber, kidney beans, clover, tomato, asparagus, peas and lawns, and when they are extremely numerous, they may also attack weeds, such as groundsel, wild oats, buttercups, etc.
Millipedes (order Diplopoda) also contains some plant eating species, including *Polydesmus angustus*, which attacks the seeds of wheat, peas and kidney beans, though it can also parasitize carrots, onions, pansies and anemones and the aerial parts of artichokes, strawberries and potatoes, or *Blaniulus guttulatus*, which attacks strawberry plants, potato tubers, squashes, cucumbers, peas, cauliflower, etc. Others, such as *Cylindriolus teutonicus, C. frisius,* and *Schizophyllum sabulosum*, are damaging to

some crops, such as lettuces, potato tubers, alfalfa, carrots and sugarbeets.

4.3.3. Insects
(class *Insecta*)

Because insects are such important plant pests this chapter is relatively long. For readers who are not experts on insects, the length of the description of the morphology and physiology of the insects may seem excessive and even boring, but understanding the insects is essential in order to correctly use the methods available to combat them.

The class *Insecta* contains almost 70% of the known species of animal. More than 700,000 species have been formally described, and these include large species, such as the butterfly *Erebus agrippina*, whose wingspan may reach 280 mm, and extremely small ones, such as some beetles of the family *Ptiliidae* which are less than 0.25 mm long. The insects are arthropods that breathe through tracheae, with a body that is clearly divided into three distinct parts, the **head**, **thorax** and **abdomen**. The head bears a single pair of antennae, one pair of mandibles and two pairs of maxillae, the second pair of which are fused. The thorax has three

pairs of legs and generally has 1 or 2 pairs of wings.

Only some insects, **phytophagous** insects, feed on plants, and only they are of importance in plant pathology. Many other insects are of no importance to human beings, and are of only biological interest. Some insects are of great economic importance, such as honeybees, which pollinate fruit trees, and without which there would be no fruit.

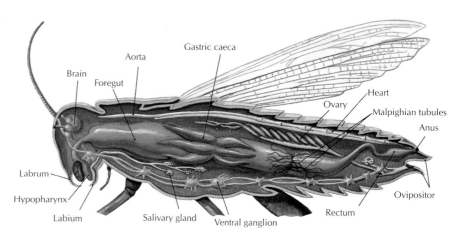

Internal organs of an insect

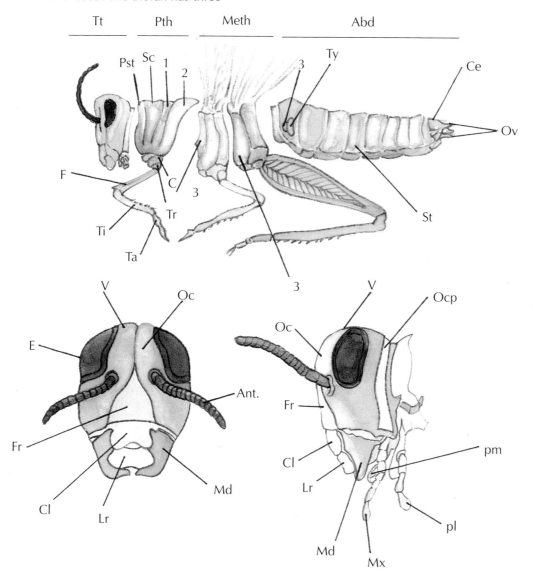

External anatomy of a grasshopper (Order Orthoptera)
Tt: head; Pth: protothorax; Meth: metathorax; Abd: abdomen; Aa: front wing; Ap: rear wing Ce: cerci; F: femur C: coxa; Ov: ovipositor; Pst: prescutum; Sc: scutum; 1: scutellum 2: postscutellum 3: spiracle Ta: tarsus; Ti: tibia Tr: trochanter Ty: tympanum (According to Herms)

Two details of the head of an insect with typical mouthparts. The front view of the head (left) shows the single pair of antennae (ant), the pair of compound eyes (CE), clypeus (Cl), the frons (Fr), the vertex (V) and three ocelli (sing: ocellus) or simple eyes. The side view shows the mouthparts in more detail: a pair of mandibles (Md), a pair of maxillae (Mx), a maxillary palp (pm), and labium (pl), a labrum (Lr) and the rear of the head, called the occiput (Ocp). (Taken from Snodgrass)

Other insects, mostly carnivorous ones, eat the phytophagous species and so these species can be used against harmful insects without having to use chemical pesticides.

The following is a brief look at the internal and external morphology of insects, but readers requiring greater detail should consult a reference work on animal physiology.

• External morphology

The **tegument** covers the insect's entire surface, mouth, the foregut and hindgut, the tracheae, the genital ducts and the glands opening to the exterior. The tegument consists of three parts: the **cuticle**, the **epidermal cells**, the **hypodermis** and the **basal membrane**. The cuticle consists of two parts, the external **epicuticle**, which is usually colorless, 4 μ thick or less, and a **procuticle**. The procuticle is secreted by the hypodermal cells and has a uniform structure.

The tegument consists mainly of the inert material chitin. Chitin is colorless and only dissolves in strong acids. In addition to chitin, the cuticle contains some lipids, polyphenols and proteins, and it acts as the insect's armour, protecting it from external agents. This is of great importance when discussing contact insecticides, because for the insecticide to reach the animal's body it must be soluble in lipids or be dissolved in some liquid that can dissolve in fats, waxes and water. The tegument forms a more or less rigid **exoskeleton**, consisting of a series of rings or **segments** that are articulated by flexible **intersegmental membranes**.

The **head** generally consists of the **cephalic capsule**, a pair of **antennae**, a pair of multi-faceted **compound eyes**, 3 **ocelli** or simple eyes and the **mouthparts**. The cephalic capsule of the primitive insects correspond to the typical head found in members of the order Orthoptera (grasshoppers, etc.). The antennae may vary greatly in form. They house a battery of sense organs and are of vital importance to the insect. They consist of three segments: the first segment, the **scape** is generally thickened; the second segment, the **pedicel** varies greatly in shape; the remaining segments form the **flagellum**.

The mouthparts of the adult insect are of great importance in their classification and identification and when studying the damage they cause to plants. Insects may have **chewing**, **licking**, **piercing** or **sucking** mouthparts.

• **Chewing mouthparts** are the most primitive, and are found in the Orthoptera (grasshoppers, etc.) and the Coleoptera (beetles). They consist of a pair of large mandibles moved by powerful muscles that cut and grind the food; a pair of **maxillae** that helps when chewing the food, each with a **maxillary palp**; a **labium** or lower lip, formed by the fusion of two parts similar to the maxillae, and two internal parts, the **epipharynx** and the **hypopharynx**. All these mouthparts are protected at the front by the **labrum**.

• **Piercing mouthparts** are typically found in the orders Heteroptera (blood-sucking bugs, etc.) and Homoptera (plant-sucking bugs, etc.). The mandibles and maxillae are greatly elongated. They have evolved into sharpened **stylets** that are closely joined to each other, forming two channels that are unequal in diameter. The insect pumps out saliva through the smaller one (**salivary channel**) into the wound and sucks up the mixture of saliva and cell sap through the **food channel**. In the order Thysanoptera (the thrips), only the left mandible is present (the right mandible is absent or vestigial) together with a pair of maxillae that are housed in a triangular conical beak on the ventral surface of the head.

• **Licking mouthparts** are characteristic of the order Diptera (the true flies, or two-winged flies). The mouthparts form a **proboscis**. In the houseflies, the labium is broader at the tip and forms a sort of suction pad that allows the fly to suck up liquids and to take up very small food particles directly. In the family Tabanidae (horse-flies, etc.), the labium is elongated and forms a channel that houses the other mouthparts. The piercing labium is fused to the epipharynx, forming a cutting blade.

• **Sucking mouthparts** are found in the order Lepidoptera (moths and butterflies). The mandibles and the labium are atrophied, and the very large

Different species of piercing insects (Courtesy of KOPPERT B.V.)

Frankliniella occidentalis

Aphis gossypii

Macrosiphum euphorbiae

Myzus persicae

Adult stage of
Cacyreus marshalli,
the geranium miner.
The caterpillar
burrows within
geranium stems,
causing very severe
damage.
(Photos courtesy of
the Department of
Agriculture,
Stockraising and
Fisheries of the
Generalitat of
Catalonia).

maxillae are joined together forming a tube that can coil up on itself, a sucking proboscis.

The **thorax** of the adult insects consists of three segments, the **protothorax**, the **mesothorax** and the **metathorax**. In almost all insects, each segment bears a single pair of legs. In most winged insects, the **tergum** consists of a large plate, the **notum**, and a narrow rear plate, the **postnotum**, that is in contact with the intersegmental membrane. The pleura consists of a front plate, the **episternum**, and a rear plate, the **epimere**, forming together the **pleural suture**.

The **legs** consist of 5 segments, the hip or **coxa**, the **trochanter**, the **femur** or thigh, the **tibia** and the **tarsus**, which typically has five segments and is tipped with one or two "claws" called **unques**. Near these claws there are often several small organs, which vary greatly in shape: **arolia** (singular: **arolium**) are small rounded pads located between the nails; the **pulvilli** (singular: **pulvillus**), which are sometimes very narrow, are located below the claws and on either side of the arolium or a large hair; the **empodium** is a small thickened nodule or a large hair located at the tip of the tarsus.

The membraneous **wings** are sited on the upper rear of the thorax, and are not present in all orders of insect. Insects lacking wings in the adult form are members of the sub-class **Apterygota** (the wingless insects), while the insects with wings belong to the sub-class **Pterygota**, which all have wings or have evolved from winged forms. The wings are crossed by many thickened longitudinal veins; there are also transversal or cross veins that divide the wing into cells.

The wings are more or less triangular; the front is the **costal edge**, the rear edge is the **inner edge** and the lateral edge is the **outer edge**. Winged insects usually possess two pairs of wings. When at rest, the rear wings are "folded", and may

even totally fold up so they are totally protected by the front wings.

The wings are of great importance when trying to identify insects. In some insects the males have wings while the females are wingless. Other orders lack the rear wings, for example, the true flies, in which the rear wings are vestigial **halteres**. In ants only some castes have wings, while in other orders the front wings are not used to fly but form a hard covering that protects the rear wings.

The **abdomen** consists of a maximum of eleven segments called **uromeres** which are joined to the terminal segment, the **telson**. In the lower orders, the eleventh segment forms the **supra-anal plaque** or **lobe**, also known as the **epiproctum**, and the eleventh **sternite** consists of the **paraprocta**. Most insects have 10 segments or less. The rear segments may fold within each other; the first segment, mainly the **sternite**, is atrophied or absent. The abdomen bears appendages in some orders of insect: the most important are the paired forcep-like **cerci** in the family *Japygidae* and in the order *Dermaptera* (the earwigs), which are located on the eleventh segment. The **genital shield** consists of appendages called **gonopods**, on the eighth and ninth segments in the female, and on the ninth segment in the male. A complete gonopod contains a small **coxa** that terminates in a distal stylet and an internal prolongation called the **gonapophysis**; the genital orifice is frequently sited on the membrane located behind the eight or ninth segment. In most insects, the male only differs from the female in the genital shield. In some species, there are major differences between the sexes, **secondary sexual characteristics**, as opposed to the **primary sexual characteristics**, the reproductive organs. The males may look very different from the females.

Panorpa

Parnassius apollo

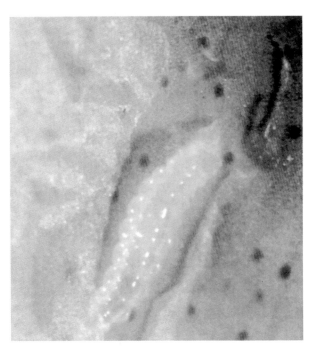

The larvae of
Phyllocnistis citrella,
the citrus leaf miner.
The larva burrows
galleries in the leaves
of citrus trees, and
causes severe damage
in plantations.
(Photos courtesy of
the Department of
Agriculture,
Stockraising and
Fisheries of the
Generalitat of
Catalonia).

A colony of larvae of the pear psyllid (Psylla piri). This is a phytophagous insect with piercing mouthparts. Its larvae and adults cause deformations in pear leaves, and when the attack is severe, the plantation may be defoliated. (Photos courtesy of the Department of Agriculture, Stockraising and Fisheries of the Generalitat of Catalonia)

This is called **sexual dimorphism**. The most varied organs may show sexual dimorphism. The mouthparts of the males may be much larger than those of the females; the compound eyes are larger in the males of some species of true flies than in the females; the antennae of mosquitoes are different shapes in the males and females. The females of several species of butterfly lack wings or only have atrophied wings, while the males have normal wings; female scale insects and fireflies lack wings and look different from the males. In butterflies, the colors of the males are often more conspicuous than those of the females. Depending on the environmental conditions during the larva's development, the coloration of the adults may vary and the insect is then said to show **seasonal dimorphism**.

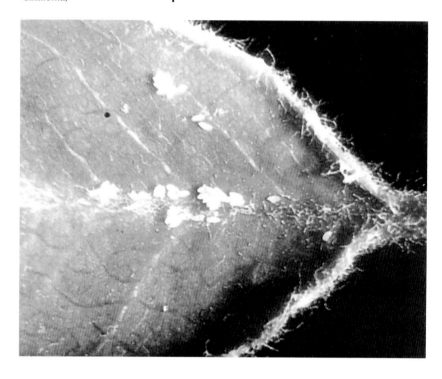

Biological pest control (Courtesy of KOPPERT B.V.) Two wasps that parasitize leaf miners like Liryomoza bryoniae.

Dacnusa sibirica

Dyglyphus isaea

• Internal anatomy

The **digestive system** varies greatly in length, as it may be straight or highly looped. It consists of three parts: the foregut or **stomodeum**, the midgut or **mesenteron** and the hindgut or **proctodeum**. The foregut and hindgut are covered in chitin, and digestion only occurs in the midgut. The food, between being eaten and being expelled from the anus, passes through several processes little different from those in the human being. The food particles pass through the **mouth**, the **oesophagus** to the **stomach** and then to the **gizzard**. They continue to the midgut which forms a relatively broad cavity (sometimes with ramifications) where digestion takes place. Finally, the food enters the hind gut and proceeds to the anus. Several species of insects contain symbiotic organisms in their digestive system that help digest the food and are vital for the host. These organisms are usually bacteria or fungi.

The **excretory apparatus** of insects consists of specific organs that vary from order to order. These include the **Malpighian tubules**, parts of the intestine that play an excretory role, and there are also lip glands that play the same role. The Malpighian tubules discharge into the junction of the midgut and the hindgut, frequently next to the **pyloric sphincter**. Different orders of insects have different numbers of Malpighian tubules, the thrips having just two, while the members of the Orthoptera may have up to 200. These tubes consist of groups of cells, called **nephrocytes**, that accumulate waste products. The excretory system also includes poison glands and the special glands that secrete silk.

The **circulatory system** consists of the dorsal vessel and the accessory pumping chambers. The **dorsal vessel** is a very long organ that runs from the hind end to the head: the front part, in the thorax, is known as the **aorta**. The rear part is the pumping organ, the **heart**, and is often divided into chambers; the wall of the heart is perforated by **ostia** (singular: **ostium**). The heart beats regularly at a rate of 12 to 150 beats per minute, depending on the order. The beat rate may vary depending on the developmental stage, how active the insect is, its age, its physiological status and the temperature. Accessory pumping chambers are located at the base of the wings, antennae and legs. The **hemolymph**, the insect's "blood", transports nutrients to the different organs of the body and collects their waste products. The blood contains different types of free cells, **hemocytes**, many of them phagocytes.

The **respiratory system** is highly developed and consists of the **tracheae**, which ramify throughout the body as **tracheoles**. The tracheae and tracheoles are stiffened by an epithelial layer that secretes a cuticle, often with spiral thickening. Many active species have air sacs, swellings of the respiratory system, with thick tracheae, or formed by the fusion of several tracheae. The tracheae open to the outside at the spiracles, which are arranged in pairs on the thorax and abdomen, one on either side. There are several types of respiratory apparatus: in the **holopneustic system**, the most primitive, all the spiracles are functional and there is one spiracle on either side of the mesothorax and the metathorax, as well as on the first 8 abdominal segments, making a total of 10 pairs. In the **hemipneustic system**, one or several pairs of spiracles are non-functional. In **apneustic** systems, all the spiracles are closed or have disappeared and respiration is by diffusion through the tegument, or using gills (in some aquatic insects). Gills are outgrowths of the tegument with thin cell walls into which oxygen can diffuse from the water. The gills are located within the rectum in many aquatic insect larvae.

The **muscle system** includes the muscles the insect needs to move. The muscles are usually greyish or translucent, except for the muscles of

Apis mellifica

the wings, which are yellow, orange or brown. Most of the muscles are **striated** and are divided into two groups: the abundant **skeletal muscles**, which are attached to a support, and the **visceral muscles**, which are circular or longitudinal.

The **nervous system** consists of nerve cells, **neurons**, which have a very long body with thin highly branched ramifications, called **dendrites**, and a long filament, the **axon**. Nerves consist of a group of many axons. There are two types of neuron, **sensory neurons** and **motor neurons**. The nervous system consists, as in vertebrates, of chains of neurons running from one end of the body to the other. The connection between two consecutive neurons is through the axon of one and the dendrites of the other; there is no continuity, the two neurons are only in contact. The point where the axon of one neuron is in contact with the dendrites of the next nerve is called the **synapse**. The nervous system consists of three closely coupled systems: the central nervous system, the sympathetic system and the peripheral nervous system.

The **central nervous system** consists of a double chain of ganglia joined to each other by longitudinal **connective** fibers and transverse fibers called **commissures**. The central nervous system consists of three parts: the **brain**, the **subesophageal ganglion** and the **ganglionic chain**. The **sympathetic system** is divided into three parts: the **stomato-gastric** system, which is directly connected to the brain and acts directly on the foregut, the midgut and the heart; the uneven-numbered ventral nerves are the nerves controlling the spiracles; and the **caudal sympathetic system**, which consists of nerves that originate in the last abdominal ganglion and supply the reproductive organs and the reargut. Several hormone-secreting glands release into the stomato-gastric system: a pair of oesophageal ganglia, known as the **corpora cardiaca** (singular: corpus cardiacum) are located below the oesophagus and behind the brain. The **corpora allata** (singular: **corpus allatum**) are joined to the **corpora cardiaca** and together

they control the initiation of molting ("ecdysis") and metamorphosis.

The **peripheral nervous system** consists of bipolar or multipolar neurons connected to sensory hairs or are arranged on the surface of the muscles and intestine.

The **reproductive system** is at the rear of the abdomen. In males, it consists of a pair of **testes**, which are more or less ovoid, that do not have suspensory filaments and consist of follicles; in many insects, the **peritoneal** covering of the follicles is highly developed and forms a covering for the two testes, the **scrotum**. The sperm are released into the two **vas deferens** (the male sexual ducts), which vary greatly in length, and stored in a **seminal vesicle** that may or may not be shared, then into the shared **ejaculatory duct** and finally into the copulatory organ, the **aedeagus** (the "penis"), whose shape may vary greatly and is an important character used to tell different species apart.

The **female reproductive system** consists of two ovaries, each consisting of more than 2,400 ovarioles. In general, an ovariole is an elongated tube with a **terminal filament** or **suspensory ligament**, that is attached to the internal wall of the tegument, the adipose tissue or the pericardial diaphragm; the **germarium** also contains the undifferentiated cells and sometimes also nurse cells; the **vitellarium** contains the developing eggs and sometimes the nurse cells. The ripe eggs are released into the oviduct and then into the **vagina** which sometimes has a **copulatory sac**; there is also a **spermatheca** or **seminal receptacle** where the sperm are stored, and one or two pairs of **accessory glands**, that secrete into the vagina. Their sticky insoluble secretion is deposited around the eggs and protects them from damage, as well as sticking the eggs to the tissue on which they were deposited. In some insects, the secretion is deposited around several eggs, forming a structure known as the **ootheca**.

Biological pest control (Courtesy of KOPPERT B.V.)

Scolia flavifrons, *a large hymenopteran*

The adult of Zeuzera pirina *or leopard moth. It generally attacks pears, but also attacks apples and other fruit trees. The caterpillar mines galleries in the tree's trunk and stems, often causing its death. (Courtesy of the Department of Agriculture, Stockraising and Fisheries of the Generalitat of Catalonia)*

• Biology

The insects can overwinter as eggs, larvae, pupae or adults and may hide on or in the soil, under dead leaves or stones, on or under the bark of trees, etc. The adults emerge relatively early in spring, depending on the species and the climatic conditions. Each species of insect only renews activity when the temperature has risen above a temperature, the **development threshold temperature**. The adults generally feed for a few days and then mate. The female lays her eggs on the plant and usually dies shortly after laying the eggs.

The length of the incubation period depends on the climatic conditions; below a given temperature, which differs from species to species, the **minimum temperature**, the egg does not grow. At higher temperatures, the incubation period is shorter, but very high temperatures delay the embryo's development. This growth ceases above a certain temperature called the **maximum** and the embryo dies if the temperature is raised above this to the **lethal temperature**. The temperature at which development is shortest is called the **optimal temperature**. Other factors also influence how long incubation takes, such as humidity, light, etc.

The larvae of many species are able to leave the plant on which they hatched and move to other plants. Their growth, like that of the egg, depends on the climatic conditions. When they have finished growing, they turn into pupa ("pupate") and then adults. Some insects only have a single generation a year. The cherry fruit fly, for example, overwinters in the soil as the pupa; the adult appears in early May, feeds and mates; the eggs are laid on the maturing cherries; the larvae develop within the fruit, and finish their development 20 to 25 days later; they then drop to the ground and pupate at a depth of 3 to 5 cm; they remain as pupae until the following May.

Other insects have several generations a year. The cabbage white butterfly (*Pieris brassicae*) overwinters as the chrysalis; the butterflies appear in the first fortnight of May and lay their eggs on wild or cultivated members of the cress family; the caterpillars eat the leaves of these plants until they develop into chrysalids in June. A few days later, from these chrysalids emerge a new generation of butterflies, which lay eggs throughout July; if temperatures are high, the caterpillars grow fast, and become chrysalids in August in temperate zones. The different generations are counted starting from the egg and not from the adult; in the case of the cabbage white, the butterflies that appear in spring belong to the overwintering generation; their eggs are the starting point for the first generation.

The cockchafer (*Melolontha melolontha* L.)

Examples	Order, and type of mouthparts in adult	Main chara	
		Rear wings	R... wi...
	DIPTERA (flies) Suckers or able to pierce and suck	membranous	transfo... into ap... haltere...
	HETEROPTERA (true bugs) Suckers	hardened at the base and membraneous at tip	memb...
	HOMOPTERA (plant bugs) Suckers	generally membraneous and transparent	mem...
	THYSANOPTERA (thrips) Suckers	wings with narrow long fringed (which may be absent)	narr... long
	NEUROPTERA (lacewings, etc.) Chewers	usually highly reticulate and membraneous	in g... high... and

* The insect shown in drawing

Classification and characteristics of the main types of insects. Note the list includes both harmful and useful insects.

Meta-morphosis	Representatives		Representatives		Main characteristics			Order, and type of mouthparts in adult	Examples
	Useful	Harmful	Harmful	Useful	Meta-morphosis	Rear wings	Rear wings		
ete	Syrphidae (hoverflies), Tachinidae (parasitic flies)	gall midges, cabbage maggot, Mediterranean fruit fly (Ceratitis)	mole crickets, locusts*, cockroaches	crickets	gradual	membranous and folded in fan	hardened and cover the front wings	ORTHOPTERA chewers	
al	anthoco-ridae (anthoco-rid bugs), Miridae (plant bugs)	chinch bugs, pear tiger	"white grubs", Colorado potato beetle	ground beetles, lady bugs	complete	membranous and single functional pair	chitinous elytra that protect the front wings	COLEOPTERA (beetles) chewers	
al		aphids*, scale insects, psyllids, or jum-ping plant lice	wood-wasps, sawflies	chalcid wasps, bracoind parasitic wasps*, ichneu-mon parasitic wasps, bees	complete	membranous: the wings are lacking in the workers of some groups	membranous, larger than front wings	HYMENOPTERA (bees, wasps, etc.) Chewers whose mouthparts may form a sucking proboscis	
al		thrips*	codling moth*, noctuid wood-boring moths	silkworm	complete	membranous and covered in scales	membranous and covered with scales	LEPIDOPTERA (butterflies and moths) suckers	
ete	lacewings *Raphidia**								

A/ Leaf miners of ornamental plants, such as Liryomiza brioniae a major problem in almost all crops. Their larvae typically burrow galleries in the leaves of the plants they attack.
(Courtesy of KOPPERT B.V.)

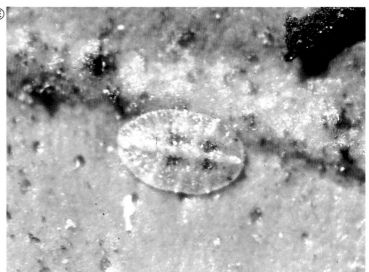

B/ Typical symptoms in an apple leaf of the galleries made by leaf miners (Cemiostoma). SHELL recommend Nomolt® (15% teflubenzorol).
C/ Damage caused by Liryomiza brioniae in an ornamental plant.
D/ The aphids are the best known members of the order Homoptera, and attack a wide variety of plants. Detail of a stem attacked by aphids.
(Courtesy of BASF, S.A.)
E/ SCHERING sells OLEANOL 83 FOR SUMMER. This is a summer mineral oil that is a very effective treatment for citrus scale.

larvae show negative phototaxis (they move away from light). Depending on the source of the stimulus, insects show other forms of taxis, such as **thermotaxis** (stimulation by temperature), **geotaxis** (stimulation by gravity), **anemotaxis** (response to the flow of air), and **chemotaxis** (a response to a chemical stimulus). They may all be positive or negative, depending on the species and its developmental stage.

The life cycle of insects varies greatly. This is because the rate of growth varies from species to species, and because growth sometimes halts for a relatively long period,

is an insect that attacks broadleaf trees but can also affect some fruit trees. It only produces a single generation in three years, and the adults appear in the first fortnight of April. The females lay their eggs in the soil. The eggs hatch in July and feed on the root hairs. The first molt takes place in September or October and then they burrow into the soil, and do not appear until April of the next year. They continue feeding on the roots of crop plants, and continue this alternation for three years. In the spring of the third year, they feed on plant roots until July, when they burrow to a depth of 30 cm and pupate; the adult has formed by August, but does not emerge until spring of the fourth year.

The conduct of the lower animals is conditioned by different types of **taxis**, the term for the animal's nervous system's response to a physical, chemical or biological stimulus, which leads first to orientation and then movement. The movement may be positive (**positive taxis**), when the animal moves towards the source of the stimulus, or it may be negative (**negative taxis**), when it moves away from the stimulus.

As an example, the adult common fly (*Musca domestica*) shows positive phototaxis (it moves towards light) but the

a phenomenon called **diapause**. Some insects go through all their developmental stages ("instars") without any halts as long as environmental conditions are suitable. If temperature, humidity or light are below the minimum or above the maximum **growth threshold**, or if the food supply is insufficient, growth halts, but restarts when the conditions are favorable again. This is called a growth halt.

In other insects, however, development may stop sharply even when growing conditions are excellent. This halt is called **diapause** and may occur at any time of year and in any developmental stage. Diapause may occur in eggs, larvae, pupae and even in adults. Insects showing these pauses are called **heterodynamic** and those that do not **homodynamic**. Many factors may trigger diapause, such as the changes in the photoperiod, excessive heat, food quality, availability of water, and any variation in any external factor that may affect the insect's growth.

4.3.3.1. Classification

Insects are essentially classified on the basis of their wing structure, i.e., their position and veins.

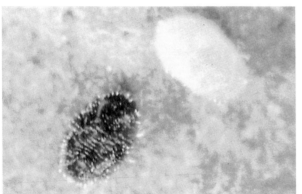

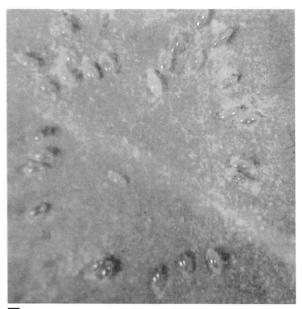

Greenhouse whitefly
1/ Adult
2/ Recently laid eggs
3/ Normal pupa (white) and one parasitized by Eucarsia formosa.
4/ Infested tomato leaf
5/ Fungal growth on tomato fruit
(Photos courtesy of the Department of Agriculture, Stockraising and Fisheries of the Generalitat of Catalonia)

January	February	March	April	May	June	July	August	September	October	November	December
		+	++	++	++	+++	+++	+++	++	++	+
o	o	oo	oo	oo	ooo	ooo	ooo	ooo	oo	oo	o
–	–	– –	– –	– –	– – –	– – –	– – –	– – –	– – –	– –	–
•	•	•	••	••	•••	•••	•••	•••	••••	••	•

Adults +

Eggs o

Larvae –

Pupae •

Some insects attack ornamental trees in villages and towns. **Corythuca ciliata** *is a small heteropteran bug that attacks plane trees.*
1/ Adults and larval colonies on the underside of the leaf.
2/ Adults in winter diapause
3/ Adult
4/ Larva
5/ Leaves showing symptoms
(Photos Courtesy of the Department of Agriculture, Stockraising and Fisheries of the Generalitat of Catalonia)

The wingless insects are considered to be primitive and form a subclass of their own, the **Apterygota**. The winged insects are considered advanced and form the other subclass, the **Pterygota**, including most of the phytophagous insects and plant pests considered here.

Insect taxonomy is very complex, and the huge number of species means that taxonomic levels unknown in other groups have to be used. These unusual divisions include **subclass**, **section**, **sub-order**, **superfamily**, and **subfamily**. You should be familiar with the ones most important to farmers: the orders Orthoptera (grasshoppers), Coleoptera (beetles), Hymenoptera (ants and bees), Lepidoptera (butterflies), Diptera (true flies), Heteroptera (true bugs), Homoptera (aphids) and Thysanoptera (thrips). On page 311, there is a table describing these groups with the most important characteristics of each one and a representative drawing.

4.3.3.2. Life cycles

• **Life cycles**

In order to understand insect life cycles, it is necessary to make some preliminary comments on their different methods of reproduction. Most insects are **oviparous**, that is to say, they lay eggs, either individually or in packets, but other forms of reproduction, such as viviparity and parthenogenesis, are found in the orders of insects that attack crops, while other reproductive methods occur in insects that are not relevant to plant pathology, such as paedogenesis (reproduction by larvae), polyembryony and hermaphroditism.

• **Viviparity**. There are three types of viviparity: **ovoviviparity**, in which the eggs hatch soon after being laid. This has been observed in thrips, cockroaches and beetles. In **adenotrophic viviparity**, the larvae hatch within the uterus where they continue to grow, thanks to a secretion from special glands, until the larvae are almost completely developed; this has only been observed in a few true flies in which the larval stage is very short, and rapidly enters the pupa stage. In **pseudoplacental viviparity**, the embryo develops in a swelling of the vagina of the mother insect from an egg without a yolk and usually without an eggshell; the embryo is nourished by a special organ similar to the mammalian placenta. This occurs among the aphids and some cockroaches.

• **Parthenogenesis**. Parthenogenesis is when an egg develops without fertilization by a sperm. In general, the fertilized eggs, the zygotes, are diploid, which means they have two sets of chromosomes (2n), one from the father (in the haploid sperm) and one from the mother (in the unfertilized egg). The eggs produced by parthenogenesis receive all their chromosomes from the mother and are diploid (2n). There are three types of parthenogenesis:

• **Haplodiploidy** in which the females lay two types of egg. After fertilization, the eggs are diploid and grow into females, while the unfertilized haploid eggs grow into males. This occurs in the Hymenoptera and Thysanoptera (thrips).

• **Obligate parthenogenesis**. There are no males at all, or they are extremely uncommon; the eggs are all diploid. This occurs in some thrips and members of the Phasmatidae (stick insects).

• **Cyclic parthenogenesis**. This is frequent in aphids (greenfly, etc.). Parthenogenetic reproduction and sexual reproduction alternate in response to a set of factors. In most cases, several generations are born by parthenogenesis in the spring; the sexual generation appears in the fall as the photoperiod diminishes.

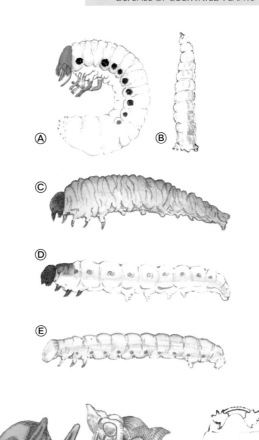

Larvae of different insects that show complete metamorphosis
A) beetle larva
B) fly larva
C) hymenopteran
D) and E) caterpillars
(According to Bovey)

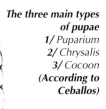

The three main types of pupae
1/ Puparium
2/ Chrysalis
3/ Cocoon
(According to Ceballos)

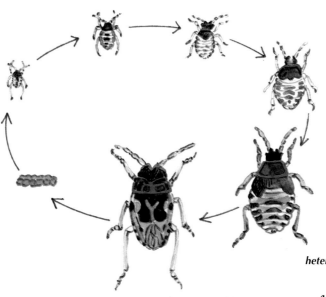

Life cycle of a heteropteran showing gradual metamorphosis (Horcias nobilellus Berg)

• Embryonic development

The eggs are generally laid shortly after mating, but may be laid several weeks, or even months, later. The eggs may be laid in one or two batches or scattered over several months. The number of eggs laid by insects varies enormously. The sexually reproducing females of some aphids only lay a single egg, while butterflies almost always lay more than a hundred. Wasps may lay 20 to 30,000 eggs and the termites found in hot countries may lay hundreds of thousands. The number of eggs laid is usually very high at the beginning of the egg-laying period and gradually declines. The rate of egg laying depends on the food supply, the humidity and the light.

The **eggs** may be deposited on the ground or on the surface of the water, but they are usually stuck to the surface of a plant by a sticky substance secreted by the accessory glands. They may be laid individually or in groups or even in a structure, an ootheca. Some insects have special egg-laying organs to deposit the eggs in the soil (locusts), on plant tissues (thrips) or within other animals (parasitic wasps). The eggs are often slightly elongated, but they may be barrel-shaped, spherical, conical or disc-shaped. After an incubation period whose length depends mainly on the temperature, the egg hatches and the larva emerges.

The **larvae** are tiny when they hatch and grow through a fixed number of **molts**. An insect's rigid exoskeleton cannot stretch, and so the animal has to shed it and grow a new, larger, one. This usually happens 4 to 6 times, and each form is called an instar. The first stage larvae may vary greatly in appearance:

1) They may closely resemble the adult: in the final molt, the genitals mature and become clearly visible. The adult forms are always wingless, and are members of the subclass Apterygota.
2) All other insects have wings in the adult form (and are members of the sub-class **Pterygota**, which means "winged insects"), except for a few parasitic species. This gives rise to two different situations: the first larval stage may basically resemble the adult, but without wings, which develop gradually over the series of molts. At the end of the final larval stage, the animal completes the changes to the adult form, and this is called **gradual metamorphosis**. The larvae have the same type of mouthparts and the same biology as the adults;

Life cycle of a butterfly, the cabbage white (Pieris brassicae L.), showing complete metamorphosis

A/ Larvae of the Colorado potato beetle (Leptinotarsa decemlineata) *(Courtesy of SHELL)*
B/ Some warm-blooded animals also eat plants. The photo shows a meadow mouse.
C/ Adult Colorado potato beetles (Photo courtesy of SHELL)

the animal does not have a long period of inactivity before the adult emerges. This is thus not a complete and drastic metamorphosis, and is called **gradual metamorphosis** or **incomplete metamorphosis**, and the insects are said to be hemimetabolous.

The adult may, however, be totally unlike the larva, with different mouthparts, with compound eyes and it may in fact be totally unlike the larva. This type of larvae generally passes through 5 or 6 molts. It then ceases to feed and develops into the pupa, which is almost always totally immobile. Within the pupa the larval tissues breakdown and are used to create the adult organs. This total change is called **complete metamorphosis**, and in these insects the wings only appear in the adult: these insects are said to be **holometabolous**. The diagram on page 315 shows larvae of insects undergoing complete metamorphosis and it is a guide for identifying which of the main orders a plant pest belongs to, which is an essential first step

towards identifying it and thus learning how to combat it.

In insects undergoing complete metamorphosis, the final molt (pupation) before turning into the imago (adult) is a special stage when the insect is most vulnerable, the **pupa**. Before pupation, the larvae usually search out a site offering some protection from their natural enemies and the weather. Some insects pupate in the soil or within a case made of earth mixed with saliva.

Others make **cocoons** consisting of strands of silk, such as hymenopterans and lepidopterans. There are three types of pupa, the puparium, the chrysalis and the cocoon (see drawing page 315). In the puparium, pupation occurs within the exoskeleton of the final larval stage. In the chrysalis, the adult body parts may be visible underneath a hard outer covering. This is found among lepidopterans. The cocoon is a silken sheath that the pupa spins around itself, and is barrel shaped.

4.3.3.3. Symptoms

Whitefly is one of the most terrible problems for farmers, both in the greenhouse and in outdoor crops. To control whitefly, BASF recommends Lancord, a mixture of methomyl and cypermethrine.

The oriental fruit moth (Graphiolita molesta) *causes severe damage to peach shoots and fruits, and to those of other fruit trees.*
1/ Adults
2/ Caterpillars
3/ Entry hole 4/ Eggs
5/ Damage to shoot
(Photos courtesy of the Department of Agriculture, Stockraising and Fisheries of the Generalitat of Catalonia)

There are so many insects that it is hard to make generalizations about the damage they cause. Leaf-eating species can cause large-scale defoliation, recognizable by the bite marks on the leaves. Examples include: the small phytophagous leaf roller moth (*Tortrix viridiana*) of evergreen and deciduous oaks, which can cause major defoliation, attacking even the acorns and bark; and many other small insects, whose larvae eat the leaves of ornamental plants, such as gerberas and chrysanthemums, burrowing galleries in the leaf parenchyma, such as *Liriomyza trifolii Burgess*, commonly known as the American gerbera miner.

Other insects attack the roots and feed on the roots, causing the plant's general appearance to deteriorate greatly. Insects that attack roots are very common and include the mole cricket (*Gryllotalpa gryllotalpa Lat.*), which attacks the roots of onions and many other agricultural plants, and the cockchafer (*Melolantha melolantha L.*) which attacks the roots of fruit trees and forestry plantations.

1

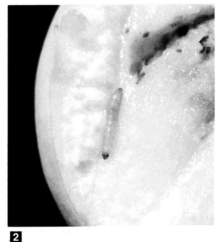

3

4

2

5

The larvae of many insects burrow galleries in the stem of the plant, totally destroying it. This is clearly shown by the leopard moth (*Zeuzera pyrina L.*), whose larvae burrow upwards within the pear's trunk, again leading to total destruction. Insects attacking horticultural crops include the artichoke stem borer (*Hidroecia xanthenes Germar.*). Other insects that bore into forest trees, such as pines, are often very small but very numerous, such as the bark beetles (family *Scolytidae*) *Tomicus pinperda L.*, *T. minor Hart.* and *T. destruens Woll.*, which excavate tunnels in tree trunks. The entry holes are clearly visible, and the tree generally shows small patches of gummosis, produced as a defense. The insects that spend part of their life cycle within a plant are known as **endophytic**.

A variety of other insects attack fruit causing economic damage of greater or lesser severity, but in all cases insect attack results in decreased fruit quality. This is true of the oriental fruit moth (*Grapholita molesta Busk*), whose larvae mainly attack peaches, or the codling moth of apples and pears (*Cydia pomonella L.*), which burrows into the fruit and eats them. A common pest of vines is the grape moth (*Lobesia botrana Schiff*), whose larvae burrow into the grapes through small wounds that they make with their mouthparts. This insect is often the vector of other parasites that enter the plant through the same wound, such as the fungi that cause the grapes to rot, like *Botrytis cinerea P.*
There is more information on insects and insect damage in chapter 8, which briefly discusses plant pests, classified by the plant they attack, such as fruit trees, extensive or intensive crops. This chapter also contains many photographs, drawings, graphs and useful information on insects and the damage they cause.

The birds are the most important vertebrate consumers of insects, but also of seeds and fruit, so that it is difficult to say whether they are beneficial or prejudicial to agriculture.
1/ Lark (Alauda arvensis)
2/ Crow (Corvus corax)
3/ Golden oriole (Oriolus oriolus)

The mammals include some insectivorous species, such as the hedgehog (Erinaceus europeus). Other species eat plants, such as chamois, rabbits, hares and wild boar, and constitute a rich fauna and their population levels should be maintained in balance with agricultural needs.

4.4. VERTEBRATES

The vertebrate classes that cause serious damage to plants are the **birds** and the **mammals**. The other classes (fish, amphibians and reptiles) are carnivorous or do not come into contact with plants, as in the case of fish. Birds and mammals are warm-blooded, homoiothermic, a physiological specialization that makes them relatively independent of the environment they live in. As their effect on crops is not very important, we shall not go into the details of their morphology or physiology.

In the case of birds, it is extremely hard to draw a line between ones that are beneficial and ones that are prejudicial. The starling (*Sturnus vulgaris*) is beneficial to agriculture because it eats a huge number of insects, but it is prejudicial if you only consider what it eats in the spring and fall, when it consumes fruit, such as grapes and cherries. Many birds are omnivorous, and so they can eat animals or plants, depending on the time of year. The sparrow (*Passer domesticus*) also feeds on insects but can be very prejudicial when it feeds on recently sown seeds and on fruit. Other birds, such as herons (*Ardea spp.*) eat small mammals, which is positive for agriculture because they eliminate the populations of rodents that gnaw the collar and roots of some plants. Some flock-forming birds may cause serious damage if they arrive in very large numbers.

The **mammals** also include phytophagous species, especially among the **rodents**. These animals eat plants almost exclusively, using their powerful incisors to cut the plant tissues, and they can attack a very wide range of crops. In addition to this damage, it is worth mentioning that rodents, especially rats and mice, live in the same habitat as human beings and often feed on stored plant products, such as flour, rice, legumes, etc. One of the most notable features of rodents is their rapid reproduction, and the fact that their populations may show great fluctuations in the number of individuals. In the meadow mouse (**Microtus arvalis**) it has been observed that the population fluctuates in cycles of 10 to 12 years. When the population peaks, invasions of crops and the resulting damage are a terrible problem for agriculture. These cyclic variations have been studied, but their causes have not been identified.

Another rodent, the mole (*Talpa spp.*), feed on the insects it finds in the subsoil when digging its galleries with its powerful claws. The mole does eat plants, but it damages crop roots when digging galleries in its search for larvae and adult insects to eat. On the whole, however, the mole is considered beneficial because it eats a large number of insect pests and its burrows improve soil aeration. The same is true of other animals such as hares (*Lepus spp.*) and rabbits (*Oryctalagus cuniculus*).

Some large mammals may cause local damage to crops, especially in sparsely populated areas near forests. These animals, including the European wild boar (*Sus scrofa*), the red deer (*Cervus elaphus*), the roe deer (*Capreolus spp.*) or badgers (*Meles taxus*), cause noticeable damage to crops, but their ecological importance and the income they may generate as game means that they should be maintained, but at a level compatible with agricultural activity.

Drawing of a thread-counter. A small magnifying glass like this makes it much easier to detect insect pests and identify them.

5. PREVENTIVE MEASURES

After this discussion of the physiological disorders and parasitic diseases of plants and the agents causing them, such as viruses, bacteria, fungi and animals, it is time to discuss how to eradicate them. In general, as in all areas of agricultural practise, prevention is better and cheaper than cure. Two types of measures can be taken against plant disorders and diseases: the first consists of **preventive measures** and the second of **control methods** to get rid of pests. Preventive methods seek to prevent these disorders and diseases from occurring, while control methods seek to eradicate the causes of the problem.

This chapter deals with the preventive methods that can be used to ensure there is little need to apply chemical pesticides, etc., and the measures available to control and treat pests and diseases of cultivated plants are dealt with in the next chapter. Sometimes the division between prevention and cure is blurred but when there are few or no disorders or diseases, then **preventive** methods are used. When pests and diseases exceed a certain **critical level**, **curative treatment** is required.

The best way to prevent pests and diseases

in a crop is simply to carry out a regular and thorough visual check. Farmers often apply unnecessary treatments to protect their crops from attack, when it would be cheaper in time and money to check the crop regularly. There is now a general tendency to avoid using chemical products unless they are strictly necessary. Instead, you are recommended to check the crop for pests every day.

Many agrochemical distributors give away small thread-counters or hand lenses to their clients, and they are very useful indeed to the farmer. They are small instruments (see drawing) that consist of a folding (or fixed) support and a magnifying lens, and which are often used by stamp collectors to look at their stamps. A small magnifying glass like this is useful when checking a plant really thoroughly for parasites.
Inspection is perhaps the most effective method, as harmful agents can be detected immediately. Once they have been detected, the farmer has to decide how to combat them in order to eliminate them as quickly as possible, at the beginning, not when the infestation is so severe that it may be very expensive, or impossible, to resolve. In most countries, government bodies linked to the official agricultural services regularly publish information bulletins that advise when is the best time to treat against a given pest, and provide a list of appropriate products. This could have been included in the next chapter, but is included here because it is very useful warning information.

Preventive methods may be **cultivation techniques** or **biological**, **physical** or **mechanical** methods. Agricultural or cultivation methods are ancestral methods passed down from father to son, and are essentially mechanical or physical. Biological methods, including the natural resistance of some plants to certain phytophagous insects, relies on a biological potential, while physical and mechanical methods are usually simple logical ways of protecting the crop.

5.1. CULTIVATION METHODS

Cultivation techniques are long established and many are typical of a region or district where they have long been practised. Many developed long ago on the basis of accumulated experience rather than scientific theory. It has been confirmed beyond all doubt that they are effective and they can be highly recommended.

• **Crop rotation**. Crop rotation is a common practise in large scale cereal cultivation, and it is based on the idea of not cultivating he same crop species on the same plot for more than one year. If the same crop is cultivated every year, its pests and parasites, such as fungi, bacteria and nematodes, have such an abundance of food that populations can grow almost exponentially. This leads to what is called "**soil fatigue**", when the soil becomes incapable of continued production, or the plants are spindly and stunted. Furthermore, cultivating the same species year after year on the same site may lead to impoverishment of any one of the nutrient elements, meaning that nutrient deficiencies are more and more likely to appear. The microelements reveal this problem most clearly, especially if the crop in question requires a lot of a particular microelement.

The idea of the **fallow period** is to leave the soil uncultivated for a year to "rest", causing the populations of harmful bacteria, fungi and nematodes to decline substantially. Leaving ground fallow is not common practise in horticultural crops, and this implies large costs for the farmer in the purchase of chemical treatments, because the problem of pests and diseases gets worse year after year, and so far more pesticides have to be used than if rotation was practised. There is also the added problem that many pests start to evolve pesticide-resistant strains, meaning that you can not use the same product year after year, but have to buy different products, which may well be new on the market and thus much more expensive.

Weeding. Weeds may well allow crop parasites or pests to complete their life cycle, such as black-stem rust of wheat, which needs the barberry to complete its life cycle (see page 291). Other animals, such as insects and especially aphids, use weeds as secondary hosts. Mechanical or chemical weeding disrupts these animals' life cycle and prevents their populations from increasing. In addition to getting rid of the weeds, it is worth removing and burning stubble, as it houses the different overwintering stages of many insects.

• **Plowing**. The different methods of plowing combat the underground stages of some insect pests, such as white maggots (*Plyphylla fullo L.*) and wire worms (*Agriotes spp.*). Plowing not only mechanically destroys the pests, but may also reduce their number indirectly by causing changes in the soil's structure, moisture content or temperature that are unfavorable for the insects.

• **Fertilizer application**. Correct application of fertilizer to crops is a good cultivation technique. It ensures the crop plants are healthy and this means they are less vulnerable to pest attacks. Fertilizer application should be balanced, with the correct proportions of nitrogen, phosphorus and potassium (N-P-K), secondary elements and microelements for

Correct fertilizer application ensures healthier plants that are resistant to pests. (Courtesy of SCHERING)

each crop species. It has been shown that excessive application of nitrogen and potassium leads to an increase in attacks by phytophagous mites. Clearly everything that helps to keep the plant in good health also helps to defend it gainst pests and diseases, as a healthy plant is a resistant plant. Healthy well-fed plants are also more resistant to climatic agents like frosts, hailstones, heat waves, cold spells, etc.

5.2. BIOLOGICAL METHODS

One preventive method of dealing with soil diseases caused by viruses, bacteria, fungi and nematodes is to graft the chosen variety onto rootstocks that are resistant. It is also possible to choose species or varieties that are resistant to the attack of some insect pests, such as phylloxera (*Phylloxera vastatrix*), a terrible pest of the vine, which devastates European varieties but has little effect on North American varieties. There are two types of plant from an agricultural perspective, monobionts and dibionts. Monobionts consist of a single organism, and includes most agricultural crops such as the vetch (*Vicia sativa*) and the Windsor bean (*Vicia fava*). These two crop plants belong to the same genus (*Vicia*) but are two different species. They are referred to as *Vicia spp.*

Dibiont plants consist of two different organisms joined together by a graft. This is done to obtain the best features of the rootstock and the fruiting variety. Most fruit trees and some ornamental plants are dibionts. In general it is only possible to graft a plant onto a rootstock of the same species, or a different variety of the same species. In fruit trees, for example the almond (*Prunus amygdalus*) can be used as graftstocks for commercial varieties of almond, peach (*Prunus persicae*), apricot (*P. armeniaca*), plum (*P. domestica*), etc. The purpose of grafting is to use rootstocks **resistant** to pests and diseases (nematodes, bacteria and fungi) and to graft onto them varieties

Pests often become resistant to a given chemical pesticide (the individuals that survive and breed are selected to be resistant). This leads to the appearance of strains resistant to the product. The only solution is to use a different product.

that are more profitable. The second section of this work, which deals with fruit trees in more detail, provides more information on grafts, the species used as rootstocks, the varieties and the pests they are resistant to.

As pointed out in previous chapters, not all plants show the same resistance to a single pest, due to the genetic differences between species and between varieties within a species. These differences are based on genetic mutations, as in all other living things and viruses, and on their recombination in sexual reproduction. Thus, **resistance** in plants may be defined as the sum of hereditary qualities possessed by a plant that determine how much damage a pest causes. In agricultural practise, this means a variety that can produce a larger and better harvest than ordinary varieties, with the same population density of insects.

There is a range of resistance to an insect within the varieties of a host species or group of species. These differences are not due only to genetic factors, because environmental factors may modify the degree of resistance. A variety is said to be **immune** when it has never been attacked at any stage of growth by a pest species that can attack other varieties of the same plant. A plant variety is said to be highly susceptible when it is attacked more than the other varieties of the species. Between these extremes a variety may be **highly resistant**, show **medium resistance** or be **susceptible**.

The term **pseudoresistance** is used for apparent resistance, the result of transitory characters in the susceptible host plants; this may be because the insect population is lower than normal, because the insect population peaks out of phase with the sensitive stage of the plant (for example, in early ripening varieties) or because the plant is unusually vigorous due to a favorable climate or very fertile soil.

In vitro cultivation in laboratories using **genetic manipulation** can produce plants that are resistant to a given pest or disease. These plants are sold as "genetically modified" and are certified to have certain properties. They are usually much more expensive than the typical local varieties, but are worth the extra cost in some circumstances, especially when the traditional crop has started to decline, either due to lack of resistance to certain diseases or because it is less productive than the new hybrids.

Plants can be cultivated in vitro in biological laboratories in order to select plants that are resistant to viruses, fungi, nematodes, etc. (Photo courtesy of SCHERING)

5.3. PHYSICAL OR MECHANICAL METHODS

• **Measures against frosts**. It may sometimes be worth leaving a layer of plant cover (weeds) on the soil. In orchards this is done to prevent frosts. A soil with plant cover will always be 1 or 2°C warmer than the same soil without any plant cover. If there are reasons to fear a late frost in spring or an early frost in fall, when the fruit is nearly ripe, the oldest preventive measure is to burn sawdust or wood, and nowadays oils and even kerosene are used in burners and heaters. Another system is to prevent the plant's temperature from falling below a critical point by spraying water directly onto the crop. The water deposited on the leaves releases energy when it freezes due to the low temperatures. When a gram of water freezes, it releases 80 calories. Part of this heat may increase the plant's temperature, and the rest is lost by irradiation.

• **Measures against sun damage**. If the weather forecast suggests very sunny weather and rising temperatures immediately after intense cold in winter, some young fruit trees should be protected with a layer of straw on the side exposed to the morning sun. It is also possible to lime the trunks which has a similar effect.

• **Measures against hail**. Anti-hail cannons, which seek to disperse the cloud or to move it away from the crop, have been used, but are not very successful.

• **Measures against wind and saline winds**. In very windy sites, hedges can be planted as windbreaks on the edges of crops to prevent wind damage as far as possible. These windbreaks are usually tall trees, such as cypresses, poplars, etc. In crops near the sea, windbreaks can be planted to prevent wind-blown salt from damaging crops.

• **Preventive measures against incorrect use of agricultural chemicals**. Taking great care in the choice, dose, mixtures and application of all agricultural chemicals will prevent later toxic effects on crop plants. The reader should refer to the corresponding sections of this section that list the precautions to be taken with pesticides, herbicides, fungicides, etc.

• **Preventive measures against possible nutrient deficiencies**. Before symptoms of deficiency of any nutrient element occurs, it is worth systematically applying fertilizer to crops, trying to supply all the elements in the proportions and quantities needed by the crop.

• **Measures against diseases caused by pruning**. Pruning fruit trees is normally carried out in winter and produces different types of wound in the plant. No precautions need be taken with small wounds, as they will heal over. As a preventive measure, however, larger wounds should be treated with a balsam or putty so they are not left exposed to the open air, when they would be entry points for bacteria and fungi. There are special preparations (usually mercury compounds) that are bactericidal and fungicidal and are sold ready for use.

• **Preventive measures against viruses**. All the plant parts used for vegetative reproduction,

such as scions for grafting, cuttings, layers, stolons, tubers, bulbs, rhizomes, etc., must be taken from healthy plants, so that they do not spread any viral diseases to the new plants. You are also recommended to destroy completely any infected plants, so that animal vectors can not transfer the infection to other plants.

• **Preventive measures against nematodes**. Some simple and easily applied rules should be enough to prevent zones free of nematodes from being contaminated. Nematodes can be transported in the soil adhering to plant roots, tubers, seeds, shoes, animals' hooves, etc, or in the earth on the wheels of machinery. It is good practise to thoroughly wash everything that may be infected before moving it on to any other crop.

• **Measures against disease in greenhouses**. Measures to prevent pests and diseases in greenhouses include **disinfection** of **soil**, **pots** and **tools**. In high-yielding crops, and especially ornamental plants, the same crop can be grown year after year in the greenhouse if the soil is suitably disinfected. Products to disinfect soil are discussed in more detail in the chapter on Soils, and the section dealing with methyl bromide explains that it is the most effective method of controlling bacteria, fungi and nematodes, as well as the precautions to be taken when using it. To some extent, soil disinfection prevents "soil fatigue", and allows successive crops of the same species on the same site, as long as appropriate fertilizer is applied. Apart from soil disinfection, sterilizing the **pots** and **tools** is also very important, and this point is discussed in more detail in the section on Greenhouse Cultivation.

A field of sunflowers with a windbreak of poplars in background. Windbreaks reduce the effect of the wind on the crop.

Pruning wounds above a certain size should be covered with putty or balsam so they are not left exposed to the air, as this would allow disease to enter.

The high temperatures and humidity in tropical habitats allow a wide range and large number of plants to grow.

Very few plant species can grow in arid zones, because they are hot and dry.

On the highest peaks, the low temperatures make plant life impossible.

6. CONTROL OR CURATIVE METHODS

As commented in the previous chapter, the division between prevention and cure is blurred, and it is often hard to determine when a preventive measure becomes a curative measure. This chapter lists the control methods available, though it includes some methods that could just as well be included in the section on preventive methods.

The control methods available to combat pests and diseases can be classified into abiotic methods, biotic methods and integrated pest management. **Abiotic means** are those that are not of organic origin, and they can be divided into **physical** or **mechanical** means and **climatic effects**. **Biological methods** use living organisms to combat pests, basically by using pheromones and predators. There is a third way, **integrated pest control**, and the section discussing it seeks to provide the individual farmer with the proven guidelines needed to bring together all these methods and draw up a calender that fits in with individual needs.

6.1. ABIOTIC METHODS

Abiotic means include three methods of combatting pests: **physical** methods, as opposed to **chemical methods**, and **climatic** effects. From the mid 19th century onwards, when laboratory synthesized pesticides began to be used, it was thought that chemical methods were the definitive answer. For many years pesticides, etc., were developed at the expense of other less spectacular methods, such as preventive cultivation techniques, physical and mechanical methods, and climatic effects. Since then, however, it has become clear that chemical methods have their disadvantages.

Many insects and mites that used to be no more than minor problems are now major causes of damage to crops, mainly because their natural enemies have been eliminated by the use of non-specific pesticides. Another drawback of the use of pesticides is that it has selected resistant strains of pests (the result of selection and adaptation to the environment), and as a result some pesticides that used to be very effective are now not very useful. Their effectiveness has declined because the insects that were intended to be destroyed have developed resistant strains. This problem has meant that the agricultural pesticide industry has regularly had to launch new products because the old products have become ineffective. The **latest generation** of new products are usually much more expensive than traditional insecticides, and so the farmer must spend much more money on pesticides, etc., often making the crop uneconomical.

However, non-chemical methods alone are not enough to protect agricultural production, which is needed to feed a growing human population. As we shall see in the section on **integrated control**, the latest tendencies in the struggle against plant pests are aimed at a mix of methods together with a rational use of chemical products. Other **biological methods** have been developed with some success, and they hold great promise for the future.

6.1.1. Climatic effects

Climatic effects, i.e., weather conditions, are beyond human control, but may be very effective at killing insects. They may occur as irregular and local accidents, eliminating large numbers of insects, but you should not rely on this. Harsh winters, for example, kill some insects and mites that are poorly protected, or which overwinter in an active form (larva or imago). Other insects, such as the caterpillar of the goat moth (*Cossus cossus L.*), can withstand temperatures as low as –20°C. The adult of the garlic moth (*Acrolepia assectella Zell.*) overwinters in the open air in southern Sweden. In general, many insects and mites can withstand temperatures of –40°C to –50°C for a few minutes, though they die after thirty minutes or so.

In any case, very low or very high temperatures slow down the animal's metabolism, greatly reducing the damage they cause. A growth halt in diapause may represent a reduction of two or more generations within the insect's active period, implying a substantial reduction of the damage caused that year. The minimum and maximum **lethal temperatures** are the temperatures that cause the pest to die within a short period of time. Extreme temperatures have a greater effect if they arrive suddenly than if the temperature change is gradual.

Violent rains may wash the eggs of some dipteran leaf miners (*Pegomyia spp.*) off the plants they were laid on. The adults of some species are washed off fruit trees in the first violent storms in late August in the Mediterranean area. This mainly affects species of homopteran, such as aphids, and the red spider mite (*Panonychus ulmi Koch*). If the rain is hard enough, they drown in the water on the soil surface after they are washed off the trees.

Abnormal climatic conditions sometimes cause great indirect mortality. A cold spell in spring

As an example of the effect of the weather, strong rains may clean trees of mites, greenflies and other parasites. (Photo courtesy of SCHERING)

after the eggs are laid may delay the start of plant growth and cause the larvae to starve. If summer is unusually hot or cold, the date at which the different generations appear may shift, meaning that the forms not usually exposed to cold will be exposed to it (for example, the caterpillar instead of the chrysalis), and this may cause great mortality.

Moisture is another question that is not strictly climatic, but is a major environmental factor that may greatly reduce insect populations. As pointed out when discussing nematodes, they require some moisture in the soil for their growth. In very dry summer periods, insect attacks on plants may be greatly decreased. Other animals, such as snails and slugs, may be devastating in very wet springs, but their effect is almost negligible in very dry conditions. When environmental humidity is excessive for them or when there is a severe drought, some insects and mites can enter diapause, and this may mean that the farmer need not apply pesticides as often.

6.1.2. Physical or mechanical methods

This is perhaps the area where it is hardest to distinguish between preventive and curative methods. The reader may think that the methods outlined are rather old-fashioned, but you should remember that for years they were the only ones available, and unfortunately in many of the poorer areas of the world they are still the only ones available. In more developed countries, direct methods to kill insects are increasingly uncommon because of the high cost of labor.

• **Methods against viruses, bacteria and fungi.** Some fungal and bacterial diseases can be transmitted by seeds, and the seeds can be disinfected by immersion in hot water. If sick plants have been detected (on a regular visual inspection) that are thought to be due to viral or bacterial attack, they should be uprooted immediately and destroyed. This prevents the disease from infecting the rest of the crop. Fungi require moisture to grow, and anything that lowers the relative humidity of the air or soil may slow down their growth. This is relevant in sites that tend to waterlogging, where mechanical measures can be taken to install drainage so the water can drain away. In the case of parasitic fungi, such as black-stem rust of wheat (*Puccinia graminis*), the seeds can be disinfected by immersing them in hot water.

• **Physical methods against nematodes and mollusks**. Nematodes are sensitive to temperatures, so seeds, seedbeds and topsoil can be disinfected by heat. **Heat treatment** is also possible in ornamental bulbs and in strawberry runners. Some greenhouse plants can be heat treated during their winter rest period. Mollusks can be removed from the plants and destroyed. Though this procedure is rarely used in large fields, it is suitable for small family vegetable plots covering a few tens of square meters.

• **Physical methods against insects**. The mechanical methods formerly used to combat insects (collection by hand, traps and baits, removal of cork from the tree trunks, mechanical destruction of eggs, protective barriers and ditches, etc.) are now only of limited interest and are only viable for small cultivated areas. Heat is still used to treat wheat and rice against grain weevils (*Sitophilus*), which attack cereal grains stored for consumption as such or for making bread.

• **Physical methods against vertebrates**. Physical or mechanical methods are still widely used against vertebrates. These include traps for small rodents, mesh or synthetic fiber netting to protect crops from starlings, nets over the soil to protect recently sown corn from rooks and crows, and finally protecting the bark of fruit trees from gnawing by hares with wickerwork or wire.

6.1.3. Chemical methods

Pesticides, **insecticides**, **anti-parasitic agents**, etc. were initially produced in laboratories, but they are now all produced in large chemical factories.

Two types of chemical product are used, fertilizers, on the one hand, and pesticides and weedkillers on the other. The nutrients applied include all the different formulations, granulates and powders, soluble ones for application in irrigation water, foliar feeds, with or without microelements, etc. Fertilizers were dealt with in detail in the section on Soils.

Pesticides can be divided into **bactericides**, **fungicides**, **helicides**, **nematicides**, **acaricides** and **insecticides**, which eliminate bacteria, fungi, snails, nematodes, mites and insects respectively, together with **herbicides** (weedkillers). The general term **agricultural chemical** refers to products produced by chemical synthesis, and includes pesticides and weedkillers.

Agricultural chemical products may under certain circumstances have toxic effects on plants. When a mistake is made in dosing, and excess is applied, the plant may be damaged. Toxic effects on plants are frequently due to mistakenly mixing incompatible products, using machinery that has not been thoroughly cleaned or the mistake of using machinery used in herbicide application for other purposes.
The label of any product of this type must contain the percentage of the **active ingredient** (*A.I.*).

It must also specify the product's formulation, what crops it can be used on, what pests it combats, its toxicology, the legal safety period and any other relevant features, and the following is a more detailed discussion of each of these concepts, using a practical example.

• The **product name**. The trade name of the product, which identifies it and distinguishes it from the other products on the market, must be at the top of the technical and commercial explanation on the label of every agricultural chemical.

List of the most common abbreviations denoting the formulations of agricultural chemicals

Abbreviations	Definition
EC	Emulsionable concentrate
GB	Granulated bait
SCr	Soluble crystals
WWE	Wax water emulsion
WSE	Wax solvent emulsion
PH	Plant hormone
UG	Ultradispersable grains
GR	Granule
SL	Suspension-forming liquid
CLS	Concentrated liquid suspension
EL	Emulsionable liquid
FL	Fumigant liquid
OL	Oily liquid
SL	Soluble liquid
XL	Other liquid formulations
MC	Microcapsules
ME	Microemulsion
MGr	Microgranules
EP	Emulsionable powder
WP	Wettable powder
SP	Soluble powder
XP	Other types of powder
CS	Colloidal suspension
SCS	Soluble crystalline solid
SS	Supersaturated solution
ST	Soluble tablets
FT	Fumigating tablets
SP	Soluble pastilles
ULV	Ultra low volume
ULVL	Ultra low volume liquid

*Example of a label
from an agricultural
insecticide
(Vademecum. Liñán)*

Commercial name: Manufacturer:
DOMINEX 10   BASF Española, S.A.
Active ingredient and type of formulation:
10% de alfa-cipermethrin emulsionable liquid (EL).
Product description:
Contact insecticide
Characteristics and authorized applications:
- Decidous fruit trees: against "Psylla", greenfly, codling moth and mining caterpillars.
- Citrus fruit: against aphids and prays.
- Olive trees: contra prays (only anthophagous generation).
- Horticultural crops like strawberries, potatoes and others; against aphids, beetles, heliothis, plusia and other grubs.
- Cotton and safflower: against aphids, heliothis, earias, pink maggot and other grubs.
- Flowers and ornamentals: against aphids, heliothis and other caterpillars.
- Elms and plane trees; against galeruca, wood-boring moths, corituca and as treatment in cut pine against wood-boring scolytid moths.
- Poplars, willows and alders; against paranthrene.
- Untilled land: against grasshopers.
- Pine woods: in treatments against the nest of the processionary moth.
Dose and method of use:
Apply in a normal spray at 0.01-0.015% in fruit trees, safflower, elm, against the processionary moth and paranthrene; 0.03-0.04% in other crops and 0.1% for cut wood.
Toxicology:
Human: Noxious Nx. Land: A: Water: C. Against bees: C.
Safety period:
For harvesting or entry of livestock: 2 days.

• **The manufacturer**. The manufacturer and/or distributor is in the final analysis responsible for the product's quantity, quality, and the technical specifications, doses and method of use recommended on the label.

• **The active ingredient** is the compound that is effective against the disease or pest to be combatted. An agricultural chemical never contains 100% active ingredient. Therefore the percentage of the active ingredient must figure on the label next to the percentage composition of active ingredient. A label (see example) might say "*10% alpha-cipermethrin*". This means that 10% of the product is alpha-cipermethrin, and the other 90% is a bulking agent.

• **The type of formulation**, the physical and chemical state of the product (the active material plus the excipient). This is usually expressed in capital letters and the most common ones are listed on the previous page. The abbreviations used may vary from country to country, but their meaning is the same.

• **The product description** tells us if it is an insecticide, an acaricide or a herbicide, etc. From the plant's perspective its method of action may be **systemic** or by **contact**. **Contact** products act when they come into contact with the animal or plant to be eliminated, but do not enter the plant. **Systemic** products enter the plant and destroy the pathogens from within. The product may act on the pathogen by: **ingestion**, when the animal eats the product which kills it; by **inhalation**, when the pathogen absorbs the product through the respiratory system; and by **contact**, when the product acts by crossing through the pest's chitinous exoskeleton.

A product may also be **polyvalent** or **specific**. A polyvalent product against pests and diseases acts upon a wide range of parasites, eliminating or controlling many of them. A highly polyvalent agent may act against a wide range of organisms of the same group (the classic fungicides based on copper or sulfur) or many different taxonomic groups (such as methyl bromide, which acts against fungi, nematodes, insects, vertebrates and weeds). Specific products act against a single organism or a small group of organisms.

• The **crops** and the **pests** and **diseases** for which the Department of Agriculture, or Ministry of Agriculture, in each country has authorized the use of a given active ingredient, and the percentage specified for use.

• The **dose** and the **method of use** recommended by the manufacturer and authorized by the agricultural authorities. In some cases the label specifies on which stage of the plant's life cycle ("phenological stage") the product should be used. In fruit cultivation, some products used against the overwintering forms of insects and mites can only be applied when the tree has no leaves and is in its winter rest, as the product is toxic to the green parts of plants. The **phenological stages** are the name for the different stages the plant passes through over the course of the year, and their names correspond to the first letters of the alphabet. In the apple, for example, the phenological stages are defined starting from observation of the buds. The winter buds are in the first, **A**, stage. Stages **B** and **C** correspond to the buds after they swell in the spring. The C_3 and **D** stages correspond to the appearance of the flower buds, and finally, the **E** and E_2 stages are when the sepals have separated enough for the petals to be seen. Some authors use further letters to continue to describe the fully formed fruit.

• **Toxicology**. Weedkillers and pesticides must be registered in the *Official Register of Agricultural Chemicals* and are given a toxicological classification. This classification varies over time as new products appear on the market. Information on this subject can be found in the regular official publications and on the labels the manufacturer must put on every product. In general, the toxicological

a:	resting winter bud	**d₃:**	emergence of flower bud and opening of calyx
b:	bud swelling (upper)	**e:**	swelling pink corolla
c:	swollen bud (lower)	**f:**	swollen pink corolla
c:	growing bud (upper)	**f₂:**	full flowering
c₃:	swollen bud (lower)	**g:**	set fruit and shedding of petals
c₃:	growing bud	**h:**	growing fruit
d:	appearance of flower bud		

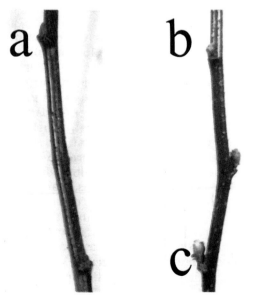

Phenological stages of the QUINCE

According to R. Dolcet

Apple leaves heavily attacked by apple scab (Venturia inaequalis) (Photo courtesy of SHELL)

classification is divided into three parts. The first part usually deals with its toxicity to human beings, the second part with its effect on the soil flora and fauna and the third part refers to its toxicity for aquatic fauna and flora. A fourth section gives a measure of the product's effect on bees, which are useful insects. In Spain, the toxicity is indicated with the letters **A**, **B**, **C** and **D** (from least to most toxic).

Some insecticides also kill mites or at least depress their numbers. Other insecticides, such as pyrethrins, have led to mites evolving resistant strains, and should be used with caution, especially when treating a mixture of insects and mites. The next section lists the main agricultural chemicals on the market. This is obviously only a brief summary, as the farmer, should stay up to date with new products as they appear, technical advances, either by directly consulting the technical service of the

• **Safety period**. The label must show the safety period, the time that must elapse between applying the product, at the authorized dose and in the authorized conditions, and harvesting or the entry of agricultural livestock.

• **Other information of interest**. The label also includes other important details, such as whether the product is inflammable, how hygroscopic it is, whether it is volatile, explosive, corrosive or if it is highly irritating. On a practical level, it is common for farmers to mix two or more products together to avoid the need for a second application. It is habitual to mix insecticides and fungicides together. It is very important to realize that not all agricultural products can be mixed together, as some mixtures may have a toxic effect on the crop plants. The product label should specify the precautions to be taken in this respect, that is to say the products it can be mixed with and those it cannot be mixed with. The reader should look at the table on page 327 showing an example of a label of an everyday agricultural chemical, defining the product's characteristics.

trade distributor or consulting the official bulletins from the different levels of administration, by performing tests with the new products or even better, in all three fields at the same time.

6.1.3.1. Antiviral agents

Unfortunately there are no chemical products that are effective against viruses in crop plants. The only measures that can be taken are preventive. Only virus-free plants should be planted, and cultivation techniques should aim to ensure they do not become infected. Most viral plant pathogens are not transmitted by seeds, and so plants grown from seed can normally be considered to be free of viruses. However, the French or green bean (*Phaseolus vulgaris*) and some fruit trees of the genus *Prunus* are exceptions to this rule, and seeds should only be taken from plants that are free of virus.

The vegetative organs used to propagate plants (graft scions, cuttings, layers, etc.) are the main means by which viruses are transmitted. In specialized reproduction and genetic selection laboratories, plants that are certified to be free of virus can be obtained by **in vitro** cultivation techniques.

Certified virus-free plants are usually more expensive, but their yield, vigor and production are also greater. Over time viruses undergo mutations that enable them to attack plants that were genetically immune.
In some cases, when it is not possible to find plants free of virus, they can be produced by heat treatment. This consists of keeping actively growing infected plants at a temperature of 37-38°C, and this not only causes the symptoms to disappear, but also eliminates the virus. The best way to protect healthy crops is to eliminate the vector organisms, using a nematicide on the soil if nematodes are the vector, or if the virus is spread by insects or mites, the use of an insecticide-acaricide.

6.1.3.2. Bactericides

There are few effective chemical methods of combatting bacterial diseases. As in viruses, the best thing is to prevent them by using resistant varieties, eliminating vectors and heat treating the seeds.
Some chemical products have been found to be relatively satisfactory as insecticides. Copper hydroxides, oxychlorides, oxinates and sulfate are all effective bacteriostatic agents (i.e., they halt bacterial growth). They have long been used as fungicides and are now sold as fungicides-bactericides. Recently produced chemicals, essentially similar to the antibiotics used in human medicine, have also been successful, for example **Kasugamycin**. It is sold as a systemic fungicide-bactericide, and has good preventive and curative results against endoparasitic fungi and bacteria.

6.1.3.3. Fungicides

Fungal diseases are caused by infection by fungi, and are basically combatted by the preventive measures described in the previous chapter together with mechanical actions. One of the most important methods of preventing fungal diseases is to grow resistant varieties. In very rainy springs or under special circumstances, these methods may not be enough and chemical products (fungicides) have to be used.
Fungicides can be divided into two distinct categories **Contact fungicides** and **systemic fungicides**. **Contact fungicides**, such as copper or sulfur compounds, can be dissolved and applied as **mixtures** to the plants or fruit or applied as a powder to the plants. Contact fungicides have a preventive action, as they kill the fungus when its spores germinate, before the mycelium can grow into the plant. Contact fungicides can be applied to the soil to disinfect it, while seeds or bulbs (and other parts that are to be propagated) can be dipped in a freshly prepared mixture. Treatment with these products must be repeated regularly as the

A tomato plant treated with a copper compound as a preventive measure against fungus attack. The bluish tinge to the leaves and fruit is typical of copper compounds. (Courtesy of BASF, S.A.)

Pear scab (Venturia pirina) (Courtesy of ICI-CELTIA)

plants and fruit grow, in order to prevent them from being attacked by fungi.

Truly **systemic fungicides** do not exist. Some products show some systemic action, and can kill the fungus once its mycelium has penetrated the leaf parenchyma. They are also known as **curative** or **eradication** fungicides, though this definition is rather misleading as truly systemic fungicides have not yet been developed. Studies have shown that these products increase the leaf epidermis's resistance to attack by the mycelium, and this prevents the fungus from spreading and further infecting the plant, but the leaves that have been attacked will never recover.
This is especially important in the case of the plants that are cultivated for their ornamental foliage, such as *Euonymus*, which are at great risk of attack by the powdery mildew *Oidium euonymi-japonici Sacc*. **Powdery mildew** forms whitish powdery patches on the leaves of the euonymus. Even if the leaves are treated with the most effective anti-mildew agent on the market these patches will never go away.
Copper products significantly increase the thickness of the leaf tissue, **captan** causes a slight increase, while **maneb** and **zineb** cause no increase at all. The following is a list of the main groups of commercial fungicides.

• **Sulfur compounds**. Sulfur is usually sold as **pure sulfur** and in formulations with different particle sizes. It is usually applied as a powder and is effective against the powdery mildews if applied at a temperature of 16-18°C or higher. Other sulfur derivatives, such as **calcium-sulfur mixture**, **wettable sulfurs**, **colloidal sulfurs**, **alkaline polysulfides** and **barium polysulfide** are effective against powdery mildew and scab in fruit trees. As long as they are not mixed with insecticides, they are not dangerous to bees and can even be applied during flowering. Wettable sulfurs and colloidal sulfurs are applied dissolved in treatment mixtures and can usually be mixed with other pesticides.

Detail of a vine leaf attacked by the downy mildew of vine (Plasmopara viticola). *(Courtesy of SCHERING)*

• **Copper compounds**. **Bordeaux mixture** has for a long time been the typical mixture for treatment with copper. Bordeaux mixture is a mixture of a solution of copper sulfate with hydrated lime or a suspension of slaked lime. If used alone copper sulfate would severely burn the plants because it is very acidic, but mixed with a calcium compound it forms a solution that is hardly soluble in water, an alkaline adhesive colloid. It is possible to make your own Bordeaux mixture, but because this does not work out much cheaper, we recommend buying one of the formulations on the market. Bordeaux mixture is the typical fungicide used against downy mildew of vine (*Plasmopara viticola*), but it is also used against copper-sensitive fungi in extensive and horticultural crops.
Copper oxychlorides were created by the chemical industry many years ago to try and resolve some of the problems associated with Bordeaux mixture. These drawbacks, due to the alkalinity of the mixture, are that the treated plants may suffer a growth halt or completely stop growing, and the leaves may be burnt. Copper oxychlorides are neutral, and there are two types on the market: the copper oxychlorides, with commercial preparations containing up to 50% copper, while the **copper-calcium oxychlorides** only contain about 35% copper.
The oxychlorides are used in silviculture, viticulture, in extensive crops and in horticultural crops against copper-sensitive fungi, such as stone fruit shot hole (*Stigmina carpophila*), fruit scab (*Venturia sp.*), peach leaf curl (*Taphrina deformans*), downy mildew of vine (*Plasmopara viticola*) and potato blight (*Phytophthora infestans*). Copper-calcium oxychlorides are mainly used to combat potato blight and sugarbeet cercospora leaf spot (*Cercospora beticola Sacc.*). There are other copper compounds on the market, such as **copper oxide**, **basic copper carbonate** and **basic copper sulfates** and **copper powders**. The different forms of copper sulfate are formulations in which the chloride has been replaced by sulfate, eliminating the possible toxic effects of the chloride ion. Copper products for powdering are used in intercalary powdering in viticulture and against fungal diseases in horticulture. They consist of different copper compounds to which inert materials and bulking agents have been added. They are usually complemented with elementary sulfur so a single treatment also combats mildew.

• **Organic fungicides**. Organic fungicides are now replacing the old copper and sulfur compounds, and are more and more common in stores. They are rarely more effective than the classic fungicides, but have the advantage of being less toxic to plants, and are also compatible with most of the pesticides on the market. Mixed fungicides are also on sale, consisting of a mixture of organic fungicides, together with a copper or sulfur compound, and they are called **organo-copper** or **organo-sulfur** fungicides. Only the best known products are discussed here, but the number of products is increasing every year and one should try to keep up to date with the technical advances occurring in this field.

Captan is also known as orthocide. This highly stable fungicide is not very soluble in water, is not volatile and is highly persistent. It shows good compatibility with most agricultural chemicals, and is not very toxic to human beings and other warm blooded animals, but is very toxic to fish. It is effective against scabs, mildew, white rot of grape (*Coniella diplodiella*), monilia leaf blight of seed and stone fruit (*Monilia sp.*), cherry leaf spot (*Blumeriella jaapii*), stone fruit shot hole, apple anthracnose canker (*Pezicula alba*) and cherry anthracnose canker (*Apiognomonia erythrostoma*) and grey rot of some horticultural plants and vines (*Botrytis cinereae*).

Dinocap or **karathane** was initially used as an acaricide, but it was later found to be very effective against powdery mildew. It not very soluble or volatile, and shows little toxicity to humans and animals, but should not be mixed with alkaline products. It is very effective against apple and apricot powdery mildew

replacing copper compounds in combatting fungal diseases in extensive cultivation.

Phaltan is also known as **folpet**. There are many copper compounds on the market with formulations containing different percentages of folpet. It is very similar to captan, in both its chemical composition and fungicidal properties.

Tiram is also known by the name **TMTD**. This carbamate does not contain a metallic ion, unlike zineb, maneb and mancozeb. It is hardly toxic to humans or bees, but can provoke an allergic reaction in some people. It is effective against scab, grey rot, monilia leaf blight, peach leaf curl, and shot hole, but it should not be applied to fruit intended for preserves. It can be used as a disinfectant fungicide for the seeds of several horticultural plants and extensive crop plants.

Zineb is an organic fungicide and is one of the many thiocarbamates. It is a salt of zinc and is very insoluble in water. It is not very toxic to plants or human beings. When used alone, it is effective against most fungal diseases, such as scab, rusts, powdery mildews, leaf curl and shot

Snails may consume a large amount of plant material in favorable weather conditions.

(*Podosphaera*), peach and rose powdery mildew (*Spaherotheca pannosa*), vine powdery mildew (*Uncinula necator*), powdery mildew of currants and gooseberries (*Microsphaera grossulariae*) and cucumber powdery mildew (*Erisyphe cichoreacorum*). It is a good substitute for sulfur in the treatment of sulfur-sensitive varieties.

Mancozeb is a mixture of zineb and maneb, and contains both zinc and magnesium. It is effective against apple scab (*Venturia inaequalis*) and pear scab (*Venturia pirina*) and potato and staghead downy mildew of tobacco (*Peronospora tabacina*).

Maneb has the same chemical formula as zineb, except that the zinc has been replaced with magnesium. Its fungicidal effect and mode of action are very similar to those of zineb (described below), but maneb is more effective against potato blight and tobacco mildew. This product is increasingly

hole. It can be sold as a mix with copper salts, which makes it effective against powdery mildew of vine and potato. If it is mixed with nickel salts, it is an ideal fungicide to combat cereal rusts (*Puccinia spp.*).

Ziram is mainly used against peach leaf curl and scabs of pear, apple and peach.

6.1.3.4. Snail and slug killers

The chemical products used to kill gastropods such as snails and slugs are called **snail and slug killers**. Their active ingredient is metaldehyde. Most of the products on the market are sold as granulates containing 5% metaldehyde that are scattered on the soil. This product is toxic to human beings, land animals and the aquatic fauna, and should be handled with great care. Metaldehyde acts by contact and by ingestion, and is an effective insecticide against orthopterans, such as the mole cricket (*Gryllotalpa gryllotalpa*) and the migratory locust (*Locusta migratoria*). It is sold as bait and should be distributed over the surface of the entire cultivated area, but especially in the areas where the snails are often present, such as in wet sites, near walls and undergrowth.

6.1.3.5. Nematicides

The best way to combat nematodes is not with chemical products, but the use of cultural practises, such as fallow, rotation, disinfecting organs for use in propagation by heat therapy, cultivating species and varieties that are resistant to nematodes, not transporting infected soil and so on. These methods ensure that the soil nematode population does not exceed acceptable levels, but the results are sufficient, though they do not eradicate the nematodes. The chemical **nematicides**, some of them highly effective, can be considered to be **soil fumigants** and **disinfectants**, rather than nematicides. Most of them are toxic to plants, and so they can not be used when the crop is in the field. Many of these products also show insecticidal, helicidal and even herbicidal action. They are sold as solids, liquids or gases, though gases are the most effective, as they reach all the pores in the soil.

Depending on whether they are solids, liquids or

from using them without the corresponding official permission.

The (chemically) weakest agents are the least effective, but they are the ones that destabilize the soil least. The most effective ones are the most toxic and act against the broadest spectrum of pests, but they usually have a highly destabilizing effect on the soil. To finish, remember that nematicides are expensive, and they are only for use in horticulture and floriculture, mainly in greenhouses. In the section on Soils in this book, within the part on soil disinfectants, there is a list of the most important nematicides or disinfectants, its mode of action, how to use them and the organisms they combat. As a reminder, the main agents used are **methyl bromide** and **chloropicrin**, **dichloropropane** or DD, **dichloropropene**, **dazomet** and **metam-sodi** or VAPAM.

6.1.3.6. Insecticides and acaricides

These are chemical products used to combat insects, mites and to a lesser extent millipedes and centipedes. Insects and mites are the most numerous parasites of cultivated plants in terms of their populations and the damage caused, and thus cause the most problems for farmers. As already pointed out, these animals can multiply very quickly, as they go through many generations a year. There is a wide range of different insect species, not all of them pests of plants, as insects may parasitize or predate other insects.

Metam-Sodium is a soil desinfectant used to control nematodes; it is also effective against weeds at high doses. It should be used in greenhouses and the soil must be covered inmediately after application. (This product is manufactured by BASF, S.A.)

The larvae of some beetles and butterflies cause serious damage to horticultural crops. (Photo courtesi of DOW ELANCO)

gases, nematicides can be applied in three different ways: injection into the soil, dissolved in irrigation water or blown under plastic sheeting. They can only be used when the site is moist, and at a suitable ambient temperature, and may have to be irrigated afterwards to seal in the gas. Some of these products are extremely toxic and can only be used by specialized technicians, as individuals are usually banned

Due to the wide range of insects and mites that exist, agricultural research laboratories have developed different formulations to deal with all the possible cases of struggle against these pests. The products on the market are usually liquids (emulsions and suspensions) and less often powders to treat the aerial part of the plant. Products to treat parasites in the soil may be liquids, granulates or powders that are directly applied. For greenhouses and disinfection of bulbs, seeds and stores in general, there are gaseous products in the form of aerosols, and there are solid products for dusting or to be dissolved, for the preventive disinfection of seeds. Their mode of action means they are effective by inhalation, contact or ingestion. To describe the action of a product, it is sometimes said to be a "shock" treatment. And people use expressions like "It's a good shock insecticide" or "It's a very effective shock acaricide". These expressions mean that the product is fast-acting, effective or even spectacularly effective against a given pest, but its persistence and posterior action are much lower, as these products decompose very quickly in contact with the soil or the plant.

Depending on their form of nutrition, plant pests may be sensitive to totally different classes of products. For insects with chewing mouthparts, an insecticide sprayed on the leaves is usually enough. However, for insects with piercing type mouthparts, systemic insecticides usually give better results, because they enter the plant cells, are transported in the sap, and are sooner or later absorbed by the insect. Insecticides and acaricides can be classified into chemical families, or to put it another way, they can be divided on the basis of their chemical characteristics. A description of all the pesticides to kill insects and mites on the market would be beyond the scope of this work, and so we have made a selection of the most representative pesticides of each type, either because they are well-known products of proven effectiveness, or because they are recent developments and have given very encouraging results in tests.

• **Natural insecticides**. Natural insecticides were the first to be used, as they are derived from naturally-occurring substances or organisms. The best-known natural insecticides include **nicotine** and the natural pyrethrins. Nicotine, an alkaloid obtained from the tobacco plant, is a good shock insecticide against aphids and leaf-miners, though it also shows good action against many insects. It does not persist for long in the environment, and so it is not considered to disrupt the ecological balance. The flowers of some species of pyrethrum, mainly *Chrysanthemum cinerariaefolium*, contain insecticidal substances, the natural pyrethrins, that act on contact and cause very rapid

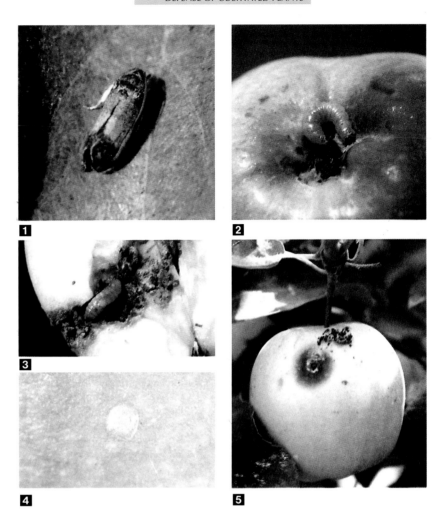

paralysis of the pest. These pyrethrins only last a very short time, as they decompose rapidly, both within the insect's body and on the treated plant.

• **Mineral oils**. The most common mineral oils are **petrol oils** or **mineral oils**, **anthracene oils**, **dinitrocresol** (DNOC), **dinitrobutylphenol**, **yellow oils**, and other petroleum oils formulated with organophosphate insecticides. These oils are derived from the fractional distillation of crude oil. Two basic types of oils are prepared for the market, **winter oils** and **summer oils**. **Winter oils** are thicker and contain more unsaturated hydrocarbons than **summer oils**, which are less viscous than winter oils and contain a greater proportion of saturated hydrocarbons. The winter oils (**yellow oils**) used in fruit cultivation can only be applied during the fruit tree's winter rest to kill the overwintering stages of insects and acarid mites, because they are highly toxic to growing plants. The summer oils (**white oils**) are applied in spring, and are very effective against mites and the larvae of many insects. Both winter and summer oils act by forming a film on the tree's bark, killing the eggs and overwintering forms of mites and insects by asphyxiating them.

Monocrotophos is an organophosphate insecticide that is very effective against the codling moth (Cydia pomonella) which attacks apples and pears.
1/ Adult
2/ Caterpillar on fruit
3/ Caterpillar within fruit
4/ Egg
5/ Exit hole
(Photos courtesy of the Department of Agriculture, Stockraising and Fisheries of the Generalitat of Catalonia).

DNOC is a very effective herbicide but is highly toxic to human beings. It is also used for chemical thinning of flowers. Mineral oils can be formulated with other insecticides such as organophosphates. In these combined formulations, the oil increases the effect and persistence of the insecticide that is added.

• **Synthetic organic insecticides and acaricides**. There are two large groups of synthetic insecticides: the **chlorates** and the **organophosphate**. The organophosphate group includes the **systemic insecticides**, which enter the plant, circulate in its sap and spread throughout the plant.

insecticide, like the other chlorate insecticides, but it is also toxic when ingested. It is also effective against the tarsonemid mite of strawberries and some eriophid mites. It has to be used at temperatures above 16°C to get the best results. This product is almost innocuous to bees.

Lindane is a polyvalent insecticide whose high stability confers great protection on the treated plant. It penetrates through the insect's cuticle very easily and acts by contact and when eaten. It is also used in soil treatments against soil white grubs (*Phyllopertha horticola*) and wire worms (*Agriotes spp.*). Note that lindane is

Cabbage leaves devastated by the larvae of the cabbage white butterfly (Pieris brassicae). A synthetic pyrethrin like cipermethrin will control this lepidopteran. (Courtesy of SHELL)

Apple leaf attacked by red spider mite. (Courtesy of SHELL)

• **Chlorate products**. These insecticides were synthesized long before the phosphates. Each molecule contains several atoms of chlorine, and though they are less toxic than the phosphates, they are more persistent in the soil. In many countries their use has greatly decreased because of their persistence in the soil and the plant, and their tendency to accumulate in the adipose tissue of vertebrates. The main active ingredients are endosulfan and lindane. **Endosulfan** is a relatively volatile sulfur chlorate compound, but less volatile than lindane. It acts mainly as a contact

so persistent in the soil that root vegetables, such as carrots, should not be planted for three years after application. Its high toxicity to bees means it can not be recommended as an insecticide.

• **Organophosphates**. **Organophosphates** is the name for the derivatives of phosphoric acid, thiophosphoric acid or dithiophosphoric acid. They act by blocking the insect's nerve impulses, by acting on cholinesterase, a substance playing a key role in the insect's nervous system. It is also highly poisonous to

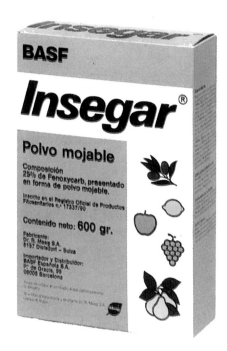

INSEGAR is an insecticidal product manufactured by BASF, S.A., and is a growth inhibitor, GrI. Its action prevents the larvae from reaching the adult stage by preventing their metamorphosis.

also effective as a contact insecticide and by inhalation. This is especially useful against aphids, as these insects often shelter on the underside of the leaf, and are hard to reach with contact insecticides.

The product may remain active for some time within the plant before it breaks down; sometimes it may turn into breakdown products that are more effective than the original compound. Absorbtion is greater in old leaves than in young leaves, greater on the underside of the leaf than its top, and is greater in warm weather than cold. There are many active ingredients on the market that are generically known as organophosphates and more will surely appear in the future. The best known are probably **acephate**, **demethoate** and **monocrotophos**.

- **Acephate**. Acephate is a systemic organophosphate that mainly acts by ingestion. It is especially effective against lepidopteran caterpillars, and is very effective against *Adoxophyes reticulana* and other caterpillars. It is also very effective against some aphids, such as peach and hop greenfly (*Myzus sp.*), but it is not very effective against aphids (*Aphis spp.*). Acephate is highly toxic to bees.

- **Dimethoate**. This is not an extremely toxic insecticide, and it breaks down within the plant to produce breakdown products that are more toxic than the original product. It gives excellent results against sucking insects, *Hoplocampa*, codling moth, *Yponomeuta*, Mediterranean fruit fly (*Ceratitis capitata*) and the sugarbeet fly (*Pegomyia betae*). A safety period of six weeks must be left before harvesting.

- **Monocrotophos**. Its characteristics are similar to the acephate and dimethoate. It is a systemic organophosphate insecticide that is highly persistent. It is effective against the codling moth (*Cydia pomonella L.*) and peel caterpillars, though its use should be avoided in sensitive apple varieties, such as *Golden Delicious, Red Delicious, Ananas pippin* and *Miagold,* on which it has a toxic effect and causes red patches. It should not be used during the flowering period and when bees are active, as it is toxic to bees.

• **Carbamates**. They form a separate group of highly effective insecticides that includes **aldicarb, carbaryl, carbofurane, diazocarb, ethiophencarb** and **methomyl.**

- **Aldicarb**. This insecticide is effective against insects, mites and nematodes, and is usually applied as granules. When applied to the soil, it is absorbed well by the roots, and is considered much more toxic than parathion.

- **Ethiophencarb**. Ethiophencarb belongs to the carbamate group. It is a systemic insecticide that is very effective against aphids, though it is not very effective against the woolly aphid of apples (*Eriosoma lanigera Hausm.*). It is systemic, acting by ingestion, contact and inhalation, and its toxicity to wildlife is relatively low.

warm-blooded animals, including human beings, acting in the same way as on insects. Some of the many active ingredients in this group are: **methyl azinphos, chlorphenvinphos, chlorpiriphos, diazinon, dichlorvos,** or DDPV, **phenitrothion, phonophos, malathion, parathion, methylparathion,** etc.

- **Chlorfenvos**. This organophosphate acts by contact and by ingestion. It is applied as a powder or as a granulate to the soil, and is highly effective at controlling the cabbage fly (*Hylemya brassicae*), the Colorado potato beetle (*Leptinotarsa decemlineata*) and the citrus scale (*Pseudococcus citri*).

- **Dichlorvos**, also known as DDPV. This insecticide was originally used to control the housefly, but is now widely used against many pests of fruit and vegetables.

- **Parathion**. This was one of the first organophosphate insecticides. It is almost insoluble in water and persists for several days on treated plants. It acts against insects by contact, and has some degree of systemic action, as it enters the cells of the plant tissues. Treatment should be carried out at a temperature of more than 15°C, and parathion should be used with caution when there is also a problem with the red spider mite, as this insecticide causes an imbalance that favours the proliferation of the mite. Great care must be taken when applying parathion as it tends to accumulate in the fats of warm-blooded vertebrates, including human beings.

• **Systemic organophosphates**. This groups shows the greatest systemic action as they dissolve in the sap and are translocated throughout the plant in the woody vessels and sapwood. This has obvious advantages when treating sucking insects, such as aphids and psyllids, because the product is most effective when ingested by the pest, though it is

- **Methomyl.** This was first synthesized in laboratories in America, and is a contact insecticide with some systemic action, and shows some effectiveness against nematodes. It is used against caterpillars, aphids, vine pyralid moth (*Sparganothis pilleriana Schiff.*) and *Adoxophyes*. It is very toxic and it is rapidly broken down within the plant.

• **Acaricides.** Several of the products cited as insecticides show good activity against mites, though resistant strains of mite have evolved due to the use of organophosphates and this has forced the chemical industry to synthesize new ones. Strains resistant to some of these new products have already evolved. It is impossible to give general guidelines for the application of acaricides, as they depend on the species of mite (especially the red spider mites), on the geographical area and the plant species to be treated. The main active ingredients on sale include **amitraz**, **bromopropylate**, **cihexastan**, **dicophol**, **fenbutestan**, **propargyte** and **tetradiphon**.

- **Bromopropylate.** This acts basically by contact, though it is more effective by inhalation. It is not very effective against mite eggs, but gives good results against larvae and adults. It has a slow but longlasting effect, and is effective against red spider mites and eriophid mites in fruit cultivation and viticulture. It partially respects the bee flora, shows little toxicity against terrestrial fauna and stays on the treated tissues for a long time.

- **Cihexastan.** This acaricide is effective against resistant yellow and red spider mites. It shows good persistence, though it does not kill the eggs; it is not dangerous to the useful fauna. It order to get round the fact that it does not kill the eggs it is sold mixed with dicophol or with tetradiphon.

- **Tetradiphon.** This is effective against mite eggs and larvae when applied in summer, and also sterilizes the adults without harming the useful fauna. Unusually, it is slightly absorbed by the leaf parenchyma, meaning it has some effect against mites on the underside of the leaf.

• **Various.** This group includes three active ingredients that are so important they can not be left out, but do not fit in the above groups. The first is **methyl bromide** sold mixed with **chloropicrin**. This soil disinfectant is discussed in detail in the section on soils, and is a good nematicide as well as insecticide. Soil disinfection with methyl bromide kills all insects at all stages.

- **Diflubenzon.** This insecticide acts by ingestion, acting on the metabolic pathways leading to chitin synthesis, thus disturbing the successive molts of the larvae. It has no effect on the

Cotton scale insect

eggs or adults, only killing the larvae. This and other similar products are called **growth inhibitors**, or **Grl**. They are not specific and affect all the insects they reach. In silviculture it is used mainly against the codling moth, leaf miners and the pear Psylla (*Psylla piri L.*).

- **Pyrethroids**. The pyrethroids are a group of active ingredients that are becoming more and more abundant. They are chemical modifications of the natural pyrethrins mentioned above, and show the advantages over the natural forms of being more persistent and not decomposing as easily. They act by contact and are highly polyvalent, killing most of the insects they contact. They have the advantage of not being toxic to human beings or terrestrial wildlife, being effective as a shock treatment and having no residual effects. They have the disadvantage of being highly toxic to aquatic wildlife, and it is thought that it favors the proliferation of mites, especially the red spider mite. Synthetic pyrethrins are usually divided into first generation and second generation depending on when they were first manufactured. The first generation includes the active ingredients cipermethrin, fenvalerate and permethrin, while the second generation includes alpha-cipermethrin, deltamethrin and sphenvalinate.

6.1.3.7. Products against vertebrates

The products used to prevent damage by vertebrates can be divided into two groups: those used to kill or repel birds and those used against rodents.

• **Birds**. There is a general tendency to use methods that repel birds or scare them away rather than methods that kill them, because the products that kill birds do not distinguish between harmful species and beneficial ones. Remember that many species of bird are protected by regulations, national laws and international treaties because of their value. Many preventive methods get good results, such as mechanical barriers to protect crops, scarecrows, shiny metal sequins, the silhouette of predator species, plastic strips over the crop, etc. Electrical acoustic devices have been used to scare crows and starlings. There are no chemical products that can be used to kill birds. In the past, very lethal products were used, but they did not distinguish between harmful species and beneficial ones. These products are now totally banned by the legislation of the European Union (E.U.) and many countries. Repellent products are only permitted for use with seeds before sowing, so that they are unattractive to seed-eating birds. One example is **anthraquinone** which repels crows, which is sold to repel birds in general, and members of the crow family in particular. It is normally formulated as a powder,

Device to scare birds and animals

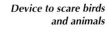

Preparation and placing of STORM® ratkiller baits, whose active ingredient is fluobenzuron. This ratkiller is distributed by SHELL.

which can be added to the seeds together with the disinfectants when dry. If a wet disinfectant is used, it is applied first, the seed is dried, and the product with anthraquinone is then applied.

• **Mammals**. As already mentioned, moles cause the most damage to crops, while other species such as red deer, roe deer and badgers cause significant damage in agriculture, but their ecological importance and value as game animals mean that their populations should be kept at a level compatible with agricultural activities. The fight against moles, voles and other rodents presents a series of problems due to their way of life. Their populations fluctuate, it is hard to find products that do not poison the entire ground fauna and the rest of the food chain, and their customs and underground way of life all make it difficult to eliminate them, or to keep their levels down to levels compatible with agriculture.
The methods available can be divided into three groups: **fumigants**, **poisons** and **traps**. The poisonous chemicals used to prepare traps and fumigant mixtures against rodents are technically known as **rodenticicides**. The rodenticides on the market may be inorganic, such as **zinc phosphide**, or organic, such as **coumarin** derivatives. They are anticoagulants, and are frequently formulated as powder, paraffin-impregnated blocks, blocks of colored fat, etc. Other products include **coumachlorine**,

The photo shows the use of a sex pheromone to confuse the oriental fruit moth (Grapholitha molesta) which attacks the peach. In this case, the strategy is to release so much pheromone into the air that the males can not find the females, thus preventing mating.
(Photo courtesy of BASF, S.A.)

diphacinone and **warfarin**, which are the base of most of the ratkillers sold.

6.2. BIOTIC METHODS

After looking at the abiotic curative methods, such as the effect of the weather, the physical or mechanical methods, and the chemical products, we shall now turn our attention to the biotic methods available. In one way or another, these use living organisms to control parasites of plants. Biotic methods can be divided into two large groups, those using the positive or negative forms of **taxis** described in the fourth chapter on the description of insects, and the methods using the specific **natural predators** of certain phytophagous insects.

6.2.1. Taxis

Insect behavior, and that of animals in general, consists of responses to external stimuli such as light, temperature, gravity, chemicals, etc., and all these stimuli can be used to influence the behavior of plant pests and change it in a way favorable to agriculture. Light traps, which make use of the fact that many insects that fly in the evening or at night are attracted to lights, are very effective and provide very useful information about the dates of appearance of some species, such as the codling moth. The behavior of diurnal insects can be influenced by colors. Yellow attracts greenflies and others pests of rapeseed, and this means you can catch them and see how they are developing, and thus decide on the best moment for treatment.

The use of the different types of taxis and their success in combatting plant pests was very limited until the discovery of insect pheromones. **Pheromones** are volatile compounds emitted by insects to communicate with each other. Pheromones play many different roles, and may signal insects to **group together** or **disperse**, or they may act as alarms (signalling an immediate threat and causing them to flee or adopt defensive postures) or be used to **mark a trail** (normally towards a food source), but we are most interested in the **sexual pheromones**, which play a vital role in bringing the two sexes together.

The **sex pheromones** are compounds released by female insects to attract males. These pheromones are **sexual hormones**, and once they were synthesized in the laboratory it became possible to make traps to attract the males and catch them with a sticky substance. This operation has many benefits; it catches lots of males, meaning they can not mate; the males are sexually "confused" when large amounts of the pest's pheromones are released,

and they can not locate the females and mate. Pheromones can also be used to attract insects away from crops towards uncultivated areas, by placing the traps well away from the crops.
So far, however, the most important use of this method is that it is possible to count, using simple sampling techniques, how many of any particular insect are present. Knowing if an insect is present in an area, and roughly how many, makes it possible to draw up its "flight curve", the change in its abundance in the ecosystem over time. Some government bodies supply farmers with these "flight curves", which can be used to determine the **optimum moment** for chemical treatment.

Pheromone trap to catch adults of the grape moth (Lobesia botrana) (Photo courtesy of DOW ELANCO)

List of the sexual pheromones on the market (Courtesy of KENOGARD)

Product	Applications
ADOXAMONE	*Adoxophyes (capua) reticulana F.R.* (apple, pear, etc., peel grub)
ANAMONE	*Anarsia lineatella Zell* (peach twig borer moth)
AGROTIS IPSILON	(black cutworm)
AONIDELLA	*Aonidiella auranti* (California red scale)
ARCHIPS PODANA	*Archips podana* (leafroller)
ARCHIPS ROSANUS	(filbert leafroller)
CERATITIS	*Ceratitis capitata Weidman* (Mediterranean fruit fly))
CLYSIA	*Clysia ambiguella* (grape bud moth)
CEW -I-	*Heliothis armiguera Boddie* (tomato budworm)
CODLEMONE	*Laspeyresia pomonella* (codling moth)
EVETRIA	*Evetria (Phyacionia) buoliana* (pine apical bud tortrix)
FUNEMONE	*Lapeyresia (Grapholita) funebrana Tr.* (plum codling moth)
GOSSYPMONE	*Pectinophora gossypiella Saund* (pink bollworm)
DISPARMONE	*Lymantria dispar* (gypsy moth)
GRAPEMONE	*Lobesia botrana Schift* (grape moth)
LITHOCOLLETIS	*Lithocolletis blancardella* (Spotted tentiform leafminer of apple and pear)
ORFAMONE	*Laspeyresia (Grapholita) molesta Buck* (eastern peach moth)
OSTRAMONE	*Ostrinia (Pyrausta) nubilalis Hb.* (European corn borer)
PANDEMIS LIMITATA	(threelined leafroller)
PHTHORIMAEA	*Phthorimaea operculella* (potato tuberworm)
PLUSIA	*Plusia gamma L.* (garden noctuid moth)
PRAYS	*Prays oleae F.* (olive prays)
PRONUMONE	*Tortrix (Cacoecia) pronubana H.* (carnation tortrix)
QUADRASPIDIOTUS	*Quadraspidiotus* (San José scale)
RHAGOLETIS	*Rhagoletis pomonella* (apple maggot) *Rhagoletis cingulata (R. cerasi)* (cherry fruitfly)

Encarsia formosa This parasitic wasp lays its eggs within the larvae of whitefly.

Phytoseiulus persimilis This predatory mite feeds on the adults and eggs of the red spider mite.

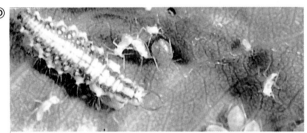

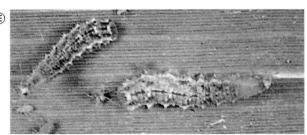

A/ SPIDEX-PLUS® is a registered trademark of KOPPERT. It consists of vermiculite mixed with the predatory mite Phytoseiulus persimilis, *and is used in the biological control of the red spider mite.*
B/ The larva of the ladybug is a very effective biological control agent, as it eats lots of aphids and mites. (Courtesy of RHÔNE-POULENC)
C/ The adult ladybug is a great consumer of greenflies. Each adult, on average, eats about 9,000 greenflies. (Photo courtesy of RHÔNE-POULENC)

This optimum period or moment is usually when the insect's population is at its maximum or the stage when the insect is most sensitive to insecticides.

6.2.2. Biological control

Like all living things, crop pests and diseases have their own parasites and predators. **Biological control** consists of using the **natural predators** and **parasites** of crop pests in order to reduce the number of chemical treatments used. It is now accepted that the massive use of pesticides, especially persistent ones, may disturb the ecological balance due to the long-term pollution of the ecosystem. Biological control is being investigated by many experts on ecology who disagree with the massive use of pesticides.

Biological control seeks to destroy harmful organisms by using their **enemies**, and is based on the use of the many helpful organisms that exist. This idea might seem simple and easy to put into practice, but in

fact there are many difficulties due to the complex and subtle interactions between species.

The organisms attacking the pests of crops may be animals, bacteria, fungi or viruses. Pathogenic microorganisms such as virus, bacteria and some fungi cause disease or death in many crop pests. *Bacillus thuringiensis Berliner* causes paralysis in the larvae of many lepidopterans, leaving them incapable of feeding. Most of the **parasites** and **predators** of crop pests are insects or mites. One of the best known predators is the ladybug (*Coccinella septempunctata*), which eats many greenflies, in both its larval stage and the adult stage. Parasites of insects include **entomophagous** parasites, which lay their eggs within the host insect and whose larvae eat the host from within, feeding on it until they reach the adult stage, when they emerge from the host, causing it to die. There are also many vertebrates that are predators of insects, including birds, bats, lizards, toads, etc. Birds of prey are very useful for agriculture because they eat many small rodents. Biological control has certain clear disadvantages for the farmer. First, the parasites and predators of crop pests have to be identified and then encouraged by

avoiding the use of broad-spectrum pesticides, which indiscriminately kill crop pests and their predators, and finally the crop must be inspected regularly for pests and their number assessed, as must the population of their predators. Below is a list of the parasites and predators of some of the insects and mites that attack plants. Biological control methods have already shown some encouraging successes, but it will be some time before biological control is totally effective.

(From previous page)
D/ Lacewing larva. This lacewing (Order Neuroptera) eats large numbers of aphids and mites. (Photo courtesy of RHÔNE-POULENC) E/ The larvae of hoverflies (Order Diptera) are very useful in biological control because they eat many mites. (Photo courtesy of RHÔNE-POULENC) F/ Bug larvae are among the most polyphagous of all insects. They eat the eggs and nymphs of Psylla and of mites. (Courtesy of RHÔNE-POULENC)

Pests		Antagonistas		Plant attacket by pest
Icerya purchasi Mask.	Cottony cushion scale	*Rodolia Cardinalis* Muls.	Coccinelid beetle	Oranges, lemons and ornamental plants
Eriosoma lanigerum Hausm.	Woolly apple aphid	*Aphelinus mali* Hald.	Chalcid wasp	Fruit trees
Grapholitha molesta Busck.	Oriental fruit moth	*Macrocentrus ancylivorus* Roh.	Braconid wasp	Peach
Archips rosanus L.	Leafroller	*Trichogramma cacoeciae* March.	Chalcid wasp	Fruit trees
Panonychus ulmi Koch.	Red spider mite	*Typhlodromus pyri* Scheuten	Phytoseid mite	Fruit trees, vine
Tetranychus urticae Koch.	Yellow spider mite	*Phytoseiulus persimilis* Athias-Henriot	Phytoseid mite	Greenhouse plants
Quadraspidiotus perniciosus Comst.	San José scale	*Prospaltella perniciosi* Tow.	Chalcid wasp	Fruit trees
Pseudaulacaspis pentagona Targ.	White peach scale	*Prospaltella berlesi* How.	Chalcid wasp	Mulberries, peaches
Several lepidopteran		*Bacillus thuringiensis* Berl.	Bacteria	Fruit trees

List of effective biological control agents (According to Bovey)

Bacillus thuringiensis is a bacteria that is very effective biological agent against the processionary caterpillar, a pest of pines.

In integrated control, all the different methods must be used: cultivation techniques, agricultural chemicals, etc.

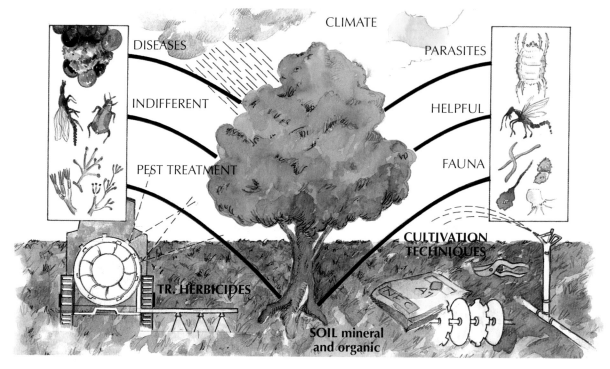

CLIMATE

DISEASES

PARASITES

INDIFFERENT

HELPFUL

PEST TREATMENT

FAUNA

CULTIVATION TECHNIQUES

TR. HERBICIDES

SOIL mineral and organic

Biological control (Courtesy of KOPPERT B.V.)

A/ Aphidius colemani. *This parasitic wasp can detect greenfly even when their population density is very low.*

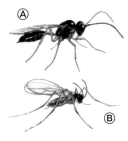

Ⓐ

Ⓑ

B/ Aphidoletes aphidomyza. *This insect lays its eggs in aphid colonies.*

C/ Orius laevigatus. *This is an excellent predator of the larvae and adults of different species of thrips.*

Ⓒ

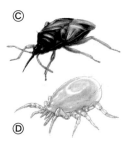

Ⓓ

D/ Amblyseius cucumeris. *Another good predator of thrips. It is a mite.*

6.3. INTEGRATED PEST CONTROL

It is now accepted that plant protection should be based on a more ecological approach that uses all the different methods available. Chemical methods provoke the appearance of resistant strains of crop pests and cause chemicals to accumulate in the soil and enter the food chain and reach human beings, and these disadvantages mean that their use, at least on a large scale, is now being questioned.

In general terms, you are recommended not to treat with chemicals until you have tried all the other easily performed methods. In the first place, try the **cultivation techniques** discussed in the previous chapter. As far as possible, you should cultivate **resistant species and varieties**. Once the crop is in the field, you should carry out regular **visual checks** in order to detect any pest in its early stages. You should also bear in mind the **climatic conditions**, as these may make many treatments unnecessary. You should also follow carefully all the information produced by the government dealing with the **flight curves** of pests measured using **pheromones**, which pests must be reported to the authorities, and the correct ways to use the recommended chemical products.

If there is information available, you should also check which pests may attack the crops you are growing, and which parasites and predators are available to deal with them. As soon as the first pests are detected, it is a good idea to treat the affected area with products that are not excessively toxic and do not leave residues. Finally, when there is no better option, treatment with chemicals should be carried out using the recommended doses and application methods, and choosing

pesticides that, as far as possible, do not kill off the useful insects.

Of course dealing with so many different parameters makes it difficult to decide what measures to take. These new integrated control techniques are based on the technical support of agricultural engineers working with a team of professional plant pathologists, chemists, toxicologists, physiologists, entomologists, ecologists, geneticists and meteorologists. All these professionals are needed to find the most suitable solution to a given problem.

6.4. PESTICIDE APPLICATION TECHNIQUES

Throughout this work, we have discussed the different formulation in which agricultural chemicals are sold (liquids, solids, granulates, powders) and the application methods suitable for these different formulations (dissolved in a mix, powdering, etc.). This range of possibilities is because of the different pests to be controlled, their stage of growth, the time of year, the phenological state of the plant to be protected, etc. Two questions should be borne in mind: the first is the type of machinery necessary to carry out the treatment, and the second is the precautions that must be taken when using the product, because even when using the most inoffensive products you must follow proper procedures in order to avoid accidents and poisoning yourself.

6.4.1. Characteristics of the devices

The most frequent case is a solid or liquid that has to be dissolved in water so it can be sprayed onto the affected plants. **Sprays** range from simple backpacks for treatments of small areas to hydraulic tanks powered by a tractor. There are even simpler versions of these, without a fan, in which the product is simply applied by pressure from a water pump, as well as more sophisticated versions, with fans that disperse the product a long distance from the tank, thus allowing the treatment of large areas in a short time and using little labor.

For products to be applied by **powdering**, there is special machinery that is simpler to use than the systems above, as the product does not need to be dissolved. They do however have some disadvantages, as they may be more toxic to the operator and less effective and persistent than liquid treatments. The only powders normally applied to growing plants are broad spectrum preventive fungicides, which normally show little or no toxicity.

In the large areas cultivated in the Americas it is quite common to use **helicopters** and

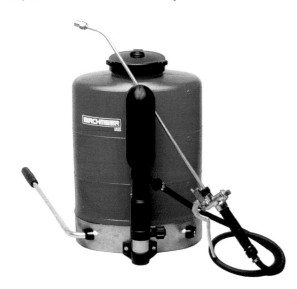

cropduster planes to apply treatment, which may be as liquids or powders. One of the advantages of this procedure is that it causes no mechanical damage to the crop or soil, as no machinery enters the field. The disadvantages of aerial treatment include the fact that it is hard to restrict the area treated. This method can only be used in large areas of crops, and traditional methods have to be used in small areas.

The products used to kill nematodes, **soil disinfectants**, were dealt with in the section on Soils of this work, together with the machinery needed and how to apply them. To recapitulate, gaseous products have to be applied under plastic sheeting, while solids or liquids can be applied at depth by suitable plowing techniques that are hydraulically activated using mechanical traction. It is the farmer who must chose which of the wide range of machinery available is most suitable. This choice should be made on the basis of the size of the area to be treated, the nature of the crop, the presence or absence of

associated crops and the site's relief. Finally, you should carry out a serious study of the economic viability of the machinery to be purchased, if you are starting with a new site or replacing the old machinery. It does not make much sense to buy a very powerful tractor - which is expensive to buy and maintain - if you are only working two or three hectares. One viable solution is to purchase a high quality machine as a cooperative with other farmers, as this means your investment will pay itself off.

Maintenance of the machinery is essential if it is to last. Agricultural machinery often represents a major expense for an agricultural exploitation and it should be maintained properly, in order to ensure its useful life is as long as possible. Throughout the growing season, the machinery used should be regularly cleaned. **Sprays**, **nozzles**, **tubing** and the **tank** should all be rinsed thoroughly several times with clean water after each use. It is also worth regularly lubricating the **pistons** and **joints**. When the growing season has finished, the machinery should be cleaned more thoroughly. All the leaves and branches, etc., that may have got caught in the in the fan, hydraulic arms, wheels, etc., should be removed. In addition, all the parts that normally come into contact with the products (nozzles, sprays, etc.) should be washed with soapy water, and the machinery should be removed to a dry covered site, if possible with the tires deflated.

Hydraulic tanks activated by mechanical traction make excellent application possible using just the pressure of the spray nozzles. (Photo courtesy of DOW ELANCO)

This backpack spray is especially suitable for use in small plots. It is produced by the BIRCHMEIER company.

Application of dichloropropene at depth to control nematodes (Heterodera sp. and Meloidogyne sp.) (Courtesy of SHELL)

6.4.2. Operator protection

Essential precautions to be taken when applying agricultural chemicals. Taken from the International Agricultural Chemicals Manufacturers Group

Below is a copy of a poster produced by the International Agricultural Chemicals Manufacturers Group which shows very clearly the essential precautions to take when applying agricultural chemicals. They are essentially common sense, but you should not underestimate how toxic some of these chemicals are, especially before they are diluted. Bear in mind that if a liter of product has to be diluted in 250 liters of water, then before dilution it is 250 times more toxic than when diluted. In other cases, the concentrated solution may be a thousand times more toxic than the mixture that is applied, and so you should be particularly careful when handling the concentrated products.

The safety measures are common sense: use gloves, appropriate clothing, eye protection when dealing with irritant products, use a gas mask when handling very volatile and toxic products, do not eat or smoke during the application, ensure the toxic substance never comes into contact with the skin (especially before dilution), and shower after applying the product, etc.

Take advice from a technician on pests and the chemical products to be used.

During transport, keep the products isolated from any other goods and all foodstuffs.
Agricultural chemicals must be stored separately from all other goods.

Agricultural chemicals must be stored in a secure place that is totally out of the reach of children.

Read the instructions on the container carefully and take advice before using an agricultural chemical.

When handling the product, follow the instructions on dosage and wear appropriate clothing.

Do not allow children or animals to approach.

Burn or bury the empty containers.

6.4.3. Storage of the products

Storage of agricultural chemicals should follow several basic and easily applied rules. A cool and dry site should be chosen exclusively for their storage. This should be a small room or storage space that other people, especially children, can not access and if possible it should be kept under lock and key. The products must be stored in the original container, and should never be stored in other containers and never, under any circumstances, in containers for food or drinks, such as bottles of wine, beer, soda water or lemonade. Most products have a lid that incorporates a measure to help in preparing mixtures. These measures are usually graduated and should only be used for this purpose. Agricultural chemicals can be stored for some time, but most only last for two or three years, and after this period their results are not as reliable. When a product has passed its expiration date (which must be stated on the label), or you have decided that it is not as effective as hoped, its contents should not just be dumped down a drain or into a river, stream, lake, etc. In agricultural areas, the government often provides special containers to deposit any leftover product that the farmer has decided to get rid of.

Never use the empty containers for any other purpose and never, never, to store water or food.

Do not spray agricultural chemicals into the wind.

Never use defective equipment or materials

Do not pollute the environment by inadequate use of agricultural chemicals.

After applying the product, wash your hands and face before drinking, eating or smoking.

Do not allow children to apply agricultural chemicals

If your clothing is contaminated with any product, change clothes and wash or shower immediately.

If anyone is poisoned, call a doctor immediately and show him or her the label.

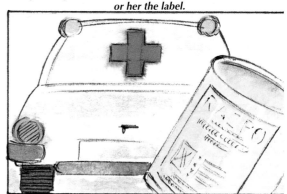

A/ A cornfield weed:the corn poppy (Papaver rhoeas) (Courtesy of SCHERING)
B/ A weed of cultivated ground: fat hen (Chenopodium album) (Courtesy of SCHERING)
C/ A weed: charlock (Sinapis arvensis L.) (Courtesy of SCHERING)
D/ A biennial weed: Queen Anne's lace (Daucus carota L.) (Courtesy of SCHERING)
E/ The Canadian thistle (Cirsium arvense Scop.) is a invasive perennial weed. (Courtesy of SCHERING)
F/ An example of a stoloniferous perennial weed: creeping buttercup (Ranunculus repens L.) (Courtesy of SCHERING)

7. WEED CONTROL

When we call a plant a "weed" we are expressing a human opinion, that it is a nuisance, because it is growing where it is not wanted. The term "weed" has no real botanical meaning, because botany is the science studying plants on the basis of their physiology and anatomy, and their classification into species, genera and families.

Farmers or agricultural technicians subjectively classify plants as "useful" or "weeds" depending on whether they are a problem to the crop being grown. Clearly the unwanted plants growing in a crop diminish the crop's growth, as they compete with it for space, light, soil nutrients, water, etc.

One of the reasons why the term "weed" is not very useful is that the same species of plant may be a crop or a weed depending on the circumstances. The cultivated broad bean can be a "weed" if a cereal is grown on the same site the following year. Within a single family, and even within a single genus, there may be both crops and weeds.

This is clearly shown by the oats, with cultivated oats (Avena sativa) and wild oats (Avena fatua). A single species of plant, such as buckwheat, may be considered a weed in one country or region and a crop in another, and buckwheat is considered a weed in warmer countries but cultivated in colder countries, such as Poland.
"Weeds" can be divided into invasive weeds, cornfield weeds, weeds of cultivated ground, etc. There are two main groups, the woody weeds and the herbaceous ones, which can be divided into annuals, biennials and perennials.

Annual weeds complete their life cycle within a single year. Common annual weeds include the corn poppy (Papaver rhoeas), fat hen (Chenopodium album), charlock (Sinapis arvensis), etc. These plants grow in crops like cereals, sugarbeet, flax, cotton, etc. They are fast-growing and short-lived. Some grow so fast that in fields of winter cereals in temperate zones, they even flower in autumn or very early in spring. Most germinate and flower when good weather arrives, yet manage to flower and disperse their seeds before harvest time.

Biennial plants complete their life cycle in two growing seasons. They may germinate in spring or fall, but they do not flower or seed until the following year. This group is not very large, but the most representative ones include the wild carrot, Queen Anne's lace (*Daucus sp.*), *Adonis sp.* and some thistles of the genera *Cirsium*, etc. Perennial herbs flower and fruit for several years in a row. They generally have dispersal mechanisms in addition to producing seeds. The most widespread perennial weeds include those that grow from rhizomes, such as the Canadian thistle (*Cirsium arvense*), Bermuda grass (*Cynodon dactylon*), Johnson grass (*Sorghum halepense*), sedges (*Cyperus rotundus*), etc. Stoloniferous plants produce stolons, long trailing shoots that grow underground or at ground level, sprout adventitious roots at the leaf nodes and produce a new plant. One example is the creeping buttercup (*Ranunculus repens*). Bulbous plants are perennial and produce a short shoot, often swollen or covered in scales, called a bulb (or corn), that can often reproduce vegetatively to produce new bulbs and new plants. Invasive bulbous plants include the wild garlic *Allium vineale* or the wood-sorrel (*Oxalis sp.*), the most widespread in temperate zones.

In general, the rhizome-producing and stoloniferous weeds are the hardest to eliminate, as they can reproduce vegetatively and thus spread faster. The highest numbers of perennials are found in meadows and fields under grass, though some ubiquitous species invade any crop.

7.1. DAMAGE CAUSED BY WEEDS

Weeds can cause large losses in agriculture. Some crops are very sensitive to their competition. Weeds can cause four main types of damage.

• **Weeds have a negative effect on crop growth**. Weeds may depress crop growth in two ways, by straightforward competition for the factors of production or due to some antagonistic effect. Weeds actively compete with crops for nutrients, water, light and air.

a) Nutrients. Weeds usually grow fast and strong, absorbing many of the nutrients applied to the soil, and especially the nitrates.

b) Water. Weeds also absorb a lot of water from the soil, and water is essential for crop productions. Water is the medium for the transport of all the mineral nutrients absorbed by the roots. Water is always circulating through the plant, and most is lost by transpiration. If less water is circulating, nutrient transport is slower, causing a decrease in photosynthesis and thus leading to smaller plants.

A/ Colchicum autumnale is a member of the lily family (Liliaceae) that is toxic to livestock. (Courtesy of SCHERING)
B/ The corn cockle (Agrostema ghitago L.) is dangerous to human health if its seeds are mixed in with those of cereals. (Courtesy of SCHERING)

Ⓐ

Ⓑ

Adequate control of weeds in cultivated ground prevents severe competition from invasive plants. (Courtesy of PROBELTE)

c) **Light**. Light plays an essential role in plant growth. The plant absorbs carbon dioxide from the air, and with the energy of sunlight, it photosynthesizes organic matter that is both the plant's source of energy and the means to build its bodily structure. So, as in the case of water, a decrease in light intensity leads to decreased vegetative growth. In extreme cases, when there are many weeds in a crop, this competition for light leads to

elongated stems that are fragile, thin and lack chlorophyll.

d) **Air**: air is also essential for plant growth, and plants may compete for carbon dioxide. Lack of carbon dioxide results in decreased vegetative growth.

e) **Allelopathy**. Allelopathy is when a weed (or other organism) releases some substance that inhibits the germination or growth of another plant. Germinating seeds and the remains of some plants (roots or buried aboveground parts) may have an inhibitory effect on other plants. This inhibition is because some plant tissues contain phytotoxic substances that are released into the soil by decomposition, or which are secreted from their root hairs, and disturb to a greater or lesser extent the growth of nearby plants. Flax and radishes are crops that suffer badly from allelopathy, while cereals are among the least affected.

• **Weeds and their seeds reduce a crop's value**. Some weeds are toxic and can cause poisoning, sometimes even death, when they are consumed in sufficient quantities by livestock. Plants that are highly toxic to livestock include *Colchicum autumnale*, *Ranunculus acris* and *Senecio jacobaea*. In processed products, such as flours and meals, some weeds can be toxic to livestock and humans, such as the corn cockle (*Agrostema ghitago*).

• **Weeds make plowing and harvesting more difficult**. If weeds are not regularly removed from the crop and surrounding area, many secondary problems may arise, such as problems when performing necessary tasks, problems when harvesting, etc. Weeds also tend to grow in irrigation and drainage channels, impeding the flow of water. When installations are abandoned for a long time, cleaning up these channels may be very expensive.

• **Weeds are hosts and shelters for insects and diseases**. A farmer may fulfil all the chemical treatment plans recommended by the technician, but they may not be effective, because insects and other pests living on the

Key to classifying weeds ("Herbicides and their use" by L.D. & J.G.)

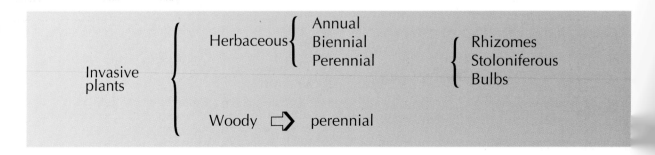

weeds at the edges of the fields, paths and neighboring fields can rapidly reinfest the treated field. A frequent case that often reaches crisis proportions is when a farmer's plot is next to abandoned fields full of the remains of previous crops and especially of weeds. Using agricultural chemicals to resolve a problem of this type is almost certain to fail. Farmers often eliminate the weeds from neighboring abandoned fields in order to protect their own crops.

Species	Seeds / plant
Poppy	50,000-60,000
Matricaria	45,000
Cirsium arvense	20,000
Plantago lanceolata	2,500-13,000
Queen Anne's lace	1,200-11,000
Chrysanthemum	1,500-25,000
Groundsel	3,000-20,000
Tussilago	5,500
Raphanus	4,500
Capsella bursa-pastoris	4,500
Sinapis	1,200-4,000
Galium	1,100
Stellaria media	500

7.2. SPREAD

The spread of weeds in crops is extraordinarily fast. Weeds produce many seeds and propagate so fast it is tremendously difficult to control them. There are two main reasons for this rapid propagation. Weeds produce large numbers of seeds. You need only look at the table above showing the number of seeds produced per plant to see they produce huge numbers of seeds within a year; a single poppy, for example, can produce 50,000 to 60,000 seeds within a year. In natural conditions, most of these seeds would die, but plowing the soil means that many can germinate and produce new plants.
Weeds spread so fast due to their high seed production and to the fact that their seeds do not all germinate at once, but in batches over a long period of time, and they are also very resistant to the methods available to destroy them. The relative impermeability of the seeds allows them to survive prolonged desiccation, and to survive burial at depth with little air. This impermeability renders the seeds dormant and they cannot germinate even when they are in optimal conditions of moisture and heat. Some seeds retain their ability to germinate for a long time, and some can survive several decades. Some seeds show poor germination in the year they are produced, germinating best after 3 to 5 years. As a result, every soil contains large numbers

of all sorts of seeds. Kosmo found that there may be between 1,700 and 34,000 seeds in the top 25 cm of soil of a square meter of ground.

7.3. CONTROL METHODS

Methods, usually mechanical ones, were developed in antiquity to control weeds growing in crops. There are many manual tools to get rid of weeds, such as the hoe, grub hoe, small weeding hoes, etc., though they are rather laborious. Manual weeding of this type may be the best solution in very small plots, but is quite unsuitable for crops of any size.

At the beginning of this century, after the 19th century industrial revolution, the first chemical products to kill weeds came into use. These products are generally known as herbicides or weedkillers, and a herbicide is any plant that kills plants. Chemical herbicides are now so important that current agricultural production would decrease sharply if they did not exist. Their importance has led us to devote the entire next chapter to them.

There are other methods of getting rid of weeds, such as mechanical devices to remove them with motor traction (see photo), which

The average number of seeds produced by a plant during its life (annual or biennial). (According to Berhault & Long)

The GARD company builds agricultural machinery, including a wide range of weeding machinery.

makes it possible to remove the weeds from a large area in a short time at a low cost. Hydraulic devices powered by a tractor, such as drills, disc harrows, etc., are all examples of this type of machinery. The machinery turns over the topsoil, burying the weeds, and so this method can only be carried out on suitable crops, such as fruit trees (between the rows), but it is not appropriate for fields of cereals.

Many technicians prefer these methods, when applicable, to chemical herbicides, because they have the advantage of incorporating organic matter into the soil and mean it is not necessary to apply chemical products to the soil, with the consequent problems of environmental pollution, especially when dealing with herbicides leaving residues in the soil that are highly persistent.

Other methods of eliminating weeds are also very important, such as laying plastic (often black) over the crop with holes for the crop plants to grow. The black plastic prevents the weeds germinating, as it blocks the light and they can not photosynthesize. This is usually done in small sites, such as greenhouses. The method is expensive, but it saves costs on weeding in the long term. In Spain it is commonly used in strawberry crops.

The next section deals in more detail with the different types of herbicidal products. The farmer must decide on the most appropriate method of removing weeds on the basis of the type of crop, the season of the year, the type of weeds to be eliminated and the abundance of each species, the type of herbicide available, the surface to be cleared, etc. Whenever possible use mechanical weeding methods, and only use chemical herbicides when really necessary.

7.4. USE OF HERBICIDES

The use of chemicals to get rid of weeds is becoming more and more common. This is partly because labor is expensive and may be in short supply, and is partly due to constant progress in organic chemistry and the development of new herbicides.

7.4.1. Classification

We can divide herbicides three different ways, by the objective, by the product's method of action and by the time of application.

• **Classification by objective**

Nonselective herbicides kill all plants they come into contact with. A **selective** herbicide is a product that kills specific species of weeds, causing little or no damage to crops. Herbicides may be **physically** or **physiologically** selective. Physical selective action is when the product's penetration into the plant depends on the plant's anatomical features (for example, leaves covered with a cuticle impermeable to the product). Physiological selective action is when the absorbed product affects different species of plant in different ways; some may react strongly, whilst others are unaffected by the product. Selective herbicides may be **soil herbicides** or **foliage-applied**.

There are several types of selective herbicides in the marketplace. Selective herbicides may act against dicotyledons (dicots,) or against monocotyledons (monocots). Herbicides with a selective action against dicot weeds do not harm monocots (at the correct dose). 2,4-D is a well-known example of an herbicide that can be used in cereal crops, because it eliminates dicots but does not harm monocots (such as cereals). There are herbicides with the opposite selective action, killing monocots without harming dicots, and these translocation herbicides or plant hormone herbicides are now widely used in horticulture.

These definitions are not absolutely precise, and should not be taken too rigorously. A nonselective herbicide may be selective at lower doses (e.g. Monuron); and to the contrary, a selective herbicide may become a total herbicide if the normal dose is exceeded.

Diagram showing different application methods and modes of action of herbicides.

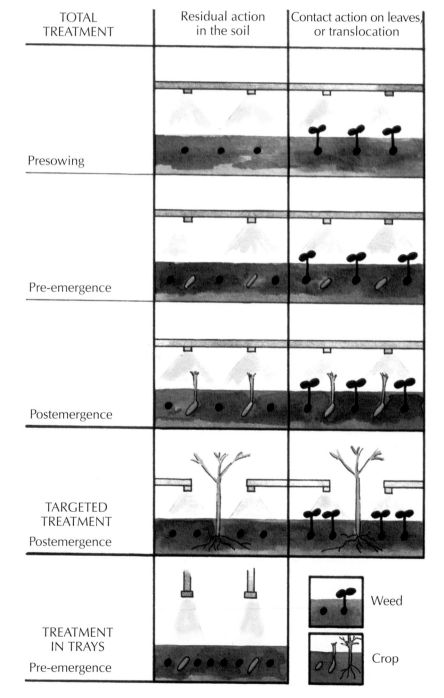

TOTAL TREATMENT	Residual action in the soil	Contact action on leaves, or translocation
Presowing		
Pre-emergence		
Postemergence		
TARGETED TREATMENT Postemergence		
TREATMENT IN TRAYS Pre-emergence		Weed / Crop

A plant is never totally resistant to a selective product if the limits indicated on the label are exceeded. Once again, we would like to point out just how important it is to respect as precisely as possible the doses recommended on the product label. This is even more important when dealing with selective herbicides, because if you apply too high a dose you may cause irreparable damage to a crop.

• **Different methods of action**

a) A **contact herbicide** acts when it comes into contact with plants or any part of a plant, killing the area affected.

b) A **translocation herbicide**, or **systemic herbicide**, or internal action herbicide, is a product that is absorbed by the part of the plant that is treated and then exerts its toxic effect on another part of the plant. They are also known as hormone weedkillers, or plant hormone weedkillers, and are very similar to the natural plant hormones; at higher doses they act as herbicides, killing the plant.

• **Different times of application**

a) A **presowing** or **preplanting** herbicide is applied after preparing the soil but before sowing the seeds.

b) A **pre-emergence** herbicide is applied after sowing the crop seeds but before they germinate.
These two classes are subdivided into **contact** herbicides and **residual** ones.
Contact herbicides are products that kill the weeds they fall on, but their toxic action does not last for a long time, as they decompose into harmless substances or evaporate.
Residual herbicides are products that remain in the soil for long enough to continue killing the weeds the moment that they germinate; these products are not toxic to crop plants or they break down into harmless substances before the crop germinates.

The accompanying drawing shows the moment to apply and how each herbicide works. Before sowing the crop, the farmer can choose a residual preplanting herbicide to eliminate the weeds as soon as they germinate, or a contact preplanting herbicide that will eliminate the weeds when they have germinated. During the pre-emergence period residual or contact herbicides can also be used, and they may even be mixed together, as the crop has not germinated and will not be affected.
Pre-emergence and contact products are usually applied shortly before the crop plant germinates, in order to kill as many of the germinated weeds as possible, whilst the pre-emergence products with a residual action

Invasive weed: Capsella bursa-pastoris (Courtesy of SCHERING)

are applied before this. The timing, dosage and application method should all be described in detail on the product's label.

c) Post-emergence herbicides are applied after weeds and crops have germinated.
It is important to know which category the product you want to use belongs to, as this detail alone gives a general idea of the right conditions and way to use the product. Thus a contact selective herbicide for post-emergence treatment should be applied to very young weed seedlings; a herbicide absorbed by roots should be distributed in a large quantity of water so that it will reach the roots rapidly, and should also be applied when the plant is growing actively, when root absorption and the rising of the sap are very active.

7.4.2. Main herbicides

As we have already pointed out, this chapter does not aim to discuss all the chemical herbicides in the marketplace. New synthetic chemical products are launched every year in the hope of replacing long-established products. The manufacturers are sometimes successful, but in many cases the new herbicides are only slightly different from the previous ones. The following is a list of the classic herbicides, their characteristics, how they act, their persistence in the soil, etc.

The farmer or technician has to adapt to the new products that appear on the market. We have a few simple general recommendations. You should always select products that are properly labelled, with the manufacturer's name, the method, dose and time of year when the product should be applied, and you should check that the label indicates clearly which crops it can be used on and which weeds it will eliminate. We also recommend you carry out a small trial before applying the product to the whole crop, especially in the case of selective or plant hormone weedkillers, in order to check they have the desired effect. This type of test should be carried out on a small area of the crop, and they are a very good idea as unknowable factors, such as temperature, wind, etc., may influence the product's effectiveness.

Weedkillers are usually classified on the basis of their chemical family. In order to consult the properties of a particular active ingredient, you need only refer to the section dealing with all the members of its family, to find a description of the product, its method of action, and its selective action. Remember that not all members of the same chemical family have exactly the same characteristics. If we are looking for information on a particular active ingredient and it is not in the list, you should look in more specialized reference works or the manufacturer's technical details. The entries in the following list are ordered as follows: the chemical name comes first, and is followed by the active ingredients forming the family.

Cornfield weed:
Fumitory
Fumaria officinalis L.
(Courtesy of
SCHERING)

7.4.2.1. Hormone weedkillers

• **Chemical name 2,4-D. 2,4 dichlorophenoxy acetic acid.**

Active ingredients: **2,4-D salts**

This is a hormone weedkiller that is absorbed through the leaves. It is very effective against dicot weeds and selectively favors monocots. While it can be absorbed by the roots, it is usually applied postemergence to the leaves. This herbicide is translocated and causes growth disorders and increased respiration, which exhaust the plant's reserves. It also acts secondarily on the uptake and metabolism of nitrogen, phosphorus and potassium (N, P, K), by inhibiting the growth of the shoots and favoring the appearance of tumors. It shows physiological selectivity in favor of grasses; lower doses are used on cereals, while higher doses are used on grazing land (or lawns) or for special applications. 2,4-D should be applied when the temperature is greater than 12°C, and should not be applied on windy or overcast days, or when it looks like it will rain. 2,4-D should not be used near dicot crops, as they are very sensitive to it.

Active ingredient: **2,4-D esters**

The product, its method of action and selectivity are all very similar to those of 2,4-D, except that the same precautions as for 2,4-D salts should be observed, but with even greater care, because the lighter (ethyl, propyl, etc.) esters are highly volatile.

• **Chemical name 2-methyl-4-chloro-phenoxyacetic acid.**

Active ingredient: **MCPA**

The product, its method of action and its selectivity are all very similar to those of 2,4-D. Like 2,4-D, it is absorbed through the leaves and by the roots, though its action on some species may be greater or lesser than that of 2,4-D. The product is not as affected as 2,4-D by washing by the rain, and its action is slower and longer-lasting.

• **Chemical name: 2-4-5-trichlorophenoxy acetic acid.**

Active ingredient: **2,4,5-T**

2,4,5-T is less toxic to plants than 2,4-D, but it is more effective against woody plants. It is less volatile than 2,4-D, and like all the plant hormone weedkillers it is translocated throughout the plant. It is mainly absorbed through the leaves, it can be applied as a powder or mixed with water or gas-oil for localized treatment in uncultivated ground to kill woody plants. It is often used mixed with 2,4-D, as a selective weedkiller acting against dicot species, and you are recommended to carry out tests in advance to assess its effect on the monocot crop species.

• **Chemical name: 2-(2 methyl-4-chlorophenoxy acid)**

Active ingredient: **MECOPROP (MCPP)**

This acts in the same way as the previous weedkillers, causing increased and disordered growth. It shows good selectivity in favor of cereals and against dicot weeds, even in spring. It is absorbed through the leaves and is translocated within the plant to the apical meristems. This agent is especially suitable for controlling *Stellaria, Convolvulus, Galium* and *Polygonum*. The best time to apply is when the weeds are showing peak growth.

• **Chemical name: 2,4-dichlorophenoxypropionic acid**

Active ingredient: **DICHLORPROP**

Like 2,4-D, etc., this is a systemic herbicide, one that acts internally. It is mainly absorbed through the leaves and acts by translocation towards the apical meristems. Its properties are very similar to MECOPROP, and it is even more effective against *Polygonum* and other weeds that are hard to kill using other conventional weedkillers. Like the preceding weedkillers, the results are best if applied when the weeds are actively growing. It should not be applied on very windy days, as plants in surrounding fields may be very sensitive: these include vegetables of all types, potato, cotton, Windsor beans, fruit trees and tobacco.

There are other systemic herbicides on the market that are mixtures of those already discussed, and there are other brand new chemical formulations, which are not discussed here. To find out more about how to use them,

the dose, application, etc., you should read the product specification on the manufacturer's label, or look it up in more specialized books.

7.4.2.2. Carbamates

• **Chemical name: Isopropyl-N-phenyl-carbamate**

Active ingredient: **PROFAM**

This is a greyish white powder, only slightly soluble in water, that the plant absorbs through its leaves and/or roots. It is more effective against grasses than against dicots, i.e., it acts against monocots rather than dicots. It acts by blocking cell division in grasses during the seedling stage, and does not act against fully grown weeds. It is effective against *Stellaria* and *Portulaca*.

• **Chemical name: N(chloro-3-phenyl) isopropyl carbamate**

Active ingredient: **CHLOROPROFAM**

This product is insoluble in water, but dissolves in organic solvents. It is absorbed by the roots and acts mainly on young plants. It blocks cell division, causing the formation of cells with many nuclei. It persists on average for eight weeks, but less at higher temperatures. Like all the herbicides that act on roots it works better when the soil is moist, and it is worth irrigating after the application to fix it into the soil.

7.4.2.3. Urea derivatives

• **Chemical name: 3(4-chlorophenyl) 1,1, dimethylurea**

Active ingredient: **MONURON**

MONURON is sold as a white wettable powder that shows low solubility in water and is very persistent in the soil. This is a nonselective herbicide with a very broad range of action. It is recommended for use on uncultivated sites, as a pre-emergence herbicide against weeds. It acts better in wet sites. The dose should be decreased on very light soils and increased on soils with high levels of organic matter. After the application, you should not till.

• **Chemical name: 3(3,4-dichlorophenyl) 1,1, dimethylurea**

Active ingredient: **DIURON**

DIURON is mainly used as a preemergence weedkiller against weeds growing in crops like cotton, olive, citrus fruits, vines, seed fruit, etc. A film of the product forms over the soil and destroys the seedlings as they germinate. It is preferentially used in spring and fall. It acts mainly against annual weeds, and has little or no effect against perennial weeds. However, its use can be recommended when

Invasive weed: the scarlet pimpernel (Anagallis arvensis) (Courtesy of SCHERING)

the vine, olive, etc., are only three or four years old.

• **Chemical name: 3(3,4 dichlorophenol) 1 methoxy 1 methyl urea**

Active ingredient: **LINURON**

Linuron is not very soluble in water, and is essentially used as a pre-emergence weedkiller because it has little effect on the leaves and acts mainly on the roots. It is mainly effective against annual weeds, and is suitable for killing weeds in crops like potatoes, onions and vines. Like all the other urea-derived weedkillers it should not be used on very sandy soils or ones containing a lot of organic matter, and it persists in the soil for one to three months.

7.4.2.4. Triazines

• **Chemical name: 2-chloro,4,6, bisethylamino-s-triazine**

Active ingredient: **SIMAZINE**

This herbicide has a very stable molecule that is almost insoluble in water, and is only slightly soluble in organic solvents. It is almost exclusively absorbed by the roots, and acts by blocking photosynthesis and sugar production. It is more effective on wet soils that have previously been watered. It acts against recently germinated seedlings, when the roots start to absorb water and nutrients from the soil, causing them to die back from the young tissues and leaf edges. As it is a preemergence weedkiller, it must be used on cleared ground. The soil should be broken down to get rid of large clods, and the treatment can be total or applied in strips. Use smaller doses on sandy sites, and larger ones on clay soils and ones with a lot of organic matter. Rain, or watering, after

Polygonum convolvulus (*Courtesy of SCHERING*)

application increases the effect.

• **Chemical name: 2-chloro, 4-ethylamine, 6-isopropylane-s-triazine**

Active ingredient: **ATRAZINE**

This weedkiller is very stable, and a little more soluble in water and organic solvents than simazine. It is mainly absorbed by the roots, but can also be absorbed through the leaves. It does not require such wet conditions as simazine and it can be used as a preemergence or postemergence weedkiller, as long as the weeds are still in the early stages of growth.

7.4.2.5. Amides

• **Chemical name: 2 chloro,2'-6'diethyl-N (methoxymethyl) acetanilide**

Active ingredient: **ALACHLORE**

A preemergence weedkiller that acts between germination and growth of the first internode. It is selective and favors corn, beans, peas, rapeseed and other members of the cress family (Cruciferae) in general. It is not very effective against weeds that are members of the Cruciferae of the dock family (Polygonaceae). It is best applied before germination of the crop and the weeds. It must be applied on soil that is in good tilth and relatively warm; rain or watering after application favour its absorbtion.

• **Chemical name: N isopropyl-2-chloracetanilide**

Active ingredient: **PROPACHLORE**

A weedkiller that mainly acts against annual weeds. It persists six to eight weeks in the soil. It acts when crops and weeds germinate, and acts selectively in favor of perennial members of the Cruciferae, members of the lily family (Liliaceae), corn and legumes. It is recommended for use as a preemergence weedkiller before the germination of the crop and weeds. Rain or watering after application increase the product's effectiveness.

7.4.2.6. Quaternary ammonium compounds

• **Chemical name: 1,1'dimethyl 4,4'bipyridylium dichloride**

Active ingredient: **PARAQUAT**

This is a yellow crystalline product that is soluble in water. It breaks down in alkaline media and is corrosive, and this makes it very useful on sites with basic soils, because it decomposes when it comes into contact with the soil. Though it acts systemically through translocation, its external effects are similar to contact weedkillers. It is absorbed through the leaves and rapidly translocated throughout the plant, and the product then oxidizes, which leads to the appearance of superoxides that kill the plant cells. It is very effective against invasive weeds, including grasses. As it is considered to be a broad-spectrum nonselective herbicide, you should take care not to wet the green parts of the crop or the neighbouring crops, and it obviously should always be applied as a postemergence weedkiller.

• **Chemical name: 1'1 ethylene-2,2-bipyridylium dibromide monohydrate**

Active ingredient: **DIQUAT**

The only difference between diquat and paraquat is that diquat is less effective against grasses. Its description, method of action and selectivity are all similar to paraquat.

7.4.2.7. Aniline derivatives

• **Chemical name: 4 trifluoro-2,6 dinitro N.N. dipropyl methylaniline**

Active ingredient: **TRIFLUORALINE**

This is a crystalline solid that is insoluble in water and breaks down in sunlight, meaning that it should be worked into the soil after application. It persists in the soil and acts immediately after the seeds germinate. It does not act against fully grown weeds, but acts as a postemergence weedkiller against both dicot and monocot weeds. It acts selectively against members of the Malvaceae (mallows), Cruciferae (cresses) and Compositae (daisies, etc.). Because this product has to be incorporated into the soil, the dose applied should be lower in sites that are dry farmed or soils with a lot of organic matter. As it is considered a preplanting herbicide, the crop should not be sown within 12 months of application.

• **Chemical name: N-butyl-N-ethyl-2,6-dinitro-4-trifluoro-methylaniline**

Active ingredient: **BENFLUORALINE**

Benfluoraline is very similar to trifluoraline, differing in its selectivity, which is greater for certain crops. Its method of action and effects are also very similar to trifluoraline. On irrigated sites in dry areas, cereals and onions should not be sown within ten months of treatment, while sorghum, corn, oats, sugarbeet and spinach should not be sown within twelve months. The product should be worked into the soil as soon as it is applied.

7.4.2.8. Halogenated fatty acids

• **Chemical name: Sodium trichloroacetate**

Active ingredient: **TCA**

This is a white powder or granulate that is very soluble in water. It mainly acts when absorbed by the roots, but acts to some extent through the leaves. It is not a very selective weedkiller, but it is most effective against annual and perennial grasses and water plants, making it especially suitable for crops of alfalfa, sugar cane, sugarbeet,

A weed: Urtica urens *(Courtesy of SCHERING)*

cotton and uncultivated ground. It causes affected plants to yellow at the tips and in the young leaves and this spreads to the entire plant. On clay soils or ones with a lot of organic matter, it persists longer in the soil. Take special care when applying near crops of cabbages, turnips, broccoli, etc., as all the members of the cress family (Cruciferae) are very sensitive to this product.

• **Chemical name: Sodium 2,2, dichloropropionate**

Active ingredient: **DALAPON**

Dalapon is mainly absorbed by the roots, though it is also absorbed by the leaves. When absorbed by the leaves, it moves slowly to the roots and several days usually elapse before the effects are noted. It is a very effective weedkiller against aquatic plants, and is very suitable for use when cleaning drains and irrigation channels. It is highly soluble in water, and so does not usually cause problems with residues in soils. It is normally used on uncultivated sites and in channels as a total weedkiller, though at lower doses it can be used as a selective herbicide to favor alfalfa.

7.4.2.9. Others

This group includes a series of widely used weedkillers that are so important that they can not be left out. They belong to very different chemical families and do not fall into the groups discussed above.

• **Chemical name: 3-amino,1,2,4-triazole acid**

Active ingredient **AMINOTRIAZOLE**

Aminotriazole is sold commercially as a water-soluble crystalline powder that is insoluble in petroleum oils, ether or acetone. It is mainly absorbed through the leaves, but is also absorbed by the roots. Its most notable feature is that it is rapidly translocated, and it may be absorbed by the cortex in branches less than three years old. It acts by inhibiting photosynthesis and speeding up the plant's respiration. Its average persistence in a loamy soil is four or five weeks. It is considered a nonselective herbicide, as it shows little selectivity. It is also very effective against Bermuda grass (*Cynodon dactylon*). It is best applied in spring, when the weeds are growing fastest. Take care not to spray the leaves or young stems of crop plants. Do not use when there are associated crops in the plantation.

• **Chemical name: 1,4-dimethyl-2,3,5,6 tetrachlorophthalate**

Active ingredient: **DCPA (CHLORTAL)**

This a nonselective weedkiller that is mainly absorbed by the roots, though in grasses it is also absorbed by the coleoptile. It is a preemergence weedkiller against weeds, and does not affect those that have already germinated. It gives very good results against *Portulaca, Amaranthus, Poa* and *Stellaria*. Before application, the crop should be well buried so the weedkiller does not affect it, and rain or watering after application is desirable.

SUMINISTROS ADARO, S.A. supplies a wide range of protective products for application of pesticides, etc., such as masks, suits, helmets, goggles, gloves, etc.

• **Chemical name: 3,6-dichloro-2-2-methoxy-benzoic acid.**

Active ingredient: **DICAMBA**

This water-soluble weedkiller acts by absorption through the leaves and it is then translocated throughout the plant. It has some effect when absorbed by the roots, and it does not act as fast as the hormone weedkillers, though it is more effective. It is often used mixed with other weedkillers such as mecocrop or MCPA. It is excellent against *Galeopsis tetrahit, Polygonum spp., Stellaria, Matricaria, Spergula, Fumaria* and *Rumex*. It is especially suitable for grasslands like meadows, lawns and cereal

fields, though in cereals it should be applied after tiller production and before stalk production. When it is mixed with hormone weedkillers or bought pre-mixed, it is essential to observe the precautions for both active ingredients.

7.4.3. Conditions for use

We have already looked at the damage caused by weeds, how quickly they spread, the available methods to combat them and the main chemical weedkillers on the market. This section deals with how these chemicals should be applied.

7.4.3.1. Toxicological classification

Weedkillers, like insecticides and all pesticides, must be registered in the national *Official Register of Agricultural Chemicals* and they are given a toxicological classification. This classification is changed from time to time as new products appear on the market. We shall not go into the details, because they vary from country to country.

Information on this matter can be found in the regular publications of the Ministry of Agriculture and the manufacturer's label on the product. In general, toxicological information is divided into three sections: the first deals with its toxicity to human beings, the second with its effect on the soil flora and fauna, and the third with its effect on the aquatic flora and fauna.

The label also includes other important details, such as whether the product is inflammable, how hygroscopic it is, whether it is highly volatile, explosive, corrosive or highly irritant. A scale from high to low is normally used to indicate how toxic a product is.

7.4.3.2. Precautions

The precautions to take obviously depend on the type of toxic product. Some precautions should, however, be taken with all products.

The products should always be stored in the original container and should never under any circumstances be transferred into containers where they may be mistaken for foodstuffs or drinks. They should be stored in a room or shed that is only used to store herbicides and other agricultural chemicals, and they must never be stored in the same space as human foodstuffs or animal feeds.

As we have said several times in this chapter on herbicides, **before applying any chemical, read the product's label and instructions very carefully**. The manufacturer is legally responsible for compensation for any damage to crops caused by use of their product. Legal cases of this type are long and expensive, because the manufacturer usually alleges the product was not used correctly; application of an incorrect dose, treating unauthorized crops, application to kill weeds other than those specified, etc. Furthermore, in most cases it is difficult to show which is responsible for the damage, the person applying the product or the product. Bear in mind that when farmers have scrupulously respected all the conditions for application, there have been very, very, few cases where the product used has caused serious damage, because the instructions for use are usually the result of long and expensive trials, and so the risk from correct application is minimal.

Do not use empty weedkiller containers for other purposes. The best thing to do with them is to destroy them or deposit them in the special containers for their recycling (if these are available). Other general precautions are: ensure the product does not come into contact with your skin, do not eat or drink during the treatment, wear loose and comfortable clothing, wash it thoroughly after use and take a shower when you have finished the application.

7.4.3.3. Machinery

We strongly recommend having one set of machinery that is only used for applying weedkillers, and another set that is used for all the other agricultural chemicals. This machinery, whether a pump powered by the tractor or a simple backpack spray, should

always be washed after use and its components rinsed thoroughly in running water. It should also be regularly inspected to ensure the correct functioning of the pumps, filters, washers, nozzles, levers, faucets, etc. This cleansing should be even more thorough after using hormone weedkillers, as they are more persistent than contact weedkillers.

7.4.3.4. Dosage

The most common method of applying herbicides is by dissolving them in water and spraying the solution onto the soil or onto the weeds. After checking the dose at which the product is to be applied, fill the backpack or tank with half the water required, then add the full quantity of the product (using a measuring device, usually supplied with the product), then close the backpack or tank and mix it up by shaking. You must then add the rest of the water to make it up to the full dilution, and mix it up again.

7.4.4. The weedkiller's behavior in the soil

We consider it necessary to emphasize the behavior of the weedkiller in the soil, though those who have read this far will have a relatively clear idea of the problem. The soil residues of agricultural chemicals have become a cause for great concern among environmentalists.

Contact weedkillers leave the least residues, as they usually break down as soon as they come into contact with the soil. Presowing and preplanting weedkillers last several days in the soil, so that they can eliminate the weeds as they germinate. These weedkillers may accumulate in the soil or in the water table if they are soluble, polluting both of these ecosystems.

As far as possible, we recommend the use of herbicides that are the most environment-friendly, such as the contact weedkillers, which breakdown on contact with the soil, and the weedkillers that leave residues should only be used when really necessary. The technical details specifying the residues the weedkiller leaves can be found in the manufacturer's label on the product.

It is important to mention the major role of organic matter in the behavior of residual weedkillers in the soil. The organic matter acts like a sponge absorbing and retaining the product in the soil. We suggest you reduce the dose slightly if a soil analysis has shown more than 2% organic matter. In dry farming cereal fields the percentage of organic matter in the soil is usually very low, but in intensively cultivated vegetable plots and in gardens the soil usually contains larger amounts of organic matter.

The photo shows weedkiller application in a cereal field using specialized machinery. The arms jut a long way out on both sides of the tractor, meaning the tractor does not have to make as many journeys. (Photo courtesy of BASF, S.A.)

The technical staff of PROBELTE offers free advice on any matter related to crops pests and diseases.

8. PESTS AND DISEASES OF SPECIFIC CROPS

Many of the pests and diseases affecting crop plants have already been discussed in the text. This chapter deals with some of the agents causing plant diseases, arranged by the plants that they affect. The first section deals with the pests and diseases of woody plants, and the second section with those of herbaceous plants. The physiological disorders, diseases and pests of the apple alone would fill several dozen pages, far beyond the scope of this work. This discussion only deals with the most common pests or diseases of crops in temperate areas. The farmer or technician seeking to find the cause of something that is not described in this work should seek advice from a technician, a biological laboratory or commercial supplier, or refer to specialized works on plant pathology, which can be found in bookshops and libraries.

Many of the larger commercial suppliers offer the services of their technical department to the farmer. The staff of these departments are usually commercial technicians who understand plant disorders and diseases and they seek to assess the farmer, starting by determining the cause of the problem and advising which product would be most suitable to treat it. These technicians are often (and rightly) accused of being partisan, as they usually recommend the products of the company they work for, even if this is not the most appropriate. In all fields there are excellent and mediocre professionals, and agricultural technicians and agronomic

engineers are no exception: many of them have excellent reputations and are totally honest.

It is also possible to contract a technician as part of the staff of the farm or as a self-employed consultant for specific cases. To identify diseases you can use the services of a private laboratory. If their fees are too expensive at a given moment, for example when starting a new farm, you can request the services of an official laboratory, which are usually cheaper. Wherever possible we have included photographs of the pests and diseases that attack plants as a guide for the farmer when dealing with these problems.

8.1. PESTS AND DISEASES OF WOODY PLANTS

This chapter deals with the agents attacking fruit trees, including vines and cane fruits. The damage can be classified by whether it is caused by a non-biological agent (physiological disorders) or by a living agent (virus, bacteria, nematodes, gastropods, millipedes and centipedes, mites, insects or vertebrates). The farmer is bound to come across other pests of fruit trees that are not dealt with here, but a thorough description of all these pests and diseases would be far beyond the scope of this work.

8.1.1. Abiotic agents

• **Chlorosis**. In many fruit trees, such as the peach and apple, chlorosis may be due to many causes, though the external symptoms are always a loss of color in the leaves, a lack of chlorophyll, and the appearance of yellow and reddish colors. In the worst cases the plant suffers necrosis. Chlorosis may be due to a lack of iron, nitrogen, magnesium, manganese, zinc or to an excess of lime in the soil, which may cause the immobilization of the assimilable iron.

• **Bitter pit**. This disorder is common in apples, where it causes corky tissue to form in some parts of the fruit, greatly reducing its quality and value. Though it is still being studied, it seems that bitter pit is probably due to excessive feeding with nitrogen. Severe pruning and the production of thick fruit are also prejudicial factors. Alternating wet and dry periods, especially late in the season, and excessive evaporation in dry hot weather all favor the disorder.

• **Scald.** This only occurs in cold storage chambers. It initially takes the form of irregular dark patches on the skin, variable in size, which progressively spread without showing clearly defined limits from the healthy parts. The underlying flesh is sometimes darkened or altered, and the pigment cells are dead. The organoleptic qualities of the fruit are unchanged but its commercial value is greatly diminished, and it can only be sold to the food processing industry to be made into jam, etc. To avoid bitter pit, which is due to the fruit's development and the substances produced by its own metabolism, you should treat the fruit as a living thing and not harvest it too early or too late.

8.1.2. Biotic agents

• **Mosaic.** Pear and apple mosaic virus is a disease that occurs all over the world and can reduce an apple tree's productivity by 30%. It affects other closely related fruit trees like peaches, apricots, cherries and quinces. The symptoms vary greatly depending on the species and variety, but are normally chlorotic patches that are pale yellow, bright yellow or light green. All that can be done is to control the animal vectors and to cultivate resistant species and varieties.

• **Canker.** Canker or **collar and root gall** mainly affects peaches, though it is also frequent in other fruit trees. It is caused by a

bacteria called *Agrobacterium tumefaciens*, which also affects other crop plants (sugarbeet) and ornamental (dahlias). The tumors usually take the form of bumpy growths, ranging from the size of a pea to a full-grown cabbage. These growths are initially white and soft, but they lignify rapidly, turning hard and dark. The plant dies because these bumps physically prevent the flow of sap in the roots, or because the tumor accumulates an excess of hydrocarbons absorbed from the plant. The bacteria is hard to eliminate, and all that can really be done is to adopt good cultivation techniques, such as crop rotation, combatting the bacteria's animal vectors, as well as the disinfection of seedling roots before final planting, using a **mercury** product like those used to disinfect seeds.

• **Apple scab** is a fungal disease caused by *Venturia inaequalis*. Another species, *V. pirina* attacks pears. The scab attacks all the fruit tree's vegetative parts and takes the form of greenish brown or dark brown patches. It appears in spring, on the leaves and the upper surface of the leaf blade. The patches are slightly translucid to begin with, and as they increase in size they turn olive green and appear to become hairy. The leaf blade is generally deformed. The inner area of the

The filbert leafroller (Archips rosanus) is a moth whose caterpillars cause serious damage to pears and apples.

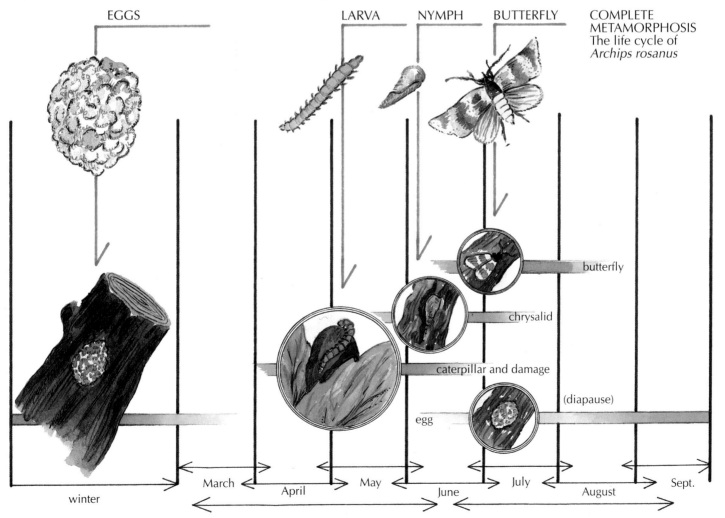

EGGS

LARVA   NYMPH   BUTTERFLY

COMPLETE METAMORPHOSIS
The life cycle of *Archips rosanus*

butterfly

chrysalid

caterpillar and damage

(diapause)

egg

March   May   July   Sept.

winter   April   June   August

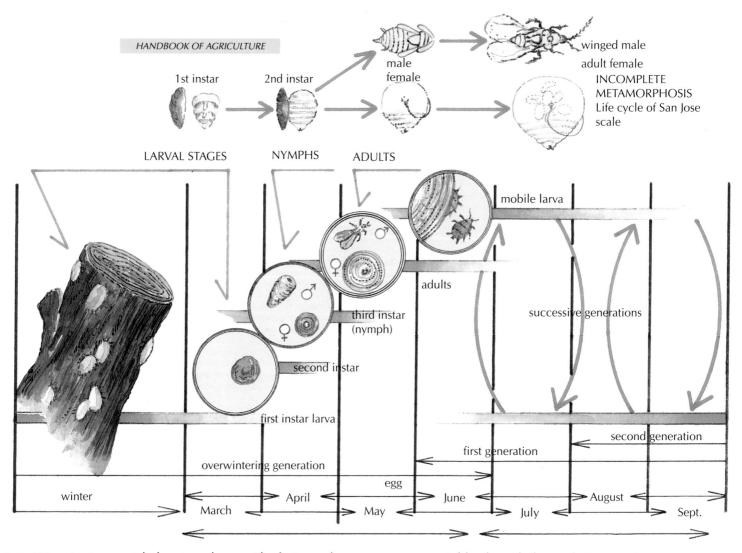

winged male

adult female

INCOMPLETE
METAMORPHOSIS
Life cycle of San Jose
scale

1st instar   2nd instar   male   female

LARVAL STAGES   NYMPHS   ADULTS

mobile larva

adults

third instar
(nymph)

second instar

first instar larva

successive generations

second generation

first generation

overwintering generation

egg

winter   March   April   May   June   July   August   Sept.

Typical life cycle of a scale insect (Quadraspidiotus perniciosus Comst.). Note that the males and females are very different (sexual dimorphism).

patch then turns brown. The fruit are also attacked by the fungus, causing premature fruit fall. If the attack occurs once the fruit is relatively large, surface patches form that turn a darkish color and sometimes have a reddish tinge on the edges. The only methods available to deal with this disease are to try and prevent it, and they are effective as they persist on the tree a long time after the treatment. Copper and sulfur products are used, though if you wish to avoid their possible toxic effects on the plant, you should use synthetic fungicides, such as **zineb**, **mancozeb**, **captan**, etc.

• **Powdery mildew**. Powdery mildew of apples and of pears does not affect all varieties of these trees to the same extent. The presence of this fungal disease can be detected in winter. When carrying out pruning, you may note that the shoots are covered with a whitish grey mycelium and the lateral and apical buds are unusually narrow and sharp. The first infections appear in spring and the affected stems and leaves are covered with a whitish floury growth. As the leaves grow, they become narrow and rigid, and frequently curl upwards. Good cultivation techniques, such as pruning, can get rid of much of the infected wood. Harsh winters decrease the population of mycelia within the tree. As in all fungal diseases, chemical treatments are most effective when applied as preventive measure; **sulfur compounds** are

suitable, through the synthetic organic fungicides, such as **dinocap**, **binapacril**, **quinomethionate**, **triphorina**, etc.

• **Mildew.** Another very widespread fungal disease is **downy mildew**. This mainly affects fruit trees, though it can attack other plants, such as ornamental conifers and horticultural crops like endives and strawberries. It is caused by *Phytophthora spp.* and leads to rotting of the trunk that sooner or later causes the tree's death. There are two forms, depending on which part of the trunk is affected. The fungus may either spread upwards from the graft junction, attacking the cultivated variety, or it may destroy the roots, only attacking the rootstock. The disease does not attack all varieties, as some are resistant to it. Wet soils favor the spread of the fungus, which infects the roots deep in the soil; the foliage turns reddish in late summer and falls prematurely.

When the rot is restricted to the aboveground part, the annual shoots are short, the foliage acquires a bronzed tinge or turns chlorotic and the fruit is small and ripens badly. The fruit can also be attacked by the downy mildew, and it is possible to detect darkish patches with undefined edges. Once it has broken through the epidermis, the fungus enters the flesh and soon reaches the core. When the apple is completely rotten, it turns a brown color, streaked yellow, green or garnet, depending on the variety and how ripe it is. The rotten apple remains hard to the touch. Copper compounds are effective at combatting fungal attack of the aboveground part, as long as treatment is preventive. There is no known compound that is really effective against rotting of the roots, though a new fungicidal active ingredient has had some success as a preventive and even as a curative treatment. This is phosethyl-Al, aluminium phosethyl, which appears to have some systemic action. After a series of applications, it seems to enter the plant, reaching the roots and manages to slow the action of the fungus.

• **Brown rot.** Brown rot of fruit, is caused by different species of the fungus *Monilia* (or sometimes *Monilinia*). It attacks both seed and stone fruit, and also attacks (to some extent) other plants like gooseberries, hazels and vines. Some brown rots penetrate the pistil of the flower in spring, when the spores are dispersed by the wind. The flowers wither and then the branches and the rest of the plant is infected. Other species of *Monilia* attack the plant entering through wounds caused by hailstones, snails and slugs, insects, small vertebrates, etc. The best way to combat brown rot is preventive treatment with fungicides, especially when brown rot can be expected, namely after a strong hailstorm.

• **Nematodes.** The free-living parasitic nematodes are the main agents transmitting viruses, and this is the main damage they cause. Endoparasitic nematodes live within the plant and feed on its cells. They mainly belong to the different species of the genus *Pratylenchus* and their main effect is causing soil fatigue. This leads to greatly decreased production of little commercial value. The best method to prevent it is by crop rotation, and in the case of fruit trees they should not

be cultivated for at least ten years and other cereal or horticultural crops should be grown instead.

• **Codling moth.** *Laspeyresia pomonella L.* is a moth whose caterpillars feed on the fleshy fruit of many fruit trees. The damage it causes decreases the fruit's value. The adult is a small moth that flies in the evening or at night, and

lays its eggs on the leaves in spring or directly on the fruit in summer. The eggs hatch to produce small larvae that feed on the fruit. The caterpillar penetrates into the fruit by making a small hole in the epidermis and then burrows into the fruit, forming spiral galleries. It reaches the core and eats the seeds and then attacks the rest of the fruit, feeding on the pith.

Several methods are available to combat codling moths. The classic method is to apply preventive chemical treatments from spring to fall. A cheaper and more economical solution is to use **pheromones** to determine the **flight curve** in order to decide when is the best time to apply it (thus saving money by reducing the number of treatments). Or you can carry out regular visual inspections and treat as soon as you detect the first larvae on the fruit. There are many different types of chemical insecticides

that successfully kill this moth. Some, such as the organophosphates, can kill the eggs, larvae and adult (imago). The safety period before harvesting must be respected if they are applied near harvest time, as the fruit produced is for direct human consumption.

• **Leafrollers.** The leafroller (*Archips rosanus L.*) is another insect that attacks many different species of plants. It is especially destructive to pears and apples, and can affect 80% of the harvest if weather conditions favor its spread. The leafroller only has a single larval period in the spring and a single flying period. It overwinters as eggs which are clearly visible on the bark of the thicker branches. They hatch relatively late (from mid April onwards), from before the apple flowers until the time when the fruit is setting.

*Several aphids attack the peach, such as the whitefly (**Pseudaulacaspis pentagona Targ**).
1/ Male scales
2/ Male scales on stem and fruit
3/ Females overwintering
4/ Females and eggs
5/ Intense attack on peach
(Photos courtesy of the Department of Agriculture, Stockraising and Fisheries of the Generalitat of Catalonia)*

*Adult of the rice stem borer (**Chilo suppressalis Walker**). The insect's larva burrows galleries within the young shoots of rice plants. It may cause serious economic damage. (Courtesy of the Department of Agriculture, Stockraising and Fisheries of the Generalitat of Catalonia)*

A/ Characteristic symptoms of collar rot in fruit trees caused by Phytophthora cactorum. (Courtesy of RHÔNE-POULENC)

B/ Ornamental conifers attacked by Phytophthora. RHÔNE-POULENC recommends Phosethyl-Al to eradicate it.
C/ Adult of Lobesia botrana Schiff *(Courtesy of DOW ELANCO)*

plant tissues, either the wood or the fruits. Its pernicious action is more due to the toxic substances that it injects into the host tissues than to the amount of the plant's sap it removes when feeding. The affected parts turn a very distinctive violet-red color; if these patches occur on the fruit, they lose their value and can not be sold.
Preventive chemical treatments include treating the tree with **winter oils** and **oil preparations** before the buds open. This is often not enough, and treatment with organophosphates is needed, such as **phosalone** and **methidathion**, when the mobile larvae hatch, during the month of June.

• **Psylla**. The pear psyllid (*Psylla pyri*) is a sucking insect closely related to the aphids,

They damage the flower buds but especially the leaves, which they roll up like a cigar (thus they are called leafrollers). In some circumstances they may also attack the fruit, feeding on the pulp; the fruit is rarely shed, but finishes its growth with major deformations. The nymph stage occurs on the organs of the attacked plant, followed shortly after by the adult, a small butterfly about 20 cm long that shows slight sexual dimorphism. In July it lays it eggs on the bark of the branches, and the eggs do not hatch until the following spring. Checking the frequency of the groups of eggs during winter makes it possible to decide whether chemical treatment is necessary. When the eggs hatch, they can be treated with a contact organophosphate insecticide, such as **chlorpiriphos** or **quinalphos**.

• **Peach greenfly**. This mainly attacks the peach and is very rare on other species of the genus *Prunus*. The eggs overwinter on the branches of the peach tree. Hatching takes place from late February onwards and the aphids feed first on the flowers and then on the leaves. They cause the flowers to abort, the leaves to roll up and the shoots to wither. Around the end of May, the winged forms of the aphids fly to secondary hosts, which include the potato, tobacco, sugarbeet, cabbage, etc. In addition to attacking the plants, this insect is one of the main vectors transmitting viruses from one plant to another. You can carry out an initial shock treatment in early spring with a good pyrethrin insecticide, such as alpha-cypermethrine. If this is not effective, use more persistent insecticides, such as **acephate**, **dimethoate**, **ethioncarb** or **pyrimicarb**.

• **San Jose scale**. This scale insect's scientific name is *Quadraspidiotus perniciosus Comst.*, and it was discovered in San Jose (California) in 1873. The San Jose scale attacks many different plants. It now has more than 150 host species, though its favorites are the apple, pear, plum and several ornamental species. This insect has a shield (a "scute") that covers its body in the immobile stages of its life cycle. The larva underneath this scute is firmly attached to the

and their flattened yellowish brown larvae form relatively large colonies on shoots, inflorescences, leaves, and sometimes on the young fruit of the pear. Piercing by the adults and larvae causes leaf deformations or premature leaf fall, and reduces the vigor of heavily infected shoots. Effective application of insecticides against psyllid bugs may require two phases. A first treatment with synthetic pyrethrins, such as **phenvalerate** is usually effective as a shock treatment against the first and second generations in spring. But if *Psylla* damage to pear trees is detected during the summer, it may be necessary to carry out a second or third treatment with more persistent products, such as **azinphos**, **methydithion**, **amitraz**, **difluobenzuron**, etc.

• **Red spider mite**. Of all the mites that attack plants, the red spider mire (*Panonychus ulmi*) and the yellow spider mite (*Tetranychus urticae*) cause the most damage to cultivated plants. The red spider mite mainly attacks apple trees, but it also attacks other fruit trees. It causes a characteristic loss of color in the leaves of apples, which turn darkish or lead grey and are unable to photosynthesize. The adults of the red spider mite can be seen with the aid of a hand lens. These mites overwinter as eggs, which hatch in spring to produce the first larvae. Attack by the larvae is usually detectable in mid April, but the most important damage is caused by the adults, which appear from early May onwards. If no specific measures have been taken, great damage can be caused during the hottest days of August, when the eggs lay two different types of eggs: the summer eggs, which give rise to new generations of phytophagous mites, and the winter eggs, which enter diapause until the following spring.

The strategy to follow is complex and requires several different types of acaricides. The first essential treatment is to apply **mineral oils** before the shoots open and just before the eggs hatch, which greatly reduces potential future populations. The second treatment is with fungicides that are effective acaricides, such as **dinocap** or **mancozeb**, which may be effective in controlling the mite populations in July. The final alternative, if the red spider mite populations have not been controlled, is to treat in August with the latest generation of acaricides, such as **pyridiben**, which is fast acting, effective and highly persistent.
The yellow spider mite is a highly polyphagous species that attacks many different species. It has been found on 200 different host species, including the hop, cotton, vine, carnations and fruit trees. The yellow spider mite overwinters as the adult female under the barks of trees or other rough surfaces. The generations follow each other very quickly and each female lays a large

number of eggs, and this is why dense populations suddenly appear. The parasite remains active until the fall, when the first frost send the females rushing in search of shelter. The products used are similar to those used against the red spider mite, with treatment every twelve days, and taking the precaution of changing the active ingredient from one application to the next in order to avoid the appearance of resistant strains.

8.2. PESTS AND DISEASES OF HERBACEOUS CROPS

Horticultural plots and cereal fields exemplify intensive and extensive cultivation, two distinct approaches to cultivation. These cultivation methods are essentially different, as is their economic profitability. From the plant pathology perspective, horticultural crops and cereals are all herbaceous plants, unlike fruit trees, which are woody plants. Because horticultural crops are essentially herbaceous they are attacked by pests and diseases that do not attack fruit trees, as their leaves are high in the air, their trunks are lignified and because most herbaceous plants are only actively growing for a short period of the year (often only a few months).

8.2.1. Abiotic agents

• **Fertilizer application**. Applying fertilizers is of great importance for herbaceous plants, because they have to grow, flower and ripen in a relatively short growing period (less than one year). Balanced and sufficient feeding is thus essential from the moment the seedling is transplanted into the field, or from the moment it is sown, in the case of cereals. Many non-parasitic disorders cause malformations due to nutrient deficiencies or imbalances. Fertilizers, the balance of nutrients necessary for plant nutrition and the disorders caused by deficiency are all dealt with in detail in the section on Soils in this volume. To recap, unbalanced fertilization of herbaceous plants may cause major disorders in their growth. In cereals, for example, excess nitrogen relative to phosphorus and potassium leads to growth of very weak stems, the grain ripens late and badly and the harvest quality is inferior. If nitrogen is lacking, the plant produces few tillers and late, the stems are fragile and short, and often turn reddish, and all this leads to a decreased yield.
In vegetables it is also very important for fertilizer application to be balanced. If we accept that nitrogen is largely responsible for the growing parts of the plant (stem, leaves, etc.) and potassium and phosphorus for the formation and quality of fruit and tubers, these three elements must present be in the right proportions for the plant to be in balance, so that different parts of the plant do not grow more than others, which would lead to decreased product quality and quantity. Each plant has its own requirements for this balance of nutrient elements, and to ensure optimum production these quantities must be respected.

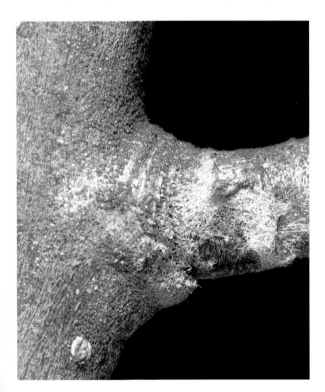

Severe attack of red spider mite on the trunk of an apple tree (Photo courtesy of SCHERING)

Ring-shaped lesion caused by the tomato wilt virus

The sucking insect Frankliniella occidentalis (Thysanoptera) is especially harmful in greenhouse crops, as it is a vector for many diseases. (Photo courtesy of AGREVO)

8.2.2. Biotic agents

• **Viruses**. Viruses are endemic diseases of many horticultural crops, such as the onion, leek, garlic and many other members of the lily family (Liliaceae). The symptoms are shown in the leaves and flower-stalks (scapes), which are flattened, showing long yellow streaks alternating with green ones that are paler than the leaves of unaffected plants. The infected plants tend to wilt as a result of losing turgidity. The leaves are also attacked by secondary rots and seed ripening is endangered. As we have already seen, there are no products that can eradicate viruses. All that can be done is to take preventive measures, such as growing resistant species and varieties, eliminating the animal vectors, and destroying each and every infected plant, in order to ensure the virus spreads no further. Other types of virus affect almost all

known herbaceous plants. Viruses such as the mosaic viruses of tobacco, pea, cucumber and alfalfa cause widespread diseases, whose symptoms are similar to those described, and they can only be combatted using preventive measures.

• **Bacteria**. *Erwinia carotovora* causes a bacterial disease known as **black rot** or **soft rot** of potato tubers. This disease appears in late spring in potato fields. The base of the affected stems show total rot of the tissues, which later turn black; the leaves are upright, then curl up, turn yellow and wither. If the disease attacks before the buds have sprouted, they do not sprout, but if it attacks when the potato is actively growing, the disease does not prevent the potato from growing fully, though they may wither in the space of a few days. The affected tubers turn dark and soft, and are sources of infection if they are left in the soil. The best methods of combatting this disease are to plant healthy uninfected tubers, disinfect the site or rotate crops.

• **Fungal root diseases**. Many soil-living fungi, including the genera *Phytium*, *Phytophthora*, *Botrytis*, *Fusarium*, *Rhizoctonia*, etc., cause different rots in the roots of vegetables and cereals, though some also attack fruit trees. The symptoms of these diseases are rotting and blackening of the roots, greatly weakening the plant or causing it to die. The affected part is covered with a blackish brown powder (*Thielaviopsis*), a pinkish cottony mass (*Fusarium*), black bodies, (*Colletotrichum*), and a felt-like layer of dark or bluish filaments with dark nodules (the *sclerotia*). The sclerotia remain alive in the

Olive branch with larvae and adults of the olive scale (Saissetia oleae Bern.) (Courtesy of the Department of Agriculture, Stockraising and Fisheries of the Generalitat of Catalonia)

soil for several years and can cause new infections. The different methods of combatting root rots are crop rotation (in the case of cereals) and using seeds freed of fungi by heat treatment. In horticultural crops, the site can be disinfected, especially in greenhouses, with fumigants such as **steam** or **methyl bromide** (see the section on Soils).

• **Aboveground fungal diseases**. Many fungi can produce rots in the aboveground parts of vegetables, such as potato blight (*Phytophthora infestans*), which also attacks the tomato and its fruit. Where this disease has been detected, crops should be rotated, avoiding planting potatoes after tomatoes or tomatoes after potatoes. Preventive chemical fungicides, such as **copper compounds**, and synthetic organic fungicides like **maneb** and **mancozeb** are suitable as part of a strategy for chemical prevention of mildew.
Another group of fungi that affect the aboveground parts of crops is the rusts (*Puccinia spp.*). The most important are black rust, dark and yellow rust of wheat and brown rust of barley. There are also rusts that attack other vegetables, such as asparagus (*P. asparagi*), chicory (*P. cichorii*), leek (*P. porri*), etc. Other fungi that affect cereals include the smuts, including oat smut (*Ustilago avenae*), corn smut (*U. zeae*), wheat smut (*Ustilago tritici*) and barley smut (*U. hordei*). The most important step in combatting fungal diseases of cereals are cultivation techniques like crop rotation, growing resistant varieties, disinfecting seeds with hot water, eliminating weeds, etc. Classic copper-based or synthetic chemical fungicides can be applied, but they should always be used as a preventive measure.

Millipedes and mollusks. Several species of millipede belonging to the families Blanniuidae and Julidae damage crops. They normally eat decomposing plants or animals, but they sometime eat the roots or tubers of crop plants. They are particularly damaging to germinating seeds and strawberries. Seedling peas and beans are also vulnerable to attack by millipedes when bad weather delays germination. The preventive methods available basically use physical and mechanical methods to destroy these animals. You should get rid of any objects serving as shelters and breeding sites for the millipedes, such as detritus, decomposing plant remains, old planks, sacks, protective matting or straw, etc. It is also possible to use some chemical agents; coating seeds with lindane-based insecticide provides some protection; some organophosphate insecticides give acceptable results when applied as a liquid treatment to the soil, though they are not totally satisfactory. Another type of insecticide, granulates such as **methiocarb**, can be used as baits when applied to the soil.

Mollusks, i.e., **snails** and **slugs**, can cause great damage when they are abundant in intensive crops, ornamental and even in extensive crops. They can be eliminated with chemical products based on **metaldehyde** in the form of baits. They can be applied in strips along vegetable crops or throughout the entire crop.

• **Nematodes**. The genera of nematodes attack a wide range of plants. There are three main types: the ectoparasites (*Pratylenchus*), the gall-forming nematodes (*Meloidogyne*) and the nematodes of stems, bulbs and roots, such as *Ditylenchus dipsaci*. The vegetables affected by these nematodes, such as wheat, potato, sugarbeet, tobacco, alfalfa, tomato and many others all show the same symptoms: stunted plants, rotting of the roots, low production, etc. There is also a clear increase in the number of viral and bacterial diseases in fields infested with nematodes. Cultivation techniques like crop rotation, seed disinfection and growing resistant species and varieties are the best and most widely used methods. In horticultural and ornamental crops, where it is economically feasible, the soil can be disinfected with agents like **steam** or chemicals such as **dazomet**, **dichloropropane**, **dichloropropene**, **chloropicrin**, **methyl bromide**, etc.

• **Insects**. One of the best known is the **Colorado potato beetle** (*Leptinotarsa decemlineata*), whose adult form has quite unmistakable yellow and black stripes. The adult insects overwinter in the soil, at a depth of 25 cm to 40 cm. They appear in spring after a rainfall, when the soil temperature reaches 14°C at the level where the insects are overwintering. From the moment they emerge, the adults devour the leaves of the young potato plants. The females start laying eggs shortly after mating. The elongated ovoid eggs are orange yellow, 1.5 mm long, and are attached in packets of 10 to 30 on the underside of the potato leaves. The eggs are laid from May to August and hatch in six or seven days. The tiny yellow larvae feed on the leaves of the potato plants, undergo three molts and finish growing in 15 days. When they have finished growing, the larvae burrow into the soil to a variable depth, where they pupate,

A/ Adult and eggs of the red spider mite of apple (Panonychus ulmi) (Courtesy of BASF, S.A.)

B/ Larvae of the wireworm (Agriotes sp.). BASF manufactures and sells a product based on phonophos, which effectively eliminates these insects from the soil.

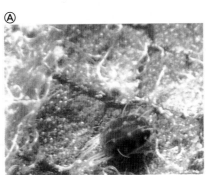
Ⓐ

Ⓑ

emerging on the surface ten days later as the adult form. It is easily controlled using the chemical products available, and organophosphates like **azinphos**, **methidathion** and **phosalone**.

• **Wire worms**. There are many species of wire worm, but perhaps the best known and representative are the species of *Agriotes*. They are the larvae of click beetles (*Elateridae*) that live in the soil and feed on the roots of many horticultural crops and cereals. The adults have a stout body and are dark in color, with designs on their wings (4 to 5 cm span) that can be used to identify them. After mating the adults lay many eggs on the underneath of low-growing plants, mainly weeds. The larvae eat the foliage during the night and stay in their burrows in the soil by day. The plants that are most damaged are lettuces, tomatoes, endives, cabbages, turnips, radishes, sugarbeet, spinach, carrots, celery, potato, tobacco and cereals. It is good practice to eliminate weeds in crops in order to get rid of the areas where the adults can lay the eggs. Good results can also be achieved by applying a granulated

chemical insecticide such as **phonophos** to the soil as a preplanting or preemergence treatment.

• The **mole cricket** (*Gryllotalpa gryllotalpa*) is a cricket (an orthopteran). It shows a clear preference for deep, fresh and loose soils. It does not thrive on stony shallow soils. The adult insect is a typical dark velvety color and is 4 to 5 cm long. One of its most notable features is that their front legs have been profoundly modified into digging forelegs. It is essentially nocturnal, and it lives in the soil where it digs tunnels and galleries to reach the plant roots it feeds on. During the day it remains inactive deep in the soil. Mating takes place in early summer and the females lay their eggs in the soil at a depth of 20 cm. The eggs hatch to give rise to larvae that resemble the adult forms. They undergo three molts and then in winter they disperse and bury themselves even deeper (up to 1 m deep) to overwinter. They renew activity in spring and the first symptoms of damage appear soon afterwards. The life cycle lasts two years. Poisonous baits are normally used to get rid of them, such as 1% **chlorpyriphos** granulates. When applying, take care it does not come into contact with the crop plants.

When buying horticultural seedlings it is very important to choose healthy plants, as this will make later treatments unnecessary. (Courtesy of GEL•BO PLANT)

BIBLIOGRAPHY

BONNEMAISON, L.
Enemigos aminales de las plantas cultivadas y forestales
Barcelona: Occidente
1st edition. Volumes I,II and III. 1964

BOVEY, R., BAGGIOLINI, M., BOLAY, A., BOVAY, E., CORBAZ, R., MATHYS, G., MEYLAN, A., MURBACH, R., PELET, F., SAVARY, A. & TRIVELLI, G.
La defensa de las plantas cultivadas
Barcelona: Omega
2nd edition, revised and extended.

DEPARTAMENT D'AGRICULTURA, RAMADERIA I PESCA
Butlletí d'avisos fitosanitaris
Barcelona: Edicions de la Generalitat de Catalunya. 1987

DETROUX, L. & GOSTINČAR, J.
Los herbicidas y su empleo
Barcelona: Oikos-Tau, 1st edition. 1967

DUALDE, V.
Biología
Valencia: Gráficas Ecir, 1st edition. 1983

DURÁN, S.
Replantación de frutales
Barcelona: Aedos. 1st edition. 1976

EDIN, M., LICHOU, J., SAUNIER, R., TRONEL, C.
Le cerisier
Paris: Edited by J. Granier - Ctifl. 1990

ENCICLOPÈDIA CATALANA
Gran Enciclopèdia Catalana
Barcelona: Reprint of 2nd edition, 1990.

FARBENFABRIKEN BAYER AKTIENGESELLSCHAFT
«Bayer» Pflanzenschutz Compendium II
Leverkusen: 1964

GONZÁLEZ, J.
La contaminació: Bases ecològiques i tècniques de correcció
Barcelona: Edicions del Servei del Medi Ambient de la Diputació Provincial de Barcelona. 1978

LIÑÁN, C.
Vademécum de productos fitosanitarios
Madrid: Ediciones Agrotécnicas. 1998

MINISTERIO DE AGRICULTURA
Guía de aplicación de herbicidas
Madrid: Publicaciones de Capacitación Agraria
1st edition. 1971

PAPE, H.
Plagas de las flores y de las plantas ornamentales
Barcelona: Oikos-Tau. 1977

PLANAS MESTRES, J.
Elementos de biología
Valladolid: Sever-Cuesta. 1975

ROS, J., NADAL, J., LLIMONA, X., ARTECHE, A. & REGUERO, M.
Enciclopedia de la Naturaleza y del Medio Ambiente
Barcelona: Primera Plana. 1992

STRASBURGER, F., NOLL, F., SCHENCK, H. & SCHIMPER, A.F.W.
Tratado de Botánica
Barcelona: Marín, 6th edition, 1981

VIVES, J.M., GINÉ, J.
Control de plagues mitjançant feromones sexuals
Magazine CATALUNYA RURAL I AGRÀRIA. Nº 5 PP. 32-34.
Barcelona: Edita Servei de Protecció del Vegetals. September, 1994

4

Agricultural techniques
in extensive crops

1. INTRODUCTION

One long-standing division of agricultural exploitation methods is into **extensive** and **intensive** cultivation. Unlike extensive crops, intensive methods seek to obtain high levels of production from small areas of ground, by making the best use of the site. To do this the farmer does not leave a fallow year and irrigates and fertilizes the soil so that it can produce crops continuously. Cereals are the typical example of an extensive crop, while market gardens and truck farms exemplify the intensive approach. Large areas of ground were formerly sown with cereals because vegetables could not be grown for climatic or soil reasons. Intensive horticultural crops require greater care, have to be cultivated in the most fertile areas with good climatic conditions, and in general fetch higher prices on the market.

The great technical and industrial advances (machinery, pesticides, fertilizers, etc.) in agriculture over the last fifty years has made this division into extensive and intensive cultivation rather outdated, because technology has made it possible to cultivate vegetables on a large scale, while some cereals, such as corn, may be even more profitable when cultivated on relatively small sites.

The extensive crops can be divided into four main groups, **cereals**, **pulses**, **industrial crops** and **grasslands**. Cereal production has met human food requirements since antiquity, and they are now so important that much of this section is devoted to them. Pulses, such as beans, peas, lentils and peanuts, are considered here as an alternative within the rotation of cereals, because they fix atmospheric nitrogen into the soil. Though they are not dealt with specifically, the reader can extrapolate the tasks recommended for cereals to these legumes. **Industrial crops** (rapeseed, coffee, cotton, etc.) are currently of great importance. These plants vary greatly in their biology and their needs and have to be dealt with separately. The last chapter in this section deals with the distinctive features of the cultivation of each of these crop plants.

Meadows are areas of grass for animal food and ornamental lawns, and they are cultivated as monocultures for many years, or more commonly a mix of species. The many cultivated grasses include species of *Agrostis, Bromus, Cynodon, Festuca, Lolium* and *Poa*. Growing a lawn is totally unlike growing a field of cereals, as the aim is different. Growing meadow grasses is a separate matter and is not dealt with here.
Cereals are grown for their seeds ("grains"), which are rich in starch and can be eaten by humans and animals. There is ancient evidence showing that Neolithic humans cultivated cereals:

they have played a very important role in human history. Cereals played a major role in human adoption of permanent settlement, as well as the development of most civilizations. The civilizations of Asia grew and developed thanks to rice. The main crop of the Pre-Columbian civilizations of the Americas was corn, and in Babylon and Egypt they were wheat and barley.

In the Mediterranean, the Romans held great festivals at sowing time and harvest time in honor of the goddess of agriculture, Ceres, who gave her name to the word

cereal. The cereals are now the most important of the all crop plants due to their high yields, high food value and ease of storage.

Most of the world's cereal production is for human consumption as a foodstuff (directly eating the grain or elaborated products or byproducts); a large amount is also fed to animals (fodder, feeds, etc.). Apart from direct consumption, much of the grain produced is used by food processing industries, such as mills, pasta-makers, bakeries, biscuits, factories that produce starch, dextrose, gluten, etc., in addition to production of beer and distilled beverages. In general, cereal grains are very rich in starches ($\approx 80\%$), moderately rich in proteins ($\approx 9\text{-}16\%$) and low in lipids ($<5\%$). They also contain mineral salts and vitamin B_1. Their high starch content means that cereals meet 50% of the world's protein and energy needs. If we also consider the animal feeds and fodder consumed in the production of meat, milk and eggs, something like 75% of humanity's protein and energy needs are met by cereal crops. Cereals are not however only important for their food value, but also for their other characteristics, including their ease of cultivation and harvest, ease of transport (as they are concentrated dry foodstuffs) and storage, and their adaptability to different climates.

It is also important to point out that the production and consumption of cereals is closely linked to the wealth and development status of each region. Economic developments tends to lead to greater cereal consumption, but is not the only factor, as the dietary habits and potential production of each country also play a role. In the poorest countries, almost all cereal production is consumed by humans. As living standards rise the first step is to replace the cereals considered secondary for human consumption (millet, sorghum, corn, etc.) by wheat. Later, there is a tendency to reduced *per capita* consumption of wheat accompanied by an increase in meat consumption. Meat production, however, requires an increase in production of secondary (fodder) cereals, with part of the wheat crop used as fodder. This means that cereal production tends to show consistent growth because in technically advanced developed countries cereals are grown for feeds rather than for human consumption, and even when production is greater than local demand, the grain can be exported.

List of the main extensive crops

	NAME	SPECIES NAME
CEREALS	Rice	*Oryza sativa*
	Wheat	*Triticum aestivum; T. durum*
	Corn	*Zea mays*
	Oat	*Avena sativa*
	Canary grass	*Phalaris canariensis*
	Buckwheat	*Fagopyrum esculentum*
	Millet	*Panicum milliaceum*
	Barley	*Hordeum vulgare*
	Two-rowed barley	*Hordeum vulgare distichum*
	Spiked millet	*Pennisetum typhoides*
	Rye	*Secale cereale*
	Sorghum	*Sorghum bicolor*
	Triticale	*Triticale hexaploide*
PULSES	Peanut	*Arachis hypogaea*
	Chick-pea (Garbanzo)	*Cicer arietinum*
	Vetch	*Vicia ervilia*
	Windsor bean	*Vicia faba*
	Fenugreek	*Trigonella foenum-graecum*
	Chickling vetch	*Lathyrus sativus*
	Lentil	*Lens esculenta*
	Vetch	*Vicia monanthos*
	Lupines	*Lupinus sp.*
	French bean (green bean)	*Phaseolus vulgaris*
	Garden pea	*Pisum sativum*
	Soybean	*Glycine max*
	White vetch	*Vicia sativa*
INDUSTRIAL CROPS AND OTHERS	Sugarbeet	*Beta vulgaris*
	Hemp	*Cannabis sativa*
	Safflower	*Carthamus tinctorius*
	Oilseed rape	*Brassica napus oleifera*
	Cotton	*Gossypium hirsutum*
	Sunflower	*Helianthus annuus*
	Flax	*Linum usitatissimum*
	Castor-oil plant	*Ricinus communis*
	Tobacco	*Nicotiana tabacum*

2. THE USE OF CEREALS

The **grain** is the main product of cereals and the basic reason for their cultivation, though it is accompanied by a by-product, **straw**, that may also be profitable. All the remains of the entire plant are normally used as fodder, though this is of secondary importance. There are two main ways of using cereals. The traditional method in areas where livestock following seasonal migration methods spent the winter was to graze the cereal during the tiller phase and use the grain produced later, while the second method was to use the entire plant as fodder (as straw or silage) in a relatively advanced stage of the growing cycle. An in-depth study of cereals for animal feed must be considered from a global perspective within the group of animal fodder crops as a whole.

2.1. USE OF THE GRAIN

As pointed out in the introduction, the average cereal grain consists largely of starch, with some proteins and lipids. The component showing the greatest variation between different grains is the **fiber** content, because of the difference between cereals with a covering and cereals without one. Fiber is present in the largest proportion in oat, rice and millet, while corn contains the least. The protein content is highest in wheat and lowest in corn. The highest levels of lipids or fats are found in oats and corn. Starch is the main reserve found in cereal seeds, especially in the seed's endosperm (more than 70% of the seed's dry weight in corn and sorghum).

From a food or technological perspective, the protein content is also extremely important. Proteins can be divided into four groups on the basis of their solubility in different solvents, but we can divide them into two categories, the **functional proteins** located in the protoplasm of the grain cells (albumins and globulins) and the **reserve proteins** (prolamins and glutens), which accumulate in the external cells of the endosperm.

Dietary fiber is the name for the plant food fraction that consists mainly of cellulose. It is not digested by humans but is a rich food source for herbivores, as they can digest it.

The Rollex 1220® is a Cambridge type roller sold by the VÄDERSTAD company. Note that it is weighed down with stones to ensure it reaches deeper.

These reserve proteins are the most important as foodstuffs, and in wheat they consist of **gluten**.

Cereal grains show a consistent food composition but they often show differences that are important when they are used. These differences may due to the genetic differences between varieties, such as the variation in the starch reserves in corn. They may, however, be due to environmental factors that influence the plant's physiology, for example in parched wheat grains the percentage and composition of the starches is modified. The composition of the different proteins and their proportions are genetically determined, but they are also influenced by environmental conditions.

Cereal grains were first used basically as human foodstuffs. After this the range of possible uses was extended to other applications, such as a **raw material** for food industries and as animal **feed**.

Depending on their use, we can distinguish between **primary cereals** (wheat and rice, and to a lesser extent, rye, corn, millet and sorghum) for human consumption, and **secondary cereals** (corn, barley, oats, etc.) which are mainly fed to animals. As the reader will have noticed, corn and sorghum, for example, are classified as a primary grain and a secondary grain, as they are used both as a human foodstuff and an animal feed. There is a third group, barley and corn, that are the basis of industrial processes.

On page 378 there is a list of the main extensive crops, with their English and scientific names, divided into these three main groups, the cereals, the pulses and the industrial crops. This list includes most of the extensive crops, though some are only cultivated on a small scale and are now of limited importance.

In terms of human nutrition, two main points should be taken into account, the **nutritional value** of the products and their **technical properties**. Cereals play a vital role in human nutrition. In many countries, cereals are the staple foodstuffs, while in richer countries they are only a complement, and for this reason we should mainly concern ourselves with their organoleptic and technical qualities (bread quality, appearance, taste, etc.),

Clod-crushing harrow, sold by VÄDERSTAD.

and not their other properties, such as whether their proteins are balanced or are deficient in any one of the essential *amino acids*. From a nutritional perspective it is important to consider which part of the cereal is going to be consumed. Clearly whole grain rice or wholemeal flour play a different role in the diet from other common foodstuffs such as biscuits.

In terms of the technological properties of the cereal grains, the main concepts are the idea of **industrial yield** (the percentage of flour/wheat, grits/corn or hard wheat etc.) and the characteristics of the **final product** (how good the flour is for making dough and baking bread, how suitable the rice grain is for precooked meals, etc.).

In rich countries, most cereal production is used for **animal feeds**, and this is true for wheat as well as the secondary cereals. Ruminants (cows, horses, sheep, etc.) can be fed on most of the different cereal species, but birds and herbivores that are not ruminants are much more demanding in their requirements for essential amino acids.

A third group is the cereals put to industrial usages. Apart from the straightforward processing required to turn wheat into flour for making bread, the preparation of precooked rice or feed production, there are other more complex processes such as **starch extraction** and **production of fermented beverages** such as beer and spirits. These processes mainly use two cereals, corn and barley.

Starch obtained from corn grains has several uses, including starch as such, which is used as a component in many food products for human or animal consumption, as well as in paper production, production of some glues and it is also used as the basis of some industrial processes in the textile, pharmaceutical and chemical industries. Malting is the clearest example of fermentation of cereal grains in beverage production, and is the basis of beer production. Malt is barley that has started to germinate and has then been dried rapidly, and has had the germ and the radicle removed, so that the grain retains only the seed's starch and enzymes. These enzymes hydrolyze the starch which can then be fermented into alcohol. Whisky, gin, vodka and other distilled drinks are based on the fermentation of the sugars found in cereals.

Spring-tooth harrow (Courtesy of VÄDERSTAD)

Amino acids are the structural units of proteins, i.e., the molecules proteins are built from.

To measure the levels of proteins, lipids, carbohydrates or cellulose, the sample must first be dried to remove the water. Once it has been dehydrated and analyzed, the results are expressed as percentages of the *dry weight*.

2.2. THE USE OF STRAW

Straw is a byproduct of cereal cultivation and is used to **feed livestock**, as **bedding** in stables and is also **mixed into soil** as a source of organic matter and to improve future harvests. Straw contains very few proteins, minerals or vitamins, but does supply energy in the form of cellulose (often more than 50% of its **dry weight**). The nutritional value of cereal straw is low and straw should be supplemented with nitrogen-containing materials. Different types of straw show great variation in their digestibility and in the amount eaten. These two conditions depend on the cereal species, the harvesting conditions and the animal's diet as a whole.

Straw makes good bedding for stabled animals because it can absorb a lot of water and is a good insulating agent. Straw can be used as an organic fertilizer either by working it directly into the soil or after use as bedding in stables. In order for the straw to be incorporated into the soil, it should previously be shredded (an option on some types of harvester). It can be worked into the surface layers and this should be accompanied by a supplement of 5-10 kg N per tonne of straw (refer to the section on Soils and Fertilizers; C/N ratio). Except when the stubble and straw from the previous crop should be burnt for reasons of crop hygiene, straw should not be burnt as this is wasting a potential benefit that can be used for other purposes, as we have seen.

Shredding the straw and stubble of the previous harvests is the best way of ensuring the next crop has a good level of organic matter in the soil. The SH 160®, sold by JF - FABRIKEN - J. FREUDENDAHL A/S, cuts and shreds the straw so it can then be worked into the soil.

3. THE PLANT

3.1. BOTANICAL CLASSIFICATION

Cereals can be classified into **botanical groups** or by the use they are put to. All the cereals belong to the grass family, the *Gramineae*, which is divided into subfamilies, such as the *Festucideae* (the winter cereals, wheat, barley, oats, rye), the *Panicoideae* (the summer cereals, corn, sorghum and millet) and the *Oryzoideae* (rice). The only exception is buckwheat, which is not a grass but a member of the dock family (*Polygonaceae*), which is not discussed in more detail because it is not very important. Not all authors agree on how the cereals should be classified, differing in the genera they locate in each tribe and subfamily, but B.N. Smith and W.V. Brown produced a scheme classifying the 198 cultivated species of the grass family within 6 subfamilies and 47 tribes. All the cereals cultivated in Spain are members of three subfamilies and seven tribes, as shown in the accompanying table.

3.2. MORPHOLOGY

Grass morphology, and plant morphology in general, seeks to describe the external parts of the plant, and to understand the role the different organs play.

3.2.1. The vegetative parts

A grass plant is in fact a clump consisting of a set of shoots called **tillers**. The first drawing on the following page shows the clumps of two grasses, wheat and oats. The tillers can be thought of as separate biological units, though they are not totally independent. When a tiller is completely developed, it produces a long flowering stem (the **culm**) that bears leaves and an inflorescence at its tip.
The stem consists of **nodes**, which are meristematic zones from which the internodes extend and where the leaves are borne. When the plant is mature, the **internodes** may be hollow (as in barley, oats and some types of wheat) or solid (as in corn, sorghum and durum) or hard (wheat). The leaf consist of two parts, the **sheath**, the lower part which encloses the stem and protects the bud at its base, and the leaf blade or **lamina**, the upper part, which is flat and long and varies in size depending on the species, the variety and the site it occupies on the stem.
In many grasses at the junction of the lamina and the sheath there is a **ligule**, which is a membraneous structure and the **auricles**, claw-like projections on either side of the base of the lamina. The presence or absence of ligules and auricles, and their shape and characteristics, are of importance when trying to distinguish between grasses that are not in flower (and identify them). The tillers sprout from the buds in the axils of the leaves and undergo two types of growth, when they grow out of the sheath and when they grow within the sheath.

Subfamily	Tribe	Species	Common name
ORYZOIDEAE	Oryzeae	*Oryza sativa*	rice
FESTUCIDEAE	Aveneae	*Avena sativa*	oat
	Hordeae	*Hordeum vulgare*	barley
		Secale cereale	rye
		Triticum aestivum	wheat, bread wheat
		Triticum durum	durum wheat
	Phalaridae	*Phalaris canariensis*	Canary grass
PANICOIDEAE	Paniceae	*Panicum miliaceum*	millet
		Panicum italicum (= *Setaria italica*)	foxtail millet
		Pennisetum typhoides	spiked millet
	Andropogoneae	*Sorghum bicolor*	sorghum
	Maydeae	*Zea mays*	corn

Botanical classification of the most widely cultivated cereals (Smith & Brown)

Corn (Zea mays) with male and female flowers

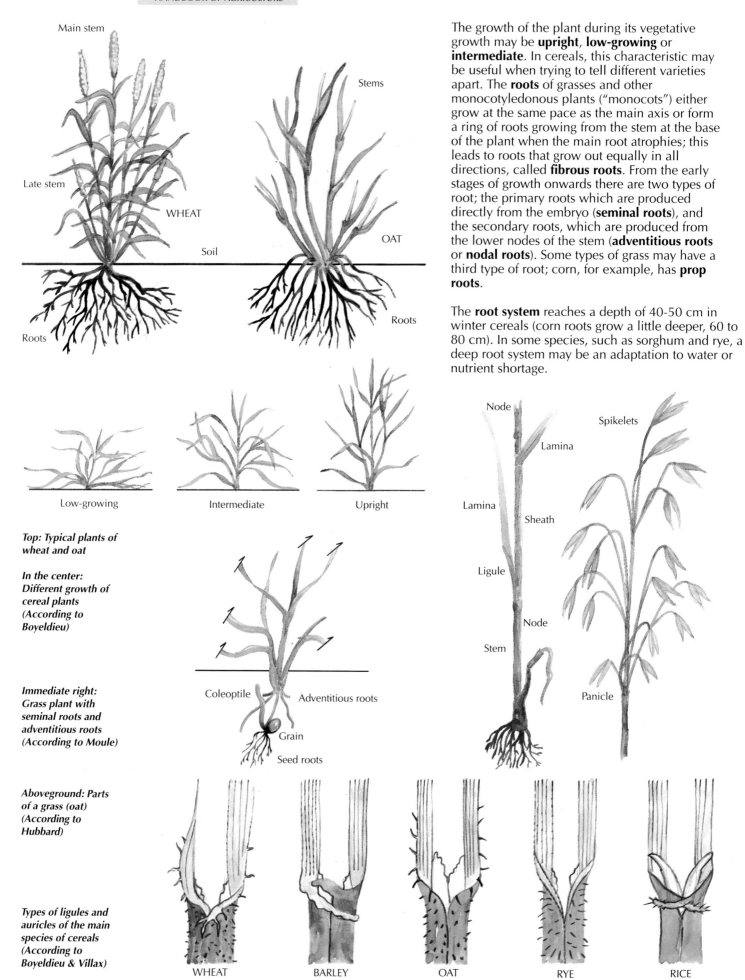

Main stem

Late stem

WHEAT

Soil

Roots

Stems

OAT

Roots

The growth of the plant during its vegetative growth may be **upright**, **low-growing** or **intermediate**. In cereals, this characteristic may be useful when trying to tell different varieties apart. The **roots** of grasses and other monocotyledonous plants ("monocots") either grow at the same pace as the main axis or form a ring of roots growing from the stem at the base of the plant when the main root atrophies; this leads to roots that grow out equally in all directions, called **fibrous roots**. From the early stages of growth onwards there are two types of root; the primary roots which are produced directly from the embryo (**seminal roots**), and the secondary roots, which are produced from the lower nodes of the stem (**adventitious roots** or **nodal roots**). Some types of grass may have a third type of root; corn, for example, has **prop roots**.

The **root system** reaches a depth of 40-50 cm in winter cereals (corn roots grow a little deeper, 60 to 80 cm). In some species, such as sorghum and rye, a deep root system may be an adaptation to water or nutrient shortage.

Low-growing

Intermediate

Upright

Node

Spikelets

Lamina

Lamina

Sheath

Ligule

Node

Stem

Panicle

Top: Typical plants of wheat and oat

In the center: Different growth of cereal plants (According to Boyeldieu)

Immediate right: Grass plant with seminal roots and adventitious roots (According to Moule)

Coleoptile

Adventitious roots

Grain

Seed roots

Aboveground: Parts of a grass (oat) (According to Hubbard)

Types of ligules and auricles of the main species of cereals (According to Boyeldieu & Villax)

WHEAT

BARLEY

OAT

RYE

RICE

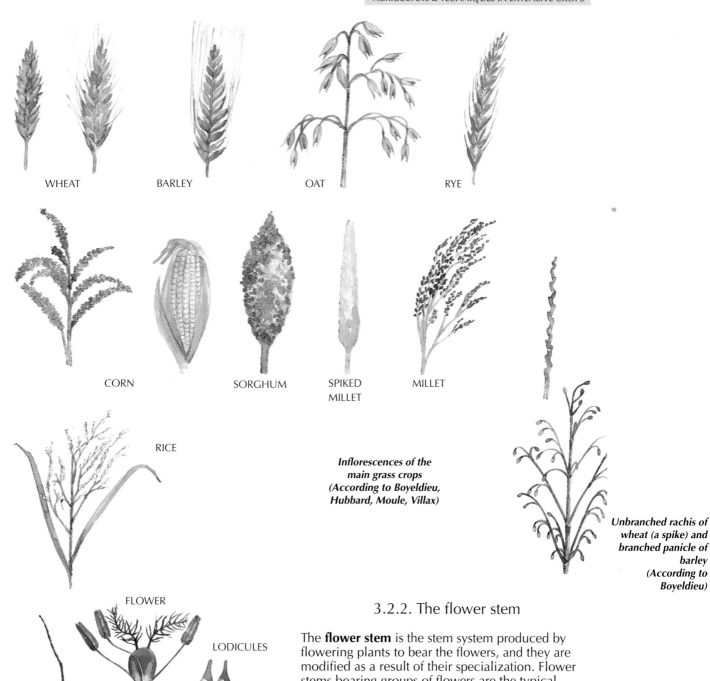

WHEAT   BARLEY   OAT   RYE

CORN   SORGHUM   SPIKED MILLET   MILLET

RICE

Inflorescences of the main grass crops (According to Boyeldieu, Hubbard, Moule, Villax)

Unbranched rachis of wheat (a spike) and branched panicle of barley (According to Boyeldieu)

FLOWER

LODICULES

GRAIN   PALEA

LEMMA

GLUME   GLUME

SPIKELET

3.2.2. The flower stem

The **flower stem** is the stem system produced by flowering plants to bear the flowers, and they are modified as a result of their specialization. Flower stems bearing groups of flowers are the typical reproductive unit of the grasses, and are called culms. The basic morphological unit of the grass flower stem is the **spikelet**. A spikelet consists of one or more flowers borne on a spike and protected at the base by two bracts, the **glumes**. There are two bracts that protect the reproductive organs of each flower, the upper is called the **palea** and the lower the **lemma**. These bracts may remain attached to the grain after threshing. The glumes may have bristles, **awns**, which are found in some varieties of wheat, barley, oats, etc. Each flower consists of three **stamens**, an **ovary** and two small scales at the base of the ovary, the **lodicules**. Grasses may have one of two types of inflorescence, either a **spike**, when the spikelets are directly joined to the spike (or **rachis**) which is unbranched, as in wheat and barley, or a **panicle**, which is branched, and bears the spikelets on the branches of the flowering stem, as in oats, rice, sorghum, etc.

Spikelet of barley broken down into its parts (According to Hubbard)

Transverse section of a spikelet (A) and detail (B) of a grass flower and its floral diagram (Taken from Boyeldieu)

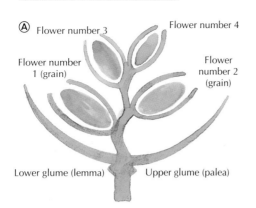

Ⓐ Flower number 3
Flower number 4
Flower number 1 (grain)
Flower number 2 (grain)
Lower glume (lemma)
Upper glume (palea)

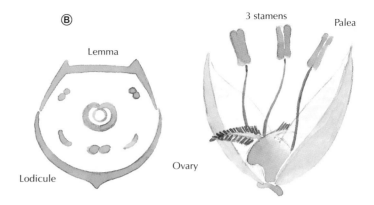

Ⓑ
3 stamens
Palea
Lemma
Ovary
Lodicule

3.2.3. The grain

The grains that are borne in inflorescences are examples of a caryopsis. They are nut-like indehiscent fruits with the seed coat united to the ovary wall. There are two types of caryopsis, the **naked caryopsis** and the **covered caryopsis**. A naked caryopsis is not covered by glumes or bracts (such as wheat, rye and corn), while a **covered caryopsis** is within the glumes or bracts, such as barley, oat and rice. The grain's structure is that of a monocot nut, with a large **endosperm**, containing the food reserves and a small compact **embryo**. The embryo consists of the **plantlet**, with its radicle protected by the **coleorrhiza**, the **plumule** (protected by the **coleoptile**) and the cotyledon, which is rich in starch. Much of the seed's protein is in the aleurone layer surrounding the endosperm.

Sections of a wheat grain (According to Boyeldieu)

3.3. THE GROWTH CYCLE

The life of a grass consists of a series of stages, phases or periods that make up its growth cycle. The concepts of stage, phase, period and cycle refer to progressively longer periods of time in the plant's life. The **stage** refers to a specific moment in the plant's development (germination, flowering, etc.); the **phase** is the time between two stages: the **period** consists of a set of phases in which the plants perform a given function (growth, ripening, etc.) and the **life cycle** is the sum of all the periods that occur in the plant's life. However, different works on this subject use the term period and phase indistinctly.

Different types of cereal grains (According to Boyeldieu & Villax)

Cereals, and annual grasses in general, die after the plant has flowered, seeded and completed its life cycle. The following pages are an introductory description of the different periods of the cereal's life cycle and they all follow the same structure; a description of the phenomena that occur, the factors that influence the different periods and their agronomic consequences.

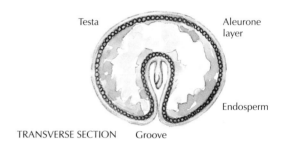

Testa
Aleurone layer
Endosperm
TRANSVERSE SECTION   Groove

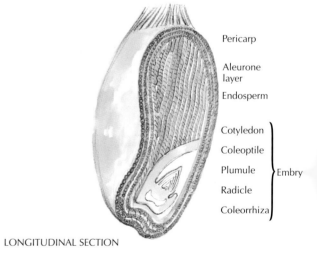

Pericarp
Aleurone layer
Endosperm
Cotyledon
Coleoptile
Plumule   } Embry
Radicle
Coleorrhiza
LONGITUDINAL SECTION

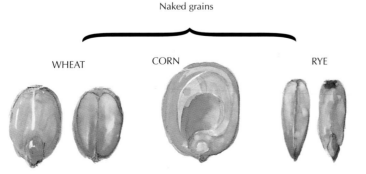

Naked grains
WHEAT
CORN
RYE

Covered grains
BARLEY
OAT
RICE
SORGHUM

PERIOD	PHASES	MOST IMPORTANT STAGES
VEGETATIVE	Germination Seedling growth Tillering	Germination 1 leaf, 2 leaves,... 1 tiller, 2 tillers,...
REPRODUCTIVE	Apical change Flower differentiation Meiosis and fertilization	State A, State B Spike formation, flower opening
RIPENING	Intense cell division Deposition of starches and proteins Drying out of grain	Milky grain Pasty grain Glassy grain

Table showing the most characteristic periods, phases and stages into which the life cycle of a grass is divided. (Taken from Pujol, M.)

3.3.1. Germination

The **vegetative period** is the part of the cereal's life cycle from germination to just before flowering; that is to say, from when the plant starts growth to the moment when it starts the morphological changes related to reproduction. The vegetative or growth period of the plant is thus asexual.

In the **germination phase**, the seed absorbs water, swells and produces the **coleoptile** and the **coleorrhiza**, which will give rise to the seminal roots. When the coleoptile emerges above the soil, germination has finished. Successful germination depends on a large number of factors. Soil moisture must be at least 35-50% for the seed to germinate. Cereals vary in their requirement for warmth (for winter cereals the temperatures needed for germination are lower, and for summer cereals they are higher).

Life cycle of a grass (wheat) (According to Boyeldieu)

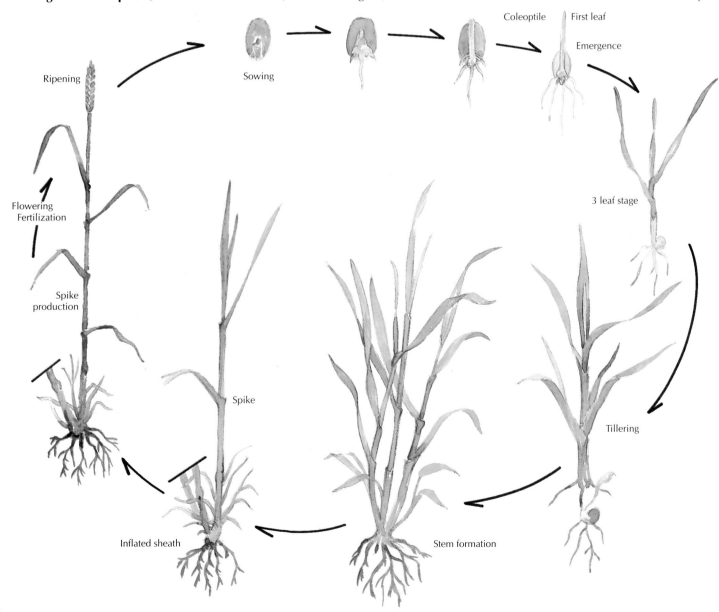

Ripening

Sowing

Coleoptile First leaf

Emergence

3 leaf stage

Flowering Fertilization

Spike production

Spike

Tillering

Inflated sheath

Stem formation

A/ *Influence of sowing depth on germination (According to Gillet)*

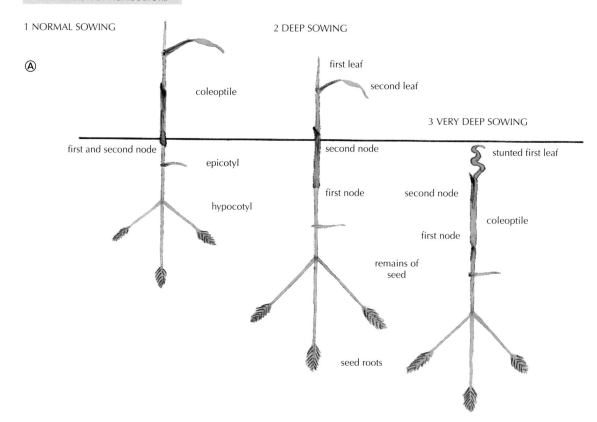

Ⓐ

1 NORMAL SOWING

2 DEEP SOWING

coleoptile

first leaf

second leaf

3 VERY DEEP SOWING

first and second node

second node

stunted first leaf

epicotyl

first node

second node

hypocotyl

coleoptile

first node

remains of seed

seed roots

B/ *Germination is faster at higher temperatures (According to Moule)*

Ⓑ

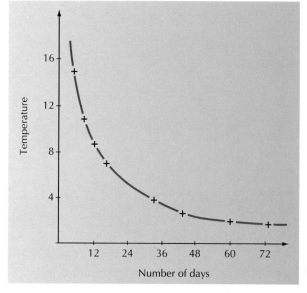

Ⓒ

germination

emergence

first leaf

coleoptile

coleoptile

coleorrhiza

seed roots

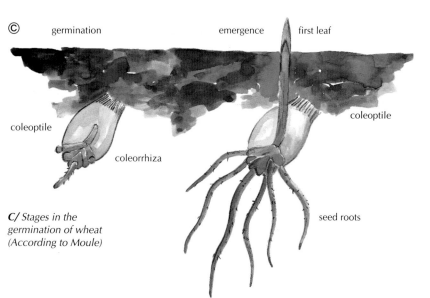

C/ *Stages in the germination of wheat (According to Moule)*

Yet germination is generally quicker at higher temperatures; the correct depth of sowing is about 4-5 cm (the deeper the seed is sown, the later the seedling emerges and the weaker and more vulnerable it is). It is important always to sow recently harvested seed, and check before sowing to see that it germinates well. To ensure the seed emerges correctly, it is important not to sow it too deep.

3.3.2. Growth

The germination phase is followed by the **growth phase**. The coleoptile is pierced by the emerging first leaf and then withers away. The first leaf grows and is followed by the second, third and fourth. When the plant has four leaves it normally produces its first tiller and then finishes the first development phase. The apical meristem regulates the plant's vegetative growth and is called the shoot tip. The apical meristem produces the leaves and the other meristematic tissues, as when the leaves are produced they bear a lateral meristem in their axil, and this will behave like the previous meristem and give rise to a new tiller. This means that there is a group of undifferentiated tissues in the soil that will give rise to all the later tillers.

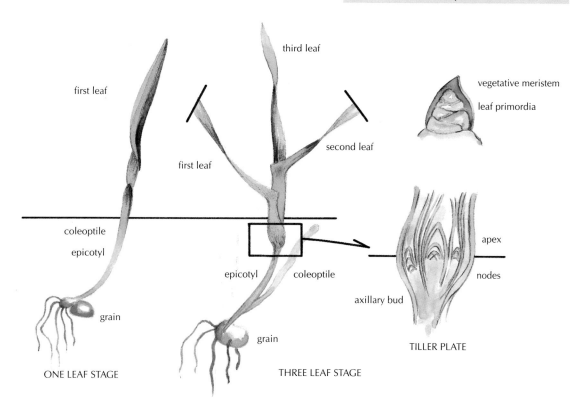

third leaf

first leaf

vegetative meristem

leaf primordia

first leaf

second leaf

coleoptile

epicotyl

apex

epicotyl   coleoptile

nodes

axillary bud

grain

grain

TILLER PLATE

ONE LEAF STAGE

THREE LEAF STAGE

This is known as the **tiller plate**. This meristematic zone is joined to the original grain by an underground stem called the **epicotyl**.

The essential factor for effective growth in this phase is the temperature. There is a linear relationship between the sum of the degrees of temperature and the production of a new leaf. This variation called the **heat integral**, ranges from 100 to 200°C and depends on the species. Other less important factors, such as light and nitrogen availability, accelerate the appearance of new leaves, slightly modifying the sum of temperatures needed. The epicotyl, like an umbilical cord between the aboveground part and the roots, may be attacked by insects or damaged by the cold. This is another reason to avoid sowing too deep.

The next phase is the **tillering phase**. The tiller is the basic unit of the grass's production, and tillers generally behave as independent plants. Each tiller develops from a lateral meristem in the **axil** of the leaf, and it grows new leaves. It becomes hard to tell the **main stem** from the tillers. The adventitious roots are produced when the tillers emerge, or shortly before. The ability to produce tillers is characteristic to each variety and depends on the cereal species. The tiller production phase finishes when the **reproductive phase** starts, due to the crop plants competing for light, space, water, etc.

Tiller production is not only affected by the species and variety but also by other factors such as the sowing time, due to the influence of the temperature, the availability of nitrogen which limits competition and favors tiller production, and the negative relationship between the depth of sowing and tiller production. As the tillers are borne in the axil of the leaves, if there are limits on the numbers of leaves, this will also reduce the number of tillers that can grow.

The most important factor in tiller production is light, but for it to have an effect it must reach the ground level, where the buds are located. Thus, when the sowing density is too high, little light reaches the soil level and the plants only produce a few tillers. High temperatures are unfavorable to tiller production, while rather low temperatures favor the formation of new leaves and new tillers. If the lighting conditions are optimal, an input of nitrogen

*Influence of sowing
depth on tiller
production
(According to
Boyeldieu)*

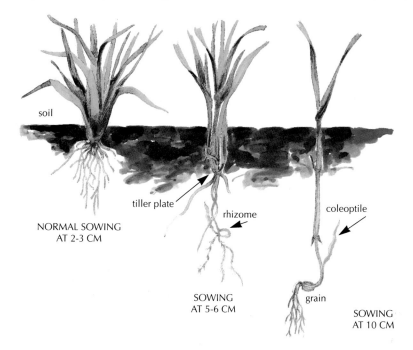

soil

tiller plate

rhizome

coleoptile

NORMAL SOWING
AT 2-3 CM

SOWING
AT 5-6 CM

grain

SOWING
AT 10 CM

Different stages in the reproductive period, also known as the spike formation stage in cereals

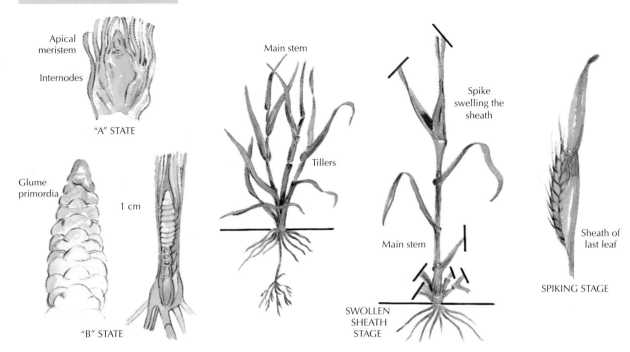

Apical meristem

Internodes

"A" STATE

Glume primordia

1 cm

"B" STATE

Main stem

Tillers

Spike swelling the sheath

Main stem

SWOLLEN SHEATH STAGE

Sheath of last leaf

SPIKING STAGE

favors tiller production. In certain circumstances, not only may new tillers fail to appear but the ones that have appeared may even die. This occurs in situations of great competition within the crop or it may occur when the reproductive period begins. Low sowing densities, if the variety and conditions are suitable, are compensated by high tiller production. In some circumstances tiller production can be induced by a complementary input of nitrogen.

3.3.3. The reproductive period

The beginning of the **reproductive period**, also known as **spike formation**, is marked by the moment when the vegetative meristem turns into a reproductive meristem. This is difficult to detect visually, as there are no externally visible changes. The morphological changes can only be seen in a cross-section of the tiny apical meristem. A series of changes take place in the apical meristem and stem during this period. The apex starts the branching that will lead to the inflorescence ("A" stage), this is followed by appearance of the glume primordia ("B" stage) and the flower begins to differentiate, which continues to meiosis (gamete formation). The stem elongates, slowly to begin with but very quickly once the apical meristem has reached the "B" stage. Stem growth later slows down and eventually stops. The stem's growth is due to elongation of the internodes, the leaves raising, which are arranged in two vertical rows running from the bottom of the stem to the top.

As the stem extends upwards, it raises the reproductive meristem, which is still protected by the sheaths of the youngest leaves, each of which is longer than the previous one. Eventually, only the sheath of the last leaf protects the inflorescence, and as the flower stalk swells it bursts through the

sheath. This is when the gametes are produced in meiosis. As the sheath shows little or no growth while the stem and inflorescence continue growing, the inflorescence emerges from the sheath and can be seen. This is the **spike stage**, which corresponds to the moment when the tip of the inflorescence emerges from the sheath.

The stem continues to grow after spike formation, a lot in some cereals but much less in others, depending on the variety and the species. When the stem has finished growing, flowering takes place; the glumes open, the stamens release the pollen, and the stigmas emerge. Fertilization occurs within a few hours, and is the **final stage** of the reproductive period. Other physiological events occur in this period; tiller production is inhibited and some tillers may even die; growth of new roots is inhibited and even the existing roots reduce mineral uptake; the plant's reserves are translocated and accumulate, in the cereals, in the sheaths and internodes.

The rate at which the different phases of the reproductive period occur is very important for grain production; if the flower stalk emerges too early, the spikes may be exposed to the cold, but if it emerges too late the ripening grain may be affected by excess heat or water shortage. The wide range of varieties of a single species that exists can adapt to a wide range of environmental conditions before and during the growth of the inflorescence.

3.3.4. Ripening of the grain

Throughout this period, the plant is photosynthesizing and translocating these reserves to the growing grain. In cereals the translocation of the reserves to the grain leads to the death of the plant.

This period consists of three important phases; a phase of intense cell division, when the grain increases in dry weight and absorbs a lot of water. After the cell division phase the grain has grown to its final shape, but it is still green. This is the **milky grain stage**. The second phase is an increase in the carbohydrate and protein content, the result of translocation to the grain from the rest of the plant. In this phase the weight of the water in the grain remains the same, but its dry weight increases greatly. At this moment, the grain cannot be squashed between the fingers, but it can be pierced with a fingernail; this is the **pasty grain stage**. In the third and final stage, the **desiccation stage**, the grain rapidly loses water, and its water content may easily decrease from 40% to 14% or less by the end of this period. In this phase, the grain successively goes through the **semi-hard**, **hard** and **glassy** stages.

The most important factors affecting the ripening of the grain are the temperature and the water supply. If sufficient water is available, the length of each phase depends on the temperature and the sunshine. This value is expressed as the sum of temperatures and is known as the **heat integral** (a value for the heat integral has been measured for each variety). Water shortage combined with high temperatures may cause the disorder known as **"shrivelling"**. Translocation of reserves to the grain represents a critical moment in the plant's water requirements. **Physiological maturity** is reached at the **pasty grain stage**, though the grain still contains too much moisture to be harvested and stored, and so you should wait till the desiccation phase to ensure the water content does not exceed 12-13%.

It is easy to harvest winter cereals with a suitable water content, but this is not true for summer cereals because they are normally harvested in the fall, a rainy season, meaning that some additional form of drying is also necessary.

How well early varieties and species of winter cereals adapt to each site is directly related to the conditions in which desiccation occurs. The oat, for example, needs mild conditions with some moisture because its second phase is very long, while barley adapts better than wheat to dry conditions because this phase is faster, among other reasons.

3.4. ECOLOGY

One notable characteristic of cereal crops is that they adapt very well to all the world's different agricultural environments. This adaptation is largely based on the variations shown by the different species of cereals, as well as the considerable variations between different varieties. The next section deals with the two main environmental factors that cereals adapt to, the climate and the soil.

3.4.1. Adaptation to the climate

The main climatic factors that may affect the crop's growth are the **temperature regime** (extreme summer or winter temperatures) and the **rainfall regime** (how much rain falls). In terms of temperature, cereals can be divided into two groups, the **cryophilic species** which require the cold and grow best in temperatures that are not excessively high, and the **thermophilic species**, which require high temperatures (summer cereals).

Winter cereals can be limited by the climate in the following ways:

1) When it is too cold to germinate; sowing in fall must be performed in suitable conditions.

2) When the plants cannot withstand the winter cold. Barley is the most sensitive cereal, while rye is the most cold-resistant (–18°C to –20°C). Wheat is more resistant than barley, but there are great differences between varieties in both species, with barley ranging from –8°C to –14°C, and wheat from –8°C to –16°C. The plant's resistance to the cold does not only depend on the lowest temperature reached but also on whether the temperature falls gradually or very sharply.

3) When there are not enough hours of cold for complete ripening. The differences between varieties are very important, as some varieties require 50 or 60 days with temperatures between 0°C and 5°C, while others hardly require any cold period at all. This point is of great importance when choosing a variety that will grow well in given zone.

4) When flowering is affected by late frosts or by excessively high temperatures, which might decrease pollination.

Summer cereals may be affected negatively by the following climatic conditions:

1) The minimum temperatures necessary for germination to occur. Corn is less demanding than millet and sorghum. Adaptation is achieved by sowing later and using varieties that have a shorter growing season.

2) Low temperatures during the crop's growth. Corn stops growing if the temperature falls below 8°C. Rice and sorghum show similar halts in growth, but at higher temperatures.

3) Excessively high temperature during the flowering period. Corn shows declines in yield if temperatures exceed 35°C, but sorghum, rice and millet do not.

4) The frost-free period marks the limits and the length of the growing season. At temperatures of –2°C or –3°C corn and sorghum freeze to death. If these frosts are frequent

Using assisted fertilization techniques it is possible to produce pure strains of wheat and barley, used in obtaining improved varieties. (Courtesy of ICI SEEDS)

in spring, sowing should be delayed; if frosts occur in fall, the ripening grain has to reach the pasty grain stage before the first frost. The availability of varieties with a very short growing season means that corn can now be cultivated in colder climates than before.

The different cereal crops can be arranged in order of their **water requirements**, from greatest to least: rice, corn, oat, wheat, sorghum and barley. Their water requirements vary as it depends on the type of soil, on how the cereal uses the water and the level of evaporation. When sufficient water is available the water consumed per tonne crop produced is about 300 m^3 for sorghum, 350 m^3 for corn, 500 to 550 m^3 for wheat, 600 m^3 for oats and rye, and 650 to 700 m^3 for rice. Their demand for water is not equal throughout their life cycle, and dry conditions are most negative during germination, spike formation and to a lesser extent during the ripening of the grain.

Most cereals adapt relatively well to soil water shortages and the yield decreases until conditions are again favorable. Rice is an exception as it requires a lot of water at all times.

3.4.2. Adaptation to the soil

Cereals are not very demanding in their soil requirements, but they all have their special characteristics. Wheat, for example, needs deep soils with a good water retention capacity; barley adapts well to light shallow soils; rye and oats can grow on acid nutrient-poor soils, while corn requires a lot of organic matter. The percentage content of mineral elements of crop and straw produced are similar in all the cereals (see table below).

Mineral composition of the main cereals expressed as % nutrient of grain (Taken from Boyaldieu)

3.5. VARIETIES

Cereal varieties may be of three types, **populations**, **pure strains** or **hybrids**. In order to understand the importance of hybrid varieties in cereal production we must introduce some of the basic concepts of genetics.

All living things bear, in the DNA of their cells, the genetic information that they will transmit to their descendants and which will determine their characteristics. Cereals can be divided into two groups on the basis of how they are fertilized; some are **self-fertilizing** (wheat, barley, oats, rice and sorghum) while **cross-fertilizing** species require pollination (corn, rye). Self-fertilizing varieties are usually **pure strains**, as they fertilize themselves. Cross-fertilizing plants, which always cross with other plants, give rise to **mixed populations**. Self-fertilizing plants, such as wheat, maintain their characteristics almost unaltered, as there is no genetic exchange between them. In other cereals, such as corn, which is cross-pollinated, the seeds obtained from the harvests are mixtures. **Populations** are varieties that consist of a set of individuals that share a number of characteristics, mainly those relating to adaptation to the environment; to the contrary, however, they are relatively heterogeneous in their other agricultural characteristics. Using assisted fertilization techniques, it is possible to obtain **pure strains**. A variety is considered a pure strain when all the individuals are genetically identical and homozygous. A series of individuals are homozygotic when their zygotes, or embryos, contain the same genetic information, and will give rise to individuals that are identical to their parents. A pure line becomes stable, generation after generation, if there is no mutation or recombination of genetic material with that of other varieties. Some pure strains of cereals on sale are resistant to some fungal diseases.

Hybrid varieties are derived from crossing two pure strains and they have the advantage of showing

Crop	Product	N	P_2O_5	K_2O	CaO	MgO
Wheat	Grain	1.90 %	1.00 %	0.5 %	0.2 %	0.20 %
	Straw	0.50 %	0.25 %	1.2 %	0.5 %	0.15 %
Barley	Grain	1.50 %	0.85 %	0.7 %	0.4 %	
	Straw	0.50 %	0.20 %	1.2 %	0.6 %	
Oat	Grain	1.80 %	0.90 %	0.7 %		
	Straw	0.60 %	0.40 %	1.7 %		
Rye	Grain	1.40 %	1.00 %	0.6 %		
	Straw	0.45 %	0.30 %	1.2 %		
Corn	Grain	1.50 %	0.75 %	0.5 %	0.3 %	0.10 %
	Straw	0.70 %	0.20 %	1.8 %	0.5 %	0.25 %

heterosis, or **hybrid vigor**. In hybrid varieties, all the individuals of the population are identical but heterozygous, meaning that they cannot produce offspring identical to themselves. The commercial varieties of corn and sorghum are hybrids. Pure lines of self-fertilizing plants can be maintained indefinitely, generation after generation, if the fields are kept free of other plants. Synthetic varieties may become unbalanced due to the selective effect of the environment on the individuals making up the initial population and they may lose productive potential. Finally, note that hybrid varieties do not breed true, that is to say their offspring show great variation, unlike their parents.

3.5.1. Characteristics of the varieties

When it comes to choosing variety, the farmer should choose the one offering the highest profit, that is to say, the **highest gross production**. The gross production is obtained by multiplying the yield in kg/ha by the sale price in money/kg. The price will depend on the product's **quality**, and the yield will depend on two factors, the variety's capacity or **potential** and its **regularity**.

3.5.1.1. Productivity

Productivity is defined as the production capacity of a cereal variety in ideal cultivation conditions; it is the same as the **maximum yield** the variety can produce. It is impossible to find the **absolute value** of the productivity for each variety. Tests based on comparing varieties can only measure the relative productivity of different varieties, or the different productivity a single variety in different areas. It is however possible to determine the factors due to the varieties, as some varieties show greater capacities than others, such as in the number of tillers produced.

A single variety, unless it is affected by incidents during its growth, can compensate for each factor; if the sowing density is low, more tillers will be produced and the spikes will be larger and the grains heavier. When certain limits are exceeded, however, it is not possible to compensate. Knowledge of the characteristics related to the yield, such as the variety's ability to produce tillers, spike size, etc., is important when deciding which cultivation techniques are most suitable.

All modern commercial varieties have a very high potential production. They differ in the reliability of the yield, and in their ability to adapt better to one site than another. To assess this ability to adapt it is necessary to carry out tests comparing the varieties. To sum up, the varieties' adaptation to a specific geographical area depends on two factors: its **growth rate** and its **resistance** to adverse environmental conditions.

3.5.1.2. Growth rate

The growth rate is defined by two characteristics: **earliness** and the **rhythm of alternation** (only in winter cereals). Early ripening (length of cycle in corn and sorghum) is one of the basic characteristics of any variety, because varieties that are too early (in a given area) may be affected by late frosts when forming the spike, while varieties that are too late may be affected by the heat (shrivelling) and may in any case lack water to terminate their cycle. In corn and sorghum, the length of the period is directly related to the productive potential of the different varieties and must be adapted to each zone.

The **degree of alternation** of the varieties of winter cereal represents the cold requirements of each variety. These needs condition the moment of sowing. and suggest which varieties may grow well in a given area and moment.

3.5.1.3. Resistance to environmental factors

Varieties vary greatly in their resistance to environmental factors. These characteristics can not be considered separately from the environment, as resistance to the cold is not important in a hot climate, and resistance to shrivelling is not important in irrigated crops. Resistance to fungal diseases is more important in some places than others, and resistant varieties should only be considered in sites where fungus is a problem.

Cold resistance is not very important in temperate-cold countries, as many species of cereal can be cultivated in very cold climates, such as northern Europe.

Spiked millet, like corn, is an out-breeding plant. Commercial varieties are usually hybrids bred in experimental fields, where the male flowers are prevented from pollinating the female ones, so that pollination can be controlled. (Courtesy of ICI SEEDS)

Remember that spring varieties have little or no resistance to the cold. When cultivating corn, it may be worth choosing varieties resistant to cold in their early stages, as this makes it possible to sow earlier in zones that would otherwise be late. **Resistance to flattening** is a very important characteristic of a variety, as this negatively affects production. Certain new varieties with a low growth form largely avoid this disorder. Resistance to flattening depends more on the plant's growth habit than on its genetic resistance. Obviously, the lower the plant, the more resistant it is to flattening. Barley is more sensitive than wheat.

Resistance to shrivelling may be of great importance in hot regions. Some varieties show much greater resistance to shrivelling than others.

Disease resistance is one of the most important factors in obtaining homogeneous harvests. If certain diseases are a problem in a given area, it is better to grow resistant varieties than to plan a series of later fungicidal treatments. Obtaining varieties resistant to different fungal diseases is one of the main aims of plant breeders.

The **pest resistance** of different cereal varieties is of less importance than resistance to fungal diseases, and geneticists have not selected many pest-resistant varieties, the reason why chemical methods are mainly used to combat them. There are some varieties that have been selected for their resistance to nematodes.

3.5.1.4. Quality

In countries where the sale price of cereals is fixed in advance, or when the price conditions do not refer to grain quality, the farmer may think that quality is irrelevant. In countries where liberalization of the cereal grains market has led to greater interest in quality, the farmer has had to chose varieties of greater quality than traditional varieties. The following quality factors should be mentioned:

• Some **physical characteristics** of the grain, such as its specific weight (the greater the weight, the greater the yield of flour), moisture content (excess moisture may be penalized by a lower price) and color (which is important in corn).

• The **technological suitability** of the grain, such as its suitability for making bread (a factor related to the variety), the percentage protein in barley for beer production, or the yield after milling in rice.

• The **fodder value** depends on the variety, too. Factors like the percentage of cellulose in oats, the amount of starch in barley, the amount of proteins in wheat and corn and the levels of the amino acid lysine in corn are all important in cereals that are intended for animal feed production.

3.5.2. The varieties available

In order for any cereal variety to be sold in Spain and the other EU countries, it has to be registered in the list of varieties of the Ministry of Agriculture. It can only be registered in this list after the Instituto Nacional de Semillas y Plantas de Vivero (The National Institute of Seeds and Greenhouse Plants, INSPV) has carried out the relevant tests over 2-3 years, and accepted it. Acceptance and registration of a variety in this Register shows that it is a stable uniform population that shows certain intrinsic advantages.

These lists of varieties often do not provide enough information on cultivation in certain areas with a given climate. The farmer may then turn to the local trials performed by local bodies (together with the INSPV), and we would also recommend consulting the technical reports regularly published by the companies that sell these varieties. There are hundreds of varieties of wheat, corn and barley, and there are dozens of varieties of rice, sorghum and oats. Describing all these varieties is far beyond the scope of this work, and the reader should refer to other publications.

4. CULTIVATION CYCLES

Depending on the length of the growing cycle, cereals can be divided into **winter cereals** and **summer cereals**. On the next page there is a calender of the sowing and harvesting dates of the main cereal crops cultivated in Spain, showing the range of average dates for sowing and harvesting in Spain. The broken line shows the percentage of harvests completed by a given date. Thus, for example, sowing six-rowed barley peaks on July 15, the date when 55% of all the barley cultivated in Spain is sown. Likewise, 48% of the barley harvest is collected on November 15. This range in the sowing and harvesting periods is mainly due to the large number of species and varieties on sale, and offers the farmer an excellent chance of drawing up an individual calendar for sowing, harvesting and crop rotation.

4.1. WINTER AND SUMMER CROPS

Winter cereals can be sown from the fall onwards (October-November) until the end of winter (February-March), depending on the zone and on the cold requirements of the varieties grown. The crop is harvested in the summer, from June onwards, and the harvest may last until August or even September.

These crops are better adapted to the climate of Spain than the summer crops, because in most cases they do not need to be irrigated. Also bear in mind the savings due to the fact that the grain does not have to be dried after the harvest.

The species and the many varieties differ in their sowing and harvesting times, and this means the farmer can divide the tasks better. This diversity of sowing dates also allows greater control of weeds, by making the sequential rotations of crops more flexible. The harvest can also be programmed so it is staggered (i.e., not all at once), though it takes place over a shorter period than sowing; the winter barleys ripen first, followed 8-15 days later by spring barleys and wheats. By modifying the sowing dates of the cereals, it is possible to extend the harvest period, thus reducing the risks of accidents (wind, hailstones) and most importantly, makes it possible to sow the following crop (double harvests).

The summer cereals are sown in spring (April-May) and are harvested in fall (September-November). The main "summer" cereals are rice, sorghum and corn. As shown in the accompanying graph, sowing is mainly from early spring to early summer. Corn and sorghum can be combined with winter cereals and can be sown at either of two times. The **first harvest** should be sown as soon as the soil temperature permits (April-May). The **second harvest** sowing is performed after the winter cereals (June-July). The end of the cycle, in both cases is usually around the time of the first fall frosts. In order to fit the growing cycle within these parameters, early or late varieties should be chosen that are apt for the local climate.

4.2. ALTERNATIVES AND ROTATIONS

Cereals almost always occupy much of the cultivated area of a farm. They are often the only crop cultivated, and then we talk of **totally cereal alternation**. The classic form of cereal cultivation is based on a **fallow period**. The fallow period is a rest that allows a plot to recover its fertility. Between fallow period and fallow period, different species (alternatives) are cultivated in a **crop rotation**: after the rotation of crops, the site is once more left fallow. In a crop rotation, the previous crop is known as the **preceding crop**, while the next crop is called the **following crop**.

Some of the reasons why a crop's potential production decreases are now understood; nutrient impoverishment, a gradual increase in diseases and soil pests, increasing levels of inhibitory substances produced directly by the crop itself, etc. In the past it was not understood why land "became exhausted",

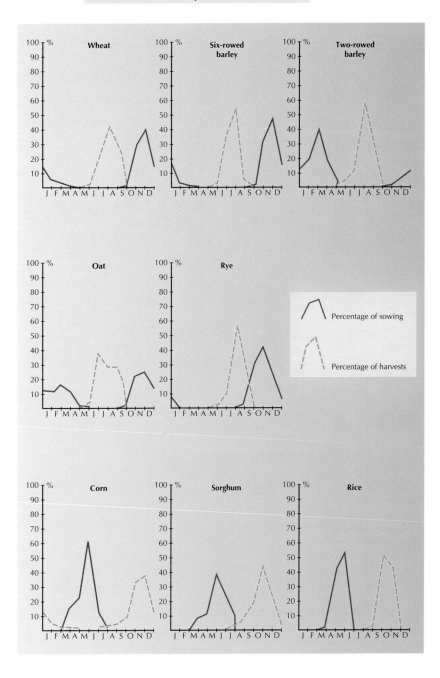

but fallow periods were planed, and this allowed the plot to recover some of its initial production potential. A cereal **monoculture** is a plot consisting of a single cereal species. Some crops can be grown as monocultures, but not all. Land used for growing rice, for example, is not usually suitable for other crops, meaning that monoculture is almost inevitable. If high and balanced nitrogen fertilizer is applied, corn can be cultivated on the same site for many years. In rice and corn the yield does not decline, maintaining a constant level of production. Barley can be grown on the same site for several years, but the yield is slightly lower each time it is planted. Wheat, unlike rice, corn and barley, can not be sown for more than two years in a row on the same site, because there is a large decrease in yield in the third year.

Calender of sowing and harvesting for the main cereals grown in Spain, according to the General Technical Secretariat of the Ministry of Agriculture

For the above reasons, each cereal crop has adopted a different role within contemporary agriculture. In order to intensify wheat production in dry farmed plots, the fallow period was abandoned and weedkillers and fertilizers came into use; yet these changes were not enough to maintain the productive level of the wheat, and so people in many areas opted to abandon wheat and grow barley instead, because it is not so sensitive to the effects of repeated cultivation, as we have already seen.

Wheat is still cultivated on the best arable land and rotations are still respected using a great variety of alternative crops. Thanks to fertilizers, especially to nitrogen fertilizers, areas that are irrigated have seen a drastic simplification of the number of alternative crops, with the farmers tending to grow monocultures of corn. The advantage of monoculture is that it allows specialization and makes production simpler, thus reducing production costs. Yet it has major disadvantages as work is concentrated at a few moments of the year, making the other months unproductive; business risks are not diversified: the costs of production increase, as expenses on pesticides, herbicides and

An example of crop rotation on a plot over eight years (Taken from M. Pujol)

Cultivation period	Crop
From November year 1 to June year 2:	winter cereals
From September year 2 to May year 3:	fodder cereals
From June year 3 to September year 3:	corn
From November year 3 to June year 4:	legumes
From November year 4 to June year 5:	winter cereals
From August year 5 to April year 6:	meadow grasses
From June year 6 to October year 6:	sorghum
From November year 6 to June year 7:	legumes
From November year 7 to June year 8:	winter cereals
From July year 8 to September year 8:	vetch
From November year 8 to June year 9:	fodder cereals

fertilizers have to increase in order to maintain the same productivity year after year.

4.2.1. Alternative crops

Almost all crops can be grown as alternatives to cereals, though some guidelines have to be respected. Wheat and corn are the crops that make best use of favorable preceding crops (such as legumes or a fallow year), and so they are normally placed at the beginning of the rotation. Barley and oats, however, make best use of the soil's lower fertility as they have lower production potential, and it is counterproductive to sow them in fertile soils as excess nitrogen may further increase their tendency to flattening. A preceding crop of sorghum is negative for all the straw cereals, especially wheat. The order of a crop rotation must also be logical. It is not, for example, a good idea to plan to follow a corn crop with a wheat crop, even though it is a good alternative, because there might not be enough time to prepare the soil and sow.

Pulses, such as vetch or beans, have for a long time been the crops that alternate with cereals. They have the great advantage of fixing atmospheric nitrogen into the soil (performed by the symbiotic bacteria living on their roots) and preventing the buildup of the pests, diseases and weeds associated with cereal crops. Legumes show many inconvenient features, however, such as their low productivity and the difficulty of mechanizing certain aspects of their cultivation, which have to be carried out by hand, greatly increasing the production costs. Another alternative to grain cereals are meadow grasses and fodder crops. Both aim to produce feed for animals, meaning that the farm must either have its own livestock or some way of selling this produce, which is not always possible. In dry farming crops, where water is a limiting factor, the number of possible alternative crops is much less, as in dry farming only winter cereals can be programmed. To finish, in some areas vetch is cultivated solely to be applied as a green manure. It is incorporated into the soil without ever being harvested, and raises the content of organic matter for the next crop.

To the left is an example of a crop rotation over eight years, including most of the crops grown in rotations. Of course it may not be possible to put all of these alternatives in practise, but it is useful as a guideline for a real case.

Legumes are a good preceding crop for wheat and barley, as the nitrogen-fixing Rhizobium *bacteria that grow on their roots supply nitrogen to the soil. The photo shows a crop of alfalfa (*Medicago sativa*). (Photo courtesy of EUROLIGO)*

5. SOIL PREPARATION

Soil preparation may start immediately after harvesting the preceding crop, and finishes at the time of sowing. When winter cereals are preceded by winter cereals, soil preparation is carried out in the summer months, from the moment the crop has been harvested in late July until the moment the next crop is sown, in late November. The tasks involved in soil preparation vary depending on the type of soil, its moisture level, the tools used and the sowing method chosen.

The aims of soil preparation for a new sowing are to ensure that the plants can thrive at the planned density, and to help the germination and growth of the crop sown. To sum up: **preparatory work** seeks to create a topsoil layer that is fine and compact enough for the seed to absorb water, and porous enough for the seed to obtain the oxygen it needs. This artificial soil structure should also allow the soil to warm up. When all these factors have been achieved, the soil is a **seedbed**. Before the preparatory surface plowing, a set of deeper operations are normally carried out in order to plow in the previous crop. This is known as **profile plowing**.

The soil humidity will determine the type of plowing to be performed before sowing. When the soil **lacks moisture**, it should be pressed together and the seed should be sown deeper. If there is **sufficient moisture**, then the main objective is to sow the seed at a homogeneous depth to ensure good tiller production. When the site is **excessively wet**, it is not a good idea to water after sowing, as this may cause the seed to rot. Some other points should be borne in mind when preparing the soil: in dry farming conditions, the structure attained after preparatory plowing should ensure maximum water storage and the best possible development of the cereal's roots, and plowing should eliminate as many weeds as possible.

Winter cereals are among the least demanding crops in terms of their soil preparation requirements. To begin with, this is because they are sown in a rainy season, when the seed does not require a lot of water, and grows rapidly because it does not grow very deep. So when preparing the soil for winter cereals the aim is to achieve an even tilth with small surface lumps (to prevent the formation of a surface crust) with a more compact layer below the seedling.

Summer cereals (corn and sorghum) require a different preparatory strategy. Evaporation is greater in summer and the soil surface soon dries out, so the emergence of the sorghum is more difficult as it must break through a hard surface crust. Special care should be paid to the preparation of the soil profile below the seedling, because these seedlings have a strong root system that needs to grow as deep as possible so the plant can have a good reserve of water.

5.1. SOIL PREPARATION STRATEGIES

The different ways of preparing the soil for growing **winter cereals** range from the classic and traditional system using the **plow** to the most simplified methods of **direct sowing** with no plowing. One or other is chosen for different agrarian and economic reasons, including optimization of water use, reducing the costs of cultivation, reducing the time needed for sowing and improving soil structure.
In dry farming conditions, the main strategy in soil preparation is to create and maintain the moisture conditions needed by the plants, and to plow the soil. When conditions are otherwise, that is to say, there is no problem of water shortage (irrigation or good rainfall), the limiting factor is time (and not water). In those cases where the preceding crop has been harvested and

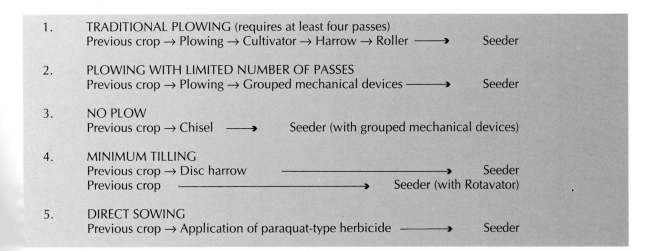

1. TRADITIONAL PLOWING (requires at least four passes)
    Previous crop → Plowing → Cultivator → Harrow → Roller ⟶ Seeder

2. PLOWING WITH LIMITED NUMBER OF PASSES
    Previous crop → Plowing → Grouped mechanical devices ⟶ Seeder

3. NO PLOW
    Previous crop → Chisel ⟶ Seeder (with grouped mechanical devices)

4. MINIMUM TILLING
    Previous crop → Disc harrow ⟶ Seeder
    Previous crop ⟶ Seeder (with Rotavator)

5. DIRECT SOWING
    Previous crop → Application of paraquat-type herbicide ⟶ Seeder

Soil preparation systems for cereals (Taken from Pujol, M.)

Deep plowing helps soil recovery and is essential to combat weeds and the residual effects of some herbicides, such as the triazines. It is also a useful way of incorporating chemical fertilizers and organic material deep into the soil.

the next crop should be sown rapidly, the simpler soil preparation methods, and even direct sowing, are justified.

On the previous page there is a diagram showing the possible systems of soil preparation for cereals. The diagram shows three methods differing in the depth of the plowing:

• **Deep plowing**, plowing and further work on the soil to prepare the tilth for sowing. This is the classic technique and is still practised by many farmers.

• **Surface plowing**, limited to a layer about 8-10 cm deep.

• **Direct sowing**, with the aid of special seeders that make it possible to work on hard ground.

Deep plowing helps the soil structure to recover, and is essential to combat the residual effects of some herbicides (**triazines**) and invasive weeds. It is also useful because it incorporates chemical fertilizers, as well as organic matter, deep into the soil. Other types of machinery can be used to get the same results as plowing. An example of this is the chisel, which is not widely used in Spain, or the different types of subsoil plows, which may show clear advantages when plowing sites where water is a limiting factor. An option that is an alternative to plowing is to get rid of the remains of the last crop with a relatively shallow disc harrow and then plow deep with the subsoil plow. The two together allow the root system to grow deep and leave the soil

prepared to withstand the storms of summer; it also means that normal plowing can be carried out in fall, even if it has not rained. It is normal when cultivating winter cereals to carry out **surface plowing** with a disc harrow or **rotavator**, as even though the soil is not plowed deep, merely preparing the seed bed is enough for these crops, especially if some organic matter and fertilizers can be incorporated into the surface layer when plowing, and measures are taken to eliminate weeds.

Finally, **direct sowing**, realized after the application of a total herbicide that does not leave toxic residues (such as **paraquat**), has the advantage of reducing the labor force required and the working time needed. In France, the normal practice of direct sowing of winter and spring cereals shows that the productive yield of traditional hoeing compared to direct sowing, is similar for the two different methods of site preparation. The application costs are roughly the same, but the direct sowing method saves money on of fuel and labor, but it does increase the expenditure on herbicide and machinery (a specialized seeder is very powerful and expensive). Direct sowing is a very interesting method if the soil has a good deep structure, and if weeds do not present a serious problem.

5.2 PREPARATORY PLOWING

The direct sowing method did not arise until it became possible to combat weeds with contact weedkillers.

Yet traditional preparatory plowing and crop alternation already offered an adequate solution for weeds. Immediately after the end of the cereal harvest, the field should receive a preliminary pass with a disc harrow to remove and bury the remains of the previous crop. This is done for three reasons: so summer rain can infiltrate into the soil better, to encourage weeds to germinate and to incorporate the stubble of the previous crop into the soil as an organic fertilizer for the following crop. This operation is very important and should be carried out before plowing.

If the farmer opts for **deep plowing**, the depth of the subsoil plow should be chosen on the basis of the soil type and the tractor power available (the deeper the plowing, the more tractor power is needed). The main purpose of this operation is to allow oxygen to enter deep into the soil. This is usually followed by **dunging**, and the organic matter is generally incorporated into the soil with a disc harrow.

Plowing, if it is carried out, should be done well before sowing, so that the soil is not too loose for the cereal's root system to grow, but leaving enough time to bury as many germinated weeds as possible when sowing.

After plowing, the **deep fertilizer** is usually

applied, preferably consisting of the fertilizer units of phosphorus and potassium. These two macronutrients, unlike nitrogen, show low mobility (they are insoluble) and should be applied close to the roots. The respective fertilizer units are often calculated for the crops in an annual rotation as a whole, and not for a particular crop. The "Law of the Minimum" postulates that: "*The size of the yield obtained is determined by the element present in the least quantity in relation to the crop's need*". This law shows the "solidarity of the fertilizer elements", that is to say, a shortage of a single essential element affects production, even if all the other

elements are present in sufficient quantities. In cereals this is true for phosphorus and potassium which, unlike nitrogen, are not determining factors in the volume of production even though their contribution is essential.

On the basis of the preceding points, the fertilizer units of the phosphorus and potassium to be incorporated with the disc harrow are determined on the basis of the following guidelines:

• For **soils rich** in phosphorus and potassium, the units extracted by the crops should only be replaced, in order to maintain the soil's fertility. The average extraction by cereal crops are of the order of 10 kg/ha of P_2O_5 and about 5 kg/ha of K_2O per 100 kg of grain produced, and these figures should be increased to take the soil factors into account (its pH, calcium levels, soil class, leaching, etc.) in order to reach a dose of fertilizer of about 60-80 kg/ha of each.

• For **soils impoverished** in phosphorus or potassium, deep fertilizer application should seek to correct this and to increase the levels of reserves. In this case, the application should be 30-50% greater than for rich soils, though this should be calculated for each soil.
The final phase is to pass the roller, which is done for several reasons. Before sowing, passing the roller presses the soil together so the soil profile is not so loose and helps penetration by the root system once the seed has germinated; this operation is called preparing the **seedbed**. After sowing, passing the roller brings the seed into closer contact with the soil, but this is a delicate operation, because depending on the soil's texture, this may cause formation of a surface crust, which would prevent the emergence of the seedling.

Left: Burying the stubble of the previous crop is a good agricultural practise as it enriches the soil with organic matter (Machinery sold by RABE WERK GmbH+Co.)

Below: Before passing the disc harrow, dung should be scattered onto the plot. Dunging may be a problem in relatively large sites. The AV 6000 ® model of the JF-FABRIKEN J. FREUDENDAHL A/S does this task very effectively.

6. SOWING

6.1. SOWING DATES

The most suitable sowing date for each crop species is shown in the diagram in chapter four of this section, which deals with the winter and summer cycles of cereals. As you will remember, there is a very broad range of possibilities for the date of sowing. This is due to the large number of commercial varieties of each crop species and their individual adaptations to different climates. The reader wishing to know the sowing date of a particular variety should consult specialist works, the catalogues of the seed distributors or the information bulletins of the Ministry of Agriculture.

6.2. DENSITY

There are two distinct tendencies in the sowing density of **winter cereals**. Some authors enthusiastically defend very dense sowing of wheat, while others suggest that less dense sowings are more productive. In fact most cereal species can be sown at a wide range of densities, mainly because of their ability to produce abundant tillers. Those supporting lower sowing densities cite the case of barley, where similar production levels have been achieved with a density as low as 50 plants/m^2 and as high as 800 plants/m^2.

As already pointed out in chapter three of this section, the cereal plant's ability to produce tillers depends on the sowing time and depth, the light, the temperature, soil nitrogen availability, and of course on the species and variety. It is thus difficult to give a figure for the sowing density, but as a guideline, production is thought to be optimal with about 450-500 spikes/m^2. Starting with the number of **spikes/m^2**, the number of plants needed per m^2 can be calculated in function of the variety and the other factors mentioned above. When the **planting density** (plants/m^2) has been decided, it is possible to calculate the **sowing dose**, the weight of seed needed (in kg/ha), which depends on the **number of seeds** and the **weight per 1,000 seeds**.

The number of seeds needed, the sowing density, will depend on the desired density of plants and the variety's **percentage germination**. The percentage germination is influenced by both the seed's ability to germinate and the germination conditions. The weight per 1,000 seeds is another factor that should be borne in mind. Because seeds vary in density, especially after the arrival of many new varieties, the same volume of two varieties of seeds may differ greatly in weight, and the same weight may consist of a different number of seeds. A new method was then established to evaluate the content of a given container, the weight per 1,000 grains of the cereal in question.

In general, products used to disinfect seeds for human consumption are much less toxic than those used to disinfect seeds for resowing. Treated seeds for resowing are obliged to include a coloring agent (generally reddish) in their formulation, in order to prevent mistakes and accidents.

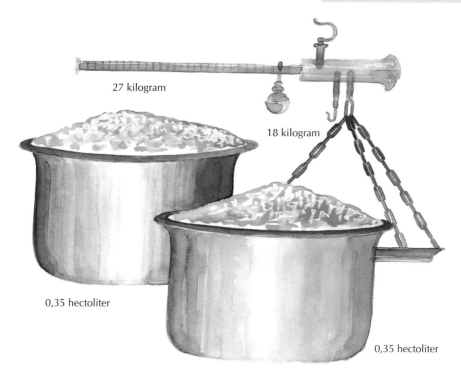

27 kilogram

18 kilogram

0,35 hectoliter

0,35 hectoliter

A given volume of two different varieties of seed often weighs a different amount.

Unlike winter cereals, summer cereals do not show this flexibility in the sowing dose, as they are much more demanding. Corn and sorghum have to be studied separately, and the reader should refer to the last chapter in this section, where these two crops are discussed in more detail.

6.3. SEED DISINFECTION

Disinfection of crop seeds is a very old and widespread practice that gives excellent results against **fungal diseases**. There are two types of fungicide: **contact fungicides** and **systemic fungicides**.

The most important chemical contact fungicides are the **organomercury compounds maneb, mancozeb** as well as **copper oxyquinolate**. Organomercury compounds are the most effective products against leaf blight of grasses (**Helminthosporium**) and against glume blotch (**Septoria**), but they may have a negative effect on seed germination if an excessive dose is applied. These products may affect germination negatively if they have been stored for some time in relatively moist conditions, and in addition they are highly toxic to humans and animals. Maneb and mancozeb are also effective but have the advantage of not affecting germination. Oxyquinolate is a product widely used to disinfect wheat seeds and it is very effective against

BECKER GmbH u. Co. KG sells this combined sowing equipment, which can treat the seed with fungicide just before sowing it.

The effectiveness of the main fungicides used in preventive disinfection of cereal seeds. Reproduced from "Les maladies des céréales", ITCF, 1983.

DISEASES / DOSE (grams of active ingredient per 100 kg of grain)	ORGANO-MERCURY COMPOUNDS	MANCOZEB		MANEB		COOPER OXYQUINOLATE	CARBOXIN		THIABENDAZOLE	ETHYRMOL
	3	96	160	80	160	30	75	100	100	650
BUNT	+++	+++	+++	+++	+++	+++	–	+	+++	–
NAKED SMUTS	–	–	–	–	–	–	++	+++	+	–
COVERED SMUTS	+++	++	+++	++	+++	+		+++	+++	
ERGOT	–	–	–	–	–	–	–	–	–	–
FUSARIUM Fusarium nivale	+++	+++	+++	+++	+++	++		+	++	–
Fusarium roseum	++	+	+	+	+	–		+	+++	–
HELMINTHOSPORIUM	+++	++	+++	++	+++	+	+	+	+	–
DOWNY MILDEW	–	–	–	–	–	–			–	++ (+)
STEM ROT	–	–	–	–	–	–			–	–
PARASITIC FLATTENING	–	–	–	–	–	–			–	–
RHYNCOSPORIUM	–	–	–	–	–	–			–	–
RUSTS	–	–	–	–	–	–			–	–
SEPTORIA	+++	+++	+++	+++	+++	++		+	+++	

Legend: +++ Highly effective; ++ Adequality effective; + Not very effective; – Ineffective.

bunt (*Tilletia*), **glume blotch** (*Septoria*) and *Fusarium nivale*, though it is not effective against the different forms of smut.

The **systemic fungicides** include the following active ingredients **carboxin, ethyrmol, thiabendazole** and **triadimenol**. These fungicides are both preventive and curative, unlike the contact fungicides, whose action is only preventive. Systemic fungicides are said to be systemic because they penetrate into the seed. They are also more persistent than the contact fungicides. Carboxin is especially active against the smuts that spread within the plant. Ethyrmol is only active against onion downy mildew before flowering; the seed should therefore be treated with a product with a broader action.

The reader should refer to the final chapter in this section which specifies for each crop the fungal diseases that may affect the seeds and the products that should be used to treat them. Before choosing a commercial product, the risk of losses due to fungal diseases should be assessed, in order to find out if the cost of the product will be compensated by the increase in production. The many fungicides on sale include formulations combining fungicides with insecticides and even with substances to repel birds.

6.4. SOWING THE SEED

Sowing the seed does not raise many problems. The most common practice is to sow **in lines**, which is technically better than scattering the seed (broadcasting). If the site is small, it is possible to use the traditional system of broadcasting by hand, but if the site is larger this can be done using machines, such as **frontal seeders** and **centrifugal fertilizer appliers**. Broadcasting the seed (by hand or using machinery) may lead to problems when trying to sow at a precise dose, as the seed is irregularly distributed. There are special sowing devices on the market to help to sow in a straight line, ensuring the seed is distributed with an adequate regulation of the distance between the rows, the number of grains per linear meter in each row and the depth of sowing. This type of machinery makes it possible to adjust the distance between the different rows.

A reference value for the separation between rows is 15 cm, though the optimum distance varies for each species and variety. An equidistant distribution has the advantage of letting light pass during vegetative growth, but it should be borne in mind that this distribution has the disadvantage of greatly increasing competition between plants in the same row.

The best **sowing depth** depends on the soil type and the moisture content; in wet soils, a depth of 3-4 cm is sufficient, but in light or dry soils, it is better to sow deeper at 5 or 6 cm or more. Seed should not be sown deep systematically, and this option should be kept in reserve for when a water shortage is feared during the germination period. It is important that the sowing depth should be **regular** (all the seeds should be at the same level), as this helps uniform emergence.

The frontal linear seeder sold by MASKINFABRIK A7S model SD 977®, allows sowing with an oscillation of 3 to 6 meters between rows and can sow 21 to 47 rows per pass.

7. TENDING THE CROP

Tending is the set of tasks that must be carried to ensure a successful harvest. In a broad sense, the work to be done begins with the preparation of the site and lasts until the harvest and removal of the stubble. In the strictest sense, and as the previous chapters dealt with soil preparations and sowing, this section deals with the work to be done for each crop in its growth cycle, dealing with harvesting later.

7.1. CULTIVATION TASKS

Though the operations that must be performed during cultivation vary from crop species to crop species, all the winter cereals require more or less the same treatment. The tasks that normally have to be done when cultivating **winter cereals** can be divided, by their timing, into four groups corresponding to the four stages of the crop's growth, from germination to harvest.

- Soil preparation
- Sowing
- Post-sowing operations, **surface** fertilizer application, treatments, irrigation, etc.
- Harvesting

Each group includes a series of tasks, as shown in the table on this page and the next. The first is a table outlining the tasks to be carried out for winter cereals, and also showing when they should be performed in a temperate climate, and how long they take. It is important to know how long they take, as this data makes it possible to organize the work and to calculate the economic costs of the crop. The time taken by these tasks (expressed in hours/hectare), is only a guideline. The farmer should work out how long they will take on the basis of personal experience, the machinery used, the

availability of labor, etc. For spring cereals, the tasks to be done are similar, but they are not done at the same time, as shown in the second table.

Soil preparation for sowing, must include cutting down the previous crop, passing the subsoil plow, dunging, plowing and levelling the soil. Sowing consists of **applying fertilizer at depth** and sowing the seed. The operations after sowing include passing the **roller**, surface application of fertilizer and the corresponding treatments with weedkiller. The final stage is harvesting, which consists of collecting the straw, **baling** it and **removing** the bales from the plot. The tasks to be done for spring cereals are similar to those described.

The essential difference between these cereals and the **summer cereals** (corn, sorghum and rice) is that summer cereals usually require irrigation, as they are cultivated in summer. The plot should be irrigated after sowing, and irrigation requirements should be calculated for each crop on the basis of their individual characteristics (see Chapter 13). In the case of corn, this requires both irrigation and a series of complementary tasks included in the section on post-sowing operations, which are described next.

• Thinning out
In small plots where corn is still sown by hand, thinning out must also be done by hand. This operation consists of uprooting the excess plants when they are 30-40 cm tall. This operation greatly increases the cost of cultivation, but is not necessary where a precision pneumatic sower has been used.

• Tassel removal
It is common practise to remove the tassel from the corn after fertilization, which can be recognized because the pistils on the cob wither. This is usually done to use the tassels as green feed for livestock. This practise cannot be recommended as it reduces the final yield, because it encourages early ripening of the grain

Cultivation tasks		WHEN PERFORMED	TIME REQUIRED (hours/hectare)
SOIL PREPARATION	Cut down	25 July - 15 August	1.5
	Subsoil plowing	15 August - 30 September	2.5
	Dunging	25 July - 30 September	9.0
	Plowing	1 September - 15 October	3.2
	Levelling		1.5
SOWING	Deep fertilization	15 October - 30 November	0.75
	Sowing		1.20
POST-SOWING OPERATIONS	Roller	15 October - 15 November	1.00
	Surface fertilization	15 February - 15 April	0.75
	Weedkiller treatment	15 March - 15 April	1.00
HARVESTING	Harvest	25 June - 15 July	1.70
	Bale straw	30 June - 20 July	0.80
	Collect bales	30 June - 30 July	4.50

Note: not all plots are dunged every year.

Example of a calender of agricultural tasks for winter cereals and the time needed to perform them (Taken from Gorche)

Example of a calendar for agricultural tasks for spring cereals and the time needed to carry them out (Taken from Gorche)

Cultivation tasks		WHEN PERFORMED	TIME REQUIRED (Hours/hectare)
SOIL PREPARATION	Cut down	1 December - 15 December	1.5
	Subsoil plowing	15 December - 30 December	2.5
	Dunging	1 December - 30 January	9.0
	Plowing	1 January - 30 January	3.2
	Levelling	1 February - 15 February	1.5
SOWING	Deep fertilization	1 February - 30 March	0.75
	Sowing		1.20
WEEDKILLER TREATMENT		1 April - 30 April	1.00
HARVEST	Harvest	10 July - 20 July	1.70
	Bale straw	10 July - 30 July	0.80
	Collect bales		4.50

Note: Some soil preparation tasks (cutting down the previous crop, subsoil plowing) can be carried out before the dates indicated. Weedkiller is not always applied.

at the expense of the final weight. It is better to dedicate a small area of the plot to producing fodder corn, which will more than compensate for the fodder obtained from the tassels.

7.2. WEEDING

Mechanical weeding is only carried out in corn (of all cereal crops), as it is sown in straight lines and produces almost no tillers, meaning that machinery can be used along the rows. Mechanical weeding should start when the corn has at least four leaves, when it has produced enough roots. The first weeding should not be delayed because the first stages of corn's development are very sensitive to competition from weeds. Weeding should be continued in order to keep the soil free of weeds, but it should only be superficial, so as not to damage the small roots.

Graph showing the accumulated absorbtion of nitrogen by wheat, in the stems and leaves, grain and husks

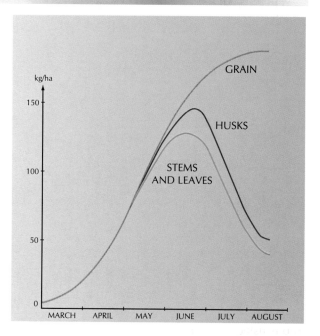

7.3. SURFACE FERTILIZER APPLICATION

Surface fertilizer application complements the deep fertilizer application described in the section on soil preparation. The deep fertilizer provides phosphorus and potassium, elements that show little mobility in the soil, but **surface fertilizer application** supplies the fertilizer units of nitrogen in distinct stages. The plant's total nitrogen needs are normally met by three applications, the first third together with the deep fertilizer application, and the other two thirds as surface applications during the growing season. Nitrogen is the main factor influencing yield in cereals, but this fertilizer has many effects, not all of them positive. As it has a dramatic effect on vegetative growth, it may unbalance the plant's growth. Excess nitrogen at the wrong moment may cause a decrease in the final yield, just as shortage of nitrogen when it is needed will decrease production. The nitrogen compounds are not applied entirely at depth because, as already

mentioned, nitrogen is a mobile element and can easily be washed from the soil by rainfall. Organic nitrogen (released by breakdown of organic matter) can be stored in the soil, but it can only be gradually absorbed by the plant, at the rate that microorganisms break it down and release it into the soil. In order for these microorganisms to break the organic matter down, the soil environment must meet certain conditions, such as warmth, moisture, and this often means the plant's requirements do not correspond to those of the microbial activity.

In winter cereals, for example, the release of organic nitrogen often occurs too late (temperature rise in spring). For corn, however, part of the nitrogen may be released at the time of germination and growth, and this should be taken into account when calculating the dose of fertilizer. These are the reasons why the nitrogen fertilizer is divided into more than one application, in order to adapt it to the crop's needs and to reduce losses by leaching.

The nitrogen requirements of wheat have probably been studied more thoroughly than those of any other crop. The growing wheat plant absorbs little nitrogen until it starts to produce tillers, when absorption starts to increase to its peak value (up to 2 kg of nitrogen per hectare per day) during the spike production phase. After flowering, nitrogen uptake decreases to zero, as the grain accumulates the nitrogen stored within the plant (nutrient translocation).

7.3.1. The effect of nitrogen on production

A crop plant's yield is determined by the variety, the sowing dose, etc., but applying nitrogen-containing fertilizers can increase the yield in favor of the farmer. In the case of wheat, for example, nitrogen has an effect on tiller production, allowing the farmer to induce more tiller production(and therefore production), if the rainfall conditions allow, by applying more nitrogen. To the contrary, if spring is too dry, the nitrogen applied can be decreased in order to partially inhibit tiller production, which makes the plant more drought-resistant.
Some authors coincide in stating that one of the factors determining the number of grains in the spike is good nitrogen nutrition during stem production. Nitrogen supply during stem and spike production increases the grain's weight and its protein content, something of special importance in durum wheat.

7.3.2. Nitrogen dose and distribution

The **total dose** of nitrogen required by each crop is listed in Chapter 13 of this section, which discusses each cereal in more detail. Calculating the total dose requires knowledge of many technical factors, which cannot be dealt with in detail here. To mention a few, these factors include the crop's needs, the plant remains left after the harvest, the assimilable mineral nitrogen left in the soil, the percentage of nitrogen that is mineralized, etc. Mineralization depends on the soil type, the crops cultivated in the past and, above all, the level of organic matter in the soil. The dynamics of nitrogen in the soil are dealt with in the section on soils and fertilizers.
When estimating the amount of nitrogen required it is very important to take into account whether the straw from the previous crop was buried. If this was done, the microbial decomposition will require a nitrogen supplement of 35 kg/ha. If this is not done, the microorganisms will compete with the crop for the nitrogen, and this may seriously reduce the harvest. We have already pointed out that the **distribution** of nitrogen over the growing cycle can be carried out partly (40-60 fertilizer units) with the fertilizer

application at depth, though it is not always necessary. Looking back to the graph of the absorbtion of nitrogen, you can see that the greatest demand is from tiller production to stem production, and then decreases, showing that nitrogen fertilizer applied in late fall, just before sowing, will only be used until tiller production, if it is not lost by leaching. This is why many agronomists think the nitrogen should only be applied on the surface, at three precise moment's of the plant life, i.e., tiller production, stem production and spike production.

Good harvests require the correct application of nitrogen.

Photo courtesy of SHELL

8. IRRIGATION

Cereals use the available soil water very effectively, and they often resist dry periods very well. In many areas cereals are dry farmed, among other reasons because the lack of irrigation means that the land can not be used for any other purpose. There are, however, some very good reasons for installing irrigation in plots where cereals are cultivated; in the first place because in extremely dry conditions not even the cereals best adapted to water shortage can produce a crop; in the second place, if the distribution of rainfall is highly irregular, it is not possible to fix the calender for cultivation because water has to be supplied in case of need (even if it does not rain); and in the third place because if production is high (applying suitable technology), a relative lack of water is the main factor limiting production.

grain quality, but a water shortage may cause its value to decrease.

• The other winter cereals are not usually irrigated.
The winter cereals, which grow when evaporation is low and rainfall is high, differ greatly from the summer cereals, which grow at a time that is hotter and when less rain falls. Maintaining the classification into winter and summer cereals:

• For **winter cereals**, taking wheat as the most productive and most demanding crop, irrigation should be considered as a complement in case the rains fail. Many authors disagree on the moment to apply water in wheat, though it seems certain that during the germination of the grain, a few days before stem production and during spike production, are the three moments when

Rice is cultivated in flooded soils and needs more water than any other cereal.
(Courtesy of the Department of Agriculture, Stockraising and Fisheries of the Generalitat of Catalonia)

As a guideline, one should always ask whether the cereal's production measured in economic terms (which is always lower than that of more intensive crops) justifies the high costs of installing irrigation.

Not all species of cereal show the same sensitivity to water shortage or respond the same way. Furthermore, the life cycle of the different crop species and varieties may adapt better or worse to the climatic conditions of the cultivation site. Agronomists make the follow distinctions between cereals:

• Rice is cultivated on flooded soils and is the cereal that requires the most water.

• Corn is a typical summer cereal needing irrigation. Except on very wet areas, it can not be cultivated without artificial inputs of water.

• Sorghum requires less water than corn. Though it is preferentially cultivated with irrigation (it is a summer cereal), it is frequently cultivated in dry farming areas with sufficient rainfall.

• Whea, because of its great importance as a winter cereal and its high economic productivity, is irrigated (when necessary) in irrigated areas because, for this crop, water is a factor limiting production: fertilizers and technology are applied to wheat to improve

water shortage leads to decreased production. Barley, compared to wheat, needs less water at the end of its life cycle, and so less irrigation is recommended.

• In **summer cereals**, corn is clearly different from both sorghum and rice. Irrigation of rice crops is special, and is considered in the last chapter of this work. Sorghum and corn have similar growth cycles but their water needs are clearly different, and the farmer should opt for sorghum if there is not enough water available for corn, but should always chose corn whenever there is enough water for it to grow, as corn is of greater economic value and is more profitable.

8.1. IRRIGATION SYSTEMS

Not all of the irrigation systems available are suitable for extensive crops. Drip irrigation, for example, is used in greenhouses for flower production, and even in the open air for intensive horticultural crops, but is totally impractical for cereal crops because of its high construction and maintenance costs. The irrigation systems used in extensive crops can be divided into two groups: the different ways of flooding the surface that have been used since antiquity, **flood irrigation** (also known as **gravity irrigation**) and irrigation by artificial raindrops, **sprinkler irrigation**.

*The main irrigation
systems:*
A/ by furrows
B/ by ridges
C/ as a sheet of water
D/ by strips

8.1.1. Surface irrigation

• Strip irrigation

In strip irrigation, the strips receive water through one of the shorter sides. The water flows slowly as a sheet throughout the irrigation period and is channelled by longitudinal ridges of soil. The lowest edge remains open. The ridges are interrupted a few meters before the end of the strip and the excess water flows into a drainage channel. This system can be recommended for large areas of alfalfa, grazing land and cereals, but it has the disadvantage of only being suitable in a narrow range of soil profiles. If the slope is greater than 6‰, small, recently germinated seedlings may be washed away. In these cases, a provisional irrigation system should be adopted until the crop is firmly rooted. The strip must be horizontal (from side to side) to ensure uniform distribution of the water throughout the width of the strip. Strip irrigation is a cheap, simple and effective method which usually has advantages over other irrigation methods in large areas of crops. On sandy soils, the water must circulate quickly to prevent deep percolation; on clay soils, to the contrary, the water should circulate slowly so it can be absorbed by the soil.

• Irrigation by furrows

In the preceding system, the irrigation water covers the entire plot and gradually filters downwards into the soil. In furrow irrigation systems only part of the soil directly receives water, and the rest is wetted by lateral movement of water. Relatively large amounts of water flow along the furrows, but they are small in comparison with strip irrigation. This method is commonly applied in intensive horticultural crops, and has a series of advantages.

On the one hand, plants with a trailing growth habit do not get wet, thus preventing the spread of fungal diseases. On clay soils the earth does not form a crust and only cracks partially, thus reducing water losses. It permits use of a smaller volume of water, thus reducing the risk or soil erosion, and this makes it suitable for use on relatively steep slopes. This irrigation system has to be used in crops that are planted in lines close to each other, such as potatoes, pulses and corn. One variation on furrow irrigation is **irrigation by corrugation** in which the furrows are shallow and are supplied by tubes running at right angles to the slope of the site, and located at the highest level on the plot.

• Sheet irrigation

The preceding methods make the water flow over the surface of the site throughout the irrigation period, with a volume of flow that barely exceeds the amount that infiltrates into the soil. When the site is almost flat, with a slope of less than 1.5‰, it is difficult to make the water flow in a thin sheet, and so a volume of flow greater than the soil's permeability is used. The water accumulates on the surface and does not flow for much of the infiltration time. It is also known as **flood irrigation**. This system has the disadvantage of causing the soil to settle excessively. This is the typical method of irrigation for rice crops.

8.1.2. Sprinkler irrigation

Sprinkler irrigation appeared around 1930, when advances in metalworking made it possible to construct lightweight conduits and ingenious distribution systems. The idea is that the crop is watered by raised sprinklers, creating an "artificial rain". When planning this system it is not necessary to take into account the site's slope and soil class; it is suitable for flat and sloping sites, as well as very permeable (sandy) sites and compact (clay) ones, because the quantity of water supplied by this irrigation method can be precisely measured.

A sprinkler irrigation system basically consists of the following items:

• A pumping system to raise the water pressure. In a small plot a simple electric pump is enough, while large sites require a complex high power system.

• A network of tubing to take the water to the hydrants, the water outlets in the plot where the user connects the irrigation equipment. The water reaches each hydrant at a known pressure and rate of flow.

• A network of distribution tubing to channel the water through the plot to be watered. This requires both main branches or feeder channels, which distribute the water throughout the plot, and the **lateral channels**, branches of the main branches and which channel the water to the sprinklers.

• The sprinklers are devices that emit the water in a form similar to rain.

A detailed description of all the elements of an irrigation system would be far beyond the scope of this work, but a brief consideration of sprinklers is needed, as they are the most important elements of the entire system. They break down the jet of water into very small droplets and spread them evenly over the site. There are several types, such as **perforated tubing**, **rotating sprinklers** and **non-rotating sprinklers**. Perforated tubing and non-rotating sprinklers are used in gardening, greenhouses and small horticultural exploitations.

In agriculture, the most widely used ones are **rotary sprinklers**, with one or two nozzles. The sprinkler turns on its axis, powered by the water pressure, and thus waters a circular area. Sectorial sprinklers, used on the edges of plots and near the rows, do not irrigate a complete circle to avoid wasting water. Rotary sprinklers can be classified in several ways, though the most relevant to the farmer is the pressure they work at.

• **Low pressure** devices work at pressures of less than 2 kg/cm^2. They usually have a rate of flow of less than 1,000 l per hour and they are installed at spacings of less than 12-15 m. They are used in gardening, small vegetable plots and to water fruit trees. These sprinklers can also be used to protect plants from frost (see part 4: Damage to cultivated plants).

• **Medium pressure** devices work at pressures of 2-4 kg/cm^2, and have a flow of 1,000 to 6,000 l/hour. They can water an area of 12 to 24 m^2 of crop and are used on a wide variety of soils and extensive crops.

• **High pressure** devices work at pressures of more than 4 kg/cm^2 with a volume of flow of more than 6,000 l/hour. This category includes irrigation cannons. As the volume of flow is so high, they have some disadvantages, including the fact that they are severely affected by strong gusts of wind and produce very heavy drops that may damage some crops and some soils.

In addition to the differences in working pressure, the farmer must also take two more factors into account when deciding which sprinkler system to buy: the **special features of the crop** and the **functional characteristics** of the different sprinklers.

• The total volume of irrigation water, its distribution and drop size are all factors to bear in mind when buying sprinklers. The type of crop, its growth habit and its ground cover all condition the type of installation. Some very fragile crops (flowers, some horticultural crops, etc.) need very gentle sprinkling with very small droplets and uniform levels of watering, and this means the sprinklers have to be close together. In tall crops (such as corn, sunflowers, etc.), you should water at a medium or high rates. In grass meadows, sprinklers with a large range should be used to reduce costs.

To find the **functional characteristics** of the different types of sprinklers, as well as those of the different irrigation systems that we will discuss next, the farmer should consult the different technical and publicity leaflets from commercial suppliers, whose technical staff can usually offer clients the most appropriate solutions for their special case. Sprinkler irrigation systems can be divided into two main groups, **stationary systems** and **mechanized systems**. Stationary systems, unlike mechanized ones, do not move over the soil during irrigation.

8.1.2.1. Stationary systems

Stationary irrigation systems may be semi-fixed or fixed, and fixed systems are also known as **total coverage** systems. **Semi-fixed stationary systems** consist of distribution branches, which are fixed, and the lateral branches, which are mobile. Thus the main PVC (polyvinylchloride) or fibrocement tubing is buried in position and the water outlets are then connected. Each outlet is connected to a mobile aluminium lateral branch, which can be adjusted to a given number of irrigation positions (see figure A). Buried **total coverage systems** consist of a PVC or fibrocement general tube system which bears the secondary tubing (of PVC). These in turn bear the tertiary tubes (made of high density polyethylene), to which the sprinklers are fixed. Irrigation is carried out simultaneously with several lines of sprinklers, as shown in figure B. The arrangement of the different tubes in these two irrigation systems may vary from manufacturer to manufacturer, but the underlying idea is the same.

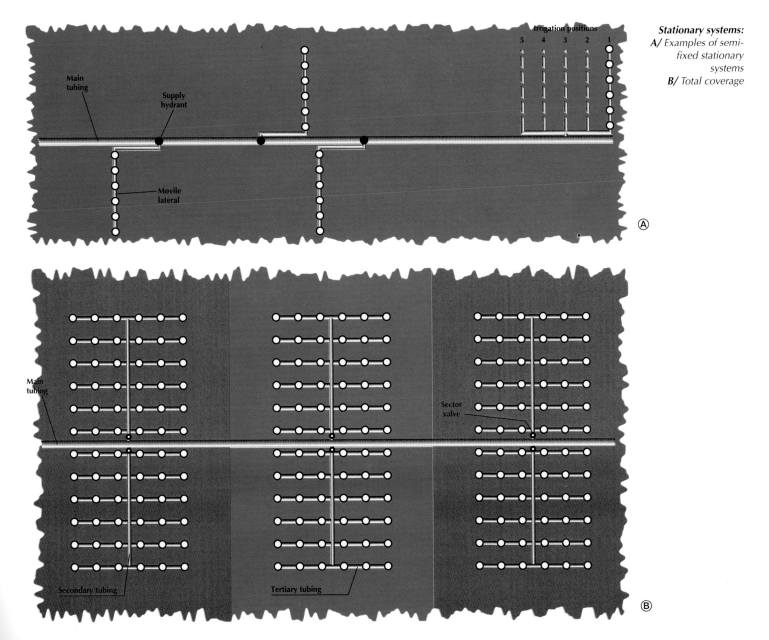

Stationary systems:
A/ Examples of semi-fixed stationary systems
B/ Total coverage

The Pivot from the Agrocaja company is an irrigation machine consisting of a metal structure and is treated with anti-corrosion products it is an effective way of ensuring higher yields per hectare irrigated.

The motorized irrigation cannon supplies water to crops, and can water strips more than 100 m wide and 500 m long. (Courtesy of AGROCAJA)

Photo upper right: the photo shows the large area reached by mechanical structures of the pivot type. (Photo courtesy of AGROCAJA)

Photo lower right: This frontally advancing lateral has a structure similar to the pivot. It is suitable for irrigating rectangular sites. (Photo courtesy of AGROCAJA)

articulated arms, each one supported by a metal tower. All the stretches are kept in alignment by means of sensors that act on the motor system. The length of the irrigation equipment can vary from 50 m to 800 m, meaning it is possible to select a model suitable for any size of farm.

The **frontally advancing lateral** system has a similar structure to the pivot system. It consists of an irrigation device mounted on supports that move and irrigate rectangular areas of soil. The water supply is taken directly from an irrigation channel that runs parallel to the edge of the plot, or from a flexible tube that is connected to water outlets sited on the edge of the field. In comparison with the pivot system, this has the advantage of watering a larger percentage of the ground area, as the areas irrigated are rectangular rather than circular, and most cultivated fields are rectangular.

8.1.2.2. Mechanized systems

Mechanized irrigation systems are relatively complex machines that are moved over the site during irrigation. The **motorized irrigation cannon** consist of a long distance, high-volume, spray (a "cannon") mounted on a frame or on rollers and connected to the water supply by a hose. The irrigation equipment always waters backwards, watering the soil it has passed over, so that it is always moving on dry ground. In one irrigation position, strips more than 100 m wide and up to 500 m long can be irrigated.

The **pivot** is a machine consisting of a metal structure that supports tubes bearing nozzles. The machine rotates around a fixed point (the pivot point) where it receives the water and electricity supply, and where the controls are located. The area watered is circular. An irrigation system of this type consist of several

9. AGRICULTURAL MECHANICS

This chapter seeks to introduce the farmer to the complex world of agricultural machinery, and lists the main machines suitable for basic agricultural tasks. Even so, the reader will have to refer to more specialized works for certain specialized but widely used widely machinery, as they are beyond the scope of this work. These include combine harvesters and special machinery for harvesting specific crops.

Combine harvesters are now widely used in cereal cultivation in minimum tilling systems. Minimum tilling consists of dispensing with many of the traditional forms of tillage (raising, passing the cultivator, disc harrow, roller and seed sower). The same tasks are now carried out by sets of implements that are linked to the tractor and can plant a new crop in a single pass, or a few passes. Reducing the tilling is done to save labor, energy and time. There are many implements that combine different types of plow, for example a spike harrow combined with a roller, or a raising plow combined with rollers and a clod-crushing harrow. There are also drillers-sowers that prepare the seedbed and then sow the seed in a single pass. Obviously these minimum tillage systems rely on the use of chemical weedkillers.

Technological advances have made it possible to build very sophisticated agricultural implements. Perhaps the most remarkable advances have been in the machinery for collection of agricultural products, such as harvesters for corn grain, grapes, cotton, potatoes, sugar beets, fruit and vegetables, etc.,

as well as the different harvesters for loading, transporting and preserving the produce. There is also specific machinery for the preparation and distribution of animal feeds, as well as highly sophisticated installations for milking cows.

Whole-cycle onion harvesters (Courtesy of SLUIS & GROOT SEEDS)

Above: complete equipment for harvesting grapes and loading and transporting them. The body is raised so it can collect the grapes while passing over the vines. (Photos courtesy of the Department of Agriculture and Stockraising of the Generalitat of Catalonia)

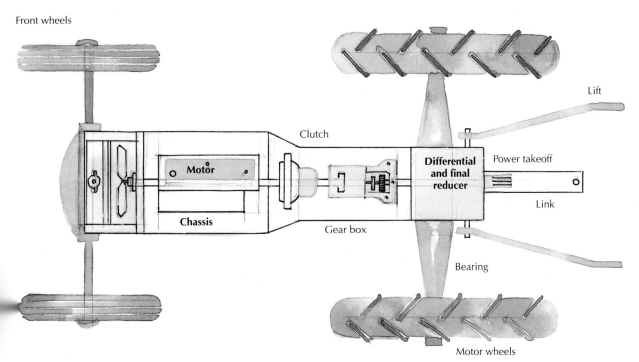

Front wheels

Clutch

Motor

Chassis

Gear box

Differential and final reducer

Power takeoff

Lift

Link

Bearing

Motor wheels

Diagram of the internal and external parts of a tractor

9.1. TRACTORS

The tasks a tractor can do, divided by the method of operation.

A tractor is a motor vehicle adapted to carry out a variety of agricultural tasks. It is self-propelled and can also pull or power the different implements used in the different types of tillage. Tractors have four wheels, the rear pair being much larger and providing the power, while the front pair's main purpose is to steer the vehicle.
Tractors normally use an internal combustion engine like those in cars in lorries. They all consist of a chassis, a motor and a transmission system that moves the wheels. A tractor consists of the following elements:

• The **hydraulic lift** is the device that makes it possible to raise or lower the implements connected to the tractor, so the tractor can manoeuver better and transport the devices.

• The **chassis** is a very tough metal frame supporting the tractor's essential mechanisms.

— Stationary   { By pulley / By power takeoff / By hydraulic equipment

— Transport
— Dragging
— Pushing

— Combined   { Transport and power takeoff / Dragging and power takeoff

• The **differential gear** is the set of gears that allows the two rear motor wheels to turn at different speeds, so it take bends more easily.
• The **steering** is the machinery that guides the tractor in the direction chosen by the driver. It moves the front wheels, which is why they are called the steering wheels.
• The **clutch** is the device that connects or disconnects the turning movement of the motor to the gear box.

The different types of work that a tractor can carry out
A/ Stationary work with pulley connected to the power takeoff, i.e., thresher, storage in silos
B/ Stationary work with the power takeoff, i.e., irrigation pumps, feed mills
C/ Stationary work with the hydraulic equipment, i.e., grain elevators
D/ Transport, i.e., trailers
E/ Dragging, i.e., moldboard plows, disc harrows
F/ Pushing, i.e., tractor loader, bulldozer
G/ Combined transport and supplying power, i.e., balers, dung distributors
H/ Combined dragging and supplying power, i.e., subsoil plows, vibrator, drill

(A)

(B)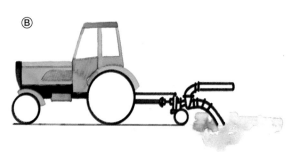

(C)

(D)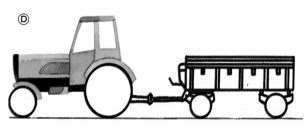

(E)

(F)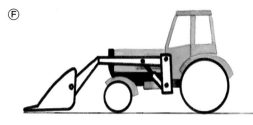

(G)

(H)

• The **link** is the element that makes it possible to connect machines or implements to the tractor. There are two types of link: a pulling bar with a single attachment point to pull implements, and the three-point linkage connected to the hydraulic lift, for suspended or semi-suspended machines or devices.

• The **brakes** are the mechanical elements that allow the driver to slow or stop the tractor.

• The **motor** is the device that transforms the energy released by the combustion of the fuel into mechanical energy, as a rotational movement.

• The **bearings** are the axes responsible for transmitting the movement from the differential to the wheels, through the final reducer.

• The **pulley** is a mechanism intended to transmit movement, by means of belts, to certain machines. It is normally connected to the power takeoff, and this moves the belts.

• The **final reducer** is the mechanism responsible for reducing, after the gear box, the speed of rotation of the wheels that increases the traction force.

• The **wheels** are the elements that are in contact with the soil and allow the tractor to move.

• The **power takeoff** is an axis with a grooved tip that is moved by the motor and which serves to move or power the different implements that can be attached to the tractor.

The different tasks that the tractor can carry out can be divided into four groups, **pulling implements**, **dragging**, **pushing** and **transmitting other movements**. On the previous page there is an outline of the tasks that can be carried out by a tractor, as well as some drawings illustrating each type of application.

9.1.1. Characteristics

• **Motor**
Normally tractors have a diesel motor and require diesel fuel. Most tractor motors have four cylinders, though the most powerful have six or even eight cylinders. The distribution, that is to say, the set of mechanisms regulating the entry and exit of gases from the cylinder, is like that of any other internal combustion engine. It consists of a valve lifter, a camshaft, cams, push-rod, rocker, valves (outlet and inlet) and manifolds (entrance and escape) (inlet and outlet).

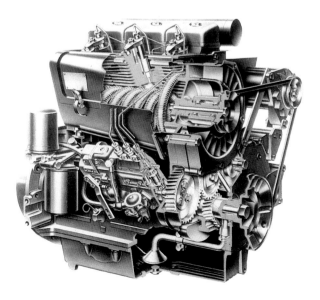

Nowadays, tractor cabins are as comfortable as those of a family car. This model is manufactured by XAVER FEND & Co.

A view of the cabin of a different tractor, the 390 model produced by XAVER FEND & Co. All the devices connected to the tractor can easily be handled from within the cabin.

Some tractors are equipped, at the front, with a hydraulic lift system, with its own power takeoff. The photograph shows a detail of the front coupling system on a tractor.

The Farmer line sold by XAVER FEND & Co. includes the 250 V® model, (50 horsepower). The illustration is a schematic diagram of the model's engine.

The weight of the tractor and the difficulty of working on flooded soil are among the main problems when cultivating rice. This can be resolved with the help of improved rear wheels, skeleton wheels, which hardly sink into the mud, meaning much less damage to the crop and the soil structure. (Courtesy of HARDI INTERNATIONAL A/S)

DIN or *Deutsches Institut für Normung*. This is an official German government body created in 1926 and responsible for standardizing industrial measures, sizes, tolerances, qualities and systems. Its scope is very broad, and for the materials it covers its norms are internationally accepted.

SAE are the initials of the *Society of Automotive Engineers*. This US body is responsible for establishing the technical standards for the automobile industry, and like the German DIN standards, its norms are internationally accepted.

• The tractor's power

The power is one of the characteristics specified by the manufacturer and the power specifications must figure in the vehicle's instruction manual. The power may be expressed according to German standards (**DIN norms**) or to American ones (**SAE**). For a given motor, the power expressed in DIN terms is always less than in SAE terms. This is because the DIN norms consider that the motor is powering all the mechanisms at the same time (water pump, fan, dynamo, etc.), while the SAE norms suppose the motor is not powering these mechanisms. The SAE norms tell us the complete power of the motor, which is logically greater than the DIN norm. The power is normally expressed in horsepower. Some times the SI units (Système International) is used and the power is expressed in kiloWatts (kW), and they can be converted using the following factors:

1 HP = 0.736 kW, or 1 kW = 1.36 HP

• Air filters

Air filtering systems are very important, because tractors often work in very dusty situations (fields, dirt roads, etc.), and so they need a really good air filter. It is usually located at the front (where there is less dust in the air), and may be based on an oil-bath or a dry system.

• Fuel supply

The engine's fuel supply is typical of diesel motors. The fuel is pumped from the tank through the fuel filters to the fuel injection pump. This pumps the fuel to the injectors, and then into the cylinder. The mixture of diesel fuel and air is compressed by the piston until combustion occurs. The excess fuel is then returned to the fuel tank.

• Cooling the motor

The combustion of the diesel fuel makes the engine very hot and so it requires an adequate cooling system. Current tractors are either air-cooled or water-cooled. Cooling by air is carried out by a single turbine that pumps air from the outside over all the cylinders to cool them down. Water cooling systems are based on a radiator that distributes the water through the jackets of the cylinders and within the block; the hot water is returned to the radiator, where it is cooled by means of a fan.

• Oil and lubrication

As in all internal combustion engines, the engine must be lubricated with oil, so the parts do not wear each other away. An oil's most important technical characteristic is its viscosity, which is measured all over the world using the SAE norms. The higher the SAE value, the higher the viscosity. High viscosity oils (SAE 20, 30, 40 or 50) are used in winter, while the lower values correspond to less viscous oils (SAE 5, 10 or 20) which are used in winter. The farmer can now choose **multigrade oils** whose physical properties are modified by additives to give them a wider range on the SAE scale. In any case, the right oil for the tractor must be clearly specified in the instructions for use and maintenance that come with the tractor.

In addition to the engine, other moving parts of the tractor must be lubricated for the same reason, to prevent the wearing away of the gears that are in contact during operation. The gear box, differential and hydraulic lifting device all contain mechanisms that need to be greased. In tractor maintenance it is very important not to confuse the oils used to lubricate the engine (with SAE values from 5 to 50) with those used to lubricate gears, whose SAE value is much higher (between 75 and 250).

• Clutch

The movement produced in the engine by the combustion of the diesel fuel is transmitted to the crankshaft and from there through the clutch to the rear wheels, the ones that provide the traction. The rotary movement produced by the motor is transmitted, through the steering wheel to the clutch, then to the gearbox, from there to the differential and from there through the axle shafts and the final reducer, to the rear wheels.

The role of the clutch is to connect or disconnect the movement of the engine to the gearbox. When the clutch pedal is in the normal position (disengaged), the clutch transmits the movement of the engine to the gearbox. When you tread the pedal, the clutch ceases to transmit this movement.

• Gearbox
The main role of the gearbox is to adapt the tractor's forward speed to the power required for a given task, so that the engine's power is used best. Clearly a higher speed will develop less force than a lower one, and vice versa (at a low speed the tractor can apply more force than at a high one). Agricultural tractors now have a wide range of speeds, so they can perform a wide range of functions.

• Differential
If the two axles joined to the wheels were joined to each other in the center, when the tractor tried to take a corner, the wheel with the shortest path would skid, as each of the wheels has to travel a different distance, and thus rotate a different number of times. This problem is resolved by the differential, which allows the two wheels to turn at different speeds, making it easier to take corners, as the turns lost by the wheel on the inside of the turning circle are gained by the one on the outside.

• Final reducer
In agricultural tractors, despite the reductions undergone by the movement in the gear box and the differential gear, the turning speed of the output from the differential is too high for the low working speed needed by a tractor. Thus the turning velocity of the bearings must be reduced, which is done by placing what is called the **final reducer** between the differential and the rear wheels.

• Electric circuits
All tractors have a similar electrical installation, though each one has its own distinctive features. The instruction manual for the tractor almost always include a diagrammatic drawing of the electrical installation, and it is important to know how to interpret it in order to make it easier to locate any electrical fault that occurs. There are normally four types of electric circuit in a tractor: the **fixed light circuit** (ordinary lighting, the hazard beacon and the indicator lights); the **control circuits** (oil pressure gauge, thermometer, fuel gauge and heaters); **loading and starter circuits** (battery, dynamo, control of load, and starter motor); and the **maneuvering circuits** (steering indicator lights, brake light and horn).

• Wheels
Most tractors have the classic type of wheel, that is to say the metal wheel supports an air-filled tube that is covered by a hard rubber tire that grips the soil. The front and rear wheels are very different. The rear, or traction, wheels are very large and usually bear a V-shaped tread pattern to get a good grip on the ground. The front, or guide, wheels are small with longitudinal grooves, and they determine which direction the tractor goes.

• Steering
There are two different types of steering system depending on the mechanism used to steer the front wheels, which as pointed out, are the ones that determine where the tractor will go. **Mechanical steering** is when all the force necessary to turn the front wheels is supplied by the tractor driver. **Power-assisted steering** is when the force of the tractor driver is increased by a double acting pump, coupled to the steering levers, that receives oil under pressure from a pump powered by the crankshaft. In **hydraulic power steering** when the tractor

An example of an implement performing several functions used in minimum tillage techniques. Called the Cultirota®, it is sold by BECKER GmbH u. Co. KG, and prepares the seedbed in a single pass.

driver moves the steering wheel, this acts directly upon the valve box, and this is joined to a piston, with oil pumped under pressure, that finally transmits the force to move the steering levers, and so almost no effort is required from the driver. Finally, **hydrostatic power steering** is when there are no levers joining the steering wheel and the front wheels, which are turned by the work of a double acting piston, whose oil under pressure is injected by a pump and controlled by a valve box that is activated by the steering wheel.

• Brakes

Like cars, tractors have two sorts of brakes. The brakes, as such, are used to slow down the tractor when it is in motion, and to stop it, which may be drum brakes (hydraulic) or disc brakes (friction). There is also a hand brake, which is used to immobilize the vehicle once it has come to a halt. The hand brake may use the same mechanism as the brake pedal, suitably immobilized by means of a latch or by a mechanism called the belt brake.

9.1.2. Connecting the implements

There are three types of connection for the implements; the **pull bar**, the **hydraulic lift** and the **power takeoff**. In the past tractors had a fourth type of connection, a **pulley**. The lateral pulley used belts to power some devices such as threshers and storers. Not many tractors have a pulley now, as it can be connected to the power takeoff and screwed to the rear of the tractor. Pulleys are hardly used because most machines carrying out stationary tasks are adapted to be powered directly by the tractor's power takeoff, meaning that using a pulley is unnecessary.

• Pull bar

Located at the rear of the tractor, the pull bar is a

long hard strong bar with a hitch at the tip. This hitch is used to couple the implements to be towed, such as disc harrows, moldboard plows, trailers, etc.

• Hydraulic lift

The hydraulic lift consists of three bars located at the rear of the tractor with a hitch on each one. These bars make it possible to raise or lower "suspended" or "semi-suspended" implements, and they form what is known as a **three-point linkage**, now found on almost all tractors.

The three-point linkage consists of two rigid pulling arms, each joined to the tractor by a ball-

and-socket joint at the tip. At the other tip, each has another ball-and-socket joint to attach the implement firmly. These arms are joined to two other short hydraulic arms that can move the implement in or out.

The third connection point is an extensible bar known as the **third point** which is joined by a ball-and-socket joint to the tractor's frame. At the other end there is another ball-and-socket joint to attach to the device. The third point is extensible because there is a central tube with two

nuts at the tips, into which are screwed two bolts joined to the ball-and-socket joints. The implements connected to the three-point link lie in a plane and can be raised for transport, or left at ground level resting position to carry out tasks. Certain implements, in addition to the three-point linkage are connected to the power takeoff to power their movement.

• The **power takeoff**
The power takeoff powers the internal mechanisms of some implements coupled to the tractor. The power takeoff is not connected to the implements that are simply powered by dragging, such as disc plows or moldboard plows. It is however very useful and is essential to power, for example, rotary cultivators, whose rotary movement is derived from the power takeoff. The power from the takeoff is independent of the tractor's speed, and it can be used when the tractor is stationary (though of course the engine must be turned on). The power takeoff is internationally standardized so that any implement can be fitted to the tractor. The turning speed of the power takeoff is standardized at 540 r.p.m. (revolutions per minute) and 1,000 r.p.m., and so the manufacturer of a given trailer must calculate the pulleys and gearings for their machine to function correctly. The farmer has to make the tractor work at the r.p.m. established by the manufacturer for the implement, so it works at the right speed.

9.1.3. Types of tractor

In general, all tractors share the features described above and the only difference is the power they generate. Each agricultural implement's technical instructions specify the power it requires. When deciding how powerful a tractor is needed, it is important to bear in mind the individual characteristics of the farm: the soil class, the size of the farm, the type of crop, etc. Suppliers of agricultural machinery produce catalog where the farmer can find a great deal of useful information on the different types of tractors, their characteristics, and the most

suitable model for each crop and size of field, the payment instalments, its useful life, etc.
The more powerful the tractor, the more expensive it will be, and so it is foolish to buy a really powerful tractor for a small farm, because it is very expensive, and it may even be difficult to handle. Having said that, the farmer must take the future into account. If the farmer intends to expand activities within the near future, it may be necessary to buy a powerful tractor. It is also worth considering purchasing a powerful tractor as a cooperative with other farmers, which may work out very well for all concerned.
Standard tractors with rear-wheel drive are suitable for almost all soils and crops. However four-wheel drive may be needed in some circumstances, such as very steep sites, or if the work is to be done on very uneven sites in forests. A really complete tractor has a lever to switch from two-wheel drive to four-wheel.
For some very specific tasks, such as the operations involved in rice cultivation, the wheels are replaced by two continuous tracks of metal plates, one on either side of the tractor, which the tractor moves on. They are known as **caterpillar tractors** or **crawler tractors**. Tractors with normal rear-wheel drive make use of about 60% of the engine's power, but caterpillar tractors make use of almost 90%, mainly because the caterpillar tracks have a large surface area in contact with the soil.

To the left: Detail of the pull bar of a tractor. It is used to hook implements onto the tractor.

Many of the new models of tractor allow you to attach implements to the front or rear of the tractor: they have a power takeoff and three-point linkage at the front and at the back, and so they can, for example, power a mower at the front and a fodder collector at the rear. In order to power two devices at the same time, the tractor must have at least 75 horsepower. The photo shows a model in the 300 Range®, produced by XAVER FEND & Co.

Many tractors have large rear wheels so they can get a better grip on the ground and make best use of their power. (Courtesy of XAVER FEND & Co.)

9.2.1. Scrub and weed removers

Plant removers are specialized machines to remove the plants from a site so it can cultivated. In general these devices are suspended or semi-suspended from the tractor and powered by the tractor's power takeoff. A 45 to 50 horsepower tractor is usually enough power a machine of this type. A wide range of machinery of this type is on sale, the most common ones use a **horizontal axis**, a **vertical axis** or **rollers**.

The main differences between them are in the mechanisms by which they eliminate different types of plant cover. The first distinction to draw is between trees and shrubs with stout trunks, and herbaceous plants. The plant cover to be eliminated will give us an idea of the type of machinery needed. Obviously machinery to remove trees and shrubs will have to be much stronger and will need a more powerful tractor than one to eliminate a layer of herbaceous plants.

Three earthmoving implements coupled to a powerful tractor. At the rear a device for digging trenches, and at the front a tractor loader and an angled blade bulldozer, an angledozer.

9.2. SOIL TREATMENT MACHINERY

The next chapters describe some of the most common implements used to prepare the seedbed, but before this we should mention some preliminary tasks. On some occasions, the ground is being cultivated for the first time, or a plot that has not been cultivated for several years is to be recovered in which case, it will be necessary to remove the plant cover and uproot any scrub. On other occasions a change in the crop grown or modernization of cultivation techniques means the stones have to be removed. We may also find that new cultivation techniques or irrigation systems require the site to be levelled. Many of these aspects of transforming a site are closer to civil engineering than to agriculture.

In the corners that the tractor cannot reach such as under trees or vine stems, the farmer can use a small device like this manual pneumatic weeder. This type of tool is typical of professional gardeners but is also widely used in agriculture. It is sold by M.A.I.B.O. s.r.l.

In general, these machines are too expensive for most medium or small farms. These devices usually consume a lot of power, and it may well be worth buying one through a cooperative or farmer's association, especially when you consider that you will probably only want to use it once or twice. This chapter, which we could call Preliminary Operations, briefly discusses how to prepare the soil so the future crop will grow well; equipment to **clear scrub**, **stone removers**, **earthmoving equipment**, and **equipment to install drainage**.

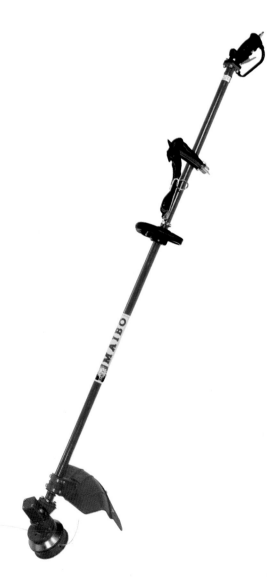

Agricultural machinery to remove plants can uproot woody plants from the soil and can also cut up the bits of plant in their path. On sites where there are a large number of large trees, machinery that is closer to **forestry** than agriculture is needed. The occasional extraction of very large tree stumps can be done with a **front loaders** or **rear digger**. For this type of work, there is also a specialized machine, a **stumper**; there is also a **stump shredder** to break the stumps into small pieces so they can be incorporated into the soil as organic matter.

9.2.2. Stone removers

Since antiquity stones have been removed from fields and small horticultural plots. It was performed by hand and was very laborious, and so they were only removed from very small plots. Technological advances, as well as the need for stone-free soil in order to get the best results from modern agricultural implements (seeders, harvesters, etc.), have all led to the development of mechanical stone removers. Stone removers are machines to remove or eliminate the stones from a plot of ground. The effectiveness of these machines depends on the type and size of the stone present. The size and topography of the plot to be cleared are also relevant, as these machines are large and unmaneuverable, and are hard to get to all corners of the site. Normally, to ensure the ground is free of stones, the stone remover should be passed every two or three years. This removes the stones to a depth of between 20 cm and 35 cm, depending on the type of site and the power of the tractor used.

Stones can be removed from a site in three different ways. They can be collected and placed in a hopper or trailer and then dumped somewhere else outside the plot. These are called **stone collectors**. A second method is to align all the stones in strips on the soil surface and then collect them using a front tractor loader. The third option is to use special machinery to crush the stones and return them to the soil, crushing the stones.

The simplest models of stone collectors can be used on tractors of only 40-50 horsepower. Stone crushers require a much more powerful machine (80-100 horsepower). The farmer should not expect them to perform miracles, as no stone crusher is going to be able to break up boulders weighing half a tonne. In plots where there are only a few large stones, it may be possible to remove them with a powerful frontloader, or smash them to pieces with dynamite and then collect the stones. This type of machinery adapts well to purchase by a cooperative, as when all the fields have been treated, the machine is no

An example of a rotary horizontal axis cutter. The implement is suspended and powered by the tractor's power takeoff. It cuts and shreds all sorts of plants, including small shrubs. Sold by JOSKIN, S.A.

Forestry is the science dealing with forests and their rational exploitation.

longer needed, except when subsoil plowing brings more stones to the surface.

9.2.3. Earth-moving equipment

Earth-moving equipment is more associated with civil engineering than agriculture, but the use of front loaders, rear diggers, levelling harrows, etc., is relatively common on farms. Digging an irrigation or drainage ditch, levelling a plot and clearing a path are all normal operations, even though they are not very frequent. The farmer should decide whether to buy the machinery or rent it on the basis of factors like farm size and the amount of work to be done.

Front tractor loaders solve many of the problems that may arise on a farm. The photo shows a device that can be coupled to the 380 GTA® tractor, both sold by XAVER FEND & Co.

Small front loaders can be coupled to any tractor, however small it is. The photo shows the M18-050®, sold by Agrostroj Prostějoc, and is only rated at 20 horsepower.

On the right: The shape of the different types of bucket on front loaders is adapted to the materials to be handled (Taken from Bernat, C.)

9.2.3.1. Front tractor loaders

There are basically two types of front loaders. Some are independent self-powered machines that are used for tasks requiring a lot of power, such as loading earth, rubble, etc., while others are intended to be coupled to the front of a tractor, even a low-power tractor (40-50 horsepower) for handling cereals, dung, and for storage in containers. The most important features of the loader are: its capacity, its maximum loading and unloading height, its unloading system, that is to say, whether it has a double hydraulic system to allow unloading in any position, and the precise shape of the bucket, which depends on what it will be used for.

PRANEDA B.V. sells tractor loaders that are easily connected to the front of the tractor. The photos, from top to bottom and left to right, show different stages in fitting it onto the tractor.

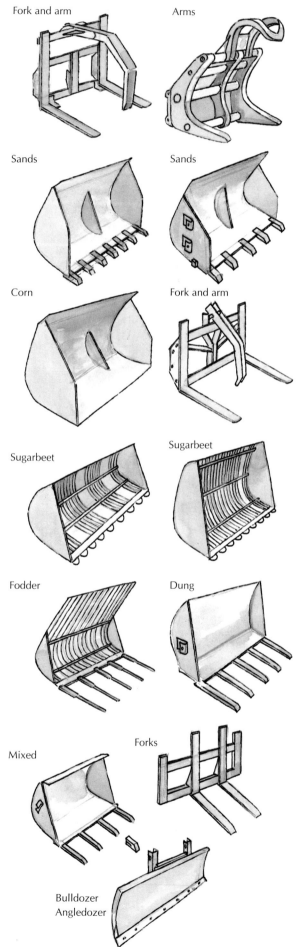

Fork and arm

Arms

Sands

Sands

Corn

Fork and arm

Sugarbeet

Sugarbeet

Fodder

Dung

Mixed

Forks

Bulldozer
Angledozer

9.2.3.2 Levelling harrows

Levelling harrows are tractor-drawn devices that load earth into a deposit or "cup" that can be unloaded rapidly or gradually. They are also known as scrapers, diggers or levellers, depending on what they do. There are also high-powered self-propelled machines, called **motorlevellers**, that are used in civil engineering.

When the levelling harrow is dragged forwards, it scrapes the soil with a knife, so that the loosened earth enters the harrow. Once full, it can be unloaded somewhere else. On farms levelling harrows are used to level plots, to make or maintain tracks, or to make small retaining walls or dams. They may even be used to bank up earth (earthing up olive trunks, for example). Tractors with 50 to 100 horsepower can power levelling harrows with a working capacity of 0.5-3 m³ of soil.

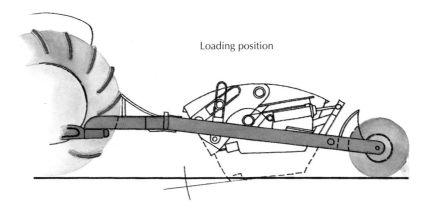

Loading position

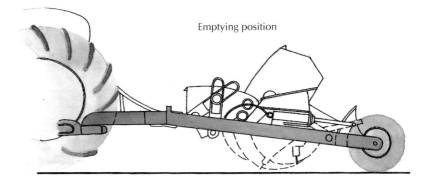

Emptying position

Diagram of the levelling harrow in the working position (loading) and emptying (Taken from Bernat, C.)

9.2.3.3. Levelling devices

Levelling devices essentially push a blade forward, and if the blade is perpendicular to the direction it is moving, it is called a **bulldozer**, while if the blade is at an angle, it is an **angledozer**.

These devices are usually mounted on caterpillar tractors, but a normal wheeled tractor with more than 60 horsepower can also be fitted with a blade. These machine can do things like fill in ditches, get rid of small irregularities in the site, improve rural paths, knock down small walls, etc.

9.2.3.4. Diggers

When setting up an irrigation system, it is often necessary to dig ditches to bury the irrigation tubing or to improve the drainage. This is generally contracted out to a specialist company, but purchasing machinery of this type may be worthwhile in some circumstances.

The photo shows an angledozer, which is different from a bulldozer as the blade is at an angle to the tractor's forward movement. The angledozer in the photo is connected to the tractor's power takeoff and is clearing the road by pushing the snow to one side. (Courtesy of BRENIG GREENMASTER LANDTECHNIK GmbH)

When it is necessary to dig a trench, diggers that can be fitted onto a tractor are very useful. There are many agricultural implements of this type, which vary in the scope of work they are intended to do. Sold by CECCATO BENITO.

When it is necessary to dig a small trench, for example to install a buried surface irrigation system, a device like that shown in the photos can be used. Fixed to a 50 to 100 horsepower caterpillar tractor, it can dig a trench 15 to 25 cm deep and 15 to 25 cm wide. This product is from GAMBETTIBARRE, s.r.l.

9.3. TILLING THE SOIL

Tillage is the set of mechanical operations carried out on the site (on the surface and at depth) in order to ensure crop seedlings and plants grow better. Human cultivation is the main cause of a series of imbalances in the soil structure, which have to be remedied so that the soil can maintain its fertility. The weight of the tractor often compacts the soil at depth, causing it to lose its structure: merely practising monoculture and leaving the soil bare during certain periods of time is enough to imbalance it.

Right: Plowing with a moldboard restores the soil's open structure, buries weeds, eliminates the larvae of some phytophagous insects and undesirable weedkiller residues. It also allows incorporation of fertilizers to the soil and ensures the soil humidity is evenly distributed. (Courtesy of BRENIG GREENMASTER LANDTECHNIK GmbH)

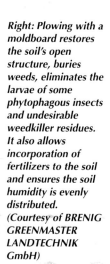

Diggers are mounted on caterpillar tractors. The digging implement is a bucket on an articulated arm at the end of a boom, and the whole unit is powered hydraulically. The depth and reach of the rear digger depends on the size of the boom and the articulated arm, which in turn depend on the power and weight of the tractor they are mounted on. These devices can be fitted onto wheeled tractors with 75-80 horsepower or more, but in these cases, they must have supports that can be fixed into the ground, in the working position, to increase the base supporting the digger. Normally, the base of the rear digger is interchangeable, and different models can be used to dig different widths of trenches or can be changed to deal with different types of soil. The **trench digger** always has some advantages over diggers for tasks of this type, as long as there are no obstacles in its path. They are usually self-powered vehicles that dig the ditch as they advance, by means of an endless chain with blades and a mechanism to remove the earth. The width of the ditch, and its depth, can be varied by alternating the size of the blades and the angle of attack of the digging mechanism.

These different forms of tillage open up the soil (recovering the micropores and macropores); they bury weeds and their seeds, the larvae of some phytophagous insects and unwanted agricultural chemical (weedkiller) residues; plowing allows incorporation of fertilizers into the soil, ensures the soil humidity is evenly distributed with the soil layer and, finally, makes it possible to create ridges, furrows, etc. Yet tilling has its disadvantages. The best known and most widely studied is the formation of a hard layer of soil just below the level the plow reaches. Some machines, such as the moldboard plow, compact the soil below the blades. This hard and relatively impermeable compacted layer is called the **plow pan**. Tilling devices connected to a tractor to carry out these different tasks can be classified by **how they work**, or **how they are connected**, or **what they do**. On the next page there is a classification of these implements by their method of functioning, that is to say if they are powered by the tractor's power takeoff or not. Depending on how they are connected to the tractor, these devices can be:

• **Dragged**
They are not connected to the tractor's power

Unpowered	Non-turning plows	subsoil plow scarifier chisel	
	Turning plows	moldboard disc	reversible non-reversible
	Harrows	teeth discs clod-breaker	
	Cultivators	fixed teeth flexible teeth vibrating teeth	
	Rakes Rollers		
Powered (by tractor p.t.o.)	Drills	transversal axis (normal drills or rotavators) vertical axis (rotary hoes)	
	Diggers		
	Powered-teeth harrow	oscillating rotary	
	Powered plows	rotary moldboard powered ridger powered discs	

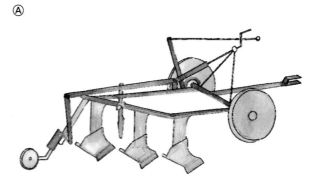

Ⓐ

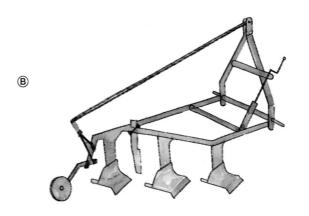

Ⓑ

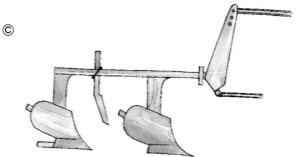

Ⓒ

Left: classification of implements by how they work

Different ways of attaching the implement to the tractor, taking the moldboard plow as an example.
A/ Drawn
B/ Semi-suspended
C/ Suspended

takeoff, and are not supported by the tractor. They are only joined at one point of contact, and can easily be hitched or unhitched.

• Suspended
These devices are fully supported by the tractor when they are in the raised position. They are connected by three hydraulic points, and their working depth and direction can be controlled by the tractor.

• Semi-suspended
They are the same as the preceding group, but rest on a set of rear wheels, meaning they are much heavier.
Depending on the type of work they perform, there are the following types:

• Primary or raising devices
They are used to prepare the soil before sowing and usually reach deep into the soil (20 to 35 cm).

• Complementary or secondary implements
These work on the surface (15 cm), such as burying stubble, weeding between rows, preparing the seed bed, etc.

• Special devices
These implements are designed for very special tasks, such as airing the soil at great depth, digging trenches, work in vineyards, machines to make ridges or break them up, etc.

9.3.1. Plows

• Drawn plows
Drawn plows have greater freedom of movement with respect to the tractor, meaning the work done is more regularly. Hitching and unhitching the device to the tractor is very quick, and they can be connected to any tractor, especially the most powerful ones. They do, however, have some problems. The purchase price is higher than that of suspended and semi-suspended models, as they have to include a series of additional systems, such

as wheels to regulate the depth, a raising system, etc. They require more space to maneuver, which reduces the total yield of the work; movement along roads, etc., is slower; they do not increase the tractor's adherence very much, especially when compared to suspended plows.

• Suspended plows
The main advantages of suspended plows are: they are cheaper than an equivalent dragged plow; the manoeuvering space is smaller, meaning they are easier to use; road travel can be at the tractor's maximum speed; and they improve the tractor's adherence to the soil. Their disadvantages include the fact that counterweights must be placed at the front of the tractor, as suspended plows may overturn the tractor when they are raised.

• Semi-suspended plows
Semi-suspended plows are in between suspended and dragged plows in their maneuverability when working, on roads, in price and their possible improvement of the tractor's adherence.

1/ Thanks to their design, some models of chisel-type plows can have a disc harrow attached behind them. This means that in a single pass it can plow the soil and prepare the seedbed. This is called minimum tilling. Courtesy of Brenig GREENMASTER LANDTECHNIK GmbH.

2/ This model of chisel plow has retractile teeth, so it can work around the stones and obstacles it meets in the subsoil. Manufactured by BRENIG GREENMASTER LANDTECHNIK GmbH.

3/ This model of subsoil plow with two teeth, produced by BRENIG GREENMASTER LANDTECHNIK GmbH, can open up and air the soil to a depth of 80 cm.

*4/ Different shapes for the teeth of a subsoil plow. **A** and **B** need less traction power.*

5/ The subsoil plow is used to air the soil at depth and break up the plow pan that may be created by repeated use of surface plows. This model is manufactured by BRENIG GREENMASTER LANDTECHNIK GmbH.

6/ Chisel-type scarifying plows, like the one shown in the photo, do not work as deep as subsoil plows, but are less heavy and do not require such a powerful tractor.

9.3.1.1. Subsoil plow and scarifier

These plows do not turn the earth, that is to say, they do turn the soil over and invert the layers, and they are used to prepare and improve the ground. This type of machinery is for working the soil at depth, and so its arms are made of very tough heavy materials. They meet special requirements, such as deep plowing, breaking new ground, installing drainage, removing stones, etc., though they are also often used to open the soil up, to air the soil at depth, and to break the plow pan that may have been formed by repeated plowing to the same depth.

Subsoil plows may have a variety of different types of teeth, as shown in the drawing below. The ones labelled A and B need 25% less traction power than C because of their angled design. Chisel-type scarifying plows, can work at a shallower depth than subsoil plows, and are often less heavy.

Ⓐ   Ⓑ   Ⓒ

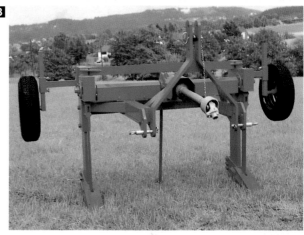

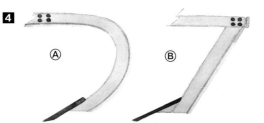

9.3.1.2. Moldboard plow

In cereals, the moldboard plow is the most typical and the most widely used by farmers. It does not work as deep in the soil as the subsoil or chisel plows, and so it can turn over the arable layer and break it down. Its use increases the soil's porosity, giving the soil a greater water retention capacity and it can also bury the remains of the previous crop. It is also effective against some insects, as it buries the larvae of many soil-living insects deep in the soil.

The moldboard consists of three distinct parts: the **share**, the **landside** and the **moldboard**. The purpose of the share is to start the process of breaking the soil, the landside finishes this breaking process and starts to overturn the furrow slice, and the moldboard completes the turning of the soil. There are three different types of moldboard on the market: **cylindrical**, **universal** and **cambered**. For the uses of each one, as well as their technical

specifications, the reader will have to refer to the publicity leaflets of the many commercial suppliers.

9.3.1.3. Disc plow

The main problem with moldboard plows is the high friction between the soil and the parts of the moldboard in contact with the soil. The working parts are rapidly worn away, especially on sandy soils. If there are many stones in the soil, they may deform and damage moldboard plows. Disc plows greatly reduce this friction and the problem of obstacles.

Disc plows consist of round concave discs which rotate on axles that are joined to the frame. These axes are at an angle to the direction the tractor is moving, and are also at

an angle to the horizontal. The soil cut by the disc presses against the disc, causing it to turn, removing and raising the furrow slice. When it has reached a certain height, a scraper directs the slice and it falls to the base of the furrow, thus turning the soil.

Disc plows are especially suitable for heavy sticky soils where the problem is getting the soil off the turning surface, and in soils where a plow pan has formed. Disc plows are also suitable for excessively stony soils, as the disc can roll over the obstacle instead of trapping it, as happens with moldboard plows. It is also suitable in soils where only the topsoil has good texture and structure, and where turning the soil is not a good idea. It can also be recommended for very hard or abrasive soils, because the disc offers less resistance to the soil (hard soils) and suffers less abrasion in sandy soils (which are highly abrasive).

In the photograph, a combined reversible moldboard plow and roller. Machinery sold by RABE WERK GmbH+Co.

Depending on the length of the work, there are many different types of moldboard plow. The photo shows a double reversible moldboard plow, sold by Angostroj Prosetějoc. As it is highly maneuverable, it can be used in small fields.

Another model of disc plow with scrapers, sold by JOSKIN, S.A.

Disc plow with scrapers that clean the disc during the plowing. Machinery sold by RABE WERK GmbH+Co.

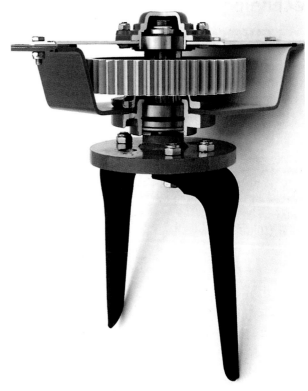

9.3.1.4. Fixed and reversible plows

When the share of a plow enters the soil and cuts a continuous slice of soil, the slice breaks into different layers due to the formation of primary cracks and then into fractions (secondary cracking), which breaks it up, and this breaking is completed when it is turned over by the moldboard. The turned furrow slices rest on the preceding slice (see diagram) and breaks down leaving a furrow as shown below. As the plow is not symmetrical, the work done is not symmetrical, meaning that work in one direction will be the opposite of work done in the other direction. The two are equal but opposite and are known as **furrows** and **back furrows**.
With fixed plows, the furrow slice is always turned over in the same direction. With a reversible plow, the slice can be turned to one side or another, and this makes it possible to plow continuously, turning the tractor and plow at the end of the row, from one end of the plot to the other.

Turning over the furrow slice
s) share
d) working depth
w) width of the share
α) angle of turning

In fact the prisms of earth are deformed, creating the furrows typical of plowing. The illustrations show results that are:
A/ correct
B/ too high
C/ too low

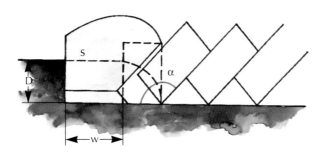

9.3.2. Rotary cultivators

The first brand of rotary cultivator was the **Rotovator** and many books and articles call all machines of this type rotovator, but this term is a trademark rather than the name of a type of machine. **Rotary cultivator** is a better name for them. They work the ground with rotating knives ("tines") powered by the tractor. This section includes another similar device, the **mechanical hoe** or **digger**. Rotary cultivators are widely used in intensive crops and orchards. They can perform a wide range of tasks, but to sum up in a single pass they air the soil, breaking it down into smaller particles and thoroughly mixing it up. In intensive crops, the rotary cultivator may complement plowing as the final step in seedbed preparation, or it may be the only preparation, completely replacing tilling. Because it only works the surface, the rotary cultivator has the disadvantage of frequently leading to the formation of a subsoil crust, the pan plow. Unlike the preceding plows, the rotary cultivator is powered by the tractor's power takeoff, and is connected by the cardan shaft. There are two types of rotary cultivators, depending on their plane of rotation: **transversal** and **vertical** ones. Transversal ones, which rotate in the direction the tractor is moving, are the most common; vertical ones, also known as **rotary hoes**, are little used, even though they advance very evenly because their teeth are always in contact with the soil.

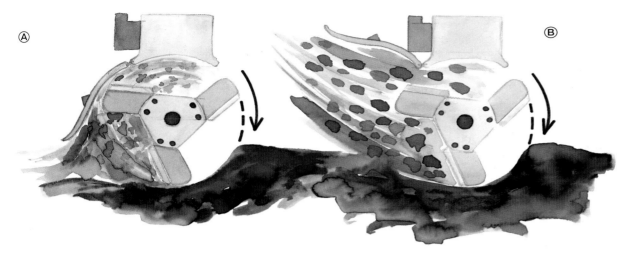

Ⓐ Ⓑ

The modus operandi of a transversal rotary cultivator. Depending on the position of the hood and the ratio of the speed of advance to the speed of rotation, this gives:
A/ fine powdering: the hood is lowered and the blade speed/speed of advance is high
B/ coarse shredding: the lid is raised and the blade speed/speed of advance is low

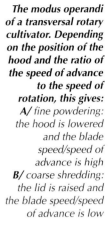

In the vertical rotary cultivator, rotary hoe or rotary toothed harrow, the teeth rotate vertically, and are in constant contact with the soil. Depending on whether the tines are large or small, it is called a rotary hoe or rotary harrow. This is the Cyclotiller model, produced by MASCHINENFABRIK RAU GmbH.

The working depth of the rotary cultivators can be regulated using the tractor's hydraulic system, but is usually between 12 and 15 cm, with heavy-duty machines reaching a depth of 25 cm. The usual width worked is between 1.40 and 1.80 m, but there are models on the market as narrow as 90 cm and as wide as 280 cm. These devices are connected to the tractor's power takeoff, so they can work on their own (they are not just dragged by the tractor). The **rotor** can be changed at will by changing the tractor's gear. In simpler tractors with no gearing for the hydraulic takeoff, it can be done by changing a set of gears. The normal rotation speed is between 140 and 250 r.p.m.

To ensure a fine tilth, the rotor should be turning as quickly as possible, the tractor should be advancing slowly and the rear hood should be lowered, while if a coarser tilth is wanted, the rotor speed should be slow, the tractor should go faster, and the rear hood should be raised. The ideal is to advance slowly (1-2 kmh) with the rotor at a slow speed. When the speed is low, less power is required, and this saves fuel. It also helps to maintain the soil structure, avoiding the formation of a plow pan. It also reduces wear to the blades.

Left: There is a wide range of models of rotary cultivators on the market. The model in the photo is manufactured by the FERRI ROMOLO company. It is small and can be powered by a medium-low powered tractor of only 35-50 horsepower.

The R2®.model from FERRI ROMOLO is suitable for tractors with 55-80 horsepower. It works a wide strip (3.35 m) and can do the work in just a few passes.

When it is necessary to work the soil deeper, rotary cultivators with larger blades can be used. Of course they need a high-power tractor, 100-150 horsepower. Sold by FERRI ROMOLO.

The rotary axle of diggers does not bear the knives typical of rotary cultivators, but shovels like the tip of a hoe. This type of machinery has the advantage that it does not lead to formation of a subsoil crust, plow pan. Courtesy of FERRI ROMOLO

9.3.2.1. Diggers

Mechanical diggers are widely used in horticultural crops as they have a major advantage over rotary cultivators, namely that they do not cause the formation of a subsoil crust. The blades or tines are powered by a crankshaft axle through the articulated quadrilaterals, meaning that its movement is oscillatory, and thus very similar to manual hoeing. There may be 3, 4 or 6 arms, the width worked may be 1-2 m, the depth 20-25 cm and the speed 1-1.5 km. Its power consumption is comparable to that of a rotary cultivator, but the deeper it digs the more power it consumes.

9.4. DEVICES FOR COMPLEMENTARY TASKS

Different types of cultivator arm
A/ rigid with spring
B/ flexible flat steel
C/ flexible with leaf spring
D/ flexible spiral

This section includes the devices that are not used to remove the previous harvest, but to prepare the seedbed for sowing the next crop. They do things like dunging, incorporation of mineral fertilizers, levelling the soil, breaking up clods, getting rid of weeds, etc.

9.4.1. Cultivators

This model of cultivator with spirally sprung flexible arms can prepare a seed bed. Manufactured by OTTAVIOLI CONSTRUCTEUR.

Cultivators are used for many purposes; weeding, breaking up clods, preparing a fine surface tilth, and incorporating herbicides and pesticides into the soil. They consist of a series of arms, tipped with teeth that can vary

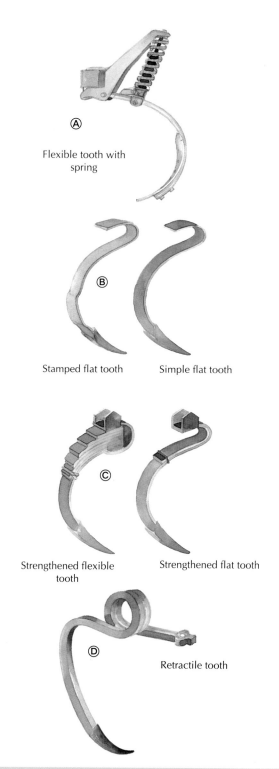

Ⓐ
Flexible tooth with spring

Ⓑ
Stamped flat tooth   Simple flat tooth

Ⓒ
Strengthened flexible tooth   Strengthened flat tooth

Ⓓ
Retractile tooth

The photo shows a device, consisting of three implements combined together, that is excellent for minimum tilling techniques. First, the chisel-type scarifying teeth penetrate the soil; then a cultivator with spring teeth breaks down the clods, and finally a roller prepares the seedbed. A single pass leaves the soil ready for sowing. Manufactured by BRENIG GREENMASTER LANDTECHNIK GmbH

greatly in form and which can generally be mounted at variable distances on a frame that essentially consists of a set of bars.
The different types of cultivators can be classified on the basis of the arm type, i.e., rigid, rigid with spring, flexible, flexible with spring, spiral flexible, etc. Perhaps the most important difference between the different types of cultivator is the different shape of the teeth, that is to say, the type of work they do.

Starting with the broadest, they are:

• **Grubhooks**, Ⓐ, are used to eliminate weeds.

• **Bankers**, Ⓑ, are used to make ridges.

• **Diggers**, Ⓒ, are similar to grubhooks, but narrower.

• **Weeders**, Ⓔ, are like the preceding ones, but narrower. They are used to remove weeds.

• **Scarifiers**, Ⓕ, are strong and tough, so they can work to a great depth.

• **Regenerators**, Ⓓ, are very narrow and sharp. They are used to regenerate meadows, turf, etc. This device is widely used to improve turf in sports fields.

9.4.2. Harrows

The main use of harrows is to prepare the seedbed, that is to say, to thoroughly break up the soil to prepare it for sowing the seed. Unlike the knives of cultivators, which only exert a frontal pressure on the soil, harrows are designed to exert frontal and lateral pressures, and so they break the soil down better. Like plows, harrows are tractor-drawn and are not connected to the tractor's power takeoff. The most important types are **rotary cross harrows**, **spike-toothed harrows** and **disc harrows**.
Rotary cross harrows consist of rotating axes with many sharp points arranged in several different horizontal axes, which moves over the earth as it is drawn by the tractor. They are often used to break clods after tilling in the previous crop, and both these operations are often carried out at the same time. The axles are supported by bearings, so they never jam. There are two types of rotary cross harrow, **star harrows** and **blade harrows**.
Spike harrows consist of a more or less rigid frame bearing many spikes perpendicular to the ground that break the topsoil by cutting it to a given depth. Spike harrows may be **rigid**, **articulated**, **reticulate** or have **mobile spikes**.

A double disc harrow with four gangs. This model is the Centor 50 A ® produced by the RAU-JEAN DE BRU company.

Above left: A cultivator with very narrow cutting teeth, these implements are also known as regenerators. It is widely used to restore the turf in sports pitches. Manufactured by WHIRLWIND HOLLAND B.V.

Left: Different types of cultivator teeth

Different types of spikes used in harrows

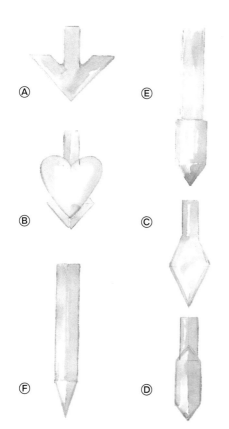

Ⓐ
Ⓔ
Ⓑ
Ⓒ
Ⓕ
Ⓓ

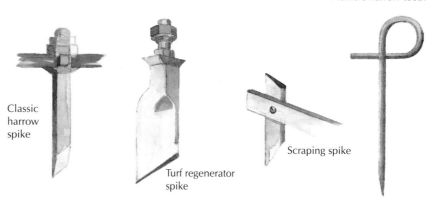

Flexible harrow tooth

Classic harrow spike

Turf regenerator spike

Scraping spike

Croskill disc rollers have rounded bumps on the side and this makes them wider than the diameter of the axis supported, so they can move freely up and down and turn at different speeds.

The marker roller in the photograph is the SKE ® model from RABE WERK GmbH+Co. Seed sowers can be fitted on at the same time.

Disc harrows consist of a set of concave discs, with a smooth or grooved edge, mounted on two or four horizontal axles and which can rotate freely supported on bearings at an angle to the direction the tractor is advancing. They are used in tasks as important as removing the previous crop, complementary tasks and minimum tillage systems. The number of discs mounted on each axle varies from 4 to 15, and the most common size is between 510 and 610 mm. Depending on the arrangement of the disc harrow's axles, we can distinguish between **simple harrows**, **double harrows** and **eccentric harrows**.

9.4.3. Rollers

Rollers are cylindrical instruments that are drawn in the same direction as the tractor. They are usually between 40 and 60 cm in diameter and roughly 2 m wide, and weigh 600-800 kg. They are used for several purposes, including consolidating the topsoil after sowing, in order to reduce the porosity of the topmost layer so the seedlings are in closer contact with the soil particles. Rollers may be used for other purposes, such as breaking clods, making the soil uniformly firm for sowing and consolidating the topsoil to avoid moisture loss. The main types of roller are **smooth**, **undulating**, **marker**, **Cambridge**, **Crosskill** and **subsoil** (see accompanying drawing).

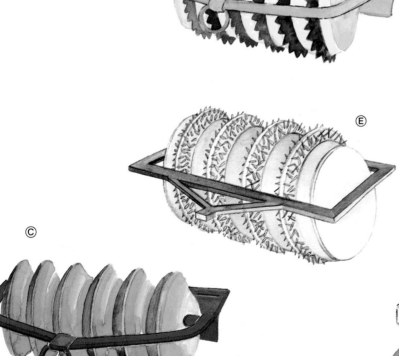

Different types of roller
a) smooth
b) undulating
c) marker
d) Cambridge
e) Crosskill
f) subsoil

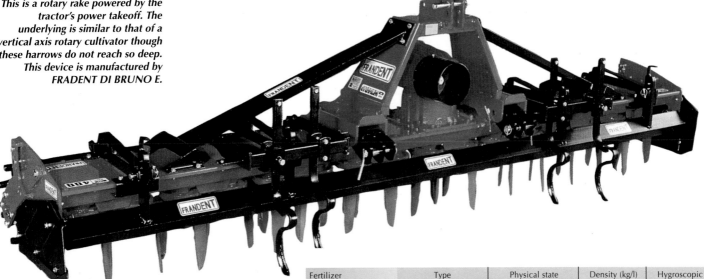

This is a rotary rake powered by the tractor's power takeoff. The underlying is similar to that of a vertical axis rotary cultivator though these harrows do not reach so deep. This device is manufactured by FRADENT DI BRUNO E.

Fertilizer	Type	Physical state	Density (kg/l)	Hygroscopic
Nitrogen-containing:				
– Ammonium	Ammonium sulfate	Crystal	1.08	Weakly
	Calcium cyanamid	Powder or granulate	1.00 - 1.05	Not
	Ammonium phosphate	Crystal	1.00	Not
	Urea	Granulate	0.71	Not
– Nitrates	Calcium nitrate	Crushed	1.00	Highly
	Sodium nitrate	Crystal	1.25	Weakly
	Potassium nitrate	Crystal	1.00	Weakly
– Ammonium nitrate	Ammonium nitrate	Granulate	1.25	Highly
Phosphates:	Superphosphate	Powder	1.00	Weakly
	Slag	Powder	2.00	Weakly
	Natural phosphates	Powder	1.28	Not
Potassium-containing:	Potassium chloride	Crystal	1.00	Weakly
	Potassium sulfate	Crystal	0.90	Weakly

9.4.4. Rakes

This group includes a wide variety of implements that are used to level the top layer of soil. They also remove any weeds present, break any crust in the soil and turn the topsoil into a fine tilth, thus improving its aeration. They used to be made by farmers out of heavy boards with nails, or from a set of iron beams soldered together.

9.5. FERTILIZER APPLIERS

The special machinery for applying fertilizer to the soil has recently undergone great changes due to the need to make highly localized applications of fertilizer. Applying the fertilizer close to the plant seeks to obtain the highest yield from the crop with the smallest expenditure. Fertilizer appliers have thus become more technically advanced and specialized, so there are now forms suitable for applying any sort of fertilizer to any crop. Obviously applying solid fertilizer is very different to applying a liquid or gaseous one, distributing evenly throughout the site is not the same as distributing in strips or at depth, and fertilizing horticultural crops is unlike fertilizing orchards or cereals, etc.

The fertilizers used on crops are usually solid. The application doses may vary between 50 kg/ha and 1,200 kg/ha. The form of application will depend on the specific conditions in each case: soil, climate, crop, etc. Solid fertilizers may be **granulates**, **crystals** and **powders**, with particle sizes of 0.5-5 mm, 0.2-1 mm and 0.001-0.1 mm respectively. One important factor about a fertilizer is how easily it is dispersed: the greater the particle size, the less likely it is to be blown away. It is also important to know how hygroscopic a fertilizer is; the more water it absorbs, the harder it will be to disperse evenly, as it will clump.

Solid fertilizers are widely used in the following circumstances and crops: in turfs, applied in lines; in strips, together with other cultivation tasks; applied at the time of sowing; distribution on the surface after harrowing, or mixed with earth and before sowing; applied at depth with the subsoil plow; distribution on the surface and before plowing; or introducing the fertilizer into the base of the furrows made by the teeth. Liquid and gaseous fertilizers are much less widely used. Liquid fertilizers are usually applied in solutions where the fertilizer is simply dissolved, or in colloidal suspensions, whose formulations allow much higher concentrations. **Ammonia** is the most widely used gaseous fertilizer. It is transported as a pressurized liquid, but turns into a gas at normal atmospheric pressure, after it is injected into the soil.

Physical characteristics of the main mineral fertilizers

To the left the rakes sold by RABE WERK GmbH+Co can break any subsurface crust and open up the topsoil.

Fertilizers with very fine particles (dusty powders) are easily blown away by the wind. Fertilizers applied in larger grains are heavier, and so they are less likely to be blown away when discharged from a fertilizer applier.

Centrifugal fertilizer distributor in use (Photo courtesy of BASF, S.A.)

The BS 1400® is a rotary plate fertilizer applier. It is manufactured and distributed by COMERCIAL VICON, S.A.

9.5.1. Appliers for solid fertilizers

A good fertilizer applier of this type should ideally fulfil the following conditions. The fertilizer should be evenly distributed; it should be made of materials resistant to corrosion, as many fertilizers are caustic; it should resolve in some way the problem that many fertilizers are hygroscopic, which leads to clumping and this makes even distribution more difficult. In general, they consist of a hopper in which a given amount of fertilizer is deposited, and a system to distribute it onto the site. There are five main types of fertilizer appliers with different distribution systems:

• Gravity distributors
Gravity distributors are suitable for granulates and powdery fertilizers, which fall by their own weight. The working width varies from 1.75 m to 5 m. The hoppers usually have a capacity of 50-100 kg per meter width treated, and they may have some type of mixing system to prevent clumping. To avoid corrosion problems choose models in stainless steel or plastic. They normally have wheels and are dragged by the tractor, though some are suspended with a three-point linkage. There are the following types of mechanisms in fertilizer appliers: **endless screw, tooth, roller, chain, rotating plates, moving base** and **central hopper**.

• Pneumatic distributors
The fertilizer is blown by a jet of air along a bar or ramp with nozzles. They are suitable for applying any type of fertilizer, including powders. There are different models on the market varying in width from 5 to 15 meters. The mineral fertilizer is introduced into the jet of air from a doser, which consists of a toothed roller or a cell wheel powered by the tractor. This system ensures very even distribution, even at very low doses.

This model of centrifugal fertilizer, from COMERCIAL VICON, S.A., ensures uniform distribution of the fertilizer, but only to the left and right, and not forwards and backwards. This helps keep the implement's mechanisms clean, as the fertilizer is not scattered over them.

• Centrifugal distributors
These are very suitable for the distribution of granulated fertilizers whose granules are mechanically scattered over the soil. They consist of a hopper beneath which is the distribution mechanism, from where the fertilizer is scattered. They have a capacity for 300-700 kg of fertilizer, and the hopper includes a removable shaking mechanism. Their working width varies from 8 to 14 m. This type of machine can be used to sow seed by scattering. These fertilizer appliers are generally suspended from the tractor's three-point linkage and powered by the tractor, though they sometimes also have wheels so they can be attached to small tractors or trailers.

• Line distributors

These devices apply the fertilizer in a narrow strip next to, or ideally underneath, the plants to be given fertilizer. They are best when used on crops grown in rows a considerable distance apart (tomato, corn, etc.). They are frequently used combined with weeders between rows of cereals, or with the sowing of the seed. Deep application models have cutting teeth behind which the fertilizer is deposited, which falls through thick flexible tubes or onto a ramp formed by the tooth itself.

• Line fertilizer appliers

They are the conventional type of hopper with an inner axle of shovels on an endless screw that propels the fertilizer through outlet tubes onto each line, or in one or two strips by means of screens. In some cases, these implements have to be coupled to a deep line distributor whose teeth have to be removed.

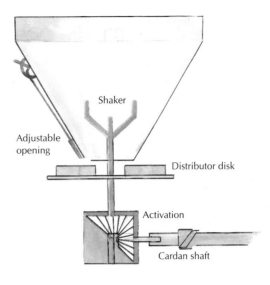

Diagram of a centrifugal fertilizer applier

To the left:
Diagrams of different methods of distributing fertilizer:
A/ endless screw
B/ grates
C/ roller
D/ chain
E/ rotary plates
F/ moving base

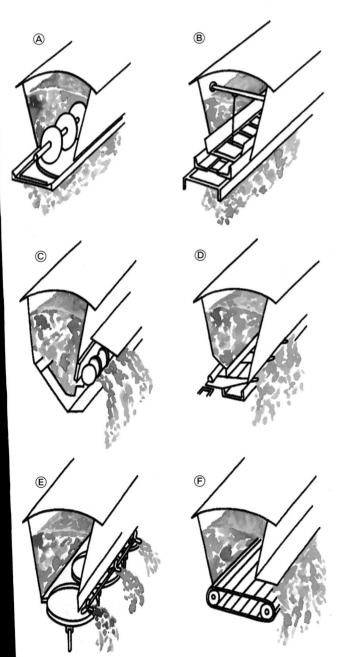

9.5.2. Appliers for liquid fertilizers.

Using liquid fertilizers is very good for the crop (they are easy to assimilate and the dose is easily controlled), though it has the disadvantage that devices to apply liquid fertilizers are expensive and take a long time to pay for themselves. There are two types: those used to inject gaseous fertilizers, which have to work at high pressure (12-18 bars) injecting the liquid ammonia into the soil (where it turns into a liquid at normal atmospheric pressure), and those intended for the application of nutrient solutions of soluble minerals (urea, ammonia, soluble nitrates, etc). In both gaseous and liquid applications, the soil must be relatively fine tilth so the fertilizer can mix with the soil water in order to be absorbed by the plant. If this is not so, the plant does not make best use of the fertilizer.

There is now special machinery to distribute liquid fertilizers on the soil surface, which do not require complex devices to inject it into the soil. Many of the formulations of soluble fertilizer on sale can be applied in this way.
Distributed by COMERCIAL VICON, S.A.

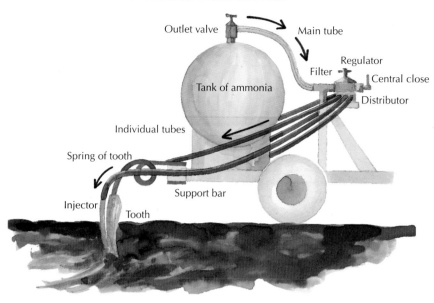

Outlet valve

Main tube

Regulator

Filter

Central close

Tank of ammonia

Distributor

Individual tubes

Spring of tooth

Injector

Tooth

Support bar

Fertilizer applier suitable for applying gaseous fertilizers. Liquid ammonia is injected into the soil at depth, where it turns into a gas, and its nitrogen can then be used by the plants.

There are now soluble fertilizers or colloidal suspensions of all the nutrients needed by plants, including the micronutrients, such as boron, iron, copper, zinc, etc. This type of fertilizer can also be applied using a drip irrigation system or by sprinklers.

9.6. SEED SOWERS

A conventional pneumatic seeder from the SOLA, S.L. company. The top of the range, the EURO 888® can sow a strip 4 m wide with a separation between rows of 12 cm. The hopper can take a load of 660 kg of seed.

There is a wide range of different seeders, machines adapted to the different methods of sowing that are commonly practised. Their purpose is to put the seeds into the soil without damaging them. The sowing method used depends on the crop grown; each crop has its own characteristics and has its own sowing requirements. The purpose of the different sowing methods, and thus seeders, is to establish the best density of plants, with an economically and agriculturally suitable planting frame. The density of the plants, their planting frame, is determined by the type of crop, the soil class and its fertility, the availability of water (dry farming or irrigation), and the cost of the work that has to be carried out at the different densities (thinning, weeding, cultivation, harvesting, etc.).

Photo of a pneumatic line seeder. These precision devices make it possible to control the rate the seed is sown very accurately. Sold by HASSIA MASCHINENFABRIK GmbH.

Plants can be sown in two ways. **Entire surface cultivation** is practised when it is not necessary to till during most of or all the plant's growth period, for example in winter cereals. **Line cultivation** is when the crop is sown in lines so that machinery can pass freely later in cultivation.

The differences between crops means there are the following different sowing methods:

• **Broadcasting** (scattering)
The seeds are distributed randomly over the surface of the ground.

• **In continuous rows**
The seed is sown randomly along furrows. The rows may be 1 cm wide or as much as 6-8 cm, in strips or bands.

• **In discontinuous rows**
As above, but sowing a group of seeds at a given distance within each row.

• **Separately**
As above, but only sowing a single seed at a time.

9.6.1. Types of seeders

• **Scatterers**
Scatterers are appropriate for small seeds and especially for meadow grasses. They are very simple, and there are two different types, **centrifugal** and **free dischargers**. Centrifugal scatterers are similar to centrifugal fertilizer appliers, and can often be adapted to distribute any sort of seed. Free dischargers let the seed fall by gravity and, at the rear, usually have a spike harrow or roller to bury the seed.

Other seed scatterers, based on completely different ideas, are used to **sow from the air** and in order to **sow in water**. Aerial sowing uses small planes and the same equipment as used for aerial application of fertilizer. Wet seed scatterers are mostly used to sow meadow grasses on slopes at the side of

| Distribution over the maximum width of strip (without accessories) | Distribution in strips (special accessory) | Distribution on both sides (central screen) | Distribution on one side only. | Distribution in a strip (total screen) |

highways and freeways. This special machinery is hooked up to the tractor, and the seed is mixed with fertilizer, water and adhesive, and then sprayed out.

• Seed drills
Seed drills have to perform the following operations: open the furrow where the seed is going to be deposited (using furrowing teeth, circular blades, etc.); dose and deposit the seed in the furrow, then bury it, and press the soil down (to favor successful germination). The hopper containing the seed should be easy to fill, empty and clean, with a capacity of at least 100 liters of seed for each meter width to be sown. These seed drills are classified on the basis of their dosing mechanism, and the following are sold: **toothed wheels**, **grooved roller**, **centrifugal** and **pneumatic**.

• Seeder units
Seeder units are high precision seeders that deposit roughly the same number of seeds each time, at equal distances along the length of a line. Single grain drills are similar but only deposit a single seed. This type of machine can be adapted from one method of sowing to another simply by changing the distributor plates. The term seed drill is usually kept for devices that sow larger seeds (corn, cotton and pulses).

These machines, modified to obtain a single-seed type drill, are used to sow very small seeds (sugarbeet, vegetables, etc.). They allow great control over the number of plants per hectare, meaning that less seed is needed, mechanization is easier (more regular rows), less labor is needed (unnecessary thinning) and conditions for harvesting are optimal.

Precision drills can be adapted to any type of seed by adjusting the mechanical parts. It is possible to regulate the sowing rate, the distance between seeds, their depth and the distance between the rows.
Other types of machinery can also be attached to these drills, such as sprays to apply weedkiller and implements to apply fertilizer at depth.

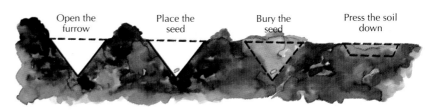

| Open the furrow | Place the seed | Bury the seed | Press the soil down |

9.7. PLANTERS AND TRANSPLANTERS

Some authors consider potato cultivation to be intensive, while others consider it extensive. It was included in the first section of this work as an intensive crop. In some zones potato cultivation is highly mechanized and covers large areas of ground. The potato tuber is vegetatively reproduced, and is not grown from seeds, unlike most other crops. Small potato tubers ("seed potatoes") have to be planted in the field, and this requires specialized machines, called **planting machines**, that are different from seed drills.

The seed potato tubers must be sown at the same depth all over the site and at the same distance within rows and between rows. The planting machine has to open the furrow, introduce the seed potato, replace the earth in the furrow and press it down. Depending on the variety, the planting density can vary between 30,000 and 60,000 seed potatoes per hectare, with a planting frame that is usually 30 x 75 cm. The potato should be planted as shown in the diagram on the following page, with the potatoes at the original level of the site and covered with a layer of at least 5 cm of earth.

Above: The different ways of distributing with a centrifugal seeder. It can also be used to apply fertilizer in the same ways.

The sequence of operations performed by line seed drills.

An automatic potato planter with chain of scoops. Manufactured by HASSIA MASCHINENFABRIK GmbH.

A/ The correct position of the potato in the ground. Planting machines ensure the planting depth is uniform over the entire site.

B/ Other sprays attached to the tractor allow application of pre-emergence weedkillers. Manufactured and distributed by Hardi International A/S

There are three different types of planting machines, depending on how the potatoes are supplied. **Manual planters** are very suitable for small surfaces, and require an operator for each line to be sown. They are not very fast, as their speed is limited by the fact an operator can only handle about 120 potatoes/hour. **Automatic** planters do not require a labor force and are specially designed for large areas; they have a chain of buckets that scoop the potatoes from the hopper to the furrow. These planters have an error-correcting device that consists of a mechanical or electronic sensor that detects the passing of the potato, and sends a signal when a scoop is empty. **Semi-automatic** models are similar to automatic ones, except that they require an operator to fill any scoops that are empty.

AA/ Detail of the scoops of the potato planter shown on the previous page. (HASSIA MASCHINENFABRIK GmbH)

Dividing a single drop with a diameter of 400 μ into drops with diameter of 200 μ gives a total of 8 drops. This means the product covers the plant better and is more likely to reach its target. (Courtesy of HARDI INTERNATIONAL, A/S)

ⒶⒶ

On the following page, upper left: **C/** Some crop protection machines are high enough to apply pesticides to crops that are actively growing, thus causing minimum damage. Machinery manufactured by GAMBETTIBARRE, s.r.l.

The purpose of transplanting machines is to transplant seedlings grown in seedbeds, and they are mainly used in horticulture and nurseries, etc. They are usually hand supplied, and work at a rate of 0.4-2.0 kmh, depending on the number of plants per hectare. The seedling is introduced manually into the machine, and the machine then places the plant upright, presses earth around the base, and in some cases irrigates the base of the furrow. They work to a depth of 20 cm or more, in rows 25-75 cm apart, and at a distance within the row that may vary between 15 and 120 cm. As they use interchangeable parts these machines can be used to transplant root crops, bulbs, seedlings, etc.

Potato planters can often be adapted to transplant plants. The machines that transplant rice are a special example of this type of machine. The mechanism consists of needles or fingers that separate a group of plants that are arranged on a tray and then introduced into the mud. As happens with the planters, the planting depth and distance can be regulated.

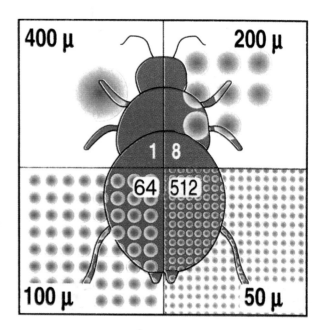

9.8. CROP PROTECTION MACHINERY

This section deals with the machinery used to protect plants, to apply products like pesticides and herbicides. These treatments seek to cover the plant with the product as evenly and continuously as possible, in order to prevent further growth of weeds, or the spread of pests and diseases. The reader should refer to the fourth part of this work which defines and discusses the concepts of active ingredients, wetting agents, bulking agents, etc., and gives guidelines on the maintenance and cleaning of these devices. The different methods of applying liquid products produce drops of different sizes, and of course the smaller the diameter of the drops the larger the number of drops, and this increases the area that can be covered with a given amount of product.

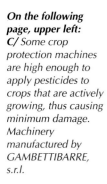

E/ The TS 3082® spray from HARDI INTERNATIONAL allows treatment between rows of fruit trees. The small drop size produced by this type of machine increases the intrinsic effect of the insecticide or fungicide.

D/ This spray is used to apply insecticides when the crop's roots and collar are most sensitive to attack. Manufactured and distributed by HARDI INTERNATIONAL A/S.

Combination 1958 and 1973
1969

1963

1972

1967

Changes in spray technology over the years

Depending on the type of product to be applied, there are two main types of machinery. **Sprays** apply liquids in the form of drops. **Dusters** distribute solid products in the form of powders. Depending on the size of the drops or particles, these devices distribute liquids with **sprays** that emit a jet, **atomizers**, **nebulizers** and **rotary** or centrifugal models.

9.8.1. Sprays

When pressure is applied to the treatment liquid, it is ejected from the nozzle in the form of drops, and the distance the drops travel varies with their size (the smallest drops are stopped first by air resistance). The diameter of these drops ranges from 150 to 450 μ. Sprays can be used to apply a wide range of pesticides and fertilizers, etc. They can also perform highly localized treatments, such as applications to fruit trees in winter. Per unit time they need less energy than atomizers, and their price is lower. The large size of the drops these devices emit means that a larger amount of product has to be used per hectare in comparison with other treatment machines. The large drop size also means that the product can not reach all the parts of the plant, and the inside of the plant often does not receive any product.

Working with a backpack pneumatic atomizer. Manufactured by WHIRLWIND HOLLAND B.V.

Backpack pneumatic nebulizer. It provides enough pressure to treat fruit trees at some distance above the ground. Manufactured by E. ALLMAN & COMPANY Ltd.

9.8.2. Atomizers

Atomizers work in a way similar to sprays, but differ in that just before the liquid is ejected, there is a fan that emits a high speed jet of air that breaks the drops into tiny droplets (this is called atomization). These droplets have a diameter of between 50 and 150 μ. Atomizers have one major advantage over sprays, the drops are much finer and thus cover a large surface area. Depending on the droplet size, there are two types, **hydropneumatic atomizers** and **pneumatic** ones.

The droplet size is much smaller in atomizers than in sprays, as the drops are smaller, and this means that less liquid has to be used (50 l/ha or less), and this increases the efficiency of the time spent, because less time is wasted filling the container. Losses due to runoff from the aboveground parts of the plant are reduced, precisely because the drops are so fine. The number of impacts on the plant increases, favoring the action of fungicides. The product is deposited on both the upper and lower surfaces of the leaf.

One problem that should be mentioned is that because they are so fine, the drops may be

blown away by the wind and onto neighbouring crops, and this may damage them if the product used is toxic (such as copper fungicides). Atomizers use more energy than sprays, and their purchase price and maintenance costs are also greater. In dry conditions with high evaporation, the drops may evaporate before they reach the plant, and this can be avoided by adding a product to the mix to reduce evaporation.

9.8.3. Nebulizers

The idea behind the nebulizer is very different to sprays and atomizers, as the drops are transported (rather than propelled) using hot gas or steam. This creates a fog of gas combined with tiny drops of the product to be applied, and this can reach all the parts of the plant. There are two types of fogs, depending on the particle size produced.

Nebulizers create a dense fog that is light and floats in the air, and which contains tiny drops of the product to be applied. Because it is so finely divided, it reaches all the surfaces of the plant. Swingfog, sold by TECTRAPLANT, S.l.

This atomizer is sold by HARDI INTERNATIONAL and is used to apply insecticides and fungicides in nurseries of forest and fruit trees.

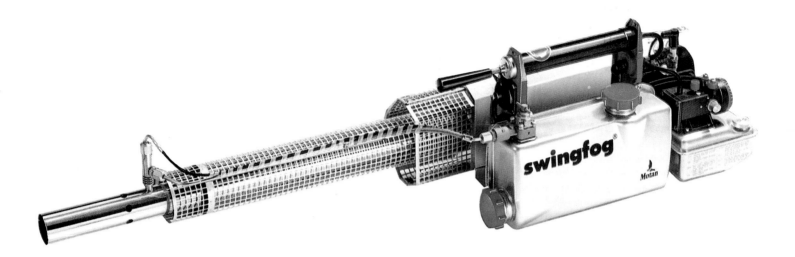

Coarse or **wetting** fogs consist of drops between 20 and 50µ, while **fine** or **dry** fogs have a drop size below 20 µ.
These devices are designed for treatments inside greenhouses, as in the open air the droplets are very easily blown away by the wind, even in very calm conditions. The manufacturers recommend them by saying that a fog is bound to reach the entire plant, and is thus better than any other type of spray or atomizer. The main disadvantage is the high purchase price and maintenance costs.

Machinery to protect crops in greenhouses: a nebulizer. Swingfog ® model from TECTRAPLANT, S.L.

9.8.4. Centrifuges

In rotary or centrifugal sprays, the treatment mixture descends due to gravity through a rotary disc (7,000-10,000 r.p.m.), giving a particle size of 170 µ (LV or low volume) and 50 µ (ULV, ultra low volume). This system is mainly used in manually operated sprays with batteries powering a 7 W electric motor.

Manual centrifugal spray, specially designed for ultra low volume (ULV) weedkiller

Duster for small areas. The GR 75® model of the GR di GAMBERINI REMO, has a capacity of 75 kg and is suspended when it is connected to the tractor, and is powered by the tractor's power takeoff.

This duster is from the same company as the one above, but has a larger capacity (up to 200 kg in the largest model). It requires a 15 horsepower tractor, and is connected to its power takeoff. It can disperse any pesticide powder to a distance of up to 8 m on either side.

On the right: A set of injector blades or teeth. They are especially suitable for the deep application of nematicides and insecticides.

9.8.5. Dusters

Dusters are machines that distribute the active ingredients on a stream of air. The main problem of this technique is that the powders do not adhere well to the plants, meaning that they do remain on the plant for long. For this reason, it is worth dusting crops early in the morning when the plants are covered in dew. Large areas should be treated at night, when there is little wind, and when soil warming does not cause updrafts that cause the product to drift away from the soil.

The dust's adherence to the plant depends on the atmospheric conditions, and may be improved by wetting techniques or electrostatically charging the particles, though these methods are still experimental.

The main advantage of dusters is that they do not need water for application (no tanks, transport or deposits, etc. are needed). They have the disadvantage that treatments cannot be applied on windy days, as they might have a toxic effect on neighbouring crops.

9.8.6. Machines for soil treatments

Some soil parasites, such as fungi and nematodes, cause serious damage to crop plants (see section four: Defense of cultivated plants). They can be combatted with physical or chemical methods, and both require suitable machinery.

Soil sterilization by heat is an ancient method of disinfection that eliminates many nematodes, some fungi and weed seeds in the soil. When a substrate has to be disinfected for use in horticulture or gardening, the problem is relatively simple, as all that needs to be done is to put the substrate in a suitable container and heat it with an electric heater element.
When soil has to be disinfected *in situ*, the soil is directly treated in small patches by injecting steam into the soil. Steam sterilization devices are rake-shaped or rectangular cushions with teeth that are inserted into the soil to the desired depth. The jets of steam emerge at a temperature of approximately 110°C. The temperature should not be higher than this to prevent biological or chemical damage to the soil.

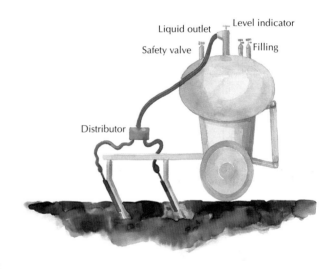

Chemical disinfection of the soil is performed using products that may be solid or liquid. Solid products applied using **localized drills** may be microgranulate or powder herbicides or insecticides, and they are usually applied when the seed is sown. The application of liquid products uses the vapour they give off as a fumigating agent, as it kills insects, fungi, nematodes, and weeds. The product is distributed using different devices, such as **injectors**, **injector blades** and **injector teeth**. This machinery injects the liquid at depth, an operation that has to be performed under sheeting because the products are highly toxic.

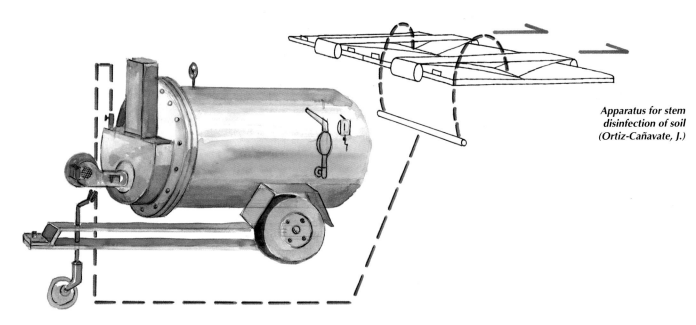

Apparatus for stem disinfection of soil (Ortiz-Cañavate, J.)

9.8.7. Machinery for aerial treatments

Low volume (LV) duster mounted on plane (cropduster) Courtesy of K. BRZICA.

Aerial treatments are carried out using small planes or helicopters, though helicopters are much more expensive per hour (two or three times more expensive). Small planes go faster, which is positive in large areas, and they can also carry a larger load. Helicopters show some advantages, as they can land and take off from very small sites (20 m² is enough) and even on waterlogged sites; they are safer and more maneuverable, and also distribute the product better because of the air blown down by their blades.

Modern agricultural airplanes have a power of 125 to 240 kW, and can take a load of 300 to 750 kg. They generally work with low volumes (LV) applying a dose of 30 to 60 l/ha with a best drop size of 180 µ. If special technology is available, the drop size can be reduced to work at ultra low volume (ULV),

in which case a dose of 0.5 to 10 l/ha is used. Wind is an important factor limiting aerial treatments, as application should not be performed if the wind speed exceeds 4 m/s, especially when using products that may damage neighboring crops, such as copper-based fungicides.

Helicopters adapted to carry out aerial treatments allow highly localized application of herbicides or pesticides. Courtesy of SANDOZ.

10. HARVESTING

Harvesting the crop and storing it for use are the main aims of cultivating crops. The harvest is collected when it is ripe, and it then has to be stored to feed humans or animals over the rest of the year. Harvesting has always been one of the most important of human agricultural activities; in prehistory, mankind learned to harvest before learning to cultivate.

Rotary disc reaper for harvesting fodder (Courtesy of COMERCIAL VICON, S.A.)

*Above: Growing sugar beet or fodder beet requires specific machinery. In this case, the trailer to transport the crop and a front tract loader suitable for handling it.
Sold by
XAVER FEND & Co.*

Reaper-harvester for fodder corn. This is the MH 90S® model manufactured by COMERCIAL VICON, S.A.

10.1. GENERAL PRINCIPLES

The grain is physiologically ripe when it reaches the pasty grain stage, but because it then contains about 40% water, harvesting must be delayed until it is drier. This does not represent any problem when part of the production has to be kept apart as seed for the next crop, as grain that has dried on the plant can still germinate normally.

The moisture content of harvested grain may thus vary from 40-45% to 10-12%. Harvesting when moisture content is higher means that the harvest can be earlier in order to plant a new crop quickly or to avoid risks. Harvesting when the moisture content is very low avoids the cost of drying the grain before storage. The following are guidelines indicating whether the harvest should be earlier or later, but each crop has its own guidelines:

• Corn should be harvested when the moisture content is 25-30%, because if it is too dry the grains break when the harvester strips the grains from the cob.

• Rice is best harvested with 18-21% moisture content. This means the yield is higher in the industrial processes of removing the husk, or "hull", and cleaning.

• The other cereals should be harvested as dry as possible, in order to avoid having to dry them (saving expense).

Harvesting is now carried out by harvesters, which separate the grain from the straw. The grain is taken to storage and the straw is left in the field for later baling and collection, or grinding and burial, or to be burnt. In the case of corn, the whole cobs can be collected using harvesters that separate the cobs from the rest of the plant. This used to done by hand.

10.2. COLLECTING FODDER

Fodder plants, those grown as food for livestock, can be eaten by the livestock directly in the field or stored in silos as a reserve for the fall and winter. The first step to take in storage is to reap the plant. This used to be done with **scythes** and **sickles**,

In the past implements like scythes and sickles were used to harvest fodder and cereal grains.

which was dangerous for the operators. Reaping devices had a sharpened blade and were used to separate the plant from the soil. They are no longer used because of the high cost of labor and the time needed to complete this task. These tools are now only used in very small sites or in underdeveloped countries.

Around 1822, an animal-drawn machine was invented that cut grass, but the definitive design was only introduced in the middle 18th century when the first cutting bars were introduced in the United States. **Reapers** can be divided into two groups, **alternating** and **rotary**, depending on the way the cutting parts move. Alternating reapers have two parts, one of them mobile while the other may be fixed or mobile and serves as a blade for the other to cut against. The two blades thus act like the two blades of a pair of scissors.

Rotary reapers are based on a different method of cutting. Cutting occurs when a blade rotating at high speed around an axis hits the stem of the plant. Some authors recommend rotary reapers when harvesting fodder for hay, because it cuts the plant in several different places and this favors natural drying in the field. In alternating reapers, harvesting is more irregular, and so it is recommended for use when harvesting fodder crops that are intended to produce new shoots.

When the fodder has been cut, it should be left on the soil until it is dry enough to be stored. Special machines, such as cutter-shredders, turning conditioners and rakes, and hay liners, are then used in order to speed up the haymaking process by ensuring drying of stems and leaves is uniform, and this avoids premature drying out of the leaves and leaf shedding.

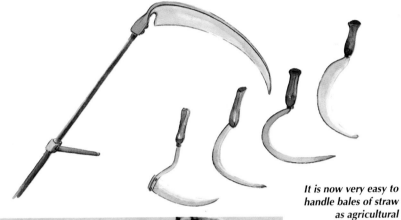

It is now very easy to handle bales of straw as agricultural machinery has undergone such great advances. This machinery is sold by PRANEDA B.V.

Detail of the pneumatic arm handling a bale. Sold by PRANEDA B.V.

A) The HSR 200 R® from the JF-FABRIKEN-J. FREUDENDAHL A/S company in action. It is a line-forming rake with a cylindrical mill.

B) The CM 1900® rotary reaper is a very practical machine that makes it possible to dry straw in a short time. It is sold by the JF-FABRIKEN-J. FREUDENDAHL A/S Company.

Reaper-shredders cut the grass along the course it follows, starting when the grass is first cut and continuing until it is beyond the reach of the blades. This machinery makes it possible to produce fodder with less than 50% moisture content. **Conditioners** are machines that break, tear and crush the stems and leaves of the plants by squashing them between two blades. This seeks to make as many cracks as possible so that the humidity is lost as quickly as possible. **Rakes** and **liners** consist of several devices that fulfil a range of tasks, such as arranging the fodder in rows, scattering it, airing it and turning it. There are different types of rake, i.e., **cylindrical mill**, **chains** and vertical axis **rotary discs**.

Ⓐ

Ⓑ

Ⓒ

Ⓓ

C) JF-FABRIKEN-J. FREUDENDAHL A/S sells a special machine for cutting and collecting fodder. Depending on the working width required, there are 3 models, the FH 1100® (1.1 m), the FH 1300 ® (1.3. m) and the FH 1450 (1.45 m).

D) A chain rake in use. This model from JF-FABRIKEN-J. FREUDENDAHL A/S arranges the fodder in lines and turns it over, treating a row up to 4 m wide at a time.

The first animal-drawn machine to cut grass was invented in 1822.

The rotary line-forming rake from JF-FABRIKEN-J. FREUDENDAHL A/S can form neat lines from rows as wide as 3 m.

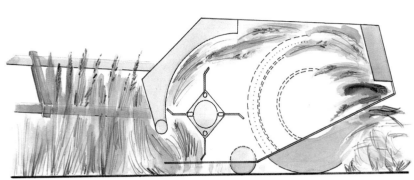

How a rotary harvester works. The plant is cut by impact.

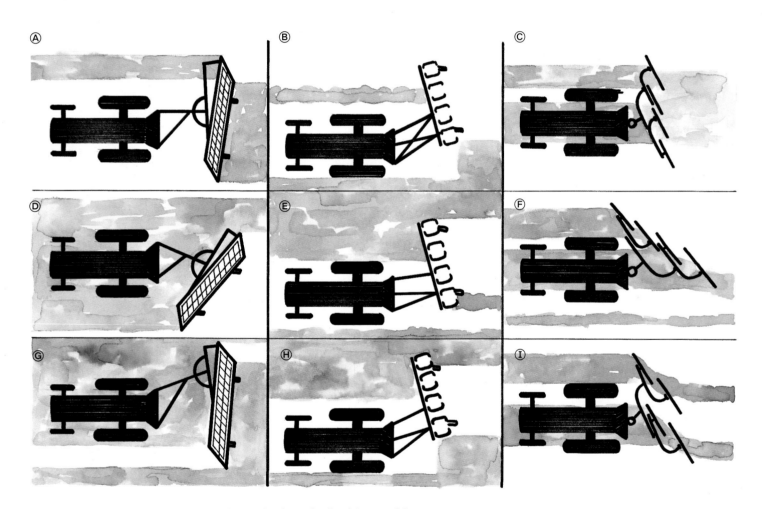

Ways of working with rakes: A/ mill; B/ chains; C/ discs, scattering

Ways of working with rakes: D/ mill; E/ chains; F/ discs, forming lines

Ways of working with rakes: G/ mill; H/ chains; I/ discs, turning

Once the fodder has dried in the field, it must be collected, cut, baled and stored in a silo. Describing all the different types of machinery used in these operations and the many types of silos is beyond the scope of this work. The reader should refer to books on agricultural machinery, which provide more detailed information on these machines and what they do, the conditions for use and their technical specifications.

Disc rake

10.3. HARVESTING GRAIN

The **chaff** is the powder or fine dust produced when grain is threshed.

The international convention is that *hUTH* represents operator working hours.

The same principles underlie harvesting the grain as cutting fodder, though harvesters must also have some sort of **threshing system** (to separate the grain from the chaff) and **collection system** (either grain or straw). The first **reaper-bunchers** were developed in the US in the late 19th century, and were animal drawn.

Reaper-bunchers are still used in combination with **threshers**. They are derived from the first machines of this type to be commercialized and, as their name suggests, they both cut the cereal and tie it into bunches. This simple type of machinery may be driven by the tractor's power takeoff or it may be self-powered by coupling it to a motocultivator. Machines of this type are known as **motorized reaper-bunchers**.

Self-powered harvester. The grain is stored within the harvester, while the shredded straw is ejected.
SAMPO ROSENLEW Ltd.

organ thresher, whose concave grain separating cylinder separates the chaff from 90-95% of the grains. This figure is also known as the **threshing efficiency** and each machine has its own value.

The machinery now used to harvest the grain is known as a **harvester**. A harvester is just a reaper-buncher and a thresher combined together in a single machine. The **reaping**, **threshing**, **winnowing** and **seed sorting** are all carried out at the same time, and this greatly reduces the labor force required. Even in the 19th century, manual sowing and harvesting a hectare of wheat represented 1,400 hUTH, but nowadays large areas of cereals are highly mechanized and the work performed is only about 10 hUTH.

The first harvesters were built in California and Australia and used animal traction. They were not connected to machinery until the first steam powered tractors were developed in the 1880s. In the early 20th century tractors started to use petrol. The first self-powered harvesters did not go on sale until 1938, in the United States. As their name indicates, **self-powered harvesters** are not powered by being dragged by a machine (such as a tractor). Two types of harvesters are now on sale, **self-powered** and **drawn**.

This model of self-powered harvester (right) is manufactured and sold by SAMPO ROSENLEW Ltd., and has a grain storage capacity of 4.2 m³. Every now and then, the grain must be unloaded into the trailer.

10.3.1. Drawn harvesters

Drawn harvesters may be powered by the tractor's power takeoff or by an engine of their own. The self-powered models are all **longitudinal**, that is to say the route of the ripe grain within the harvester is from front to rear without changing direction. There are still some drawn harvesters that are **longitudinal-transversal**, and this seeks to make the machine shorter but broader. The main advantage of this type of harvester is its low cost, which makes them worthwhile for small areas. Their main drawback is their low mobility.

The first thresher was built in Scotland in the 18th century. A thresher is a more complex machine than the reaper-buncher, because it not only cuts and ties but also separates the grain from the **chaff**, dust and other seeds. For threshing, the seed should contain 20-22% moisture, but for storage its moisture content should not exceed 15%. Perhaps the most important part of this type of machinery is the

Self-powered harvesters first thresh the plant then emit the finely divided straw onto the soil. This can then be plowed into the soil. The photo shows the 2060® model manufactured by SAMPO ROSENLEW Ltd.

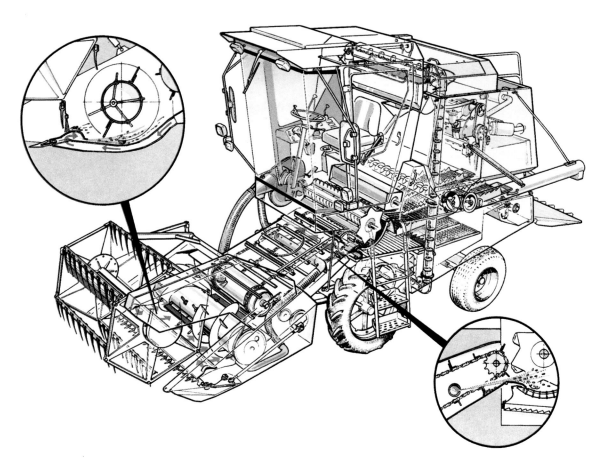

Self-powered harvester. Automatic harvesters do not need to be connected to the tractor's power takeoff. (Courtesy of SAMPO ROSENLEW Ltd.)

10.3.2. Self-powered harvesters

Self-powered harvesters usually have three speeds forward, and reverse. Each speed can be gradually and continuously regulated at a ratio of 1:2 or even 1:2.5, using a continuous trapezoidal belt change. Thus when the engine is at full power there is a continuous variation of forward speeds between 2 kmh and 16 kmh. Harvesters are used to harvest cereals (wheat, barley, oats, rye, corn, sorghum and rice). Harvesters can also be used to harvest other types of seeds, such as sunflowers, rapeseed, safflower, and pulses (lentils, vetch, beans, peas, chickpeas, etc.). All that has to be done is to vary the distance separating the concave from the degraining cylinder, and the number of revolutions per minute the cylinder turns. As already pointed out, five operations are carried out within the harvester, **reaping** and **supplying** the cutting platform, **threshing and separating the grain** from the chaff in the beaters, **cleaning** the grain in the sieves and finally **storing and unloading** the grain.

The 2000® series, consisting of the 2045®, 2050®, 2055® and 2060®, are manufactured by SAMPO ROSENLEW Ltd. The only difference between them is their working capacity, the width of the cutting bar and their power. The photo shows 4 of the 2060 model harvesting at the same time.

To the right: Integrated equipment for harvesting apples. Manufactured by MUNCKHOF.

A small complete equipment for collecting corn for fodder. The tractor powers the reaper, which ejects the fodder into the trailer. Sold by JF-FABRIKEN-J. FREUDENDAHL A/S.

10.3.3. Harvesting equipment

Nowadays there are integrated machines that can collect, pack and store a wide range of crops. There are harvesters for cotton, corn grain, sugarbeets, carrots, fruit, vegetables, etc. These whole-cycle machines perform a wide range of functions in order. As an example, a machine to harvest sugarbeets has to perform the following operations: separating the leaves from the root, uprooting it (from the soil), arranging the roots in lines on the soil, collecting the root from the soil, cleaning it by

removing any earth, loading it and transporting it. For each of these tasks there is a mechanical device that can be regulated independently of the others.

The market for agricultural machinery includes **gatherers**, which are harvesting machines that perform many functions at the same time. The farmer may opt to buy separate machines, each one of them performing just a single task (the result of the tasks performed by each machine is the same as the result of using a single harvester. Choosing between a **gatherer** or separate machines should be done on the basis of the farm's size and special conditions. The separate machines are cheaper, and it is possible to choose between a wide range of equipment and specifications (with a width of 1-6 rows) and they also perform each operation very well, because that is exactly what the machine is designed to do. They are also easy to use, without complex rules on how to use them, and they can be used on small farms.

In their favor, harvesting equipment has a high working yield, especially self-powered models with two or three rows. They are expensive to buy, but this can be split if the purchase is made by an association of farmers with common interests. These machines have to work a large area of ground to be really worthwhile. They cannot be recommended for irregularly shaped small sites, as they are difficult to manoeuver.

11. STORING THE GRAIN

11.1. RECEIVING THE GRAIN

After harvesting, the grain has to be stored in conditions that preserve it in good condition without deterioration. During storage, the germination rate of cereal grains increases, and in the case of wheat grains, the percentage yield of bread flour also increases. In general, however, the grain deteriorates because it undergoes changes, even though it is dormant.

In grain silos and stores, the grain performs **respiration**, meaning that it loses weight and creates heat. If there is little oxygen in the store but moisture and temperatures are suitable, anaerobic fermentation may occur (lactic or acetic ferments).

- **Enzymatic degradation**
Under certain storage conditions, the grain may find circumstances suitable for germination, which causes changes in its nutrient reserves, and thus a loss of quality.

- **Attacks by macroorganisms**
Some butterflies, moths and beetles can attack dry stored grain as long as it is warm enough for their growth. They may have come from the field or from within the store. Note that some insects can attack whole grains, while others can only attack broken grains, and these insects can also attack flour. Some large animals, such as rodents and birds, can cause serious damage in granaries.

- **Attack by microorganisms**
Bacteria may attack stored grain. Bacterial attack is frequent when the grain has been stored with too high a moisture content (≈ 21%), meaning

that the relative humidity within the granary (especially if it is poorly ventilated), may easily reach 90%. Other organisms, such as **fungi** may arrive from the field (fungi need 20% moisture content in the grain), or they may be from within the granary (some fungi can attack dry grains).

To sum up, the main factors affecting grain storage are its **temperature** and **humidity**. Look at the graph of temperature against humidity on the next page, which shows the different agents that may affect stored grains in different conditions. The **physical integrity** ("wholeness") of the grain is important given that its coverings protect it from some types of attack. Dust, bits of straw and other seeds are foci concentrating moisture and creating conditions suitable for infectious fungi to grow, and so it is worth trying to ensure the grain is as **clean** as possible before storing it. **Cleaning the machinery** that manipulates the grain is also extremely important. A few rotten old grains in the harvesting machinery, etc., may infect the entering harvest with many diseases. **Disinfecting granaries** is also very important, especially killing all the insects, as an incorrectly disinfected silo can cause the entire crop to be lost.

Grain silos allow automatic control of the internal moisture and temperature conditions, and this means they can store the grain in the best conditions.

11.2. WHOLENESS, CLEANING, DRYING AND COOLING

The physical integrity of the grain is ensured by paying adequate attention to the working conditions of the harvester and in the later handling of the grain. The most suitable machines to transport the grain to storage are **rubber conveyor belts**. **Endless belts** are more or less suitable, but **pneumatic tubes** are considered the least suitable method.

Complete grain drying, cleaning and storage plant. Manufactured by INDUSTRIAS LUIS PERIS, S.A.

A general diagram of the storage of grains, in function of the silo's temperature and moisture. (Adapted from Burges & Borrell)

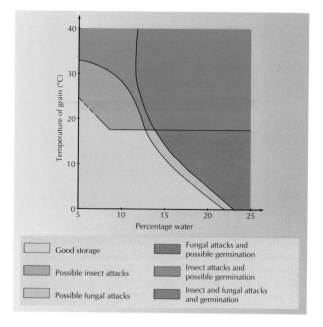

Good storage	Fungal attacks and possible germination
Possible insect attacks	Insect attacks and possible germination
Possible fungal attacks	Insect and fungal attacks and germination

Aluminium installations for grain drying. Drying machines made of aluminium are the best for installation in the open air, because they resist rusting. This model is manufactured by INDUSTRIAS LUIS PERIS, S.A.

The Rotoclean® from the company INDUSTRIAS LUIS PERIS, S.A. is a cleaner for all sorts of cereals, including corn and rice. It is well known for its high capacity when treating loads with a very high water content.

Cleaning the grain to be kept as seed for future crops requires the product to be extremely pure, often more than 99.9% purity (depending on the species). The cleaner Delta Super Universal 108® from INDUSTRIAS LUIS PERIS, S.A., equipped with a good sieving and draft system, can work with a very wide range of seeds, such as cereals, horticultural crops, meadow grasses, flowers, etc.

Cleaning the grain starts with the harvester, meaning that the harvester must be correctly adjusted, though the results obtained also depend on the initial state of the grain. It is relatively easy to clean the grain later by sieving it. This operation is normally carried out on the farm using simple machinery (manual sieves). There are now more complex machines that have a higher capacity and do the job better.

Drying the grain may give rise to added costs, considerably greater than cleaning it, and so you are strongly recommended to let the grain dry in the field wherever possible. Drying it to a suitable moisture level is one of the most important factors in ensuring good conservation. Rice is normally dried on **threshing floors**, still using traditional manual

methods. Manual drying operations represent a large increase in the costs of production, precisely because they are performed manually. Corn, like sorghum, can be dried naturally (see chapter thirteen) in **cage-silos**, which consist of an iron frame raised 50 or 60 cm above the soil and solidly attached to the foundations. It is however more frequent to dry corn in artificial drying installations. The other cereals, which ripen and are harvested in drier conditions, usually dry out thoroughly in the field.

If environmental conditions are favorable (moderate temperature and low relative humidity), the grain can be dried out at ambient temperature, if it is suitably stored in isolation. The seed's greater or lower moisture content depends on the species of seed and the environmental relative humidity. A third factor, the air temperature, also plays an important role. To dry moist grain, it should be left in contact with air, and the warmer and drier the air, the better. Hot air is most effective, as this lowers the relative humidity.

If the grain is heaped, it behaves differently. The layers of grain act as an insulating agent, reducing the exchange of air between the interior and the exterior, and thus slowing down the grain's loss of moisture. Thus it is not a good idea to pile warm moist grain and hope it will dry out suitably; it should be turned to air it, or it should be artificially ventilated. A final operation, cooling the grain, simply reduces its temperature. This removes the excess moisture from the grain (as cold air is drier than wet air). Before storage, the grain must pass through a forced ventilation system to lower its temperature to 15-16°C in summer, and 5-10°C in winter.
Artificial drying of the grain can be carried out in special **drying installations**, immediately after harvest or later on. The hot air should not be

hotter than 80°C, so as not to damage the grain. When the grain is intended to be kept as seed for the next crop, the temperature should not exceed 43°C, or the seed's germination will be negatively affected. One useful drying method is installations combining hot air and cold air. The seed is treated with hot air until its moisture content has decreased to 18-20%, and is then ventilated strongly with cold air for a couple of months in order to complete the drying. This method has several advantages, such as the reduction of costs, increasing the capacity of the drying installation and making handling the grain easier, as it is easier to handle when dry. When the grain is cold and dry, it stores much better. It is however important to check on the stored grain's temperature, as this may rise due the several factors. If it rises, the grain should be cooled down again by forced ventilation.

11.3. STORING THE GRAIN

Maintaining low temperatures using the procedures described in the previous section greatly limits the reproduction of the insects that attack the grain. Once the grain has been stored in the best moisture and temperature conditions, there is a further job to be done, chemical treatments against **insects** in the granary. This is done at two levels:

• A preliminary **chemical treatment** of the granary, in order to eliminate any resident insects that might infect the new batch of grain.

• **Treatment of the grain** when it is put into storage, or within the silo.
The farmer can choose between a wide range of products (solids, liquids and gases) that are suitable for this purpose. There are certain restrictions on which agents can be used. Products that leave residual toxicity should not be used when the grain is intended, whether directly or indirectly, for human or animal consumption. Those products that may negatively affect germination should be rejected if it is to be kept as seed for the next crop. Some highly toxic disinfectants (categories C and D), must be applied by specialized technical staff and under specified safety conditions. The reader can refer to the accompanying table detailing the possible active materials that can be used and their toxicological categories.
Other animals, such as **rodents** and **birds**, have to be fought using mechanical and preventive measures, with results that are not totally satisfactory. These methods include installing grilles over the windows, sealing their runs or holes with broken glass, etc. Other traditional systems, such as the use of cats and rat traps, and more modern methods, such as chemical products, are not 100% effective either.

• **Ventilation**. In order to prevent rodents and birds entering granaries, the windows could all be sealed, but this is not a good idea as if the silo is sealed, the grain's temperature and moisture content increases, and this is not

The capacity of the silos on sale ranges from as little as 10 m³ to more than 10,000 m³, like the ones in the photo. (Courtesy of BALLARINI SOCAMA, S.p.A.)

CLASI-FICATION	TREAT GRAIN	TREAT EMPTY SILOS	ACTIVE INGREDIENT
A	YES	YES	BROMOPHOS
A	YES	YES	MALATHION
A	YES	YES	METHYL-PYRIMPHOS
A	YES	YES	PYRETHRINES
A	NO	YES	FOXIM
A	NO	YES	METHYL CHLORPYRIPHOS
C*	YES	YES	CARBON DISULFIDE
A	NO	YES	METACRIPHOS
C*	NO	YES	DICHLORVOS
D*	YES	YES	PHOSPHAMINE
D*			
D*	YES	YES	METHYL BROMIDE

*Insecticides registered and authorized by the Spanish Ministry of Agriculture for the treatment of stored grain intended for human and animal consumption. * Manipulation by authorized companies only.*

good for its storage. Granaries should thus be well ventilated, but windows and other openings should be suitably protected to prevent animals entering. The raised granaries still found in the Galicia region of Spain are typical of how granaries used to be. They are raised above the soil to prevent moisture damaging the stored cereals, and designed for good ventilation for the same reason. Nowadays, airtight metal silos meet the same demands by means of complex controls for moisture (relative humidity) and temperature. These installations are expensive, but the investment may well be worthwhile for agricultural cooperatives, as their high storage capacity means they can store several different harvests at the same time.

Previous page. Threshing floors are areas of clean firm ground where cereals can be threshed.

Protecting the crop is one of the most important factors in obtaining good harvests. Getting rid of weeds is essential to meet your production targets. The photo shows a special machine to weed cereal crops. The bar is 24 m wide and has flexible teeth, meaning that there is almost no damage to the crop.

Next page: which phenological (growth) stage of the cereals to apply the different active ingredients of herbicides. Adapted from a work produced by the Generalitat of Catalonia's Plant Protection Service.

Weeding corn. This model from HATZENBICHLER is extremely versatile and can be used to weed, to apply fertilizer in rows, and as a distributor.

12. ACCIDENTS, WEEDS, DISEASES AND PESTS

The agents that can damage cereal plants can be divided into **abiotic** and **biotic** agents. Abiotic agents include all the different types of damage caused by non-living agents, such as the temperature, frosts, wind, etc. Biotic agents include all the different types of damage produced by living agents, such as weeds, fungi, nematodes and insects.

12.1. ACCIDENTS

Temperatures below 0°C may cause freezing and damage to the cereal plants. As the temperatures fall, the water in the cells is displaced from the cell and the plant dies of dehydration. Resistance to the cold varies and is characteristic for each species and variety. Rye for example is much more resistant to cold than wheat, and so it can be cultivated in the colder areas of northern Europe, although the seedlings of winter cereals are usually most resistant to cold when they only have three or four leaves, and this resistance declines after the fifth leaf is produced. During the flowering period, temperatures below 16°C prevent fertilization (the flowers are **shed**).

Excess **humidity**, mainly on clay soils, may cause **root asphyxia**. Asphyxia is especially dangerous because it encourages the anaerobic bacteria that may cause the roots to rot. Other microorganisms, the aerobic ones that fix atmospheric nitrogen, may die if there is insufficient oxygen in the soil, meaning that less nitrogen will be available to the plant. Excessive rainfall causes leaching of nitrates to deeper layers in the soil, where it is unavailable to the plant.

Excess **heat** may cause **shrivelling**. Ripening is the last phase of the growth period and this is when starch accumulates in the grain. This starch is produced by photosynthesis, which continues in the last leaves and in the spike.

These sugars and proteins are translocated towards the spike when the grain is ripening. If temperatures are very high and the wind is strong and parching, mobilization of the last water resources required for the translocation of sugars and proteins is prevented, and shrivelling occurs, leaving the grains wrinkled because they cannot accumulate all the reserves they need.

Flattening is the name of an accident that is common in cereals, when the plant's stem is bent down to the ground. This leads to defective flowering, meaning that the grains are small and malformed. **Non-parasitic flattening**, already described as an accident, may occur on fertile soils as a result of unbalanced application of fertilizer. Applying excess nitrogen relative to phosphorus and potassium causes the plants to grow excessively long, fragile and prone to flattening. It may also be due to an imbalance between the plant's nitrogen supply and its carbon metabolism. There are other causes for flattening, such as excessively dense sowing, heavy rainfall and persistent winds, or even strong gusts of wind. Non-parasitic flattening should not be confused with parasitic flattening, which is described later. Varieties differ in their resistance to flattening.

Flower shed may also occur, and this is when fertilization does not take place, and may be due to an imbalance in the application of the macronutrients nitrogen, phosphorus and potassium.

12.2. WEEDS

Combatting weeds in cereal crops is basically done by herbicides, in both winter and summer cereals. Until recently cereal crops were weeded by hand when the crop was growing. In the last 25 years, herbicides have made a decisive contribution to the spread of cereal monoculture and to increasing crop yields, but the growing use of the many products on the market leads to several problems:

• The cost of the products is relatively low, and is more than compensated by the increase in production, but it does increase the costs of production.
• Just how these herbicides are broken down in the soil is often not fully understood.
• Sometimes the products are incorrectly and indiscriminately used by the farmer, totally ignoring the recommended doses and without knowing what effect it will have.
• Forms resistant to the products used gradually arise by evolution.
• Some varieties of cereal are sensitive to some weedkillers.

For all these reasons, the alternative and traditional methods of getting rid of weeds are still very relevant, and at least in terms of prevention their use should always be considered. Suitable preparation of the site can eliminate weeds that germinate early, and rotating crops by alternating with fodder cereals on which different herbicides can be used, are the most widely used traditional systems to combat weeds.

Herbicides can be applied at different stages in the cereal's growth: on sowing (presowing or pre-emergence), when the vegetation starts growing or during the growing period. Choosing to apply weedkiller at the time of sowing implies that you are certain that cornfield weeds will rapidly invade the field and that later control will not be possible or effective. It is justified in very intense production systems.
This type of treatment is common in corn and sorghum. In winter cereals, however, it is better to wait until the weeds have germinated naturally and to apply the weedkiller at the most appropriate moment, as long as it is still possible to enter the field in order to carry out the post-emergence treatment of the crop.
Regardless of the moment it is applied, the herbicide should be chosen on the basis of the weeds that are growing and their state of growth: it is best to treat weeds as soon as possible. This also depends on the species and variety of cereal, the soil type, the rainfall and on the temperatures expected after it is applied. After considering all these points, the herbicide should also be chosen on the basis of its price, that is to say, the cost of the product should be compared to the profits expected from its application.

As a general rule, when choosing which

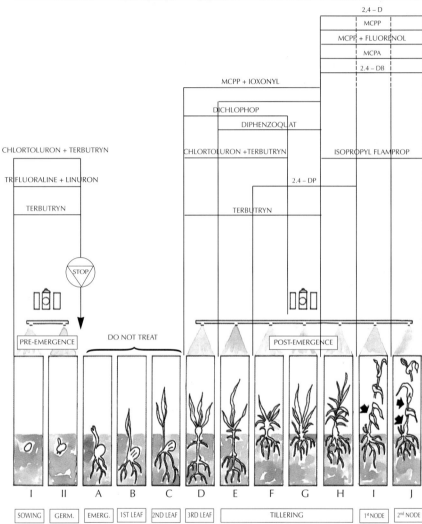

I	II	A	B	C	D	E	F	G	H	I	J

| SOWING | GERM. | EMERG. | 1ST LEAF | 2ND LEAF | 3RD LEAF | TILLERING | 1st NODE | 2nd NODE |

herbicide to apply, we suggest you consider each of these points in order. First of all, identify the weeds present and their state of development, then select suitable chemical products, taking care not to chose one that may damage the crop, and, of course, respecting all the manufacturer's instructions on the dosage and how to apply the product. Once you have chosen the herbicide, it should be applied in the most effective way to the weeds and with as little effect as possible on the cereal. At the same time, you should take into account the risk of toxic effects on neighboring crops, and the safety of the person applying the treatment.

Many implements can be coupled to the tractor to carry out the mechanical weeding frequently mentioned in the text, especially when the seedlings are young and the tractor can move along the rows. The photo shows a device produced by HATZENBICHLER weeding between young corn plants.

To ensure the greatest effect against the weeds to be treated, never apply at a dose lower than that recommended by the manufacturer. Pre-emergence treatments should be applied when the soil is well prepared and with enough moisture for the plant to absorb the weedkiller. Post-emergence treatments should be applied at a moment when the temperature favors active weed growth. To finish, the weedkiller should always be applied as evenly as possible over the whole field.

To ensure the cereal is damaged as little as possible by the weedkiller, the dose recommended by the manufacturer should never be exceeded. If a particular weed resists a particular weedkiller, it is better to use a herbicide with a different active ingredient than to increase the dose. The weedkiller should not be applied when the cereal is less resistant, for example if it has not reached or has passed the developmental stage specified by the manufacturer, if the cereal is not healthy for some other reason, and when weather conditions are unfavorable (frosts, ice or very dry conditions).

12.3. FUNGAL DISEASES

Fungal diseases are having an increasing effect on the yield of **winter cereals**. Many agricultural experts agree that the current increase in fungal disease is due to the simplification of cultivation practises and the use of higher yield varieties. Diseases like powdery mildew and glume blotch (**Septoria**) were relatively unimportant in the recent past, but now cause serious economic losses in cereal crops, and this has meant that farmers have to apply fungicides as a matter of course in the most extensive cultivation systems.

In **summer cereals**, however, fungal diseases are less important. In corn, for example, damage caused by **rust** or **powdery mildew** (*Ustilago maldis*) is of little economic importance. In rice, diseases like *Piricularia oryzas*, *Helminthosporium oryzas* and even *Sclerotium oryzas*, have not yet caused serious attacks. However, the effect of the different diseases can and does vary from area to area.
Fungal diseases of cereals may lower the **crop yield** and the **harvest quality**. When they affect the yield, they lower the number of stems and tillers (**parasitic flattening, powdery mildew** and **Fusarium wilt**, etc.), depress the fertility of the spike (**rusts, bunt, Helminthosporium**, etc.) and decreases the weight per 1,000 seeds. Fungi, such as the different **Fusarium wilts** attacking the spike, *Sclerotium oryzas*, **parasitic flattening, powdery mildew** and the different **rusts** all cause **shrivelling**. They all disturb the water supply and the translocation of reserves to the grain.

The harvest quality can be affected by:

• Reduction of the **unit seed weight**, as a result of the diseases that attack the sheathes, leaves, stems or glumes: **parasitic flattening, fusarium wilt, glume blotch, rusts**, etc.

• When **powdery mildew** attacks **malting barley** it increases the number of small grains, ruining the harvest.

•**Rusts** cause a decrease in the **protein content** of the seed.

• **Rusts** also cause a decrease in the **meal yield** and the **percentage of glassy grains** in hard wheats.

Some of these fungal diseases can be dangerous to human beings, such as the consumption of grains affected by ergot (*Claviceps purpurea*), the cause of the illness called **ergotism**, or the consumption of derivatives with a high percentage of grains affected by **Fusarium**. Combatting these diseases requires three different strategies. First, preventive cultural techniques must be carried out, then the seed must be disinfected before sowing. If these do not work, then chemical products should be used.

12.3.1. Preventive cultivation techniques

The progress and spread of diseases and their effect on the yield of cereals depends on a number of factors. The parameters determining whether a crop will be attacked include the weather in the year in question, the type of soil, the varieties sown, the rotation of species that is performed, the cultivation techniques, the date and the density and depth of sowing. That is to say, the cultivation techniques used have an influence, not just the environmental factors. Combatting these diseases must begin with appropriate cultivation techniques. There is now a tendency to ignore these former practises because they do not always give the desired results. Cultivation techniques and practises may play an important role by favoring or preventing the growth of parasitic fungi by increasing or decreasing the cereals' resistance to fungal attack. Some fungal diseases cannot yet be chemically controlled because there are no products available, and diseases like **stem canker** (*Ophiobolus graminis*) or ergot (*Claviceps purpurea*) can only be combatted by good cultivation techniques.

Growing genetically selected **resistant varieties** is a good preventive method, but it does have its drawbacks. Mutations often appear in the fungus allowing it to attack previously resistant varieties.

The frequent repetition of the same crop on the same site favors these diseases. **Crop rotation** and **alternation** are good preventive cultivation practises to keep down fungal population levels. Thus, converting from traditional extensive agricultural practises to an intensive model tends to cause an increase in fungal diseases.

The **sowing date** may have a different effect depending on the type of disease: early sowing in fall favors parasitic flattening and rusts, while plants sown late in spring tend to be more affected by powdery mildew. **Dense sowing** favors the transmission of parasitic fungi by the roots, at the same time as creating a moist microclimate that favors the growth of fungal diseases on the leaves. **Deep sowing** delays the emergence of the cereal and makes it more vulnerable to diseases.

Excessive or late **nitrogen supply** favors the development of young tissues and increases the plant's vulnerability to disease. **Potassium** seems to increase a cereal's resistance to some diseases. In general, a well-nourished plant shows greater resistance to all diseases, as long as its nutrition is not unbalanced, and nitrogen, potassium and phosphorus are administered in the right proportions.

12.3.2. The main diseases and combatting them

In the last chapter of this section, the reader will find detailed information on the diseases affecting each crop, together with the main products used to treat them. As a guide we include two tables. The one on the next page is a list of the fungal diseases, the crops affected and the part of the plant showing the symptoms. The table on this page is a key that can be used to identify which type of disease you are dealing with. The growth of these fungi, and the time they should be treated, depend on the weather

and even the microclimate of each site. The treatment dates can be consulted on the bulletins published by different administrative bodies.

Preventive treatments to **disinfect seeds** (see chapters 6 and 11) can prevent many diseases from appearing in crops. This treatment effectively controls the **smuts** of wheat and barley, **Helminthosporium** in wheat, **bunt** in wheat, **Fusarium** and **Septoria**, which affect most cereals when they are germinating, as well as **powdery mildew** of barley.

It is also possible to carry out **post-emergence treatment**. Fungicides were first used against parasitic fungi of cereals in Europe in 1970. They give good results against the diseases that affect the collar of the plant, such as **parasitic flattening** and the different forms of **Fusarium**, and also against the fungi that attack the stems and spikes, such as **Fusarium**, **Septoria**, **powdery mildew**, and yellow rusts and dark rusts. In barley, this treatment also combats **Rhynchosporium** effectively, as well as the diseases that attack wheat.

12.4. NEMATODES

Nematodes are the worst pest of cereals in Spain, especially the species *Heterodera avenae*. The fourth part of this work contains a thorough description of nematodes (their biology, their

Diagram classifying the fungi that way attack cereals, showing the parts affected and the scientific name of the fungus. Reproduced from "El cultivo del trigo y la cebada" (Cultivation of wheat and barley), by the Spanish Agricultural Extension Service.

List of fungal diseases, the crops they affect and the part of the plant that they affect. Taken from "Les maladies des céréals", ITCF, 1980

FUNGAL DISEASE (Hongos parásitos)	CROPS AFFECTED				PARTS AFFECTED							
	WHEAT	BARLEY	OATS	RYE	Seedlings	Roots	Base of stem	Nodes	Sheath (Leaves)	Blade (Leaves)	Spike	Grain
BUNT (Tilletia caries)	*											*
NAKED RUSTS (Ustilago sp.)	U. Tritici	U. Muda										*
COVERED RUSTS (Ustilago sp.)		U. hordei	U. avenae									*
ERGOT (Claviceps purpurea)	*	*	*	*								*
FUSARIUM (Fusarium nivale and F. roseum)	*	*	*	*	*	*	*	*	*		*	*
HELMINTHOSPORIUM (Helminthosporium sp.)		H.gramineum H. Teres							*	*		
POWDERY MILDEW (Erysiphe graminis)	f. sp. tritici	f. sp. hordei	f. sp. avenae	f. sp. secalis					*	*	*	
STEM CANKER (Ophiobolus graminis)	*	*	*	*		*	*				*	
PARASITIC FLATTENING (Cercosporella herpotrichoides)	*	*	*	*			*		*		*	
RINCOSPORIOSIS (Rhynchosporium secalis)		f. sp. hordei		f. sp. secalis					*	*		
BROWN RUST (Puccinia sp.)	P. tritici	P. hordei		P. dispersa						*		
RUST (Puccinia coronata)			*							*		
YELLOW RUST (Puccinia striiformis)	f. sp. tritici	f. sp hordei		f. sp. secalis						*	*	
BLACK-STEM RUST (Puccinia graminis)	f. sp. tritici	f. sp. hordei	f. sp. avenae	f. sp. secalis					*	*	*	
GLUME BLOTCH (Septoria sp.)	S. nodorum S. tritici	S. passerinii	S. avenae	S. secalis	*		*		*	*	*	

Key: * = crop or plant part affected

life cycles, the different species they attack and the damage they cause). Briefly, they are small worm-shaped animals that live in the soil or within the roots and stems of the plants they are feeding on. They eat a wide range of plants, including cereals, fruit trees, vegetables and flowers grown in greenhouses.

Strategies to combat nematodes in cereal farms have to be based on indirect methods that reduce the damage as far as possible. In general, these are preventive measures, but there are some chemical disinfectants that can be used. Alternation of crops usually gives good results, so avoid planting cereals after cereals, and choose a totally different crop instead, such as sugarbeets. In affected sites, it may be worth delaying sowing so this critical period of the crop plant's life does not coincide with the peak activity of the larvae after they hatch from their cysts. Combatting weeds is also important, as they are also hosts for nematodes and contribute to increasing the population of nematodes. Cereals are most sensitive to nematodes in the early stages of their life cycle, meaning that a surface application of nitrogen at the seedling stage may be a good idea, as it speeds up the seedling's growth. Treatment to kill nematodes, in extensive estates, can only be recommended for very small patches, as the cost is not covered by the profits made from an extensive exploitation. As an alternative, it is possible to lower the usual dose and extend the treatment to the entire plot, so that the plant is protected at the most sensitive stage.

12.5. INSECTS

A very large number of insects can cause serious damage to cereal crops, but insect pest populations generally remain more or less stable within limits that suppose tolerable economic losses. On the next page there is a key that the reader can work through to identify the different pests depending on the symptoms that can be detected on the different parts of the plant. The fourth part of this work describes in detail the different insects that attack crops, their life cycles, the damage they cause to plants, etc., and should be read if you want to learn more about pests.

There are many species of **soil insects**, the most important ones including wire worms (*Agriotes sp.*), the cockchafer (*Melolontha melolontha*), black cutworms (*Scotia ypsilon*) and turnip moths (*S. segetum*) and crane flies (*Tipula spp.*). The populations of larvae of these insects, which live in the soil and feed on the roots of cultivated plants, may increase if the soil is not turned by plowing and if crop alternation is not practised. The first step to combat these insects is to protect the seed with an insecticide, or later by means of insecticide treatment of the soil, either locally or throughout the plot. Applying insecticide to an entire plot is expensive, and can only be justified if the attack is very heavy.

Wheat is attacked by two similar bugs, *Aelia rostrata* and the senn bug (*Eurygaster maurus*).

These are migratory insects that spend the winter in high areas and in spring move to cultivated fields, where they reproduce. The adult forms that hatched in the spring are the forms that damage the cereals. The marks of their bites on the grain are very distinctive, and they inject a substance that destroys the gluten and gives the flour an unpleasant taste and smell flour an unpleasant taste and smell. If 5% of grain is affected by bites, it is rejected by mills. They can be combatted by cultivation methods such as growing early varieties so that they can mature before the adult forms appear, or by using classic insecticides such as **trichlorphon**, **carbaryl**, or **malathion**.

Recently there have been increasing references in the technical literature to **aphids** as a pest attacking cereal crops, something that was unknown until recently. Their presence in crop fields is due to the ecological imbalance partly caused by massive and indiscriminate chemical treatments (which destroy their natural enemies) and partly because the cultivation cycle of the crops has changed, with the introduction of new varieties, earlier sowing, etc.

Aphids are a pest showing high adaptability to the environment, because they can multiply fast when they find a suitable host (they characteristically live on a set of different plants), and because they can reach new sites if they cannot find enough food in one place. Aphids can also adapt to local climatic conditions, overwintering as eggs in very cold zones, or as parthenogenetic adults in zones with mild winters. The most important of the 15 species of aphids that attack cereals in Europe are: the variable green *Sitobion avenae F.*; the light green *Metopolosiphum dirhodum Wlk.*; and the greenish brown *Rhopalosiphum padi L.* They may cause damage to the cereals directly, or indirectly as vectors of diseases such as viruses. Other insects, such as the cereal totrix moth (*Cnephasia pumicana*), cause damage to the stem, causing abortion of the spike. This moth is a specific pest of wheat and barley. Treatment with a **malathion-type** insecticide during stem formation is sufficient to control this pest.

The pests of corn include corn borers (*Pyrausta*) and *Sesamia*, which destroy the inside of the stem, and are dreaded by farmers. In Spain, there are two main pests of rice, the rice stem borer (*Chilo suppressalis*) and the rice stinkbug (*Eusarcoris inconspicuus H.S.*).

12.6 BIRDS

Sparrows, crows, starlings, herons, etc., may be serious pests of all cereals, whether at the time of sowing (as they can eat the seed and seedlings) or just before harvest (as they can eat the ripening grain). At sowing time, treatment with repellents may be sufficient. There is no really effective way to protect harvests from seed-eating birds.

The two types of wheat bug: Aelia rostrata (above) and the senn bug (below). (Taken from Bonnemaison)

Key to identify the main animals that attack cereals, giving the plant part affected, the symptoms and the animal that is probably to blame. Taken from "The Cultivation of Wheat and Barley", by the Spanish Agricultural Extension Service

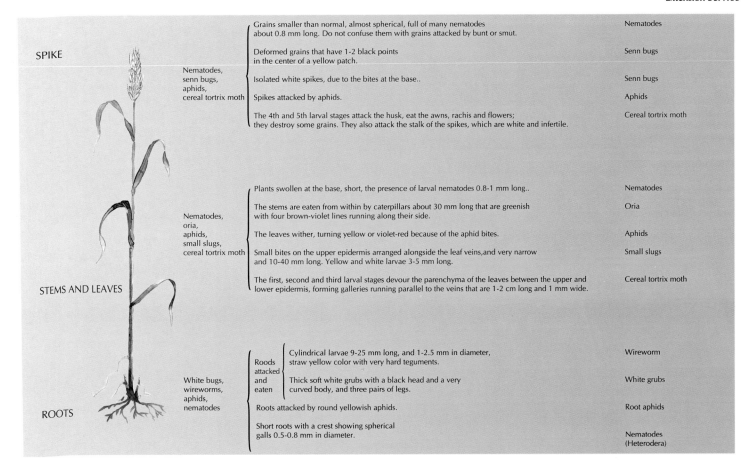

Plant part	Animals	Symptoms	Animal to blame
SPIKE	Nematodes, senn bugs, aphids, cereal tortrix moth	Grains smaller than normal, almost spherical, full of many nematodes about 0.8 mm long. Do not confuse them with grains attacked by bunt or smut.	Nematodes
		Deformed grains that have 1-2 black points in the center of a yellow patch.	Senn bugs
		Isolated white spikes, due to the bites at the base..	Senn bugs
		Spikes attacked by aphids.	Aphids
		The 4th and 5th larval stages attack the husk, eat the awns, rachis and flowers; they destroy some grains. They also attack the stalk of the spikes, which are white and infertile.	Cereal tortrix moth
STEMS AND LEAVES	Nematodes, oria, aphids, small slugs, cereal tortrix moth	Plants swollen at the base, short, the presence of larval nematodes 0.8-1 mm long..	Nematodes
		The stems are eaten from within by caterpillars about 30 mm long that are greenish with four brown-violet lines running along their side.	Oria
		The leaves wither, turning yellow or violet-red because of the aphid bites.	Aphids
		Small bites on the upper epidermis arranged alongside the leaf veins, and very narrow and 10-40 mm long. Yellow and white larvae 3-5 mm long.	Small slugs
		The first, second and third larval stages devour the parenchyma of the leaves between the upper and lower epidermis, forming galleries running parallel to the veins that are 1-2 cm long and 1 mm wide.	Cereal tortrix moth
ROOTS	White bugs, wireworms, aphids, nematodes	*Roots attacked and eaten:* Cylindrical larvae 9-25 mm long, and 1-2.5 mm in diameter, straw yellow color with very hard teguments.	Wireworm
		Thick soft white grubs with a black head and a very curved body, and three pairs of legs.	White grubs
		Roots attacked by round yellowish aphids.	Root aphids
		Short roots with a crest showing spherical galls 0.5-0.8 mm in diameter.	Nematodes (Heterodera)

Detail of a sugarbeet crop (Courtesy of BASF, S.A.)

harvesting, and a final section detailing the product's main uses.

13. THE MAIN EXTENSIVE CROPS

This chapter provides a description of the main intensive crops, which deserve a section of their own because they are of such great economic importance and are so widely cultivated in many countries. A thorough description of just one of these crops, wheat, would occupy dozens of pages. Readers wanting further information should refer to more specific works, where they will find far more information. We have tried not to miss any of the more important aspects of cultivation of each crop.

In addition to cereals, extensive crops include other plants such as rapeseed (a member of the cress family - Cruciferae), some legumes (such as soybeans), sunflowers (daisy family - Compositae), and other species, such as coffee, cotton and tobacco, which are grown on a large scale all over the world. The text considers the crop's growth cycle, the different varieties, its place in a crop rotation, its cultivation requirements, details on fertilizer application, sowing, irrigation, herbicides, damage suffered, diseases, pests,

13.1. WHEAT

Like all other cereals, wheat is a herbaceous monocotyledonous plant (monocot) belonging to the grass family (Gramineae). All the different species and varieties of wheat belong to the genus *Triticum*. The different types of wheat are classified on the basis of the number of chromosomes in their adult cells into diploids (2n = 14 chromosomes), tetraploids (2n = 28) and hexaploids (2n = 42). Wheat has a fibrous root system, with most of the roots at a depth of less than 25 cm, though they may reach a depth of 150 cm. Their growth depends on variables like the variety, the developmental stage, the soil texture, the depth of the water table, the weather during germination and the seedling's growth, etc. At the beginning of the vegetative phases, the stem is within a mass of cells, the **tillering node**. This stem has axillary buds that will give rise to the tillers. In the **stem formation** phase, the elongating stem bears 7 or 8 leaves, each sheathed until the next internode. In almost all varieties of wheat, the stem is solid to begin with and then becomes hollow as it grows, except at the leaf nodes where the stem remains solid.

The leaves are strap-shaped with parallel venation and a pointed tip. The flower spike forms in the apical meristem of the tiller. When tiller production has finished, the spike starts to rise from the stem, at the same time as the stem elongates in the stem formation phase. When the stem's growth has finished, the fully-formed spike appears within the sheath of the last leaf. This is the **spike formation** phase. A spike consists of a cental stem called the **rachis**, which bears spikelets on both sides, always less than 25 in total. These little spikes, spikelets, are borne directly on the rachis and overlap each other. Each spikelet contains several flowers, but only 2-5 of them (depending on the variety) are fertile. The **spikelet** consists of two **bracts** called **glumes**. Above the glumes and attached to a peduncle (flower stalk) is the lower bract, which bears in its axil a flower, which has an upper bract. These bracts are known as **scales**.

The flowers are fertilized before they open,

Diagram showing the growth cycle of the winter cereals: the seeds are sown in fall, the seedlings pass the winter without growing actively, and grow actively when spring arrives. Harvesting is in late spring or early summer, depending on the variety cultivated.

Plowing and harrowing   Sowing   Winter   Growth   Harvest

First, the site is tilled and harrowed. This delays the growth of weeds, and allows the wheat to be sown earlier. Seed used to be sown by hand but this is now totally mechanized.

It is beneficial if the soil is protected from frost in winter by an insulating layer of snow. A cold winter with little snow is usually followed by a poor harvest. When the snow melts and the sun warms up the soil, the seeds start to grow actively.

Harvesting is a delicate operation that is highly dependent on good weather, as if the grain is wet when it is harvested, it deteriorates rapidly in the grain silos.

Wheat in flower

When the soil is warm enough and the seed has absorbed enough water to mobilize its reserves, a small root appears, called the **radicle**, that is covered by a protective layer of cells called the **coleorrhiza**, and at the same time the **coleoptile** is produced, which is a sheath covering the **plumule**. The radicle grows using the food reserves contained in the grain and then starts to absorb the minerals dissolved in the soil solution. The grain grows towards the surface, at the same time as new roots appear. When the coleoptile emerges above the soil, the first true leaf appears. When this is half grown, the second leaf appears. By this stage the plant has 5 or 6 rootlets. Soon afterwards, the tip of the third leaf can be seen, and through the transparent coleoptile a stem (the **rhizome**) can be seen, ending in a swelling that starts to expand until it forms a tillering node. When the fourth leaf is about to appear, tiller production (tillering) occurs, and this is the formation of secondary stems from the rhizome. The tillering node thickens until it is as large as four or five nodes together. Each one of them will give rise to a new leaf, and in its axil bears an axillary bud that will give rise to a secondary stem. The greater or lesser production of tillers depends on the variety of wheat, the weather and the soil's nitrogen content.

The **stem formation** phase is when a herbaceous stem turn into stems with a flower spike at the tip, while the development of the herbaceous stems halts or stops. This is followed by the **spike production** phase, when the plant shows the greatest growth. Wheat synthesizes an estimated 3/4 of its total dry weight between the tillering phase and flowering. The final phase is the **ripening period**, when starch reserves accumulate in the grain.

meaning that wheat is **self-fertilized**. Wheat varieties can thus maintain their original characteristics over many generations. In order to cross two different varieties of wheat, special artificial fertilization techniques have to be used, such as **emasculation**, removing the stamens from the plant that is to be the seed parent. The flower gives rise to a single fruit, the **grain**, which contains an **embryo**, called the **germ** or "wheatgerm", and food reserves.

13.1.1. Wheat's growth cycle

The growth cycle of wheat consists of three different phases. The first phase is from sowing to stem formation, and is the **vegetative period**. The second phase is the **reproductive period** and lasts from stem formation to the end of spike production. The third and final phase, the **ripening period**, lasts until the grain is harvested and consists of the ripening of the grain.

If the grain to be sown is stored in silos at a moisture content of less than 11%, it should retain its ability to germinate for about 10 years, though you are normally recommended not to store it for more than two years, because its ability to germinate starts to decline. Once it is in the soil, the grain germinates when it has absorbed 25% of its own weight in water (though it can retain as much as 65%). It then emerges from dormancy and starts growing actively. The best temperature for germination is 20-25°C, though the first roots sprout in winter varieties when the temperatures reach 3-4°C. In addition to the temperature, the soil humidity and depth of sowing should also be taken into account. As a general rule, wheat germinates in 12 to 15 days when the soil moisture content is about 60% of the field capacity (see the section on Soils and Fertilizers), as long as the seed is not sown too deep.

Mature spikes of beardless wheat. (Courtesy of AGRO LORIN)

13.1.2. Varieties

Wheat varieties are classified on the basis of the cereal's cultivation period into **winter wheats** or **long cycle varieties**, **spring wheats** or **short cycle varieties**, and **alternative varieties**. The difference between these groups is based on how long a growing period they need. In order to grow to completion, each variety requires a given amount of heat, and this is measured by the sum of differences between the average temperature on each day and the **vegetative 0**. This quantity of heat is known as the **heat integer** and varies in different varieties. Winter wheats require from 1,900 to 2,400°C, while spring wheats require 1,250 to 1,550°C. It may seem illogical that the heat integer of the spring varieties is lower than that of the winter varieties, but this is because the spring varieties have a short life and thus accumulate less hours of heat (because they are in the soil for a shorter time) than the winter, or long cycle, varieties. **Winter** or **long cycle** varieties need a relatively longer period of time in order to produce their spike. This is why they are sown in fall or early winter. There are different degrees of earliness within this group, both for spike formation and grain ripening. Long cycle varieties are suitable in areas with cold winters, where sowing should be performed in October or November; it should not be delayed later than the first ten days of December because this might lead to irregularities in the ripening of the grains. They can produce many tillers and need a winter rest period. There are enough different varieties, varying in their resistance to disease, productivity and adaptation to the soil, etc., to meet almost any requirement.

Spring varieties or **short cycle varieties** do not require a winter rest period in the soil. They are usually planted at the end of winter, and are frequently sensitive to low temperatures. They produce very high quality flour for making bread, and this partly compensates their tendency to produce few tillers, and consequent low productivity. They are not as widely appreciated as the long cycle varieties, because of the risk involved; they should be sown in December or January in temperate areas, and may thus suffer the effects of late frosts in spring.

The **alternative varieties** show the best features of the winter and spring wheats. They can be sown late at the end of winter, and then ripen like the long cycle or winter varieties. The alternative varieties include **early**, **medium** and **late** varieties. The late alternative varieties must be sown early (November-December), while the early alternative varieties can be sown in December or January or even February in cold zones.

Choosing the most suitable variety in each case is an important and complex decision, as many factors must be taken into account. It is worth seeking advice from a local professional, reading the bulletins from the Ministry of Agriculture and the technical information from commercial suppliers, and this should be enough to correlate factors like climatic conditions (rainfall, distribution of rainfall over the year, extreme temperatures, humidity, etc.), soil characteristics (depth, soil class, organic matter content, pH, pF, drainage, nutrient content, etc.), the crop's position in the rotation and the purpose for which the crop is grown. As a function of the characteristics of the varieties available, it is possible to select a variety with a given production, quality, earliness, genetic resistance to diseases and accidents, and tiller production.

The genus *Triticum* contains several species, many of them cultivated since antiquity. The Romans, for example, cultivated durum (hard) wheat (*T. durum*). The different varieties of wheat are classified on the basis of their grain characteristics into **soft wheats**, **hard wheats** and **hybrid** varieties obtained by breeding and selection. Each of these groups consists of a large number of varieties (including both winter and spring varieties), each with its own special characteristics.

A thorough description of every single variety of wheat would be far beyond the scope of this work, but this text is accompanied by two lists, the first with the most common varieties of hard wheat, and the second with the most common varieties of soft wheat. Commercial suppliers and official selection and

Variety name	Registration date
Abadía	14-8-76
Alaga	2-12-74
Aldeano	12-12-84
Anento	3-03-88
Antón	25-04-84
Ardente	26-11-85
Benor	6-08-80
Bidi-17	2-12-74
Bonzo	25-04-84
Camacho	11-03-81
Castronuevo	12-12-84
Cibeles (T-343)	3-02-82
Cocorit	7-09-77
D-104	2-12-74
Esquilache (TD-253)	2-12-74
Ferox	26-11-85
Jabato	15-04-87
Jaguar	12-12-84
Jiloca	6-08-80
Kidur	6-08-80
Mexa	6-08-80
Mundial	12-11-83
Nita	1-07-81
Nuño, Yavaros C-79	12-12-84
Páramo (TD-330)	7-03-83
Peñafiel	12-12-84
Pingüino	7-09-77
Randur	6-08-80
Roqueño (TD-332)	3-02-82
Safari	6-08-80
Senatori Capelli	2-12-74
Tejón (TD-335)	15-09-82
Valgera	15-09-82
Valnova	3-02-82
Vitrón	12-11-83

Current list of hard wheat varieties registered with the Spanish National Register of Varieties of the National, Institute of Seeds and Greenhouse Plants

Variety name	Registration date	Variety name	Registration date	Variety name	Registration date
Abel	6-08-80	Cárdeno	25-04-84	Navarro-105	2-12-74
Ablaca (T-331)	3-02-82	Cargifaro	12-11-83	Nivelo	15-04-87
Aboukir	12-12-84	Cartaya	12-12-84	Noroit	6-08-90
Adalid	12-12-87	Cascón	2-12-74	Novisad-7000	26-11-85
Adonay	12-12-87	Castán	7-09-77	Oroel	11-03-81
Albares	12-11-82	Chamorro	26-11-85	Orso	2-12-74
Alcalá	25-04-84	Champlein	2-12-74	Osona	15-04-87
Alcazaba	15-04-87	Cocagne	12-11-83	Pané-247	2-12-74
Alcázar	30-10-78	Compadre (T-85)	2-12-74	Partizanka	30-10-78
Alcotán	15-09-82	Corzo	2-11-81	Pavón (F-76)	15-09-82
Alfori	12-12-87	Costal	17-09-83	Pesudo (T-336)	16-11-82
Alejo	1-08-86	Diego	26-11-85	Pilos	1-07-81
Alto	1-07-81	Dollar	12-12-87	Pirón	15-09-82
Alud	1-08-86	Don Antonio	7-09-77	Pistou	15-09-87
Amiro	12-12-87	Echo	12-12-84	Potam-70	15-09-82
Amón	12-12-87	Emilio Morandi	3-02-82	Prinqual	3-02-82
Anza	2-12-74	Escualo (GV-1)	1-07-81	Pursang	7-03-82
Apuesto	12-12-87	Estrella	2-12-74	Recital	25-04-84
Aragón-03	2-12-74	Festín	12-11-83	Rex	2-12-74
Aranda	7-03-83	Fiel	15-09-82	Rinconada	1-07-81
Arcole	3-02-82	Florence Aurora	2-12-74	Sansa	6-08-80
Ardec	1-07-81	Fondo	3-02-82	Sevillano	12-12-84
Arganda	12-11-83	Forton (T-326)	16-11-82	Shasta	3-02-82
Argelato	2-12-74	Frandoc	26-11-85	Siete Cerros	2-12-74
Ariana	2-12-74	Garant	12-12-87	Silver	15-09-82
Artal	15-04-87	Gaucho	15-04-87	Sion	30-10-78
Asteroide	12-11-83	Glauco	12-12-84	Splendeur	2-12-74
Astral	2-12-74	Golo	15-09-82	Super X	3-02-82
AT-14	2-12-74	Hardi	2-12-74	Sureño (T-338)	3-02-82
Autonomía	2-12-74	Hugo	12-12-87	Taba	15-09-82
Azor	1-07-81	Impeto	2-12-74	Talento	7-09-77
Azulón	26-11-85	Inia-66	2-12-74	Tanori	14-08-76
Bastión	7-09-77	Itrio	16-11-82	Tarot	1-07-81
Bellido (T-323)	3-02-82	Jupateco	4-11-76	Tauro (King-33)	1-07-81
Betres	3-02-82	Lachish	3-02-82	Tavares	2-12-74
Beuno	16-11-82	Lozano	7-09-77	Titán	26-11-85
Bolero	12-11-83	Maestro	3-02-82	Top	12-12-84
Boulmiche	2-12-74	Manero	25-04-84	Tornado	1-07-81
Bracero (T-346)	2-05-83	Mara	2-12-74	Tramontana	15-04-87
Cabezorro	2-12-74	Marca	6-08-80	Triana	12-12-87
Cajeme-71, sin. Bluebird-4	2-12-74	Marius	6-08-80	Vakon	15-09-82
Candeal Arévalo	2-12-74	Montcada	2-12-74	Yafit	7-09-77
Capitole	2-12-74	Montserrat	1-12-74	Yécora, sin. Bluebird-2	2-12-74
Carat	12-11-83	Nacozari (M-76)	15-09-82	Yenca	15-09-82

improvement laboratories must register the new varieties in the Ministry of Agriculture's **Official Register of Varieties**. Note that all the varieties are accompanied by the date they were registered. Artificial fertilization techniques are used to breed new varieties that bring together features from both parents. These new varieties are known as **hybrid** varieties. The cost of producing seeds is 2.5 to 3 times that of traditional seeds, but their production is usually 15% or more greater than varieties obtained by traditional selection methods.

13.1.3. Wheat's place in rotations

A plot of ground that is not cultivated one year is said to **lie fallow**. The following year, the ground is plowed to bury all the weeds that have germinated. The weeds also provide organic matter and fertilize the soil. If a chemical fertilizer has been applied at depth, the soil is now ready to be sown with a new crop.

Wheat is usually planted after a fallow year, and in general wheat should not be planted after wheat.

13.1.4. Cultivation requirements

In terms of the soil, wheat grows well on deep open soils. On these soils the root system grows well. Slightly acid sites (pH ≈ 5.4-7) are best, though wheat plants tolerate more alkaline soils. It is best if the rainfall is evenly distributed over the course of the year, and should be more abundant in spring and less abundant in summer. Wheat has been shown to grow well with only 300-400 mm of rain a year, as long as the soil is good (neither too much clay nor too much sand) and the rain is evenly divided, though annual rainfall of 500 to 600 mm is preferable. The soil class is a very important factor when evaluating water requirements; soil with a lot of clay may be negative in winter as it will retain too much moisture; excessively sandy soils may suffer water shortage in spring.

Current list of soft wheat varieties registered with the Spanish National Register of Varieties of the National, Institute of Seeds and Greenhouse Plants

13.1.5. Applying fertilizer

For information on wheat's requirements for fertilizer, consult the first section of this work which outlines the general principles of correct fertilizer application, which we shall not discuss here. Remember that phosphorus and potassium should be applied at depth, buried with the remains of the previous crop before sowing. They are usually applied as ternary compound fertilizers, which also contain nitrogen. Nitrogen should be supplied regularly during growth, especially during the critical phases, such as tiller production, stem production and grain ripening. Thus three separate applications of fertilizer should be carried out over the wheat's growing period: the first is applied at depth with all the fertilizer units of phosphorus and potassium, and one third of the units of nitrogen; the remaining nitrogen is divided into two applications over the course of growth, preferably during tiller production and stem formation, and is applied to the surface.

Table showing the average number of fertilizer units to apply, on the basis of the expected wheat production.

Expected wheat production (t)	Units N	Units P_2O_5	Units K_2O
1	35	25	25
2	70	50	50
3	105	75	75
4	140	100	100
5	175	125	125
6	210	150	150

The fertilizer units supplied are directly proportional to the expected production. The accompanying table shows the number of fertilizer units necessary to achieve a given yield (in metric tons). However, when calculating the total requirement of nitrogen you should take into account the amount of organic matter present in the soil (an additional input of nitrogen), and the preceding crop, because if it was a legume it will have fixed between 60 and 150 kg of nitrogen per hectare of soil. It is also important to take into account that dunging represents a significant input of nitrogen. To finish, remember how important it is to fertilize the plant with a balanced mixture of nitrogen, phosphorus and potassium, as nutritional imbalances may have serious effects on the plant, severely affecting productivity in wheat.

13.1.6. Sowing

Before sowing, the site must be **prepared**. This is done with heavy disc harrows to bury the plant remains from the preceding fallow, and a second pass to apply the fertilizer at depth. Before sowing it is worth disinfecting the seed

Basagran® is a selective post-emergence herbicide used to control dicot weeds in cereal crops, especially in rice. It is manufactured by BASF, S.A.

with an appropriate fungicide (**benomyl, methylthiophanate, copper oxychlorides,** etc.). The quantity of seed varies between 60 kg per hectare on infertile dry farmed sites and 250 kg per hectare on very fertile irrigated sites. In any case, the amount of seed will depend on the annual rainfall, the variety sown, the method and time of sowing, and the preparation of the site.

The seed can be sown by hand (scattering) or using a specialized seed sower. If machinery is used, the distribution of the seed is more regular and the amount of seed used can be reduced. It is normal for fixed sowers to sow at a distance of 17 to 18 cm between rows.

13.1.7. Irrigation

In areas where irrigation is possible, if fall is dry it is worth irrigating either before or after emergence. Stem production is a period of intense water uptake, and the wheat should be irrigated then, too. It is also worth irrigating during spike formation, though this may be dangerous in zones where high temperatures favor fungal diseases, especially rusts. In temperate climates, when spring is excessively hot and dry, the wheat should be watered twice in March, the first time at the beginning of the month, and the second just before spike production.

Sprinkler irrigation is becoming more and more important, at the cost of traditional methods, because to a large extent it avoids damaging the plant.

13.1.8. Herbicides

For information on weeding, the reader should consult the third section of this work, which lists the most common herbicides and how to apply them. Herbicides can be applied pre-plantation, pre-emergence in

13.1.9. Accidents, diseases and pests

Many **accidents** can cause physiological disturbances to the wheat (see chapter 12). **Temperatures** below 0°C may cause frosts that affect the plants. Resistance to low temperatures varies from variety to variety, though wheat plants are generally most resistant to the cold when they have three or four leaves, and this resistance declines after production of the fifth leaf. During flowering, temperatures below 16°C prevent fertilization (**flower shed**). Excess **moisture**, especially on clay soils, may cause **root asphyxia**. Excess **heat** may cause **shrivelling**.
Wheat may also suffer from **flattening**. Flower production is then defective and the wheat grains are small and deformed. Reaping and harvest are also harder. Non-parasitic flattening, already mentioned as an accident, may occur on fertile sites due to unbalanced fertilizer application.

Flattening may also occur as a result of excessively dense sowing, heavy persistent rain and even strong wind. Resistance to flattening varies from variety to variety. **Flower shedding** may also occur in wheat when fertilization fails occur, which may be due to an imbalance in the application of the three macronutrients nitrogen, phosphorus and potassium.

Fungal diseases are of great importance in cultivation of cereals, especially wheat. The life cycle of a rust (*Puccinia graminis*), was described in detail in the third section of this work. They give rise to pustules on the leaves and spikes of cereals. They disrupt photosynthesis, nutrient uptake and metabolism, and thus reduce yields. They attack the water-conducting vessels in the stem and reduce sap transport. The pustules release a huge number of wind-dispersed spores that spread the disease to nearby plots. Fungi should be controlled using preventive methods, such as growing resistant varieties, although the use of chemical fungicides, such as **triademephon** and **butryzol**, as a preventive treatment is also appropriate.

There are other genera of harmful fungi, such as *Helminthosporium*, *Septoria*, powdery mildews (*Erysiphe graminis*), bunt (*Tilletia*), rust of wheat (*Ustilago tritici*), stem canker (*Ophiobolus graminis*), *Fusarium* and **parasitic flattening** (*Cercosporella herpotrichoides*) are frequent in wheat fields. Though they differ in their details, their effects on the cereals are very similar, and these can be summed up as a general decrease in production and in a reduction in flour quality when they are widespread in the crop. In addition to the preventive measures mentioned above,

Invasive weeds: rough pigweed (Amaranthus retroflexus) Courtesy of SCHERING)

Weeds: sheep sorrel (Rumex acetosella) Courtesy of SCHERING)

wheat or when the crop is actively growing. There are new products on the market that eliminate monocot weeds like the wild oat (*Avena fatua*), but do not damage other grasses, such as wheat. An example of this is **trialate**, a carbamate herbicide. This should be applied immediately after sowing the wheat (pre-emergence), and should never be applied three or four days after sowing.

Other herbicides, such as the hormone weedkillers selective against dicots, can be applied when the wheat is actively growing. But they have the disadvantage of being highly volatile, and use has to be restricted on windy days because it often affects neighbouring crops. Hormone weedkillers like **MCPA**, **MCPB** and **2,4-D** are used a lot, though they do not kill some dicot weeds, such as *Matricaria chamomilla, Gallium aparine* and the corn poppy (*Papaver rhoeas*).

such as the use of resistant varieties and treatment with chemical fungicides, it is worth disinfecting the seeds before sowing with a product like **benomyl**, **methylthiophanate** or benomyl mixed with **copper oxyquinoleate**. You are strongly recommended to remove all stubble (as it is the second host for many fungi) and to follow a systematic rotation of crops (as some fungi are specific for one species and do not attack others).

Nematodes may also attack wheat, and nematodes of the genus *Heterodera* are common in wheat fields, and their effects can be noted in localized patches; the wheat plants affected lose their characteristic green color and turn a darker color. Controlling nematodes and the methods used are described extensively in section three of this work, which deals with plant pathology and how to resolve these problems.

Pests of wheat include the insects of the genera *Aelia* and *Eurygaster*, the senn bugs of wheat, which are a tremendous problem for farmers in the zones where they are common. These bugs attack the spikes and feed on the grain which shrivels and becomes deformed. They have an effect on the end quality of the flour produced rather than a simple reduction of the yield. These insects inject an enzyme that destroys the gluten, meaning that the resulting flour is of lower quality. A regular visual check of the wheat field is the best way to detect them. The bugs can be detected when the sun rises on the spikes, on the side facing the rising sun. The **agrochemicals** that can be used to kill them include **dimethoate**, **malathion**, **trichlorphon**, etc. These treatments should be applied at dawn and not late in the day, because these insects shelter from the heat by going down to the soil and its much harder to reach them with the product.

Aphids or greenfly are also sap-feeders that attack wheat. They mainly attack the leaves and spikes, and when they are abundant they may cause serious reductions in grain yield. Insect-eating insects are available as a biological method of controlling them, and they noticeably reduce the aphid populations. The ladybug (*Coccinella septempunctuata*) and the lace wing *Chrysopa vulgaris* are examples. The weather may also get rid of many aphids, such as heavy rainfall in late winter-early spring. When rain is intense, 90% of the winged dispersal forms of the aphids are destroyed, greatly reducing dispersal to other fields. The chemical products mentioned above, such as **dimethoate** (used against senn bugs), are suitable for controlling aphids. Other types of insects may also damage wheat production, including wasps, flies, beetles, thrips, etc.

Grain weevils (*Sitophilus granarius*) are

among the worst insects attacking wheat. They live in granaries and cause great damage to the wheat stored there. They are small almost black insects 4 or 5 mm long, and the females lays her eggs within the grain. The egg hatches within a few days and the small larva takes a month to finish its growth within the grain, which it totally destroys. Preventive measures include storing the grain with a low moisture content (less than 11%), as weevils thrive in moist conditions. It is also common to disinfect the grain and the granary with **aluminium phosphide**, which is extremely toxic to humans when inhaled, but is highly effective against grain weevil eggs, larvae and adults if it is applied correctly. Other insects that attack stored grain include butterflies and moths such as *Citrotoga cerealella*, *Tinea granella*, and flour beetles (*Tribolium*) and cadelle beetles (*Tenebroides*). Aluminium phosphide can be used as an insecticide to combat them, too.

13.1.10. Harvesting

All wheat is now harvested using self-powered harvesters. It is common practise to leave the gathered wheat in piles on the threshing floor so it can dry, especially if the wheat is to be stored rather than for immediate delivery. Even if the wheat contains less than 11% moisture when it is harvested, further drying is worthwhile to avoid attack by weevils if stored moist. In general, wheat collected with harvesters usually looks unattractive because of the bracts attached to the grain (the **husks**) and to the presence of broken grains and other impurities. It is worth passing the grain through a cleaning machine. During storage, the store should be dry and well ventilated, and measures should be taken to prevent insects and small rodents entering through the windows.

Ripe dry spikes of barley; this is the time to harvest. (Courtesy of AGRO LORIN)

13.1.11. Use

Wheat grain is usually measured in **hectoliters** (100 liters, hl), and 1 hl weighs 76-80 kg/hl. In very unusual cases with very productive varieties grown in excellent conditions, it may weigh 80 kg/hl. In general, the better the wheat, the higher its weight per hectoliter, its **specific weight**. Wheat grain is usually made into bread, and the **milling yield** becomes important. This is the percentage yield of bread flour obtained by milling the grain. For white flour this yield varies between 70% and 75% (the rest being the germ, bran, etc., which are not included in white flour). There are different ways of calculating this yield, such as the *Chopin index*, the *Zeleny index* and the *Hagberg index*.

Barley is a plant with a low growth form, due to its short stem, and its root system is more superficial than that of wheat. In the early stages of its development, barley is usually less upright than wheat, and has narrower leaves that are a lighter green color. At the point where the leaf blade separates from the stem, at the end of the leaf sheath, there are two auricles that cross over on the opposite side of the stem, with a short toothed ligule tight against the stem. The flowers have three stamens and a pistil with two stigmas. Like wheat, fertilization occurs before the flowers open, so barley is **self-fertilizing**. This means that varieties maintain their characteristics over many generations. Except for the "huskless" varieties, the barley grain is a caryopsis with the husk attached.

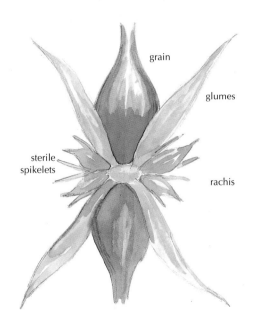

A cross section of the spike of four-rowed barley is square; a cross-section of six-rowed barley is hexagonal.

13.2. BARLEY

Barleys are members of the genus *Hordeum* (Gramineae), which contains several species and many varieties. The spikelets are directly joined to the rachis and overlap each other. The glumes are elongated with sharp edges, and the bracts are attached to the grain, except in "huskless barley". The bracts are prolonged as an arista. In general, barleys are classified on the basis of how many florets in each spikelet produce grain, and thus how many rows there are on each side. If all three florets bear grain, the plant is six-rowed barley (*Hordeum hexastichum*). If the central floret aborts, and only the two lateral ones bear seeds, this is four-rowed barley. If the central floret ripens and the two laterals abort it is two-rowed barley, or malt barley (*H. distichum*).

A cross section of the four-rowed barley spike is square; a section of six-rowed barley is hexagonal. The spike of two-rowed barley is flattened and the spikelets are attached alternately on opposite sides and perpendicular to the plane of flattening.

13.2.1. Varieties

In order to choose the most suitable variety for your growing conditions three of the **variety's characteristics** must be taken into account: the **productivity**, the **regularity of yield** and the grain **quality**.

The productivity is measured on the basis of the variety's production capacity in dry farming conditions in mediocre cultivation conditions (little fertilizer, poor soil, etc.). Obviously, when barley can be irrigated the variety

Current list of six-rowed barley registered with the spanish National Register of Varieties of the National Institute of seeds and Greenhouse Plants

Variety name	Registration date
Ager	15-05-74
Albacete	15-05-74
Alicia 2	27-10-80
Almunia	15-05-74
Alsekal	7-09-77
Ao 1	15-05-74
Astrix	15-05-74
Barbarrosa	27-10-80
Begoña	25-04-84
Berta	15-05-74
Bosquet	3-02-82
Carla	1-08-86
Cerro	15-05-74
Criter	16-11-82
Dacil	16-11-82
Dobla	6-08-80
Ejea	6-08-80
Gerbel	27-06-81
Hatif de Grignon	15-05-74
Hexa	27-06-81
Hop	14-08-76
Mikado	27-06-81
Monlon	15-05-74
Motan	16-11-82
Nimphe	15-05-74
Nuevede	3-02-82
Plaisant	16-11-82
Precoz Lepeuple	15-05-74
Ribera	6-08-80
Robur	6-08-80
Saphir	14-08-76
Steptoe	27-06-81
Sutter	27-06-81
Tabaida	12-12-84
Tatiana	16-11-82
Tecla	25-04-84
Tina	30-10-78
Vegal	27-10-80

	Registration date
Cebito	15-09-82
Claret	6-08-80
Cleo	5-01-88
Cobra	15-04-87
Copelia	6-08-80
Cresta	15-04-87
Custa	27-06-81
Elixir	26-11-85
Evasem	27-06-81
Fitamara	15-09-82
Flika	26-11-85
Gabriela	12-11-83
Georgie	15-05-74
Goldmarker	3-03-82
Hassan	15-05-74
Helena	26-11-85
Hellas	15-05-74
Icare	3-03-88
Igri	3-02-82
Ingrid	15-05-74
Iranis	7-03-83
Irene	26-11-85
Janes	27-06-81
Júpiter	6-08-80
Koru	6-08-80
Klaxon	12-12-84
Kristina	15-05-74
Kym	6-08-80
Logra	15-05-74
Lud	15-05-74
Maika	27-06-81
Marta	26-11-85
Melusine	1-08-86
Menuet	2-11-81
Miranda	7-09-77
Mogador	16-11-82
Melum	6-08-80
Murieta	3-02-82
Nepri	27-06-81
Nett	3-02-82
Ofelia	12-11-83
Olivia	12-11-83
Osa	12-11-83
Osiris	12-12-84
Pamela	12-12-84
Pallas	15-05-74
Pastel	15-04-87
Patty	26-11-85
Patrik	6-08-80
Paula	17-09-83
Pen	27-06-81
Pirouette	27-06-81
Polka	6-06-80
Porthos	7-09-77
Priver	12-11-83
Procer	15-05-74
Pronta	14-08-76
Regatta	5-01-88
Reinette	12-11-83
Sandra	6-08-80
Sonja	15-09-82
Tipper	26-11-85
Trait D'Union	15-05-74
Troubadour	7-03-83
Unión	15-05-74
Varsa	27-06-81
Vasy	27-06-81
Viva	12-12-84
Volare	7-09-77
Welam	3-02-82
Wisa	15-05-74
Zaida	26-11-85
Zephir	7-09-77

Current list of two-rowed barley registered whith the spanish National Register of Varieties of the National Institute of seeds and Greenhouse Plants

Variety name	Registration date
Abacus	15-05-74
Abundancia	26-11-85
Adorra	27-06-81
Almudena	26-11-85
Alpha	4-11-76
Alva	6-08-80
Amaltea	12-12-84
Angélica	1-08-86
Apex	5-01-88
Arabella	15-04-87
Aramir	27-06-81
Araya	17-09-83
Artemis	15-09-82
Atem	3-02-82
Athos	7-09-77
Aurea	12-11-83
Aurore	15-05-74
Avatar	3-02-82
Avera	3-03-88
Bacchus	12-11-83
Beka	15-05-74
Berac	30-10-78
Brava	2-11-81
Camelot	5-04-87
Cameo	15-04-87
Carina	15-05-74

can be expected to yield more.

The most important factors affecting yield regularity are **earliness, flattening, cold resistance, resistance to pests and diseases** and **quality factors**. Earliness is a determining factor in zones where late frosts are frequent in spring. In these cases, the earliest varieties possible should be chosen. An earlier variety will also be more drought resistant. In rainy years on fertile soils varieties that are resistant to flattening should be chosen, as it may significantly reduce barley yields. Cold resistance is another factor that varies from variety to variety. Short-cycle varieties are usually more sensitive to the cold. There are varieties of barley that are resistant to different fungal diseases, which are normally the same as those affecting wheat. Quality is another important factor when choosing which variety to grow. Long-cycle barleys have a higher protein content and are usually fed to animals, while short-cycle varieties contain less protein and are malted for beer production.

Like wheat, there are very many varieties of barley on the market, and this work cannot describe them all. There is an accompanying table of commercial varieties of two-rowed and six-rowed barleys. Four-rowed barleys are not included because they are little grown.

13.2.2. Cultivation requirements

Farmers usually consider that barley requires less water than wheat, but this is not completely true. Its transpiration coefficient is higher than that of wheat, but because barley has a shorter growing cycle, the total quantity of water absorbed from the soil over the growing cycle is slightly lower. Its water needs are concentrated at the beginning of its growth, an advantage over wheat because this lowers the risk of shrivelling before harvest.

Barley prefers fertile soils, though more than adequate harvests can be obtained on shallow stony soils as long as its water requirements are met during early growth. It grows poorly on clay soils, doubtless because they drain badly in winter and are easily waterlogged, and on these clay soils wheat should be sown instead. Barley grows well over a wide range of soil pH values, but better on alkaline soils and it can tolerate soils with high calcium levels.

13.2.3. Appliying fertilizer

Barley requires a lot of nutrients during the first phase of its growth. Barley naturally tends to flattening, and so the nitrogen input should be calculated very carefully. Even more care should be taken with the dose of nitrogen when growing two-rowed barley (malt barley), as excess nitrogen favors protein synthesis and reduces the value of the grain for the brewing industry. The average extraction per hectare and metric tonne of barley produced are roughly:

$$26 \text{ kg N}$$
$$20.5 \text{ kg P}_2\text{O}_5$$
$$25 \text{ kg K}_2\text{O}$$

Using these data for extraction it is possible to calculate the nutrients required by a barley crop that is expected to produce about 3,000 kg/ha. It will thus be necessary to supply the soil with approximately 80, 80 and 80 fertilizer units of N, P_2O_5 and K_2O. These quantities have been calculated bearing in mind the possible washing of nitrogen and some degree of retrograde fixation of phosphorus. Part of the nitrogen can and should be applied to the soil surface, but this has to be carried out early, as the plants' needs are highest during early growth and later application of N increases the risk of flattening. These calculations should also be adjusted to take into account the previous crop, the dung incorporated, the results of soil analysis, etc.

13.2.4. Sowing

As for wheat and all cereal crops in general, **soil preparation** is very important to ensure the soil is not too fine a tilth before sowing, and this is done by passing the disc harrow or roller, or both, a few times. Many authors recommend sowing at a rate of 120-160 kg/ha. The best time to sow is so that the plant does not spend the coldest part of winter in too late a stage of its development. In temperate zones, the best time is fall, when the seasonal rains have not yet started. Like wheat, barley can be sown by hand or by machine, which is more predictable and requires a little less seed.

13.2.5. Irrigation

Barley should only be irrigated during the stem formation period, as if barley is watered during the spike formation period, it may suffer flattening or be attacked by rust.

13.2.6. Herbicides

Herbicides are not normally used on fodder crops for animal consumption, as the farmer prefers not to lose the botanical diversity of the plot for the coming years, so it can continue to complement the livestock's diet. Herbicides should be used on barley intended for industrial use (brewing,) as they ensure maximum productivity. When herbicides are used, the same active ingredients should be used as for wheat. Depending on the product, they can be applied pre-sowing, pre-emergence or during active growth.

Weeds: sheep sorrel (Rumex acetosella) (Courtesy of SCHERING)

Invasive weeds: garden sorrel (Rumex acetosa) (Courtesy of SCHERING)

13.2.7 Accidents, diseases and pests

By far the worst **accident** affecting barley is flattening, which has already been described. The **fungal diseases** affecting barley include the fungi mentioned in the section on wheat, species of the genera *Puccinia, Ustilago* and *Septoria*. Other fungi are held to be specific to barley, such as *Helminthosporium gramineus*, which causes elongated longitudinal dark violet patches in late spring. Heavy attacks may halt the plant's growth and prevent the grain in the spike from ripening normally. It is worth disinfecting the barley grains that are to be sown, as this will prevent later problems with fungi. The fungicides to be used are the same as those listed for wheat.

To the right: Enlarged detail of oat inflorescence (Photo courtesy of BASF, S.A.)

To the left: Funbas® is a systemic fungicide sold by BASF, S.A. to combat powdery mildew, rusts and Rhyncosporium in wheat and barley.

The **pests** that attack barley include the senn bugs that attack wheat (*Aelia* and *Eurygaster*), but they cause less damage, or less economic damage, than to wheat. Firstly because these insects reduce the quality of the bread flour made from wheat, and this is not very important in the case of barley, and secondly because barley forms spikes and matures before wheat and is thus not so susceptible to attack from these bugs. Grain weevils also attack barley, and they should be controlled in the same way as for wheat.

13.2.8. Harvesting

Harvesting is performed using the same machinery as for wheat, which can be used to harvest all the different cereal crops.

13.2.9. Use

Barley can be used as fodder for direct consumption by livestock in the field. For fodder it is often sown together with legumes, such as vetches. Both the grain and straw are also stored for fodder to feed stabled livestock.

Barley's most important use for human consumption is in making beer. Two-rowed barleys, malt barleys, are the best for making beer. They do not contain a lot of protein, and this favors high quality beers. They should not receive excessive nitrogen fertilizer during their growth as this favors increased protein content. Malt barley usually shows at least 95% germination and the grain has to be harvested when it is thoroughly ripe and dry.

13.3. OATS

Oat (*Avena fatua*) is a grass with a pseudofibrous root system that is larger than that of wheat or barley. The elongated leaves are flat, and there is a ligule at the junction with the stem, though it lacks auricles. The oat can be distinguished from barley as it is a more bluish color, and barley is lighter green. The flower spike of the oat consists of a bunch of spikelets with two or three florets on long flower stalks. This type of inflorescence is called a **panicle**. The fruit is a **caryopsis**, with the husks attached. Like wheat it is self-fertilizing, though some flowers open their glumes and bracts before the stamens and pistils ripen, which means some cross-pollination takes place, and this leads to the degeneration of selected varieties.

13.3.1. Varieties

As for wheat and barley, there are too many varieties of oats for a full description here. The reader can find more information on the characteristics of each variety in the leaflets published by the companies distributing the variety and in the bulletins of the Ministry of Agriculture. Some of the more common varieties cultivated in Spain are **Previsión**, **Coker 227**, **Sol II**, **Blancanieves**, **Cóndor**, **Moyencourt** and **Blenda**.

13.3.2. Place in rotations

Most authors think oats do best as the second or third crop in a rotation. Because it is tough, oats should be grown after wheat or barley as a second or third crop. On poor sites, where neither wheat or barley is normally sown, oats are sown directly after the fallow year.

13.3.3. Cultivation requirements

Water is one of the main factors limiting the production of oats. When spring is very rainy, production is highest, as this plant transpires abundantly and needs a lot of water. It can however be damaged by waterlogging on excessively clay sites. It grows well on slightly acid sites (pH from 5 to 7). Oats are not very demanding in their soil requirements, but very limy soils are best avoided.

13.3.4. Applying fertilizer

As described in the first part of this work, which deals with the application and dosage of fertilizers, one approximate way of calculating the fertilizer needs of a plant is based on the crop's **extraction**. These extractions are calculated on the basis of complex analyses carried out in the laboratory, and are the average amounts of the different nutrients that a given crop plant has extracted from the soil. The oat's average extraction of nutrients per hectare and per metric tonne of yield is:

$$27.5 \text{ kg N}$$
$$12.5 \text{ kg } P_2O_5$$
$$30.0 \text{ kg } K_2O$$

On the basis of these extractions, it is possible to calculate that the needs of a crop of oats that is expected to give a production of 2,500 kg/ha will be approximately 70, 32 and 75 kg of fertilizer units of N, P_2O_5 and K_2O respectively. These amounts correspond to the replacement needs, but correction coefficients should be applied if the amounts of nutrients in the soil are known (soil analysis), if a good dose of dung has been applied before sowing (less fertilizer is required) or if the preceding crop was a legume (less nitrogen is needed).

Duplosan KV® is a selective herbicide that kills dicot weeds and is recommended for use in cereal crops (wheat, barley and oats).
It is manufactured and distributed by BASF, S.A.

13.3.5. Cultivation

Before sowing, **site preparation** should be carried out as for barley. In general little care is given to oat crops, and the yields are not very impressive. If more attention is paid to soil preparation and applying fertilizer, this crop can offer relatively high production when the spring rainfall conditions are favorable. The amount of seed sown can vary widely, but a normal dose is 100 to 150 kg/ha. Cultivation is similar to that of barley, though herbicide use is not common. Harvesting is usually by self-powered harvester.

13.3.6. Accidents, diseases and pests

Oats have show many features in common with wheat and barley, and is attacked by **rust** (*Ustilago avenae*) which behaves similarly to wheat **rust**. The rusts that are specific to oats include *Puccinia coronifera*, whose spores are a bright orange color and which may cause serious damage. Oats are attacked by other fungi, such as **black rust**, several types of **Fusarium**, **powdery mildew**, **black rot**, **parasitic flattening** and **Septoria**. The value of the oat harvest is low and it is rarely worth applying fungicides, but it is

*Cornfield weeds:
curled dock (**Rumex
crispus**)
(Courtesy of
SCHERING)*

worth disinfecting the seeds before sowing,
using the same products as mentioned for
wheat. If you decide to use a fungicide against
rust, use well-proven fungicides like **maneb,
mancozeb, benomyl**, etc.
Stored oat grain, like wheat and barley, is
attacked by grain weevils, though the damage
is much less severe.

13.3.7. Use

Most of the oats produced are used in
manufacturing animal feeds. The grain contains
a lot of vitamin E, making it very suitable for
working and breeding animals. It is also widely
used as a green fodder when grown together
with vetches or barley for livestock
consumption. For human consumption, it is
used in production of alcohol and alcoholic
drinks, as well as in diet products.

13.4. RYE

*Right:
Weeds:
Mugwort (**Artemisia
vulgaris**)*

Rye (*Secale cereale*) has a fibrous root system
similar to wheat, though larger. It is considered
to be one of the toughest cereals. The stem is
long and flexible and the leaves are narrow.

Like barley, the spikelets are not borne on a
stalk but directly on the rachis, with a single
spikelet at each notch of the rachis. The glumes
are elongated and sharp at the tip and the
bracts are hairy in the center and prolonged to
form a long awn. The spike is very thin and
long, and each spikelet bears three flowers of
which one is usually aborted.

13.4.1. Varieties

In Mediterranean zones, wheat is cultivated
more than rye, and so few varieties of rye are
known in Spain. Those grown are mainly from
northern Europe, such as **Petkus** from Germany
and **Royal** and **Varne** from Poland.

13.4.2. Cultivation requirements

Rye is undemanding in its soil requirements,
but grows better in the acid sandy soils and
cold climates of northern Europe than in Spain.
It grows well on shallow soils in mountainous
areas. Because it tolerates the harsh northern
winter well, rye is grown instead of wheat in
countries with cold climates, such as Germany
and Poland, where rye bread is a substitute for
wheat bread.

13.4.3. Cultivation

As rye is cultivated on poor soil, a poor yield is normally expected, and so little or no effort is put into cultivation. Although most land under rye receives no fertilizer, you are recommended to apply 20-40 units of N, 70 to 80 units of P_2O_5, and 70 fertilizer units of K_2O.

As rye is cultivated in cold zones, it must be sown very early. It is generally sown before the first rains in fall. The seed is sown at a dose of 100 to 120 kg/ha.

13.4.4. Accidents, diseases and pests

Rye is attacked relatively frequently by **nematodes.** They can be combatted by growing resistant varieties and rotating crops.

The **fungal diseases** attacking rye include **rust** (*Puccinia graminis secalis*), **leaf rust** (*Puccinia graminis recondita*) and **yellow rust** (*Puccinia striiformis*). **Ergot** (*Claviceps purpurea*) is another fungus that especially attacks rye though it also attacks other members of the grass family, such as meadow grasses, wheat, barley and rice. The ergot fungus produces an alkaloid toxic to human beings that is used in the pharmaceutical industry. Rye seed should be treated with **organo-copper** compounds, as it is sensitive to *Fusarium nivale*.

Some **insects**, cephid woodwasps, such as *Cephus pygmaeus* and *Trachelus tabidus*, can bore into rye plants and also wheat.

Cornfield weeds: herb mercury (Mercurialis annua) (Courtesy of SCHERING)

13.4.5. Use

Rye is widely cultivated in the cold zones of northern Europe, where it is used to make a dark bread that lasts longer than white bread made from wheat. The area under rye is decreasing every year all over the world.

13.5. CORN

Corn (*Zea mays*) is another member of the grass family, like all the preceding cereals. It originated in the Americas and first reached Europe (where it is known in English as maize) after the discovery of the Americas. The first classification of corn was made in the United States, and the English names of the different types are internationally used. This classification is based on the shape and structure of the grains produced:

• **Dent corn** has a depression in the crown of the kernel.
• **Flint corn** has a hard smooth grain.
• **Sweet corn** is sugary, not starchy.
• **Soft corn** has a lot of soft starch.
• **Popcorn** is heated dry to burst it before consumption.
• **Pot corn** is corn on the cob.
• **Waxy corn** is the type of corn used industrially to produce starch. It is also known as industrial corn.
• **Coyote corn** is uncultivated corn, the corn that germinates in uncultivated fields.

Invasive weeds: Black henbane (Hyoscamus niger) (Courtesy of SCHERING)

13.5.1. Growth cycle

Corn's growth cycle begins with germination, which takes six to eight days, and is the period from sowing till the coleoptile emerges into the light. Once the corn has germinated, the **growing period** starts and a new leaf is produced every three days in normal growing and weather conditions. Twenty days after germination, the plant should have five or six leaves, and is in full leaf within five or six weeks. The **flowering** phase begins when the male tassel emerges and starts to disperse pollen, and when elongation of the styles occurs. The pollen is emitted over a period of eight to ten days, depending on the temperature and the availability of water.

Fruiting begins when the ovules are fertilized by the pollen, and this has finished when the styles of the cobs turn brown. The cob is full-sized within three weeks of fertilization, and the grains containing the embryo are fully formed. The grains then fill with a milky substance containing a lot of sugars, that is turned into starch as the grain ripens. A month and a half after fertilization, at the end of the eighth week, the grain is **physiologically mature** and reaches its maximum dry weight. At this stage moisture content is usually about 33%. Due to the low environmental moisture and high temperatures, the grain continues to dry until it reaches **commercial ripeness**.

13.5.2. Varieties

In corn, hybrid varieties are of great importance. The reader should refer to section 3.5 "Cereal varieties" for information on the relevant basic concepts of genetics: pure strains, homozygote and heterozygote, heterosis (hybrid vigor), hybrid varieties, etc.

Field of corn

The most widely cultivated are dent corn and flint corn. The glassy corns have grains with less flour, with a lower starch and higher protein content. Waxy corn is only grown for its high starch content. Most of this type of corn is grown in North America, and this has led the EU to try to promote waxy corn cultivation in the warmer areas of Europe, such as Spain and Italy.

Corn is a very vigorous plant, with a fibrous root system that grows deep and fast. The stems often reach four meters and the broad leaves clasp the stem. Corn is monoecious, i.e., the male and female flowers are borne on the same plant, though in different places. The male flowers are borne in a panicle (usually called a **tassel**) terminating the main stem. The female flowers arise in the axils of some leaves and are all borne in rows on a spike surrounded by bracts ("shucks"). This spike (usually known as the **cob**) bears the long brush-like styles, which are known as **"silks"**. The cob consists of a central axis which bears several hundred grains. The axis represents about 15-30% of the cob's weight. The flowers may be fertilized by pollen from the plant's own tassel or by pollen from other plants. When pollination is by pollen from other plants, it is common for some of the grains to differ in color.

Classification of the hybrid varieties of corn on the basis of how early they are.

Cycle	Name	Days from germination to physiological maturity
100	Ultra-early	< 80
200	Very early	80-90
300	Early	90-100
400	Semi-early	100-105
500	Semi-early	105-110
600	Medium cycle	120-125
700	Semi-late	125-130
800	Late	130-140
900	Very late	140-150
1000	Ultra-late	> 155

Corn produces male and female flowers on the same plant and can thus self fertilize, but 98% of corn pollination is **cross pollination**. Thanks to the wind, corn plants can fertilize each other, and not themselves.

CYCLE 200

Name	Country of origin	Type of hybrid	Registration date
Adour-250	France	HD	1975
Caldera-535	USA	HD	1981
Dea	USA	HS	1982
DK-216	USA	HD	1982
Eperón	France	HS	1988
G-4082	USA	H3L	1977
Hórreo-270	Spain	H3L	1982
Hórreo-330	Spain	H3L	1981
Inra-260	France	H3L	1974
Luisa-PR-3949	USA	HS	1987
Minedor	USA	HS	1988
Misión-201	Spain	H3L	1987
M-280	France	HD	1986
PS-271	Spain	H3L	1984
PS-272	Spain	H3L	1987
PS-290	Spain	H3L	1981
PX-7	USA	H3LE	1984
SN-96	France	H3L	1984
Trial	France	H3L	1988

CYCLE 300

Name	Country of origin	Type of hybrid	Registration date
AE-260	USA	HS	1988
AE-325	USA	HS	1988
DK-498	USA	HS	1988
Eva	USA	HS	1981
G-350	USA	HD	1974
H-81501	USA	HS	1987
LG-15	France	HD	1974
Melisa PR-3704	USA	HS	1988
Montejo	Belgium	HS	1982
M-379	France	HDE	1982
Pau-360	France	H3L	1984
PS-366	Spain	H3L	1984
PX-20	USA	HS	1977
PX-9283	USA	HS	1988
Rebeka PR-3803	USA	HS	1986
Vulcano	USA	HS	1988
XL-312	USA	H3LE	1983

CYCLE 400

Name	Country of origin	Type of hybrid	Registration date
AE-431	USA	HS	1988
Altón	USA	HS	1986
Cantaleso LG-18	France	HS	1985
Danika	USA	HS	1987
Demar	USA	HS	1988
DK-222	USA	HD	1978
Domino-440	Spain	H3L	1983
Domino-450	Spain	HS	1981
PS-431	Spain	HD	1974
PX-9292	USA	HS	1988
RX-39	USA	H3L	1974
Sabrina PR-3707	USA	HS	1984
Valeria PR-3540	USA	HS	1986
Volga PR-3475	USA	HS	1986
XL-32A	USA	HS	1981

CYCLE 500

Name	Country of origin	Type of hybrid	Registration date
A-335	USA	HD	1974
Adour-534	France	HS	1978
Bremta	USA	HS	1986
Calvi	France	HS	1988
Damon	USA	H3LE	1983
Golf	USA	HS	1987
G-4408	USA	HS	1980
G-4524	USA	HS	1982
Lenor G-4441	USA	HS	1986
Orellana	Belgium	HS	1982
Pamela PR-3471	USA	HS	1988
Potro-577	France	HS	1986
PR-3551	USA	HS	1984
PS-551	Spain	HD	1974
P-3536	USA	H3LE	1981
P-3543	USA	H3L	1978
Sonar	USA	HS	1985
XL-32AA	USA	HS	1980

CYCLE 600

Name	Country of origin	Type of hybrid	Registration date
Acturus W-4.000	USA	HS	1986
AE-707	USA	H3LE	1980
Comet	USA	H3LE	1983
Cortes	USA	HS	1982
Gheppio	France	HS	1985
Granada	USA	H3L	1988
Isora PR-3380	USA	HS	1985
Logos DK-636	USA	HS	1988
Luana PR-3377	USA	HS	1984
Matador	Germany	HS	1981
Meteor	USA	HS	1984
Miejour LG-57	France	H3L	1985
Nelson	USA	HS	1988
Ortis G-4597	USA	HS	1988
Palma PR-3352	USA	HS	1985
Palomar	USA	HS	1982
Pizarro	Belgium	HS	1982
PN-9635	USA	H3L	1982
Remo	USA	HS	1988
SNH-731	USA	HS	1986
Mistral	—	HS	—
Pantera	—	HS	—
SC-874	—	HS	—
XP-7.286	—	HS	—

CYCLE 700

Name	Country of origin	Tipe of hybrid	Registration date
Adour-640	France	HS	1976
AE-664	USA	HS	1986
AE-703	USA	HS	1974
AE-750	USA	HS	1986
AE-7020	USA	H3L	1983
Agus LG-2661	France	HS	1986
Albufera	USA	HS	1984
Flamingo	USA	HS	1987
Furia PR-3297	USA	HS	1986
Futuro M-8556	USA	HS	1988
G-4507	USA	HS	1977
Ivana PR-3181	USA	HS	1988
Max	USA	HS	1984
Monteverde	Spain	HS	1986
Mundial	USA	H3LE	1983
M-650	France	H3LE	1983
Nella PR-3198	USA	HS	1984
Nepris	France	HS	1987
Polaris	USA	HS	1984
PS-71	Spain	HS	1985
PS-734	Spain	HD	1974
PX-74	USA	HS	1979
PX-675	USA	H3L	1982
PX-9540	USA	HS	1988
Río Bravo	USA	HS	1984
Roxalis	France	HS	1988
RX-90	USA	HS	1978
Valdivia	USA-Belgium	HS	1988
XL-72	USA	HS	1974
XL-72AA	USA	HS	1981
XL-75A	USA	H3LE	1980
Bianca	—	HS	—
Itala	—	HS	—
Mérida	—	HS	—
Pianosa	—	HS	—

CYCLE 800

Name	Country of origin	Type of hybrid	Registration date
AE-8004	USA	HS	1981
Alios	—	HS	—
Amanda P-3186	USA	HS	1984
Aneto-810	Spain	HS	1987
Badajoz	USA	HS	1988
Celina PR-3124	USA	HS	1988
G-4647	USA	HS	1985
G-4727	USA	HS	1982
G-5050	USA	H3L	1976
M-770	France	HS	1983
Molto	—	HS	—
Paolo DK-711	USA	HS	1987
Prisma G-4730	USA	HS	1986
P-3183	USA	HS	1980
RX-904 Mincio	USA	HS	1980
Virax G-4754	USA	HS	1986
XL-365	USA	H3LE	1974

List of current varieties of corn classified by earliness. Numbers from 200 to 800 are assigned to each growth cycle, with 200 being the earliest and 800 the latest.

The grains on the cobs of corn do not all give rise to plants belonging to the same variety but to different varieties, and thus give rise to populations rather than authentic varieties. Using complicated systems of emasculating the stamens and fertilizing the flowers, **pure lines** can be obtained in corn, and when they are crossed they give rise to hybrid varieties. These varieties consistently show higher yields than the old open pollination varieties, normally 25-35% more, and maybe even more.

In an international meeting organized by the Food and Agriculture Organization of the United Nations (FAO), it was agreed to classify hybrid varieties of corn into ten groups on the basis of their earliness. This table is on page 472, and assigns the cycles of these groups a number ranging from 100 to 1,000, that is to say from ultra-early to ultra-late. The table also shows the days between germination and physiological maturity. The introduction of two new indices is being considered in order to create a clearer classification of corn. The **base index** or **earliness index** is the number of temperature units above 6°C needed between sowing and the date by which 50% of the pistils have appeared. The other, the **maturity index** is the number of temperature units above 6°C needed from the date the styles elongate until the grain reaches 33% moisture content, when it is considered to have reached physiological maturity.

In order to choose the most suitable varieties for each case and climate, the characteristics of each variety should be carefully consulted in the publicity leaflets of the commercial suppliers and in the bulletins of the Ministry of Agriculture. In general terms, short-cycle varieties manage to avoid the negative effects of fall frosts, because they are harvested early. This has the added advantage that the crop is

dried naturally on the same plot, eliminating the costs of artificial drying. Short-cycle varieties tend to show less productive potential, meaning that the varieties chosen should have the longest cycle compatible with the regional climate.

On the previous page there are seven tables with all the varieties on sale in Spain classified by their earliness and numbered 200 to 800, in accordance with the guidelines mentioned above.

13.5.3. Place in rotations

After fallow, corn can follow any other crop. On irrigated ground, and as a second crop, corn should be planted after wheat or beans.

13.5.4. Cultivation requirements

Corn is a summer cereal, as it requires a lot of warmth. Corn should not be sown below 10°C, and it requires 15°C for germination, and 18°C for flowering, though the ideal temperature in the growing phase is between 24 and 30°C. The introduction of varieties with different cycle lengths has allowed corn cultivation to spread to colder zones. Short-cycle varieties can be cultivated in cooler climates, though their yields are lower.

Corn adapts well to a wide range of soils, preferring a neutral or slightly acid soil pH (pH 6-7). Perhaps the only limitation of this type is in excessively limy and alkaline soils, where availability of some micro-elements may be blocked. Corn should be grown with irrigation or in zones with high rainfall, as it requires a lot of water in the flowering period.

13.5.5. Applying fertilizer

Corn's average extraction of the main macro-elements N-P-K per metric tonne are 25 kg N, 11 kg P_2O_5 and 23 kg K_2O. As a guideline, for every tonne of expected production you should apply the following amounts of fertilizer:

30 kg N
15 kg P_2O_5
25 kg K_2O

Depending on the expected grain production, the farmer should multiply these fertilizer units by the appropriate factor. If the expected production is 9,000 kg (9 tonnes), the fertilizer units given above should be multiplied by 9. As in the preceding crops, these quantities should be corrected in accordance with the results of soil analyses and the levels of assimilable potassium and phosphorus they show. Factors that affect the dose of nitrogen required should also be taken into account, such as recent dunging or if the preceding crop was a legume.

Nitrogen is largely absorbed by corn in the period from just before flowering until 25 to 30 days after flowering, when its nitrogen needs are greatest. When the plant lacks nitrogen, the leaf tips turn yellow, spreading along the central vein in a "V"-shape: the overall appearance of the plant is mediocre, its vigor diminishes, the leaves are small and the tips of the cobs lack kernels.

The period of maximum phosphorus requirement in corn coincides with that of nitrogen. Phosphorus shortage usually affects the setting of the seed of the cob, leading to malformed kernels and some rows show rudimentary kernels.

When the plant suffers from potassium shortage in the early stages of its life, the young plants turn yellow or greyish-yellow, sometimes with yellowish stripes or patches. The tips and edges of the leaves wither, appearing scorched or burnt. Later, potassium shortage lowers the plant's resistance to flattening and fungal attack.

Magnesium deficiency can be detected because the plant has yellowish stripes along the veins, and frequently shows a purple coloration on the underside of the lower leaves. The cobs are also affected by magnesium shortage and are smaller than those on well-nourished plants.
On very acid soils, there may be a deficiency of assimilable boron. In these cases, cobs are wrinkled on the side facing the stem, and the rest of the cob is unaltered.

In corn, grain production and quality depend more on the fertilizer applied than in any other cereal crop. Nitrogen should be applied ten or fifteen days before flowering to ensure sufficient protein in the grain and a correct level of production. All the fertilizer units of phosphorus and potassium should be applied at depth together with one third of the nitrogen. The remaining two thirds of the nitrogen should be applied to the surface, once at the time of thinning and the rest a month later.

13.5.6. Sowing

The site must be **prepared** before sowing. The object is to prepare a tilth that is deep but not too loose. Soil preparation also seeks to eliminate surface weeds, break down clods of earth and level the soil.

According to several authors it is possible, and even preferable, to sow late in February in relatively warm areas, as even if there are late spring frosts, the corn plant is not damaged. Sowing in February has other advantages. The early stages of the plant's growth escape attack by soil insects, etc; pollination is early, before the hottest temperatures; the grain also ripens at a time when temperatures are not excessively high; less irrigation is required; the crop can be harvested early, which is useful if you intend to plant another crop in autumn. Treatment against red spider mite is also usually not necessary. In general, sowing is carried out when the soil temperature has risen above 10°C.

When calculating the sowing density there is an important agricultural guideline to bear in mind. If the crop is sown too densely, production is lower than expected (due to competition between plants). If the sowing density is low, productivity per plant is high, but the total productivity of the plot is lower, because the high production per plant does not compensate for the lower number of plants. In corn, this means that if the density is high, the cobs are small, and if the density is low, then the cobs are large, but this does not compensate for the low density. The best sowing density varies from variety to variety. There are thus varieties that can be grown at high densities without decreasing the total production of the plot.
Corn is often sown using **precision pneumatic sowers**, which leave a space between rows of plants. The distance between the rows and between the plants in the rows (i.e., the **planting frame**) is a priority question that has to be chosen on the basis of the characteristics of the variety cultivated. As already mentioned, you should consult the technical information from the suppliers and the official bulletins of the Ministry of Agriculture to obtain specific information on the characteristics of a given variety. As a guideline, sowing with a precision sower at a dose of 100,000 plants per hectare and a germination rate of 85-90% will give a real density of 85-90,000 plants/ha. This rate might be suitable for some varieties, in a good climate and with irrigation, but in dry farming the sowing rate should be much less.

The grain should not be buried too deep. It should be sown no deeper than 2-3 cm in wet clay soils or 8-10 cm in sandy soils, because they dry out more easily. The pneumatic sower covers the seed with a thin layer of soil, which should not be more than 3-5 cm. When many of the seeds fail to germinate, further seed should not be sown, as second sowings do not do well. Remove the rest of the crop and sow again with a different variety with a shorter cycle (because you will have lost a few days).

13.3.7. Irrigation

It is negative for corn to suffer periods of water shortage, because the stomata close and photosynthesis is reduced, meaning the final yield is lower. Water shortage is especially negative during the flowering period, and may decrease the harvest by 30%. The crop can be irrigated on the surface with a sheet of water or using special machinery. If you chose semi-fixed sprinklers, should be sown further apart (some rows) to make lanes for the mobile tubing to move along, in order to make irrigation easier.

Dicot weeds	
Latin name	Common name
Amaranthus retroflexus	Rough pigweed
Chenopodium sp.	Goosefoot
Convulvulus sp.	Bindweed
Sonchus sp.	Sow-thistle
Solanum nigrum	Black nightshade

Monocot weeds	
Latin name	Common name
Digitaria sanguinalis	Crabgrass
Echinochloa crus-galli	Barnyard grass
Cyperus sp.	Sedge
Phalaris canariensis	Canary grass
Poa annua	Annual bluegrass
Setaria sp.	Bristle grass (foxtails)
Cynodon dactylon	Bermuda grass
Sorghum halepense	Wild sorghum

number year after year, even if weedkillers are used. This is clearly shown by weeds like sedges (*Cyperus*) and *Sorghum*. Above is a list of the most common weeds of corn.

Mechanical weeding between the rows is essential in the first stages of cultivation, as this is when weeds compete most with the corn. Only the soil surface should be disturbed so as not to damage the corn's roots. Mechanical weeding should be combined with application of chemical weedkillers. Whether to use chemical weedkillers or weed mechanically depends on factors like the size of the plot, the climate, the phenological stage of the crop, the types of weed present and their of developmental stage, etc.

Most herbicides used on corn crops are pre-emergence herbicides. The herbicide should be applied after sowing and before the crop

13.5.8. Herbicides

As explained in the third section of this work, weeds compete directly with the crop for nutrients and space. This competition is especially severe when the plants are young, before their root system fully occupies the soil. Where the same crop is grown year after year, the most resistant weeds increase in

of pollen shedding and style elongation. If there are frosts before the cob is fully ripe, before all the sugars in the grain have been transformed into starch, the process is totally disrupted, and the grain remains soft and is much more difficult to dry thoroughly, because when the frost finishes, the plant puts its last efforts into transporting water to the grain.

Fungal diseases affecting corn include **corn smut** (*Ustilago maydis*). This disease is characterized by the appearance of large tumors on the stem and even on the leaves. Under the epidermis of the affected area a black powder can be seen. This powder consists of the chlamydospores of the fungus, which are dispersed by the wind. When they come into contact with water they form a promycelium that produces basidiospores, which germinate and penetrate into the plant, where the mycelium develops among the tissues, causing tumors to form in different points. Some hybrid varieties on the market are resistant to this fungus, while others are highly sensitive. In zones where this fungus is common, the farmer should choose resistant varieties.

Some authors point out that good preventive results are obtained by disinfecting the seed with the systemic fungicide **carboxyn**. Other fungal diseases, such as *Helminthosporium* or the rust *Puccinia sorghi,* are also frequent in corn and should be combatted using preventive disinfection of the seed with **maneb**, **mancozeb** etc., or by applying **carbendazime**, **maneb**, **tridemorph** etc., against *Helminthosporium*, while rust should be treated with **oxycarboxine**, **propiconazol**, **triademenol**, etc.

Some deleterious **mites**, such as the red spider mite (*Tetranychus*), causes serious damage to corn in hot zones, especially in the months of September and August. Visual checks on the weed *Solanum nigrum* are an effective way of detecting this mite in crops: it infects this weed before corn plants. Once populations of the red spider mite have been detected, a semi-preventive treatment can be carried by dusting sulfur, and this has the advantage of not killing the useful insects. When there is a serious infestation of red spider mite, you can treat with acaricides such as **tetradiphon** and **diphocol** or a commercial mixture of the two.

Insects can be classified according to the damage they cause. Soil insects, such as **wireworms** (*Agriotes lineatus*), *Anoxia villosa*, **cockchafers** (*Melolantha melolantha*), *Tropinota irta*, etc., **cutworms** (*Agrotis segetum*) and **leatherjackets**, the larvae of the crane fly *Tipulia oleracea*, all have different habits and life cycles, but they affect the stem and roots of the corn plant in similar ways. Preventive measures can be taken before sowing, such as plowing the fields, disinfecting the soil with granulated insecticides such as **dazomet**, or with **carburophan**, **chlorpyriphos**, **phonophos**, etc., at the time of sowing. Once the crop is well established, curative treatments

Weeds: musk thistle (Carduus nutans) (Courtesy of SCHERING)

has emerged above ground. The active ingredient remains in the top few centimeters of soil, eliminating the weeds and leaving the ungerminated corn untouched. For more information on herbicides, active ingredients, application methods, doses, etc., consult the third section of this work, as well as the supplier's technical specifications for each product and the technical bulletins from the different official bodies. The chemical herbicides most widely used on corn crops are triazines (**simazine**, **atrazine**, **terbutryn**, etc.), amides (**metholachloro** and **propachloro**), thiocarbamates (**eptan** and **butylate**) and anilines.

Other herbicides, such as the selective hormone weedkillers that eliminate dicots but leave monocots unharmed, should be used with precaution after the crop seedlings have emerged. They should only be applied to the earliest developmental stages of the plant, before the corn plant has grown four or five leaves, and as long as the ambient temperature is not above 20°C or extremely cold. Selective hormone weedkillers such as **2,4-D** and **MCPA** are frequently used in corn crops. When the corn has reached a height of 50 cm contact herbicides can be used, such as **paraquat**, as long as care is taken to ensure the product does not wet the plant when applied.

13.5.9. Accidents, diseases and pests

Sometimes in dry conditions the leaves turn yellow, all of the leaves on the plant turning yellow at the same time, unlike what happens when this is due to nitrogen deficiency. If the temperatures are extremely high, there may be problems at the time should be applied based on synthetic pyrethrins (**protenophos**, **tetrachlorvinphos**, **trichlorphon**, etc.), or on **methyl-pyriniphos** or **trichlorphon** baits.

Other insects such as **corn borers** (*Sesamia nonagrioides* and *Pyrausta nubilalis*), moths whose larvae bore within the corn's stem. Both insects overwinter as caterpillars within the galleries they bore in the stem, where they pupate. The adult moths emerge in May and the eggs are laid in sites sheltered from the sun and rain, usually on the underside of the leaf. When the eggs hatch, the larvae that emerge eat the corn's leaves, and only later bore into the plant. Corn sown in June and July is worst affected and this is when preventive treatments should be applied. The many methods available include biological ones, such as **Bacillus thuringensis** or chemical ones, such as **chlorpyriphos**, **diazinon**, **phenitrothion**, **trichlorphon**, etc.

One of the worst lepidopterans to attack corn is budworm (*Heliothis*). Like the previous insects, it is phytophagous, first attacking the leaves and then the grain on the cobs as they form. This drastically reduces the harvest, making this butterfly one of the worst pests of corn. **Endosulphan**, **chlorpyriphos**, **deltamethrin** and **methamidophos** are some of the insecticides used to control it.

drying machinery. **Driers** are used to lower the moisture content to 13-14%, the optimum for storage in silo. Driers normally use a current of hot air to evaporate the moisture from the grain. In the past this specialized drying machinery did not exist and grain was stored in special granaries whose design ensured the grain was well-aired, a feature common to many traditional designs. Harvesting the corn includes **breaking off** the cob, **removing the bracts**, **removing the grain** and **shredding the stems**. Harvesters for wheat, etc., can be adapted to harvest corn, but if it is economically possible it is better to acquire a **harvester** for corn. In a single pass this machine gathers the cobs, removes the bracts and separates the grain.

Glassy corn intended for flour production should be harvested when the moisture content is less than 14%, as otherwise it would lose proteins when it is being dried. It would be better to bury the crop remains after harvesting, in order to add organic matter to the soil, but they are frequently piled together and burnt, due to the difficulty of incorporating them into the soil. There is now special machinery, shredders, that can shred the crop remains for later incorporation into the soil using normal methods. Corn has the highest potential yield of all the cereals, and on irrigated ground it can reach 15,000 kg of grain per hectare.

A wide range of possible farm machinery is available to farmers. The photo shows a fodder corn harvester with a single tooth which can be powered by a low power tractor, just 45 horsepower. (Courtesy of AGROMÁQUINAS Y REMOLQUES, S.A.)

13.5.10 Harvesting

When 50-75% of the shucks (bracts) covering the cobs have turned yellow the corn is considered to have reached **physiological maturity**, meaning it can be harvested. However, the corn still contains too much moisture. When corn is grown in hot climates, the cobs can be left on the plant to dry out. In zones where the ripening period is rainy, the cobs have to be harvested while they still contain too much moisture, and so they must be dried using special

Some varieties of corn have attractive multi-colored cobs, and are sold as ornamentals. Selected and sold by VILMORIN.

13.5.11. Use

Corn was brought to Spain shortly after Columbus discovered the Americas, and from there spread throughout Europe and the Mediterranean Basin. Corn adapts well to a range of climates and soils and so it is cultivated on all five continents in a strip ranging from the hot zones in temperate climates to wet tropical zones.

Corn production is mainly for livestock feed; it is very nutritious and is one of the basic components of feeds for fowl and hogs. For use in feed the corn is ground and the meal contains proteins, phosphorus, some calcium, fats, pro-Vitamin A, etc. To complement corn-based feeds, the ground cob (ground with the grain), barley, oats, etc., are added. The crop remains of corn can be fed as a green fodder to livestock. Part of the corn produced is used industrially as a source of starch, some flours for human consumption and to obtain cooking oil.

Glassy corns are used for several purposes in the food industry, such as breakfast cereals, baby foods, popcorn, etc. In addition to these uses, the corn stems can be collected and dried and made into paper pulp. The stripped cobs are an excellent fuel: two tonnes of dried cobs is roughly equal to a tonne of coal. The stripped cobs are also used in some industrial processes. They are used for example in the production of furfural, a compound that is much used in the chemical industry. Some varieties have cobs bearing grains of several different colors, and are grown as ornamentals.

13.6. SORGHUM

Sorghum (*Sorghum vulgare)* is a member of the grass family (*Gramineae*). On deep permeable sites, its root system can reach a depth of 2 m. In Spain sorghum produces male and female

Detail of sorghum inflorescence. (Courtesy of ICI SEEDS)

inflorescences on the same plant, that is to say they bear stamens and pistils in the same flower. In Sudan, however, dioecious sorghum plants have been found. The sorghum plant can reach a height of 1-2 meters, the inflorescences are panicles and the seeds, or kernels, are spherical or oblong and about 3 mm in diameter, and reddish or yellowish black.

13.6.1. Varieties

The different varieties of sorghum can be divided be the length of their growing cycle into four major groups, **late**, **medium**, **early** and **very early**. Below is a complete list of the varieties on sale in Spain and their cycles.

Detail of a precision corn sower

List of sorghum varieties registered with the Spanish authorities

Variety name	Country of origin	Cycle	Date of registration
Aralba	France	Late	1984
Argence	France	Late	1985
A-28	USA	Very early	1981
Bartol	USA	Early	1985
Bravo E	USA	Medium	1989
Bravo M	USA	Medium	1986
Cetrero	USA	Medium	1983
Corral	USA	Medium	1985
Dorado DR	USA	Early	1985
Dorado E	USA	Early	1974
Double TX	USA	Late	1974
D-55	USA	Medium	1981
Eneka-1580	USA	Early	1984
E-59	USA	Medium	1980
Granador	USA	Early	1983
G-550	USA	Medium	1986
G-1400	USA	Early	1986
G-1516 BR	USA	Late	1987
H-7910	USA	Late	1985
Hazera-226	Israel	Very early	1977
Hazera-610	Israel	Medium	1974
Hazera-6078	Israel	Medium	1983
HW-5445	USA	Early	1987
NK-121	USA	Very early	1979
NK-180	USA	Early	1979
Pilos-708	Germany	Late	1988
PRB-864	USA	Early	1983
PR-8239	USA	Medium	1986
PR-8244	USA	Medium	1985
PR-8416 A	USA	Medium	1982
PR-8515	USA	Medium	1988
PR-8680	USA	Early	1983
PR-8686	USA	Very early	1989
P-8501	USA	Medium	1979
Regulus-705	Germany	Medium	1988
Tamaran	USA	Early	1989
TE-Y-45	USA	Medium	1983
Topaz	USA	Semi-late	1977
Veloz-701	Germany	Early	1988

13.6.2. Place in rotations

On irrigated sites, sorghum can follow wheat or beans as a second crop. As a first crop, it can be followed by any other crop.

13.6.3. Cultivation requirements

Sorghum grows well on alkaline soils, especially **sweet sorghum** which requires calcium carbonate in the soil. In soils with suitable pH values, the sucrose content of the stems and leaves is higher. It grows best on good deep soils that are not too heavy. It shows some tolerance of saline conditions.

In comparison with corn, sorghum is more resistant to dry conditions, and needs less water than corn to synthesize a kilogram of dry matter. If conditions are very dry, sorghum's growth halts, but it can renew growth when water is available again. In terms of temperatures, sorghum can withstand low temperatures in its early growth, like corn. Sharp drops in temperature at the time of flowering may lead to reduced yields. Sorghum tolerates excessively high temperatures better than corn. If it is growing in relatively cool soil, the flowers are not shed in very hot weather. Temperature may be a factor limiting germination. The seed does not germinate below 12-13°C, and so sorghum should be sown roughly three or four weeks

after corn. Sorghum does not begin to grow actively until the temperature exceeds 15°C, and its optimum temperature is around 32°C.

13.6.4. Applying fertilizer

Sorghum produces more when irrigated than when dry farmed. Irrigated sorghum might be expected to produce about 7,000 kg of grain per hectare. The fertilizer requirements, in fertilizer units, are:

200 kg N
100 kg P_2O_5
150 kg K_2O

The units of nitrogen are calculated taking into account the average nitrogen losses due to leaching and the loss of phosphorus due to retrograde fixation. As in the crops already discussed, these doses should be recalculated if the soil is particularly rich in potassium and/or phosphorus or if the soil is especially rich in organic matter.

13.6.5. Sowing

Before sowing, the site should be **prepared** as for corn. The previous crop can be dug in deeply, then a final plowing and passes of the cultivator to remove weeds. When the site is prepared to receive the seed, it is sown,

Weeds:
Field sow-thistle (Sonchus arvensis) *(Courtesy of SCHERING)*

To the right: cornfield weeds: Buxbaum's speedwell (Veronica persica) *(Courtesy of SCHERING)*

usually beginning fifteen to thirty days after the normal sowing date for corn in the same region. The sowing density is not excessively important because sorghum can compensate for a low density by producing tillers, unlike corn. A density of 20 to 30 plants/m² and a distance between rows of 20-60 cm is recommended by most authors. As a guideline you can assume a quantity of 15 kg/ha of seed. The seed should not be buried too deep, no more than 2-4 cm. Sorghum seed can be sown successfully with the same mechanical sowers as discussed for wheat or those used for sowing corn, after adapting the types of disc to the size of the sorghum grain.

13.6.6. Irrigation

Sorghum should be watered five times over the course of its growing period. Sorghum can withstand drought well, but adequate watering ensures better harvests. Sorghum's critical period for water is from the moment the panicle appears in the leaves of the stem tip until the end of the milky grain stage. The stems are short and so it is easier to water by sprinkler than corn and requires less labour. When it is uncertain how much water will be available to supply to the crop, it is better to opt for sorghum over corn, as sorghum requires less water.

13.6.7. Herbicides

In general pre-emergence treatments are preferable to treatment with selective weedkillers. Chemical products, such as **atrazine** and **atrazine + terbutryn** are absorbed by the leaves and roots, and should be applied after sowing, as a pre-emergence herbicide. Other weedkillers, such as selective herbicides against dicot weeds should be used post-emergence, taking due precautions to protect neighboring crops. Widely used active ingredients, such as **MCPA** and the esters of the salts of **2,4-D**, are examples of selective weedkillers suitable for treating actively growing sorghum.

13.6.8. Accidents, diseases and pests

All the fungal diseases that can attack corn can attack sorghum, and the same methods can be used to treat them. Some varieties of sorghum are highly susceptible to some agricultural chemicals, and it is a good idea to consult the technical services of the supplier before using any pesticide.

13.6.9. Harvesting

Like most cereals, sorghum is now harvested using a harvester. The grain should not be stored with a moisture content of more than 15%. If it is to be stored for a long time, the grain's moisture content should exceed 12%.

13.6.10. Use

Most sorghum produced is used as fodder for livestock, though the whole grain can also be fed to animals. Sorghum has a lower energy content than corn. Sorghum contains less fats but more protein than corn. Care should be taken when livestock are allowed to graze sorghum, because the tips of the plant contain high levels of a glucoside that releases cyanide ions when it is broken down by enzymes (diastases), and cyanide is extremely toxic. The glucoside accumulates in the tender shoots during the plant's growth, but disappears after flowering; the levels of glucoside increase with frosts, high temperatures and dry conditions. Irrigation helps to lower the level of this glucoside.

Sorghum is easy to sell as it is a raw material for the industrial production of starch, dextrine, dextrose, oils, etc.

13.7. RICE

Rice (*Oryza sativa*), like all cereals, is a member of the grass family (Gramineae). It is an annual herbaceous plant, and its stem consists of an upright spike that can reach a height of more than 1 meter. The leaves are alternate, sheathed, lineal and rough to the touch, and 5-10 mm wide.
The whitish green flowers are arranged in spikelets, and form a long narrow panicle at the end of the stem that weeps downwards

Rice is cultivated on the waterlogged soils of deltas and marshes and in paddies. (Courtesy of AGRO LORIN)

Variety name	Registration date
Bahía	7-05-74
Balilla	7-05-74
Balilla X Sollana	7-05-74
Balstirsia AC	7-05-74
Belle Patna	7-05-74
Betis	21-09-82
Blue Belle	7-05-74
Blue Bonnet	7-05-74
Bomba	7-05-74
Bombón	7-05-74
Bond	4-08-87
Colina	7-05-74
Delta	7-05-74
Francés	7-05-74
Girona	7-05-74
Gulfmont	18-05-88
Indio	4-08-87
Italpatna	7-05-74
Júcar	24-10-78
Lebonnet	4-08-87
Lemont	18-05-88
Liso	7-05-74
Matusaska	7-05-74
Nano X Sollana	7-05-74
Newbonnet	4-08-87
Niva	24-10-78
Pegonil	7-05-74
Ribe	7-05-74
Ribello	7-05-74
Rinaldo Bersani	7-05-74
Rubino	9-04-86
Senia	9-04-86
Sequial	7-05-74
Skybonnet	18-05-88
Starbonnet	7-05-74
Tebonnet	4-08-87
Tebre	9-04-86
Thaibonnet, L-202	4-08-87
Thainato	4-08-87
Veneria	9-04-86

Current list of the rice varieties registered in Spain

after flowering. Each spikelet bears a single flower and has a glume with two small valves. The grain, which is a caryopsis a few millimeters long, consists of an **embryo** and the **endosperm** within a surface bran layer rich in proteins and thiamine (vitamin B_1), that is a whitish brown color, and a tough cellulose-rich **husk** or **hull**. **White rice** has been milled to remove both husk and bran, while **brown rice** retains the bran layer.

13.7.1. Growth cycle

In many zones of Asia, rice is still cultivated using traditional systems.

Once the rice grain has germinated, the roots, stem and leaves appear. The **panicle**, or **flower spike**, starts to form thirty days before the spike emerges and seven days later it is already 2 mm long. Flowering takes place on the same day as the spike emerges or on the following day, late in the morning.

not very saline, the water reaching the lowest paddy is usually pumped back up to higher levels. When the water cannot be reused because it is too saline, it should be directly discharged.

Rice needs a minimum temperature of 10-13°C to germinate, but germinates much better in seedbeds at a temperature of 30-35°C. It does not germinate at temperatures of more than 40°C. Once it has germinated it grows well at temperatures between 7°C and 23°C. If temperatures are higher than 23°C, the plant grows very quickly, but its tissues are too soft and they are very vulnerable to fungal diseases. Rice needs a minimum of 15°C for flowering, and the optimum is about 30°C. Pollination is negatively affected if the weather is rainy and temperatures are low during flowering. When the grain is ripening, it is best if the nights are cool, as if they are

| Site preparation | Planting | Ripening | Harvesting |

The paddy is flooded. The soil is prepared with hoes and plows pulled by water buffalo, creating a layer of mud at the bottom. The banks separating the paddies retain the water and act as paths.

The rice plants are germinated in seedbeds and planted by hand in the mud.

When the spikes emerge, the field is drained and the plants finish growing on soil that is dry. The grains are then harvested with the traditional sickle.

13.7.2. Varieties

On the previous page there is a list of varieties of rice that are frequently cultivated in Spain. Each one has its own characteristic growth form, color, leaf form, stem, type of spike and grain length and shape. Refer to more specialized works for more details on each variety's earliness of spike production and ripening, its productive capacity, its resistance to accidents such as flattening and shedding the grain, resistance to diseases and pests, etc.

13.7.3 Cultivation requirements

In sharp contrast to all other cereals, and all other crops in general, rice has to be grown in fields ("paddies") that are flooded most of the time. The site is divided into plots separated by large levees that retain the water. Each levee has some sluices that allow water to flow from the paddies with higher levels to those with lower levels. When the water is

too hot, the plant's respiration increases, and this burns up many of the sugars photosynthesized during the day, which does not favor good ripening of the grains. Fifteen days before spike emergence, when the spike is growing rapidly, the plants are especially sensitive to adverse environmental conditions.

13.7.4. Applying fertilizer

It is normal practise when cultivating rice to apply fertilizer at depth with organic matter in addition to chemical fertilizers. In restored unproductive plots, a dose of 50 to 70 tonnes of organic matter (dung, city waste, etc.) per hectare is correct: when applying a maintenance organic fertilizer, the dose is about 10 t/ha.

On the basis of the extractions of N-P-K from the soil, it is possible to calculate the average nutrient requirements per metric tonne of rice produced. The values are

21 kg N, 11 kg P_2O_5 and 18 kg K_2O. If the expected yield is about 6 tonnes per hectare, then as a guideline we would expect to apply the following quantities of fertilizer:

$$125 \text{ kg N}$$
$$90 \text{ kg } P_2O_5$$
$$90 \text{ kg } K_2O$$

The potassium supplied should be based on potassium sulfate and never on potassium chloride, because rice is grown in deltas and marshes in river mouths where the water already contains a lot of salt (sodium chloride), meaning that applying fertilizer with potassium chloride may cause salination due to excess chloride ions. To meet rice's fertilizer requirements, **ammonium sulfate**, **superphosphate**, **potassium sulfate**, and even **urea** and **ammonia** can be applied.

As in the crops described above, the fertilizer units of potassium and phosphorus should be applied before the crop is planted. They are incorporated when the soil is dry, using a disc harrow or a cultivator. They can also be incorporated when the soil is flooded, **fertilizing the mud**. Suspensions of liquid fertilizers are used more and more in rice cultivation. Some authors consider that nitrogen absorption is greater and better for the plant when soluble fertilizers are applied, as they favor vigorous germination, a crop that is a few days earlier and of course they are easier to apply.

Depending on the stage of cultivation, its requirements of nutrient elements vary. Nitrogen and potassium absorption is greatest during the tiller production period. The requirements of phosphorus, magnesium and calcium all peak at the end of the tillering period. Most nutrients are absorbed before grain formation, and after this nutrient uptake declines to almost nothing. To finish, note that the absorption of nutrients is directly proportional to the root mass of the plant, and this also depends on the circulation of irrigation water to ensure the water is well aerated.

13.7.5. Sowing

Before sowing, the site should be prepared, and this is very important for rice, as it is such an unusual crop. The site has to be thoroughly levelled. The slope should be less than 1‰ and they can even be made totally flat (0 gradient), in which case the paddies can be up to 15 ha. The technical advances in agricultural measuring machinery mean that levelling a site should now be performed using laser instruments. Though lasers are expensive, they can be bought with other farmers as a cooperative. Using lasers means levelling can be done with great precision, with minimum expense and in a short time.

At the end of winter, the previous crop is **raised** with a moldboard plow or a **hook cultivator**. In order to break up the soil thoroughly, it may be necessary to carry out a couple of double crossed passes with the **scarifier**.

The seed can be sown directly (**direct sowing**) or germinated in nurseries and then transplanted to the definitive site. Rice is sown when the water is less than 5 cm deep, and the grain is not covered after sowing. Sowing can be performed by hand, by scattering, using special machinery, or by plane on large sites. Whichever system is used, the seed should be sown when the water is clear and the mud has settled to the bottom. The amount of seed to be used varies from variety to variety and depends on the variety's ability to produce tillers. After tillering, there should be about 250-300 stems/m^2 in thick-stemmed varieties with short panicles. Thin-stemmed varieties with long loose panicles should have 300-350 stems/m^2, which requires sowing seed at about 130-150 kg/ha.

When the seed is sown with specialized sowing machines, these are equipped with tubes that float on top of the water. Where the large size of the area cultivated justifies it, sowing seed by light aircraft has to be done on a calm day and from a height that does not cause the seed to sink into the soil, as if it is buried too deep it does not germinate successfully. Depending on the local climate, rice should be sown between the second fortnight in April and the end of May.

Facet® is a herbicide to control Echinochloa spp. in rice crops. Distributed by BASF, S.A.

Basagran M® is a selective weedkiller that controls dicot weeds and sedges in rice crops. Sold by BASF, S.A.

13.7.6. Irrigation

The flow of water needed for rice cultivation is very large, between 2 and 4 liters per second per hectare, depending on the soil and climate. When irrigating, it is important to ensure the water reaches the correct height in relation to the plant's development. In the first stages of growth after germination, the water level should be high, as this has several advantages; it protects the seedlings from frosts, it prevents growth of weeds, it prevents breakdown of certain herbicides (if they are used), and it also prevents surface movement of the water due to the wind from uprooting the seedlings. In the later stages, the water should be kept at a height that allows the leaves to rise above the surface.

Weeds: Sedge **(Juncus compressus)** *(Courtesy of SCHERING)*

The water should be renewed regularly, so it remains well oxygenated, and the paddy is usually allowed to dry out after the end of tillering until the panicle starts to form. This is done, among other reasons, to reduce the risk of flattening, prepare the plant for the fruiting period, to defend the rice from the many algae that compete with it, and to take advantage of the moment to apply selective contact hormone weedkillers.

Weeds: Crabgrass **(Digitaria sanguinalis)** *(Courtesy of SCHERING)*

13.7.7. Herbicides

One of the problems of growing rice is the presence of algae in the paddies. The algae compete with the rice and may be a serious problem, as they can make it impossible to plow, weed, etc. They are usually eliminated by placing lumps of copper sulfate (or other algicide) in the channels. Where copper sulfate is used, this represents an additional input of sulfate and copper to the crop, meaning that the fertilizer dose should be recalculated.

The weedkiller treatments to be performed in paddies can be divided according to the weed to be controlled. One of the invasive weeds whose presence and competition is most negative is **barnyard grass** (*Echinochloa crus-galli,* Gramineae). The herbicides **molinate** (formulated as a granulate or as an emulsionable liquid), **thiocarbacyl** (emulsionable liquid) or **thiobencarb** (granulate) can be applied before sowing the rice and before the germination of the *Echinochloa,* though the same products can also be used after sowing the rice and before or after germination of the *Echinochloa.* It is also possible to carry out a post-emergence herbicide treatment of the *Echinochloa* before the weed has produced its third or fourth true leaf. This is usually done with formulations based on **propanil**, and it should be applied with a backpack spray or by cropduster light aircraft, as the wheels of heavy machines cause serious damage to the young rice plants. We strongly advise the reader to read the product label carefully to ensure the product is applied at the right dose and in the correct way.

The classic selective hormone weedkillers, such as **2,4-D**, **2,4,5-T**, **MCPA** and **2,4,5-TP**, have long been used to control dicot weeds and some monocots, such as members of the water plantain family (Alismataceae) and sedges (Cyperaceae). As they cause serious damage, their use for this purpose has been banned in Spain by the Ministry of Agriculture. Only a few not very volatile forms of **MCPA** are still authorized. **Benthazone**, a recently produced herbicide, is considered to be highly selective against rice weeds, as it eliminates the same weeds as the hormone weedkillers, and it is hardly volatile meaning it represents little risk to neighboring crops. Compared to classic hormone weedkillers, **benthazone** has the advantage that it can be applied at any stage of the rice plant's growth, unlike the other selective weedkillers, which all have to be applied between 55 and 85 days after sowing, i.e., in the period between the end of tiller production and the phase when the spike is in the sheath. Selective weedkillers for use in rice crops often cause damage to the rice crop. To avoid this, farmers often mix a small amount of N-P-K foliar feed in with the herbicide in order to reduce the negative effect of the herbicide.

Other weeds, such as the perennial water couch grass (*Paspalum distichum*) and broad-leaved pondweed (*Potamogeton natans*) are unfortunately resistant to these herbicides. When weeding was performed manually, these plants were easily controlled, but they have now become very serious weeds of rice crops. They can be eliminated before the rice germinates by treatment with **paraquat**, but after germination weeding must be done by hand. In order to eliminate these weeds from the dividing ridges, apply **paraquat** or even **dalapon** or **glyphosate** in highly localized applications, taking great care the product does not come into contact with the crop.

13.7.8. Accidents, diseases and pests

• Accidents

Irrigation water for rice should not contain more than 1 gram of sodium chloride per liter. When this level is reached or exceeded, the rice starts to suffer from the salinity. Another common accident of rice crops is **flattening** to which rice plants are especially vulnerable: excess nitrogen usually increases rice's risk of flattening. At some stages of rice's growth it may turn yellow due to nitrogen shortage, and this should be remedied by applying a supplement of 100 kg/ha of nitrate, to be applied after draining the paddy. The nitrogen is then absorbed by the plant within 24 hours.

• Diseases

Fungal diseases can usually be prevented by treatment with a commercial mix of **carboxin + thiram** to disinfect the seed. If the grain is

Weeds: Barnyard grass (Echinochloa crus-galli) (Courtesy of SCHERING)

soaked in a solution of these active ingredients, this greatly reduces later problems during the crop's growth from fungi, such as *Rhizoctonia* and *Helminthosporium*. Other previously mentioned preventive measures, such as growing varieties resistant to a specific fungal diseases, avoiding excess nitrogen fertilization, applying a correct mix of the three macro-elements, destroying the remains of previous harvests, etc., all favor the rice's health and help to prevent fungal diseases.

Perhaps the worst disease of rice is the **stem rot** caused by *Pericularia oryzae*, which attacks the rice's panicles and leaves. The worst attacks are those at the basal node of the spike, where a necrotic zone forms above and below the collar, preventing sap flow and tissue growth. It can be treated with Bordeaux mixture, but to be effective this has to be repeated, which is expensive and reduces the final yield. Other fungicides can be used, such as **triclazol** or some mercury products, but the mercury compounds have the disadvantage of being toxic to some varieties of rice.

• Pests

The insects that attack rice include the rice stem borer (*Chilo suppressalis*), and the black *Spodoptera littoralis*, hemipteran bugs such as the rice stinkbug (*Eusarcoris sp.*), some dipteran flies such as grubs attacking seedlings (members of different families) and less frequently, aphids. They have different life cycles and customs but all damage the rice plant. The stem borer bores galleries in the stems, the stinkbug reduces the crop's commercial value, *Spodoptera* defoliates the leaves. Chemical insecticides such as **phenitrothion**, **malathion**, **carbaryl**, **trichlorphon**, etc., are authorized to control these pests. The rice weevil (*Sitophilus oryzae*) is a phytophagous insect that feeds on stored grain. All the points about disinfecting the grain and silo in the section on wheat apply to rice.

13.7.9. Harvesting

Harvesting is carried out with harvesters that are essentially based on the same concepts as those used for other cereals, except that they have to use caterpillar tracks. When the rice starts to form the grain, irrigation is stopped. By this stage the grain should be hard enough to resist chewing between the teeth. Depending on the climate and the time of harvesting, the moisture content of the rice leaving the harvester may vary. Ideally the grain should contain less than 14% moisture. When the grain contains 25-30% moisture it should be dried in the sun or in a drying machine.

The best quality rice with the highest value retains the highest number of whole grains after harvesting. If the grain contains a lot of moisture or if it was harvested very early, drying is essential, but this causes many of the grains to split. After harvesting, the stubble is burnt and the mud is churned up using special wheels.

Rice forms part of the culinary tradition of Europe and the Americas. (Courtesy of GROUPE ROULIER)

13.7.10. Use

Rice, after wheat, is one of humanity's most important staple foodstuffs. This is shown by the world's high production of rice, which is exceeded only by wheat. In the very densely populated areas of Asia, far more rice is grown and eaten than wheat, and it is the main and sometimes the only staple foodstuff for a large population. Brown rice is an excellent foodstuff, but white rice contains little protein and vitamin B_1, and is an unbalanced foodstuff. In Europe and the Americas, rice forms part of the culinary tradition of every country. Its high starch content means it is used as the basis of many products of the food industry, and it is also used in beer production, in order to make a more alcoholic drink.

Rice straw, mixed with other substances, has many applications, including use in the manufacture of some types of high quality paper, use as packing for delicate materials (for tiles, glass, porcelain, etc.) and even as fuel. The residues from dehusking the grain are fed to livestock.

13.8. SUNFLOWERS

The sunflower (*Helianthus anuus, Compositae*) was brought to Europe in the days of Columbus, and for several centuries was only cultivated as an ornamental. In the early 19th century, sunflower oil was extracted for the first time by a Russian farmer using a small press. Since then its cultivation for oil has spread rapidly throughout Europe, especially the Slav countries. It is now cultivated in many countries around the world, such as Spain, France, Argentina, etc.

The sunflower is an annual plant with a main **root**, unlike the fibrous root system of cereals. The root grows down 50-70 cm and is often longer than the stem. The plant then produces secondary roots that colonise a large area. The root mass is greatest when flowering occurs, when the lateral roots reach 40 cm from the main root and to a depth of 30 cm. The size of the roots and the volume they exploit depends largely on the water available in the soil; in dry farming the roots grow deeper in search of moisture. When there is rain during the growing period, **adventitious** roots grow that rapidly occupy the topsoil.

The stem varies from 60 to 220 cm. When the **capitulum** or flower head of the sunflower

Sunflower plantation (Courtesy of EUROLIGO)

Name	Country of origin	Type of variety	Year of registration
Adalid	Spain	HS	1983
Adalid E.	Spain	HS	1985
Albasol	France	HS	1988
Algazul	Spain	H3L	1982
Alhama	Spain	HS	1983
Alhama E.	Spain	HS	1985
Alhama S.	Spain	HS	1985
Alhama-12	Spain	HS	1984
Almansur S.	Spain	HS	1985
Arbung N-123	Spain	HS	1983
Arbung E-353	Spain	HS	1983
Arbung F-253	Spain	HS	1987
Arbung G-133	Spain	HS	1983
Arbung L-233	Spain	HS	1986
Arbung P-113	Spain	HS	1983
Arbung V-183	Spain	HS	1986
Ariflor	France	H3L	1987
Arocha	Spain	H3L	1982
Calera	USA	HS	1988
Cepsola	Spain	HS	1984
Cerflor	France	H3L	1983
Enano	USA	HS	1987
Fantasia-2	USA	HS	1986
Fantasia-3	USA	HS	1986
Fantasia-4	USA	HS	1987
Flo-328	Rumania	HS	1986
Florasol	USA	HS	1981
Florida	USA	HS	1984
Florida-2	USA	H3L	1985
Florida-2000	Spain	H3L	1985
Floril	USA	HS	1988
Girador	Spain	H3L	1985
Girospan-70	USA	HS	1985
Girospan-80	USA	HS	1985
Heliandalus	Spain	HS	1984
Helioespaña F-223	Spain	HS	1988
Helioespaña R-173	Spain	HS	1988
Hysun-33	Austria	HS	1987
HS-891	USA	HS	1983
Lidia	France	HS	1985
Lotus-915	Germany	HS	1988
Luzsol	France	HS	1987
Maribel	USA	HS	1988
Mirasol	USA	HS	1979
Monro-45	Spain	HS	1987
Montenuovo	Spain	HS	1986
Numa-1	Spain	H3L	1987
Numa-6	Spain	H3L	1988
Numa-12	Spain	H3L	1987
Numa-17	Spain	HS	1988
Odil	USA	HS	1988
Orosol	USA	HS	1985
Osuna HS-101 C	Spain	HS	1977
Osuna HS-105 C	Spain	HS	1977
Pemir	Spain	Polin. libre	1981
Peredovick	Spain	Polin. Libre	1974
Riosol	USA	HS	1985
Rustiflor	France	H3L	1983
SH-21	Spain	HS	1984
SH-25	USA	HS	1976
SH-26	USA	H3L	1984
SH-31	Spain	HS	1983
SH-222	Spain	HS	1985
SH-3322	Spain	HS	1987
SH-3422	Spain	HS	1986
SH-3822	Spain	HS	1986
Sirio	USA	HS	1988
Solre	USA	HS	1987
Sunbred-257	USA	HS	1988
Sunbred-285	USA	HS	1988
Sungro-372 A	USA	HS	1979
Sungro-380	USA	HS	1979
Sungro-386	USA	HS	1988
Tesoro	USA	HS	1983
Tesoro-92	Spain	H3L	1987
Texas	USA	HS	1983
Toledo-2	Spain	HS	1984
Toledo-8	Spain	HS	1984
Toledo-9	Spain	HS	1984
Toledo-55	Spain	HS	1984
Tomejil	Spain	HS	1981
Topflor	France	HS	1983
Tornasol	USA	HS	1986
Ulises	Spain	HS	1986
VYP	Spain	H3L	1985
VYP-60	Spain	H3L	1986
VYP-70	Spain	H3L	1988

Current list of the sunflower varieties registered in Spain

is mature, the tip of the stem droops and turns towards the sun, meaning that the capitulum is not held flat. This stem inclination varies from variety to variety, and it protects the seed from scorching and the attacks of seed-eating birds. In general every stem bears a single capitulum, though in recent improved varieties the stem may branch and each secondary stem will bear a **secondary capitulum**.

The sunflower's **leaves** are large, have long petioles and number between 12 and 40. The basal leaves, i.e., the first two or three pairs of leaves, are opposite, while the following leaves are borne alternately. Leaf color varies from dark green to almost yellow, and like the number of leaves, depends on the variety. The flower-head, or capitulum, of the sunflower can be 10-40 cm wide and is an inflorescence bearing many individual flowers on a discoid receptacle. In the daytime, capitula that are not fully ripe follow the track of the sun across the sky and remain horizontal during the night. At the end of ripening, the capitula remain pointing towards the sunrise. The capitula bear two types of flowers, **ray florets** and **disc florets**. The yellow ray florets (the "petals") do not usually produce seed and are arranged in a row around the edge of the capitulum (the "flower"). The disc florets are tubular, hermaphrodite and produce seed. They are arranged on the disc in spirals ("Fibonacci spirals") radiating out from the center.

The sunflower is an outbreeding plant (with cross-fertilization), because the female part of the disc floret matures after the male flower, and there is a genetic incompatibility system that prevents self-fertilization. The flowers are mainly pollinated by insects, and to a lesser extent by the wind. The **seeds** contain some protein and carbohydrates, but a lot of oil. Sunflower seed containing 9% moisture and 2% impurities, may yield 44% of its weight in oil, which corresponds to 50% of the dry weight. The shells can also be used, as they contain 1.6-6% oil, as well as lots of cellulose and other carbohydrates.

year after year on the same site. This question is not very relevant in those areas with a hot climate where the crop's growing period is during relatively dry periods. In many areas of Spain, the crop can be repeated every two years, and serious damage due to the action of these fungi has not yet been detected. Sunflowers are normally successful in Spain when grown in an alternation of wheat-sunflower-wheat-sunflower, though the following seem to have shown they are more productive: wheat-chickpea-wheat-sunflower or wheat-beans-wheat-sunflower. Under irrigation, sunflowers seem to do well when they follow corn.

13.8.3. Cultivation requirements

• **Temperature**. The sunflower is a plant from hot climates, and grows well at high temperatures. For germination the seed requires a temperature of at least 5°C for 24 hours. It can easily withstand temperatures of only 6-8°C as a seedling, but once the fifth leaf has appeared it requires more warmth. The optimum temperatures for growth are 25-30°C, but it grows well (though slowly) at 13-14°C. The heat integer needed for growth is 1,600-2,000°C, taken as the sum of daily temperatures above 5°C. Insufficient heat may prevent ripening by affecting active growth and flowering.

• **Light**. Light is essential for good plant growth. During the growth period the photoperiod (the length of light each day) is important, speeding up development and leaf formation or slowing it down. When the plant has started to differentiate the receptacle of the flower, the photoperiod becomes less important, and other aspects of the light become more important, such as its strength.

• **Moisture**. Together with the temperature and light, water is one of the main factors affecting the sunflower's production. It consumes large quantities of water, both during the active growth period and in the formation and growth of the seeds. In places like Spain, where rainfall in dry farming conditions is insufficient, soil water resources play a vital role in production. Only in cold wet regions with little sunshine are light and temperature the main factors. Several authors state that the rainfall in fall-winter, together with the plant's deep root system and ability to use the water accumulated in the soil, is enough for the critical period of maximum water needs, which is during the formation of the capitulum.

The sunflower's capitulum, an inflorescence with two types of flowers, is borne at the end of the stem. Depending on the variety, the stem may be 60-120 cm tall.

Sunflower capitula (Courtesy of AGRO LORIN)

13.8.1. Varieties

Until the early 1970s, the sunflower varieties grown in Spain were from Russia and Eastern Europe. Since then, official bodies such as the **INIA** (National Agricultural Research Institute) and the different seed companies joined together to breed more productive hybrid varieties of sunflower, with the added advantage that some of them are resistant to mildew, the worst disease affecting the sunflower. The varieties of sunflower can be divided into two large groups, those grown for **oil** and those grown for **human consumption**. On the previous page there is a list of the sunflower varieties registered with the Spanish authorities.

13.8.2. Place in rotations

Sunflowers should be rotated with surface rooting plants like cereals, because the sunflower obtains water at some depth. Many agronomists agree that it is a good idea to wait 4-6 years before sowing sunflowers again. Among other reasons, they argue that the populations of certain pernicious fungi, such as *Sclerotinia sclerotiorum* and especially **mildew**, increase when sunflowers are sown

• **Soil**. Sunflowers prefer sandy-clay soils with a water table near the surface that are rich in organic matter and permeable. Soils with extreme characteristics are not suitable, for example, strongly acid or alkaline soils, stony soils, saline soils, etc.

13.8.4. Applying fertilizer

Many studies have shown that the sunflower has the highest levels of nutrients (as % weight) at the beginning of flower production, a moment that is considered critical, as it coincides with the greatest absorbtion of nutrients. As the plant grows, the concentrations (expressed as percentages of the plant's total weight) decline. During seed ripening nitrogen and phosphorus are translocated from the vegetative parts to the reproductive organs. This explains why the nutrient content of the vegetative parts is lowest after the capitulum has formed.

It is also important to take into account that the sunflower, unlike other crops, has a large root mass that favors uptake of fertilizer applied at depth. Many trials have shown that applying fertilizers which are not very mobile, such as phosphorus and potassium, to the soil surface does not increase seed production.

In order to work out the correct dose of fertilizer, and when to apply it, we must distinguish between the two main cultivation systems, dry farming and irrigation. In Spain, fertilizer is not applied to dry farmed sunflowers, as experience has shown that the factor limiting production in dry farming is water and not fertilizer, and this means that money spent on fertilizers does not lead to increased production.

In irrigation, when water is not the limiting factor, it seems to have been shown that proportionally higher doses of nitrogen increase seed production per hectare, as well as the diameter of the capitulum, while the oil content decreases. Similar results have been obtained for increased doses of phosphorus and potassium. The results of many trials show that on irrigated land between 50 and 100 fertilizer units of N, P and K should be applied per hectare, with half the nitrogen being applied at depth together with the phosphorus and potassium, while the other half of the nitrogen should be applied to the surface three or four weeks after the seedling emerges. The first application of fertilizer should be at depth, or better still near the seed, especially on sites poor in phosphorus and/or potassium.

Boron is the micronutrient most likely to be deficient for sunflower crops, as this crop has a high boron requirement. The first symptom of boron deficiency is when the young leaves turn a reddish-brown color at the base and edges, becoming distorted and curling downwards, and more rigid than normal. As the effects get worse, necrotic zones appear at the edges and tip of the leaf, and the entire leaf may sometimes suffer necrosis, especially near the top of the plant. To confirm a boron deficiency, samples of the plant tissues should be taken to a specialist laboratory. If the results are equal to or less than 37 ppm of boron, you are recommended to apply boron as a foliar feed at a rate of 0.5-1 kg/ha at the start of flower formation. When the crop has been cut, it is worth applying corrective fertilizer at a rate of 1-2 kg of boron per hectare.

13.8.5. Sowing

The site should be **prepared** for sowing. Depending on the region, site preparation should include the normal tasks, such as plowing the remains of the previous crop or fallow into the soil, subsoil plowing, applying weedkiller, plowing to incorporate fertilizer at depth, and to prepare the seedbed; though in order to save labor, a "no-till" system may be chosen. Choosing one or the other system, or the intermediate "minimum tilling" system, should be done on the basis of the soil characteristics, the weed population and the availability of labor to perform the work. For more information on soil preparation systems refer to chapter 5.

Remember that sunflowers root deep and so the soil should be prepared to some depth. You are recommended to plow to a depth of 35-40 cm, as this eliminates many weeds. On sites that harden when the tilth is broken up, the best thing is subsoil plowing, as this prevents water loss.

Sowing can be carried out with the same sowers as cereals, after modification, though they break some of the seeds and are not as precise as the use of specialized sunflower seed sowers, **single grain sowers**. These sowers have to be adjusted to the size of the sunflower seed, by ordering plates with a suitable hole size to ensure the seeds fall one at a time. Sunflower seed germinates at temperatures above 5°C, but you are recommended not to sow until the soil temperature reaches 10°C, as at lower temperatures germination and growth are very slow. In warmer zones, sunflowers are usually sown in March-April, but in colder zones they are sown in May-June.

The **sowing depth** depends on the temperature, moisture and the soil class. In excellent conditions, the sowing depth should be 1-3 cm, but in areas with low rainfall the seed should be buried to a depth of 6-8 cm.

To calculate the sowing **density** or dose, the following data should be taken into account: the planting frame, i.e., the number of plants desired (plants/ha), the number of seeds per kilogram, and the percentage germination of the variety (which must be above 85% by law). The number of seeds in each kg varies depending on its size:

• Extra large: at least 8,700 seeds/kg
• Large: at least 10,000 seeds/kg

Weeds: stinging nettle (Urtica dioica) (Courtesy of SCHERING)

13.8.6. Irrigation

On irrigated ground, sunflowers should be watered when the capitulum has reached the size of an artichoke. About 50-60 liters of water/m^2 should be given. A second watering, which is also very important, of about 60-80 liters/m^2 should be given in the flowering period. If irrigation is by sprinkler, this watering should be delayed as it may disturb pollination. Sunflowers should be watered a third time, for ripening, at the end of flowering. This helps to plump the seed. If the soil is not in good tilth for germination, the crop can be watered before sowing, at a dose of 20-25 l/m^2, and this favors successful germination.

13.8.7. Herbicides

Weeding is frequently carried out using herbicides, though mechanical weeding is also common. When pre-sowing herbicides are not used, the weeds should be removed when the site is prepared. The seed should be sown immediately after the seedbed is prepared, to ensure the sunflowers emerge at the same time as the weeds, and not after them as this would lead to increased competition with the crop. Mechanical weeding is also frequent, using a cultivator after germination, when the seedling has 4-6

- Standard: at least 13,400 seeds/kg
- Medium: at least 19,800 seeds/kg.

Supposing 20% of the plants that germinate are lost as a result of accidents, such as attacks by birds, rodents, soil insect larvae, etc., only 80% of the plants will grow. The formula to determine the sowing density is:

$$\text{Sowing dose} = \frac{\text{Final number of plants}}{\text{Number of seeds/kg} \times 0.85 \times 0.80}$$

As an example, a farmer wanting a final density of 55,000 plants/ha and sowing large seeds should use a dose as follows:

$$\text{Sowing dose} = \frac{55,000}{10,500 \times 0.85 \times 0.80} = 7.7 \text{ kg/ha}$$

In dry farming, the desired density is between 40,000 and 70,000 plants/ha. On irrigated land, the number is of course much higher. It is normally recommended to sow a little denser than calculated in order to thin and to ensure the best final density and distribution of the plants. The sunflower emerges above ground about 15 days after sowing and it grows very slowly to begin with. You must wait until the rows are clearly visible before passing the cultivator.

Weeds: dandelion (Taraxacum officinale) (Courtesy of SCHERING)

true leaves, when the plants are more resistant to breakage and burial.

When dry farming sunflowers, it is common to pass the cultivator once or twice before the plants grow so much that they block its passage. If the pre-sowing herbicide application was successful, then it is better not to weed mechanically, as working the soil increases water loss. Once the crop is growing well, it smothers any weeds.

When using chemical weedkillers, it is important to distinguish between those that are applied in **pre-sowing** and those applied in **pre-emergence**. The herbicides to be used before sowing the crop should be incorporated into the soil at a depth of 6-8 cm with two passes of the cultivator. Products like **trifluoraline**, **ethafluoraline** and **dinitramine** are widely used. Products to be applied after sowing, but before the emergence of the crop and weeds, should be applied within two days of sowing. Some, such as **alachloro** are highly selective against grasses while having little effect on the sunflower. Other pre-emergence weedkillers include **terbutryn**, **linuron**, **fluorchloridone** and some commercially prepared mixtures.

13.8.8. Accidents, diseases and pests

• Accidents

Accidents due to weather and climatic conditions are the most important agents affecting the sunflower's growth. If there is heavy rain at the time of pollination, they wash the pollen off the styles and also prevent bees from flying, with negative effects on fertilization. Sunlight can dry out the pollen, making it less effective at pollination.

Necrosis of the bracts of the capitulum occurs where temperatures are very high, and is thus considered to be heat damage caused by the sun. Geneticists are trying to breed varieties of sunflower with capitula that droop after ripening, as this would not leave them exposed to the sunlight.

• Diseases

The viruses that affect the sunflower include **sunflower mosaic virus**. It is transmitted by infected seeds or by some insect vectors. Affected plants show a mosaic of colorless patches.

Fungal diseases include a pernicious mildew (*Plasmopara helianthi*) that attacks all parts of the plant at any stage of growth, causing **dwarfism**. The leaves of the affected plants show a mosaic of chlorotic patches which contrast with the normal green of the healthy tissue. If the infection occurs during germination, the plants eventually die without forming a capitulum. If infection occurs during active growth, development halts and the capitula produced are small with only a few small seeds. There are no chemical products to combat the infection, so the only methods to combat it are preventive. Possible preventive methods include use of seeds that are certified disease free, not sowing sunflowers again on affected plots for six or seven years, cultivating resistant varieties, etc.

Another very harmful disease is **grey mold** of sunflower (*Sclerotinia sclerotiorum*). This disease thrives in conditions of heavy rainfall and low temperatures, and can attack sunflowers at any stage of their development, though they are most sensitive when they are young or during capitulum formation. The tissues attacked by the fungus go soft, turn brown and rot, causing the seedlings to wither and die. In relatively humid conditions, the affected parts are covered with a fine white layer, the fungal hyphae. There are no resistant varieties on the market, and the only way to combat the disease is not to repeat sunflowers on the same site for six or seven years if severe attacks of the fungus have been detected.

Attacks by other fungal have also been reported, but they produce less damage. They include **grey mold** (*Botrytis cinerea*), **sunflower rust** (*Puccinia helianthi*), *Septoria helianthi*, *Alternaria helianthi* and *Verticillium dahlias*.

• Pests

Sunflowers are attacked by two types of pests, those living in the soil and those that attack the aboveground parts. The **soil insects** should be treated at the same time as seed is sown, using microgranulate distributors attached to the sower. Granulated insecticides, such as **chlorpyriphos**, **carbofuran**, **phonophos**, **bendiocarb**, etc., are very effective against cutworms, wireworms and millipedes and centipedes. On irrigated sites, there are usually late attacks of wireworms (*Agriotis segetum*), and this should be treated with synthetic pyrethrins, and applied in the evening because *Agriotis* is a nocturnal insect.

Perhaps the most important **insect** to attack the aboveground parts is the **sunflower moth** (*Homoesoma nebulella*), whose larvae are 3-16 mm long and attack the plant during or after flowering. They eat the flowers, the pollen and often leave bitemarks on the seeds or perforate them.

There is also an unusual flowering plant that attacks the sunflower, the broomrape *Orobanche cumana*. It has little chlorophyll and like other broomrapes is a root parasite of other plants. It causes a general decrease in growth and the sunflower can not form the capitulum properly. The capitulum is small and the seeds are empty. The only method to combat broomrape is to grow resistant varieties.

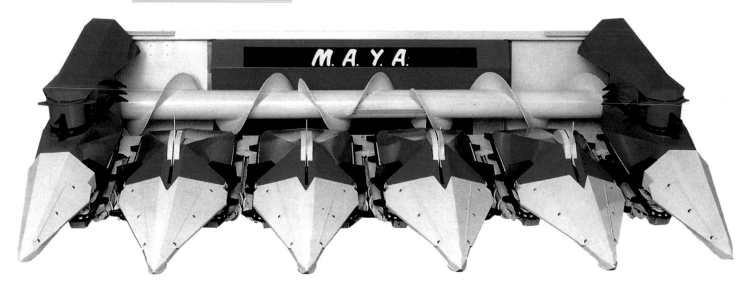

Detail of the cutting bar of a harvester for both sunflowers and corn. Manufactured by M.A.Y.A., S.A.

13.8.9. Harvesting

When the seed ceases to increase in dry weight and oil content it has reached **physiological maturity**. The seed has a moisture content of about 30%, but is still not ready for harvesting. The best time to harvest in order to ensure good storage is when the grain contains only about 9% moisture. Harvesting should be soon after reaching this level, as there may be losses because the seeds are shed from the capitulum.

Sunflowers can be harvested with the harvesters used for cereals, after adapting them. Bats are fitted to the cutting bars, and the drum wheel is generally removed. The velocity of the grain removing cylinder should not exceed 500 rpm. The distance between the cylinder and the concave plate should be 25-30 mm at the entrance and between 12 and 18 mm at the outlet. The speed of the harvester should be lower than when harvesting cereals. When sunflowers have to be harvested in wet weather because there is no other alternative, the plants can be treated by dusting with magnesium chlorate at a dose of 20-30 kg/ha.

13.8.10. Use

About 70% of the world supply of oils and fats comes from plants. Animal oils and fats account for about 20% and industrial and marine oils for about 10%. Sunflower is the second most important oil crop after soya. The most important flour as a source of protein is soy flour, followed a long way behind by cotton, rapeseed and sunflower.

Sunflower oil is one of the best vegetable oils, because of its energy content and digestibility, and is the most butter-like of the oils. Sunflower oil does not, however, increase cholesterol and phospholipids in the blood in the way butter does, and is recommended by many nutritionists as a preventive measure against arteriosclerosis and cardiovascular disease.

Sunflowers are an important source of protein in animal feeds, and is also grown as fodder, which is harvested when flowering starts and is then stored.

Lesser but still important uses include consumption of toasted sunflower seeds and honey production. The whole seed of the varieties is toasted and salted, and eaten. Honey production is also important, as bees can produce 20-30 kg of high value honey from a hectare of sunflowers.

13.9. RAPE

The world's main producers of rape are India and China, where it is grown on about 3,000,000 ha. It is also grown on large areas in other countries, such as Canada, France, Sweden, Germany, Slovakia and England. In Spain it is not grown very widely. In the late 1980s, rape was grown on about 6,785 ha in Spain, 95% of them dry farmed, and the remainder irrigated.

Rape (*Brassica napus* var. *oleifera,* Cruciferae) is also known as colza. It has a tap root that grows deep into the soil, and the secondary roots ramify well and colonize the soil efficiently, especially if the tap root encounters an obstruction as it grows down. The stem can easily reach 1.4-1.8 m. The lower leaves have petioles, but the upper leaves are lanceolate and entire.

The flowers are small and yellow. They have four sepals arranged in a cross, six stamens and the pistil. The flowers are borne in bunches at the end of the stem. The fruit is a pod (a *siliqua*) 5-6 cm long containing 20-25 seeds. The length of the pods and number of seeds vary from variety to variety. The seeds are spherical, 2-2.5 mm in diameter, and when ripe they are dark brown, reddish or black.

A kilo of seeds contains 300,000-400,000 seeds.

13.9.1. Life cycle

Depending on the climate, rape can be sown in fall or in spring, depending on whether it is a winter variety or a summer variety. After 10-20 days, germination occurs and the two cotyledons (seed leaves) appear. The first 6-8 leaves all arise from a basal rosette at ground level, before the plant produces a stem. In this phase it appears to halt growth, but the roots are growing actively in the subsoil. The plant is ready for the next phases, **stem formation**, **flowering** and **fruiting**. Three or four months after sowing, when temperatures increase with the arrival of spring, the plant produces a stem, and this is when the plant shows its greatest vegetative growth. Flowering starts about 20 days after stem formation, starting at the top and followed by the lower flowers, and lasts for thirty or forty days.

Rape plantation (Courtesy of EUROLIGO)

Rape crop (Courtesy of AGRO LORIN)

Fully developed rape (Courtesy of HARDI INTERNATIONAL A/S)

Name	Country of origin	Cycle	Year of registration
Anouk	Spain	Long	1986
Arenal	Germany	Long	1984
Brutor	France	Short	1976
Cresor	France	Short	1978
Doral	Germany	Long	1985
Duplo	Germany	Short	1981
Gulliver	Sweden	Short	1978
Hércules	Sweden	Long	1985
Husky	Spain	Long	1986
Kabel	Spain	Short	1986
Karat	Sweden	Short	1986
Ledos	Germany	Long	1981
Liberia	Germany	Short	1985
Librador	Germany	Long	1983
Liquita	Germany	Long	1984
Lirama	Germany	Long	1983
Liraspa	Alemania	Short	1983
Malpa	Spain	Medium	1983
Navafría	Spain	Long	1986
Niklas	Sweden	Short	1985
Quinta	Germany	Medium	1978
Rafal	France	Long	1981
Vigor	Germany	Long	1984

List of the varieties of rape registered in Spain

13.9.2. Varieties

Rape was once cultivated in the cereal zones
of the northern hemisphere. New varieties
have been developed that grow well in both
hemispheres. Rape is divided into two groups,
long cycle and **short cycle** varieties. The long
cycle or **winter** varieties are in the field for
nine of ten months, and **halt growth** in winter.
The short cycle, or **spring** varieties, develop
rapidly after germination and fruits within five
or six months. On the previous page there is a
list of the varieties currently authorized in
Spain.

13.9.3. Place in rotations

Rape usually precedes cereals, which have
surface roots. Burying the remains of the rape
stems and roots greatly improves the soil's
organic matter and nutrient content. Rape is
usually given intensive fertilizer applications,
so the reserves of macronutrients are not
depleted for the following crop.

13.9.4. Cultivation requirements

• **Temperature**. Rape used to be thought of as
a crop plant of cold northern climates. The
new varieties now available can adapt to all
the different climates. From germination to
the rosette stage, the temperatures should not
fall below -2°C or -3°C, but in the rosette

stage it can tolerate temperatures as low as
–15°C. Low temperatures from rosette
formation until stem formation favor root
growth and are not a problem. During the
flowering period, high temperatures can be a
problem, as they may cause the plant to
shorten its cycle and the seed to fail to
develop correctly.

• **Water**. Rape is a plant of dry conditions,
especially in its early stages. It tolerates dry winters
well, and heavy rain or waterlogged sites are
detrimental. In spring, the abundant rains favor
flowering and seed set. 400 mm of rainfall well
distributed over the year is enough for its
development.

• **Soil**. It grows well on deep soils that are
slightly acid and well drained. It does not
tolerate alkalinity above pH 7.7. It can
tolerate some degree of salinity.

13.9.5. Applying fertilizer

In Spain, rape is the alternative to sunflowers on
infertile sites that are irrigated, and so the maximum
yield does not usually exceed 1,500 kg/ha. Each
tonne of crop extracts 66 kg N, 37 kg P_2O_5 and 36
kg K_2O. As recommended for other crops, all the
fertilizer units of phosphorus and potassium, 50-60
kg of both, should be applied at depth, together with
part of the 50 kg/ha units of nitrogen. The rest of the
nitrogen fertilizer, to which rape responds very well,
should be applied on the surface when the rosette
phase is finishing and the stem is beginning to grow,

at a far higher dose than extraction (about 70-90 additional fertilizer units). Rape requires a lot of sulfur, and so the potassium fertilizer included in the fertilizer at depth should be in the form of potassium sulfate, not potassium chloride.

13.9.6. Sowing

As rape has a tap root that grows deep, the soil should be prepared to a greater depth than for other crops. The seed is small, and so special care should be taken that the soil is in fine tilth with no clods, and this is done by passing the disc harrow a couple of times to get rid of the clods. The roller should also be passed so the soil is flat.

Depending on the local climate, the sowing date should aim for the plant to have reached the rosette stage (six to eight true leaves) before the harshest cold of winter. It is worth waiting till the autumn rains to get rid of the weeds, and then sow the seed, though this should be done before November 10.
Sowing can be carried out with the same devices as for cereals, sealing alternate openings to leave a space between lines of 34-36 cm. Usually about 6-8 kg of seed is sown per hectare. The seed should be sown no deeper than 1 cm, as otherwise germination may not be successful.

13.9.7. Herbicides

The presence of weeds causes a decrease in production and in the quality of the oil produced. If there are many weeds, this also makes harvesting more troublesome. This crop can not be weeded mechanically because the rows are too close together for the cultivator to pass. Weeding should thus be done using herbicides. The normal pre-sowing herbicides are **trifluoraline**, **napronamide** and commercial mixtures, which are incorporated into the soil. In pre-emergence, **trialate** is normally used and should also be incorporated into the soil. When the crop reaches a height of 20 cm, post-emergence, **aloxidin** can be applied.

13.9.8. Accidents, diseases and pests

• Diseases
Different species of *Alternaria,* commonly known as rape **black spot** cause small necrotic patches on the leaves that are surrounded by a lighter halo. These patches then increase in size and form circular necrotic patches 5-12 mm in diameter. The development of this disease, and of other fungal diseases like rape **black stem** (*Phoma lingam*), requires a relatively high temperature and humidity. The methods of combatting are usually preventive, such as destroying the remains of previous crops, eliminating wild

members of the cress family, increasing the distance between the rows on the plot and growing resistant varieties.

• Pests
Several pests attack rape. The rape weevils (*Ceutorrhynchus napi, C. picitarsis* and *C. assimilis*) respectively attack the stem, terminal bud and the pods. They differ in their characteristics, but they all cause similar damage to the plant, bites and stem deformation, destruction of the terminal bud and the pods, and leading to late and irregular flowering. These animals are often not treated, as the level of attack does not cause a significant reduction in the harvest. If they have to be treated, insecticides like **endosulphan**, **phosalone** and **lambdacyhalotrin** can be used.

The other insects that attack rape can be controlled with the same insecticides. These include the gallwasp (*Dasyneura brassicae*), cress tipworm (*Meligethes sp.*), rape flea beetle (*Psyllodes chrysocephala*) and cabbage flea beetle (*Phylotreta sp.*).

13.9.9. Harvesting

The long cycle varieties are harvested in May-June in warmer zones, and in June-July in colder ones. The short cycle varieties are harvested between late August and early September. The best time to harvest is when the seeds in the pods halfway up the stem turn from red to a darkish or even black color. Harvesting should not be left after this, as there is the risk the pods will split and shed their seeds.

The WR 322® model reaper conditioner is manufactured by COMERCIAL VICON, S.A. It is ideal for reaping and conditioning of crops grown for fodder, such as rape, flax and other species. The fodder can be harvested at a given height, and so a height can be chosen when the plant contains little erucic acid or glucosinolate.

13.9.10. Use

Rapeseed, the seed of the rape, contains a lot of oil. In the countries where it is cultivated, it is used for the same purposes as the other oils from oilseeds. It is used in production of margarine, the prepared foods industry and for human consumption. After extraction of the oil, the cake left represents about 60% of the seed weight. This oil is also used in metallurgy to temper metals.

A flour is extracted from the rapeseed that is used for animal feeds for ruminants, hogs and fowl, and it is perfectly nutritious. In countries where a lot of rapeseed is grown, it is given as fodder to livestock, and is usually stored.

Rapeseed oil and flour contain erucic acid and glucosinolate, respectively. These two substances can be poisonous. Laboratory experiments with rats have shown that erucic acid can cause deformation of the adipose tissue of the muscles of the heart. Glucosinolate at high doses can cause damage to the thyroids. Genetic selection has led to varieties free of these substances. When a variety contains no erucic acid it is said to be "zero", and if it is free of glucosinolate too it is said to be "zero-zero".

13.10. SOYA

Soya or soybean (*Glycine max*) is an annual plant belonging to the pea family (Leguminosae). It reaches a height of 40 to 100 cm and has trifoliate leaves, with white or purple butterfly-like flowers, and its fruit, a legume, contains three or four seeds. The seed is normally spherical, the size of a pea, and yellow. The different parts of the plant, the leaves, stem and pods are all fuzzy (they have downy hairs).

13.10.1. Varieties

More than 50% of the world's soya crop is produced in the United States, where soya has been classified into ten different groups on the basis of when they ripen and the length of their growing cycle. These groups are numbered from 00 to VIII. The flowering of soya is controlled by the changes in day length, the photoperiod. Therefore, in addition to the temperature, moisture and soil conditions, this point should also be borne in mind when deciding when to sow. There are almost three thousand varieties of soya, with growing periods ranging from ninety days to almost two hundred, and with different day length requirements. Accompanying this text is a list of the soya varieties registered in Spain.

13.10.2. Place in rotations

Soy is a legume whose roots support colonies of *Rhizobium* bacteria that fix atmospheric nitrogen, and so this plant improves the soil. For this reason, soya can be sown as a second harvest after a winter wheat. The nitrogen provided by this legume is excellent for the germination and later growth of wheat sown in fall, though many farmers plant winter wheat as the first crop after fallow.

13.10.3. Cultivation requirements

• **Temperature**. Soya's growth slows down when the temperature is below 10°C, and halts below 4°C. It can resist temperatures of –2°C to –4°C without dying. When the temperature exceeds 38°C growth halts. The best temperatures are 15-18°C for sowing and 25°C for flowering.

• **Light**. Soya is sensitive to the changes in day length (it is a short-day plant, meaning that for flowering to occur the day length must have shortened to the length the variety requires).

• **Water**. Over its growing period, soya needs at least 300 mm of water, which can be in the form of irrigation water on irrigated ground, or as rainfall in the moist temperate zones, where rainfall is normally sufficient. As a general rule, soya can be grown on cool soils that are also suitable for growing corn and Windsor beans.

Soya plantation (Courtesy of EUROLIGO)

Variety	Group	Year of registration
Akashi	II	1988
Amsoy	II	1974
Azzurra	I	1987
Beeson	II	1974
Calland	IV	1974
Canton	I	1988
Clark 63	IV	1974
Chipewa 64	I	1974
Furia	I	1988
Futura	II	1988
Gallarda	II	1986
Katai	III	1988
Kawetanya	III	1984
Kawevera	III	1984
Kingsoy	II	1985
Panter	IV	1987
Soimira	0	1987
Soinova	II	1987
Turchina	II	1987
Williams	III	1977

The pod of the soya bean (Glycine max, Leguminosae). (Photo courtesy of ICI SEEDS)

Current list of varieties of soya bean registered in Spain. The different varieties are classified into groups on the basis of their growing cycle.

Weeds: black nightshade (Solanum nigrum) (Courtesy of SCHERING)

• **Soil**. Soya is not very demanding in terms of soils with high nutrient levels, and it is often considered to be an alternative for unfertile soil that is unsuitable for other crops. It grows best in neutral or slightly acid soils. Good yields are obtained on soils with a pH between 6 and neutral (pH 7). It is very sensitive to waterlogging, and so it cannot be recommended for clay soils with a tendency to waterlog. If the site is relatively flat, it should be levelled, so the water does not form pools and cause waterlogging. However, soya requires a lot of water, and on sandy sites it should be watered frequently. It shows some resistance to salinity.

13.10.4. Applying fertilizer

Many authors recommend supplying a **minimum of nitrogen** to soya crops, because the *Rhizobium* bacteria supply all the nitrogen the plant needs, with the added advantage that it is made available gradually. Even so, we recommend applying some nitrogen, 25-30 fertilizer units/ha, before sowing, and if some patches show yellowing, apply a further 50-60 fertilizer units/ha. This supplement should be applied to the surface and is best applied in the two weeks before flowering, as this is when the plant's needs are greatest. If it is applied later, it may reduce the harvest.

Phosphorus should be applied at depth before sowing at a rate of about 100 fertilizer units/ha. But some points should be taken into account. If the remains of the previous harvest are incorporated into the soil, bear in mind that soya contains a lot of phosphorus, and this represents a complementary input of the nutrient to the soil. Nitrogen-fixing bacteria need a lot of phosphorus in the soil to grow well, and so this factor should be checked continuously, and complementary inputs made if necessary. In serious cases of phosphorus deficiency, a supplement near the

roots may be necessary, of about 20-30 fertilizer units of phosphorus, just when the first roots are forming.

Soya does not show any period of critical need for potassium, though potassium absorption is greatest during the vegetative growth period, and decreases when the seeds start to form and finishing 15-20 days before they mature. If laboratory analysis shows potassium deficiency in the soil, then about 100 fertilizer units of potassium, the same as for phosphorus, can be applied.

Soya requires the other fertilizer elements and micro-elements. Soya requires a lot of sulfur (to produce a sulfur-containing amino acid), and so if ternary fertilizers are applied they should contain potassium sulfate rather than potassium chloride. Magnesium is not particularly vital for soya, but it is important to know the interrelation between magnesium and nitrogen, or if there is deficiency of magnesium due to excess potassium in the soil (ion antagonism).

Micro-nutrients such as molybdenum, iron, zinc, etc., may also cause deficiencies. These have to be measured in the laboratory and corrected with special fertilizers. For more information on possible deficiencies and how to correct them, refer to the first section of this work, where nutrient deficiencies, their symptoms and possible corrections are discussed in detail.

13.10.5. Sowing

The site should be prepared before sowing. If it is grown as a second harvest, the seed should be sown as rapidly as possible after harvesting the winter cereal, because a delay of just a single day may represent a loss of production of 15 kg//ha or more. If soya is grown as a second crop, the stubble of the preceding crop should be burnt or buried, and the soil should then be watered to create a good tilth. The disc harrow should be passed to break up soil clods, followed by the cultivator, and then the seed is sown. If soya is the first crop, the previous crop

Weeds:
Common chickweed
(Stellaria media)
(Courtesy of
SCHERING)

should be raised before these tasks, in order to ensure the roots can grow well.

Before sowing it is normal to inoculate the seed with the soya's own species of nitrogen-fixing bacteria, *Rhizobium japonicum*, as this means less nitrogen is necessary. This is a tricky operation, and you should consult the technical reports from the commercial suppliers.

Sowing time is conditioned by the fact that soya is a "short-day" plant and flowering does not occur until the day length has decreased to the length the variety in question requires. As sowing soya as a second crop is always done after the short days, it is very important to sow as quickly as possible, because any delay leads to smaller harvests. Soya should never be sown after July 10, and in colder areas it should always be sown before this. The minimum temperature for the seeds to germinate is 9-10°C.

The best **sowing depth** is 2-4 cm, though in very loose soils, where the seed might desiccate before emerging above ground, it can be sown as deep as 7 cm, but never deeper. The sowing density, using machinery and in rows 50-60 cm apart, is usually 45-50 plants/m². The amount of seed required is normally 140-160 kg/ha.

13.10.6. Irrigation

Soy should not be watered abundantly, in order to avoid waterlogging. In addition to the watering before sowing, mentioned above, soya should be irrigated every ten days, a minimum of seven times more. Once irrigation has been started it should not be stopped, as this causes a serious decrease in production. Many authors consider that soya's water requirements increase after flowering, meaning that the plant should be regularly irrigated from then till harvesting.

13.10.7. Herbicides

As soy does not smother weeds, herbicide use is almost essential. The most widely used are **trifluoraline**, **ethalfluoraline**, **alachloro** and **linuron**. They are applied pre-sowing and can be incorporated into the soil with the disc harrow. Each one has its own characteristics, volatility, degradation by organic matter, residues, etc., and you should consult technical documentation from the supplier before using any of them. It is also possible to apply post-sowing herbicides, using a commercial mixture of **alachloro** and **linuron**. These are herbicides that can be applied dissolved in irrigation water applied by sprinkler, immediately after sowing. If herbicides have not given the desired results, you can weed between the rows manually while the plants are still young and it is possible to pass between them.

13.10.8. Accidents, diseases and pests

• Diseases

Fungal diseases affecting soya include the soil fungi *Pythium, Rhizoctonia* and *Fusarium,* which attack and destroy the seedling's roots. Soya seedlings are usually especially sensitive, and this is why just after the seedlings emerge, there are usually yellowish patches in the field caused by these fungal diseases. One good way to combat them is to disinfect the seed before sowing with the same products (**tiram, captan**, etc.) mentioned above, though this disinfection has the disadvantage of killing the nitrogen-fixing bacteria soya depends on. The experts can not agree on the best answer to this problem, but it seems the best answer is to inoculate the seed with a higher dose of *Rhizobium* in order to counteract the bactericidal effect of the fungicides. Soya is resistant to one soil fungus, *Verticillium,* which seriously attacks cotton, and so soya is considered a good alternative to cotton.

• Pests

Soya can also be attacked by the red spider mite (*Tetranychus*). A semi-preventive treatment is to dust with sulfur, and this method has the advantage of not harming the beneficial insects and providing a further source of sulfur for the crop. When mite infestation is serious, acaricides can be used, for example **tetradiphon, dicophol,** or a commercial mixture.

The soil insects that attack the roots of the soya include the **black screwworm** (*Spodoptera littoralis*), which has nocturnal larvae that eat and devastate its roots. There are many insecticides available to treat them, such as **phoxim, methomyl, acephate, monocrotophos, cypermethrin**, etc. Other types of insects can attack soya. These include lepidopterans like *Laphygma exigua* and *Heliothis,* and homopterans like the greenhouse whitefly (*Trialeurodes vaporariorium*), aphids (*Aphis*) and plant bugs. They can all be controlled with organophosphate insecticides like **diazinon, trichlorphon, chlorpyriphos, lindane, tetrachlorvinphos**, etc.

13.10.9. Harvesting

When the soya's pod matures, it turns from green to dark. When ripening starts, the leaves start to turn yellow and are shed, leaving only the pods on the plant. Soya can be harvested with a harvester, but if the correct precautions are not taken, there may be significant losses due to seed breakage. Harvesting should not be left too late, as the pods would split and shed their seeds,

causing losses.

The soya bean's moisture content at the time of harvesting is usually about 15%, bearing in mind that after passing through the harvester, its moisture content will be reduced to about 13%. This is however still considered too high, and the beans should be cleaned and dried. This removes the plant remains attached to the beans; the beans are dried in the sun or using special machinery, ensuring that the hot air used is no hotter than 60°C.

13.10.10. Use

Work in the United States to select improved soya beans has made it the most important oil producing plant in the world. The soya bean contains about 16% oil, which is used in human foodstuffs or for industrial uses. In terms of human nutrition, soya flour can be made into meal with a high content of digestible proteins (44% to 50%). Another use for milled soya is production of lecithin, which is used in the production of margarine, chocolate, sweets, etc.

Soya flour contains all the essential amino acids, and is one of the main components of industrial feeds for livestock. In Spain, soya flour is used in 20% of the animal feeds manufactured. It is also used as a green fodder crop, adding a rich source of protein to the livestock's diet.

13.11. COTTON

Cotton (*Gossypium spp.*) is a member of the mallow family (Malvaceae). The genus contains about 45 species, all of them subtropical in origin, and includes annuals, biennials and perennials, and herbaceous, shrub and tree species. Three species are widely cultivated for their cotton.

Cotton has a taproot and secondary roots that grow roughly horizontally. They both colonize the soil relatively deeply, and so cotton is considered to be deeprooting. In deep soils with good drainage, the roots can reach a depth of two meters, but in shallow badly drained soils, they only reach a depth of 50 cm. Cotton has an upright main stem, with secondary stems branching from the main stem, whose growth varies from species to species. The leaves are palmate and the flowers have a calycle, with five white or yellow petals, four bracts, many stamens and a single pistil. It is a self-fertilizing plant, though some flowers open before fertilization and produce hybrid seeds. The seedpods, or bolls, are ovoid capsules, elongated or spherical, green with red patches, and there are six to ten

(percentage harvest in first picking), from not very early to very early. In cold zones it is worth growing early varieties, and to harvest as much of the cotton as possible in the first picking, before the first rains fall.

Cotton cultivation and manufacturing involve both the farmer and then industrial processing in cotton gins, and so cotton varieties are classified on the basis of a number of characteristics, some of greater importance to farmers and others to industry. The farmer is most interested in a variety suitable for mechanical harvesting and disease resistance (especially to the *Verticillium* fungus). Cotton gins are interested in the variety's yield of fiber, its quality and uniformity.
Cotton 's quality is measured on the basis of its **appearance** and **length**. The **appearance** refers to physical characters like color, shine, impurities and preparation (fiber texture). The fiber **length** is rated on a four-point scale, from long to short.

Detail of cotton capsule, the boll, (Courtesy of BASF, S.A.)

on each flower. The seed hairs on the outer skin of the seeds are the fiber we call cotton.

13.11.1. Varieties

Current list of varieties authorized in Spain

The most widely cultivated varieties of cotton belong to three species. **Egyptian cotton** or long fiber cotton (*Gossypium barbadense*); **American cotton** or medium fiber cotton (*G. hirsutum*); and **Indian cotton** or short fiber cotton (*G. herbaceum*). The cotton fiber is calibrated on the basis of its length and diameter. The fibers of long fiber cotton are 32-34 mm long and have a diameter of 15 μ. The medium fibers of *G. hirsutum* are 20-25 μ in diameter and 24-34 mm long. The short fibers of *G. herbaceum* have a diameter of about 25 μ and are 23 mm long. On this page there are two tables, the first a list of the varieties cultivated in Spain, and the second table classifies these varieties according to their earliness

Cotton varieties classified by earliness (percentage total harvest in first picking) (Taken from Guerrero)

Variety name	Registration date
Acala SJ-1	7-05-74
Acala SJ-2	25-03-82
Alba	1-06-88
Coker 210 (synonym of Carolina Queen)	7-05-74
Coker 208	31-07-84
Coker 304	25-03-82
Coker 310	7-05-74
Coker 312	9-02-80
Coker 315	1-06-88
Jerez	3-07-80
McNair 220	31-07-84
Palma 76	31-07-84
Promese	7-05-74
Stroman 254	25-03-82
Tobladilla 100	3-07-80

13.11.2. Place in rotations

The cotton seedling is extremely delicate, and successful germination is the most tricky aspect of cotton cultivation. Cotton germinates much better in wheat stubble or after corn than after sugarbeets. Sugarbeet residues are a medium for the growth for fungi that rot the seed or the seedlings after germination.
The same is true for irrigated potatoes, and so it is not a good idea for cotton to follow potatoes, probably for the same reason as given for sugarbeets. Cotton can be grown on the same site for several years in a row, as long as there are no problems with *Verticillium* fungus.

First group Not very early	Second group Relatively early	Third group Early	Fourth group Very early
Acala GC-510	Coker 304	Tabladilla 100	Jaén
Acala SJ-2	Coker 310	Tabladilla 13	
Deltapine Acala 90	Coker 208	Tabladilla 16	
Coker 312	McNair 220	Promese	
Coker 315	Jerez		
Vered	Palma 76		
Stoneville 506			

13.11.3. Cultivation requirements

• **Temperature**. Cotton originated in sub-tropical climates and requires high temperatures. It has a heat integer of 3,500°C for early varieties, and 4,000°C for long cycle varieties. The seed does not germinate below 14°C, but if it germinates in warmth and the temperature then falls, the seedling dies. For the fruits to ripen and open, the plant requires a lot of light and warmth.

• **Water**. The soil should be in good tilth for germination, as if the soil is not moist enough, the seed fails to germinate, but if the soil is too wet, it rots. If harvesting is delayed and the plant is affected by fall rains, pollination and fruit set are negatively affected and the crop is ruined; it still requires water during this period, but in the soil not the air. In the 30 days before flowering, cotton is very sensitive to dry conditions.

• **Soil.** Cotton roots require deep permeable soil for their respiration. It is prejudiced by acidic soils, and requires a neutral or alkaline soil, though it does not tolerate excess calcium. Vegetative growth is very good on clay and silty soils, if they are fertile, but the cycle is prolonged, and many of the bolls do not fully ripen, and flowering takes place over a long time.

13.11.4. Applying fertilizer

On the basis of the quantities of N-P-K removed from the soil by the cotton, it is possible to calculate the average need for nutrients for each tonne of crop produced. The numerical values are 60 kg N, 25 kg P_2O_5, and 48 kg K_2O. If the expected production on irrigated land is roughly 3-4 tonnes/ha, the following figures are guidelines

for how much fertilizer should be applied:

$$120\text{-}180 \text{ kg N}$$
$$120 \text{ kg } P_2O_5$$
$$80 \text{ kg } K_2O$$

As we have explained for other crops, the fertilizer units of nitrogen should be divided into three applications. The first should be 70-80 fertilizer units at depth, 50-60 units after thinning and another 50-60 units in the second half of June. The fertilizer units of phosphorus and potassium should be applied at depth before sowing. Soils do not normally show shortages of potassium, but if they do, the dose of potassium should be increased to 120 units.

13.11.5. Sowing

Before sowing, the site should be prepared. Because cotton has a tap root, the soil should be plowed to a depth of 35-50 cm, using a moldboard plow or better still a subsoil plow. This should be followed by the scarifier, and the soil should then be levelled with the roller. It is normal to disinfect the site with **aldicarb** before sowing. In general the disinfectant product is applied with the sower at a dose of 5-10 kg active ingredient/ha. This treatment protects the young seedlings from aphids and other insects. This product is an insecticide-nematicide that has a systemic effect within the plant and acts by contact within the soil.
We have already said that **sowing** the cotton plant is a very delicate operation, and that successful germination depends on it. For emergence, the soil must be in good tilth, and this is achieved by watering with sprinklers. The seed should be watered before sowing, not after, otherwise a surface crust forms that makes the seedling's emergence harder: the soil temperature falls, and this delays germination and favors fungal diseases such as *Rhizoctonia*.

The sowing date also raises important problems. If sowing is late, when the temperature is high and there is no risk of late frosts, germination is much better, but because there are less days between germination and the fall rains, the correct ripening of the bolls is harder. In general, cotton is sown in early April, when the ambient temperature has reached 13°C. It is important to note that good germination depends on the soil temperature (and thus on the soil texture) and not on the air temperature. If the soil is sandy, it will warm up quicker and can be sown earlier, while on clay soils the situation is the opposite.

Sowing is normally carried out using semi-precision or precision seeders. The rows are normally sown 70 cm apart if they are going to be harvested by hand, and 96-100 cm apart if they are to be harvested mechanically. About 35 kg of seed is used per hectare, giving a density of plants of 125,000-150,000/ha. The seed should be sown about 3 cm deep, and never more than 7 or 8 cm deep. The cotton gins supply disinfected seed, meaning the farmer does not have to disinfect it.

13.11.6. Irrigation

In the first forty days after germination, cotton needs little water, only 2.5 l/m^2/day. After this, water needs increase to 6 mm/m^2/day after 40 days. In the third period, about 60 days after emergence, water needs peak at 10 l/m^2/day. In the last 40-50 days of the growing cycle, water needs decline from 6 mm to 2.5, until ripening has finished. The critical period for water needs is the three weeks after the opening of the first flowers.

If the first irrigation is delayed as long as

Weeds: Wild chamomile (Matricaria chamomilla) (Courtesy of SCHERING)

possible, this helps good root development. When the stem turns a characteristic reddish color, the first irrigation should be carried out. If this is not done, growth halts and the plant never makes up the lost time.

The best time for the first irrigation is determined by the soil type. If it is sandy, the plant will need water earlier. In the opposite case, a clay soil, the first irrigation can be applied later. When harvest time approaches, you should stop watering about 20 days before harvest, to stop vegetative growth and encourage faster ripening.

Cotton can be irrigated using gravity or sprinklers, and the use of drip systems is becoming increasingly common.
Drip irrigation is mainly used in soils tending to saline, with a shallow saline water table. Drip irrigation is also highly recommended for sites with excessive slopes and for very sandy soils.

13.11.7. Herbicides

The active ingredient **trifluoraline** is normally used on cotton crops as a pre-sowing herbicide, and is incorporated by passing the disc harrow or cultivator. This herbicide does not kill the weed black nightshade (*Solanum nigrum*), and if present **fluomethuron** should be used. It should be applied immediately after sowing, as a crop pre-emergence herbicide. If suitable machinery is available to apply the herbicide without wetting the cotton plants, it can be applied post-emergence between the rows. If the cotton's leaves are splashed, they show a very typical chlorosis and a slight growth halt. In addition to these two, there is a wide range of active ingredients that can be used, including pre-sowing, pre-emergence and post-emergence herbicides, which can be applied when the crop is actively growing.

In the early stages of cotton's growth, if herbicides have had the desired effect, a moderate weeding is often performed by hand. If the area cultivated is large, especially on irrigated ground, the use of chemical fertilizers is frequently alternated with passing the cultivator between the rows of cotton. Cultivators have the advantage of bringing the deeper soil into contact with the sun, causing the soil as a whole to warm up, which improves the conditions for the crop's growth.

13.11.8. Accidents, diseases and pests

The worst accident caused by human intervention is applying too much nitrogen, and in cotton this causes an increase in the number of vegetative branches (to the detriment of flowering ones), decreased resistance to pests (such as the red spider mite and *Verticillium*), and a lengthening of the crop's growing cycle. The plants may also be affected by cotton **wilt**, which usually occurs when temperatures during germination were low, and is more common on heavy soils, because they are colder and wetter. The affected plants topple over when they emerge and they will never right themselves, so they should be uprooted and replacements sown.

• Diseases

The most frequent fungal disease is produced by *Verticillium alboatrum*, a soil fungus that attacks the germinating cotton seed. Many agronomists consider that the "fall" discussed above is only caused by this fungus, but "fall" can also be a purely abiotic accident. In general, when temperatures at germination are optimum, damage by this fungus is hardly noticeable, and it only has a serious effect in colder years. Like most other parasitic soil fungi, *Verticillium* attacks increase rapidly if cotton is grown on the same site year after year. If this fungus attacks, other crop species should be grown and cotton should not be planted for several years. Another preventive measure is to grow resistant varieties.

Cotton can also be attacked by fungi of the genus *Fusarium*. This is another fungus that lives saprophytically in the soil on the remains of previous crops, and infects the collar of the cotton seedling when it germinates. *Fusarium* attacks the vessels of the plant, causing the roots to rot and the plant to wither and die.

• Pests

One of the most important pests attacking cotton is the **red spider mite** (*Tetranychus*). You are advised to check the cotton regularly for the first symptoms of this mite, which usually appears on the sides of paths, water channels or sites where there are weeds (underneath electricity pylons, etc.), and then spread to the crop.

If you take this into account and try to form a barrier around these foci of infection, this is enough to prevent most attacks by red spider mite, and it has the advantage of not eliminating the mite's predators, as the field as a whole is not treated. **Pyrethrins** should not be used against red spider mite, as these products kill all the insects that prey on it, and cause its population to grow even faster. Specific acaricides like **abamectin**, **propargite**, **tetradiphon**, **dicophol**, etc., are widely used.

The insects that attack cotton include the aphid *Aphis gossypii*, which can be combatted with specific insecticides like **thiomethon**, **malathion**, etc. Another insect, the lepidopteran *Platyedra gossypiella* or **pink grub**, causes serious damage to the cotton seeds on which it feeds. Frequent use of insecticides like **flucitrinate**, **cifluthrin** or **phenvalerate** eliminates these insects. Another

lepidopteran, the budworm *Heliothis*, mentioned in the section on sunflowers, also attacks cotton and destroys many of the capsules. In addition to the insecticides listed for controlling *Heliothis* in the sunflower, there are others, such a **biphentrin**, a commercial mixture of the active ingredients **prophenophos + cypermethrin**. The first signs of attack by pink grubs can be detected in June, the presence of *Heliothis* is detected in July, and another lepidopteran causes serious damage in August, *Earias insulana*. In addition to the insecticides listed for treating the pests above, you can also use **endosulfan** and **chialotrin**, etc.

13.11.9. Harvesting

The flowering of cotton is staggered, and harvesting is therefore also staggered. Cotton used to be harvested manually. 50% of the cotton harvest in Spain is now mechanized. In dry farming, cotton ripening is more uniform and harvesters are used that strip all the bolls off the plant at once, regardless of their ripeness. In dry farming, fiber harvesters are used, of which there are two types: **drum harvesters** and **endless chain harvesters**. They are very heavy machines but are highly manoeuverable.

13.11.10. Use

The main use of cotton is as a source of fibers for the textile industry. From the point of view of human consumption, the fluff obtained by carding is used in production of cotton wool and in the manufacture of cushions, felt, thread for making rope, wicks for lamps, cooking gloves, dressings, etc. In the chemical industry, the fluff

PIX ® is a product of BASF, S.A., that inhibits the unwanted vegetative growth of cotton and favors growth of the reproductive organs.

is used in the manufacture of plastics, lacquers, cinema film, cellophane, etc.

The seed contains 18-20% edible oil, and can be milled to produce meal, or can be made into a seedcake suitable for swine, cattle and fowl. The cake contains a lot of protein, but also contains a toxic alkaloid called **gossypol**. Seedcake free of this toxic compound can now be produced.
The seed case can be fed to animals, and can also be used as a fuel.

13.12. TOBACCO

The first reports of tobacco in Europe were brought by Columbus, after his discovery of the Americas (1492). The American Indians of the Caribbean and other areas smoked tobacco leaf, either by inhaling the smoke through a stick full of shredded tobacco leaf, or rolling the leaves into the form of a cylinder that they lit at one tip (the Spanish "conquistadors" called this primitive cigar a "tizon"). The first known graphic representation of someone smoking tobacco is in a Mayan relief from the Palenque site, which historians date from the 6th or 7th century AD.

Tobacco plant in flower. This is the moment to remove the flower in order to favor leaf growth

Tobacco spread to the rest of Europe from the Iberian Peninsula, and by the 16th century tobacco was smoked in Italy and the Netherlands. It was known in England due to contacts with sailors from Portugal or from the northern coastline of Spain, and was also reported by Drake and Hawkins after their direct contact with the North American Indians.

Tobacco (*Nicotiana tabacum*) is a dicotyledonous plant ("dicot") and is a member of the potato, or nightshade, family (Solanaceae). The leaves are lanceolate, alternate and petiolate. The flowers have a reddish corolla and are born in a terminal panicle; it is a self-fertilizing flower because its flowers self-fertilize before they open. The fruit is a capsule containing very small white seeds. It is a perennial plant that can live for several years in its countries of origin, though it is always cultivated as an annual. It is a large plant, easily reaching a height of 1 or 2 meters. It has a penetrating root system, though most of the roots remain near the surface on rich soils.

13.12.1. Varieties

There are many different forms of the species *Nicotiana tabacum*, and many grow wild in their place of origin. Geneticists are working to create improved varieties with lower concentrations of nicotine and tar, that are more productive and show greater resistance to fungal diseases. All the cultivated varieties can be placed in four groups:

• **Havaniensis type**
This type is represented by the varieties **Vuelta abajo**, **Java** and **Sumatra**. They are medium-sized plants that bear 20-25 leaves, that are initially almost horizontal to the stem, with elliptical light green leaves. The inflorescence is open, with spaced and almost horizontal lower branches. The flowers are small with sepals that are tipped with a point and adhere to the corolla tube. The petals are red or scarlet.

• **Braziliensis type**
This type is represented by the variety **Brasil-Bahía**. It is a medium-sized plant, with light green leaves that are twice as long as they are broad. They turn shiny brown with a reddish tinge when they are dried. The inflorescence does not rise very far, with unbranched almost horizontal stems that are close together. The flowers are vertical, with a not very clear point at the tip of the sepals, which adhere to the corolla tube. The corolla is short with pink or red petals.

• **Virginia type**
This includes both **Virginia** and **Kentucky** type tobacco. The plants are tall and strong with a thick stem, and very short internodes at the base of the plant. The lanceolate leaves are 3 times as long as they are broad. The leaves are dark green and turn copper-brown when dried. The large flowers have triangular petals and are pink or reddish.

• **Purpurea type**
These plants have a cylindrical stem, with leaves that are borne horizontally and internodes that shorten from the base to the top of the plant. The leaves are petiolate with a wavy edge, and are yellow-green turning light brown when dried. The inflorescence is small and flattened. The flowers point downwards, and have long, tipped, sepals. The corolla is a dirty-white color at the base, gradually turning purple towards the tip. The varieties **Oriente** and **Sumatra** are representative of these varieties.

13.12.2. Cultivation requirements

Tobacco is originally from tropical and subtropical climates, and grows best in hot wet climates. The best growing areas are between 45°N and 30°S. However, it is cultivated between 60°N and 40°S, and it is grown in Canada and Brazil, in Belgium and South Africa.

• **Temperature**. When the temperature is constant and humidity is high, tobacco leaves transpire little water, resulting in greater vein growth and thinner leaves. It requires temperatures of 13-15°C for germination, but requires higher temperatures for optimum growth, 18-28°C. The temperature should not fall below 14°C or rise above 32°C for good crop growth, as growth halts below 14°C and above 32°C.

• **Light**. Light intensity should also be constant, without large variations, as this leads to thinner leaves, which are of better quality. Some authors consider that stronger light increases the amount of nicotine in the leaves. In the years when there are less hours of sunshine during the crop's growing period, tobacco leaves contain less nicotine.

• **Moisture**. Two types of moisture are relevant here, environmental humidity and soil moisture. Tobacco thrives in conditions of high environmental humidity. In these conditions the leaves are thinner and contain less nicotine. The humidity conditions are thus suitable in coastal zones, though the crops have to be some distance from the sea, as salt deposition on the leaves is highly detrimental. Yet very high relative humidity is prejudicial, because it favors the growth of fungal diseases. In dry climates, the leaves are smaller and contain more alkaloids, which is not much appreciated, as the tendency is towards tobacco with less nicotine and tar. When growing, the plant tolerates relatively dry conditions better than excessive soil moisture.

• **Soil**. Tobacco's roots need a deep permeable soil where they can respire well. That is to say, it does best on loamy or sandy loam soils, with enough organic matter to ensure fertility. The best pH values are slightly acid or neutral for light tobacco (Burley and Virginia) and neutral or slightly alkaline for dark tobacco, whether for cigars or cigarettes. Tobacco should not be sown on soils rich in chlorides, and thus should not be sown on salty soils.

13.12.3. Applying fertilizer

When applying fertilizer, there is an important distinction between plants destined for cigarette production and those intended for cigars. Excess protein in the leaves, the result of excessive fertilization with nitrogen, causes a reduction in the **sugar/protein** ratio in the leaf. Tobacco for cigarettes should produce a sweet, acid, smoke. For cigars, however, they should have an alkaline taste and aroma, meaning that the **sugar/protein** ratio should be lower. As a result, when growing tobacco for cigars nitrogen can be applied over the entire growing season, but if the tobacco is for cigarettes, and especially if it is Virginia type tobacco, the nitrogen should be supplied in a way that ensures that at harvest time there is almost no nitrogen left in the soil for the plant to absorb.

Another general point is that the quantities of each of the nutrients, and especially the micronutrients, applied to the crop have to be carefully calculated, because an excess or shortage of a given element may greatly affect the organoleptic qualities of the manufactured product. The amount and timing of fertilizer application depend on the type of soil, the variety of tobacco that is cultivated, climatic factors and other agronomic factors that you should consult with a technician, or if this is not possible, with government agencies or directly with the local commercial suppliers.

• **Nitrogen**. As tobacco is cultivated for its leaves, its vegetative parts, nitrogen is essential for tobacco's growth. The leaves contain between 1.5 and 4% nitrogen, depending on the type of tobacco, but it is very important that the nitrogen supply is balanced, just as in other crops. Excess nitrogen causes tobacco to produce leaves with softer tissues that cure and burn badly. A deficiency of nitrogen leads to small leaves and thin stems, and the lowest leaves turn pale green. The nitrogen has to be applied in small doses, which the plant can use better.

We have repeatedly stressed how important it is for fertilizer application to be balanced. In tobacco, more units of phosphorus and potassium should be applied than of nitrogen. A well balanced fertilizer for tobacco should be in the proportions 1 N; 1.5 P_2O_5; 2.5-3 K_2O (1:1.5:2.5). On clay soils, 80 % should be applied at depth when preparing the seedbed, and the remaining 20% to the surface and then incorporated using the cultivator. On loose or sandy soils, 30% of the nitrogen should be incorporated before sowing, 40% during crop growth and the remaining 30% with the earthing up.

• **Phosphorus**. Phosphorus accelerates the ripening of the leaves. Tobacco for cigarettes is best with high sugar levels in its tissues when harvested, and requires higher applications of phosphorus. Tobacco for cigars does not need so much phosphorus, and excess phosphorus leads to leaves that are fragile and rubbery, that burn badly and give a darkish ash. Phosphorus deficiency symptoms appear when leaf levels fall below 0.3%. The leaves turn a bluish green because the proportion of chlorophyll increases, and this is especially negative for Burley type light tobacco. The optimum phosphorus concentration for tobacco intended for cigars is about 0.6%, while for cigarettes it can be up to 1%.

• **Potassium**. Potassium is a very important element affecting the tobacco's quality. The potassium salts present in the leaf make it burn well, and help to maintain the acid/base balance of the plant's tissues within the most desirable limits. Potassium inhibits the production of proteins, and also acts as a catalyst in the formation of sugars in the leaves. Fertilizer rich in potassium increases the **sugar/protein** ratio, which is very favorable in tobacco intended for cigarettes.

• **Microelements**. Chloride ions have a very negative effect on the tobacco's quality. When they exceed 1.1% of the tissue weight, the tobacco's burnability decreases. Above 2%, the burnability of the final product is unacceptable. If it exceeds 4%, leaf ripening may be delayed by fifteen days. This means that the potassium fertilizer applied should be in the form of potassium sulfate, never potassium chloride, in order to ensure the crop has as little chloride ion in the soil as possible.

Other important secondary nutrients, such as calcium and magnesium, are important in producing high quality tobacco. Their levels depend on the greater or lesser levels in the soil. The normal level of calcium in tobacco leaves should be between 3 and 6%, while that of magnesium should be between 1 and 2%. Other microelements, such as boron, iron, manganese, etc. can also affect the final leaf quality, and thus the organoleptic qualities of the final product.

13.12.4 Sowing

• **The seedbed**. Tobacco seed is very small, and the seedling is very delicate and vulnerable during germination and emergence, and because it is often necessary to advance the ripening of the tobacco plant, tobacco is always germinated in seedbeds.
The seedbeds should be no wider than 1.20 m and should contain between 250 and 500

plants/m^2. The small seedlings should be protected from the winds using suitable windbreaks made of canes or artificial materials, such as plastics. The seedbeds can be established on a cold bed or a warm bed. Ordinary seedbeds are made by mixing the soil to a depth of 20-30 cm. with thoroughly rotted dung, and adding a top layer of 6-8 cm of good mulch well mixed in a 2:1 ratio with earth. Warm seedbeds are similar to cold ones, but when they are prepared a layer of fresh dung should be added just above the drainage and below the earth. This way, the organic fertilizer rots, releasing some warmth which is good for the tobacco seedlings.
It is a good idea to disinfect the seedbeds 20 days before sowing to protect the seedlings from fungi (*Fusarium, Pythium*, etc.) and from nematodes. This is frequently done using **metam sodium** and **dazomet**. The dose, and how and when to apply, are discussed in the fifth section of this work, the section on soil disinfection. At the time of sowing, the seed is usually mixed with flour, sand or ash to make it easier to handle, because the seeds are so small. They are sown by hand and watered immediately afterwards. The seed is sown at a rate of 1/4 to 1/2 g of seed per m^2. It is also worth applying some fertilizer in order to speed up germination and growth, using a soluble fertilizer dissolved in water. This fertilizer usually contains superphosphate, nitrate, potassium sulfate and some form of chelated iron, also in the form of the sulfate. It is now normal practise to cover the seedbed with plastic or glass as if they were small greenhouses. This protects the plants against possible frosts and other adverse factors, such as rain, wind, etc., as well as increasing the temperature for germination. In addition, it is important to keep a careful eye on the health of tobacco seedlings, and treati them when necessary.

• **Soil preparation**. When the definitive site is prepared, this should be well washed and with the topsoil as a fine tilth. In both dry farming and irrigation, the stubble of the previous crop should be raised, followed by scarification and then a pass of the disc harrow, finishing with the cultivator and if necessary passing the roller.

• **Planting**. If the tobacco is planted very densely, the yield per plant is smaller. If less plants are grown per hectare, the production per plant increases. The amount of leaf harvested will depend on the planting frame, and also on the height at which the flower is removed. If the flower is to be cut low, the density of plants can be increased. The variety grown is also important, as some varieties can be grown at higher densities without a decrease in productivity. The amount of nicotine is another important factor; the higher the planting density, the lower the number of roots, meaning the nicotine content is lower. Higher planting density also leads to thinner leaves, meaning the final product is of higher quality.

The following data can be used as a guideline. Under irrigation, the tobacco is usually planted on ridges in rows 1 to 1.2 m apart, and the plants in each row are 50-60 cm apart. If it is going to be necessary to carry out further tasks with the tractor, service lanes should be left free for machines to carry out treatments. Once the final site has been prepared, the tobacco seedlings should be transplanted, using a planter, and taking great care not to damage the roots. There are now planting machines that save a great deal of manual labor. These mechanized planters are attached to the tractor, and the most complete ones open the hole, plant the seed ling, water it and then apply fertilizer, all in a single pass.

13.12.5. Cultivation

Tobacco requires a series of tasks intended to increase leaf yield and quality.

• **Deleafing**. When the plant is about 40 cm tall, it has 10 or 12 leaves. At this stage it is worth removing the lowest leaves, those in contact with the soil, because they compete with the other leaves for nutrients, but are not suitable for producing quality tobacco, because of the attached soil, dirt, etc.

• **Earthing up**. After removing the lower leaves, the plant is usually earthed up, which is raising the level of earth at the base of the plant, which encourages formation of new roots and makes the plant more vigorous. This operation is usually carried out with the cultivator and then finished off manually with a hoe.

• **Flower removal**. Tobacco is grown for its leaves, not for its flowers or fruit, and so the terminal flower panicle is normally removed, so it does not consume the nutrients it requires for its growth. When strong, thick and tough leaf is wanted, the flower stem is removed low, removing several of the upper leaves together with the flower bud, shortly after it forms. This is done when the entire plant is harvested, and it is desirable for all the leaves to mature at the same time. When the object is fine tobacco, it does not matter if the leaves do not all ripen at the same time. A larger number of leaves are left on the plant, and flower stem removal is left till later.

• **Shoot removal**. When the flower stem is removed, this encourages the plant's lateral buds, in the axils of the leaves, to grow. They too should be removed. This was traditionally done by hand, but there are now chemicals that give the same results.

• **Removing sidestems**. In tobacco cultivation, it is very important to remove any sidestems that may appear during the plant's growth. This is also done to remove competition that may affect the quality of the leaves.

13.12.6. Irrigation

In wet rainy zones, tobacco can be dry farmed, and the plant need only be watered when the seedling is planted, after transplanting and in especially dry years.

The irrigation water used on irrigated tobacco crops should contain no chloride or as little as possible. Water three times when planting: when preparing the site before transplanting, immediately after transplanting and when replacing failures. Before removing the flower stem, water should be given if necessary, but remember that excess water always reduces the quality of the leaves. After removing the flower stem, the plants are not usually watered except in very dry zones, where the crop is watered about 20 days before harvesting.

13.12.7. Herbicides

Weeding is frequently carried out by hand, or using machinery when a planting frame has been chosen that allows mechanical weeding. It is also possible to use herbicides to eliminate weeds.

The active ingredients of the herbicides most commonly used in tobacco cultivation are applied pre-emergence of the weeds. **Benfluoraline** is considered to be selective against tobacco weeds, but treatment should be performed six weeks before transplanting. Other herbicides, such as **diphenamide** and **methobromuron**, can be applied several weeks after transplanting the tobacco, as they favor tobacco by selecting against monocot and dicot weeds in pre-emergence.

13.12.8. Accidents, diseases and pests

• **Accidents**
Damage can be caused to tobacco leaves by sharp and drastic changes in temperature, environmental humidity, soil humidity and light, and this causes the plant to take up more chloride ions, which negatively effects the quality of the final product, because high concentrations of chloride reduce the burnability of the tobacco leaf. During harvesting, if the picked leaves are exposed for a long time to the sun, they may be **scalded**, which makes the leaves totally useless. Quality is also diminished if the leaf is picked before the dew has evaporated (the leaves remain moist and lose some of their quality). Excess nitrogen leads to an increase in the leaf concentration of nicotine, nitrates and ammonium, leading to lowered assimilation of phosphorus and potassium. When excess nitrogen fertilizer is applied, there is a measurable decrease in the levels of phosphorus and potassium in the plant's tissues.

• **Diseases**
• *Viral diseases*. There are at least 9 viruses that cause mosaic disease in tobacco. The most common one affects the leaves, causing regularly distributed light green patches; sometimes there are relatively clear bumps. Like all

viruses, the tobacco viruses are transmitted by insect bites, especially by thrips and aphids. Viral damage greatly reduces the value of the leaves, because the tobacco will not burn well. There are no ways of treating viral attacks, and the only way to deal with them is to practise preventive cultivation methods. The affected plants should be uprooted and burned, all tools and implements should be cleaned and disinfected, greenfly should be treated, and these measures usually give acceptable results.

After harvesting, the tobacco should first be dried at ambient temperature.

• *Fungi.* Fungal diseases frequently attack seedbeds, and the tobacco seedlings are very sensitive. Parasitic fungi, such as seedbed **powdery mildew** (*Pythium debaryanum*), tobacco black root rot (*Thielavia basicola*), **Fusarium wilt** (*Fusarium sp.*), blue mold (*Perenospora tabacina*), etc., rot the roots, collar or aerial parts of the tobacco seedlings. Preventive measures include not using acidic mulches, destroying the remains of previous tobacco harvests, using guaranteed and disinfected seed, etc., and all give good results in preventing attacks by these fungi. They can also be combatted with curative or preventive chemical treatments, such as **methyl thiophanate**, **mancozeb**, **methaxyl**, etc. Tobacco crops are frequently attacked by **blue mold**, the most important disease damaging tobacco crops. It can be controlled by dusting with the fungicides **zineb**, **probineb** and **methalaxyl**. Tobacco is also frequently attacked by mildew (*Erysiphe cichoreacorum*), *Thielavia basicola* and the

The green fruit of the coffee, the bean, which later turns bright red.

different wilts caused by species of the genus *Fusarium*.

• **Pests**
The animals that attack the seedbeds include gastropods, such as the slug *Agriolimax agrestis* and the snail *Helix hortensis*, which should be controlled with **metaldehyde** baits. There are also some soil insects, such as **ants** and the **mole cricket** (*Gryllotalpa gryllotalpa*), which

should be controlled with baits of **barium fluorosilicate**, **magnesium phosphide**, **sodium arsenite** or **lindane**. These are highly toxic products, and extreme caution should be taken when handling, storing and applying them.

When the tobacco has been transplanted into the field, it may be attacked by other insects, such as wireworms (*Agriotes segetum, A. lineatus*), thrips, aphids, etc.

13.12.9. Harvesting

Choosing the right time to harvest tobacco is very important, as this determines the leaf color and how it changes during curing. When the entire plant is harvested, the lower and middle leaves should be showing signs of maturity,

even if it has barely started in the upper leaves. When the leaves mature, yellow patches appear at the edges that gradually get larger until they cover most of the leaf blade. The edges curl, the leaves become brittle, breaking easily when bent and they acquire a characteristic shine and bumps. Each type of tobacco has to be harvested in its own special way. **Dark tobacco** is usually harvested as the entire plant, leaving them on the soil, where they undergo a first airing, and then they are transported to the drying sheds. **Cigar wrapper tobacco** has to be harvested leaf by leaf. After choosing the mature leaves, the leaves from the bottom third of the plant are removed first, which are the first to mature, shown by the leaves turning gradually yellow. Later, the leaves are removed from the middle third, and finally, from the upper third.
Light tobacco is **flue-cured** in an artificial atmosphere, and should be harvested when the leaves are completely mature. They are collected leaf by leaf, selecting the most mature ones on the plant and those that are yellowish green. In general, the plants are harvested five times in all. **Burley** type tobacco

and **dark** fire-cured tobacco have to be harvested when they are still not fully mature. They do not have to be fully mature to be harvested. These types of tobacco are usually harvested as whole plants, rather than leaf by leaf.

If the area cultivated is large, the farmer requires special machinery for collection. You can choose between **semi-automatic harvesters** that require one or more operators, or **automatic** ones, which save a great deal of manual labor. It is also possible to adapt side-cutting cereal harvesters to do this. This system requires the plants to be collected by hand after they have been cut, but has the advantage that you do not need to buy a new machine for the harvest.

13.12.10. Curing

To turn tobacco leaf into a product that can be smoked, the leaf has to be cured and fermented. The fermentation is usually performed by the companies buying the tobacco, and the farmer need not worry about it. Curing is almost always carried out by the farmer, though it is possible to pass the leaf to a company specialized in drying tobacco. When the leaf is collected, it contains 80-90% water. Over the course of curing, it should lose 65% of its weight in water, and at the end should contain only 15-25% moisture. There are several ways to cure tobacco: **air curing**, **flue curing**, **sun curing** and **fire curing**. Each type of curing has specific systems of handling the leaf, regularly checking the temperature and humidity levels. These processes are too complex to include in this work.

13.12.11. Use

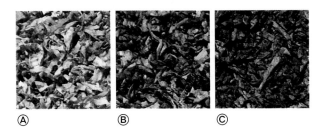

Ⓐ　　　　Ⓑ　　　　Ⓒ

As is well known tobacco, in all its different forms, is essentially grown for smoking. Cigarettes, cigars, pipe tobacco, chewing tobacco, snuff, etc., are all found in most tobacconists. There is commercial distinction between **light tobacco** and **dark tobacco**, i.e., the color of the leaf after it has been cured and fermented. The internationally accepted commercial classification of tobacco is:

• Dark air-cured tobacco
These tobacco are used to make dark tobacco and for the shredded tobacco used as filler for some types of cigar. They are also used in some types of pipe tobacco. They are large

leaf tobacco that change on curing to a color between dark cinnamon and mahogany. They are known as **Maryland**.

• Dark air-cured tobacco for cigar wrapper and binder
They are medium to large leaf tobaccos that are used, because of their thinness and good physical properties (elasticity, resistance, burnability) for the outer layers (wrapper) or inner layers (binder) of cigars.

• Light air-cured tobacco
These tobaccos contain a lot of nicotine and absorb a lot of the artificial additives used in manufacturing. The different commercial types of product made with this tobacco contain a lot of additives, and many physicians consider it the most harmful way of smoking tobacco. They are large leaf tobaccos and after curing they are conditioned by heat, and are then manufactured after a relatively long natural process. They are known as **Burley** type tobacco.

• Light flue-dried tobacco
Like all light tobaccos, they contain high levels of nicotine and tar. The leaf is light green and turns lemon yellow or orange during the long curing process they require. When used alone, they give rise to English-type mild cigarettes. When they are mixed with other types of tobacco, the cigarettes are still light in color but stronger in every other way; they are called American-type mild cigarettes. They are also used in some mixes to be smoked in a pipe.

• Aromatic tobacco
Aromatic tobaccos are also known as Turkish tobaccos. They are small plants with small light yellow leaves that contain little nicotine, and are normally very aromatic. They are sun-cured and are made into pipe tobacco and mixed light cigarettes.

• Fire-cured
Fire-cured tobaccos are used in the manufacture of snuff, and they are also processed for use as pipe tobacco and as filling for cigars. These are the tobaccos typical of Virginia and Kentucky. The leaves are cured over a hot fire, and turn a dark brown color.

• Homogenized tobaccos
What are called homogenized tobaccos are used to manufacture products that are low in tar and nicotine. First, a paste is made of the leaves or the entire plant, using processes similar to the production of paper pulp. This paste is then subject to physical, chemical or biological processes to obtain a less harmful product. It is also more economically profitable for the manufacturer, because the entire plant is more thoroughly used, even though the processing is expensive.

Types of tobacco
A/ Virginia type tobaccos are mild and aromatic. They vary in color from light yellow to golden orange. When fire-cured, they turn dark brown. (Courtesy of MAC BAREN TOBACCO COMPANY)
B/ Burley type tobaccos are mainly cultivated in the US states of Kentucky and Virginia. They are light air-cured tobaccos. (Courtesy of MAC BAREN TOBACCO COMPANY)
C/ Latakia tobacco. Aromatic tobaccos are also known as Turkish tobaccos and are generally used in mixes with other tobaccos, to enhance their aroma and taste. (Courtesy of MAC BAREN TOBACCO COMPANY)

Coffee plantation

In the United States, the tobacco industry uses the following classification for cured leaves.

- **Class I**. *Flue-cured*. These are the tobaccos cured in an artificial atmosphere.
- **Class II**. *Fire-cured*. They are cured using fire.
- **Class III**. *Air-cured*. The leaves are cured in the air.
- **Class IV**. *Cigar-filler*. The leaves are used for the filling the cigar.
- **Class V**. *Cigar-binder*. The leaves are used for the layer below the wrapper of the cigar.
- **Class VI**. *Cigar-wrapper*. The leaves are used for the outer layer of the cigar, the wrapper.

Coffee plant in a Brazilian plantation

An even wider range of names is used within the industry. Thus, for example, the classes used in the United States are further subdivided, from the perspective of their industrial uniformity, on the basis the leaf thickness, the part of the plant that is selected (the central part of the stem, the intermediate leaves, low leaves, etc.), according to their quality, and there is even a classification of the residues or leaf fragments used for other less important purposes.

The tobacco industry in Spain also has its own classification of tobacco leaf. Burley and Virginia type tobaccos are the most widely cultivated in Spain, and the leaf is classified on the basis of the position of the leaf on the plant and its quality. If considering cultivating this member of the nightshade family for the first time, the farmer should consult the local agricultural services in order to obtain information on the most suitable varieties for each zone, and the different classifications applicable in each region.

13.13. COFFEE

The coffee bean is bright red when ripe (Photo courtesy of A. GOSTINCAR for this publication)

The coffee grown now is produced by plants of two species of the genus *Coffea, C. arabica L.* and *C. canephora P.* Smaller areas of the following are also cultivated, *C. liberia B., C. abeokutal C.* and *C. dewevrei W. & D.* Coffee (*Coffea arabica*) is originally from the highlands of Abyssinia (Ethiopia). It spread first

to Arabia, and from there to the other Islamic countries (Egypt, Syria, Turkey, etc.). It was first imported into Europe in the 16th century by Venetian merchants, and was introduced to Asia from Africa. In the late 17th century, it started to be grown in Dutch colonies like Dutch Guyana, and from there it spread to Colombia and the Antilles, later spreading to Brazil and throughout the Caribbean. *C. canephora* is considered an African species, like *C. arabica* as its different varieties are widely cultivated in Africa.

C. arabica is cultivated in Africa, in the highlands of Kenya, Tanzania, Uganda, Malawi, Congo, Cameroon, Ethiopia, Rwanda and Madagascar. In Asia, it is grown in the highlands of Arabia (Yemen), India, the Philippines, Indonesia, Vietnam and Laos. In the Americas, the highlands of Mexico, Guatemala, Honduras, San Salvador, Nicaragua, Costa Rica, Panama, Colombia, Venezuela, Ecuador, Paraguay and Peru are all famous for their quality coffee. It is also cultivated at lower altitudes in South America in Brazil, the Caribbean islands, etc. *C. canephora* is cultivated in Africa in the lowlands of the Congo, Angola, Ivory Coast, Equatorial Guinea, and at medium altitudes in Cameroon, Uganda, and Tanzania. In Asia it is cultivated in lowlying and medium altitude zones of Indonesia, India, the Philippines, New Caledonia, etc.

Other species of *Coffea*, such as *C. liberica, C. excelsa* and *C. abeokutae*, are still cultivated on a very small scale in different areas of Africa (Central Africa Republic, Ivory Coast), Asia (Indonesia) and South America (Surinam).

13.13.1. Botany

- **C. arabica**. This is the best known species and the most widespread, and is what most people consider coffee. It is a small self-fertile tree of the madder family (Rubiaceae) that grows eight to ten meters tall in the wild. The cultivated varieties of coffee are usually smaller, because they have been selected for better quality to the detriment of their vigor, and this makes cultivation easier (agricultural tasks, treatments, harvest, etc.). The leaves are evergreen, slightly upturned and 10 to 15 cm long, and

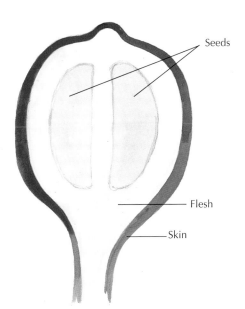

Seeds

Flesh

Skin

Longitudinal section of a coffee bean

and 4 to 6 cm wide, dark green, shiny above and with a high caffeine content (1.26%). The tree has a taproot reaching a depth of 0.3-0.5 m. On deep soils they can reach a depth of 1 m. The root mass often extends through 12 to 15 m³.

The small star-shaped flowers are white and very fragrant. They are borne in the axils of the leaves. The fruit are berries similar in size to a cherry, but yellow, and are called coffee beans. They contain two seeds (the true seeds) that are elongated and hemispherical with a longitudinal groove and covered by a relatively tough membrane. The seed, from which the coffee is made, is yellowish grey or slate grey, with a bluish tinge, and varies from variety to variety, as does seed size and shape. Most seeds are 10 mm long, 6-7 mm wide and 3-4 mm thick. They weigh 0.15-0.20 g. After drying, the coffee grains contain about 1.5% caffeine.

• **C. canephora** is self-sterile, unlike *C. arabica*. It is an evergreen shrub that reaches a height of 8 to 12 m. The branches are long and twisted, the leaves are large (20-35 cm long, and 8-15 cm wide), oblong and upturned. The inflorescences are borne in the axils and consist of three whorls, each with fifteen or thirty white fragrant flowers, and a corolla consisting of five to seven petals. The ovoid fruit are 8 to 16 mm long. The bean is red when it is ripe and contains two ovoid seeds with one face flattened, and varying in size but generally small. In general, varieties of *C. canephora* grow more vigorously and are more productive than *C. arabica*. It is also more resistant and is less sensitive to several diseases. Though the organoleptic qualities of the coffee produced are different from *C. arabica*, they are increasingly appreciated by consumers.

13.13.2. Varieties

Many varieties of coffee are natural mutations ("sports") of existing varieties or crosses. Over the years, wherever coffee has been introduced, new varieties have arisen that are well adapted to local climatic conditions, where they produce large and high quality yields. In general, varieties of *C. arabica* are cultivated in South America, while varieties of *C. canephora* are cultivated in Africa and Indonesia. These two species account for 98% of the world's coffee production. The other 2% comes from several species, whose cultivation is highly localized and which are rapidly being replaced by *C. canephora var. robusta*.

• **C. arabica var. typica L.** Introduced into Brazil at the end of 18th century from French Guyana, it gave rise to the coffee grown in Brazil before the spread of the modern selected varieties that are now dominant.

• **C. arabica var. amarella Ch.** This variety originated in Bahía (Brazil) and is a mutation. It is a large tree with large leaves, long internodes and large fruit. The seeds are larger than those of *C. arabica var. typica*. This variety is grown in Brazil and other South American countries.

• **C. arabica var. Bourbon Ch.** This variety is thought to be a recessive mutation and to have arisen on Réunion Island. It has been introduced all over the world. Its most typical characteristic is that its growth form is smaller than that of *C. arabica var. typica*, but its vegetative growth is denser, as a result of the heavy branching of nearby nodes. The seed is also larger than that of the typical species. Many plantations in Brazil consist of selected strains of this variety, which is greatly appreciated for its high quality and production.

• **C. canephora var. robusta**. This is the most widely cultivated variety in the world, and accounts for 90% of all the plantations of *C. canephora*. Compared to *C. arabica*, it is more vigorous, more productive and shows greater resistance to some diseases. Its cultivation has spread all over the world, not just throughout Africa (Ivory Coast, Congo, Cameroon, Uganda, Angola, etc.) but also throughout the far east (India, Indonesia, etc.) and Oceania (New Caledonia, etc.).

• **C. canephora var. kouillou**. This variety, similar to **robusta**, grows wild in Congo, Ivory Coast, Gabon, etc. This form varies greatly and differs from **robusta** in that the leaves are more oblong and the fruit and seeds are smaller. It is cultivated in Ivory Coast side by side with **robusta**, and has the advantage of resisting dry conditions better. It is now grown in Madagascar and Guinea too.

13.13.3. Cultivation requirements

• **Temperature**.*C. arabica* is better able to tolerate variations in temperature than the other species. It can withstand, for a short time, temperatures only a little above 0°C or as high as 30°C.

Ⓐ

Ⓑ

Ⓒ

A/ Mature coffee beans. They are the fruit collected from the coffee plant.
B/ Coffee seeds after extraction from the bean. This is the Blue Mountain "caracolillo" variety, which is unusual because each bean only has a single seed, instead of two.
C/ Coffee after toasting. This is the normal appearance of the coffee we buy.

C. canephor cannot withstand low temperatures as well as *C. arabica*. The damage takes place when temperatures fall to 8-10°C, and the coffee plants die long before the temperatures reach 0°C The temperature is one of the most important factors limiting coffee. The optimum temperatures are between 22-26°C, without sharp changes.

• **Light**. In its natural habitat, the coffee grows in shady or partly shaded places, such as forest clearings, gallery forest along watercourses, on river banks, etc. It requires shading from the earliest stages until the plant is considered adult, and it gradually adapts to direct sunshine.

• **Moisture**. After temperature, rainfall is the second factor limiting coffee cultivation. This is not just the total rainfall over the year, but its distribution over the course of the year. Coffee grows best with 1,500-1,800 mm of rain, with a rainfall regime that includes some less rainy, or relatively dry, months that should coincide with the rest period just before peak flowering. *C. canephora* adapts well to very abundant rainfall, over 2,000 mm per year. In addition to rainfall, atmospheric humidity is also very important. If it is very high, the plants transpire less and thus need less water.

• **Soil requirements**. Deep well-drained soils are best for growing coffee, as they have a taproot and lateral roots that spread over a large area of soil to obtain nutrients. In shallow soils, the roots usually only colonize the top 30 cm of subsoil, meaning that fertilizer supplements are often necessary to get good production. In their natural habitat, they grow well in highly acid soils (pH 4.0 to 5.0) with a great deal of humus, though they grow well in soils close to neutral (pH 7).

13.13.4. Applying fertilizer

The doses of fertilizer to apply to coffee plants vary depending on the soil class. They should be larger on sandy soils, which lose more nutrients by leaching, and smaller on clay soil. The annual surface fertilizer should be divided into three doses, using complex ternary fertilizers or simple traditional ones. The nitrogen and potassium supplied should be in the form of sulfates, as coffee does well with supplementary sulfur. The plant requires little phosphorus, and it should only be applied once a year.

As a guideline, the fertilizer units of N-P-K required should be applied as ammonium sulfate (21% N), dicalcium phosphate (40% P_2O_5) and potassium sulfate (50% K_2O) required for sandy soils:

Quantities given per m²

• **March**
50 g ammonium sulfate = 10.5 FU or g N
50 g bicalcium phosphate = 20 FU or g P_2O_5
40 g potassium sulfate = 20 FU or g K_2O

• **July**
30 g ammonium sulfate = 6.3 FU or g N
40 g potassium sulfate = 20 FU or g K_2O

• **October**
30 g ammonium sulfate = 6.3 FU or g N
40 g potassium sulfate = 20 FU or g K_2O

For clay soils, the quantities are:

• **March**
30 g ammonium sulfate = 6.3 FU or g N
50 g bicalcium phosphate = 20 FU or g P_2O_5
30 g potassium sulfate = 15 FU or g K_2O

• **July**
20 g ammonium sulfate = 4.2 FU or g N
40 g potassium sulfate = 20 FU or g K_2O

• **October**
20 g ammonium sulfate = 4.2 FU or g N
30 g potassium sulfate = 15 FU or g K_2O

13.13.5. Sowing

The seeds should be hardened off before being planted in the final site. First a seedbed should be prepared on flat soil with rich light soil. Care should be taken to site the seedbed near a source of water (river, pool, well, etc.). The soil is dug up to a depth of 30 cm, and dung or well rotted organic matter should be incorporated. Preparation finishes with a final pass with the disc harrow in order to break down any remaining clods. The site is then divided into rectangular plots 1.2 m wide and 60 cm apart. Furrows should be made in each plot for the seed, and using a planting frame of 8 x 4 cm. That is to say each row is 8 cm apart and the seeds in the row are 4 cm apart. The seed should not be sown too deep (no more than 2-3 cm). The seed should be placed on its flattened side, with the groove downwards.

The coffee seedlings should germinate successfully within six weeks. The coffee seedlings are very delicate. They cannot withstand direct sunshine, and it is often necessary to construct some sort of roofing to protect them. They need a lot of water, and they should be watered frequently. The seedbed should be weeded to prevent weeds smothering the seedlings, and they should be checked frequently for phytophagous insects.

Three weeks after successful germination, the seedling's two cotyledons ("seed leaves") are well developed. They should now be transplanted to the nursery.

The nursery should be close to the seedbed. It should be dug up to a depth of 40 cm and dung incorporated, the clods broken up and fertilizer mixed in. Similar plots should be prepared, with a width of 1.2 m and paths 60 cm wide. The seedlings are transplanted to the nursery at a much lower density, so the planting frame should be 30 cm between sowing, and the holes for the new plants should be 30 cm apart and 15 cm deep. When transplanting, reject plants with twisted or diseased roots. Take great care not to twist or damage the main root. Firm the soil down around the transplanted seedling but do not bury the collar.

While it is in the nursery, the seedling should be watered, protected from direct sunshine with roofing like that in the seedbed, and treated with pesticides if attacked by pests. The shading should be removed gradually so the young plants can adapt to the sunshine. The saplings can remain in the nursery for 4 to 5 months if the intention is to plant them in their definitive site in the next rainy season, or for a year if they are to be planted the following year. Their definitive site should have deep soil. If the site has never been used to grow coffee before, it is worth leaving some trees to protect the coffee trees from the sunshine and to protect the soil from erosion by rain.

In sites where there is serious soil erosion, the plant remains from clearing the plot should be left in the soil, as they provide organic matter. Legumes can also planted among the coffee trees. This plant cover prevents erosion and supplies nitrogen to the soil, but they should be removed after three years, as they may compete with the crop in the soil for nutrients.

The young coffee plants should be transplanted to their definitive site with a 3 x 3 m planting frame, which gives a density of 1,000 plants per hectare. The holes to plant them in should be cubic, 50 cm x 50 cm x 50 cm. They should be transplanted in the rainy season, and when they are transplanted you should remove some of the plant's leaves to reduce its transpiration. You should also take great care not to break the main root. Fertilize the base of the hole and protect the seedling from the sun for the next few days. Water the saplings immediately after transplanting them.

13.13.6. Vegetative cycle

The life of the coffee plant is in three stages. The first stage, the growing phase, starts with seed germination, and finishes with maturity, and this takes from four to seven years depending on the species and environmental conditions. The second period is the productive period, usually considered to be fifteen to twenty years, sometimes more. The third

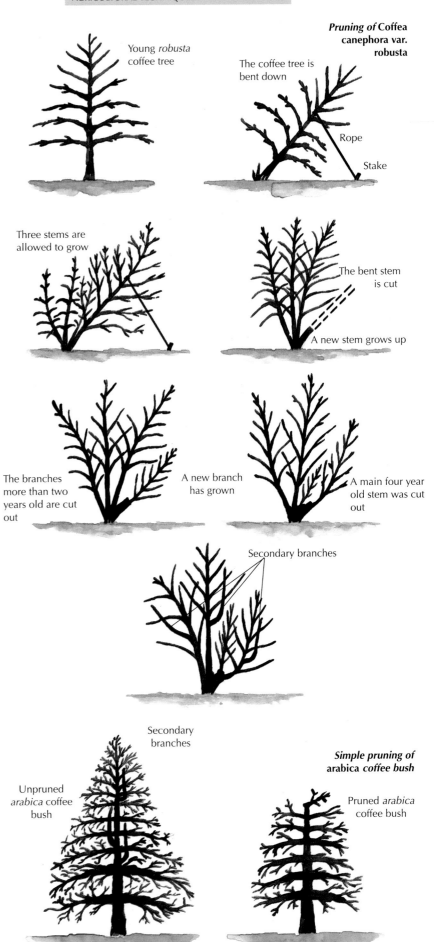

Pruning of **Coffea canephora var. robusta**

Young *robusta* coffee tree

The coffee tree is bent down

Rope

Stake

Three stems are allowed to grow

The bent stem is cut

A new stem grows up

The branches more than two years old are cut out

A new branch has grown

A main four year old stem was cut out

Secondary branches

Secondary branches

Unpruned *arabica* coffee bush

Simple pruning of arabica *coffee bush*

Pruned *arabica* coffee bush

period is physiological decline, which ends in the death of the plant. From an agricultural perspective, the third stage is of no commercial interest, as the yield is declining, and so the plant should be uprooted.

13.13.7. Pruning

In its first years, the plant produces many branches, and the first year's production of beans is usually very abundant. However, the coffee bush can age rapidly, and it has to be pruned to rejuvenate it and maintain its production. If it is allowed to grow unpruned, it soon becomes large (10 to 15 m), and this makes harvesting very difficult.

The trees of the species *C. arabica, C. liberica, C. abeokutae* and *C. dewevrei* only have a single trunk. This is very easy to prune. The apical bud, the tip, should be cut to ensure the tree does not exceed a height of 1.5-2 m. Then the branches near the base of the trunk are cut. Only the thickest and best branches borne on the trunk are left. All the small branches that are borne on the trunk should be cut out, and all the dead, withered or sick branches should be cut too.

The coffee plants of the species *C. canephora* have several main stems. The stretches of stem that bore flowers last year will not flower and only the new terminal buds produced from the old branches will produce beans. Pruning these varieties is rather more complicated. The main stem has to be bent from the vertical with a rope tied to the ground. The bent trunk will give rise to new upright stems that will bear new fruit, and finally the former main stem must be cut. All the smaller branches should be removed, as they will not bear fruit, and any diseased, withered or dead branches should also be removed.

In both cases, the bean production of very old trees declines, and they have to be rejuvenated as follows. In the rainy season all the main stems are cut, except for one. Fertilizers are applied to encourage the appearance of new stems. When these new stems have developed sufficiently, the old stem is cut, leaving the stump, and more fertilizer is applied.

13.13.8. Irrigation

In the regions where the dry season is long and harsh, irrigation is needed. This is common in Arabia (Yemen), Mysore (India), Kenya, Brazil, etc. Coffee can be irrigated by gravity or by sprinkler. In the first case, the water is transported to the higher parts of the plots, and then flows downwards from coffee plant to coffee plant, along channels that follow the slope of the site. The main disadvantage of this method is that it causes surface erosion, often made worse by heavy washing of the soil.

Sprinkler irrigation has the advantage of not causing erosion on steep sites, though it has the disadvantage that installation is expensive, requiring a pump, network of channels, irrigation devices, etc. It also saves water, about 20-40% with respect to gravity irrigation. It can also be used for irrigation of coffee grown on flat sites, where gravity irrigation would be very difficult. Sprinkler irrigation also creates a moist microclimate in the coffee plantations, which is very good for the foliage and the plant in general.

Water may have to be supplied to the coffee plants to make up for insufficient rainfall during the critical fruit formation period. That is to say, coffee must be watered immediately after flowering. When rainfall is very low, irrigation may help the bushes to resist the long seasonal dry periods better.

13.13.9. Herbicides

In their definitive location, weeds usually grow between the rows of coffee plants, and so weeding should be continued. In many areas, coffee plantations are still weeded by hand. It can be performed using machinery, taking care to avoid damaging the trunk or roots of the plants, or using herbicides. Products that are persistent in the soil such as **diuron** and **simazine** are used as preventive herbicides at a dose of 1 kg/ha. The use of herbicides to eliminate weeds after they germinate, such as **2,4-D**, **MCPA** and **dalapon**, preserve the soil structures, and means that they are preferable to traditional manual or mechanical methods.

The elimination of certain perennial grasses is a serious problem, as the use of **dalapon** at high doses may damage the young coffee bush. Books on agriculture recommend using this and other herbicides, such as **diuron**, **linuron**, **triazine**, etc., though they should be applied with all the due precautions, as the coffee bush's response to the different active ingredients depends on many local factors, the weed species, the rainfall, the soil's characteristics, etc.

The coffee berry borer (Stehanodenes hamjei)

13.13.10. Accidents, diseases and pests

• Accidents

The main accidents affecting coffee bushes are due to excess light and heat. When the temperature rises above 30°C, coffee is negatively affected, especially if the air is dry. Continued transpiration dehydrates the tissues, the foliage withers, turns blacks and falls. The bush makes use of the next rain to restore its foliage, and so the production is considerably reduced. Low temperatures, close to 0°C, are very negative, especially for the leaves and buds. If low temperatures persist, they may damage the trunk, creating entrances for fungal diseases.

Water requirements are as discussed in the section on cultivation requirements. Below 800-1,000 mm average annual rainfall, even if they are evenly distributed over the course of the year, growing coffee is unreliable and the production may vary greatly. Irrigation can make up for part of this shortfall or uneven distribution of rainfall. Nutritional imbalances may also seriously affect production; a uniform chlorosis of the foliage indicates nitrogen deficiency (it is easy to detect visually). Other deficiencies (potassium, phosphorus, iron, zinc, magnesium, etc.) are harder to detect and soil samples should be sent to a specialist laboratory for analysis.

• Diseases

Coffee is affected by several fungal diseases, but most damage is caused by **coffee leaf rust** (*Hemileia sp.*) which attacks the branches and leaves. The fungal mycelium invades the leaf tissues, revealing its presence by the appearance of small yellow translucent spots that join up to form patches that occupy a relatively large part of the leaf blade. The center of the patch frequently shows necrosis. On the underside of the leaf a fine yellow powder can be detected, the fungus's fruiting bodies. The diseased leaves, whose chlorophyll has been destroyed, are shed and the tree enters decline and eventually dies. The most normal way of combatting this is to stop growing *C. arabica* and grow varieties less sensitive to this fungus, such as *C. canephora*, *C. liberica*, etc.

Tracheomycosis (*Fusarium xylaroides*) is another fungal disease, which attacks the branches, stems and trunk, causing the death of the tree. Preventive measures must be taken, such as cutting off infected branches and burning them, and disinfecting pruning wounds and any other damage by dusting with **Bordeaux mixture**.
Other diseases attack the fruit, reducing its quality, for example **anthracnose** (*Colletotrichum coffeanum*). This takes the form of sunken patches on the fruit, which may appear at any stage of their development. The fruit darken, turn black and rot, and are then shed. This disease can be fought by repeated applications of copper-based compounds.

Other pernicious fungi affect the roots and collar, such as **white rot** (*Leptoporus lignosus*), **dark rot** (*Phellinus lamaoensis*) and **black rot** (*Rosellinia bunodes*). Almost the only measure that can be taken is good cultivation practise, such as uprooting affected patches and burning them, not replanting with coffee for a period of three or four years on plots affected by the fungus, and correct soil problems before replanting (drainage, fertilization, etc.).

• Pests

The insect pests attacking coffee trees can be divided into groups depending on the part of the plant that they attack. One of the most destructive animals that attacks the cherry beans is the berry borer (*Stephanoderes hamjei*). Originally from Africa, this insect now attacks coffee plantations all over the world. The adult is 1.5 cm long and a blackish brown color and a body recalling a hedgehog, because it is covered in dark bristles. The legless larvae are white and crescent-shaped. The adult female lays the eggs in deep holes in the fruit. The larvae mine galleries in the young seeds and feed on their reserves. All the blackened cherries should be removed and burnt, as should those that have fallen to the soil. The pesticides that can be used include **lindane**, **parathion** and **endosulfan**. The parasite may be present in harvested grain, and it attacks stored grain if contains more than 12.5% moisture.

The fruit may also be attacked by dipteran flies, the **fruit flies**. There are three species that attack coffee *Ceratitis capitata*, *Pterandrus fasciventris* and *Trirhithrunis coffeae*. A moth attacks the berries, *Thiliptoceras octoguttalis*. Other pests of flowers and fruit include bugs (hemipterans) like *Volumnus obscurus*, *Lygus sp.* and the coffee bug (*Antestiopsis lineaticolis*).

Leaf and stem **miners** are also frequent, and include *Bixadus dierricola*, *Anthores luconotus* and *Apate monachus*. Their larvae mine within the stem, and treatment consists of removing the affected branches. If the entire plant is infested it must be uprooted and burnt. They can be treated with chemicals like organophosphates. To finish, there are soil insects that attack the roots, such as the African mole cricket (*Gryllotalpa africana*) or the cricket *Brachtripes membranaceus*, that cause serious damage to the roots they feed on. They can be treated with granulated insecticides such as **lindane**.

13.13.11. Harvesting

When harvesting cherry beans for sowing to produce new plants, only ripe beans should be collected. The pulp is then removed and the grains are left to dry in the shade. If they are left in the sun, the grains dry too fast and the embryo may die. The seeds should not be kept for more than two weeks. If they are stored for longer, the embryo may die and the seed will not germinate.

Harvesting for consumption has to be carried out three or four times over the course of the year. Only the red beans (the "cherrus") are harvested from the tree, and not the green immature ones. Coffee yields range from 100-500 kg/ha in Africa to much higher yields in South America, sometimes more than 1,000 kg/ha.

After the harvest, great care must be taken in storing the cherries, to ensure that they do not rot. The most common way is to scatter them on dry ground that is well protected from rainfall. Once the beans have dried, the pulp has to be removed from the grain. This can be done by hand, or using special machinery to remove it. After this, the coffee seeds are winnowed, when any pulp or skin still attached to the grain is removed. Special machinery is available to do this, but it is usually done by hand, using a sieve. After cleaning the grain it should be dried in single layers on trays placed somewhere dry. The seeds should be turned frequently.

Variety of long peanut. They contain 3 or 4 seeds.

13.13.12. Use

Coffee is widely drunk as a beverage all over the world. The organoleptic characteristics of the coffee obtained from *Coffea canephora* are less appreciated than those of *C. arabica*, which contains more caffeine. Caffeine occupies an important place in the pharmacopoeia. It is mainly used as a cardiac tonic. It also has physiological effects on the nervous, muscular and circulatory systems. It excites brain activity and is an effective diuretic. Coffee can be sold as grains or already ground, natural roasted or sugar-roasted, but some coffee is used industrially, in the preparation of instant (freeze-dried) coffee.
On the sites where coffee is grown the fruit, which represents 75% of the fruit's dry weight, is used as organic fertilizer and fodder for animals.

13.14. PEANUTS

The peanut is probably originally from Brazil. It is grown for its oil-bearing fruit, and the plant is also used as fodder. The largest areas of peanuts are in Asia (China, India, Myanmar [Burma] and Indonesia), Africa (Senegal, Sudan, Nigeria and Zaire), North America (US) and Latin America (Brazil and Argentina).
The peanut (*Arachis hypogaea*) is a member of the pea family (*Leguminosae*), and of the subfamily Papilionatae. It is an annual

herbaceous plant, usually growing to a height of 40-65 cm. The stems are branched at the base, and depending on the variety may be prostrate, semi-upright or upright. They are slightly hairy and are often square in cross-section. The peanut produces a taproot normally 15-20 cm long. The leaves are alternate, and compound with four equal leaflets, which are oval, hairless, entire and relatively matt on the underside.

The yellow flowers are hermaphrodite, with a butterfly-shaped corolla and bearing bracts at the base. The way the peanut produces its fruit is most unusual. After fertilization and flower shed, a meristem at the base of the ovary produces a slender, sturdy organ, the **peg**, with the ovaries in its pointed tip. This organ shows positive geotropism, i.e., it grows down into the soil. When it is a few cm below the surface it stops growing downwards, and the pod starts to develop. The peanut thus ripens underground.

The fruit is an elongated indehiscent pod that usually contains between two and four seeds. The pods, or shells, are netted on the outside, and contracted between the spaces occupied by the seeds. The seeds are covered by a yellowish, white or reddish "skin".

13.14.1. Varieties

Four varieties of peanut are cultivated in Spain, mainly in the Valencia region. The two-seeded peanut, or **short peanut**, the two or four seeded peanut, or **long peanut**, the **Moruno** and **Palma**. The short peanut has one or two seeds within each shell and is the most productive of all, with the seeds accounting

Detail of a peanut plant. It is a legume

for 78% of the total weight of the pod (only 22% shell). The long peanut is the most appreciated for eating as such. It looks better than any of the other varieties and its market price is also higher. The fruit of the **Moruno** peanut are similar in pod color to the two-seed variety, but the pod is much larger and the "netting" is larger. The Palma variety was formerly cultivated in the Balearic Islands, but is now cultivated along the coastline of Valencia. The aboveground parts of this peanut resemble those of the long peanut, but the fruit are more similar to the short peanut.

Apart from the varieties cultivated in Spain, there other international varieties, such as *Virginia, Rouge de Ludima, Improved Spanish, Jaune Paille, Tennessee Red*, etc. The productivity of each variety depends a lot on the type of soil and the regional climate; furthermore, market demand influences whether to cultivate one variety or another. Different authors agree that the best way to cultivate peanuts is to begin by choosing varieties that are known to grow well locally, and then to cultivate small plots of the new improved varieties that are on the market.

13.14.2. Place in rotations

Peanut cultivation, due to its high nutrient requirements, is considered to impoverish the soil. It should therefore not be repeatedly grown on the same site, as the crops are much lower in the second and third year.
In general, the peanut is a suitable crop to follow other legumes, which supply nitrogen to the soil, and to follow crops, such as potatoes, that are given abundant dung and fertilizer. If peanuts are grown after cereals, fertilizer has to be applied to restore nutrients.

13.14.3. Cultivation requirements

• **Temperature** and **humidity**

Peanuts prefer a hot dry climate: excess moisture or abundant rains damage the plant, leading to small, poor quality, harvests.

• **Soil requirements**

Peanuts prefer fertile open soils. They should not be grown on compact soils, partly because it is hard for the peg to grow into soil and partly because the fruit harvested would be very irregularly shaped. On rather heavy soils, the *Moruno* variety may be the best. On medium consistency open soils, the two-seeded variety would be expected to do best.

The long-seeded variety should be sown on loose soils, as its larger fruit may not develop well in compacted soils or open ones. Furthermore, the peanut is harvested by **uprooting** (explained below) because the flower stalks (peduncles) of this variety are weaker, and many of the fruit are separated from the stem and remain in the soil. So when this variety is sown in hard soil, it should be harvested mechanically or using a hoe.

13.14.4. Applying fertilizer

The peanut requires a lot of mineral nutrients, but does not usually respond very well to fertilizers applied while it is growing. Applying fertilizer can thus be omitted when peanuts follow a crop requiring a lot of fertilizer, such as potatoes. Applying fertilizer can also be omitted when peanuts follow other legumes (beans, vetches, etc), because they incorporate nitrogen into the soil. Because the peanut plant is removed entirely from the soil, with the roots, stems and leaves for use as fodder, growing peanuts impoverishes the soil in organic matter and nutrients, and so the peanut should be considered a plant that exhausts the soil.

If peanuts are grown after crops that receive little fertilizer (cereals), fertilizer should be applied at depth. All varieties do better on soils rich in calcium because the tough shell contains a lot of calcium, and so the fertilizer applied at depth should contain phosphorus in the form of calcium superphosphate. The units of nitrogen should be split into two or three applications, between germination and flowering. Nitrogen should not be applied after flowering, as this has a negative effect on the subsequent ripening of the pods in the ground.

The following are guidelines for the amount of fertilizer to apply, per hectare: 300 kg of ammonium sulfate, 500 kg of calcium superphosphate and 200 kg of potassium sulfate. If these binary fertilizers contain 21% N, 18% P_2O_5 and 50% K_2O, then the amounts to be applied will be:

$$63 \text{ kg N}$$
$$90 \text{ kg } P_2O_5$$
$$100 \text{ kg } K_2O$$

Note that the amounts of nitrogen recommended are lower than those of the other elements. As mentioned previously, legumes form a symbiosis with bacteria that can fix atmospheric nitrogen, and this must be taken into account when calculating how much nitrogen fertilizer should be applied. Peanuts rapidly show symptoms of deficiencies of secondary elements and micronutrients, and is even used in some laboratories as a test plant for deficiencies. It frequently suffers from shortage of calcium, magnesium, iron, boron, copper, zinc, etc. For information on deficiencies, refer to the fifth section of this work.

13.14.5. Sowing

• Soil preparation
After harvesting the previous crop, if the earth is very solid, it is worth carrying out a shallow pass of the moldboard plow, and leaving the soil exposed to the sun for a few days. If it is cultivated after the potato or any other crop that does not leave the ground very compact, this need not be done, and the cultivator can be passed directly. A few days before sowing, water the soil to make sure it is in good tilth. Then prepare the seedbed by passing the disc harrow, the roller and then a rake. As mentioned in the chapter on site preparation, some of these passes can be omitted, depending on the state of the soil.

The furrows should be made when the soil is

in good tilth, in order to ensure the peanut seed does not suffer from too much or too little water. If cultivation is not mechanized, the furrows are usually 50 to 55 cm apart, and the seeds should be sown at a distance of 20-25 cm in each row. If cultivation is mechanized, wider spaces have to be left between rows so that the tractor can circulate.

• Sowing
Sowing is by drill, depositing two seeds at a time in the open furrows; the furrows are then filled in, and the rake is passed to firm and level the soil. The seed should be buried at a depth of 6 to 8 cm. If two or three seeds are deposited at a time, this represents about 80 to 100 kg/ha of shelled peanuts. The seed can be sown by mechanical sowers that sow one or more rows at a time. The seed should be shelled before sowing. Shelling can be done by hand or by machine, but requires care, as any damage to the peanut is an entry point for fungal diseases causing rots.

The best time to sow depends on the local climate. In temperate zones, it is normally in early May and it should not be left much later than this in order to ensure that the peanut does not ripen during the autumn rainy period. When the rains come before the fruit are ripe, they prevent harvest of the peanuts, which turn black, reducing their quality. In some circumstances, the peanuts may even germinate before they are harvested. Germination occurs at temperatures of about 12°C, and they take about twelve days to emerge above ground.

13.14.6. Irrigation

Once the plant has started flowering, about a month and a half after sowing, the plant is watered for the first time. Watering is delayed to force the plant to flower, so it produces its fruit as soon as possible. This means that the pegs can grow into the soil more easily, as the plant is still very short when it flowers; it also means that a higher percentage of fruit will be totally ripe, and the harvest will therefore be more uniform.

Fifteen days later, the plant is irrigated a second time, and this is followed by two further waterings, at intervals of fifteen days. It is not a good idea to give any more water until just before harvesting, when the crop is watered for the last time, if necessary, to bring the soil into a good tilth to make uprooting the plant easier. The peanut thus usually receives four or five waterings, not including the one given before preparing the site.

13.14.7. Herbicides

The **banking plow** is widely used in peanut cultivation. This tool, unlike the plows described in the section on machinery, is symmetrical and displaces the earth on both sides. Eight or ten days after the plants have emerged above ground, it is worth carrying out a pass with this tool between the rows, which gets rids of the weeds and breaks the soil lumps, and helps the later banking up of earth around the plants' stems. Fifteen days later, earthing up is repeated with the same type of plow, but at greater depth.

The second pass with the plow gets rid of weeds, prepares furrows for irrigation at flowering time, and more importantly breaks the soil down to a fine tilth so that the pegs do not meet problems when they grow into the soil. Banking up the soil around the peanut plants brings the earth to the pegs, and thus increases the harvest.

Chemical herbicides can also be used. The active ingredient **benthazone** is widely used as a herbicide post-emergence of crop and weeds. It selects against the weeds of the peanut, and should be applied when the peanut has produced two of three compound leaves.

13.14.8. Accidents, diseases and pests

• **Fungal diseases**
The most important fungal disease of peanuts is **Cercospora** infection, caused by different species of *Cercospora*. The symptoms are small circular patches no less than 1 mm in diameter, that are pale or colorless to begin with, and then turn dark brown or black. It affects production directly or indirectly in different ways depending on when the plant is attacked. If infection is when the plant is young, it is defoliated. If it is late, the infection attacks the fruit directly. Attacks by this fungus can be prevented by simple cultural methods: not repeating the crop two years in a row keeps the population of he fungus down to acceptable levels; keep the soil free of weeds, many of which are also attacked by the fungus; preventive treatment with classic fungicides like **Bordeaux mixture**, **copper oxychloride** and **copper** (I) **oxide** is effective, though organic ones, such as **zineb**, **ziram**, **maneb** and **captan**, can also be used.

There are other less important fungal diseases, such as powdery mildew (*Erysiphe cichoriacearum*), which also attacks, tobacco, and also the melon and other members of the squash family (Cucurbitaceae). It can be controlled with **dinocap**. Rots that attack the collar are also important, such as *Fusarium, Sclerotium, Rhizoctonia, Diplodia,* while *Penicillium, Aspergillus* and others attack the fruit. These are usually dealt with by cultivation methods like sowing disinfected seed, growing resistant varieties, or even disinfecting the soil before planting the crop.

• **Pests**
The red spider mite (*Tetranychus telarius*) causes serious damage. It grows on the underside of the leaves it feeds on and emits fine threads. It can be controlled by dusting with sulfur, though more effective acaricides are now used, such as **naled**, **phosalone**, **tetradiphon** and **dicophol**.

Different species of leafhopper (*Empoasca*) can attack peanuts, eating the plant and causing serious damage. Organophosphate insecticides, such as **methyl parathion** control them effectively. Other insects, such as the tobacco thrip (*Thrips tabaci*), can also attack the peanut. The leaves show a range of bitemarks and deformations due to the bites of both the larvae and adults. Common insecticides like **lindane**, **carbaryl** and **diazinon** are widely used to treat them.

Because the peanut's fruit grows underground, special attention should be paid to any attack by soil insects. The insects already discussed for other crops, cutworms, screwworms, wireworms, etc., can damage a large number of the fruit. They can be controlled using modern granulated or powder insecticides that are made specially to be incorporated into the soil. Pesticides like **trichlorphon** and **lindane** are very frequently used.

13.14.9. Harvesting

The best time to start harvesting the peanuts is shown by the general status of the plants, which turn yellowish and the lower leaves start to wither. The harvest should not be left for long after this because the plants are uprooted whole, and any rot in the neck region may cause the stem to break and leave the fruit in the ground. As long as the aboveground parts remain green, harvesting should be delayed so that as many fruit as possible can ripen fully, as this means the final yield will be higher. About forty days elapse between peg formation and fruit ripening.

If harvesting is delayed and the autumn rains arrive, the peanut seed may start to germinate, and this can be detected by the presence of split pods with the emerging roots of the seedlings. This should be checked visually by uprooting a plant to inspect the seeds. The two-seed varieties are simply uprooted by pulling, as can four-seed varieties in very open soil. In harder soils, four-seed varieties should be harvested by hoeing. It is now possible to harvest the crop using the same machines as used to harvest potatoes.

After uprooting the plants, they are usually left on the ridges for two or three days to dry out, when the plants and fruit lose 60-65% of their weight in water. The plants are then threshed to separate the fruit from the plants. The fruit are then taken to the threshing ground to finish drying, and they are fully dry when the seeds rattle within the pods. They are then winnowed to remove any remaining earth, plant remains and empty fruit. The peanuts are now ready for storage or sale.

13.14.10. Use

Peanuts contain a great deal of energy (5.9 cal/g), calcium (1 mg/g) and phosphorus (4 mg/g). They can be eaten directly, though they are normally processed industrially before consumption. One of the most common methods is to roast and salt them. They can also be fried in oil and then salted. The peanut can be salted, sugared, roasted, etc.

Peanuts can also be ground to make a meal, consumed by human beings and to make a cake used in making glues for binding and sticky paper.
Oil is also extracted from peanuts, whose cotyledons contain a lot of oil. Peanut butter is made by roasting, bleaching and finely grinding the fruit.
A protein is also obtained that is used to make a textile fiber with properties similar to wool (*sarelon*).
Peanuts are also used in industrial chocolate production.

Tea plantations in Japan. Mount Fujiyama is visible in the background

Peanuts are used in animal feeds as peanut cake for feed production; after threshing, the plants are often fed to animals.

13.15. TEA

The zones where tea grows naturally are in India (Assam valley) and China (Yang-Tse region), and its wide acceptance and increasing consumption led to tea cultivation spreading to many other countries. Ceylon, now Sri Lanka, is one of the world's major tea producers, as the island's temperatures and climates are suitable for tea. The British introduced tea cultivation and processing into Ceylon in the 19th century to meet the growing demand. It gradually spread to different regions all over the world. It spread from Ceylon to South America, where it was taken to areas of Brazil and Argentina with suitable soils. There are now 42,000 ha of tea cultivated in Argentina, producing 41,000 tonnes a year. The area under tea in Brazil is much smaller (less than 5,000 ha), but its production is relatively much higher, more than 10,000 tonnes a year.

In Africa the main producer countries are Kenya (50,000 ha) and Malawi (20,000 ha). Other producer countries, such as Peru, Ecuador, Cameroon and Congo, have invested in tea production, and though it is not a basic product of their economy, the results cannot be ignored and the yields are acceptable. Another country with high production is Russia, where work to adapt tea to the local climate has been successful.

The tea plant belongs to the tea family (Theaceae), and consists of two species. *Thea sinensis* is originally from China, while the tea from India (Assam valley), is *T. assamica*. In the wild the tree can reach a height of more than 10 m, and so in plantations the shoot tips are removed to prevent excessive growth, leaving the plants just 2 m tall in order to make collecting the leaves easier. When pruned this way, the tea plant is roughly shaped like an upside down cone with its tip pointing towards the trunk, which is straight and bears the main branches. These primary branches, which are not straight due to the presence of relatively separated nodes, will bear the secondary branches, which all grow to a similar size, giving the tree a harmonious and well-balanced appearance.

The leaves, whose color varies from variety to variety, are lanceolate, petiolate, serrated and alternate. Each one has a central vein, which is tough and bumpy on the underside, and bears the other veins, which are also strengthened, and this gives the leaf its typical wavy look. When they are young (leaf age is a very important characteristic in tea), the leaves are covered with a fine down that disappears as they grow.

The flowers are the same size as those of the cherry, and are very similar to their close relatives, *Camellia*, though they are not as robust and a purer shade of white. Single flowers are borne in the axils, or in groups of two or three, and generally have

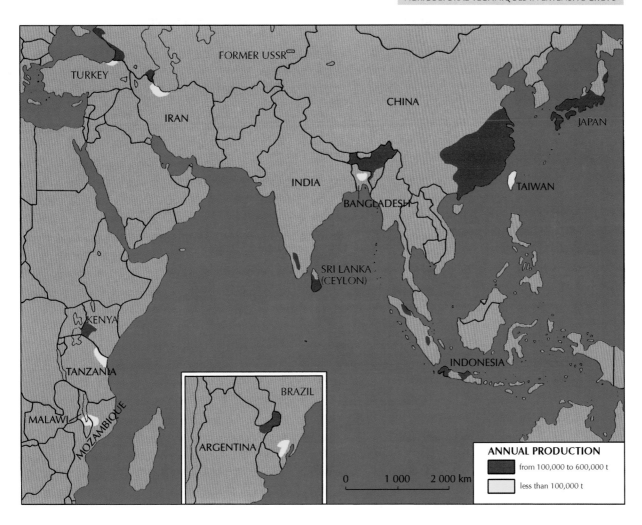

Map of the world's main tea producers

ANNUAL PRODUCTION

from 100,000 to 600,000 t

less than 100,000 t

six to nine petals, depending on the variety. Flowering usually takes place at the beginning of the winter period. The fruit is a three-lobed fruit, each lobe containing a bitter-tasting seed that if eaten causes abundant salivation and nausea.

13.15.1. Cultivation requirements

• Temperature and humidity

The effective culture of tea requires temperatures that vary between 18 and 21°C, strong but not prolonged sunshine, and a high level of humidity. Thus in tropical zones it must be grown at altitudes of more than 600 m. This all depends on the species, as *T. sinensis* resists low temperatures (10°C) much better than *T. assamica*, which is more adapted to a tropical climate.

• Soil requirements

The soil has to be rich in organic matter and with a loamy texture in order to retain moisture without getting waterlogged, as the roots require good aeration. In excessively clay soils, the roots rot due to excess moisture, with negative effects on leaf color and quality, and reducing the organoleptic qualities of the resulting infusion. The most suitable sites for tea cultivation are mountain slopes and steep valleys, where the water can drain rapidly due to gravity.

13.15.2. Sowing

• Soil preparation

Before sowing the seed, the soil should be suitably prepared by removing the forest to get rid of the trees and vegetation that might compete with the crop. This should be done in the driest season, so that the soil is prepared to receive the rain when they arrive. On some occasions it may be worth leaving some trees to provide protection for the crop against excess sun, and more importantly against the wind, which is very damaging to the delicate leaves and to the site itself, desiccating them both.

Once the forest has been removed, it is worth installing good drainage, especially if the plantation is not on a slope. Then prepare ridges with raised lanes on both sides to make it easier for the laborers to carry out the different operations needed in tea cultivation and harvesting. Finally, a suitable irrigation system should be installed to ensure a constant supply of water for the plantation even when there is little rainfall.

• Sowing

Once the site has been prepared, the seed can be sown. The seed used should be rigorously selected, in order to select the best

In many zones of Asia, tea is still harvested by hand. A plantation in Sri Lanka.

quality specimens, and reject all those that are not vigorous and healthy enough to produce strong plants. It is also very important to detect and eliminate all the seeds that may be damaged by fungal diseases, as they may ruin the entire plantation if they spread through the soil to the other plants.

In China, the seed is sown in holes 30 to 40 cm deep. The hole is filled with fertilizer or dung, the seed is sown at a depth of about 5 cm, and is then covered with a layer of topsoil. A hectare of ground can hold up to 8,000 plants, though the figure varies from country to country. After sowing, if the soil is ideal and contains enough humus, no more chemical fertilizer need be applied, though it is important to check their nutrient status. If needed, the necessary amounts of nitrogen, phosphorous, potassium, calcium and magnesium should be applied.

13.15.4. Harvesting

In Asia, tea is harvested by hand, though in the former Soviet Republics (Azerbaijan, Georgia, etc.) mechanical harvesters are used. These are tractor-drawn devices, that cut and blow the leaves into the corresponding trailers. The irregularity of the sites where tea is usually cultivated makes mechanization difficult. When the tea plant is in full production, at an age of three years, the leaves are harvested several times a year.

13.15.5. Use

Depending on how the leaves are treated, there are three types of tea, **black tea**, **green tea** and **oolong tea**. All three are derived from the same leaves, but they are processed in different ways.

Black tea is the most appreciated in the West and in much of the world. The leaves are first desiccated (they are exposed to open air or put in a draft of hot air). About twenty hours later, the leaves are put into rotary presses, where the cells are broken to release the sap and enzymes. At this stage the tea leaves acquire their characteristic curly shape and start to change color. The leaves are pressed together when they leave the machine, and have to be separated with mechanical sieves. The smallest are the first to pass through the sieve, and the leaves are graded as they separate.

In Japan, pruning is mechanized, and this saves a lot of labor.

13.15.3. Pruning

Tea has a useful life of 25-40 years, starting production in its third year when the leaves can start to be collected. This is the best time to carry out the first pruning, which aims to stimulate the production of the young leaves that produce the best tea. After this pruning is carried out every two and a half years. The second pruning is performed three months after the first and does not produce high quality leaves, but is necessary to favor the continued production of young shoots that, after the new harvest, will reach the highest level of quality and can be harvested at an interval of two or three weeks until the plant's rest period. The annual growth halt is easy to recognize because the terminal bud ceases to grow.

The leaves are then fermented. The leaves are placed on glass or cement in a cool moist atmosphere. After three hours the leaves turn a coppery color. The next stage in the process is **firing**, which consists of treating the fermented leaves to continuous blasts of hot air. This stops the fermentation and ensures the black color, that will later color the water in the teapot. Finally, the leaves are freed of impurities, then selected and classified.

Before they are packed, the leaves are divided into two grades, *broken* and *leaf* (whole).
The *broken* grade includes the smallest leaves, both those that were originally small and those that were broken in the different phases of processing, though this detail has nothing to do with their quality. The difference between the two grades is that *broken* tea produces a darker and stronger infusion than *leaf*. The broken leaves are then divided into *Broken Orange, Broken Pekoe, Broken Pekoe Souchong, Fannings* and *Dust*. The whole leaf grade is used to classify the largest leaves into *Orange Pekoe, Pekoe* and *Pekoe Souchong*.

Green tea is more appreciated in some countries, such as Algeria, Morocco, China and Japan, and is different because it is processed differently. The ferments and enzymes that remain in the leaf of black tea after withering, are destroyed in green tea, not by means of oxidation or previous exposure to fans, but by steam treatment before curling which maintains their green color, and will later give a bitter and more astringent taste. The process of producing green tea is thus not as complex as that for black tea. The leaves are manipulated less and do not break. This is why green tea is not graded on the basis of size, but by leaf age and the treatment undergone.

Oolong tea is obtained by a method halfway between that for green tea and black tea. To produce oolong tea the fermentation process has to be interrupted, so the leaves are only half fermented. It is then treated with steam like green tea. The resulting tea has a dark brown color and can be treated in different ways to enrich its aroma and taste.

As is well known, tea is usually drunk as a hot infusion, though it is also used for other purposes than as a drink. Tea is an excellent tonic for all skin types. Rubbing the skin with a piece of cotton soaked in cold tea and then leaving it to dry helps to revive the epidermis. If a cup of tea is used in the final rinse after washing the hair, it recovers gloss and color; brown hair becomes lighter and gets highlights, while dark brown and black hair become more shiny.

Industrial processes exist that produce caffeine-free tea. Soluble extracts of tea are prepared by evaporating the water from an infusion. There are now cold drinks on the market that are based on tea, either alone or mixed with lemon or other flavoring agents.

13.16. COCOA

The Europeans first learned of cocoa when they reached the Americas. The indian people living between the Isthmus of Tehuantepec (Guatemala) and the Darién mountain range (Panama) used cocoa as both a food and as currency. In South America, in the basin of the River Magdalena and other tributaries of the Amazon, the cocoa's pulp was used to make an

Leaves and fruit of cocoa tree (Photo courtesy of A. GOSTINČAR for this publication)

alcoholic drink. From the 16th century onwards, cocoa cultivation spread to the islands of the Caribbean and in the 17th century it spread to the Philippines. In the 19th century, cocoa spread to Ceylon (Sri Lanka), Java and Brazil, from where it spread to Africa. The most important producers are now Ivory Coast, Brazil, Nigeria and Ghana. It is cultivated throughout the tropical zones, including Oceania.

Cocoa (*Theobroma cacao*) or cocoa tree, is a member of the Sterculiaceae family. It reaches a height of 5-7 m in plantations (where it is pruned) and 8-10 m in the wild. The evergreen leaves are alternate, petiolate, bright green, smooth, hard and oblong. The plant produces taproots. The small flowers are white or pink and are borne in small bunches on the trunk and larger branches (i.e., it is cauliflorous). The cocoa tree's fruit is a pod 20 cm long and 8 cm wide. Its surface is hard and warty, with five or ten longitudinal grooves, and is yellow or orange. The mature pods contain 30-40 seeds. The flattened seeds are in five rows and surrounded by a sticky pulp.

Cocoa tree pods

Seeds

Shell

Pulp

Longitudinal section of a cocoa pod

13.16.1. Varieties

There is a wide range of cultivated cocoa, and the different varieties can be divided into three main groups. **Criollo** cocoas have elongated fruit that are yellow or red, with a sharp tip and large seeds, and have long been cultivated in Venezuela and Central America. This variety is not very productive, but the cocoa it produces is highly appreciated. The **Amazon** cocoas have rounded fruit that are smooth with shallow grooves, flattened grains and short yellow cotyledons. The last group, the **Trinidad** cocoas vary greatly. The pods may be short or long, red or yellow, and intermediate between the other two types. This means that they probably originated as hybrids between criollo and Amazon cocoas.

13.16.2. Cultivation requirements

• **Temperature, light and moisture**
Cocoa is a tropical lowland plant, with an

Left: Costa Rica, a cocoa treatment and sales point. Photo courtesy of A. GOSTINČAR

Right:
Fruit of the cocoa cultivated in Central America

optimal average annual temperature of about 25°C, and which requires the average of the minimum daily temperatures to be greater than 15°C. It needs average annual rainfall of more than 1,200 mm, and the dry season should be less than three months long. It needs shading, especially when young. This shade may be natural, the result of leaving trees when clearing the jungle, or artificial, by planting bananas (*Musa sp.*), cassava (*Manihot esculenta*), erythrina trees (*Erythrina*), etc.

• **Soil requirements**
Cocoa trees almost always adapt to any type of soil, and can tolerate a wide range of pH values, though the optimum is about pH 6.5. As they produce a taproot, they prefer deep soils, where they produce best.

13.16.3. Applying fertilizer

The cocoa plant is a tree, and so it does not make much sense to apply fertilizer throughout the grove, meaning that each tree should be fertilized independently. On sandy sites, you can choose a ternary 1.3-1-1.5 (13:10:13) fertilizer. On more clay soils, apply a fertilizer with less units of nitrogen and more of phosphorus: 1-1.25-1.5 (12-15-18). This is because the soluble units of nitrogen will not be washed as easily as on a sandy soil, and because it is worth increasing the supply of phosphorus so that the roots can grow better on a hard soil that is difficult to penetrate.

The quantity to be applied should be 250 g per tree applied twice a year (April and September) when the cocoa tree is still young, and increasing to 500 g per tree, also in two applications, when the trees are adult.

13.16.4. Sowing

Cocoa seeds can be sown directly, or plants from nurseries can be planted. The seed are sown in rows 2.5 to 3 m apart, with a distance of 2.5-3 m between plants in each row. This planting frame will give a density of 1,000 to 1,600 plants per hectare.

The seed should be sown, or seedlings from nurseries transplanted, when the tropical rainy season starts. The plants need a lot of water, especially just after sowing. They should be watered frequently. When the plantation is young, weeding should be performed four or five times a year. When the trees have grown, their shade prevents the growth of weeds, meaning that weeding need not be so frequent.

13.16.5. Accidents, diseases and pests

Some types of virus affect the cocoa plant, though they have not yet been thoroughly described. The most important **fungal diseases** are **black pod** of the pods (*Phytophthora palmivora*), and the **witches' broom** caused by *Marasmius perniciosus*. The plant may also be attacked by several rot fungi. Insects like **capsid bugs** and **miners** bore into the plant's branches, fruit and trunk, and they can be controlled by applying any organophosphate pesticide.

13.16.6. Harvest

The cocoa produces its first flowers two years after planting. These flowers should however be removed so as not to exhaust the young plant. The first pods can be harvested in the fourth year. Only pods that are completely ripe should be harvested, and this can be recognized by the characteristic yellow or red color. The green pods are left to be harvested later. Harvesting is by hand and is performed every fifteen days.

13.16.7. Use

The cocoa seed, or "bean", is a complete food, as it contains fats, sugars, proteins, and mineral salts (especially phosphorus and potassium). It is a stimulant because it contains theobromine and traces of caffeine. Theobromine is used pharmaceutically in the production of stimulants and diuretics. Cocoa is the basic ingredient of chocolate.

To produce cocoa butter, from which chocolate is manufactured, the grains are left in the open air, stored in boxes where they are subject to a preliminary fermentation, encouraged by the fruit's acid and sugar-rich pulp, which favors the growth of a complex microbial flora. This fermentation takes five to seven days in Amazon cocoas and two to three days in criollo varieties. The seeds start purple and are reddish when it is over. After fermentation, the cocoa seeds are sun-dried (5 to 10 days) and then gently roasted. The shell is removed with fans. Finally the cocoa is ground in **ball mills**. The grinding generates a lot of heat. This melts the fat and turns the cocoa into a semi-liquid paste.

One byproduct of cocoa paste production is **cocoa butter**. This is the fat contained in the cocoa seed. It is partially separated by pressing the paste obtained from the grains. It is a solid mass that melts on the tongue, yellowish white in color, and smelling and tasting of cocoa. Cocoa butter derivatives are used in the pharmaceutical industry as excipients in suppositories and lotions or in sticks of lip balms.

13.17. SUGARCANE

Sugarcane plantation

Saccharum officinarum, the sugarcane, is a perennial member of the grass family (Gramineae). It has a thick rhizome that bears many upright stems that may be 2 to 6 m tall, and 2 to 8 cm in diameter. These stems consist of a sugary tissue, sappy and spongy, from which the sugar is extracted. The leaves are narrower and longer than those of the common cane, and are borne more densely. The terminal panicle is pyramidal and consists of many purple or reddish flowers.

This plant is only known in cultivation, and is the main source of sugar. Historians consider it was first cultivated in India, apparently more than 3,000 years ago.

Detail of sugarcane cultivation in Madeira

Gaps between rows allow irrigation and the pass of implements

From there it spread to Egypt (641 AD), to the Iberian Peninsula (755 AD), and then to tropical and subtropical zones throughout the world. The world's main producers of sugarcane, starting with the largest, are Brazil, India, Cuba, China, Mexico and Pakistan.

13.17.1. Cultivation

Sugarcane requires a hot wet climate. The best soils are alluvial plains near the sea, especially silty soils rich in humus and calcium. Sugarcane can be multiplied by seeds, but it is usually propagated by cuttings. They can be planted between July and September, depending on the region's climate. The most important fertilizer to supply is potassium, which plays an important role in sugar production. The canes are ripe about three months after flowering, ten or fifteen months after planting. When the plant is physiologically ripe, the stems turn golden or purple, and the plant's sap is very thick and sticky.

13.17.2. Weeds and insects

The different weeds that appear in sugarcane plantations (*Rumex, Digitaria, Echinochloa,* etc) can be treated with the active ingredient **asulam**. This is a systemic herbicide that can be absorbed by the roots or leaves, and should be applied in pre-emergence or post-emergence of the crop. In general, legumes are sensitive to this herbicide, which does not harm sugarcane. The crop is frequently attacked by **aphids**. They can be controlled with any of the commercial formulations of the systemic insecticide **ethiophencarb**, which acts by contact or ingestion.

13.17.3. Harvesting

To harvest sugarcane, the canes are cut down to the ground and then sent to a mill for pressing. The canes contain 80% sap, and this contains about 20% sugar. The rhizomes produce new stems the following year, and the plants can be grown on the same site for four or five years, and produce several harvests during this period. After this, the

plant should be removed, as cane production decreases. There are many varieties of sugarcane, but they are usually classified into four groups by the color of their leaves and stems. There are green, white or yellow, streaked and red canes.

13.17.4. Use

The main sugar in human foods is sucrose, mainly obtained from sugarcane and sugar beets. It is a pure easily digested sugar with a sweet taste and with no nutritional value other than as a source of energy, because all the vitamins, minerals and proteins are removed in the extraction and purification process. When the sugar is extracted from the sugarcane, the sucrose is accompanied by impurities such as organic acids, amino acids, etc. The sugar industry technology extracts and purifies the sucrose. The first stage in obtaining sugar from the canes is to mill and press the canes. The extracted juice is diluted in water; it is then cleaned with calcium (which precipitates the impurities). It is then carbonated (to get rid of the excess calcium), filtered, bleached, concentrated by evaporation (to crystallize the sucrose), and finally centrifuged to separate the sugar from the liquid impurities that do not crystallize.

Sugar is used as a complementary ingredient in foods and drinks, both directly and in the huge number of food products with added sugar. There are many different manufacturing processes, and they produce different sugars for human consumption.

These include: **demerara sugar**, obtained from the second production, whose color varies from light yellow to dark brown, depending on the amount of mixture left in the crystals; **semi-refined sugar**, which is molded into rectangular prisms; **flowers of sugar**, the first sugar obtained and very pure; **rock candy sugar**, reduced to crystals by clarification and slow evaporation; **caster sugar**, obtained by crushing the crystals and sieving them (also known as icing sugar); and **muscovado sugar**, the second production from the cane; and **white sugar**, or **refined sugar**, the purest form, obtained in refineries.

BIBLIOGRAPHY

ARNAL, P.V., LAGUNA, A.
Tractores y motores agrícolas
Madrid: Mundi-Prensa 2nd edition, 1989

BERNAT, C.
Maquinaria para agricultura y jardinería
Barcelona: AEDOS 1st edition, 1980

BERNAT, C., MARTÍ, R. & PLANAS, S.
Lèxic de maquinària agrícola
Generalitat de Catalunya. Departament
d'Agricultura, Ramaderia i Pesca. 1st edition.
Barcelona: 1993

BONNEMAISON, L.
Enemigos animales de las plantas cultivadas y fo-
restales
Barcelona: Ediciones de Occidente.
Volumes I ,II y III, 1964

BOVEY, R., BAGGIOLINI, M., BOLAY, A., BOVAY, E., COR-
BAZ, R., MATHYS, G., MEYLAN, A., MURBACH, R., PELET,
F., SAVARY, A. & TRIVELLI, G.
La defensa de las plantas cultivadas
Barcelona: Omega 2nd edition, revised and exten-
ded, 1984

CORNEJO, J. & GARCÍA, C.
El cacahuete
Madrid: Publicaciones de Extensión Agraria
2nd edition, 1973

COSTE, R.
El café
Barcelona: Blume 1st edition, 1969

DELOYE, M. & REBOUR, H.
El riego
Madrid: Mundi-Prensa 1st edition, 1967

DEPARTAMENT D'AGRICULTURA, RAMADERIA I PESCA
Butlletí d'avisos fitosanitaris
Generalitat de Catalunya.
Barcelona: 1987

DETROUX, L. & GOSTINČAR, J.
Los herbicidas y su empleo
Barcelona: Oikos-Tau, 1st edition, 1967

DUALDE, V.
Biología
Valencia: Gráficas Ecir 1st edition, 1983

ENCICLOPÈDIA CATALANA
Gran enciclòpedia catalana
Barcelona: Reprint of 2nd edition, 1990

FARBENFABRIKEN BAYER AKTIENGESELLSCHAFT
«Bayer» Pflanzenschutz
Leverkusen: *Compendium II,* 1964

FUENTES, J.L.
Técnicas de riego
Madrid: Instituto Nacional de Reforma y Desarrollo
Agrario, 1992

GUERRERO, A.
Cultivos herbáceos extensivos
Madrid: Mundi-Prensa 5th edition, 1992

LANZARA, P. & PIZZETI, M.
Guía de árboles
Barcelona: Grijalbo. 4th edition,1979

LIÑÁN, C.
Vademecum de productos fitosanitarios
Madrid: Agrotécnicas, 1998

MINISTERIO DE AGRICULTURA
Guía de aplicación de herbicidas
Madrid: Publicaciones de Capacitación Agraria.
1st edition, 1971

U.N.O.
El cacao
Roma: Spanish edition FAO, 1980
El café
Roma: Spanish edition FAO, 1980

ORTÍZ-CAÑAVETE, J.
Las máquinas agrícolas y su aplicación
Madrid: Mundi-Prensa 4th edition, 1993

PLANAS MESTRES, J.
Elementos de biología
Valladolid: Sever-Cuesta, 1975

PUJOL, M.
Conceptes de morfologia i biologia de les
gramínies
Published by the author.
Capellades (Barcelona): 1982
Els cereals: Generalitats
Published by the author.
Capellades (Barcelona): 1983

PUJOL, M., COMAS, J.
Quadern de pràctiques
Published by the author.
Capellades (Barcelona): 1983

ROS, J., NADAL, J., LLIMONA, X., ARTECHE,
A. & REGUERO, M.
Enciclopedia de la naturaleza y
del medio ambiente
Barcelona: Primera Plana, 1992

SERRAT, M.
Sendas del té
Barcelona: CPC España 1st edition, 1986

STRASBURGER, F., NOLL, F., SCHENCK, H. &
SCHIMPER, A.F.W.
Tratado de botánica
Barcelona: Marín. 6th edition,1981

5

Horticulture

13. "IN VITRO" CULTIVATION IN HORTICULTURE 646
BIBLIOGRAPHY —— 648

1. INTRODUCTION

Horticulture is generally thought of as an exploitation system using small areas of crop and low levels of mechanization, though there are extensive and industrial forms of horticulture.

Another important characteristic of intensive horticulture is rotation. These follow each other in rapid succession, and it is common for two or three crops to be grown on a single plot in one year.

In horticulture it is very important to thoroughly plan the sale of the produce in advance, identifying sales routes, because the produce is highly perishable.

It is worth having suitable storage facilities, and treatment after harvest is extremely important to maintain the produce in good condition for sale.

The main types of horticultural exploitation can be classified into:

• **Intensive horticultural exploitations**, where the area cultivated is not very large. They are rarely more than a hectare, and are highly specialized in a small number of crops. They normally aim to produce vegetables to be eaten fresh.

• **Forced or accelerated horticultural exploitations**, on equally small areas. Their aim is to produce very early or very late produce, and they rely on tunnels and greenhouses to create suitable climatic conditions for the crop to grow.

• **Extensive horticultural exploitations** cultivate larger areas than the preceding ones. They alternate horticultural crops with extensive crops like cereals and legumes. In these exploitations, the period between crops is longer than in the preceding forms, but the use of machinery is much greater.

• **Industrial horticultural exploitations**. Their main aim is to produce crops for industrial usage, and the produce may be tinned, frozen, made into jam, dehydrated, etc.

1.1. OPEN-AIR CULTIVATION

The climate is the average weather conditions at a particular place over a long period of time.

The climate is the most important limiting factor, especially for crops grown in the open air.

The environment of the horticultural exploitation is defined by the following factors:

• **Sunlight**. Sunlight is the source of energy for plants. Energy from sunlight powers carbon dioxide fixation, synthesis of organic matter and all the plant's vital functions. Plants need light to grow.

• **Temperature and frost regime**. Plants need warmth to grow. Each crop plant requires a given set of temperature conditions, which determine its growth and future production.

In general, plants grow more at higher temperatures, until the temperature reaches an optimal value, where growth is greatest, and further increases do not speed up growth, but eventually cause it to halt. The same happens as temperatures decrease.

Of all the components of climate, the temperature is perhaps the factor that most affects the growth and development of the plant, as it affects almost all its biochemical processes, such as photosynthesis, respiration and transpiration.

It is important to know the details of the regime of low temperatures and frosts of the site where the exploitation is located, as frost may completely destroy crop plants.

• **Moisture**. Moisture influences a wide range of physiological processes, often interacting with the temperature. Sites located in deep valleys that are always wet should not be chosen for horticulture, as production is uncertain and yields are low.

Other factors to bear in mind in open air horticulture, apart from the climate, are:

• **The soil**. The soil is the environment where the plant lives and grows. The relationship between the soil and the plant is thus very close.

The farmer can modify, to some extent, the arable topsoil, and thus improve growing conditions for the crop. Correcting the factors limiting crop growth is not always economically worthwhile for the horticulturist.

It is possible to greatly improve soil fertility by applying fertilizer, and to improve the soil structure by adding organic matter and by suitable tilling practises.

To make irrigation easier, the site should be flat with a slight slope. An excessive slope causes the soil to wash away, erosion, and is very inconvenient to till.

• **Altitude** The altitude limits the horticultural uses of some areas. An increase of 100 m in altitude represents a fall in temperature of 1°C, and maybe as much as 6°C.

• **Proximity to the sea**. This mainly influences the climate, making it milder in summer and winter. However, coastal sites may suffer greatly from salination and salty winds, which cause serious damage to the plants.

• **Dominant winds**. Strong winds, whether warm or cold, moist or dry, are bad for plants, and limit the growth of horticultural crops.

In areas with strong winds, if it is not possible to vary the orientation of the crop, protective structures are often needed. The most widely used structures for protecting crops from the wind are espaliers and windbreaks.

• *Espaliers.* These are light trellis structures that protect the row of plants. They consist of small canes or wooden stakes that support a material such as cardboard, plastic or straw.

• *Windbreaks.* Windbreaks are physical barriers that slow down the wind, and are sited between the wind and the crop to be protected. Windbreaks may be living, such as hedges, or inanimate.

Living windbreaks, hedges, have the disadvantage of competing with the crop.
Other materials used as windbreaks include cane and plastic meshing fixed on stakes. They are all permeable to the wind.

You are not recommended to construct brick walls, or other barriers that are totally impermeable to the wind, as they create turbulence that has a negative effect on the crop to be protected.

In addition, the greater the resistance to the wind, the stronger and more substantial the structure has to be, and this obviously means it will be more expensive.
Finally, a windbreak of a given height totally protects the plants in an area from its base to a distance 3 to 5 times its height.

• **Orientation**. Within each plot, the crops must be sited so that they make best use of the sunshine. Plots sited on slopes receiving the most sunshine produce earlier crops.

To obtain produce between fall and spring, when the heat required is greatest, the furrows in the plot should be aligned on an east-west axis, so that all the plants receive the same sunshine.

During the months of greatest heat or when the plant requires less heat, the furrows should be aligned north-south in order to reduce the risk of scorching and burns.

1.2. PROTECTED HORTICULTURAL CROPS: MULCHING, TUNNELS AND GREENHOUSES

Nature does not always provide the best environmental conditions for plant growth and development. During the colder seasons, it is too cold for many vegetables to grow.

These unsuitable environmental conditions and the farmer's interest in increasing the total crop and lengthening the production period, has led horticultural undertakings to develop different techniques and to create special installations for vegetable production.

Obviously, the profitability of these techniques and installations varies in relation to the economic and productive possibilities of each specific crop, and with the characteristics of the site in question.

These installations improve the plant's environmental conditions, especially during the most vulnerable periods of its growth, when the crop could not be cultivated in the open field without some type of protection.
To sum up, the aim is to "protect" the crop from unfavorable climatic conditions.
The protective methods used in vegetable production vary greatly. The most widely used are:

• **Mulching the soil**. Mulching is covering the soil around the crop with any of a range of materials, such as plastic sheets, plant remains, straw, leaves, etc.

This technique has several advantages:

- Harvests are earlier
- Production is higher
- The quality of the produce is higher
- The soil temperature is more stable
- Soil moisture is conserved
- Erosion is prevented
- Weeds are smothered

• **Tunnels**. The plastic tunnels used in horticulture are semi-cylindrical constructions of semicircular arches of different widths that are covered by a sheet of flexible transparent plastic. Their purpose is to produce earlier crops, with the advantages this implies, such as greater production and a better price.

• **Greenhouses**. A greenhouse is an enclosed site

where a favorable environment is created for the plants at a time when the conditions outside would not permit their cultivation in the open air.

The greenhouse's structure may be made of wood or metal, and the covering may be glass, rigid plastic, or sheets of flexible plastic.

The main aims of greenhouse production are:

- To obtain produce out of season when the climatic conditions are unfavorable for cultivation in the open air
- To increase the production and yield
- To improve the quality of the commercial produce obtained

2. PROPAGATION OF HORTICULTURAL PLANTS

Plants can be propagated in two ways:

• **Sexually**, by seed
• **Asexually**, or **vegetatively**, by dividing the organs of the plant

Suitable plant organs include:
• *Tubers*, such as the potato and the chufa sedge. They are sections of underground stems that are swollen with food reserves.
• *Rhizomes* like the asparagus. They are stems that grow horizontally underground.
• *Bulbs*, such as garlic. They are also underground stems, but they have been greatly modified, shortened, and surrounded by thick fleshy leaves, which are arranged around the growing point and protect it.
• *Tuberous roots*, such as the sweet potato. They are roots that are swollen with food reserves.

• *Offshoots*, such as those produced by the artichoke. They are lateral shoots of the plant that produce new plants.
• *Stolons*, like the strawberry. They are stems produced by the plant that can grow roots at their tip, at some distance from the parent plant, and then produce a new plant.

There are other forms of vegetative propagation, such as **layering**, **cuttings** and **grafting**. These methods are used in fruit tree cultivation and for ornamental plants.
Sexual reproduction by seed prevents the spread of many diseases, such as viral infections. In crops grown for their seed, the greatest care should be taken to ensure high standards of hygiene, rejecting all the plants that show signs of pest attack or diseases, and collecting the seed when it is ripe.

Hybridization is a technique widely used in vegetable cultivation. It consists of crossing two different varieties or pure-breeding strains of a crop plant, producing offspring called the F_1 generation. The F_1 generation shows hybrid vigor (it is more vigorous than the parent varieties), and they are more uniform in production time, quality, etc. If the F_1 generation hybrids are crossed with each other, this gives the second generation, the F_2 generation, but this does not show the F_1 generation's uniformity or hybrid vigor. Thus when buying hybrid seed, buy from recognized suppliers.

Vegetative propagation is based on the regeneration of an entire plant from part of a plant. The individual obtained is exactly the same as the individual it was taken from, as there has not been any type of sexual reproduction.

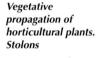

Vegetative propagation of horticultural plants. Stolons

The parts of the plant that can be vegetatively propagated include the stem, roots, leaves and buds. The organ to be used depends on the plant to be propagated.

Vegetative propagation has a series of advantages over sexual propagation. It is possible to reproduce seedless forms and to propagate slow-growing plants faster. But it has the major disadvantage that the plant material deteriorates over the course of time, giving rise to weak individuals that accumulate more and more diseases.

"In vitro" propagation techniques, based on tissue culture, are now very widespread in horticulture (See chapter 13).

To the left:
stolons
To the right:
offshoots

offshoots

2.1. PURCHASING SEEDS

There are two ways for the farmer to obtain seeds. The first is to use seeds from their own crop, and the second is to purchase them.

Seed that is purchased should be certified to fulfil all the desired agricultural requirements, and they make higher yields possible.

To obtain information on the use of selected seeds or new varieties, consult the agricultural extension services where you can find lists of recommended varieties.

Certification systems guarantee, because they are officially controlled, that the various operations involved in seed production and handling were carried out in accordance with the approved technical regulations. The seeds that have been obtained in accordance with these regulations are

inspected and analyzed, and then labelled, sealed and officially classified into different categories.

Seed must meet the following requirements before it can be sold:

• It has to correspond to a variety registered in the Official List of Commercial Varieties. This list contains the names and characteristics of the varieties that have shown their agricultural value, and available as certified seed.

• Production has to be carried out in accordance with a controlled system in which, for each species, the categories of seeds accepted for certification are fixed.

• Seeds for sale have to fulfil certain minimum conditions of purity, germination and freedom from disease.

• The seeds have to be sold in officially sealed and labelled packets. The label must include the name of the body responsible for certification, the name of the variety, the category of the seed, and other relevant data which varies from species to species.

• The officially sealed packet may also have a label from the producer that specifies the producer's name and address, together with the species, variety, category, purity and percentage germination.

Vegetative propagation of horticultural plants: tubers

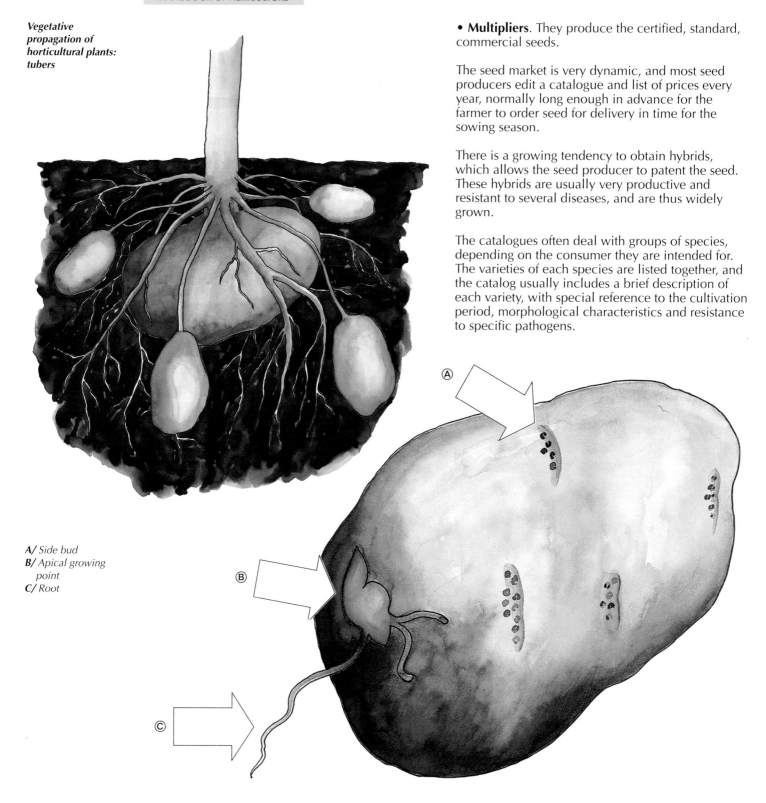

A/ Side bud
B/ Apical growing point
C/ Root

• **Multipliers**. They produce the certified, standard, commercial seeds.

The seed market is very dynamic, and most seed producers edit a catalogue and list of prices every year, normally long enough in advance for the farmer to order seed for delivery in time for the sowing season.

There is a growing tendency to obtain hybrids, which allows the seed producer to patent the seed. These hybrids are usually very productive and resistant to several diseases, and are thus widely grown.

The catalogues often deal with groups of species, depending on the consumer they are intended for. The varieties of each species are listed together, and the catalog usually includes a brief description of each variety, with special reference to the cultivation period, morphological characteristics and resistance to specific pathogens.

Seed producers can be divided, depending on the category of the seed obtained, into:

• **Obtainers**. They produce original or parental material by means of genetic selection and improvement. These are the pre-base seed from which base seeds are obtained.

• **Selectors**. They produce base seeds starting from the parent material.

They also contain other important information, such as the number of seeds per unit weight, the sowing dose and crop rotation programs.

The commercial bodies that sell their seeds on the international market normally have an office in each country. The company either has distributors in each country or its own direct sales system.

2.2. THE MAIN PROPERTIES OF SEEDS

For sale as quality seed, the seed must fulfil a whole series of requirements. They have to be free of all types of pest and disease that may affect the growth of the seedling after germination. Other properties to take into account are purity, germination rate, vigor, dormancy, size, caliber and specific weight.

1/ and 2/ Tuberous roots
3/ Rhizome
4/ Bulb

A/ Flower bud
B/ Leaf
C/ Fleshy leaf
D/ Bud
E/ Stem
F/ Root

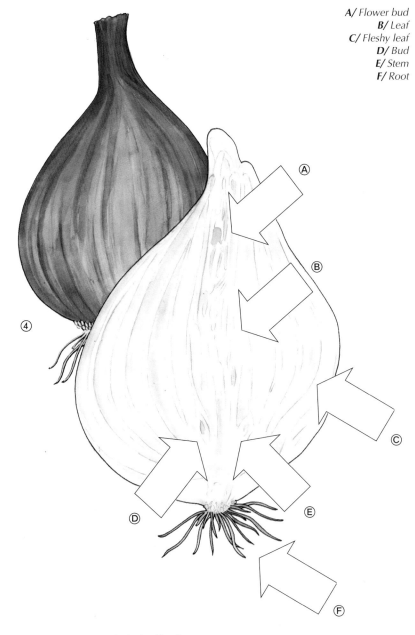

2.2.1. Purity

The purity is the number of pure seeds in the total number of seeds. It is measured as a percentage, and the remainder corresponds to stones, gravel, broken seeds and seeds of other plants.

A purity of 97% means that 97 out of every 100 seeds are of the desired plant, and the other 3% consists of unknown bodies.

Different models of machinery for planting horticultural crops

2.2.2. Germination

This refers to the seed's ability to germinate normally in suitable conditions and within a specified period of time, something that varies from species to species. That is to say, it refers to the seed's ability to grow into a seedling in normal germination conditions.

The seed's ability to germinate after harvesting varies from species to species and declines as it gets older. The longevity or life of the seed reflects how well it retains its ability to germinate after several years.

EU regulations for the sale of some horticultural seeds (INSPV, 1986)

2.2.3. Vigor

This refers to the seed's ability to germinate in unfavorable conditions.

Vegetable	Specific purity (% weight)	Germination of pure seed or glomerules (% weight)	Maximum content of seeds of other species (% weight)
Asparagus	96	70	0.5
Bean	98	75	0.1
Borage	97	65	0.5
Broad bean	98	80	0.1
Broccoli	97	70	1.0
Brussels sprouts	97	75	1.0
Cabbage	97	75	1.0
Cardoon	96	65	0.5
Carrot	95	65	1.0
Cauliflower	97	70	1.0
Celery	97	80	1.0
Chinese cabbage	97	75	1.0
Coffee chicory	95	65	1.5
Cucumber	98	80	0.1
Eggplant	96	65	0.5
Endive	95	65	1.0
Endive or Brussels chicory	95	65	1.5
Fennel	96	70	1.0
Kale	97	75	1.0
Kohlrabi	97	75	1.0
Leek	97	65	0.5
Lettuce	95	75	0.5
Melon	98	75	0.1
Onion	97	70	0.5
Parsley	97	65	1.0
Pea	98	80	0.1
Pepper	97	65	0.5
Radish	97	70	1.0
Red cabbage	97	75	1.0
Runner bean	98	80	0.1
Savoy cabbage	97	75	1.0
Scorzonera	95	70	1.0
Spinach	97	75	1.0
Squash	98	75-85	0.1
Sugarbeet	97	70(gl)	0.5
Swiss chard	97	70(gl)	0.5
Tomato	97	75	0.5
Turnip	97	80	1.0
Watermelon	98	75	0.1
Zucchini	98	75	0.1

gl = glomerule

2.2.4. Dormancy

When a seed is in favorable conditions for germination, and does not do so, even though it is alive, then it is said to be dormant. Some horticultural plants, such as celery, lettuce or water cress, show this problem. Some time has to elapse between harvesting the ripe seed and germination. The seed may take a long time to emerge from its natural dormancy, and so artificial ways of breaking dormancy have to be used.

Dormancy is due to several causes:

• **Structural causes**: the thick, or hard, seed shell physically prevents water from entering.
• **Chemical causes**: the presence of substances that inhibit germination (germination inhibitors).

There are many different processes to break dormancy, such as low temperature (treatment) for short periods of time (24-48 hours), illumination with red light in the case of lettuce, applying gibberellic acid to potatoes and celery, etc.

The use of one or other method depends on the species of plant to be treated and the type of dormancy.

2.2.5. Size, caliber and specific weight

The size of the seed is partly determined by genetic factors, but other factors also play a role, such as the growing conditions of the seed parent and the position on the plant that the seeds were taken from. Seeds taken from the shaded base of the plant are usually smaller than those taken from parts with more light.

Seed size is expressed as the weight per 1,000 seeds. The caliber is the uniformity of size. This should be as uniform as possible, especially for precision seeder units, where this factor is very important.

Plant species	Purity %	Ability to germinate (years) (in normal conditions)	Number of seeds/kg	Weight per liter, in gm
Asparagus	99	5	40,000	
Broccoli	98	4	345,000	670
Celery	97		2,540,000	480
Cucumber	99	5 - 6	35,000	500
Chinese cabbage	98	4	350,000	500
Eggplant	99	5 - 6	250,000	500
Endive	95	3	690,000	295
Fennel	90	3 - 4	200,000	
Leek	98	2	400,000	550
Lettuce	98	3	1,100,000	425
Melon	99	5 - 6	30,000	360
Parsley	97	2	760,000	500
Pea	99	3	200 - 500	700 - 800
Pepper	98	3 to 4	175,000	
Radish	98	4	85,000	685
Runner bean	99	3	700 - 1,000	800 - 850
Spinach	97	4	115,000	510
Swiss chard	97	4	75,000	255
Tomato	99	3 - 4	300,000	200 - 300
Watermelon	99	5	10,000	470
Zucchini	99	5	7,000	330

Useful data on seeds

Ekengårds potato planter, manufactured by OY JUKO, LTD

3. SOWING

Sowing depth (in cm), which depends on the seed size

The seed to be sown must fulfil a series of requirements. It must have been stored properly, retaining an ability to germinate greater than the legally required minimum, and must not have exceeded the number of years it is considered to retain its ability to germinate.
In order to store seed correctly until the moment of sowing, it should fulfil the following storage conditions:

• The seed should be stored somewhere dry and cool, in the dark or in deep shade, at as low a temperature as possible. Increases in temperature and humidity may cause the seed to germinate.
• The seed should not be kept in hermetic flasks or in plastic sacks if the seed is not thoroughly dry.
• It should be treated with insecticides.
• It should not be left for longer than the period of time it is considered to retain its ability to germinate, as the germination rate declines over the years.
To carry out sowing, the soil should be suitably moist and warm. Each species has its own minimum requirements for temperature and moisture for germination to occur. The best time to sow is determined by the weather, if the seed is sown in the open air, and by the soil type and characteristics. As a general guideline, to find the best time to sow you should pay great attention to accumulated local experience.
The sowing depth is related to thickness of the layer

Below left: germination temperatures
Right: time taken for germination at optimum temperature

of earth over the seed required for successful germination. This varies from species to species, depending mainly on the size of the seed, though it also depends on other factors, such as the soil conditions, the weather and the time of year. The seed should be shown at a shallower depth if the soil is wet or compact or if weather is unsasonably cold and deeper if the soil is dry or very open, or in the hot season.
Small seeds can germinate successfully even if they are simply left on the soil surface, though it is always better to cover them with a fine layer of soil

Crop	TEMPERATURE (°C)		
	Minimum	Optimum	Maximum
Asparagus	6-8	20-25	40
Broccoli	8	18-25	30-35
Cabbage	—	20-30	35
Cabbage	5	25-30	35
Cardoon	4	20-30	35
Carrot	5	20-30	35
Cauliflower	5	20-30	35
Celery	8	18-25	30
Cucumber	12	30-35	35
Eggplant	15	20-30	35
Endive	3	15-20	30
Fennel	12	20-25	40
Gourd	10	20-30	44
Leeks	8	15-20	35
Lettuce	4	15-20	30
Melon	13	28-30	45
Onion	4	20-30	35
Parsley	6	18-25	35
Pea	6	14-25	30
Pepper	13	20-30	40
Radish	10	20-25	35
Runner bean	12	15-25	30
Spinach	5	15-25	30
Sugarbeet	5	25-30	35
Swiss chard	5	18-22	35
Tomato	10	25-30	35
Watermelon	13	25	40
Zucchini	10	20-30	40

Plant species	In germinator (days)	In soil (days)
Artichoke	8	12 - 15
Asparagus	—	20 - 30
Broccoli	3	5 - 7
Cabbage	3	3 - 10
Cabbage	3	5 - 7
Cardoon	8	10 - 21
Carrot	6	7 - 21
Cauliflower	3	4 - 10
Celery	10	15 - 22
Chicory	3	5 - 8
Cucumber	4	6 - 8
Eggplant	7	8 - 10
Endive	5	8 - 12
Gourd	4	8 - 10
Onion	8	10 - 20
Leek	6	12 - 14
Lettuce	7	6 - 8
Melon	4	9 - 12
Pea	5	6 - 15
Parsley	10	15 - 25
Pepper	6	7 - 9
Radish	4	5 - 6
Runner bean	5	5 - 8
Spinach	5	5 - 7
Sugarbeet	3	3 - 14
Swiss chard	7	8 - 10
Tomato	5	6 - 10
Watermelon	5	7 - 9
Zucchini	4	8 - 12

Maximum % humidity		4-10°C	21°C	27°C
	Beetroot	14	11	9
	Cabbage	9	7	5
	Carrot	13	9	7
	Celery	13	9	7
	Cucumber	10	9	7
	Gourd	11	9	8
	Lettuce	10	7	5
	Onion	11	8	6
	Pepper	10	9	7
	Runner beans	15	11	8
	Spinach	13	11	9
	Turnip	10	8	6
	Tomato	13	11	9
	Watermelon	10	8	7

Storage conditions to store some seeds for 1 year

in order to defend them from sharp falls in temperature and from attacks by birds.

Sowing too deep delays the seedling's emergence above ground, and this means it is less vigorous, and raises the added risk that the seedling may not emerge above ground if it exhausts its food reserves in the process.

As a general rule, you are recommended to cover the seed with a layer of soil as deep as the diameter of the seed.

Germination is a process that starts when the seed enters conditions suitable for germination and lasts until the seedling produces its first true leaves and is established as a new plant.

Water plays a vital role in the germination process. Once the seed has absorbed sufficient water, the embryo's root and stem start to swell and grow, and they eventually break the seed case. They grow by using the reserves accumulated in the seed. These developmental and growth processes are all chemical reactions that are activated by water and which occur faster at higher temperatures.

A whole series of environmental factors influence germination.

• **Moisture**. In general, as soon as the seed absorbs water, it swells and germinates if the other environmental factors are suitable. The amount of water needed for germination varies from species to species. In general, excess moisture is bad for germination, as the seedling cannot get enough air, and this encourages diseases.

• **Temperature**. Every species has its own temperature requirements, though germination is normally faster at higher temperatures. In general, plants develop better if the temperatures vary than if they are constant.

• **Oxygen**. The seed requires oxygen for germination. The deeper the seed is sown, the less oxygen is available to it.

• **Light**. Not all species need light to germinate, but some species germinate faster in the light, such as celery, eggplant and endive.

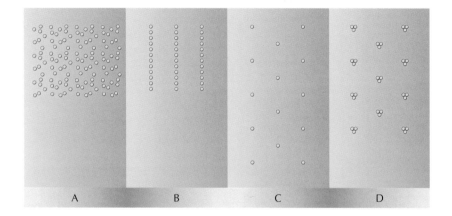

3.1. SOWING SYSTEMS

The most widely used sowing systems are:

• **Scattering**. The seeds are scattered by hand, as uniformly as possible, over the entire area to be sown. This system is the most widely used in seedbeds, where regular spacing is not so vitally important, as the plants are going to be pricked out within a short time.

It is difficult to ensure the seeds are evenly distributed when scattering, especially if the seeds are very small. Very small seeds can be mixed with an inert filler such as fine sand, and if you use sand of a color different to the soil, this also makes it possible to check the regularity of the seed distribution.

• **Continuous sowing**. This is sowing a continuous line of seeds into a row prepared for sowing. It can be used for sowing in the final site, or in seedbeds. The young plants in each row will have to be thinned out in order to leave them at the best distance from each other.

• **Discontinuous sowing**. This is sowing a seed, or several seeds, at a regular and defined distance along the row prepared for sowing. This system is widely used, because it cuts down the number (and cost) of seeds used and reduces the labor required for thinning.

Different systems of sowing
A/ Scattering
B/ In rows
C/ Single seed at a time
D/ Several seeds at a time

Mini-Tex transplanter designed for motocultivators

Precision seed sowing is now extremely important, as it ensures significant reduction in the seed used precisely because the excess do not need to be removed. It can only be used if the site conditions are suitable. To sow seed in this way, the machinery has to be calibrated for the seed, or pelleted seed has to be used.

Pelleted seed is prepared industrially so it can be used in precision seeders. Pelleted seeds are covered with a material that rapidly breaks down when it comes into contact with the soil water.

3.2. PREPARING THE SEED

There are several operations to encourage successful germination and speed it up:

Seed disinfectants

• **Seed disinfection**. This is done to prevent all the pests and diseases that may affect the seedling after germination, and to prevent soil insects from eating the seed. Before sowing, the seed is treated with an insecticide-fungicide mix. In general, seeds are all disinfected by the seed company selling them, and so the farmer need not disinfect them.

Product	Action	Presentation	Dose	Toxicity
Captan	Fungicide	Liquid	2.5-6 cc/Kg	AAC
Carboxyn	Fungicide	Wettable powder	1.5-2 g/Kg	AAC
Copper	Fungicide	Dustable powder	2 g/Kg	AAB
Ethridiazol	Fungicide	Liquid	100 cc/Hl	BBA
Malathion	Insecticide	Dustable powder	0.1 g/Kg	A--
Mancozeb	Fungicide	Liquid	2.5-3.5 cc/Kg	AAB
Maneb	Fungicide	Wettable powder	2.5-3.5 g/Kg	BBC
Maneb 40% + Lindane 10%	Fungicide and insecticide	Wettable powder	3-4 g/Kg	BBC
Maneb 40% + Lindane 20%	Fungicide and insecticide	Wettable powder	1.5-2.5 g/Kg	BBC
Methalaxil	Fungicide	Liquid	6 cc/Kg	AAA
Quintocene	Fungicide	Wettable powder	2-5 g/Kg	BAA
TCMTB	Fungicide	Dustable powder	2-5 g/Kg	BAC
Thiram (TMTD)	Fungicide	Liquid	3-5 cc/Kg	BBB
Thiabendazole	Fungicide	Wettable powder	1.5 g/Kg	AAA

In the toxicity table, the first letter corresponds to the danger to human beings, the second to the danger to the terrestrial fauna, and the third to its toxicity to the aquatic fauna; A represents low danger, B represents some danger, C represents danger, and D represents grat danger.

The diseases that can affect the seedling, known as **damping** off, all have similar symptoms and are all caused by fungi, such as *Pythium*, *Rhizoctonia* and *Phytophthora*.

• **Soaking**. This consists of soaking the seeds for 5 or 6 days. This has the disadvantage that prolonged immersion in water may prevent seed respiration, causing the seed to die.
• **Pre-germination**. This is done to encourage the rapid germination of some vegetables. It has the advantage that it saves seeds and means they germinate more evenly. It also speeds up the

seedling's growth and reduces the risk of disease in the seedbeds. Several techniques are used to achieve this.

• Putting the seeds in a heated germinator at 25°C.
• Immersing the seeds in warm water (30°C) for 6-12 hours. They are then removed, dried, treated with fungicide and placed on a moist cotton cloth, which is left somewhere suitable until germination.

It is worth pre-germinating melons, watermelons, zucchini, cucumbers, tomatoes, eggplant and sweet peppers.

3.3. SOIL PREPARATION FOR SOWING

Soil preparation depends on the type of crop to be sown, some crops being sown directly into the soil while others are sown in seedbeds.

3.3.1. Direct sowing

The crop plant spends its entire life in the site where it was sown. This is why it is so important to plant the seeds in soil that meets the plant's requirements. Directly sown species include broad beans, runner beans, melons, cucumbers, carrots, radishes, etc.
The sowing system used can be continuous lines, for vegetables with small seeds, such as spinach, lettuce and carrots, or discontinuous lines for larger seeds, such as peas or beans.
The soil must contain the water the seed needs. The seed is sown at the appropriate depth using one of the sowing systems mentioned above, then covered with soil and watered.
If the seed is sown in sand-covered soil, it is necessary to water a few days in advance, so the soil is wet enough when the seed is sown.
If the crop is to be watered with an installed drip irrigation system, the rows follow the lines of the water distribution tubing.

3.3.2. Sowing in seedbeds

The young plants are going to be pricked out and then transplanted after they germinate. They may be transplanted into the open air or into greenhouses. The seedlings may be pricked out as bare roots without any earth, or with a ball of earth around their roots, depending on the system used.
The seedbed can be a small plot in the open air on a favored site, where the seed can be sown. Seedbeds have to be well orientated, well aired, protected from the dominant winds and easy for the farmer to tend. These plots may or may not be protected.
A seedbed can be protected by building a small structure of plastic sheeting, or tunnels, to protect the plot from bad weather.

Cold frames are another protection system. They are rectangular structures that vary in length (but are usually 1.3-1.5 m wide), with the rear higher than the front, with a sloping cover of some transparent material, oriented to trap as much sunshine as possible and allow the rain to drain off.

Cold frames can also be heated electrically, and are then called "hotbeds". This is done by installing a heating cable in a bed of sand at the base of the cold frame, which is connected to the power supply and a thermostat at the level of the plant roots. The

Plants with a good root ball ready to be transplanted

The soil or substrate need only have good physical properties in order to favor successful germination; open and cool, but not cold, so can warm up easily, and which it does not present any significant resistance to the emerging seedling and its growing roots.

1/ Osmocote. Long-lasting mineral fertilizer used for horticultural seedlings. It can be mixed with the substrate or scattered over the surface.
2/ The seedling is removed from the compartment of the tray with its root ball intact
3/ Seedtrays after germination
4/ Greenhouse producing seedlings for sale

heater cable is covered by sand and then the substrate for the plant to grow in.

The seed can be sown directly in the substrate inside the cold frame, or in containers that are then placed inside the cold frame. These containers are usually pots or polystyrene trays, which may or may not be divided into separate compartments.
Seedbeds have the advantage that many plants can be grown in a very small space, and so they can be given a lot of care and a favorable environment.

To sow in trays: fill the container with the substrate, water it a little, and sow one seed per compartment in compartmented trays, or scatter seed over undivided ones. Sieve a little of the substrate over the surface and press it down around the seed to ensure it is in good contact with the substrate. Finally, water by sprinkler and place somewhere with suitable temperature and moisture conditions. Using seedbeds with compartments, or sowing in small pots, produces seedlings with an intact ball of soil around the roots. This method ,

The seedbeds have to be kept free of weeds as these will cause problems very quickly. Weeding must be carried out early, as cultivated plants are more fragile and delicate than invasive weeds which compete with them for the root space.

3.4.1. Thinning out

Thinning out is usually performed when seed is sown directly into the final site rather than into seedbeds. Seedlings from seedbeds are normally pricked out, and do not need to be thinned.

Thinning out consists of removing excess seedlings that have germinated, in order to avoid competition for light, nutrients, root space, etc., and thus leave the number of plants considered best.

The soil has to be moist and in good tilth to prevent disturbing and damaging the roots of the plants that are not removed. This operation should be performed twice, 7 days apart, to ensure none are overlooked. After the second thinning out, only the plants that are going to be cultivated remain.

3.4.2. Pricking out

Seedlings grown in seedbeds have to be pricked out, normally into trays with divisions or into small pots. This is performed in between germination and transplanting the crop into its definitive site.

This procedure means the seedling is larger and more balanced, with a stronger root system because it has not been competing for root space. Because the seedlings are transplanted with a good root ball and not just bare roots, there is no halt in growth.

Pricking out should be done with great care, avoiding damage to the seedling's roots when removing them from the soil.

The seedlings should be pricked out when they have 2 or 3 true leaves, using a spatula ("dibble"), taking care that the roots bear some earth, and they are then placed in their new home.

When the seed germinates, it produces a root and leaves. The first leaves are the seed-leaves, or cotyledons, and they differ in shape from the true leaves that are produced afterwards.
A/ True leaves
B/ Seed leaves, or cotyledons
C/ Root

avoids delays in growth and failures due to poor root growth, and it also avoids growth halts after transplantation, as the seedling is transplanted without any roots being broken. Another method is to prick out the seed from the seed bed into the compartments of trays. This ensures better root development, as they are not competing with each other in the same space.

Highly specialized installations use special germination chambers, where the trays are incubated after sowing in the best conditions for germination. After germination they are removed to a latticework shelter or to greenhouses to continue growth, and can then be transplanted to their definitive site.

3.4. TREATMENTS AFTER GERMINATION

After sowing, the soil should be kept cool and moist by watering gently, or if this is not done, by covering with plastic or temporary screens.

Pricking out
A/ Water the seed tray
B/ Tap the tray to separate the earth from the sides
C/ Remove the seedling using a dibble
D/ Hold it gently by the leaves
E/ Plant in a pot or a divided tray, and then water

(A)

(B)

(C)

(D)

(E)

Pricking out is normally performed with a dibble (shaped like a spatula). First, make a hole in the soil in the new container and then put the roots in it. Then press the dibble towards the plant at an angle, in order to gently compress the earth and make it adhere to the roots.

The new container can be a tray with compartments or individual small pots. In both cases, the substrate is prepared as for sowing. The roots should be exposed to the air for the shortest time possible so they do not dry out. After pricking out, water abundantly and give the same care as for a seedbed.

Some time later, the seedlings will have grown enough roots and leaves to be transplanted to their definitive site.

How to remove a seedling with its entire root ball.

To the right: placing the plant in the soil
A/ Enough soil to cover the roots
B/ The depth depends on the size of the roots
C/ The collar should be at ground level
D/ The roots should be in the normal position

3.4.3. Transplanting and planting

After transplanting, the seedlings are installed in their final site, where they will grow. Transplanting is moving the seedling from one site to another, and planting is placing the seedling in the ground where it will grow. Planting is more than just placing the seedling in the soil, as other plant organs can be planted to grow vegetatively.

Before removing the plants from the tray divisions or small pots and planting them in the soil, they should be watered thoroughly so the plants can be removed without damaging the roots.

The more soil there is on the roots, the better, as they will root better in the new soil.

The best time to transplant is in the evening, as the seedlings are less likely to dry out in the hours after planting.

It is very important that the soil should be moist enough when the seedlings are planted, as this ensures fast root growth.

Planting into the ground is best on a flat site, though some species are grown on raised tables and others on ridges, and are transplanted either with a root ball or as bare roots. Some plants are cultivated in this way because they have problems if they come into direct contact with water. To make raised tables, ridges, etc., planters and planting machines can be used.

To vertical

Too slanted

Correct

If the seedling is planted on a ridge, the tool is inserted halfway up the ridge and the seedling is introduced into the hole, taking care that all the plants are on the same side of the ridge.

The plants have to be treated with care so their roots do not break, and they have to be placed in an upright position. The roots have to be completely covered, but must not be bent or twisted, and the plant's collar should not be buried.

To finish, the earth that was removed should be replaced and pressed down around the roots, and the seedlings are then watered.

When planting vegetative organs like bulbs, tubers, etc., first make a furrow and then plant the tubers, etc., at a regular distance. Cover with the earth removed from the furrow and press it down lightly. If the soil is sufficiently loose and spongy, planters can be used instead of making furrows.

When planting into sand-covered soil, first remove the layer of sand, then plant the seedling, and replace the sand. To finish, water the plants.

Automatic planting is normally used in extensive horticultural crops, as it considerably reduces labor costs.

Correct planting

Mechanical planter

Too shallow

Too deep

Correct

4. SOIL PREPARATION

The main purpose of soil preparation is to create the most favorable conditions for the plant from the moment it is sown till the end of its production cycle.
Good soil preparation is a precondition for good production.
The better the soil is prepared, the better the plant will grow in it. Suitable preparation can greatly improve the soil's characteristics, increasing its volume, its fertility and its permeability to air and water. Thoroughly breaking the clods aids the growth and spread of roots. Loose, spongy, soils are well aired and allow the flow of water, and this means that watering and fertilizer application are much more effective.

Plowing may be deep, ordinary or surface.

• **Deep plowing**. This is done to a depth of 30 cm. It is especially positive in compact soils, where it improves drainage. It increases the depth of the topsoil and thus increases the volume the roots can grow in.
It is done with a subsoil plow (if you do not want to turn the soil), or with a moldboard plow (if you want to turn the soil).

• **Ordinary plowing**. This is carried out between two consecutive crops to incorporate fertilizer into the soil and to ensure the soil is in good tilth, and is normally to a depth of 15-20 cm.

• **Surface plowing**. This is plowing the top 10-15 cm of soil with disc harrows or cultivators. It is done to break the surface crust and kill weeds.

In general, plowing to prepare the site should be performed at least 15-25 days before planting or sowing.
It should be done when the soil moisture is optimal. If the soil is too moist, plowing is harder and the soil forms clumps, while if the soil is very dry, its structure breaks down into large clods and dust.

In general, soil preparation for cultivation requires the following steps:

• Incorporating organic matter into the soil, whenever dunging is necessary.
• Plowing with a moldboard or disc harrow to a depth of 30 cm.
• Disinfecting the soil if necessary.
• After the period recommended by the disinfectant manufacturer has elapsed, plow the soil again, to air it.
• Apply mineral fertilizer at depth.
• Pass the drill or cultivator over the surface to ensure there are no clods and the surface is flat.
• Prepare the soil for planting or sowing, which depends on the characteristics of the crop to be grown. This is done by making flat areas, ridges or flat-topped ridges.

4.1. BREAKING THE TILL PAN AND DISINFECTION

Any subsurface crust should be broken when a plot is brought under cultivation for the first time. Any subsurface crust that forms should also be broken up every 4-5 years on plots that are permanently cultivated. This breaks the till pan that forms in soils that are always plowed to the same depth.
Breaking this till pan is done by plowing at depth, perhaps more than 40 cm. The aim is to break any crust that has formed deep in the soil, and thus improve drainage.
This operation makes the soil looser and this means the roots can grow better and take up more water. The soil is also better drained and healthier, as its aeration improves, favoring microbial growth and their release of soluble nutrients.

Breaking the pan is done with a turning plow, whenever the subsoil is better than the topsoil, though sometimes the soil is mixed. If the subsoil is of lower quality, a subsoil plow should be used that breaks the pan but neither turns nor mixes the soil.

Soil disinfection is performed to prevent the problems caused by excessive accumulation of parasites and pests, as they endanger the plant's life and yield.
The main diseases and pests that may accumulate in the soil are insects, nematodes, fungi, bacteria and viruses.

Several different techniques can be used to get rid of these pests:

• **Indirect actions** such as appropriate cultivation techniques, growing resistant varieties, leaving the soil fallow, rational crop rotation systems, and using seeds, bulbs and tubers that have been disinfected and are guaranteed free of diseases.

• **Direct action**, using physical or chemical methods. Physical techniques are based on the sterilizing power of heat applied in different ways, such as steam. The most widely used chemical methods of soil disinfection are methyl bromide, dazomet and metam-sodium, all of which are very effective biocides.

4.2. HARROWING AND RAKING

This operation is done to level the site and break up clods before the crop is sown or planted. In an established plot, it breaks any surface crust and uproots the weeds.
It is done with different types of harrow, small motor cultivators, rakes or rake-scarifiers.
Raking is also performed after sowing medium-sized seeds, in order to cover them properly.

Cultivators break the surface crust in order to improve the infiltration of irrigation water.

4.3. FIRMING

Firming destroys any clods of earth that the harrow or rake has left. It also ensures the soil is homogeneous and is in good contact with the seeds once they have been sown.

The seed's contact with the soil provides it with enough moisture to start germinating. Firming the soil prevents the formation of air pockets around the seeds and increases the area of contact, meaning that it is easier for the seeds to take up the water they need for germination, which is faster, regular and more uniform.

Firming is done with rods or iron rollers, hand powered on small sites or with larger tractor-drawn cement rollers on larger sites.

4.4. WEEDING

Moisture and good soil preparation favor the growth of the crop but also growth of the weed seeds present in the soil.

These undesired plants cause direct damage to the crop, because they compete for space, light, nutrient elements and water. They also transmit pests and diseases.

They also cause a direct decrease in production, and decrease the quality of the product harvested.

Weeds multiply and grow rapidly, as they are usually more vigorous than the crop and compete with it, and may eventually smother it.

Weeding seeks to eliminate the weeds growing among the crop. This can be done by cultivation methods or by using chemicals.

Cultivation methods include weeding by hand with suitable tools, harrowing, or passing the cultivator.

In order to control weeds, you should follow a rotation that alternates deep plowing and shallow plowing, and exposing the soil to the sun, which kills the uprooted weeds by desiccation. For the same reason, it is best to weed on a hot day, when it is most effective.

Weeding is often combined with earthing up. Weeding has to be carried out throughout the crop's growth, not just during the preparation of the soil.

Weeding can also be performed with chemical weedkillers. These are normally applied between two consecutive crops, but the high degree of overlap between horticultural crops makes this a problem.

Weedkillers can be total or selective.

• **Selective weedkillers** are only toxic to some plants, that is to say they kill some plants selectively. The plants eliminated are dicots, while the monocots are left unharmed.

• **Total weedkillers** kill all sorts of plants.

The way the herbicide is to be applied depends on the chemical nature of the product:

• Pre-sowing products are applied before sowing
• Pre-emergence products are applied before the crop's emergence above the ground
• Post-emergence products are applied after the crop's germination or when it is already established

They are classified on the basis of their method of action, distinguishing between systemic weedkillers, which eliminate the entire plant including the root, or contact ones, which only kill the part exposed. When using any herbicide you must follow the manufacturer's instruction carefully.

One indirect way of preventing the appearance of weeds in crops is to grow certified seeds, bulbs or tubers, and to create favorable conditions for their rapid growth, so that from the moment they start growing they can grow faster than their competitors and smother them.

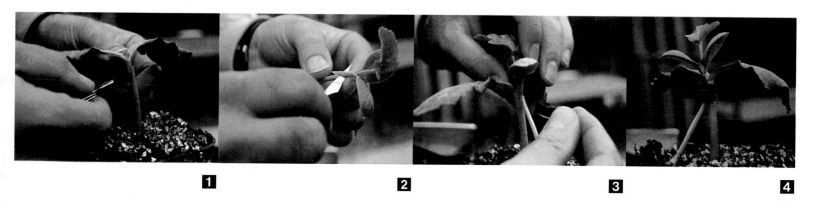

Approach graft
1/ Slanted cut in the stock plant.
2/ Cut in the scion.
3/ Joining the two together.
4/ Fixing together with a graft clip.

Scion graft
5/ Cut in the scion.
6/ Vertical cut in the stock plant.
7/ Joining the two.
8/ Fixing together with a graft clip

5. TENDING THE CROP

Throughout the production cycle, the plant requires a series of specific treatments and attention that are not the same for each species, as they depend on the special characteristics of each crop.
The cultivation methods needed vary from one species to another, but they all have the same objective, to ensure the plant's growth and production is the best possible. Some of these cultivation methods are essential if acceptable yields are to be achieved.

5.1. PINCHING OUT

Pinching out is the removal of the growing tip, the terminal bud of the plant's main stem, in order to prevent vertical growth and to favor the development of side branches or fruit formation.
Pinching out is performed on tomatoes, melons and eggplants.

The secondary stems, branches, leaves, etc., may also be pinched out in order to shape the plant as desired.
This is done in tomatoes, melons, eggplants and cucumbers.
Pinching out can favor the spread of viral infections through a crop if the necessary hygienic precautions are not taken.

The tool used should be disinfected after pinching out one plant and before treating the next. This can be done using two tools, and while one is being used the other can be immersed in a liquid disinfectant.
Sometime leaves are removed from the plant, in order to get rid of dead, old or unhealthy leaves. Sometimes a large number of leaves, even healthy ones, may be removed.

Very lush growth creates an excessively humid environment that favors the spread of fungal diseases. Removing some leaves improves aeration and decreases humidity. Leaves may also be removed from lush plants to let the light reach the fruit, which would otherwise ripen late or unevenly.
Removing leaves should be carried in the hours of least heat, so the leaf loss does not affect the plant seriously.

Leaf removal should not be excessive, as in some cases it may cause growth disorders that affect the quantity or quality of the produce, and in other cases it may lead to burns on parts previously protected from the sun.
Thinning of fruits can be included in this group of cultivation operations. It is done to remove defective fruit lacking commercial value, or in order to improve fruit quality, by improving the plant's vigor.

5.2. STAKING, TRAINING AND TYING

Some horticultural species have a tendency to climb and need supports to climb properly. The farmer takes advantage of this to ensure better growth of the plant, and thus higher production.
Other advantages shown by this type of cultivation include more air and light for the plant, meaning that flowering and fruit set are better. As the fruit receives more warmth, harvest is earlier.
The fruit are healthier, as contact with the soil is avoided, and cultivation operations, eg. for example, weeding and pruning, are easier.
Their vertical growth means they make better use of the soil area, and the yield per unit area increases substantially.

Several systems of support are used:

• **Stakes**. They are usually upright pieces of wood, cane or plastic driven into the ground next to the plant.

• **Espaliers**, which consist of several stakes joined together to form a trellis.

The crops that require staking are tomatoes, peppers, cucumbers, melons and eggplants.
When crops are staked, in the early stages of growth it is necessary to guide the plant's growth on the stakes, and this is called **training**.

The plant's stem may be **tied** to the stake or espalier. Tying is normally done with raffia, string or plastic. Plastic binds the stem firmly but does not strangle it, as its elastic nature means it can cede as the plant grows.
The leaves may also be tied to the main stem, for example in lettuce or endive, to blanch the inner leaves.

5.3. PRUNING AND EMASCULATION

Pruning aims to remove broken, dead, diseased or excess branches, in order to improve the plant's health and save nutrients which will instead be devoted to increased fruit production and quality. The first pruning is called training, and it seeks to shape the plant's future growth. Pruning is also frequently carried out to favor **flowering**, **fruiting** and **renewal**.

Some vegetables can produce fruit without fertilization of the female flowers by pollen. These are called **parthenocarpic fruit**, and include the cucumber and the squash.
When these plants' male flowers emit their pollen, this spreads and pollinates the female flowers. Fertilization is not uniform, causing uneven or malformed fruit to grow. To prevent this from occurring, the male flowers are removed, and this is called **emasculation**.
To ensure the growth of the parthenocarpic fruit and to prevent fertilization, emasculation should

The tie should support the stem, which should not rub against the stake as this would cause friction damage.

be performed every 2 days from the beginning of flowering until it ceases.

There are now hybrids on the market that only produce female flowers, and thus do not need to be emasculated.

5.4. EARTHING UP AND BLANCHING

Earthing up consists of piling up soil around the collar or base of the plant, and it is done for different reasons in different crops.

In general, earthing up favors the growth of tubers, such as the potato and sweet potato, and of adventitious roots (for example, in the zucchini, red pepper and tomato), which increase the plant's mechanical stability and water uptake.

It also acts as a support for the base of plants like corn, beans or peas, and protects the collar of plants from frosts (cabbage and cauliflower).

Earthing up is also done to blanch vegetables, that is to say, to produce plants with a white collar or white leaf stalks, such as leeks, celery and cardoons.

Blanching seeks to prevent chlorophyll formation in certain edible parts of the plant, which remain white and soft and taste better. Blanching can be performed by earthing up soil, but it can also be done by tying the leaves to the main stem or by placing sleeves of black plastic aroud the plant to prevent light passing.

Concept	Color		
	Transparent	Smoky grey	Opaque black
Transmission of sunlight	80%	35%	0%
Heat absorbtion	Low	Medium	High
Possibility of preventing frosts	Some	Little	None
Early harvests	High	Medium	Low
Crop yields	Medium	Medium	High
Weeds	Abundant	Scarce	None
Useful life of plastic	Less	Medium	Greater

5.5. MULCHING

In some cases, it is worth protecting the soil with some type of covering. This is known as **mulching**, and the mulch used can vary widely, from plant matter to plastics.

The idea of mulching is to cover the soil surface with a layer of straw, dry leaves or peat, i.e., plant matter. The covering should be uniform, and should be applied when temperatures are low. It should be between 6 and 8 cm thick.

Mulching is done for several reasons:

• To prevent the growth of weeds, thus reducing weeding operations.
• To keep the soil structure open and spongy, thus favoring rapid absorption of rainwater and good air circulation.
• To reduce evaporation of water, and make the soil water reserves last longer.
• To reduce temperature variations in the soil, which protects the roots.
• To incorporate organic matter for later crops.
• To reduce the need for weeding and earthing up.

In general, the crops that are mulched have a long growth cycle, such as tomatoes, peppers, eggplants, zucchini, cucumbers and melons.

Mulching with plastic consists of placing a sheet of polyethylene (or polythene, PE) or polyvinylchloride (PVC) over the soil surface.

	Transparent PE	Opaque black PE	Smoky grey PE
Useful life	In seasonal crops	In 1 to 3 year crops	In seasonal crops and 2 year crops
Weeds	On weed-free sites treated with herbicides	On sites infested with weeds	On sites not heavily infested with weeds
Temperature	In cold zones at risk from frost	In hot zones not at risk from frost	In hot and cold zones
Aim	When early crops are wanted, rather than higher yields	When higher yields are sought, rather than an early crop	When seeking an earlier crop and a higher yield
Disadvantages	Problems with the weeds that grow under the plastic, which grow fast and may raise the sheeting	They may cause burns to the plants where they are in contact with the sheeting	

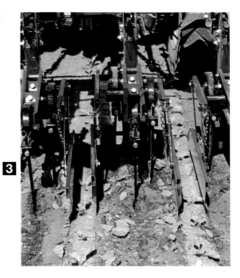

Three types of plastic are suitable for mulching, transparent, smoky grey and opaque black, each with its own characteristics.

Mulching with plastic has a series of advantages:

• **Earlier harvest**. The plastic increases the soil temperature, even at night, and this increases root activity and uptake.
• **Protecting the root system** against low temperatures.
• **Avoiding the need for weeding and earthing up**, meaning that the soil maintains its structure, and does not form a surface crust, something that impedes the infiltration of water.
• **Conserving soil water**. Mulching reduces the loss of soil humidity by evaporation, and the soil is left uniformly moist.
• **Better use is made of fertilizers**. They are not washed deep into the soil by irrigation water.

Mulching with plastic is widespread in cultivation of the tomato, eggplant, pepper, zucchini, cucumber, strawberry, melon, watermelon, lettuce, endive, lettuce, Swiss chard and celery.

5.6. TILLING

The tasks carried out during the crop's growth are referred to as cultivation. This includes harrowing, scarifying, passing the cultivator, etc.

Harrowing is also included among the tasks to

prepare a site. It is also done to complete soil preparation before sowing, and to bury medium-sized seeds after sowing.
Harrowing is also done to break any surface crust, because this is an obstacle to seedling emergence and to gas exchange between the soil and the air.

Passing the scarifier is done for the same reasons, but is more vigorous. It seeks to till the soil to a depth of 10 cm in order to break any surface crust and improve the flow of water and air. It can also be done to incorporate organic matter to the soil.

Tilling is typically performed during crop growth to conserve soil moisture. This consists of a very gentle surface tilling that breaks the surface crust formed after irrigation, creating an even and smooth soil horizon that prevents the capillary rising of water and protects the soil from evaporation. This operation should be performed carefully, because if large clods are left in the soil, the action performed will be totally ineffective.
This is performed after every watering from the start of the crop's growth until the moment when the vegetation covers the soil. This is done with the cultivator, a toothed rake or any manual tool.
The beneficial effect of passing the cultivator can be attributed to the fact that it acts to increase soil aeration, thus speeding up microbial activity and therefore leading to faster mineralization of organic matter, as well as increasing nitrate uptake.

1/ A rolling plow applied to sugarbeets.
2/ 2-row rolling plow used to earth up potatoes.
3/ Sugarbeet weeder with 6 rows for young plants.

6-row machine to weed sugarbeets, with manual steering as an additional facility.

The best time to weed is when the potato plant has reached a certain size. The weeds are uprooted and left to dry out on the soil surface.

A corn weeder can be used to weed sunflowers because the large teeth allow wider rows.

6. APPLYING FERTILIZER

Fertilizers are applied to supply the soil with the mineral nutrients that plants need and which the soil cannot supply, either because they are not present or because they are present in a form that cannot be assimilated.

Organic fertilizers are also applied to improve the soil's structure, texture and other physical properties. In horticultural cultivation, nutrient inputs are very important, not only in terms of their quantity but also in striking the right balance between the different nutrient elements.

Only some elements, the nutrients, are used by plants. This means it is important to know how the main nutrient elements act.

Horticultural substrate

• Nitrogen
Nitrogen acts on plant growth and on the amount of chlorophyll synthesized.
Nitrogen deficiency leads to a general weakening of the plant and a sharp reduction in its yield and production. The leaves also turn yellow due to a lack of chlorophyll.
Excess nitrogen causes the plant to grow very vigorously, giving rise to a series of problems like late ripening, greater sensitivity to disease and to changes in temperature and humidity.

• Phosphorus
Phosphorus is required by the plant throughout its entire growing cycle. It is extremely important at flowering time and during fruit formation. It is also important for growth of the root system, as well as advancing flowering and favoring earlier harvests.

There is a wide range of different substrates premixed by the supplier.

• Potassium
Potassium forms part of plant tissues, especially growing tissues, and it is also required for chlorophyll synthesis. In general it increases the plant's resistance to water shortage, as it lowers transpiration. It also increase the plant's resistance to low temperatures, as the concentration of salts, i.e., nutrient ions, in the cell sap increases.

• Calcium
Calcium favors plant growth in general. It is important in fruit formation and ripening.

• Sulfur
Sulfur together with nitrogen and phosphorus is vital for protein synthesis. It also favors a balanced microbial flora in the soil.

• Magnesium
Magnesium is required for production of chlorophyll, the molecule that performs photosynthesis, and it also helps phosphorus uptake.

• Microelements
Unlike the nutrients discussed so far, the microelements are only required in very small amounts. This does not mean they are unimportant, however, as deficiency of any one can cause serious problems to the plant, and even its death. The microelements include iron, manganese, boron, zinc, copper and molybdenum.

It is difficult to give general guidelines for applying fertilizers, because the plant's nutrient requirements change over the course of its life cycle. Successful fertilizer application does not only depend on applying the correct amount of fertilizer to the soil but also on applying it at the moment when the plant needs it.
In general terms, nitrogen should be applied to the soil shortly before the beginning of the

main growth period, and also in greater doses throughout cultivation of the crop.
Phosphorus and potassium are necessary for the plant's early growth, and throughout its life as surface applications of fertilizer.

Kg/ha	N	P_2O_5	K_2O
Carrot	150	90	400
Cauliflower	200	80	250
Celery	200	150	500
Cucumber	150	80	300
Lettuce	80	40	200
Melon	90	40	200
Onion	90	40	120
Pepper	200	60	300
Potato	175	60	300
Tomato	250	90	400

Other factors that have to be taken into account if the fertilizer application is to give good results:

• The fertilizer should have the right balance of nitrogen, phosphorus and potassium.
• The best time to apply does not depend just on the crop but also on the state of the soil.
• The soil has to be well aired, well drained and has to have a good structure and retain water well.
• The use of productive varieties, which should be appropriate for the area in which the crop is cultivated.
• Weeds have to be controlled, as do the pests and diseases that may attack the crop.
• Following a suitable crop rotation.

In order to know how much fertilizer to apply there are some guidelines for different types of horticultural crop plants. Leaf crops should be given a balanced 1-1-1 fertilizer, which means that you should apply a fertilizer with the same amount of nitrogen (N), phosphorus (P_2O_5) and potassium (K_2O). For a 1-1.5-2 ternary fertilizer, you should choose a formulation with one part N, one and a half P_2O_5 and two of K_2O.

The following are guidelines for the nutrient balance of the fertilizers to apply to horticultural plants.

• Horticultural crops grown for their leaves: 1-1-1.

• Horticultural crops grown for their tubers: 1-1-1.

• Horticultural crops grown for their bulbs: 1-1.5-2.

• Horticultural crops grown for their fruit: 1-2-2.

There are several ways of finding out how much of a nutrient to apply to the soil for a crop:

• Direct observation of the crop. This is a subjective method that requires a lot of previous experience.
• Soil analysis.
• Measuring the quantities of mineral nutrients extracted by the crop in previous years.

6.1. FERTILIZER APPLICATION TECHNIQUES

The fertilizer application technique most widely used in horticultural crops is to apply fertilizer at depth before sowing or planting, at the same time as the other soil preparation tasks, applying organic matter and mineral fertilizers based on nitrogen, phosphorus and potassium.
It is important to apply abundant organic matter. This is done to obtain a good soil structure and to maintain it for as long as possible. You should apply between 2 and 5 kg of organic matter per square meter, depending on the soil's organic matter content.
Fertilizer should be applied to the soil surface several times during the plant's life cycle, using a mineral fertilizer, and depending on the requirements of the crop in question. It is important that the fertilizer be placed at the correct depth for it to be taken up by the roots. The roots grow more and better where they can find moisture and nutrients.
Fertilizers should not be simply scattered on the soil surface, as this would cause the roots to grow up to the soil surface, and this would leave them vulnerable to water shortage.
There are two different types of fertilizer application, at depth and on the surface. Application at depth seeks to raise soil fertility to an optimum level before the crop starts growing, while surface application seeks to meet the plant's nutrient requirements over the course of its productive cycle.

6.1.1. Applying fertilizer at depth

This is carried out before planting or sowing the crop.

Main nutrients extracted by some horticultural crops

Composition of some organic fertilizers

	% Nitrogen	% P$_2$O$_5$	% K$_2$O	% Organic matter	pH
Chicken droppings	1.1-4	0.5-3	0.5-2	50-75	Basic
Cattle dung	0.5-0.7	0.2-0.3	0.5-0.6	30	Acid
Sheep dung	1-2	0.7-1	1-2.5	60	Acid
Horse dung	0.6-1	0.2-0.7	0.6-0.8	30	Acid
Goat dung	2.7	1.7	2.8	60	Acid
Rabbit dung	2	1.3	1.2	50	
Dried blood	13	1.5	1	80	Acid
Cotton seed cake	6.5	3	1.5	80	Acid
Worm compost	2-3	2-3	2-3	50	
Alfalfa straw	1.5	0.3	1.5	82	
Cereal straw	0.6	0.2	1.1	80	
Wool remains	0.8	1.2	—	—	

The product should be distributed uniformly over the entire soil surface, either by hand or using machinery.

This can be done using gravity distributors if the fertilizer is not granulated, or using centrifugal distributors for granulated fertilizers.

Once the fertilizer has been spread, it has to be worked into the soil. This is done by tilling, to a depth that depends on the species to be grown. The fertilizer is applied at depth when preparing the site. This is the usual method of applying organic fertilizers, phosphate fertilizers and potassium fertilizers, as well as nitrogen-containing fertilizers based on urea or ammonium.

6.1.2. Surface fertilizer application

Surface application consists of applying small amounts of fertilizer at precise moments in the crop's production cycle.

This is based on the idea of anticipating the plant's needs, foreseeing what nutrients it is going to consume, and aims to ensure the soil's fertility remains constant.

Once the crop is well established, the fertilizer can be applied in strips parallel to the rows of the crop, or in discrete doses between plants, and always at some distance from the plant, so the fertilizer does not burn the roots but is accessible to them.

Working fertilizer applied in strips into the soil is performed by first passing the moldboard harrow to open the soil. The fertilizer is then applied, and then the cultivator is passed in order to mix it all together. If the fertilizer is applied in discrete doses, the first thing to do is to dig the holes where the fertilizer will go. They are made with a planter and should be a distance of about 25 cm from the plant. They should be about 10 cm in diameter, and their depth

varies depending on the crop being grown and its root system.

After placing the fertilizer in the holes manually, the hole is covered and pressed down so that no air pockets are left.

This method is frequently used to apply nitrogen-containing fertilizers based on nitrates.

6.2. TYPES OF FERTILIZER USED

To start we should distinguish between organic fertilizers and chemical (or mineral) fertilizers.

6.2.1. Organic fertilizers

Organic fertilizers consist of organic matter or dung, mainly from stockraising farms.

A tonne of dung supplies 100 kg of humus to the soil, and this is the main source of nitrogen.

In horticultural crops, the dung used should be well rotted, and applied at a rate of 30 to 50 t/ha, depending on the species being cultivated, the expected production, the application system and the rotation being followed.

The dung should be spread over the soil surface and buried immediately afterwards to prevent it being washed by the rain, which would cause loss of part of the nutrients. It should be incorporated into the soil when preparing the site.

In general, organic fertilizers have the following advantages:

• It lightens heavy or clay soil.
• It increases the soil temperature by absorbing more sunlight.
• It increases the soil's ability to retain water and nutrients.

- It supplies large amounts of nitrogen.
- It favors microbial activity in the soil.

The disadvantage of organic fertilizers is that they may cause increased attack by certain pests or diseases, as well as favoring weeds.

Green mulches are harvest remains that are incorporated into the soil. This improves the soil's fertility and its physical state.
Compost is produced by the rotting of plant remains and is also used as an organic fertilizer. It improves physical conditions in the soil, and is normally used (mixed with soil) in seedbeds.

6.2.2. Mineral fertilizers

Dung supplies large amounts of organic matter, but little nitrogen, phosphorus or potassium, and so it has to be complemented with mineral fertilizers. Chemical fertilizers are not a replacement for organic fertilizers, as they do not improve the physical conditions in the soil.

The advantages of mineral fertilizers are:

- They supply consistent amounts of the nutrient elements
- They are simpler and more convenient to spread
- Fertilizer doses are more precise
- They can be dissolved in the irrigation water
- They are easy to obtain in the market

The main groups of chemical fertilizers are those that contain nitrogen, phosphorus and potassium.

- **Nitrogen-containing fertilizers** have to be applied before the period when the plant is growing most. They are applied in small amounts over the course of the plant's growing cycle, in order to ensure they are used as efficiently as possible.

The nitrogen-containing fertilizers can be divided into:

- *Nitrate fertilizers.* Nitrates are very soluble in water, and are thus easily lost in drainage water. They are easily absorbed by roots, and so they are applied to the surface.
- *Ammonium fertilizers.* Nitrogen is better retained in the soil as ammonium, and it is then oxidized to nitrate by microbes. Plants can thus make use of ammonium for longer than nitrates.
- *Ammonium-nitrate fertilizers* contain part of their nitrogen as nitrates and part as ammonium, for example ammonium nitrosulfate.
- *Amides.* The nitrogen is present in a form plants cannot take up, but it is converted to an assimilable form and released by the action of soil microbes.

- **Phosphate-containing fertilizers** are applied at depth, in sufficient quantity to meet the plant's needs over its entire growth cycle.
- *Superphosphates* contain 16-18% assimilable phosphate. They are soluble in water and easily absorbed. In acid soils that are rich in iron and aluminium, phosphorus becomes insoluble.

- **Potassium-containing fertilizers** are normally required in large amounts. They have to be applied in several fractions: the first at the time of planting,

Different products from S.E.S.

when potassium is applied with the other fertilizers at depth, and it is then applied to the surface during the crop's growth.

Fertilizers may be simple, compound or complex.

- **Simple fertilizers** supply a single nutrient element.
- **Compound fertilizers** contain more than one nutrient element, but they have not reacted together to form a single fertilizer compound.
- **Complex fertilizers** consist of a single fertilizer compound containing more than one nutrient element. They are the most widely used because they are well-balanced in nitrogen, phosphorus and potassium. They also use less storage space and store well.

A fertilizer's strength is normally expressed on the factory label as three numbers, which are the fertilizer units of each element present expressed as a percentage.
It is important to grasp that the fertilizer units are expressed in nitrogen-containing fertilizers as nitrogen (N), in phosphate fertilizers as P_2O_5, and in potassium fertilizers as K_2O.
Thus a 12-12-17 fertilizer contains, per 100 kg, 12 kg of N, 12 kg of P_2O_5 and 17 kg of K_2O.

Sometimes in addition to these three figures, the label of a complex fertilizer may have another number, and the label should specify clearly which element it refers to. In general, the fourth figure refers to magnesium.
Finally, fertilizers can also be classified by their presentation or their physical state, i.e., solid, liquid or gas. Solids include forms like powder, granulates and crystals.

6.3. APPLYING FERTILIZER IN IRRIGATION WATER

Soluble fertilizers can be applied in irrigation water simply by dissolving them in the irrigation water, which is then distributed by the installed irrigation system. Liquid fertilizers or highly soluble ones can be used, as long as they do not react with the salts in the irrigation water and are not corrosive to the irrigation installations and devices.
This system allows the fertilizer to be applied to the crop in fractions, giving control over the moment of application, and thus diminishing the danger of the accumulation of salts and saline residues.
Combining fertilizer application with irrigation favors nutrient uptake by the roots, even fertilizer application and requires less fertilizer to be used, as it is only applied to a very specific portion of the soil, where the roots are growing, and not to the soil as a whole.
This system also has the advantage of saving labor, as the distribution can be mechanized using the automatic regulation of the irrigation system.
The disadvantage of combining fertilization with irrigation is that the plant's root system is small, because it does not need to explore a large volume of soil. The plants are also more susceptible to deficiency of a mineral element and shortage of fertilizer, especially deficiencies of micro-elements, which are taken up in smaller quantities because the roots do not explore so much soil.
To use this system properly you must know the amounts of nutrients extracted from the soil by the crop, and the rate of absorption, which depends on the climatic conditions and the plant's stage of development. This is needed to establish the balance, or ratio, between the water requirements and the nutrient requirements, the dose and the best time for application.

Pneumatic bar for low volume irrigation. It supplies 80 to 250 liters per hectare.

All the different sorts of soluble fertilizers can be dissolved in irrigation water, as long as they do not give rise to problems by clogging the outlets of the irrigation system with insoluble precipitates. You should avoid using mixtures of fertilizers that cause insoluble precipitates to form; do not combine fertilizers supplying calcium or magnesium with ones containing sulfates or phosphates, in order to prevent them forming insoluble precipitates.

The nitrogen can be supplied in the form of commercially available solutions or by using urea, which dissolves easily, or potassium nitrate if you wish to supply potassium at the same time.
It is more difficult to apply phosphorus fertilizers in irrigation water, as its solubility is low, and it may be precipitated out by the calcium in the irrigation water.
Micro-elements are applied in the form of chelates to avoid their reacting with the salts dissolved in the water as this would mean the roots could not absorb them.
The water reaching the plants should contain no more than 1-2 g/l of dissolved salts, in order to prevent salts accumulating in the soil, and its pH should be between 6 and 6.5. If the irrigation water has a pH of 7 or more, this may cause precipitation of some minerals, and not only does this mean that the plants cannot absorb the nutrients but also causes severe problems due to blockages in the piping of the irrigation system. If the pH is below 6, the water may cause corrosion of the metal components of the irrigation system.
A conductivity meter can be installed to measure the salinity and sound an alarm when it exceeds an established maximum value. If the pH of the irrigation water is excessively alkaline, it can be corrected by adding some sulfuric acid or nitric acid

The fertilizers can be added to the irrigation water using one of the following systems:

• **Tank fertilizer**. This is a deposit treated with anticorrosive agent, that varies in volume, connected in parallel to the main conduit. The fertilizers are placed in the tank, and using a series of cocks, part of the flow of water enters the tank and dilutes the fertilizer and is then returned to the irrigation system. This functions on the basis of pressure differences. It has the disadvantage that the fertilizer is not very evenly distributed, because the concentration of fertilizer in the irrigation water is not constant, as the concentration of the stock solution varies with time.
• **Fertilizer injectors**. These are units in which the concentrated fertilizer solution is placed within a deposit, and from there it is injected into the irrigation supply by means of some sort of pump.
• *Electric doser*. This is an electric pump that takes the fertilizer from a deposit using suction and then injects it into the irrigation system. This device does not adjust fertilizer input in proportion to any changes in the volume of water.
• *Proportional injection electric doser*. This basically consists of a meter to measure the flow of water, and every time the pre-established volume is exceeded it sends an electric signal to a sensor that tells the

Mineral element	Fertilizer	Content	Use
Nitrogen	Ammonium nitrate	33% N	Rapid supply of nitrogen (Surface application)
	Ammonium calcium nitrate	20-30% N	
	Calcium nitrate	15% N	
	Sodium nitrate	16% N	
	Ammonium nitrosulfate	26% N	
	Ammonium sulfate	21% N	Average nitrogen supply (Apply at depth)
	Calcium cyanamide	20% N	
	Urea	46% N	
	Ammonium	—	
	Urea-sulfur	24% N	Slow nitrogen supply (Apply at depth)
	Isobutylidenurea	32% N	
	Ureate	29% N	
Phosphorus	Natural phosphate or rock phosphate	25-40% P_2O_5	Slow phosphorus supply (Apply at depth)
	Slag	15% P_2O_5	
	Calcium phosphate	65% P_2O_5	
	Simple superphosphate	18% P_2O_5	
	Triple superphosphate	45% P_2O_5	
	Phosphoric acid	50% P_2O_5	Supply as liquid (In irrigation water)
	Superphosphoric acid	76% P_2O_5	
Potassium	Potassium chloride	60% K_2O	Apply at depth
	Potassium sulfate	50% K_2O	

pump to add more fertilizer, injecting the appropriate volume of stock solution. The solution is always the same strength, whatever the volume of flow.
• *Hydraulic doser*. The pump has a chamber with a piston. When the chamber fills it sucks fertilizer from a deposit, and it is emptied as it injects fertilizer into the water supply.

The most widely used simple fertilizers

Above: lateral irrigation with fertilizer

Irrigated vegetable plot

7. IRRIGATION

Horticultural crops need large amounts of water. Vegetables consist of about 90% water. The yield of the crop depends to a large extent on the availability of water in the soil.

Water is essential for sugar production and to maintain healthy cells. It is the transport medium for nutrients within the plant and for the substances the plant produces, as well as being a primary reagent in many basic physiological processes. Cell turgor, and thus the turgor of the entire plant, also depends on water.

Most of the water taken up by the roots evaporates from the leaves in the process of transpiration, and only 5% of the water taken up is retained in the plant tissues. The rate of transpiration is directly related to the ambient temperature.

The water evaporated by the leaves is replaced by water that is taken up by the root system, which contains the mineral elements present in the soil solution.

When more water is lost by transpiration than is taken up by the roots, a water deficit arises within the plant, and this has a negative effect on growth and thus yield.

The effect of this shortage depends on the duration and intensity of the **water stress**, and also on the plant's stage of growth. Intense water stress, even for a short period, may reduce the plant's growth more than a moderate but longer lasting water stress. This is especially negative when it occurs during the more delicate stages of the plant's life cycle, such as flowering, flower induction and bud differentiation. Wilting occurs when the plant is subject to long-lasting or intense water stress. The plant loses water and therefore rigidity and turgor. Air movement is another factor increasing the plant's transpiration of water.

Applying water to the soil is the best way to meet the crop's water requirements, and this is done through irrigation.

When irrigating, it is important to take into account how much water is already present in the soil. There are several different ways of measuring the moisture levels. One of them is for the farmer to observe the different signs and symptoms shown by the soil and the plant, such as whether the soil is visibly dry or the crop is wilting.

It has been clearly shown that the plant should be irrigated before the plant suffers intense water stress, and above all before the soil moisture level declines to the plant's wilting point.

The decision to irrigate can be taken on the basis of previous experience or of methods that directly measure the soil moisture levels. This is done with devices to measure humidity, such as a tensiometer. The tensiometer measures the soil's pressure, that is to say, the pressure the roots have to exert in order to absorb water from the soil. The greater this pressure is, the less water there is in the soil, as the roots have to apply more pressure; and to the contrary the lower the pressure is, the greater the moisture present and the less force is required.

A tensiometer consists of a porous ceramic capsule that is filled with water and buried in the soil at the depth to be measured. This depth should be the depth the crop's roots grow most actively. The capsule is connected to a manometer, a pressure gauge, by means of a tube of water. The porous ceramic is permeable to water but not to air.

The soil exerts pressure on the water in the column, which causes the height of the water to fall below the soil level, showing the negative pressure which can thus be measured.

You should place two tensiometers at different levels within the soil in order to know more about the moisture status of the soil.

The tensiometer scale goes from 0 to 100. 0 indicates the highest possible water content and 100 indicates total absence of water. The readings mean the following:

- From 0 to 10: the soil is saturated
- From 10 to 30: the soil is at field capacity
- From 30 to 60 : usable water is present
- More than 60: water is lacking

It is not only important to know how much water is present in the soil, but also to know how much water the crop requires at each stage in its life cycle. In general, you should water abundantly but not often. Meager irrigation loses a lot of water through evaporation and causes formation of a surface crust that prevents the infiltration of water in future irrigation.
In hot periods it is best to water in the evening. This means the plants can make best use of the water, because less is lost through evaporation. The temperature of the irrigation water should be about 20°C, because if it is any lower it may cause the crop's growth to halt.

7.1. THE IRRIGATION SYSTEMS USED

The systems for distributing the water may vary greatly, as they depend on the nature of the site, the requirements of the crop being grown, the cultivation techniques being used, the cost of labor, the availability of water and the possibility of automating the site.

All the many different types of irrigation system in use can be divided into three groups:

• Gravity
• Sprinkler
• Installed systems

7.1.1. Irrigation by gravity

This is also known as **surface irrigation**. This is the traditional method and is widely used for unprotected horticultural crops.
Irrigating using gravity does not require any energy to distribute the water, because the water is distributed by gravity, even if energy is required to bring the water to the site.
The water is applied to the soil surface at the highest point on the site, and it then descends along the gradient.
The general advantages of this system are that installation costs are low, it is highly effective at washing salts from the soil and a large volume of the soil available to the plant is wetted.
The disadvantages of irrigating by gravity are that the water is distributed irregularly over the soil surface, much of the irrigation water is lost, soil fertility is lost and the site requires more preparation, implying greater labor expenses.
Irrigation systems based on gravity can be divided into two systems, irrigation as a sheet or along furrows.

• **Flooding or sheet irrigation**. More water is applied than can infiltrate into the soil, so that a continuous sheet of water forms that floods the soil and gradually infiltrates into the soil.

This system is suitable for permeable sites with little or no slope. It has the disadvantage of causing excessive soil compaction, and thus leading to problems of soil aeration.

• **Irrigation along channels**. This consists of causing a sheet of water to circulate along the furrow in the earth between two ridges within the field, thus ensuring the water can infiltrate sideways and downwards.
This can be done on sites that have been adequately levelled. It is important that the soil is perfectly level so that the water can flow without problems, but without causing soil erosion. The best gradient for the furrows is between 0.5% and 1.5%.

Irrigation by gravity avoids wetting the plant directly, but this system does have a series of disadvantages; the distribution of water is not uniform, the water cannot be dosed, more labor is necessary, and it may give rise to problems of poor aeration due to compaction of the soil, and also cracking if the soil dries out.

The Tardiente canal (Huesca, Spain)

7.1.2. Irrigation by sprinkler

This supplies the water as an artificial rain. The system consists of the following components:

- **The pumps**, which supply the volume of water and the pressure needed for irrigation.
- **The distribution network**, which consists of tubing of different diameters. The tubing should be lightweight and easily connected. The main tube bears the water to a network of secondary tubes where the sprinklers are located.
- **The sprinklers** are the components that disperse the water. They do this in a variety of ways, depending on the sprinkler type used and the crop's needs. They may emit a fan, parallel jets or a single jet of water. They may also vary in their angle of inclination, and be fixed or rotatory. The most widely used ones are rotary and produce a jet that can reach 10 to 25 m and with a volume of flow of 0.5 to 7.5 m^3 of water per hour at a pressure of 2 to 3.5 kg/cm^2.
- **The outlet valves**, whose purpose is to prevent the system dripping after irrigation has finished.

Rotary microspray (340° of arc is irrigated). Manufactured by RAIN-BIRD

A field of sunflowers being watered by sprinklers.

When installing a system of this type, you must take into account the nature and structure of the soil to be irrigated. The maximum amount of water to be given is limited by the speed the water can infiltrate into the soil. If more water is applied, the soil will become waterlogged.

The advantages of a sprinkler irrigation system are that:

- It saves water and labor.
- It does not cause movements of soil and can be used to irrigate sites that have not been levelled.
- Distribution of the irrigation water is uniform and controlled.
- It reduces excessive washing of the soil.
- Irrigation can be automated and programmed.
- Fertilizers can be applied in the irrigation water.
- It can be used when there is the risk of a frost.

However, a sprinkler system also has a series of disadvantages:

- More energy is needed.
- Water is lost by evaporation, especially in dry environments when the water is emitted as very fine drops.
- The water is unevenly distributed on windy days.
- It favors the spread of some diseases.
- It causes problems in taller-growing crops.
- It interferes in fertilization, especially in crops grown for their fruit.
- The installation cost is high.

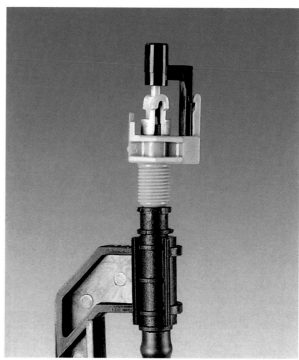

7.1.3. Installed irrigation systems

Installed irrigation supplies very small doses of water to the plant, with a lower volume and higher frequency. This maintains the soil at an optimum level of humidity for the crop to grow.

Installed irrigation systems are classified by the way they emit the water:

- **Microsprinklers**. The water is emitted from a rotary sprinkler, which is moved by the pressure of the irrigation water.
- **Diffusors**. Diffusors are fixed and emit the water as a very fine spray that is distributed evenly over the surrounding circular area.
- **Drip** or **trickle** irrigators release small volumes of water directly into the soil. They form a permanently wet area around the plant's roots called the **moist bulb**.

The installed irrigation system consists of the following components:

- The irrigation nozzle
- The water distribution system
- The emitters

7.1.3.1. The irrigation nozzle

This consists of a set of components that serve to control the volume of flow, the pressure, the filtering and the addition of fertilizer to the water that is going to be supplied to the crop.

• **Flow and pressure**. They are controlled by means of volumetric valves or time-regulated solenoid valves, and by pressure regulators or manometers. The manometers are located at both the inlet and the outlet of the filters and show the loss of flow that occurs within. When there is an excessive pressure difference between the inlet manometer and the outlet manometer, this means that there may be problems due to obstructions and that an overhaul and thorough cleaning are necessary.
At the tip of the irrigation nozzle, after the filters, there is a meter to measure the water flow towards the distribution system.

• **Filtering**. The filters are responsible for removing the impurities from the irrigation water, to avoid problems due to blockages forming in the nozzles. The volume of flow indicates the size and number of filters that should be used.
The filters may be located before the fertilizer application equipment or after it, though you are advised to put it before, in order to avoid problems due to corrosion caused by the fertilizers.

There are three types of filters:

• *Hidrocyclones*. They eliminate suspended silt and heavy metals. They are upside-down truncated cones, larger at the top. The water enters at the side, the suspended particles are moved by centrifugal force to the periphery and sink under the action of gravity to the base, where there is an outlet for sediments, while the water exists through the center at the top.
• *Mesh filter*. This consists of a frame of some corrosion-resistant material, within which there is a cartridge consisting of one or more stainless steel meshes, fine enough to retain any impurities that may have got past the sand filters.
It is usually located after the fertilizer application equipment.
• *Ring filters*. The mesh filters can be replaced by ring filters, which consist of a set of pressed porous discs.

The type of filter to be used is determined by the quality of the water used, with all three types of filter used together for very dirty water, a sand filter and a mesh one for normal water, and just a mesh filter for very clean water.

• **The fertilizer application equipment**. This has already been dealt with in the section of applying fertilizer in irrigation water.
• **The pumping station**. This provides the water with the pressure needed to make it flow. The characteristics of the equipment are imposed by the volume of flow required and the output pressure of the filters, and the losses in filtering. These losses are normally estimated to be 5 m^3 for each filter in series.

Ⓐ
Ⓑ

Sectorial/complete circle sprinklers
A/ With a 25° angle of emission
B/ With a 27° angle of emission

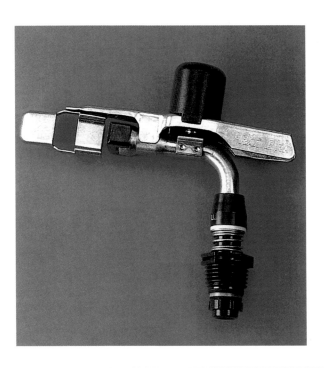

1/2 full circle sprinkler.
Manufactured by
RAIN BIRD

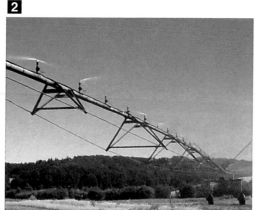

1/ Mobile sprinkler system
2/ Mobile pivot system
3/ Fixed sprinkler cannon system
4/ Mobile sprinkler system with rollable hose
5/ Emerging jet of water

7.1.3.2. The distribution network

The distribution network consists of a set of tubes that channel the water from the irrigation hydrant to the nozzles.

The tubing may be of fibrocement or galvanized iron, though they are normally made of PVC or polythene, especially the secondary tubing and the nozzle supports.

A main tube bears water from the irrigation hydrant, and then divides into primary tubes and then secondary tubes, and the network is completed with the tubes leading to the nozzle supports.

7.1.3.3. Emitters

They are the devices that emit the water.

There are several types:

• **Tricklers**. There are many different models on the market, and new ones appear every day. They may be turbulent or laminar, with a vortex-type orifice, with a long channel, with microtubes, self-compensating, multi-outlet, self-cleaning, etc. They provide between 4 and 6 l/h with the pressure of a 10 m column of water.

• **Diffusors**. These do not have any turning mechanism. They emit the water at some pressure, after projecting it against a fixed deflector. The flow is less than 150 l/h.

• **Micro-sprinklers**. The water is pulverized by rotary mechanisms and then emitted. The volume of flow is between 20 and 150 l/h, with a range of less than 6 m.

7.1.3.4. Advantages and disadvantages of an installed irrigation system

This type of irrigation system has the following advantages:

- It increases the crop's production.
- It saves water, labor and fertilizers.
- It gives greater control of soil humidity.
- Any sort of soil can be irrigated, however poor it is.
- The installation does not require major earthworks.
- Fertilizer can be applied with the irrigation water.
- It reduces the appearance of weeds.
- It allows the whole system to be mechanized by using an irrigation program.

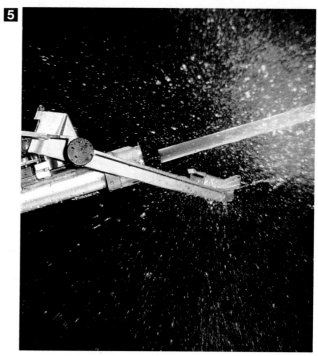

The most important disadvantages are the following:

- The high investment required for the equipment.
- Problems due to blockages in the devices emitting the water, which have to be specially cleaned.
- The use of soluble fertilizers, which are more expensive.
- The mineral salts concentrate in the "wet bulb", which may give rise to salinity problems with the crop.

7.2. QUALITY OF IRRIGATION WATER

The most important characteristics of the irrigation water, the ones conditioning its use in irrigation, are: the temperature, the quantity and nature of the suspended materials present, and the concentration and composition of the dissolved substances. Vegetables do not only need a lot of water, but also need high quality water, and this mainly depends on its origin.
Water can be supplied by extraction from wells, rivers, lakes or aquifers. Substances in suspension are typically found in surface waters and may be organic or mineral in origin.

The water used in irrigation may be derived from:

• **Rainwater**. This is good quality water, well aired and free of salts, generally rich in nitrogen and suitable for all sorts of crops.

• **Spring water**. These are generally cold, unaerated and they vary greatly in their chemical composition, which depends on the nature of the local soil.
• **Water from wells**. These are cold, unaerated and normally contain a lot of mineral salts.
• **Water from lakes or pools**. These waters are rich in organic matter, but are not aerated and are rather acid.
• **Industrial discharges**. These can only be used if the suspended particles are of animal or plant origin. Discharges from the chemicals or metallurgical industries must be rejected because of the risk that they may be toxic to the crop.

In general, irrigation water should contain no substances that are harmful to plants, it should not be cloudy or silty, and it should not be colder than the soil it is applied to.
The temperature of the water should be as close as possible to that of the soil the root system is growing in, and the difference should never exceed 10°C.
It is important to know which salts and elements are dissolved in the water before using it. The amount of dissolved salts, the salinity of the water, can be found by measuring the water's electrical conductivity (E.C.), as the higher the amount of dissolved salts, the greater the conductivity. The conductivity is expressed in microsiemens/cm^2, and when multiplied by the factor 0.64 this gives us the total number of milligrams of salts per liter.

Removable turbulent maze tricklers. They are especially suitable for vegetables, orchards and vineyards. Manufactured by TWIN DROPS IBÉRICA, S.A.

8. APPLICATION OF PLANT GROWTH REGULATORS

8.1. INTRODUCTION

In horticulture, plant hormones are widely applied to many crops, and this is done for a wide range of specific and very different reasons.

Plant growth regulators are plant hormones. Hormones are organic substances synthesized by the plant that can activate, inhibit or modify the plant's growth even at very low concentrations.

Great care must be taken when using plant hormones, as there are several factors, such as the environmental conditions, the species and dose applied, and the plant's susceptibility or tolerance, that can all influence the crop's response to the application and give rise to undesired results.

You must know in advance the plant's response to the growth regulator, the dose to apply, the plant's susceptibility or tolerance, the correct environmental conditions and the levels that may be toxic.

Their main action is to speed up or slow down certain phases of the plant's development. They are used in vegetable crops, ornamental plants and cut flowers for the following reasons:

• To promote rooting of cuttings.
• To improve germination and root formation in germinated seeds.
• To promote growth of the root system after transplanting and pricking out of young plants.
• To increase and advance flowering.
• To encourage formation of parthenocarpic fruit, i.e., fruit formation without fertilization by the pollen grain.

When applying plant growth regulators, the following concepts should be borne in mind:

• A single product need not have a beneficial effect on all plants in terms of a given physiological process, such as flowering, fruiting, etc.
• A single product need not be equally advantageous in two different physiological processes, even within the same plant.
• Plant growth regulators should not be mixed with other products, such as agricultural chemicals or fertilizers, as the growth regulator may be made ineffective, or it may even have a synergistic effect, increasing the power and effect of the hormone.
• They should be applied in the coolest hours of the daytime, the best time being late afternoon. Do not apply after heavy dew or during rainfall, because when the hormone is diluted with water, its concentration, and thus its effect, is lower.
• The dosage has to be very carefully worked out and applied. The manufacturer's instruction should be obeyed thoroughly. If the dose is too low, we will not get the desired results, but if the dose is too high it may be toxic to the plant.
• The product should only be applied to healthy plants. The soil should be supplying them with all the moisture and nutrients they need. If these

demands are not met and the soil can not meet the additional new demands, when flowering or fruiting is induced, we may cause serious damage to the plant that it will never recover from.
• After application, there should be abundant water in the soil, because the rapid growth of the fruit means that consumption is greater, and water shortage may cause malformations and unbalanced plant growth and development.

8.2. TYPES OF HORMONES AND THEIR EFFECTS

The effects of the plant growth regulators are mainly related to the stimulation of the roots, increasing flowering, fruit ripening, and in general, the growth and development of the plant and its organs.

Not all these regulators have the same effects on the same physiological processes. The hormones most widely used in horticulture are the **auxins**, **gibberellins**, **cytokinins** and **other substances**.

8.2.1. Auxins

Auxins are chemically related to indole acetic acid (IAA). They are mainly characterized by their ability to activate growth. They generally have a feminizing effect on flowers, that is to say, they increase the number of female flowers.

Auxins are used:

• To improve fruit set, for example in the tomato and eggplant.
• To improve fruit growth in unfavorable climatic conditions, mainly low temperatures.
• To favor the rooting of cuttings in plants like the carnation and artichoke.

When an effect on the fruit is desired, auxins are applied to the flower buds at the specified dose, as an excess can cause malformation or empty fruit.

The auxins include the following:

- Chlorophenoxyacetic acid
- Chlorophenoxypropionic acid
- Naphthylacetic acid (NAA)
- Naphthoxyacetic acid
- Indolebutyric acid (IBA)
- Indole acetic acid (IAA)

To promote rooting, IBA, IAA or NAA are normally used.
To induce flowering auxin derivatives are normally used, such as dichlorophenoxyacetic acid (2,4-D) or sodium naphthaleneacetate.

8.2.2. Gibberellins

The gibberellins are all chemically related to gibberellic acid. They are mainly characterized by their influence on stem elongation, and thus the growth of the plant. In general, they have a masculinizing effect. They alter the proportion of male and female flowers, increasing the percentage of male flowers. The main gibberellin is gibberellic acid.

Gibberellins are used:

• To force vegetative growth in unsuitable conditions, in the case of celery and lettuce.
• To keep the plants growing when temperatures start to fall, for example in peppers, eggplant, strawberry and lettuce.
• To obtain earlier flower heads in the artichoke.
• To break dormancy in some vegetative organs, such as potato tubers, and also in seeds, such as celery, watercress, lettuce, etc.
• To induce greater stolon formation in strawberries.
• To increase the number of male flowers in the cucumber.
• To replace the cold requirement or long-day requirements of many plants for flower production.
• To induce parthenocarpy, the formation of fruit without fertilization, especially in plants on which auxins do not have this effect.

8.2.3. Cytokinins

Cytokinins are chemical derivatives of adenine. They are mainly characterized by their ability to influence cell division. They are very useful for *in vitro* tissue culture, making it possible to maintain plant tissues alive and to stimulate cell division.

They are used:

• To induce parthenocarpy in some fruits.
• To stimulate the formation of buds in leaves separated from the parent plant.
• To stimulate water loss by transpiration in some plants.
• To stimulate tuber formation in potatoes.

8.2.4. Other substances

This group includes several substances that are used in horticulture for their effects on plants.

They are:

• **Abscisic acid**. This can inhibit many plant growth phenomena. It also inhibits chlorophyll synthesis and induces the leaf stomata to close, thus reducing transpiration.
• **Ethylene** (ethene). This affects plant growth, leaf fall and ripening of some fruit.
• **Tolylphthalam**. This substance increases flower formation in tomato crops. It is sold under the name *Tomapar.*
• **Etephon**. This acts on fruit ripening and coloring. In peppers and tomatoes, it brings ripening forward and allows fruit color to become more intense and uniform. It increases the resistance of small tomato plants to the cold, and favors the formation of female flowers in the cucumber.
Etephon is used to induce and regulate flowering in

the pineapple. It is also used together with gibberellins to break the dormancy of some seeds. It is sold under the name *Fruitel* or *Ethel 48.*

• **Daminozide** and **chlormequat**. These are growth retardants, substances that reduce growth. Daminozide is used in flowers for cutting and in ornamental plants to bring flowering forward.

Blocks of different substrates for rooting cuttings

Chlormequat or chlorocholine chloride is sold as *Cycocel* or *CCC* and regulates flowering, both in flowers for cutting and horticultural crops. In melons, it reduces vegetative growth, and promotes greater fruit set and higher production. In potatoes and sweet potatoes it induces earlier and more abundant tuber formation. In ornamental plants, it is used to obtain dwarf plants which are of great commercial appeal.

• **Maleic hydrazide**. This is a compound that acts by inhibiting the plant's growth and by delaying the sprouting of bulbs, tubers and rhizomes during storage, though other substances, such as methyl naphthaleneacetate or IPC, are used. In celery it is used to prevent the plant bolting (flowering prematurely).

• **Anti-transpiration agents**. These substances lower very high rates of transpiration, thus reducing water losses, and when applied after collection they prevent the fruit losing weight. This group includes Di-1-p-menthene, sodium alginate and oxythylene-dodecanol.

9. ALTERNATION AND ROTATION IN HORTICULTURAL CROPS

9.1. INTRODUCTION

In horticulture, crops are continually extracted from the soil in order to obtain the greatest quantity of produce from the soil in the shortest possible time. And in order to obtain the maximum yield from the soil, the intention is that the soil is always growing a crop, and so it requires abundant fertilizer and water.

To increase the soil's yield even further, crops can be grown together (associations). This is based on the idea that the adult plants of one species protect the young plants of the next crop.

When the first crop has finished growing and is harvested, the young plants take over.

As a general rule, repeatedly growing the same crop on the same plot leads to a clear fall in yields. Yields are high to begin with, but decline over the years as the same crop is repeated year after year.

The most effective solution to this problem is not to increase the doses of fertilizer, as the soil is "exhausted". This condition is known as "soil fatigue", and occurs when a site is cultivated continuously and intensively. The most important causes of soil fatigue are:

• Secretion by the plants of substances that accumulate in the soil. Some of these substances are toxic when they reach high concentrations in the soil.

• The impoverishment of some levels of the soil, with a decrease in the amounts of certain nutrient elements.

• Imbalances in the soil microorganisms, due to the repetition of a crop. This repetition acts to select in favor of certain organisms in the soil.

• An increase in some pests, including insects, mites, fungi, weeds, bacteria, viruses and above all nematodes, because if the same plant is grown year after year its natural enemies can increase and intensify their attacks.

For all these reasons, the repeated cultivation of the same crop species only raises problems. The solution is to follow a well-planned alternation of crops.

There is a difference between **rotation** and **alternation**.

• **The rotation** is the set of crops that follow each other on the same plot aver a fixed number of years.
• **Alternation** is the simultaneous cultivation of crops in the different parcels making up the site. It is annual in nature and overall it forms the rotation.

The main purpose of rotation and alternation is to get higher yields and to improve the quality of the harvest, while preventing soil impoverishment by the crops and the accumulation of diseases and pests.

9.2. BASIC RULES FOR ROTATIONS AND ALTERNATIONS

The basic rules to follow are general in nature and sometimes they are not enough to ensure optimum yields.

Rules for rotations and alternations

	Tomato	Pepper	Eggplant	Cucumber	Melon	Watermelon	Zucchini	Runner bean	Strawberry	Lettuce	Asparagus	Chinese cabbage	Garlic	Onion	Spinach	Sugarbeet
Goes well after	Leek Lettuce Onion Runner bean Pea	Runner bean Broad bean Pea Onion Leek		Tomato Pepper Eggplant Runner bean Celery Lettuce	Tomato Pepper Eggplant Runner bean Celery Lettuce	Tomato Pepper Eggplant Runner bean Celery Lettuce	Tomato Pepper Eggplant Runner bean Celery Lettuce		Zucchini Runner bean Pea Broad bean Lettuce *	Cucumber Melon Watermelon Zucchini Tomato Celery Pepper Eggplant Carrot		Cucumber Melon Onion Watermelon Zucchini Tomato Pepper Eggplant Leek Asparagus	Potato Cabbage Spinach	Cereals Legumes	Swiss chard Sugarbeet and other unweeded crops	Beans Broad beans Peas Cereals
Does not go well after	Melon Watermelon Cucumber Pepper Eggplant Potato	Tomato Eggplant Potato		Melon Watermelon Cucumber Squash	Watermelon Cucumber Zucchini	Watermelon Cucumber Zucchini	Watermelon Melon Cucumber Zucchini	Pea Broad beans Spinach	Tomato Eggplant Peppers Sugarbeet	Endive Endive Runner bean Asparagus Turnip Chinese cabbage Cabbage	Garlic Onion Leek	Cabbage Radish Broccoli Pea		Cabbage Sugarbeet		
Goes well before		Lettuce Bean Pea														
Does not go well before						Watermelon Cucumber Zucchini			Broad bean Pea				Garlic Onion Leek			
Observations to bear in mind in the alternation	Do not repeat crop	Should not be repeated within 3 years. Requires a lot of organic matter.	It has deep roots. It exhausts the soil. It eliminates many weeds.	It improves the soil. Do not repeat within 3 years.	It improves the soil.. Do not repeat within 3 years.	It improves the soil. The root system is superficial. Do not repeat within 3 years	Greedy plant. Eliminates many weeds.	Improves the soil. Fast-growing Surface root system.	Improves the soil. Surface roots.		Greedy plant. Surface rooting.				Do not repeat within 3 years.	Do not repeat within 3 years.

* and any plant with deep roots

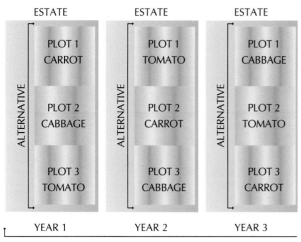

FAMILY	CROPS
Bean (Leguminosae)	Runner bean, pea and broad bean
Cucumber (Cucurbitaceae)	Melon, watermelon, cucumber and zucchini
Cress (Cruciferae)	Chinese cabbage, cabbage, radish, broccoli, cauliflower and turnip
Daisy (Compositae)	Lettuce, endive, green endive, artichoke and cardoon
Goosefoot (Chenopodiaceae)	Swiss chard, sugar beet and spinach
Lily (Liliaceae)	Onion, leek, asparagus and garlic
Nightshade (Solanaceae)	Tomato, pepper, eggplant and potato
Rose (Rosaceae)	Strawberry and raspberry
Umbellifer (Umbelliferae)	Celery, fennel, parsnip, parsley and carrot

To the right: some crops and the botanical families they belong to.

To the left: the difference between rotation and alternation

ESTATE / PLOT 1 CARROT / PLOT 2 CABBAGE / PLOT 3 TOMATO — YEAR 1

ESTATE / PLOT 1 TOMATO / PLOT 2 CARROT / PLOT 3 CABBAGE — YEAR 2

ESTATE / PLOT 1 CABBAGE / PLOT 2 TOMATO / PLOT 3 CARROT — YEAR 3

ALTERNATIVE

3 YEAR ROTATION

• After plants with deep roots, you should grow plants with surface roots, and vice versa, in order to make best use of the different layers of soil.
• After plants that require a lot of dung, you should grow plants that prefer well-rotted dung. These should in turn be followed by plants requiring abundant mineral fertilizers.
• A crop that requires several different cultivation operations and deep plowing should be followed by one that requires a weed-free soil.
• In general, two consecutive crops should have needs that are as different as possible.
• A plant that exhausts or impoverishes the soil should be followed by one that improves or enriches the soil, such as legumes.
• A crop with a long cycle should be followed by a fast-growing crop with a short cycle.
• Plants should not follow members of the same botanical family. They should be grown at least 3 years apart.
• If you cannot wait three years, the soil should be disinfected and fertilized.

9.3. FACTORS AFFECTING THE CHOICE OF CROP ROTATION

When planning a crop rotation, many factors must be taken into account. They can be divided into economic and agricultural factors:

• Economic factors:

The more intensive the estate, the greater the investment required. Several economic factors intervene when choosing a crop rotation:

• *The availability of labor.* This is generally one of the largest expenses. Sometimes, skilled labor is required.
• *Availability of capital* to meet the costs of production of the horticultural crop in question.
• *The possibility of selling the produce*, that is to say, being sure that the produce can actually be sold.
• *Means of transport*
• *Availability of cold chambers or refrigerated transport* if the crop requires it.

• Agricultural factors

The most important factor determining the crop rotation is the climate (temperature, rainfall, wind regime, etc.)
Another important factor is the availability of organic matter. Some crops require a lot of dung, and if this is not produced on the same estate and it cannot be obtained at a satisfactory price, this should be taken into account when planning the crop rotation.
The nature, depth and physical properties of the soil are also determining agricultural factors when choosing a crop rotation.

9.4. ASSOCIATED CROPS

As already pointed out, the association of crops is based on the simultaneous production of two different crop species, that is to say, they are grown in the same soil at the same time. The young plants grow together with the adult plants, and are protected by them.
When they reach the end of their life cycle, the adult plants are collected, and the young plants continue growing and take their place.
The following is a list of some rules for the successful association of different crops in accordance with their horticultural characteristics.

• **The tomato** is a difficult plant to associate. Irrigation and fertilization play a crucial role in tomato cultivation, and so any plant to be grown in association must adapt well to these factors. Tomatoes should not be grown with cucumber, melon or zucchini.

• **Peppers** should be associated with a crop of runner beans, as long as the planting frame is made slightly larger. It is not recommended that grow peppers by grown with eggplant, melon or cucumber.

• **The eggplant** can be grown with runner beans, but not with tomato, pepper, melon, cucumber or zucchini.

Time course of an intensive rotation over 4 years

• **For the cucumber**, one of the most important requirements is moisture. The long cucumber can be associated with fast growing plants cultivated for their leaves, such as the lettuce. The short cucumber can be associated with runner beans. Neither short nor long cucumbers should be associated with the tomato, melon, watermelon or zucchini.

• **The melon** cannot be associated with crops that do not like a lot of water. It can be associated with the runner bean, as long as it is the main crop, and it should not be associated with cucumber, zucchini, tomato or eggplant.

• **Zucchini** can not be grown in association, due to its rapid and vigorous growth, which covers the entire ground area.

• **The watermelon** like the zucchini, covers the entire ground and so it cannot be grown in associations.

• **The runner bean**, if it is a dwarf variety, can be grown with most other crops.

• **The strawberry** can not be grown in association with other crops because of its dense planting frame.

• **Chinese cabbage** is fast-growing and has a surface root system, meaning it can be associated with other crops.

• **Lettuce**, for the same reasons as Chinese cabbage, can be grown with other crops. Lettuce is also very adaptable to different climates.

9.5. TYPES OF CROP ROTATION

Two different types of rotation exist, varying in the degree of overlap between the crops:

• **Intensive rotation** is when many harvests follow one another in a short period of time. This is done on small sites with a small number of very specialized crops.
In these crop rotations, it is vital to establish a logical plan for the succession of different crops, as the exploitation of the soil is so intensive.

Associated crops

The tables show examples of rotations for 4 or 5 years.

• **Extensive rotation**. When the degree of overlap between two successive crops is not very high. It is often reduced to a single crop per year. When planning an extensive rotation, crops other than horticultural species are occasionally introduced, such as wheat (see example).

	Tomato	Pepper	Eggplant	Cucumber	Melon	Zucchini	Watermelon	Runner bean	Strawberry	Chinese cabbage	Lettuce
Tomato	✕										
Pepper	—	✕									
Eggplant	NO	NO	✕								
Cucumber	NO	NO	NO	✕							
Melon	NO	NO	NO	NO	✕						
Zucchini	NO	NO	NO	NO	NO	✕					
Watermelon	NO	NO	NO	NO	NO	NO	✕				
Runner bean	—	YES	YES	YES	YES	—	—	✕			
Strawberry	NO	NO	NO	NO	NO	NO	NO	NO	✕		
Chi. cabbage	YES	YES	YES	YES	YES	NO	NO	—	NO	✕	
Lettuce	YES	YES	YES	YES	YES	NO	NO	—	NO	—	✕

On the left: time course of an intensive rotation over 5 years

On the right: time course of an extensive rotation over 10 years

Left diagram (intensive rotation over 5 years):

YEAR 1 — SPRING, SUMMER, FALL, WINTER — SPINACH / TOMATO / CABBAGE/GREEN ENDIVE

YEAR 2 — SPRING, SUMMER, FALL, WINTER — PEAS/BEANS / CAULIFLOWER

YEAR 3 — SPRING, SUMMER, FALL, WINTER — LETTUCE

YEAR 4 — SPRING, SUMMER, FALL, WINTER — LETTUCE / ARTICHOKE

YEAR 5 — SPRING, SUMMER, FALL, WINTER — SPINACH

YEAR 1 — SPRING, SUMMER, FALL, WINTER — TOMATO / CELERY

YEAR 2 — SPRING, SUMMER, FALL, WINTER — MELON / CAULIFLOWER / ONION

YEAR 3 — SPRING, SUMMER, FALL, WINTER — SWEET POTATO

YEAR 4 — SPRING, SUMMER, FALL, WINTER — PEA

YEAR 5 — SPRING, SUMMER, FALL, WINTER — STRAWBERRY / LETTUCE

Right diagram (extensive rotation over 10 years):

YEAR 1 — SPRING, SUMMER, FALL, WINTER — CUCUMBER / SPINACH / WHEAT

YEAR 2 — SPRING, SUMMER, FALL, WINTER — RUNNER BEANS

YEAR 3 — SPRING, SUMMER, FALL, WINTER — TOMATO

YEAR 4 — SPRING, SUMMER, FALL, WINTER — GARLIC / CARROTS / WHEAT

YEAR 5 — SPRING, SUMMER, FALL, WINTER

YEAR 6 — SPRING, SUMMER, FALL, WINTER — ARTICHOKES

YEAR 7 — SPRING, SUMMER, FALL, WINTER

YEAR 8 — SPRING, SUMMER, FALL, WINTER

YEAR 9 — SPRING, SUMMER, FALL, WINTER — WHEAT / CAULIFLOWER

YEAR 10 — SPRING, SUMMER, FALL, WINTER — PEPPER

10. HARVESTING HORTICULTURAL PRODUCE

The best moment to harvest varies from crop to crop and directly influences the later quality of the product.

The period of time when the crop should be harvested depends on which part of the crop is used. Crops grown for their fruit are harvested over a shorter period of time than those grown for their leaves, roots or tubers.

The location of the point of sale is a further factor to take into account when deciding when to harvest. The further the sales point is from the farm, the earlier harvest should be, as far as the crop permits. This is done so the product is fully ripe when it reaches the market, attaining full maturity at the point of sale.

The optimum degree of maturity for sale always depend on the taste and requirements of the market where the product is sold.

In general, harvest time is determined by the size or weight of the usable part of the crop or by the compactness of the head, depending on the crop in question.

10.1. GENERAL RULES FOR HARVESTING

The following are some general guidelines for harvesting horticultural produce:

• The product should not be harvested if it is wet or if the weather is very hot, as fruit collected when it is sunwarmed ripens more quickly.
• It is best to collect the produce early in the morning, the coolest hours of daylight, when the produce is most turgid and in best condition to withstand transport and handling.

Vegetable harvester

• The least possible damage should be caused to the produce in the harvesting process, and so it should be handled very carefully.
• When harvesting, a preliminary selection should be made, rejecting damaged or substandard produce.
• Any diseased or damaged parts should be removed from the plant and surrounding soil, so they do not act as foci of infection.
• After harvesting, the produce has to be protected from desiccation, especially in the hottest months of the year.
• The time taken in harvesting and handling the produce to prepare it for sale should be as short as possible.
• In the period before harvesting you should take great care when applying agricultural chemicals like pesticides, and always, always, respect the safety period specified by the manufacturer. If it is necessary to apply a chemical treatment near to an area about to be harvested, it should be treated after the harvest, and never before.

10.2. HARVESTING SYSTEMS

Harvesting is normally the horticultural operation requiring the most labor. There is therefore a growing tendency to mechanize harvesting.

Most horticultural produce for fresh consumption is picked by hand. There are some products, such as bulbs, roots and tubers, that can be harvested mechanically without too many problems.

There are almost no mechanical systems for harvesting the crops that are grown for their leaves, though several companies are beginning to promote machinery for this purpose.

The general tendency is to use more rational manual harvesting systems. This trend is particularly clear in the crops grown to be eaten fresh, which are less suitable for mechanization than horticultural crops grown for industrial use.

This trend to greater rationalization consists of breaking down the operations involved in harvesting so that they can be carried out simultaneously by different groups of people. These are operations like cutting the produce from the plant, removing the leaves, cutting the tip off, packing in the field or carriage to some means of transport, and some of these depend on the characteristics of the produce being harvested.

This leads to a reduction in labor costs and a reduction in harvesting time.

CROP	WHEN TO HARVEST
Swiss chard	60-75 days after sowing
Chicory (salad)	Single leaves: 10-15 cm Entire plant; minimum weight 200-300 g
Garlic (leaf)	Stems 1-2 cm in diameter
Celery	Minimum weight per plant: 300 g
Watercress	5-6 weeks after sowing
Spring Onions	Base diameter: 8-15 mm
Brussels sprouts	Sprouts 2 cm in diameter
Endive (green)	Minimum weight: 200-300 g
Spinach	Leaves: 40-60 days after sowing Entire plant: at least 10 leaves
Leek	Minimum diameter: 25-30 mm 5-6 months after sowing
Borage	50 days after sowing
Parsley	80-90 days after sowing
Asparagus	Green asparagus: when 12-15 cm tall White asparagus: when it reaches soil level
Broccoli	Minimum inflorescence diameter: 10 cm
Cauliflower	Minimum inflorescence diameter: 11-13 cm
Garlic	Minimum bulb weight 25 g
Artichoke	Minimum flower weight: 100 g Minimum diameter: 6 cm
Eggplant	Long varieties: minimum diameter 4 cm and minimum length 10 cm Globose varieties: minimum diameter 6 cm
Zucchini	45-60 days after sowing Minimum diameter of fruit 4 cm, and minimum length 15 cm
Cardoon (thistle)	Minimum weight per plant: 1.5 kg
Onion	Minimum diameter 5-6 cm and minimum weight 50 g
Strawberry	90-120 days after planting
Pea	90-120 days after sowing
Fennel	Minimum weight of stem base: 150 g
Runner bean	65-100 days after sowing
Cucumber (gherkin)	Minimum length 6 cm, and minimum diameter 2 cm 60-70 days after planting
Pepper	Minimum fruit weight: 30 g 70-100 days after planting
Watermelon	75-100 days after sowing
Tomato	90-100 days after planting Minimum weight: 60 g

There is also mechanical help for all sorts of harvesting chores, and these reduce the time lost in journeys to and fro. They include:

• Tractors with a trailer, which can be loaded directly with the produce.
• Forklift tractors, which can take the produce from the field to the transport vehicle.

• Conveyor belts, which can take the produce from the field to the transport vehicle.

To finish, you should use containers or packaging that do not cause any damage to the produce and are strong enough to prevent damage by crushing.

11. STORAGE OF HORTICULTURAL PRODUCE

Below:
Machine to process
horticultural produce

After harvesting, a series of operations should be performed to ensure good storage, which depend on the produce and its intended use.
If the intended use of the produce is for the grower's own consumption, little handling is necessary, but if it is for sale it is important to distinguish between the handling in the field and that carried out later in wholesale horticultural and fruit markets.

In the field, manipulation is mainly cleaning, selecting and calibrating the produce, and it may also be packed. From there it is taken to the chambers for cold treatment, and it will then be dispatched for sale.

Other produce, after harvesting, is taken directly to the wholesale center, where all the operations needed to make the produce ready for sale are carried out.

Ethylene production
by some vegetables

VEGETABLE	ETHYLENE PRODUCTION
Swiss chard, artichoke, celery, Brussels sprout, red cabbage, cauliflower, endive, spinach, green endive, asparagus, strawberry, leaf and root vegetables, potato	VERY LOW
Eggplant, squash, melon cucumber, pepper, watermelon	LOW
Tomato	MEDIUM
Cantaloupe melon	HIGH

11.1. POST-HARVESTING OPERATIONS

There is a whole series of operations to carry out after harvesting the produce: cleaning, selection, calibration, weighing and packing.
These operations differ from one crop to another.

11.1.1. Pre-cooling or pre-refrigeration

This procedure consists of rapidly cooling the produce after harvest.
The temperature the produce is cooled to depends not only on the produce being treated but also on how near or far the sales point is, and on the cooling method used.
The temperature for vegetables is between 2°C and 10°C and the length of cooling depends on the method used, the packing used and the initial temperature of the produce.
Pre-cooling seeks to delay ripening and ensure longer storage, while at the same time reducing the fresh weight of the produce and diminishing the effects of desiccation.
Pre-cooling has other positive effects, such as reducing microbial damage and improving product quality.

The most common pre-cooling techniques are:

• **Cold water**. This is based on spraying or submerging the produce in water at 0°C for 15-20 minutes.

• **Wet air draft**. This is done by placing the produce in a cold wet draft of air.

• **Vacuum treatment**. This consists of placing the produce in a controlled vacuum for 15-35 minutes, depending on the produce.

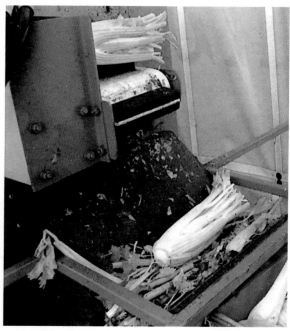

11.1.2. Selection

Selection consists of sorting the sellable vegetables into different groups, and removing any that show anomalies.

The selection criteria depend mainly on the produce. Some products are selected by their color, shape, ripeness or sugar content. The selection methods are usually visual, though in some cases there are specialized machines to do the job.

Storage conditions for some vegetables in a cold chamber

PRODUCE	TEMPERATURE (°C)	HUMIDITY (%)	TIME (days)
Swiss chard	0	90-95	10 - 15
Garlic	-1 - 0	70-75	180 - 240
Artichoke	-1 - 0	85-95	10 - 40
Celery	0	90-95	30 - 90
Eggplant	5 - 10	90-95	10 - 20
Broccoli	0	95	7 - 15
Zucchini	0 - 4	80-90	60 - 90
Onion	-1 - 0	70-75	120-240
Cabbage	0	90-95	20 - 90
Brussels sprout	-1 - 0	90-95	20 - 35
Cauliflower	0 - 1	90-95	20 - 40
Endive	0 - 2	90-95	15 - 20
Green endive	0 - 1	90-95	15 - 30
Asparagus	0 - 1	90-95	10 - 30
Spinach	0	90-95	7 - 15
Strawberry	-0,5 - 0	85-90	3 - 10
Pea	0 - 1	85-95	7 - 20
Fennel	0 - 1	90-95	30 - 60
Runner bean	0 - 4	85-95	7 - 15
Lettuce	0 - 1	90-95	7 - 20
Melon	7 - 10	80-85	10 - 20
Cantaloupe melon	0 - 7	90	10 - 15
Turnip	0	85-90	120 - 150
Potato	2 - 10	90	120 - 170
Cucumber	7 - 10	90	7 - 15
Pepper	7 - 10	90	7 - 30
Leek	0	85-90	40 - 50
Radish	0	90-95	8 - 20
Sugar beet	0	90-95	30 - 100
Watermelon	2 - 4	85-90	14 - 25
Green tomato	10 - 15	90-95	7 - 20
Ripe tomato	2 - 10	90-95	7 - 15
Carrot	0	90-95	60 - 150

*Assembly line
processing
vegetables*

11.1.3. Cleaning

Horticultural produce must be cleaned before packing and sale.

Cleaning the produce includes removing the leaves and washing off any remaining soil.
The most widely used methods of washing are immersion and spraying with water to which some antiseptic has been added.
After being cleaned with water, the produce must be thoroughly dried before packing to prevent rotting.

Other produce, such as peppers and eggplant, should be cleaned without water, using cloths or rags.

11.1.4. Calibration

Calibration is selecting the produce by size, usually the diameter but sometimes the length. This can be done manually or using specialized machinery. There are specific regulations for the calibration of each type of horticultural produce. Documentation on this subject can be bought from the Ministry of Agriculture.

11.1.5. Weighing and packaging

Weighing is done using electric scales, leaving a margin in the weight to compensate for the losses that may occur in transport to the sales point.
The produce used to be put directly into in wooden or cardboard packaging, but there is now a wider range of materials that can be used to pack the product:

• Waxed paper in the bottom of the box.
• Individual protection, which may be cardboard, for melons, or plastic, for lettuces.
• Plastic bags or mesh bags, with a defined weight of produce.
• Trays covered with a plastic sheet, for endives.
• Small transparent plastic boxes ("punnets"), for strawberries.
• Plastic boxes that can be stacked, for tomatoes.

As a general rule, these types of packaging should fulfil all the following characteristics: protect the produce, allow good aeration and make transport and handling easier.

For transportation, it is important that the packages should be easy to stack on pallets. Thus the size of the packaging has to be an exact division of the size of the standard pallet. Another important point is that the packaging has to bear all the relevant information.

The outside of the packaging should clearly state the crop species and variety, as well as its origin and class, category, color, caliber and total weight.

Integrated system to process vegetables in the field

11.2. STORAGE

Some horticultural produce is perishable, and its long-term storage requires it to be placed in a cold chamber. The temperature inside varies from –1°C to 10°C, depending on the species, as some crops, like cucumbers, peppers and eggplant, are damaged by low temperatures.

11.2.1. Techniques

There are a series of rules to bear in mind when storing produce in the cold chamber.

• **Material**. Only place sound healthy produce in the cold chamber.
• **Stacking**. The packaging containing the produce should be stacked in layers on the pallets, to a maximum of 7 boxes. They should not be touching the walls of the chamber, and be at a distance of at least 5 cm from the wall. They must also be a distance of at least 1 m from the fans ventilating the cold chamber.
• **Distribution**. The stacks of pallets should be arranged in rows so that the cold air can circulate, and so the produce can be removed easily.
•**Planning**. Different types of produce should not be stored in the same cold chamber, because they normally have different requirements. Another reason why produce should not be mixed is that during the ripening process some vegetables release ethylene, and this may have a negative effect on other vegetables, causing them to age and deteriorate.

Good storage depends on the correct size of the cold chamber, its total insulation and the correct arrangement and stacking of the produce. The doors should be opened and the lights turned on as little as possible, so as to avoid external inputs of heat.
The possibility of using cold chambers with wet air systems means it is possible to obtain a relative humidity of 98-99%, meaning that the produce stores longer and better.

11.3. TRANSPORT

The storage conditions should be the same during transport, but of course this is much harder to achieve.

Most produce is transported overland in the following ways:

• **Trucks with natural ventilation**. They lack any refrigeration system. They are mainly used to transport non-perishable produce to nearby markets.
• **Insulated trucks**. These do not have any refrigeration system, but the truck's frame is insulated.
• **Cooled trucks**. They have some cooling substance, such as dry ice (solid carbon dioxide), liquid nitrogen or even ice deposits.
• **Refrigerated trucks**. They have their own refrigeration system, powered by the truck's engine.

12. CHARACTERISTICS OF THE MAIN VEGETABLE CROPS

12.1. CROPS GROWN FOR THEIR ROOTS AND TUBERS

12.1.1. Turnips

12.1.1.1. General concepts

The turnip (*Brassica napus*) belongs to the cress family (Cruciferae), and originated in Europe and Central Asia.
It is a biennial, but cultivation only lasts one year. Its growth cycle takes 50-60 days in early varieties and 70-100 days in late varieties.
The seeds are dark red and rounded, and retain their ability to germinate for 4-5 years.
The root system is swollen and partly white and partly reddish. The many varieties can be divided into several groups on the basis of the turnip's shape.

Turnip. Varieties:
A/ Flattened
B/ Cylindrical
C/ Globular

Ⓐ

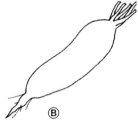
Ⓑ

Ⓒ

• **Elongated or cylindrical varieties:**

Virtudes	Nantes
Mantillo	Nantes strain Candia
Fuencarral	

• **Round flattened varieties**:

Bola de nieve	De Nancy
Milan Red	Shogoin
Bola de oro	White Round
Just Right	Supertop Bency
Black Round	Ping Pong

12.1.1.2. The plant's requirements

• **Climate and temperature**. The turnip requires a cool moist climate. High summer temperatures have a negative effect. There are some varieties that tolerate mild frosts.

• **Water**. Turnips require a lot of water. One effect of

Varieties of turnip:
Milan Red.
Courtesy of SEMILLAS VILMORIN

Nutrient value per 100 g edible produce	
Proteins	1.12 g
Lipids	0.24 g
Sugars	7.77 g
Fiber	1 g
Vitamin B$_1$ (thiamine)	60 mcg
Vitamin B$_2$ (riboflavin)	50 mcg
Vitamin C (ascorbic acid)	61 mcg
Calcium	46 mg
Phosphorus	50 mg
Iron	0.5 mg
Energy value	39 calories

Climatic requirements		
CRITICAL TEMPERATURES	Freezing point	-5 - -8°C
	Zero growth	3 - 5°C
	Optimal growth	15 - 18°C
	Maximum for growth	25 - 30°C
GERMINATION	Minimum temperature	5 - 8°C
	Optimum temperature	20 - 25°C
	Maximum temperature	30 - 35°C
MOISTURE		HIGH
LIGHT		LOW

dry conditions is that the plant flowers prematurely (it "bolts").

• **Soil**. It requires an average texture and good drainage. It can not tolerate waterlogged soil, but likes a cool soil that retains water.
The best soil pH value is 6-6.9. Excessively limy soils cause growth of stringy roots that have an unpleasant taste.

• **Soil extractions**. The nutrient extraction per hectare of turnips is:

100 kg N
60 kg P$_2$O$_5$
100 kg K$_2$O

• **Applying fertilizer**. It is sensitive to fresh dung. The dung should be added to the soil during the previous crop.

Fertilizer at depth per hectare:

40 kg N
128 kg P$_2$O$_5$
164 kg K$_2$O

Surface fertilizer per hectare: 75 kg N

• **Deficiencies**. The turnip is sensitive to boron deficiency.

12.1.1.3. Soil preparation and sowing

The turnip requires a loose soil. The soil should be deep plowed, and the rotovator passed to break the soil down evenly.
Once the site has been prepared, furrows should be made 30-40 cm apart.
The seed should be sown at the end of summer or in early fall, or alternatively in spring, depending on the desired harvest time. The seed should be sown at a depth of 2-3 cm, and it is usually sown in continuous rows.

12.1.1.4. Cultivation and harvest techniques

• **Thinning**. The distance between plants should be 10-25 cm.
• **Weeding**. Weeds can be eliminated manually or by using selective weedkillers.
• **Earthing up**. This is mainly done to avoid the negative effects of frosts.
• **Harvesting**. Turnips are harvested manually, pulling the leaves with the help of a hoe and uprooting the plant. For very large areas, harvesting machinery is used.
• **Sale**. After harvesting, the following operations have to be performed: leaf removal, washing, selection and calibration. Sale is in bunches or in boxes or small sacks.
• **Storage**. Turnips can be stored in cold chambers at 0°C and a humidity of 85-95% for 4-5 months.

12.1.1.5. The most common pests, diseases and physiological disorders

• **Pests**
- *Cress leaf miner*. The larvae mine galleries within the leaves, and the adults eat the leaves.
- *Cress caterpillar*. Leaf-eaters.
- *Miners*. Mine galleries at the base of the stem.
- *Greenfly*. Cause the plant as a whole to turn yellow and the leaves to curl up.
- *Snails and slugs*. Leaf-eaters.

• **Fungal diseases**.
- *Downy mildew*. Produces yellow blotches on the upper side of the leaf and grey mealy areas on the underside.
- *Damping off*. Causes the seedling to collapse at the soil level.
- *Rhizoctonia*. Produces reddish zones in the collar.

• **Viral diseases**
- *Cauliflower mosaic virus*. This virus also attacks the turnip. The veins remain dark green against a generally chlorotic leaf.
- *Turnip mosaic virus*.

• **Physiological disorders**
- *Boron deficiency*. This causes the inside of the root to rot totally.

12.1.2. Potato

12.1.2.1. General concepts

The potato (*Solanum tuberosum*) is a member of the nightshade family (Solanaceae).
It appears to have been cultivated in three different South American centers of origin in Peru, Bolivia and southern Chile.
It is not only used as a human foodstuff, but is also fed to animals.

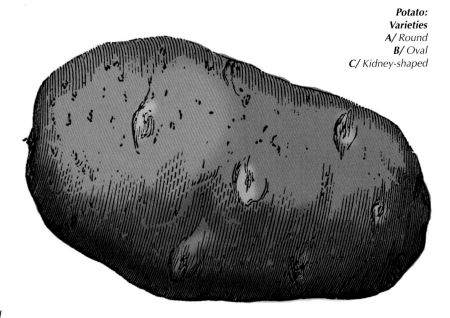

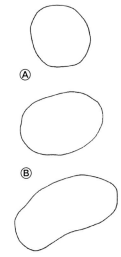

Potato:
Varieties
A/ Round
B/ Oval
C/ Kidney-shaped

Ⓐ

Ⓑ

Ⓒ

Nutritional information per 100 g of foodstuff	
Proteins	1.56 g
Lipids	0.25g
Sugars	19.83 g
Fiber	1.34 g
Vitamin C (ascorbic acid)	10-40 mg
Vitamin B₁ (thiamine)	100 mcg
Vitamin B₂ (riboflavin)	30 mcg
Calcium	8 mg
Phosphorus	56 mg
Iron	0.7 mg
Energy value	72-80 calories

Climate requirements		
CRITICAL TEMPERATURES	Freezing point	–2°C
	Zero growth	6 - 8°C
	Optimum growth	15 - 18°C
MOISTURE		HIGH
LIGHT		MEDIUM

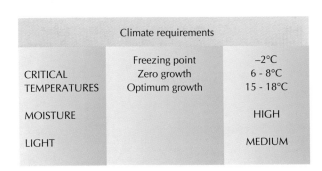

The potato is a perennial herbaceous plant that produces underground stems that bear the tubers. The tubers are oval or round swellings of the underground stems.

The tubers bear a few scale leaves with the buds, the "eyes", in their axils.

The entire potato plant can produce a poisonous alkaloid, solanin, which is produced in the tuber after direct exposure to sunlight. The tuber is the only part of the potato plant that is edible.

The potato tuber is dormant, and this should be borne in mind when planting. To break its dormancy, the tuber can be immersed for an hour in any one of several different solutions.

There are a huge number of varieties of potato on the market. They are classified on the basis of the length of their crop cycle, and include the following:

• **Early varieties**, with a crop cycle of 90 days. These include:

White varieties:

> Arran-Banner
> Kennebec
> Blauka

Yellow-flesh varieties:

> Bintje   Aura
> Belladonna   Spinta

• **Semi-late varieties**, with a crop cycle of 120 to 150 days.

White-flesh varieties:

> Olalla   Gelda
> Turia   Majestic

Yellow-flesh varieties:

> Claudia
> Desirée
> Heidi

• **Late varieties**, with a crop cycle of 150 to 200 days

White-flesh varieties:

> Victor
> Up-to-date

Yellow-flesh varieties:

> Alava   Goya
> Alfa   Baraka

12.1.2.2. The plant's requirements

• **Climate and temperature**. The potato requires a temperate climate, with cool nighttime temperatures, as they favor tuber formation. It is sensitive to late frosts.

• **Water**. The critical period for water supply is the period between the beginning of tuber formation and flowering.

• **Soil**. It requires a deep light soil rich in organic matter.

The potato is resistant to salinity, and can tolerate pH values as acidic as 5.5-6.

• **Extractions**. The amount of nutrients extracted varies with the yields obtained. In general, a hectare of potatoes extracts:

> 200 kg of N
> 50 kg P_2O_5
> 200 kg K_2O

• **Applying fertilizer**. 20-30 t/ha of well shredded dung should be applied.

Fertilization at depth per hectare:

> 80 kg N
> 70-100 kg P_2O_5
> 200-300 kg K_2O

Surface application of fertilizer: 40-60 kg N

• **Deficiencies**. It does not have any special need for boron, but cannot grow well if there is little magnesium in the soil.

12.1.2.3. Soil preparation and planting

The potato requires a homogeneous soil in good tilth and well aired. After deep plowing, the site should be harrowed several times.

The potato is propagated vegetatively by division, by planting tubers or small parts of the tuber (with at least one bud). Certified disease-free "seed" potatoes should always be planted.

Planting can be manual, or mechanical if the site is very large.

Make furrows 50-70 cm apart, plant at a depth of 7-8 cm and leave a distance between plants of 30-40 cm.

The time to plant depends on the variety and its crop cycle.

- Extra-early cycle: early/mid fall
- Early cycle: late fall/mid winter
- Medium cycle: early spring/mid spring
- Late cycle: late spring/ early summer
- Very late cycle: mid summer

12.1.2.4. Cultivation techniques and harvesting

• **Passing the roller**. To firm the soil around the newly planted tubers.

• **Mechanical weeding** to break any pan and create a good tilth, as well as to kill weeds.

• **Weeding**. Either manually or with selective weedkillers.

• **Earthing up**. This should be done when the plants are 15-20 cm tall.

• **Mulching**. This is optional. It consists of covering the soil with a transparent polythene film. This should be 200 galgas thick and have holes for the

crop plants to emerge through. The aim is make the plant sprout earlier. It is removed when the plant is 25-30 cm tall.
• **Harvest**. Harvest is performed when the clumps begin to wither. Before harvesting, all the aboveground parts of the plant have to be destroyed, either mechanically or chemically.
Harvesting is either manual, with the help of a hoe, or mechanical. The machines available range from normal hoes for uprooting to self-powered complete harvesters that uproot, clean and bag the potatoes.

The harvesting period will depend on when the potatoes were planted:

- Extra-early cycle: mid/late winter
- Early cycle: early to late spring
- Medium cycle: from early to late summer
- Late cycle: early/mid fall
- Very late cycle: mid/late fall

• **Sale**. In mesh bags or sacks.
• **Conservation**. The storage sites should be well insulated from changes in temperature and have good ventilation. The storage temperature is 4-6°C at a humidity of 85-90%.

Sometimes, to prolong storage, anti-sprouting agents are applied to prevent or delay sprouting.

12.1.2.5. The most common pests, diseases and physiological disorders

• **Pests**
- *Colorado potato beetle*. Both the larvae and the adults eat potato leaves voraciously.
- *Potato moth*. The larvae mine galleries in the tuber.
- *Wireworms*. Eat holes in the tubers.
- *Greenfly*. They weaken the plant.
- *Cutworms*. They eat the top of the root.
- *Black screwworms*. Leaf-eaters.
- *Red spider mite*. General weakening of the plant.
- *Nematodes*, such as potato cyst eelworm produce cysts on the root system, in addition to general weakening of the plant.

• **Fungal diseases**:

- *Mildew*. This causes patches to form that are brown on the upper surface of the leaf and greyish on the underside.
- *Alternaria*. This causes semi-dry patches on the leaves, with clear edges.
- *Fusarium*. This attacks the tubers damaged during handling, causing them to rot during storage.
- *Rhizoctonia*. This causes the aerial part of the plant to collapse and black pustules to form on the tubers, which then rot.
- *Potato scab*. They cause a wet rot, and stem gangrene.

• **Bacterial diseases**:
- *Bacterial infections*. They cause soft rots, brown rots and gangrene of the stem.

• **Viral diseases**:
- *Viral diseases* are very important in potato

cultivation, as they cause the vegetatively propagated varieties to deteriorate.

• **Physiological disorders**
- *Frosts* may partially or totally destroy the leaves. If very intense, frost may even damage the tubers.
- *Greening*. The tuber turns green when it is exposed to light, and this is accompanied by solanin formation.
- *Scorching*. If the environmental temperature is high and exposure to light is prolonged, the tuber turns bronze-green and the cells below the bleached area die.
- *Legginess*. The sprouts are long and thin. This is due to excessive heating of the tuber.
- *Strings of tubers*. The tubers are small and in a row. This is frequent in late varieties due to several successive interruptions to tuber formation.
- *Cracks and hollowness*. These are due to excessively brusque changes in environmental factors, such as water and temperature, and excess nitrogen availability in the later phases of growth.
- *Excess lenticel production*. The tuber produces many wart-like lenticels due to excess nitrogen supply.
- *Colored patches within the tubers*. They vary in color and texture, and may be blackish or reddish, and elongated. They are due to many different causes.

12.1.3. Radishes

12.1.3.1. General concepts

The radish (*Raphanus sativus*) is a member of the cress family (Cruciferae).
Small radishes are thought to have originated in the Mediterranean region, while large radishes are from China and Japan.
Radishes are highly diuretic and contain a lot of vitamin C.
The seeds are brownish red and rounded, and retain their ability to germinate for 3-5 years.
The root is thickened and may be round or elongated. Radishes may be red, yellow or black, and all have a relatively peppery taste.

The many varieties can be broadly divided into three groups with different biological cycles:

• **Short cycle varieties**. Between 25 and 30 days. These are the most widely grown and are greatly appreciated in the marketplace. They include:

Radishes: Varieties
A/ Round
B/ Intermediate
C/ Long

Ⓐ

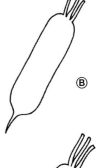

Ⓑ

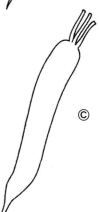

Ⓒ

*Intermediate variety of radish: **Dix Huit Jours**. Courtesy of SEMILLAS VILMORIN.*

Cherry Belle Rota Marteau
Fakir Novired Delog pont
Gandry Carnaval II Flamboyant
Round Scarlet Saxa Kiva
Matador Red-devil

Nutritional composition per 100 g of foodstuff	
Proteins	0.86 g
Lipids	0 g
Sugars	2.44 g
Vitamin A	30 IU
Vitamin B_1 (thiamine)	30 mcg
Vitamin B_2 (riboflavin)	20 mcg
Vitamin C (ascorbic acid)	24 mcg
Calcium	37 mg
Phosphorus	31 mg
Iron	1 mg
Energy value	14 calories

Climate requirements		
CRITICAL TEMPERATURES	Freezing point	-2°C
	Zero growth	6°C
	Minimum for growth	8°C
	Optimum growth	18 a 22°C
	Maximum for growth	30°C
GERMINATION	Minimum temperature	16°C
	Optimum temperature	20 - 25°C
	Maximum temperature	30 - 35°C
MOISTURE		HIGH
LIGHT	Not very demanding	MEDIUM

• **Medium cycle varieties**. Between 40 and 45 days. Larger than short cycle varieties, and include:

Bamba
Golo
White Round
Stuttgart Giant White

• **Long cycle varieties**. Between 100 and 110 days. They are the largest. They include:

Round Thick Black
Long Thick Black

12.1.3.2. The plant's requirements

• **Climate and temperature**. The radish requires a cool moist environment. If temperatures are too high, the taste is more peppery. Short cycle varieties are the most sensitive to frosts.
• **Soil**. The radish requires cool rich soils with a lot of organic matter. It does not tolerate soil salinity well.

• **Soil extractions per hectare**. These differ from variety to variety, but are around:

80-110 kg N
40-60 kg P_2O_5
100-200 kg K_2O

• **Applying fertilizer**. The radish requires an input of well-rotted manure together with the fertilizer applied at depth. The amount is usually between 10 and 20 t/ha.

Deep fertilizer application per hectare:

50 kg N
48 kg P_2O_5
117 kg K_2O

Fertilizer is not applied on the surface because the radish has such a short growing cycle.

• **Deficiencies**. It is sensitive to shortages of boron.

12.1.3.3. Soil preparation and sowing

Soil preparation for growing radishes is similar to that described for turnips.
Radishes can be sown on levelled beds 130-150 wide, or in furrows about 40-50 cm apart. The separation between the rows should be 20 cm and the seed is sown at a depth of 2 cm. The short cycle varieties can be sown throughout the year, bearing in mind that winter sowings will grow more slowly.

12.1.3.4. Cultivation and harvest techniques

• **Thinning**. The distance between plants should be 10-20 cm, depending on the size attained by the variety. Short cycle radishes can be grown only 5 cm apart.
• **Weeding**. Either manual or using selective herbicides.
• **Harvesting**. This should be done before the roots hollow. In small sites, harvesting is manual, and on large sites it is done by machinery.
• **Sale**. Once harvested, the leaves are removed, and the roots are washed, calibrated and packaged. They are either sold in bunches or in bags.
• **Storage**. They can be stored in a cold chamber at 0°C and a humidity of 90-95% for 3-4 weeks.

12.1.3.5. The most common pests, diseases and physiological disorders

Radishes are attacked by the same pests and diseases as turnips, but also by:

• **Pests:**
- *Ants*. They mainly attack at the moment of germination.

• **Physiological disorders:**
- *Hollow root*. This effect may be due to several

causes, such as excess ripening, frosts or imbalances in soil moisture.

- *Split root*. This occurs when the soil texture is not smooth enough.

12.1.4. Sugar beet

12.1.4.1. General concepts

Sugar beet (*Beta vulgaris*) is a member of the goosefoot family (Chenopodiaceae).
It originated in Europe. It can be eaten fresh, boiled

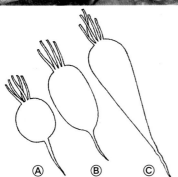

or as a preserve. It is also the source of a red coloring agent, **betacyanin**, used in the food industry.

Sugar beet is a biennial that produces its swollen root in its first year of growth, and flowers in its second year. Its cultivation cycle is usually 210-215 days, though some varieties can be harvested after only 90-100 days.

The seeds retain their ability to germinate for 3-5 years. The shape of the root varies from elongated to round to slightly flattened. The color varies from red to yellowish on the outside and from red to pale pink on the inside.

The varieties are grouped on the basis of their shape.

• **Elongated varieties**. They are 30-40 cm long and include:

> Long Red Virtudes
> Long Covent Garden
> Cylindra

• **Round and flattened varieties.** These are most appreciated in the marketplace, and thus the most widely grown. They include:

Egyptian Red	Monopoly
Round Red	Egyptian Flattened
Detroit	Claudia
Bikores	Red Clapaudine
Boltardy	Red Short Top

Nutritional composition per 100 g of foodstuff	
Proteins	1.6 g
Lipids	0.1g
Sugars	9.9 g
Fiber	0.8 g
Vitamin A	20 IU
Vitamin B$_1$ (thiamine)	0.03 mg
Vitamin B$_2$ (riboflavin)	0.05 mg
Niacin	0.4 mg
Vitamin C (ascorbic acid)	10 mg
Calcium	16 mg
Phosphorus	33 mg
Iron	0.7 mg
Sodium	60 mg
Potassium	335 mg
Energy value	43 calories

Climatic requeriments		
CRITICAL TEMPERATURES	Freezing point	-5 - -7°C
	Zero growth	5 - 7°C
	Optimum growth	22 - 25°C
	Maximum growth	30 - 35°C
GERMINATION	Minimum temperature	5 - 8°C
	Optimum temperature	20 - 25°C
	Maximum temperature	30 - 35°C
MOISTURE		MEDIUM
LIGHT		MEDIUM

12.1.4.2. The plant's requirements

• **Climate and temperature**. It requires a mild moist climate, though it adapts well to other climatic conditions. The young plants are the most sensitive to cold temperatures, and die at temperatures of –3°C or less.

• **Soil**. The sugar beet requires an open, light, deep soil, which should be as homogeneous as possible and without stones or gravel. It resists salinity well within a pH range of 6-8.

• **Nutrient extraction from the soil**. The extraction of nutrients per hectare is:

On the left:
A variety of sugar beet
for the table:
Egyptian Red
Courtesy of SEMILLAS VILMORIN.

Sugar beet.
Varieties
A/ *Spherical*
B/ *Cylindrical*
C/ *Long*

84 kg N
45 kg P_2O_5
168 kg K_2O

• **Applying fertilizer**. The organic matter should be applied well in advance.

Fertilizer application at depth, per hectare:

35 kg N
80-100 kg P_2O_5
150-200 kg K_2O

Surface fertilizer per hectare: 70 kg N

• **Deficiencies**: It is sensitive to a shortage of boron.

12.1.4.3. Soil preparation and planting

Soil preparation for growing sugar beets is the same as for growing carrots.
In temperate zones, the seed can be sown from late winter until the end of spring.
The seed is sown in rows 35-40 cm apart and at a depth of 2-3 cm.
Sugar beet seeds need pre-germination treatment of immersion in warm water for several hours before sowing.

12.1.4.4. Cultivation and harvest techniques

• **Thinning**. The distance between plants should be 20-30 cm, and they should be thinned when they have 4-5 leaves.
• **Weeding**. Manually or with a selective herbicide.
• **Harvest**. They are harvested when the root has reached a diameter of 3-6 cm, depending on the requirements of the market, and their weight is 100-200 g, and this is between midsummer and early fall.
Harvesting can be manual or mechanized. If it is mechanized, the leaves should be removed before harvest.
• **Sale**. After the harvest, the beets are washed, any remaining leaves are removed, they are calibrated and placed in trays, 4 to 15 at a time, and covered with transparent plastic.
• **Storage**. They can be stored in a cold chamber at 0°C and 90-95% humidity for 1-3 months.

12.1.4.5. The most common pests, diseases and physiological disorders

• **Pests:**
- *Sugar beet fly*. It mines galleries in the leaves.
- *Sugar beet plantlouse*. This eats the leaves.
- *White grubs*. The larvae attack the roots.
- *Aphids*. They generally weaken the plant and cause the leaves to curl up.
- *Screwworm*. This eats the leaves.
- *Cutworms*. They eat the collar of the root.
- *Nematodes*. They attack the roots.

• **Fungal diseases:**
- *Cercospora*. This causes circular necrotic patches.

- *Rhizoctonia*. Causes the roots to rot.
- *Sugar beet mildew*. This causes yellow patches on the edges of the leaves, and greyish cottony fibers appear on the underside.

• **Viral diseases:**
- *Sugar beet mosaic.*

• **Physiological disorders**:
- *Bolting* (premature flower production)
- *Boron deficiency*. This causes cracks in the roots and rotting of the interior.

12.1.5. Carrot

12.1.5.1. General concepts

The carrot (*Daucus carota*) belongs to the family Umbelliferae. It originated in Asia Minor, where it can be found growing wild.
It is a biennial plant that lays down reserves in its swollen tap root during its first year of growth, and produces its flower spike in the second year. In less than 12 months it completes both cycles, though the crop cycle can be reduced to 3-8 months, depending on the variety.
The small seeds are dark green with 2 asymmetric faces and bearing curved spines at the ends. They retain their ability to germinate for 3-4 years.
Root size and color are very important when classifying carrots. The most important commercial varieties are orange red, while yellow ones do not sell well.

They are divided by size into:

• **Short varieties**. Less than 10 cm long. They include:

Nancy Red
Dutch Short
Guérande Short
Paris Market
Flakko

Varieties of carrot:
Nantes.
Courtesy of SEMILLAS VILMORIN

• **Intermediate varieties**. 10-20 cm long. They include:

Primato	Amsterdam
Nantes	Karaf
Forto	Tantal
Express	Halle semi-long
De Chantenay	Guérande blunt
Foram	Nandor

• **Long varieties**. Longer than 20 cm. They include:

Micolor	Flacoro
Bercoro	De Colmar
Saint Valéry	Danro

12.1.5.2. The plant's requirements

• **Climate and temperature**. Excessively high or low temperatures have a negative effect on the coloring of the roots, which are paler. Carrots can withstand mild frosts. Low temperatures, at certain times in their growth cycle, may cause the plant to "bolt" (flower prematurely).
• **Water**. Carrots require a lot of water.
• **Soils**. Carrots prefer light, open and deep soils that are well drained and retain moisture well. Compact or stony soils are not suitable. Carrots are sensitive to salinity. They grow best in soils with a pH between 5.5 and 6.8.
• **Nutrient extraction from the soil**. The extraction per hectare differs from variety to variety, but is generally about:

$$150\text{-}160 \text{ kg N}$$
$$60\text{-}100 \text{ kg P}_2\text{O}_5$$
$$250\text{-}500 \text{ kg K}_2\text{O}$$

• **Fertilizer application**. 20-25 t/ha of well-rotted dung should be applied as a source of organic matter.

Fertilization at depth per hectare:

$$40 \text{ kg N}$$
$$30 \text{ kg P}_2\text{O}_5$$
$$150 \text{ kg K}_2\text{O}$$

Surface application: 2 applications of 40 kg of N and 30 kg of P_2O_5 per hectare. The potassium should be applied in fractions between application at depth and 1 or 2 surface applications.

• **Deficiencies**. It is sensitive to boron deficiency.

12.1.5.3. Soil preparation and planting

Carrot cultivation requires good soil preparation. The soil should be deep plowed, incorporating the fertilizers to be applied at depth, and then the surface is tilled as many times as needed to leave the soil fine and loose.
The seed should be sown in spring or early summer in furrows 30-40 cm apart, where 2 rows are sown 25-30 cm apart. The seed is sown at a depth of 1 cm.

Carrot seeds need a pre-germination treatment of immersion in warm water at 20°C for three days.
The seed is sown in continuous rows. For a really large estate, precision seeders can be used.

12.1.5.4. Cultivation and harvest techniques

• **Thinning**. Carrots are thinned twice, 10 days apart, the first time when the seedling has 3-4 leaves. After thinning, the plants should be 6 to 10 cm apart, depending on the variety being cultivated.
• **Weeding**. Either manual or using selective weedkillers.
• **Harvesting**. Carrots can be harvested manually or using machinery. The machines used have toothed discs that cut the leaves off and then uproot the plant.
• **Sale**. After harvesting, the carrots are washed and divided into 2 or 3 categories. This calibration can be manual or mechanical. They are then placed in plastic bags for sale. They are also sold as bunches, with the leaves attached.
• **Storage** . Carrots can be stored in cold chambers at 0°C, with a humidity of 90-95%, for 2-3 months.

12.1.5.5. The most common pests, diseases and physiological disorders

• **Pests:**
- *Carrot fly*. The larvae mine galleries in the root.
- *Wireworms*.
- *Cutworms*. They eat the base of the seedlings.
- *Aphids*. They cause general yellowing of the plant.
- *Nematodes*. They produce swellings and deformations in the roots.

• **Fungal diseases:**
- *Root black rot*. This causes damage that is covered by a blackish mold near the top of the root.
- *Alternaria*. This produces dark patches at the edges of the leaves.
- *Cercospora*. This produces semicircular patches on the leaves that turn a dark grey color.
- *Mildew*. This causes patches to form that are yellow on the upper side and grey on the underside of the leaf.
- *Powdery mildew*. The leaf is covered by a whitish powder.

• **Physiological disorders:**
- *Drought*. The roots are tough and stringy.
- *Cracking of the roots*. This is caused by sharp changes in soil moisture.
- *Cleft roots*. This is the result of inadequate soil preparation, especially in stony soils.
- *Boron deficiency*. This causes gummy patches in the root which turn brown.
- *Bolting* (premature flower production).

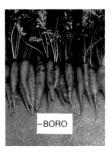

Carrots. Differences in caliber and quality in boron deficient soil, and after treatment

12.1.6. Other crops

12.1.6.1. Sweet potato

The sweet potato (*Ipomoea batatas*) is a member of the convolvulus family (Convulvulaceae). It is originally from the tropical Americas and is a human foodstuff, though it is also fed to animals. It is rich in carotene and vitamin C.
The sweet potato produces an annual climbing stem, with a growing cycle of 120-150 days, depending on the variety.
The large tubers contain the plant's reserves and are the edible part.

The most widely cultivated varieties are:

Malaga sweet potato
American red sweet potato
Centennial
Jaspers
Catameco

12.1.6.1.1. The plant's requirements

• **Climate and temperature**. It requires a hot moist climate, and tolerates hot conditions well, though it does not cope well with sharp changes in temperatures. Frosts may completely destroy the plant.
• **Moisture** is important, especially during the early stages of growth.
• **The soil** has to be light and cool, with good drainage. It shows average resistance to saline conditions and is highly resistant to acidic conditions, growing in soil with a pH of 5.5-6.

Climatic requirements		
CRITICAL	Freezing point	0°C
	Zero growth	10°C
TEMPERATURES	Minimum for growth	12-14°C
	Optimum growth	21 - 24°C
	Maximum for growth	35°C
MOISTURE		HIGH
LIGHT		MEDIUM

• **Nutrient extraction** per hectare is estimated to be:

70 kg N
20 kg P_2O_5
110 kg K_2O

• **Fertilizer**. 20-25 t/ha of well-rotted dung should be applied, ideally during the previous crop.

Fertilizer application at depth:

25-50 kg N
60-120 kg P_2O_5
50-150 kg K_2O

Surface application per hectare:

40 kg N
50-100 kg K_2O

12.1.6.1.2. Soil preparation and planting

The soil has to be open and in good tilth, so first the soil should be deep plowed, incorporating the fertilizer at depth at the same time, and then tilling the surface as many times as needed to leave the soil in perfect growing conditions.
The sweet potato is propagated vegetatively, using a small part of a tuber with a bud, just like the potato.
The sweet potato is planted in spring in furrows 60-90 cm apart and at a depth of 12-15 cm. The distance between plants should be between 35 and 50 cm.

12.1.6.3. Cultivation techniques

• **Earthing up**. This should be carried out 50 days after planting.
• **Weeding**. It is important to control weeds during early stages of growth. As the aboveground parts grow they cover the soil surface.
• **Pinching out**. To control the plant's spread.
• **Harvesting**. This is in fall, when the leaves start to turn yellow. Harvesting can be manual or by machinery. Like the potato, all the aboveground parts have to be removed before harvesting if machinery is going to be used.
• **Sale**. After harvesting, it is cured to prevent rotting of any wounds suffered during harvesting and handling. The roots are placed in ventilated chambers at a temperature of 27-29°C and a humidity of 85-90% for 5-7 days.
• **Storage**. Sweet potatoes can be stored at a temperature of 11-15°C and a humidity of 80-85% for several months.

12.1.6.4. The most common pests, diseases and physiological disorders

• **Pests:**
- *Screwworms*
- *Wireworms*
- *White grubs*
- *Aphids*
- *Nematodes*

• **Fungal diseases:**
- *Fusarium*. This attacks the vascular system causing progressive yellowing of the leaves and a black rot of the roots.
- *Sweet potato dry rot.*

• **Viral diseases:**
- *Sweet potato mosaic*
- *Sweet potato corky heart.*

12.1.6.2. Parsnips

The parsnip (*Pastinaca sativa*) is a member of the umbellifer family (Umbelliferae). It originated as a wild plant in the temperate zones of Europe. It is a biennial that produces a whitish root. During its first year of growth it produces the thickened root, and it flowers the following year. Its cultivation cycle is only 4-5 months.
The seeds are flattened and retain their ability to germinate for 1 or 2 years.

The best known varieties are:

> Long White
> Early Round
> Caranchueca
> Tender and True
> Avon Resister
> Ofen Ham
> White Gem
> Hollowcron
> Improved

It requires a moist temperate climate, and its temperature requirements are very similar to those of the turnip.
The soil should be cool, light, deep and rich in organic matter.
Per hectare, parsnips extract about:
> 96 kg N
> 28 kg P_2O_5
> 125 kg K_2O

The input of rotted organic matter should be between 20 and 30 t/ha.

The application of fertilizer at depth should be:

> 60 kg N
> 80 kg P_2O_5
> 144 kg K_2O.

The surface application should be two doses of 25 kg/ha N.
The seed is sown in late winter/early spring in furrows 40-50 cm apart, at a depth of 1-2 cm.
The crop requires thinning, leaving the seedlings 20-25 cm apart when they have 2-3 leaves.
Harvest is in late summer/early fall.
The pests and diseases that attack the parsnip are more or less the same as those that attack the carrot.

12.1.6.3. Chufa

Chufa (*Cyperus esculentus sativus*) is a form of the orgeat sedge, or yellow nut sedge (*Cyperus esculentus*), and is a member of the sedge family (Cyperaceae). It originated in the Nile valley, and is grown for its tubers, known in English as chufa, earth almonds and tiger nuts. The tubers are used to make a delicious and refreshing summer drink called "**horchata**" in Spanish, and similar to "orgeat" or barley water.
It is a perennial plant, that only grows actively in the summer, with a more or less rounded root system.
The tubers are borne at the tips of the roots. They are brown on the outside and white within.
The size and shape of the tuber depend on the soil texture, and it prefers loose, open, soils. It is planted in the middle of spring, in furrows 50-60 cm apart and with a distance between plants of 10-15 cm.
It is not very demanding in terms of fertilizer, and the only labors performed are weeding and earthing up the furrows.
The tubers are harvested in fall, and the aboveground parts have to be removed first.
After harvest, the tubers must be washed, dried and bagged in sacks for sale.
The most common pest is the chufa miner which mines galleries within the plant. The fungal diseases that may affect it include *Fusarium* and *Rhizoctonia*.

12.1.6.4. Cassava

Cassava (*Manihot esculenta*) is a member of the spurge family (Euphorbiaceae). It is also known as manioc, mandioc and yuca.
It originated in South America, and the most important producers are Brazil, Zaire, Indonesia, Thailand, and Nigeria.
It is used as a human foodstuff and fed to livestock. It is also a source of a range of products like alcohol, starches and **tapioca**.
Tapioca is obtained by a complex cooking process that eliminates the toxic cyanide the plant contains, allowing it to be eaten.
Cassava is a small "shrub" that produces tubers that can be eaten like potatoes. The tubers are spindle-shaped, 30-70 cm long, and 5-15 cm in diameter. They may weigh up to 2 kg.

The latest generation of parsnip varieties:
De Guernesey. *Courtesy of SEMILLAS VILMORIN*

The skin of the tuber is yellowish white and the flesh inside is white, yellow or pink.
There are two groups, which differ in their taste and cyanide content:

- bitter cassava
- sweet cassava, which is much more widespread

The planting time varies from zone to zone, and the tubers are harvested 9-10 months after planting.

12.1.6.5. Yams

The name yams is used for many species of the genus *Dioscorea*, which are all members of the family Dioscoreaceae. They have many names in English, including **Chinese yam** and **air potato**.
It is a typical plant of tropical and subtropical zones, but some species have adapted to cultivation in more temperate climates. Yams are a major foodstuff in many African and Asian countries. It is said to be beneficial for arthritis.
It is a climbing plant with a well-developed root system that produces the edible tubers.
The growing cycle lasts 6-7 months and it requires deep, fresh, soils with a lot of organic matter.
It is propagated vegetatively, like the potato and sweet potato, starting from a tuber or any part bearing a bud.
Yams are planted in early spring, in furrows 80-110 cm apart with a distance of 40-80 cm between plants.
It does not require a lot of cultivation, and the most important operation is staking the stems.
Harvesting is in fall/early winter, before the first frosts. Good aeration is important for storage.
Its worst enemies are tropical ants and nematodes. The most common fungal diseases affecting it include *Rhizoctonia* and *Alternaria*.

12.2. CROPS GROWN FOR THEIR BULBS

12.2.1. Garlic

12.2.1.1. General concepts

Garlic (*Allium sativum*) is a member of the lily family (Liliaceae).
It originated in southwest Asia and southern Europe. It has diuretic, purgative and antiseptic properties, as well as stimulating the appetite. It is used as both a cooking herb and a flavoring. It contains a highly effective bactericide, **allicin**. The distinctive smell of garlic is due to an essential oil based on allyl sulfide compounds.
It is a perennial bulb that rarely produces

Nutritional composition per 100 g of foodstuff	
Proteins	4 g
Lipids	0.5 g
Sugars	20 g
Vitamin B$_1$	0.2 mg
Vitamin B$_2$ (rivoflavin)	0.11 mg
Niacin	0.7 mg
Vitamin C (ascorbic acid)	9-18 mg
Calcium	10-24 mg
Phosphorus	40-195 mg
Iron	1.7-2.3 mg
Potassium	540 mg
Energy value	98 calories

Climatic requirements		
CRITICAL TEMPERATURES	Freezing point	-5°C
	Zero growth	5°C
	Minimum for growth	6°C
	Optimum for growth	10 - 20°C
	Maximum for growth	35°C
SPROUTING	Minimum temperature	6°C
	Optimum temperature	20 - 22°C
	Maximum temperature	30°C
MOISTURE		HIGH
LIGHT		MEDIUM

flowers in temperate climates, and never produces seed. Its cultivation cycle is 4-5 months for spring varieties and 7-8 months for the fall varieties. It sprouts 12-15 days after planting.
The bulb consists of individual bulblets, called cloves, each one covered by a protective papery husk, which varies in color. The bulb is covered by a membraneous skin, the tunic, that is usually whitish.
The number of cloves varies from variety to variety (between 3 and 14). A garlic bulb may

weigh 30-100 g, and some varieties may even reach 200 g.

The different varieties of garlic can be divided into three groups:

• **White garlic**. They are usually consumed dry. They are stronger growing, with good levels of production and they store well.

White garlic	Ronda White
Fino de Chinchón	Chinchón White
Pardo rocambola	Thermidrome
Canary garlic	Messindrome
Cuenca White	

• **Pink garlic**. The tunic surrounding the bulb is reddish-pink. These varieties do not store as well as white ones, and should thus be consumed first.

Provence Red
Lemosin round garlic
Early Pink
Pedroñera dark garlic
California Late
California Early
Chonan
Caçador
Lavinia

• **Dark garlic**. The protective tunic around the bulb is a dark, purplish, color. The comments on storage for pink garlics also apply.

Bañolas
Yegen
Cuenca Red
Castro Red
Germidour
Créole

12.2.1.2. The plant's requirements

• **Climate and temperature**. It is a robust plant of temperate climates, and can withstand low temperatures.

• **Soil**. It requires light, open and permeable soils with a pH of 6-7. It can tolerate slight soil acidity.

• **Extractions**. Garlic's nutrient extraction per hectare is about:

50 kg N
15 kg P_2O_5
30 kg K_2O

• **Fertilizer**. The site should not be dunged immediately before planting garlic.

Fertilizer application at depth per hectare:

50 kg N
50-60 kg P_2O_5
60-70 kg K_2O

Surface fertilizer application per hectare:
50 kg N

You are also advised to supply some sulfur in the

fertilizer applied at depth.

• **Deficiencies**. Garlic is sensitive to shortage of zinc, boron and molybdenum.

12.2.1.3. Soil preparation and planting

Soil preparation is similar to that for the onion. Garlic is propagated vegetatively from the cloves, which are planted in late fall or early winter.

Planting may be:

• **flat**, with rows 20 to 30 cm apart and a distance of 10-15 cm between plants.
• **in furrows** 40-60 cm apart, with 2 rows in each furrow and a distance of 10-25 cm between plants.

The planting depth is 4-5 cm. The soil should be watered before planting so it is in good tilth for planting. The pointed tip should be pointing upwards.

Planting can be manual or using planting machinery, in which case the teeth have to be accurately calibrated.

12.2.1.4. Cultivation and harvest techniques

• **Tilling.**
• **Weeding**. Manual or using herbicides.

Onions:
Varieties
A/ For pickling
B/ For salad
C/ and D/ Bulbs

Ⓐ   Ⓑ

• **Earthing up**, to limit the rows of garlic in the furrows.
• **Harvesting**. In early summer, when the bulbs have reached a diameter of 40-45 mm and weigh 25-75 g. Harvesting can be manual, though large sites are now beginning to be mechanically harvested.
Once the bulb has been uprooted, it is left to dry on the soil.
• **Sale**. They can be sold as loose bulbs, in mesh bags or in boxes.
• **Storage**. When the bulbs are sufficiently dry they can be stored in well aired sites. They can easily resist temperatures of –10°C. They can be stored in cold chambers at –1 to 0°C with a humidity of 70-75% for 6-8 months.

12.2.1.5. The most common pests, diseases and physiological disorders

These include those mentioned in the following section on onions, and also:

• **Pests**:
- *Garlic weevil*. The larvae cause serious damage to the bulbs and can also attack onions.
- *Nematodes*. Nematodes are serious pests of garlic, causing stunted growth, general

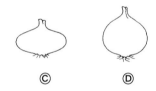

Ⓒ   Ⓓ

Varieties of onion:
Babosa
Courtesy of SEMILLAS
VILMORIN

yellowing of the plant and asymmetric leaf growth.

• **Viral diseases**:
- *Garlic mosaic*.

12.2.1.6. Other crops

• **Leaf garlic**:

The plant is collected whole, including the leaves, when the bulb is still forming, and eaten as such immediately after collection. This is just the same as normal cultivation of garlic, but harvesting earlier. For this reason, the planting frame can be denser.

12.2.2. Onions

12.2.2.1. General concepts

The onion (*Allium cepa*) is a bulb and a member of the lily family (Liliaceae).
It is originally from Asia, and is a tonic and easily digested foodstuff, a diuretic, and said to be effective against rheumatism and to have an aphrodisiac effect. It is used fresh, stored, pickled and even dehydrated. Some onion extracts are also produced.

The onion is a biennial plant that produces a bulb during its first year of growth and flowers in the following year. Its growing cycle is 100-200 days, depending on the variety.
The seeds are round and black and only retain their ability to germinate for 1 year.

All the parts of the onion have a characteristic smell, which is due to the presence of sulfur-containing volatile oils.

The onion is a crop with many varieties that tend to be closely adapted to their local conditions, and they are thus adapted to this particular climate.

In general:
- At latitudes lower than 35°, the varieties that adapt best are short-day varieties.
- For latitudes between 35° and 38°, the medium varieties adapt best, and
- For latitudes above 38°, long-day varieties grow best.

There are very many varieties, including:

• **Spanish varieties**

Babosa
French White
De la reina
Sangre de buey
Lanzarote
Spanish White
Fuentes Large White

Nutritional composition per 100 foodstuff	
Proteins	0.5-1.6 g
Lipids	0.1-0.6 g
Sugar	6-11 g
Vitamin A	Traces
Vitamin B_1 (thiamine)	0.03-0.05 mg
Vitamin B_2 (riboflavin)	0.02 mg
Vitamin B_6	0.06 mg
Inositol	90 mg
Vitamin C (ascorbic acid)	9-23 mg
Vitamin E	0.2 mg
Phosphorus	27-73 mg
Calcium	27-62 mg
Iron	0.5-1 mg
Potassium	120-180 mg
Sulfur	61-73 mg
Magnesium	16-25 mg
Iodine	0.03 mg
Energy value	20-37 calories

Late Large White
De Grano
Colorada de conserva
Morada de Zalla
Morada de Amposta
Molina Red
Colorada de Figueras
Lérida White
De Liria

• **French varieties**:

Pompey White
De Malakoff
Brunswick Red
Hyper
Hygro
Superba

• **Italian varieties**:

Large Italian Silver White
La Rocca Dark Giant

• **English varieties**:

Big Ben
Granex

• **American varieties**:

Gornet
Texas Grano
Crystal Wax
White Lisbon
White Knight

• **Dutch varieties**:

Mirato
Carolus

• **Japanese varieties**:

Hayate	Senshu yellow
Hi-keeper	Top-keeper
Hi-ball	Tropic-Ace
Keep-well	Esquino
Buffalo	Dragon Eye

Climatic requirements		
CRITICAL TEMPERATURES	Freezing point	3°C
	Zero growth	5°C
	Minimum for growth	7°C
	Optimum growth	12-23°C
	Maximum for growth	45°C
GERMINATION	Minimum temperature	2-4°C
	Optimum temperature	20 - 24°C
	Maximum temperature	40°C
MOISTURE		MEDIUM
LIGHT		MEDIUM

12.2.2.2. The plant's requirements

• **Climate and temperature**. Bulb formation and ripening both require high temperatures.
• **Water**. Sharp fluctuations in soil humidity may cause the bulb to split.
• **Soil**. Onions need a medium to light soil that is well drained.
The onion can tolerate a little salinity, but does not tolerate acidity.
• **Soil extractions**. For a hectare, these are estimated to be:

$$80\text{-}100 \text{ kg N}$$
$$30\text{-}40 \text{ kg P}_2\text{O}_5$$
$$100\text{-}140 \text{ kg K}_2\text{O}$$

• **Fertilizer**. Onions need a lot of nitrogen in the first phases of growth. If organic matter is applied, this should be done when the preceding crop is in the field. If this is done, only a small amount of well-rotted dung should be applied.

Fertilizer application at depth per hectare:

$$40\text{-}100 \text{ kg N}$$
$$70\text{-}100 \text{ kg P}_2\text{O}_5$$
$$100\text{-}150 \text{ kg K}_2\text{O}$$

• **Deficiencies**. It has an average requirement for boron, and needs a lot of sulfur and calcium.

12.2.2.3. Soil preparation and planting

Onions do not require any deep plowing, but need a homogeneous and loose topsoil layer.

The seed can be sown directly or in protected seedbeds.

The seed can be sown directly, and seedlings transplanted, into:

• **Flat sites** in 3-5 m raised tables and with the rows 15-20 cm apart.
• **Furrows**, which should be 40-60 cm apart and planted with 2 rows.
• **Earthed up furrows** 100-120 cm apart, each furrow with 4 rows of onions. This is the most widely used system.

The best time to sow and transplant varies from variety to variety.

• **Early varieties** are sown in late summer and transplanted in late fall.
• **Medium varieties** are sown in mid/late fall and transplanted in mid/late winter.
• **Late varieties** are sown in winter and transplanted in spring.

Acoustic device to scare birds, etc,. Manufactured by DAZON B.V.

12.2.2.4. Cultivation and harvest techniques

• **Thinning**. Thinning should be performed if the seed is sown directly, and should leave 15-20 cm between plants.
• **Weeding**. In this crop, it is important to keep the soil free of weeds. This can be done manually or using weedkillers.
• **Harvest**. The bulbs are ready to be harvested when the first leaves wither.

The time to harvest depends on the variety cultivated.

• **Early varieties**: mid spring
• **Medium varieties**: late spring, early summer
• **Late varieties**: mid summer

Onions are harvested when they have reached a diameter of 5-12 cm and a weight of 50-300 g.
Harvesting is manual, though it is now tending to be mechanized, at least partly. Once the onion has been uprooted, it is left on the soil for 8-10 days to dry out. The leaf tops can be cut off to speed up the drying process.
• **Sale**. Once dry, the onions are gathered and classified by size into different calibers. Onions are usually sold in mesh sacking or in boxes of wooden laths.
• **Storage**. To ensure good storage, the onions should be treated with an anti-sprouting agent 15 days before harvest. Onions should be stored in well-aired premises and well protected from any moisture.
They can be stored in cold chambers at a temperature of 0-2°C and a humidity of 75-85% for 4 to 6 months, depending on the variety.

12.2.2.5. The most common pests, diseases and physiological disorders

• **Pests:**
- *Mole cricket.* This mainly attacks the seed beds.
- *Onion thrip.* This attacks the leaves, causing bites, loss of color and deformations.
- *Onion fly.* The larvae (maggots) mine galleries in the bulb, damaging them.
- *Onion miner.* The larvae mine galleries in the leaves.
- *Wireworms.* They attack the roots and bulbs, causing damage.
- *Nematodes.* They attack the bulb and the roots.

• **Fungal diseases:**
- *Onion powdery mildew.* This causes elongated patches on the leaves, and may eventually look like a burn.
- *Onion smut.* To begin with, silvery lesions can be seen, and these then turn into sooty pustules on the outer layers of the bulb.
- *Onion anthracnose.* This causes blackish patches on the outer scale leaves, especially in white onions.
- *Onion rust.* This produces small reddish-brown pustules on the leaves.
- *Soft rot.* This causes areas of the bulb to rot and the leaves to wither, causing the plants to collapse and die.
- *White rot.* Cottony growths appear on the collar when the temperature is above 8°C and the humidity is high.

• **Bacterial diseases:**
- Several bacterial diseases cause onions to rot.

• **Viral diseases:**
• *Onion mottle virus.* This causes a mosaic which is accompanied by crinkling of the leaves.

• **Physiological disorders:**
- *Scorching*. Burns during drying in the field caused by excessive sunshine.
- *Cracked and double bulbs*. These are caused by sharp variations in soil humidity.
- *Bolting*. This may be provoked by several causes, such as sowing too early, and climatic and cultivation factors, e.g., excess nitrogen, excessive irrigation, etc.

12.2.2.6. Other forms of production

• **Spring onions**. These are normally harvested 1.5-2 months after planting, before the bulbs have ripened. They are sold in bunches of 4-6 plants, and only the leaf tips are removed.

The varieties grown include:

> White Lisbon
> Ishikura

• **Onions for pickling**. The seed is sown directly and very thickly in spring. As they are sown densely, small bulbs (2-2.5 cm) form within 75-80 days of planting. These varieties are grown in early spring and harvested in mid summer, and are pickled.

There are several special varieties, such as:

> Paris silver skin
> Berletta

12.2.3. Leeks

12.2.3.1. General concepts

The leek (*Allium porrum*) is a member of the lily family (Liliaceae).
It originated in Europe and western Asia, and is greatly appreciated for its flavor in cooking. Not only the bulb is consumed but also the adjacent leaf parts, which are blanched.
The leek is a biennial plant, with a single elongated bulb. It flowers in its second year of growth, but is only grown 4-5 months.

Nutritional composition per 100 g foodstuff	
Proteins	2.2 g
Lipids	0.3 g
Sugars	11.2 g
Fiber	1.3 g
Vitamin A	40 IU
Vitamin B$_1$ (thiamine)	0.11 mg
Vitamin B$_2$ (riboflavin)	0.6 mg
Niacin	0.5 mg
Vitamin C (ascorbic acid)	17 mg
Calcium	52 mg
Phosphorus	50 mg
Iron	1.1 mg
Sodium	5 mg
Potassium	347 mg
Energy value	52 calories

The seeds are similar to those of the onion, but a little smaller and darker. They retain their ability to germinate for 2-3 years.

Leeks are classified on the length of the white leaf bases:

• **Long leeks:**

> Gennevilliers long   Argenta
> Colonna   Large American Flag
> Mezières long   Kilima
> Abel   Helvetia
> Bulgarian long

• **Semi-long and short leeks:**

> Rouen thick
> Malabare
> Electra
> Platina
> Arcadia
> Kajack
> Blizzard

Climatic requirements		
CRITICAL TEMPERATURES	Freezing point	2°C
	Zero growth	5°C
	Minimum for growth	7°C
	Optimum growth	13 a 24°C
	Maximum for growth	35°C
GERMINATION	Minimum temperature	4°C
	Optimum temperature	20 - 24°C
	Maximum temperature	30°C
MOISTURE		MEDIUM
LIGHT		MEDIUM

Varieties of leek:
Colonna.
Courtesy of SEMILLAS SLUIS & GROOT

Different sacks for bagging agricultural produce. Manufactured by BENITEX, S.A.

12.2.3.2. The plant's requirements

In general, the leek's requirements are similar to those of garlic and the onion.

• **Climate and temperature**. It requires a moist temperate climate, and can resist the cold.
• **Water**. Try to ensure that there are no sharp variations in soil humidity during the leek's growth.
• **Soils**. It prefers soils with average textures that are deep, rich and cool, though the leek is very adaptable. It shows little resistance to soil acidity, and it is best if the soil is not excessively alkaline, either. The soil pH should be between 5 and 6.1.

• **Extractions per hectare:**

85-100 kg N
40-60 kg P_2O_5
100-120 kg K_2O

• **Fertilizer:**

Fertilization at depth per hectare:

50 kg N
80-100 kg P_2O_5
150-170 kg K_2O

Surface application per hectare:

50 kg N

12.2.3.3. Soil preparation and planting

The soil is prepared in the same way as for growing onions.
The seed can be sown directly or in seedbeds. In the seedbed the seed is sown by scattering, and covered with 3-4 mm of soil.

After two months the seedlings are transplanted into furrows 25-30 cm apart, with a separation between plants of about 15 cm.
The seed can be sown directly into furrows, as above, or in lines 30-40 cm apart.

12.2.3.4. Cultivation and harvest techniques

• **Thinning**. This is necessary if the seed is sown directly. The plants should be left 10-15 cm apart.
• **Weeding**. Manual or with selective weedkillers.
• **Pinching out**, so that vegetative growth is not excessive.
• **Earthing up**. This is done to blanch the base of the plant. Banking up should be done 20-30 days before collection.
• **Harvesting**. This is performed 4-5 months after sowing, when the plant is 20-25 cm long (35 cm for long varieties) and is 2-3 cm in diameter.
• **Sale**. After harvesting, the leeks must be cleaned, any yellow leaves removed and the roots trimmed.
They are sold in bunches or loose. They are also sold in polythene bags containing several leeks.
• **Storage**. Leeks can be stored at 0-1°C at a humidity of 90-95% for 1 to 3 months.

12.2.3.5. The most common pests, diseases and physiological disorders

They are generally the same as those that attack the onion.

12.2.4. Other crops

12.2.4.1. Welsh onions

Welsh onions (*Allium fistulosum*) are also known as ciboule, and they are members of the lily family (Liliaceae).
The plant is perennial and it is similar in shape to the onion, but is smaller with an elongated and less distinct bulb. They can be used as a flavoring or in salads.

It does not require a very good climate, and it can adapt well to a wide range of soils.

It is sown directly in spring/summer at a distance of 20 cm, or in seedbeds.
The crop is harvested 3 months after sowing.

12.2.4.2. Chives

Chives (*Allium schoenoprasum*) is a member of the lily family (Liliaceae).

The leaves are used to flavor several dishes and also in salads. The leaves are cut 1 cm above soil level.
The plant is a perennial similar to the onion, but the leaves and bulbs are much smaller.
It is not very demanding in its soil requirements. It prefers moist soils, but tolerates low temperatures and relatively dry conditions.

12.3. CROPS GROWN FOR THEIR STEMS

12.3.1. Asparagus

12.3.1.1. General concepts

Asparagus (*Asparagus officinalis*) is member of the lily family (Liliaceae). It is thought to have originated in southern Europe and in Asia. It can be eaten fresh or preserved. It is a diuretic.
The asparagus is a perennial with an underground fleshy root system shaped like a platform (a "crown"). The young shoots are eaten.
These shoots are produced at the base of the crown, and remain white while they are growing underground. As soon as they emerge into the light they start to turn green.
The seeds retain their ability to germinate for 5 to 8 years

The best known varieties in Spain include:

Aranjuez White
Zaragoza White
Navarre Purple
Navarre White

Nutritional composition per 100 g foodstuff	
Proteins	1.62-1.79 g
Lipids	0.11-0.25 g
Sugars	0.37 g
Fiber	0.81-1.04 g
Vitamin B$_1$ (thiamine)	25 mg
Vitamin B$_2$ (riboflavin)	170 mg
Vitamin C (ascorbic acid)	30 mg
Calcium	20 mg
Phosphorus	60 mg
Iron	1 mg
Energy value	26 calories

The best known varieties in other countries include:

Limbras	Larc
Spaganiva	Diane
Connover's Colossal	Minerva
Loella	Junon
Darbonne Nº 4	Mira
Argenteuil	Aneto
Martha Washington	Bruneto
Dutch White	Desto
Rosa Chérault	Cito

12.3.1.2. The plant's requirements

• **Climate and temperature**. Asparagus adapts well to very different climates, including tropical, subtropical and temperate climates.
• **Water.** It resists dry conditions well.
• **Soil**. Asparagus occupies the same site for many years, and so the site has to be carefully chosen and well prepared. It requires a light, deep and cool soil, with good drainage to prevent waterlogging. The plant is sensitive to root asphyxiation, but resistant to salinity and grows best at a pH of 7.5-7.8.
• **Extractions**. Per hectare, asparagus extracts:

86-100 kg N
28-34 kg P$_2$O$_5$
90-106 kg K$_2$O

It has a high requirement for boron.

• **Fertilizer:**
In the first year, apply 50-60 t/ha of dung.

Fertilizer applied at depth, per hectare:

96 kg P$_2$O$_5$
250 kg K$_2$O

Fertilizer applied on surface:

3 applications of 50 kg N per hectare.

In the second year, fertilizer should be applied in winter, at a rate of 15 t/ha of dung.

Climatic requirements		
CRITICAL TEMPERATURES	Optimum growth	18-25°C
	Sprouting by crowns	11 - 13°C
GERMINATION	Minimum temperature	6 - 8°C
	Optimum temperature	20 - 25°C
	Maximum temperature	35 - 40°C
MOISTURE		MEDIUM
LIGHT		MEDIUM

Fertilizer applied at depth:

60-90 kg P_2O_5
100-180 kg K_2O

Surface fertilizer application:

100-200 kg N
60-100 kg P_2O_5
150-250 kg K_2O

From the third year onwards, 15 t/ha of dung should be applied every year.

Fertilizer applied at depth:

100-200 kg N
60-100 kg P_2O_5
150-250 kg K_2O

Surface fertilizer application:

100 kg N

You should also apply 30-40 kg of borax every 2-3 years.

12.3.1.3. Soil preparation and planting

The soil has to be very thoroughly prepared, as the asparagus is going to remain there for about 10 years. The soil should be deep plowed several times, and harrowed several times to leave the soil well broken down.
Asparagus is propagated vegetatively by small parts of the rhizome, or from one-year-old crowns grown from seed.
The seed is sown in seedbeds between late winter and the end of spring, either in continuous rows or individually, and at a depth of 3-4 cm.

When fall arrives, the seedling's aboveground parts wither, and are pruned. It remains like this throughout the winter, and in spring, the small rhizomes are transplanted to their final site. The seedbed phase lasts 1 year.
The crowns are transplanted into the base of ditches or furrows.

• **Ditches**. These should be 30-35 cm deep and 180-220 cm apart. The crowns should be 40-50 cm apart.

• **Furrows**. These should be 20-30 cm deep and 150-200 cm apart. The distance between the crowns should be 30-40 cm.

When it is time for transplanting, reject the rhizomes that are in bad condition or weigh less than 20 g. For greater reliability, you can buy the crowns from commercial suppliers. If they are home grown, be sure to disinfect them thoroughly before planting. They should be planted at a depth of 10 cm, and as the shoots are produced, earth should be banked up around the newly planted crowns.

In fall, once the aboveground parts have withered, they are all cut off and some earth is removed from above the rhizomes. Fertilizer is applied at depth in winter, and covered with new topsoil, and any plants that have failed should be replaced. After this, the rhizomes should be earthed up again.
The first shoots cannot be collected until the third year, but the first real harvest will be in the fourth year.

12.3.1.4. Cultivation and harvest techniques

• **Earthing up**. This consists of piling earth on the newly planted rhizome, so that the shoots grow underground and can be harvested white.

*Planting aparagus
A/ in ditches
B/ in furrows*

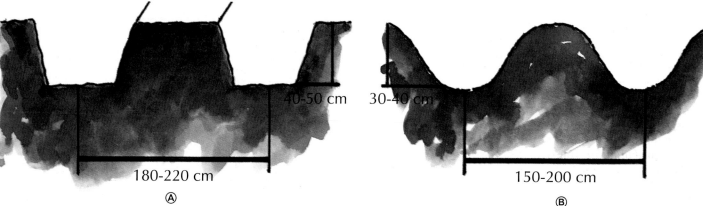

40-50 cm   30-40 cm

180-220 cm
Ⓐ

150-200 cm
Ⓑ

• **Weeding**. This can be done manually or using selective weedkillers. The herbicide is mainly applied between the rows.

• **Vegetative growth**. Once the first shoots ("tips") have been harvested, other shoots are allowed to grow fully so that the aboveground parts can produce the reserves needed to produce the next harvest.

• **Pruning**. Remove the aboveground parts completely once they have withered.

• **Mulching**. This can be carried out after earthing up and application of herbicides, and consists of placing a plastic sheet, normally transparent, on the soil. This is done to warm up the soil and increase its water retention. It also brings the harvest forward by 2-3 weeks.

• **Harvesting**. No asparagus can be harvested until the fourth year of cultivation, and yields peak between the sixth and tenth year.
The harvesting period lasts from early spring to early summer, lasting 2 to 3 months, with a daily check for new tips. Harvesting is performed manually using a special tool with a broad semi-circular end.
After harvesting, the asparagus should be removed from the soil as quickly as possible, because they dry out and turn stringy. If they are to be eaten fresh, they should be pre-cooled and washed. If they are to be preserved, they are washed, boiled and then bottled for sale.

• **Sale**. If they are sold to be eaten fresh, they are normally sold in bunches held together by a rubber band.

• **Storage**. Asparagus can be stored in cold chambers at 2-3°C and a humidity of 95%.

12.3.1.5. The most common pests, diseases and physiological disorders

• **Pests:**
- *Asparagus beetle* (*Crioceris*). This is an insect whose adult and larvae attack the foliage and young stems, especially of the young asparagus plants, leaving them sticky and black.
- *Field fly*. The larvae burrow into the tips, causing them to collapse.
- *Asparagus fly*. The larvae burrow downwards within the tips to the rhizome. The stems wither and eventually die.
- *Cut worms* (*Agrotis*).
- *Aphids*. They attack the foliage and severe attacks may weaken the rhizome.

• **Fungal diseases:**
- *Violet root rot* (*Rhizoctonia*). This disease takes several years to reveal itself. One of the symptoms is the production of stunted, thin and rather hard tips. The roots are covered by a purple-violet fungus, and as the disease progresses, the plant withers and dies.
- *Asparagus rust*. This causes reddish patches on the stems, weakening the aboveground parts and causing them to wither and die.
- *Asparagus Fusarium*. This destroys the vessels in the root tissues and may kill the plant.
- *Cercospora*.
- *Sclerotinia*.

• **Physiological disorders:**
- *Tip rust*, or *false rust*. This gives rise to long reddish streaks along the stems, and is the result of a metabolic disorder in cold moist weather.
- *Dieback of the young shoots*. They wither when they are about to branch. This is attributed to several causes, such as boron deficiency or insufficient water absorption.
- *Autumn shoot production*. When conditions are favorable in the autumn, shoots may be produced, but this is detrimental to the next spring's crop, as these shoots are using up the reserves laid down for the next spring's shoots. To prevent this, avoid watering in this period, and apply fertilizer in the winter.

12.3.1.6. Other crops

• **Green asparagus**. Production costs are lower for green asparagus, because it does not need to be blanched by making ridges. The planting density of green asparagus can be higher than in asparagus for blanching, 125-150 cm between rows and 30 cm between plants. One important factor is to grow a variety with bracts that are held tightly against the stem, something that not all varieties show.

The most suitable varieties for growing in this way are:

California Green
Jacq. Ma. green
UC-157

• **Cultivation in tropical and sub-tropical zones**. In these climates, asparagus starts producing much earlier, normally in the first year after planting. As there is no winter growth halt, this has to be induced by heavy pruning.
One system used is known as **main stem**. Over the course of the cultivation, some stems are allowed to grow vegetatively so that they can accumulate new reserves for the production of new stems.
These stems are cut back regularly. Though harvest may be continuous over the course of the year, there are some periods when the stems are left to grow and not harvested. In these areas, asparagus can not be cultivated for as long as in temperate climates, as the plants get "worn out".

Green varieties of asparagus imply lower production costs. Courtesy of SEMILLAS SLUIS & GROOT

12.4 CROPS GROWN FOR THEIR LEAVES

12.4.1. Chard

12.4.1.1. General concepts

Chard (also known as Swiss chard) is a form of beet.
Its latin name is *Beta vulgaris* var. *cicla*, and
it is a member of the goosefoot family
(Chenopodiaceae).
It is normally eaten boiled, and favors intestinal
transit and helps digestion.
It is a biennial plant that flowers in its second year of
growth. The seeds retain their ability to germinate
for 6 years.

*Varieties
of chards: **Paros.**
Courtesy of
SEMILLAS
SLUIS & GROOT.*

Nutritional composition per 100 g foodstuff	
Proteins	2.4 g
Lipids	0.3 g
Sugars	4.6 g
Fiber	0.8 g
Vitamin A	6.500 IU
Vitamin B$_1$ (thiamine)	0.06 mg
Vitamin B$_2$ (riboflavin)	0.17 mg
Niacin	0.5 mg
Vitamin C (ascorbic acid)	3.2 mg
Calcium	88 mg
Phosphorus	39 mg
Iron	3.2 mg
Sodium	147 mg
Potassium	550 mg
Energy value	25 calories

The varieties are divided on the basis of their
production cycle, which is usually 3-4 months. They
include:

- **Spring-summer varieties:**

    Cutting green
    Narrow white-stalk green
    White-stalk Bressane green
    White-stalk Ampuis green
    Lyon yellow
    Cutting yellow
    White silver
    Hawaii

- **Fall-winter varieties:**

    Broad white-stalk green
    White-stalk Nice green
    Giant Fordhook
    Paros

Climatic requirements		
CRITICAL TEMPERATURES	Freezing point	-5°C
	Zero growth	5°C
	Minimum for growth	7°C
	Optimum growth	15 - 25°C
	Maximum for growth	30 - 35°C
GERMINATION	Minimum temperature	5°C
	Optimum temperature	18 - 22°C
	Maximum temperature	27 - 33°C
HUMIDITY		MEDIUM
LIGHT		LOW

12.4.1.2. The plant's requirements

- **Climate and temperature**. It requires a moist
temperate climate, and the plant is sensitive to frosts.

- **Water**. Chard requires constant soil humidity.

- **Soils**. It requires an open, rich, cool soil. It does
not resist soil acidity well. It is resistant to salinity.
The best pH for its growth is between 6 and 8.

- **Fertilizer application**. 20-30 t/ha of well rotted
organic matter should be applied.

Fertilizer at depth per hectare:

    30-40 kg N
    40-60 kg P$_2$O$_5$
    80-100 kg K$_2$O

Surface fertilizer application: it requires 3
applications of 30-40 kg N.

12.4.1.3. Soil preparation and planting

Deep plow, then surface plow a couple of times to
prepare the soil.

The seed can be sown throughout the year, except in the months with low temperatures. The spring/summer varieties are sown in late winter/early spring. The fall/winter varieties can be sown from the middle of summer to early fall. The seed can be sown directly or in seedbeds, at a depth of 2-3 mm.

If the seedlings are transplanted, this is done 30-40 days after sowing, when the plants have 5-6 true leaves, and they are transplanted into furrows 40-60 cm apart.

12.4.1.4. Cultivation and harvest techniques

• **Thinning**. If the seed is sown directly, the seedlings must be thinned to leave a distance of 30 cm between plants.

• **Weeding**. This can be done manually or using herbicides.

• **Harvest**. The crop is harvested 2.5-3 months after sowing, when the plant should weigh between 750 g and 1 kg. Harvesting can be manual and repeated several times, cutting off the largest leaves each time.

• **Sale**. After collection, the leaves are washed and bundled.

• **Storage**. Chard can be stored in cold chambers at 0°C and a humidity of 90% for 10-12 days.

12.4.1.5. The most common pests, diseases and physiological disorders

These are the same as those that attack the sugar beets, though miners, aphids, *Cercospora*, snails and slugs are all more serious pests to chard.

12.4.2. Celery

12.4.2.1. General concepts

Celery (*Apium graveolens*) is a member of the umbellifer family (Umbelliferae).

It originated in the Mediterranean and is said to purify the blood, and has diuretic properties. It is eaten fresh and cooked as a flavoring in stews, casseroles, soups, etc.

Celery is a biennial plant that flowers in its second year of growth. The seeds retain their ability to germinate for 5 to 9 years. Its cultivation usually lasts between 110 and 125 days.

The many varieties are divided into 2 groups:

• **Green varieties**

  Easter full green
  Elne green
  Utah
  Florida
  Florimart
  Slow Bolting
  Pascal
  June-Belle
  Lepage green
  Avonpearl

*Varieties of celery: **Pascal**. Courtesy of SEMILLAS SLUIS & GROOT*

• **White or golden varieties:**

  Full golden
  Barbier golden
  Carthom Blanching
  Jason
  Celebrity
  Chatteris
  Golden Plume
  Golden Spartan
  Avon Resister
  Dore Chemin

Nutritional composition per 100 g foodstuff	
Proteins	0.5-2 g
Lipids	0.1-0.5 g
Sugars	1-1.2 g
Fiber	0.7-2.7 g
Vitamin A	0-120 IU
Vitamin B$_1$ (thiamine)	0.02-0.05 mcg
Vitamin B$_2$ (riboflavin)	0.02-0.04 mcg
Vitamin B$_5$	0.3 mcg
Vitamin B$_6$	1.54 mcg
Vitamin C (ascorbic acid)	0.2-1.5 mg
Vitamin E	0.45 mcg
Iron	0.3-0.5 mg
Phosphorus	27-65 mg
Chlorine	137-183 mg
Iodine	0.012 mg
Magnesium	3-40 mg
Manganese	0.16 mg
Potassium	160-400 mg
Sulfur	15-20 mg
Sodium	96-240 mg
Energy value	5-22 calories

12.4.2.2. The plant's requirements

• **Climate and temperature**. Intense frosts causes hollow cavities to form in the leafstalks.
• **Water**. It requires a lot of water, but an excess can be prejudicial.
• **Soil**. It requires deep loamy soils that are well drained and rich in organic matter. The optimum pH value is between 6.8 and 7.2. It is sensitive to salinity.
• **Soil extractions**. Celery's nutrient extraction per hectare is about:

$$130 \text{ kg N}$$
$$50 \text{ kg P}_2\text{O}_5$$
$$200 \text{ kg K}_2\text{O}$$

• **Fertilizer**: Apply 25-30 t/ha of dung.

Fertilizer to be applied at depth:

$$50 \text{ kg N}$$
$$100 \text{ kg P}_2\text{O}_5$$
$$150 \text{ kg K}_2\text{O}$$

Surface application of fertilizer: 2-3 inputs of 30-40 kg N can be applied.

• **Deficiency**. Celery is sensitive to boron deficiency, and so you are advised to include some borax when applying the fertilizer at depth.

Climatic requirements		
CRITICAL TEMPERATURES	Freezing point	0°C
	Zero growth	8°C
	Minimum for growth	9-10°C
	Optimum growth	18 - 25°C
	Maximum for growth	30°C
GERMINATION	Minimum temperature	5°C
	Optimum temperature	15 - 25°C
	Maximum temperature	30°C
HUMIDITY		HIGH
LIGHT		LOW

12.4.2.3. Soil preparation and planting

The soil should be plowed deep a couple of times and also surface plowed several times to ensure the site's drainage is very good.
The seed is sown in spring, after all risk of frost has passed. Sow the seed in seedbeds protected from the sun, and ensure they are never short of water.
The seed requires pre-germination treatment, maintaining them moist at 20°C for 2-3 days. They should be transplanted after 2-3 months when the seedlings have 4-5 leaves, selecting those that are as uniform as possible and 10-12 cm tall.

The seed can also be sown in compressed peat "pots", meaning that the seedling's rootball remains intact.
You are advised to clamp the seedling after transplantation to reduce water loss by transpiration and to help it root well.
They should be planted in furrows 35-40 cm apart with a distance of 15-20 cm between plants. If the celery is going to be earthed up to blanch it, the rows should be further apart.

12.4.2.4. Cultivation and harvest techniques

• **Weeding**. Using selective weedkillers.
• **Blanching**. The plants should be earthed up one month before harvesting.
• **Harvesting**. This is normally by hand in summer and fall.
• **Sale**. Once harvested, the top is cut off, the plant is washed and bunched, either in boxes or in plastic bags.
• **Storage**. Celery can be stored in cold chambers at 0-1°C at a humidity of 90-95% for several weeks.

12.4.2.5. The most common pests, diseases and physiological disorders

• **Pests:**

- *Cutworms* (*Agrotis*). They mine galleries in the aboveground parts of the plant and destroy them.
- *Screwworm*. This attacks the seedbeds.
- *Psylla*. The larvae mine galleries in the collar of the plant, causing a halt in growth and a general yellowing.
- *Celery fly*. The larvae mine galleries in the leaves.
- *Aphids*. They cause the leaves to curl up, and a general yellowing of the plant.
- *Snails* and *slugs*.
- *Nematodes*. The plant's growth is slow, the leaves are yellow and deformed, and there are swellings and bumps on the roots.

• **Fungal diseases:**

- *Sclerotinia*. This causes the heart of the plant to rot, causing all the leaves to droop and wither.
- *Septoria*. This causes light brown patches with black points at the tips in the leaves, which curl up and wither.
- *Cercospora*. The symptoms are similar to Septoria.
- *Celery mildew*. This causes general drooping and withering of the leaves.
- *Soil fungi*. These include *Pythium*, *Fusarium* and *Rhizoctonia*. They cause damage at the level of the collar and to the plant's root system.
- *Botrytis* or grey rot.

• **Bacterial diseases:**

- *Pseudomonas*. This is a consequence of high humidity and low temperatures. It causes irregular brown patches with a yellow border on the leaves.
- *Erwinia*. This causes a soft, watery rot, with a rapid and general decline of the plant.

• Viral diseases:
- *Celery mosaic virus.*
- *Cucumber mosaic virus.*

• Physiological disorders:
- *Black heart.* This causes necrosis and drooping of the young leaves in the center of the celery plant. It is caused by factors like high temperatures, imbalances in soil humidity, excess nitrogen fertilizer, calcium deficiency or excess soil salinity.
- *Boron deficiency.* This causes dark brown lines to form along the leaf veins.
- *Magnesium deficiency.* The old leaves of the plant turn yellow.
- *Hollow leaf stalks.* This may be due to a period of frost or to over-maturity.
- *Bolting* (premature flowering). This is caused by a period of two weeks with temperatures below 10°C when the plant is young.

12.4.3. Cabbages

12.4.3.1. General concepts

The cabbage (*Brassica oleracea*) is a member of the cress family (Cruciferae). There is a form with reddish (or purple) leaves known as "red cabbage".

It originated in Europe and is considered to be easily digestible.

Nutritional composition per 100 g foodstuff	
Proteins	2-4.2 g
Lipids	0.2 g
Sugars	5-7 g
Fiber	0.8-1 g
Vitamin A	130 IU
Vitamin B$_1$ (thiamine)	0.05 mg
Vitamin B$_2$ (riboflavin)	0.06 mg
Vitamin C (ascorbic acid)	50-60 mg
Niacin	0.3 mg
Calcium	50-60 mg
Phosphorus	30-50 mg
Iron	0.5-0.9 mg
Sodium	20-25 mg
Potassium	240-260 mg
Energy value	24-31 calories

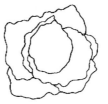

Cabbages. Varieties.
A/ Spring
B/ Summer
C/ Winter

Cabbage is a biennial plant that forms a bud surrounded by very tightly packed leaves in its first year, and flowers the following year. Its cultivation cycle is between 5 and 6 months.
The round seeds retain their ability to germinate for 3-4 years.
The varieties are classified into two main groups depending on the type of leaf, and then divided by the time of harvest. The spring/summer varieties grow faster than the fall/winter varieties.

• Smooth-leaved varieties:

- *Spring/summer harvest:*

Hornet	Enterprise
Early Bacalan	Brunswick
Large Corazón de Buey	Express
San Dionisio	Delphi
Small Corazón de Buey	Tucana
Golden Acre	Holanda
Enkhuizen Glory	
Golden Cross F$_1$	
Minicole	

Fall/winter harvest:

Stonar	Aranjuez cabbage
Erdeno	Lares
Wondergen	Junes Star
Taurus	Rodolfo
Tucana	Hitoma
Murciano	Red Savoy
Red Autoro	

• **Curly leaved cabbages, also known as Savoy cabbages:**

Spring/summer harvest:

Saint Jean	Easter
Julius	Siete semanas de verano
Savoy King	Court Hâtif
Estibal	San Juan
Marcelino	Reglo
King of Milan	

Fall/winter harvest:

Tarvoy	Conquest
Late Mars	Virtudes
Spivoy	Large Christmas
Winter King	Hamasa
Ice Queen	Savoy
Reglo	Novum
Vertos tight	Havro

Varieties of cabbage:
Bacalan.
Courtesy of SEMILLAS VILMORIN

Climatic requirements		
CRITICAL TEMPERATURES	Freezing point	-10 a -15°C
	Zero growth	3 - 5°C
	Minimum for growth	6°C
	Optimum growth	13 a 18°C
	Maximum for growth	30°C
GERMINATION	Minimum temperature	5 - 8°C
	Optimum temperature	20 - 25°C
	Maximum temperature	30 - 35°C
HUMIDITY		HIGH
LIGHT		LOW

Iceprince	Ostara
Precursor	

12.4.3.2. The plant's requirements

• **Climate and temperature**. The cabbage prefers cool environments, but adapts to a wide range of climates. In terms of temperatures, the spring/summer varieties are resistant to high temperatures, while the fall/winter varieties resist low temperatures, and some can even resist –10°C.

• **Water**. It is very sensitive to dry conditions. It needs a constant level of humidity in the soil.

• **Soils**. It requires soils with an average texture, cool and rich, with good drainage. It does not grow well on acid soils and is relatively resistant to salinity.

• **Extractions**. This depends on the variety grown and the yield, but for a hectare of cabbages they can be estimated as:

$$200\text{-}300 \text{ kg N}$$
$$85\text{-}100 \text{ kg P}_2\text{O}_5$$
$$250\text{-}500 \text{ kg K}_2\text{O}$$

• **Applying fertilizer**. Apply 25-30 t/ha of dung.

Fertilizer application at depth:

$$70\text{-}100 \text{ kg N}$$
$$65\text{-}85 \text{ kg P}_2\text{O}_5$$
$$150\text{-}200 \text{ kg K}_2\text{O}$$

Surface fertilizer application:

$$50 \text{ kg N}$$

The winter varieties grow larger. The doses of fertilizer should thus be a little higher than for the spring/summer varieties.

• **Deficiencies**. Cabbage requires a lot of boron and does not grow well in soils with little manganese.

12.4.3.3. Soil preparation and planting

Together with the fertilizer at depth, deep plow once and surface till several times to leave the soil ready for the crop.

The seed is sown in seedbeds, by scattering, and covered to a depth of 2-3 mm.
The time to sow depends on the crop's cycle. Fall/winter varieties are sown between mid spring and the end of summer.

Transplant to the definitive site 40-50 days after sowing, when the plants are 15-18 cm tall and the stem has reached a diameter of 4-5 mm.

The planting furrows should be 70-80 cm apart for large varieties and 50-69 cm for normal and Savoy varieties. The distance between plants should be 60-70 cm for large varieties and 40-50 cm for normal and Savoy cabbages.

12.4.3.4. Cultivation and harvest techniques

• **Weeding.** Either manual or with herbicides. Take care when using weedkillers, as cabbage tends to suffer toxic effects.
• **Earthing up**. A little, 25-30 days after transplanting.
• **Harvesting**. When the leaves around the head are tightly pressed, and the head weighs between 2 and 3 kg.

Harvesting is manual, or using harvesting machinery on very large areas. The harvest time varies depending on the variety's growth cycle. The fall/winter varieties are harvested from early fall to early spring. The spring summer varieties are harvested from mid spring to early summer.

• **Sale**. Once harvested, the outer leaves are removed, and the cabbages are wrapped in polythene bags and packed in wooden or plastic boxes.
• **Storage**. Cabbages can be stored in a cold chamber at 0-1°C at a humidity of 85-90%.

12.4.3.5. The most common pests, diseases and physiological disorders

• **Pests:**
- *Cabbage fly.* The larvae mine gallery at the base of the stems.
- *Leaf miners.* They mine galleries in the leaves.
- *False hernia of cabbage.* This forms galls at the base of the stem, within which are the larvae.
- *Cress flea beetle.* The larvae mine galleries within the leaves, and the adults eat the leaves.
- *Ash greenfly of cabbage.* The plant as a whole turns yellow and the leaves droop.
- *Cabbage bug.* Yellow patches form on the leaves, caused by its bites.
- *Cabbage white butterfly.* The larvae eat the leaves.
- *Cress moth.* They mine galleries in the leaves.
- *Cabbage noctuid moths.* The larvae eat the leaves.
- *Black screwworm.* The larvae eat the leaves.
- *Cut worms.* They eat the stems of recently transplanted seedlings.
- *Snails and slugs.* They attack the leaves.
- *Nematodes.* They cause damage to the roots.

• **Fungal diseases:**
- *Clubroot.* Growth is stunted and large bumps

appear on the roots. It attacks most frequently on acid soils.
- *Cress mildew.* Yellowing of the leaves, with the appearance of grey mycelium on the underside.
- *White rust of cresses.* This causes deformations in the plant, as well as whitish pustules.
- *Pythium.* This attacks the seedbeds.
- *Rhizoctonia.* This causes deformations in the collar and the roots.
- *Black stem.* This causes rotting at the junction of the collar and the roots, and necrotic patches on the stems and leaves.
- *Mycosphaerella.* This produces round corky patches on the old leaves ("leaf spot").
- *Alternaria.* This produces irregular patches on the leaves.

• **Viral diseases:**
- *Round black patch virus.*
- *Cauliflower mosaic virus.* This causes a mosaic on the leaves.

• **Physiological disorders:**
- *Bolting.* Premature flowering can be caused by, among other factors, exposure of the young plants to low temperatures for a period of time, or by periods of drought.

12.4.4. Brussels sprouts

12.4.4.1. General concepts

Brussels sprouts (*Brassica oleracea* var. *gemmifera*) is a variety of cabbage and a member of the cress family (Cruciferae).

They cabbage grows an elongated main stem, the head of which is removed. The plant then develops small lateral buds, the sprouts. It originated in Belgium.
They are mainly consumed fresh, though they are also used industrially.

It is a biennial plant that produces a main stem that may reach a height of 1 m, and which bears the buds, called sprouts.
The varieties are classified by their cultivation cycle, which is between 5 and 8 months:

• **Early varieties**. With a cultivation cycle of 150-170 days:

    Parsifal
    Topscore
    Oliver
    Camelot
    Jade Cross
    Long Island Improved
    Peer Gynt
    Acropolis
    Silvestar
    Predora
    Titurel
    Lancelot
    Alcazar
    Goldmine

The photo shows the hybrid Brussels sprout variety **Roger**. *Courtesy of SEMILLAS SLUIS & GROOT.*

Nutritional composition per 100 g foodstuff	
Proteins	4.7-4.9 g
Lipids	0.4-0.5 g
Sugars	7.5-8.3 g
Vitamin A	33-550 IU
Vitamin B_1 (thiamine)	0.1 mg
Vitamin B_2 (riboflavin)	0.1-0.16 mg
Vitamin C (ascorbic acid)	68-100 mg
Calcium	36-38 mg
Phosphorus	50-80 mg
Iron	1-1.5 mg
Energy value	45-53 calories

• **Medium varieties**. With a cultivation cycle of 170 to 200 days:

Giant Fall Magis   Thor
King Arthur   Merlon
Citadel   Anagor
Lunet   Bengalor
Camelot   Golfer
Prince Askold   Fripostar

• **Late varieties**: With a cultivation cycle of 200 to 250 days:

Sigmundo
Fortress
Belfort
Seven Hills
Asmer Hermer
Herka
Erwin
Rampart
Arctic

12.4.4.2. The plant's requirements

• **Climate and temperature**. It is a robust plant that prefers cool moist climates. It is very resistant to cold temperatures but not to hot ones.
• **Water**. It requires good soil humidity.

Climatic requirements		
CRITICAL TEMPERATURES	Freezing point	-15°C
	Zero growth	3 - 5°C
	Minimum for growth	6°C
	Optimum growth	16 - 18°C
	Maximum for growth	30°C
GERMINATION	Minimum temperature	6 - 8°C
	Optimum temperature	18 - 25°C
	Maximum temperature	30 - 35°C
HUMIDITY		HIGH
LIGHT		LOW

• **Soils**. It requires average-textured soils that are rich but not with a lot of nitrogen.

- *Extractions per hectare:*

200-400 kg N
56-90 kg P_2O_5
280-430 kg K_2O

• **Fertilizer application**: An excess of nitrogen fertilizer increases leaf production, and the sprouts are not so tight, and are not of good quality. Between 25 and 30 t/ha of dung should be applied.

Fertilizer application at depth, per hectare:

20-30 kg N
50-80 kg P_2O_5
100-200 kg K_2O

Surface fertilizer application:

20-30 kg N

• **Deficiencies**: The plant is sensitive to boron and molybdenum deficiency in the soil.

12.4.4.3. Soil preparation and planting

Soil preparation is the same as for cabbages.
The seed should be sown between late spring and mid summer.
The seed can be sown directly or into seedbeds, at a depth of 2-3 mm.
The seedlings are transplanted 40-50 days after sowing, when they are 14-18 cm tall, in rows 60-70 cm apart, with a distance of 40-50 cm between plants.

12.4.4.4. Cultivation and harvest techniques

• **Thinning**. If the seed is sown directly.
• **Pinching out**. The top of the plant is cut off a month before harvesting, to ensure production is more uniform.
• **Leaf removal**. Especially if harvesting is manual.
• **Weeding**. Manual or using selective herbicides.
• **Harvesting**. Harvesting may be manual or mechanized. Because of the time taken harvesting manually, Brussels sprouts are increasingly harvested by machinery.
Harvest time varies depending on the cultivation cycle of the variety grown. In early varieties harvest is in early to mid fall. In medium varieties it is in mid fall, and in late varieties it is between late fall and mid winter.
• **Sale**. After harvesting, the sprouts are selected, calibrated and packed in boxes or trays covered with plastic, or in sacks.
• **Storage**. Brussels sprouts can be stored in cold chambers at 0-1°C with a humidity of 85-90%.

12.4.4.5. The most common pests, diseases and physiological disorders

Brussels sprouts are attacked by the same pests as their close relative, the cabbage.

• Physiological disorders:
- *Loose sprouts.* This may be due to several causes, such as high temperatures or excessive fertilization with nitrogen.
- *Boron deficiency.* The sprouts are loose and there are gummy patches.
- *Internal browning of the sprout.* This happens if the sprouts are left on the plant too long.

12.4.5. Endive

12.4.5.1. General concepts

Endive (*Cichorium endivia*) is a member of the daisy family (Compositae). Its taste is slightly bitter and it is considered to stimulate the appetite. It is mainly used in salads. It originated in eastern India.

Endive is a biennial plant that flowers in its second year.
It does not form a tight head. The seed is larger than the lettuce's, and retains its ability to germinate for 3-5 years. Its cultivation cycle lasts for 3 to 4 months.

Nutritional composition per 100 g foodstuff	
Proteins	1.7 g
Lipids	0.1 g
Sugars	4.1 g
Fiber	0.9 g
Vitamin A	3,300 IU
Vitamin B$_1$ (thiamine)	0.07 mg
Vitamin B$_2$ (riboflavin)	0.14 mg
Niacin	0.5 mg
Vitamin C (ascorbic acid)	10 mg
Calcium	81 mg
Phosphorus	54 mg
Iron	1.7 mg
Sodium	14 mg
Potassium	294 mg
Energy value	20 calories

Two varieties are cultivated:
• Cichorium endivia var. **crispa** (curly-leaved or narrow-leaved endive). The leaves are very finely dissected and divided. There are winter/fall varieties and spring/summer varieties.

Winter varieties:
> Cabello de Angel
> Prat curly
> Winter curly

Fall varieties:
> Wallonner Frisan
> Ruffec
> Ruffec-Armel strain

Spring/summer varieties:
> Summer double curly
> Pavia
> Summer fine - Anjou strain
> Rouen Fine
> Yellow heart - Dabis strain

Varieties of endive:
Ruffec.
Courtesy of SEMILLAS VILMORIN

• Cichorium endivia var. **latifolia** (Batavian, or broad-leaved, endive). These varieties have lobed or toothed leaves. There are also winter/fall and spring/summer varieties.
Winter:
> Anjou Cornet
> Round green
> Full Heart
> Giant orchard Maral

Fall:
> Giant orchard
> Round green full heart
> Brevo
> Elna
> Solera

Spring/summer:
> Bouclée tight
> Italian fine-curled
> Malan

Climatic requirements		
CRITICAL TEMPERATURES	Freezing point	-6 - -8°C
	Zero growth	5°C
	Minimum for growth	6°C
	Optimum growth	15 - 20°C
	Maximum for growth	25 - 30°C
GERMINATION	Minimum temperature	6°C
	Optimum temperature	14 - 16°C
	Maximum temperature	30°C
HUMIDITY		MEDIUM
LIGHT		MEDIUM

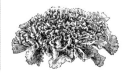

12.4.5.2. The plant's requirements

• **Climate and temperature**. In general, they are sensitive to low temperatures.
• **Soils**. It requires soils with an average texture that are not too open. It can tolerate soil acidity.
• **Extractions**. The plant's extractions per hectare are:

$$89 \text{ kg N}$$
$$40 \text{ kg P}_2\text{O}_5$$
$$227 \text{ kg K}_2\text{O}$$

• **Dunging**. Apply 15-20 t/ha of well-rotted dung.

Fertilizer application at depth per hectare:

$$30\text{-}40 \text{ kg N}$$
$$50\text{-}60 \text{ kg P}_2\text{O}_5$$
$$100\text{-}120 \text{ kg K}_2\text{O}$$

Surface fertilizer per hectare:

30-40 kg N halfway through growing cycle.

12.4.5.3. Soil preparation and planting

Soil preparation is similar to that for the lettuce.

The seed can be scattered in seedbeds, or sown directly into the site, in continuous rows, discontinuous rows, or using pelleted seed. The sowing depth should be 2-3 mm. If the seed is sown in seedbeds, transplant when the seedling is about 8 cm tall.

Before sowing, the seed needs a pre-germination treatment of immersion in water at 20°C for 2 days.

*New highly productive varieties of spinach: **Viroflay**. Courtesy of SEMILLAS VILMORIN*

The cultivation furrows should be 40-50 cm apart and the plants should be 25-40 cm apart.

When to sow depends on the growing cycle of the endive. The many varieties include:

- Spring/summer varieties: sow in early/mid winter
- Fall varieties: mid/late spring
- Winter varieties: late summer/early fall.

12.4.5.4. Cultivation and harvest techniques

• **Thinning**. If sown directly, thin to leave 25-40 cm between plants, depending on the variety.
• **Weeding**. Manual or using herbicides.
• **Blanching**. The leaves are tied together 15-20 days before harvest to blanche the inner leaves.
• **Harvesting**. Manual or mechanized.
• **Storage**. Endives can be stored in cold chambers at a temperature between -1°C and 1°C at a humidity of 90% for 15 to 30 days, the same as lettuce.
• **Sale**. Depending on the variety, they may weigh from 500 g to 2 kg.

12.4.5.5. The most common pests, diseases and physiological disorders

These are the same as for the lettuce, though bolting can also be caused by low temperatures.

12.4.6. Spinach

12.4.6.1. General concepts

Spinach (*Spinacia oleracea*) is a member of the goosefoot family (Chenopodiaceae). Spinach can be eaten raw, boiled or fried. It is widely used in industry and its high iron content makes it effective against anemia. It originated in Asia.
It is a biennial dioecious plant, meaning that the male flowers are borne on male plants, and the female flowers on female plants. The seeds retain their ability to germinate for 4 years.
The varieties are classified on the basis of their production cycle, which is 50 to 80 days long.

• **Fall/winter varieties:**

Viking
Winter Giant
Viroflay Monster
Early Hybrid
Andros
Samos
Grandstand hybrid
Hybrid Palona
Virkarde Polka
Marathon
Universal
Astí Curly
Califlay
Roga

Nutritional composition per 100 g foodstuff	
Proteins	2.3-3.7 g
Lipids	0.4-0.6 g
Sugars	3-3.6 g
Fiber	.8 g
Vitamin A	9,420-13,000 IU
Vitamin B_1 (thiamine)	1.1-1.5 mg
Vitamin B_2 (riboflavin)	2-2.7 mg
Vitamin B_6	4 mg
Vitamin C (ascorbic acid)	59-116 mg
Vitamin E	1.7 mg
Calcium	81 mg
Phosphorus	55-86 mg
Iron	3-10 mg
Magnesium	37 mg
Potassium	774 mg
Zinc	1.2 mg
Sulfur	306-600 mg
Manganese	8.5 mg
Energy value	26 calorías

• **Spring/summer varieties:**

> Pavana
> Bloomsdale
> Market Wonder
> Capella
> Symphonie
> Estivato
> High Pack
> Hybrid Summic
> Rhapsody
> Hybrid Indian Summer
> Lagos
> Protekta
> Spinoza
> Vital
> Butterfly

12.4.6.2. The plant's requirements

• **Climate and temperature**. It requires a cool climate and cannot withstand high temperatures.
• **Soil**. It requires deep, rich, loamy soils with good drainage. It can not tolerate excessively alkaline or acidic soils. The best pH is 6.5. It is resistant to salinity.
• **Extractions**. The spinach's extraction of nutrients per hectare is about:

> 60-90 kg N
> 27-40 kg P_2O_5
> 100-230 kg K_2O

• **Fertilizer application**. It does not grow well after recent application of organic matter, and so the dung should be applied to the previous crop.

Application of fertilizer at depth:

> 40-60 kg N
> 40-60 kg P_2O_5

> 100-150 kg K_2O

Surface application of fertilizer:

> 45-60 kg N

12.4.6.3. Soil preparation and planting

To prepare the soil, it should be deep plowed once and surface plowed several times to leave the soil in good tilth.
The seed is sown directly in continuous rows if the spinach is for industrial use. The seed is sown at a depth of 2 cm.
The seed requires a pre-germination treatment of immersion in water for 12 hours. The seed should be sown in mid/late summer for production in fall/winter, and at the end of winter for production in spring/summer.

12.4.6.4. Cultivation and harvest techniques

• **Thinning**. The plants should be left 5 to 15 cm apart, depending on whether production is extensive or for industrial use. Thin when the plants have 4-5 leaves.
• **Weeding**. Manual or with selective herbicides.
• **Harvesting**. This can be done manually cutting the largest leaves a few at a time, or by harvesting the entire plant, or it can be done mechanically if the spinach is for industrial use.
Spinach is harvested 2 or 3 months after sowing.
• **Sale**. If it is to be consumed fresh, the spinach should be washed to eliminate any remaining soil. For sale fresh, the spinach can be sold as loose leaves or as whole plants in bunches.
They should be transported in cold chambers to ensure they reach the sales point in good condition. Spinach for industrial production should be transported immediately, without any preliminary washing.
• **Storage**. Spinach can be stored in cold chambers between –1°C and 1°C at a humidity of 90-95% for a period of 2-4 weeks.

Climate requirements		
CRITICAL TEMPERATURES	Freezing point	-5°C
	Zero growth	5°C
	Minimum for growth	6°C
	Optimum growth	15 - 25°C
	Maximum for growth	30°C
GERMINATION	Minimum temperatures	5 - 7°C
	Optimum temperature	15 - 18°C
	Maximum temperature	25 - 30°C
HUMIDITY		MEDIUM
LIGHT		MEDIUM

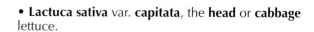

12.4.6.5. The most common pests, diseases and physiological disorders

• Pests:
- *Beetroot fly.* This mines galleries within the leaf.
- *Aphids.* General weakening and curling up of the leaves.
- *Agrotis* (cutworms). They eat the collar of the root.
- *Snails* and *slugs.* They eat the leaves.

• Fungal diseases:
- *Downy mildew of spinach.* This causes yellow patches on the leaves, with grey mycelium on the underside.
- *Cercospora.* This causes round patches with a yellow edge.
- *Botrytis* or grey rot.
- *Pythium.* This causes necrosis of the roots and a halt in growth.

• Viral diseases:
- *Cucumber virus* ("spinach blight"). This causes a yellowing and deformation (puckering) of the leaves.
- *Beetroot mosaic virus.* This causes light patches with black points.

12.4.7. Lettuce

12.4.7.1. General concepts

Lettuce (*Lactuca sativa*) is a member of the daisy family (Compositae). It is the most widely used leaf vegetable in salads. Other species of *Lactuca* grow wild throughout the temperate zone.
It is a biennial plant with more or less rounded leaves and seeds with a feather-like pappus. The seed retains its ability to germinate for 4-5 years. The productive cycle may be 2 to 6 months, depending on the variety and the time of year.
The lettuce includes two different varieties that are widely cultivated, cos and cabbage lettuces.

• Lactuca sativa var. **longifolia**. This is known as the **Cos** (or **romaine**) lettuce, which is elongated and does not form a heart.

Winter:

Pink long cos
Green long cos
White-seed blond cos
Narrow-leaved blond mule's ear
Winter Density
Romaserra
Gorrión
Colibrí
Gaviota
Romance

Spring/summer:

Green Three Eyes
White Three Eyes
Romea
Romabelle
Romaverde

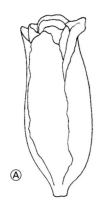

Lettuce: Varieties
A/ *Cos, or romaine, lettuce*
B/ *Head, or cabbage, lettuces*

Varieties of lettuce:
Trocadero
Courtesy of SEMILLAS SLUIS & GROOT

• Lactuca sativa var. **capitata**, the **head** or **cabbage** lettuce.

Winter:

- *Crisp-head types:*

Winter Galician	Red Batavia
Caravan	Grenobloise
Salinas	Coolguard
Winter Marvel	Bowl rouge
Lydia	Red salad rouge

- *Butter-head types:*

Trocadera black seed	Corine
Trocadera white seed	Platine
Ravel	Sabine
Hamlet	

Nutritional composition per 100 g foodstuff	
Proteins	0,8-1,6 g
Lipids	0,1-0,2 g
Sugars	1,2-2,1 g
Vitamin A	300-2,600 IU
Vitamin B$_1$ (thiamine)	0.07-0.1 mg
Vitamin B$_2$ (riboflavin)	0.03-0.1 mg
Vitamin B$_5$	0.3-0.5 mg
Vitamin C (ascorbic acid)	5-24 mg
Calcium	13-36 mg
Iron	1.1-1.5 mg
Magnesium	6-16 mg
Phosphorus	25-45 mg
Potassium	100-400 mg
Sodium	5-10 mg

Spring/summer:

- Crisp-head types:

Batavia rubia	Maravilla de verano
Batavia Flavia	Calmar
Iceberg type	Climax
German Batavia	Montemar
Naples cabbage-type	Vanguard

- Butter-head types:

Queen of May	Summer
Lucía	Queen
Aprilia	Cyria
Toria	Oresto
Amerika	Tessy
Rigoletto	Reskia
Orfeo	Aurelia
4 seasons wonder	Capitán
Faust	Augusta
Oberto	Nadia
Clarion	Verpia

12.4.7.2. The plant's requirements

• **Climate and temperature**. It prefers moist temperate climates. High temperatures negatively affect head formation and cause bolting.
• **Soils**. It prefers rich well-drained loamy soils that are not too moist. It resists slight salinity and is sensitive to soil acidity. The optimum pH for lettuce is 6.7-7.4.
• **Extractions**. These depend on whether the crop is grown in the open air or in a greenhouse, where extractions are higher (the figures are per hectare):

$$50\text{-}100 \text{ kg N}$$
$$20\text{-}50 \text{ kg P}_2\text{O}_5$$
$$120\text{-}250 \text{ kg K}_2\text{O}$$

• **Fertilizer application**. Apply 15-20 t/ha of well-rotted dung.

Fertilizer application at depth:

$$30\text{-}40 \text{ kg N}$$
$$30\text{-}50 \text{ kg P}_2\text{O}_5$$
$$100\text{-}150 \text{ kg K}_2\text{O}$$

- Surface fertilizer application: Three applications of the same quantity of nitrogen as applied at depth: once after thinning, another 15 days later and the last one when head formation starts.

12.4.7.3. Soil preparation and planting

The soil should be deep plowed and harrowed until it is a fine tilth. The seed needs a pre-germination treatment of immersion in water at 20°C for 48 hours.
The seed can be sown in seedbeds or directly into the field.

Climatic requirements		
CRITICAL TEMPERATURES	Freezing point	-6°C
	Zero growth	6°C
	Minimum for growth	14 - 18°C
	Optimum growth	5 - 8°
	Maximum for growth	30°C
HEAD FORMATION	Optimum day temperature	10 - 12°C
	Optimum night temperature	3 - 5°C
GERMINATION	Minimum temperature	3 - 5°C
	Optimum temperature	15 - 20°C
	Maximum temperature	25 - 30°C
HUMIDITY		MEDIUM
LIGHT		MEDIUM

In seedbeds the seed can be scattered, at a rate of 1 g seed per m^2 and at a depth of 3-4 mm. The seed can be sown in trays, pots or pressed peat pots. This means that the seedlings maintain their rootball intact, meaning that transplanting losses are lower than with seedlings with no root ball.

30-40 days after sowing, the lettuces can be transplanted into rows 50 cm apart, or in tables 80-100 cm apart and with 2 rows, and distance between plants (in both cases) of 25-30 cm. The seed can be directly sown into furrows or table as above, but can be sown as continuous rows, discontinuous rows or using pelleted seed, which requires precision seeders.

The time to sow depends on the productive cycle, and is as follows:

- Winter cycle: mid/late summer.
- Spring cycle: early/mid winter.
- Summer cycle: early/mid spring.
- Fall cycle: early/mid summer.

12.4.7.4. Cultivation and harvest techniques

• **Thinning**. After direct sowing, thin to leave the plants 25-30 cm apart. Thin when the plants are 6-8 cm tall.
• **Weeding**. As often as needed, either manually or using herbicides.
• **Blanching**. The leaves are held together (with a rubber band or string) 8-12 days before harvesting. The inner leaves are blanched, especially in varieties that do not form a head (cos or romaine lettuces).
• **Harvest**. Lettuce is normally harvested manually, though it can also be mechanized.

Depending on the productive cycle:

- Winter cycle: late fall/late winter.
- Spring cycle: early/late spring.
- Summer cycle: mid/late summer.
- Fall cycle: early/late autumn.

• **Sale**. After harvesting, the outer leaves are

removed, and the stem is cut below the leaves. The lettuces are then calibrated and put in polythene bags for sale.

• **Storage**. In the most suitable conditions, 0-1°C and 90-95% humidity, lettuces can be stored 15-30 days.

12.4.7.5. The most common pests, diseases and physiological disorders

• **Pests:**
- *Larvae*. Eat the leaves.
- *Cutworms* (*Agrotis*). They attack the stems of young plants.
- *Wireworms*. They attack the roots.
- *Greenfly*.
- *Whitefly*. In greenhouses.
- *Leaf miners*.
- *Slugs and snails*.
- *Nematodes*. The plant's growth is stunted and cysts form on the roots.

• **Fungal diseases:**
- *Lettuce mildew*. Yellow patches form between the veins on the leaves, which are covered by a grey mycelium and then wither.
- *Powdery mildew*. The leaves are covered by a whitish mycelium.
- *Sclerotinia*. This causes a soft rot at the base of the plant.
- *Pythium*. This attacks the base of the collar, causing the aboveground parts to die.
- *Rhizoctonia*. This causes a rot of the junction of the collar and root.
- *Botrytis* or *grey rot*.

• **Viral diseases**. This produces a mosaic of light and dark green in the leaves.
- *Cucumber mosaic virus*. Stunted growth and general yellowing.

Watercress

- *Lettuce thick vein virus*. This causes swellings on the leaves and the veins turn yellow.

• **Physiological disorders:**
- *Bolting*. This is due to high temperatures.
- *Poor head formation*. Due to deficient fertilizer or excessive fertilization with macronutrients.
- *Lettuce tipburn*. Brown patches appear and the edges of the leaves dry out. This seems to be related to calcium deficiency.

12.4.8. Other crops

12.4.8.1. Sorrels

The name *sorrel* is applied to several species of the genus *Rumex*, the most common being garden sorrel (*Rumex acetosa*), and they are all members of the knotweed family (Polygonaceae).
It is normally used as a flavoring, but its continued consumption is recommended for those with kidney stones, as they are rich in oxalic acid.
Sorrel is a perennial plant that requires moist sites that do not receive a lot of direct sun. The plantation lasts for 10 years but has to be resown every three years.
The seed is sown from late spring to late summer, and harvest begins 3 months after sowing.

12.4.8.2. Watercress

Watercress (*Nasturtium officinale*) is a perennial aquatic plant and a member of the cress family (Cruciferae).
The leaves have a peppery flavor and they are eaten in salads or to flavor other foods. It is said to be cleansing and is rich in vitamin C.
The plantation lasts for 2 years. In terms of climate, it can not tolerate low temperatures, requiring water at a temperature of at least 10°C and with little lime.
The seed is sown in spring, and harvesting begins after 2 months. After washing, they are sold in bunches. They can be stored at a temperature of 1°C with a humidity of 95%.

12.4.8.3. Kale

Kale (*Brassica oleracea* var. *acephala*) is a form of cabbage, and it is a member of the cress family (Cruciferae).
The leaves are used in salads and to make a few dishes, and to feed livestock. A form with broader leaves is known as **collard**.
Their climatic and soil requirements are similar to those of winter cabbages, as are their growing cycles and cultivation techniques.

12.4.8.4. Borage

Borage (*Borago officinalis*) is a member of the borage family (*Boraginaceae*).
It originated in the Mediterranean region and is mainly eaten boiled. It has diuretic and expectorant properties.

It is a very robust annual plant that prefers deep cool loamy soils rich in organic matter. The seed is sown in mid summer, with a separation of 25-30 cm between plants. Harvesting begins in late fall and finishes in early winter.

12.4.8.5. Cardoon

The cardoon (*Cynara cardunculus*) is a member of the daisy family (Compositae). It originated in the Mediterranean region and is usually eaten boiled.
It is a perennial plant cultivated as an annual. The seeds retain their ability to germinate for about 5 years.
In terms of climatic requirements it grows well in temperate climates and is sensitive to frosts. It prefers light deep soils rich in organic matter, with a neutral or slightly alkaline pH. The seed is normally sown in summer, but can be sown in spring if the winters are very severe. It is important to blanch, which is done one month before harvest.
Harvesting is staggered, beginning in fall and ending in winter.
The produce is placed in plastic bags that are placed in boxes for sale. In a cold chamber at 0°C and 90-95% humidity, cardoon can be stored for up to 2 months.

12.4.8.6. Fennel

Fennel (*Foeniculum officinale*) is a member of the umbellifer family (Umbelliferae). Fennel has soothing, anti-inflammatory and diuretic properties, and is used to flavor meals and salads.
It is a biennial plant whose seeds retain their ability to germinate for 3-4 years. Its cultivation cycle lasts about 4-5 months.

In terms of its climatic requirements, it is sensitive to low temperatures and droughts, which induce bolting. It prefers a deep, cool, loamy soil rich in organic matter. It is also tolerant of soil acidity.
The seed is sown in early summer, and at the end of summer the seedlings are transplanted into furrows 30-40 cm apart.
Harvesting is staggered from late fall to late winter. After harvesting the outer leaves are removed, the fennels are calibrated and placed in bags or boxes.
Fennel can be stored in cold chambers at 0-1°C and 90% humidity.

12.4.8.7. Parsley

Parsley (*Petroselinum crispum* syn. *P. sativum*) is a member of the umbellifer family (Umbelliferae).
It originated in the Mediterranean region and is mainly used to flavor meals.

Automatic spinach harvesters. Manufactured by DE PIETRI

It is a biennial plant that flowers in its second year. The seeds retain their ability to germinate for 2 years. Its cultivation cycle lasts 3 months. There are smooth-leaved and curly-leaved varieties.
In terms of its climatic requirements, it is sensitive to frosts, drought and strong winds. It prefers deep, cool, loamy soils rich in organic matter.
The leaves can be harvested several times, from 3 months after sowing, or in a single harvest, in which case the entire plant is removed.
The leaves are sold in bunches or in plastic bags.
Parsley can be stored at 0-1°C and 85-90% humidity for 1-2 months.

12.5. CROPS GROWN FOR THEIR FLOWERS

12.5.1. Artichoke

12.5.1.1. General concepts

The artichoke (*Cynara scolymus*) is a member of the daisy family (Compositae).
It originated in the Mediterranean region and is eaten both fresh and tinned or frozen.

A liquid extract can be made from artichokes that is an ingredient in some aperitif drinks.
The artichoke is a perennial plant, whose flower heads are surrounded by a series of fleshy bracts. The flower heads (capitula) are the edible part of the plant, and if they are not collected at the right moment, the bracts open and the deep purplish flowers emerge.

Artichoke seeds retain their ability to germinate for 6 to 12 years, but they are normally propagated vegetatively by cuttings, as plants grown from seed vary greatly in their production.
They are generally grown for a period of 3-4 years, depending to a large extent on the climate. There is normally a period of 5 to 7 months between planting and the first harvest.
Artichoke varieties are usually classified on the basis of the color of the flower head, which is either green or violet.

• **"White" or light green heads:**
Great Britain thick
Tudela White
Callosinas Madrileñas
Paris Green
Aranjuez
Monquelina
Green Globe
Getafe

• **Violet heads:**
Provence violet
Early violet
Tuscany violet

Nutritional composition per 100 g foodstuff	
Proteins	2.59 g
Lipids	0 g
Sugars	6,72 g
Vitamin A	270 IU
Vitamin B₁ (thiamine)	180 mcg
Vitamin B₂ (riboflavin)	10 mcg
Vitamin C (ascorbic acid)	5 mcg
Calcium	50 mg
Phosphorus	90 mg
Iron	0.5 mg
Energy value	38 calories

Climatic requirements		
CRITICAL TEMPERATURES	Freezing point	-4°C
	Zero growth	5°C
	Minimum for growth	6 - 8°C
	Optimum growth	18 - 25°C
	Maximum for growth	30°C
HUMIDITY		MEDIUM
LIGHT		MEDIUM

Long Sicilian violet
Purple Globe

12.5.1.2. The plant's requirements

• **Climate and temperature**. Artichokes are sensitive to frosts. The aboveground parts may die completely if the temperature falls to –2°C, and the root system at a temperature of –10°C. Very high temperatures also cause growth and production to halt. Hot dry winds in the early stages of growth may cause serious damage.
• **Water**. It is a plant that cannot can not grow well with excess soil moisture.
• **Soil**. It does not require a very good soil, but does not do well on sandy soils. It prefers loamy clay soils with good drainage. It is resistant to salinity and the optimum pH for growth is between 7.3 and 7.6.
• **Extractions**. The extractions per hectare are estimated to be:

220-230 kg N
50-100 kg P_2O_5
500-750 kg K_2O

• **Fertilizer application**: Apply 30-40 t/ha of dung.

Fertilization at depth per hectare:

100-150 kg N
120-170 kg P_2O_5
120-250 kg K_2O

Surface application of fertilizer: 3 applications of 75 kg/ha of N.

12.5.1.3. Soil preparation and planting

Soil preparation is important when growing artichokes as the plants will remain in the soil for several years. The site should be deep plowed twice and harrowed several times to get the soil into good condition.
The most common method of vegetative propagation is the division of a rhizome. After they are disinfected, the rhizomes are planted in early/mid summer in furrows at a distance of 80-150 cm (depending on their size) and with a distance of 80-100 cm between plants. The rhizomes are planted at a depth of 5-6 cm.

12.5.1.4. Cultivation and harvest techniques

• **Replace any rhizomes that have not sprouted**.
• **Earthing up**. This is normally done after the second watering.
• **Weeding**. This is normally done with selective herbicides.
• **Pruning**. The plant is cut back when it has ceased producing in its first year, when the plant starts to wither.
• **Harvest**. The time to harvest depends mainly on the climate, but usually lasts for a period of 2-3 months.
In temperate climates production lasts from mid fall to early spring.
In climates with cold winters and mild summers production lasts from late spring to mid fall.
In climates with relatively cold winters and hot summers, there are two periods of production; the first lasting from fall till the temperatures drop sharply, and the second from the period temperatures rise again until it is too hot, when production is halted.
The flower head is harvested with up to 10 cm of its stalk, and the flower head should have a diameter of at least 6 cm.
• **Sale**. Artichokes for fresh consumption are placed in boxes lined with waxed paper. Those for industrial use are first boiled, and then tinned, bottled, etc.
• **Storage**. Artichokes can be stored in a cold chamber at 0-1°C at 90-95% relative humidity for 15 to 30 days.

12.5.1.5. The most common pests, diseases and physiological disorders

• **Pests:**
- *Screwworms*. This leaf eater attacks at night.
- *Cutworms (Agrotis)*. They destroy the collar of the plant, causing it to wilt. They attack at night.
- *Artichoke miner*. The larvae penetrate the leaf veins and form galleries in the stems and may reach the flower heads.
- *Aphids*. General yellowing of the plant and drooping of the leaves.
- *Artichoke flea beetle*. The larvae devour the leaf from within.
- *Artichoke apio* and *artichoke fly*. Their larvae mine galleries in the leaves and heads.
- *Artichoke fungus beetle*. The larvae devour the leaves.
- *Red mite, snails* and *slugs*.

• **Fungal diseases:**
- *Powdery mildew*. A grey mycelium forms on the underside, while the upper surface goes yellow.
- *Lettuce mildew*.
- *Artichoke blight*. This causes patches around the nerves of the old leaves.
- *Ascocyta*. This causes circular black patches to form at the tip of the bracts surrounding the heads.

• **Bacterial diseases:**
- *Artichoke oily patch*. This causes oily patches to form around the bracts of the heads.

• **Viral diseases:**
- *Yellow mosaic virus*
- *Ringspot virus*
- *Curly dwarf virus*

• **Physiological disorders:**
- *Frosts* cause blackish patches to form on the bracts around the heads, which wilt.
- *Scalding*. The young plant dies, due to watering during the hours of daylight.

Storage of artichokes with stacking for pallets

Cauliflower varieties:
Snowball.
Courtesy of SEMILLAS
VILMORIN

12.5.2. Cauliflower

12.5.2.1. General concepts

The cauliflower (*Brassica oleracea* var. *botrytis*) is a variety of cabbage and is a member of the cress family (Cruciferae).
It originated in the eastern Mediterranean and it is mainly eaten fresh, though it is now also sold deep-frozen.
The cauliflower is a biennial plant, and the edible part is the tightly packed inflorescence that is produced before flowering.
The seeds retain their ability to germinate for 5-8 years.

The varieties are classified on the basis of their cultivation cycle.

• **Short cycle varieties**. Their growing cycle lasts for three months:

Olga	Lawyna
Alpha	Catalina
Erfurt	Orco
Early Snowball	Talmira
Canberra	Everest
Kangaroo	Selandia
Delira	Snowball
Soro	Florablanca
Lago	

• **Medium cycle varieties**.
Their growing cycle lasts 3-4 months:

Snowcap	Christmas
Italian Giant	Parnas
Prebaco Early	Kibo-tardo
Rubaco	Christmas turkey

• **Long cycle varieties**: Their growing cycle lasts 4-6 months.

Odin winter cauliflower	Armando Mirado
Atlas late cauliflower	Lent late
Angers late	San José cauliflower

Nutritional composition per 100 g foodstuff	
Proteins	2.48 g
Lipids	0.34 g
Sugars	4.55 g
Vitamin A	90 IU
Vitamin B₁ (thiamine)	110 mcg
Vitamin B₂ (riboflavin)	100 mcg
Vitamin C (ascorbic acid)	69 mcg
Calcium	22 mg
Phosphorus	72 mg
Iron	1.1 mg
Energy value	32 calories

Climatic requirements		
CRITICAL TEMPERATURES	Freezing point	-10°C
	Zero growth	3 - 5°C
	Minimum for growth	6°C
	Optimum growth	16 - 18°C
	Maximum for growth	30°C
GERMINATION	Minimum temperature	6 - 8°C
	Optimum temperature	18 - 25°C
	Maximum temperature	30 - 35°C
MOISTURE		HIGH
LIGHT		LOW

12.5.2.2. The plant's requirements

• **Climate and temperature**. It requires moderate temperatures and moist environments. Excessively dry winds are prejudicial.
• **Soils**. It requires soils with a light texture and good water retention. They are relatively resistant to salinity and need a soil pH of 6-6.5.
• **Extractions**. Average extraction per hectare is estimated to be:

$$175\text{-}200 \text{ kg N}$$
$$60\text{-}80 \text{ kg P}_2\text{O}_5$$
$$200\text{-}250 \text{ kg K}_2\text{O}$$

• **Fertilizer application**. Apply 30-40 t/ha of dung.

Fertilizer application at depth per hectare:

$$60 \text{ kg N}$$
$$80 \text{ kg P}_2\text{O}_5$$
$$200 \text{ kg K}_2\text{O}$$

Surface fertilizer application: 2 application of 100 kg N/ha.

• **Deficiencies**. It is sensitive to shortage of boron and molybdenum in acid soils.

12.5.2.3. Soil preparation and planting

Soil preparation should include a deep plowing and several surface plowings.

The seed is sown between mid spring and mid summer, depending on the cycle of the variety sown.
The seed can be sown into seedbeds or into peat pots. The seedlings are transplanted when they have 5-6 leaves and are 15-20 cm tall, about 40-50 days after sowing.

The seedlings are transplanted into furrows 60-80 cm apart, with a distance of 40-60 cm between plants, depending on the variety and the length of its cycle. On very large sites, the seeds can be sown directly using precision seeder units.

12.5.2.4. Cultivation and harvest techniques

• **Replace dead seedlings after transplanting**.
• **Thinning**. If the seed was sown directly.
• **Earthing up**. Once the plant is well rooted in the site.
• **Weeding**. With selective herbicides.
• **Protecting the inflorescences**. Once the leaves have opened out, one of them should be bent back to protect the inflorescence ("curd") to prevent yellowing.
• **Harvesting**. Cauliflowers are harvested manually 5-10 times, depending on the production cycle.
- *Short cycle varieties*. Harvesting lasts from summer to mid fall.
- *Medium cycle varieties*. Harvest lasts from mid fall to mid winter.
- *Long cycle varieties*. Harvest lasts from mid winter to early spring.

The moment to harvest is when the curd has reached its greatest size, just before the flower opens. Specialized machinery can also be used that only harvests heads that are larger than a selected minimum size.
• **Sale**. Once harvested, the heads are selected and packed in boxes lined with waxed paper, or in polythene bags. They may be sold with their leaves or without.
• **Storage**. Cauliflowers can be stored perfectly at 0-1°C and humidity of 85-90% for 15-20 days.

12.5.2.5. The most common pests, diseases and physiological disorders

The pests and disease that attack the cauliflower are the same as those that attack the cabbage.

The following are the most important **physiological disorders**:

- *Appearance of bracts within the head*. This may be due to a sharp increase in temperatures, either after flower induction or in the first stages of its development.
- *Premature head formation* (i.e., before the plant is fully grown. This means the heads are small and malformed. This may be due to excessively low temperatures during the early stages of the plant's growth.
- *Premature flowering*. This may be due to high temperatures during flower formation.
- *Grey dots on the head*. Burns caused by the action of the sun's rays on drops of dew.
- *Boron deficiency*. This provokes corky patches in the leaves and blackish patches on the head.
- *Molybdenum deficiency*. This causes anomalous leaf growth. The heads that form are very small.

Field of cauliflowers

12.5.3. Other crops

12.5.3.1. Capers

The caper (*Capparis spinosa*) is a member of the caper family (Capparidaceae).

It is a perennial plant with a climbing habit, that grows wild in dry areas of the Mediterranean region. The part consumed is the flower bud before it opens, and it is usually consumed pickled or preserved in vinegar.
It begins its development in spring and flowers in summer, and all the aboveground parts die back when winter arrives.

Cultivated capers can live for 40 years, but they do not start production until the third year after planting, peaking at 5 years.

In terms of its climatic requirements, it is sensitive to water shortage, but is also sensitive to frosts and excess moisture.

The most important cultivation techniques to use are earthing up, once when the plant starts growing, once in summer and again in fall, after cutting back. Harvesting starts in late spring and lasts throughout the summer.

Once the capers have been harvested, they are cleaned, selected and then spread out to air dry. They are then treated industrially.

12.5.3.2. Broccoli

Broccoli (*Brassica oleracea* var. *italica*) is very closely related to the cauliflower, and is sometimes called winter cauliflower. It is a member of the cress family (Cruciferae).

The varieties are classified on the basis of the color of the inflorescence:

• **Violet varieties**:

Christmas Purple
Sprouting
Early Purple Sprouting
Late Purple Sprouting
Purple Sprouting

• **White varieties**:

Early White Sprouting
Late White Sprouting

White variety of broccoli.
Courtesy of SEMILLAS SLUIS & GROOT

• **Green varieties:**

Express Corona
Green comet
Italian Sprouting
Corvet
Romanesco
El Centro

Its climate, temperature, water and soil requirements are similar to those of the cauliflower. Broccoli is sown in spring and transplanted in summer.

The main inflorescences are harvested first, causing lateral inflorescences to grow and these are harvested later. White and violet varieties are harvested in spring and summer, while the green varieties are harvested in fall.

They are sold in plastic bags or in "punnets" covered with a sheet of plastic. Broccoli can be stored in a cold chamber at 0°C and 90-95% humidity for up to 3 weeks.

The pests and diseases that attack broccoli are the same as those that attack the cabbage.

The inflorescences may turn yellow if the cold storage conditions are not correct.

12.6. CROPS GROWN FOR THEIR FRUIT

12.6.1. Eggplant

12.6.1.1. General concepts

The eggplant, or aubergine (*Solanum melongena*), is a close relative of the potato and a member of the nightshade family (Solanaceae).

It originated in India and China and is mainly eaten fried or roasted.

The eggplant is usually grown as an annual, and its fruit is a fleshy berry, varying greatly in shape and color.

The seeds retain their ability to germinate for 5-6 years, and its growing cycle is 100-125 days.

The fruit may be white, violet, dark purple, black or streaked.

Nutritional composition per 100 g foodstuff	
Proteins	1.2 g
Lipds	0.2 g
Sugars	3.1-5.6 g
Fiber	0.9 g
Vitamin A	10-30 IU
Vitamin B₁ (thiamine)	0.04-0.05 mg
Vitamin B₂ (riboflavin)	0.05 mg
Viamin C (ascorbic acid)	5 mg
Calcium	12-15 mg
Phosphorus	26-37 mg
Sodium	2 mg
Iron	0.4-0.7 mg
Potassium	214 mg
Energy value	25 calories

Climatic requirements		
CRITICAL TEMPERATURES	Freezing point	0°C
	Zero growth	10 - 12°C
	Maximum for growth	40 - 50°C
	Optimum growth	22 - 27°C
	Minimum for growth	13 - 15°C
NIGHTTIME TEMPERATURE	Optimum	17 - 22°C
GERMINATION	Minimum temperature	15°C
	Optimum temperature	20 - 25°C
	Maximum temperature	35°C
FLOWERING	Optimum temperature	20 - 30°C
HUMIDITY		MEDIUM
LIGHT		HIGH

12.6.1.2. The plant's requirements

• **Climate and temperature**. The plant is very sensitive to frosts, but is not damaged by high temperatures.
• **Water**. It requires a lot of water.
• **Soils**. It requires deep rich soils that are sandy-clay in texture and with good drainage. It is not very resistant to salinity.
• **Extractions**. The extractions per hectare are estimated to be:

$$50 \text{ kg N}$$
$$50 \text{ kg P}_2\text{O}_5$$
$$100 \text{ kg K}_2\text{O}$$

• **Fertilizer application**. Apply 40-50 t/ha of dung.

Fertilizer application at depth:

$$50\text{-}60 \text{ kg N}$$
$$120\text{-}150 \text{ kg P}_2\text{O}_5$$
$$250\text{-}300 \text{ kg K}_2\text{O}$$

Surface fertilizer application: 2-3 applications of 30-40 kg N.

These quantities are increased for greenhouse crops.

12.6.1.3. Soil preparation and planting

Soil preparation is the same as that for growing tomatoes.
The seed is sown in seedbeds or in peat pots and then transplanted when the seedlings are 12-15 cm tall into furrows that are 100-130 cm apart, and with a separation of 70 cm between plants. The seeds need a pre-germination treatment of storage in wet conditions at 20-22°C for 5-6 days.

The cultivation periods and productive cycles are the same as those of the pepper.

Variety of elongated eggplant. Courtesy of SEMILLAS SLUIS & GROOT.

The different cultivated varieties are classified on the basis of the fruit shape:

• **Round:**

Smooth round purple	Burpee Hybrid
Smooth round violet	Blacknite
Ribbed round purple	Reina Negra
Black Beauty	Black-bell
Giant New York	Agora
Bonica	Galine
Mission Bell	

• **Elongated:**

Early long black	Eras
Early long violet	Barn
Long black	Prelane
Long purple	Linda
Caminal	Solara
Long Purple	Sultana
Bari vedette	

Eggplant is widely grown in greenhouses and plastic tunnels.

12.6.1.4. Cultivation and harvest techniques

- **Replace transplanted seedlings that die**.
- **Earthing up**.
- **Pruning**. 4-5 branches are left on the main stem, and the plant has a trailing growth habit. Pruning is only to eliminate internal branches
- **Staking**.
- **Weeding**.
- **Protection**. Similar to that for the tomato.
- **Plant growth regulators**. These are used to increase fruitset.
- **Thinning out**.
- **Harvesting**. This is done by hand, when the fruit is ripe and is fully plump.
The fruit must be handled carefully, as it is sensitive to mechanical damage.
Harvesting times are the same as for the pepper.
- **Sale**. After harvesting, the fruit are selected and the fruit's stem is cut back to a length of 2-3 cm. The fruit are then placed in boxes lined with waxed paper.
- **Storage**. Eggplant fruit can be stored in cold chambers at 4-5°C for a period of 10-12 days.

12.6.1.5. The most common pests, diseases and physiological disorders

The pests, diseases and physiological disorders affecting the eggplant are effectively the same as those affecting the tomato.
The plant may be attacked by the Colorado potato beetle, or suffer excessive growth of the branches, accompanied by deformations of the flowers, anomalous fruitset and irregular and deformed fruit. This is caused by lack of light in the first stages of growth accompanied by high humidity.

12.6.2. Zucchini

12.6.2.1. General concepts

The zucchini (*Cucurbita pepo*), or courgette, is a relative of the marrow and a member of the cucumber family (Cucurbitaceae).

It originated in central America and is mainly eaten boiled or fried.
The fruit are generally elongated and cylindrical and the skin is normally smooth.
There are three different groups of varieties, differing in the color of the skin:

- **Green-skinned:**

Long green	Elite
Diamante	Tala
Black Princess	Majestic
Zucchini	Maya
Aristocrat	Albina
Senator	Calista

- **Yellow-skinned:**

Seneca	Goldbar
Lemondrop	Gold Slice

- **White-skinned:**

Early white marrow	Neu
Medium long	Lemondrop
Long ribbed	

Zucchini plant in flower. Courtesy of SEMILLAS SLUIS & GROOT

Nutritional composition per 100 g foodstuff	
Proteins	1.76 g
Lipids	0.11 g
Sugars	2.14 g
Vitamin A	100 IU
Vitamin B₁ (thiamine)	60 mcg
Vitamin B₂ (riboflavin)	40 mcg
Vitamin C (ascorbic acid)	20 mg
Calcium	18 mg
Phosphorus	21 mg
Iron	0.6 mg
Energy value	17 calories

12.6.2.2. The plant's requirements

- **Climate and temperature**. In general it is less demanding than the melon or cucumber.
- **Soils**. It requires average texture soils, that are deep, rich in organic matter and well drained. It is somewhat resistant to soil salinity and the optimum soil pH is 5.5-6.5.

Symptoms of boron deficiency in the strawberry.
A/ Normal flower
B/ Flower showing boron deficiency
C/ Poor root growth, with excessive secondary root growth
D/ Seeds located at the tip of the strawberry accompanied by deformed fruit
E/ Malformation due to deficient pollination
Courtesy of BORAX ESPAÑA, S.A.

• **Extractions**. The extraction per hectare is estimated to be:

$$110 \text{ kg N}$$
$$160 \text{ kg P}_2\text{O}_5$$
$$90 \text{ kg K}_2\text{O}$$

• **Fertilizer application**. Apply 30-40 t/ha of dung.

Fertilizer application at depth:

$$60\text{-}80 \text{ kg N}$$
$$60\text{-}80 \text{ kg P}_5$$
$$100\text{-}120 \text{ kg K}_2\text{O}$$

Surface fertilizer application: Apply 20-25 kg nitrogen 3 times.

Climatic requirements		
CRITICAL TEMPERATURES	Freezing point	-1°C
	Zero growth	8°C
	Maximum for growth	35°C
	Optimum growth	25 - 30°C
	Minimum for growth	10°C
GERMINATION	Minimum temperature	10°C
	Optimum temperature	20 - 30°C
	Maximum temperature	40°C
HUMIDITY		MEDIUM
LIGHT		HIGH

12.6.2.3. Soil preparation and planting

Deep plow several times and then harrow several times.
The seed can be sown directly into the site, or in protected seedbed. The seed requires a pre-germination treatment, consisting of placing the seeds in water at 18-20°C for 24 hours, and then keeping them moist at 22-25°C for 2-3 days.
Whether the seed is sown directly or is transplanted, the furrows should be 100-120 cm apart with a separation between plants of 80-100 cm.

The sowing period depends on the growing cycle:

- *Extra-early cycle*. The seed is sown in protected conditions in mid fall.
- *Early cycle*. The seed is sown in tunnels in mid winter.
- *Medium cycle*. The seed is sown early spring.
- *Late cycle*: The seed is sown in mid summer.

12.6.2.4. Cultivation and harvest techniques

• **Thinning**. When the seed is sown directly.
• **Pruning**. Remove the secondary stems or leaves, if growth is abundant.
• **Staking**.
• **Weeding**.
• **Harvest**. The zucchini is collected when it is still small and tender, 40-70 days after sowing, and the harvest is staggered.

The harvest time depends on the plant's growth cycle:

- *Extra-early cycle*. Harvest starts in mid winter.
- *Early cycle*. Harvest starts in mid spring.
- *Medium cycle*. Harvest begins in late spring.
- *Late cycle*. Harvest starts in fall.

• **Sale**. When handling the fruit, remember that its skin is sensitive to blows and scratching. Zucchini are sold in boxes lined with waxed paper.
• **Storage**. Zucchini can be stored in cold chambers at a temperature of 4°C and a humidity of 80% for a period of 30 days without any problems.

12.6.2.5. The most common pests, diseases and physiological disorders

It is in general a resistant plant to most pests, diseases and physiological disorders. It is affected by the same pests as the other members of the cucumber family.

12.6.3. Strawberry

12.6.3.1. General concepts

The many different types of strawberry are all forms of the species *Fragaria*, and members of the rose family (Rosaceae).
Some forms originated in the forests of Europe, but there are now very many modern crosses and hybrids.
Strawberries are mainly consumed fresh as a dessert, but can be used for many other purposes, such as jams and spreads.
The plant is perennial, but is cultivated as an annual, with a cultivation cycle of only 90-180 days.
The strawberry is not a true berry, but a set of seeds on the swollen end of the stem, the receptacle, which is red and more or less conical.

The strawberry is not a berry but a set of fruits (the seeds) borne on the swollen end of the stem (the receptacle).
Courtesy of SEMILLAS VILMORIN.

The best-known varieties are divided into 2 groups:

• Varieties giving more than one crop a year:

Small fruit:

Caen Giant	Queen of the Valleys

Large fruit:

Geneva	Rabunda
Humi-Gento	Revada
Ostara	

• Varieties giving a single harvest per year. All produce large fruit:

Early:

Alyso	Surprise
Missionary	Toro
Sequoia	Favette

Medium:

Belrubí	Tioga
Fresno	Vola
Gorella	Aiko
Senga	

Late:

Coupil	Talisman
Tago	Bogotá

Nutritional composition per 100 g foodstuff	
Proteins	0.5-0.9 g
Lipids	0.1-0.4 g
Sugars	5-10 g
Vitamin A	60 IU
Vitamin B$_1$ (thiamine)	0.03 mg
Vitamin B$_2$ (riboflavin)	0.07 mg
Niacin	0.6 mg
Vitamin C (ascorbic acid)	20-70 mg
Iron	1 mg
Sodium	1 mg
Potassium	164 mg
Calcum	21 mg
Phosphorus	21 mg
Energy value	37 calories

12.6.3.2. The plant's requirements

• **Climate and temperature**. Strawberries adapt well to cool and hot climates. It is resistant to frosts, except for the flowers.
• **Water**. It is important to maintain the soil humidity levels.
• **Soils**. It requires loose soils, well drained, with a medium texture. The optimum pH value is between 5.5 and 6.5.
It is sensitive to lime and soil salinity.
• **Extractions**. The extractions per hectare are estimated to be:

$$108\text{-}135 \text{ kg N}$$
$$52\text{-}70 \text{ kg P}_2\text{O}_5$$
$$190\text{-}218 \text{ kg K}_2\text{O}$$

• **Fertilizer application**. Apply 30 t/ha of well-rotted dung.

Fertilizer application at depth, per hectare:

90 kg N
120 kg P_2O_5
180 kg K_2O

Surface fertilizer application: A total of 80 kg of nitrogen in several applications.
• **Deficiencies**: It is sensitive to chlorosis due to shortage of available iron.

Climatic requirements		
CRITICAL TEMPERATURES	Freezing point	-3 - -5°C
	Zero growth	2 - 5°C
	Optimum day temperature	15-18°C
	Optimum night temperature	8 - 10°C
ROOTING	Minimum temperature	10°C
	Optimum temperature	18°C
	Maximum temperature	35°C
RIPENING	Optimum day temperature	18 - 25°C
	Optimum night temperature	10 - 13°C
HUMIDITY		MEDIUM
LIGHT		MEDIUM

12.6.3.3. Soil preparation and planting

The soil should be deep plowed several times, and the harrow then passed several times to ensure that the soil is levelled.
Strawberries are propagated by stolons. Plants are only grown from seed to obtain new varieties. Strawberries are very sensitive to viral diseases, and so it is very important only to propagate from healthy material. You are advised to purchase small plants from specialist nurseries that guarantee the plants are virus-free.

There are two ways of planting:

• **With cold-stored plants**. The stolons are uprooted in early winter, the old leaves are removed and the plants are disinfected. They are then placed in plastic bags, and then in a cold chamber, at a temperature of –1°C and a humidity of 90%, for 6-8 months. They are planted in their final site in late spring/early summer, after hardening (acclimatizing) them outside the chamber for 12-24 hours.

They are planted in furrows 50-60 cm apart, or in 100-120 cm tables with 2 alternating rows. In both cases, the distance between plants is 30 cm.

• **With fresh plants**. The plants are obtained from stolons separated from the parent plant, and they are planted in mid/late fall.

This system enters production earlier, but the yield is lower.

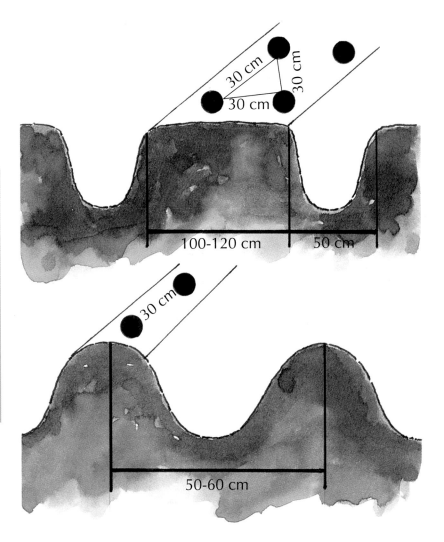

Planting distances when cultivating strawberries

12.6.3.4. Cultivation and harvest techniques

• **Replacing any transplanted plants that die.**
• **Removing the flowers after planting in summer.**
• **Removing the stolons that form in early/mid fall.**
• **Earthing up.**
• **Weeding.**
• **Pruning in fall.**
• **Mulching**. This is done to ensure the fruit does not get dirty through contact with the soil, and also to obtain earlier crops.
• **Plastic tunnels for forced crops**.
• **Harvesting**. Strawberries are harvested manually, at the stage of ripeness determined by the distance from the market. For distant markets, you should harvest when half of the fruit are ripe, and for nearby markets when 3/4 are ripe.
• **Sale.** Once they have been harvested, they are classified and placed in punnets covered with transparent plastic, and they are then put in boxes if they are to be eaten fresh.
• **Storage**. They can be stored in a cold chamber at a temperature of -1–0.5°C and a humidity of 85-90% for 10 days.

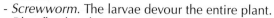

12.6.3.5. The most common pests, diseases and physiological disorders

• **Pests:**
- *Screwworm.* The larvae devour the entire plant.
- *Blue flea beetle.*
- *Strawberry snout weevil.* This causes the flower buds to wilt and dry out.
- *Greenfly.* They cause a general yellowing and curling up of the leaves.
- *Red spider mite.* The leaves lose color and turn yellowish.
- *Attacks by nematodes.*

• **Fungal diseases:**
- *Strawberry stem mildew.* This disease attacks the vessels and causes the edges of the old leaves to wither and dry out.
- *Strawberry Verticillium.* This is a disease of the vessels that causes drooping and withering of the edges of the old leaves.
- *Powdery mildew.* A white powder appears on the underside of the leaves.
- *Botrytis,* or *grey mold.*
- *Red core* ("Lanarkshire disease"). Reddish patches form on the leaves.
- *Anthracnose.*

• **Viral diseases:**
- *Strawberry mottle virus.* This causes inter-vein chlorosis, together with mottling of the leaves.
- *Yellow edge virus.*
- *Curly mosaic virus.* Causes deformations of the leaves, accompanied by the appearance of yellow patches.
- *Crinkle virus.* Causes the leaves to pucker.

• **Physiological disorders:**
- *Scalding of the fruit.*
- *Strawberry tipburn.* This is caused by excess moisture, and the tips of the young leaves are deformed and die ("burnt").

12.6.4. Melon

12.6.4.1. General concepts

The melon (*Cucumis melo*) is a member of the cucumber family (Cucurbitaceae). It seems to have originated in Asia. It is mainly eaten fresh, and it is also used to prepare some sweets.
The part eaten is the fruit. The fruit's flesh is white, yellow or orange, while the rind is green, yellow or orange, and may be smooth or ribbed. They are normally ovoid or spherical.

The seeds retain their ability to germinate for 5-6 years and their production cycle lasts for 4-5 months.

The best-known varieties of melon are:

• **Cantaloupe:**

Charentais	Cosmos
Doublon	Athos
Vedentrais	Savor
Cavaillon	Vedor
Bellegarde	Laguna
Jivaro	Romeo
Alaska	Trapio
Galia	Delta

• **Musk melons:**

Elongated large
Villaconejos
Reticulate fragrant green

• **Winter melons:**

Winter Olive
Provence winter melon
Golden Beauty

Cantaloupe-type variety of melon: **Chanterais**. Courtesy of SEMILLAS VILMORIN

Nutritional composition per 100 g foodstuff	
Proteins	0.6-1.2 g
Lipids	0.1 g
Sugars	6.2-10 g
Fiber	0.1-0.2 g
Viamin A	483-4,000 IU
Vitamin B$_1$ (thiamine)	0.04-0.08 mg
Vitamin B$_2$ (riboflavin)	0.01-0.02 mg
Vitamin C (ascorbic acid)	19-47 mg
Niacin	0.4-1 mg
Calcium	5-11 mg
Iron	0.2-0.5 mg
Phosphorus	7-50 mg
Energy value	26-41 calories

• **For open-air cultivation:**

Green tendral	Piel de sapo
Amarillo oro	Hidalgo
Amarelo	Sapiel
Rochet	Meloso
Piñonet	Cuper

• **For greenhouse cultivation:**

Amarillo oro	Geisna
Amarelo	Ogen
Cuper	Overgen
Rochet	Jivaro
Hidalgo	Galia
Piñonet	Makdimon
Piel de sapo	Marina
Meloso	Biga
Sapiel	

12.6.4.2. The plant's requirements

• **Climate and temperature**. Melons require warm temperatures. They are sensitive to frosts and to excessively hot weather. Temperatures above 35-40°C cause burns in the fruit.
• **Water**. The plant is drought-resistant and very moist conditions are not good for it.
• **Soils**. It requires a loose, deep soil rich in organic matter that retains water well. It is moderately resistant to salinity and the optimum soil pH is between 6 and 7.
• **Extractions**. The extraction of nutrients per hectare are estimated to be:

50 kg N
20 kg P_2O_5
100 kg K_2O

• **Fertilizer application:** Apply 20-40 t/ha of dung.

Fertilizer application at depth, per hectare:

50-100 kg N
60-130 kg P_2O_5
100-150 kg K_2O

Climatic requirements		
CRITICAL TEMPERATURES	Freezing point	1°C
	Air temp. for zero growth	13 - 15°C
	Soil temp. for zero growth	8 - 10°C
	Air temp. for optimum growth	18 - 24°C
	Soil temp. for optimum growth	18 - 20°C
GERMINATION	Minimum temperature	13°C
	Optimum temperature	28 - 30°C
	Maximum temperature	45°C
FLOWERING	Optimum temperature	20 - 23°C
RIPENING	Optimum temperature	25 - 30°C
HUMIDITY		MEDIUM
LIGHT		HIGH

Surface fertilizer application:

50-60 kg N
60-80 kg K_2O

When cultivating melons in a greenhouse these quantities should be substantially increased.

• **Deficiencies**. It is sensitive to deficiencies of magnesium, boron, manganese and molybdenum.

12.6.4.3. Soil preparation and planting

The site should be deep plowed several times, and then harrowed several times to break the soil down to a tilth.

Planting melons

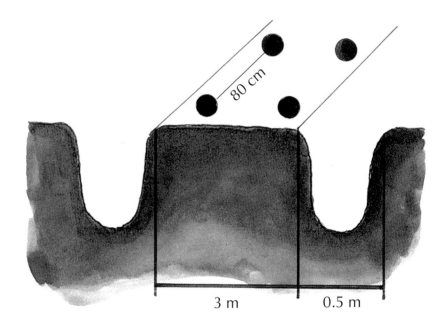

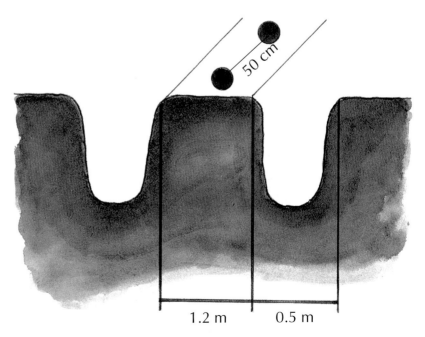

Symptoms of boron deficiency in sugar beet

The seed is sown directly in the site, and can be sown on 3 m raised tables with two rows on each, or on 120 cm tables with a single row of plants. The distance between the plants should be 50-80 cm.

Seed is sown in protected seedbeds in the case of the extra-early varieties. The seed should be given a pre-germination treatment of immersion for 24 hours in water at 18°C, after which they are drained and kept moist for 2-3 days at 25-35°C.

The seedlings are transplanted about 2 months after sowing.

The best time to sow depends on the variety's cultivation cycle:

- *Extra-early cycle*. The seed is sown at the end of fall, normally in protected seedbeds.
- *Early cycle*. The seed is sown in late winter/early spring.
- *Medium-late cycle*. Seed can be sown over the spring period.

12.6.4.4. Cultivation and harvest techniques

• **Thinning**. If the seed is sown directly, leave a distance of 50-80 cm between plants.
• **Pinching out**. Pinch out the main stem when it has 5-6 leaves, and pinch out the side branches when they have 5-6 leaves.
• **Earthing up**.
• **Weeding**. Use herbicides with great caution, as they may be toxic to melons.
• **Staking**. This is normally only done in greenhouses.
• **Thinning the fruit**. Leave 1 fruit per branch, and 5-6 per plant.
• **Protection**. Use plastic tunnels or grow in a greenhouse.
• **Harvesting**. It is very important to harvest at the right moment, as the melon's sugar content does not increase after harvesting.
The time to harvest is determined by the cultivation cycle and usually lasts about 2 months.

- *Extra-early cycle*. Harvesting begins in mid spring.
- *Early cycle*. Harvesting begins in late spring.
- *Medium-late cycle*. Harvesting starts in early summer.

• **Sale**. Once harvested, the melons are selected by size, and then placed in boxes lined with wood or cardboard.
• **Storage**. Melons can be stored in cold chambers at a temperature of 5-10°C (2°C for cantaloupe melons) at a humidity of 75-85% for 15-50 days depending on the variety.

12.6.4.5. The most common pests, diseases and physiological disorders

• **Pests:**
- *Melon aphid*. This secretes honeydew that fungi can grow on.
- *Aphids*. Weakening and general yellowing of the plant, and curling of the leaves.
- *White fly*. Especially in greenhouse cultivation.
- *Melon vacanita*. An insect that gnaws the leaves.
- *Melon galeruca*. The adult attacks the leaves, while the larvae attacks the leaves and roots.
- *Wireworms*. The larvae attack the root system.
- *Cutworms (Agrotis)*. The larvae devour the base of the plant's stem.
- *Thrips*. They eat the leaf and cause weakening of the plant.
- *Leaf miners*.
- *Red spider mite*.
- *Snails and slugs*.
- *Diseases caused by nematodes*. They normally cause cysts on the roots or malformations.

• **Fungal diseases:**
- *Fusarium* and *Verticillium* wilts. They both damage the plant's vascular system, causing general yellowing and wilting of the plant.
- *Stem rot*. Due to high humidity.
- *Melon anthracnose*. This gives rise to round darkish-black patches all over the plant.
- *Powdery mildew*. Patches of grey powder form, causing the leaves to wither.
- *Gourd mildew*. This causes patches on the edges of the leaves, which wither.
- *Alternaria*. This causes patches to form that are similar to those observed in the tomato.
- *Septoria*. Points appear on the leaves.
- *Botrytis* or *grey rot*.

• **Bacterial diseases:**
- *Gourd oily patch*.
- *Bacterial wilt*. This causes a dramatic halt in growth.

• **Viral diseases:**
- *Cucumber mosaic virus*.
- *Melon veining virus*. This causes reddish patches and necrosis of the veins.
- *Watermelon mosaic virus*. The leaves become chlorotic, and are stunted and swollen.
- *Zucchini mosaic virus*.

• **Physiological disorders:**
- *Scalding*. White patches form that are the result of excessively strong sunshine and high temperatures.
- *Splits in the melon*. This is due to water imbalances.

12.6.5. Cucumber

12.6.5.1. General concepts

The cucumber (*Cucumis pepo*) is a member of the cucumber family (Cucurbitaceae).
It originated in tropical Africa. It is mainly eaten in salads, though it is also pickled as "gherkins".
The cucumber is an annual plant. The fruit are elongated, more or less cylindrical, and have a watery white flesh within a skin that may be green, yellow or white.
The seeds retain their ability to germinate for 5-6 years and their cultivation cycle lasts 70-90 days.

Cucumber varieties are divided into 2 large groups:

• Cucumbers to be eaten fresh:

Long green English
Calahorra Factum medium long
Market medium long
Marketer
Brillante
Sporu
Pica
Saticoy
Princesa
Toska
Brillant
Palmera
Belcanto
Pepinex
Virel
Bambina
Monique
Pollex
Aries
Astrea
Amazona   Sandra
Uniflora   Jason
Farbio   Noval

• Cucumbers for pickling ("gherkins"):

Cornichon
Paris green
Rhineland early
Wisconsin
Chipper
Addis
Kobus
Calypso
Explorer
Levo
Ceto
Score
Parifin
Carolina
Ginor
Tagor
Levina
Olimpia

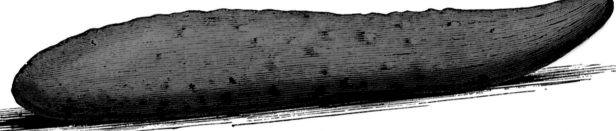

12.6.5.2. The plants's requirements

Cucumber variety to be eaten fresh:
***Marketer**.*
Courtesy of SEMILLAS VILMORIN

In general, its climatic requirements are similar to those of the melon.
• **Soil**. It requires soils with a medium sandy-clay texture, in good tilth, cool and rich in organic matter, and with good drainage. It is somewhat resistant to salinity, and the optimum soil pH is between 6 and 7.2.
• **Extractions**. The extractions per hectare are estimated to be:

60 kg N
40 kg P_2O_5
80 kg K_2O

• **Fertilizer application**. Apply 10-35 t/ha of dung.

Fertilizer application at depth, per hectare:

50 kg N
100-150 kg P_2O_5
100-200 kg K_2O

Surface fertilizer application: 2 applications of more than 50 kg of nitrogen.

For greenhouse cultivation, these amounts should be substantially increased.

• **Deficiencies**. It is sensitive to magnesium deficiency.

Nutritional composition per 100 g foodstuff	
Proteins	1.6 g
Lipids	0.2 g
Sugars	2.4 g
Fiber	1.2 g
Vitamin A	250 IU
Vitamin B_1 (thiamine)	30 mcg
Vitamin B_2 (riboflavin)	40 mcg
Vitamin C (ascorbic acid)	8 mg
Oxalic acid	27 mg
Phosphate	33 mg
Potassium	200 mg
Calcium	16 mg
Sulfur	12 mg
Magnesium	12 mg
Energy value	17 calories

Climatic requirements		
CRITICAL TEMPERATURES	Freezing point	-1°C
	Zero growth	10 - 12°C
	Soil temp. for zero growth	20 - 25°C
	Night temp. for optimum growth	18°C
SOIL TEMPERATURE	Minimum temperature	12°C
	Optimum temperature	18 - 20 °C
GERMINATION	Minimum temperature	12°C
	Optimum temperature	30°C
	Maximum temperature	35°C
HUMIDITY		MEDIUM
LIGHT		MEDIUM

12.6.5.3. Soil preparation and planting

Soil preparation is similar to that for the melon. The seed of early and extra-early varieties is sown in seedbeds or in small peat pots. The seed should be given a pre-germination treatment of soaking in water at 18°C for 1 day, and then drained and left moist at 25°C for 2 days.

When sowing directly or transplanting make 200-240 cm raised tables with two lines of plants, or 80-100 cm tables with a single row of plants. The distance between plants should be 50-60 cm.

Cucumber variety for pickling, "gherkin":
Paris green.
Courtesy of SEMILLAS VILMORIN

The time to sow depends on the production cycle:

- *Extra-early cycle.* The seed is sown in late fall and then transplanted at the beginning of winter.
- *Early cycle.* The seed is sown in late winter.
- *Medium cycle.* The seed is sown in early spring, normally directly into the final site. This cycle is mainly used for gherkins.

12.6.5.4. Cultivation and harvest techniques

- **Thinning**. If the seed is sown directly.
- **Pinching out**. To regulate production.
- **Earthing up**.
- **Weeding**.
- **Staking**.
- **Emasculation**. Remove the male flowers to prevent the formation of abnormal fruit.
- **Harvesting**. The individual fruit should be collected when it is ripe.
Gherkins are collected when they are 6-8 cm long and about 2 cm in diameter.
The time to harvest depends on the cultivation cycle, and may last 30 to 60 days.
- *Extra-early cycle.* Harvesting begins at the end of winter.
- *Early cycle.* Harvesting begins in late spring.
- *Medium cycle.* Harvesting begins in early summer.

- **Sale**. Avoid damage and blows to the fruit as they may cause it to rot later. Cucumbers are sold in wooden or cardboard boxes.
- **Storage**. Cucumbers can be stored in cold chambers at 3°C and 80% humidity for 30 days.

12.6.5.5. The most common pests, diseases and physiological disorders

In general, the pests and diseases that attack the cucumber are the same as those that attack the melon.

12.6.6. Peppers

12.6.6.1. General concepts

The pepper or garden pepper (*Capsicum annuum*) is one of the many plants known as peppers. It is a member of the nightshade family (Solanaceae).

Planting cucumbers

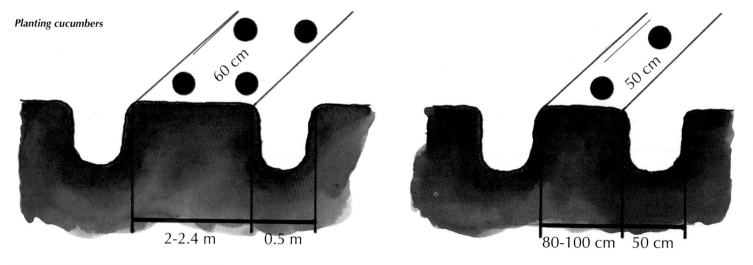

60 cm

2-2.4 m   0.5 m

50 cm

80-100 cm   50 cm

It originated in South America and is used a spice or flavoring for many dishes, and as a coloring agent.

The pungent taste of some varieties of pepper is due to a substance called **capsaicin**, which has digestive and diuretic properties.

The pepper is a biennial herbaceous plant, bearing a fruit that may be red, yellow or green, and varying in shape; square, elongated, round, rectangular, etc.

The seeds retain their ability to germinate for 3-4 years. There are three main groups of varieties:

• **Sweet peppers**, mainly grown in greenhouses:

Valenciano
Cornicabra
Lamuyo
Gedeon
Argos
Sonar
Toledo
Clovis
Jerico
Vidi
Latino
Apolo
Pacific
Dulce
Italiano
Lipari

• **Varieties for paprika production**. These varieties are sweet and ball-shaped.

• **Hot or chili peppers**. They are elongated and very widely cultivated in South America:

Cascabel   Jalapeño
Piquin   Caloro
Long thin Cayenne Serrano chili

Peppers are now widely grown in greenhouses and in tunnels of plastic.

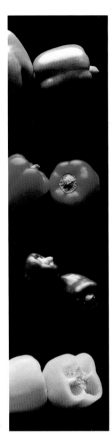

The fruit of the pepper may be red, yellow or green.
Courtesy of SEMILLAS SLUIS & GROOT.

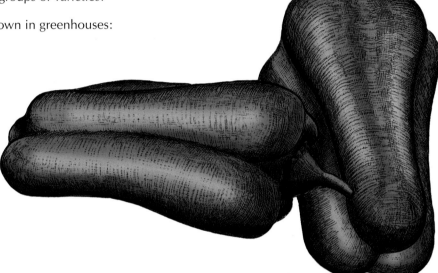

Nutritional composition per 100 g foodstuff				
	Sweet green	Sweet red	Hot green	Hot red
Proteins	1.2 g	1.4 g	2.3 g	2.3 g
Lipids	0.2 g	0.3 g	0.2 g	0.4 g
Sugars	4.8 g	7.1 g	9.1 g	15.8 g
Fiber	1.4 g	1.7 g	1.8 g	2.3 g
Vitamin A	420 IU	4,450 IU	770 IU	21,600 IU
Vitamin B$_1$ (thiamine)	0.08 mg	0.08 mg	0.09 mg	0.1 mg
Vitamin B$_2$ (riboflavin)	0.08 mg	0.08 mg	0.06 mg	0.2 mg
Niacin	0.5 mg	0.5 mg	1.7 mg	2.9 mg
Vitamin C (ascorbic acid)	128 mg	204 mg	235 mg	369 mg
Calcium	9 mg	13 mg	10 mg	16 mg
Phosphorus	22 mg	30 mg	25 mg	49 mg
Iron	0.7 mg	0.6 mg	0.7 mg	1.4 mg
Sodium	13 mg	—	—	25 mg
Potassium	213 mg	—	—	564 mg
Energy value	22 calories	31 calories	37 calories	65 calories

Climatic requirements		
CRITICAL TEMPERATURES	Freezing point	-1°C
	Zero growth	10°C
	Minimum for growth	15°C
	Optimum for growth	20 - 25°C
	Maximum for growth	30°C
NIGHTTIME TEMPERATURE	Optimum	16 - 18 °C
GERMINATION	Minimum temperature	13°C
	Optimum temperature	25°C
	Maximum temperature	40°C
FRUITSET	Minimum temperature	18 - 20 °C
	Optimum temperature	25°C
	Maximum temperature	35°C

12.6.6.2. The plant's requirements

• **Climate and temperature**. It is sensitive to frosts and excessively hot temperatures.
• **Water**. Between 50-70% humidity. Lower humidity levels affect it negatively.
• **Soil**. It requires deep, open rich and well-drained soils.
• **Extractions**. The estimated extraction per hectare is:

$$200 \text{ kg N}$$
$$50 \text{ kg P}_2\text{O}_5$$
$$270 \text{ kg K}_2\text{O}$$

• **Fertilizer application**. One application of 30-40 t/ha of dung.

Fertilizer application at depth, per hectare:

$$100 \text{ kg N}$$
$$90\text{-}150 \text{ kg P}_2\text{O}_5$$
$$200\text{-}300 \text{ kg K}_2\text{O}$$

Surface fertilizer application, per hectare:
4 applications of 40-50 kg of nitrogen, and one application of some K_2O.
When cultivated in greenhouses, these quantities are increased.

12.6.3.3. Soil preparation and planting

Soil preparation is similar to that for tomato cultivation.
The seed is sown in seedbeds or in peat pots, and depending on the time of year, etc., they may need to be protected.
The pepper seeds require a pre-germination treatment in which the seeds are stored moist for 7 days at 18-22°C.
Crops grown for industrial usage are normally sown directly.

The sowing time varies depending on the cultivation cycle:
- *Extra-early cycle*. The seed is sown from late summer onwards, in order to transplant it into a greenhouse in mid fall.
- *Early cycle*. The seed is sown in mid fall and transplanted to tunnels in mid winter.
- *Medium-late cycle*. The seed is sown at a time that means it can be transplanted into the open air without any protection.

12.6.6.4. Cultivation and harvest techniques

• **Thinning**. If sowing is direct.
• **Replacing failures**. If transplanted.
• **Earthing up**.
• **Training pruning**. To eliminate shoots low on the plant.
• **Staking**.
• **Weeding**.
• **Protection**. Using espaliers like those for the tomato, or plastic tunnels.
• **Thinning the fruit**. If the quantity of fruit prevents the plant's normal growth.
• **Application of plant growth regulators**. To encourage early flowering and better fruitset.
• **Harvesting**. Harvest when the fruit is fully ripe.
The time to harvest depends on the crop cycle:
- *Extra-early ctcle*. Harvesting starts in mid winter.
- *Early cycle*. Harvesting starts in mid spring.
- *Medium-late cycle*. Harvesting lasts throughout the summer.

• **Sale**. Once harvested, the peppers are selected, cleaned and packed in boxes for sale.
• **Storage**. Peppers can be stored in cold chambers at 0°C and 85-90% humidity for 30-35 days.

12.6.6.5. The most common pests, diseases and physiological disorders

The pests, diseases and physiological disorders are almost exactly the same as for the tomato. There may also be problems of flower fall and fruit shed due to very high temperatures accompanied by low humidity levels.

12.6.7. Watermelon

12.6.7.1. General concepts

The watermelon (*Citrullus lanatus* also known as *C. vulgaris*) is a member of the cucumber family (Cucurbitaceae).

It originated in tropical Africa and it is eaten fresh. The watermelon is an annual plant with fruit that are berries, normally globose, containing pinkish or red flesh. The rind is normally smooth and may be light or dark green.

Nutritional composition per 100 g foodstuff	
Proteins	0.5 g
Lipids	0.2 g
Sugars	6.4 g
Fiber	0,3 g
Vitamin A	590 IU
Vitamin B$_1$ (thiamin)	0.03 mg
Vitamin B$_2$ (riboflavin)	0.03 mg
Niacin	0.2 mg
Vitamin C (ascorbic acid)	7 mg
Calcium	7 mg
Phosphorus	10 mg
Iron	0.5 mg
Sodium	1 mg
Potassium	100 mg
Energy value	26 calories

Variety of watermelon: **Sugar Baby**. *Courtesy of SEMILLAS VILMORIN*

The seeds are flattened and vary in color from white through brown to black, and retain their ability to germinate for 5 years.

Watermelons are classified on the basis of their crop cycle into:

• Early varieties with a cycle of 75-80 days:

Spherical fruit:

> Sugar Baby
> Black Pearl
> Panonia
> Rocio
> Fabiola
> Valentina
> Rubin

Elongated fruit:

> Striped Klondyke
> Prince Charles

• Medium-late varieties with a cycle of 90-110 days:

Spherical fruit:

> Pileña
> Sayonara
> Ali
> America Sweet

Elongated fruit:

> Fairfax
> Congo
> Charleston Gray

12.6.7.2. The plant's requirements

In general, the plant's climatic requirements are the same as those of the melon. The plant is sensitive to frosts and resistant to dry conditions.

• Soils. It requires loamy-sandy soils rich in organic matter, well aired and best with a pH between 6 and 7.4.

• Extractions. The extractions per hectare are estimated to be:

> 50 kg N
> 15 kg P$_2$O$_5$
> 65 kg K$_2$O

• Fertilizer application. Apply 25-30 t/ha of dung.

Fertilizer application at depth:

> 30 kg N
> 90-120 kg P$_2$O$_5$
> 100-125 kg K$_2$O

Surface fertilizer application: 3 applications of 20-30 kg of nitrogen.

• Deficiencies: It is sensitive to shortages of magnesium in the soil.

Climatic requirements		
CRITICAL TEMPERATURES	Freezing point	0°C
	Zero growth	11 - 13°C
	Optimum for growth	23 - 28°C
	Optimum for flowering	18 - 20°C
GERMINATION	Minimum temperature	13°C
	Optimum temperature	25°C
	Maximum temperature	45°C
HUMIDITY		MEDIUM
LIGHT		HIGH

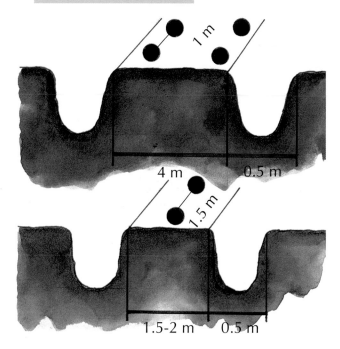

*12.6.7.3. Soil preparation
and planting*

**Planting tomatoes.
Tomato. Varieties**
A/ *Smooth skin*
B/ *Ribbed*

Soil preparation is similar to that of the melon. The seed of early varieties can be sown in protected seedbeds. Before sowing, the seeds require a pre-germination treatment of being soaked for one day and then drained and stored moist at 25°C for 2 days.
Seed should be directly sown and seedlings transplanted into 4 m raised tables with 2 rows, or 1.5-2 m tables with a single row. The distance between plants should be 1-1.5 m.

The planting times as well as the cultivation cycles, coincide with those of the melon.

*12.6.7.4. Cultivation and harvest
techniques*

- **Thinning**. If seed is sown directly.
- **Pruning**. Leaving 3 main branches.
- **Removing any defective fruit**.
- **Earthing up**.
- **Weeding**.

- **Staking**. Especially when grown in greenhouses.
- **Harvesting**. The harvesting times coincide with those of the melon.
- **Sale**. In wooden or cardboard boxes.
- **Storage**. Watermelons can be stored in cold chambers at 2-4°C and 85-90% humidity for 25 days.

*12.6.7.5. The most common pests, diseases
and physiological disorders*

They are the same as those discussed under the melon.

12.6.8. Tomato

12.6.8.1. General concepts

The tomato (*Lycopersicon esculentum*) is a member of the nightshade family (Solanaceae).

It originated in the Americas and is grown as an annual, though it may live several years. The aboveground plant is covered with hairs and has a distinctive smell.
The tomato's fruit is a globose berry, normally red when ripe, with a surface that may be smooth or ribbed.

The seeds retain their ability to germinate for 3 to 6 years.
There are 3 different production cycles:

- *Short cycle*, 90-110 days
- *Medium cycle*, 100-120 days
- *Long cycle*, 110-125 days.

There is a huge number of tomato varieties on the market, and so it is worth investigating in detail before deciding which variety to grow. Below is a list of tomato varieties classified by their use:

- **To use fresh:**

Candela	Zafiro
Marmande	Amatista
Cuarentena	Feria
Catala	Valenciano
Moneymaker	Montfauet
Vemone	Pyros
Quatuor	Amigo
Monita	Lucy
Motabo	Montecarlo
Monalbo	Fandango
Precador	Flamingo
Fusor	Bonset
Etna	Meltine
Cintra	Corindo
Tobol	Turmalina
Sonato	Dombo
Tango	Dombito
Carmelo	Dombelo
Diego	Melody
Primset	Dario
Bonset	Hymar
Robin	Maindor

• **For industrial use:**

Earlired	Redstone
San Marzano	Slumac
Roma	Napoli
Romano	Julimac
Ace	Río Grande
Arno	Hypeel
Campbell	Petomech
Heinz	Pearson
Mecano	Roforto
Ventura	Walter
Romulus	California
Esla	Ropreco

Tomatoes are widely cultivated in greenhouses, because this is profitable. Tomatoes may be cultivated in simple plastic tunnels or in highly sophisticated greenhouses.

12.6.8.2. The plant's requirements

• **Climate and temperature**. It requires a hot climate, with some change between day and night. Low temperatures seriously affect the plant.
• **Water**. Environmental humidity is important, especially at the time of pollination, and should ideally be 55-60%.

Nutritional composition per 100 g foodstuff	
Proteins	1 g
Lipids	4 g
Sugars	0.2 g
Vitamin A	1,700 IU
Vitamin B$_1$ (thiamine)	0.1 mg
Vitamin B$_2$ (riboflavin)	0.02 mg
Niacin	0.6 mg
Vitamin C (ascorbic acid)	2.1 mg
Calcium	13 mg
Phosphorus	27 mg
Iron	0.5 mg
Sodium	3 mg
Potassium	244 mg
Energy value	23 calories

Climatic requirements		
CRITICAL TEMPERATURES	Freezing point	-2°C
	Zero growth	10 - 12°C
	Minimum for growth	15 - 17 °C
	Optimum for growth	20 - 24°C
	Maximum for growth	30°C
SOIL TEMPERATURE	Minimum	12°C
	Optimum	20 - 24 °C
	Maximum	34°C
GERMINATION	Minimum temperature	10°C
	Optimum temperature	25 - 30°C
	Maximum temperature	35°C
FLOWERING	Day temperature	23 - 26 °C
	Night temperature	15 - 18 °C
RIPENING	Optimum temperature	15 - 22 °C
MOISTURE		MEDIUM
LIGHT		HIGH

*Tomato variety for industrial use: **Roma**. Courtesy of SEMILLAS VILMORIN*

The tomato plant is sensitive to insufficient and excess water.
• **Soils**. It requires siliceous-clay loose, deep, well-drained soils that are rich in organic matter. The best pH value is between 6 and 7. It requires a lot of calcium, potassium and magnesium.
• **Extractions**: The extractions vary depending on the variety cultivated and yield obtained.

$$150\text{-}300 \text{ kg N}$$
$$55\text{-}96 \text{ kg P}_2\text{O}_5$$
$$250\text{-}680 \text{ kg K}_2\text{O}$$

• **Fertilizer application**. When cultivating in the open air, a single application of 30 t/ha of dung.

Fertilizer application at depth, per hectare:

$$60 \text{ kg N}$$
$$80\text{-}100 \text{ kg P}_2\text{O}_5$$
$$200\text{-}250 \text{ kg K}_2\text{O}$$

Surface fertilizer application, per hectare:

$$3 \text{ applications of } 90 \text{ kg N}$$
$$1 \text{ application } 20 \text{ kg P}_2\text{O}_5$$
$$90 \text{ kg K}_2\text{O}$$

For greenhouse cultivation, these quantities have to be increased and divided into a larger number of applications, in order to increase production.

12.6.8.3. Soil preparation and planting

To prepare the soil, first pass the subsoil plow, then harrow several times to leave the surface a fine tilth. You are recommended to apply an insecticide together with the fertilizer applied at depth, in order to prevent attacks by soil larvae.
The seed is either sown in seedbeds or in small pots, discs of pressed peat, to ensure the seedlings have a good root ball. In any case, the seedbeds should be protected by tunnels covered with plastic sheeting. You are recommended to give a pre-germination treatment of keeping the seeds moist at 20°C for 5-6 days.
The seedlings should be transplanted into furrows 80-120 cm apart, leaving a distance of 30-50 cm between plants. If the tomatoes are for industrial use, the seed can be sown directly, with a smaller distance between furrows and a separation between plants of 25-35 cm.

The time to sow depends on the cultivation cycle of the variety grown.

- *Extra-early cycle*. The seed is sown in early fall and the seedlings transplanted into a greenhouse at the end of fall.
- *Early cycle*. The seed is sown in the middle of fall, and the seedlings transplanted in the middle of winter to the open air, with protection.
- *Medium cycle*. The seed is sown at the beginning of winter, and the seedlings transplanted once the danger of frosts has passed.
- *Late cycle*. The seed is sown at the end of spring and the beginning of summer, and the seedlings transplanted at the end of summer.

12.6.8.4. Cultivation and harvest techniques

- **Thinning**. If the seed is sown directly.
- **Replacing failures**. If the seedlings are transplanted.
- **Earthing up**. This is done four weeks after transplanting.
- **Pruning**.
- **1st training pruning**. Leave a single main stem if early production is desired, or 2-3 main branches for normal cultivation.
- **2nd training pruning**. Remove the side shoots and the old leaves.
- **Pinching out**. Pinch out the terminal buds on the main stems. This restricts the plant's vertical growth.
- **Staking**. To ensure the plant grows upright, and the fruit does not come into contact with the soil. This is done using canes, wooden boards, or belts and wires in greenhouses.
- **Weeding**. Use selective herbicides.
- **Encouraging fruitset**. This is done to achieve early fruitset, and uses several techniques.
    — *Application of plant growth regulators to the inflorescences. The substances used are normally auxins, and cause the fruit to develop parthenocarpically. They are applied by wetting the inflorescences once or twice. Take great care to apply the correct dose, as an excess may be toxic to the tomato.*
    — *Mechanically shaking the inflorescences. This method is widely used in greenhouses. Shaking the inflorescences means that more pollen is released.*
- **Protection**. To protect the plant against cold winds and mild frosts, especially in early varieties. This is done by placing espaliers in the soil at an angle in the side of the furrow opposite the plant.
- **Harvesting**. The fruit is harvested when it is ripe. The fruit may be any of 3 different colors:

- Mature green, a light green.
- Turning, which is turning to red
- Ripe red, which is deep red.

If the tomatoes are to be eaten fresh, they are harvested manually, several times, and collecting those that are green or turning. If the tomatoes are for industrial use, harvesting can be mechanized, and collection is when the fruit are deep red.

The time to harvest depends on the cultivation cycle:

- *Extra-early cycle*. Harvesting starts in the middle of winter.
- *Early cycle*. Harvesting starts at the beginning of spring.
- *Late cycle*. Harvesting is from late summer to mid winter.

- **Sale**. After harvesting, the tomatoes are selected, washed and placed in boxes for transport to the market.
- **Storage**.
- *Ripe red*. They can be stored at a temperature of 5°C and 95% humidity for 10-15 days.
- *Ripe green* or *turning*. They can be stored at 10-12°C for 30 days.

12.6.8.5. The most common pests, diseases and physiological disorders

- **Pests:**

- *Tomato grub*. The larvae make holes in the tomatoes, when they penetrate them.
- *Cutworms* (*Agrotis*). The larvae attack the base of the stem, causing the plant to wilt.
- *Screwworms*. The larvae damage the fruit and leaves.
- *Wireworms*. The larvae attack the root system.
- *Leaf miner*. They mine galleries within the leaf.
- *Whitefly*. They especially attack greenhouse tomatoes, weakening the plant.
- *Aphids*.
- *Tomato bug*. Causes patches to form and deformations of the leaf.
- *Tomato mites*. They cause shiny brown patches to form on the underside of the leaf, and the leaf then withers and falls.
- *Red spider mites*. The upper surface of the leaf turns yellow.
- *Diseases caused by nematodes*. Nematodes normally attack the roots of the tomato, causing yellowing and generally stunted growth.

- **Fungal diseases:**

- *Vascular problems*. This can be caused by *Fusarium* and *Verticillium* wilts. They cause general necrosis of the roots and wilting of the plant.
- *Tomato mildew*. This causes yellow patches on the leaves, which then wilt.
- *Alternaria* of tomato. This causes dark round patches with a yellow edge.
- *Cladosporium* of tomato. A grey mold forms on the underside of the leaves. This mainly occurs in greenhouses.

- *Tomato anthracnose.* This causes circular necrotic patches on the leaves and fruit.
- *Tomato powdery mildew.* This causes yellow blotches on the leaves, which then die.
- *Botrytis.* This causes a soft rot of the fruit. This usually only occurs in greenhouses.

• **Bacterial diseases:**

- *Stem canker.* This causes rapid wilting of the leaves and splitting of the stems.

• **Viral diseases:**

- *Tomato mosaic virus.* This causes a yellowish mosaic on the leaves, leading to a halt in growth and irregular ripening of the fruit.
- *Thread mosaic virus.* This causes deformations in the leaves, as well as mosaics on the leaves.
- *Cucumber mosaic virus.*

• **Physiological disorders:**

- *Apical necrosis.* Caused by irregular watering or excessive salinity.
- *"Cracking" of the fruit.* This is mainly the result of irregularities in water supply, or the combination very high temperatures with low humidity.
- *Hollow fruit.* This is caused by poor pollination and fertilization, or excessive doses of plant growth regulators to promote fruitset.
- *Toxic effects.* Those caused by an excess of plant growth regulators may cause malformations in the fruit.
- *Scalding.* Whitish patches form over watery tissues. This is due to excess sunshine.
- *Shedding of flowers or fruit.* This is due to very high temperatures accompanied by low humidity. It may also be the result of low temperatures.

12.6.9. Other crops

12.6.9.1 Gourds

Gourds, also called squashes, are members of several different species, all of them members of the cucumber family (Cucurbitaceae).
They originated in Central and South America.
They are mainly used to make desserts, though the seeds are also eaten as such.

The best known varieties of gourd are:

Spanish green
Dulce de Horno
Buttercup
Cabello de Angel
Totanera

In terms of their climatic requirements, the plant requires a lot of warmth, and it is sensitive to low temperatures. It needs a rich cool soil in good tilth. It does not tolerate excess soil moisture, and the best pH value is 6.

The seed is sown early in spring, and the crop is harvested 6 months later, normally in early fall.
Before storage, they should be dried in the sun. They can be stored at 8-12°C for several months.
The pests, diseases and physiological disorders are similar to those that affect other members of the cucumber family (Cucurbitaceae).

12.6.9.2. Okra

Okra (*Hibiscus esculentus*) is a member of the mallow family (Malvaceae). It originated in tropical Africa, and may be eaten boiled or fried.
Okra is an annual plant that bears elongated capsules up to 30 cm long. Its cultivation cycle is 80-140 days.
In terms of its climatic requirements, this tropical plant requires a tropical climate and is very sensitive to low temperatures. It requires a clay soil rich in organic matter.

12.6.9.3. Papaya

The papaya or pawpaw (*Carica papaya*) is a member of the family Caricaceae.
The papaya is dioecious, meaning that there are male plants bearing male flowers and separate female plants with female flowers. Only the female plants bear fruit.

Varieties of gourd:
Buttercup.
Courtesy of SEMILLAS VILMORIN

Peas attacked by different insects

The fruit is a spherical berry with smooth skin, green or yellow on the outside, and yellow-orange within. It requires a tropical climate, and is very sensitive to low temperatures. The soil should be rich in organic matter and have good drainage.

12.6.9.4. Pineapple

The pineapple (*Ananas comosus*) is a member of the bromeliad family (Bromeliaceae).
It produces an inflorescence at the tip of its stem that will turn into the fruit. The "fruit" consists of a large number of small fused fruit, covered by a tough skin, and known as the pineapple.
Fruiting starts 1 or 2 years after planting.
In terms of its climatic requirements, it cannot withstand low temperatures, and it requires open, cool, well-drained soils rich in organic matter.

12.7. CROPS GROWN FOR THEIR SEEDS

12.7.1. Peas

12.7.1.1. General concepts

The pea (*Pisum sativum*) is a member of the legume family (Leguminosae).
It originated in Europe. The seeds may be dried or eaten fresh, but horticultural production is for fresh consumption.
The fruit is a legume, a pod, containing the more-or-less spherical seeds, which may be smooth or wrinkled. The seeds retain their ability to geminate for about 3 years.
The cultivation cycle lasts 3-5 months. Peas are classified into 3 groups on the basis of their growth:

• Dwarf varieties, for industrial use:

Alaska	Allegro
Arabal	Asterix
Coronet	De Grace
Frescoroy	Frila
Heralda	Hybris
Negret	Nugget
Oreste	Orfeo
Profino	Proval
Recette	VilVence
Almirex	Oberon
Aubine	Precovil
Fabina	Rag

Varieties of pea. Courtesy of SEMILLAS VILMORIN

Gloria de Quimper	Voluntario
Kalife	

• Climbing varieties, to be eaten fresh:

Lincoln	Herald
Onward	Rey de los Carouby
Carouby de Maussane	Rey de las Aldot
Cuerno de las Aldot	Miragreen
Gradus	

• Harvests:

Express generoso	Senador
Serpette	Teléfono
Alderman	

Nutritional composition per 100 g foodstuff	
Proteins	6.3 g
Lipids	0.4 g
Sugars	14.4 g
Fiber	2 g
Vitamin A	640 IU
Viamin B$_1$ (thiamine)	0.35 mg
Vitamin B$_2$ (riboflavin)	0.14 mg
Niacin	2.9 mg
Vitamin C (ascorbic acid)	27 mg
Calcum	26 mg
Phosphorus	116 mg
Iron	1.9 mg
Sodium	2 mg
Potassium	316 mg
Energy value	84 calories

Climatic requirements		
CRITICAL TEMPERATURES	Freezing point	-3 - -4°C
	Zero growth	5 - 7°C
	Minimum for growth	10°C
	Optimum growth	16 - 20°C
	Maximum for growth	35°C
GERMINATION	Minimum temperature	6°C
	Optimum temperature	14 - 25°C
	Maximum temperature	30°C
HUMIDITY		MEDIUM
LIGHT		HIGH

12.7.1.2. The plant's requirements

• Climate and temperature. It requires a moist temperate climate. It is sensitive to frosts and to very high temperatures.
• Soils. It requires a soil with an average texture, light, cool and with good drainage. The optimum pH value is between 6 and 7.
• Extractions. The extractions per hectare are estimated to be:

$$125 \text{ kg N}$$
$$45 \text{ kg P}_2\text{O}_5$$
$$90 \text{ kg K}_2\text{O}$$

•**Fertilizer application**. Like all the legumes, peas form nodules on their roots that are home to symbiotic *Rhizobium* bacteria that fix atmospheric nitrogen. Peas require a moderate application of well-rotted dung.

Fertilizer application at depth, per hectare:

20-30 kg N
50-80 kg P_2O_5
100-140 kg K_2O

12.7.1.3. Soil preparation and planting

The soil should be deep plowed once, then harrowed several times to ensure it is loose and in good tilth.
The seed can be sown directly into the final site, or in protected seedbeds for the early varieties.
Whether the seed is sown directly or into seedbeds, the seedlings are placed in furrows at a distance of 80-90 cm for climbing varieties, and 50-60 cm for dwarf varieties.
The plants should be 70-80 cm apart in climbing varieties and 40-60 cm for shorter ones.
The time to sow depends on the variety's growing cycle, and is between fall and the following spring. The varieties may be extra-early, early, medium and late.
The dates are similar to those given for the runner bean.

12.7.1.4. Cultivation and harvest techniques

• **Weeding**. Manually or with herbicides.
• **Thinning**. If the seed is sown directly.
• **Staking**. In the climbing varieties.
• **Harvesting**. The varieties grown to be eaten fresh are picked by hand, while those intended for industrial use are generally harvested mechanically.
• **Sale**. Varieties grown to be eaten fresh are sold in transparent polythene bags.
• **Storage**. Peas can be stored in cold chambers at 10°C and 85% humidity for 20 days.

21.7.1.5. The most common pests, diseases and physiological disorders

• **Pests:**
- *Whitefly*. In greenhouses. They cause general weakening of the plant.
- *Greenfly*.
- *Sitona* weevil. The larvae destroy the nodules on the roots.
- *Pea weevil*. It mines galleries in the pods and destroys the seeds.
- *Red spider mites*. Its effects are similar to those on the runner bean.
- *Pea thrip*. Their bites cause deformations of the leaves, which turn a silvery color.
- *Leafminers* (*Agromyzidae*). The larvae mine galleries in the leaves.
- *Snails and slugs.*

• **Fungal diseases:**
- *Pea anthracnose*. This causes brown patches with a yellowish center on both leaves and pods.

- *Pea rust*. This causes patches that are yellow on the leaf blade and brown on the underside.
- *Diseases of the plant's collar*. The attack is similar to that in the runner bean.

• **Viral diseases:**
- Pea virus I. This causes swellings on the leaves, accompanied by a characteristic mosaic.
- Pea virus II. This causes a green-yellow mosaic on the leaves.

12.7.2. Broad beans

12.7.2.1. General concepts

The broad bean (*Vicia faba*) is a member of the legume family (Leguminosae).
It originated in the Mediterranean area. The unripe and tender seeds are the part consumed, either fresh or preserved.
The broad bean is an annual plant whose roots bear nodules containing *Rhizobium* bacteria that can fix atmospheric nitrogen.
The fruit is a pod, a legume, which contains the seeds. The seeds retain their ability to germinate for 4 years.

The best known varieties of broad bean are:

Agua dulce	Muchamiel
Comprimo	Ramillete
Granadina	Segureña
Mahon	Beryl
Primabel	De Sevilla
Reina Mora	Histal
Aranjuez	Aquitaine early
Sicilian Common	Reina Blanca
Windsor bean	

*Varieties of broad bean: **Agua dulce** Courtesy of SEMILLAS VILMORIN*

Nutritional composition per 100 g foodstuff	
Proteins	7.9 g
Lipids	0.4-0.7 g
Sugars	17-20 g
Fiber	2.75 g
Vitamin A	200 IU
Vitamin B$_1$ (thiamine)	0.3 mg
Vitamin B$_2$ (riboflavin)	0.18 mg
Niacin	1.80 mg
Vitamin C (ascorbic acid)	25 mg
Calcium	105 mg
Potassium	1,390 mg
Phosphorus	600 mg
Magnesium	240 mg
Copper	3 mg
Iron	2 mg
Energy value	25 calories

Climatic requirements		
CRITICAL TEMPERATURES	Freezing point	-5°C
	Zero growth	6 - 8°C
	Minimum for growth	8 - 10°C
	Optimum growth	18 - 22°C
	Maximum for growth	35°C
GERMINATION	Minimum temperature	7°C
	Optimum temperature	12 - 20°C
	Maximum temperature	30°C
HUMIDITY		MEDIUM
LIGHT		MEDIUM

12.7.2.2. The plant's requirements

• **Climate and temperature**. It is a plant that grows in the fall and which is negatively affected by high temperatures.
• **Water**. The plant is sensitive to lack of water.
• **Soil**. It requires clay soils with good water retention. The best pH values are between 6.5 and 7.5.
• **Extractions**. Extraction is estimated to be:

$$120 \text{ kg N}$$
$$30 \text{ kg P}_2\text{O}_5$$
$$80 \text{ kg K}_2\text{O}$$

• **Fertilizer application**. If the soil is poor, apply 10-15 t/ha of dung well in advance.

Fertilizer application at depth:

$$20\text{-}30 \text{ kg N}$$
$$65\text{-}80 \text{ kg P}_2\text{O}_5$$
$$90\text{-}150 \text{ kg K}_2\text{O}$$

12.7.2.3. Soil preparation and planting

The soil should be prepared between the end of summer and the beginning of fall, though it may be extended into the fall. In areas with very cold winters, the seed is sown in spring.

Before sowing the seed, you are advised to carry out a pre-germination treatment of immersion in warm water for 24 hours.

The seed is sown directly into the final site, in furrows 50-60 cm apart, with a distance of 30-40 cm between plants.

12.7.2.4. Cultivation and harvest techniques

Cultivation is simple, and is generally limited to weeding and earthing up.

• **Harvesting**. Broad beans to be eaten fresh are normally collected by hand, collecting when the pod reaches 3/4 of its normal size.
• **Storage**. They store best at a temperature of 0-1°C and 85-95% humidity.

12.7.2.5. The most common pests, diseases and physiological disorders

• **Pests:**

- *Sitona* weevils. The larvae destroy the root nodules and the adults attack the leaves.
- *Blackfly of beans*. The plants turn yellow and the leaves curl up.
- *Pea thrip*. Similar to the attack on the pea.
- *Weevils*.
- *Snails and slugs*.

• **Fungal diseases:**

- *Bean mildew*. This causes patches at the edges of the leaves, which then wilt.
- *Bean rust*. Causes small brown patches on the underside of the leaf.
- *Sclerotinia*. Similar to that described for the runner bean.
- *Parasites*
- *Broomrape* (*Orobanche*). This is a root parasite which attacks the roots and takes their nutrients.

12.7.2.3. Runner beans

12.7.3.1. General concepts

The runner bean (*Phaseolus vulgaris*) is a member of the legume family (Leguminosae).

It originated in Central America. It can be consumed in two ways; the seed (mature or immature) may be eaten, or the entire immature pod may be eaten. In horticulture, runner beans are grown for their pods, while in extensive cultivation they are grown for the beans.

The runner bean is an annual plant whose fruit is a legume, which may be white, green or streaked. The seeds retain their ability to germinate for 3-4 years.

The varieties are divided into 2 main groups:

• **Low-growing varieties**, which are mainly grown for industrial use. They are usually cultivated in association with other vegetable crops.

- Yellowish-white pod:

Constanza	Monte de Oro
Kinghorn	Gamalan
Findor	Orbane Saxa Gold

- Green pod:

Arian	Astra
Belna	Blulect
Constant	Eagle
Gitana	Groffy
Kora	Lit
Radar	Ranger

Nutritional composition per 100 g foodstuff	
Proteins	1.9 g
Lipids	0.2 g
Sugars	7.1 g
Fiber	1 g
Vitamin A	600 IU
Vitamin B$_1$ Thiamine)	0.08 mg
Vitamin B$_2$ (riboflavin)	0.11 mg
Niacin	0.5 mg
Vitamin C (ascorbic acid)	19 mg
Sodium	7 gr
Potassium	132 mg
Calcium	56 mg
Iron	0.8 mg
Phosphorus	4.4 mg
Energy value	32 calories

Skil	Strike
Athena	Processor
Bountiful	Reginel
Garrafal	Tivoli
Janus	

- Streaked pod:

Faraybel	Marbel

• **Climbing varieties**, which are mainly grown to be eaten fresh. They are normally grown in greenhouses.

- Yellowish-white pod:

Cascada	Aragon Butter

Climatic requirements		
CRITICAL TEMPERATURES	Freezing point	1°C
	Zero growth	8 - 10°C
	Minimum for growth	10 - 12°C
	Optimum growth	18 - 30°C
	Maximum for growth	35 - 40°C
GERMINATION	Minimum temperature	12°C
	Optimum temperature	15 - 25°C
	Maximum temperature	30°C
FLOWERING	Minimum temperature	12 - 15°C
	Optimum temperature	15 - 25°C
	Maximum temperature	30 - 40°C
HUMIDITY		MEDIUM
LIGHT		MEDIUM

Rhine Gold	Oro de Verna
Torrente de Oro	San Fiacre
Algerian Garrafal	

- Green pod:

Bertina	Potomac
Esmeralda	Zondra
Buenos Aires, or Parade Lindra	Meru
Diamond	Perona
Fenomeno	Promo
Helda	Climbing Garrafal de Oro
Kadi	Climbing Pevir
Perfeccion	Remo

- Streaked pod:

Buenos Aires Gimenez	Prague Coco

12.7.3.2. The plant's requirements

• **Climate and temperature**. It requires a hot climate. Sharp changes in temperature cause the pods to form twisted and stringy.
• **Water**. The runner bean requires a lot of water.
• **Soils**. It requires loose, open, soils with good drainage. It grows well on soils with a pH

*Variety of climbing runner bean: **Buenos Aires**. Courtesy of SEMILLAS VILMORIN*

Harvester for potatoes, onions, carrots and chicory

between 5.5 and 7, and is sensitive to salinity and to lime.
• **Extractions**. It is estimated to extract, per hectare:

$$120 \text{ kg N}$$
$$35 \text{ kg P}_2\text{O}_5$$
$$90 \text{ kg K}_2\text{O}$$

• **Fertilizer application**. Runner beans can obtain some nitrogen from the air as well as from the soil, because they have nitrogen-fixing *Rhizobium* bacteria in nodules on their roots. All legume vegetables can fix nitrogen.
In general, dunging should consist of a single application of 15-20 t/ha of well-rotted dung.

Fertilizer application at depth, per hectare:

$$50\text{-}60 \text{ kg N}$$
$$60\text{-}80 \text{ kg P}_2\text{O}_5$$
$$125\text{-}150 \text{ kg K}_2\text{O}$$

• **Deficiencies**: It is sensitive to deficiency of magnesium, manganese and zinc.

12.7.3.3. Soil preparation and planting

Soil preparation requires a deep plowing, followed by the surface tilling needed to ensure the soil is in good tilth.
The seed can be sown directly, or into protected seedbeds for early varieties.
Seed should be sown and seedlings transplanted into furrows 50-60 cm apart (low growing varieties) or 80-100 cm apart (climbing varieties). The distance between plants should be 30-40 for low growing plants and 65-70 cm for climbers.
The sowing time depends on the cultivation cycle of the variety.

- *Extra-early cycle*. The seed should be protected and sown in late fall/early winter.
- *Early cycle*. The seed is sown between mid and late winter.
- *Medium cycle*. The seed is sown early in spring, once the risk of frost has passed.
- *Late cycle*. The seed is sown in early or mid summer.
- *Extra-late cycle*. The seed is sown in late summer/early fall.

12.7.3.4. Cultivation and harvest techniques

• **Thinning**. If the seed is sown directly.
• **Mulching**. If early cycle varieties are sown directly into the final site, the soil should be mulched and holes left for the seeds to emerge.
• **Weeding**. Manually or with herbicides.
• **Supports**. Only in the climbing varieties.
• **Harvesting**. Low growing plants for industrial use are harvested mechanically, while runner beans to be eaten fresh are harvested manually.

The harvest period depends on the variety's cultivation cycle:

- *Extra-early cycle*. Harvest starts late in winter.
- *Early cycle*. Harvest starts in mid spring.
- *Medium cycle*. Harvest starts at the end of spring.
- *Late cycle*. Harvest starts at the end of summer.
- *Extra-late cycle*. Harvest starts at the end of fall.

• **Sale**. Runner beans are sold in plastic bags (varying in size).
• **Storage**. Runner beans can be stored at 2-4°C and a humidity of 85%.

12.7.3.5. The most common pests, diseases and physiological disorders

• **Pests:**

- *Bean seedfly*. This causes germination to be irregular, and leaves the plant susceptible to fungal attack.
- *Leafminers*. The larvae mine galleries within the leaves.
- *Screwworms*. Leaf eaters.
- *Aphids*. General yellowing of the plant and drooping of the leaves.
- *Greenhouse whitefly*. They cause general weakening of the plant.
- *Red spider mite*. The leaves and pods turn yellow then red.
- *Weevils*. They attack the seeds stored for sowing.
- *Snails and slugs*.
- *Diseases caused by nematodes*. They cause cysts to form on the roots, and the aboveground parts turn yellow and their growth is stunted.

12.7.4. Other crops

12.7.4.1. Sweet corn

Sweet corn (*Zea mays*) is a member of the grass family (Gramineae).
It belongs to the same species as corn, but has a much higher sugar content.
It is mainly eaten fresh, but is also preserved and frozen.
The most important varieties, classified by their cultivation cycle, are:

- **Early varieties** (less than 75 days):

    Aztec
    Goldcrest
    Comanche
    Marcross

- **Early varieties** (75-85 days)

Apache	Bonanza
Comet	Cherokee
Guardian	Merit
Snowbelle	

- **Medium varieties** (more than 85 days)

    Lobelle
    Market
    Lochief
    Valley

The effect of boron deficiency on the potato and corn cobs ready for harvest

- **Fungal diseases:**

- *Foot and root rot.* This can be cased by several fungi, such as *Rhizoctonia*, *Fusarium*, and *Pythium*.
- *Anthracnose.* Black patches form near the leaf veins.
- *Sclerotinia.* This causes patches at the base of the plant that give rise to soft rots.
- *Bean rust.* This cause small patches that are yellow on the upper surface and dark on the underside.
- *Botrytis* or *grey mold.*

- **Bacterial diseases:**

- *Bean oily patch.* This causes oily patches to form on the leaves and pods.

- **Viral diseases:**

- *Common bean mosaic virus.* Causes a mosaic of light and dark green to form on the leaves.
- *Yellow mosaic virus.*
- *Young plant necrosis virus.*
- *Pod formation necrosis virus.*

- **Physiological disorders:**

- *Pod aging or discoloring.* This occurs in the cold storage of produce from old plants.
- *Stringy pods.* This is due to low temperatures.
- *Shedding of flowers and young pods.* Caused by very high temperatures and very low humidity.

The seed is sown in spring in furrows 70-90 cm apart, with a distance between plants of 20-25 cm. Harvesting may be manual or mechanical, and is when the seeds are fully developed. Once they have been collected, the surrounding bracts ("shucks") are removed and the cobs are placed in trays covered by a sheet of transparent plastic.
They can be stored at 0°C and 85-90% humidity.

*New varieties of sweet corn, such as **Dynasty F₁ Hybrid**, are constantly appearing on the market. This is a selection from SEMILLAS SLUIS & GROOT.*

13. "IN VITRO" CULTIVATION IN HORTICULTURE

"In vitro" cultivation is the cultivation is sterile conditions of plants, seeds, and plant embryos, tissues and cells.
"In vitro" literally means "in glass", because this type of experiment is performed in glass vessels, such as test tubes.

The main features of "in vitro" cultivation are:

• It is on a very small scale, with a relatively small area.
• Environmental conditions such as humidity, temperature and nutrition are optimized.
• All the microorganisms (fungi, bacteria and virus) as well as pests (insects and nematodes) are excluded.

"In vitro" cultivation has to be aseptic, meaning that the tissues have to be sterilized in advance and removed without contaminating them, and the cultivation conditions must ensure the tissue cultures are not exposed to any possible contamination.

This technique has given rise to great expectations, but it is only economically worthwhile for a small number of plant species to be grown this way for purposes of production.

This technique requires a large initial investment and has to be performed by specialized technicians. The running costs are also high, including laboratory materials, electricity and a skilled labor force.

Every living plant cell is capable of reproducing the entire plant from which it was taken. That is to say, a whole plant can be grown from any portion of tissue or single cell, or even a pollen grain.

This growth depends on a series of factors, such as

Five stages in the micro-propagation process
Stage 0: *Growing and preparing the parent plant*
Stage 1: *Placing and establishing the tissues or organs in sterile conditions*
Stage 2: *Growth phase*
Stage 3: *Rooting of the micro-cuttings in sterile conditions*
Stage 4: *Adaptation of the micro-plants to normal cultivation conditions*

Stage 1

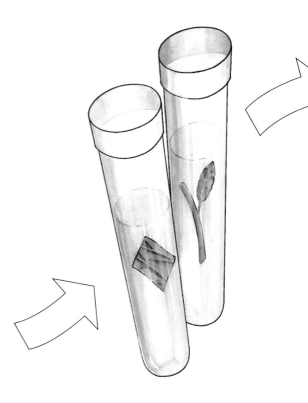

Stage 0

the species, the culture media and the environmental conditions, and this gives rise to a series of problems when using micro-propagation techniques.
The method most widely used in horticulture is meristem culture, and it is mainly used to obtain plants that are free of virus. The meristem is the tissue at the stem tip where the cells continue to divide and thus grow.
The meristem has to be removed from the parent plant in totally aseptic conditions, in special "laminar air flow" chambers designed to ensure a sterile environment.

The tissue is sterilized by immersion in a solution of sodium hypochlorite and is then washed several times in distilled water.
The meristem is then placed in a test tube containing a suitable culture medium. This culture medium consist of water, mineral salts, carbohydrates, vitamins, hormones and a gel, usually **agar-agar**, to solidify the ingredients.

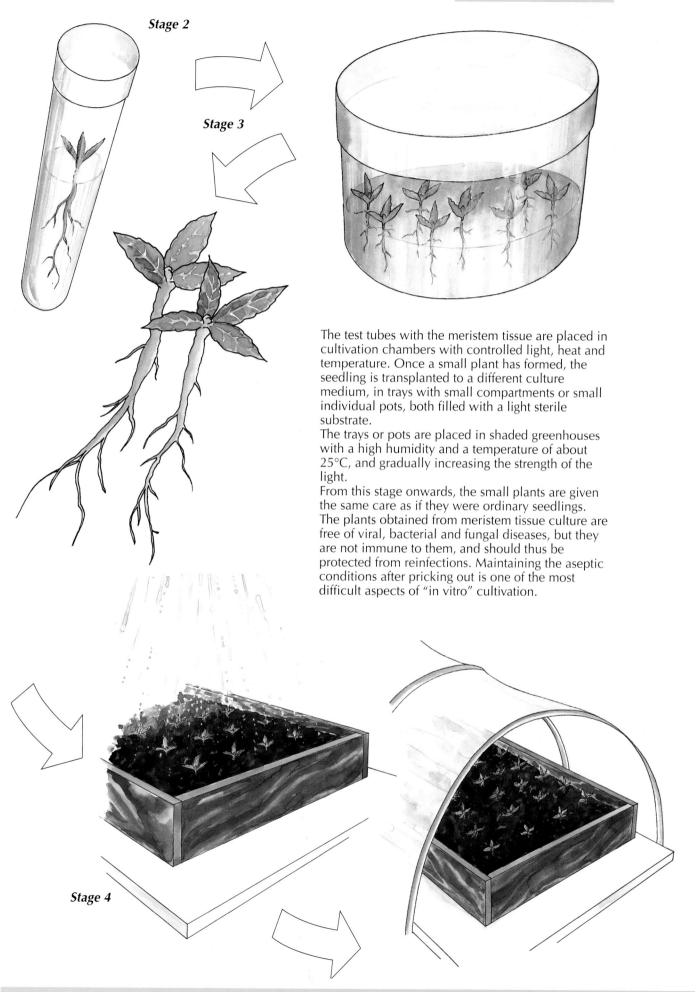

Stage 2

Stage 3

The test tubes with the meristem tissue are placed in cultivation chambers with controlled light, heat and temperature. Once a small plant has formed, the seedling is transplanted to a different culture medium, in trays with small compartments or small individual pots, both filled with a light sterile substrate.

The trays or pots are placed in shaded greenhouses with a high humidity and a temperature of about 25°C, and gradually increasing the strength of the light.

From this stage onwards, the small plants are given the same care as if they were ordinary seedlings. The plants obtained from meristem tissue culture are free of viral, bacterial and fungal diseases, but they are not immune to them, and should thus be protected from reinfections. Maintaining the aseptic conditions after pricking out is one of the most difficult aspects of "in vitro" cultivation.

Stage 4

BIBLIOGRAPHY

ALSINA, L.
Horticultura especial
Sintes. Les Fonts de Terrassa 1980

AMORÓS, M.
Horticultura
Dilago 1980

ANDERLINI, R.
El cultivo del tubérculo
Mundi-Prensa. Madrid 1970

BESNIER ROMERO, FERNANDO.
Semillas. Biología y tecnología
Mundi-Prensa. Madrid 1989

BISBAL, A.
El espárrago
Spanish Ministry of Agriculture. Madrid 1975

CASALLO, A. Y SOBRINO, E.
Variedades de hortalizas cultivadas en España.
Spanish Ministry of Agriculture. Madrid 1965

CASSERES, E.
Producción de hortalizas
Sucesores de Herrero Hermanos. México 1971

CUISANCE, PIERRE.
La multiplicación de las plantas y el vivero.
Mundi-Prensa. Madrid 1988

DARPOUX, R. Y DEBELLEY, M.
Plantas de escarda
Mundi-Prensa. Madrid 1969

DOMÍNGUEZ, F.
Plagas y enfermedades de las plantas cultivadas
Dossat. Madrid 1965

DOMÍNGUEZ VIVANCOS, A.
Abonos minerales
Spanish Ministry of Agriculture. Madrid 1978
Tratado de fertilización
Mundi-Prensa. Madrid 1984

EDMOND, J.B.
Principios de horticultura
C.E. Continental 1981

FERNÁNDEZ CUEVAS, A.
Horticultura intensiva
Spanish Ministry of Agriculture. Madrid 1968

FERSINI, A.
Horticultura práctica
Diana. México 1976

GARCÍA ALONSO, C.R.
El ajo. Cultivo y aprovechamiento
Mundi-Prensa. Madrid 1990

GEORGE, R.
Producción de semillas de plantas hortícolas
Mundi-Prensa. Madrid 1989

GRACIA LÓPEZ, C.
Mecanización de los cultivos hortícolas
Mundi-Prensa. Madrid

GUERRERO, A.
Cultivos herbáceos extensivos
Mundi-Prensa. Madrid 1977

HARTMANN, H.T.
Propagación de plantas. Principios y prácticas
CECSA. México 1984

HUME, W.G. Y KRAMP K.V.
Producción comercial de cebollas y guisantes
Acribia. Zaragoza 1971

JACOB, A. Y VON VESKÜLL,L.H.
Fertilización, nutrición y abonado de los cultivos tropicales y subtropicales
Euroamericanas. México 1976

MARGARA, JACQUES.
Multiplicación vegetativa y cultivo in vitro
Mundi-Prensa. Madrid 1988

MAROTO, J.V.
Elementos de horticultura general
Mundi-Prensa. Madrid 1990
Horticultura especial herbácea
Mundi-Prensa. Madrid 1986

MELA, P.
Cultivos de regadío
Agrociencia. Zaragoza 1971

MESSIAEN, C.M.
Las hortalizas
Blume. Barcelona 1979
Enfermedades de las hortalizas
Oikos-Tau. Vilassar de Mar. (Barcelona) 1967

NAMESNY, ALICIA.
Post-recolección de hortalizas
Ediciones de Horticultura. Reus 1993

SARLI, A.E.
Tratado de Horticultura
Hemisferio Sur. Buenos Aires 1980

SERRANO CERMEÑO, ZOILO.
Cultivos hortícolas enarenados
Spanish Ministry of Agriculture. Madrid 1974
Prontuario del horticultor
Gráficas Murgis. Almería 1985

TAMARO, D.
Manual de horticultura
Gustavo Gili. Barcelona 1968

TURCHI, A.
Horticultura práctica
Aedos. Barcelona 1981

OTROS:
Las semillas hortícolas en la CEE
Spanish Ministry of Agriculture.
Madrid 1986.

6

Cultivation in greenhouses

1. INTRODUCTION

1.1. GENERAL POINTS

The main difference between greenhouse cultivation and open air cultivation is that the environment is controlled to provide the most favorable environment for plant growth.

The geographical distribution of plants is basically determined by the environmental temperature, together with other lesser factors such as light intensity, water and nutrient availability. But it is the temperature that determines whether the plant thrives or merely survives.

The greenhouse modifies the plant's environment to create the desired conditions. Greenhouse cultivation is essentially a special case of small-scale protected horticulture.

Protected cultivation is when the plant's environment is controlled to some extent for part or all of the plant's life cycle, and this is also known as "forcing".

Protected cultivation is not only based on modifying the environment, but includes other aspects, such as specialized cultivation techniques, combining fertilizer application with irrigation, using agricultural chemicals, different sowing times, etc.

The most important form of protected cultivation is the greenhouse, a space that is enclosed or surrounded by a metal or wooden structure covered by glass or transparent plastic, within which the crop is grown in controlled conditions. Greenhouses are equipped with a heating system to provide warmth at the most critical moments. They may also have some sort of backup lighting system, as well as other devices to regulate environmental factors within the greenhouse, such as cooling systems to cool excessively high temperatures, to control humidity levels, the carbon dioxide concentration, etc.

Greenhouses also save water for the farmer. Some crops only use half as much water in a greenhouse as if they were growing in the open air. Crop yields are also 3 to 5 times greater than crops grown in the open air. There are also special varieties that have been selected to give really high yields in greenhouse conditions.

Further advantages of greenhouse cultivation are that strict

control of crop pests and diseases is possible, and 2 or 3 crops a year can be obtained from the site.

Greenhouses are often divided by the temperature within:

• **Cool greenhouses**. Where the minimum (the nighttime) temperature is between 5°C and 8°C.
• **Warm greenhouses**. Where the minimum temperature is between 10°C and 14°C.
• **Hothouses**. Where the minimum temperature is between 16°C and 20°C. They include specialized greenhouses for propagation.

A series of internal and external factors affect the temperature within the greenhouse:

• **External factors** include the external temperature and humidity, together with the wind, the intensity of the sunshine and the heat emitted by the sun and the earth.
• **Internal factors** include the properties of the covering material used as a roof, the soil temperature and humidity, the ambient temperature and humidity, the rate of turnover of the air within the greenhouse and its movement within the greenhouse, the level of evapotranspiration and the condensation of water vapor.

The aims of greenhouse cultivation can be summed up as follows:

- To protect the crop from unfavorable climatic events and factors, such as wind, rain, frost, hail and drought.
- To cultivate crops when external conditions are unsuitable for the plant to grow, flower and fruit. This makes it possible to obtain early crops and extra-early crops that fetch a higher price at market, because they arrive before the normal harvest period, and are thus scarce.
- To prolong the production period when temperatures start to fall, thus obtaining late produce, which fetches a higher price for the same reason as early produce.
- To increase the total amount produced. This is because the crop receives more care and is grown in more favorable environmental conditions, and the production period is longer. This gives a higher yield and makes better commercial use of the plot.
- To improve crop quality so the final product can compete on the marketplace.

2. PROTECTIVE STRUCTURES

2.1. TYPES OF PROTECTION

Early sowing and the demand for early or late produce have led horticulturalists to develop protective systems suitable for the crops that they grow. These special systems also increase yields and profits.

Crop protection installations vary greatly, not only in their structure and the materials used but also in how far they modify the environmental conditions.

2.1.1. Tunnels

The materials most widely used in tunnels are plastic sheeting for the roof supported by an iron or wooden structure. They can be combined in many different ways, depending on the crop being grown.

Tunnels are normally between 50 and 150 cm wide and 40-60 cm high, and they may vary greatly in their length.

The supports are usually arches and may vary greatly in their shape.

The distance between supports depends on the size of the tunnel, and the longer the tunnel, the smaller the distance between supports should be. Bear in mind that the bigger the tunnel is, the more warmth it traps.

The different types of tunnel are:

• **Fixed tunnels**. These consist of small iron hoops covered by plastic sheeting. The hoops are fixed into the earth, and the plastic sheeting is held down at the sides and at the end. This type of tunnel isolates the environment within from that outside, and lacks any ventilation. It may thus get very hot inside, and there may be a lot of condensation of the water vapor in the air. They should only be used for short periods of the crop's growth cycle.

• **Aired tunnels**. These are used in crops that spend a longer period of their crop cycle in the tunnel. They consist of a tunnel with two series of arches. One series supports the plastic sheeting, while the other series is placed over the sheeting, and keeps the tunnel in contact with the soil so it can be lifted to air the inside of the tunnel. A more complicated system that can be used is to construct frames about 5 m long, placing sheets of plastic with a system of lifts so one side of the tunnel can be raised to ventilate the interior.

• **Perforated tunnel**. This is based on using perforated sheets of plastic that allow a continuous but limited exchange of air between the inside of the tunnel and the exterior.

Tunnels showing different ways of fixing the plastic in the soil and the hoops
A/ Tunnel with the plastic held down by the soil
B/ Tunnel with plastic held down by wooden planking alongside
C/ Tunnel held up by wooden stakes in ground and with support cable running lengthwise
D/ Tunnel held down by iron structures.
E/ Tunnel with double tensing arches
F/ Tunnel with wire to tense the sheeting
G/ Tunnel with crossed wires to tense the sheeting

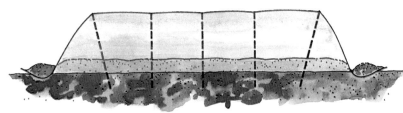

Ⓐ

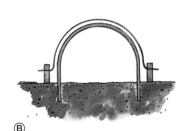

Ⓑ

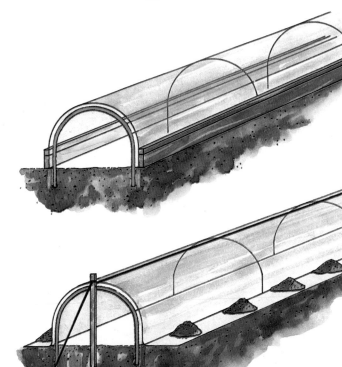

Ⓒ

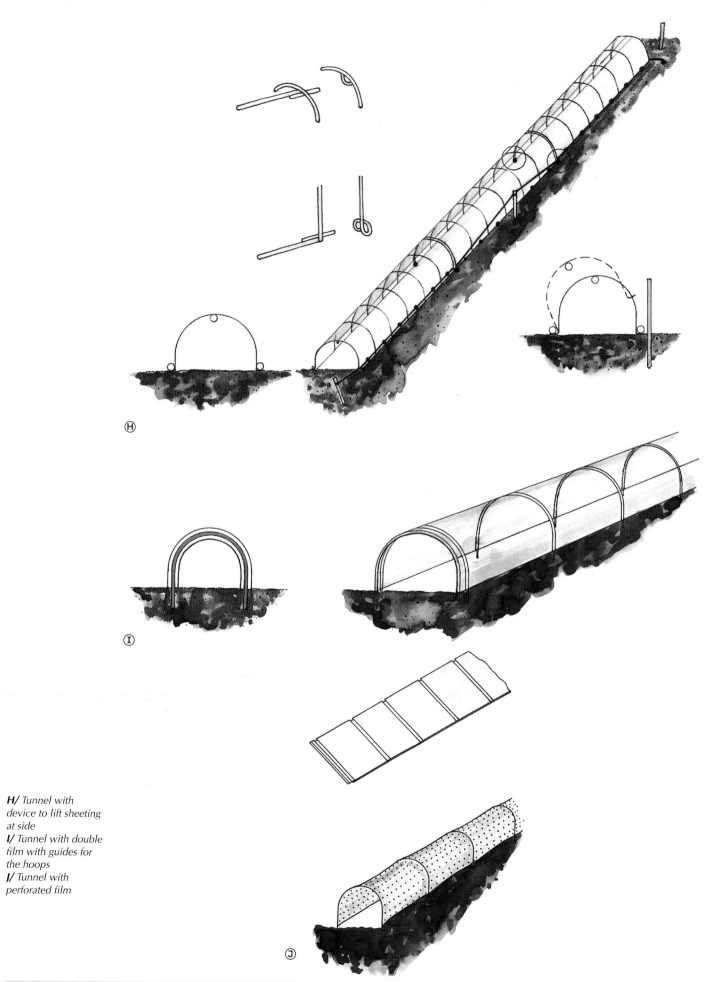

H/ Tunnel with
device to lift sheeting
at side
I/ Tunnel with double
film with guides for
the hoops
J/ Tunnel with
perforated film

2.1.2. "Cloches"

Cloches (French for "bell") were originally bell-shaped glass covers used in early crops to protect individual plants from the risk of frost. They have now been replaced by plastic tunnels.
The transparency of the cloche depends on the material it is made of. They are usually made of PVC (polyvinyl chloride), though some are still made from glass. They are usually truncated cones, shaped like a little greenhouse.

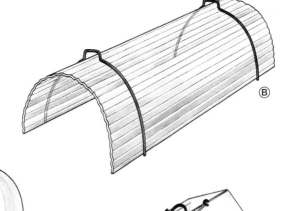

Cloches
A/ Glass bell-shaped cloche
B/ Sheets of corrugated plastic that are bent and anchored with wire hoops
C/ Glass panes joined by clips to form roof
D/ Glass joined by clips to form a "tent"
E/ Tunnels consisting of strips of plastic that run over some sheets and under others.
F/ Plastic panes on wire frames

2.1.3. Cold frames and seedbeds

Cold frames are shallow rectangular boxes consisting of walls and a roof. The walls are higher at one end than the other and the side walls are at a slant (see D) and support and permit adjustment of a roof, which may be a sloping sheet or have a central ridge (see A).

These constructions are often 1.2 m x 1.8 m. They can be built along the sides of a greenhouse, in order to make use of some of the heat it loses.
The rear of the cold frame is higher than the front so that the glass cover slopes downward so that rainwater does not accumulate, and to make best use of the light.
The cover of the cold frame may be wood and glass, wood and plastic, iron and glass or iron and sheets of rigid plastic. One variation is to use sliding doors with a lightweight metal frame and with a plastic cover. The purpose of cold frames is to protect the plants that will be grown in the open air after they are removed from the greenhouse. They are also used for the propagation of plants, both germination and the rooting of cuttings.
Cold frames may or may not have heating, and if they are heated they are called hotbeds. Hotbeds have a built-in system of heater elements to heat the substrate. This is connected to a thermostat to control the temperature.

Cold frames
A/ Aluminium with glass sides
B/ Lightweight plastic
C/ With sliding glass
D/ With glass "window"

2.1.4. Greenhouses

The greenhouse is the only protection system that allows a crop to be grown totally out of season. Depending on the requirements of the crop plant and its profitability, there is a huge range of structures, structural materials and systems to control the environmental conditions.

The basic differences between different types of greenhouse lie in the structural materials used and the covering materials used. The way the two are combined determines the type of greenhouse. The most widely used structural materials are wood,

The materials used for the covering are glass and plastic, as either rigid or flexible sheets.

Greenhouses normally have either a wooden structure covered in plastic or a metallic structure covered in glass or rigid plastic.

Wooden greenhouses are usually 5-8 m wide, 2.5-3 m high in the centre and 1.6-1.8 m high at the sides. The highest point should not be more than 3 m above the ground to avoid problems from strong winds. In general it is not a good idea for a single greenhouse to be more than 10 m wide, and to

Commercial greenhouse, with a flattened curved plastic roof on a metal frame

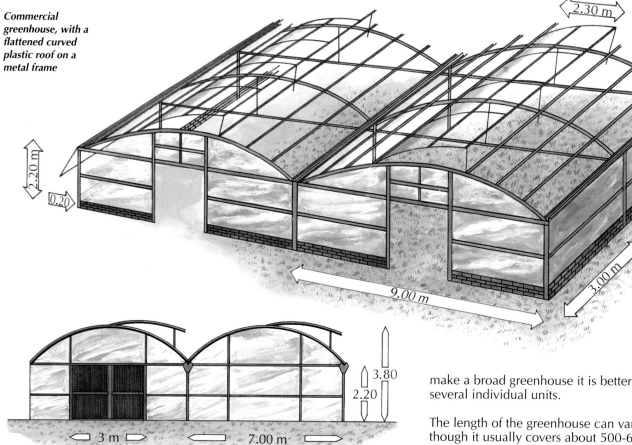

galvanized iron, and in some cases or perhaps just some parts, aluminium.

On the basis of the shape of their roof, greenhouses can be divided into:

- Flat-roofed greenhouses
- Greenhouses with a central ridge and two sides, a gable roof
- Greenhouses with curved roofs

Each of these types shows many variants. Greenhouses with gable roofs may be asymmetric, saw-shaped or pagoda type with overhead ventilation. Greenhouses with curved roofs may have semi-circular arches, flattened semi-circular ones or elliptical ones shaped like the handle of a basket.

make a broad greenhouse it is better to join together several individual units.

The length of the greenhouse can vary greatly, though it usually covers about 500-600 m². Industrial greenhouses built of metal can cover up to 1,500 m². To obtain the maximum light in the greenhouse, the roof should have two slopes, though their shape and angle may vary greatly. A trip around Spain reveals the diversity of construction types used, and the tendency to a local design in each region. In the very dry Mediterranean coastal province of Almería, greenhouse cultivation is very widespread, the most common model being the "PARRAL" (literally a vine arbor). This is a large flat-roofed greenhouse, with a structure built of eucalyptus wood. It reaches a height of about 2.5 m, and is open at the sides and ventilated. It is covered by plastic sheets that last about 2 years and which sometimes include agents to make them more effective insulation against heat. They are fixed to the structure by a mesh system of galvanized wires. This type of greenhouse is not suitable for areas with a normal or high rainfall, but it is very cheap,

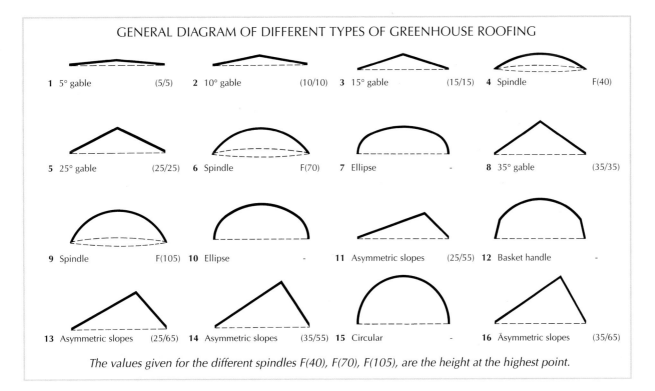

GENERAL DIAGRAM OF DIFFERENT TYPES OF GREENHOUSE ROOFING

1 5° gable (5/5) **2** 10° gable (10/10) **3** 15° gable (15/15) **4** Spindle F(40)

5 25° gable (25/25) **6** Spindle F(70) **7** Ellipse - **8** 35° gable (35/35)

9 Spindle F(105) **10** Ellipse - **11** Asymmetric slopes (25/55) **12** Basket handle -

13 Asymmetric slopes (25/65) **14** Asymmetric slopes (35/55) **15** Circular - **16** Äsymmetric slopes (35/65)

The values given for the different spindles F(40), F(70), F(105), are the height at the highest point.

probably the main reason why it is so widespread. It is also very resistant to the wind, and it adapts well to make use of the site, as they can be joined together to form large units, possibly covering more than 1 hectare.

In the relatively wet coastal Maresme region (northeastern Spain), the typical greenhouse is built from chestnut wood and covered with a gable roof of sheets of plastic.

The most important aspect of greenhouse design is to make best use of the area's natural conditions, by choosing a suitable shape, orientation, the slope of the roof and the material it is made of, and of course the type of crop to be grown has to be taken into account.

The main aim of greenhouse design is to use the minimum materials necessary to achieve the desired increase in production, and at a cost the farmer can afford.

2.2. COVERING MATERIALS

The material used to cover a greenhouse is of great importance, because it modifies the environmental conditions of the enclosed area.

The optical and thermal properties of the material used determine the difference between the microclimate created within the greenhouse and the external conditions.

The roofing material to be used depends on a series of criteria or indicators that interact with each other, and these help when choosing the material.

Selection criteria:

- The agricultural response, earliness of the crop, production and quality
- The rate of aging, i.e., the useful life, of the product

- The greenhouse's structure and the way the covering is to be attached

The choice should be made on the basis of the material's cost, its effectiveness as an insulating agent, and the type of crop that is going to be grown under it.

Glass roofs have now largely been replaced by plastic, and this has represented a major advance in greenhouse construction.

In addition to the classic gabled greenhouse with a central ridge and two glass sides, it is now possible to construct greenhouses with curved roofs, thanks to the use of new flexible materials. This has also meant that greenhouse structures now have to bear less weight, and are thus cheaper.

The materials used to cover greenhouses can be classified as follows:

Glass

Plastic:

• **Rigid plastic panes**

- Polyester, corrugated or smooth
- PVC (polyvinyl chloride) panes
- Methyl polymethacrylate ("Plexiglas")

• **Flexible plastic sheets**

- Polyethylene (PE), either normal, long-life, insulating or infrared
- Vinyl acetate-ethylene copolymer (V.A.E.)
- Polyester film
- Reinforced PVC, with or without nylon mesh

2.2.1. Materials

2.2.1.1. Glass

- It transmits light well, not modifying its spectral composition.
- It transmits about 90% of the light, so the levels of light within the greenhouse are almost as high as outside.
- It is almost totally opaque to the long wave radiation emitted by the plants and soil at night, and so the heat loss at night is less than in greenhouses covered with plastic.
- It is an effective insulation against heat.
- It is not degraded by environmental factors like heat, humidity, acids and blanching products, it has a very long useful life and does not deteriorate with age.
- It does not burn.

A greenhouse with glass roofing

Glass has a series of disadvantages, namely:

- Glass is much heavier than plastic, and thus requires a much stronger and more substantial, and therefore more expensive, support structure.
- It is very fragile, and may even be broken if it vibrates in the wind. If it is not firmly fastened to the structure, it can easily be broken by strong gusts of wind.

Glass is now being replaced by plastic as a roofing material for greenhouses, though it still has to be used in very cold climates or for crops that require a stable high temperature. The glass used is almost always "sheet" glass. It is usually between 2 mm and 6 mm thick, and the panes are about 60 cm x 60 cm. The crystalline glass that is used in houses is hardly used, as the light it transmits has an intense effect on the plants.

The sheet glass used is polished on just one side and left rough on the other, with the rough side on the inside of the greenhouse. Thus one side receives the sunlight which passes through the glass and is then scattered in every direction.

2.2.1.2. Plastic materials

Plastics for use in greenhouse construction have recently undergone major developments, as they have shown how useful they are for these purposes.
In general the plastics used are polymers, and may be used as films, i.e., flexible sheeting, or rigid plaques (panes). Independently of its width, it is common for flexible plastic sheeting to be classified on the basis of its thickness, which is normally measured in gauge units. 100 gauge is equal to 0.025 mm.

The transparency to different types of radiation of some materials used in different types of roofing.

MATERIAL	Ultra-violet radiation 300 nm -380 nm	Visible light 380 nm -760 nm	Sunlight 300 nm- 2.500 nm	Infra-red radiation 2.500 nm- 26.000 nm
3 mm glass	0.53	0.87 - 0.90	0.85	0.00
1 mm polyester	0.15	0.85 - 0.93	0.6 - 0.7	0.00
3 mm methyl polymethacrylate	0.68	0.85 - 0.93	0.73	0.00
1 mm PVC pane	0.72	0.77 - 0.80	0.75	0.00
0.15/0.20 mm PVC film	0.72	0.80 - 0.87	0.82	0.28
0.1/0.15 mm low density polythene	0.68	0.70 - 0.85	0.80	0.73
0.1/0.15 mm Vinyl acetate - ethylene copolymer (V.A.E.)	0.68	0.70 - 0.85	0.80	0.60 - 0.75

However, plastic sheeting is usually sold by weight, so that it is very important to know the density of a given film.

Which type of plastic material you should choose depends not only on its cost but also on the support structure to be used.

Roofing plastics should have all the following properties:

• **Transparency**. They should let as much light pass as possible. The amount of light they absorb depends basically on three factors:

- *The absorbtion of light by the plastic.* Different materials absorb greater or lesser amounts of the light passing through.
- *Reflection.* Some of the light does not enter the plastic, because it is reflected from the outer surface, and this depends on the angle of incidence of the sunshine and how reflective the material used is.
- *Scattering.* The light scatters when it passes through the material, and this means that the light is more evenly distributed.

Scattering	Reflection
Greatest	
Glass	Polysterene
Polyester with	Polythene
glass fiber	Polyester with
PVC panes	glass fiber
PVC sheets	Methyl
Ethylene	polymethacrylate
vinyl acetate	PVC
Polythene	
Least	

• **Transmission of infrared radiation**. The roofing material should not let all the heat emitted by soil and plants during the night escape as long wave radiation. This is called thermal insulation, and only some plastics show this property. They are relatively opaque to long wave radiation, and thus reduce or eliminate thermal inversion effects, improving conditions for the plant.

Plastic sheeting is considered to be thermal insulation if less than 20% of the long wave radiation that is emitted can escape.

- *Heat retention.* It should not let the heat that has accumulated within escape.
- *Effectiveness as insulation.* The temperature difference between the inside of the greenhouse and the exterior.
- *Lightweight.*
- *Flexible.* It should be possible to fit to any shape greenhouse.
- *Seal.* It should seal well, to prevent heat escaping.
- *Useful life.* This depends on the following factors:

Table: The characteristics of the main materials used to cover greenhouses

Below: Greenhouse with side and top ventilation

Characteristics	FLEXIBLE		RIGID			
	Polyethylene (0.08 mm)	PVC (0.1 mm)	Corrugated PVC (1-2 mm)	Methyl polymethacrylate (4 mm)	Stratified polyester (1-2 mm)	Glass (2.7 mm)
Density	0.92	1.3	1.4	1.18	1.5	2.40
Refraction index	1.512	1.538	—	1.489	1.549	1.516
% age stretching to destruction	400-500	200-500	50-100	low	low	none
Resistance to cold and heat	−40°+50°C	−10°+50°C	−20°+70°C	−70°+80°C	−70°+100°C	
Useful life	2 years	2-3 years	High	High	High	Very high
Transparency % (380-760 nm)	70-75	80-87	77	85-93	70-80	87-90

The higher the value for the density, the more rigid the material is, the worse its mechanical properties are, and the more fragile it is at lower temperatures. Only low density sheeting is used in agriculture. The method normally used to assess the degree of aging of a film is to measure the stretching to destruction, and see how much this has decreased with respect to new film. It is generally accepted that a film whose stretching to destruction value has not declined to 50% of its initial value is in condition to be used. (Spanish Norm U.N.E. 53-165)

Plastic greenhouses with curved roof and overhead ventilation

- *Ambient light intensity.* The stronger the sunlight, the greater the degradation caused by ultraviolet light.
- *The orientation of the sheeting's exposure to the sunshine.*
- *Treatment with inhibitors.* If the roofing material is treated with anti-oxidants and products that inhibit the action of ultra-violet light, the roofing lasts longer.
- *Thickness of the sheet.* The thicker it is, the longer it lasts.

- *The type of structure and how the plastic is attached.*
- *The wind regime.*

POLYETHYLENE (PE)

PE is derived from coal or oil. This is the plastic that is most widely used for forcing crops in tunnels or greenhouses. It is so widely used because it is the cheapest of the many polymers on the market. PE can be fitted to almost any type of structure because it is very flexible, but it is degraded faster than other materials. Exposure to sunlight causes degradation of the material, especially in the spring-summer period. Even if additives are added to increase its resistance, exposure to strong sunshine means that it only lasts a single year.
It also shows good resistance to wind and hail, and it is also resistant to tearing and splitting. It is not degraded by the sorts of chemicals likely to be used in a greenhouse. It is a poor conductor of heat and causes a good increase in the internal temperature.
When placing and fixing the plastic to the support structure, it is important to place it correctly, as poorly fixed plastic may split, thus reducing its useful life.

• **Main properties of PE:**

> Absorbtion........... 5-30%
> Reflection 10-14%
> Scattering............. low
> Transparency 70-85%

There are two types of PE, low density PE (LDPE) and high density PE (HDPE). HDPE is more rigid and more fragile than LDPE at low temperatures. LDPE is less resistant to breakage and tearing. HDPE is more resistant than flexible PVC, but less resistant than other plastics. LDPE weighs less per unit area for the same thickness. In general, PE does not turn as dark as PVC and polyester, and it is easier to cut and stick.

• **Types of PE:**

- *Normal PE:*
Normal PE is relatively transparent to the infrared radiation emitted by the soil at night. It allows about 70% to escape and this may even give rise to a thermal inversion; when the temperature outside is 0°C to –3°C, it may be even colder inside. Its useful life is one year, and only 10 months if the sunshine is strong and prolonged.
- *Long-life normal PE:*
It is similar to normal PE, but includes additives to protect the plastic, and its useful life is 2 or 3 years, depending on the intensity of the sunshine and the wind regime.
- *Long-life insulating PE:*
This is relatively opaque to the infrared radiation emitted at night, allowing only 18% to escape, and this prevents any possible thermal inversions. The plastic includes additives to scatter the light and to prevent dripping due to condensation.

There is a wide range of different thicknesses and widths on the market. PE is sold by weight and it is important to know the thickness of the sheets. The thickness is measured in gauges, and the PE is sold by the reel, in widths that vary from 80 cm to 12 m. The thickness may vary from 200 to 1,200 gauge, that is to say from 0.05 mm to 0.3 mm. A 200 gauge sheet of PE weighs 46.7 g/m^2, a 400 gauge sheet weighs 93 g/m^2, and an 800 gauge sheet weighs 187 g/m^2.

PVC (POLYVINYL CHLORIDE)

This rigid material is obtained by reacting acetylene and hydrogen chloride. Plastifying agents have to be added during manufacture to produce flexible sheets. It is little used in Spain (only about 2% of all greenhouse roofs use PVC).
There are several types of PVC on the market, including rigid panes, flexible sheets, semi-flexible sheets reinforced with nylon mesh or linear polyester.

- *Colored PVC*. Blue: this is for crops that grow horizontally. The plant stems are shorter, meaning that the weight of the leaves, roots and stems is greater. It is also used in seedbeds and in crops grown for their leaves and tubers.
Red: Red PVC is recommended when growing tomatoes, peppers, strawberries and watermelon and for flower production.
- *Black PVC*. This is used for mulching and when blanching vegetables.
- *Reinforced PVC*. This is a sheet consisting of two transparent sheets joined together, with a strengthening mesh between them.

Flexible PVC allows 90% of sunlight to pass, and rigid panes allow 80% to pass. It absorbs about 5% of the light in flexible sheets and 5% to 10% in rigid sheets. It reflects 5-8% of incoming light, and scatters the light less than polyester and more than polyethylene.
It causes a greater greenhouse effect than PE, because it conducts less heat than PE.

Upper right : "Chapel" type greenhouse with wooden structure

• Other types of PVC:

- *Transparent PVC*. Its transmission of sunlight is comparable to PE and VAE.
- *Translucent PVC*. It allows the same radiations to pass as transparent PVC, but it diffuses the light more, and this means the light is more equally distributed.
- *Photo-selective PVC*. This selectively transmits the wavelengths of light that are of most importance for the plant and favor its growth.

It thus retains more heat at night, and thus prevents any possible thermal inversion. It retains 80-90% of the heat, while PE retains 10-15%. It is however more expensive than PE.

Little humidity condenses on PVC. Its resistance is similar to that of PVC. It is less flexible than PE at low temperatures, but flexible PVC is a little more resistant to tearing and breakage than PE.

Panes of rigid PVC are more resistant than

Upper left and below: Greenhouses covered with PVC

PE and a little less than polypropylene. PVC is more resistant to oxidation and the action of chemicals than PE. One disadvantage of PVC is that dust sticks to the surface more than on other materials.

PVC has a longer useful life than PE. It ages less quickly, considering aging as loss of transparency and greater fragility and tendency to break. This is due to chemical changes caused by the heat and light in the presence of oxygen.

PVC has a useful life of 2 to 3 years. It varies in thickness between 0.05 mm and 3 mm (200 to 1,200 gauge) and it is sold in sheets up to 8 m wide. The density of flexible PVC is between 1,200 and 1,400 kg/m^3. 1 m^2 of 100 gauge PVC weighs between 33 and 35 grams. Rigid PVC is sold in panes 7 mm thick. It has a useful life of 6 years and does not require any special care when it is installed.

Plastic mesh used to shade the greenhouse

its physical resistance but by the fact that it becomes less transparent over time.

If the pane is not protected on the outside, it is rapidly worn away by environmental agents, and it may turn opaque in a few years of use. If it is protected, this wearing away is slowed down, but not the yellowing of the plastic.

PMM (POLYMETHYL METHACRYLATE)

This an acrylic material derived from acetylene. It is sold as acrylic glass or Plexiglas. There are two types, colorless and translucent white.

It is very transparent, transmitting 85-92% of the sunlight. It lets almost all the ultra-violet rays pass. It shows almost no scattering of the light.

It is highly opaque to the long wave radiation emitted at night, and is highly resistant to breakage, tearing and aging. But because it is not a very hard material, it is easily scratched by sharp objects, causing its optical properties to deteriorate.

It has a longer useful life than polyester panes. It is manufactured in rigid panes up to 2 m wide. Its density is 1,190 kg/m^3, and 1 m^2 of PMM 3 mm thick weighs approximately 3.5 kg.

Long-life PMM is produced by adding anti-oxidants and agents to inhibit the effects of ultra-violet light. Thermal PMM also includes agents to reduce emission of infra-red radiation.

POLYESTER

The most widely used polyester plastic is polyester reinforced with fiber glass, in the shape of rigid panes.

It is very transparent, transmitting 70-80% of the sunshine, reflecting 5-8% of the light and absorbing 15-20%. It is also very good at scattering the light. It is opaque to long wave radiation, the heat emitted at night, and in this aspect is very similar to glass. It is not recommended for greenhouses that are growing plants that will later be planted in the open air, because it absorbs all the ultra-violet radiation.

Polyester panes adapt well to any type of structure because they are highly flexible. They are also quite resistant to breakage. They expand very little when they heat up.

To the right: Cross-strut support made of galvanized tubing. (Manufactured by AGRIMEC)

They last for 8-15 years, depending on how the pane is attached. Its useful life is not limited by

POLYPROPYLENE

Polypropylene transmits about 83% of the incoming light, and scatters about 77% of this, and transmits about 48% of the long wave infrared radiation.

Its density is 910 kg/m^3 and it is sold in 500 gauge sheets.

Its useful life is 3 years, and when it is manufactured it is given a special anti-stick treatment so that dust does not accumulate on the surface. It is easy to handle when fixing it to the structure.

This material shows some porosity, which has some advantages, such as avoiding the condensation of humidity, but dripping rainwater can cause problems if the roof is not steep enough.

VINYL ACETATE-ETHYLENE COPOLYMER

This is a co-polymer, a product obtained by polymerizing a mixture of different plastics. In this case, VAE is a mixture of vinyl acetate with ethylene. A range of products can be added during the process to improve the material's useful life.

Depending on the percentage of vinyl acetate, there are various types of VAE, and their characteristics are similar to those of polyethylene sheeting if the percentage of vinyl acetate is low (around 6%) and closer to PVC if the percentage is high (around 18%).
In general, VAE sheets are more effective as insulation than thermal PE.
It is more flexible and more resistant to impacts than PE, but much less resistant to scratching.
It is more transparent to sunlight than PE and a little less than PVC, and it diffuses the light more then PE.
800 gauge VAE has a useful life of 2 years, and 400 gauge VAE has useful life of 1 year.
You are recommended not to use VAE with a high acetate content in sites with very high light intensity and high temperatures, because this material expands a lot with heat, which gives rise to sagging where rainwater accumulates and to breakages caused by the wind.

In general, VAE is very elastic, and so it requires a supporting structure or fabric to prevent excessive expansion due to the heat, because once it has stretched, the sheet never returns to its original size and it remains loose.
The density of VAE copolymer is 0.926 to 0.937, and 1 m^2 of 100 gauge sheet 2.5 mm thick weighs 23.42 g.

Overhead windows in glass greenhouse

2.2.2. Some general points about roofing and light. The greenhouse effect

Transparency is the most important feature to bear in mind when choosing roofing materials to ensure favorable growing conditions.

The behavior of sunlight inside a greenhouse
Part of the radiation is reflected back into space, part is absorbed by the material and the rest enters the greenhouse

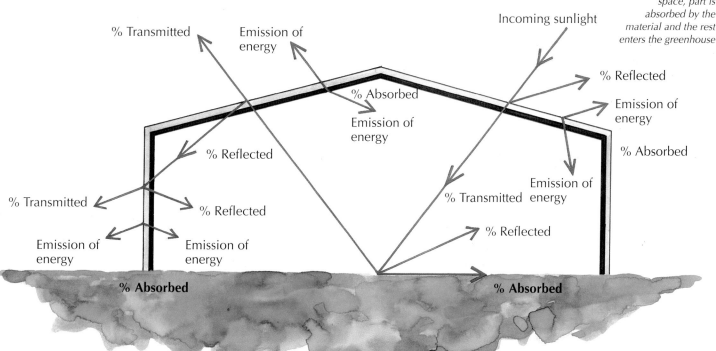

A good covering material should have the following optical properties:

• To reduce as little as possible the visible light that powers photosynthesis.
• Not to alter the spectral composition of the light. This means that any material to be used as roofing for a greenhouse should possess the following characteristics:

* As transparent as possible to sunlight and short wave radiation (380 to 3,000 nanometers).
* As opaque as possible to the long wave radiation (more than 3,000 nanometers) emitted by the soil and plant cover at night.

The combination of these two factors leads to what is called the "greenhouse effect". The intensity of this phenomenon depends on the greenhouse's ability to warm up during the hours of sunshine and keep warm during the night. The roofing material's ability to transmit short wave radiation varies depending on the angle of incidence of the sunshine on the greenhouse's roof. This has consequences on the greenhouse's design, because depending on the latitude where it is to be built, the installation's design must have an angle of incidence for the sunlight that does not lead to a reduction of light transmission.

Opening and closing mechanisms using rack and pinion systems

In general terms, it can be said that different flexible films trap the heat inside to the same extent, maintaining a temperature inside roughly 2-2.5°C warmer than the outside temperature.

It should be pointed out that these values are guidelines and the real values may be higher or lower, depending on the environmental conditions, such as the temperature, the cloud cover, the wind regime, rainfall, etc.

If you use glass or panes of plastic the temperature difference between the inside and the outside may be 4°–5°C, depending on the weather outside.

2.3. STRUCTURAL MATERIALS

2.3.1. The general requirements of the structure

The materials have to be resistant and ensure the structure is stable. Eliminate everything that is superfluous, as it increases the weight and reduces the incoming light.
The most widely used materials are wood, steel and aluminium.

Wood has the disadvantage of casting a deep shadow on the plants. The planks supporting the roof occupy much of the area of the greenhouse.

2.3.2. Materials

2.3.2.1. Wood

Structures made with posts, beams and poles are still very common, especially when the intention is to spend as little money as possible on a structure, most of which can be replaced anyway.
Wood is the cheapest material and has the lowest thermal conductivity, but it is not very airtight. Wood casts more shade within the greenhouse, as there have to be more supports, and their diameter is greater.

Resistant wood should be used for the structure. The wood used is usually either untreated eucalyptus, or old railways sleepers treated with creosote or aluminium salts. Before installing them, treat with preservatives and then varnish them. The greater the care taken when preparing the wood, the longer it will last.

2.3.2.2. Steel

The great advantage of steel is that it does not need any internal supports, and this gives greater freedom of movement and manoeuvrability for machinery.
The greater distance between the structural components and the fact that they are smaller in cross-section means that there is more light, and this is of course good for the crop.
Steel can also bear heavier loads than wood and has a greater ability to dissipate heat by conduction. This heat loss is more than compensated by the fact that it is possible to make the structure much more airtight. A better seal saves fuel for heating. It is also possible to install a better designed airing system that is easier to use.

The disadvantages include the high initial cost and the need for annual maintenance. Corrosion is a problem, and so the steel has to be well protected, either by painting it every year or by using galvanized material when constructing the greenhouse.

2.3.2.2. Aluminium alloy

This has characteristics similar to those of steel. Aluminium is more resistant to corrosion than steel and can be used to build very complex structures, but it is very expensive, and soldering one piece to another is very difficult.

2.4. LOCATION AND SITE

The greenhouse's size, shape and layout depend basically on the farmer's needs. The type of greenhouse chosen should be a response to the desired purpose, the available capital and the features of the site.

To the left: Wood and glass greenhouse

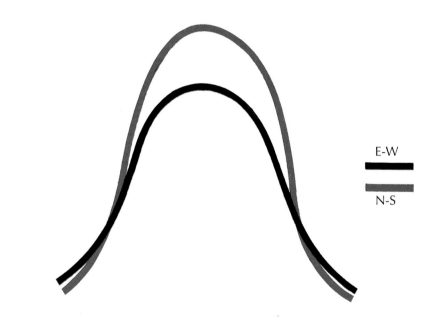

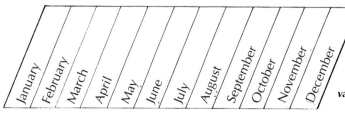

The annual variation in light received by a greenhouse
The graph shows the annual variation in the strength of the sunshine received by a greenhouse with a symmetric gable roof (35°/35°) with an east-west orientation (black line) or a north-south one (red line).

The external characteristics are the first conditioning factors that have to be defined. These conditioning factors are largely determined by the climatic conditions of the site, but are also influenced by the soil's chemical and physical properties, the availability of irrigation water and its quality, and many other factors such as the possibility of connecting to an electricity supply and access to the site.

The greenhouse's site should be well away from obstacles that cast shade.

The amount of light received by a greenhouse in relation to its orientation and the type of roof

When choosing a spot, attach great importance to the soil type and the site's microclimate, paying great attention to changes in the temperature and humidity at the site over the course of the day and year, the period it is free from frost, as well as the light intensity, the hours of daylight when it receives direct illumination, and the wind regime.

Take special care when siting the greenhouse to choose a spot far from the shade of buildings or trees, especially in winter when the sun is at its lowest in the sky. Take care to avoid the shade of nearby mountains or hills. Avoid sunken areas of ground, because cold air collects in these sites in winter, putting them at risk of frosts.

Choose a site where there is some breeze to help ventilation and choose a site with good soil, avoiding soils that are compacted or have bad drainage.

Of course it is hard to find a site that meets all these conditions, and at the same time, but they should all be taken into account when choosing a site.

The orientation of the greenhouse is influenced by the division of parcels on the plot and by the direction of the dominant winds.

The orientation should also be chosen to trap as much sunlight as possible in the winter, and this factor is related or combined with the shape of the greenhouse's roofing.

An east-west orientation has the advantage of obtaining more sunlight in the period between the autumnal and vernal equinoxes (i.e., between September 22 or 23 and March 20 or 21 in the northern hemisphere) than a north-south orientation. Likewise, curved roofs trap more sunlight than flat roofs or gable roofs, as they favor the entry of the most sunlight at the most effective angle.

Asymmetric constructions should have the same orientation as symmetric ones, with their longest side facing the south (in the northern hemisphere). You are normally advised to lay out the crops at right angles to the longest axis of the greenhouse, north-south, to ensure the plant gets the most sunlight over the course of the day. This avoids the shadows cast by the roof beams and shiny reflections from bars.

If you are considering a series of greenhouses side by side, the individual greenhouses should have a north-south orientation, as an east-west orientation may cause them to shade each others.

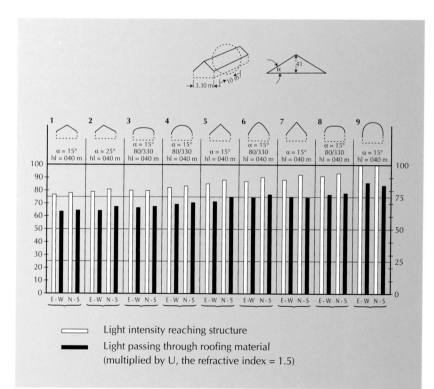

Light intensity reaching structure

Light passing through roofing material
(multiplied by U, the refractive index = 1.5)

2.4.1. Establishing the areas of shade

The first thing to do is to find out the lowest angle of the sun over the horizon, at the winter solstice. Then bolt together two pieces of wood or rulers. Using a simple protractor (drawing A), fix the two pieces of wood at the correct angle. They should be about 30 cm long. Then place the bottom piece of wood on a spirit level flat on the soil on the planned site of the greenhouse. Face the device south (in the northern hemisphere) and ensure it is perfectly level, if necessary using supports to ensure the spirit level is correctly levelled (with the bubble in the center).

The upper piece of wood is now at the angle of the sun at its lowest in midwinter. Looking straight along the upper piece of wood you can see if there are any trees or buildings that are going to cast a shadow on the greenhouse in midwinter.

Summer

The sun's path across the sky

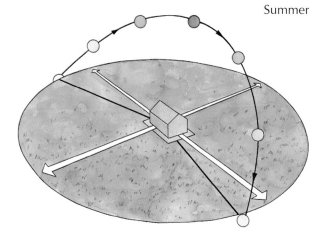

Winter

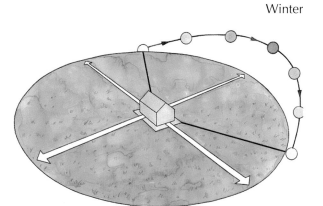

The apparent position of the sun varies greatly from winter to summer and this variation should be borne in mind when choosing the location and the type of greenhouse.

In the northern hemisphere, in winter only the south side of the greenhouse receives direct sunshine; in summer, the two ends are also lit, in the morning and in the afternoon.

Measuring the areas of shade using a calibrator of angles
A/ To check if a site will be shaded, find out the lowest angle of the sun over the horizon in mid winter. Then bolt together two pieces of wood. Using a protractor to determine the angle, fix the two pieces together. Tighten the bolt.
B/ Place the lower piece of wood on a spirit level in the planned site for the greenhouse. Point the device due south, taking care that it is perfectly flat.
C/ The upper arm is now pointing to the lowest position of the midday sun (on the winter solstice). Looking along the upper piece of wood you can see which trees or buildings will cast a shade on the site you have chosen.

2.5. CONSTRUCTION

There are four main factors to take into account when building a greenhouse.
• The covering materials should transmit as much light as possible.
• The structure should be stable.
• The area covered should be large enough to use machinery, if appropriate.
• Low cost.

The exact area the greenhouse is going to occupy should first be marked on the site. This should be selected carefully and marked before starting to build the greenhouse. Use greenhouse floorplans

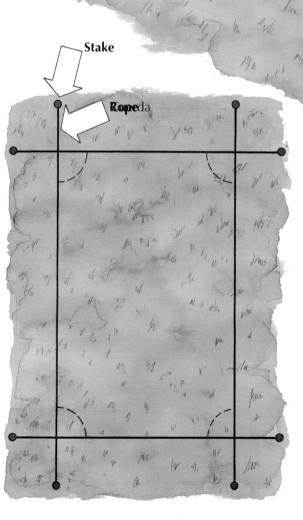

Marking out the site
Mark the position of one side of the greenhouse with two stakes and tense a rope between them. Make sure the stakes are at the same level. Then measure a right angle, using a set square, to establish the position of one of the load-bearing walls. Repeat the operation for the remaining angles. Check that the 8 stakes are at the same level.

Stake

Ropeda

that are adjusted to the most accepted measures. The line each wall is going to follow should be marked, as should the site of each of the main structural supports, such as wooden poles or metal support structures.
After marking the greenhouse's first wall, you must establish the correct angle for the other walls. Each one of these lines should be marked using a series of stakes fixed into the ground beyond the point where the lines cross. Thus the line of each wall is marked by a rope between two stakes, and the corners are where the ropes intersect. That is to say, the ropes cross, and the intersections mark the corners of the greenhouse.

It is important that the site be flat and well drained. If the site is sloping or very uneven, the first thing to do is to level it. The soil should not be compacted when levelling as excess pressure may damage the soil structure and cause drainage problems and the loss of fertility.

Use the spirit level all the time to ensure the base is perfectly level. This level is of course marked by the ropes between the stakes. Any mistake in levelling at this stage will lead to serious problems when it comes to building the structure and fixing the roofing. Levelling mistakes may have effects in the medium and long term, as the structure will be subject to stresses and strains that will eventually cause many problems.

If cement has been used to anchor the structure, it should be left to set for 48 hours, and longer in cold areas. But if concrete is used after building the structure, take the precaution of only using it on sunny days with stable temperatures.

The same is recommended for the installation of pre-fabricated modules that have not completely set, and which also require a structure reinforced with concrete. Time is of great importance. If the cement does not fully set, it will suffer the effects of environmental factors, especially the wind, which will eventually get in, thus negating the basic idea of the greenhouse.

2.5.1. Basic construction calculations

The structure supporting a greenhouse should be able to support its own weight, and also possible additional weights, such as stakes for the crop, extra loads due to snowfall, the force of the wind, any automated devices, etc.

The greenhouse must support permanent and incidental loads:

• **Permanent loads** include the weight of the roofing material, the weight of the structures supporting the roof, the weight of the structure itself, as well as any other permanent loads acting permanently on the structure and any installed loads, such as tubing.
• **Incidental loads**, such as snow or the action of the wind.

PERMANENT LOADS

The weight of the material for the roofing varies depending on the material used. The weight is about 7 kg/m² for glass 6 to 8 mm thick, and between 1 and 5 kg/m² for rigid plastic. This weight can be added to, or be considered together with, the weight of the support structure.

In the absence of a detailed calculation, the following are guidelines for different types of greenhouse:

The load tolerated by a greenhouse depends on its shape and the construction materials

TYPE OF GREENHOUSE	AVERAGE WEIGHT (kg/m²)
- Wooden greenhouse	8-26
- Steel greenhouse with rigid or flexible plastic roofing	4-10
- Aluminium greenhouse with glass roof	5-8
- Steel greenhouse with glass roof	8-14

SNOW LOAD

The following factors must be taken into account:

• The areas where greenhouses are built have mild climates, and thus low snowfall.
• The nature of the roofing material favors the snow slipping off.

• The heat within the greenhouse causes the snow to melt when it comes into contact with the roofing material.
• If a lot of snow accumulates, it breaks the roofing material, when it exceeds the load-bearing capacity.

It is reasonable to adopt, for the snow load, a value of 25 kg per horizontal square meter.

THE ACTION OF THE WIND

The action of the wind depends on the height of the structure and its situation within the local relief. The situation is considered to be exposed along coasts, on ridges and in valleys. The additional load due to the wind depends on the structure's shape and the angle the wind approaches.

For purposes of calculation, the action of the weight is compared to a static weight that is proportional to the value of W in kg/m^2.

$$W = V^2 / 16$$

where V is the velocity of the wind in meters per second.
Given that these greenhouses are rarely higher than 6 m, it seems justified to suppose a fixed value for the dynamic pressure, regardless of its location, a value of 50 kg/m^2.
The foundations transmit the loads acting on the greenhouse to the soil, and so the load-bearing capacity of the soil is also important.

The weight of the foundations should be at least 3 times greater than the pressure of the wind within the greenhouse, after deducting the weight of the greenhouse.

$$Gw / (Wp - Fw) \geqslant 3$$

where Gw is the weight of the greenhouse
Wp is the pressure of the wind
Fw is the weight of the foundation

In greenhouses with plastic sheet roofing, the action on the foundation should not be less than 1.5 times the pressure created by the wind within the

greenhouse, according to the relevant Spanish government regulations.

2.5.2. Construction using wood

Some of the specific requirements for building a wooden structure are:

• The posts should have a wooden block at their base that is as wide as the post (or nearly so), roughly 30-50 cm, and is buried to act as an anchor when the post is fixed onto it. It should be at least 30 cm deep to prevent variations caused by the settling of the different support posts, and especially to prevent the wind uprooting it from the soil. This is the only join that is nailed.
• The rests of the joins between posts, beams and struts use nuts, washers and bolts, to prevent the structural weakening that would be caused if nails were used, basically a consequence of the movement caused by the wind.
• To avoid rotting caused by soil humidity, char the surface of the posts to a height of about 50 cm above ground level.

The roof is normally made of plastic film, like the other surfaces. For the roof, a wire-based protection has to be made. This protection should consist of a grid network of squares about 30 cm x 30 cm for the lower part, which supports the plastic film, in order to prevent rainwater causing the film to sag. This net should run from one side of the top of the greenhouse to the other, following precisely the way the plastic (that it will support) is going to be fixed and so it should be firmly attached.

The upper part of the protective network consists of cables that cross from one side of the greenhouse to the other above the plastic, which they keep stable against the push of the wind from below, giving the fragile plastic roofing a permanent stability as long as is permitted by the quality of the plastic.

The structure should be arranged so that the end of each wire can be tightly fixed in its correct place. This installation should be equipped with side channels to collect the rain water and channel it to a deposit where it can be stored for later use.

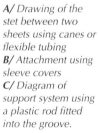

A/ Drawing of the stet between two sheets using canes or flexible tubing
B/ Attachment using sleeve covers
C/ Diagram of support system using a plastic rod fitted into the groove.

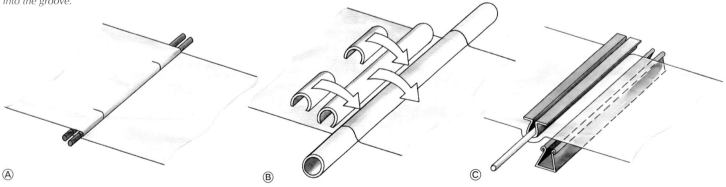

Ⓐ   Ⓑ   Ⓒ

• On wood, place a wooden lath or plastic strip above it. Fix with a tack.
• By winding the sheeting around a cane, spool or other cylindrical object. It is then tied with a galvanized wire and attached where necessary.
• Using galvanized tensor cables or wires in a grooved slit.
• Using metal and plastic eyelets.
• Placing mesh above and below the plastic sheeting.

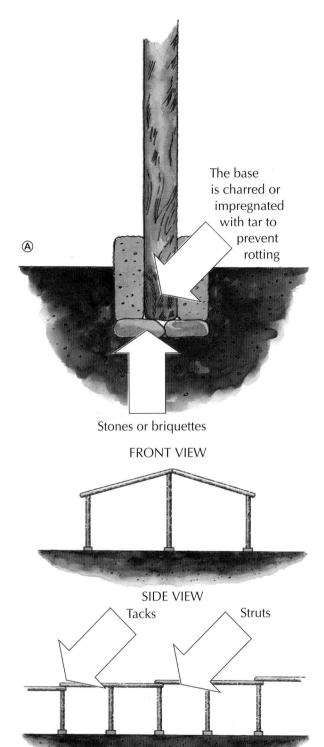

The base is charred or impregnated with tar to prevent rotting

Stones or briquettes

FRONT VIEW

SIDE VIEW

Tacks   Struts

Front view of a wooden greenhouse and the arrangement of pillars within it.
A/ The installation of wooden pillars
B/ The position of the struts on the side wall of a wooden greenhouse and the arrangement of rafters in the roof

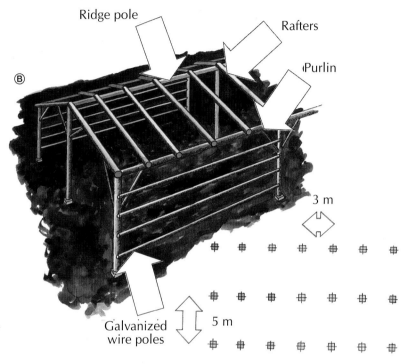

Ridge pole   Rafters   Purlin

3 m   5 m

Galvanized wire poles

2.5.3. Metal greenhouses

Nowadays all metal greenhouses are built with tubing, whether they have a flat roof, a gable roof or are a multiple tunnel. They are very easy to assemble, using galvanized clips for all the connections.
To make the supports more solid, blocks of concrete in the form of the base of a pyramid 20 cm x 30 cm x 20 cm are used. They have a hole in the top for the pole, to make it more stable. The pole is fixed with wire through the pole and fixing it underneath the block.
Metal greenhouses can be covered by plastic sheeting, which is fixed in the same way as for wooden structures, or with sheets of rigid plastic. These panes are fixed using hooks, bolts and washers.
If glass panes are used, they are attached to the structure using polyvinyl-based glues that do not need maintenance. They may also be fixed using iron springs, thus eliminating the need for glue.

When handling glass, bear in mind that installation is dangerous on moist days, due to the constant risk that it may slip and fall. Furthermore, the sealing agents used, such as mastic, do not adhere well in moist conditions.

When placing the plastic sheeting on the roof, bear in mind that these materials expand and contract when the temperatures change, and so the sheeting should be fixed during the hottest hours of the day. It may be necessary to protect the roll of film from the heat, as this would cause it to expand. It should not be stretched to avoid stresses at the points of attachment when the weather gets cold.
The many ways of attaching the plastic to the structure include:

3. THE INFLUENCE OF ENVIRONMENTAL FACTORS ON GREENHOUSES

The local climate determines which type of greenhouse to install and its orientation. The most important thing affecting the crop's growth is the microclimate within the greenhouse, and this is clearly largely determined by the outside environment, but the microclimate within the greenhouse is substantially different merely because it is covered.

3.1. TEMPERATURE

Of the many environmental factors, the temperature has the most impact on the plant's growth. The temperature is the most important environmental factor that is regulated in greenhouse cultivation.
There is an optimum temperature for plants to carry out each of their functions. Above or below this temperature, these functions become more difficult. The farmer who installs a greenhouse does not only wish to prevent temperatures from falling to critically low

values, but also to maintain temperatures at the levels that are most beneficial to the crop's respiration, transpiration, photosynthesis, germination, growth and fruiting.

Low temperatures can cause a range of damage to plants. Their proteins may be denatured by the cold, and when temperatures fall below –4°C ice crystals start to form in the cells, causing dehydration of the cell and splitting of the cell membranes.

When temperatures are too high, the cytoplasm starts to coagulate and the cell dies. The plant stops growing long before the heat causes it to die.

Exceeding the plant's limits of tolerance leads to its death. Even if they do not actually die, they will be left inactive and useless.
However, bear in mind that some difference between the daytime and nighttime temperature is necessary for good growth.

The essential aim of greenhouse cultivation is to maintain a temperature suitable for plant growth.

Air temperatures (°C) (Tesi, 1969)
The columns are:
- Lethal minimum temperature.
The temperature at which irreversible damage occurs to the plant with the possibility of death if it lasts a long time.
- Minimum biological temperature:
The temperature at which the plant halts vegetative growth.
- Optimum temperature:
The optimum temperature for the plant's growth.
- Maximum biological temperature.
Above this temperature the crop's physiology is disordered, and its growth and production totally halts.

SPECIES	Lethal minimum temperature	Minimum biological temperature	Optimum temperature Night	Optimum temperature Day	Maximum biological temperature
VEGETABLES:					
Tomato	0-2	8-10	13-16	22-26	26-30
Cucumber	0-4	10-13	18-20	24-18	28-32
Melon	0-2	12-14	18-21	24-30	30-34
Squash	0-4	10-12	15-18	24-30	30-34
Runner bean	0-2	10-14	16-18	21-28	28-35
Pepper	0-4	10-12	16-18	22-28	28-32
Eggplant	0-2	9-10	15-18	22-26	30-32
Lettuce	(–2)-0	4-6	10-15	15-20	25-30
Strawberry	(–2)-0	6	10-13	18-22	—
FLOWERS					
Carnation	(–4)-0	4-6	10-12	18-21	26-32
Rose	(–6)-0	8-12	14-16	20-25	30-32
Gerbera	0	8-10	13-15	20-24	—
Chrysanthemum	—	6-8	13-16	20-25	25-30
Gladiolus	0-2	5	10-12	16-20	25-30
Tulip	—	4-6	12-18	22-25	—
Iris and daffodil	—	3-5	8-15	15-20	—
Lily and freesia	—	6-8	10-16	18-24	30-34
Cyclamen	—	2-4	12-18	20-22	—
Calla lilies	—	—	10-13	14-20	—
Azalea-Rhododendron	—	6-8	12-14	14-20	—
Poinsettia	0-4	8-10	18-20	20-25	26-28
Gloxinia	—	—	18-20	20-25	—
Primrose and Calceolaria	—	—	18-20	20-25	—
Pelargonium	—	6-10	14-16	20-25	26-30
African violet	—	10-12	16-20	20-24	—
Kalanchoe	—	—	15	20-25	—
Hydrangea	—	—	10-18	20-25	25-27
Gardenia	(–8)-0	—	15-17	21-23	—
Hothouse orchid	—	—	16-18	18-21	28-30
Warm greenhouse orchid	—	—	13-16	16-18	23-25
Cool greenhouse orchid	—	—	10	13-16	18-22
Croton, Ficus	—	—	15-20	23-24	35-40
Philodendron, Anthurium Dieffenbachia	—	—	20-23	25-30	35-40

Crop	Germination			Growth		Flowering	
	Minimum	Optimum	Maximum	Night	Day	Night	Day
Tomato	10°	25-30°	35°	13-16°	18-21°	15-18°	23-28°
Pepper	13°	25-30°	40°	16-18°	20-25°	18-20°	25°
Eggplant	15°	20-25°	35°	17-22°	22-27°	18-20°	20-30°
Runner bean	12°	15-25°	30°	16-20°	18-30°	15-20°	20-25°
Strawberry	10°	18°	35°	10-15°	18-25°	8-10°	15-18°
Cucumber	12°	30°	35°	18-22°	20-25°	18-22°	20-25°
Zucchini	10°	20-30°	35°	20-25°	25-35°	20-25°	22-30°
Melon	13°	28-30°	40°	20-24°	25-30°	18-22°	20-23°
Chrysanthemum	—	—	—	16-18°	18-22°	13-15°	15-17°
Carnation	—	—	—	10-12°	22-22°	10-12°	20-22°
Rose	—	—	—	10-12°	20-25°	14-16°	24-25°

Critical temperatures for some crops

Different plants require different temperatures and anyone seeking to cultivate them in the best conditions should concentrate on obtaining the best results, despite the atmospheric conditions in each season of the year. Carried to extremes, this makes it possible to ignore the external calender.

The temperature should not be fixed at the minimum value given, as this will only ensure the plants survival, allowing them to grow a little, but not to achieve their entire potential.

The temperature is controlled using natural or artificial systems to supply heat, and by using ventilation, lighting, shading and creating and maintaining appropriate levels of humidity.

Before using this equipment, it is important to understand the needs of the crop you intend to grow. Most plants can grow within a temperature range that we have divided into four levels, though there is no strict division:

- Cold regime: 4.5°C to 0°C.
- Cold and stable regime: 4.5°C
- Warm regime: 10°C
- Hot regime: 16°C

If the temperature falls below 4.5°C the crop will probably be damaged. This is not a clearly defined limit, but a generalization that is worth bearing in mind.
The minimum nighttime temperature in winter for greenhouses is thus 10°, and the peak is 16°C. Cold frames can be kept warmer.

The temperature defines the limits between which a plant species can survive, grow and achieve its peak production. Production depends on the interaction of the temperature with other factors such as light, water and

Recording thermometer

nutrients. Thus for each species there is a minimum and maximum temperature below or above which the plant dies. Between these two limits lies an optimum temperature for the plant's growth and development.

In order to understand the effect of temperature on crop plants and how to control it, you have to understand a series of concepts:

Shading in greenhouses buffers the temperature within, making it more stable.

• **Zero vegetative**. This is the temperature below which the plant does not grow.

• **Critical temperatures**. These are the minimum or maximum temperatures below which or above which the plant is damaged.

• **Optimum temperature**. This is the temperature at which, all other environmental factors being equal, the plant grows best.

• **Minimum temperature**. This is the temperature below which the plant cannot perform a given phase of growth, such as flowering or fruit production.

• **Maximum temperature**. This is the temperature above which a given phase of growth is clearly negatively affected.

• **Sum of day-degrees**. All plants have overall heat requirements to grow successfully. The sum of the day-degrees is an indirect way of quantifying this temperature requirement. It is the sum of the differences (one for each day of growth) between the average daily temperature and the temperature for zero vegetative growth in the crop species. This system of accumulated heat units only provides a broad guideline for planning the crop.

3.1.1. Soil temperature

The plant requires a suitable soil temperature to grow well. This allows the roots to grow well, and ensures that the salts in the soil are soluble. A suitable soil temperature supports good microbial growth, and the microbes mineralize organic matter and thus return minerals to the soil.

The soil heat is obtained from the sunlight and the heat accumulated by the air within the greenhouse. The soil can be artificially warmed by installing a heating system. Absorption and

Tomatoes growing in a shaded greenhouse

accumulation of heat by the soil are indirectly affected by humidity levels, the greenhouse's orientation, height and the type of roofing material.

The soil class has a major effect on its temperature. Sandy soils retain little heat, meaning that they warm up and cool down relatively rapidly. Clay soils, loamy soils and those rich in organic matter undergo much smaller changes in temperature, because they contain much more water than sandy soils. As a generalization, dark soils absorb more heat than light-colored soils.

3.1.2. Temperature-related phenomena

TRANSPIRATION

If the temperature within the greenhouse is high and the plants do not receive enough water, they wilt. This is particularly damaging, even irreversible, in young plants.

THERMOPERIODICITY

Thermoperiodicity, or thermal periodicity, is the phenomenon by which crops grow best when the nighttime temperature is lower than the daytime temperature. Within limits, the plant grows better if the night is cooler than the day, rather than if the temperature is maintained constant.

It is best if there is a temperature difference of 8-10°C for optimal growth, though this of course also depends on the daylength, the light intensity and the temperature.

VERNALIZATION

Some plants are adapted to environments with cold winters, and have evolved systems to ensure that they do not flower until the cold period has passed. Their flowering is thus induced by low temperatures, and this phenomenon is called "vernalization".

In temperate zones where this type of plant is cultivated, it is important to control the temperature, as there is a risk that a period of unusually low temperatures may cause premature flowering which decreases the product's quality.

FROSTS

When the air temperature falls below 0°C, this is called a frost, and the longer it is and the colder it is, the more severe it is said to be. The dew-point is the temperature at which the water vapor in the air condenses. If the air's relative humidity is high and the temperature then falls, the dew-point will be a relatively high temperature, and when the temperature falls below this value, dew will start to form (the water condenses from a gas to a liquid).

When the temperature falls below 0°C, there is a further change of state (from liquid to solid) and the dew is then deposited on the plant's leaves as ice crystals, commonly known as hoarfrost.

If the relative humidity of the air is low, however, the dew point may be at a temperature lower than 0°C. In this case, a fall in temperature below freezing point may not be accompanied by the deposition of ice crystals but can still cause serious damage to the plants cultivated. A killing frost without hoarfrost is

known as a "black frost", because the plant wilts very clearly, and the parts affected turn black.

TEMPERATURE INVERSION

Altitude is a major factor influencing the temperature, and the higher the altitude the lower the temperatures. This is because atmospheric pressure decreases with altitude, and when the air expands it cools down. The temperature falls by between 0.6°C and 1.0°C for every 100 m increase in altitude.

3.2. LIGHT

The sunlight reaching the earth is divided into 3 types, long wave ultra-violet light, visible light, and short wave infrared.

• **Ultra-violet radiation** has a wavelength between 300 and 400 nanometers (nm). UV light influences normal plant growth, and is responsible for the deterioration of the roofing plastic. Most UV radiation is absorbed by the ozone present in the upper atmosphere.

• **Visible light** consists of the light with a wavelength between 400 nm and 760 nm. It is essential for the plant's growth, and flowering and germination depend on sufficient light. The part of the electromagnetic spectrum that includes visible light is called PAR (photosynthetically active radiation). This band of radiation is where the plant's photosynthetic pigments, the different forms of chlorophyll, absorb most.

• **Infra-red radiation**, with a wavelength of 760 nanometers to 2,500 nanometers, is responsible for warming the soil and the plants.

Type of light	Plant physiology	Germination	Stem growth	Leaf size	Photosynthesis	Rooting
UV	Near	X	X	X		X
UV	Far	Restrictive effect			N	
VISIBLE	Violet					
VISIBLE	Indigo	O	G	O	O	O
VISIBLE	Blue					
VISIBLE	Green	N	N	N	G	N
VISIBLE	Yellow					
VISIBLE	Orange	O	O	G	O	G
VISIBLE	Red					
INFRA-RED	Near	Needed for heating				
INFRA-RED	Far	Needed to conserve heat				

X = Bad   O = Optimum   G = Good   N = Normal

The physiological effects of light on plants (J. Bueno Abella, in "Fertilización", issue 93)

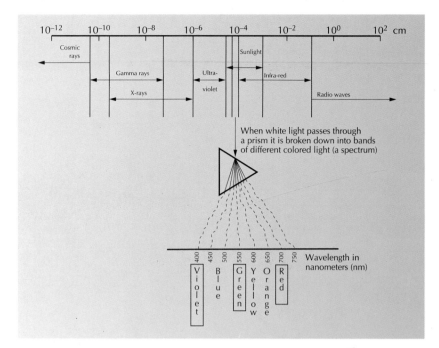

The electromagnetic spectrum

When white light passes through a prism it is broken down into bands of different colored light (a spectrum)

Wavelength in nanometers (nm)

The soil and plants absorb heat during the day, and emit it at night as infrared radiation with a wavelength of more than 2,500 nanometers. The purpose of the plastic roofing materials on the greenhouse is to trap this heat by preventing the infrared radiation from escaping. How much radiation the plastic traps and how much it lets pass determines how effective the plastic is as an insulating agent.

The intensity of the sunlight basically determines the light levels within the greenhouse. The strength of the sunlight in turn depends on environmental factors (varying with the latitude and the position of the sun over the course of the day), the characteristics of the greenhouse's structure and, most of all, the roofing.

The strength of the sunshine can be considered the "potential" light intensity within the greenhouse while the "real" value corresponds to the fraction of the radiation that actually enters the greenhouse, and which is determined by the characteristics of the roofing and the presence of clouds in the sky.

Not all plants require the same light to achieve the maximum rate of photosynthesis. In many plants, too much light may cause a series of negative phenomena, such as decreased net photosynthesis, which may decline to zero, and the accumulation of intermediate substances, catabolic processes, etc.

3.2.1. Phenomena related to light

PHOTOSYNTHESIS

In photosynthesis, organic compounds are synthesized by the reduction of carbon dioxide using energy absorbed by chlorophyll from sunlight.
The carbon dioxide is reduced with hydrogen obtained by splitting the water molecule, and results in formation of carbohydrates and the release of oxygen. This reaction cannot take place without light. It can be represented by the equation:

$$6\ CO_2 + 6\ H_2O \xrightarrow{\substack{\text{Sunlight} \\ \text{(674,000 calories)}}} C_6H_{12}O_6 + 6\ O_2$$
$$\text{Carbohydrates}$$

General transmission curve for different wavelengths of sunlight and infrared (black body) radiation.

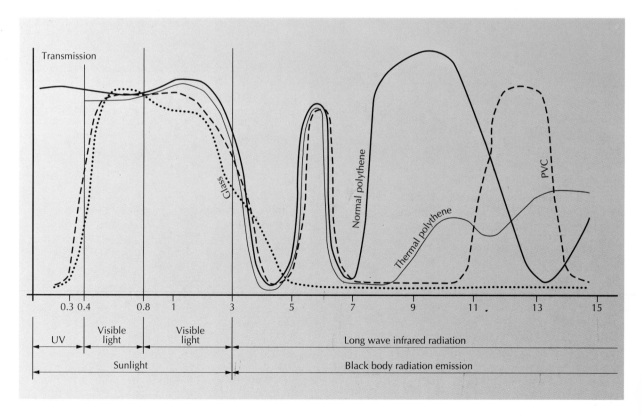

Transmission

Glass

Normal polythene

Thermal polythene

PVC

0.3 0.4   0.8   1   3   5   7   9   11   13   15

UV | Visible light | Visible light | Long wave infrared radiation

Sunlight | Black body radiation emission

PHOTOPERIODISM

This is a physiological phenomenon by which the flowering of many plants occurs in response to the lengthening and shortening of the periods of light and dark, and especially the duration of the dark period.

Some plants require long days to flower, others require short days, and some plants are neutral or unaffected by the length of the day period. Flowering is not induced in long-day plants until the daylength exceeds a critical period (such as 14 hours daylight). In short-day plants, flowering is not induced until daylength has shortened to less than a critical value. Day-neutral plants are insensitive to day length, and do not flower in response to shortening or lengthening days.

PHOTOMORPHOGENESIS

The spectral composition of the light has a major effect on how plants grow.

Ultraviolet radiation has an unfavorable effect on the plant's shape, leading to large leaves but short stunted plants.

Infrared radiation has little influence as such in growth; however, the heat effect these radiations cause does have an influence.

Illumination with red light causes excessive elongation of the stems and the formation of small leaves.

PHOTOTROPISM

This is the phenomenon by which plants grow towards the source of the light.

This is because the light acts on the formation or breakdown of the plant hormones called auxins, which are responsible for cell elongation and division. Auxins are not produced in the illuminated part, but the less illuminated areas do produce auxins. Thus, the part of the stem exposed to the light grows less than the shaded part, and the stem then bends towards the source of the light.

3.3. HUMIDITY

Each crop plant has its own humidity requirements for best vegetative growth in a greenhouse. When the humidity is much higher or lower than these limits, the plants suffer disorders that lead to decreases in yield, and the plant may even die. The humidity of the air varies greatly and depends strictly on the temperature, and the relative humidity may decrease from 60% to 25% if the same air is warmed from 10°C to 25°C. For the same amount of water in the air, the environment is moister at lower temperatures, and less humid at higher temperatures.

The amount of moisture present in the air within the greenhouse is directly proportional to the moisture content of the soil, that is to say, the amount of water that the site has stored and retained, and also to the water needs of the crop throughout its growing cycle.

The environmental humidity has a major influence on transpiration and plant growth. Low environmental humidity leads to dehydration of the plant. Dry environmental conditions always represent a check to growth, and this may be irreversible, meaning that watering the plant will not be enough for full recovery. When humidity is excessive, growth slows down, even if temperatures are ideal.

It is necessary to know the rate of evapotranspiration corresponding to the quantity of water that the plants and the soil contribute to the environment.

Plants grow towards the light

If the plant needs more water than is available in the soil, the plant rations the water by closing the stomata, halting carbon dioxide uptake at the same time as water loss, and this obviously limits photosynthesis. If the water supplied by the plant and the soil to the environment is sufficient, the stomata remain open.

Over the course of the day, the plant's water requirements increase, and the stomata close when the water supply is insufficient, thus slowing down photosynthesis just when the light is most intense.

The moisture content of the air is limited by the air temperature, and the maximum moisture content is greater at higher temperatures. When the moisture content increases beyond this, it reaches the saturation point, when the air contains so much water vapor that it can absorb no more, and the water condenses. This saturation point increases with the temperature.

THE QUANTITY OF WATER VAPOR REQUIRED TO SATURATE 1 CUBIC METER OF AIR, AT DIFFERENT TEMPERATURES	
Temperature in °C	Grams of water per m^3 of saturated air
–10°	2.2
–5°	3.2
0°	4.9
5°	6.8
10°	9.3
15°	12.7
20°	17.2
25°	22.8
30°	30
35°	39
40°	51

Air generator with inlet conduct

Above:
Roof installation of evaporation conditioner with vertical outlet

Below:
Installation at ground level, evaporation conditioner with roof outlet (Manufactured by GESTIÓN, ESTUDIOS Y REALIZACIONES, S.A.)

of just these two elements (50% carbon and 40% oxygen).

Carbon is absorbed from the air in the form of carbon dioxide, and in photosynthesis it is fixed in organic molecules that the plant then uses to make all the other compounds in its bodily structure.

Photosynthesis depends on light. On hot sunny days the plants in a greenhouse photosynthesize intensely, and the concentration of carbon dioxide within the greenhouse is usually lower than outside, and may be insufficient for the crop's needs, and thus become a limiting factor.

3.3.1. Humidity-related phenomena

CONDENSATION OF WATER VAPOR

Condensation of water vapor on the plastic sheeting used in tunnels and greenhouses may increase the "greenhouse effect", by blocking the escape of the infrared radiation emitted by the soil during the night, thus making the roofing more effective as insulation. Yet if this condensation leads to dripping, this may damage the plant(s) it falls on.

TRANSPIRATION

Transpiration is the process by which plants lose most of the water absorbed by their roots, almost all of which (99%+) is transpired. When temperatures increase, transpiration also increases, and more energy is used in the evaporation of water. This cools down the plant's leaves, and thus regulates the plant's temperature.

When the humidity is very high and normal transpiration cannot occur, if the roots are absorbing a lot of water, some plants can eliminate the excess water through their stomata as tiny drops of liquid.

EVAPOTRANSPIRATION

Evapotranspiration is the sum of the water lost by transpiration from the plant and the evaporation from the soil, and can be quantified for the real conditions for a unit surface area and a given period of time. This phenomenon thus depends on the crop and the weather conditions and the soil type.

3.4. CARBON DIOXIDE

Carbon and oxygen are essential to plants, roughly 90% of a plant's dry weight consisting

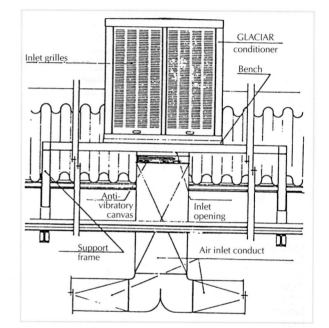

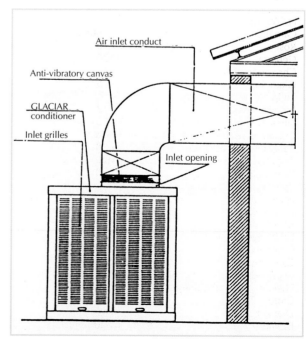

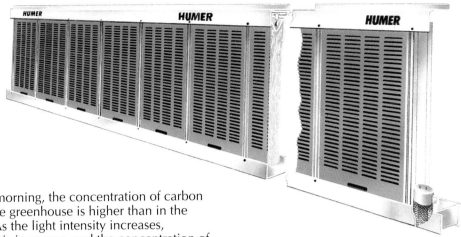

In the early morning, the concentration of carbon dioxide in the greenhouse is higher than in the outside air. As the light intensity increases, photosynthesis increases, and the concentration of carbon dioxide rapidly falls to a very low level (200 ppm).
In normal conditions, the concentration of carbon dioxide in the air is over 300 ppm.

Note that in winter, on cloudy days, the concentration of carbon dioxide is lower than on sunny days with a clear sky, as on cloudy days the environmental conditions mean the greenhouses are closed all day long, and all the carbon dioxide within is absorbed by the plants, and cannot be replenished from outside.

Therefore, shortage of carbon dioxide must be taken into account as a possible factor limiting plant growth in the months of winter, together with light. The level of carbon dioxide is above all related to the sunlight and the outside temperature.

In the summer months, high temperatures within the greenhouse make it necessary to open the windows. The level of carbon dioxide, which may have declined, then returns to normal.

Optimum levels of CO_2, relative humidity (%), substrate temperature and light intensity (Tesi, 1969)

DN = Day-neutral
LD = Long-day
SD = Short-day

Species	Optimum substrate temperature (°C)	(CO_2) (ppm)	Relative humidity (%)	Light	
				Light intensity (in lux)	Length (hours)
VEGETABLE					
Tomato	15-20	1,000-2,000	55-60	10,000-40,000	DN
Cucumber	20-21	1,000-3,000	70-90	15,000-40,000	LD
Melon	20-22	—	60-80	—	LD
Pepper	15-20	—	65-70	—	LD
Eggplant	15-20	—	65-70	—	LD
Lettuce	10-12	1,000-2,000	60-80	12,000-30,000	LD
Strawberry	12-15	—	60-70	—	SD
FLOWERS					
Carnation	15-18 (rooting)	500-1,000	70-80	15,000-45,000	DN
Rose	15-18	1,000-2,000	70-75	Full sunshine	DN
Gerbera	18-20	—	60-70	Full sunshine	DN
Chrysanthemum	18	400-1,200	60-70		SD
Gladiolus	10-15	—	60-70	Full sunshine	LD
Tulip	8-12	—	70-80	Full sunshine	LD
Iris and daffodil	10-13	—	60-70	Full sunshine	LD
Lily and freesia	10-15	—	60-70	Full sunshine	LD
Cyclamen	14-16	—	60-70	Semi-shade	DN
Azalea-Rhododendron	15-18	—	80-95		LD
Begonia	18-20	—	60-70	Semi-shade	LD
Poinsettia	18-20	—	60-70	Full sunshine	SD
Primrose and Calceolaria	—	—	60-75	Full sunshine	LD
Pelargonium	—	1,000-2,000	60-70	Full sunshine	LD
African violet	20-22	—	70-80	5,000-20,000	
Kalanchoe	—	—	60-70	Full sunshine	SD
Hydrangea	18-20	—	70-80	Full sunshine	LD
Gardenia	19-22	—	—	Full sunshine	LD
Cymbidium	10-14	—	80-90	15-30,000	DN
Cypripedium	10-14	—	80-90	15-30,000	DN
Phalaenopsis and Cattleya	16-18	—	80-90	15,000	SD
Croton, Ficus	21-21	—	80-90	Full sunshine	
Dieffenbachia	18-20	800-1,200	85-95	12,000-15,000	
Bromeliads	18-20	—	80-90	Semi-shade	

4. HEATING AND COOLING

The increasing shortage of energy means that studies aiming to reduce energy consumption are more and more common.

Of all the climatic variables that can be controlled within the greenhouse, the temperature is perhaps the most important, especially in countries with cold climates, though all the other variables must also be taken into account, such as humidity, light and carbon dioxide concentrations.

Unheated greenhouses may not be warm enough for the crop to achieve the desired levels of production. It is thus normally necessary to install heating to prevent temperatures falling so low they kill the plant, and to ensure an optimum temperature for the crop.

The greenhouse's geographical location and the crop species grown in it are the main factors determining

Cooling the greenhouse by opening the windows

• **Convection**. This is the transfer of heat by movement of a heated fluid (such as a gas or liquid). This may be:

- with the external environment
- with the environment inside the greenhouse
- with the soil
- through a wall that is not airtight

• **Conduction**. This is the transfer of heat through a material. This may be:

- through the soil inside
- through the wall of the greenhouse

To combine all these energy exchanges into a single mathematical expression is complex, given that many factors intervene, and they all influence each other, as already pointed out.

Finding out the values of the individual processes composing the greenhouse's energy balance makes it possible to calculate the energy input required to maintain a daytime-nighttime temperature regime suitable for the crop species.

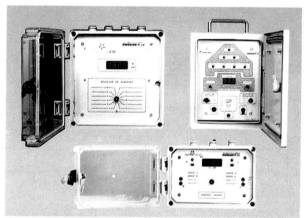

Upper right: Devices to monitor environmental parameters

how much energy is needed to maintain a suitable microclimate within the greenhouse.

The energy exchanges between the interior of the greenhouse and the exterior are complex and influence each other, involving all three different mechanisms of heat exchange, radiation, convection and conduction. These energy transfers can be summed up as follows:

Upper right and center: Measurement systems used in temperature control (Manufactured by SERDIA)

• **Radiation**. Energy is transferred without being transmitted by any intermediate material. The energy may be:

- derived from the soil, the atmosphere, the environment and the vegetation
- emitted towards the environment by the structure and by the roof.

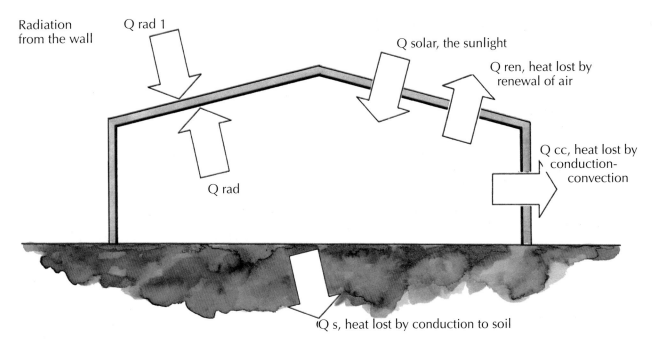

Radiation from the wall
Q rad 1
Q solar, the sunlight
Q ren, heat lost by renewal of air
Q rad
Q cc, heat lost by conduction-convection
Q s, heat lost by conduction to soil

Energy exchanges in the greenhouse

4.1. DAYTIME ENERGY BALANCE

The energy balance of the air in the greenhouse, for the daytime period with sunshine, is defined as the state of balance between the energy inputs and outputs, and can be simplified and represented as follows:

Accumulated energy = energy inputs - energy losses

The initial equation is:

$$Q(Kcal/h) = Q\ solar - (Q\ cc + Q\ rad + Q\ s + Q\ ren)$$

where:

- Q is the nett gain or loss within the greenhouse.
- Q cc is the heat lost by conduction-convection. The losses of heat through the roofing to its surface, and due to the temperature difference between the inside and the outside.
- Q rad is the heat lost by radiation. This depends on the surface, the covering material's coefficient of heat transmission, and on the temperatures inside and outside the greenhouse.
- Q s is the heat lost by conduction to the soil. The heat lost to the soil is influenced by the soil's composition and its moisture content.
- Q ren is the heat lost through the renewal of air, and depends on the number of times per hour the air of the greenhouse is renewed.
- Q solar is the energy input from the sun.

The term Q solar indicates an energy gain and the values that are deducted from it represent energy losses. Thus, if Q is positive, this means that the greenhouse's energy balance is positive and it is warming up, while if the value of Q is negative, the greenhouse is losing heat, and getting colder.

Calculating the values for the nighttime period, when Q solar is zero, gives the heating capacity to install in the greenhouse.

4.2. TEMPERATURE CONTROL

4.2.1. Heating

The purpose of building a greenhouse is to create an artificial microclimate more favorable than adverse external conditions. This is done by maintaining temperatures and humidity levels close to the optimum for the plant's growth.

Heating installation using hot water pipes

Mobile shading systems to control the temperature

Crops can be grown in a greenhouse at any time of year, including the summer, but the most profitable crops are those produced when low external temperatures prevent successful open air cultivation. In this case, in order to grow crops in these critical cold periods, it is necessary to warm the air and soil in the greenhouse to levels that allow the plants to grow.

Heating the greenhouse is of great importance, both to prevent damage by unforeseen and exceptionally cold conditions and to provide the plants with an optimal growing environment.

As a guideline, the energy input required can be estimated at 10 Kcal/h per square meter of greenhouse floorspace per degree Celsius of temperature difference between the greenhouse and the outside.

There are some simplified formulae to calculate the heat input required for the greenhouse. One formula advanced by several authors is based on bringing together all the losses into a single expression of the type:

$$Q = Kr \cdot S(Ti - To)$$

where:

- Q is the heat required in kilocalories per hour
- Kr is a constant with the following values:
  - glass: 5
  - 400 gauge polyester: 6 to 6.5
  - double chamber polyester: 2.7
  - normal polyester: 4.5 to 5
  - iron: 6.5
  - polythene: 6
- S, the surface area of the walls and roof in square meters
- (Ti - To), the temperature difference, that is to say the desired minimum temperature inside the greenhouse (Ti) minus the lowest minimum temperature outside (To) over the course of the year.

The resulting value is increased by 10% to 15% to give a rough outline of the number of Kcal/h required to heat the greenhouse.

A worked example:

We have a greenhouse covering a surface area of 1,000 m^2, and we wish to raise the temperature inside about 10°C above the temperature outside. The total surface area of the (polyethylene) walls and roofs is estimated to be 1,500 m^2. The number of kilocalories per hour required to heat the greenhouse can be calculated using the formula above. Thus:

$$Q = 6 \text{ Kcal/h/m}^2/°C \times 1,500 \text{ m}^2 \times 10°C$$

$$Q = 6 \times 1,500 \times 10 = 90,000 \text{ Kcal/h}$$

Increasing this by 10% gives:

$$Q = 99,000 \text{ Kcal/h, or roughly 100,000 Kcal/h}$$

Every square meter of the area of the greenhouse requires:

$$100,000/1,000 = 100 \text{ kcal/h/m}^2$$

4.2.2. Heating systems

The heat lost by the greenhouse is compensated by the installation of a heating system used to control the temperature within the greenhouse.

A greenhouse heating system must fulfil the following conditions:

• **Reliability**: it must be reliable as a failure during the cold period may seriously affect the plant's growth.

• **Effectiveness**: it must be capable of maintaining the fixed temperature throughout the greenhouse.

• **Flexibility**: That is to say the heater unit should perform other functions, such as producing steam to disinfect soil.

The basic criteria for selecting the heating system are:

• The crop is to be grown in the greenhouse (its temperature and water requirements)
• The structural features of the greenhouse
• The surface area of the greenhouse
• The cost of the installation
• The availability and cost of different types of fuel
• Operating costs
• The installation's guarantee
• The installation's useful life
• The zone's climate
• The ease of supply
• Purchasing facilities

The air in the greenhouse can be heated in several different ways. The systems used vary tremendously, from rudimentary woodburning heaters to highly sophisticated centrally controlled automatic heating systems.

Heating systems can be divided into:

• Hot air systems
• Hot water systems
• Electric heating systems

Electric heater systems are not widely used in greenhouses, because other systems are cheaper, even though automated electric heating systems eliminate the need for labor. Electricity is worth using to heat the greenhouse soil, if the energy is used at night at the cheaper nighttime tariff. Greenhouses, especially small greenhouses, can be heated with special heaters called storage heaters.

Installing storage heaters is relatively expensive. However using offpeak electricity at the nighttime tariff is a significant saving. A storage heater consists of:

• A heating device consisting of a set of resistances.
• A storage device made of some heat-resistant material, which may reach a temperature of 1,000°C.
• A device to recover the accumulated heat

A fan sucks air from within the greenhouse, and then circulates it along channels within the heat-resistant material. The hot air emerges from the top of the heater. The installation is of course controlled by a thermostat.

Heaters range from small models similar to domestic heaters, consuming 1 or 2 kilowatts, to industrial models consuming hundreds of kilowatts.

Temperature regulation is based on thermostats, which may be set to a precision of +/– 1°C. The most widely used type is the ambient thermostat. This is based on the variation of the vapor pressure of a liquid within a membrane.

The precise location of the thermostat is very important, and the air around it must faithfully reflect the average ambient temperature of the greenhouse. It can be placed two thirds of the distance between the heater and the far end of the greenhouse, not too close to the flow of warm air, and taking care not to locate it where it will be heated by direct sunlight. You should consider installing a second thermostat to give an alarm signal when the temperature falls below a set limit.

Hot air generators (Manufactured by GER)

4.2.3. Air heating

Air heating systems can be divided into two groups:

• **Radiator heating systems** heat the surrounding layers of air by convection and irradiation. These systems may be based on hot water or steam.

• **Hot air systems** in which the air is heated within a device and is then blown by a fan into the greenhouse. They are hot air generators and aerotherms.

4.2.3.1. Centralized heating by hot water

One of the most widely used heating systems in greenhouse cultivation is to heat the inside by circulating hot water.

Water can accumulate a lot of heat. It absorbs the heat when it passes through a heating device consisting of a boiler and a burner, and it then distributes the heat as it circulates around the greenhouse. The water returns to the boiler again after it has lost part of its heat. The heater may use fuel-oil, gas oil C or natural gas.

Thermosiphon principle

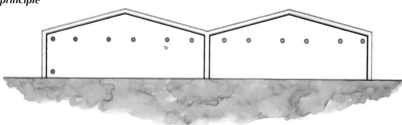

Hot water heating system
A/ Burner
B/ Boiler
C/ Pump
D/ General tubing
E/ Valves generating water flow
F/ Air thermostat
G/ Radiant tubing

These heating installations require the following components: boiler, burner, fuel deposit, tubing, pumps, automatic mechanisms and accessories. Nowadays most hot water heating systems are driven by pumps that circulate the water, but there are also older installations that rely on the

thermosiphon principle. This system requires an adequate difference in height between the greenhouse and the boiler.

A thermosiphon works as follows: when the water is heated, its density decreases and it rises within the tubing. As it cools down, it increases in density and descends under the action of gravity towards the boiler, thus completing the cycle.

Hot water systems using the thermosiphon effect are no longer used, having been replaced by systems that pump hot water. If a pump is installed at the outlet from the boiler, an expansion vessel is also needed to allow for the expansion in volume of the water in the circuit, otherwise the tubes would split.

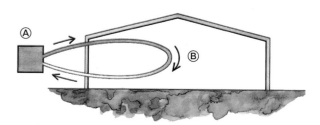

The water absorbs heat when it passes through the boiler (A), and diminishes in density, rising up through the tubing and releasing the heat on its path (B). In the greenhouse it cools down and thus increases in density, and then falls back to the boiler, where it starts the cycle again.

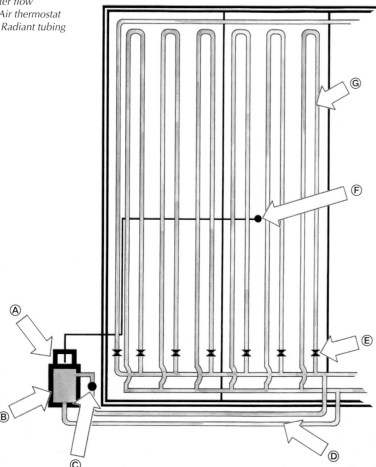

The use of a pump to drive the water means that narrower tubes can be used, with a diameter of 10 to 2.5-5 cm. The tubing must be installed at the right height for the heat to be distributed evenly throughout the greenhouse. This also reduces the number of obstacles and the area shaded.

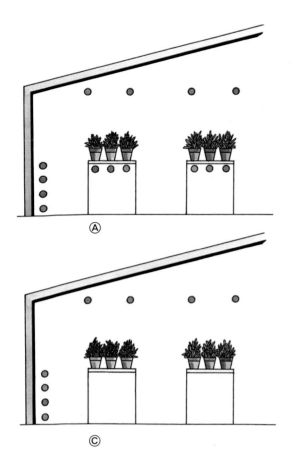

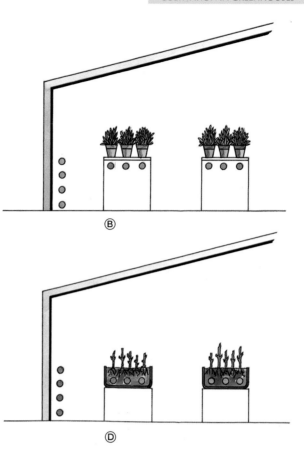

Diagram of different ways of distributing the heating pipes. Simultaneous heating of the air in the greenhouse and the air around the plant pot.
D shows the rooting of cuttings on a heated bench.

Cultivation benches with heating

The arrangement of the tubing in the greenhouse may vary greatly, though they are generally installed around the perimeter and follow the line of the guttering above the growing plants.

Installing the heating tubing at ground level is, however, becoming increasingly common. This creates a more favorable microclimate at the level the plant is growing, and avoids the formation of a layer of hot air at the top of the greenhouse.

Figures A, B, C and D show plant cultivation in pots with heating of the air in the greenhouse as a whole and that around the pots. Diagram D refers to rooting cuttings.

Metal tubing is normally used to transport hot water to the greenhouse, but plastic tubing, such as polypropylene and polyethylene, is increasingly common.

In these tubes, the water circulates at a temperature between 35°C and 45°C. This temperature regime has given rise to a new layout for the tubing in the greenhouses where the plants are cultivated on benches.

The metal tubes may be in the air and/or mounted on the greenhouse walls, while plastic tubing runs along the benches and is in contact with the pots with the plants.

Heating by hot water is worthwhile in greenhouse cultivation of ornamental plants.

It has the following advantages:

• The system is easy to use
• It does not dry out the air too much
• Maintenance costs are low
• The system has some thermal inertia (resistance to sudden changes)

Cross-section of a bench with heating for pot plant cultivation
A/ Polyethylene sheet
B/ and E/ Aluminium sheeting
C/ Porexpan
D/ 20 mm heating tube

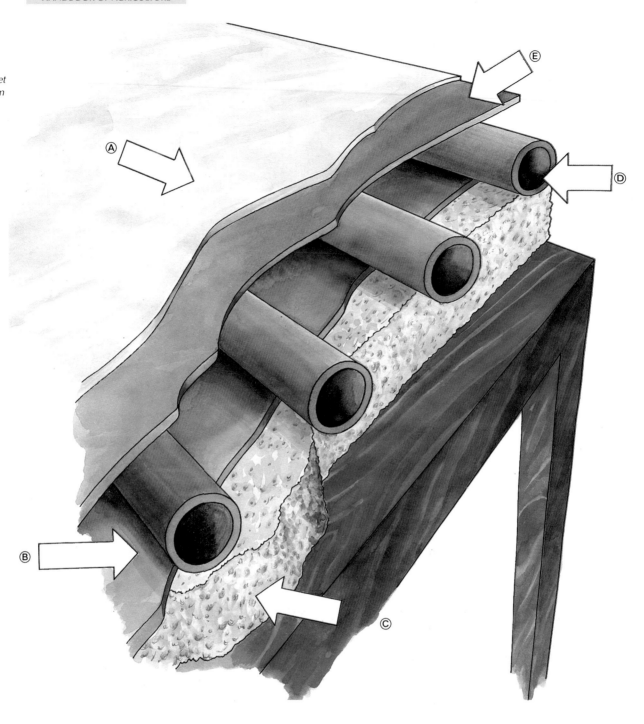

Thermal inertia means the installation's ability to maintain the warmth after it stops working. The slower the installation cools down, the greater its thermal inertia.

Its high thermal inertia has the following disadvantage: there is a long delay between turning the system on and reaching the desired temperature. The cost of installation is also high.

4.2.3.2. Heating by steam

The difference between this and the preceding system is that the water emerges from the boiler as steam under pressure. This steam derived from the boilers or high pressure

generators is channelled at medium or low pressure along the tubing and throughout the greenhouse. The heat is lost over the course of its path as it condenses, meaning that it returns to the boiler as liquid water.

This system has very little thermal inertia, and it sometimes shows very unequal distribution of heat, and it also dries out the environment more than heating by hot water.

4.2.3.3. Heating by hot air

Heating with hot air consists of passing air over a heat source, which may be direct or by means of a heat exchanger, and then emitting it into the greenhouse.

The heating system installed can use hot air generators. This system offers two possibilities:

• The hot air passes directly from the generator into the greenhouse, where it diffuses, and the waste gases are discharged through a chimney.

• The hot air is mixed with the discharge gases and circulated inside the greenhouse, and is then eliminated by means of special tubes that discharge it outside.

The generator can be placed inside or outside the greenhouse. If placed inside, it requires an air supply sufficient to ensure combustion, and it must have a chimney to discharge the exhaust gases outside the greenhouse.

If the device is placed outside, this can be done in a way so it can heat more than one greenhouse.

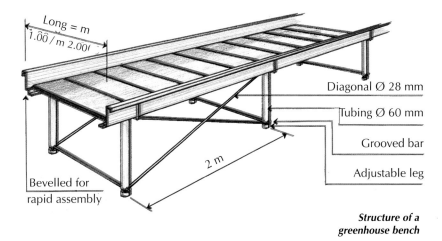

Structure of a greenhouse bench

Power Kcal/hour	Volume m³/hour/20°C	Fan motor in HP	Chimney diameter (mm)	Fuel-oil consumption, in Kg/h
80,000	5,550	2	180	9.4
200,000	13,700	2 x 2.5	300	23.5
250,000	17,200	2 x 3	350	29.5
300,000	20,800	2 x 3	350	35.4
400,000	27,500	2 x 4	350	47
550,000	38,000	2 x 4	400	64.8
640,000	44,100	2 x 5	400	72.5
800,000	55,400	2 x 7.5	400	94

Technical characteristics of hot air generators

In both cases, the hot air can be blown into tubing consisting of perforated plastic sheets placed about 1.5 m above ground level to distribute the heat more uniformly and create greater air turbulence.

The advantages and disadvantages of hot air heating systems are:

• **Advantages:**

- The installation costs are lower than hot water heating systems.
- It does not get in the way when working inside the greenhouse.
- The movement of the hot air makes it possible to eliminate the condensation on the roof in plastic greenhouses.
- In summer, the generators can be used to lower the temperature by blowing air from outside the greenhouse into it.
- As they have little thermal inertia, once turned on they soon raise the temperature inside the greenhouse.

• **Disadvantages:**

- The turbulence created within the greenhouse leads to increased heat losses.
- In case of malfunction, the temperature falls rapidly, as the system has little thermal inertia.
- The running costs are greater than in hot water heating systems.
- Heat distribution is not uniform. This can be resolved by using the distributor tubes mentioned above, i.e., by placing perforated plastic tubing over the hot air outlets that then run all over the greenhouse.

Fuel-oil is the fuel normally used in these systems, which usually also have an electric ignition system activated by a thermostat.

4.2.3.4. Aerothermal heating

Aerotherms, or hot air furnaces, are devices that use hot water or steam generated in boilers or generators to heat the surrounding air, and then propel it into the greenhouse by means of powerful fans. This is a mixed system halfway between

heating by hot water and heating by hot air.
The water or steam from the central boiler reaches
the aerotherm installed inside the greenhouse
through an iron tube; the aerotherm itself consists of
radiator panels and a fan.

The water or steam heats the radiator and returns to
the boiler through another tube. The water may
move within the tube by the thermosiphon effect, or
be pumped.

This system has the disadvantage that the heat is
badly distributed if the different elements
making up the system are not installed
absolutely correctly. The aerotherm is suspended
within the greenhouse and the air may be blown
horizontally or vertically.

The hot air leaves the aerotherm at a defined
temperature, and its temperature decreases the
farther it moves from the aerotherm. To get around
this problem, it is necessary to work with a large
volume of air that passes through the aerotherm
slowly (roughly 1 m/second at 30°C).

4.2.4. Heating the soil

Artificially heating the soil is expensive. It is
not worthwhile in most greenhouse crops,

when the costs are weighed against the benefits
obtained.

Soil heating is only used when cultivating a small
area of a highly profitable crop, such as some
flowers for cutting, some ornamentals, and for the
propagation of many different types of plant
(seedbeds, cuttings, etc.).

4.2.4.1 Heating the soil in cold frames

Electrically heated cold frames ("hotbeds") consist of
the following layers of different materials over the
base:

• Drainage
• Insulating material that also functions as drainage,
such as glass wool
• A layer of sand 5 to 10 cm deep
• Electric cables arranged in parallel lines (the
cables must not cross or touch each other).
• Sand to cover the cables (which are protected
from the plants by a metallic net).
• A layer of earth for the plants or the seeds to be
germinated

The installation can be regulated automatically
or manually. The length of heating can b
controlled manually by a switch. If it is

***Diagram of a soil
heating system***
A/ *Boiler*
B/ *Electricity supply*
C/ *Return*
D/ *Pump*
E/ *20 cm*
F/ *1/2" or 3/4" PE
tube loops*
G/ *Lines of tubes 40
cm apart*
H, I, *and* ***J/*** *Main
heating tubes, one
supply and two
returns*

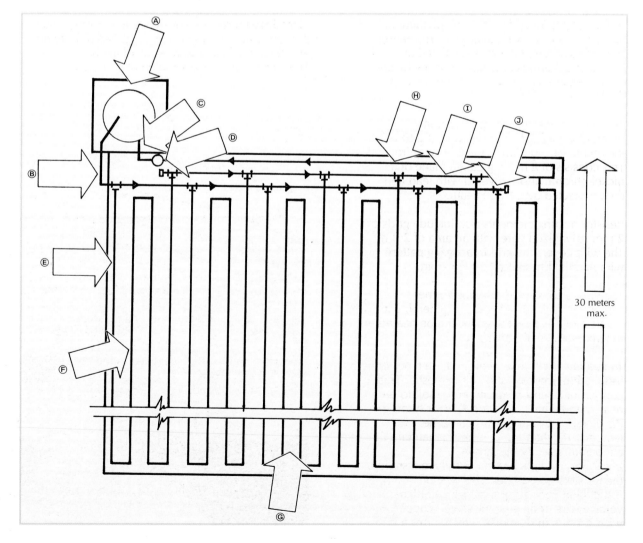

30 meters
max.

automatic, it is controlled by a thermostat installed in the soil.

4.2.4.2. Heater elements and grids

There are now electric systems to heat the soil. Heat produced in a heating element (an electrical resistor) warms up the surrounding soil. The soil temperature is regulated by means of a thermostat which turns the heater off to save energy. The temperature is set by the farmer.

This type of installation can function at different voltages. When working at 220 volts, the complete installation must be acquired. This consists of a heating cable, made of a nickel-chrome alloy, covered with lead and then protected by a PVC sheath.

Resistor heating elements also include grids of 2 mm galvanized wire with an area of 2-3 m^2. The wire covers the grid in a zigzag pattern, with parallel lines about 20 cm apart.

The grid is connected to a transformer supplying power at 40 volts. In general, with resistors of this type the wires can be heated to temperatures of 50°C.

To regulate the temperature in these types of installation, there must be a thermostat. This consists of a sensor located in the soil to be heated and which is responsible for turning the electric current on and off as the temperature changes, in order to maintain the temperature set by the farmer.
If very low voltage heaters, 24 V, are installed, the resistors are simpler. This type of installation consists of a galvanized high-voltage wire or an enamel-coated copper wire 3.4 mm in diameter, which forms a grid

above the soil to be heated; these wires can take a supply of 40-50 watts.

There also has to be a main transformer with an output at 24 volts. The cable joining the transformer to the grid must be copper, with a cross-section of 80 mm^2 or more. The connections have to be very good, to ensure good contact. There must also be a thermostat to regulate the temperature.

4.2.4.3. The tubing network

Heating the soil by a hot fluid requires a heat source and a network of tubes to transfer the heat. The heat source can be the same if the greenhouse is heated by hot water or steam. The most suitable material for the tubing is plastic, either polyethylene (PE) or PVC, which can work perfectly with water at 60°C, and can even tolerate temperatures of 100°C for a short time.

There are two types of tubing system:

• **Buried tubing**: to heat the soil in a greenhouse where the crop is grown in the soil. The tubes are buried at a depth of 20 cm, and are fixed in place by supports. The lines of tubes are placed 50 cm apart.

• **Tubing at soil level**: this is used to heat the soil or concrete benches on which the containers are placed. The tubes are covered with 4-5 cm of sand, beneath which are panels of expanded polystyrene 20-30 cm thick, to ensure none of the heat is lost. The water in the tubes should not be hotter than 50°C.

Traditional heating system based on heated tubes suspended from the ceiling

Electric lights to increase photosynthesis

Ventilation systems

Above: Details of heating system with piped hot water and a ventilated radiator

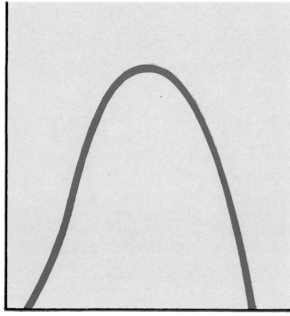

Growth

Temperature in °C

Growth curve
Graph of root growth at different temperatures

4.2.4.4. The advantages of heating the soil in greenhouses

As the heat is applied to the base, the air temperature of the greenhouse is much more uniform than with the traditional heating by hot tubes hanging from the roof. There is almost no vertical temperature gradient, and so the greenhouse roof is not at risk of overheating when trying to raise the temperature on which the level the plants are growing.
Water at 30-40°C can be used to heat the soil,

allowing use of heating systems like geothermal energy, industrial waste heat and low temperature solar power installations. Heating at a relatively low temperature is also convenient in installations that burn any type of fuel, as low temperatures lose less energy than high temperature ones.

It has been shown that it is to some extent possible to reduce the air temperature below that considered optimum, as long as the root temperature is increased at the same time.

If the soil is warm, this increases the thermal inertia of the greenhouse, and so if there is a power cut, for example, the temperature remains near the desired value for some time.

4.2.5. Choosing the heating system and power supply

The main points to bear in mind when choosing a heating system are:

• Hot-air installations are cheaper than central heating or hot water heating systems, but they have a shorter useful life because they are turned on and off relatively frequently.
• Centralized hot-air heating for several greenhouses is more expensive than if it is installed separately in each one.
• The temperature within the greenhouses varies less if it is heated by hot water than if it is heated by hot air, and so in crops that are sensitive to temperature changes, you are recommended to use central heating based hot water or steam.
• In the case of a failure, a greenhouse heated by hot air cools down rapidly, while one heated by hot water keeps warm for much longer, because water has a very high thermal inertia.
• Heating by hot water or by aerotherms creates less obstacles within the greenhouse than heating by hot water using tubing as radiators, and so mechanization is easier, and less shade is cast on the crop.
• Water condensation is a more serious problem in greenhouses heated by hot water, and this increases the risk of fungal diseases, while the flow of air in greenhouses heated by hot air has a positive effect on the plant's health.
• The outlet for the hot air has to be properly positioned otherwise it may desiccate nearby plants.
• In summer, the fans of the hot air system can be used to cool the greenhouse, normally by 3-5°C.
• If the heating system is used for many hours over the course of the year, it is worth installing a highly efficient system even if the initial investment is higher. If the installation is not going to be used for many hours a year, a low cost installation is better, even if it is less energy efficient.

The fuels used are distilled from petrol, and vary in

their physical and chemical properties, such as viscosity, flash point, freezing point, water content, sulfur content energy content and yield.

The sulfur content is very important and varies in petrol from different sources.
The energy content is the amount of heat obtained from complete combustion of a unit weight of the fuel.
The energy crisis obliges us to find new sources of energy, including use of solid fuels instead of petrol. Materials to be used as fuels should have a suitable energy content, a low sulfur content and produce little ash.

most plants, as long as it does not produce excess sulfur derivatives.

• Propane and electricity are not recommended for use in hot-water systems with an energy requirement of more 100,000 kcal/h. Below 500,000 kcal/h, use gas-oil and above this value, use gasoline.

• Heating a greenhouse by burning propane inside it has the advantage that the exhaust gases do not need to be discharged, because they contain no sulfur derivatives, unlike other solid and liquid fuels.

The characteristics of the most widely used fuels

Fuel	Energy content Kcal/kg	Energy yield	Boiler efficiency	Installation costs	Maintenance costs	Preheating requirement
Liquids						
- Light fuel oil	10,600	80%	70%	Low	Medium	No
- Gasoline C	11,000	85%	75%	Low	Medium	No
- Heavy fuel oil	10,000	80%	70%	Medium	High	Yes
Gas						
- Propane gas	12,000	90%	80%	High	Low	No
- Butane gas	11,800	90%	80%	High	Low	No
- Natural gas	9,500	90%	80%	High	Low	No

Burning solid fuels, such as coal, peat or residues, presents greater heating problems than use of liquid or gas fuels.
Direct supply and combustion make the system less flexible to the greenhouse's requirements. Excess heat during the day often means the windows have to be opened.
Another disadvantage of this type of heating is that the fuel has to be stored, and thus requires a storage space considerably larger than that required for liquid fuel, which can be stored in buried cisterns.
The cost of installation of solid-fuel burners is greater than that for liquid fuels, and before deciding it is important to be sure that the fuel will be available for a long time after completion.

| Fuel | Generators using: | | |
	Hot air Direct combustion	Hot air Heat exchanger	Hot water
Petrol	1.10 1/h	1.12 1/h	
Gas-oil	1.00 1/h	1.26 1/h	
Propane	0.90 Kg/h	0.91 Kg/h	
Natural gas	1.05 m/h	1.05 m/h	
Light fuel oil		1.25 Kg/h	1.25 Kg/h
Heavy fuel oil		1.26 Kg/h	1.50 Kg/h

Fuel required to produce 10,000 kcal/hour

Some points about combustion:

• You are advised to install several different heating devices (boilers, heaters, generators, etc.), not just a single device, so that there is some backup in case of mechanical failure.

• Heating systems that burn fuel within the greenhouse must have an outlet for the exhaust gases, otherwise they may cause severe damage to the plants, especially ornamentals and flowers. However, the carbon dioxide produced by combustion is very positive for

• When it is necessary to heat the greenhouse during the daytime, burning propane gas within the greenhouse produces carbon dioxide and water vapor, and this is positive because carbon dioxide is used in photosynthesis and the water maintains the environmental humidity. Propane-burning installations are also cheaper.

• Most fuels require an electricity supply to move the fuel. This is not necessary for propane.

4.3. PROTECTIVE TECHNIQUES

Protective technique should ensure that the greenhouse absorbs as much sunshine as possible while reducing its heat losses.

4.3.1. Insulation

There are a series of measures that reduce the heat lost by the greenhouse, and thus the cost of heating it.
Insulation can be improved by:

• Improving the seal and making the greenhouse more airtight. Poorly adjusted roofing, or a poor seal around doors and windows, may cause the temperature to fall when cold air enters from outside.

This has the disadvantage that a perfect seal would cause the level of carbon dioxide within the greenhouse to fall, causing photosynthesis to decline. It may also create excessively humid conditions within the greenhouse, with the risk of fungal diseases.

These problems can be resolved by good ventilation.

• Placing tunnels along the rows of crops within the greenhouse. This reduces the heat loss, but the plant photosynthesizes less because it is covered by two layers of material.

In cold periods, plastic tunnels can be installed in an unheated greenhouse. They can raise the temperature within the tunnel by 5-7°C with respect to the temperature in the open air when this is in the range –5°C to 5°C.

The structure of the greenhouse itself is used to support these tunnels. At a height of 1.5 m above the soil, a galvanized wire frame is prepared that is supported by the uprights of the greenhouse structure. Plastic sheeting is placed on this frame, with the sides reaching down to the soil level. This forms large tunnels that are as wide as the separation between the uprights of the greenhouse.

Installation of an insulation panel and how it works

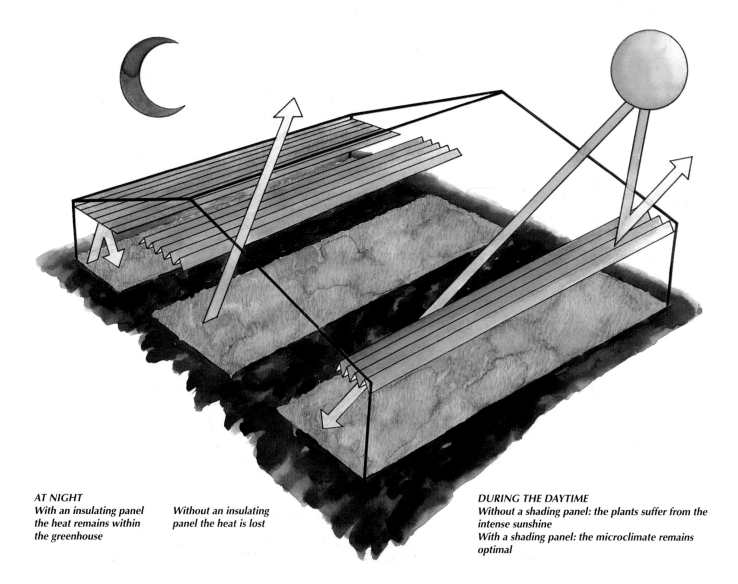

AT NIGHT
With an insulating panel the heat remains within the greenhouse

Without an insulating panel the heat is lost

DURING THE DAYTIME
Without a shading panel: the plants suffer from the intense sunshine
With a shading panel: the microclimate remains optimal

To the left:
Shading inside a
greenhouse

Smaller tunnels can be built that consist of a sheet of plastic material supported by arches. The arches are made of 3-4 mm diameter galvanized wire, and each arch is fixed into the ground at right angles to the crop rows, ensuring that all the arches forming the tunnel are in a straight line, at the same height, and at the same distance from each other. The arches at the two ends of the tunnel are fixed into the soil leaning outwards, so that they offer some resistance when the sheeting is tensed.

To attach the sheeting, bury one end in the sand, then make furrows along both sides of the arches, in order to bury the sides of the sheet, then place the sheet over the arches, covering the edges at the same time with sand.

• Use insulating panels within the greenhouse. These can be made of many different materials, such as polyethylene or polyester, sometimes partly covered with metal. They are unfolded at night and act as barriers to the heat emitted as long wave infrared radiation from the soil. They are also used as insulation, and can be used to shade the plants during the day.

Other techniques used to reduce heat loss from greenhouses include:

• Installation, in some situations, of external windbreaks to protect the greenhouse and to reduce conduction-convection losses and those caused by the turnover of air.

• Installing the greenhouse in the most suitable orientation, in order to ensure as much sunlight as possible.

4.3.2. Double roofing

Double roofing consists of placing a sheet of thin plastic inside the greenhouse adjacent to the panels or sheeting of the greenhouse's roof. An air chamber forms between the two layers that prevents loss of

To the right:
Shading systems,
activated by a light
meter.

the heat emitted as radiation by the soil and plants, and which stores the natural heat accumulated during the day for a longer time.

This double roofing prevents temperature inversions. In order to install this interior sheeting the structure of the greenhouse is used. The better the seal around the sheet and the fewer points it has in contact with the roof, the better the insulation will be. For this technique to be effective, there has to be an air layer between the two layers.

Double roofing is most effective as insulation if the distance between the two layers is 1.8-3.75 cm. Using a double roof, which may be the same material as the roof or different, reduces heat losses. The disadvantage is that the light intensity within the greenhouse is lower (at least 15% lower, depending on the material used).

4.4. COOLING AT HIGH TEMPERATURES

Greenhouses are cooled in hot weather for two main reasons: in order to make use of the greenhouse throughout the year, and because high temperatures have a negative effect on plant growth. The following methods can be used to lower the temperature within the greenhouse:

- Cooling the roofing:
- Shading systems, such as liming and meshing

- Cooling the roof with water:
- Ventilation
- Making the air inside circulate
- Cooling it by evaporating water

- Cooling system
- Mist system

4.4.1. Shading systems

The reason for installing shading is to reduce the intensity of the sunshine reaching the greenhouse, and thus cause the temperature to fall. This is the most widely used system because it is very simple and easy to carry out, though it does not lower the temperature sharply.
Shading systems may be static or mobile. Static systems cast a constant shade and cannot be adjusted. Mobile shading systems make it possible to adjust the intensity of the sunlight in function of the environmental conditions within the greenhouse.

Shading system based on whitewashing the greenhouse

Black polyethylene mesh used inside the greenhouse for shade

4.4.1.1. Static shading systems

WHITEWASHING

One of the most widely used shading systems is to paint the greenhouse walls white with whiting or quicklime. Whiting (calcium carbonate in the form of ground chalk) is more easily washed away, and so is more appropriate in areas with little rainfall.
To make whitewash, mix 15 to 20 kg of commercial whiting in 100 L of water. Add "sizing" to make it stick better to the walls. The mixture can be sprayed using any of the machines used for spraying agricultural chemicals.
Apply 2 or 3 times to ensure the window is sufficiently opaque. This regulates the light intensity better at higher light intensities and temperatures. To get rid of the whitewash, use a 10% solution of ammonium sulfate, taking great care not to drip any onto the crop, which would be damaged.
The effect of whitewashing on the air temperature depends on the type of greenhouse, and the effect is greater in a sealed greenhouse than in a freely ventilated one.
The main disadvantages of whitewashing are that the degree of shading cannot be adjusted and that non-uniform applications cause differences in the intensity of light reaching the plants.

FIXED MESHES

The most widely used meshes are woven polyethylene raffia or polypropylene raffia, though other materials can also be used. They may be any color, though they are usually black or white, and used separately or in combination. There is a wide range of meshes on the market, each with its own value for light transmission and reflection, and porosity to air. The meshes on the market are 30-90% opaque (they reduce the light intensity by 30-90%).
There is a second type of polyester raffia with a shiny aluminium coating, aluminized mesh. They have the advantage that they reflect part of the sunshine, instead of absorbing it, and this leads to lower temperatures.

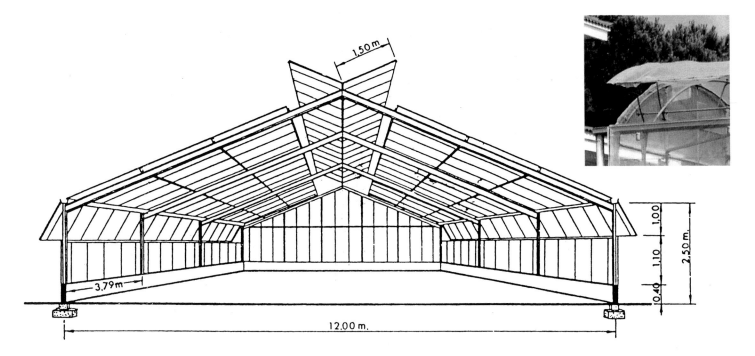

Whenever possible, the shading mesh should be installed outside the greenhouse, leaving an air chamber between the greenhouse roof and the mesh. This reduces the mesh's useful life and installation is more complex, but makes it more effective at reducing the temperature. If the mesh is inside the greenhouse, it absorbs the sunshine and emits it as heat, which has to be removed by ventilation. Mesh outside the greenhouse can lose heat to the outside air.

Meshes must also be permeable to air, as they often reduce the gas exchange between the vegetation and the external environment. Shading and ventilation have to be planned together.

Temperatures in greenhouses shaded with mesh inside or outside

TYPE OF GREENHOUSE	AVERAGE TEMPERATURE
- Open air	33.0°C
- Unshaded greenhouse	46.6°C
- Greenhouse with 45% shading (outside)	40.8°C
- Greenhouse with 45% shading (inside)	50.5°C

4.4.1.2. Dynamic shading systems

MOBILE MESHES

The use of fixed mesh has a major disadvantage. In the early morning and late afternoon, and on cloudy days, the shading is excessive, and the plant's photosynthesis is limited.

If there is a mechanism to draw the mesh or to open and close it as the light changes, the sunlight is used more effectively.

The basic elements needed to move the mesh are a rotary axis motor, steel cables that roll around the axis and move the mesh, a set of pulleys and a device to measure sunlight intensity inside the greenhouse.

4.4.2. Cooling the roof with water

Some greenhouses have a watering system installed on the roof to create a film of water that flows down the walls. This is an effective way of reducing peak temperatures, and hardly reduces the light intensity. If a colorant is added to the water, it absorbs more light, and thus causes a greater reduction in light levels within the greenhouse, and thus also lowers the temperature.

The water used should contain little calcium carbonate (it should be "soft water"), as this is deposited on the glass and forms encrustations on the structure, and also blocks the nozzles emitting the water.

This system has the disadvantage of requiring a lot of water. Even if the installation recycles the water, the water consumption is always excessive, as a lot is lost by evaporation.

4.4.3. Ventilation

The air in the greenhouse must be renewed, it must turn over, as the air influences the environment created inside the greenhouse. Ventilation not only changes the air temperature, it also changes the levels of humidity, carbon dioxide and oxygen. Ventilation may be natural or mechanical.

4.4.3.1. Natural ventilation

Natural ventilation is based on the fact that hot air weighs less than cold air and thus rises.

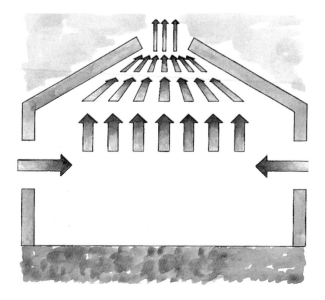

Natural overhead ventilation

The opening and closing of the windows can be more or less totally mechanized. The side windows can be opened by turning a crank handle to roll the plastic sheet down or roll it up again. One commonly used method of opening and closing the windows is to use a rack and pinion system. This mechanism consists of a gear with teeth (the pinion) that mesh with a bar with teeth (a rack) connected to the windows. Thus when the pinion is turned, the rack rises or falls, opening or closing the window. This can be done manually or using an electric motor.

Opening and closing the windows can also be regulated automatically using an electric motor controlled by a thermostat activated by temperature changes inside the greenhouse. When the temperature rises above the established limit, the electric circuit is closed and the motor opens the windows. The windows are closed when the temperature falls below the lower limit set on the thermostat.

4.4.3.2. Mechanical ventilation

Using fans allows more precise control of the turnover of air, and thus the temperature within the greenhouse. The air flow is produced by extractor fans that expel the hot air from the greenhouse, which is then replaced by air from the outside.

To maintain a temperature in the greenhouse about 5°C warmer than that outside, the air has to turn over between 40 and 60 times per hour, though in summer it is difficult to lower the temperature to acceptable levels.
In order to ensure the air flow is uniform, the distance between ventilators should not be greater than 7.5 m, and at the air outlet there should a distance of at least twice the diameter of the

Air circulation in a greenhouse with overhead ventilation

Natural ventilation is simply opening the windows and letting the air flow in and out. The surface area of the windows open in the greenhouse depends on the greenhouse's size, and the type of windows installed.

The windows may be in the side walls or in the roof, and those installed in the roof are called overhead windows. A greenhouse may have side windows or overhead ones, but it is best to have both.
If the greenhouse only has side windows, they should represent 20% of the greenhouse's surface area, and the same is true for overhead windows. If both are used, the overhead windows should represent 10% of the greenhouse area, and the side windows 15%.

Natural ventilation in a greenhouse with a curved roof opened by a rack and pinion system

ventilator that is free of obstacles. The fans should be placed on the greenhouse roof if obstacles prevent installation on the side walls. They should also be sited in the same direction as the dominant winds in summer.

4.4.4. Making the air circulate inside

Moving the air inside the greenhouse makes the temperature more uniform and also increases the evaporation of water from the soil and the transpiration from the plants, and thus tends to increase the humidity of the air and lower the temperature.

If the greenhouse is heated by a hot air system, the air can be made to circulate by turning on the fans but not the heating.

4.4.5. Cooling by evaporation of water

This type of cooling is based on the fact that when liquid water evaporates to vapor, it absorbs heat from its surroundings and thus lowers the temperature.

4.4.5.1. Cooling system

The cooling system consists of blowing outside air through a continuously wetted panel and into the greenhouse. This cools the incoming air and saturates it with humidity.

The panel consists of a porous material that is saturated with water by a watering device. The water enters at the top of the panel, and is collected by a drain at the base. The porous panels are located on one side or the front of the greenhouse, and the ventilator-extractor fans are sited at the opposite end.

The ventilator-extractor fans consist of six radial blades attached to a central hub, and a 1.5-2 horsepower motor able to move 200 to 400 m^3 of air per minute. The air inlet includes a protective system, and aluminium shutters with an automatic system to ensure the intake is held open when the fan is in use and to close it when the fan is not in use, thus maintaining the greenhouse airtight.

6-blade fan used in the cooling system, (Ventigran Series X model, manufactured by GER)

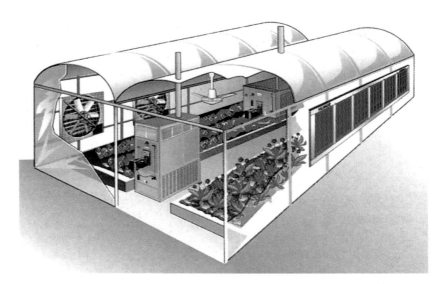

The cooling system uses fans with blades and refrigeration modules. The evaporation of water increases the air's humidity. (Ventigran edge fan, Ventigran extractor fan, wind hot air generator, and Humer refrigeration modules. Manufactured by GER)

poorer quality, and unlike panels that use wood shavings, they do not need auxiliary support structures. As time passes, wood shavings become compacted and leave spaces where the air can flow without absorbing humidity.

In order for the system to be effective, the

RELATIVE HUMIDITY OF OUTSIDE AIR IN C°	TEMPERATURE OF AIR OUTSIDE (°C)				
	20	25	30	35	40
20	16	19	22.5	26	11.8
30	13.9	17.5	21.8	25.5	29.2
40	15.5	10.9	24	28	32.2
50	17	22.2	26.2	30.7	35

The cooling panel is usually made of metallic fiber, within which are wood shavings or sawdust or some cellulose-based material. These are more expensive but are better, as they can be used with water of

greenhouse should be airtight, so that the incoming air all passes through the cooling panels.
The lower the humidity of the external air, the more effective the cooling system is.

Misting inside a greenhouse

The unused water that is not evaporated in the panels returns to the deposit where the water is stored. The deposit houses the pump to supply water to the top of the panels. A supply of water is also necessary, as about 40% of the water entering the panels evaporates.

Other points to bear in mind include:

• An area of 0.35 m² of panels should be installed for every 1,000 m³/h of air that is renewed.
• The panel should be located on the greenhouse's north wall, so that it casts less shade in midwinter.
• The distance between the screen and the extractor fan should be 20-25 m, and never greater than 40 m.
• To be most effective, the air should turn over 40-60 times per hour.
• The panel should not be more than 2.5 m high or less than 0.5 m tall, to ensure water flow is regular.

Below left: Misting system that can also be used to treat against pests

• There should be 3 extractor-fans for every 1,000 m² of greenhouse roof.
• The distance between adjacent fans should not be more than 10 m.
• For every 1,000 m² the water distributor pump requires 2 horsepower.

The system is operated automatically by a thermostat.

This opens and closes contacts that turn the fans and water pumps on or off, when the temperature rises and falls. Anyway, the system's maintenance may be expensive.

Below right: Pulsfog misting system. It is possible to mist from the door, as long as the greenhouse has fans inside to distribute the mist. (Manufactured by Dr. STAHL + SOHN)

At the moment, it can only be recommended in high-value crops or those that have to be cultivated in periods of excessive heat. It has also given good results in plant propagation, encouraging rooting.

4.4.5.2. Mist system

The mist system is based on the creation of an artificial mist inside the greenhouse. Misting is the distribution in the air of a very large number of particles of water with a drop size of about 10 micrometers. Because the particles are so small, they remain suspended in the air within the greenhouse for long enough to evaporate without ever wetting the plant.

The system consists of an installation of high-pressure tubes bearing water to nozzles called "misters". They contain the most delicate component of the system, the nozzle. The quality of the installation depends on the mister's nozzle.

The nozzle receives the water under pressure, breaks it down into tiny drops and disperses them at a short distance, an operation that requires the input of energy. Depending on the energy supply, the nozzles may be hydraulic, gas, centrifugal or use heat.

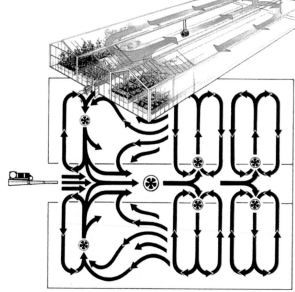

The Fog Turbomatic misting system allows stationary misting. It is equipped with a fan to distribute the air. (Manufactured by Dr. STAHL + SOHN)

The most widely used misting systems are:

4.4.5.3. High pressure nozzles

These nozzles are connected to tubes able to support a working pressure of 60 kg/cm². The nozzle is designed so that the jet of water hits an obstacle as it emerges, and disperses into a cone of small drops, 95% of them smaller than 20 micrometers. One nozzle is normally installed for every 6-8 m² of greenhouse. Each nozzle evaporates about 5 l/h of water.

The main elements of the installation are:

The Ribertec is an ultra low volume electronic misting system (Manufactured by TECTRAPLANT, S.A.)

• High pressure pump, motor and tubing.
• Water treatment injectors. If the water comes from a surface well or a deposit, 0.5 ppm chlorine should be added to stop algae and bacteria growing.
• High pressure mesh filters.
• Mister nozzle
• Control equipment: a thermostat to control the temperature and humidistat arranged in series to turn the device off when the humidity exceeds a pre-established value.

4.4.5.4. Low pressure nozzles

The system consists of the same components as above, but at a working pressure of only 3-6 kg/cm².

Different types of nozzle can be used:

• Openings about 10 micrometers wide from which small drops of water emerge. They are the simplest nozzles but are the most easily blocked.
• Ultrasonic humidifier. The current of compressed air is directed against a round hollow resonator opposite the water outlet. The water crosses a field of waves and disperses into drops 10 micrometers or less in diameter. Ultrasonic misting produces a very fine dry mist, but is very expensive.
• Cheaper nozzles with acceptable quality.

They use a mixture of air at 6-8 kg/cm² and water at 3-5 kg/cm², mixing them inside the body of the nozzle.

The installation is less expensive, but an air compressor is needed. The current of air also cleans the nozzle and prevents dripping when the water supply is disconnected.

4.4.5.5. Air humidifiers

Air humidifiers consist of moving parts. Some models use centrifugal force to produce small drops of water and a fan to spread them through the greenhouse. The devices have a capacity to turn about 40 l to 200 l of water into a mist. The quality is lower than in other misting systems and it requires more maintenance, but the cost is lower.

4.5. REGULATING THE HUMIDITY

The ambient humidity in the greenhouse depends basically on the water present in the soil, the humidity of the external environment and the temperature.
The ambient humidity in the greenhouse is derived from evapotranspiration, ignoring any gains or losses due to air entering from the exterior.

The following methods can be used to correct low humidity:

Below:
Controlling humidity
is essential when
growing young plants
and seedlings

• Maintaining soil humidity by:

- Irrigation: give as much water as is suitable for the crop, taking care not to overwater
- Pools of water: small deposits of water distributed over the greenhouse soil.

To the right:
Technical
characteristics of a
humidifier

• Increasing the soil water content by:

- Sprinkler irrigation: for crops that require high humidity. Water in the hours when the humidity levels in the greenhouse are lowest.
- Cooling system
- Mist system

• Lowering the temperature by:

- Ventilation
- Moving the air
- Reducing the light intensity

Excess humidity is difficult to remedy, especially when due to high external humidity, such as when there is fog at night and dew. The measures that can be taken to reduce humidity are:

• Forcing the entry of dry air from the outside
• Avoid excess soil moisture by watering correctly and not mulching.
• Increasing the temperature by heating.

4.5.1. Humidifiers

The atmosphere can be made more humid by using humidifiers placed in the most appropriate places within the greenhouse.
The purpose of the devices is to regulate the changes in the humidity of the greenhouse's atmosphere over the course of the day.
These systems work automatically using remote control systems, with a thermostat that records and maintains the necessary level of humidity in the greenhouse's atmosphere. This device is called a humidistat.
Humidifiers are centrifugal-type emitters that can pulverize 7-40 l/h of water.
The air in the greenhouse can also be humidified by coupling a special humidification system to the hot air generators used for heating.

– Supply of atomized water	15 l/h
– Volume of air	3,000 m³/h
– Motor	
(three-phase or single phase 220/380)	0.5 CV
– Water pressure	not needed
– Water supply tubing	
(internal diameter)	10 mm
– Outlet tubing	
(internal diameter)	22 mm
– Approximate weight	25 kg

4.5.2. Controlling condensation on the greenhouse roof

Condensation on the inside walls of the greenhouse depends on the environmental humidity, the roofing material and on the temperature inside and outside.

The air inside the greenhouse need not be saturated for condensation to form on the walls. All that is needed is for the outside temperature to be a few degrees centigrade lower than that inside.

The roofing material greatly influences the amount of condensation, because depending on its surface tension, it will retain more or less of the sheet of condensed water. In the case of polyethylene, the condensed water does not trickle down the sloping walls, but forms drops that get larger and larger and then fall as drips.

To control condensation, ventilate to reduce the humidity inside the greenhouse, or heat to increase the temperature of the greenhouse roof.

4.6. ARTIFICIAL LIGHTING

It is really very difficult to supply the plant with all the light it needs over the course of its cultivation cycle by artificial means. It is complicated and is not worthwhile in most cases, because it needs a lot of electricity, and the installation cost is high.

Artificial lighting is used at certain very specific moments of the plant's productive cycle, for example to induce or inhibit phenomena such as flowering, leaf or flower pigmentation, and stem elongation. It is also used in companies specialized in a single operation, such as propagation.

4.6.1. Lighting techniques

The lighting techniques used in a greenhouse depend basically on their purpose. They are divided into:

• **Photoperiodic lighting**. Artificial lighting to alter the plant's photoperiod.
• **Supplementary illumination**. This is in addition to the natural lighting and aims to increase net photosynthesis.

4.6.1.1. Photoperiodic illumination

Artificial light is used to modify the plant's photoperiod.

Two different lighting techniques can be used:

• Lengthening the daylength, by prolonging the hours of daylight either before dawn or after sun set. This implies two different periods of lighting, the first being the period of high intensity natural light, and the second being the extension period. The extension period only requires low levels of artificial light. Normally, the light intensity is about 10-15 W/m².
• Interrupting the dark period separating one day from the next. This also means there are two light periods, the high intensity natural illumination and the short low intensity light period interrupting the dark period.

The quantity and quality of the light required is different for the two types of light treatment. Extending the daylength is used to induce flowering in long-day plants (plants that need a day more than 12 hours long to flower).

The second type of lighting, interrupting the dark period, is used to inhibit flowering in short-day plants (plants that need a daylength of less than 12 hours to flower). This increases the vegetative growth in a period when flowering would be induced before the plants were large enough.

Interrupting the dark phase can be used for other purposes, such as preventing leaffall in deciduous ornamental trees or to ensure better rooting in cuttings of ornamental plants.

Interrupting the night period is extremely effective in short-day plants. This type of treatment shows very little effect on long-day plants.

Extending the daytime should be done using incandescent lamps, so that it can be effective for long-day plants and even for short-day plants. Incandescent lamps cost less, cast less shadow in the greenhouse during the day, and they are almost as energy efficient as white fluorescent lighting.

Artificial light could be applied during the nighttime to increase photosynthesis, but the yield achieved would in no way compensate for the electricity used. The light intensity needed for a plant to photosynthesize normally is 500-1,000 watts/m².

Illumination with electric lights

4.6.1.2. Supplementary lighting

For much of the winter natural light levels are insufficient to ensure adequate plant growth, and so supplementary artificial light is necessary to increase photosynthesis.

This supplementary lighting is different from photoperiodic lighting because it is applied at the same time as the natural light, and different types and strengths of light sources are used.

4.6.2. Types of light source

The type of light source to be used depends on its purpose. Use a light source with an emission spectrum as similar as possible to sunlight, and whose light emission is uniform over the course of its useful life.

The technical characteristics to bear in mind are:

- The spectral distribution of the light emitted
- The cost of installation and of electricity consumption
- The power required
- The efficiency of the light source, i.e., the percentage of the electricity converted into light

4.6.2.1. Incandescent lamps

Incandescent lamps emit a lot of red and infrared radiation, meaning that they emit a lot of heat. They are used to induce flowering in long-day plants. If they are used as a light supplement, they may cause stem elongation.

4.6.2.2. Fluorescent tubes

Fluorescent tubes are not widely used in greenhouses because they cast a lot of shade. They can be used as additional lighting and to interrupt the night period.
They emit little infrared radiation, and thus little heat.

Types and characteristics of light sources used in greenhouses (T. Sante Beltramelli)

Characteristics	Type of lamp			
	Incandescent	Mercury vapor	Incandescent and mercury vapor	Fluorescent lighting
Light produced	6-7% visible light Red and infrared (a lot of heat)	16% visible light and ultraviolet	Mixed	25% visible light Mixed, dominated by blue and red
Power	3 W/m²	150-200 W/m²	—	—
Efficiency in producing light	10%	90%	30%	90% (emits little heat)
Useful life	1,000 hours	3,500 hours	2,000 hours	3,500 hours
Application	Large greenhouses; bringing flowering forward or delaying it	Normal plant growth	Bringing flowering forward	Normal plant growth
Observations	Low installation cost	Pay attention to the type of light	High cost of use	Weak light intensity: place 3-4 in a battery

Types of light source, the power needed and the height of the light source (According to J. Miranda de Larra y de Onis)

Purpose	Type of lamp	Power needed W/m²	Length of lighting	Height
Supplementary lighting to increase photosynthesis	a) Large mercury fluorescent lamps b) Fluorescent tubes	50-100	4 to 10 hours	0.8 - 1.5 m
Growth in sunlight. Growth chambers or artificial environments	a) Large mercury fluorescent lamps b) Fluorescent tubes c) Incandescent (seed trays)	200-1,000	16 hours (long days) 10 hours (short days)	1 m
Supplementary lighting to obtain long days	a) Fluorescent tubes b) Incandescent (in some cases)	5-50	5 hours	1 m
Forcing bulbs	a) Fluorescent tubes b) Incandescent	100	12 hours	0.8 - 1.5 m
Forcing flowering by artificial lighting alone	a) Fluorescent tubes b) Incandescent	25-100	10-12 hours	—
To interrupt the dark period	a) Fluorescent tubes b) Mixed	25	1-3 hours (total length of flashes))	—

4.6.2.3. Mercury vapor lamps

Mercury vapor lamps are used when the crop needs intense light. The spectral distribution of the light emitted is well balanced and is suitable as supplementary lighting.

They are also used to increase photosynthesis and for photoperiodic induction. They do not emit a lot of red and infrared radiation.

4.6.2.4. Sodium-vapor lamps

They have a restricted emission spectrum that may produce unbalanced growth by the plant. They are used in supplementary illumination.
They are very effective at producing visible light, with an efficiency of more than 25%, they emit little heat and their useful life may be as long as 7,000 hours.

4.7. SOLAR GREENHOUSES

Solar power is one of the most attractive alternative energy sources, because it is not going to run out and is non-polluting.

At the latitude of Spain there are roughly 2,000-3,000 hours of sun in the summer and about 1,000 in the winter.

There have been many studies into using solar power in greenhouses, but very few installations actually use it. In southern Europe, solar power is more widely used for domestic water heating.

Most installations using this free source of energy have high yields. Solar installations have low running costs, but the initial investment may be quite large. For the installations to be worthwhile, the devices must trap sunshine and emit it as heat inexpensively and simply enough to be suitable for use in the countryside.

There are two ways of trapping solar power to heat a greenhouse:

• The greenhouse itself can act as the collector, either because the collector forms part of the greenhouse's structure, or because the collector is inside the greenhouse.
• By installing solar panels on the outside of the greenhouse.

4.7.1. Collectors within the greenhouse

The simplest solar collector system within the greenhouse consists of a series of transparent polyethylene tubes full of water alongside the rows of crop plants. This absorbs the warmth of the sun during the day, heating the water in the bags, which slowly releases the heat at night.

As a whole, the system is cheap to install and maintain. If the crop plants do not shade the bags as they grow, the system increases nighttime temperatures by 4-6°C, and so it is also an effective method of preventing frosts.

This system only works in small greenhouses, and is not suitable for large ones. It is most suitable for seedbeds or greenhouses specialized in propagation.

4.7.1.1. The angle of the walls

The south-facing wall receives the most sunlight (in the northern hemisphere) and so should be glassed. To trap as much sunlight as possible, the roof should be at an angle of 60°. The choice of angle also depends on the lighting system and the structure of the greenhouse.

The sun's position varies over the course of the year. Thus the angle of the glass should be the average of the sun's position in the different months. Selecting the angle is technically complex, but in practise it can be obtained by adding 20° to the latitude (in degrees) of the greenhouse's site.

The north wall receives much less sunshine than the south-facing one. To achieve complete insulation and reduce heat losses, this wall should be a non-glass material. One way of doing this is by introducing polythene between two plastic sheets, and painting the inner side white so that the radiation entering the glass wall is reflected back into the greenhouse. In this type of greenhouse, the side walls are small and the sun only shines on them for a short period, and so their contribution to heating the greenhouse is very small.

Artificial lighting favors photosynthesis

An array of panels reflecting sunlight onto the solar furnace on top of the tower in the center

4.7.1.2. Insulation

The non-glass walls can be insulated by placing materials inside the brick walls. The area of glass must be clean in order for as much radiation as possible to enter. At night, the glass surfaces lose heat.

To solve this loss of heat, a double wall can be installed, though this has the disadvantage of reducing light intensity within the greenhouse. Another method is to cover the glass surfaces at night with opaque blinds, which, when they are opened in the morning, can be adjusted to act as parabolic reflectors to increase the light intensity within the greenhouse. The opening and closing mechanism can be automatic.

4.7.2. Collectors outside the greenhouse

Conventional flat collectors are expensive and occupy a lot of space, meaning they reduce the area of crop that can be grown. This can be resolved by locating them on a slope or on wasteland. The collector is most effective if it is pointed towards the midday sun, and when the plane of the collector is perpendicular to its rays.

A conventional system to trap and distribute heat from sunshine consists of the following parts:

• External collectors
• Storage system
• Heat distribution system

A conventional system to trap and distribute heat from solar energy
A/ Distribution system within the greenhouse
B/ Storage system
C/ Conventional heating system, as back-up
D/ Network to transfer heat to the storage system
E/ Insulation
F/ Transparent roofing
G/ Solar collector

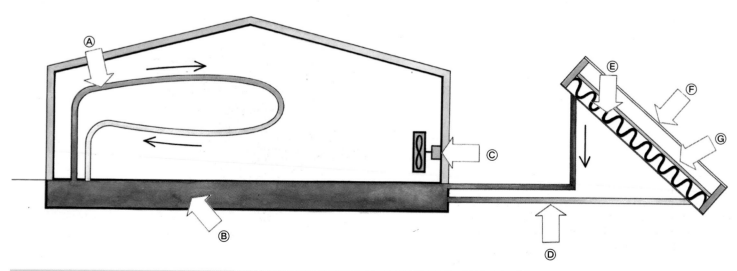

4.7.2.1 Collectors

Flat solar collectors consist of 3 components:

• The absorbing surface, which absorbs the sunlight and where it is converted into heat. The absorbent material is usually black (as black absorbs more radiation) and made of metal: copper, aluminium or galvanized iron. Polycarbonate plastics or EPDM (ethylene-propylene-diene monomer) rubber can also be used, and this has the advantage that it is cheaper and resists corrosion.

• A cover, which should be transparent to short wave radiation and opaque to the long wave radiation that transmits heat. The material used is glass, which allows the maximum transmission of sunlight towards the absorbent material. Polyethylene can also be used, which is cheaper, but it is not as good and degrades faster.

• Thermal insulation on the side and rear of the collector to reduce heat losses.

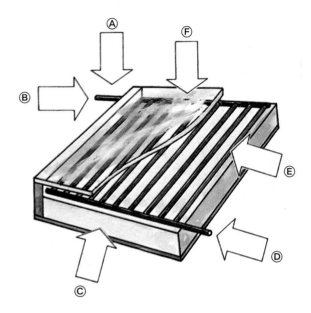

Diagram of three types of flat solar panels
A/ 1″ tube
B/ Water flow
C/ 3″ insulation
D/ 1″ tube
E/ 1-2″ tubes soldered to the collector panel
F/ Glass
G/ Water flow
H/ Water collector
I/ 3″ insulation
J/ Corrugated iron soldered and riveted to the base of the collector
K/ Glass
L/ Air flow
M/ Blackened corrugated aluminium or steel

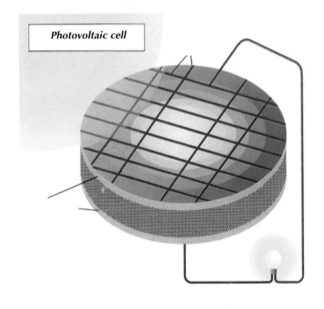

Photovoltaic cell

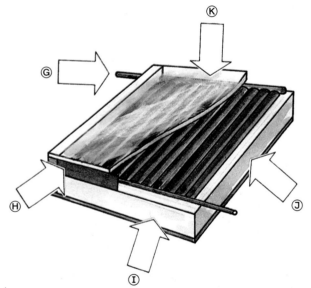

4.7.2.2. Storage system

The heat trapped has to be stored if it is to be used later. The storage capacity should be large enough to compensate for the heat lost at night. This requires use of a substance with a high specific heat capacity (such as water). Stones can also be used, though their specific heat capacity is only one third that of water. They are usually placed on the on the greenhouse floor, if this is possible with the crop being grown.

4.7.2.3. Heat distribution

The heat is distributed within the greenhouse by a network of tubes. Water heated by solar energy circulates at 40°C, while conventional heating by hot water uses water at 80°C, and so a larger surface area is needed for heat exchange, meaning a greater length of tubing is required.

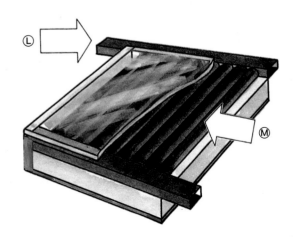

Diagram showing installation of a drainage tube
A/ - A 50 cm layer of topsoil
B/ - Sheet of perforated plastic
C/ - 30 cm gravel and sand
D/ - 10 cm diameter drainage tube 15 cm below plastic sheet

5. CULTIVATION SYSTEMS IN GREENHOUSES

Greenhouse cultivation techniques are based on the same general principles as the rest of agriculture, but with some minor adaptations. It is based on principles and techniques long used in traditional crops, and all that has changed is that some techniques have improved greatly over many years of application.

These improved techniques are essential for cultivated crops, as otherwise all the benefits of creating an artificial microclimate inside the greenhouse would be wasted.

A badly thought out cultivation system or operation limits production, and so these aspects deserve special attention, and may have to be adapted for different types of greenhouse.

Mechanization of cultivation, mainly in small greenhouses, should be considered in advance

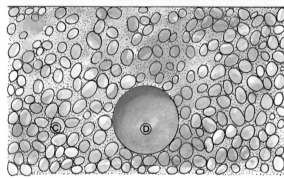

Seedlings a few days after germination, ready to be pricked out

because of the special nature of the machinery that has to be used.

5.1. CULTIVATION IN THE SOIL

The same agricultural techniques are used as in open air cultivation. The only real difference is that the crops are protected, and this gives greater control of the microclimate and all the advantages of greenhouse cultivation.
Good results require a good quality, deep, well-aired and well-drained soil rich in organic matter.

The greenhouse soil should be levelled with a mild

slope of about 2‰ and be 100 cm deep so that the crop's roots have a large volume of soil to grow in.

The greenhouse's drainage is also important, not only to avoid waterlogging, but to ensure good washing of the salts applied as fertilizers and those present in the irrigation water.
To install good drainage, dig trenches 50 cm wide and 75 cm deep, with a slope of 5‰. Cover the bottom with sand to a depth of 5 cm, and lay a 10 cm diameter drainage tube. Cover the tube with 15 cm of coarse sand, then place a sheet of perforated plastic on top and fill in the trench with topsoil.
If the soil is good, growing greenhouse crops directly in the soil requires little investment.
If the original soil is dreadful or if the plants to be grown require a very special soil, the soil can be replaced by artificial soils based on substrates, or other cultivation systems can be used.

Planting bare-root plants or one with rootballs in the soil consists of the following steps:

• Make a hole with the dibber halfway up the ridges, on the appropriate side for the season of the year.
• Introduce the plant into the hole, ensuring that several centimeters of the plant's stem are below the soil level. All the plants should be planted at the same height along the ridge.
• The soil is pressed down around the roots, filling in the depression created when the plant was placed in the hole.
• Once all the plants in each furrow have been planted, the soil is watered. Take care to ensure that the water level in each furrow reaches the level of the line of plants.

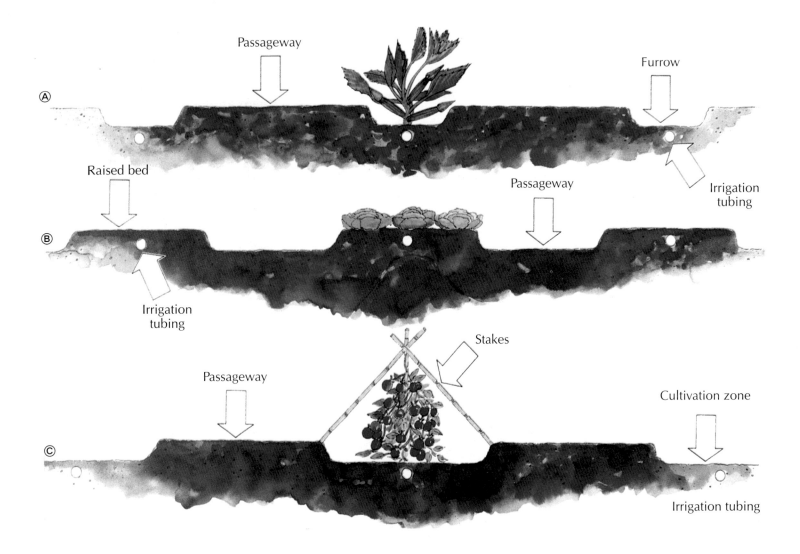

Passageway

Furrow

Ⓐ

Raised bed

Passageway

Irrigation
tubing

Ⓑ

Irrigation
tubing

Stakes

Passageway

Cultivation zone

Ⓒ

Irrigation tubing

*Diagram of some
greenhouse
cultivation systems*

When cultivating on the flat, planting is no problem. The main disadvantage of cultivating the soil in the greenhouse is that it is very difficult to get rid of weeds, pests and diseases, once they have been established. Soil sterilization to a shallow depth is not effective and the reinvasions may be even more serious because the natural balance has been disrupted.

A further disadvantage is that poor quality water may lead to serious salination of the soil.

5.2. CULTIVATION IN BEDS

Cultivation in beds is based on dividing the site into plots, so that the growing area is clearly bounded.

These divisions are usually 150-200 cm wide and are separated by lanes 40-50 cm wide. These beds can be divided by their structure:

• Fixed beds at soil level or slightly raised, with or without a base
• Fixed or mobile raised beds

5.2.1. Beds at ground level

This cultivation system can be applied with good results in greenhouses as well as in the open air, depending on the type of crop to be grown. Greenhouses are more suitable for bed cultivation for a number of technical and agricultural reasons.

The site to be divided into beds has to be deep, without any impermeable layers, as the irrigation water has to circulate well.

Site preparation starts with plowing the soil to a depth of 50 cm, incorporating some well-rotted organic matter at the same time, then placing thin slabs that separate the cultivation plots. The slabs can be up to 25 cm tall and may be made of various materials, such as wood, fibrocement, hard plastic, bricks or prefabricated concrete units.

The beds can be filled with substrate well mixed with the soil, or the soil can be totally replaced by substrate to improve growing conditions. The substrates used in the beds are normally mixtures based on topsoil, sand, peat, organic fertilizers, mulches and other materials.

Peat is used in nearly all the mixtures, because its excellent physical properties favor both water retention and good aeration.

If peat is going to be added to the earth in the raised bed, the quantity added should be 3-8 kg/m^3, depending on the original physical characteristics of the soil.

Different types of beds at ground level
A/ Perspective diagram of a bed without a bottom. The passageway is covered in earth and edged by slabs

0.3-0.4

Ⓐ
0.4 m
1 m
0.4 m

B/ Beds without a bottom, bounded by L-shaped units that form a bottom to the passageway and the edge of the bed

0.3-0.4

Ⓑ
0.4 m
1 m
0.4 m

C/ Beds with bottom and passageways made of prefabricated units

0.3
0.4

Ⓒ
0.4 m
1 m
0.4 m

Upper right: Greenhouse cultivation with trickle irrigation system

The bed is usually 30-40 cm deep, and must be well-drained, to ensure any excess water drains away. Cover the base with a layer of 10 cm of gravel and sand, and then cover this with the soil for cultivation.

The passageways should be covered with slabs, in order to prevent spreading diseases and insulate the soil or substrate as much as possible. If the raised beds do not have a bottom, U-shaped prefabricated concrete units can be placed upside down. Not only do they cover the passageway, but they also form the side walls of the raised beds.

5.2.2. Raised beds

These are beds raised about 80-100 cm above the soil level, with the same wall height and passage width as for beds on the ground.

These raised beds may be made of wood or galvanized iron.

Mobile benches have recently been introduced into this type of cultivation. Their main advantage is that they make better use of the space in the greenhouse than fixed benches.

The surface covered by the crop is 85-90%. The main disadvantage is the high installation cost.

5.2.3. The system's advantages and disadvantages

Growing in beds makes it possible to control the physical conditions of the substrate and its nutrient content, as well as irrigation and any pests, diseases, etc. The soil can be easily sterilized using steam.

One variation of cultivation in raised beds is cultivation with a closed bottom.

This completely isolates the soil, preventing many problems, such as possible infection from the soil.

The soil in a raised bed with a bottom can be replaced by a mixture of substrates, though substrates are mainly used for crops grown in pots. The raised bed is the most functional system for growing ornamental and flowering plants.

Use of the greenhouse area is more rational and more profitable, speeding up and facilitating all the different cultivation operations, as they only have to treat the soil that is in the raised bench and not the entire area of the cultivated site.

Watering can also be easily automated by installing tubing with trickle outlets along the sides of the beds, which can be turned on and off by timers, or by tensiometers placed in the soil and connected to the pump system. Local irrigation systems can also be installed.

The disadvantage of benches without a bottom is that the crop is in direct contact with the soil beneath, and may thus be attacked by pests or diseases. The cost of installation is also higher than if the crop is grown in the soil.

5.3. SANDING

Sanding is also used in open air cultivation, but it is much more advantageous in a greenhouse. It is considered to be an excellent technique to make best use of the greenhouse.

The initial cost is high, and so it is only used to cultivate crops that generate high profits, either because production is higher or they fetch a higher price at market. Sanding is used in greenhouses and intensive truck farms in hot climates.

Sanding consists of covering the soil surface with a layer of about 8 cm of dung and then a 10-12 cm layer of sand.

The soil has to be free of stones, absolutely flat and rich in nutrients. Levelling the soil is important, as otherwise water would accumulate in the depressions and damage the plants growing there, and plants on the higher areas would not receive all the water they need.

The idea is that the root system can grow in the dung and the nutrient rich topsoil, and the sand acts as an anchor and support, also providing several other advantages for the crop.
The water and nutrients accumulate in a small volume of soil, and so it is important that they be supplied continuously. This is why there is a lot of interest in local irrigation and combining fertilizer application with irrigation. This is in fact a semi-hydroponic system.

This system can grow plants in excellent conditions for 3-4 years as the cultivation operations are restricted to the top layer of soil. After this, the soil has to be regenerated to recover the fertility it has lost. This set of operations is known as retrenching.

Retrenching consists of three phases:

• Removing the sand
• Applying fertilizer to the soil at depth and replacing the layer of dung
• Replacing the layer of sand that was removed

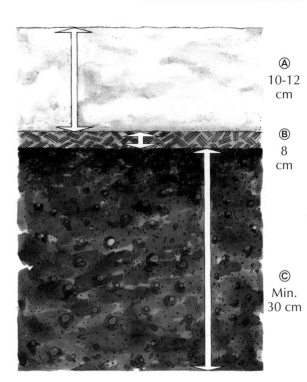

The different layers in a sanded crop
A/ Sand. Protective and supportive layer
B/ Dung. Nutrient layer
C/ Soil. The original topsoil

Ⓐ 10-12 cm

Ⓑ 8 cm

Ⓒ Min. 30 cm

Most of the roots grow in the dung and topsoil

The most important advantage of this system is that it is possible to cultivate very poor quality, or saline, soil, and utilize irrigation water with a high salt content. This is because the sand tends to absorb the salt and prevent it from reaching the top soil.
This desalination effect is based on the fact that less water rises through the soil due to capillary action, as there is less evaporation, and also because the microbial activity of the soil is greater and more carbon dioxide is released, which makes the salts more soluble and so they are washed away in the irrigation water.

5.3.1. Applying the sand

First the site must be prepared by removing stones to a depth of 75 cm. It is then levelled to a slope of 4‰ on the shortest side and 2-3‰ on the longest side. The slope should not be greater than this or the sand will be washed away, and the slope should not be any lower or the water would flow too slowly and infiltrate the soil, causing waterlogging. The excess water drains off horizontally, and this is made easier by the layer of sand.

This simple technology makes it possible to cultivate even salt-sensitive plants on salty soils with saline water.

Right below:
Strawberries irrigated
by trickle irrigation

After levelling the site, it is plowed several times, harrowed or treated with a cultivator, and then finished with the rotary cultivator to break the clods and smooth the soil.
Fertilizer should be applied at depth with these preparatory treatments.
Sanding begins with spreading the dung, which should be firm but not too warm or moist. The dung is dumped in piles on the soil and then spread as evenly as possible with shovels.

Below:
Tomatoes grown in
bags

Once the dung layer has been spread, the sand is distributed on top of it. This is usually unloaded as piles, and can then be spread with shovels, like the dung.

The crop plants are transplanted deep to encourage root growth in the dung layer and the topsoil. Once the crop has been harvested, all remains of the plant must be removed to keep the surface of the sand layer clean. When uprooting the plants take care to pull as little soil as possible into the sand layer.

Below right:
Seedbeds for
vegetables

The crops most frequently grown this way, and the ones giving the best results, are tomatoes, peppers, eggplant, zucchini, melons, watermelons, cucumbers and runner beans, and, among tropical crops, pineapples and pawpaws.

5.3.2. Materials

The main material needed for sanding is sand, but not all types of sand are suitable. It has to be siliceous sand that is loose, clean and ideally with a particle size between 0.2 and 3 mm, according to the guidelines of the International Soil Association. The sand does not last indefinitely and should be replaced after 20-25 years, depending on the cultivation conditions and the state of the sand.

The other important material used is the dung. This plays a vital role, as it is the source of most of the nutrients the plant needs. Not all sorts of dung are suitable, and the best is horse dung, and dung that is too potent should not be used. You absolutely must not use dung from animals fed with leftover vegetable scraps as this dung is a serious source of disease. In general, the dung should be sterilized before use.

The dung has to be well-rotted and should not be too wet: it should be loose and not form balls. Apply between 8 and 10 kg per square meter. It acts as a nutrient-rich layer between the soil and the sand. Part of the dung is buried and the rest spread on the soil surface.

5.3.3. Retrenching

Retrenching is the series of operations and tasks performed in sanded soil in order to regenerate it. This improves the soil's aeration and also makes it easier to sterilize the soil and to apply fertilizer at depth. First, the sand is watered and the surface raked to open it up, and it is collected in strips, leaving lanes between every 2 furrows. Take care not to mix the sand with the dung or the soil, and to keep the sand as clean as possible.

Surface plow and deep plow the lanes formed in the same way as when applying the sand the first time, applying fertilizer at depth, levelling and dunging. Once the lanes have been treated, the sand remaining in strips is transferred to the lanes, and the newly exposed strips are treated in the same way. The old sand is then washed and replaced, adding new sand to a uniform height of 10-12 cm throughout the site.

5.3.4. The system's advantages and disadvantages

The advantages of sanding are:

• The main advantage is that excellent crops can be grown on poor quality or saline soils, using irrigation water with a high salt content. This systems diminishes the problem of salinity because the sand creates a barrier to the loss of water by evaporation (thus stopping the rise of salts to the surface layer by capillary action), and because the sand's limited ability to retain water means that irrigation water with a relatively high salt content can be used without causing problems of salt accumulation in the soil.
• It increases the number of harvests, as planting the next crop does not require the previous crop to be plowed in or preparatory tilling.
• The presence of the sand layer reduces moisture loss from the soil by evaporation, meaning that the plants need less water, and less often.
• Better growing conditions, as the sand warms up rapidly in the sunlight, and this warms up the soil and the roots which retain this warmth even when the sand cools down. The fact that the roots are warmed means that harvests are earlier. Some light is also reflected from the sand onto the leaves, thus contributing to photosynthesis.
• Less plowing, as only the surface layer of sand needs treatment.
• It prevents soil compaction and cracking.
• It makes best use of the soil surface, as the plants do not have to be grown on ridges but can be grown on the flat, and this means more plants can be grown per unit surface area.

The main disadvantages are:

• The high cost of starting the system and the high cultivation expenses, due to the very high price of quality sand.

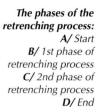

The phases of the retrenching process:
A/ Start
B/ 1st phase of retrenching process
C/ 2nd phase of retrenching process
D/ End

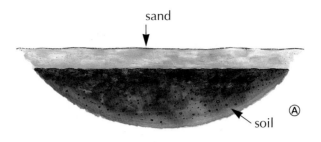

sand
soil
Ⓐ

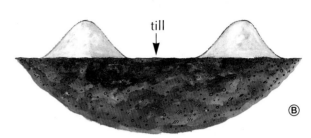

till
Ⓑ

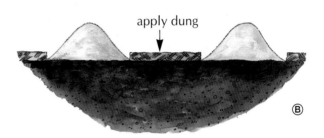

apply dung
Ⓑ

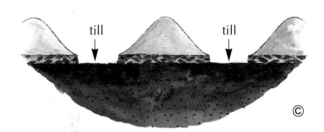

till   till
Ⓒ

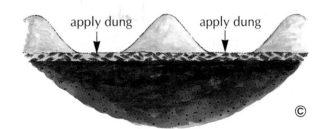

apply dung   apply dung
Ⓒ

Ⓓ

The characteristics of water as a base for a nutrient solution. The salt concentrations should be lower than the values given.

• Water rich in suspended matter cannot be used, as it would silt up the sand.
• The sand is rapidly invaded by weeds, which are difficult to eliminate because this has to be done manually or with herbicides, which increases the expense.
• Soil pests (nematodes, screwworms, wireworms, etc.) can increase. The soil has to be disinfected every 3-4 years for cultivation to remain profitable, because it creates such a favorable environment for pests. The soil is not plowed for 3-4 years, which would normally get rid of many pests.
• Fungal diseases can spread fast. The plant stems are inevitably wetted during irrigation, and they remain in contact with the water for a long time. This creates a moist microclimate at a temperature suitable for the outbreak and spread of fungal diseases.
• Sand is an excellent medium for planting or sowing, but has a major disadvantage. If it dries out, it heats up and may damage or kill the plant. Thus when sowing seed or planting seedlings into sand, the sand must be kept cool, until the plant has "hardened off" and can resist this damage.
• The heat limits the growth cycle of some crops. The sand gets too hot in the summer, and so these crops cannot be grown in the summer.

SALT OR ELEMENT	CONCENTRATION
Sodium chloride	50 ppm
Free chlorine	5 ppm
Boron	2 ppm
Fluorine	2 ppm
Manganese	2 ppm
Sulfates	20 ppm
Magnesium	10 ppm
Iron	1 ppm

Hydroponic cultivation techniques are based entirely on control of the composition of the nutrient solution. The nutrient solution is the source of all the nutrients available to the plant, and must meet all the plant's needs. Only those nutrients required for optimum growth are added to the solution. The nutrient solution must also be at the appropriate pH, and must allow good aeration of the roots.

There are two different types of hydroponic cultivation:

• Soilless cultivation, those that do not use soil or an equivalent inert substrate.
• Hydroponic cultivation in an inert substrate such as perlite, rock wool, sand, etc.

5.4.1. The nutrient solution. Preparation and handling

The nutrient solution must contain all the elements required for the plant's nutrition in a suitable form and at a suitable concentration.

The nutrients are all dissolved in water, and the water must fulfil certain conditions if it is to fulfil this objective. It should be as pure as possible, and it should have a total salt concentration of less than 200 ppm. The best thing to do is to use rainwater, or water from a well.

Excess calcium in the water, even if it is present in proportions higher than the plants needs, does not usually cause any problems.

When preparing the nutrient solution it is very important to know and understand the characteristics of the water you are using, as this may cause the solution to immobilize nutrients or introduce new elements.

There are two methods of preparing nutrient solutions. The first is to prepare a concentrated stock solution for each nutrient element;

Roots submerged in nutrient solution. HUMATOR is a product manufactured by SICOSA, and is a concentrate of humic acids.

5.4. HYDROPONIC CULTIVATION

Hydroponic cultivation consists of replacing the soil with a natural or artificial substrate, that provides the plant with what it normally finds in the soil, namely anchorage, water, air, and nutrients.
Hydroponic cultivation is thus not limited to cultivation in water, but includes cultivation in inert media such as Perlite, tuff or expanded clay, and are fed by nutrient solutions.
This has the advantage over soil that the substrate used can be chosen in accordance the crop's requirements.

Chemical formula	Chemical name	Molecular weight	Elements supplied	Solubility ratio of solution in water	Cost	Other data
A) Macroelements						
•KNO_3	Potassium nitrate (Saltpeter)	101.1	K^+, NO_3^-	1:4	Low	Highly soluble, very pure
•$Ca(NO_3)_2$	Calcium nitrate	164.1	Ca^{2+}, $2(NO_3^-)$	1:1	Low-medium	Highly soluble, but forms a greasy slick that has to be removed from the nutrient solution.
•$(NH_4)_2SO_4$	Ammonium sulfate	132.2	$2(NH_4^+)$, SO_4^+	1:2	Medium	These compounds can only be used in very good lighting conditions or to correct deficiencies.
$NH_4H_2PO_4$	Ammonium dihydrogen phosphate	115.0	$NH_4^+H_2PO_4^-$	1:4	Medium	
NH_4NO_3	Ammonium nitrate	80.05	NH_4^+, NO_3^-	1:1	Medium	
$(NH_4)_2HPO_4$	Ammonium phosphate	132.1	$2(NH_4^+)$, HPO_4^{2-}	1:2	Medium	An excellent salt, highly soluble and pure, but very expensive
KH_2PO_4	Potassium dihydrogen phosphate	136.1	K^+, $H_2PO_4^-$	1:3	Very expensive	
•KCl	Potassium chloride	74.55	K^+, Cl^-	1:3	Expensive	This should only be used in cases of K deficiency, and when sodium chloride is not present in the solution.
K_2SO_4	Potassium sulfate	174.3	$2K^+$, SO_4^{2-}	1:15	Cheap	Its solubility is very low, but it dissolves in hot water
$Ca(H_2PO_4)_2H_2O$	Calcium dihydrogen phosphate	252.1	Ca^{2+}, $2(H_2PO_4^-)$	1:60	Cheap	It is very difficult to dissolve
$CaH_4(PO_4)_2$	Triple superphosphate	Variable	Ca^{2+}, $2(PO_4^{3-})$	1:300	Cheap	Its solubility is so low it can only be applied dry, and it cannot be used in nutrient solutions.
•$MgSO_4.7H_2O$	Magnesium sulfate (Epsom salts)	246.5	Mg^{2+}, SO_4^-	1:2	Cheap	An excellent salt, cheap, highly soluble and pure
$CaCl_2.6H_2O$	Calcium chloride	219.1	Ca^{2+}, $2Cl^-$	1:1	Expensive	Highly soluble, excellent in correcting calcium deficiency, but it should only be used if there is no NaCl in the nutrient solution
$CaSO_4.2H_2O$	Calcium sulfate (gypsum)	172.2	Ca^{2+}, SO_4^-	1:500	Cheap	Very insoluble, it cannot be used in nutrient solutions.
H_3PO_4	Phosphoric acid Orthophosphoric acid	98.0	PO_4^{3-}	Concentrated solution of acid	Expensive	Very good at correcting P deficiencies
B) Microelements						
•$FeSO_4.7H_2O$	Iron (II) sulfate	278.0	Fe^{++}, SO_4^{2-}	1:4		
$FeCl_3.6H_2O$	Iron (III) chloride	270.3	Fe^{+++}, $3Cl^-$	1:2		
$FeEDTA$	Iron chelate (Sequestrene) (10.5% Iron)	382.1	Fe^{++}	Highly soluble	Expensive	The best source of iron; it is dissolved in hot water
•H_3BO_3	Boric acid	61.8	B^{3+}	1:20	Expensive	The best source of boron; it is dissolved in hot water
$Na_2B_4O_7.10H_2O$	Sodium tetraborate (borax)	381.4	B^{3+}	1:25		
•$CuSO_4.5H_2O$	Copper sulfate	249.7	Cu^{++}, SO_4^-	1:5	Cheap	
•$MnSO_4.4H_2O$	Manganese sulfate	223.1	Mn^{++}, SO_4^-	1:2	Cheap	
$MnCl_2.4H_2O$	Manganese chloride	197.9	Mn^{++}, $2Cl^-$	1:2	Cheap	
•$ZnSO_4.7H_2O$	Zinc sulfate	287.6	Zn^{++}, SO_4^-	1:3	Cheap	
$ZnCl_2$	Zinc chloride	136.3	Zn^{++}, $2Cl^-$	1:1,5	Cheap	
$(NH_4)_6Mo_7O_{24}$	Ammonium molybdate	1,163.9	NH_4^+, Mo^{6-}	1:2,3 Highly soluble	Relatively expensive	
$ZnEDTA$	Zinc chelate	431.6	Zn^{++}	Highly soluble	Expensive	

the second is to add the salts directly to the water, as long as the compounds used do not show any incompatibility. To prepare solutions with simple compounds, they should be dissolved separately before they are mixed.

To mix them together, the most soluble and most acidic salts are added first, and then the others, in this order,

- magnesium sulfate
- calcium dihydrogen phosphate
- potassium nitrate
- calcium sulfate

Summary of the fertilizer salts used in hydroponic cultivation

To the right:
Roots growing
submerged in
HUMATOR nutrient
solution,
manufactured by
SICOSA. This can be
applied a little at a
time, allowing better
regulation of the
plant's nutrition

To the left: Some
nutrient solutions
that can be used. The
value for balance
refers to N:P:K

1) Hoagland's solution

Potassium nitrate	540 g
Calcium nitrate	90 g
Calcium dihydrogen phosphate	140 g
Magnesium sulfate	130 g
Iron sulfate	14 g
Manganese sulfate	2 g
Borax	1.7 g
Zinc sulfate	0.8 g
Copper sulfate	0.6 g
A total of 919.1 g of solutes in 1,000 l of water	
Approximate concentration	0.1 %
Balance	1:0.83:2.84

2) Turner & Henry's solution developed at the New Jersey Experiment Station

Calcium nitrate	2,430 g
Ammonium sulfate	150 g
Potassium dihydrogen phosphate	285 g
Magnesium sulfate	570 g
A total of 3,435 g of solutes in 1,000 l of water	
Approximate concentration	0.34 %
Balance	1:0.36:0.25

3) Ellis and Swaney's California Experiment Station»

Calcium nitrate	720 g
Potassium nitrate	660 g
Ammonium dihydrogen phosphate	120 g
Magnesium sulfate	520 g
A total of 2,020 g of solutes in 1,000 l of water	
Approximate concentration	0.2 %
Balance	1:0.29:1.42

This solution can be complemented 2,5 g of iron sulfate, 1-2 of manganese sulfate, 1.4-2.8 g of boric acid, 0.1-0.2 g of copper sulfate and 0.1-0.2 g of zinc sulfate to supply the microelements needed.

4) Meier-Schwart's spring and fall solution

Chilean nitrate	750 g
Triple superphosphate	350 g
Potassium sulfate	700 g
Calcium sulfate	200 g
Magnesium sulfate	450 g
Iron sulfate	5 g
Borax	20 g
Copper sulfate	0.1 g
Zinc sulfate	0.2 g
Manganese sulfate	1 g
A total of 2,476,3 g of solutes per 1,000 l of water	
Approximate concentration	0.25 %
Balance	1:1.1:2.9

5) Kiplin-Laurie's Ohio State Experiment Station Solution

Potassium nitrate	608 g
Ammonium sulfate	110 g
Calcium dihydrogen phosphate	282 g
Calcium sulfate	1,219 g
Magnesium sulfate	511 g
A total of 2,725 g of solutes in 1,000 l of water	
Approximate concentration	0.27 %
Balance	1:1.39:2.73

6) Steiner's solution

Calcium nitrate	900 g
Potassium nitrate	440 g
Ammonium sulfate	10 g
Double superphosphate	250 g
Patentkali	400 g
Magnesium sulfate	150 g
Borax	10 g
Manganese sulfate	5 g
Zinc sulfate	0.04 g
Copper sulfate	0.04 g
Sodium molybdate	0.125 g
A total of 2,129.205 g of solutes in 1,000 l of water	
Approximate concentration	0.2 %
Balance	1:0.47:1.59

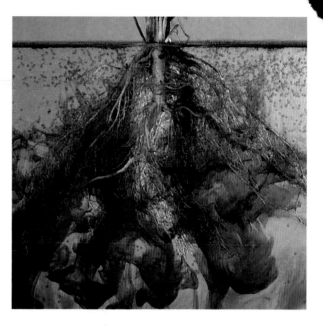

Anyway, the microelements must be dissolved last and in separate solutions. Dissolve the less soluble salts in warm water, using rainwater to avoid the chemicals precipitating out.

The pH of the culture solution has to be suitable for the plant being cultivated. Most cultivated plants tolerate a pH between 5 and 6.5. An excessively alkaline pH can lead to the precipitation of iron, manganese, phosphate, calcium and magnesium in the form of insoluble salts that cannot be assimilated by the plant, while excessively acidic solutions may lead to calcium deficiency.

The solution's pH is adjusted to between 6 and 6.5, using sulfuric acid to acidify it and potassium hydroxide to make it more alkaline, then stirred for 10 minutes and the pH is again adjusted to ensure it is stable.

There is not a single defined nutrient solution for every plant, and there are several formulas, the result of the work of many researchers.

The concentration of salts in the nutrient solution depends on three factors:

• The temperature, which depends on the time of year; the salts are more dilute in summer than in winter
• The plant's state of development; in the early phases, total salt concentration should be less than 0.1%; in full growth it can be between 0.15% and 0.4%.
• The plant's tolerance to salinity.

As time goes by, the nutrient solutions deteriorate and change, and so it is important to monitor their composition.

Analyze the nutrient solution to work out which nutrient ions have been taken up and to check the pH. It will be necessary to replace the solution from time to time so that undesirable ions from the substrate or water cannot accumulate. The solution lasts for 2-3 weeks and never lasts more than 2 months. You should check the solution twice a week. In addition, the volume of water decreases every day by 5-30%, and this has to be measured and replaced.

Reference	pH	Ca^{++}	Mg^{++}	Na$^+$	K$^+$	N as NH$_4^+$	N as NO$_3^-$	P as PO$_4^{3-}$	S as SO$_4^{2-}$	Cl$^-$	Fe	Mn	Cu	Zn	B	Mo
Knop (1865)		244	24		168		206	57	32		Trace					
Shive (1915)		208	484		562		148	448	640		Trace					
Hoagland (1919)	6.8	200	99	12	284		158	44	125	18	As required					
Jones & Shive (1921)		292	172		102	39	204	65	227		0.83					
*Rothamsted	6.2	116	48		593		139	117	157	17	8	0.25			0.2	0.016
Hoagland &Snyder (1933, 1938)		200	48		234		210	31	64	–1	As required	0.1	0.014	0.01	0.1	0.01
Hoagland & Arnon (1938)		160	48		234	14	196	31	64		0.63/week	0.5	0.02	0.05	0.5	0.05
Long Ashton Soln	5.5–6.0	134–300	36	30	130–295		140–284	41	48	3.5	5.6 o 2.8	0.55	0.064	0.065	0.50	
Eaton (1931)		240	72		117		168	93	96		0.8	0.5			1	
Shive & Robbins (1942)		60	53	92	117		56	46	70	107	As required	0.15		0.15	–0.1	0.01
Robbins (1946)		200	48		195		196	31	64		0.5	0.25	0.02	0.25	0.25	
White (1943)	4.8	50	72	70	65		47	4	140	31	1.0	1.67	0.005	0.59	0.26	0.001
Duclos (1957)	5-6	136	72		234		210	27	32		3	0.25	0.15	0.25	0.4	2.5
Tumanov (1960)	6-7	300–500	50		150		100–150	80–100	64	4	2	0.5	0.05	0.1	0.5	0.02
A. J. Abbott	6.5	210	50		200		150	60	147		5.6	0.55	0.064	0.065	0.5	0.05 0.1
E. B. Kidson	5.5	340	54	35	234		208	57	114	75	2	0.25	0.05	0.05	0.5	
Purdue A (1948)		200	96		390	28	70	63	607		20	0.3	0.02	0.05	0.5	
Purdue B		200	96		390	28	140	63	447		1.0	0.3	0.02	0.05	0.5	
Purdue C		120	96		390	14	224	63	64		1.0	0.3	0.02	0.05	0.5	
Standard (For testing Nutrient Soln)		200	96		390	28	140	63	64		1.0	0.3	0.02	0.05	0.5	
Schwartz (Israel)		124	43		312		*98	93	160							
California		160	48		234	15	196	31	64							
New Jersey		180	55		90	20.5	126	71	96							
South Africa		320	50		300		200	65								0.027
CDA A		131	22		209	33	93	36.7	29.5	188	1.7	0.8	0.035	0.094	0.46	0.027
Saanichton B		146	22		209	33	135	36.7	29.5	108	1.7	0.8	0.035	0.094	0.46	0.027
B. C. Canadá C		146	22		209	33	177	36.7	29.5	—	1.7	0.8	0.035	0.094	0.46	—
Dr. Pilgrim C		272	54.3	—	400	—	143.4	93.0	237.5	—	—	—	—	—	—	—
Elizabeth B		204	40.7	—	300	—	107.6	69.75	178.1	—	—	—	—	—	—	—
N. C. USA A		136	27.15	—	200	—	71.7	46.5	118.75	—	—	—	—	—	—	0.02
Dr. H. Resh C		197	44	—	400	30	145	65	197.5	—	2	0.5	0.03	0.05	0.5	0.02
Univ. of B. C. B		148	33	—	300	20	110	55	144.3	—	2	0.5	0.03	0.05	0.5	0.02

Table showing the nutrients present in different nutrient solutions. The composition is given in ppm.

Chemical name	Common name
Potassium nitrate KNO$_3$	Saltpeter. Do not confuse true saltpeter with Chile saltpeter (sodium nitrate - NaNO$_3$) or with "niter cake", which is impure sodium sulfate)
Sodium nitrate Na No$_3$	Chile saltpeter (see above), or sodium nitre, or nitrate of soda
Ammonium dihydrogen phosphate NH$_4$H$_2$PO$_4$	There is no special common name for this term
Urea NH$_2$CO NH$_2$	Carbamide; carbonyldiamide
Potassium sulfate K$_2$SO$_4$	Sulfate of potash
Potassium dihyfrogen phosphate KH$_2$PO$_4$	There is no special common name for this term
Potasium chloride KCl	Potash muriate, chloride of potash (which should not be confused with chlorate of potash (potassium chlorate - KClO$_3$), sylvite, or sylvine, (the natural mineral form)
Calcium dihydrogen phosphate Ca(H$_2$PO$_4$)$_2$H$_2$O	Calcium "superphosphate", (normally with a 20% P content), calcium "triple super-phosphate" (normally with a 75% P content),
Phosphoric acid H$_2$PO$_4$	Orthophosphoric acid (the U.S.P. standard is 85-88% H$_3$PO$_4$, but the commercial grade is commonly 70-75% as (H$_3$PO$_4$)
Calcium sulfate CaSO$_4$2H$_2$O	Precipitated calcium sulfate, native calcium sulfate, alabaster, gypsum

The chemical names and synonyms of the compounds most commonly used in nutrient solutions

Diagram of a hydroponic cultivation installation that incorporates the nutrient solution to the surface in the form of a fine rain
A/ The deposit of nutrient solution
B/ Culture container full of gravel or sand

It is more and more frequent for computers to be programmed to check the nutrient solutions automatically.

5.4.2. Hydroponic cultivation systems

Hydroponic cultivation systems are essentially divided into those that do not use a substrate and those that do, and then by the type of support used for the plants.

- **Cultivation in an inert substrate**
  - Hydroponic cultivation in gravel or sand
  - Hydroponic cultivation in rock wool

- **Cultivation without a substrate**
  - Cultivation in a tank of nutrient solution
  - Nutrient film technique
  - Aeroponics

5.4.2.1. Cultivation in gravel or sand

This system consists of the following components:

- A brickwork bed, 120-150 cm wide and about 30 cm deep, covered with a protective layer to insulate it
- Gravel or sand to fill the bed
- A deposit containing the nutrient solution
- A distribution system with a network of tubing
- A pump
- A drainage system that allows the water to be reused

The nutrient solution can be added to the bench by watering the surface or by subsurface irrigation, that is to say from below. If the nutrient solution is applied to the soil surface as a fine rain, the soil aeration is better.

Diagram of hydroponic cultivation in gravel or sand with sub-irrigation
A/ Rotary pump
B/ Deposit (approx. 1/3 of the bed's content)
C/ Mesh-covered intake
D/ Return
E/ Deviation to mix the solution
F/ Filter
G/ Plastic sheet
H/ Tubing to cover cables
I/ Outgoing circulation

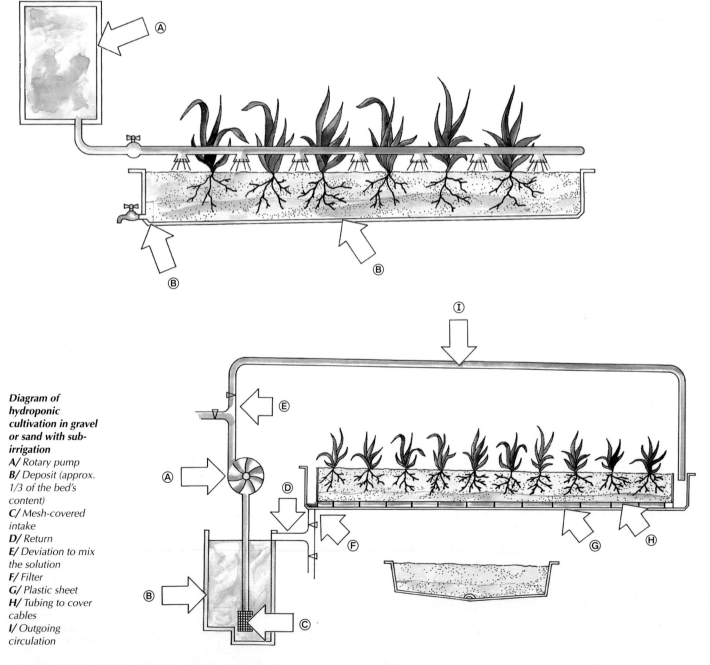

In beds where sand is used as a substrate, there are less problems with blockages of the drainage tubing by the plant roots, a real problem in gravel benches. This is because the sand's greater density favors root growth sideways rather than downwards.

This system is generally well aerated and well drained, but bear in mind that over time the sand may become compacted and show problems of aeration.

The gravel or sand in the beds can be replaced by other substrates, such as perlite, expanded clay or tuff, as long as they are suitable for this usage.

5.4.2.2. Cultivation in rock wool

This consists of placing plaques of rock wool bearing the plant on sheets of the same material arranged longitudinally. The nutrient solution may flow down these longitudinal sheets, or it may be applied locally.

Tubes can also be used to distribute the nutrient solution, and the sheets of rock wool with the plants are inside the tubes.

5.4.2.3. Cultivation in tanks

In cultivation in tanks the plant roots are suspended in a liquid medium that incorporates the nutrient solution.

This system requires the same infrastructure as gravel or sand benches, and the tank can be the same recipient as the bed, simply replacing the substrate with a liquid medium.

As a support for the plants, plastic or metal mesh is placed over the water at a distance of 7 cm above the tank.

The roots have to be in complete darkness, or algae will grow and compete with the roots.

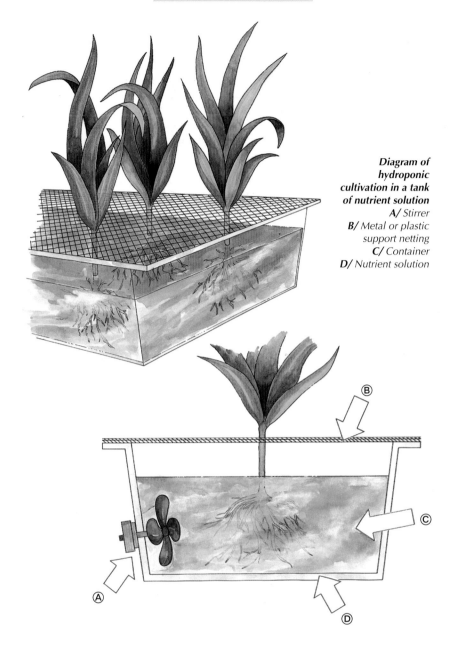

Diagram of hydroponic cultivation in a tank of nutrient solution
A/ Stirrer
B/ Metal or plastic support netting
C/ Container
D/ Nutrient solution

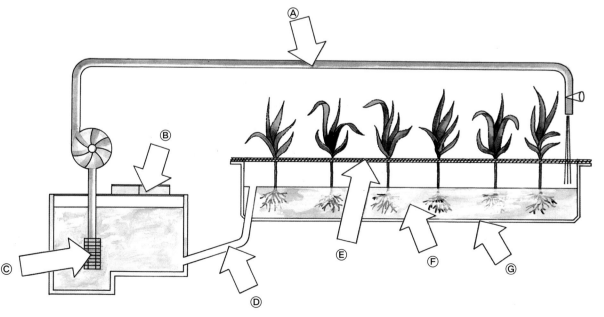

Diagram of an installation for cultivation in a solution of salts
A/ Outgoing circulation
B/ Deposit
C/ Mesh-covered intake
D/ Return
E/ Bed for the plants
F/ Nutrient solution
G/ Plastic sheeting

Detail of the preparation of the plastic channels in the NFT system
A/ *Holes for the plants*
B/ *Folding*
C/ *Nutrient film*

This system uses a relatively cheap medium, water, but this has the disadvantage of poor aeration, as well as the difficulty of supporting the plant adequately. Poor aeration can be solved by circulating the solution, but even so only a limited number of plants can grow with their roots immersed in water.

5.4.2.4. Nutrient film technique

NFT, (Nnutrient film technique), is a technique for cultivation in water in which the plants grow with their root system within a sheet of opaque plastic, through which the nutrient solution is continuously circulated.

The plastic forms tubes that are closed around each plant, and clipped together. They are usually polythene and last for up to 6 years.

At the base of the plastic, below the plant, a thin sheet of rock wool is placed as a capillary mulch. This ensures that none of the plants dries out in the early phases of growth, while leaving most of the roots exposed, allowing good aeration. The roots grow freely and are mainly located at the bottom of the plastic tube, in direct contact with the solution.

Installation using NFT hydroponic cultivation system

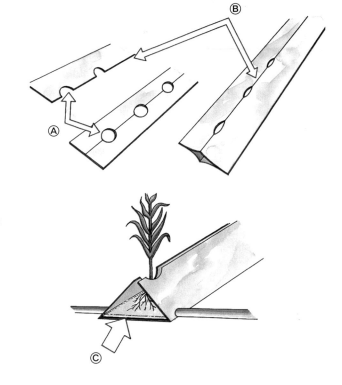

ELEMENT	SYMBOL	CONCENTRATION
Nitrogen	N	200
Phosphorus	P	60
Potassium	K	300
Calcium	Ca	170
Magnesium	Mg	50
Iron	Fe	12
Manganese	Mn	2
Boron	B	0.3
Copper	Cu	0.1
Molybdenum	Mo	0.2
Zinc	Zn	0.1

For the nutrient solution to flow adequately, the plastic must have a slope of at least 1%. The solution is collected at the end for reuse in a drainage channel that takes the excess solution to a deposit, from which it is once more pumped to the main tube, and then towards the plastic tubes.

The recommended temperature for the circulating solution is 20-25°C with a flow of 1.5 l/minute. This is controlled by monitoring the electrical conductivity of the solution. The solution is replaced every 5 days. Monitoring the pH is important, as it has a tendency to increase.

This system has the advantage of simplicity of installation, and the operations to be performed are also simple, water consumption is low and the water is well aerated.

The disadvantages are:

• Over time, water circulation is harder because the plant grows many roots.
• The roots are cold in winter and too hot in summer.
• The solution spreads diseases, making them harder to control.

5.4.2.5. Aeroponics

An aeroponic system is when the plant's roots grow within a totally dark chamber on a support structure. The nutrient solution is regularly injected into the chamber, so that it maintains a constant 100% humidity.

This systems makes the best use of the surface area available, but is hard to find on the market.

HORIZONTAL AEROPONICS

This is the cultivation of plants in rigid PVC tubes 12-15 cm in diameter, within which the nutrient solution flows from one end to the other and is collected in a deposit from where it is recovered and reused.

The plants are placed in holes 3-4 cm in diameter about 40-50 cm apart and are held in place by a plug of some spongy material.
The tubes with the plants are stacked parallel to each other in metal structures 1 m apart.

VERTICAL AEROPONICS
In this system, the plants are cultivated in tall columns, up to 2.20 m, made by piling cylindrical PVC containers 4 cm tall and 14 cm in diameter on top of each other, each one having two openings.

The columns are placed in parallel rows to give all the plants the same light and temperature conditions.
Passageways 80 cm wide are left between the rows, while the columns are arranged in alternating rows with a distance of 140 cm between the columns.

The plants are installed in the openings in the container, and the nutrient solution runs down from the top of the column and to the bottom, where the excess is collected for reuse.

The theoretical ideal concentration (ppm) of elements in a nutrient solution for use in NFT

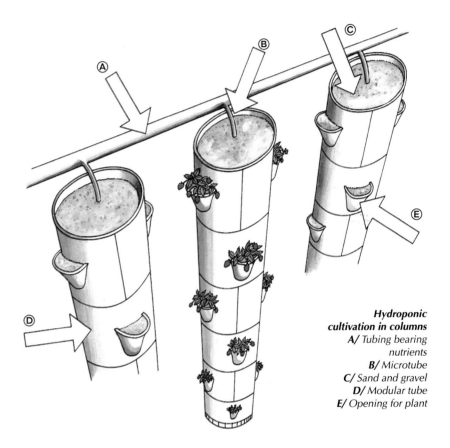

Hydroponic cultivation in columns
A/ Tubing bearing nutrients
B/ Microtube
C/ Sand and gravel
D/ Modular tube
E/ Opening for plant

5.4.3. Substrates

Hydroponic systems can be divided into those that are cultivated in a nutrient solution and those that are cultivated in a solid inert substrate.

For a substrate to be a good base for the plant roots to grow in, it has to meet the following requirements:

• It must be chemically and biologically inert;
• It must not contain toxic elements or disease organisms;
• It must have a uniform size;
• It should resist degradation over time;
• It must have a good capacity to retain water and also allow good aeration;
• It must be easy to disinfect.

The most widely used substrates in hydroponic systems are:

• **Gravel**. This is a natural substrate derived from quartz or pumice stone. Its diameter is 2 to 20 mm. Its porosity is high, it drains well and it retains little water. It lasts for several years and is normally cheap, though transport is expensive. It should be washed before use.

• **Sand**. This is a natural substrate differing from gravel in that the particles are smaller. The sizes used range from 0.2-2 mm. Quartz sand is the most suitable, but it is expensive. It has a high aeration capacity and an average retention of water. It lasts a long time, though over time it tends to present problems due to compaction. Before use, the substrate must be washed to remove any particles of earth, as otherwise this might affect the composition of the nutrient solution.

• **Perlite**. This is an inert artificial substrate, that varies in structure. The particles are white and may be 2-6 mm in diameter. It is chemically stable, with a pH of 7-7.5, but is mechanically fragile, and the particles can be broken by the root pressure. It has a very high water retention capacity and high porosity, meaning that its drainage is good. It is lightweight and only lasts a single growing season.

• **Expanded clay**. This is an artificial substrate that is expensive to manufacture. The particles have a diameter between 2 and 10 mm. It retains little water and has a good aeration capacity. It is inert and neutral with a pH between 5 and 7, and long-lasting.

• **Vermiculite**. This substrate consists of 5-10 mm flakes. It can retain a lot of water and has a good aeration capacity. It is mechanically fragile, and over time it becomes compacted. Its main disadvantage is that it is not chemically inert, and disinfection may be a problem. It has a pH between 7 and 7.2.

Different types of substrates
A/ Pumice stone
B/ Black peat
C/ Perlite
D/ Swedish peat
E/ Irish peat fraction 2
F/ Irish peat fraction 1
G/ Irish peat fraction 3
h/ Norwegian peat
I/ Russian peat
(Produced by Primasta B.V.)

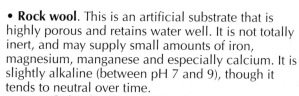

• **Rock wool**. This is an artificial substrate that is highly porous and retains water well. It is not totally inert, and may supply small amounts of iron, magnesium, manganese and especially calcium. It is slightly alkaline (between pH 7 and 9), though it tends to neutral over time.

• **Expanded polystyrene**. This substrate consists of small white balls, with a diameter of between 4 and 12 mm. It weighs very little, retains little water and aerates well. Its pH is between 6 and 6.5. With some types of disinfection, for example high temperatures, the structure may degrade.

• **Tuff**. This is highly porous, retains little water and aerates well.

5.4.4. The system's advantages and disadvantages

The main advantage of this system, in addition to saving water, is the precise control of the plant's nutrition, which results in it making better use of the fertilizers applied.

Other advantages include:

• Less pests, etc., as the crop is growing in a very good environment
• The problem of soil fatigue is eliminated, and the risk of the plant suffering as a result of water shortage declines.
• It reduces cultivation tasks
• The abundant production is of high quality

The most important disadvantages of this system are that:

• The cost of installation and maintenance are high
• It is complex and requires technical knowledge
• The plants are at risk from fungal diseases
• The shock of transplantation is worse in hydroponic cultivation than in the ground, because all the soil has to be washed off the seedling's roots.

5.5. CULTIVATION IN BAGS

In this cultivation method, the substrate is placed in sealed plastic bags. The substrate used is peat, or a mixture of peat and other materials, such as sand or perlite, to increase aeration and improve drainage.

This system completely isolates the plant from the soil and makes it easy to change the material at the end of the harvest, as well as allowing rapid installation of the following crop.

The cultivation system is completed with an irrigation system to meet the plants' water and nutrient requirements. Irrigation and fertilizer application are by trickle or local irrigation. The sacks can be placed vertically or horizontally.

5.5.1. Horizontal sacks

The sacks are placed horizontally on the site, and two separate slits are made in the upper surface, where the plants are then placed.

5.5.2. Vertical sacks

This is a variant of the horizontal sack system. The sacks form vertical tubes or columns. This system was originally based on metal barrels or drums.

Openings for the plants are made along the plastic tube. If a mineral substrate is used instead of peat, the water with the nutrient solution can be applied at the top of the column, thus achieving hydroponic cultivation.

The vertical sack system is used in strawberry cultivation using 15 cm diameter tubes that may be up to 2 m tall, and with up to 32 plants in each vertical sack, with a separation of 1 m between rows and 0.5 m within the row. This gives very high yields.

To the left:
Cultivation in horizontal sacks using the substrate Perlite

Below:
Cultivation in vertical or hanging sacks
A/ Tubing bearing nutrients
B/ Microtubing
C/ Hook
D/ Plastic sack
E/ Plants cultivated in the openings
F/ Drainage holes

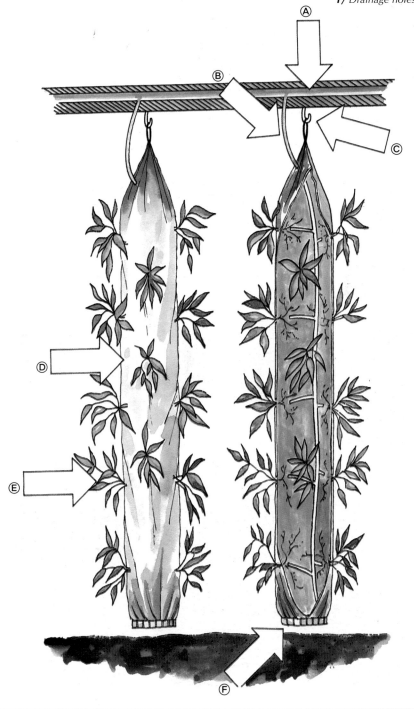

6. FERTILIZER APPLICATION

6.1. FACTORS INFLUENCING THE NUTRITION OF GREENHOUSE CROPS

Greenhouse plants grow in an artificial environment created to control all the factors affecting their growth and development. It is a fragile environment in which any mistake in handling can have dramatic effects. This is because the techniques used are increasingly sophisticated.

One of the main advantages of greenhouse cultivation is that it is possible to control the plant's nutrition, which is influenced by a series of factors that we can sum up as follows:

- Genetic factors, that is to say the species cultivated. Not all plants have the same nutrient requirements.
- Environmental factors inside the greenhouse, such as light intensity, carbon dioxide levels, temperature, humidity and available water.
- The properties of the cultivation medium, such as its chemical characteristics, texture, structure, humidity, pH and nutrient content.

All these factors interact, and the plant responds to the resulting environment.
The soil temperature is an important factor in nutrition, as there is a temperature range where the plant absorbs the most water. This temperature interval is characteristic for each species, but

The effect of soil pH on nutrient availability

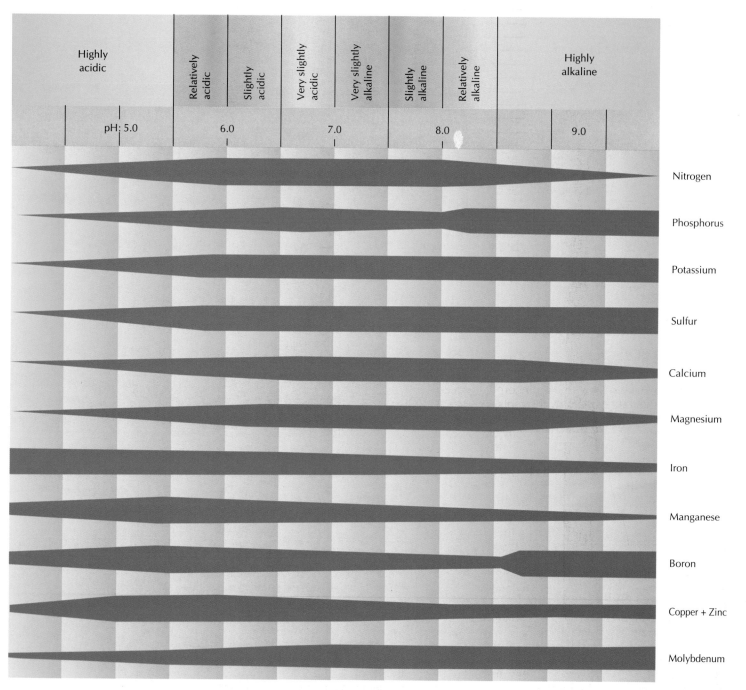

Crops	pH range								
	(4.0	4.5	5.0	5.5	6.0	6.5	7.0	7.5	8.0
Peas									
Cucumber									
Melon									
Rhubarb									
Carrot									
Beans and Lima beans									
Runner beans									
Sweet corn									
Endive									
Parsnip									
Squash									
Pepper									
Turnip									
Tomato									
Eggplant									
Asparagus									
Spinach									
Lettuce									
Celery									
Radish									
Onion									
Beetroot									
Cauliflower									
Broccoli									
Cabbage									
Brussels sprouts									

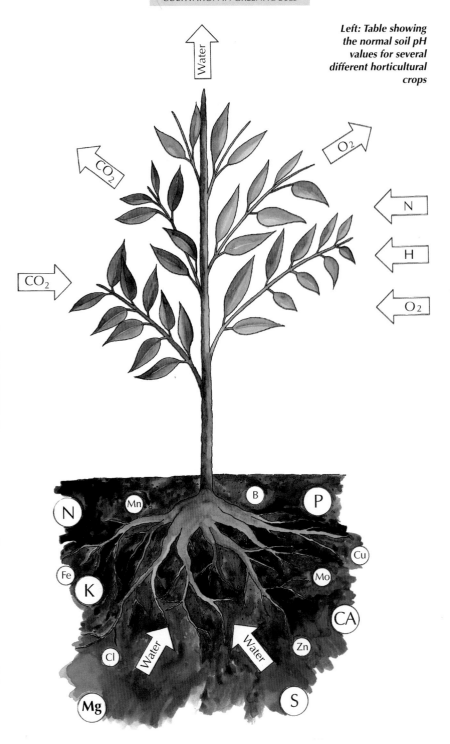

Left: Table showing the normal soil pH values for several different horticultural crops

Above: The elements used by plants

in general nutrients are absorbed best at soil temperatures of 20-25°C.
The soil pH is another factor influencing nutrient absorption, with a pH value of 6-6.5 being the best for ensuring that as many nutrients as possible are in a soluble form in the soil. Of course, every species is different, and each has its own precise pH requirements.

6.2. GENERAL PRINCIPLES OF FERTILIZER APPLICATION

Crop plants can only take up mineral elements from the soil, and not organic products, and this means that the nutrients in organic matter incorporated into the soil can only be taken up by the roots when they have been mineralized, i.e., decomposed to inorganic forms. Only a few of the many chemical elements are used by plants. Carbon, hydrogen and oxygen are obtained from the air and water, and the other elements are taken up from the soil solution by the roots.

The mineral elements are divided into macronutrients and micronutrients by the quantity in which they are needed for the plant's growth and development:

• **Macronutrients** are required in the largest quantities.
- Carbon, oxygen and hydrogen are obtained from the air and water.
-Nitrogen, phosphorus and potassium. They become scarce in the soil, and thus deserve special attention.
- Calcium, magnesium and sulfur. They are less likely to become scarce in the soil.

• **Micronutrients** are vital, but only in very small amounts.
- Iron, zinc, manganese, copper, boron, molybdenum and chlorine.

• **Nitrogen** has a major effect on plant growth and is essential for production of chlorophyll.
More nitrogen is absorbed from the soil than any other element. It can be absorbed as nitrate (NO_3^-) and ammonium (NH_4^+) ions.

Nitrogen is applied as fertilizers containing nitrates, ammonium salts or urea. Nitrate fertilizers are rapidly lost by washing of the soil, while ammonium and urea remain in the soil for longer.

It is important to understand the crop's nutrient requirements to get the best yield.

• **Phosphorus** is absorbed from the soil as $H_2PO_4^-$, and only rarely as HPO_4^{2-} or PO_4^{3-}. Phosphorus is essential in early growth (and also favors root development) and sprouting, flowering and fruiting. It is thus important to choose a fertilizer with the phosphorus in a form that is soluble enough to be easily assimilated.

• **Potassium** is important in fruit formation, increasing the fruit's weight and sugar content. It also plays a role in photosynthesis and favors root growth. It is absorbed as the K^+ ion.

• **Calcium** plays a role in several physiological phenomena in the plant and in root growth. It is absorbed as the Ca^{2+} ion.

• **Magnesium** is an essential component of the chlorophyll molecule. It is absorbed as the Mg^{2+} ion.

• **Sulfur** is essential for protein synthesis. It is absorbed as the sulfate ion (SO_4^{2-}).

• **Copper** plays a role in the formation of chlorophyll.

• **Zinc** plays a role in seed production and ripening, and in the production of plant growth substances.

• **Boron** is needed for good root growth, and for fruit and seed production.

• **Manganese** is involved in photosynthesis.

• **Iron** plays a role in several physiological processes and in chlorophyll production. It also plays a role in the plant's respiration.

Yield and nutrient extraction of a single plant species as a function of the cultivation system

• **Molybdenum** participates in protein synthesis and is essential for the plant to use the nitrogen it takes up.

• **Chlorine** plays a role in photosynthesis.

6.3. DETERMINING THE NEEDS OF GREENHOUSE CROPS

The yield from a greenhouse crop is greater than that from one grown in the open air, and the plant's extraction of nutrients is also greater. Bearing in mind that the root system in greenhouse crops has limited space in which to grow, the input of fertilizer has to be applied to the top 30 cm of soil.

In greenhouses, fertilizers have to be applied in a greater number of smaller applications to avoid the risk of soil salination. Applying excess fertilizer is a danger to watch out for. As a guideline, the doses of fertilizer to apply to the soil depend on the chemical composition of the

Species	Cultivation system	Yield per m²	Extractions g/m²		
			N	P_2O_5	K_2O
Tomato	open air	9 kg	20	7	30
	greenhouse	12-15 kg	65	10	86
Carnation	open air	6-8 flowers	50	16	40
	greenhouse	10 flowers	120	30	110

Crop	Unit production	Nutrient elements (Kg/unit production)		
		N	P_2O_5	K_2O
Tomato	10 t	30-40	8-12	40-65
Spinach	10 t	50-60	14-16	50-100
Carrot	10 t	30-50	10-12	55-65
Lettuce	10 t	30-40	10-20	60-80
Onion	10 t	25-40	10-15	30-35
Artichoke	10 t	75-80	30-35	150-160
Cauliflower	10 t	40-45	14-16	60-65

Left:
Average extraction of
nutrients by different
crops

This method of applying fertilizer is based on the idea of anticipating the plant's future needs, foreseeing what the plant is going to consume, so that the soil's fertility can be maintained constant. Maintenance fertilization should be applied when the soil contains 6-8% organic matter, 0.05-0.06% P_2O_5 soluble in water and 0.18-0.20% K_2O soluble in water.

The most convenient way of supplying the fertilizers to the crop in the greenhouse after the crop has been planted is to apply the fertilizer in the irrigation water. This is the system normally used when there is a local irrigation system.

Fertilizers can also be applied as leaf feeds. The different elements are dissolved in water and sprayed onto the aboveground parts of the plant. Foliar feeding is a fast and effective method of solving nutrient shortages, especially micronutrient deficiencies.

There are limits on how much fertilizer can be applied as leaf feeds, because high concentrations of some salts can burn the leaves. For this reason, it is normally only considered as a supplement.

Sphagnum peat
Produced by
HASSELFORS

harvests extracted, checking the plant's growth and development, and correcting any deficiencies.

To determine the nutrient elements to apply to the soil, the following methods are available:

• Observation of the crop (a subjective method that requires previous experience)
• Measuring the pH and salinity
• Soil analysis
• Measuring nutrient extraction by the crops

6.4. APPLYING THE FERTILIZER

Before planting the crop in the greenhouse, add a lot of organic matter to the soil, to ensure there is enough. There should be about 5% by weight. The input should be between 2 and 5 kg/m², depending on the soil's original content. This input seeks to ensure that the soil maintains its structure for as long as possible under cultivation.
Solid mineral fertilizers should be applied to the greenhouse soil at depth during preparatory plowing in the form of slow-release fertilizers. Once the crop has been planted, it is difficult to apply fertilizer at depth.
It is important to distinguish between fertilization at depth and maintenance application of fertilizer. Fertilization at depth seeks to raise the soil's nutrient status to an optimum level before starting cultivation, while maintenance application aims to meet the crop's nutrient requirements during the course of the year.

APPLYING FERTILIZER AT DEPTH

This is done by burying the fertilizers when preparing the soil, so that they are in the soil close to the roots.
This is the usual system of applying organic fertilizers, phosphate fertilizers and potassium fertilizers, as well nitrogen fertilizers in the form of urea or ammonium.

MAINTENANCE OR SURFACE FERTILIZER APPLICATION

The fertilizer is applied when the plant is growing in the greenhouse. The fertilizers applied usually contain nitrates.

6.5. TYPES OF FERTILIZER

Greenhouse cultivation requites more intense use of the soil. This is the main reason why more fertilizers are used, in order to meet all the crop's needs and to obtain the maximum yield.

Soil salination, due to this heavy use of fertilizers, should be avoided, and soil salinity should never exceed concentrations of 2‰ if the plant is not resistant to saline conditions.

The following points should be borne in mind when applying the main types of fertilizers:

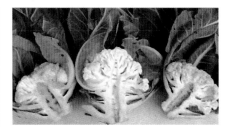

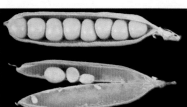

Plants showing symptoms of boron deficiency

Laboratory for biological and organic testing

• **Nitrogen-containing fertilizers**. The fertilizer should be applied before the plant's peak growth period, and depending on its growth. Applications should be small and split over the cultivation period. Bear in mind that nitrogen in the form of nitrate salts is very soluble in water and may thus be lost by drainage. Nitrogen in the form of ammonium is retained longer in the soil, until it too is turned into nitrate.

• **Phosphorus-containing fertilizers** should be applied in a single application at depth to meet the plant's needs for the entire growing period. It is especially important at sowing time and when transplanting.

• **Potassium-containing fertilizers**. They are generally needed in large amounts. Divide the total amount applied, applying part at the time of planting and the rest to the surface, depending on the plant's requirements.

Fertilizers, in general lines, are classified into simple, compound and complex. Simple fertilizers only supply a single element. Compound fertilizers contain two or more elements as different substances. Complex fertilizers contain two or more elements but in the form of a single fertilizer compound.

Technically balanced fertilizers contain nitrogen, phosphorus and potassium, and are used in formulations that vary in the proportions of the different elements, and thus meet different needs or are for use at different stages of growth.

SOLUBLE SOLID FERTILIZERS

• **Ammonium sulfate** contains 21% nitrogen in the form of ammonium, and 23% sulfur. It has a solubility of 730 g/l at 20°C.
• **Ammonium nitrate** contains 34% nitrogen, and has a solubility of 1,920 g/l at 20°C.
• **Urea** contains 46% nitrogen, and has a high solubility of 1,033 g/l at 20°C.
• **Calcium nitrate** contains 15-16% nitrogen in the form of nitrate, with a solubility of 1,220 g/l at 20°C.
• **Potassium nitrate** is an excellent product for fertilizer application in irrigation water. It contains 13.5% N and 44-46% K_2O. It is totally soluble but it has a low solubility, 316 g/l at 20°C.
• **Ammonium dihydrogen phosphate** contains 12% nitrogen and 60-62% P_2O_5, and has a solubility of 661 g/l at 20°C.
• **Ammonium hydrogen phosphate** contains 21% nitrogen and 52-54% P_2O_5
• **Ammonium polyphosphate** contains 10% nitrogen and 30% P_2O_5. It is highly soluble.
• **Urea phosphate** contains 17% nitrogen and 44% P_2O_5
• **Potassium sulfate** contains 50% K_2O and 17% sulfur. It solubility is low, 110 g/l at 20°C.
• **Potassium hydroxide** contains 50% K_2O.

LIQUID FERTILIZERS

• **Ammonia** contains 82% nitrogen in the form of ammonia.
• **Liquid calcium nitrate**. This is a solution of the solid product, containing 7% nitrogen in the form of nitrate. It is mainly used as a source of calcium.
• **Liquid magnesium nitrate**. Like liquid calcium nitrate, but it contains 7% magnesium.
• **Nitric acid**. At a concentration of 56.5% it contains 12% N.
• **Phosphoric acid**. This contain between 40% and 54% P_2O_5.

FERTILIZER APPLICATION AT DEPTH

• Slow-release sources of nitrogen:

- Ammonium sulfate
- Calcium cyanamide
- Urea
- Different forms of ammonium

• Slow-release sources of phosphorus:

- Natural phosphates, or rock phosphates, containing 25-40% P_2O_5
- Calcium hydrogen phosphate, containing 65% P_2O_5
- Simple and triple superphosphates with 18% and 45% P_2O_5 respectively

• Slow-release sources of potassium:

- Potassium chloride, containing 60% K_2O
- Potassium sulfate, containing 50% K_2O

6.6. COMBINED FERTILIZATION AND IRRIGATION

This is based on adding soluble fertilizers to the irrigation water, which is then distributed by the installed irrigation system. Liquid fertilizers or totally soluble solids can be used, as long as they do not react with the salts contained in the irrigation water and are not corrosive to the irrigation installations and devices. This system allows the total amount of fertilizer applied to greenhouse crops to be divided into small doses, controlling the time of application and thus diminishing the risk of the accumulation of salts and saline residues.

Combining fertilizer application and irrigation favors absorbtion of the nutrient elements by the roots, results in more even application of the fertilizer and uses less fertilizer, as they are only applied to a very precisely chosen part of the soil, where the roots are growing, and not to all the soil.

Another advantage of this fertilizer application system is that it saves labor, as water distribution can be regulated automatically.

Irrigation system that automatically distributes pesticides, etc. Manufactured by A. GUASTAPAGLIA

Characteristics of the main products used when applying fertilizer in irrigation water

Products	Nutrient content %				Solubility in g/l		Density	Salt index	Acid (A) or Base (B)
	N	P_2O_5	K_2O	Others	0°C	20°C			
NITROGEN-CONTAINING:									
Ammonium sulfate	21			22 S 69	700	760	110 A		
Urea	46				670	1033		75.4	85 A
Ammonium nitrate solution	33.5				1180	2190		104.7	59 A
Calcium nitrate	15			30 Ca	1020	1220		52.5	20 B
Nitric acid	13						1.36		26 A
Nitrogenated solution 20	20						1.25	57.3	
Nitrogenated solution 32	32						1.32	70.1	58 A
Magnesium nitrate	7			6 Mg				42.6	
PHOSPHATE-CONTAINING									
55% phosphoric acid	40						1.40		38 A
75% phosphoric acid	54						1.48		
POTASSIUM-CONTAINING									
Potassium sulfate			50	18 S	75	120		46.1	neutral
Potash solution			10	3 S			1.1	18.8	
BINARY AND TERNARY FERTILIZERS:									
Ammonium dihydrogen phosphate	12	61			225	400		34.2	65 A
Ammonium hydrogen phosphate	21	53				450		24	
Urea phosphate	17	44				620			
Potassium dihydrogen phosphate		51	34		148	230		8.4	neutral
Potassium nitrate	13		46		130	335		73.6	26 B
Ternary solution	4	8	12					29.1	
Ternary solution	8	4	10					29.2	
Binary solution	0	20	10					16.4	26 B
CONTAINING SECONDARY AND MICROELEMENTS									
Magnesium sulfate				16 MgO y 13 S		700			
Iron sulfate			36 Fe		155	260			
Copper sulfate			25 Cu		140	200			
Manganese sulfate			32 Mn			900			
Zinc sulfate			23 Zn			750			

Detail of motor for overhead watering system Manufactured by A. GUASTAPAGLIA

The disadvantage is that plants grown using combined fertilization and irrigation tend to have very small root systems, because they do not need to explore the soil. They are also more sensitive to deficiencies and to shortage of fertilizer, especially the microelements, which the roots cannot find in sufficient quantity in the limited space they occupy. In order to use this system, you need to know the amount of each nutrient the crop extracts and also the rate of uptake, which depends on the climatic conditions and the plant's development. This is important to strike the right balance between the plant's water requirements and its nutrient requirements, how much fertilizer to apply and when to apply it.

When applying fertilizer in irrigation water, any type of soluble fertilizer can be used, as long as it does not give rise to blockages of the nozzles of the irrigation system. Avoid mixing fertilizers that will form insoluble precipitates, and so you should not mix fertilizers containing calcium or magnesium with ones containing sulfates or phosphates, as they will form insoluble precipitates.

Chemical compatibility of fertilizer mixes

	AN	AS	AP	SN	CN	PN	PS	PC	MS	U
AN - Ammonium nitrate	C									
AS - Ammonium sulfate	C	C								
AP - Ammonium phosphate	C	C	C							
SN - Sodium nitrate	C	C	C	C						
CN - Calcium nitrate	C	O	O	C	C					
PN - Potassium nitrate	C	C	C	C	C	C				
PS - Potassium sulfate	C	C	C	C	O	C	C			
PC - Potassium chloride	C	C	C	C	C	C	C	C		
MS - Magnesium sulfate	C	C	O	C	O	C	C	C	C	
U - Urea	X	C	C	C	X	C	C	C	X	C

I: Incompatible; L: Limited compatibility; C: Compatible

Nitrogen can be supplied as commercial solutions or as urea, which dissolves easily, or potassium nitrate if you wish to supply potassium at the same time.

Phosphorus is the most difficult element to apply in combined fertilization and irrigation, because it is very insoluble, and it may well precipitate out with the calcium contained in the irrigation water. The microelements are applied in the form of chelates to prevent them reacting with the dissolved salts, which would prevent them being taken up. The water reaching the plant should contain 1-2 g/l of dissolved salts, in order to avoid salt accumulation in the soil, and have a pH between 6 and 6.5. If the irrigation water has a pH greater than 7, this may cause the precipitation of some elements, and this means the plants cannot take up enough, and also causes serious problems of blockages in the irrigation system. If the water's pH is below 6, it may corrode the metal components of the installation.

On the following page: Automatic systems to apply fertilizer in irrigation water. Manufactured by A. GUASTAPAGLIA

A conductimeter can be installed to check the salinity and sound an alarm if it exceeds a preestablished value. Nitric acid and sulfuric acid can be used to correct the pH of the irrigation water if it is excessively alkaline.

The soil has to be wetted before application, to avoid salt accumulation around the roots.

The fertilizers can be added to the irrigation water using one of the following systems:

• **Tank deposit**. The tank, which may be small or large, is treated to prevent corrosion and installed in parallel with the main conduct. The fertilizer is placed inside the tank, and by means of a set of valves, part of the volume of flow is diverted to dilute the fertilizer and is then returned to the circuit. The system works by differences in pressure. It has the disadvantage that the distribution of the fertilizer is not even, because the concentration of fertilizer in the irrigation water is not constant, as the concentration of the stock solution varies over time.

• **Fertilizer injectors**. Units containing the concentrated fertilizer solution are placed in the water tank, and inject the fertilizer into the irrigation system using different types of pumps:

- *Electric*. The pump sucks the fertilizer from a deposit and injects it into the irrigation supply. The fertilizer injected is not proportional to the volume of water.

Product	Concentration g/l	pH	Conductivity mS/cm
Ammonium nitrate	2	5.4	2.8
33.5 % N	1	5.6	0.9
	0.5	5.6	0.8
	0.25	5.9	0.5
Urea 46 % N	3	6.3	0.1
	1	5.8	0.07
	0.5	5.7	0.07
	0.25	5.6	0.05
Ammonium sulfate 21 % N	1	5.5	2.1
	0.5	5.5	1.1
	0.25	5.5	0.5
20 % N solution	1	6.4	1.3
	0.5	6.8	0.7
	0.25	6.9	0.4
32 % N solution	2	7.2	2.3
	1	7.1	1.1
	0.5	6.6	0.6
	0.25	6.1	0.3
Phosphoric acid 54 % P_2O_5	1	2.6	1.7
	0.5	2.8	1.0
	0.25	3.1	0.5
Phosphoric acid 40 % P_2O_5			
	1	2.3	1.7
	0.5	2.5	1.1
	0.25	2.7	0.6
Crystalline ammonium	1	4.9	0.8
dihydrogen phosphate	0.5	5.0	0.4
12-61-0	0.25	5.3	0.2
Phosphate-Urea	1	2.7	1.5
17-44-0	0.5	2.9	0.8
	0.25	3.2	0.5
Potassium nitrate	1	7.0	1.3
13-0-46	0.5	6.6	0.6
	0.25	6.6	0.3
Pure potassium	1	7.1	1.4
sulfate 50 % K	0.5	6.6	0.8
	0.25	6.6	0.3
Potassium solution	2	2.5	1.6
1-0-10	1	2.8	0.8
	0.5	3.0	0.5
	0.25	3.1	0.3
Liquid complex	2	2.8	1.4
4-8-12 AC	1	3.1	0.6
	0.5	3.3	0.4
	0.25	3.5	0.2
Liquid complex	3	3.1	1.2
8-4-10 AC	2	3.2	0.9
	1	3.4	0.5

- *Proportional injection electric doser.* This consists essentially of a meter measuring the flow of water, and every time the flow exceeds a preestablished value, it sends an electric signal to a sensor telling the pump to add more of the stock solution. The solution is always the same strength, whatever the volume of flow of water.

- *Hydraulic doser.* The pump has a chamber with a piston. When the chamber fills, it sucks fertilizer from a deposit and then empties as it injects the stock solution into the irrigation network.

6.7. CARBON DIOXIDE FERTILIZATION

Carbon dioxide is essential for photosynthesis. Increasing the concentration of CO_2 in the air in the greenhouse favors plant growth, increases production and improves the quality of the produce. To obtain peak yields in photosynthesis, the concentration can be increased from the 0.03% in the atmosphere up to around 0.1%. Exceeding 0.3% may be toxic to the plant.

Carbon dioxide can be added to the greenhouse air in the following ways:

• By injecting pure gas from canisters and distributing by perforated tubes. This system has the lowest cost of use and is thus the most widely used.
• By putting solid CO_2 ("dry ice") in the greenhouse, where it slowly diffuses into the air.
• By burning propane or butane.

Salinity of products at the concentration normally used in combined fertilization and irrigation

7. IRRIGATION

7.1. THE SOIL, THE WATER AND THE PLANT

The plant needs a lot of water for its growth. 80-90% of its total weight is water, depending on the species of plant.

Water is essential for sugar production and cell maintenance. It is the medium of transport for nutrients and the substances synthesized by the plant, and is the main reagent in many vital physiological processes. The cell, and thus the plant, requires water to maintain its turgid.

Most of the water absorbed by the roots is evaporated by the leaves in transpiration. Less than 5% is retained in the tissues. The rate of transpiration is directly related to the ambient temperature.

The water evaporated by the leaves is replaced by water taken up by the roots, and the mineral elements present in the soil are taken up by the plant at the same time.

When the plant loses more water through transpiration than the roots can take up, the plant suffers a water deficit, which has a negative effect on growth, and thus on yields. The effects that are derived from this shortage vary with the duration and intensity of the water stress the plant suffers, and on its growth stage. Intense water stress, even for only a short period, may reduce growth more than moderate water stress over a long time. Intense water stress is very damaging when it occurs during the most delicate stages of the plant's growth cycle, such as flower induction, flowering, or bud differentiation.

Wilting occurs when the plant is subjected to a prolonged or high intensity water stress. The plant loses water, and thus its turgor and rigidity. To limit water stress, transpiration can be reduced by lowering the temperature or shading. Increasing the concentration of carbon dioxide in the air and applying anti-transpiration agents, also reduces stomatal opening, and thus reduces the rate of water loss. Air movements also increase transpiration.

Applying water to the site is the most effective method of meeting the crop's water needs, and this is done by irrigation.

Right: Placing of tensiometers, depending on the depth the measurement is made

Determining the soil's moisture levels by observation

7.2. USEFUL WATER

The water entering in the soil occupies all the free spaces, both the large spaces (macropores) and the very small ones (micropores). The water occupying the macropores falls due to gravity, and these spaces are then occupied by air. This contributes to soil drainage by eliminating the excess water. This water is known as gravitational water.

The micropores, or capillary spaces, retain the water, and the water moves between them by capillary action.

After watering, the soil loses the gravitational water by drainage, and then it gradually loses moisture due to the evaporation of water from the soil surface, and because the plants take up water from the soil for transpiration. The joint effect of evaporation and transpiration is known as evapotranspiration, and they are

Moisture content	Organoleptic observation
Dry	Dry powder
Low (critical	Soft, does not form a ball
Medium (normal for the time of year)	Forms a ball, but softens when it is shaken several times
Good	Forms a ball that remains intact after being shaken 5 times; it sinks slightly if subject to a slight pressure.
Excellent	It forms a durable and flexible ball; it is easily flattened; a relatively large area will be depressed when pressed hard with the thumb.
Too wet	With firm pressure, some water can be squeezed from the ball of soil.

responsible for the gradual drying out of the soil.

On this basis, the soil is always in one of the following states:

• **Saturated**. This is immediately after irrigation, when the water occupies, or saturates, all the spaces between particles. This is not a good situation, as the roots are deprived of the air they need for respiration.
• **Field capacity**. This is when the soil has lost all its gravitational water. This occurs within 24 hours of irrigation. Soil at field capacity contains as much water as it can retain.
• **Semi-humid**. The soil has an intermediate moisture level, halfway between field capacity and the wilting point.
• **Wilting point**. The soil has lost so much water by evapotranspiration that the plant can no longer absorb water.

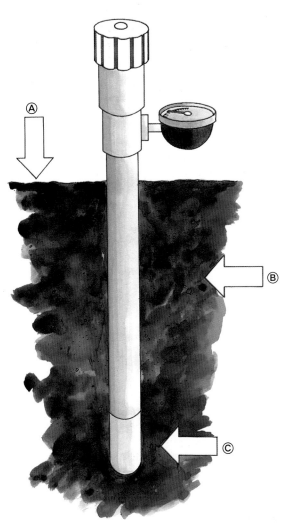

Diagram of a tensiometer
A/ Soil surface
B/ Tube full of water
C/ Porous ceramic

Left:
Automatic sprinkler irrigation system
Manufactured by
RAIN BIRD

Available water is defined as the water present between field capacity and the wilting point. The field capacity depends on the soil class, while the wilting point depends on the plant as well as the soil.

The available water, the water between these two limits, represents the quantity of water contained in the soil that can really be taken up by the plant.

Interpretation of tensiometer values to give soil moisture status

Tensiometer reading cb	Soil moisture content	Interpretation of soil moisture content
0-10	Saturated	Values like this are normal, for 24-36 hours after irrigation. If they last for longer than this, there may be drainage problems, or watering may have been excessive.
10-25	Optimal	This corresponds to optimal soil water levels; when the values are in this range, the plants do not need to be watered.
5-30	Water needed, no danger	On light soils, when the reading is about 30 cb, it needs irrigating again; on heavier soils, water when the reading reaches 45 cb.
50-70	Wilting point	The plant is in danger of irreversible wilting. The plant cannot extract the little water left in the soil.

Infiltration of water into a sandy soil and a clay soil
A/ *Sandy soil*
B/ *Clay soil*
In a sandy soil, irrigation water reaches deeper. In clay soils, the water sinks more slowly, but spreads more sideways.
The type of irrigation used must take this into account.

Lower right:
Trickle irrigation systems.
Both manufactured by RAIN BIRD

A Pressure compensator. Self-cleaning.

Water is also present as the water forming part of the different soil minerals. This is of no agricultural importance, as it cannot be released in normal conditions for use by the plant. The water retained in the soil with a force greater than the suction exerted by the roots is unavailable for use by the plant.

Numerous experiments have shown how important it is to water the plant before it suffers severe water stress, and above all before the soil moisture level declines to the wilting point.

You should water when one third or, at most, one half of the available soil water has been lost through evapotranspiration. The decision can be based on the farmer's experience, or on methods that directly measure the soil's moisture, such as tensiometers.

A tensiometer is a device to measures the soil's water pressure, that is to say, the force the roots must exert to absorb water from the soil. The greater this

levels in the soil, in order to get a better understanding of the soil's water status.
The reading of the tensiometer ranges from 0, the maximum moisture content possible, to 100, total lack of water.

The values are:

    0-10: soil saturation
    10-20: field capacity
    30-60: useful water
    60 or more: water shortage

7.3. IRRIGATION SYSTEMS IN GREENHOUSES

When irrigating a greenhouse, the aim is an almost exact dosing of the water supplied, thus saving water and allowing the use of limited water resources. The system can be automated, thus reducing labor requirements to a minimum.

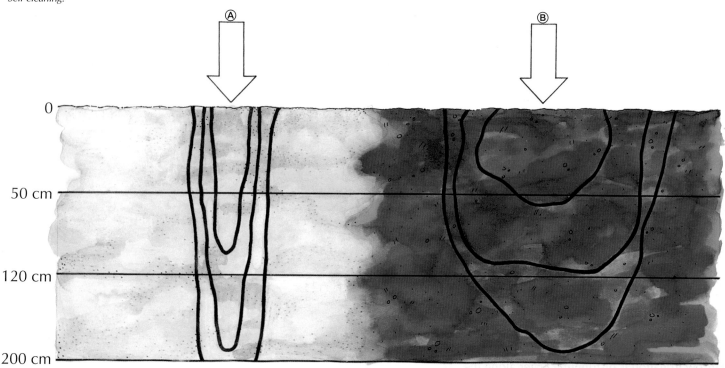

Ⓐ   Ⓑ

0

50 cm

120 cm

200 cm

High resistance plastic. Available in 3 different volumes of flow, 2 l, 4 l and 8 l per hour. Self-regulating from 1 to 3.5 Bars.

B 6 independent self-compensating outlets. Self-cleaning. Volume of flow: 4 l/h per outlet. Pressure range from 1 to 3.5 Bars. The system is supplied with one outlet open and ready for use, and the other outlets have to be opened according to need.

tension, the less water there is in the soil and the more pressure the roots must exert. And to the contrary, the lower the tension, the more moisture is present and the less pressure the plant need exert.

A tensiometer consists of a capsule of porous ceramic full of water, which is buried in the ground at the level where the humidity is to be measured. This should be the depth where root growth is most active. The capsule is connected to a manometer by a tube full of water. The porous ceramic is permeable to water, but not to air.
The soil sucks water from the capsule, causing the water in the column to fall in height below the soil level, to a depth that can be measured and corresponds to the tension.

Two tensiometers should be placed at two different

Ⓐ

Ⓑ

There are many different types of water distribution system for greenhouses, varying greatly with soil type, crop requirements, cultivation techniques, labor costs, water availability and the possibility of automating the installation.

The many different irrigation systems can be divided into three main groups:

- Irrigation by gravity
- Sprinkler irrigation
- Local irrigation

7.3.1. Irrigation by gravity

Irrigation by gravity requires no external energy to distribute the water. That is to say the water is distributed by the force of gravity alone, even if energy has to be used to get the water to the greenhouse.

The water is applied at the highest point of the greenhouse site, and flows down the gradient. Traditional gravity watering systems include sheet irrigation and furrow irrigation.

Sheet irrigation consists of covering the ground with a sheet of water by supplying more water than can filter into the soil, so that the water covers the soil, and then gradually filters into it.

Furrow irrigation is distributing water along the furrow between two ridges, meaning that the water can infiltrate sideways as well as downwards.

Irrigation by gravity has the following advantages over other systems:

• Low installation cost
• It effectively washes salts from the soil
• A large volume of soil is wetted and available to the roots.

However, it has the following disadvantages:

• The water is irregularly distributed
• A lot of the water used is lost
• The site has to be prepared
• Fertility is lost

7.3.2. Sprinkler irrigation

Sprinkler irrigation supplies the water as an artificial rain. This can be done at low pressure with a gentle short-distance jet, or at high pressure with a stronger, further-reaching, jet.

The system consists of the following components:

• Pumping group: this must provide the pressure needed and the volume of flow.
• The network of distributor tubes: this consists of a main tube that channels the water under pressure, and a secondary network that branches from the

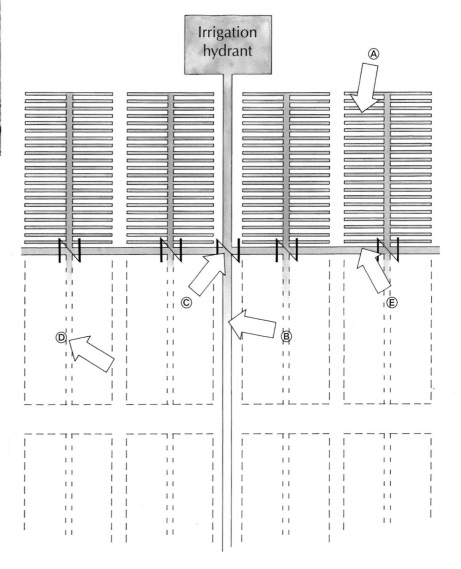

main tube and which bears the sprinkler nozzles.
• Sprinklers: they distribute the water in different ways, depending on the type used and the crop's requirements (in a fan, parallel jets, single jets, etc.). They can be fixed or rotary and may be positioned above or below the plant. Rotary models are most widely used, with jets reaching 10-25 m and a volume of flow of 0.5-7.5 m^3/h at a pressure of 2.5-3 kg/cm^2.
• Discharge valves: these serve to prevent the dripping that occurs in aerial installations after the end of watering.

*Left:
Micro-sprinkler watering system. Micro-Quick is a system of micro-sprinklers that brings together the water and energy savings of local irrigation systems in an easily assembled, ingenious and developing system.
Manufactured by
RAIN BIRD*

Diagram of distribution in an installed irrigation system
A/ Trickler tubes
B/ Main tube
C/ Stopcocks
D/ Tertiary tubing
E/ Secondary tubing
*Courtesy of
AUXILIAR
ANDALUZA DE
RIEGOS (Seville)*

When installing this system, the composition and structure of the soil to be watered must be taken into account. The maximum water input that can be applied will be limited by the speed at which the water filters into the soil. If the water supplied exceeds this amount, waterlogging will occur.

The advantages of this system are that:

• It saves water.
• Earthworks are unnecessary.
• The water is evenly distributed without having to prepare the site.
• It reduces the risk of excessive washing of the soil.
• It can be automated by a control program.
• Fertilizers can be applied directly in the irrigation water.
• It has a positive influence on the environment within the greenhouse. It reduces the temperature and maintains the environmental humidity.
• It can be used to protect from the risk of frost.

The disadvantages are that:

• More energy is consumed
• If the sprinkler drops are very small and the air is very dry, a lot of the water simply evaporates.
• It favors the development of plant pathogens.
• The cost of installation is high.

Sand filter
A, B and *C/*
Stopcocks
D and *E/* Manometers

7.3.3. Installed local irrigation

Installed local irrigation provides the plant with water in very low doses and volumes, but more often. It thus maintains excellent humidity levels in the soil for the plant.

Installed irrigation systems are classified by the way in which the water is emitted:

• **Micro-sprinklers**. The water is emitted by rotary sprinklers, that rotate due to the pressure of water.
• **Diffusors**. The water is emitted by fixed nozzles that pulverize the water and distribute it evenly in a circle around them.
• **Tricklers**. They slowly emit the water directly into the soil. They form a permanently wet volume around the plant's roots called the wet bulb.

These installations consist of:

• Irrigation hydrant
• Distribution network
• Emitters

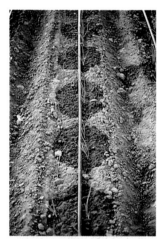

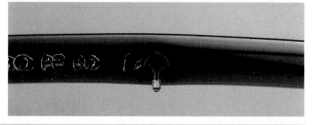

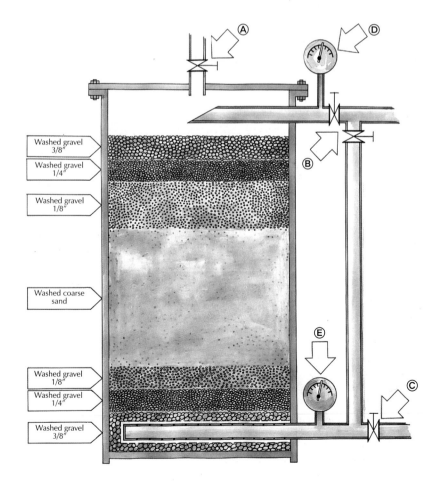

Washed gravel 3/8″
Washed gravel 1/4″
Washed gravel 1/8″
Washed coarse sand
Washed gravel 1/8″
Washed gravel 1/4″
Washed gravel 3/8″

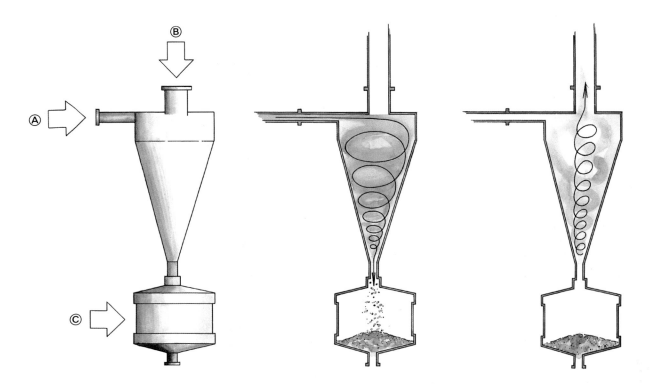

Hydrocyclone
A/ Water inlet
B/ Outlet for clean water
C/ Container for sediments

7.3.3.1. Irrigation hydrant

The components of this device control the volume of flow, pressure, filtering and the addition of fertilizer to the water being supplied to the crop.

• **Flow and pressure**. This is controlled by volumetric valves or time-regulated solenoid valves, pressure regulators or manometers.

The manometers are located at the inlet and outlet of the filters, to show the loss of load within the filters. When there is an excessive pressure difference between the inlet and outlet manometers, this shows that the filters may be blocked and should be inspected and cleaned.
At the end of the irrigation hydrant, after the filters, a water meter is installed to show the volume of water flowing to the distribution system.

• **Filtration**. The irrigation water is filtered to remove the impurities so they do not block the nozzles. The volume of flow determines the size and number of the filters required.

The filters may be located before or after the fertilizer application installation, though you are advised to place it before to avoid problems due to corrosion by the fertilizers.

There are 3 types of filter:

• **Hydrocyclones**. They eliminate suspended silt and heavy metals. They are upside-down truncated cones, larger at the top. The water enters at the side, the suspended particles are moved by centrifugal force to the periphery and sink under the action of gravity to the base, where there is an outlet for

sediments, while the water exits through the center at the top.

• **Sand filter**. This consists of a closed container which the water enters at the top and outlet collectors surrounded by gravel, on top of which there is a relatively thick layer of clean siliceous sand that acts as a filter. Reversing the flow of water is enough to drive out all the accumulated dirt. This has to be performed regularly. This filter is used to remove coarse inorganic impurities and organic impurities like algae and organic matter, and it is placed before the fertilizer equipment.
• **Mesh filter**. This consists of a frame treated with anti-corrosive agents, within which there is a cartridge consisting of one or more stainless steel meshes thick enough to trap any impurities that get through the sand filter.
The mesh filters are usually located after the fertilizer application equipment.

Tube to supply trickle irrigation system (Manufactured by TWIN DROPS IBÉRICA, S.A.)

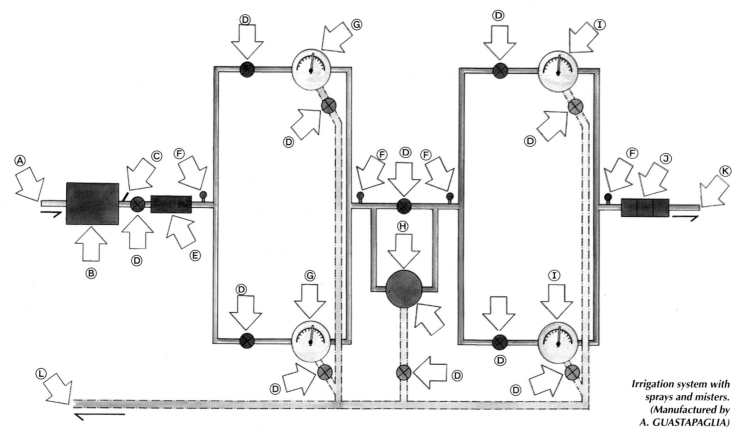

Irrigation system with sprays and misters. (Manufactured by A. GUASTAPAGLIA)

Diagram of an irrigation hydrant
A/ Water inlet tube
B/ Pumps
C/ Retaining valve
D/ Stopcock
E/ Pressure regulator
F/ Manometer
G/ Sand filters
H/ Fertilizer application equipment
I/ Mesh or ring filter
J/ Water meter
K/ Outlet to main tube
L/ Water outlet
Courtesy of AUXILIAR ANDALUZA DE RIEGOS (Seville)

Mesh filters can be replaced by ring filters. These consist of a set of pressed porous discs.
The type of filter to use is determined by the quality of the water used, using all three types for very dirty water, the sand filter and mesh filter for normal water, and the mesh filter alone for very clear water.

• **Fertilization equipment.** This was dealt with in the section on combined fertilizer application and irrigation.

• **Pumping equipment.** This supplies the pressure needed to make the water circulate. The equipment's specifications are determined by the volume of water needed, the outlet pressure of the filters and the losses due to filtering. These pressure losses are usually estimated at 5 m for every filter in series.

7.3.3.2 The distribution network

This consists of the set of tubes that bear the water from the irrigation hydrant to the nozzles.
The tubes may be made of fibrocement or galvanized iron, though they are normally made of PVC or polystyrene, especially the secondary tubes and structures bearing the nozzles.

A main tube leaves the irrigation hydrant and splits into primary tubes, and these then bear secondary tubes leading to the tubes supplying the nozzles.

7.3.3.3. Emitters

These are the devices that disperse the irrigation water, and there are many different types.

• **Tricklers**. There is a wide range of different

models on the market, and new ones are appearing all the time. They may use turbulent or laminar flow, holes, or be vortex-type, long conduct, microtubes, self-compensating, multi-outlet, self-cleaning, etc. They supply 4-6 l/h of water, with a pressure of a 10 m water column.

• **Diffusors**. They do not have any sort of rotation mechanism, and launch the water against a fixed deflector and then eject it under pressure. The volume of flow is less than 150 l/h, with a range of less than 6 m.
• **Microsprinklers.** They emit the drops of water through the air by means of a rotary mechanism. They supply 20-50 l/h, and have a range of less than 6 m.

7.3.3.4. The advantages and disadvantages of local irrigation systems

The advantages are:

• Higher production
• Very economical use of water, labor, and fertilizers
• Greater control of soil moisture levels
• The possibility of watering any type of soil, however poor it is
• The installation does not require earthworks
• Fertilizer can be applied in the irrigation water
• It reduces the appearance of weeds
• It can be automated using a control program

The disadvantages are:

• The installation is an expensive investment
• The water-emitting devices may be blocked, and so they have to be specially cleaned.
• Soluble fertilizers have to be used, which are more expensive
• Salts build up in the "wet bulb"

Watering by automatic misting with anti-pest treatments (Manufactured by A. GUASTAPAGLIA)

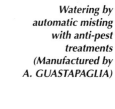

Hydraulic valve to limit flow. It is made entirely of plastic and is for use in irrigation systems. It is electrically operated by a solenoid. Rapid response hydraulic action. Easy fit. Manufacturing units, 1 or 2 inches. (Manufactured by TWIN DROPS)

Installed irrigation system for tomatoes

Sprinkler system with moving bar

7.4. THE QUALITY OF IRRIGATION WATER

The most important characteristics to monitor in the water and which most affect its quality for use in greenhouses are: its temperature, the amount and composition of the materials in suspension, and the concentration and composition of the dissolved substances.

The water supply may come from wells, rivers, lakes or aquifers. Water from surface bodies tends to contain more particles in suspension, both mineral and organic.

In general, irrigation water should not contain substances that are toxic to plants, it should not be cloudy or full of silt, and should not be colder than the soil being watered. The water should be as close as possible to the temperature of the soil occupied by the root system, and the temperature difference should never be more than 10°C.

Maximum tolerated concentrations in irrigation water

It is very important to know how much of each salt or element is dissolved in the water before using it. The amount of dissolved salts, the water's salinity, can be determined by measuring its electrical conductivity, as the conductivity is directly related to the total amount of dissolved salts. The conductivity is expressed in microSiemens/cm, and this value is then multiplied by the empirical factor 0.64 to give the total number of micrograms of salts per liter of water.

Water can be classified by its electrical conductivity into

• Not very saline water, with a conductivity value of 0-250 microSiemens/cm
• Moderately saline water, with a conductivity value of 250-750 microSiemens/cm
• Relatively saline water, with a conductivity value of 750-2,250 microSiemens/cm
• Very saline water, with a conductivity value of more than 2,250 microSiemens/cm

Water with values between 0 and 750 is suitable for irrigation, but relatively saline water causes problems and can only be used in very permeable soils and with plants that are very tolerant of salinity.

Element	For water used continuously on all soils mg/l	For use for up to 20 years on fine-textured soils, with a pH of 6 - 8.5 mg/l
Aluminium	5.0	20.0
Arsenic	0.1	2.0
Beryllium	0.1	0.5
Cadmium	0.01	0.05
Chrome	0.1	1.0
Cobalt	0.05	5.0
Copper	0.2	5.0
Fluorine	1.0	15.0
Iron	5.0	20.0
Lead	5.0	10.1
Lithium	2.5	2.5
Manganese	0.2	10.0
Molybdenum	0.01	0.05
Nickel	0.2	2.0
Selenium	0.02	0.02
Vanadium	0.1	1.0
Zinc	2.0	10.0

Bar with sprinklers

Excessively salty water causes problems because salts accumulate around the water, making it harder for the plant to absorb water from the soil, and causing water stress that may even lead to death.

Of all the elements present in the water, sodium (Na) is the most important, as it has a very negative effect on the soil. It damages the soil structure, its permeability and the velocity the water filters into the soil.

To assess the danger of this element the SAR measurement is used, the sodium absorbtion ratio (SAR)

$$SAR = \frac{Na^+}{\sqrt{\dfrac{Ca^{2+} + Mg^{2+}}{2}}}$$

Irrigation system with detachable on-line trickle labyrinth nozzle with a nominal flow of 2 l/h at a pressure of 0.9 kg/cm². Manufactured by TWIN DROPS

The concentration of the three elements is expressed in milliequivalents per liter (meq/l).

The SAR values obtained are interpreted as follows:

• SAR values of 0-10. Little danger.
• SAR values of 10-18. Medium danger.
• SAR values of 18-26. Great danger.
• SAR values of 26-30. Very great danger.

Classification of irrigation water by its content of chloride ions

Chloride index	Concentration (meq/l)	Potential risks
1	<2	None, not even with sensitive plants
2	2 - 4	Slight damage in sensitive plants
3	4 - 8	Slight damage in relatively tolerant plants
4	>8	Slight to moderate damage even in tolerant plants

For values between 6 and 10, there is the possibility of permeation in soil with tendency to contractions and expansion, while if the value exceeds 10, there is a risk of soil salination, and the higher the SAR value, the greater the risk of salination.

Bicarbonate ions must also be taken into account. At high concentrations, bicarbonates cause precipitation of calcium and magnesium ions (Ca^{2+} and Mg^{2+}) within the soil, and thus increase the concentration of sodium. The RSC (residual sodium carbonate) is used when discussing bicarbonate levels.

$$RSC = CO_3^{2-} + HCO_3^{-}) - (Ca^{2+} + Mg^{2+})$$

The four ions are expressed in meq/l.

Water with RSC greater than 2.5 meq/l is not considered suitable for irrigation, while water with a value of 1.25 meq/l is acceptable. The lower the value, the better the quality of the water.

8. FIGHTING PESTS AND DISEASES: SANITARY PROBLEMS IN GREENHOUSES

Due to the high investment required for plant production in greenhouses, it is essential to take all the measures required to ensure the crop is a success.

The appearance of disease in a plant is the result of a series of related factors. The first is that the plant has to be susceptible, i.e., the pathogen can attack it. The second is the presence of the infectious agent. The third is that the environment is suitable for the attack to progress. The environmental factors are the most important factors affecting the plant and parasite.

The modification of environmental factors inside the greenhouses creates more favorable conditions for the plants than in the open air, but they also favor the appearance of infections. The same pest species may cause a different pathology in the open air and inside a greenhouse.

It is important to understand the plant and its cultivation characteristics from a sanitary perspective, knowing which diseases and pests affect it worst, and when they are most likely to attack. This knowledge makes it possible to apply preventive treatment before the infection develops. Some characteristics of greenhouse cultivation favor the spread of many diseases and pests.

Use of misting devices to fight plant pests and diseases

• **Humidity**. High humidity levels, especially if they are very high for long periods of time, mean that diseases that are of little importance in the open air may be very serious in greenhouses.

• **The roofing material**. Plastic roofing generally shows large temperature variations between the day and night, and this favors greater condensation of humidity.

• **Cultivation techniques**. Several factors favor disease:

- *Specialization, or repeating the same crop.* There is a tendency to greater and greater specialization, reducing the number of species cultivated and following very close rotations.

Repeating a single crop favors the appearance of infectious diseases, as the pathogens can gradually increase in the crop remains that are left in the soil.

- *More intensive production.* Each crop is made to produce the greatest amount of saleable plant material per unit area in the shortest time possible.

Intensive cultivation implies a greater number of tasks, each of which may contribute to spreading diseases, especially viral infection.

- *Density.* The plants are cultivated at high densities to achieve high yields. The aboveground parts of the crop occupy much of the space in the greenhouse, maintaining high humidity and favoring contact between the pathogen and the susceptible organs of the plant.

- *Fertilizer application*. Nitrogen-rich fertilizers cause lush vegetative growth, but they also cause the plants to be more susceptible to attack by some fungi.

- *Smaller root systems*. In greenhouse cultivation, the aboveground parts of the plant are normally much more developed than the root system. This imbalance is due to the use of a local irrigation system which causes the roots to grow in a smaller volume of soil. Thus any damage to the roots rapidly affects the entire plant.

In general, keeping the greenhouse clean and tidy, and making sure there are no weeds inside it or near it, are essential to prevent possible attacks. The old leaves should be removed from the crop, and when the crop has finished, the entire plant should be removed, both the aboveground parts and all the roots.

8.1. COMBATTING THE MAIN FUNGAL, BACTERIAL AND VIRAL DISEASES

Infected soil is the main way that diseases are spread. The spread of pests and diseases is favored by the movement of earth, irrigation water and horticultural tools, and may even be spread on the farmer's shoes. The transplanting of seedlings, even if they do not have apparent symptoms, can spread infections.

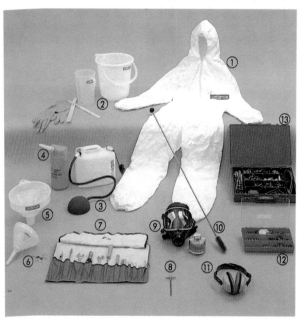

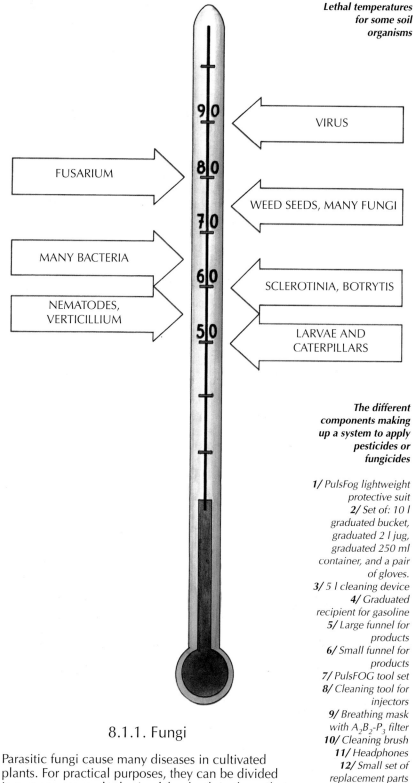

Lethal temperatures for some soil organisms

FUSARIUM

MANY BACTERIA

NEMATODES, VERTICILLIUM

VIRUS

WEED SEEDS, MANY FUNGI

SCLEROTINIA, BOTRYTIS

LARVAE AND CATERPILLARS

The different components making up a system to apply pesticides or fungicides

1/ PulsFog lightweight protective suit
2/ Set of: 10 l graduated bucket, graduated 2 l jug, graduated 250 ml container, and a pair of gloves.
3/ 5 l cleaning device
4/ Graduated recipient for gasoline
5/ Large funnel for products
6/ Small funnel for products
7/ PulsFOG tool set
8/ Cleaning tool for injectors
9/ Breathing mask with A_2B_2-P_3 filter
10/ Cleaning brush
11/ Headphones
12/ Small set of replacement parts
13/ Large set of replacement parts
(Manufactured by DR. STAHL + SOHN GmbH)

8.1.1. Fungi

Parasitic fungi cause many diseases in cultivated plants. For practical purposes, they can be divided in two groups on the basis of the depth in the soil at which they occur.

- Less than 10 cm deep. These include many fungi that cause "damping off", such as *Rhizoctonia, Phytophthora, Sclerotinia* and *Botrytis*. They all attack the plant at the level of the soil surface and infect the roots and the collar.

- More than 10 cm deep. *Fusarium* and *Verticillium*. They attack the roots and the vessels.

When using agricultural chemicals to combat pests and diseases, you must follow the doses recommended by the manufacturers

For fungal diseases to attack and spread, the following conditions must be met: the fungal spores must be present, there must be a host plant for the fungus to attack, and the temperature and humidity conditions have to be ideal for the spores to germinate and grow.
The following is a brief description of the main diseases produced by fungi, and the methods to combat them.

DAMPING OFF

This is mainly caused by *Pythium, Rhizoctonia, Fusarium* and *Botrytis*. They cause the roots and stem of the seedling to rot, and the plant collapses and dies.
The appearance of this disease is determined by the seedling's resistance and the moisture content of the soil or substrate.
It is mainly combatted using preventive techniques, by disinfecting the seedbeds, choosing healthy plant material and controlling humidity.

ROOT AND STEM ROTS

Fungal attack of the root system prevents the transport of water and nutrients to the aboveground parts, which wilt, turn yellow and die.
Stem rots are caused by *Sclerotinia* and *Botrytis*, rotting of the basal leaves is due to *Rhizoctonia*, while rots of the root system may be due to *Fusarium, Phytophthora* and *Pythium*. They are all surface fungi.

They can be combatted by treating the surface with products such as Iprodione, Prozimidone and Thiabendazole (against *Sclerotinia* and *Botrytis*) and Benomyl, Iprodione, Dithianone and Thiabendazole (against *Pythium, Phytophthora* and *Rhizoctonia*).

VASCULAR DISEASES

The symptoms consist of a general wilting of the plant, a downward curving of the lower leaves and a chlorosis that become necrotic. There are no visible symptoms on the root system, as the disease is within the plant's vessels.
This can be caused by *Fusarium* and *Verticillium*. They enter the plant through some damage to the root caused by cultivation or by nematodes or soil insects.

These diseases can be controlled by thorough soil or substrate disinfection with Dazomet or Metam-sodium, and by following long crop rotations.
There are some products with a systemic action, that is to say, that are absorbed by the plant and act within it, such as Benomyl, Thiabendazole, Thiophanate and Methylthiophanate.

GREY ROT (*BOTRYTIS*)

This is a serious problem in greenhouses. It is due to poor ventilation and excessive humidity at a temperature of around 20°C. It is highly virulent if the environmental conditions are favorable, attacking damaged organs or those weakened by other pathogens.
Patches of fuzzy ash-grey mold appear on the aboveground parts and grow to cover the entire plant.

This is mainly combatted by lowering the humidity and improving the ventilation. If the plants are attacked, lower the humidity, ventilate the greenhouse and remove and destroy all the infected leaves. The chemical products that can be used include Dichlofluanide, Iprodione, Prozimidone and Chlozolinate.

POWDERY MILDEW

This fungus attacks the aboveground parts of the plants. The symptoms are whitish patches that have a floury or dusty texture. It can grow in dry environments.
It is controlled by chemical products such as sulfur, Bupyrimate, Benomil, Dinocap, Ethyrimol, Penconazol, Triphorine, and Triadimephon.

The most important animal pests

Animal	Attack	Product	Presentation	Toxicity	Safety period in days
Nematodes	Damage to roots	Aldicarb	Granulate	CCC	45
		Carbofurane	Granulate	BBC	60
		Oxamyl	Granulate	BAB	30
		Phorate	Granulate	CCC	120
Mites		Dicophol	Liquid	AAC	15
		Propargil	Liquid	BAC	28
		Tetradiphon	Liquid	AAA	15
Thrips		Endosulfan	Liquid and wettable powder	BBC	15
		Phenitrothion	Liquid and powder to dust	BBB	30
		Lindane	Liquid and wettable powder	BBC	15
		Malathion	Powder and liquid	AAC	10
White fly	Attack the aboveground parts	Deltamethrin	Liquid	BAC	3
		Phenpropatrin	Liquid	CBC	30
		Methomyl	Liquid	CCB	7
		Permethrin	Liquid and wettable powder	AAC	7
Aphids		Acephate	Soluble powder	BAA	14-20
		Endosulfan	Liquid an wettable powder	BBC	15
		Malathion	Powder and liquid	AAC	10
		Methomyl	Liquid	CCB	7
		Primicarb	Granulate	BBB	3-14
Scale insects		Carbaryl	Liquid, wettable powder and for dusting	BBB	7
		Diazinon	Liquid and powder for dusting	BBB	30
		Phenitrothion	Liquid and powder for dusting	BBB	30
Leatherjackets	Damage to the roots system, stem and leaves	Chlorpyriphos	Liquid and wettable powder	BBC	21-30
		Lindane	Liquid and wettable powder	BBC	15
Wireworms		Methomyl	Liquid	CCB	7
		Trichlorphon	Liquid, wettable powder and powder for dusting	BBB	10
Potato beetle		Carbayl	Liquid, wettable powder and for dusting	BBB	7
		Lindane	Liquid and wettable powder	BBC	15
Weevils		Trichlorphon	Liquid, wettable powder and for dusting	BBB	10
Screwworms		Chlorpyriphos	Liquid and wettable powder	BBC	21-30
		Lindane	Liquid and wettable powder	BBC	15
		Methomyl	Liquid	CCB	7
		Trichlorphon	Liquid, wettable powder and for dusting	BBB	10
Tortrix moth larvae	Damage to the terminal buds, flowers and fruit.	Carbaryl	Liquid, wettable powder and for dusting	BBB	7
		Endosulfan	Liquid and wettable powder	BBC	15
		Phenitrothion	Liquid and powder for dusting	BBB	30
		Methyl-azinphos	Powder for dusting	BCC	28
Leaf miners	Tunnels within the leaves	Acephate	Soluble powder	BAA	14-20
		Methamidophos	Liquid	CCC	21
		Naled	Liquid	BAC	10

Adult

Eggs in dung

Larva

Pupa

The life cycle of the housefly.
The adults only live one month, and the females lay several batches of 100-150 eggs, making a total of about 2,000, which they lay in the dung of horses, hogs, cattle and human excrement.

RUST

In greenhouses, rust is not very common. The symptoms are dusty pustules that are dark orange or brown, and they are accompanied by chlorosis of the leaf, followed by its death.
It can be effectively controlled by Oxycarboxyn, Triphorine and Triadimephon

LEAF SPOT

This type of damage is caused by several diseases with similar symptoms.

- **Rose black spot**. This can be fought with Dodine, Triphorine or Captan.

- **Tomato Cladosporium**. This gives rise to fuzzy black lesions on the underside of the leaf. The disease is worse in greenhouses than in the open air, as it needs high humidity to develop. Apart from taking measures to reduce the humidity, you can treat with chemical products like Captan, Dodine Chlorthalonil.

- **Alternaria**. Round dark patches that later turn black. These patches take the form of a series of small concentric circles that are necrotic and parchment-like. This occurs in conditions of high humidity and can be treated with Chlorthalonil, Dodine, Maneb, Mancozeb and Ipriodone.

- **Septoria**. Small more-or-less circular patches that are very numerous and light grey in the center and dark brown on the edge. This group includes strawberry blight. It is combatted with the same agents as *Alternaria*.

- **Anthracnosis**. Small circular blackish patches that are slightly sunken and scattered irregularly over the leaf surface. It is controlled with the same products as *Alternaria* and *Septoria*.

8.1.2. Bacteria

Bacteria cause rots of the collar and root, and attack the vascular system, like fungi, but they are mainly combatted using preventive techniques, as curing the plant is very difficult. The soil and tools have to be disinfected.

One typical bacterial disease is bacterial canker, caused by *Agrobacterium*. In greenhouses, roses are affected the worst by this disease. Bacterial disease can also be caused by *Xanthomonas* and *Pseudomonas*. They cause small lesion on the leaves which necrose and die. They also attack the stem, causing it to rot.

8.1.3. Viruses

Plants attacked by virus are unlikely to die, but their yield is much lower and their fruit are unsaleable.
The most important virus that attack in greenhouses are spread by contact, normally by insect vectors. The most serious ones are caused by mosaic viruses.

The only way to combat viruses, like bacteria, is by preventive measures, i.e., by preventing their entrance and spread. Always use disease-free seed or buy healthy seedlings from a nursery. Disinfect the soil or substrate to be used, as well as all the tools, cultivate virus-resistant varieties, and remove and destroy any plants showing any symptom of viral disease.

8.2. COMBATTING THE MAIN PESTS FOUND IN GREENHOUSES

8.2.1. Nematodes

They are the most dangerous of all the soil-borne pests, as once they have invaded the greenhouse they attack many horticultural crops so heavily that they are hard to cultivate.
Greenhouse conditions of high humidity, relatively stable temperatures and well-aired soil all favor nematodes.

Nematodes, or eel-worms, look like very small earthworms between 0.2 and 5 mm long. They attack the root system by settling on the surface of the root and suck out the cell contents, causing galls and deformations, and this weakens the plant and halts its growth. The aboveground parts are stunted and yellow.
Nematode attack raises a second problem, as the lesions they cause to the roots allow entry of fungi, bacteria and viruses.

The soil can be disinfected as a method of preventing attack. The products suitable for treating an outbreak include Aldicarb, Carbofurane, Oxamil and Phorate.

8.2.2. Mites

Mites are distinguished by having four pairs of legs in the adult. They are very small, with a round body and they attack many crop plants. They are often found on the underside of young leaves where they feed by perforating the tissues with their piercing stylets and sucking out the cell contents.

The upper side of the leaf turns a bronze color and small chlorotic patches appear, while what look like very small spiders can be seen on the underside. The leaf dries up and the plant loses its leaves.

Mites multiply very fast. They also develop resistance to chemicals very quickly, making it difficult to treat them. You are advised to grow varieties that are resistant to this pest.
The most widely used mite-killers (acaricides) are Dicophol, Tetradiphon, Chlorphenson and Propargil, though biological control methods using natural predators of the mites are now also available.

8.2.3. Insects

Insects pass through several stages during their life cycle, and the larval stages, such as caterpillars, are the most harmful and cause the most damage, as they feed with their chewing mouthparts on any plants they can find.

In the adult state, the most harmful insects are those that pierce the plant and suck its sap, such as aphids, plant bugs, white fly and scale insects, while the adults that have chewing mouthparts (beetles) and those with sucking mouthparts (butterflies) cause little damage.
Many insects attack cultivated plants.

The most important orders of insects that attack crops are:

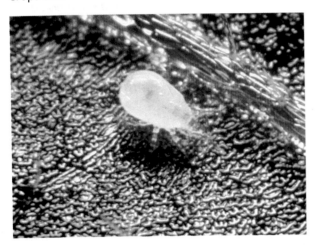

THYSANOPTERA

The thrips. They are small (1-2 mm long) with an elongated body and piercing and sucking mouthparts. They attack the young parts of the plant, causing bites that necrose and giving the leaf a silvery look.
Their bites are entrance points for fungi, bacteria and virus. They can be fought with Endosulfan, Phenitrothion, Lindane and Malathion.

HEMIPTERA

There are three main groups of interest, the white fly, the aphids and the scale insects.
The white fly is a typical greenhouse pest, as it finds conditions ideal. It settles on the underside of young leaves, preferring downy ones. It has piercing and sucking mouthparts. It not only creates access for infections, but also transmits viral diseases.

It can be treated with products like Deltamethrin, Ethrinphos, Phenpropatrin, Methomyl and Permethrin, but the eggs are hard to kill. Control using physical methods makes it possible to reduce the number of chemical treatments required and is based on capture using sticky papers.

Aphids are one of the pests that attacks nearly all greenhouse crops. They attack the young shoots and the underside of the leaves.

Their bites cause tissues lesions that necrose and cause malformations and galls. These wounds are an important entry point for other infections, especially viruses. Aphids also secrete a sugary liquid, honeydew, that favors the growth of fungi known as "sooty molds".

They are difficult to combat with chemicals, because their rapid reproduction means that resistant forms evolve rapidly. This means it is very important to treat them as soon as the first individuals appear. The most widely used chemical products are Acephate, Endosulfan, Malathion, Methomil and Primicarb.

Scale insects are almost immobile and mainly attack ornamental plants.
The insect's body is covered with a protective waxy shell, often resembling scales or cottony cushions. Like aphids they produce honeydew, encouraging the appearance of sooty molds.
They generally weaken the plant, because they feed on its sap.

They are difficult to combat chemically, as both eggs and adults are protected from insecticides by their impermeable coverings. The most widely used products are Carbaryl, Diazinon and Phenitrothion.

Left: Yellow spider mite. (Courtesy of the KOPPERT company, which sells products for biological and natural pest control.

COLEOPTERA

Many coleopteran (beetle) larvae cause serious damage in the root system or to the stem and leaves. This group includes white grubs and wireworms. The treatment consists of application of disinfectants to the soil surface and topmost layers. The products giving the best results are Chlorpyriphos, Lindane, Methomyl and Trichlorphon.

The beetle order also contains species whose adult forms damage crops, such Colorado potato beetles and weevils. They can be combatted with Carbaryl, Lindane and Trichlorphon.

Some insects are predators of other insects, and are used in biological control.

LEPIDOPTERA

The lepidopterans (butterflies and moths) have larvae known as caterpillars. The damage is caused by those living in the topmost layers of soil where they attack the collar of the plant. They emerge after dark to feed on the aboveground parts of the plant. They are known as screwworms. They can be combatted with soil disinfectants such as Chlorpyriphos, Lindane, Methomyl and Trichlorphon.

Other forms, such as the tortrix moths (the leaf roller moths) typically roll the leaves around the stems, flowers or fruit to form a nest that protects them to some extent from insecticides. Treat by dusting with Carbaryl, Methyl azinphos, Endosulfan or Phenitrothion.

DIPTERA

The KOPPERT company is specialized in products for the biological control of pests and the natural pollination of harvests.

The dipterans are the "true" flies, such as the housefly. The damage is caused by the larval forms, as many of them burrow under the leaf epidermis, forming galleries or tunnels. They are commonly known as leafminers. They can be treated with Acephate, Methamidophos, Naled and Pyrazophos.

It is also possible to try to catch all the adults before they lay their eggs.

8.3. SOME POINTS ABOUT USING AGRICULTURAL CHEMICALS IN GREENHOUSES

To cultivate a crop successfully in a greenhouse, chemicals often have to be applied at short intervals of time. Different products may be applied with a only 10 days apart. Most applications are intended to prevent the appearance of a specific disease or pest.
This intense application of agricultural chemicals may lead to serious toxicity problems. Many crops cultivated in greenhouses are for direct human consumption, and so the danger posed by the accumulation of poisonous residues should be taken into account.
In greenhouses, because there is no rain or wind, agricultural chemicals leave more residues than in the open air, meaning that they pose a greater threat. You must fully respect the legally established safety period between application and harvest,

to make sure that your edible produce does not contain poisonous levels of residues of agricultural chemicals.
It is necessary to plan a calendar of preventive treatments to ensure the crop remains free of pests and diseases. Following a rational program partly based on chemical treatments and partly on preventive measures, such as controlling the different environmental factors within the greenhouse, will mean that fewer curative treatments are necessary later.

You should use products that degrade rapidly with a short safety period. You should also assess how toxic the product is before you apply it, especially if you are using a new product or an experimental mixture. Plants grown in greenhouses are more sensitive to agricultural chemicals than those grown in the open air, because their cuticle is not so thick.
When treating, it is important to know if the product can have any sort of toxic effect on the plant species being grown. Each product persists in the plant to some extent, and this should be taken into account when programming the applications.

You must take even greater care than usual when applying treatments in a greenhouse, because the still air means that the product remains for longer in the air, thus representing a greater risk of poisoning by inhalation.
You should also consider the environmental impact of the treatment, especially on useful insects like pollination agents and the predators attacking insect pests.

8.4. SOIL DISINFECTION

Soil and substrate disinfection is essential for greenhouse cultivation, as the temperature and humidity conditions created inside are not only favorable for the plants. Together with the rich soil, they create an ideal habitat for the growth of soil organisms, many of them pathogens. Thus, the main aim of soil disinfection is to kill the organisms that may damage the crop.

The main methods used are:

• **Physical**, sterilizing the soil with steam.
• **Chemical**, using chemical products or fumigants to disinfect the soil.

8.4.1. Disinfection with steam

This method is based on the sterilizing effect of heat, and the soil is treated with hot steam. The temperature of the application should be between 80°C and 90°C and last for 10 minutes. This temperature destroys insects, mites, nematodes, fungi and weeds. You should not exceed a temperature of 100°C, as this would destroy the useful bacteria in the soil, which would be left sterile.
There is a wide range of devices on the market to generate steam. The steam produced

Fungus	Attack	Fungicide	Presentation	Toxicity	Safety period in days
Pythium	Damping off	Benomyl	Wettable powder	AAB	
Rhizoctonia		Dithianone	Liquid	AAC	14
Phytophthora		Iprodione	Wettable powder and for dusting	AAA	28
Sclerotinia	Root and	Thiabendazole	Concentrate	AAA	15
Botrytis	collar rot	Iprodione	Wettable powder and for dusting	AAA	28
		Procimidone	Wettable powder and for dusting	AAB	15
		Thiabendazole	Concentrate	AAA	15
Fusarium	Damping off. Attacks the vascular tissues	Benomyl	Wettable powder	AAB	
		Thiabendazole	Concentrate	AAA	15
Verticillium	Attacks the vascular tissues	Methylthiophanate	Liquid and wettable powder	AAA	21
		Thiophanate	Liquid and wettable powder	AAA	21
Powder mildew	Attacks the aboveground parts	Sulfur			
		Benomyl	Wettable powder	AAB	
		Bupyrimate	Liquid	AAB	15
		Dinocap	Liquid and wettable powder	ABC	21
		Ethyrimol	Liquid	AAA	21
		Triphorine	Liquid	AAA	7
Rust	Attacks the aboveground parts	Oxycarboxyn	Liquid and wettable powder	AAB	15
		Triphorine	Liquid	AAA	7
Black spot		Captan	Wettable powder and for dusting	AAC	7
		Dodine	Liquid and wettable powder	ABB	7
		Triphorine	Liquid	AAA	7
Cladosporium	Blotches on the leaves	Captan	Wettable powder and for dusting	AAC	7
		Chlorthalonyl	Liquid and wettable powder	BAC	15
		Dodine	Wettable powder	ABB	7
Alternaria		Chlorthalonyl	Liquid and wettable powder	BAC	15
		Dodine	Wettable powder	ABB	7
Septaria		Iprodione	Wettable powder and for dusting	AAA	28
Anthracnosis		Mancoceb	Liquid and wettable powder	AAB	15
		Maneb	Wettable powder and for dusting	AAB	15

In the toxicity column, the first letter denotes the danger to humans, the second the danger to the terrestrial fauna, and the third refers to the danger to the aquatic fauna, and A denotes low toxicity, B medium danger, C danger, and D is very dangerous

is distributed by means of sealed tubes to prevent escapes. If the seal is not perfect, the heat loss may be very high.

The disinfection method depends on the cultivation system being followed, and may be applied to the surface or at depth. It is most effective if it is performed in the soil with the right degree of moisture (neither too wet or too dry). To ensure this, water 5 days in advance. The application should be performed when the greenhouse is not very busy, i.e., before sowing or planting the crop. Some authors advise not starting cultivation until 4-6 weeks after disinfection.

The surface systems include the following:

• **Under canvas**. The soil to be treated is covered with a sheet of plastic, normally polyethylene, and the steam is applied under the canvas by means of a perforated tube. This can be used to disinfect soil to a depth of 25 cm.
• **Under a bell**. This is the same as the previous system, but the plastic sheet is replaced by a metal bell that covers a surface of up to 6.5 m^2.

The systems to apply steam at depth include:

• The use of drainage tubes already installed in the greenhouse soil.

The most important pathogenic fungi

• **Grid-shaped devices**. These are flat frames with extensions in the shape of perforated tubes that are introduced into the earth. If the site is not flat, the distribution of steam will be irregular.

• **Mobile plows**. The steam is distributed by means of instruments attached to a plow.

The disadvantages of disinfection by steam are that it is very expensive and that it eliminates almost all the organisms in the soil, including the useful ones. This creates a vacuum in the soil that can be occupied by pathogenic organisms.

The high temperature at which the steam is applied may also alter the physical or chemical properties of the soil or substrate, and thus negatively impact crop growth.

Steam disinfection is frequently followed by a rapid accumulation of nitrogen in the form of ammonia, which may be toxic to the plants.

Biological and natural pollination system, supplied by KOPPERT

8.4.2. Disinfection with fumigants

Fumigants are chemical products that kill fungi, insects, mites, nematodes and weeds, though their precise effect depends on the disinfectant used and the characteristics of the soil or substrate.
They are applied to soils that are not very wet, as water blocks the movement of the fumigant within the soil, as well as diluting it. Irrigate 4 or 5 days before disinfecting the soil. The soil should not have a standing crop, as the products used are normally toxic to plants.

The most widely used fumigants are:

• **Methyl bromide**. This is a broad action fumigant. It is presented in the form of a highly volatile liquid,

and so when it is applied the soil should be covered with a plastic sheet to prevent it being lost by evolution. Methyl bromide has no smell, and it is normally used mixed with the strong-smelling compound chloropicrin, so that any escape is noticed quickly. There are commercial products prepared with a mix of 98% methyl bromide and 2% chloropicrin.

Methyl bromide must only be applied by specialized personnel, because it is extremely poisonous, and you must respect the safety period before working the soil. The plastic should not be removed before 12 days.

To check if methyl bromide remains in the soil, sow some seed of watercress to see if they germinate successfully.

• **Chloropicrin**. This is a very poisonous volatile liquid that is only used in a mixture with methyl bromide.

• **Dichloropropene**. This is mainly used to treat nematodes, though it also kills wireworms and other pests.

• **Metam-sodium**. This is a broad-spectrum liquid fumigant. It has a safety period of 20-30 days, and airing starts 15 days after application.

• **Dazomet**. This is a broad-spectrum granulated fumigant. The safety period is 30 days, and the soil is aired 10 days after application.

Fumigants are normally applied by means of special devices that may work by injection or by continuous distribution. Continuous distribution devices are attached to a toothed harrow, with the fumigant vapor emerging from in front of each of the teeth.

Injectors are devices consisting of a deposit, a pump and an injector syringe. The product is pumped from the deposit into the syringe and from there into the soil.

To prevent the gaseous products from being lost by evaporation, cover the soil with a plastic sheet. If the fumigant is water soluble, it can be distributed using the irrigation system, and if it is a powder or granulate, it should be buried in the soil by plowing and then watered.

Before sowing or planting, air the soil, in order to eliminate any gases that may remain. The soil should not be plowed too deeply, taking care not to mix the disinfected soil with the underlying untreated soil.

The main disadvantages of disinfection using fumigants is the residues of the active ingredient accumulate in the soil because they do not evaporate, and this may be toxic to the following crops.

There is also an excessive accumulation of nitrogen in the form of ammonia after applying the fumigant.

GLOSSARY OF INTERNATIONAL TERMS

ENGLISH	SPANISH	FRENCH	GERMAN
A			
A horizon	horizonte A	horizon A	A-Horizont
abaxial surface of the leaf	cara superior de la hoja	surface de la feuille	Blattoberseite
abnormality	anormalidad	anomalie	Abnormalität
absorb (to)	absorber	absorber	aufsaugen
absorption	absorción	absorption	Absorption
absorption capacity	poder de absorción	pouvoir d'absorption	Absorptionsvermögen
acaricide	acaricida	acaricide	Akarizid
acarinosis of vine	acariosis de la vid	acariose de la vigne	Kräuselkrankheit der Rebe
acclimation	aclimatación	acclimatation	Akklimatisierung
achene	aqueno	akène	Achäne
acid soil	suelo ácido	sol acide	saurer Boden
acidity	acidez	acidité	Acidität
active carbon	carbono activo	charbon actif	Aktivkohle
active ingredient	principio activo	principe actif	Wirkstoff
active substance	sustancia activa	principe actif	Wirkstoff
adaxial surface of the leaf	cara inferior de la hoja	dessous de la feuille	Blattunterseite
adhesion	adhesión	adhésion	Adhäsion
adsorption	adsorción	adsorption	Adsorption
advancement of ripening	maduración acelerada	maturité hâtée	Reifebeschleunigung
adventitious bud	yema adventicia	bourgeon adventif	Adventivknospe
adventitious root	raíz adventicia	racine adventive	Adventivwurzel
aerial pest control	tratamiento aéreo	traitement aérien	Schädlingsbekämpfung aus der Luft
aerial root	raíz aérea	racine aérienne	Luftwurzel
aerobic	aerobio	aérobique	aerob
after-ripening	postmaduración	post-maturation	Nachreife
aftershoot	retoño	rejeton	Nachschosser
agricultural meteorology	meteorología agrícola	météo agricole	landwirtschaftliche Meteorologie
agrology	agrología	agrologie	Agrologie
agronomic crop	fruto del campo	produit des champs	Feldfrucht
air blast sprayer	atomizador	atomiseur	Sprühgerät
air brake	freno de aire	frein pneumatique	Druckluftbremse
algicide	algicida	algicide	Algizid
alkali soil	suelo calizo	sol alcalin	Alkaliboden
alkaline soil	suelo alcalino	sol alcalin	alkalischer Boden
alkalinity	alcalinidad	alcalinité	Alkalität
alkalization	alcalización	alcalisation	Alkalisierung
alkaloid	alcaloide	alcaloïde	Alkaloid
allelopathy	alelopatía	allélopathie	Allelopathie
allochthonous	alóctono	allochtone	allochthon
allogamy	alogamia	allogamie	Allogamie
alluvion	aluvión	alluvion	Anschwemmung
almond moth	polilla de la almendra	teigne de l'amandier	Dattelmotte
alternaria leaf spot	alternariosis de la remolacha	alternariose de la betterave	Alternaria-Blattfleckenkrankheit
alternative arable farming	cultivo alternativo	agriculture biologique	alternativer Landbau
altise	altisa	altise	Erdfloh
ammomium carbonate	carbonato amónico	carbonate d'ammonium	Ammoniumkarbonat
ammonia fertilizer	abono amoniacal	engrais ammoniacal	Ammoniakdünger
ammoniacal	amoniacal	ammoniacal	ammoniakhaltig
ammonium chloride	cloruro amónico	chlorure d'ammonium	Ammoniumchlorid
ammonium nitrate	nitrato de amonio	nitrate d'ammonium	Ammoniumnitrat
ammonium phosphate	fosfato amónico	phosphate d'ammonium	Ammoniumphosphat
ammonium sulfate	sulfato amónico	sulfate d'ammonium	Ammoniumsulfat
anaerobic	anaerobio	anaérobique	anaerob
angiosperms	angiospermas	angiospermes	Angiospermen
angleshades moth	flogofora meticulosa	phlogophore méticuleux	braune Achateule
animal excreta	deyecciones animales	déjections d'animaux	tierische Ausscheidungen
animal residues	residuos animales	déchets animaux	tierische Abfälle
anion exchange	intercambio de aniones	échange d'anions	Anionenaustausch
anion exchange capacity	capacidad de intercambio de aniones	capacité d'anions	Anionenaustauschkapazität
annual precipitation	precipitación anual	précipitations annuelles	Jahresregenmenge
annual temperature	temperatura anual	température annuelle	Jahrestemperatur
annual weather	climatología anual	conditions météorologiques annuelles	Jahreswitterung
antagonism	antagonismo	antagonisme	Antagonismus
anthela	antela	anthèle	Spirre
anther	antera	anthère	Anthere
anthropogenic soil	suelo antropogénico	sol anthropogène	anthropogener Boden
anti-slip device	dispositivo antideslizante	dispositif d'adhérence	Antischlupfvorrichtung
antigen	antígeno	antigène	Antigen
antler moth	polilla de pasto	noctuelle de l'herbe	Graseule
ants	hormigas	fourmis	Ameisen
aphicide	aficida	aphicide	Aphizid
aphid trap	trampa para pulgones	piège à pucerons	Blattlausfalle
apple blossom weevil	gorgojo del manzano	charançon du pommier	Apfelblütenstecher
apple ermine moth	arañuelo del manzano	chenille fileuse du pommier	Apfelbaumgespinstmotte
apple mildew	oidio del manzano	oïdium du pommier	Apfelmehltau
aquatic plants	plantas acuáticas	plantes aquatiques	Wasserpflanzen
aqueous	acuoso	aqueux	wasserhaltig
aqueous solution	solución acuosa	solution aqueuse	wäßrige Lösung
aquiferous	acuífero	aquifère	wasserführend
arable	arable	arable	ackerfähig
arable land	tierra de cultivo	sol arable	Ackerland
arid zones	zonas áridas	zones arides	Trockengebiete
artesian well	pozo artesiano	puits artésien	artesircher Brunnen
articulated pod	vaina articulada	gousse lomentacée	Gliederhülse
artificial rain	lluvia artificial	pluie artificielle	künstlicher Regen
asexual reproduction	reproducción asexual	reproduction asexuée	ungeschlechtliche Fortpflanzung
assimilation	asimilación	assimilation	Assimilation
atmospheric humidity	humedad atmosférica	humidité atmosphérique	Luftfeuchtigkeit
atomizer pistol lance	pistola atomizadora	lance-pistolet	Sprühpistole
autochthonous	autóctono	autochtone	autochthon
autogamy	autogamia	autogamie	Autogamie
autotrophic	autótrofo	autotrophe	autotroph
availability	disponibilidad	disponibilité	Verfügbarkeit
available for plants	utilizable por las plantas	disponibles pour les plantes	pflanzenverfügbar
awner	desbarbadora	ébarbeur	Entgranner
axe	hacha	hache	Axt
axillary bud	yema axilar	bourgeon axillaire	Achselknospe
azonal soil	suelo azonal	sol azonal	azonaler Boden

ENGLISH	SPANISH	FRENCH	GERMAN
B			
B horizon	horizonte B	horizon B	B-Horizont
bacteriae	bacterias	bactéries	Bakterien
bacterial	bactérico	bactérien	bakteriell
bacterial blight of walnut	tizón bacteriano del nogal	bactériose du noyer	Bakterienbrand der Walnuß
bacterial ring rot of potatoes	bacteriosis anular de la patata	bactériose annulaire de la pomme de terre	Ringbakteriose der Kartoffel
bactericidal	bactericida	bactéricide	bakterizid
bacteriological	bacteriológico	bactérologique	bakteriologisch
baling	prensado en balas	mise en balles	Einballen
balled plant	planta con cepellón	plante avec la motte	Ballenpflanze
band fertilization	abonado en fajas	fumage en bandes	Banddüngung
band seeding	siembra en hileras dobles	semis en bandes	Bandsaat
band sprayer	pulverizador en fajas	pulvérisateur en bandes	Bandspritzgerät
band spraying	pulverización por fajas	pulvérisation par bandes	Bandspritzung
bark	corteza	écorce	Borke
bark beetle of fig	barrenillo de la higuera	scolyte du figuier	Feigenborkenkäfer
bark grafting	injerto de corona	greffage en couronne	Rindenpfropfen
barren seed	semilla estéril	graine stérile	tauber Samen
basal liming	encalación de fondo	chaulage de fond	Grundkalkung
basalt meal	harina de basalto	poudre de basalt	Basaltmehl
base content	alcalinidad	alcalinité	Basengehalt
base exchange	intercambio de bases	échange des bases	Basenaustausch
base fertilizing	abonado de fondo	fumure de fond	Grunddüngung
base rest	residuo básico	résidu basique	Basenrest
base saturation	capacidad de cambio de bases	capacité d'échange de bases	Basensättigung
basic seeds	semilla básica	semences de base	Basissaatgut
basin irrigation	riego por submersión	irrigation par submersion	Überstauung
bastard fallow	barbecho de corto plazo	jachère partielle	Halbbrache
bean rust	roya de las judías	rouille du haricot	Bohnenrost
bean yellow mosaic	mosaico amarillo de las judías	mosaïque jaune du haricot	Gelbmosaik der Bohne
bed	cuadro	carré	Beet
bedding cultivation	capa de cultivo del suelo	billonage	Beetkultur
bedrock	roca firme	fond rocheux	Felsuntergrund
beet stem nematode	nematodo de los tallos	anguillule des tiges	Rübenkopfälchen
biennial	bienal	biannuel	zweijährig
biodynamic fertilization	fertilización biodinámica	fertilisation biodynamique	biodynamische Düngungsmethode
biological activity	actividad biológica	activité biologique	biologische Aktivität
biological pest control	lucha biológica antiparasitaria	lutte biologique antiparasitaire	biologische Schädlingsbekämpfung
biosynthesis	biosíntesis	biosynthèse	Biosynthese
biotic	biótico	biotique	biotisch
bird damage	daño causado por pájaros	dégâts causés par les oiseaux	Vogelfraß
biting insect	insecto masticador	insecte broyeur	fressendes Insekt
bitter pit	bitter pit	bitter pit	Stippigkeit
black earth	tierra negra, mantillo	terre noire, terreau	Schwarzerde, Gartenerde
black mold	fumagina	fumagine	Rußtau
blanching	blanqueo	blanchiment	Bleichen
bleaching	lavado	blanchiment	Ausbleichung
blight head	fusariosis de la espiga	furiose de l'épi	Åhrenfusariose
blue mold of citrus	moho azul de los agrios	moississure bleue des agrumes	Blaufäule der Zitrusfrüchte
bog iron ore	piedra ferruginosa	limonite	Raseneisenerz
bog soil	terreno cenagoso	sol marécageux	Moorboden
bog soil-cover cultivation	cultivo en terreno pantanoso	culture sur sol de tourbière	Moordamm-Kultur
boggy	pantanoso	marécageux	moorig
border	platabanda	plate-bande	Rabatte
boron	boro	bore	Bor
boron deficiency	deficiencia bórica	carence en bore	Bormangel
boron fertilizer	abono bórico	engrais boraté	Bordünger
botrytis	botritis	pourriture grise	Botrytisfäule
bottom gradient	gradiente del suelo	inclinaison du sol	Sohlengefälle
boulder clay	arcilla sílica	argile à silex	Geschiebelehm
bound water	agua de retención	eau de rétention	Haftwasser
bow hoe	azada de tracción	binette à tirer	Ziehhacke
brackish water	agua salobre	eau seaumâtre	Brackwasser
bract	bráctea	bractée	Deckblatt
branch	rama, ramo	branche, rameau	Ast, Zweig
bread grains	cereales panificables	céréales panifiables	Brotgetreide
broad-leaved	de hojas anchas	à feuilles larges	breitblättrig
broadcast seeder	sembradora a voleo	semoir à la volée	Breitsämaschine
broadcast sowing	siembra a voleo	semis à la volée	Breitsaat
brock	arroyo	ruisseau	Bach
brown rust of rye (of wheat)	roya parda del centeno (del trigo)	rouille brune du seigle (du blé)	Braunrost des Roggens (des Weizens)
buckwheat	trigo negro	sarrasin	Buchweizen
bud (to)	echar yemas	bourgeonner	knospen
bud dormancy	letargo de las yemas	temps de repos des boutons	Knospenruhe
budding	injerto de yema	greffage des yeux	Okulation
bulb	bulbo	bulbe	Zwiebel
bulbel	bulbo	bulbillon	Brutzwiebel
bull wheel	rueda motriz	roue motrice	Triebrad
bundle	gavilla	gerbe	Garbe
bursting of buds	apertura de las yemas	débourrement	Knospenaufbruch
C			
C horizon	horizonte C	horizon C	C-Horizont
C:N-ratio	relación C:N	rapport C:N	C:N-Verhältnis
cacao bean	grano de cacao	fève de cacao	Kakaobohne
cacao shells	cascarilla de cacao	coques de cacao	Kakaoschalen
cadmium	cadmio	cadmium	Cadmium
calcareous	calcáreo	calcaire	kalkhaltig
calcareous soil	suelo calizo	sol crayeux	karbonathaltiger Boden
calcicolous plants	plantas calcícolas	plantes calcicoles	kalkliebende Pflanzen
calciphobous plants	plantas calcífugas	plantes calcifuges	kalkfliehende Pflanzen
calcium ammonium nitrate	nitrato amónico cálcico	nitrate d'ammoniaque calcique	Kalkammonsalpeter
calcium carbonate	carbonato cálcico	carbonate de calcium	Kohlensaurer
calcium cyanamide	cianamida cálcica	cyanamide calcique	Calciumcyanamid
calcium eluviation	lixiviación cálcica	lixiviation de la chaux	Kalkauswaschung
calcium loss	pérdida de cal	perte de chaux	Kalkverlust
calcium nitrate	nitrato cálcico	nitrate de chaux	Kalksalpeter
calcium phosphate	fosfato cálcico	phosphate de chaux	Kalkphosphat
calibrated seed	semilla calibrada	semence calibrée	kalibrierter Samen
calorimetry	calorimetría	calorimétrie	Kalorimetrie
calycle	calículo	calicule	Nebenkelch
calyptra	caliptra	calyptre	Kalyptra
calyx	cáliz	calice	Blütenkelch

ENGLISH	SPANISH	FRENCH	GERMAN
capillary water	agua capilar	eau de capilarité	Kapillarwasser
capitulum	capítulo	capitule	Köpfchen
capsule	cápsula	capsule	Kapsel
carbon content	contenido en carbono	teneur en carbone	Kohlenstoffgehalt
carbonate	carbonato	carbonate	Carbonat
carrion crow	corneja negra	corneille noire	Rabenkrähe
caryopsis	cariopsis	caryopse	Karyopse
catalysis	catálisis	catalyse	Katalyse
catchpit	pozo negro	puisard	Senkgrube
cation exchange	intercambio catiónico	échange de cations	Kationenaustausch
catkin	amento	chaton	Kätzchen
cell division	división celular	division cellulaire	Zellteilung
cell tissue	tejido celular	tissu cellulaire	Zellgewebe
cell wall	pared celular	paroi cellulaire	Zellwand
cellular content	contenido celular	contenu cellulaire	Zellinhalt
cellular membrane	membrana celular	membrane cellulaire	Zellhaut
cellular sap	jugo celular	sève cellulaire	Zellsaft
cellulose	celulosa	cellulose	Zellulose
centipede	ciempiés	myriapode	Tausendfüßler
cereal grain	grano de cereal	grain de céréale	Getreidekorn
cereal leaf beetle	criocero de los cereales	criocère des céréales	rothalsiges Getreidehähnchen
cereal leaf miner	mosca minadora de los cereales	mineuse des céréals	Minierfliege der Gerste
cereal mildew	oidio de los cereales	oïdium des céréales	Getreidemehltau
cereals	cereales	céréales	Getreide
change of crops	cambio de cultivo	culture assolée	Fruchtwechsel
characteristic	característica	caractéristique	Merkmal
chemical and physical properties of soil	propriedades fisico-químicas del suelo	propriétés physicochimiques du sol	chemische und physikalische des Bodens Eigenschaften
chemical fertilizer	abono químico	engrais chimique	Kunstdünger
chemical pest control	lucha química antiparasitaria	lutte chimique antiparasitaire	chemische Schädlingsbekämpfung
chernozem	chernozem	tchernoziom	Tschernosem
chitted seeds	simientes pregerminadas	semences prégermées	vorgekeimtes Pflanzgut
chloride	cloruro	chlorure	Chlorid
chlorine	cloro	chlore	Chlor
chlorophyll	clorofila	chlorophylle	Chlorophyll
chlorosis	clorosis	chlorose	Chlorose
choke of grasses	rueca de las gramíneas	quenouille des graminées	Erstickungsschimmel
chrome	cromo	chrome	Chrom
chromosome	cromosoma	chromosome	Chromosom
citrus psorosis	psorosis de los agrios	psorose des agrumes	Psorosis der Zitrusfrüchte
clay	arcilla	argile	Ton
clay cutan	capa arcillosa	couche d'argile	Tonbelag
clay formation	formación de arcilla	formation d'argile	Tonbildung
clay fraction	fracción de arcilla	fraction d'argile	Tonfraktion
clay mineral	minerales de la arcilla	minéraux argileux	Tonmineral
clay soil	suelo arcilloso	terre argileuse	Tonboden
clayey	arcilloso	argilleux	tonig
cleaning crop	cultivo mejorante	culture améliorante	Gesundungsfrucht
cleft grafting	injerto de hendidura	greffage en fente	Spaltpropfen
climbing plant	planta trepadora	plante grimpante	Kletterpflanze
climbing wine	parra	treille	Kletterrebe
clod	terrón	grumeau	Klumpen
clod buster	desterronadora	émotteuse	Krümler
cloddy	grumoso	grumeleux	klumpig
club-root of cabbage	hernia de la col	hernie du chou	Kohlhernie
coagulation	coagulación	coagulation	Koagulation
coarse sand	arena gruesa	sable grossier	Grobsand
coarse soil	tierra basta	sol grossier	grober Boden
cobalt	cobalto	cobalt	Kobalt
cohesion	cohesión	cohésion	Kohäsion
cold resistance	resistencia al frío	résistance au froid	Kälteresistenz
collecting ditch	colector	fossé collecteur	Fanggraben
collecting drainage	drenaje de retención	collecteur de drainage	Fangdränung
colloid	coloide	colloïde	Kolloid
colloidal	coloidal	colloïdal	kolloidal
colluvium	coluvión	colluvion	Kolluvium
Colorado potato beetle	escarabajo de la patata	doryphore	Kartoffelkäfer
combined harvesting	siega y trilla	moissonnage-battage	Mähdrusch
combined seed and fertilizer drill	sembradora-fertilizadora	semoir-fertiliseur	kombinierte Drill-und Düngerstreumaschine
coming into ear	formación de la espiga	épiaison	Ährenbildung
commercial fertilizer	abono industrial	engrais industriel	Handelsdünger
commercial fruit growing	cultivo frutícola comercial	culture fruitière commerciale	Erwerbsobstbau
commercial product	preparado comercial	produit commercial	Handelspräparat
commercial seed	semillas comerciales	semences commerciales	Handelssaatgut
commercial vegetable growing	cultivo comercial de hortalizas	culture maraîchère commerciale	Erwerbsgemüsebau
common corn smut	carbón del maíz	charbon du maïs	Maisbeulenbrand
compacted horizon	capa dura	horizon compacte	Verdichtungshorizont
compacted soil	suelo compacto	sol compact	bindiger Boden
compatibility for plants	compatibilidad vegetal	compatibilité pour les plantes	Pflanzenverträglichkeit
complementary seeding	siembra complementaria	semis complémentaire	Nachsaat
compositae	compuestas	composacées	Korbblütler
composition of soil	composición del suelo	composition du sol	Zusammensetzung des Bodens
compost (to)	hacer compost	composter	kompostieren
concretion	concreción	concrétion	Konkretion
cone	cono	cône	Zapfzen
coniferous tree	conífera	conifère	Nadelbaum
consumption of water	consumo de agua	consommation d'eau	Wasserverbrauch
contact herbicide	herbicida de contacto	herbicide par contact	Kontaktherbizid
contaminated	contaminado	contaminé	befallen
continental climate	clima continental	climat continental	Kontinentalklima
conveyor dryer	secador de cinta continua	séchoir à tablier transporteur	Bandtrockner
convolculaceae	convolvuláceas	convolvulacées	Windengewächse
copper	cobre	cuivre	Kupfer
copper fertilizer	abono cúprico	engrais cuivrique	Kupferdünger
copper sulphate	sulfato de cobre	sulfate de cuivre	Kupfersulfat
core	antro	trognon	Kerngehäuse
corky	suberoso	liégeux	korkig
corn	maíz	maïs	Mais
corn ear	panocha	panouille	Maiskolben
corn field	trigal	champ de blé	Kornfeld
corn husks	espatas de maíz	spathes de maïs	Maislieschen
corn rust	roya del maíz	rouille du maïs	Maisrost

ENGLISH	SPANISH	FRENCH	GERMAN
corolla	corola	corolle	Korolle
cortical tissue	tejido cortical	tissu cortical	Rindengewebe
cotton anthracnose	antracnosis del algodonero	anthracnose du cotonier	Baumwollanthraknose
cotton worm	rosquilla negra	chenille défoliante	Baumwollraupe
cotyledon	cotiledón	cotylédon	Kotyledon
cover of mulch	cubierta de mulch	couverture de mulch	Mulchdeke
covering	cobertura	couverture	Bedeckung
cowshed manure	estiércol vacuno	fumier de bovins	Rindermist
creep	deslizamiento	reptation	Kriechen
crop area	área de cultivo	surface cultivée	Anbaufläche
crop cultivation	cuidados culturales	entretien des cultures	Kulturpflege
crop damage	daños sufridos por la cosecha	dégâts causés au récoltes	Ernteschäden
crop density	densidad de plantas	densité des plantes	Pflanzendichte
crop lifter	elevador de espigas	releveur d'épis	Ährenheber
crop losses	pérdidas de las cosechas	pertes des récoltes	Ernteverluste
crop remains	residuos de cosecha	résidus de récolte	Ernterückstände
crop rotation position	posición sucesiva en el cultivo de rotación	position dans la rotation des cultures	Fruchtfolgestellung
crop yield	rendimiento de la cosecha	rendement des récoltes	Ernteertrag
cropping	cutivo	culture	Kultur
cropping intensity	intensidad de explotación	intensité de culture	Bewirtschaftungsintensität
cross plowing	labor cruzada	labour croisé	Querpflügen
crosskill roller	rodillo crosskill	rouleau crosskill	Crosskillwalze
croton bug	cucharacha alemana	blatte germanique	Hausschabe
cruciferae	crucíferas	crucifères	Kreuzblütler
crumb formation	formación de grumos	formation de grumeaux	Krümelbildung
crumb stability	estabilidad glomerular	stabilité des glomérules	Krümelstabilität
crumb structure	estructura grumosa	structure grumeleuse	Krümelgefüge
crust	corteza	sol d'incrustation	Kruste
crusting	incrustación	incrustation	Verkrustung
cultivate (to)	cultivar	cultiver	anbauen
cultivated plant	planta cultivada	plante cultivée	Kulturpflanze
cultivation development	desarrollo de cultivos	évolution des cultures	Anbauentwicklung
cultivation frequency	frecuencia de cultivo	fréquence de culture	Anbauhäufigkeit
cultivation method	método de cultivo	méthode de culture	Anbaumethode
cultivation of slopes	cultivo de laderas	culture de pentes	Hangbau
cultivation period	distancia de cultivo	distance des cultures	Anbauabstand
cultivation test	prueba de cultivo	essai de culture	Anbauversuch
cut mowing	siega	fauchage	Schnitt
cuticle	cutícula	cuticule	Kuticula
cutter bar	barra cortadora	barre-faucheuse	Mähbalken
cutting	estaca	bouture	Steckling
cutworm	gusano gris	ver gris	Erdeule
cyme	cima	cyme	Trugdolde

D

ENGLISH	SPANISH	FRENCH	GERMAN
damage by game	daño causado por animales de caza	dégâts causés par le gibier	Wildschaden
damaging substances	substancias nocivas	substances nuisibles	Schadstoffe
dealkalination	desalcalinización	désalcalinisation	Entbasung
debris	detritus	débris	Schutt
decay	podredumbre	pourriture	Fäulnis
decomposition	descomposición	décomposition	Verwesung
deep cultivation	labor profunda con cultivador	travail profond au cultivateur	Tiefgrubbern
deep fertilizing	abonado profundo	fumure en profondeur	Tiefendüngung
deep plowing	labor de desfonde	labour profond	Tiefpflügen
deficiency disease	enfermedad carencial	maladie carentielle	Mangelkrankheit
deficiency symptom	síntoma de carencia	symptôme de carence	Mangelerscheinung
degenerate (to)	degenerar	dégénérer	degenerieren
degeneration	degeneración	dégénération	Degeneration
degeneration disease	enfermedad degenerativa	maladie de dégénérescence	Abbaukrankheit
degree of acidity	grado de acidez	degré d'acidité	Säuregrad
degree of base saturation	grado de saturación de base	degré de saturation basique	Basensättigungsgrad
degree of ripeness	grado de madurez	degré de maturité	Reifegrad
degree of saturation	grado de saturación	degré de saturation	Sättigungsgrad
delayed germination	germinación retardada	germination retardée	Keimverzug
delayed ripening	retraso de la maduración	retard de la maturité	Reifeverzögerung
denitrification	desnitrificación	dénitrification	Denitrifikation
depth of topsoil	espesor de la capa arable	épaisseur de la couche arable	Krumentiefe
desanilization of water	desalinización del agua	dessalement de l'eau	Entsalzung des Wassers
desert	desierto	désert	Wüste
desert soil	suelo desértico	sol désertique	Wüstenboden
desertification	desertificación	désertification	Desertifikation
deshiscent fruit	fruta dehiscente	fruit déhiscent	Streufrucht
desiccation	desecación	séchage	Trocknung
dibbling machine	sembradora a golpes	semoir en poquets	Dibbelmaschine
dicalcium phosphate	fosfato dicálcico	phosphate dicalcique	Dicalciumphosphat
dicotyledoneae	dicotiledóneas	dicotylédones	Dikotylen
dictyospermum scale	piojo rojo de los agrios	pou rouge des agrumes	rote Mittelmeerschildlaus
dieback of apricot	apoplejía del albaricoquero	dépérissement de l'abricotier	Baumsterben der Aprikose
differential	diferencial	différentiel	Differential
diffusion	difusión	diffusion	Diffusion
diffusion coefficient	coeficiente de difusión	coefficient de la diffusion	Diffusionskoeffizient
dig (to)	cavar	bêcher	umgraben
dioecious	dioico	dioïque	diözisch
direct drilling	siembra directa	semis direct	Direktsaat
disc ridger	aporcadora de discos	butteuse à disques	Scheibenhäufler
disc wheel	rueda de disco	roue à disque	Scheibenrad
disease control	lucha contra enfermedades	lutte contre les maladies	Krankheitsbekämpfung
disease gradient	grado de contaminación	degré de l'infestation	Befallsgrad
dispersion	dispersión	dispersion	Dispergierung
dissemination	diseminación	dissemination	Aussamung
distichous	dístico	distique	zweizeilig
distortion	torcimiento	distortion	Verdrehung
distribution rill	reguera de distribución	rigole de répartition	Verteilerrinne
ditch (to)	cavar zanjas	creuser des tranchées	Gräben ausstechen
dock sawfly	hopocampla de la acedera	tenthède de l'oseille	Ampferblattwespe
dolomite	dolomita	dolomite	Dolomit
dominant	dominante	dominant	dominant
dosage	dosificación	dosage	Dosierung
double grafting	doble injerto	double greffage	Zwichenveredlung
downy mildew of hop	midiú del lúpulo	mildiou du houblon	Hopfen-Peronospora
drag harrow	grada oscilante	herse oscillante	Grubberegge
drain	escorrentía, tubo de drenaje	écoulement, drain	Abfluß, Drän
drain (to)	evacuarse	s'évacuer	abfließen

ENGLISH	SPANISH	FRENCH	GERMAN
drain outlet	aliviadero	bouche de décharge	Dränauslauf
drainage	drenaje	drainage	Dränung
drainage network	red de drenaje	réseau de drainage	Drännetz
draining ditch	zanja de drenaje	fossé de drainage	Drängraben
dressing powder	polvo desinfectante	poudre désinfectante	Beizpuder
drop irrigation	riego por goteo	irrigation goutte-à-goutte	Tröpfchenbewässerung
drought	sequía	sécheresse	Dürre
drought resistance	resistencia a la sequía	résistance à la sécheresse	Trockenresistenz
drupe	drupa	drupe	Steinfrucht
drying	desecación	dessèchement	Trockenlegung
dryland farming	cultivo de secano	cuture en sols arides	Trockenfeldbau
dung	estiércol	fumier	Dung
dung water	purín	purin	Jauche
dung yard	estercolero	fumière	Dungstätte
duration of flowering	duración de la floración	durée de la floraison	Blühdauer
durra	sorgo durra	sorgho durra	Durra
dusting	espolvoreo	poudrage	Stäuben
dusting preparation	producto para espolvoreo	produit pour poudrage	Stäubemittel
dying	marchitamiento	dépérissement	Absterben
E			
ear	espiga	épi	Ähre
ear emergence	espigazón	épiaison	Ährenschieben
ear selection	selección de espigas	séléction des épis	Ährenauslese
early crop	cosecha temprana	récolte précoce	Frühernte
early frost	helada precoz	gelée précoce	Frühfrost
early maturing	de maduración precoz	de maturation précoce	frühreifend
earth up (to)	aporcar	butter	häufeln
earthworms	lombrices de tierra	vers de terre	Regenwurm
ecological balance	equilibrio ecológico	équilibre écologique	ökologisches Gleichgewicht
eluvial horizon	horizonte aluvial	horizon éluvial	Auswaschungshorizont
eluviation of nitrate	erosión por nitrato	lixiviation de nitrates	Nitratauswaschung
emergence	emergencia	émergence	Aufgehen der Saat
endocarp	endocarpo	endocarpe	Endokarp
endosperm	endosperma	endosperme	Endosperm
ensile (to)	ensilar	ensiler	einsilieren
environment	medio ambiente	milieu ambiant	Umwelt
enzyme	enzima	enzyme	Enzym
epidermis	epidermis	épiderme	Epidermis
epiphytes	epífitos	épyphites	Epiphyten
eradicate (to)	exterminar	exterminer	ausrotten
ergot of rye	cornezuelo del centeno	ergot du seigle	Mutterkorn
eriophyd mite	eriófido	ériophyde	Gallmilbe
erode (to)	erosionar	éroder	erodieren
eroded soil	suelo erosionado	sol érodé	erodierter Boden
erosion	erosión	érosion	Erosion
essential nutrients	nutrientes esenciales	éléments nutritifs essentiels	Hauptnährstoffe
etiolation	etiolación	étiolement	Etiolierung
European canker	chancro del manzano	chancre du pommier	Obstbaumkrebs
European earwig	tijereta	perce-oreille	Ohrwurm
European fruit scale	cochinilla ostriforme	cochenille ostréiforme	Austernschildlaus
European grain moth	falsa polilla de los graneros	teigne des grains	Kornmotte
European red mite	ácaro rojo	acarien rouge	Obstbaumspinnmilbe
European red raspberry	frambuesa	framboise	Himbeere
European shot-hole borer	barrenillo dispar	xylébore disparate	ungleicher Holzbohrer
European wheat stem sawfly	cefo del grano	cèphe des chaumes	Getreidehalmwespe
evaporate (to)	evaporar	évaporer	verdunsten
evaporation	evaporación	évaporation	Evaporation
evaporation rate	grado de evaporación	quote-part d'évaporation	Verdungstungshöhe
evapotranspiration	evapotranspiración	évapotranspiration	Evapotranspiration
evergreen	sempervirente	semper virens	immergrün
excess moisture	exceso de humedad	excès d'humidité	Überfeuchtigkeit
expanding harrow	grada-binadora	herse-bineuse	Hackegge
exuberance	exuberancia	exubérance	Wucherung
eye cutting	esqueje de yema	bouture d'oeil	Augensteckling
eye-spotted bud moth	polilla de las yemas	tordeuse rouge des bourgeons	roter Knospenwickler
F			
factorial analysis	análisis de factores	analyse des facteurs	Faktorenanalyse
fall catch drop	cultivo intermedio de otoño	culture dérobée d'automne	Herbstzwischenfrucht
fall tillage	labores de otoño	travaux d'automne	Herbstbestellung
fallow (to)	barbechar	jachérer	brachen
fallow land	barbecho	jachère	Brachfläche
family	familia	famille	Familie
fan	ventilador	ventilateur	Gebläse
farm tire	neumático agrario	pneu agraire	Farmerreifen
farming system	sistema de cultivo	système de culture	Anbausystem
feeding sugar beet	remolacha semiazucarera	betterave demi-sucrière	Gehaltsrübe
ferment (to)	fermentar	fermenter	gären
fermentation	fermentación	fermentation	Gärung
fermentation horizon	horizonte de fermentación	horizon de fermentation	Fermentationshorizont
fertile	fértil	fertile	fruchtbar
fertile soil	suelo fértil	sol fertile	fruchtbarer Boden
fertility	fertilidad	fécondité	Fertilität
fertilization	fecundación	fécondation	Befruchtung
fertilization date	época de abonado	date de la fumure	Düngungszeitpunkt
fertilization proportionment	proporciones del abonado	évaluation des besoins de fertilisants	Düngungsbemessung
fertilize (to)	fecundar, fertilizar	féconder, fertiliser	beruchten, düngen
fertilizer box	recipiente de abono	caisse à engrais	Düngerbox
fertilizer containing three macronutrients	abono de tres elementos fertilizantes	engrais ternaire	Dreinährstoffdünger
fertilizer control	control de fertilizantes	contrôle des engrais	Düngermittelkontrolle
fertilizer drilling	enterrado del abono en líneas	enfouissage de l'engrais en lignes	Eindrillen des Düngers
fertilizer salt	sales fertilizantes	sel fertilisant	Düngesalz
fertilizer silo	silo de abono	silo à engrais	Düngersilo
fertilizer use	utilización de abonos	application d'engrais	Düngereinsatz
fertilizing influence	influencia de fertilizantes	influence de la fertilisation	Düngungseinfluß
fertilizing level	nivel de abonado	phase de fumure	Düngungsstufe
fertilizing scheme	plan de fertilización	plan de fertilisation	Düngungsplan
fertilizing substances	sustancias fertilizantes	matières fertilisantes	Dungstoffe
fertirrigation	fertirrigación	irrigation fertilisante	düngende Bewässerung
fiber	fibra	fibre	Faser
fibrous	fibroso	fibreux	faserig
field	campo	champ	Acker
field capacity	capacidad de campo	capacité au champ	Feldkapazität

ENGLISH	SPANISH	FRENCH	GERMAN
field mouse	ratón de campo	campagnol des champs	Feldmaus
field vole	ratilla	campagnol agreste	Erdmaus
field work	labores de cultivo	travaux de culture	Feldarbeit
filament	filamento	filet	Staubfaden
filamentous	filiforme	filiforme	fadenartig
filed crop sprayer	pulverizador de cultivos	pulvérisateur pour cultures de plein champ	Feldspritze
filter layer	capa filtrante	couche filtrante	Filterschicht
fine sand	arena fina	sable fin	Feinsand
fine seeds	simientes de calidad	semences de qualité	Feinsämereien
fine-textured soil	tierra fina	sol fin	feiner Boden
first sowing	siembra temprana	semailles précoces	Frühsaat
five-crop farming	rotación quinquenal	rotation quinquennale	Fünffelderwirtschaft
fleshy	carnoso	charnu	fleischig
flocculation	floculación	floculation	Flockung
flood	inundación	innondation	Flut
flood (to)	inundar	inonder	überfluten
flood deposit	depósito aluvial	dépôt alluvial	Hochwasserablagerung
flood irrigation	riego por desbordamiento	irrigation par déversement	Staubewässerung
flora	flora	flore	Flora
Florida red scale	cochinilla roja floridana	pou de Floride	Florida-Schildlaus
floriferous	florífero	florifère	blütentragend
flower	florecer	fleurir	blühen
flower (to)	flor	fleur	Blüte
flower abscission	caída de las flores	coulure	Blütenfall
flower bud	yema floral	bouton floral	Blütenknospe
flowering glume	gluma de cobertura, glumélula superior	glume de couverture, glume supérieur	Deckspelze, Vorspelze
flowerless	sin flores	cryptogame	blütenlos
flume	desaguadero	rigole d'écoulement	Ableitungsrinne
fluvial sediment	sedimento fluvial	sédiment fluviatile	Flußablagerung
foil cultivation	cultivo bajo plástico	culture sous feuilles plastiques	Folienanbau
foliage	follaje	feuillage	Laub
foliage leaf	hoja vegetativa	feuille	Laubblatt
foliage plants	plantas de hoja	plantes à feuilles	Blattpflanzen
foliar fertilization	abonado foliar	nutrition foliaire	Blattdüngung
foliar nutrient	abono foliar	engrais foliaire	Blattdünger
foliation	foliación	foliation	Laubentfaltung
forage beet	remolacha forrajera	betterave fourragère	Runkelrübe
forest soil	suelo forestal	sol forestier	Waldboden
fork	horquilla	fourche	Gabel
formation of tubers	tuberización	tubérisation	Knollenbildung
fossil soil	suelo fósil	sol fossile	fossiler Boden
four-wheel drive	tracción en las cuatro ruedas	propulsion à quatre roues motrices	Allradantrieb
fowl manure	gallinaza	galline	Hühnermist
frame	cajonera	châssis	Kasten
French bean	judía verde	haricot vert	grüne Bohne
fresh manure	estiércol fresco	fumier frais	frischer Mist
fresh water	agua dulce	eau douce	Süßwasser
friable	friable	friable	bröckelig
friable soil	tierra suelta	terre meuble	lockerer Boden
frit fly	mosca frit	mouche de frit	Fritfliege
front loader	cargador frontal	chargeur frontal	Frontlader
frost protection irrigation	riego por aspersión contra heladas	aspersion antigel	Frostschutzberegnung
frost resistance	resistencia a las heladas	résistance au gel	Frostresistenz
frost-susceptible	sensible al frío	sensible au froid	frostempfindlich
fructification	fructificación	fructification	Fruchtbildung
fructify (to)	fructificar	fructifier	Frucht tragen
fruit grader	calibradora de frutas	calibreur de fruits	Obstsortierer
fruit plantation	plantación frutal	plantation fruitière	Obstplantage
fruit pulp	pulpa del fruto	pulpe du fruit	Fruchtfleisch
fruit tree	frutal	arbre fruitier	Obstbaum
fuel injection pump	bomba de inyección	injecteur	Einspritzpumpe
full ripeness	madurez completa	pleine maturité	Vollreife
fumigant	fumigante	produit fumigant	Begasungsmittel
fumigate (to)	fumigar	fumiger	begasen
fumigation	fumigación	fumigation	Begasen
fumigator	fumigador	fumigateur	Räucherapparat
fungi	hongos	champigons	Pilze
fungicide	fungicida	fongicide	Fungizid
fungus development	desarrollo de hongos	développement de champignons	Pilzentwicklung
furrow	surco	sillon	Ackerfurche
furrow drilling	siembra en surcos	semis en sillons	Rillensaat
furrow irrigation	riego por surcos	irrigation par sillons	Furchenbewässerung
furrow opener	surcador	silloneuse	Furchenzieher
fusarium wilt of cucumber	fusariosis del pepino	fusariose du concombre	Fusariumwelke der Gurke

G

ENGLISH	SPANISH	FRENCH	GERMAN
gamete	gameto	gamète	Gamet
garbage compost	mantillo de compost	compost de détritus	Müllkompost
garden chafer	abejorro de las huertas	hanneton horticole	Gartenlaubkäfer
garden cress	berro de huerta	cresson	Gartenkresse
garden slug	babosa del huerto	limace des jardins	Gartenwegschnecke
gas substitution	intercambio gaseoso	échange de gaz gelée précoce	Gasaustausch Frühfrost
gene	gen	gène	Gen
generative phase	fase generativa	phase générative	generative Phase
generative propagation	multiplicación generativa	multiplication générative	generative Vermehrung
genotype	genotipo	génotype	Genotyp
genus	género	genre	Gattung
geotropism	geotropismo	géotropisme	Geotropismus
germ	germen	germe	Keim
germ cell	célula germinal	cellule germinale	Keimzelle
germ tube	tubo germinal	alvéole germinale	Keimschlauch
germicidal	germicida	germicide	keimtötend
germinate (to)	germinar	germer	keimen
germinating bed	cama germinadora	couche de germination	Keimbett
germination	germinación	germination	Keimen
germination period	período de germinación	période de germination	Keimzeit
germination power	poder germinativo	pouvoir germinatif	Keimkraft
germination test	control de germinación	analyse de germinabilité	Keimfähigkeitsuntersuchung
germinator	germinadora	germoir	Keimapparat
glass bed	table de mantillo	châssis vitré	Frühbeet
glasshouse whitefly	mosca blanca de invernadero	mouche blanche des serres	weiße Fliege
gley soil	suelo de gley	sol à gley	Naßboden
glomerule	glomérulo	glomérule	Samenknäuel

ENGLISH	SPANISH	FRENCH	GERMAN
glume	gluma	glume	Spelze
gneiss	gneis	gneiss	Gneis
goblet training	poda en cubilete	taille en vase	Gobeleterziehung
goggles	gafas protectoras	lunettes de protection	Schutzbrille
grade	pendiente	pente	Gefälle
graft (to)	injertar	greffer	veredeln
grafting	injerto	greffage	Veredelung
grain	grano	graine	Samenkorn
grain conditioning	acondicionamiento de granos	conditionnement des grains	Getreideaufbereitung
grain dryer	secadora de granos	séchoir à grains	Getreidetrocknungsanlage
grain drying	secado de cereales	séchage des céréales	Getreidetrocknung
grain harvest	cosecha de cereales	moisson	Getreideernte
grain sieve	criba para granos	crible à grains	Körnersieb
grain storage	almacenamiento de los granos	stockage des grains	Getreidelagerung
graining	granazón	grenaison	Kornbildung
graining phase	fase de formación del grano	phase de grenaison	Kornbildungsphase
gramineae	gramináceas	graminacées	Gräser
granary weevil	gorgojo de los graneros	charançon du blé	Kornkäfer
granulated fertilizer	abono granulado	engrais granulé	granulierter Düngen
granulation	granulación	granulation	Granulierung
granule	granulado	granulé	Granulat
grape blister	erinosis de la vid	érinose de la vigne	Pockenkrankheit
grape downy mildew	mildiú de la vid	mildiou de la vigne	Falscher
grape louse	filoxera	phylloxéra	Reblaus
grape powdery mildew	oidio de la vid	oïdium de la vigne	Echter
grape spalier	parra	vigne en espalier	Rebenspalier
grape vine	vid	vigne	Rebe
grapefruit	pomelo	pamplemousse	Grapefruit
grass blade	brizna	brin d'herbe	Grashalm
grass bryobia mite	briobia gramínea	mite bryobie graminée	Grasmilbe
grass weed	mala hierba	mauvaise herbe	Ungras
grassland steppe	estepa de gramíneas	steppe de graminées	Grassteppe
gravel	grava	gravier	Kies
gravelly sand	arena gravosa	sable graveleux	kiesiger Sand
great winter moth	falena deshojadora	grande phalène hiémale	großer Frostspanner
green corn	maíz verde	maïs vert	Grünmais
green lacewing	mosca de encaje	chrysoperle	Florfliege
green leaf weevil	filobio	phyllobie	Grünrüßler
green manure	abono verde	engrais vert	Gründünger
green manuring	abonado en verde	apport d'engrais vert	Gründüngung
greendung	abonado en verde	fumure verte	Gründüngung
greenhouse	invernadero	serre	Gewächshaus
grey field slug	babosa del campo	limace agreste	graue Ackershnecke
ground moraine	limos glaciales	moraine glaciaire	Grundmoräne
ground substrate	substrato del suelo	substrat du sol	Bodensubstrat
groundwater	aguas freáticas	eaux phréatiques	Grundwasser
groundwater gauge	nivel de la capa freática	niveau de la nappe phréatique	Grundwasserpegel
groundwater training	captación de aguas subterráneas	captage des eaux souterraines	Grundwasserfassung
growing point	punto de crecimiento	point de végétation	Vegetationspunkt
growth curve	curva de crecimiento	courbe de croissance	Wachstumskurve
growth hormone	fitohormona	phytohormone	Wuchsstoff
growth inhibition	inhibición del crecimiento	inhibition de la croissance	Wachstumshemmung
growth inhibitor	inhibidor del crecimiento	inhibiteur de la croissance	Wuchshemmer
growth phase	fase de crecimiento	phase de croissance	Wachstumsperiode
grub (to)	descepar	essoucher	ausroden
grubbing	escarificación	scarification	Grubbern
gummosis of citrus	gomosis bacilar de los agrios	gommose parasitaire des agrumes	Gummosis der Zitrusfrüchte
gymnosperms	gimnospermas	gymnospermes	Gymnospermen

H

ENGLISH	SPANISH	FRENCH	GERMAN
hail damage	daños causados por el granizo	dégâts causés par la grêle	Hagelschaden
half-bog soil	suelo semiturboso	sol semi-tourbeux	anmooriger Boden
half-track tractor	semioruga	semi-chenille	Halbraupe
halophytes	halófitos	halophytes	Salzpflanzen
hand seeder	sembradora de mano	semoir à bras	Handdrillmaschine
hand sowing	siembra a mano	semis à la main	Handsaat
handrake	rastrillo de mano	râteau à main	Handrechen
hard-setting soil	suelo de asentamiento	sol se tassant difficilement	hartsetzender Boden
harmless to bees	inocuo para las abejas	inoffensif pour les abeilles	bienenunschädlich
harrow	grada	herse	Egge
harrow (to)	gradar	herser	eggen
harvest	cosecha	récolte	Ernte
harvest time	tiempo de la recolección	temps de la récolte	Erntezeit
harvesting machinery	máquinas de recolección	machines de récolte	Erntemaschinen
harvesting method	procedimiento de recolección	méthode de récolte	Ernteverfahren
harvesting operations	trabajos de recolección	travaux de récolte	Erntearbeit
hay baling press	prensa de balas	presse-botteleuse	Ballenpresse
healthy soil	suelo sano	sol sain	gesunder Boden
heat capacity	capacidad térmica	capacité calorifique	Wärmekapazität
heat conductivity	conductividad térmica	conductibilité calorifique	Wärmeleitfähigkeit
heath soil	terreno de brezal	terre de bruyère	Heideboden
heavy metal	metal pesado	métal lourd	Schwermetall
heavy soil	tierra pesada	terre lourde	schwerer Boden
heliophilous plants	plantas heliófilas	plantes heliophiles	Lichtpflanzen
hemp	cáñamo	chanvre	Hanf
herbaceous	herbáceo	herbacé	krautig
hermaphrodite	hermafrodita	hermaphrodite	Hermaphrodit
Hessian fly	mosca de Hesse	mouche de Hesse	Hessenfliege
heteroecious	heteroico	hétéroïque	heterözisch
heterotrophic	heterótrofo	hétérotrophe	heterotroph
high-volume sprinkling	aspersión de alto volumen	arrosage intensif	Starkberegnung
highmoor	turbera supra-acuática	tourbière haute	Hochmoor
hoe	azada	houe	Haue
hoeing machine	binadora	bineuse	Hackmaschine
hoeing	cava	piochage	Hacken
horizon of soil cultivation	horizonte de cultivo del suelo	horizon de culture du sol	Bearbeitungshorizont
horn meal	harina de asta	farine de corne torréfiée	Hornmehl
horse manure	estiércol de ganado caballar	fumier de cheval	Pferdemist
horticultural product	producto hortícola	produit horticole	Gartenbauerzeugnis
horticulture	horticultura	horticulture	Gartenbau
hortisol	hortisol	hortisol	Hortisol
host pant	planta huésped	plante-hôte	Wirtspflanze
hot-water dressing	tratamiento con agua caliente	désinfection à l'eau chaude	Heißwasserbeizung

ENGLISH	SPANISH	FRENCH	GERMAN
house mouse	ratón común	souris grise	Hausmaus
humic acid	ácido húmico	acide humique	Huminsäure
humic compound	compuesto húmico	composé humique	Humusverbindung
humic gley soil	terreno de prados	sol de prairie	Wiesenboden
humic soil	suelo humificado	sol humifère	Humusboden
humification	humificación	humification	Humusbildung
humus balance	balance húmico	bilan humique	Humusbilanz
humus content	contenido de humus	teneur en humus	Humusgehalt
humus earth	tierra húmica	terreau	Humuserde
humus fertilizing	suministro de humus	apport d'humus	Humuszufuhr
hybrid	híbrido	hybride	Hybrid
hybrid corn	maíz híbrido	maïs hybride	Hybridmais
hydraulic conductivity	conductividad hidráulica	conductivité hydraulique	Wasserleitfähigkeit
hydraulic lifting device	elevador hidráulico	dispositif de relevage hydraulique	hydraulischer Ausheber
hydrolisis	hidrólisis	hydrolyse	Hydrolyse
hydromorphic soil	suelo hidromórfico	sol hydromorphe	hydromorpher Boden
hydrophily	hidrofilia	hydrophilie	Hydrophilie
hydrophoby	hidrofobia	hydrophobie	Hydrophobie
hydrophytes	hidrófitos	hydrophytes	Hydrophyten
hydroponic cultivation	cultivo hidropónico	culture hydroponique	Hydrokultur
hygroscopic	higroscópico	hygroscopique	hygroskopisch
hypocotyl	hipocótilo	axe embryonal	Hypokothyl
hypogeous	hipogeo	souterrain	unterirdisch
I			
illuvial horizon	horizonte iluvial	horizon illuvial	Einwaschungshorizont
immature soil	suelo inmaduro	sol jeune	Rohboden
immunity	inmunidad	immunité	Immunität
immunity breeding	selección de especies inmunizadas	séléction d'espèces immunisées	Immunitätszüchtung
impermeable soil	suelo impermeable	sol imperméable	undurchlässiger Boden
implantation	implantación	implantation	Implantation
impurity of seed	impurezas de las semillas	impuretés des semences	Verunreinigung des Samens
in-vitro-technique	técnica in vitro	technique in vitro	In-vitro-Verfahren
incorporation of fertilizer in the soil	incorporación del abono al suelo	incorporation de l'engrais au sol	Einarbeiten des Düngers
increase	incremento	accroissement	Zuwachs
infection	infección	infection	Ansteckung
infertile soil	tierra estéril	sol stérile	unfruchtbarer Boden
infest (to)	infestar	infester	befallen
infiltration	infiltración	infiltration	Versickerung
inflorescence	inflorescencia	inflorescence	Blütenstand
inhibition of germination	inhibición de la germinación	inhibition de la germination	Keimhemmung
inoculation	inoculación	inoculation	Impfung
insect pest	plaga de insectos	fléau d'insectes	Insektenplage
insect-borne disease	enfermedad transmitida por insectos	maladie transmise par des insectes	durch Insekten übertragene Krankheit
insecticide	insecticida	insecticide	Insektizid
insectivore	insectívoro	insectivore	Insektenfresser
insects control	lucha contra insectos	lutte contre les insectes	Insektenbekämpfung
insoluble	insoluble	insoluble	unlöslich
integrated pest control	lucha integrada	lutte intégrée	integrierter Pflanzenschutz
interflow	paso de agua	débit	Wasserdurchfluß
internal drainage	drenaje interno	drainage interne	Binnenentwässerung
internode	entrenudo	entre-noeud	Internodium
interval of crop	descanso de cultivos	pause des cultures	Anbaupause
intrazonal soil	suelo intrazonal	sol intrazonal	intrazonaler Boden
inversion	inversión	inversion	Inversion
ion exchange	intercambio iónico	échange des ions	Ionenaustausch
ionization	ionización	ionisation	Ionisation
ionize (to)	ionizar	ioniser	ionisieren
iron deficiency	deficiencia de hierro	carence en fer	Eisenmangel
iron-induced chlorosis	clorosis férrica	chlorose ferrique	Eisenchlorose
irradiation	irradiación	irradiation	Bestrahlung
irrigate (to)	irrigar	irriguer	bewässern
irrigated farming	cultivo de regadío	culture irriguée	Bewässerungskultur
irrigated soil	tierra de regadío	sol irrigué	Bewässerungsboden
irrigation	irrigación	irrigation	Bewässerung
irrigation by infiltration	riego por infiltración	irrigation par infiltration	Einstau
irrigation chanel	canal de riego	canal d'irrigation	Bewässerungskanal
irrigation ditch	zanja de irrigación	fossé d'irrigation	Bewässerungsgraben
irrigation pump	bomba para riego por aspersión	pompe d'arrosage	Beregnungspumpe
irrigation reservoir	embalse de riego	réservoir d'irrigation	Bewässerungsspeicher
irrigation system	sistema de riego	système d'irrigation	Bewässerungssytem
irrigation trickling water	escurrimiento superficial	irrigation par ruissellement	Rieselbewässerung
irrigation water	agua de riego	eau d'irrigation	Bewässerungswasser
J			
Japanese beetle	escarabajo japonés	hanneton japonais	Japankäfer
junction drain	colector de entrada	collecteur d'entrée	Einmündungsdrän
june bug	escarabajo de San Juan	hanneton de la Saint-Jean	gemeiner Brachkäfer
K			
kernel fruit	fruto de pepita	fruit à noyau	Kernfrucht
knapsack duster	espolvoreador de mochila	poudreuse à dos	Rückenstäuber
knapsack sprayer	pulverizador de mochila	pulvérisateur à dos	Rückenspritzgerät
knapsack sulfur duster	sulfatadora de mochila	soufreuse à dos	Rückenschwefelgerät
knot	nudo	noeud	Astknoten
L			
labiatae	labiáceas	labiacées	Labiatae
lackey moth	oruga de librea	bombyx à livrée	Ringelspinner
ladybird	mariquita	coccinelle	Marienkäfer
lake deposit	depósito lacustre	dépôt lacustre	Seeablagerung
lamella	hojita	foliole	Blättchen
land clearing machine	desbrozadora	débroussailleuse	Rodemaschine
land planting	plantación	mise en herbe	Begrünung
land reclamation	desbrozo	défrichement	Urbarmachung
land roller	rodillo agrícola	rouleau agricole	Ackerwalze
land surveying	levantamiento topográfico	arpentage	Landesuafnahme
landslide	corrimiento de tierras	éboulement de terrain	Erdrutsch
large-scale growing	gran cultivo	grande culture	Großanbau
larval gallery	daño causado por larvas	dommage causé par un phytophage	Fraßstelle
late crop	cosecha tardía	récolte tardive	späte Ernte
late frost	helada tardía	gelée tardive	Spätfrost
late manuring	abonado tardío	fumure tardive	Spätdüngung
late vintage	vendimia tardía	vendange tardive	Spätlese
latent period	período de latencia	période de latence	Latenzzeit
lateral ramification	ramificación lateral	ramification latérale	Seitenverzweigung

ENGLISH	SPANISH	FRENCH	GERMAN
laws of yield	leyes de rendimiento	lois du rendement	Ertragsgesetze
lay fallow (to)	estar de barbecho	être en jachère	brachliegen
layer	capa, acodo	couche, marcotte	Lage, Absenker
layer (to)	acodar	marcotter	absenken
layering	acodado	marcottage	Absenken
leach (to)	lixiviar	lixivier	auslaugen
leached soil	suelo lavado	sol lessivé	ausgewaschener Boden
leaching	lixiviación, percolación	lixiviation, percolation	Auslaugung, Versickerung
leading shoot	tallo principal	pousse principale	Haupttrieb
leaf	hoja	feuille	Blatt
leaf area	superficie foliar	lime	Blattfläche
leaf axil	axila de la hoja	aisselle de la feuille	Blattachsel
leaf base	base foliar	base foliaire	Blattgrund
leaf bud	yema foliar	bourgeon foliaire	Blattknospe
leaf curl mite	ácaro torcedor de hojas	mite frisante de la feuille	Kräuselmilbe
leaf cutting	esqueje de hoja	bouture de feuille	Blattsteckling
leaf fall	defoliación	chute des feuilles	Blattfall
leaf herbicide	herbicida foliar	défoliant	Blattherbizid
leaf lamina	limbo	limbe	Blattspreite
leaf sheath	vainas foliares	gaine foliaire	Blattscheide
leaf vein	nervio foliar	nervure	Blattrippe
leaf-cutting ant	hormiga cortadora	fourmi parasol	Blattschneiderameise
leaf-edge	borde de la hoja	verticille	Blattrand
leafless	sin hojas	défeuillé	blattlos
leek	puerro	poireau	Porree
leguminosae	leguminosas	légumineuses	Leguminosen
leguminous crops	leguminosas	légumineuses	Leguminosen
lenticell	lenticela	lenticelle	Lentizelle
leopard moth	taladro amarillo de los troncos	coquette	Blausieb
lethal dose	dosis mortal	dose mortelle	tödliche Dosis
level the soil (to)	aplanar el terreno	égaliser le sol	planieren
levelling fertilization	fertilización de equilibrio	fumure de correction	Ausgleichsdüngung
liber	líber	liber	Bast
lifting by frost	descalzamiento por heladas	déchaussement par la gelée	Auffrieren
light leaf spot	cilindrosporiosis	cylindrosporiose	Cylindrosporium-Blattfleckenkrankheit
light radiation	radiación luminosa	rayonnement lumineux	Lichtstrahlung
light soil	tierra ligera	terre légère	leichter Boden
lignification	lignificación	lignification	Verholzung
lignify (to)	lignificarse	se lignifier	verholzen
lignin	lignina	lignine	Lignin
ligula	lígula	ligule	Ligula
lime	cal	chaux	Kalk
lime deposit	depósito de caliza	dépôt calcique	Kalkablagerung
lime fertilizer	abono cálcico	engrais calcique	Kalkdünger
lime marl	marga cálcica	marne calcaire	Kalkmergel
lime requirement	necesidad de cal	besoins en chaux	Kalkbedarf
lime soil	suelo calizo	sol calcaire	Kalkboden
limestone	piedra caliza	pierre calcaire	Kalkstein
liming	enmienda caliza	chaulage	Kalken
linkage drawbar	barra de enganche	barre d'attelage	Ackerschiene
linkage frame	barra porta-aperos	barre porte-outil	Werkzeugrahmen
liquid fertilizer	abono líquido	engrais liquide	Flüssigdünger
liquid manure pit	pozo de purín	fosse à purin	Jauchegrube
liquid manuring	aplicación de purín	purinage	Jauchen
litchee	litchí	litchi	Litchipflaume
lithiasis pit of pears	piedra del peral	lithiase des poires	Steinfrüchtigkeit der Birne
load platform	plataforma de carga	plateau de chargement	Ladefläche
loam	suelo franco	terre glaise	Lehm
loamy sand	arena arcillosa	sable argileux	lehmiger Sand
loamy soil	suelo limoso	sol limoneux	lehmiger Boden
local strain	variedad local	variété locale	Landsorte
locust	langosta	criquet	Wanderheuschrecke
locust control	lucha contra la langosta	lutte antiacridienne	Heuschreckenbekämpfung
lodged grain	cereales encamados	céréales versées	Lagergetreide
lodicule	lodículo	lodicule	Schwellkörperchen
loess	loess	loess	Löß
long-tailed field mouse	ratón de monte	mulot	kleine Waldmaus
long-term fallow	yermo	jachère à longue durée	Dauerbrache
loss of nutrients	pérdida de elementos fertilizantes	perte de fertilisants	Nährstoffverlust
lot	parcela	parcelle	Teilfläche
low moor	pantano bajo	tourbière basse	Niedermoor
low-volume sprinkling	aspersión de bajo volumen	arrosage léger par aspersion	Schwachberegnung
lysin	lisina	lysine	Lysin
M			
macro elements	macronutrientes	macro-éléments	Massennährstoffe
macro pore	macroporo	macro-pore	Grobpore
macrofaune	macrofauna	macrofaune	Makrofauna
macronutrients	macronutrientes	macro-éléments	Makronährstoffe
magnesium	magnesio	magnésium	Magnesium
magnesium sulphate	sulfato magnésico	sulfate de magnesium	Magnesiumsulfat
main crop	cultivo principal	culture principale	Hauptfrucht
main ditch	drenaje principal	drain principal	Hauptvorfluter
main harvest	cosecha principal	récolte principale	Haupternte
main root	raíz principal	racine principale	Hauptwurzel
maintenance cut	poda de mantenimiento	taille d'entretien	Erhaltungsschnitt
maintenance works	labores de mantenimiento	travaux d'entretien	Pflegearbeiten
malting barley	cebada cervecera	orge de brasserie	Braugerste
management of pesticides	utilización de pesticidas	utilisation de produits phytosanitaires	Pflanzenschutzmitteleinsatz
manganese	manganeso	manganèse	Mangan
manganiferous	rico en manganeso	manganésifère	manganhaltig
manganous sulphate	sulfato de manganeso	sulfate de manganèse	Mangansulfat
manual hoe	azada	houe manuelle	Handhacke
manure (to)	abonar	fumer	düngen
manure peat	turba fertilizante	tourbe fertilisante	Düngetorf
manure unity	unidad de abono	unité d'engrais	Dungeinheit
manurial requirements	exigencias de abonos	besoins en engrais	Düngerbedarf
manurial value	valor fertilizante	valeur d'engrais	Düngewert
manuring method	método de abonado	méthode de fumure	Düngerverfahren
march fly	bibio de las huertas	bibion maraîcher	Gartenhaarmücke
maritime climate	clima marítimo	climat maritime	Küstenklima
marl	marga	marne	Mergel
marling	enmargado	marnage	Mergeldüngung

ENGLISH	SPANISH	FRENCH	GERMAN
marsh	terreno bajo pantanoso	marais	Marsch
marsh plants	plantas palustres	plantes palustres	Sumpfpflanzen
maturation	maduración	maturation	Abreife
maximum yield	rendimiento máximo	rendement maximal	Höchstertrag
mechanical hoe	binadora	houe mécanisée	Maschinenhacke
mechanized	mecanizado	mécanisé	maschinell
medium early	semi-temprano	demi-précoce	mittelfrüh
medium late	semi-tardío	demi-tardif	mittelspät
medlar	níspero	nèfle	Mispel
melioration	mejora del suelo	amendement	Melioration
membrane	membrana	membrane	Membran
Mendel's laws	leyes de Mendel	lois de Mendel	Mendelsche Gesetze
meristem	meristema	méristème	Meristem
meristem cultur	cultivo de meristemas	culture de méristème	Meristemkultur
mesh	malla	maille	Masche
mesocotyl	mesocótilo	mésocotyle	Mesokotyl
mesophyll	mesófilo	mésophylle	Blattgewebe
metamorphosis	metamorfosis	métamorphose	Verwandlung
micro pore	microporo	micro-pore	Feinpore
micro-organisms	microorganismos	microorganismes	Mikroorganismen
microbial	microbiológico	microbien	mikrobiell
microclimate	microclima	microclimat	Mikroklima
microfauna	microfauna	microfaune	Mikrofauna
microflora	microflora	microflore	Mikroflora
mineral constituents	componentes minerales	composants minéraux	mineralische Bestandteile
mineral fertilization	fertilización mineral	fertilisation minérale	mineralische Düngung
mineral fertilizer	abono mineral	engrais minéral	Mineraldünger
mineral soil	suelo mineral	sol minéral	Mineralboden
mineralization	mineralización	minéralisation	Mineralisierung
mixed cropping	cultivo mixto	culture mixte	Mischkultur
mixed fertilizer	abono mixto	engrais mélangé	Mischdünger
mixed forest	bosque mixto	forêt mixte	Mischwald
mixture of varieties	mezcla de variedades	mélange de variétés	Sortenmischung
modification	modificación	modification	Modifikation
moist cereal grains	cereales húmedos	céréales humides	Feuchtgetreide
moisture metre	medidor de humedad	humidimètre	Feuchtigkeitsmesser
moisture tension	tensión del agua	tension de l'eau	Wasserspannung
mole	topo	taupe	Maulwurf
molluscicide	molusquicida	mollusquicide	Molluskizid
molybdenum	molibdeno	molybdène	Molybdän
monocotyledons	monocotiledóneas	monocotylédones	Monokotylen
monoculture	monocultivo	monoculture	Monokultur
monoecious	monoico	monoïque	monözisch
mor	sustancia vegetal en descomposición	substance végétale en décomposition	Moder
morphogenetic	morfógeno	morphogène	formbildend
morphology	morfología	morphologie	Morphologie
moss	musgo	mousse	Moos
mother plant	planta madre	pied-mère	Mutterpflanze
mouse trap	ratonera	souricière	Mausefalle
mud	fango	vase	Schlamm
mulch	abono orgánico vegetal, mulch	paillis, mulch	Mulch
mull	mantillo suave	mull	Mull
multicellular	pluricelular	pluricellulaire	mehrzellig
multinutrient fertilizer	abono complejo	engrais composé	Mehrnährstoffdünger
muskrat	rata almizclera	rat musqué	Bisamratte
mutability	mutabilidad	mutabilité	Mutabilität
mutant	mutante	mutant	Mutant
mycelium	micelio	mycélium	Pilzmycel
mycorhiza	micorriza	mycorhize	Mykorrhiza
mycosis	micosis	mycose	Mycosis

N

ENGLISH	SPANISH	FRENCH	GERMAN
natural lime	cal natural	chaux naturelle	Naturkalk
necrosis	necrosis	nécrose	Nekrose
nematicide	nematicida	nématicide	Nematizid
nematode	nematodo	nématode	Nematode
net bloth of barley	helmintosporiosis de la cebada	helminthosporiose de l'orge	Blattfleckenkrankheit der Gerste
neutral soil	tierra neutra	sol neutre	neutraler Boden
newly reclaimed land	terreno nuevo	terre de nouveau défrichée	Neuland
nickel	níquel	nickel	Nickel
night frost	helada nocturna	gelée nocturne	Nachtfrost
nitrate fertilizer	abono nítrico	engrais nitrique	Salpeterdünger
nitrate nitrogen	nitrógeno nítrico	azote nitrique	Nitratstickstoff
nitrate of soda	nitrato sódico	nitrate de sodium	Natronsalpeter
nitrate stress	carga de nitrato	charge des nitrates	Nitratbelastung
nitrification	nitrificación	nitrification	Nitrifikation
nitrogen	nitrógeno	azote	Stickstoff
nitrogen compound	compuesto nitrogenado	combiné azoté	Stickstoffverbindung
nitrogen cycle	ciclo del nitrógeno	cycle de l'azote	Stickstoffkreislauf
nitrogen extraction	extracción de nitrógeno	épuisement de l'azote	Stickstoffentzug
nitrogen fertilizing	fertilización nitrogenada	fertilisation azotée	Stickstoffdüngung
nitrogen fixation	fijación del nitrógeno	fixation de l'azote	Stickstoffbindung
nitrogen magnesia	nitrato magnésico	azote-magnésie	Stickstoffmagnesia
nitrogen phosphate	fosfato nitrogenado	engrais azoté-phosphaté	Stickstoffphosphat
nitrogen potassium	potasio nitrogenado	engrais azoté-potassique	Stickstoffkali
nitrogen-fixing bacteria	bacterias nitrificantes	bactéries nitrifiantes	nitrifizierende Bakterien
nitrogenous fertilizer	abono nitrogenado	engrais azoté	Stickstoffdünger
node of the culm	nudo del tallo	noeud de la tige	Halmknoten
non-parasitic disease	enfermedad no parasitaria	maladie non parasitaire	nichtparasitäre Krankheit
notch grafting	injerto de incrustación	greffage en incrustation	Geißfußpfropfung
nucellus	nucelo	nucelle	Nucellus
nucleus	núcleo	noyau	Zellkern
number of grains/ear	número de granos/espiga	nombre de grains/épi	Kornzahl/Ähre
nursery	vívero	pépinière	Anzucht
nutrient	nutriente	substance nutritive	Nährstoff
nutrient absorption	asimilación de nutrientes	absorption d'éléments nutritifs	Nährstoffaufnahme
nutrient balance	balance de nutrientes	bilan des substances nutritives	Nährstoffbilanz
nutrient extraction	extracción de nutrientes	épuisement des éléments nutritifs	Nährstoffentzug
nutrient humus	humus nutritivo	humus nutritif	Nährhumus
nutrient level	nivel de nutrientes	niveau de nutrition	Nährstoffspiegel
nutrient solution	solución nutritiva	solution nutritive	Nährlösung
nutrient substratum	substrato nutritivo	substrat nutritif	Nährmedium
nutrient supply	suministro de nutrientes	approvisionement en éléments nutritifs	Nährstoffangebot

ENGLISH	SPANISH	FRENCH	GERMAN
nutrition intake	absorción de nutrientes	absorption d'éléments nutritifs	Nährstoffaufnahme
nutritive salt	sal nutritiva	sel nutritif	Nährsalz
nutritive substratium	substrato nutritivo	substrat nutritif	Nährsubstrat
O			
oats	avena	avoine	Hafer
offshoot	rebrote lateral	faux bourgeon	Nebentrieb
old peat	turba negra	tourbe noire	Schwarztorf
oligotrophic	oligotrófico	oligotrophe	oligotroph
one-way plow	arado para bancales	charrue ordinaire	Beetpflug
one-way plowing	labor en tablas	labour en planches	Beetpflügen
onion yellow dwarf	abigarrado de la cebolla	bigarrure de l'oigon	Gelbstreifigkeit der Zwiebel
open ditch	zanja abierta	tranchée ouverte	offener Graben
open ditch draining	drenaje por zanjas abiertas	drainage par tranchées ouvertes	Entwasserung mittels offener Gräben
optimum of yield	rendimiento óptimo	rendement optimal	Ertragsoptimum
organic farming	cultivo orgánico	agriculture organique	organischer Landbau
organic manure	abono orgánico	engrais organique	natürlicher Dünger
organic manuring	estercolado	apport de fumier	organische Düngung
organic mass	capa orgánica	masse organique	organische Masse
organic matter	materia orgánica	matière organique	oganische Substanz
organic soil	suelo orgánico	sol organique	organischer Boden
organism	organismo	organisme	Organismus
Oriental cockroach	cucaracha oriental	blatte orientale	orientalische Schabe
original seed	semilla original	semence d'origine	Originalsaatgut
original soil	suelo natural	sol en place	gewachsener Boden
osmosis	ósmosis	osmose	Osmose
osmotic	osmótico	osmotique	osmotisch
outfall	colector de salida	collecteur de sortie	Ausmündungsdrän
outgrowth	germinación en la espiga	germination sur pied	Auswuchs
outlet ditch	zanja de desagüe	fossé d'écoulement	Abzugsgraben
outside furrow	surco lindero	enrayure	Grenzfurche
ovary	ovario	ovaire	Fruchtknoten
over-run brake	freno de inercia	frein à inertie	Auflaufbremse
overfertilization	fertilización excesiva	excès de fertilisation	Überdüngung
ovicide	ovicida	ovicide	Ovizid
ovule	óvulo	ovule	Samenanlage
ox-bow lake	agua estancada	bras mort	Altwasser
oxidation	oxidación	oxydation	Oxydation
oxidize (to)	oxidar	oxyder	oxydieren
oxigen content	contenido en oxígeno	teneur en oxygène	Sauerstoffgehalt
oxigen deficiency	deficiencia en oxígeno	manque d'oxygène	Sauerstoffmangel
oxigenation	oxigenación	oxygénation	Sauerstoffzufuhr
oxisol	oxisol	oxisol	Oxisol
P			
paddy soil	terreno granuloso	sol de rizière	Reisboden
palisade tissue	tejido empalizado	tissu palissadé	Palisadengewebe
palmette	palmeta	palmette	Palmette
panicle	panícula	panicule	Rispe
papaveraceae	papaveráceas	papavéracées	Papaveraceae
papilionaceae	papilionáceas	papilionacées	Papilionaceae
parasite	parásito	parasite	Parasit
parasitic	parasitario	parasitaire	parasitär
parasitic attack	ataque parasitario	attaque parasitaire	Schädlingsbefall
parasitic disease	enfermedad parasitaria	maladie parasitaire	parasitäre Krankheit
parasitic plant	planta parasitaria	plante parasite	Schmarotzerpflanze
parenchyma	parénquima	penrenchyme	Parenchym
parent material	materiales madres del suelo	matériel d'origine	Ausgangsmaterial
parent rock	roca madre	rochemère	Muttergestein
parental generation	generación parental	première génération	Elterngeneration
parthenogenesis	partenogénesis	parthénogénèse	Parthenogenese
particle diameter	diámetro de las partículas	diamètre des particules	Korngrößendurchmesser
particle size	tamaño de las partículas	dimension des particules	Korngröße
pathogenous	patógeno	pathogène	pathogen
pear psylla	psila del peral	psylle du poirier	Birnenblattsauger
pear scab	moteado del peral	tavelure du poirier	Birnenschorf
peat	turba	tourbe	Torf
peat bog	turbera	tourbière	Torfmoor
peat formation	transformación en turbera	transformation en tourbe	Vermoorung
peat soil	tierra turbosa	terre tourbeuse	Torfboden
pectin	pectina	pectine	Pektin
pedicel	pedúnculo	pédoncule	Blütenstiel
pedology	pedología	pédologie	Pedologie
peg-tooth harrow	grada articulada	herse articulée	Gelenkegge
penetration resistance	resistencia a la penetración	résistance à la pénétration	Eindringwiderstand
percentage of germination	porcentaje de germinación	pourcentage de germinabilité	Prozentsatz der Keimfähigkeit
percolate (to)	infiltrarse	s'infiltrer	versickern
perennial	perenne	pérenne	ausdauernd
perianth	periantio	périanthe	Perianthium
pericarp	pericarpio	péricarpe	Perikarp
permafrost	helada permanente	permafrost	Dauerfrost
permanent charge	carga permanente	charge permanente	permanente Ladung
permanent cropping	cultivo permanente	culture permanente	Dauerkultur
permeable soil	suelo permeable	sol perméable	durchlässiger Boden
pesticide	pesticida	pesticide	Pestizid
petal	pétalo	pétale	Blütenblatt
petiole	peciolo	pétiole	Blattstiel
petroleum oil	aceite mineral	huile minérale	Mineralöl
pH value	valor del pH	valeur du pH	pH-Wert
pheasant	faisán	faisan	Fasan
phenotype	fenotipo	phénotype	Phänotyp
pheromone	feromona	phéromone	Pheromon
phosphatic fertilizer	abono fosfatado	engrais phosphaté	Phosphatdünger
phosphoric acid	ácido fosfórico	acide phosphorique	Phosphorsäure
photochemistry	fotoquímica	photochimie	Photochemie
photoperiod	fotoperíodo	photopériode	Photoperiode
photosynthesis	fotosíntesis	photosynthèse	Photosynthese
phyllotaxy	filotaxis	phyllotaxie	Blattstellung
physiological	fisiológico	physiologique	physiologisch
physiology	fisiología	physiologie	Physiologie
phytochemical	fitoquímico	phytochimique	phytochemisch
phytofagous	fitófago	phytophage	phytophag
phytohormone	fitohormona	phytohormone	Wuchsstoff
phytopathology	fitopatología	phytopatologie	Phytopathologie

ENGLISH	SPANISH	FRENCH	GERMAN
phytosanitary certificate	certificado fitosanitario	certificat phytosanitaire	Pflanzengesundheitszeugnis
phytotoxic	fitotóxico	phytotoxique	phytotoxisch
phytotoxin	tóxico vegetal	poison végétal	Pflanzengift
phytotron	fitotrón	phytotron	Phytotron
pick	pico	pic	Spitzhacke
pick potatoes (to)	arrancar patatas	arracher des pommes de terre	Kartoffeln roden
pieplant	ruibarbo	rhubarbe	Rhabarber
pig manure	estiércol de cerdo	fumier de porc	Schweinemist
pigment	pigmento	pigment	Pigment
pilous	piloso	pileux	haarig
piper line	tubería	tuyauterie	Rohrleitung
pistil	pistilo	pistil	Stempel
pistillate flower	flor femenina	fleur femelle	weibliche Blüte
pith	médula	pulpe	Mark
plant ashes	ceniza vegetal	cendres végétales	Pflanzenasche
plant bed	cuadro de cultivo	planche de culture	Pflanzbeet
plant cell	célula vegetal	cellule végétale	Pflanzenzelle
plant gall	agalla	galle	Galle
plant hygiene	higiene de las plantas	hygiène des plantes	Pflanzenhygiene
plant mass	masa vegetal	masse végétale	Pflanzenmasse
plant pathologist	fitopatólogo	phytopathologue	Pflanzenpathologe
plant propagation	multiplicación de las plantas	multiplication des plantes	Pflanzenvermehrung
plant protection	protección de las plantas	protection des plantes	Pflanzenschutz
plant protection product	producto fitosanitario	produit phytosanitaire	Pflanzenschutzmittel
plant residue	residuo vegetal	résidu vegetal	Pflanzenrückstand
plant residues	residuos vegetales	déchets végétaux	pflanzliche Abfälle
plant spacing	espacio entre plantas	écartement des plantes	Pflanzenabstand
plantation	plantío	plantage	Plantage
planting	plantación	plantation	Pflanzen
planting depth	profundidad de plantación	profondeur de plantation	Pflanztiefe
planting hole	hoyo de plantación	trou de plantation	Pflanzloch
planting machine	plantadora	planteuse	Pflanzmaschine
planting row	hilera de plantación	rangée de plantation	Pflanzreihe
planting season	época de plantación	époque de plantation	Pflanzzeit
plantlet	plantón	plant	Setzling
plastic covered greenhouse	invernadero de plástico	serre en feuilles	Foliengewächshaus
plastic soil	tierra plástica	terre compacte	plastischer Boden
plate grafting	injerto de plancha	greffage en placage	Anplatten
plateau	altiplanicie	plateau	Hochebene
plow	arado	charrue	Pflug
plowing	labranza	labour	Pflügen
plowsole	planta del arado	fond de raie du labour	Pflugsohle
pocket drilling	siembra a golpes	semis en poquets	Dibbelsaat
pod	vaina	gousse	Hülse
podzol	podzol	podzol	Podsol
poisenous plant	planta venenosa	plante vénéneuse	Giftpflanze
polarity	polaridad	polarité	Polarität
polarization	polarización	polarisation	Polarisation
polder soil	pólder	polder	Polderboden
poll (to)	desmochar	écimer	gipfeln
pollen	polen	pollen	Pollen
pollen-grain	grano de polen	grain de pollen	Pollenkorn
pollen-sac	saco polínico	sac pollinique	Pollensack
pollen-tube	tubo polínico	tube pollinique	Pollenschlauch
pollinate (to)	polinizar	polliniser	bestäuben
polygamous	polígamo	polygame	polygam
polyspermous	polispermo	polysperme	vielsamig
poor soil	tierra pobre	terre maigre	magerer Boden
pore size	tamaño del poro	dimension des pores	Porengröße
pore volume	volumen de los poros	volume des pores	Porenvolumen
pore water	agua intersticial	eau interstitielle	Porenwasser
porous	poroso	poreux	porös
pot (to)	enmacetar	empoter	eintopfen
pot experiment	ensayo en maceta	expérience en pots	Gefäßversuch
potash	potasa	potasse	Kali
potash detection	detección de potasio	détection de la potasse	Kalinachweis
potassic fertilizer	abono potásico	engrais potassique	Kalidünger
potassium chloride	cloruro potásico	chlorure de potasse	Kaliumchlorid
potassium magnesia	potasio magnésico	magnésie potassique	Kalimagnesia
potassium nitrate	nitrato potásico	nitrate de potasse	Kaliumnitrat
potassium sulfate	sulfato potásico	sulfate de potasse	Kaliumsulfat
potato aucuba mosaic	abigarrado amarillo de la patata	panachure infectieuse de la pomme de terre	Aucuba-Mosaik der Kartoffel
potato buquet disease	enfermedad buquet de la patata	maladie buquet de la pomme de terre	Buckettkrankheit
potato late blight	podredumbre parda de la patata	mildiou de la pomme de terre	Braunfäule der Kartoffel
potato leaf roll	enrollamiento de las hojas de la patata	enroulement des feuilles de la pomme de terre	Blattrollkrankheit der Kartoffel
potato rot nematode	anguilulosis de la patata	maladie vermiculaire de la pomme de terre	Älchenkrätze der Kartoffel
potato sprouts	brotes de patata	germes de pomme de terre	Kartoffelkeime
pre-sowing treatment	tratamiento de presiembra	traitement en présemis	Vorsaatbehandlung
preceding crop	cultivo precedente	culture précédente	Vorfrucht
precision seed	semilla de precisión	semences de précision	Präzisionssaatgut
precision seed drill	sembradora de precisión	semoir de précision	Einzelkornsägerät
precision sowing	siembra de precisión	semis de précision	Präzisionssaat
precocity	precocidad	précocité	Frühreife
predatory insect	insecto depredador	insecte prédateur	räuberisches Insekt
premature ripening	madurez prematura	maturation prématurée	Notreife
preservation liming	encalado de mantenimiento	chaulage d'entretien	Erhaltungskalkung
pressure dewatering	drenaje por presión	assèchement par pression	Druckentwässerung
preventive measures	medidas preventivas	mesures de prévention	Verhütungsmaßnahmen
preventive pest control	lucha preventiva antiparasitaria	lutte préventive antiparasitaire	vorbeugende Schädlingsbekämpfung
prick out (to)	repicar	repiquer	pikeren
primary cultivation	cultivo primario	culture primaire	Grundbodenbearbeitung
primeval forest	selva virgen	forêt vierge	Urwald
process of growth	proceso de crecimiento	processus de croissance	Wachstumsvorgang
processed seed	semilla tratada	semence traitée	bearbeiteter Samen
propagation by cuttings	estaquillado	bouturage	Stecklingsvermehrung
prophylaxis	profilaxis	prophylaxie	Prophylaxe
protective clothing	traje protector	vêtements protecteurs	Schutzkleidung
protective mask	máscara protectora	masqué de protection	Schutzmaske
provenance of seed	procedencia de las semillas	origine des semences	Herkunft des Samens
prune (to)	desramar	élaguer	ausästen
pruning	poda	taille	Baumschnitt
pruning for cuttings	poda de plantación	taille de plantation	Pflanzschnitt

ENGLISH	SPANISH	FRENCH	GERMAN
pseudocarp	pseudocarpio	faux fruit	Scheinfrucht
pump irrigation	riego por bombas	irrigation par pompage	Pumpenbewässerung
pumping plant	instalación de bombeo	station de pompage	Pumpanlage
pure nutrient	nutriente puro	substance nutritive pure	Reinnährstoff
purification of seed	purificación de las semillas	purification des semences	Saatgutreinigung
pustule	pústula	pustule	Pustel
Q			
quicklime	cal viva	chaux vive	Branntkalk
R			
raceme	racimo	grappe	Traube
radiation	radiación	radiation	Ausstrahlung
radication	radicación	radication	Wurzelbildung
radicle	radícula	radicule	Radicula
radius of action	radio de acción	spectre d'action	Wirkungsbreite
rain factor	factor pluvial	facteur de pluie	Regenfaktor
rain water	agua de lluvia	eau de pluie	Regenwasser
raining season	estación de lluvias	saison des pluies	Regenzeit
ramification	ramificación	ramification	Verzweigung
ranunculaceae	ranunculáceas	renonculacées	Ranunculaceen
rat control	desratización	dératisation	Rattenvertilgung
rate of fertilizer application	dosis de abonado	dose d'engrais	Düngergabe
rate of propagation	tasa de propagación	vitesse de propagation	Verbreitungsgeschwindigkeit
raticide	raticida	raticide	Rattengift
raw humus	humus ácido	humus acide	Rohhumus
readiness for germination	facultad germinativa	disposition à la germination	Keimbereitschaft
reaper	cosechadora	moissoneuse	Getreidemäher
reaper-binder	segadora-atadora	moissoneuse-lieuse	Bindemäher
reaping scythe	guadaña para cereales	faux à céréales	Getreidesense
reclaim land (to)	desbrozar	défricher	urbar machen
recurrence horizon	horizonte turboso de transición	horizon tourbeux de transition	Grenzhorizont
red earth	tierra roja	terre rouge	Roterde
red spider	araña roja	araignée rouge	rote Spinne
red steele of strawberry	podredumbre roja de la raíz de la fresa	pourriture du collet et des racines du fraisier	Erdbeerwurzelfäule
redoximorphic soil	suelo redoximórfico	sol rédoximorphe	redoximorpher Boden
reduction phase	fase de reducción	phase de réduction	Reduktionsphase
regenerate (to)	rebrotar	se régénérer	nachwachsen
regeneration	regeneración	régénération	Erneuerung
regulating	regulador	régulateur	regulativ
regulation	regulación	régulation	Regulation
release lever	palanca de desenganche	levier de décrochage	Auslösehebel
relief	relieve	relief	Relief
repellent	sustancia repelente	produit à effet répulsif	Repellent
replant (to)	replantar	replanter	nachpflanzen
reproduction	reproducción	reproduction	Fortpflanzung
resistance to lodging	resistencia al encamado	résistance à la verse	Standfestigkeit
resisting spore	espora resistente	spore résistante	Dauerspore
resorb (to)	reabsorber	résorber	resorbieren
resorption	resorción	résorption	Resorption
respiration	respiración	respiration	Atmung
respiration rate	tasa de respiración	taux respiratoire	Respirationsrate
respiratory poison	veneno respiratorio	poison asphyxiant	Atemgift
respiratory root	raíz respiratoria	racine respiratoire	Atemwurzel
rewetting	rehumedecimiento	réhumidification	Wiederbefeuchtung
rhachis	raquis	rachis	Spindel
rhizome	rizoma	rhizome	Rhizom
rice plantation	arrozal	rizière	Reisfeld
rice seeding	plántula de arroz	plantule de riz	Reissetzling
ridge sowing	siembra en caballones	semis sur billons	Kammsaat
ridging	aporcadura, labor en caballones	buttage, labour en billons	Anhäufeln, Kammpflügen
ridging plow	aporcadora	butteuse	Häufelpflug
rill	reguera	rigole	Rinne
rill erosion	erosión en surcos	érosion en rigoles	Rillenerosion
ripe	maduro	mûr	reif
ripeness	madurez	maturité	Reife
ripper	escarificadora	scarificateur	Grubber
river bed	cauce	lit	Flußbett
rock	roca	roche	Gestein
rock phosphate	fosfato en bruto	phosphate brut	Rohphosphat
rodent	roedor	rongeur	Nagetier
rodenticide	rodenticida	rodenticide	Rodentizid
roll (to)	pasar el rulo	rouler	walzen
romaine lettuce	lechuga larga	laitue romaine	römischer Salat
root activity	actividad radicular	activité des racines	Wurzelaktivität
root collar	cuello de la raíz	collet de la racine	Wurzelhals
root crop cultivation	cultivo de raíces y tubérculos	culture sarclée	Hackfruchtbau
root crops	raíces comestibles	racines plantes	Wurzelfrüchte
root cutting	estaca de raíz	bouture de racine	Wurzelsteckling
root development depth	desarrollo de las raíces en profundidad	profondeur de pénétration des racines	Durchwurzelungstiefe
root hair	pelo radicular	poil radiculaire	Wurzelhaar
root mass	masa radicular	masse de racines	Wurzelmasse
root out (to)	desraizar	déraciner	entwurzeln
root system	sistema radicular	système radiculaire	Wurzelsystem
root tip	punta radicular	extrémité radiculaire	Wurzelspitze
root tuber	tubérculo radicular	tubercule radiculaire	Wurzelknolle
root-cap	pilorriza	coiffe	Wurzelhaube
rooting	enraizamiento	enracinement	Durchwurzelung
rootstock	portainjerto	porte-greffe	Unterlage
rosaceae	rosáceas	rosacées	Rosaceae
rose grain aphid	pulgón de los cereales	puceron des céréales	Haferblattlaus
rot (to)	pudrirse	se putréfier	verrotten
rotary harrow	grada giratoria	herse rotative	Kreiselegge
rotary tiller	rotocultor	machine à pailler	Mulchgerät
rotting	putrefacción	putréfaction	Verrottung
rotting on the vine	podredumbre noble	pourriture noble	Edelfäule
row cropping	cultivo en líneas	culture en lignes	Reihenkultur
row fertilization	abonado en líneas	épandage d'engrais en ligne	Reihendüngung
row seeding	siembra a chorrillo	semis en lignes	Reihensaat
runner	retoño	stolon	Ausläufer
rye	centeno	seigle	Roggen
S			
sacharated	sacaroso	sucré	zuckerhaltig
saline soil	suelo salino	sol salin	Salzboden

ENGLISH	SPANISH	FRENCH	GERMAN
salinity	salinidad	salinité	Salinität
salt accumulation	acumulación salina	accumulation de sel	Salzakkumulation
sandstone	arenisca	grès	Sandstein
sandy soil	suelo arenoso	sol sablonneux	Sandboden
sap	savia	sève	Pflanzensaft
saprophytes	saprófitas	saprophytes	Saprophyten
saturated	saturado	saturé	gesättigt
saturation	saturación	saturation	Sättigung
scald	escaldadura	échaudure	Schalenbräune
scale	escama	écaille	Schuppe
schizomycets	esquizomicetos	schizomycètes	Spaltpilze
scion grafting	injerto de púa	greffage par rameaux détachés	Pfropfen
scrophulariceae	escrofuláceas	scrofulacées	Rachenblütler
scrub clearing	desbrozo	débroussaillement	Gestrüppentfernen
scrubber	niveladora	niveleuse	Ackerschleppe
seasonal	estacional	saisonnier	jahreszeitlich
secondary effect	efecto secundario	effet secondaire	Nebenwirkung
secondary root	raíz secundaria	racine secondaire	Seitenwurzel
sediment	sedimento	sédiment	Sediment
sediment; deposit	sedimento; depósito	sédiment; dépôt	Ablagerung
sedimentary rock	roca sedimentaria	roche sédimentaire	Absatzgestein
sedimentary soil	suelo de sedimentación	sol sédimentaire	Sedimentboden
sedimentation analyse	análisis por sedimentación	analyse par sédimentation	Schlämmanalyse
seed	semilla	semence	Samen
seed bank	banco de semillas	banque de semences	Samenbank
seed breeding	selección de semillas	sélection de semences	Saatzucht
seed breeding station	centro de selección de semillas	centre de sélection de semences	Saatzuchtanstalt
seed certification	certificación de semillas	certification des semences	Saatanerkennung
seed health testing	control sanitario de las semillas	inspection sanitaire des semences	Gesundheitsprüfung von Saatgut
seed mixture	mezcla de semillas	mélange de semences	Saatgutmischung
seed producer	productor de semillas	producteur de semences	Saatgutvermehrer
seed production	producción de simientes	production de semences	Samenproduktion
seed purity	pureza de las semillas	pureté des semences	Reinheit des Saatguts
seed rate	cantidad de semilla requerida	dose de semis	Saatmenge
seed survival	viabilidad de la semilla	viabilité des semences	Samenlebensfähigkeit
seed testing laboratory	laboratorio de análisis de semillas	laboratoire d'analyse des semences	Saatgut-Untersuchungslabor
seed tray	caja de semillero	caissette à semis	Saatkiste
seed treatment	tratamiento de las semillas	traitement des semences	Saataufbereitung
seed viability	viabilidad de las semillas	viabilité des semences	Lebensfähigkeit des Samens
seed weight	peso de las semillas	poids des semences	Samengewicht
seedcorn maggot	mosca de los sembrados	mouche grise des semis	Bohnenfliege
seeder	sembradora a chorrillo	semoir en lignes	Sämaschine
seeding	planta de semillero	plante de semis	Keimling
seep into (to)	infiltrarse	s'infiltrer	einsickern
seepage loss	pérdida por percolación	perte par infiltration	Sickerverlust
selected seed	semilla seleccionada	semence sélectionnée	Zuchtsaat
selection	selección	séléction	Selektion
selective herbicide	herbicida selectivo	herbicide séléctif	Selektivherbizid
self-fertilization	autofecundación	autofécondation	Selbstung
semi-desert	semidesierto	semi-désert	Halbwüste
semi-liquid manure	licuame	lisier	Gülle
semiparasitic plants	plantas semiparásitas	plantes hémiparasites	Halbschmarotzerpflanzen
sensitive to light	fotosensible	sensible à la lumière	lichtempfindlich
sepal	sépalo	sépale	Kelchblatt
septoria leaf bloth	septoriosis del trigo	septoriose des céréales	Blattdürre des Weizens
serological test	control serológico	test sérologique	serologische Prüfung
sewage	aguas residuales	eaux résiduelles	Abwasser
sex hormone	hormona sexual	hormone sexuelle	Sexualhormon
sexual reproduction	reproducción sexual	reproduction sexuée	generative Fortpflanzung
shade (to)	sombrear	ombrager	schattieren
shade influene	influencia de la sombra	effet de l'ombre	Schattenwirkung
shallow drill	semilla plana	semis superficiel	Flachsaat
shape pruning	poda de formación	taille de formation	Erziehungsschnitt
sheep manure	sirria	fumier de moutons	Schafmist
sheet erosion	erosión laminar	érosion en nappes	Flächenerosion
shell-fruits	frutos de cáscara	fruits à coques	Schalenfrüchte
shoot (to)	brotar	bourgeonner	ausschlagen
shoot apex	ápice del vástago	pointe de la pousse	Sproßspitze
shovel	pala	pelle	Schaufel
shrivelled grain	grano rugoso	grain ridé	Schmachtkorn
sickle	hoz	faucille	Sichel
side shoot	vástago lateral	pousse latérale	Seitentrieb
sierozem	sierosem	sierozem	Sierozem
sieve	tamiz	tamis	Sieb
silicate	silicato	silicate	Kieselsäuresalz
siliceous	silíceo	siliceux	kieselsäurehaltig
siliceous soil	tierra silícica	sol siliceux	siliziumreicher Boden
silo filler	ensiladora	ensileuse	Siliermaschine
silt	limo	limon	Schluff
silt and clay	arcilla y limo	argile et limon	Abschlämmbares
silting	colmateo	colmatage	Auflandung
silver leaf disease	mal del plomo	maladie du plomb	Bleiglanzkrankheit
single cross	cruzamiento simple	croisement simple	Einfachkreuzung
single nutrient fertilizer	abono simple	engrais simple	Einnährstoffdünger
single-grain structure	estructura monogranular	structure monogranulaire	Einzelkornstruktur
skim plowing	labor superficial	labour superficiel	Schälen
skiophilous plants	plantas esciófilas	plantes sciophiles	Schattenpflanzen
slaked lime	cal muerta	chaux éteinte	gelöschter Kalk
slate	esquisto	schiste	Schiefer
slit drainage	drenaje por abertura	drainage par fossés	Schlitzdränage
slope	pendiente	pente	Hang
slope gradient	inclinación	déclivité	Hangneigung
slow-acting fertilizer	abono de acción lenta	engrais à action lente	Langzeitdüngemittel
small seed	siembra clara	semis clair	Dünnsaat
small winter moth	falena invernal	petite phalène hiémale	kleiner Frostspanner
snow mold	moho de la nieve	moisissure de la neige	Schneeschimmel
sodic soil	suelo sódico	sol sodique	Natriumboden
sodium	sodio	sodium	Natrium
sodium nitrate	nitrato de sosa	nitrate de soude	Natriumnitrat
soft wheat	trigo candeal	blé tendre	Weichweizen
soil	suelo	sol	Boden
soil acidification	acidificación del suelo	acidification du sol	Bodenversauerung

ENGLISH	SPANISH	FRENCH	GERMAN
soil acidity	acidez del suelo	acidité du sol	Bodensäure
soil adjuvant substance	substancia coadyuvante del suelo	adjuvant du sol	Bodenhilfsstoff
soil aeration	aireación del suelo	aération du sol	Bodendurchlüftung
soil air	aire del suelo	air présent dans le sol	Bodenluft
soil analysis	análisis del suelo	analyse du sol	Bodenanalyse
soil association	asociación de suelos	famille de sols	Bodengesellschaft
soil bacteria	bacterias del suelo	bactéries du sol	Bodenbakterien
soil ball	cepellón	ballot de terre	Ballen
soil capillaries	capilares del suelo	capillaires du sol	Bodenkapillaren
soil chemistry	química del suelo	chimie du sol	Bodenchemie
soil classification	clasificación de suelos	classification des sols	Bodenklassifizierung
soil color	color del suelo	couleur du sol	Bodenfarbe
soil compaction	compactación del suelo	compactage du sol	Bodenverdichtung
soil conditions	condiciones del suelo	état du sol	Bodenverhältnisse
soil conditions for cultivating	propiedades del cultivo	propriétés culturales	Bearbeitungsfähigkeit
soil conservation	conservación de suelos	conservation des sols	Bodenerhaltung
soil consistence	consistencia del suelo	consistance du sol	Bodenkonsistenz
soil constituents	componentes del suelo	constituants du sol	Bodenbestandteile
soil contamination	contaminación del suelo	contamination du sol	Bodenverseuchung
soil cover	capa vegetal del suelo	couverture végétale	Bodendecke
soil crusting	incrustación del suelo	encroûtement du sol	Bodenverkrustung
soil degeneration	degradación del suelo	dégradation du sol	Bodendegeneration
soil density	densidad del suelo	densité du sol	Bodendichte
soil depth	profundidad del suelo	profondeur du sol	Bodentiefe
soil development	desarrollo del suelo	développement du sol	Bodenentwicklung
soil disinfection	desinfección del suelo	désinfection du sol	Bodendesinfektion
soil dynamics	dinámica del suelo	dynamique du sol	Bodendynamik
soil exhaustion	agotamiento del suelo	fatigue du sol	Bodenmüdigkeit
soil extract	extracto del suelo	extrait de sol	Bodenextrakt
soil fertility	fertilidad del suelo	fertilité du sol	Bodenfruchtbarkeit
soil fertilization	fertilización del suelo	fertilisation du sol	Bodendüngung
soil formation	formación del suelo	formation du sol	Bodenbildung
soil frost	helada a ras de suelo	gelée à ras du sol	Bodenfrost
soil health	sanidad de la tierra	santé du sol	Bodengesundheit
soil impoverishment	empobrecimiento del suelo	appauvrissement du sol	Bodenverarmung
soil improvement	enmienda del terreno	amendement du sol	Bodenverbesserung
soil improvement agent	substancia mejorante del suelo	agent d'amélioration du sol	Bodenverbesserungsmittel
soil management	tratamiento del suelo	travail du sol	Bodenbearbeitung
soil mineral	mineral del suelo	minéral du sol	Bodenmineral
soil moisture	humedad del suelo	humidité du sol	Bodenfeuchtigkeit
soil mottling	moteado del suelo	tacheture du sol	Bodenfleckung
soil on pleistocene sediments	suelo diluvial	sol diluvial	Diluvialboden
soil parasites	parasitos del suelo	parasites du sol	Bodenschädlinge
soil particle	partículas de tierra	particules du sol	Bodenteilchen
soil permeability	permeabilidad del suelo	perméabilité du sol	Bodendurchlässigkeit
soil physics	física del suelo	physique du sol	Bodenphysik
soil porosity	porosidad del suelo	porosité du sol	Bodenporosität
soil preparation	preparación del suelo	soins du sol	Bodenpflege
soil profile	perfil del suelo	profil du sol	Bodenprofil
soil properties	propiedades del suelo	propiétés du sol	Bodeneigenschaften
soil protection	protección del suelo	protection du sol	Bodenschutz
soil respiration	respiración del suelo	respiration du sol	Bodenatmung
soil salinization	salinización del suelo	salinisation du sol	Versalzung des Bodens
soil sample	muestra del suelo	échantillon du sol	Bodenprobe
soil sampling	toma de muestras del suelo	prélèvement d'échantillons du sol	Entnahme von Bodenproben
soil sterilization	esterilización del suelo	stérilisation du sol	Bodensterilisation
soil structure	estructura del suelo	structure du sol	Bodenstruktur
soil temperature	temperatura del suelo	température du sol	Bodentemperatur
soil texture	textura del suelo	texture du sol	Bodentextur
soil thermal	temperatura del suelo	température du sol	Bodenwärme
soil use	utilización del suelo	utilisation du sol	Bodennutzung
soil water	agua del suelo	eau du sol	Bodenwasser
soil-cover cultivation	cultivo de cubrición	culture de couverture	Deckkultur
solanaceae	solanáceas	solanacées	Solanazeen
solar-heating	calefacción solar	chauffage solaire	Solarheizung
solibility	solubilidad	solubilité	Löslichkeit
solid dung	estiércol sólido	fumier solide	Festmist
solubility	soluble	soluble	löslich
solum	solum	solum	Solum
sooty blotch of apple	fumagina de la manzana	maladie de la suie	Rußfleckenkrankheit des Apfels
sorption capacity	capacidad de absorción	capacité de sorption	Sorptionsvermögen
sorption complex	complejo de absorción	complexe de sorption	Sorptionskomplex
source of infection	foco de infección	foyer d'infection	Ansteckungsherd
sow (to)	sembrar	semer	besäen
sow broadcast (to)	sembrar a voleo	semer à la volée	säen
sowing in the open	siembra de asiento	semis sur place	Freilandsaat
sown area	superficie sembrada	surface ensemencée	Aussaatfläche
soya bean	soja	soja	Sojabohne
spacing of drains	distancia de los tubos de drenaje	écartement des drains	Dränabstand
spadix	espádice	spadice	Kolben
species	especie	espèce	Art
species conservation	conservación de especies	conservation de l'espèce	Arterhaltung
species diversity	diversidad de especies	diversité d'espèces	Artenvielfalt
species impoverishment	depauperación de la especie	appauvrissement de l'espèce	Artenverarmung
specific gravity of soil	peso específico del suelo	poids spécifique du sol	spezifishes des Bodens Gewicht
spicula	espiguilla	épillet	Ährchen
spindle stage	estadio de huso	stade de fuseau	Spindelstufe
spindle training	poda en cono	taille en quenouille	Kegelschnitt
spontaneous mutation	mutación espontánea	mutation spontanée	spontane Mutation
sporangium	contenedor de esporas	sporange	Sporenbehälter
spore	espora	spore	Spore
sporulation	formación de esporas	sporulation	Sporenbildung
spotted wilt of tomato	bronceado del tomate	maladie bronzée de la tomate	Bronzefleckenkrankheit der Tomate
spray mixture	caldo de pulverización	bouillie de pulvérisation	Spritzbrühe
spraying	pulverización	pulvérisation	Spritzen
spread (to)	propagar	propager	ausbreiten
spread manure (to)	esparcir el estiércol	épandre le fumier	Mist streuen
spreading	esparcido	épandage	Streuen
spreading of semi-liquid manure	distribución de licuame	épandage de lisier	Begüllung
spring barley	cebada de verano	orge de printemps	Sommergerste
Spring cultivation	labores de primavera	travaux de printemps	Frühjahrsbestellung
springtail	brincacola	élater	Springschwanz

ENGLISH	SPANISH	FRENCH	GERMAN
sprinkle (to)	rociar	asperger	bespritzen
sprinkler irrigation	riego por aspersión	irrigation par aspersion	künstlich Beregnung
sprinkler with spreader	regador de chorro horizontal	arroseur horizontal	Flaschstrahregner
sprout	vástago	pousse	Sproß
sprout (to)	echar brotes	pousser	austreiben
spur pruning	poda de fructificación	taille fruitière	Fruchtholzschnitt
stabilization of soil structure	estabilización de la estructura del suelo	stabilisation de la structure du sol	Gefügestabilisierung
stable humus	humus estable	humus stable	Dauerhumus
stage of growth	estadio de crecimiento	stade de croissance	Wachstumsstadium
stage of maturity	estado de madurez	stade de maturité	Reifestadium
stagnant water	agua estancada	eau stagnante	stagnierendes Wasser
stake	tutor	étai	Stütze
stake (to)	tutorar	étayer	stützen
stalk	tallo	tige	Stiel
stamen	estambre	étamine	Stamen
staminate flower	flor masculina	fleur mâle	männliche Blüte
starch	almidón	amidon	Stärke
starting motor	motor de arranque	démarreur	Starter
stem	tronco	tronc	Stamm
stem break	encamado parasitario de los cereales	piétin-verse	Halmbruchkrankheit des Getreides
stem rot of clover	esclerotina del trébol	sclérotinose du trèfle	Kleekrebs
steppe	estepa	steppe	Steppe
sterility	esterilidad	stérilité	Sterilität
stigma	estigma	stigmate	Narbe
stimulatory effect	efecto estimulante	effet stimulant	stimulierende Wirkung
stipule	estípula	stipule	Nebenblatt
stomates	estomas	stomates	Stomata
stony soil	suelo pedregoso	terre caillouteuse	steiniger Boden
stoppage of the pipes	obstrucción de los tubos	obstruction des drains	Vertopfung des Dräns
storage decay	podredumbre de almacenamiento	pourriture d'emmagasinage	Lagerfäule
stratification	estratificación	stratification	Schichtung
straw manuring	estiércol de paja	fumier de paille	Strohdüngung
stress	estrés	stress	Belastung
stress capacity	capacidad de estrés	tolérance	Belastbarkeit
strip cropping	cultivo en fajas	culture en bandes	Streifenkultur
striped tortoise beetle	casida noble	casside noble	Schildkäfer
stubble	rastrojo	chaume	Stoppel
stunting plant	panta raquítica	plante chétive	Kümmerpflanze
subsoil	subsuelo	sous-sol	Unterboden
subsoiling	substrato	substrat	Untergrund
subterranean vole	topillo	campagnol souterrain	Kurzohrmaus
succulent plant	planta suculenta	plante grasse	Saftpflanze
sucking insect	insecto chupador	insecte suceur	saugendes Insekt
sucking root	raíz absorbente	racine absorbante	Saugwurzel
suction force	capacidad de absorción	force de succion	Saugkraft
sulfur	azufre	soufre	Schwefel
sulfur fertilizing	abonado con azufre	fertilisation de soufre	Schwefeldüngung
sulfurizer	sulfatadora	soufreuse	Schwefler
sun radiation	radiación solar	rayonnement solaire	Sonnenstrahlung
supporting crop	planta de cobertura	plante de couverture	Stützfrucht
surface compost mulch	compost de superficie	compostage en surface	Flächenkompostierung
surface drainage	drenaje de superficie	drainage superficiel	Oberflächenentwässerung
surface irrigation	riego de superficie	irrigation de surface	Oberflächenbewässerung
surface layer	capa de superficie	couche superficielle	Oberflächenschicht
surface run off	flujo superficial	ruissellement superficiel	Oberflächenabfluß
surface soil	suelo agrícola	couche arable	Bodenkrume
surface tension	tensión superficial	tension superficielle	Oberflächenspannung
surface tissue	tejido cutáneo	tissu superficiel	Hautgewebe
surface water	aguas superficiales	eaux de surface	Oberflächenwasser
surplus water	agua sobrante	eau excédentaire	Überwasser
susceptibility	propensión	sensibilité	Anfälligkeit
suspension	suspensión	suspension	Suspension
swamp	pantano	marécage	Sumpf
swelling	hinchamiento	gonflement	Quellung
swivel plow	arado basculante	charrue bascule	Pnedelflug
symbiosis	simbiosis	symbiose	Symbiose
syncarpy	infrutescencia	infrutescence	Fruchtstand
synoicous	sinoico	synoïque	synözisch

T

ENGLISH	SPANISH	FRENCH	GERMAN
take root (to)	arraigar	prendre racine	anwurzeln
take-all of cereals	mal del pie de los cereales	piétin-échaudage des céréales	Schwarzbeinigkeit des Getreides
tap-root	raíz pivotante	racine pivotante	Pfahlwurzel
tarnished plant bugs	chinche de los frutales	capside des cultures fruitières	Blattwanzen
tegument	tegumento	tégument	Samenschale
temperature drop	diferencia de temperatura	écarts de température	Temperaturgefälle
tendency tô laying down	tendencia al encamado	tendance à la verse	Lagerneigung
tendril	zarcillo	vrille	Ranke
tensile strength	resistencia a la tracción	résistance à la traction	Zugfestigkeit
tensiometre	tensiómetro	tensiomètre	Tensiometer
terminal bud	yema terminal	bourgeon terminal	Endknospe
terminal shoot	brote terminal	pousser terminale	Endtrieb
termites	termitas	termites	Termiten
terrace cultivation	cultivo en bancales	culture en terrasses	Terrassenkultur
thalamus	tálamo	réceptacle	Blütenboden
thawing layer	suelo blando	mollisol	Auftauschicht
thermophil	termófilo	thermophile	thermophil
thickness of soil	espesor del suelo	épaisseur du sol	Bodenmächtigkeit
thinning	aclareo	éclaircissage	Vereinzeln
thorn	espina	épine	Dorn
three-year rotation	rotación trienal	assolement triennal	Dreifelderwirtschaft
thresher	trilladora	batteuse	Dreschmaschine
threshing	trilla	battage	Drusch
thrips	trips	thrips	Thripse
tiller	hijuelo	talle	Bestockungstrieb
tillering	ahijamiento	tallage	Bestocken
tilth	capa de cultivo del suelo	couche du sol agricole	Bearbeitungsschicht
tin	estaño	étain	Zinn
tip (to)	despuntar	pincer	pinzieren
tobacco blue mold	mildiú del tabaco	mildiou du tabac	Blauschimmel des Tabaks
tobacco moth	efestia	teigne friande	Heumotte
tolerance limit	límite de tolerancia	seuil de tolérance	Toleranzgrenze
tomato leaf mold	cladosporiosis del tomate	cladosporiose de la tomate	Braunfleckigkeit der Tomate

ENGLISH	SPANISH	FRENCH	GERMAN
tongue grafting	injerto inglés	greffage à l'anglaise	Kopulieren mit Gegenzunge
top (to)	descabezar	étêter	köpfen
top cutting	esqueje terminal	bouture de tête	Kopfsteckling
top fertilization	abonado de cobertura	fumure de couverture	Kopfdüngung
top graft (to)	reinjertar	regreffer	umpfropfen
top soil	capa arable	couche arable	Krume
total nitrogen	nitrógeno total	azote total	Gesamtstickstoff
toxic	tóxico	toxique	giftig
toxic to bees	tóxico para las abejas	nocif pour les abeilles	bienenschädlich
toxicity	toxicidad	toxicité	Toxizität
trace elements	oligoelementos	oligo-éléments	Spurenelemente
trace elements fertilizng	abonado con oligoelementos	fertilisation d'oligo-éléments	Spurenelementdüngung
tractor	tractor	tracteur	Traktor
tractor-mounted implements	aperos colgados	outils portés sur tracteur	Anbaugeräte
trafficability	transitabilidad	praticabilité	Befahrbarkeit
trailed implements	aperos de arrastre	outils traînés	Anhängegeräte
transition soil	suelo de transición	sol de transition	Übergangsboden
translocation	translocación	translocation	Translokation
transmission by aphids	transmisión por pulgones	transmission par pucerons	Blattlausübertragung
transpiration coefficient	coeficiente de transpiración	coefficient de transpiration	Transpirationskoeffizient
transpiration flux	flujo de transpiración	flux de transpiration	Transpirationsstrom
transplant (to)	transplantar	transplanter	verpflanzen
transplanter	transplantadora	repiqueuse	Pikiermaschine
transport of substances	transporte de substancias	transport de substances	Stofftransport
tree nursery	vivero	pépinière	Baumschule
tree sparrow	gorrión	moineau	Feldsperling
tree-top	copa del árbol	cime de l'arbre	Baumkrone
treillis	espaldera	espalier	Spalier
trichogramma wasp	icneumón	ichneumon	Schlupfwespe
trim roots(to)	desbarbar	habiller	wurzeln
tropical plant	planta tropical	plante tropicale	Tropenpflanze
tropical rain forest	selva ecuatorial	forêt équatoriale	Regenwald
truncated soil	suelo truncado	sol tronqué	geköpfter Boden
tuber	tubérculo	tubercule	Knolle
tuberous chervil	perifollo tuberoso	cerfeuil tubéreux	Kerbelrübe
tuberous plant	planta tuberosa	plante tubéreuse	Knollenpflanze
turgor pressure	turgencia	turgescence	Turgor
turning of the soil	volteo de la tierra	retournement du sol	Umbruch
turnip	nabo	navet	weiße Rübe
twig canker	apoplejía del cerezo	maladie rhénane	Baumsterben der Kirsche
two-way plow	arado doble	charrue pour labour à plat	Kehrpflug
two-wheeled tractor	motocultor	motoculteur	Zweiradschlepper
ultra-violet rays	rayos utravioleta	rayons ultra-violets	ultra-violette Strahlen

U

umbel	umbela	ombelle	Dolde
umbelliferae	umbelíferas	ombellifères	Umbelliferen
underground fertilization	fertilización bajo pie	fertilisation en sous-sol	Unterfußdüngung
undersown crop	siembra bajo cobertura	semis sous couverture	Untersaat
ungrafted	franco de pie	franc de pied	wurzelecht
unicellular	unicelular	unicellulaire	einzellig
unisexual	unisexual	unisexué	eingeschlechtig
unmanured	sin abonar	non fumé	ungedüngt
useful insect	insecto útil	insecte utile	Nutzinsekt

V

vacuole	vacuola	vacuole	Vacuole
variable charge	carga variable	charge variable	variable Ladung
variety	variedad	variété	Varietät
vegetables	verduras	légumes	Gemüse
vegetation	vegetación	végétation	Vegetation
vegetation period	período de vegetación	période de végétation	Vegetationsperiode
vegetative phase	fase vegetativa	phase végétative	vegetative Phase
vegetative propagation	multiplicación vegetativa	multiplication végétative	vegetative Vermehrung
vegetative reproduction	reproducción vegetativa	reproduction végétative	vegetative Fortpflanzung
vegetative rest	reposo vegetativo	repos de la végétation	Vegetationsruhe
venom	veneno	venin	Gift
verticil	verticilo	verticile	Wirtel
verticillium wilt	verticilosis de la patata	verticillose de la pomme de terre	Pilzringfäule der Kartoffel
vessel	vaso	vaisseau	Gefäß
vine	cepa	cep	Weinstock
vine variety	variedad de cepa	cépage	Rebsorte
vineyard	viñedo	vignoble	Weinberg
virgin soil	suelo virgen	sol vierge	jungfräulicher Boden
virology	virología	virologie	Virusforschung
virosis	virosis	virose	Virose
virulence	virulencia	virulence	Ansteckungskraft
virus vector	transmisor viral	vecteur viral	Virusvektor
viscosity	viscosidad	viscosité	Viskosität
vitality	vitalidad	vitalité	Vitalität
viticulture	viticultura	viticulture	Weinbau
volatilize (to)	volatilizarse	se volatiliser	verflüchtigen (sich)
volcanic soil	tierra volcánica	terre volcanique	vulkanischer Boden
volubilate plant	planta voluble	plante volubile	Schlingflanze
volunteer plants	plantas accidentales	plantes accidentelles	sortenfremder Aufwuchs

W

wart disease of potatoes	sarna negra de la patata	gale noire de la pomme de terre	Kartoffelkrebs
waste land	erial	terrains incultes	Ödland
water (to)	regar	arroser	begießen
water balance	balance hídrico	bilan hydraulique	Wasserbilanz
water core of apples	vitrosidad de las manzanas	pommes vitreuse	Glasigkeit der Apfels
water economy	ahorro de agua	économie hydraulique	Wasserhaushalt
water erosion	erosión por el agua	érosion par l'eau	Wassererosion
water gauge	fluviómetro	indicateur du niveau de l'eau	Pegel
water inlet	acometida de agua	amenée d'eau	Wasserzuleitung
water level	nivel del agua	niveau de l'eau	Wasserspiegel
water needs	necesidades de agua	besoins d'eau	Wasserbedarf
water pipes	tubería de agua	conduite d'eau	Wasserleitung
water raising	elevación del agua	élévation de l'eau	Wasserförderung
water shoot	brote adventicio	gourmand	Wasserschoß
water soluble	soluble en agua	soluble dans l'eau	wasserlöslich
water stress	escasez de agua	carence d'eau	Wassermangel
water supply	suministro de agua	approvisionement en eau	Wasserversorgung

ENGLISH	SPANISH	FRENCH	GERMAN
water volume	caudal	débit d'eau	Wasserführung
water-bearing stratum	capa acuífera	nappe aquifère	wasserführende Schicht
watercourse	corriente de agua	cours d'eau	Wasserlauf
waterholding capacity	capacidad de fijación del agua	capacité de rétention de l'eau	Wasserbindungsvermögen
waterlily leaf beetle	galeruca del fresal	galéruque du fraisier	Erdbeerkäfer
waterlogged	impregnado de agua	imbibé d'eau	durchtränkt
waterlogging	saturación del suelo por el agua	saturation du sol par l'eau	Vernässung
weed (to)	escardar	sarcler	jäten
weed competition	competencia entre malas hierbas	concurrence des mauvaises herbes	Unkrautkonkurrenz
weed control	lucha contra las malas hierbas	lutte contre les mauvaises herbes	Unkrautbekämpfung
weed-killer	matamalezas	désherbant	Unkrautbekämpfungsmittel
weeder	escardadora	désherbeuse	Ackerbürste
well sinking	perforación de pozos	forage de puits	Brunnenbau
well water	agua de pozo	eau de puits	Brunnenwasser
wet bleaching	blanqueo en mojado	blanchiment mouillé	Naßbleichung
wet treatment	tratamiento líquido	désinfection par voie humide	Naßbeizen
wheat earworm	noctuido del trigo	noctuelle du chiendent	Queckeneule
white grub	gusano blanco	ver blanc	Engerling
wilt (to)	marchitarse	se fanner	welken
wilt of hop	verticilosis del lúpulo	verticillose du houblon	Hopfenwelke
wilted	marchito	fané	welk
wilting point	punto de marchitamiento	point de flétrissement	Welkepunkt
wind erosion	erosión eólica	érosion éolienne	Bodenverwehung
wind pollination	polinización por viento	pollinisation par le vent	Windbestäubung
wind sediment	depósito eólico	dépôt éolien	Windablagerung
wine grape	uva de vinificación	raisin de cuve	Keltertraube
winter frost	helada invernal	gelée d'hiver	Winterfrost
winter soil moisture	humedad invernal	humidité hivernale	Winterfeuchtigkeit
winter tillage	labores de invierno	travaux d'hiver	Winterbestellung
wire worm	gusano de alambre	ver fil-de-fer	Drahtwurm
wood	madera	bois	Holz
woody plants	plantas leñosas	plantes ligneuses	Holzpflanzen
X			
xerophytes	xerófilas	xérophytes	Xerophyten
Y			
yam	ñame	igname	Yamswurzel
yield decrease	disminución del rendimiento	diminution du rendement	Ertragsminderung
Z			
zinc	zinc	zinc	Zink
zonal soil	suelo zonal	sol zonal	zonaler Boden
zucchini	calabacín	courgette	Zucchini

BIBLIOGRAPHY

ALPI, A. y TOGNONI, F.
Cultivo en invernadero
Mundi-Prensa. Madrid, 1984

BOVEY, R. y OTROS AUTORES
La defensa de las plantas cultivadas
Omega. Barcelona, 1984

CADAHÍA, C.
Fertilización en riego por goteo de cultivos hortícolas
E.R.T. Fertilizantes. Madrid, 1988

DIEHL, E., MATEO J.M. y URBANO, P.
Fitotecnia general
Mundi-Prensa. Madrid, 1982

DOMÍNGUEZ VIVANCOS, A.
Abonos minerales
Ministerio de Agricultura. Madrid, 1978
Fertirrigación
Mundi-Prensa. Madrid, 1993
Tratado de fertirrigación
Mundi-Prensa. Madrid, 1984

FERNÁNDEZ CUEVAS
Horticultura intensiva
Ministerio de Agricultura. Madrid, 1968

GORDON HALFACRE, R. y BARDEN, JOHN A.
Horticultura
AGT Editor México, 1984

JIMÉNEZ, R.
Sistemas de cultivo, substratos y enarenados
Escuela Técnica Superior de Ingenieros Agrónomos
Universidad de Córdoba, 1984

LÓPEZ BELLIDO, L. y CASTILLO GARCÍA, J.E.
Horticultura mediterránea de invernadero
Escuela Técnica Superior de Ingenieros Agrónomos

Universidad de Córdoba, 1984
Nuevas tecnologías en cultivos en invernadero
Escuela Técnica Superior de Ingenieros Agrónomos
Universidad de Córdoba, 1987

LÓPEZ GÁLVEZ, J.
Curso internacional sobre Agrotecnia del cultivo en invernadero
Instituto de Fomento de Andalucía.
Almería, 1991

MAINARDI, F.
Los cultivos hidropónicos
De Vecchi. Barcelona, 1979

MAROTO, J.V.
Elementos de horticultura general
Mundi-Prensa. Madrid, 1990
Horticultura herbácea especial
Mundi-Prensa. Madrid, 1989

MARTÍN DE SANTA OLALLA MAÑAS, F. y DE JUAN VALERO, J.A.
Agronomía del riego
Mundi-Prensa. Madrid, 1993

MARTÍNEZ GARCÍA, P.F.
Características climáticas de los invernaderos de plástico
Ministerio de Agricultura. Madrid, 1978

MATALLANA, A. y MARFA PAGES, J.O.
Los invernaderos y la crisis energética
Instituto Nacional de Investigaciones Agrarias
Ministerio de Agricultura. Madrid, 1980
Invernaderos. Diseño, construcción y ambientación
Mundi-Prensa. Madrid, 1989

MEDINA SAN JUAN, J.A.
Riego por goteo
Mundi-Prensa. Madrid, 1988

MESSIAEN, C.M. y LAFON, R.
Enfermedades de las hortalizas
Oikos Tau. Vilassar de Mar, 1968

PALOMAR, F.
Nuevas técnicas de invernadero
Caja de Ahorros de Almería. Almería, 1988

PENNINGSFELD, F. y KURZMANN, P.
Cultivos hidropónicos y en turba
Mundi-Prensa. Madrid, 1975

RESH, D.
Cultivos hidropónicos. Nuevas técnicas de producción
Mundi-Prensa. Madrid, 1982

ROBLEDO DE PEDRO, F. y MARTÍN VICENTE, L.
Aplicación de los plásticos en la agricultura
Mundi-Prensa. Madrid, 1981

ROBLES, J.
Cómo se cultiva en invernadero
De Vecchi. Barcelona, 1987

SERRANO CERMEÑO, Z.
Cultivos hortícolas enarenados
Publicaciones de Extensión Agraria
Ministerio de Agricultura. Madrid, 1974
Invernadero. Instalación y manejo
Ministerio de Agricultura. Madrid, 1980
Pronturario del horticultor
Gráficas Murgis. Almería, 1985
Técnicas de invernadero
Suministros Gráficos, Sevilla, 1990

SITTA, GIORGIO
El ABC de la horticultura protegida
Mundi-Prensa. Madrid, 1988